W9-ASQ-422

Neuromechanics of Human Movement

FOURTH EDITION

Roger M. Enoka, PhD
University of Colorado at Boulder

Human Kinetics

Library of Congress Cataloging-in-Publication Data

Enoka, Roger M., 1949-
Neuromechanics of human movement / Roger M. Enoka. -- 4th ed.
p. ; cm.
Includes bibliographical references and index.
ISBN-13: 978-0-7360-6679-2 (hard cover)
ISBN-10: 0-7360-6679-9 (hard cover)
1. Kinesiology. 2. Human mechanics. I. Title.
[DNLM: 1. Biomechanics. 2. Movement--physiology. 3. Adaptation, Physiological. 4. Musculoskeletal Physiology. WE 103 E59n 2008]
QP303.E56 2008
612.7'6--dc22

2008020965

ISBN-10: 0-7360-6679-9
ISBN-13: 978-0-7360-6679-2

Acquisitions Editor: Loarn D. Robertson, PhD; **Developmental Editor:** Maureen Eckstein; **Assistant Editor:** Christine Bryant Cohen; **Copyeditor:** Joyce Sexton; **Proofreader:** Sarah Wiseman; **Indexer:** Craig Brown; **Permission Manager:** Dalene Reeder; **Graphic Designer:** Fred Starbird; **Graphic Artist:** Kathleen Boudreau-Fuoss; **Cover Designer:** Keith Blomberg; **Photographer (Cover):** Neil Bernstein; **Visual Production Assistant:** Joyce Brumfield; **Photo Office Assistant:** Jason Allen; **Art Manager:** Kelly Hendren; **Associate Art Manager:** Alan L. Wilborn; **Illustrators:** Accurate Art and Carolyn Barry; **Printer:** Sheridan Books

Printed in the United States of America 10 9 8 7 6 5 4 3 2 1

Human Kinetics
Web site: www.HumanKinetics.com

United States: Human Kinetics
P.O. Box 5076, Champaign, IL 61825-5076
800-747-4457
e-mail: humank@hkusa.com

Canada: Human Kinetics
475 Devonshire Road Unit 100, Windsor, ON N8Y 2L5
800-465-7301 (in Canada only)
e-mail: info@hkcanada.com

Europe: Human Kinetics
107 Bradford Road, Stanningley, Leeds LS28 6AT, United Kingdom
+44 (0) 113 255 5665
e-mail: hk@hkeurope.com

Australia: Human Kinetics
57A Price Avenue, Lower Mitcham, South Australia 5062
08 8372 0999
e-mail: info@hkaustralia.com

New Zealand: Human Kinetics
Division of Sports Distributors NZ Ltd.
P.O. Box 300 226 Albany, North Shore City, Auckland
0064 9 448 1207
e-mail: info@humankinetics.co.nz

To Bonny, Joel, and Seth
and
Maro, Chrissy, Katelyn, and Abby

Contents

Preface

Neuromechanics of Human Movement is intended to provide a foundation for studying how the nervous system controls the actions of muscles to exert forces on their surroundings and thereby produce movement. Because movement is constrained by the laws of physics, both the activation signals generated by the nervous system and the forces exerted by the muscles must accommodate these constraints. Accordingly, the content of the text is derived from the disciplines of neurophysiology (neuro-) and physics (mechanics) to provide a neuromechanical perspective on the study of human movement.

The text is organized into three parts. Part I focuses on Newton's laws of motion and their application to the study of human movement. This material, typically referred to in a physics text as mechanics, outlines the physical laws that define the properties of movement. The four chapters in part I examine the concepts required to describe motion, the external forces that act on the human body, the forces that exist within the human body, and the techniques that can be used to analyze movement with examples from running, jumping, and throwing. The most significant changes in this section from the third edition are that the material on electromyography has been moved to chapter 5 on electricity, new figures have been included, and the citations have been updated.

Part II introduces the essential concepts from neurophysiology needed to understand how movement is produced by the nervous system and muscles. Those parts of the human body involved in the production of movement are collectively known as the motor system. The three chapters on the motor system address excitable membranes, muscle and motor units, and voluntary movement. Chapter 5 deals with electricity, the resting membrane potential, the properties of neurons, synaptic transmission, excitation-contraction coupling, and electromyography. Chapter 6 addresses motor units, muscle mechanics, and the organization and activation of muscles. Chapter 7 examines spinal reflexes, central pattern generators, and supraspinal control of voluntary movement. Part II is completely revised from the third edition.

Part III focuses on the acute and chronic changes that can occur in the motor system in response to various interventions. The acute adjustments include warm-up effects, flexibility, muscle soreness and damage, muscle fatigue, muscle potentiation, and arousal. The chronic adaptations comprise the effects of strength and power training, the effects of reduced use, motor recovery from injury, and adaptations with aging. Although part III is similar to that in the third edition, it has been updated to include more recent concepts and citations, as well as many new figures, and the discussion on aging has been reorganized and expanded.

Because the intent of the text is to provide a scientific basis for the study of human movement, the ideas and principles are discussed in scientific terms, and more attention is devoted to precise definitions and measurements than is commonly the case in everyday conversation. There is also a presentation package and image bank that accompany this text. It contains PowerPoint slides that include most of the figures, tables, and photos in this text.

With this foundation, it should be possible to advance the study of human movement. This goal is well illustrated by the analogy from Sherlock Holmes on the following page: How could Holmes know of Watson's intentions? The answer is, of course, that he used his well-known ability to apply deductive reasoning. In a similar vein, movement can be considered the conclusion of a process, and our task, based on rigorously defined terms and concepts, is to identify the intervening steps between the starting point and the conclusion. The state of our knowledge on the neuromechanics of human movement is rather rudimentary, so I hope that you find the fourth edition of the text both a valuable resource and an inspiration for new ideas.

THE STRAND MAGAZINE.

Vol. xxvi. DECEMBER, 1903. No. 156.

THE RETURN OF SHERLOCK HOLMES.

By A. CONAN DOYLE.

III.—The Adventure of the Dancing Men.

HOLMES had been seated for some hours in silence with his long, thin back curved over a chemical vessel in which he was brewing a particularly malodorous product. His head was sunk upon his breast, and he looked from my point of view like a strange, lank bird, with dull grey plumage and a black top-knot.

"So, Watson," said he, suddenly, "you do not propose to invest in South African securities?"

I gave a start of astonishment. Accustomed as I was to Holmes's curious faculties, this sudden intrusion into my most intimate thoughts was utterly inexplicable.

"How on earth do you know that?" I asked.

He wheeled round upon his stool, with a steaming test-tube in his hand and a gleam of amusement in his deep-set eyes.

"Now, Watson, confess yourself utterly taken aback," said he.

"I am."

"I ought to make you sign a paper to that effect."

"Why?"

"Because in five minutes you will say that it is all so absurdly simple."

"I am sure that I shall say nothing of the kind."

"You see, my dear Watson"—he propped his test-tube in the rack and began to lecture with the air of a professor addressing his class —"it is not really difficult to construct a series of inferences, each dependent upon its predecessor and each simple in itself. If, after doing so, one simply knocks out all the central inferences and presents one's audience with the starting-point and the conclusion, one may produce a startling, though possibly a meretricious, effect. Now, it was not really difficult, by an inspection of the groove between your left forefinger and thumb, to feel sure that you did *not* propose to invest your small capital in the goldfields."

"I see no connection."

"Very likely not; but I can quickly show you a close connection. Here are the missing links of the very simple chain: 1. You had chalk between your left finger and thumb when you returned from the club last night. 2. You put chalk there when you play billiards to steady the cue. 3. You never play billiards except with Thurston. 4. You told me four weeks ago that Thurston had an option on some South African property which would expire in a month, and which he desired you to share with him. 5. Your cheque-book is locked in my drawer, and you have not asked for the key. 6. You do not propose to invest your money in this manner."

"How absurdly simple!" I cried.

"Quite so!" said he, a little nettled. "Every problem becomes very childish when once it is explained to you. Here is an unexplained one. See what you can make of that, friend Watson." He tossed a sheet of paper upon the table and turned once more to his chemical analysis.

Note. From The Illustrated Sherlock Holmes (p. 369) by A. C. Doyle, 1985, New York: Clarkson N. Potter, Inc.

Acknowledgments

As with the previous three editions of this book, the work is the product of a community that includes students at the University of Colorado in Boulder, current and former members of the Neurophysiology of Movement Laboratory, and colleagues who share a passion for the neuromechanics of movement. I am indebted, in particular, to critical contributions by four individuals: Professor Jacques Duchateau, who is a source of inspiration and a helpful editor; Carolyn Barry, who drew most of the new figures; Joel Enoka, who assisted with many editorial details; and Kenji Narazaki, who scrutinized much of the text for errors. To these individuals, and to those of you who assisted in other ways, many thanks.

part I

The Force–Motion Relation

Movement has long been a source of curiosity for individuals from numerous disciplines. Since the pioneering work of Aristotle (384-322 BC), Borelli (1608-1679), Marey (1830-1904), Sherrington (1857-1952), and Bernstein (1896-1996), we have known that the activation patterns used by the nervous system and muscle to produce movement are constrained by the laws of physics. To understand how the nervous system and muscles generate movement, therefore, it is necessary to appreciate the existing physical constraints. Although this text is primarily about the neural control of movement, its foundation is the principles of mechanics. This is called a **neuromechanical** focus.

Part I describes the mechanical interaction between the world in which movement occurs and the body parts that are moved. The discussion includes an introduction to the terms and concepts commonly used to describe motion, a description of the various forces that are involved in human movement, and demonstrations of the biomechanical techniques used to analyze motion. Although many of these aspects of the relation between force and motion are illustrated with a variety of numerical examples that can be bewildering, the student is encouraged to focus on the systematic application of the various equations and methods.

OBJECTIVES

The goal of this text is to describe movement as the interaction of the human body with the physical world in which we live. In part I, the aim is to define the biomechanics of human movement. The specific objectives are

- to describe movement in precise, well-defined terms;
- to define force;
- to consider the role of force in movement; and
- to demonstrate the biomechanical techniques that are used to analyze movement.

chapter 1

Describing Motion

Although it is not difficult to appreciate the aesthetic qualities or the difficulty of a movement such as a triple-twisting, backward one-and-a-half somersault dive, it is another matter to describe the movement in precise terms. The accurate and precise description of human movement is accomplished by the use of the terms *position, velocity,* and *acceleration.* Such a description of motion, one that ignores the causes of motion, is known as a **kinematic** description. These kinematic terms are often used in everyday language, but without concern for or knowledge of their precise meanings. In biomechanics, as in any scientific endeavor, the observations and principles that emerge are only as good as the concepts and definitions on which they are based. The complexity of movement makes it important, indeed crucial, that our analyses rely on the rigorous definitions of these motion descriptors. To emphasize this need, we precede our discussion of kinematics with some reminders on the essentials of measuring physical quantities.

MEASUREMENT RULES

The scientific method requires that we agree on a measurement system and that we perform calculations with sufficient accuracy to have confidence in the result. The most commonly used measurement system is the metric system. The international metric system is known as the SI system (for Le Système Internationale d'Unites), comprising seven independent base units from which all other units of measurement are derived (appendix A).

SI System

For part I of the book, we are mainly interested in the fundamental quantities of length, mass, and time and their derivatives. **Length** is measured in meters (m), with one meter defined as the length of the path traveled by light in a vacuum during a time interval of 1/299,792,458 of a second. **Mass** is measured in kilograms (kg), with one kilogram defined as the quantity of matter contained in the reference preserved at the International Bureau of Weights and Measures at Sèvres, France. **Time** is measured in seconds (s), with one second determined by an atomic clock as the duration of 9,192,631,770 periods of the radiation corresponding to the transition between the two hyperfine levels of the ground state of the cesium-133 atom.

In addition to the seven base units of measurement, there is a supplementary unit to measure angle. The unit of measurement is the radian (rad). As shown in figure 1.1, an angle is defined as the ratio of an arc length (s) to the radius of a circle (r). When the ratio has a value of 1, then the angle (θ) is equal to one radian (~57.3°). To become familiar with measuring angles in radians, recognize that a right angle is equal to 1.57 rad, that when the arm is extended the

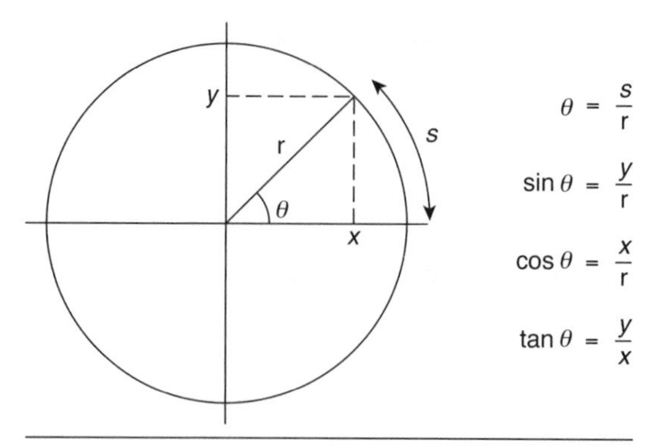

Figure 1.1 Definition of a radian.

elbow joint is at an angle of 3.14 rad, and that one complete circle is 2π rad.

Because it is convenient for numbers to range from 0.1 to 9999, prefixes (table 1.1) can be attached to the units of measurement to represent a smaller or larger amount of the unit. For example, to express long distances (e.g., marathon—26 miles, 385 yards), it is more economical to refer to thousands of meters. The prefix *kilo* (k) represents 1000; thus, 1 km = 1000 m. A marathon race, therefore, covers a distance of approximately 42.2 km. Similarly, prefixes referring to parts of a meter can be used for small distances. For example, there are 1,000,000 micrometers (μm) in one meter (1 μm = 0.000,001 m). The diameter of a muscle fiber is most appropriately expressed in micrometers, with a typical value of 55 μm.

Table 1.1 Prefixes Used With SI Units of Measurement

Prefix	Symbol	Power
yetta	Y	1,000,000,000,000,000,000,000,000 = 10^{24}
zetta	Z	1,000,000,000,000,000,000,000 = 10^{21}
exa	E	1,000,000,000,000,000,000 = 10^{18}
peta	P	1,000,000,000,000,000 = 10^{15}
tera	T	1,000,000,000,000 = 10^{12}
giga	G	1,000,000,000 = 10^{9}
mega	M	1,000,000 = 10^{6}
kilo	k	1,000 = 10^{3}
hecto	h	100 = 10^{2}
deca	da	10 = 10^{1}
—	—	1 = 10^{0}
deci	d	0.1 = 10^{-1}
centi	c	0.01 = 10^{-2}
milli	m	0.001 = 10^{-3}
micro	μ	0.000,001 = 10^{-6}
nano	n	0.000,000,001 = 10^{-9}
pico	p	0.000,000,000,001 = 10^{-12}
femto	f	0.000,000,000,000,001 = 10^{-15}
atto	a	0.000,000,000,000,000,001 = 10^{-18}
zepto	z	0.000,000,000,000,000,000,001 = 10^{-21}
yocto	y	0.000,000,000,000,000,000,000,001 = 10^{-24}

Changing Units

Sometimes it is necessary to change the unit of measurement from another system (e.g., English units) to the SI system. Appendix A lists SI units and a number of common conversion factors for this purpose, and appendix B lists selected conversion factors. To convert a quantity from one measurement system to another, treat the units as arithmetic quantities. For example, to convert your height (say 5 ft 8 in.) to SI units, get the conversion factor from appendix B and then perform the conversion:

$$(68 \text{ in.}) \bullet \left(\frac{0.0254 \text{ m}}{1 \text{ in.}}\right) = 1.73 \text{ m}$$

Similarly, to convert a speed from miles per hour (mph) to meters per second (m/s), the conversion would proceed as follows:

$$(65 \text{ mph}) \bullet \left(\frac{1 \text{ h}}{3600 \text{ s}}\right) \bullet \left(\frac{1609 \text{ m}}{1 \text{ mile}}\right) = 29.1 \text{ m/s}$$

The advantage of using these procedures is that you will be less likely to invert a conversion factor if you pay attention to canceling units. If you are not familiar with SI units, then become acquainted by remembering reference values (e.g., height, weight) rather than memorizing conversion factors. For example, to judge whether or not a movement is fast, compare the speed of the movement to the average speed (10 m/s) of a person who runs 100 m in 10 s. Similarly, remember your height in meters as a reference for distance and your body mass in kilograms as a standard for mass.

When converting units of measurement for area, remember that the conversion factor must be squared. For example, to convert cm^2 to m^2, visualize a square that has sides of 1 m in length, which means 100 cm along each side. The area of the square is 100×100 cm^2 or 1 m^2. Thus, there are 10,000 cm^2 in 1 m^2; that is, 1 $m^2 = 100^2$ cm^2. Similarly, when we are converting units of measurement for volume, the conversion factor must be raised to a power of three.

Accuracy and Significant Figures

When we measure a physical property of an object, we obtain an estimate of its true value. The closeness of the estimate to the true value indicates the **accuracy** of the measurement. The accuracy depends on the resolution of the measurement device. For example, if we measure the body mass of an individual whose actual mass is 79.25 kg, then we need a scale that can measure one-hundredths of a kilogram to get an accurate estimate. The digits in a number that indicate the accuracy of a measurement are known as the **significant figures.** The number 79.25 has four significant figures, 79.3 has three, and 79 has two. In biomechanics, it is common to use three significant figures for most measurements.

Two practices concerning significant figures are usually followed when performing calculations. First, for adding or subtracting, the number of digits to the right of the decimal point in the answer should be the same as for the term in the calculation with the least number of digits to the right of the decimal point. Similarly, for multiplying or dividing, the answer should have the same number of significant figures as the least accurate term in the calculation. Second, when calculations involve small differences, greater accuracy is required in order to estimate the difference to three significant figures. For example, if one group of subjects took 1.2503 s to perform a movement and another group took 1.2391 s, then the difference between the two groups would be 0.0112 s. To find the difference to three significant figures (0.0112), it was necessary to measure movement time to five significant figures.

MOTION DESCRIPTORS

Movement involves the shift from one position to another. It can be described in terms of the size of the shift (displacement) and the rate at which it occurs (velocity and acceleration).

The **position** of an object refers to its location in space relative to some baseline value or axis. For example, the term *3 m diving board* indicates the position of the board above the waterline. Similarly, the height of a man is specified as the distance from his feet to his head; the position of the finish line in a race is indicated with respect to the start; the third and fifth positions in ballet refer to the position of one foot relative to the other; and so on. When an object experiences a change in position, it has been displaced and **motion** has occurred. Motion cannot be detected instantaneously; rather, detection relies on the comparison of the object's position at one instant in time with its position at another instant. *Motion, therefore, is an event that occurs in space and time.*

When an object is described as experiencing a **displacement,** the reference is to the spatial (space) element of motion, that is, the change in position of the object. Alternatively, an account of both the spatial and temporal (time) elements of motion involves the term *velocity* or *speed.* Velocity tells how fast and in what direction, whereas speed defines how fast. Because it has both magnitude and direction, **velocity** is a vector quantity that defines the change in position with respect to time. It indicates how rapidly the change in position occurred and the direction in which it took place. **Speed** is simply the magnitude of the velocity vector and, as such, does not concern the direction of the displacement. Because displacement refers to a change in position, velocity can be described as the time rate (derivative) of displacement. In calculus terminology, velocity is the derivative of position with respect to time.

Figure 1.2 represents two observations, separated in time by 3 s, of the vertical position of an object above some baseline value. The change in vertical position over this 3 s period was 2 m; therefore, the rate of change in position was 2 m in 3 s, that is, 2 m/3 s or 0.67 m/s. Thus the average velocity of the object moving from position 1 to position 2 is 0.67 m/s, where m/s refers to meters per second (this unit of measurement can also be expressed as $m \cdot s^{-1}$). Stated more explicitly,

$$\text{Velocity} = \frac{\Delta \text{ position}}{\Delta \text{ time}} \tag{1.1}$$

where Δ (delta) indicates a *change* in some parameter. Graphically, velocity refers to the slope of the position–time graph. Because a line graph (such as figure 1.2) depicts the relation between two (or sometimes more) variables, *a change in the slope of the line as it becomes more or less steep indicates a change in the relation between the variables.* We determine the slope of the line numerically by subtracting an initial-position value from some final position (Δ position) and dividing the change in position by the amount of time it took for the change to occur (Δ time). **Slope,** therefore, refers to the rate of change in a variable such that the steeper the slope, the greater the rate of change, and conversely, the lesser the slope, the slower the rate of change.

Throughout this text, many concepts are presented in the form of graphs. Typically, a graph shows the relation between at least two variables. Figure 1.2 shows the relation between position and time. The relation can be indicated as the line, or data points, plotted on the graph. The main feature of a graph is to show the trend or pattern of the relation between the variables; more precise quantitative data are presented as tables or sets of numerical values. In evaluating a graph, first determine the variables involved (i.e., those on the axes) and then examine the relation between the variables. The relation between position and time in figure 1.2 is relatively straightforward; it can be represented by a single measurement, the slope of the line.

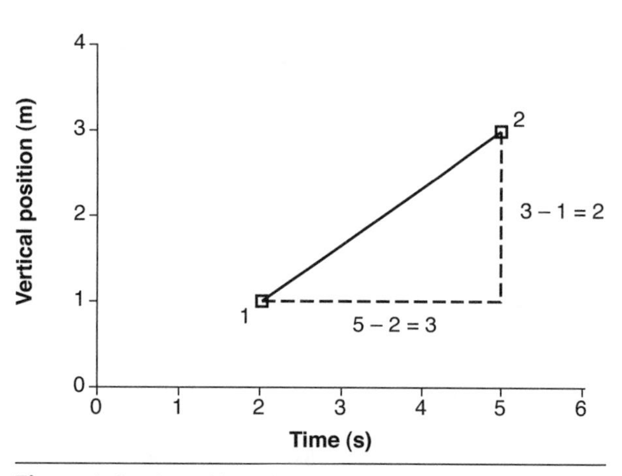

Figure 1.2 A position–time graph.

Vertical displacement can vary not only in *magnitude* (i.e., size) but also in *direction* (i.e., up-down). Figure 1.3 illustrates some of these alternatives by showing the position of an object at five instances in time. Use of equation 1.1 produces velocities of 0.75, 1.50, 0, and −1.00 m/s for movement of the object from position 1 to 2, from position 2 to 3, from position 3 to 4, and from position 4 to 5, respectively. These values are plotted in figure 1.4. Velocity increases with the steepness of the slope of the position–time graph (e.g., position 2 to 3 = 1.50 m/s vs. position 1 to 2 = 0.75 m/s) (figure 1.4). Conversely, a downward slope (e.g., position 4 to 5) indicates a negative velocity. The absence of a change in position (e.g., position 3 to 4) represents a zero velocity (i.e., no change in position). This example illustrates an important point about velocity: When the sign of the velocity value (positive, negative, or zero) changes, the movement has changed direction. Furthermore, when the direction of a movement changes, the velocity–time graph must pass through zero. Figure 1.4 indicates that an object initially moved in one direction (arbitrarily called the positive direction—note the positive slope in figure 1.3), then was stationary (zero velocity), and finally moved in the other direction (negative slope). Because velocity has both magnitude and direction, it is a vector quantity.

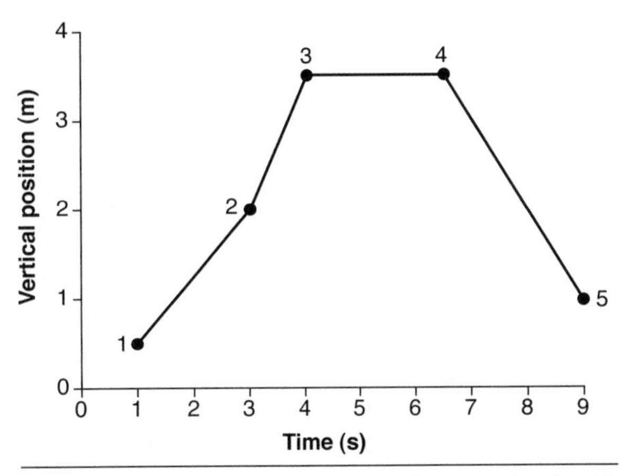

Figure 1.3 The variation in velocity associated with unequal changes in magnitude and direction in a position–time graph.

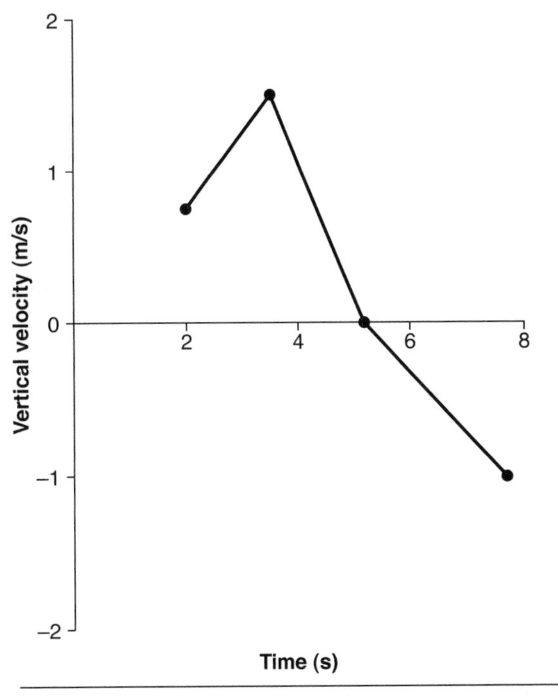

Figure 1.4 Average velocities of the displacements shown in figure 1.3.

It is not sufficient to describe motion only in terms of the occurrence and rate of a displacement. For example, a ball held 1.23 m above the ground and then dropped will reach the ground 0.5 s later. The change in position is 1.23 m, and the average velocity is 2.46 m/s (i.e., 1.23 m/0.5 s). But the ball does not travel with a constant velocity; the velocity changes over time. Starting with a zero velocity at release, the speed of the ball increases to a value of 4.91 m/s just prior to contact with the ground. This rate of change in velocity is referred to as **acceleration;** that is, acceleration is the derivative of velocity with respect to time or the second derivative of position with respect to time. The acceleration that the ball experiences while it falls is constant and has a value of 9.81 m/s^2. If the velocity of an object is measured in meters per second (m/s), then acceleration indicates the change in meters per second each second (m/s^2). Consequently,

$$\text{Acceleration} = \frac{\Delta \text{ velocity}}{\Delta \text{ time}} \quad (1.2)$$

Because velocity is defined as the change in position with respect to time, velocity can be represented graphically as the slope of the position–time graph. Similarly, acceleration can be indicated as the slope of the velocity–time graph. For example, suppose figure 1.3 was relabeled as a vertical velocity–time graph. If point 2 had the coordinates of 2.0 m/s and 3 s and point 3 had the coordinates of 3.5 m/s and 4 s, the rate of change from point 2 to point 3 (the acceleration) could be calculated with equation 1.2 as follows:

$$\text{Acceleration} = \frac{3.5 - 2.0 \text{ m/s}}{4 - 3 \text{ s}}$$
$$= 1.5 \text{ m/s/s (or m/s}^2 \text{ or m}\cdot\text{s}^{-2})$$

Similarly, the acceleration from points 1 to 2, from points 3 to 4, and from points 4 to 5 would be 0.75, 0, and –0.83 m/s², respectively. As with velocity, acceleration can have both magnitude and direction and thus is a vector quantity.

The acceleration experienced by the ball when it was dropped from a height of 1.23 m was due to the gravitational attraction between two masses, planet Earth and the ball. The force of **gravity** produces a constant acceleration of approximately 9.81 m/s² at sea level; this is usually written as –9.81 m/s² to indicate the downward direction. In general, an object acted upon by a force will be accelerated. A constant force (i.e., gravity) applied to an unsupported object produces a constant acceleration; conversely, the absence of a force means that the object is at rest or traveling at a constant velocity (i.e., no acceleration). Because acceleration can be depicted as the slope of the velocity–time graph, it should be possible to visualize the shape of the velocity–time graph when an object is accelerating and when it is not. When the object accelerates, the slope of the velocity–time graph is nonzero. Conversely, when the object does not accelerate, the velocity–time graph has a zero slope.

EXAMPLE 1.1

Kinematics of the 100 m Sprint

As an example of the relations among position, velocity, and acceleration, consider the kinematics of a person running the 100 m sprint as fast as possible. When a person performs this event, the displacement (the horizontal difference in position between the start and the finish) is 100 m. To describe the kinematics of the performance, however, we need to determine the position of the runner at various times during the race. One way to do this is to videotape the runner and measure the displacement of the hip joint. If the frame rate of the video camera is set at 10 frames per second, then we can determine the position of the runner every 100 ms during the race. If the person ran the 100 m in 10.8 s, then we would have about 109 measurements of position along the 100 m track, each at a known point in time. We can then plot these data on a position–time graph.

In a position–time graph, the change in position between any two consecutive data points represents the displacement that occurs during the selected interval. Is the displacement of the runner for each time interval during a 100 m race constant? If it were, then the shape of the position–time graph would be a straight line. The data in figure 1.5*a* show that the distance–time relation is not quite a straight line, especially at the beginning of the race. This deviation from linearity is amplified in the velocity– and acceleration–time graphs, which are derived from the position–time graph.

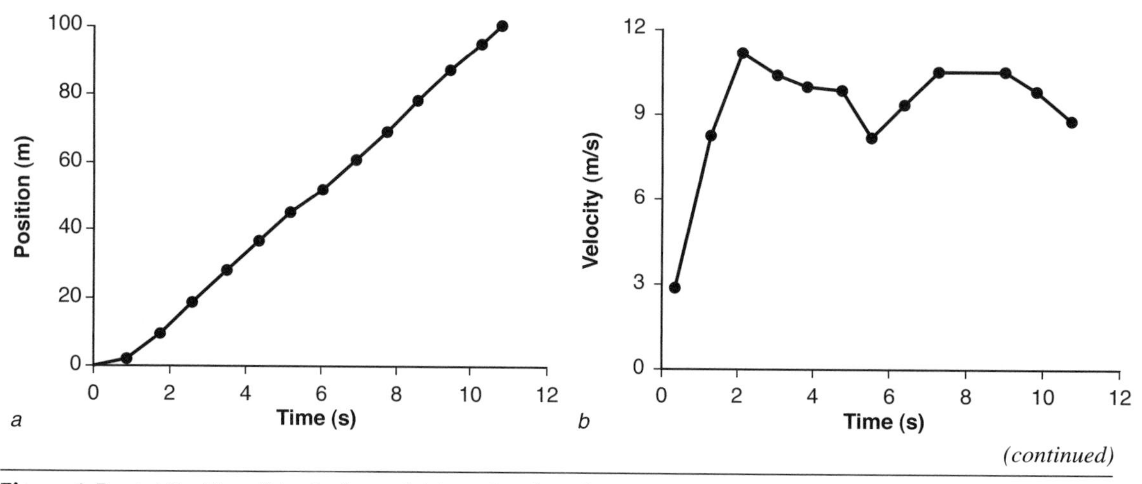

(continued)

Figure 1.5 *(a)* Position, *(b)* velocity, and *(c)* acceleration of a runner during a 100 m sprint.

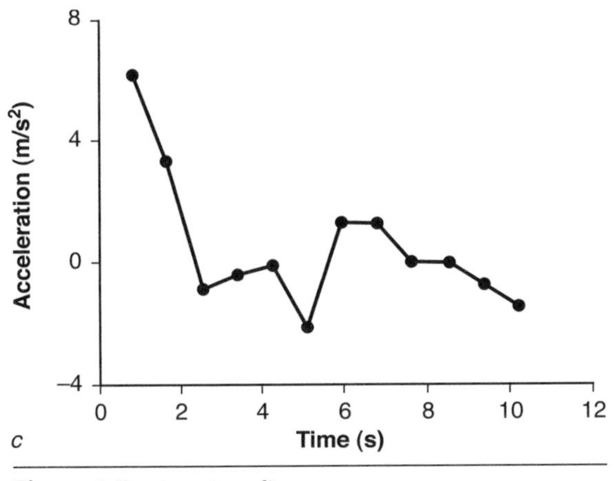

Figure 1.5 *(continued)*

From figure 1.3, we know that the slope of the position–time graph tells us about the velocity of the runner. Before calculating velocity, however, let's think about how we would expect the velocity of the runner to change during the race. It seems reasonable to expect that at the beginning of the race the runner's velocity would increase from zero to some maximum value that the runner would attempt to sustain. Look now at the position–time data in figure 1.5*a:*The slope starts off low, then increases, and finally becomes constant. With the application of equation 1.1 to each interval, we can determine the velocity–time data for the runner and then graph these data in figure 1.5*b*. Similarly, we can derive the acceleration–time graph for the runner. The expected shape of the graph can be estimated from the slope of the velocity–time graph. When velocity changes a lot in an interval, then the runner's acceleration is relatively large for that interval. Conversely, when there is no change in velocity in an interval, the runner does not accelerate. Based on equation 1.2, the acceleration–time data for each interval can be determined and graphed as in figure 1.5*c*.

CONSTANT ACCELERATION

When acceleration is constant, motion can be described with some simple equations. These equations can be used to find the velocity of an object or the distance it has moved after it has experienced a certain amount of acceleration. From the elementary definitions of velocity (equation 1.1) and acceleration (equation 1.2), we can derive algebraic expressions involving time (t), position (r), velocity (v), and acceleration (a). In these equations of motion, v_i and v_f refer to initial and final velocities, respectively, at the beginning and end of selected time intervals; r_i and r_f similarly refer to the initial and final positions for each interval; and t indicates the duration of the interval. Because acceleration is constant, the average ($\bar{a}$) and instantaneous (a) values are the same and yield the following:

$$a = \frac{\Delta \text{ velocity}}{\Delta \text{ time}}$$

$$a = \frac{v_f - v_i}{t}$$

$$v_f = v_i + at \tag{1.3}$$

As described previously, a ball dropped from a height of 1.23 m will reach the ground 0.5 s later with a final velocity of –4.9 m/s. According to equation 1.3, the variables that influence final velocity are the initial velocity (v_i) of the ball, its acceleration (a), and the duration of the fall (t). In this example, v_i is zero, a is the acceleration due to gravity (–9.81 m/s^2), and t is 0.5 s. Suppose you want to determine the final velocity of a pitched baseball as it crosses the plate. What would you need to know? As with the ball-drop example, you would need to know v_i, a, and t. The major difficulty would be in determining a, because other forces, in addition to gravity, would influence the motion of the ball and cause a to vary as a function of time. Under such conditions, we can determine a by deriving it numerically from video images of a movement. Alternatively, a can be measured directly with an instrument known as an accelerometer.

We can use a similar approach to determine how far an object will be displaced (Δr) after a given amount of acceleration. The equation is derived from our definition of velocity:

$$\text{Average velocity} = \frac{\Delta \text{ position}}{\Delta \text{ time}}$$

$$\frac{v_f + v_i}{2} = \frac{r_f - r_i}{t}$$

Substitute equation 1.3 for v_f:

$$r_f - r_i = \left(\frac{v_i + at + v_i}{2}\right)t$$

$$r_f - r_i = \left(\frac{2v_i + at}{2}\right)t$$

$$r_f - r_i = v_i t + \tfrac{1}{2}at^2 \qquad (1.4)$$

This equation indicates that the change in position (displacement) of an object depends on three variables: its initial velocity (v_i), the acceleration (a) it experiences, and time (t). This relation can be used to determine the change in position of an object during a movement. For example, consider an individual who dives off a 10 m tower; by varying the value of t from 0 to 1.5 s in 0.1 s increments, we can determine the change in position ($r_f - r_i$) for each 0.1 s interval and thereby obtain the trajectory (position–time graph) of the diver during the performance. When the initial velocity of the diver is zero, as in this example, equation 1.4 reduces to

$$r_f - r_i = \tfrac{1}{2}at^2$$

If we assume that the effects of air resistance are so small that we can ignore them, then the acceleration is simply that due to gravity and is constant during the dive. We can determine the set of position–time data that represents the trajectory of the diver by doing a number of calculations using equation 1.4 and incrementing the value of t each time by 0.1 s.

In a similar way we can derive an equation to determine the final velocity as a function of displacement (equation 1.5) rather than time (equation 1.4).

$$\text{Average velocity} = \frac{\Delta \text{ position}}{\Delta \text{ time}}$$

$$\frac{v_f + v_i}{2} = \frac{r_f - r_i}{t}$$

In this expression, t is unknown so we rearrange equation 1.4 to express t as the dependent variable [$t = (v_f - v_i) / a$] and substitute for t.

$$\frac{v_f + v_i}{2} = \frac{r_f - r_i}{(v_f - v_i)/a}$$

$$\frac{v_f + v_i}{2} = (r_f - r_i)\,\frac{a}{v_f - v_i}$$

$$2a(r_f - r_i) = (v_f + v_i)(v_f - v_i)$$

$$2a(r_f - r_i) = (v_f^2 - v_i^2)$$

$$v_f^2 = v_i^2 + 2a(r_f - r_i) \qquad (1.5)$$

When the initial velocity for an interval is zero, such as for the object beginning at rest, then the equations can be simplified:

$$v_f = at$$

$$r_f - r_i = \tfrac{1}{2}at^2$$

$$v_f^2 = 2a(r_f - r_i)$$

EXAMPLE 1.2

Penalty Kick in Soccer

When a player takes a penalty kick in soccer, the goalie stands stationary in the middle of the goal (7.32 m wide) with the ball placed on the penalty spot (11 m in front of the goal). Suppose that a player made a penalty kick such that the ball left her foot with an initial velocity of 63 mph and traveled along the ground into the goal just inside the goalpost with a final velocity of 54 mph.

A. What was the initial and final velocity of the ball in SI units?

$$v_i = 63 \text{ mph} \times 0.447 \frac{\text{m/s}}{\text{mph}} = 28.2 \text{ m/s}$$

$$v_f = 54 \text{ mph} \times 0.447 \frac{\text{m/s}}{\text{mph}} = 24.1 \text{ m/s}$$

B. What was the length of the path traveled by the ball from the penalty spot to the goal?

$$r_f - r_i = \sqrt{\left(\frac{7.32}{2}\right)^2 + 11^2}$$

$$\Delta r = 11.6 \text{ m}$$

C. What was the average acceleration of the ball between the initial and final velocities?

$$v_f^{\,2} = v_i^{\,2} + 2a(r_f - r_i)$$

$$a = \frac{v_f^{\,2} - v_i^{\,2}}{2(\Delta r)}$$

$$a = \frac{24.1^2 - 28.2^2}{2 \times 11.6}$$

$$a = -9.24 \text{ m/s}^2$$

D. How much time did the goalie have to reach the ball from the moment it left the player's foot until it entered the goal?

$$\text{Average velocity} = \frac{\Delta \text{ position}}{\Delta \text{ time}}$$

$$\frac{v_i + v_f}{2} = \frac{\Delta r}{t}$$

$$t = \frac{\Delta r}{\bar{v}}$$

$$t = \frac{11.6}{\left(\frac{28.2 + 24.1}{2}\right)}$$

$$t = 0.444 \text{ s}$$

EXAMPLE 1.3

Calculation of Velocity and Acceleration

Kinematic analyses are usually based on a set of position–time data that are obtained with a recording device such as a video camera. A video record of a movement represents a set of still images (frames) that are subsequently projected individually onto a measuring device, and the locations of selected landmarks with respect to some reference are determined. The instrument used in this procedure, called a **digitizer,** is capable of determining the *x-y* coordinates of the selected landmarks. Once we have a set of position–time data, we can use equations 1.1 and 1.2 to determine the average velocity and acceleration between each position measurement. Table 1.2 provides an example of this procedure.

Table 1.2 Calculation of Velocity and Acceleration From a Set of Position–Time Data

	Position (m)	Time (s)	Velocity (Δ position/Δ time) (m/s)	Acceleration (Δ velocity/Δ time) (m/s²)
1	0.00	0.000		
		0.050	(0.59 – 0.00) / (0.100 – 0.000) = 5.9	
2	0.59	0.100		(3.6 – 5.9) / (0.150 – 0.050) = –23.0
		0.150	(0.95 – 0.59) / (0.200 – 0.100) = 3.6	
3	0.95	0.200		(1.0 – 3.6) / (0.255 – 0.150) = –34.7
		0.225	(1.00 – 0.95) / (0.250 – 0.200) = 1.0	
4	1.00	0.250		(–1.0 – 1.0) / (0.275 – 0.225) = –40.0
		0.275	(0.95 – 1.00) / (0.300 – 0.250) = –1.0	
5	0.95	0.300		(–3.6) – [–1.0]) / (0.350 – 0.275) = –34.7
		0.350	(0.59 – 0.95) / (0.400 – 0.300) = –3.6	
6	0.59	0.400		(–5.9 – [–3.6]) / (0.450 – 0.350) = –23.0
		0.450	(0.00 – 0.59) / (0.500 – 0.400) = –5.9	
7	0.00	0.500		(–5.9 – [–5.9]) / (0.550 – 0.450) = 0.0
		0.550	(–5.9 – 0.00) / (0.600 – 0.500) = –5.9	
8	–0.59	0.600		(–3.6 – [–5.9]) / (0.650 – 0.550) = 23.0
		0.650	(–0.95 – [–0.59]) / (0.700 – 0.600) = –3.6	
9	–0.95	0.700		(–1.0 – [–3.6]) / (0.725 – 0.650) = 34.7
		0.725	(–1.00 – [–0.95]) / (0.750 – 0.700) = –1.0	
10	–1.00	0.750		(1.0 – [–1.0]) / (0.775 – 0.725) = 40.0
		0.775	(–0.95 – [–1.00]) / (0.800 – 0.750) = 1.0	
11	–0.95	0.800		(3.6 – 1.0) / (0.850 – 0.775) = 34.7
		0.850	(–0.59 – [–0.95]) / (0.900 – 0.800) = 3.6	
12	–0.59	0.900		(5.9 – 3.6) / (0.950 – 0.850) = 23.0
		0.950	(0.00 – [–0.59]) / (1.000 – 0.900) = 5.9	
13	0.00	1.000		

The 13 position values listed in table 1.2, each recorded at a different instant in time, represent the vertical path that an object traveled over a 1 s interval. The object first rose above an initial position (0.0 m) to a height of 1.0 m before being displaced by an equal amount (–1.0 m) below the original position and finally returning to 0.0 m. We calculate the velocity of the object during this motion by applying equation 1.1 to the intervals of time for which position information is available. For example, from table 1.2 we can select the intervals of

0.0 to 1.0 s, 0.0 to 0.25 s, or 0.0 to 0.1 s. If we applied equation 1.1 to each of the intervals, the average velocity for each interval would be as follows:

$$0.0-1.0=\frac{0.0-0.0}{1.0-0.0}=0 \text{ m/s}$$

$$0.0-0.25=\frac{1.0-0.0}{0.25-0.0}=4 \text{ m/s}$$

$$0.0-0.1=\frac{0.59-0.0}{0.1-0.0}=5.9 \text{ m/s}$$

Similarly, if we measure the position at the times of 0.0, 0.5, and 1.0 s, we will get the impression that the object did not move. Clearly, the smaller the intervals of time we measure, the more closely the calculated velocity will match that experienced by the object. However, economy of effort and measurement error suggest that we do not want to measure too frequently and that there must be some intermediate value. Most human movements can be measured adequately with frame rates that range from 50 to 100 frames per second, which correspond to intervals of 0.01 to 0.02 s between consecutive data points.

Table 1.2 shows velocity as determined for each interval between the position data. For example, the displacement during the first interval (0.59 – 0.0 = 0.59 m) was divided by the time elapsed during the interval (0.1 – 0.0 = 0.1 s) to yield the velocity for that interval (0.59 / 0.1 = 5.9 m/s). The calculated value (5.9 m/s) represents the *average* velocity over that interval and consequently is recorded at the midpoint in time of the interval (0.05 s). Similarly, the first acceleration value (–23.0 m/s^2), which was determined with equation 1.2, is listed at the midpoint in time (0.10 s) of the first velocity interval (0.05 to 0.15). By this procedure *the average value of velocity is determined for each position interval and the average acceleration is calculated for each velocity interval.* Thus, from a set of 13 position–time observations, we calculate 12 velocity–time and 11 acceleration–time values.

The graphical relation between a motion descriptor (e.g., position and velocity) and its rate of change has already been mentioned. Further evidence of these relations is provided in table 1.2. When position increases (positions 1 to 4 and 10 to 13), velocity is positive; when position decreases (positions 4 to 10), velocity is negative. A similar dependency exists between the slope (increase or decrease) of the velocity–time graph and the sign of the acceleration values. These relations are shown in figure 1.6.

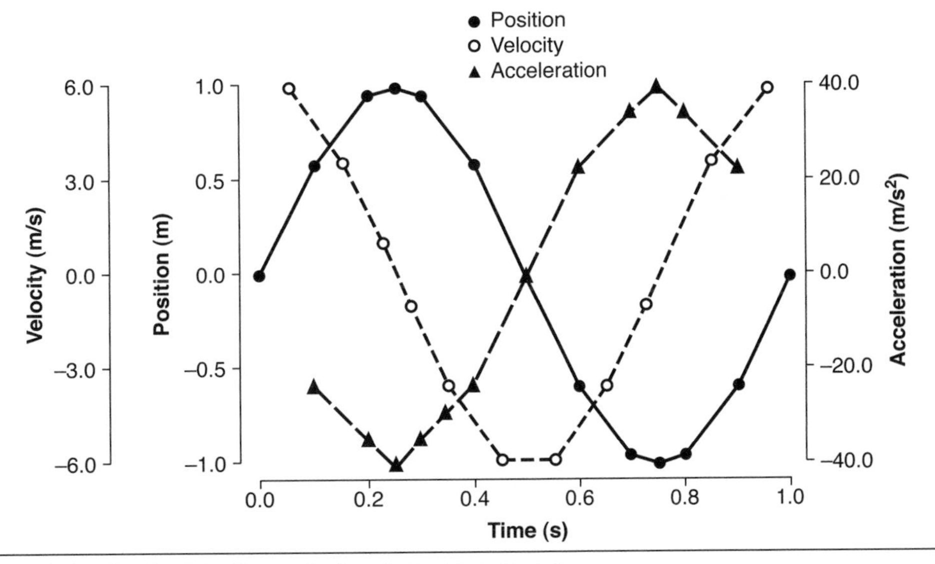

Figure 1.6 Graph of the kinematic data derived in table 1.2.

UP AND DOWN

When an object is thrown, kicked, or hit into the air or when a human jumps into the air, and air resistance is negligible, the motion of the center of mass is predictable and is characterized by a set of rules derived from the equations of motion (equations 1.3, 1.4, and 1.5). Such airborne objects are often referred to as projectiles. Projectile motion has the following characteristics:

- The effect of gravity is to cause the trajectory to deviate from a straight line into a curved path that can be described as a parabola.
- When the release height and landing occur at the same level, the time taken for the object to reach the peak of its trajectory will be identical to the time taken to go from the peak to the landing.
- The vertical velocity of the projectile (v_v) will change from an upward value (positive) at release, to zero at the peak of the trajectory (when it changes direction), to downward (negative) when it returns to the ground.
- The only significant force that the object experiences while in the air will be that due to gravity, and this will cause a vertical acceleration of -9.81 m/s^2.
- Because there is no force acting in the horizontal direction, the horizontal acceleration of the object will be zero, which means that the object's horizontal velocity (v_h) will remain constant. Consequently, the horizontal distance traveled by the object can be determined as the product of horizontal velocity and the flight time of the object.
- The flight time will depend on the vertical velocity at release and the height of release above the landing surface.

Although the trajectory of an object is parabolic, the shape of the parabola depends on the velocity—both magnitude and direction—at release. Three different parabolas are shown in figure 1.7. On the basis of the rules just listed, we can describe the velocity vector at selected instances throughout the trajectory; this is shown for the hammer throw in figure 1.7. The trajectory is characterized by a constant horizontal velocity, a positive vertical velocity on the upward phase, a negative vertical velocity on the downward phase, and zero vertical velocity at the peak of the trajectory. The vertical velocity will have the greatest absolute value at the beginning and the end of the flight. These features apply to such projectiles as a shot during the shot put, a gymnast performing a vault, a high jumper clearing the bar, and a basketball player performing a jump shot.

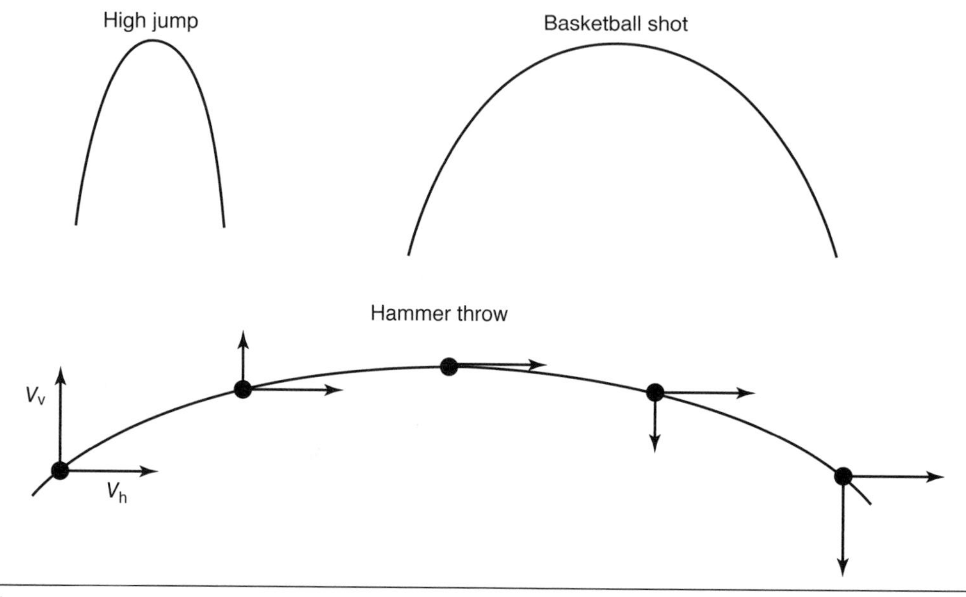

Figure 1.7 The parabolic trajectory experienced by projectiles.

EXAMPLE 1.4

Trajectory of a Ball

Let us consider a ball thrown at an angle of 1.05 rad with respect to the horizontal from a height of 2.5 m above the ground, with a resultant velocity along the line of projection of 6 m/s. For solving problems of projectile motion, it is convenient to begin with a sketch of initial conditions (see figure 1.8).

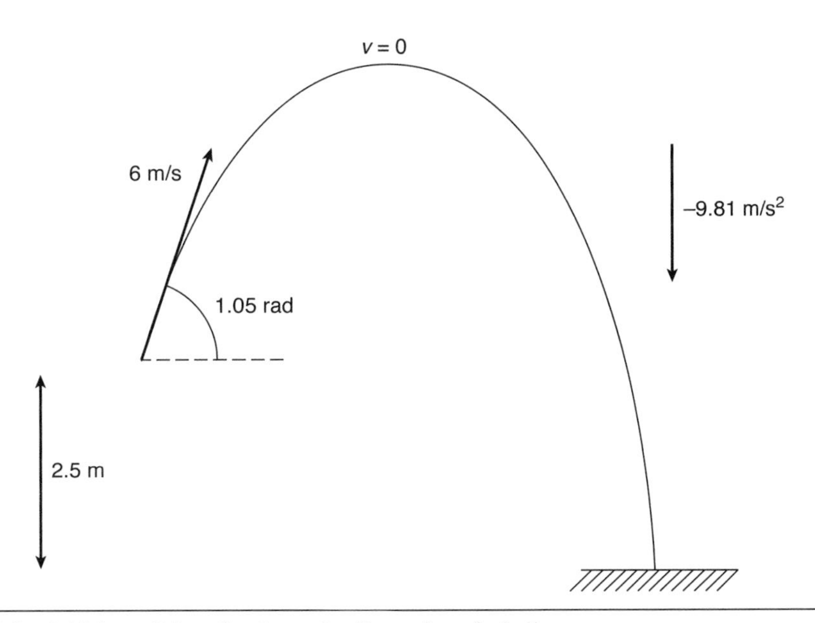

Figure 1.8 Initial conditions for the projectile motion of a ball.

A. How long does the ball take to reach its highest point? The trajectory is parabolic, so the ball will continue going up until the vertical velocity has a value of zero (where $\bar{a}$ = average vertical acceleration = -9.81 m/s^2, and the initial vertical velocity = 6 sin 1.05 = 5.2 m/s).

$$\bar{a} = \frac{\Delta v}{\Delta t}$$

$$\bar{a} = \frac{v_f - v_i}{t}$$

$$-9.81 = (0 - 5.2)/t$$

$$t = -5.2/-9.81$$

$$t = 0.53 \text{ s}$$

B. How high (vertically) does the ball get? The height reached depends on the vertical component of the release velocity plus the height above the ground at which the ball is released. We already know the following: vertical velocity at release (5.2 m/s), vertical velocity at the peak of the trajectory (0 m/s), vertical acceleration (-9.81 m/s^2), release height (2.5 m), time to reach the peak (0.53 s). We also know that average velocity = $\bar{v}$.

$$\bar{v} = \frac{\Delta r}{\Delta t}$$

$$\frac{5.2+0}{2} = \frac{\Delta r}{0.53}$$

$$\Delta r = \frac{0.53(5.2+0)}{2}$$

$$\Delta r = 1.38 \text{ m}$$

$$\text{Total height} = \Delta r + \text{Release height}$$

$$= 1.38 + 2.5$$

$$\text{Height} = 3.88 \text{ m}$$

C. What is the vertical velocity of the ball just before it hits the ground? To answer this question, we consider the second part of the trajectory, from the peak down to contact with the ground. The initial velocity is 0 m/s, and the ball falls 3.88 m from the peak to the ground while it experiences an acceleration of –9.81 m/s^2, which enables us to use equation 1.5:

$$v_f^2 = v_i^2 + 2a(\Delta r)$$

$$v_f^2 = 2 \times -9.81 \times -3.88$$

$$v_f^2 = \pm 76.1 \text{ (m/s)}^2$$

$$v_f = -8.72 \text{ m/s}$$

(Note that negative = downward.)

D. How long does it take the ball to reach the ground from the peak? Because we know v_i, v_f, and a, we can determine the time taken for the ball to reach the ground by using equation 1.3:

$$v_f = v_i + at$$

$$-8.72 = 0 + (-9.81)\,t$$

$$t = \frac{-8.72}{-9.81}$$

$$t = 0.89 \text{ s}$$

E. How long did the ball spend in flight?

$$t = t_{up} + t_{down}$$

$$t = 0.53 + 0.89$$

$$t = 1.42 \text{ s}$$

F. What horizontal distance did the ball travel (how far was it thrown)? In contrast to the previous calculations, this one uses the horizontal, as opposed to the vertical, information. Because we were given the velocity at release, we can determine the horizontal velocity at release (v_i = 6 cos 1.05 = 2.99 m/s). We also know that in the absence of significant air resistance, the horizontal velocity is constant because the ball experiences no horizontal acceleration.

$$\bar{v} = \frac{\Delta r}{\Delta t}$$

$$\frac{v_i + v_f}{2} = \frac{\Delta r}{t}$$

$$\frac{2.99+2.99}{2} = \frac{\Delta r}{1.42}$$

$$\Delta r = 1.42 \times 2.99$$

$$\Delta r = 4.25 \text{ m}$$

This example demonstrates that the time of flight for a projectile depends on the vertical velocity at release (or takeoff) and the release height. The vertical velocity at release determines the height reached by the projectile and the time taken to reach the maximal height. Because parabolic trajectories are symmetrical, the time taken to return to the release height is identical to that taken to reach the peak point. If the landing height is lower than the release height, however, then we must also consider the time taken to travel this extra distance when determining the total flight time.

When the goal of a projectile event is to maximize the horizontal displacement, it is necessary to choose the correct combination of flight time (vertical velocity + release height) and horizontal velocity. This involves selecting the **optimum** angle of release (or takeoff). The necessary angle depends on the relative positions of the release and landing heights. When the release and landing heights are identical, then the optimum angle of release is 0.785 rad (45°). But when the landing height is lower than the release height, the optimum angle of projection is less than 0.785 rad. When the landing height is above the release height, the optimum angle of projection is greater than 0.785 rad. Furthermore, as the velocity of release increases for a particular release height, the more closely the optimum angle approaches 0.785 rad. Similarly, as release height increases while the release velocity remains constant, the optimum angle (below 0.785 rad) of projection gets lower (Hay, 1993).

EXAMPLE 1.5

Takeoff Angle in the Long Jump

An experienced athlete performing a long jump for maximal distance can achieve a takeoff velocity of about 9.95 m/s. The takeoff angle (θ) used by the athlete will determine the maximal height achieved, the time spent in the air, and the horizontal distance (figure 1.9).

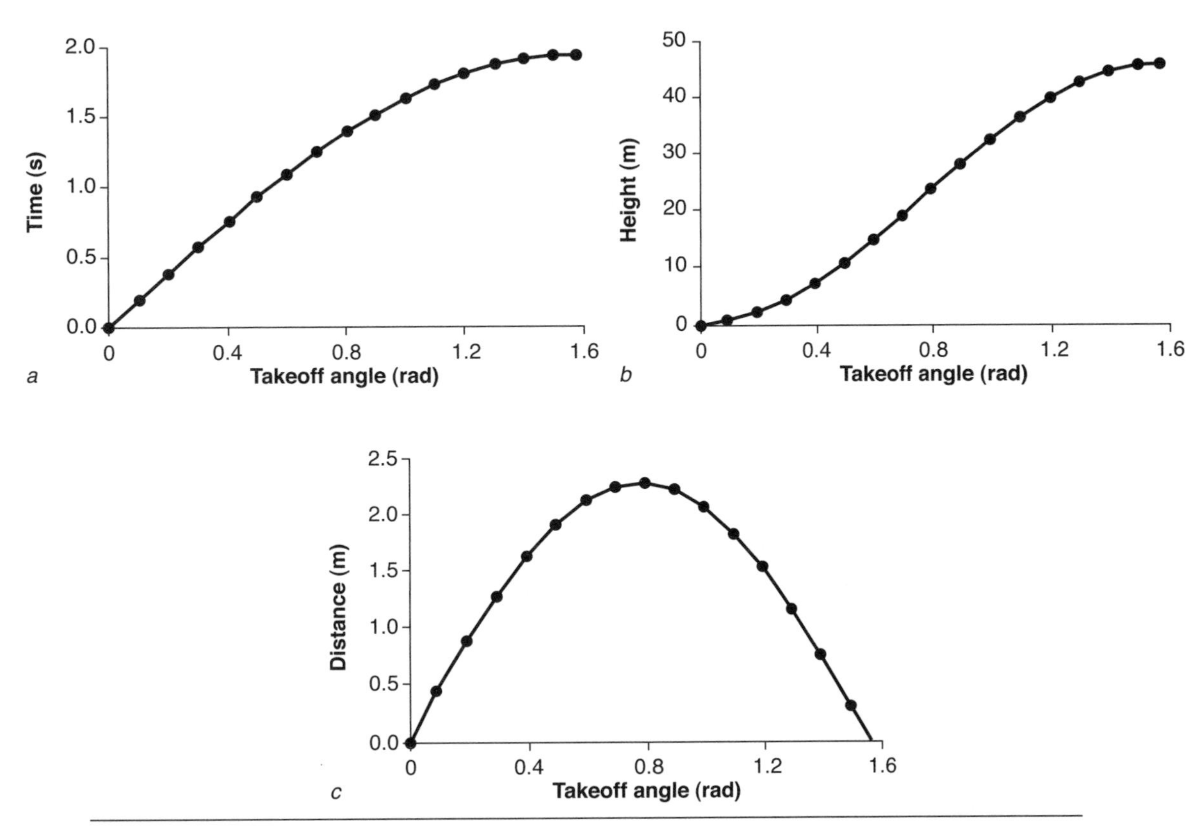

Figure 1.9 *(a)* Flight time, *(b)* vertical height, and *(c)* horizontal distance as a function of takeoff angle for the long jump.

If we assume that the takeoff and landing heights are approximately the same, then the equations for projectile motion can be rearranged to yield

$$\text{Height} = \frac{1}{2a}(v_i \sin\theta)^2$$

$$\text{Flight time} = \frac{2}{a}(v_i \sin\theta)$$

$$\text{Distance} = \frac{2v_i^2}{a}(\sin\theta)(\cos\theta)$$

where a is the magnitude of the acceleration due to gravity and v_i is the takeoff velocity. We can use these equations to plot the effect of takeoff angle on each of the variables. Figure 1.9 indicates that vertical height and flight time are maximal when the takeoff velocity is completely vertical ($\theta = 1.57$ rad). However, the distance jumped is maximal at a takeoff angle of 0.785 rad.

GRAPHIC CONNECTIONS

Because velocity and acceleration are derived from position, the three measures are related when plotted on a graph. Figure 1.10 shows these connections based on the changes in thigh angle of a runner for one stride (defined as one complete cycle, from left foot toe-off to left foot toe-off in this example). Thigh angle is an absolute angle that is measured with respect to the right horizontal, and its measurement is indicated in the upper panel of the figure (the measured angle is for the limb with the filled-in shoe). The angle is measured in *radians* (1 rad = 57.3°), the SI unit for angle.

To show these connections, we graph thigh angle as a function of time and then derive the velocity– and acceleration–time graphs. The first step is to identify the relative *minima* and *maxima* in the position–time graph. Peaks (maxima) and valleys (minima) in the curve should be noted. These points denote instants at which the rate of change has a value of zero. For a small Δt at the peak or the valley the Δr is zero, which indicates that the slope of the graph is neither upward (positive) nor downward (negative), but zero. In figure 1.10, thigh angle reaches one minimum and one maximum during the stride. From these points of zero slope, and thus zero velocity, a perpendicular line is extended to the time axis of the velocity graph to mark the locations in time at which velocity is zero. In figure 1.10 these occur at about 0.03 and 0.30 s, respectively.

The second step in the derivation is to determine the slope of the position graph between these maxima and minima. The slope of the graph will be the same (i.e., positive or negative) between

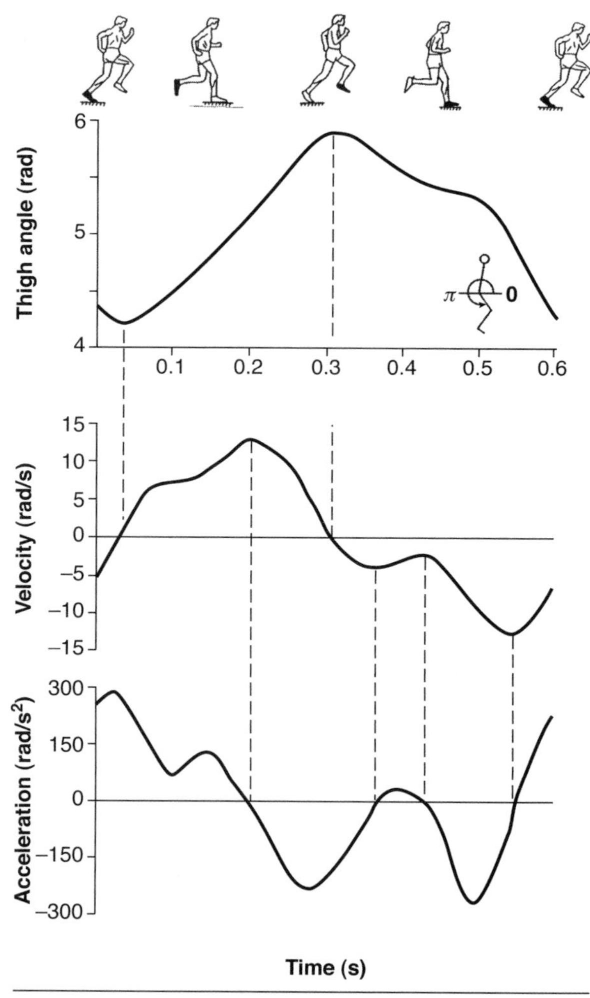

Figure 1.10 The angular velocity as a function of time can be graphically derived from the graph of angular position, and the graph for angular acceleration can be derived from the graph for angular velocity.

these points because, as points of zero velocity, they identify the location in time when the position curve changes its slope (i.e., changes direction). In each interval between the minima and maxima, the slope may become more or less steep, but it will remain either upward (positive) or downward (negative). In the thigh angle graph, there is one minimum and one maximum; there are therefore three such intervals—from the beginning of the movement to the minimum, from the minimum to the maximum, and from the maximum to the end of the movement. The slopes of the position graph associated with these intervals are negative, positive, and negative, respectively. Thus the velocity graph has values (positive or negative) similar to the slope of the thigh angle graph for each interval. For example, for the first interval, from the beginning of the movement to the minimum, both the slope of the position graph and the velocity values are negative. Because a negative velocity value is associated with a downward slope of the position graph, the negative velocities in figure 1.10 indicate a *backward* rotation of the thigh (i.e., a reduction in the measured angle). In total, the velocity graph of figure 1.10 indicates two intervals of backward thigh rotation separated in time by an interval of forward thigh rotation. The variation in the magnitude of the velocity over time indicates how the speed of this rotation varies, whereas the sign (positive or negative) indicates the direction (forward or backward) of rotation.

The derivation of the acceleration graph from the velocity graph is accomplished by the same two-stage procedure: (a) identification of the relative minima and maxima and (b) determination of the slope during the identified intervals. From figure 1.10, the velocity curve contains four minima and maxima, and thus there are four instances at which the acceleration graph crosses zero. The result is an acceleration graph that contains five intervals of alternating positive and negative values. The interpretation of an acceleration graph is generally more complicated than for position and velocity graphs. In figure 1.10, a positive acceleration indicates an acceleration of the thigh in the forward direction; during the first acceleration interval, the thigh rotates first backward and then forward (seen from the velocity graph), but throughout this interval the thigh accelerates in the forward direction. This example indicates that displacement and acceleration do not always act in the same direction. Furthermore, it is not possible to tell the direction of acceleration from the direction of a movement.

For example, consider the motion of a ball that a juggler tosses into the air and then catches. The motion of the ball is represented by a parabolic graph of the vertical position (figure 1.7) of the ball above the juggler's hand. What is the shape of the acceleration–time graph? The answer is simple. We could determine the answer using the technique shown in figure 1.10, but an object in free fall, such as a ball tossed into the air, experiences a constant acceleration of $-9.81\ m/s^2$. Although the ball moves up and down when tossed into the air, this displacement does not provide any intuitive clue about the direction of the acceleration that the ball experiences.

By the graphical technique outlined in this section, we are unable to know the precise magnitude of the rate of change in a variable. This procedure represents a **qualitative analysis;** it merely gives a positive or negative sign for the rate of change and possibly an approximate value. In contrast, table 1.2 indicates a **quantitative analysis** by which we can more accurately determine the values of the derivatives. A qualitative analysis tells us what type or kind, whereas a quantitative analysis tells us how much.

SCALARS AND VECTORS

Quantities that convey magnitude and direction are called **vectors,** for example, displacement, velocity, acceleration, force, momentum, and torque. Those variables that are defined by a magnitude only are called **scalars,** for example, mass, length, speed, time, temperature, and work. To distinguish between these two quantities, the variables that correspond to vectors are indicated in bold font throughout the text. Vectors can be represented graphically as an arrow with a certain length and direction. The position vector (**r**) in figure 1.11*a,* for example, shows that an object is located 15 m from an origin with a direction of 0.4 rad relative to the horizontal. The direction of a vector is sometimes referred to as its **line of action** (figure 1.11*b*). As long

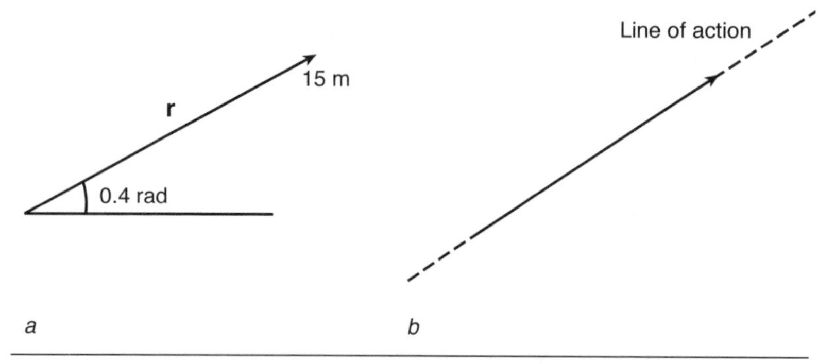

Figure 1.11 (*a*) Characteristics of a position vector include magnitude and direction, and (*b*) its direction can be indicated by a line of action.

as the magnitude of the vector remains constant, it can slide along its line of action without changing its mechanical action.

Often in biomechanics we want to determine either the net effect of several vectors or the effect of a vector in several different directions. These two procedures can be accomplished graphically with the **parallelogram law of combinations.** For example, figure 1.12 shows how we can find the net force ($\mathbf{F}_r$) acting on an object due to the action of two forces ($\mathbf{F}_p$ and $\mathbf{F}_q$). We do this by constructing a parallelogram with the vectors $\mathbf{F}_p$ and $\mathbf{F}_q$ and then drawing the diagonal of the parallelogram, which represents the resultant effect. Note that the tails (opposite the arrowheads) of the three vectors originate from the same point. The combination of two or more vectors into a single resultant vector is called **composition;** that is, the procedure is to compose the resultant. This association can be represented by the addition of the two vectors,

$$\boldsymbol{F}_r = \boldsymbol{F}_p + \boldsymbol{F}_q$$

The converse procedure, called **resolution,** is to take a vector and to resolve it into one or more components. It is often useful, for example, to deal with vector components that are perpendicular to one another, such as in the *x* and *y* directions. Given a vector **V** (magnitude and direction), its *x* and *y* components can be determined using standard trig functions (figure 1.12):

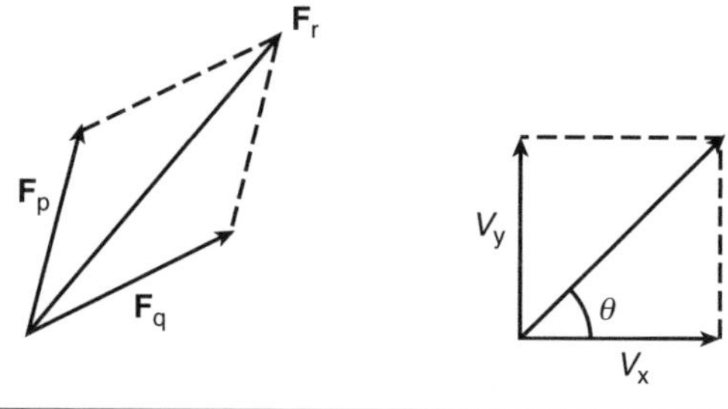

Figure 1.12 Vector parallelograms.

$$V_y = V \sin \theta \quad \text{and} \quad V_x = V \cos \theta$$

Conversely, if we are given the components V_x and V_y, we can determine the magnitude and the direction of the resultant (V) by the following relations:

$$V = \sqrt{V_x^2 + V_y^2}$$

$$\theta = \tan^{-1} \frac{V_y}{V_x}$$

EXAMPLE 1.6

Net Muscle Force

Figure 1.13 illustrates how composition is used to determine the resultant effect of coactivating different parts of the pectoralis major muscle. Figure 1.13*a* indicates the direction and magnitude of the force exerted by the clavicular ($F_{m,c}$) and sternal ($F_{m,s}$) portions of the pectoralis major muscle. Suppose that the clavicular component exerted a force of 224 N, which is directed at an angle of 0.55 rad above the horizontal, and that the sternal component had a magnitude of 251 N and acts 0.35 rad below the horizontal. To compose the resultant, we add these two components head to tail by sliding either one of the vectors along its line of action (figure 1.13*b*) and then joining the open ends (open tail to open head) to produce the resultant vector (figure 1.13*c*).

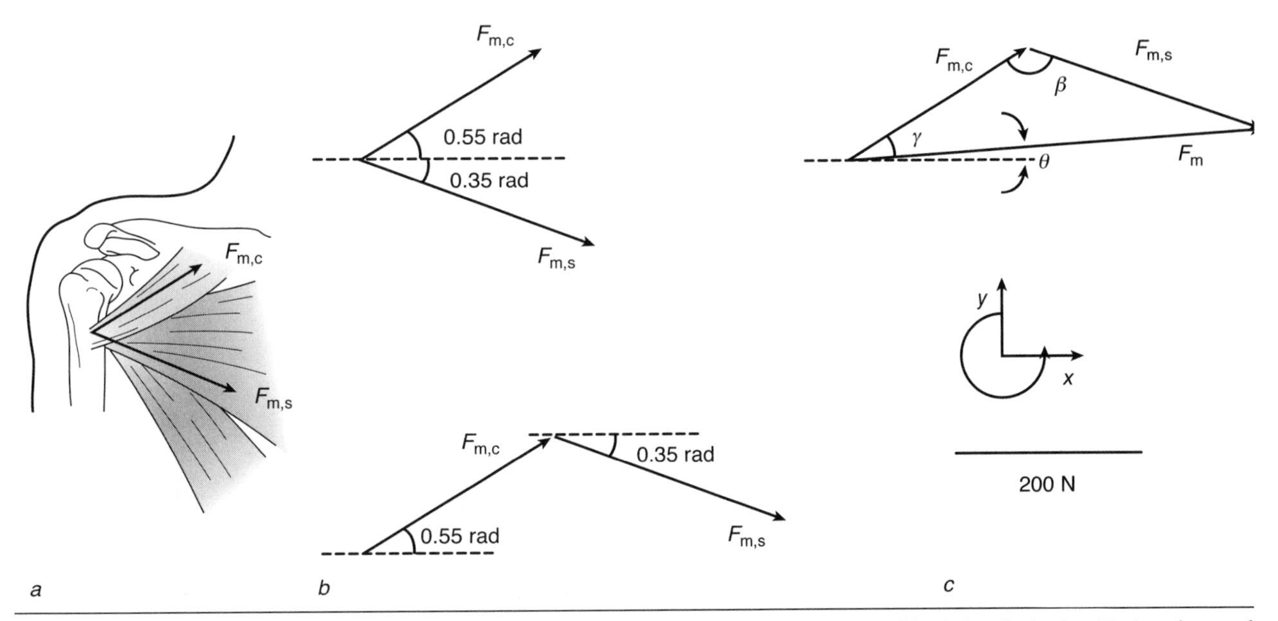

Figure 1.13 Geometric composition of the resultant muscle force (F_{m}) due to activation of both the clavicular ($F_{m,c}$) and sternal ($F_{m,s}$) components of the muscle pectoralis major. *(a)* Orientation of the two vectors; *(b)* graphic addition of the two vectors; and *(c)* calculation of the resultant vector ($\beta = 3.14 - 0.55 - 0.35 = 2.24$ rad).

The magnitude of the resultant can be obtained by applying the law of cosines, which enables us to determine the length of one side of a triangle provided we know the length of the other two sides and the angle between them. The following is the equation to determine the magnitude of F_{m}:

$$F_{m}^{2} = F_{m,c}^{2} + F_{m,s}^{2} - (2 \times F_{m,c} \times F_{m,s} \times \cos \beta)$$

$$F_{m} = \sqrt{224^{2} + 251^{2} - (2 \times 224 \times 251 \times \cos 2.24)}$$

$$F_{m} = 428 \text{ N}$$

By applying the law of sines, we can determine the magnitude of angle γ:

$$\frac{\sin \beta}{F_{m}} = \frac{\sin \gamma}{F_{m,s}}$$

$$\gamma = \sin^{-1}\left[F_{m,s} \times \frac{\sin \beta}{F_{m}}\right]$$

$$= \sin^{-1}\left[251 \times \frac{\sin 2.24}{428}\right]$$

$$\gamma = 0.478 \text{ rad}$$

Because $F_{m,c}$ is at an angle of 0.55 rad with respect to the horizontal, the direction of F_{m} is as follows:

$$\theta = 0.55 - \gamma$$

$$\theta = 0.072 \text{ rad}$$

Right-Hand Coordinate System

Movement involves the displacement of an object from one point in space to another point. To describe a movement, therefore, it is necessary to specify the location of the object at any point in time. This is typically done with a set of *xyz*-axes (figure 1.14). The *x, y,* and *z* directions are perpendicular to each other with the positive directions defined in the following way. Curl the fingers of your right hand in the direction of rotating the positive *x*-axis onto the positive *y*-axis; then the direction of your extended thumb indicates the positive *z*-axis. For this reason, this configuration of the *xyz*-axes is known as the *right-hand coordinate system.*

To specify a vector in the right-hand coordinate system, we state that it has so many units in the *x* direction, so many in the *y* direction, and so many in the *z* direction. This idea is abbreviated with use of the terminology **i, j,** and **k;** these are known as unit vectors. The symbol **i** represents the *x* direction, **j** refers to the *y* direction, and **k** indicates the *z* direction. For example, a vector described by the expression $-286\mathbf{i} + 812\mathbf{j} + 61\mathbf{k}$ has a component of 286 units in the negative *x* direction, a component of 812 units in the *y* direction, and a component of 61 units in the *z* direction. The magnitude of the vector can be determined with the Pythagorean theorem:

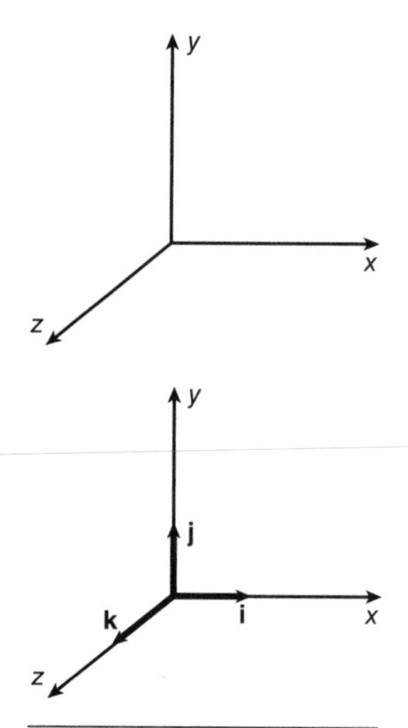

Figure 1.14 Right-hand coordinate system and unit vectors.

$$v = \sqrt{(-286)^2 + 812^2 + 61^2}$$

$$v = 863$$

Coordinate systems such as the *xyz*-axes can be fixed in space or can be fixed to an object that moves in three-dimensional space. In biomechanics, we often use both types of coordinate systems. When describing gross movements, however, we typically use an *xyz* system that is attached to the center of mass of the body with the *y*-axis going in the head-to-toes direction, the *x*-axis going front to back, and the *z*-axis going side to side. Movements can occur in any of the three planes *(x-y, x-z,* or *y-z)* or about the three axes *(x, y,* and *z).* The *x-y* plane is referred to as the **sagittal plane,** and it divides the body into right and left. The *x-z* plane is known as the **transverse plane,** and it divides the body into top and bottom. The *y-z* plane is described as the **frontal plane,** and it divides the body into front and back. The three axes are usually given functional names based on the type of movement that would be performed when a person rotated about the axis. Based on this convention, the *x*-axis is referred to as the **cartwheel axis,** the *y*-axis as the **twist axis,** and the *z*-axis as the **somersault axis.**

Vector Algebra

When describing physical systems and the interaction of their components, we often manipulate vectors in algebraic expressions. The addition and subtraction of vectors is a simple procedure in that we add or subtract the **i, j,** and **k** terms separately. For example, let us add $\mathbf{F}_p$ and $\mathbf{F}_q$:

$$\boldsymbol{F}_p = 10\boldsymbol{i} + 28\boldsymbol{j} + 92\boldsymbol{k} \qquad \text{and} \qquad \boldsymbol{F}_q = 3\boldsymbol{i} - 11\boldsymbol{j} + 46\boldsymbol{k}$$

$$\boldsymbol{F}_p + \boldsymbol{F}_q = (10 + 3)\boldsymbol{i} + (28 - 11)\boldsymbol{j} + (92 + 46)\boldsymbol{k}$$

$$\boldsymbol{F}_p + \boldsymbol{F}_q = 13\boldsymbol{i} + 17\boldsymbol{j} + 138\boldsymbol{k}$$

Multiplication, however, is a more involved procedure. Actually, there are two procedures: scalar products and vector products. (Although the procedures are more commonly known as *dot* and *cross* products, the terms *scalar* and *vector* seem more appropriate as they define the type of product.) The distinction between the two procedures, scalar (dot) and vector (cross) products, has to do with the character of the result, that is, whether it is a scalar or a vector quantity.

The definition of a **scalar product** is given by this expression:

$$\boldsymbol{d} \bullet \boldsymbol{F} = dF \cos \theta \tag{1.6}$$

The dot product (notice the dot between **d** and **F**) of the two vectors **d** and **F** is calculated as the magnitude of **d** (d) times the magnitude of **F** (F) multiplied by the cosine of the angle (θ) between the two vectors (figure 1.15*a*). This procedure gives the magnitude of **F** that is directed along **d** multiplied by the magnitude of d. The magnitude of **F** directed along **d** is shown in figure 1.15*b* as the base of the right triangle (i.e., the side that equals **F** cos θ). Thus the scalar product is just d times $F \cos \theta$.

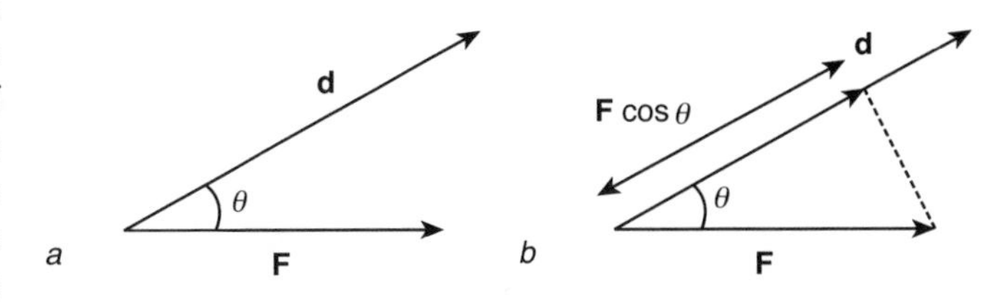

Figure 1.15 The scalar product of two vectors involves *(a)* determining how much of **F** acts in the same direction as **d** and *(b)* then multiplying the two terms.

The multiplication of vectors with the scalar product is appropriate for calculating scalar quantities. One example is the calculation of work, where work is defined as the product of force and displacement (distance). Both force and displacement are vectors. Work (a scalar quantity derived by the scalar product) is defined as the component of force in the direction of the displacement times the magnitude of the displacement, that is, the work done by a force.

Alternatively, when two vectors are multiplied and the product is a vector (magnitude and direction), the procedure is called the vector (cross) product. This is given by the expression

$$\boldsymbol{r} \times \boldsymbol{F} = rF \sin \theta \quad (1.7)$$

which reads "**r** cross **F** is equal to the magnitude of **r** (r) times the magnitude of **F** (F) multiplied by the sine of the angle (θ) between the two vectors." This states that the magnitude of a vector product is equal to the product of two vector magnitudes and that the direction is along the axis about which the rotation of vector **r** onto vector **F** occurs (i.e., cross from **r** to **F**). This means that the direction of the product is always perpendicular to the plane that contains the other two vectors and is located at the base (intersection) of the two vectors. This relation is illustrated in figure 1.16.

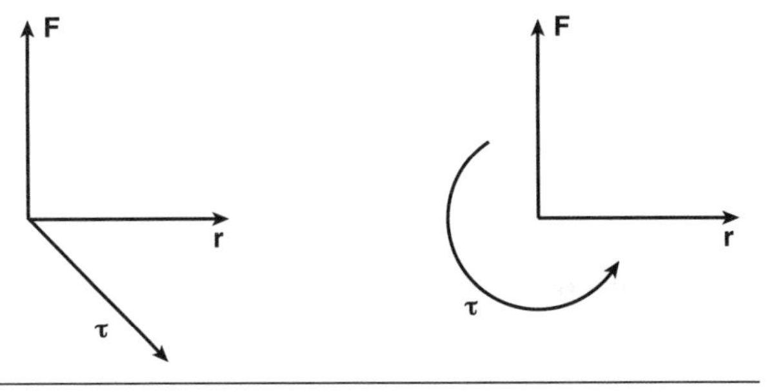

Figure 1.16 The vector product of **r** and **F** is the vector **τ**.

Vector products can involve such combinations as position and force, position and angular velocity, and position and momentum. For example, the vector product of position (**r**) and force (**F**) is torque (**τ**), as shown in figure 1.16. This product represents a rotary or angular force, the magnitude of which depends on the size of the force and the distance (moment arm) of its line of action from the axis of rotation. The direction of the torque vector is along the axis about which the object would rotate due to the action of the force. That is, torque is equal to the product of force and the moment arm. For example, if a net muscle force and the distance to its line of action were located in the plane of this page, then the direction of the net muscle torque would be perpendicular to the page. In general, the direction of such a product can be either toward or away from you and depends on the directions of the two vectors that contributed to the product. Because it is difficult to show three-dimensional directions on a page, vectors that are perpendicular to the page are often drawn as curved arrows. Figure 1.16 shows torque (**τ**) drawn as a curved arrow, which the right-hand rule indicates as a vector that comes out of the page.

LINEAR AND ANGULAR MOTION

In the preceding discussion, you may have noted that displacement has been indicated with either of two units of measurement, meters (m) or radians (rad); the distinction between the two is that of **linear** and **angular motion,** respectively. Linear motion refers to an equivalent displacement in space of all the parts of the object. Conversely, when all the parts of the object do not experience the same displacement (Δ position), the motion has included some rotation (angular displacement). A combination of linear **(translation)** and angular **(rotation)** motion in a single plane is called **planar** motion and involves rotation about a point that is itself moving. In most human movement, our body segments undergo both linear and angular motion.

A *meter,* the unit of measurement for linear motion, is defined as the length of the path traveled by light in a vacuum during about one three-hundred thousandths of a second. A *radian* is the ratio of arc length to the radius of the circle (figure 1.1). When the arc length equals the radius, the ratio has a value of 1, and the object has rotated 1 rad (57.3°). For example, consider a discus throw by an athlete whose arm length from the shoulder to the discus has a value of 63 cm. As the arm rotates about a twist axis through the shoulder joint, the discus moves in a circular path. When the discus has moved along its path 63 cm (equal to the length of the arm and thus equal to the radius of the circle), the arm and the discus have been rotated through an angle of 1 rad.

The commonly used symbols and associated units of measurement for linear and angular position, velocity, and acceleration are listed in table 1.3. As you can see, the symbols are usually Latin letters for linear terms and Greek letters for angular terms.

Table 1.3 Linear and Angular Symbols and the SI Units of Measurement

	LINEAR SYMBOL			ANGULAR SYMBOL	
	Scalar	Vector	Unit	Symbol	Unit
Position	r	$\mathbf{r}$	m	θ (theta)	rad
Displacement	Δr	$\Delta\mathbf{r}$	m	$\Delta\theta$ (theta)	rad
Velocity	v	$\mathbf{v}$	m/s	ω (omega)	rad/s
Acceleration	a	$\mathbf{a}$	m/s^2	α (alpha)	rad/s^2

Angle-Angle Diagrams

In measuring human movement, we usually graph some variable (e.g., thigh angle, ball height) as a function of time. Because human movement is accomplished by the rotation of body segments about one another, it is often more instructive to examine the relation between two angles during a movement. Such graphs, called **angle-angle diagrams** (Cavanagh & Grieve, 1973), usually plot a **relative angle** (i.e., the angle between two adjacent body segments) against the **absolute angle** of a body segment (i.e., the angle relative to a reference in the surroundings).

Figure 1.17 shows two angle-angle diagrams for part of a weightlifting movement in which the barbell was lifted from position 1 through position 10. The trunk-knee diagram shows that the movement comprises three distinct phases: (a) positions 1 to 5, slight forward rotation of the trunk and extension of the knee joint; (b) positions 5 to 8, backward rotation of the trunk and flexion of the knees; and (c) positions 8 to 10, some backward-forward trunk rotation and knee joint extension. Similarly, the thigh-ankle diagram comprises three phases: (a) forward thigh rotation and ankle plantarflexion; (b) constant thigh angle and ankle dorsiflexion; and (c) forward thigh rotation and ankle plantarflexion. Examination of the phases composed of the numbered set of positions in figure 1.17 shows the extent to which the three phases for the two angle-angle diagrams coincide; the movement is accomplished by coordinated displacements about the joints of the leg. In figure 1.17 there is a constant 10 ms interval between each dot; so the greater the distance between the dots, the greater the velocity of movement.

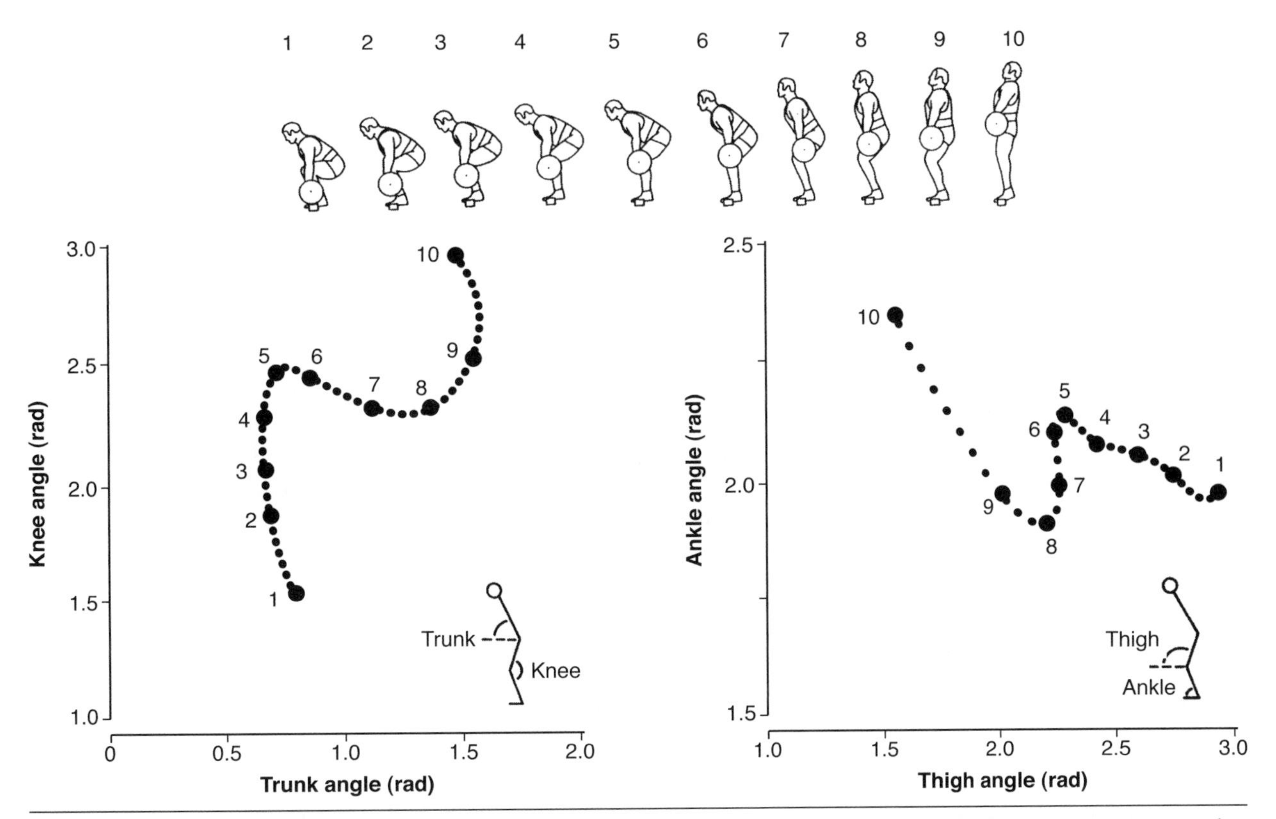

Figure 1.17 Angle-angle relations during the first part of a weightlifting event. The numbers in the diagrams correspond to the positions of the lifter indicated at the top of the figure.

Reprinted, by permission, from R.M. Enoka, 1988, "Load and skill-related changes in segmental contributions to a weightlifting movement," *Medicine and Science in Sports and Exercise* 20: 185.

Angular Kinematics

Because human movement typically involves both translation and rotation, it is necessary to know the relations between the linear and angular measures of position, velocity, and acceleration. When a rigid body of fixed length (r) rotates about a point from position 1 to position 2 (figure 1.18*a*), equation 1.8 can be used to determine the displacement (s) experienced by the end of the rigid body:

$$s = r\theta \tag{1.8}$$

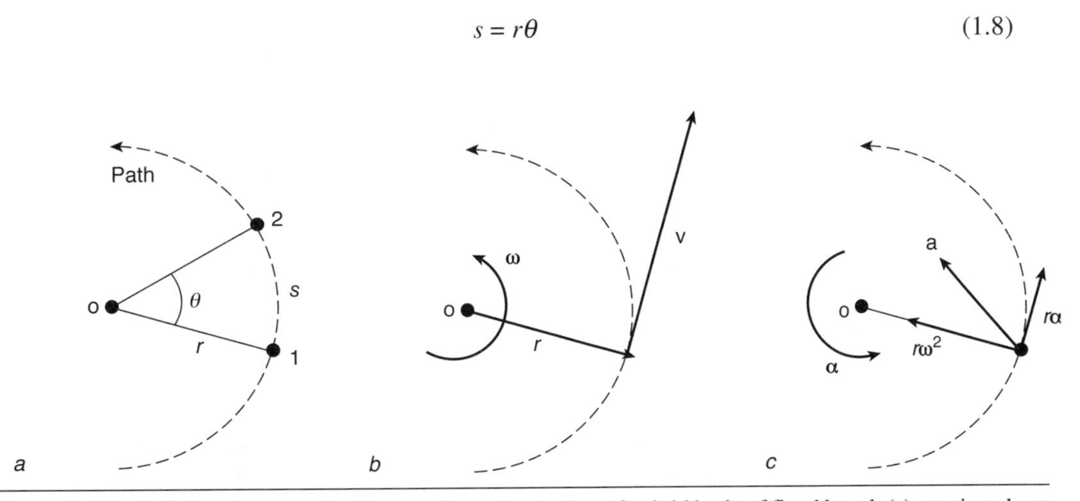

Figure 1.18 Relations between the linear and angular motion of a rigid body of fixed length (r) rotating about a fixed axis: *(a)* position, *(b)* velocity, *(c)* acceleration. The vectors $\boldsymbol{\omega}$ and $\boldsymbol{\alpha}$ are perpendicular to the page.

The linear velocity (v) of the end of the rigid body is determined as the rate of change in s with a direction that is tangent to the circular path.

$$\frac{\Delta s}{\Delta t} = \frac{\Delta(r\theta)}{\Delta t}$$

Because r has a fixed magnitude and does not change over time, this expression reduces as follows:

$$\frac{\Delta s}{\Delta t} = \frac{r\Delta\theta}{\Delta t}$$

$$v = r\omega \tag{1.9}$$

Equation 1.9 indicates that the linear velocity (v) of any point on a rigid body is equal to the product of the distance from the axis of rotation to that point (r) and the angular velocity of the rigid body (ω). For different points along a rigid body, therefore, both r and v vary. As anyone who has been part of a rotating human chain on skates knows, the person farthest from the axis of rotation experiences the greatest linear velocity. Furthermore, the direction of the linear velocity vector (**v**) is tangent to the path of the rigid body (figure 1.18*b*).

The variables **r, v,** and **ω** are vectors and have both magnitude and direction. The magnitude and direction of **r** and **v** are straightforward, as shown in figure 1.18*b*. When a vector is shown as a curved arrow, its direction is actually perpendicular to the page and is located at the axis of rotation. In figure 1.18*b,* **ω** is directed out of the page. Equation 1.9 is the scalar form of the relation between linear and angular velocity. The vector relation is indicated in equation 1.10:

$$\boldsymbol{v} = \boldsymbol{\omega} \times \boldsymbol{r} \tag{1.10}$$

Equation 1.10 states that linear velocity (**v**) is equal to the cross product (×) of angular velocity (**ω**) and position (**r**). The cross product is a vector operator that is used to multiply vectors, the result (product) of which is a vector that is perpendicular to the plane of the original vectors. The **ω** vector is perpendicular to the plane of motion.

To determine the relation between linear and angular acceleration, we need to use equation 1.10 to account for the change in both magnitude and direction of each velocity vector (**v** and **ω**):

$$\frac{\Delta \boldsymbol{v}}{\Delta t} = \frac{\Delta(\boldsymbol{\omega} \times \boldsymbol{r})}{\Delta t}$$

$$\boldsymbol{a} = \left(\boldsymbol{\omega} \times \frac{\Delta \boldsymbol{r}}{\Delta t}\right) + \left(\frac{\Delta \boldsymbol{\omega}}{\Delta t} \times \boldsymbol{r}\right)$$

$$= (\boldsymbol{\omega} \times \boldsymbol{v}) + (\boldsymbol{\alpha} \times \boldsymbol{r})$$

$$\boldsymbol{a} = \boldsymbol{\omega} \times (\boldsymbol{\omega} \times \boldsymbol{r}) + (\boldsymbol{\alpha} \times \boldsymbol{r})$$

In scalar terms, the magnitude of linear acceleration (a) can be determined.

$$a = \sqrt{(r\omega^2)^2 + (r\alpha)^2} \tag{1.11}$$

The term $r\omega^2$ accounts for the change in direction of **v,** and the term $r\alpha$ represents the change in magnitude of **v** (figure 1.18*c*). Because the direction of **v** changes during angular motion, $r\omega^2$ is never zero; but $r\alpha$ may be zero if the magnitude of **v** is constant, which occurs in uniform circular motion. Recognize also that from the scalar relation between linear and angular velocity (equation 1.9), we can derive a relation for linear acceleration for uniform circular motion ($r\alpha = 0$).

$$a = \frac{v^2}{r} \tag{1.12}$$

Sometimes the $r\omega^2$ term is referred to as the normal or radial component (a_n) and the $r\alpha$ term as the tangential component (a_t) of linear acceleration; these names indicate the direction of each component relative to the path of the rigid body. When the motion is planar (fixed axis of rotation), the lines of action of **ω** and **α** are collinear (lie on the same line) at the axis of rotation.

EXAMPLE 1.7

Kicking a Football

A kicked football leaves the foot of a punter with a vertical velocity (v_v) of 25.9 m/s and a horizontal velocity (v_h) of 14.2 m/s.

A. What is the magnitude of the resultant linear velocity?

$$v = \sqrt{v_v^2 + v_h^2}$$
$$v = \sqrt{25.9^2 + 14.2^2}$$
$$v = 29.5 \text{ m/s}$$

B. What is the direction (an angle relative to the horizontal) of the resultant linear velocity?

$$\tan\theta = \frac{v_v}{v_h}$$
$$\theta = \tan^{-1}\frac{v_v}{v_h}$$
$$\theta = 1.07 \text{ rad}$$

C. If the leg (hip-to-ankle length = 0.53 m) is straight at the moment of contact with the ball, what is the angular velocity of the leg?

$$v = r\omega$$
$$\omega = \frac{29.5}{0.53}$$
$$\omega = 55.7 \text{ rad/s}$$

D. What is the value of the acceleration component that accounted for the change in direction of the linear velocity?

$$r\omega^2 = 0.53 \times 55.7^2$$
$$r\omega^2 = 1644 \text{ m/s}^2$$

EXAMPLE 1.8

Kinematics of an Elbow Movement

An elbow extension-flexion movement is performed slowly in a transverse plane passing through the shoulder joint (figure 1.19). The movement begins with the upper arm raised to the side so that it is horizontal, with an angle of 0.70 rad (40°) between the upper arm and forearm. In one continuous movement of moderate speed, the upper arm is held stationary while the elbow joint is extended horizontally to 3.14 rad (180°) and then flexed back to the starting position (0.70 rad).

A. What is the shape of the position–time graph for this movement?

Because the upper arm remains stationary while the forearm rotates about the elbow joint, a graph of elbow angle over time should adequately describe the movement (top trace in figure 1.19).

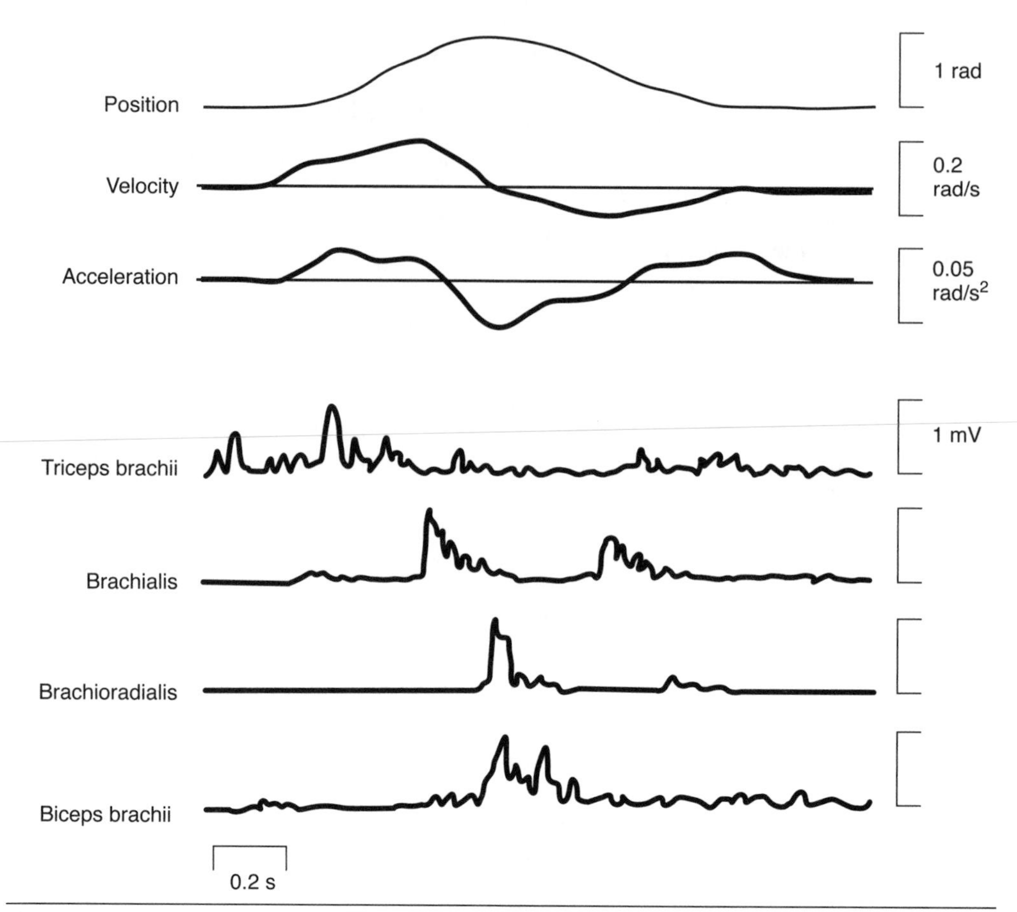

Figure 1.19 Kinematic graphs and electromyogram patterns for an extension-flexion movement of the forearm-hand segment about the elbow joint. The kinematic features of the movement are determined by the net muscle activity.

B. When is angular velocity zero?

The velocity is zero when there is no angular displacement—at the beginning and end of the movement and, for an instant, when the direction of the movement changes from extension to flexion; that is, when elbow angle is at a maximum. The velocity graph (second trace in figure 1.19) indicates that when velocity is positive (above zero), the elbow is extending; when the velocity is negative, the elbow is flexing. Thus a change in the sign of velocity (e.g., positive to negative) indicates a change in the direction of movement.

C. When is angular acceleration zero?

The angular acceleration is zero when the value of angular velocity is at its maximum and minimum, which occurs twice in figure 1.19—that is, when the slope of the velocity graph is zero.

D. Why does the acceleration graph (third trace in figure 1.19) have three phases?

As we discuss in chapters 2 and 3, the acceleration experienced by a system (the forearm-hand in this example) depends on the forces acting on the system. Because the muscles that cross the elbow joint control this movement, the three phases of the acceleration graph can be explained by the net muscle activity. For the first part of the movement, the arm accelerates in the direction of extension (positive values in this example). In the middle phase of the movement, the arm accelerates in the direction of elbow flexion. The movement concludes with a final phase of acceleration in the direction of elbow extension. As indicated by the records of muscle activity (electromyogram) in the lower traces of figure 1.19, acceleration in the direction of elbow extension is due to activation of the elbow extensor muscle (triceps brachii), whereas acceleration in the direction of elbow flexion is associated with activation of the elbow flexor muscles (brachialis, brachioradialis, and biceps brachii). Note that the angular acceleration graph cannot be predicted directly from the graph of elbow angle.

CURVE FITTING AND SMOOTHING

All measures of movement contain some error due to limitations in the technology and inaccurate measurements by the person doing the analysis (Elliott et al., 2007). Consider, for example, a ball that is tossed straight up into the air, with its trajectory recorded on videotape. The ball rises to some peak height and then falls to the ground (figure 1.20). As we have discussed previously, the vertical velocity of the ball is zero at the peak of the trajectory and the vertical acceleration is a constant $-9.81\ m/s^2$ for the duration of the trajectory. A motion analysis system, however, will incorrectly estimate that the ball experienced a nonconstant vertical acceleration (figure 1.20). The reason for this discrepancy is that the measurement of ball position by the system contains small errors that are magnified when the derivatives (velocity and acceleration) are determined. Errors in the measurement of position are compounded about 20 times by the time acceleration is calculated from the original position data.

To minimize such error propagation, processes known as *curve fitting* and *smoothing* are used to manipulate the data. Curve-fitting techniques involve deriving a mathematical function to represent a data set. Common curve-fitting procedures include polynomial functions (equation 1.13) and Fourier series (equation 1.14). Alternatively, smoothing techniques use averaging

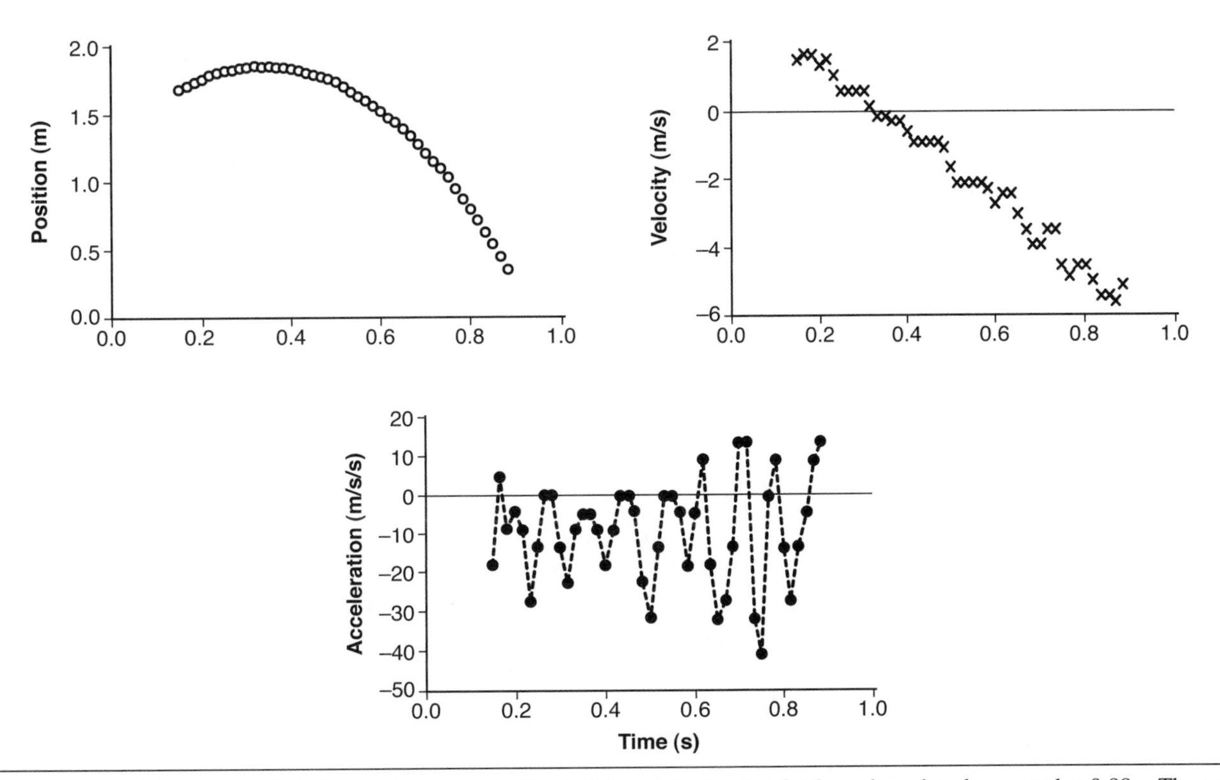

Figure 1.20 Kinematics of a ball tossed into the air. The ball leaves the hand at 0.15 s and reaches the ground at 0.88 s. The position of the ball is measured with a videotape recorder, and the position, velocity, and acceleration curves are determined by a motion analysis system.

procedures to reduce irregularities in a data set. Biomechanists often use a digital filter (equation 1.15) to smooth the data. These three techniques have the following forms:

$$x(t) = a_0 + a_1 t + a_2 t^2 + \ldots + a_n t^n \quad (1.13)$$

$$x(t) = a_0 + \sum_{i=1}^{n} \left[a_i \cos\left(\frac{2\pi t}{T} \bullet i\right) + b_i \sin\left(\frac{2\pi t}{T} \bullet i\right) \right] \quad (1.14)$$

$$x'(i) = a_0 x_i + a_1 x_{i-1} + a_2 x_{i-2} + b_1 x'_{i-1} + b_2 x'_{i-2} \quad (1.15)$$

Polynomial Functions

The simplest type of mathematical function to represent a set of data is the equation for a straight line. Data sets that can be represented by a straight line are described as linear functions. The general form of the equation for a straight line is

$$y = mx + b$$

where y is the dependent (outcome) variable, m corresponds to the slope of the line, x is the independent (predictor) variable, and b indicates the y-intercept. When m has a value of 1 and b is equal to 0, then the equation reduces to the line of identity. The effect of m is to change the slope of the line and can even be negative. The line of best fit is commonly determined by **regression analysis,** which is a statistical procedure that minimizes the differences between the line and the data.

When a straight line cannot represent a data set, then we use nonlinear functions. For example, we can use an expression in which the independent variable is raised to a power. The shape of the function depends on the power used and can be modified by adding several terms together, multiplying the terms with coefficients, and including constants in the expression. A sum of terms that are powers of a variable is known as a polynomial function (equation 1.13). For example, the function $x^3 + 3x^2 - 8x$ is a third-degree polynomial; second-degree polynomials are known as quadratic functions, and third-degree polynomials are referred to as cubic functions. As with the straight line, an appropriate polynomial expression for a set of data can be determined by regression analysis.

Polynomials are sometimes used to represent the change in a kinematic variable as a function of time. For example, a graph of position during a movement may be represented by a fifth- or seventh-degree polynomial (Wood, 1982)—that is, equations in which the terms can be raised to the fifth or seventh power. Once such a data set is represented by a polynomial, it is a simple matter to determine the velocity and acceleration graphs because this involves taking the first (velocity) and second (acceleration) derivatives of the position graph. However, fitting a single polynomial function to an entire set of position data generally does not produce sufficiently accurate derivatives (Wood, 1982). One alternative is to use several polynomial functions, each representing a different part of the data, and then combine these functions to describe the entire data set.

Fourier Analysis

Any signal can be represented by a set of sine and cosine terms; such a set is known as a **Fourier series** (equation 1.14). To understand how a combination of sine and cosine terms can be derived to represent a signal, it is necessary to review some basic properties of these functions. The functions $y = \sin x$ and $y = \cos x$ complete one revolution when x varies from 0 to 6.28 (2π) rad (figure 1.21*a*). The peak values for y are ±1 for both functions, although they occur at different locations in the cycle. Three features of these basic functions can be changed:

1. The number of cycles completed in one revolution—to vary this, a coefficient other than 1.0 is placed in front of the x term. Figure 1.21*b* shows that the expression $y = \sin 2x$ contains two cycles of the function in one revolution (2π rad). Cycle rate is expressed in terms of frequency: The completion of one cycle in 1 s would be described as a

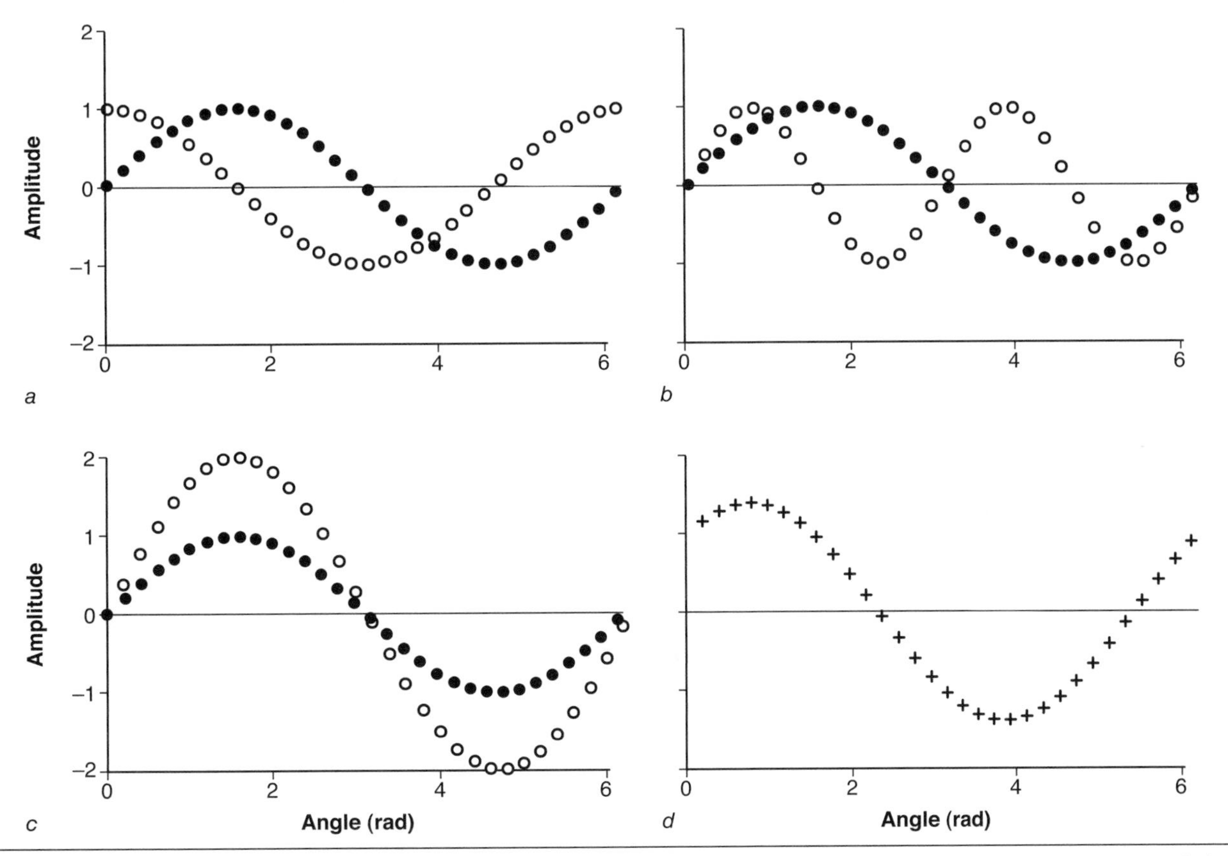

Figure 1.21 Properties of sinusoidal functions. *(a)* The basic sine (●) and cosine (○) functions complete one cycle per revolution (2π rad) with peak-to-peak values of ±1.0. However, *(b)* the number of cycles per revolution and *(c)* the amplitude can be varied by the inclusion of coefficients in the function or by *(d)* the addition of sine and cosine terms. The functions shown are *(a)* $y = \sin x$ and $y = \cos x$; *(b)* $y = \sin x$ and $y = \sin 2x$; *(c)* $y = \sin x$ and $y = 2 \sin x$; *(d)* $y = \sin x + \cos x$.

frequency of 1 Hz. Similarly, the completion of two cycles in 1 s would represent a frequency of 2 Hz.

2. The peak-to-peak amplitude of the function—to vary this, a coefficient other than 1.0 is placed in front of the sine or cosine term. Figure 1.21*c* indicates that the expression $y = 2 \sin x$ represents a function whose peak values are two times ±1.0. Similarly, when the coefficient is less than 1.0, the peak values are less than ±1.0.
3. The phase of the function—when sine and cosine functions are added, the basic function has peak values greater than ±1.0 and its phase is shifted along the *x*-axis; that is, the angles at which the function is zero are shifted along the *x*-axis (figure 1.21*d*).

The derivation of the Fourier series (equation 1.14) for a particular signal that varies as a function of time involves the following steps:

- *Mean:* Calculate the mean value (a_0) of the signal so that the sinusoidal function varies about the correct absolute value.
- *Fundamental:* Use regression analysis to obtain the single sine + cosine term that best describes how the signal varies over time during one cycle. The sine and cosine functions are evaluated at set points in time (t), which are normalized to the total time (T) it takes for one cycle to be completed. The functions can be evaluated with a constant Δt between data points (e.g., 0.1 s), as occurs when the data are obtained with a video camera.
- *Harmonics:* Derive the multiple cycles (harmonics) of the fundamental. We accomplish this by placing integer coefficients ($1, 2, 3 \ldots n$) in front of the x term (figure 1.21*b*). This is included in equation 1.14 as the i term within parentheses. The number of harmonics

(n) needed to describe a signal depends on its smoothness; the more peaks and valleys in the signal, the greater the number of harmonics that will be required for the Fourier series to match the data set.

- *Weighting coefficients:* Adjust the amplitude of each harmonic with coefficients (a_i, b_i) that alter the peak-to-peak amplitude of each sine and cosine term (figure 1.21*c*). Typically, the coefficients decrease as the harmonic number increases, which means that higher harmonics contribute less to replicating the signal. The weighting coefficients are defined as follows:

$$a_i = \frac{2}{T}\int_0^T x(t)\cos\frac{2\pi t}{T} i \bullet dt$$

$$b_i = \frac{2}{T}\int_0^T x(t)\sin\frac{2\pi t}{T} i \bullet dt$$

The amplitude of the harmonic (c_i) is then determined as the square root of the sum of squares of the weighting coefficients, and the phase (θ_i) of the harmonic is given by the arctangent of the coefficients:

$$c_i = \sqrt{a_i^2 + b_i^2}$$

$$\theta_i = \tan^{-1}\frac{a_i}{b_i}$$

- *Sum the terms:* Add the weighted harmonics to the mean value to produce an expression that comprises a series (Σ) of sine and cosine terms. For example, a series of five terms is as follows:

$$\begin{aligned} x(t) = a_0 &+ \left[a_1 \sin\left(\frac{2\pi\, i}{T}\bullet t\right) + b_1 \cos\left(\frac{2\pi\, i}{T}\bullet t\right)\right]_{i=1} \\ &+ \left[a_2 \sin\left(\frac{2\pi\, i}{T}\bullet t\right) + b_2 \cos\left(\frac{2\pi\, i}{T}\bullet t\right)\right]_{i=2} \\ &+ \left[a_3 \sin\left(\frac{2\pi\, i}{T}\bullet t\right) + b_3 \cos\left(\frac{2\pi\, i}{T}\bullet t\right)\right]_{i=3} \\ &+ \left[a_4 \sin\left(\frac{2\pi\, i}{T}\bullet t\right) + b_4 \cos\left(\frac{2\pi\, i}{T}\bullet t\right)\right]_{i=4} \\ &+ \left[a_5 \sin\left(\frac{2\pi\, i}{T}\bullet t\right) + b_5 \cos\left(\frac{2\pi\, i}{T}\bullet t\right)\right]_{i=5} \end{aligned}$$

As with polynomial functions, it is not too difficult to determine the derivatives of the sine and cosine terms with respect to time and thereby produce expressions for velocity and acceleration.

One of the important applications of a Fourier analysis is to perform a *spectral analysis of a signal.* We accomplish this by determining the frequency of the fundamental (number of cycles per second), the number of harmonics necessary to represent the signal, and the weighted amplitude of each harmonic. The outcome is the transformation of a signal from the **time domain** to the **frequency domain** (figure 1.22). For example, a sine wave at a constant frequency is represented as a single data point with a given power or amplitude in the frequency domain (f_o). Similarly, the addition of two sinusoids (the second with three times the frequency of the first) appears as two data points (f_o and $3f_o$), each with a different amplitude (peak-to-peak value), in the frequency domain. Many biological signals (e.g., the electromyogram) require many sinusoids (sine + cosine terms) in the Fourier series and therefore contain many frequencies from a lower limit (f_1) to an upper limit (f_2). A common range used with the electromyogram is 10 Hz to 500 Hz. The representation of a signal in the frequency domain is often referred to as the **power density spectrum** because it characterizes the relative amplitude (power) of each frequency over the chosen range of frequencies (figure 1.23). In general, the more peaks and

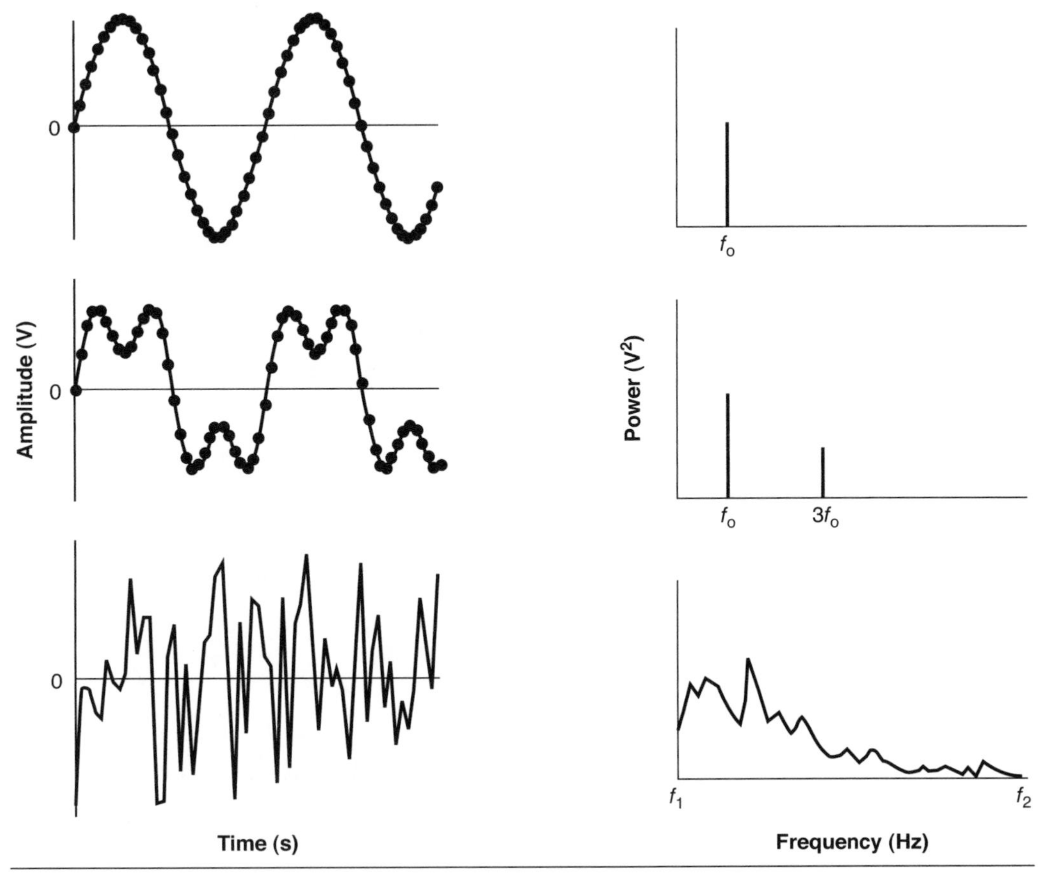

Figure 1.22 Fourier transformations of signals from the time domain (left column) to the frequency domain (right column).

Reprinted from D.A. Winter, 1990, Transformations (Fourier) of signals from the time domain to the frequency domain. In *Biomechanics and Motor Control of Human Movement* (New York: Wiley), 28.

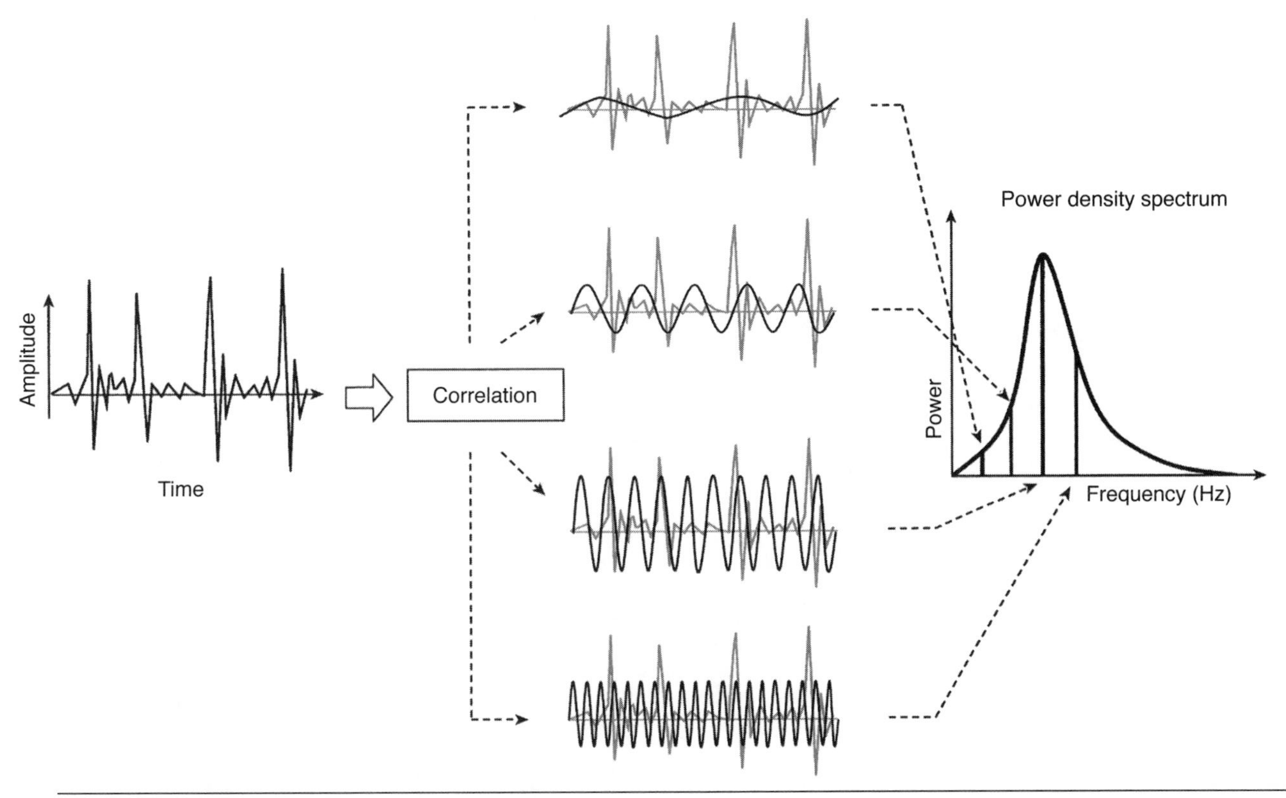

Figure 1.23 Transformation of a signal from the time domain (left) to a power density spectrum in the frequency domain (right). The middle column illustrates the amplitude (power) and frequency of four terms that contribute to the power density spectrum.

Figure developed by François G. Meyer, Ph.D.

valleys that exist in a signal, the greater the range of frequencies in the signal and therefore the greater the upper limit in the power density spectrum. A spectral analysis provides information about the rapidity of the fluctuations in the signal, which is sometimes useful to know when one is comparing the effects of various factors on biological processes.

EXAMPLE 1.9

Time and Frequency Domains

One can examine the relation between a time-domain signal and its frequency-domain version by reconstructing the time-domain signal from the components of the power density spectrum. This process is shown in figure 1.24, with the signal of interest being the hori-

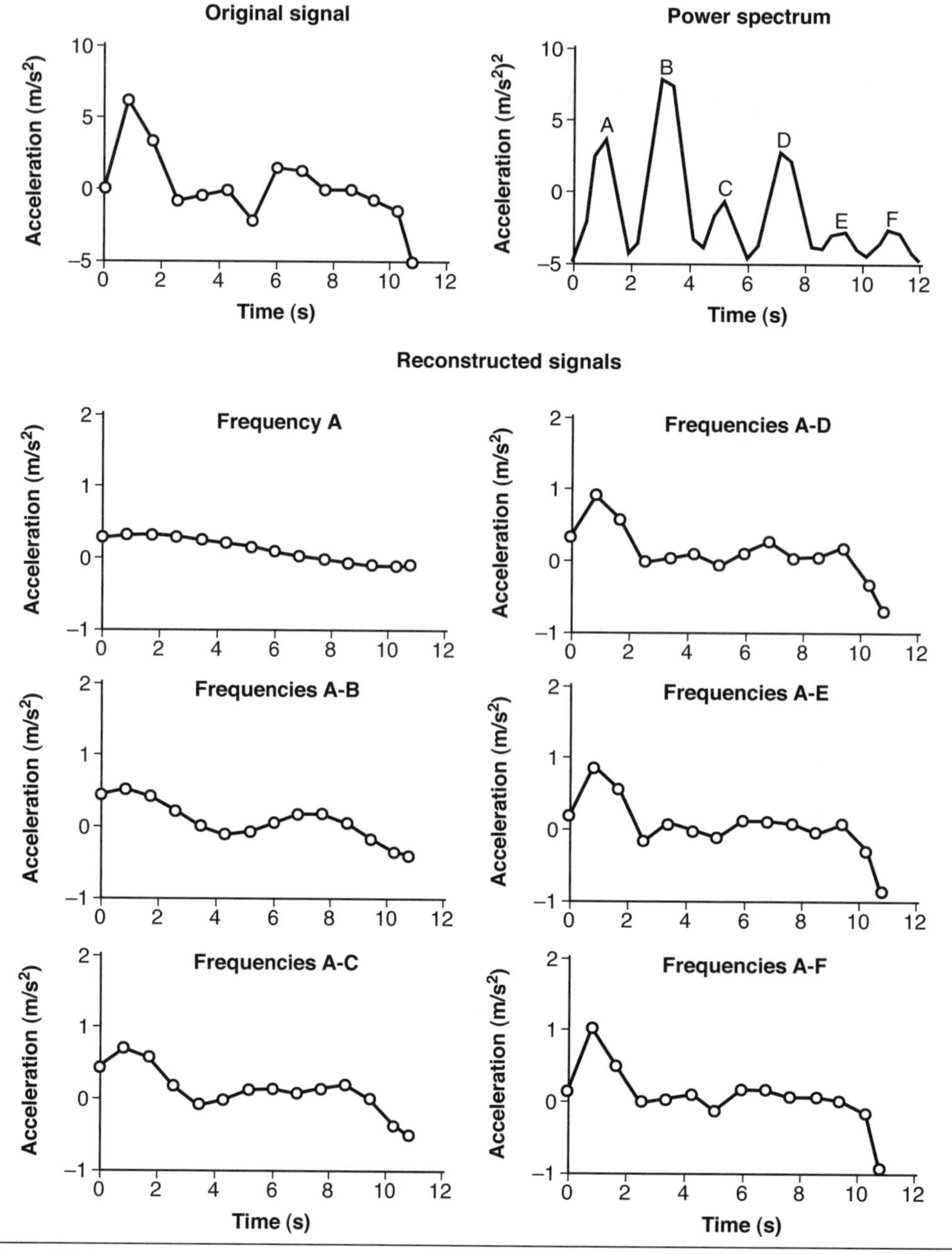

Figure 1.24 Reconstruction of a time-domain signal (acceleration) from the peaks in its power density spectrum.

zontal acceleration of a person running the 100 m sprint (figure 1.5*c*). The acceleration of the runner is shown in the upper left panel of figure 1.24, and the power density spectrum derived from the acceleration data is shown in the upper right panel. Six peaks (A-F) were identified in the power density spectrum.

The lower panels in figure 1.24 represent the time-domain signal (acceleration) as it was reconstructed from various combinations of the peaks in the power density spectrum. The graph labeled "Frequency A" indicates the amount of the acceleration record that is represented by peak A in the power density spectrum. The function in the graph was obtained using the frequency (sinusoid) and amplitude (power) of the peak (A). Similarly, the other graphs show the functions that result from summing several of the peaks (frequency + amplitude). The lower right panel represents the sum of the six peaks in the power density spectrum, which can be compared with the original acceleration graph (upper left). This analysis indicates that the acceleration data can be represented reasonably well by the first four peaks in the power density spectrum.

EXAMPLE 1.10

Comparison of Power Density Spectra

The rate of fluctuations in a record determines the range of frequencies required in the power density spectrum. Acceleration data, for example, have a greater rate of fluctuations compared with position data and thus require higher-frequency sinusoids for their frequency-domain representation. Figure 1.25 shows the acceleration and position of the index finger as a hand muscle performed an isometric contraction (top row) and a slow abduction and adduction movement (bottom row). The subject pushed against a rigid restraint during the isometric contraction, and lifted a light load with a shortening contraction of the hand muscle (first

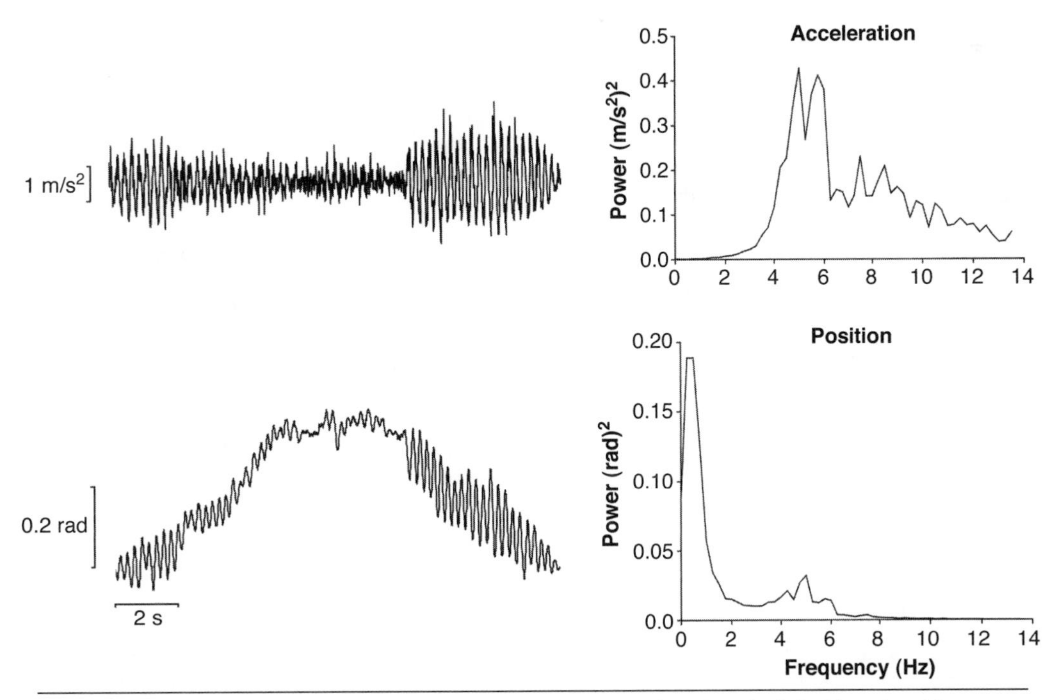

Figure 1.25 Position and acceleration of the index finger in the time domain (left column) and frequency domain (right column) during an isometric contraction (top row) and during a slow (12 s) abduction and adduction movement (bottom row).

Data from Shinohara et al. 2008.

dorsal interosseus) and lowered it with a lengthening contraction of the same muscle during the slow movement. The subject was asked to perform both tasks as steadily as possible. Despite this instruction, the tasks involved fluctuations in acceleration and position. The fluctuations in acceleration had higher frequencies than those for position.

Digital Filter

The digital filter is an example of a smoothing technique (see chapter 2 in Winter, 1990). Unlike the Fourier analysis, the digital filter does not allow us to derive an equation to represent a signal. Rather, the digital filter provides an averaging technique that reduces unwanted fluctuations in a signal. Typically, these fluctuations are the result of measurement errors and will severely contaminate the estimates of velocity and acceleration from position data.

Previously we considered the frequency content of signals (Fourier analysis) when we saw how every signal can be represented by a fundamental sinusoid and its harmonics (multiples of the fundamental). Higher-order harmonics account for the high-frequency content (sharp edges and peaks) of a signal (figure 1.25). When the position of an object is measured from a video image, the measurement includes a true estimate of the position of the landmark in addition to errors due to such effects as camera vibration, distortions in the videotape and projection system, and inaccurate placement of the cursors. These error terms (referred to as noise) are generally random and are located in the high-frequency region of a power spectrum. Digital filters represent numerical techniques that can modify the region of the power spectrum in which the noise is thought to be located (figure 1.26).

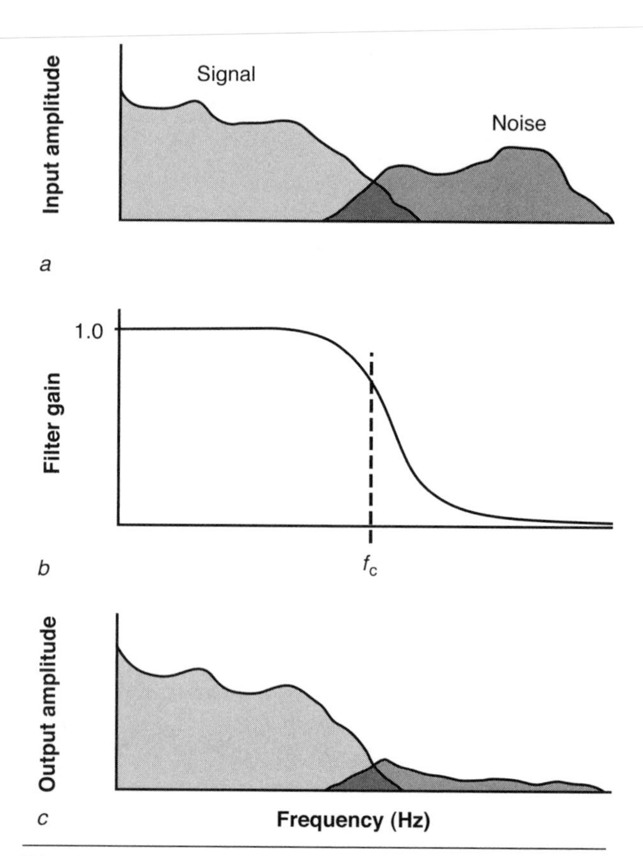

Figure 1.26 The hypothetical effect of a low-pass digital filter on the frequency content of a signal. *(a)* An original input signal contains both desired information (signal) and noise. The noise is generally located in the higher-frequency region of the power spectrum (frequency–amplitude graph). *(b)* The effect of a digital filter on the amplitude of different frequencies. At the cutoff frequency (f_c), the amplitude of the original signal is reduced to about 70% of its original value. *(c)* Power density spectrum of the signal once it has been passed through the digital filter.

The key step in the digital-filter procedure is to identify the frequency that separates the desired signal from the noise; this is referred to as the **cutoff frequency** (f_c). Unfortunately, this is not a simple matter, because the frequencies of the desired signal and noise overlap to some extent (figure 1.26*a*). The digital filter is a numerical procedure that manipulates the frequency spectrum of an input signal to produce an output signal containing a substantially attenuated frequency content above f_c. The numerical effect of a digital filter is shown in figure 1.26*b*, where the frequencies either are left untouched (below f_c) or are reduced (above f_c). The cutoff frequency is defined as the frequency where the amplitude of the output signal is reduced to 70% of its input value; that is, the power in the signal at that frequency is reduced by one-half. The hypothetical effect of this procedure is shown in figure 1.26*c* (compare the amplitude–frequency graph of the input [$x(f)$] in figure 1.26*a* to the output [$x'(f)$] in figure 1.27*c*).

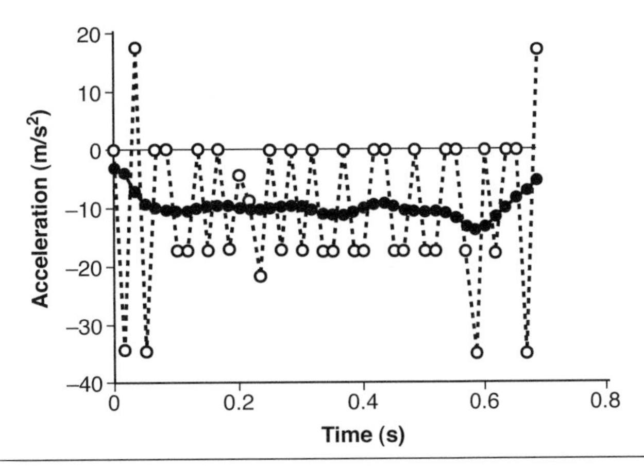

Figure 1.27 The vertical acceleration of a ball tossed into the air, as estimated from the unfiltered (○) and the filtered (●) position data.

A commonly used digital filter (Butterworth, second-order) has the following form:

$$x'(i) = a_0 x_i + a_1 x_{i-1} + a_2 x_{i-2} + b_1 x'_{i-1} + b_2 x'_{i-2}$$

where:

x = unfiltered (input) original data

x' = filtered output data

i = ith frame of data

i–1 = one frame before the ith frame of data

a_0, a_1, a_2, b_1, b_2 = filter coefficients

From this expression, it should be apparent that the output [$x'(i)$] is a weighted average of the immediate (i) and past (i–1, i–2) unfiltered data plus a weighted contribution from the past filtered output (i–1, i–2).

To implement a digital filter, we need to determine the values of the coefficients (a_0, a_1, a_2, b_1, b_2). However, this requires that we first specify f_c. Perhaps the most comprehensive procedure available to determine f_c is the **residual analysis** (Winter, 1990). This involves comparing the difference (residual) between filtered and unfiltered signals over a wide range of cutoff frequencies and choosing an f_c that minimizes both signal distortion and the amount of noise that passes through the filter. For many human movements, an f_c of 6 Hz is generally adequate. Once f_c has been determined, we can obtain the coefficients by calculating the ratio of the sampling frequency (f_s, frames per second) to f_c. For an f_c of 6 Hz and using video data obtained at 60 Hz, the ratio is 10 and the coefficients are as follows:

$a_0 = 0.06746$

$a_1 = 0.13491$

$a_2 = 0.06746$

$b_1 = 1.14298$

$b_2 = -0.41280$

The digital-filter coefficients for different f_s/f_c ratios are available in Winter (1990).

EXAMPLE 1.11

Calculating Filter Coefficients

The coefficients for a Butterworth filter can be calculated with a simple algorithm based on f_s and f_c. For example, given an f_s of 200 Hz and a desired f_c of 8 Hz:

$$S = \sin\left(\pi \times \frac{8}{200}\right) = 0.1253$$

$$C = \cos\left(\pi \times \frac{8}{200}\right) = 0.9921$$

$$SC = \frac{S}{C} = 0.1263$$

$$K = 2SC \times \sqrt{0.5} = 0.1787$$

$$L = SC^2 = 0.0160$$

$$M = 1 + K + L = 1.1946$$

From these parameters, the coefficients can be calculated:

$$a_1 = \frac{L}{M} = 0.0134$$

$$a_2 = 2\frac{L}{M} = 0.0268$$

$$a_3 = \frac{L}{M} = 0.0134$$

$$b_1 = 2\frac{1-L}{M} = 1.6474$$

$$b_2 = \frac{K-L-1}{M} = -0.7009$$

The final consideration is to correct any unwanted distortion that the digital filter introduces in the filtered data. This effect is a phase distortion that can be observed as a shift of a sine wave along the horizontal axis; if one cycle equals 2π rad, then a phase distortion of $\pi/2$ rad is equal to a shift of the sine wave by one-quarter of the cycle. This is the magnitude of the phase distortion that is introduced by a second-order Butterworth filter. To remove the phase distortion, we need to pass the data through the digital filter twice—once in the forward direction and once in the backward direction. The forward direction means beginning with the real first data point, while the backward direction means beginning with the last data point. The forward-backward filtering of data with a second-order filter results in the application of a fourth-order, zero-lag filter.

Once the position data for a movement have been smoothed with a digital filter, the derivatives (velocity and acceleration) can be determined using a numerical procedure called **finite differences.** This process involves using the smoothed position data (x) to calculate both velocity (equation 1.16) and acceleration (equation 1.17) as functions of time:

$$\dot{x}(i) = \frac{x_{i+1} - x_{i-1}}{2(\Delta t)} \tag{1.16}$$

$$\ddot{x}(i) = \frac{x_{i+1} - 2x_i + x_{i-1}}{(\Delta t)^2} \tag{1.17}$$

The effect of a fourth-order Butterworth filter is shown in figure 1.27. The vertical position of a ball tossed into the air (figure 1.20) was passed through the filter (f_c = 6 Hz), and the vertical acceleration of the ball was estimated by finite differences (equation 1.17). The actual acceleration of the ball was about -9.8 m/s^2, which was closely approximated by the acceleration data derived from the filtered position coordinates (figure 1.27).

SUMMARY

This chapter provides the foundation for the study of human movement. It examines the rigorously defined relations between position, velocity, and acceleration and establishes the concepts and definitions that are necessary to describe movement in precise terms. The study of kinematics has its foundation in physics, yet the application to human movement typically tests the limits of our understanding of these relations. Without a clear understanding of these principles, it is difficult to proceed to the next levels in the study of human movement. For this reason, the chapter repeats important concepts and presents many numerical and graphical examples of these relations. The chapter describes some tools that are used in the study of movement.

SUGGESTED READINGS

Hamill, J., & Knutzen, K.M. (2003). *Biomechanical Basis of Human Movement.* Baltimore: Williams & Wilkins, chapters 8 and 9.

Medved, V. (2001). *Measurement of Human Locomotion.* Boca Raton, FL: CRC Press LLC, chapter 4.

Winter, D.A. (1990). *Biomechanics and Motor Control of Human Movement.* New York: Wiley, chapter 2.

chapter 2

Movement Forces

Force is a concept that is used to describe the physical interaction of an object with its surroundings. It can be defined as an agent that produces or tends to produce a change in the state of motion of an object—that is, it accelerates the object. For example, a ball sitting stationary (zero velocity) on a pool table will remain in that position unless a force acts on the ball. Similarly, a person gliding on ice skates will maintain a constant velocity unless a force changes the motion. The study of motion that includes consideration of the force as the cause of movement is called **kinetics.**

LAWS OF MOTION

Isaac Newton (1642-1727) characterized the relation between force and motion with three statements, known collectively as the laws of motion. These laws, which are referred to as the laws of inertia, acceleration, and action-reaction, were originally formulated for particles but are also relevant for rigid bodies, such as human body segments (Nigg & Herzog, 1994).

Law of Inertia

A particle will remain in a state of rest or move in a straight line with constant velocity, if there are no forces acting on the particle.

More simply, a force is required to stop, start, or alter motion. This law is evident in a weightless (microgravity) environment, for example, when astronauts toss objects to one another or perform acrobatic stunts. In a gravitational world, however, forces act continuously upon bodies, and a change in motion occurs when there is a net imbalance of forces. In this context, the term body can refer to the entire human body, or just a part of the human body (e.g., thigh, hand, torso), or even some object (e.g., shot put, baseball, Frisbee).

To appreciate fully the implications of this law, it is necessary to understand the term **inertia.** The concept of inertia relates to the difficulty with which an object's velocity can be altered. Mass, expressed in grams (g), is a measure of the amount of matter composing an object and is a quantitative measure of inertia. Consider two objects of different mass but with the same amount of motion (similar velocities). It is more difficult to alter the motion of the more massive object; hence, it has a greater inertia. Because motion is described in terms of velocity, the inertia of an object is a property of matter that is revealed only when the object is being accelerated—that is, when there is a change in velocity.

According to the law of inertia, an object in motion will continue in uniform motion (constant amplitude and direction of velocity) unless a force acts on the object. This means that the tendency of an object in motion is to travel in a straight line. For example, consider a ball tied to a string and swung overhead in a horizontal plane. When the string is released, will the trajectory of the ball be a straight line or a curved line? To answer this question, we invoke Newton's law of inertia. If no force acts on an object, then the object will travel in a straight line. Once the individual releases the string, no horizontal force acts on the ball and so it will travel in a straight line. But then how can a pitched ball in baseball or softball be made to deviate from a straight line? According to the law of inertia, other forces (e.g., air resistance, gravity) must act on the ball once it is released. Remember from chapter 1 that the trajectory of a projectile will be a straight line unless there is an influence by gravity or air resistance on the object.

Because of these effects, angular motion requires the presence of a force. This force prevents an object from traveling in a straight line. Although the length of the velocity vector (its magnitude) may be constant during angular motion, its direction changes continuously. An inwardly directed force known as centripetal force causes the change in direction. Centripetal force (F_c) is defined as

$$F_c = \frac{mv^2}{r} \tag{2.1}$$

where m = mass, v = velocity, and r = the radius of the curved path. What would happen to F_c, according to equation 2.1, if the speed of the ball (v) remained constant, the length of the string (r) stayed the same, and the mass of the ball increased? F_c would have to increase to satisfy these conditions. Conversely, F_c would decrease if the mass of the ball (m) and velocity (v) remained constant while the length of the string (r) increased.

Law of Acceleration

A particle acted upon by an external force moves such that the force is equal to the time rate of change of the linear momentum.

The term **momentum** (G) describes the quantity of motion possessed by a body and is defined as the product of mass (m) and velocity (v). A runner with a mass of 60 kg moving at a horizontal speed of 8 m/s possesses a momentum of 480 kg•m/s. Thus

$$G = mv \tag{2.2}$$

and the rate of change in momentum can be written as

$$\frac{\Delta G}{\Delta t} = \frac{\Delta(mv)}{\Delta t}$$

Because m does not change in human movement, the applied force (F) is proportional to the product of mass and the rate of change in velocity ($\Delta v/\Delta t$ = acceleration):

$$F = \frac{\Delta G}{\Delta t} = \frac{m\Delta v}{\Delta t}$$

$$F = ma \tag{2.3}$$

Thus equation 2.3 is the algebraic expression of Newton's law of acceleration and states that force is equal to mass times acceleration. Conceptually, this is a cause-and-effect relation. The left-hand side (F) can be regarded as the cause because it represents the physical interactions between a system and its surroundings. In contrast, the right-hand side reveals the effect by indicating the kinematic consequences (ma) of the interaction on the system. Equation 2.3 is the most direct way to measure the force that is applied to an object.

Law of Action-Reaction

When two particles exert force upon one another, the forces act along the line joining the particles, and the two force vectors are equal in magnitude and opposite in direction.

According to the law of action-reaction, each force exerted by one body on another is counteracted by a comparable force exerted by the second body on the first. For example, when a person performs a jump shot in basketball, he exerts a force against the ground, and the ground responds simultaneously with a reaction force (ground reaction force) on the jumper. The law of action-reaction states that the forces between the jumper and ground are equal in magnitude but opposite in direction. The consequence of this interaction is that each body (the jumper and the ground) experiences an acceleration that depends on its mass ($F = ma$).

The law of action-reaction provides the foundation for many devices that are used to measure force. Examples include strain gauges, spring balances, piezoelectric crystals, and capacitors. When a force is applied to one of these devices, it is deformed microscopically and the amount of deformation can be calibrated in units of force (newtons). The signals corresponding to the deformation include stretch (strain) for a strain gauge, displacement for a spring balance, electrical charge for a piezoelectric crystal, and electric current for a capacitor. Through measurement of the deformations in response to a set of known forces, each device can be calibrated and subsequent deformations can be used to estimate the magnitude of an applied force.

FREE BODY DIAGRAM

Because human movement often involves many forces, biomechanists use free body diagrams to aid in an analysis. A **free body diagram** defines the extent of an analysis and identifies the significant forces involved in the action (Dempster, 1961; Komistek et al., 2005). It is usually drawn as a stick figure, along with a set of coordinates, with the forces indicated as arrows. The stick figure part of the diagram indicates the system to be considered in the analysis, as seen in the following examples. The actual candidate forces that can be included on a free body diagram are described subsequently.

The first example involves running (figure 2.1). Suppose that we want to determine the magnitude and direction of the force exerted by a runner on the ground. According to the law of action-reaction, we can measure the force exerted by the runner on the ground by recording the reaction force of the ground, which is known as the ground reaction force. The first step in the analysis is to draw a free body diagram, which involves specifying the system. For this analysis, we will use the runner's body as the system (figure 2.1*b*). The next step is to identify all the external forces acting on the system as arrows with the correct length (magnitude) and direction, and to label them appropriately (figure 2.1*c*). The forces shown in figure 2.1*c* represent air resistance ($\mathbf{F}_a$), weight due to gravity ($\mathbf{F}_w$), and ground reaction force ($\mathbf{F}_g$). In general, the forces to be included on a free body diagram are the weight of the system and forces arising from contact with the surroundings (e.g., ground reaction force, air resistance).

Free body diagrams show only the forces acting on the system from the surroundings and not those within the system. The free body diagram in figure 2.1*c,* for example, does not show any muscle forces across the knee joint, even though the quadriceps femoris muscle is undoubtedly contracting; this is not shown because the muscle force is internal to the system. If, however, the aim of an analysis is to examine muscle activity across a joint during movement, as we will do in chapter 3, then a different system must be defined. We accomplish this by terminating the system at the joint of interest so that we can identify the forces exerted on the joint by its surroundings. For such a system, two types of forces are identified at the joint: one representing

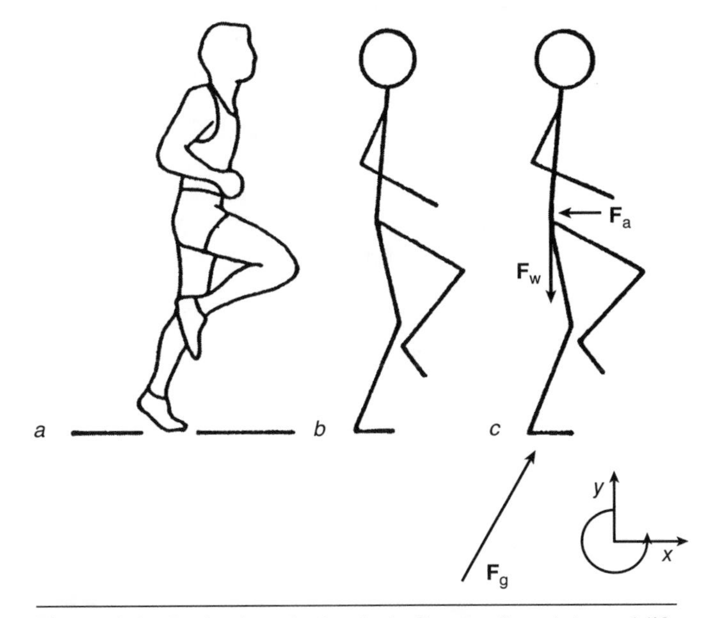

Figure 2.1 Derivation of a free body diagram from *(a)* a real-life figure to *(b)* the identification of the system and *(c)* the inclusion of the forces acting on the system from its surroundings.

the net muscle force and another denoting the effect of other tissues (e.g., joint capsule, ligaments). Let us return to the example shown in figure 2.1. To determine the magnitude and direction of the net muscle force at the instant shown in figure 2.1*b*, the appropriate free body diagram is shown in figure 2.2.

As another example of drawing a free body diagram, suppose we were interested in determining the forces acting in the lower back of a weightlifter. To determine the net muscle force, we must define the system to isolate the joint of interest. In this example, we decide on the L5-S1 intervertebral joint, and the system comprises the upper body of the lifter down to this joint (figure 2.3*b*). The forces acting on this system include weights, joint forces, and any other external forces. There are two weights, one due to the upper body ($\mathbf{F}_{w,u}$) and the other due to the barbell ($\mathbf{F}_{w,b}$). Weight vectors always act vertically downward. As in figure 2.2, the joint forces are the net muscle force ($\mathbf{F}_m$) and the joint reaction force ($\mathbf{F}_j$). The joint reaction force corresponds to the reaction of the adjacent body parts on our system. Although the direction of $\mathbf{F}_m$ is relatively straightforward in that it opposes the rotation produced by the load (i.e., the two weights), the direction of $\mathbf{F}_j$ is usually unknown and arbitrarily drawn as a compressive force acting into the joint. In addition to these four forces, it is necessary to add a force produced by the pressure in the abdomen, which is known as the intra-abdominal pressure force ($\mathbf{F}_i$). $\mathbf{F}_i$ tends to cause extension about the hip joint (Andersson et al., 1977; Eie & Wehn, 1962; Hagins et al., 2006). This reflexively controlled force is a protective mechanism and has a significant role in lifting activities (Arjmand & Shirazi-Adl, 2006; Marras et al., 1985; Miyamoto et al., 1999).

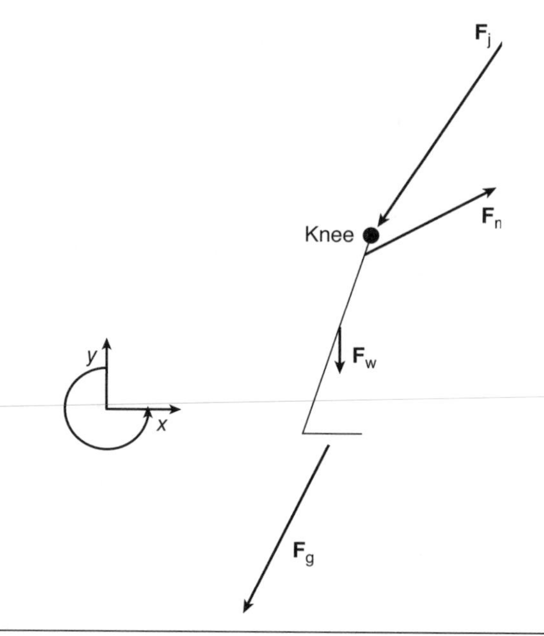

Figure 2.2 Free body diagram of a system (shank + foot) upon which four external forces are acting. $\mathbf{F}_g$ = ground reaction force, $\mathbf{F}_j$ = joint reaction force, $\mathbf{F}_m$ = resultant muscle force, $\mathbf{F}_w$ = weight of the system.

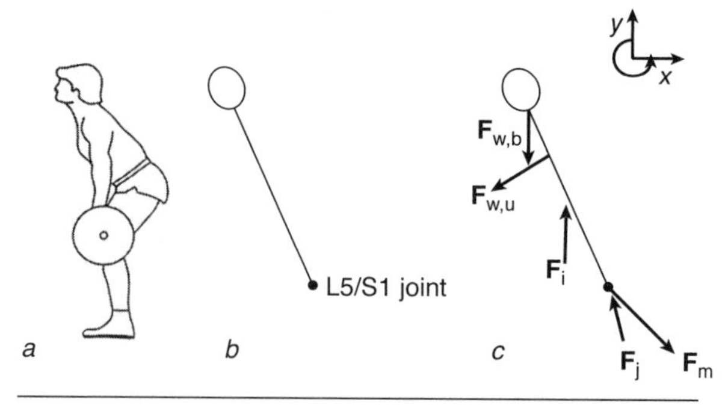

Figure 2.3 The derivation of a free body diagram to analyze part of a weightlifting movement: *(a)* whole-body diagram of the weightlifter; *(b)* system used to expose the forces across the L5-S1 intervertebral joint; *(c)* free body diagram showing the interactions between the system and its surroundings.

From these examples, you should understand why it is necessary to draw a free body diagram and how we do this. The purpose of a free body diagram is to identify the conditions associated with an analysis. There are two steps to drawing a free body diagram. First, define the system you need in order to answer the question, which usually involves drawing a stick figure of the system. Second, identify all the forces between the system and its surroundings and include these on the free body diagram as force vectors. How we know which force to include on a free body diagram is discussed in the remainder of this chapter and the next chapter.

TORQUE

All human movement involves the rotation of body segments about their joint axes. These actions are produced by the interaction of forces associated with external loads and muscle activity. Human movement is the consequence of an imbalance between the components of these forces that produce rotation. The capability of a force to produce rotation is known as **torque** or **moment of force.** Torque represents the rotational effect of a force with respect to an axis; it is the amount of rotation that a force can produce.

Torque is a vector that is equal to the magnitude of the force times the perpendicular distance between the line of action of the force and the axis of rotation. This distance is the **moment arm** and may be expressed as a scalar or a vector quantity. As a scalar quantity, it is the magnitude of the distance from the line of action to the axis of rotation. As a vector quantity, it is specified as being directed from the axis of rotation to the line of action of the force. Algebraically,

$$\boldsymbol{\tau} = \boldsymbol{r} \times \boldsymbol{F} \tag{2.4}$$

where $\boldsymbol{\tau}$ = torque, $\boldsymbol{r}$ = moment arm, and $\boldsymbol{F}$ = force. Because the moment arm is a perpendicular distance, it is the shortest distance from the line of action of the force to the axis of rotation. The direction of the torque vector is perpendicular to the plane that contains the force and moment arm vectors. As explained in chapter 1, when two vector quantities are multiplied and the result is a vector, the direction of the product is always perpendicular to the plane that contains the other two vectors; this procedure is called the vector product or cross product. Torque is a vector product.

For graphic convenience, torque is often represented as a curved arrow in the same plane as the moment arm and force vectors. We can determine the direction of the curved arrow by applying the right-hand-thumb rule, as shown in figure 2.4. This involves drawing the vectors to be multiplied so that their tails are connected and then applying the right-hand-thumb rule by rotating the extended fingers of the right hand from the direction of the moment arm (**r**) to the direction of the force ($\mathbf{F}_m$); that is, $\mathbf{r} \times \mathbf{F}_m$. The extended thumb, in this example, indicates that the product ($\boldsymbol{\tau}_m$) is directed out of the page, and this is shown as the curved arrow with a counterclockwise rotation.

Torque is always determined relative to a specific axis (as indicated by the moment arm) and must be expressed with respect to the same axis. To discuss the rotary effect of a load or a muscle force, we need to indicate the axis about which the rotation will occur. Because torque is calculated as the product of a force (N) and a distance (m), the unit of measurement is newtons times meters (N•m).

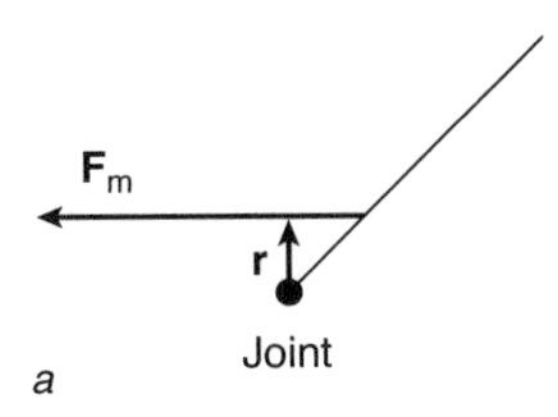

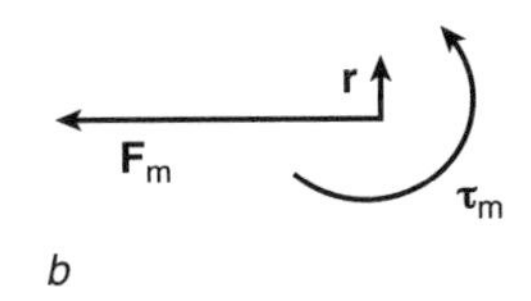

Figure 2.4 Use of the right-hand-thumb rule to determine the direction of a muscle torque ($\boldsymbol{\tau}_m$) vector. *(a)* Muscle torque is the product of a moment arm (**r**) and a muscle force ($\mathbf{F}_m$). *(b)* Realignment of **r** and $\mathbf{F}_m$ so that the tails connect and the direction of $\boldsymbol{\tau}_m$ can be determined with the right-hand-thumb rule.

EXAMPLE 2.1

Influence of Moment Arm

As an example of the role of torque in human movement, consider an individual who sits on a bench and performs a knee extension exercise (figure 2.5). The person is rehabilitating a knee injury and has a mass that can be attached to the shank to vary the difficulty of the exercise. Would it require more effort to perform the exercise if the mass was attached at the ankle or in the middle of the shank?

This exercise involves the muscles around the knee joint, which must be contracted to exert a torque that can displace the load. The magnitude of the muscle activity, therefore, depends on the load. In this instance, the load is a torque about the knee joint due to the mass of the system (attached mass + shank and foot mass) and is obtained from the following relation:

$$\boldsymbol{\tau}_l = \boldsymbol{r} \times \boldsymbol{F}_w$$

where τ_l is the load torque, **r** is the moment arm, and $\mathbf{F}_w$ is the weight of the system ($\mathbf{F}_w = mg$, where m is the mass of the system and g is the acceleration due to gravity). If you use the right-hand-thumb rule to perform the vector product (**r** cross **F**), which is the direction of τ_l? On the basis of the description in figure 2.4, point the fingers of your right hand in the direction of **r** and then curl them in the direction of $\mathbf{F}_w$, and your extended thumb will indicate the direction of τ_l. The result should tell you that τ_l is a vector pointing into the page.

Because the weight of the system does not change in this example, τ_l varies with the moment arm (**r**). So the question becomes, is **r** greater when the mass is attached at the ankle or at the midshank? At any point in the range of motion of the exercise, **r** is longer when the mass is attached at the ankle. Thus, the exercise is more difficult when the load is farther from the axis of rotation (Nourbakhsh & Kukulka, 2004).

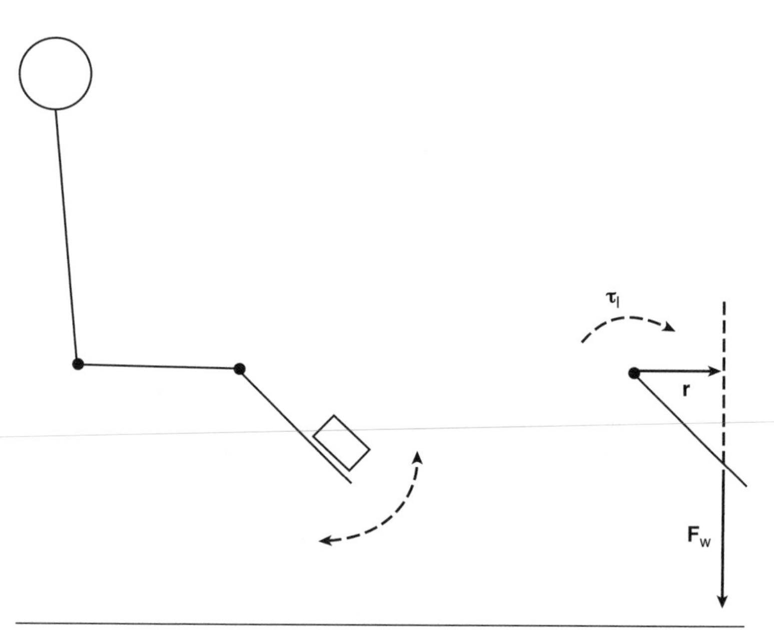

Figure 2.5 An analysis to determine the influence of the position (**r**) of a weight ($\mathbf{F}_w$) on the lower leg on the torque that it produces about the knee joint (τ_l).

FORCES DUE TO BODY MASS

Once we define a system with a free body diagram, we are able to consider the forces that are likely to exist between the system and its surroundings. In general, the forces involved in human movement are those due to body mass and those due to contact with the surroundings. These forces are distributed throughout the human body and can be estimated with techniques that we will examine in this and the next chapter.

Gravity

Newton characterized gravity in a statement known as the **law of gravitation:** All bodies attract one another with a force proportional to the product of their masses and inversely proportional to the square of the distance between them. That is,

$$F \propto \frac{m_1 m_2}{r^2} \tag{2.5}$$

where m_1 and m_2 are the masses of the two bodies and r is the distance between them.

These forces of attraction are generally regarded as negligible in the study of human movement, with the exception of the attraction between Earth and various objects. **Weight** indicates the amount of gravitational attraction between an object and Earth. As a force, weight is measured in newtons (N). Weight varies proportionally with mass—the greater the mass, the greater the attraction—but the two are separate quantities. Weight (force) is a derived variable in the SI system, whereas mass is a base unit (appendix A). Due to the law of gravitation, the weight of an object varies with its distance to the center of Earth; hence, weight is slightly less at higher altitudes than at sea level.

We can determine the magnitude of the total-body weight vector by reading the value from a weight scale. The validity of this procedure can be demonstrated with a simple analysis based on Newton's law of acceleration (equation 2.3),

$$\sum \boldsymbol{F} = m\boldsymbol{a}$$

which states that the sum (Σ = Sigma) of the forces ($\boldsymbol{F}$) produces an acceleration ($\boldsymbol{a}$) of the system that depends on the mass (m) of the system. Because gravity acts in the vertical direction (y), the analysis can be confined to this direction. The forces acting on a person standing on a weight scale include her weight (F_w), which is directed downward and indicated as negative, and the vertical component of the ground reaction force ($F_{g,y}$), which is directed upward and indicated as positive. The decision to label F_w as negative and $F_{g,y}$ as positive is quite arbitrary; however, it is essential to distinguish these differences in direction. To this end, each free body diagram should be accompanied by a reference axis (coordinate system) that shows the positive directions of each component (figure 2.6): positive vertical (y), horizontal (x), and rotation directions. The appropriate free body diagram is shown in figure 2.6, and the analysis is as follows:

$$\sum F_y = ma$$

$$-F_w + F_{g,y} = -ma_y$$

$$F_w = F_{g,y} - ma$$

Because the person is stationary, $a_y = 0$.

$$F_w = F_{g,y} - 0$$

$$F_w = F_{g,y}$$

The downward-directed weight vector originates from a point known as the **center of mass** (CM), which represents a balance point about which all the particles of the object are evenly distributed. The CM is an abstract point that moves when the body segments are moved relative to one another.

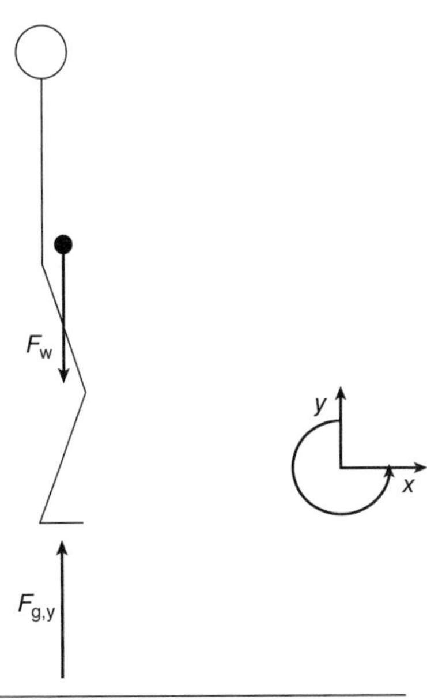

Figure 2.6 Free body diagram of the whole body.

Segmental Analysis

Like most forces encountered in the analysis of human movement, weight is a distributed force with the net effect shown at a central location, the CM. For example, a free body diagram would show the contact force between a player's head and a soccer ball as occurring at a point, although the contact is distributed over a large part of the forehead. Similarly, the CM of the object is the point about which the mass of the object is evenly distributed. Because the CM for the body moves during movement, biomechanists use a procedure known as a segmental analysis to determine its location at various instants during the movement. The approach considers the human body to comprise a set of rigid bodies (segments) and deals with them individually to determine the location of the whole-body CM (figure 2.7). A segmental analysis involves estimating the mass and CM location for each body segment and then using this information to estimate the location of the CM for the whole body. This procedure is described in chapter 3. At this point, we introduce the database on segmental masses and CMs, which represents the magnitude and location of the weight vector for the various segments of the human body.

Several groups of investigators have derived regression equations to estimate various anthropometric (measurements of the human body) segmental dimensions (Chandler et al., 1975; Chester & Jensen, 2005; Dempster, 1955; Dumas et al., 2007; Jensen et al., 1996; Matsuo et al., 1995; Pavol et al., 2002; Shan & Bohn, 2003). Some of the data from one of the most comprehensive cadaver studies (n = 6) are reported in tables 2.1 and 2.2. These data consist of regression equations to estimate segmental weights and CM locations (table 2.1) as well as actual values for segmental moments of inertia about the three axes (table 2.2).

Table 2.1 presents regression equations for estimating segment weight from total-body weight and for estimating CM location for each segment based on the measurement of segment length. For an individual weighing 750 N,

$$\text{Trunk weight} = 0.532 \times 750 - 6.93$$
$$= 392 \text{ N}$$

which accounts for 52% of total-body weight. Similarly,

$$\text{Hand weight} = 0.005 \times 750 + 0.75$$
$$= 4.5 \text{ N}$$

which represents about 0.6% of total-body weight. The segment boundaries used by Chandler and colleagues (1975) are shown in figure 2.7*a*. Table 2.1 can also be used to estimate CM

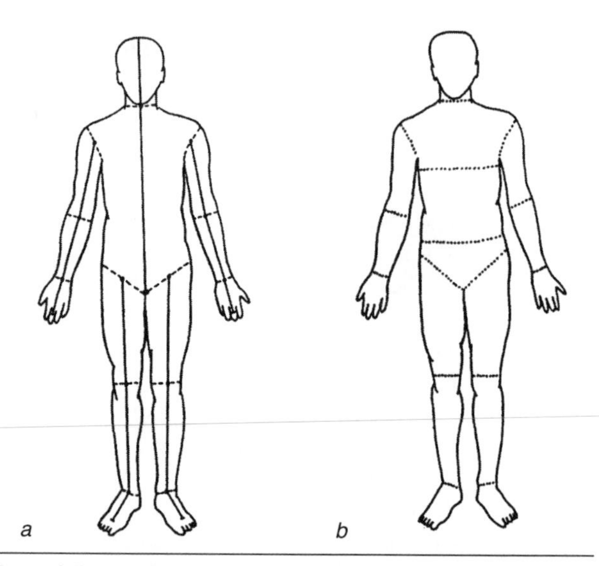

Figure 2.7 The body segment organization used in two studies: *(a)* Chandler and colleagues, 1975—the 14 segments comprised the head, trunk, upper arms, forearms, hands, thighs, shanks, and feet; *(b)* Zatsiorsky and Seluyanov, 1983—the 16 segments included the head, upper torso, middle torso, lower torso, upper arms, forearms, hands, thighs, shanks, and feet.

Table 2.1 Regression Equations Estimating Body Segment Weights and Locations of the Center of Mass

Segment	Weight (N)	CM location (%)	Proximal end of segment
Head	$0.032\ F_w + 18.70$	66.3	Vertex
Trunk	$0.532\ F_w - 6.93$	52.2	C1
Upper arm	$0.022\ F_w + 4.76$	50.7	Shoulder joint
Forearm	$0.013\ F_w + 2.41$	41.7	Elbow joint
Hand	$0.005\ F_w + 0.75$	51.5	Wrist joint
Thigh	$0.127\ F_w - 14.82$	39.8	Hip joint
Shank	$0.044\ F_w - 1.75$	41.3	Knee joint
Foot	$0.009\ F_w + 2.48$	40.0	Heel

Note. Body segment weights are estimated from total-body weight (F_w), and the segmental center-of-mass (CM) locations are expressed as a percentage of segment length as measured from the proximal end of the segment.

Table 2.2 Whole-Body Moments of Inertia (kg•m²) About the Somersault, Cartwheel, and Twist Axes

Position	Somersault	Cartwheel	Twist
Layout	12.55	15.09	3.83
Open pike (arms out to side)	8.38	8.98	4.79
Closed pike (fingers touching toes)	8.65	6.60	3.58
Tuck	4.07	4.42	2.97

location. If the length of an individual's thigh (hip-to-knee distance) is 36 cm, then the CM for that thigh would be located at 39.8% of that distance from the hip joint:

$$\text{CM location} = 36 \times 0.398$$
$$= 14.3 \text{ cm from the hip joint}$$

Because human motion is based on the rotation of body segments about one another, the distribution of mass in a segment is as important to know as its mass. The distribution of mass can be quantified by a parameter known as the moment of inertia. The **moment of inertia** is the angular equivalent of inertia (mass) and represents a measure of the resistance that an object offers to a change in its motion about an axis. The moment of inertia (I) is defined as

$$I = \sum_{i=1}^{n} m_i r_i^2 \quad (2.6)$$

where n indicates the number of elements (particles or segments) in the system, m_i represents the mass of the ith element of the system, and r_i is the distance of the ith element from the axis of rotation. Equation 2.6 is useful for systems that comprise a relatively small number of mass elements. For objects that have a continuous distribution of matter, however, the same relation should be expressed as an integral.

$$I = \int r^2 dm \quad (2.7)$$

Equation 2.7 is used for a system that has a large number of small mass elements (dm), such as the particles that comprise the human body or a body segment. The integral states that I is calculated by summing the products of two terms, the mass of each element and the square of its distance (r^2) from the axis of rotation. The mass elements refer to the parts into which the object can be divided.

Both equations 2.6 and 2.7 provide a measure of the mass distribution about an axis of rotation. The distribution of mass is usually characterized with reference to three orthogonal (perpendicular) axes that are termed the principal axes of rotation. In the analysis of human movement, these axes are often defined as the somersault (side-to-side), cartwheel (front-to-back), and twist (longitudinal) axes. Table 2.2 lists estimates of the moment of inertia about the three principal axes for the human body in different positions (Miller & Nelson, 1973). Note that the moment of inertia declines as the mass of the body is brought closer to the axis of rotation; for example, for each body position the moment of inertia is least about the twist axis.

From such measurements as those listed in table 2.2, it is possible to estimate whole-body moment of inertia about any axis of rotation used in human movement. This is accomplished with the **parallel axis theorem.** This theorem states that the moment of inertia about an axis of rotation (I_o) is equal to the sum of the moment of inertia about any parallel axis and a transfer term (md^2). In biomechanics, we often use moments of inertia that have been measured relative to the CM (I_g) and determine the value about another parallel axis (equation 2.8),

$$I_o = I_g + md^2 \quad (2.8)$$

where m represents the mass of the system and d is the distance between the two parallel axes. For example, the parallel axis theorem can be used to calculate the whole-body moment of inertia about a transverse axis (somersault axis) passing through the hands of a gymnast (71.2 kg) who is performing stunts on a high bar.

$$I_{hands} = 12.55 \text{ kg}\cdot\text{m}^2 + [71.2 \text{ kg} \times (1.02 \text{ m})^2]$$
$$= 12.55 + 74.08 \text{ kg}\cdot\text{m}^2$$
$$= 86.6 \text{ kg}\cdot\text{m}^2$$

where 12.55 kg•m^2 is the moment of inertia about the transverse axis passing through the whole-body CM in a layout position (table 2.2) and 1.02 m^2 is the distance from the CM to the hands (the axis of rotation).

EXAMPLE 2.2

Calculating the Moment of Inertia

Describe the mass distribution for a system (figure 2.8) that has four weights connected by a massless wire with its CM (*g*) located at 0.57 m from the proximal end (*o*).

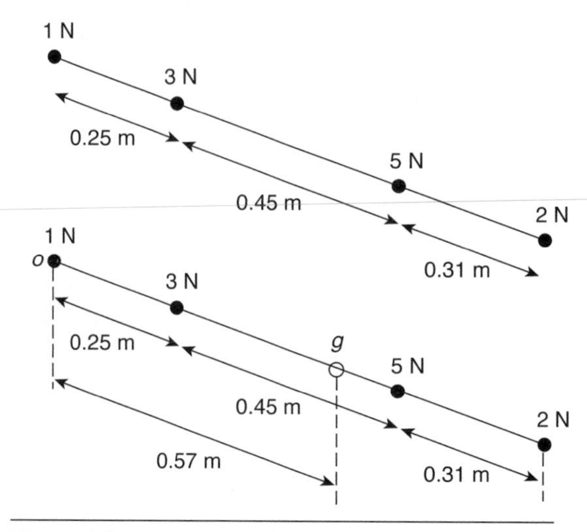

Figure 2.8 A system of four weights connected by a massless wire.

A. Use equation 2.6 to calculate the moment of inertia of the system about its center of gravity (I_g).

$$I_g = \sum_{i=1}^{4} m_i r_i^2$$

$$I_g = m_1 r_1^2 + m_2 r_2^2 + m_3 r_3^2 + m_4 r_4^2$$

$$I_g = \frac{1}{9.81}0.57^2 + \frac{3}{9.81}0.32^2 + \frac{5}{9.81}0.13^2 + \frac{2}{9.81}0.44^2$$

$$I_g = 0.0331 + 0.0313 + 0.0086 + 0.0395$$

$$I_g = 0.1125 \text{ kg} \bullet \text{m}^2$$

B. Use equation 2.6 to calculate the moment of inertia of the system about the proximal end (I_o).

$$I_o = \frac{3}{9.81}0.25^2 + \frac{5}{9.81}0.70^2 + \frac{2}{9.81}1.01^2$$

$$I_o = 0.0191 + 0.2497 + 0.2080$$

$$I_o = 0.4768 \text{ kg} \bullet \text{m}^2$$

C. Use the answer from part A and the parallel axis theorem (equation 2.8) to determine the moment of inertia of the system about the proximal end.

$$I_o = I_g + md^2$$

$$I_o = 0.1125 + \frac{11}{9.81}0.57^2$$

$$I_o = 0.1125 + 0.3643$$

$$I_o = 0.4768 \text{ kg} \bullet \text{m}^2$$

One way to determine the moment of inertia for an object such as the whole body or a body segment is to use the **pendulum method.** This involves measuring the magnitude of the moment of inertia for the object about the suspension point and then using the parallel axis theorem to calculate the moment of inertia about its CM. To use this method you must know the location of the CM and the distance between the CM and the suspension point. The pendulum method consists of suspending an object from a fixed point, setting it in motion by shifting it a few degrees from its resting position, and measuring the time it takes to complete one period of oscillation. The moment of inertia about the suspension point (I_o) is given by

$$I_o = \frac{F_w \cdot h \cdot T^2}{4\pi^2} \tag{2.9}$$

where

F_w = weight of the object,

h = distance between the CM and the suspension point, and

T = period of one oscillation.

Once I_o has been determined, I_g can be found with use of the parallel axis theorem.

In addition to estimates of mass and CM location, the cadaver studies have yielded segmental moments of inertia. Because there is low variability in these data across specimens, they are typically reported as mean values (table 2.3) (Chandler et al., 1975).

To supplement the segmental data derived from cadavers, some investigators have used mathematical modeling procedures. With this approach, the human body is represented as a set of geometric components, such as spheres, cylinders, and cones (Durkin & Dowling, 2003; Hanavan, 1964, 1966; Hatze, 1980, 1981*a, b;* Miller, 1979; Yeadon, 1990). One of the first to use this approach was Hanavan (1964, 1966). Figure 2.9*a* shows the Hanavan model, which divides the human body into 15 simple geometric solids of uniform density. The advantage of this model is that only a few simple anthropometric measurements (e.g., segment lengths and circumferences) are required to personalize the model and to predict the CM and moment of inertia for each body segment. However, three assumptions typically used in modeling body segments limit the accuracy of the estimates: (1) Segments are assumed to be rigid; (2) the boundaries between the segments are assumed to be distinct; and (3) segments are assumed to have a uniform density. In reality, there can be substantial displacement of the soft tissue during movement; the boundaries between segments are fuzzy; and the density varies within and between segments.

Hatze (1980) developed a more detailed model of the human body (figure 2.9*b*). Hatze's hominoid consists of 17 body segments and requires 242 anthropometric measurements for individualization. The model subdivides the segments into small mass elements of different geometrical structures; this allows the shape and density fluctuations of a segment to be modeled in detail. Furthermore, no assumptions are made regarding bilateral symmetry; and the model differentiates between men and women, adjusting the densities of certain segmental parts according to the value of a special subcutaneous fat indicator. The model is able to account for changes in body morphology, such as those due to obesity and pregnancy, and can accommodate children. Indeed, the model has been used to estimate the mass, CM, and moment of inertia for the limb segments of infants (Schneider & Zernicke, 1992).

Table 2.3 Segmental Moments of Inertia ($kg \cdot m^2$) About the Somersault, Cartwheel, and Twist Axes

Segment	Somersault	Cartwheel	Twist
Head	0.0164	0.0171	0.0201
Trunk	1.0876	1.6194	0.3785
Upper arm	0.0133	0.0133	0.0022
Forearm	0.0065	0.0067	0.0009
Hand	0.0008	0.0006	0.0002
Thigh	0.1157	0.1137	0.0224
Shank	0.0392	0.0391	0.0029
Foot	0.0030	0.0034	0.0007

Although acquisition of the input for the model (242 anthropometric measurements) is time-consuming, the output of the model provides accurate estimates of volume, mass, CM location, and moments of inertia for the identified body segments. Typical estimates for two men are indicated in table 2.4. One unique feature of the model is the approach used to represent the trunk and shoulder segments (figure 2.9*b*).

Another technique available to determine body segment parameters is based on imaging procedures, such as computed tomography and magnetic resonance imaging (Durkin & Dowling, 2003; Engstrom et al., 1991; Martin et al., 1989). An example of this approach is the use of radioisotopes to measure the intensity of a gamma-radiation beam before and after it passes through a body segment. The principle involves scanning a subject's

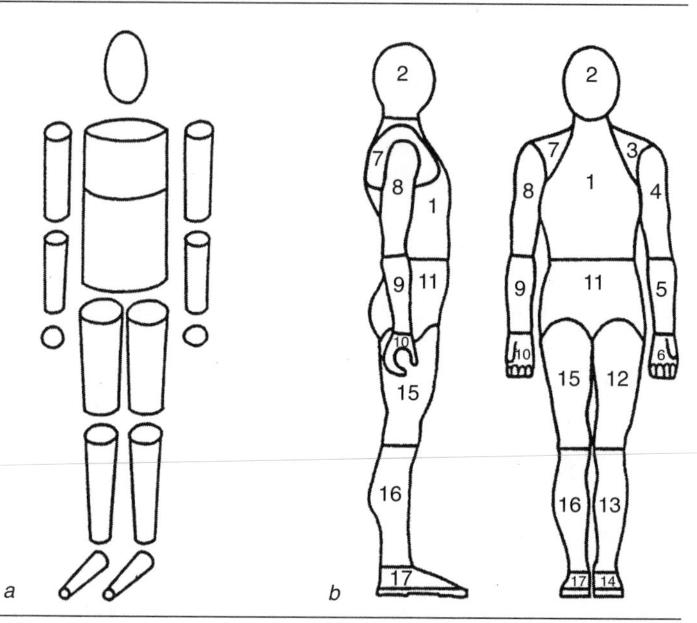

Figure 2.9 Models of the human body. *(a)* The Hanavan model of the human body; *(b)* the 17-segment hominoid of Hatze (1980). The 17 segments are 1, abdominothoracic; 2, head; 3, left shoulder; 4, left upper arm; 5, left forearm; 6, left hand; 7, right shoulder; 8, right upper arm; 9, right forearm; 10, right hand; 11, abdominopelvic; 12, left thigh; 13, left shank; 14, left foot; 15, right thigh; 16, right shank; 17, right foot.

Table 2.4 Segmental Parameter Values for Two Men Computed With the Hatze Model

	F.B. (23 YEARS)			C.P. (26 YEARS)		
Segment	**Volume**	**CM location**	I_{zz}	**Volume**	**CM location**	I_{zz}
Head-neck	4.475	0.517	0.0303	4.537	0.516	0.0337
Abdominothoracic	19.111	0.439	0.3117	19.803	0.444	0.3302
Left shoulder	1.438	0.727	0.0047	2.042	0.706	0.0080
Right shoulder	1.890	0.711	0.0071	2.121	0.699	0.0084
Left upper arm	2.110	0.432	0.0196	2.123	0.437	0.0203
Right upper arm	2.021	0.437	0.0168	2.340	0.428	0.0229
Left forearm	1.023	0.417	0.0067	1.223	0.413	0.0086
Right forearm	1.190	0.404	0.0079	1.313	0.412	0.0093
Left hand	0.453	0.515	0.0011	0.416	0.533	0.0010
Right hand	0.446	0.531	0.0011	0.417	0.524	0.0010
Abdominopelvic	8.543	0.368	0.0399	9.614	0.395	0.0541
Left thigh	8.258	0.479	0.1475	8.744	0.473	0.1653
Right thigh	8.278	0.480	0.1415	8.729	0.466	0.1702
Left shank	3.628	0.412	0.0615	3.856	0.420	0.0798
Right shank	3.686	0.417	0.0663	3.798	0.417	0.0747
Left foot	0.887	-------	0.0041	1.032	-------	0.0051
Right foot	0.923	-------	0.0042	1.055	-------	0.0051

Note. Volume = computed segment volume (L); CM Location = distance from proximal end of segment (proportion of segment length, where 0.500 indicates 50% of segment length), I_{zz} = moment of inertia about the somersault (zz) axis through the CM. Segment densities were abdomen = 1000 + 30 i_m, head = 1120, neck = 1040, and arms and legs = 1080 + 20 i_m kg/m^3, where i_m = 1 for men and 0 for women.

body and obtaining the surface density and dimensions of the body segments subjected to the radiation. Zatsiorsky and colleagues (Zatsiorsky & Seluyanov, 1983; Zatsiorsky et al., 1990*a, b*) performed this procedure on 100 men (age = 24 ± 6 years; height = 1.74 ± 0.06 m; mass = 73 ± 9 kg) and 15 women (age = 19 ± 4 years; height = 1.74 ± 0.03 m; weight = 62 ± 7 kg). The segmental data on these young adults were subsequently adjusted by de Leva (1996), and the results are listed in tables 2.5 and 2.6. The proximal ends of the segments are similar to those used by Chandler and colleagues (table 2.1) with the exceptions of the trunk and upper torso (suprasternale), midtorso (xyphion), and lower torso (omphalion). The end of the hand segment was defined as the third knuckle, between the metacarpal and the finger.

Table 2.5 Segment Length, Mass, and Center-of-Mass (CM) Location for Young Adult Women (W) and Men (M)

	LENGTH (cm)		MASS (%)		CM LOCATION (%)	
Segment	W	M	W	M	W	M
Head	20.02	20.33	6.68	6.94	58.94	59.76
Trunk	52.93	53.19	42.57	43.46	41.51	44.86
Upper torso	14.25	17.07	15.45	15.96	20.77	29.99
Middle torso	20.53	21.55	14.65	16.33	45.12	45.02
Lower torso	18.15	14.57	12.47	11.17	49.20	61.15
Upper arm	27.51	28.17	2.55	2.71	57.54	57.72
Forearm	26.43	26.89	1.38	1.62	45.59	45.74
Hand	7.80	8.62	0.56	0.61	74.74	79.00
Thigh	36.85	42.22	14.78	14.16	36.12	40.95
Shank	43.23	43.40	4.81	4.33	44.16	44.59
Foot	22.83	25.81	1.29	1.37	40.14	44.15

Table 2.6 Segmental Moments of Inertia ($kg \cdot m^2$) for Young Adult Women (W) and Men (M) About the Somersault, Cartwheel, and Twist Axes

	SOMERSAULT		CARTWHEEL		TWIST	
Segment	**W**	**M**	**W**	**M**	**W**	**M**
Head	0.0213	0.0296	0.0180	0.0266	0.0167	0.0204
Trunk	0.8484	1.0809	0.9409	1.2302	0.2159	0.3275
Upper torso	0.0489	0.0700	0.1080	0.1740	0.1001	0.1475
Middle torso	0.0479	0.0812	0.0717	0.1286	0.0658	0.1212
Lower torso	0.0411	0.0525	0.0477	0.0654	0.0501	0.0596
Upper arm	0.0081	0.0114	0.0092	0.0128	0.0026	0.0039
Forearm	0.0039	0.0060	0.0040	0.0065	0.0005	0.0022
Hand	0.0004	0.0009	0.0006	0.0013	0.0002	0.0005
Thigh	0.1646	0.1995	0.1692	0.1995	0.0326	0.0409
Shank	0.0397	0.0369	0.0409	0.0387	0.0048	0.0063
Foot	0.0032	0.0040	0.0037	0.0044	0.0008	0.0010

EXAMPLE 2.3

Performing a Segmental Analysis

Let us briefly consider a segmental analysis of the weightlifter depicted previously in the "Free Body Diagram" section. To determine the contribution to the movement made by muscles crossing the knee, we first decide how many segments are necessary to represent the lifter (figure 2.10*a*). In the movement, shown at the top of figure 1.17, the arms do not bend at the elbow, so we can consider the arms as one segment. And because the movement is confined to the sagittal *(x-y)* plane, we assume that the left and right sides of the body are more or less doing the same thing. These simplifications allow us to reduce our free body diagram from the maximum of 15 to 17 segments (based on Hanavan's and Hatze's models) down to six segments. Furthermore, the whole-body stick figure will become six separate systems, one for each segment (figure 2.10*b*); this is the essence of a segmental analysis.

Next we draw the appropriate free body diagram (figure 2.10*c*). This includes identifying the system and determining the location (CM) and magnitude of the weight vector ($F_{w,s}$). Because we want to determine the net muscle force across the knee joint ($F_{m,k}$), we use the shank as the system. Both the CM location and the magnitude of the weight vector can be estimated from the data in table 2.1. According to table 2.1, the CM for the shank is located at 41.3% of shank length from the knee joint. If the shank of the weightlifter measured 36 cm, then the CM would be situated 14.9 cm (36 × 0.413 = 14.9) from the knee. Similarly, the magnitude of the weight vector is determined as a function of total-body weight. For an 800 N weightlifter, the weight of one shank would be 33.5 N (0.044 × 800 – 1.75 = 33.5). Finally, the other forces acting on the system, which we have not described yet, must be added to the free body diagram.

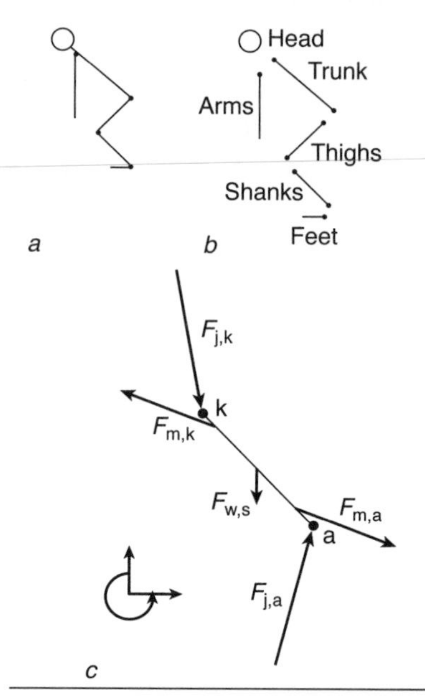

Figure 2.10 Free body diagram of a weightlifter. *(a)* Whole-body stick figure; *(b)* segmental components; *(c)* free body diagram of the shank. $F_{j,a}$ and $F_{j,k}$ = ankle and knee joint reaction forces; $F_{m,a}$ and $F_{m,k}$ = ankle and knee resultant muscle forces; $F_{w,s}$ = weight of the shank.

Wobbling Mass Model

In a segmental analysis, the body segments are usually assumed to be rigid and connected by simple joints. Because body segments are not rigid, estimated forces can contain substantial errors, especially for movements that involve significant accelerations. One can reduce these errors by representing each segment with two components: a rigid link and a soft tissue component that is referred to as the **wobbling mass** (Challis & Pain, 2008; Liu & Nigg, 2000; Pain & Challis, 2004). With this approach, the wobbling mass can be represented as a tube that surrounds a rigid cylindrical segment and is connected to it by springs and dampers.

Pain and Challis (2006) developed a wobbling-mass model to examine the forces during the first 100 ms after landing from a jump. The model comprised four segments: foot, shank, thigh, and torso. The model was constructed through estimation of segment masses and CM locations (Zatsiorsky et al., 1990*b*), moments of inertia (Challis, 1996), and the relative masses and densities of the rigid and wobbling components (Clarys & Marfell-Jones, 1986; Clarys et al., 1984). The initial position and velocity of the segments were determined from the kinematics of the subject who performed the task. The other model parameters included actuators at each of the joints and the stiffness and damping properties of the connections between the wobbling mass and the rigid segment. With this model, Pain and Challis (2006) were able to obtain a reasonable estimate of the vertical component of the ground reaction force during the first 100 ms of the landing, and found up to 50% differences in the joint forces and torques between the rigid-segment and wobbling-mass models (table 2.7). This comparison indicates that the segmental analysis of such actions as landings and foot contacts during running should consider using a wobbling-mass model to represent the mass properties of each segment (Boyer & Nigg, 2006; Gittoes et al., 2006; Zadpoor et al., 2007).

Table 2.7 Peak Joint Forces and Torques During Landing From a Jump for the Rigid-Segment and Wobbling-Mass Models

	RIGID SEGMENT		WOBBLING MASS	
Joint	Vertical force (kN)	Torque (N•m)	Vertical force (kN)	Torque (N•m)
Ankle	17.1	−370	11.1	−228
Knee	13.3	500	7.72	267
Hip	7.70	−460	5.10	−240

Inertial Force

Newton told us that a moving object moves in a straight line and at a constant speed unless it is acted on by a force (law of inertia). The object's resistance to any change in its motion is called its inertia. An object in motion can, due to its inertia, exert a force on another object. To demonstrate this effect, place one of your forearms in a vertical position (hand pointing upward) with the upper arm horizontal. Relax the muscles of your forearm that cross the wrist joint. Slowly begin to oscillate your forearm about the elbow joint in a small arc in a forward-backward direction. As the movement becomes more vigorous, you should notice that your hand begins to flail, particularly if your forearm muscles are relaxed. As you suspect, the motion of your forearm causes your hand to flail. The forearm (because of its motion) has exerted an inertial force on the hand and caused the motion of the hand; this interaction is known as a **motion-dependent effect.** This simple example emphasizes the mechanical coupling that exists between the components of a linked system, such as between our body segments.

The hammer throw provides an extreme example of the inertial force. We learned in the section on projectile motion (chapter 1) that the distance an athlete can throw an object depends on the object's speed and the angle and height at which it is released. For example, the world-record performance for a hammer throw is 86.7 m, which required a release velocity of 29.1 m/s (65 mph) if the angle of release was optimum (0.78 rad, 45°). To achieve this release speed, the athlete performed five revolutions while traveling across a 2.13 m diameter ring. The athlete applied two forces to the hammer during this procedure: (a) a pulling force that provided an angular acceleration to increase the speed of rotation, and (b) a centripetal force to maintain the angular nature of the motion. Brancazio (1984) estimated that to achieve a release velocity of 29.1 m/s, the athlete provided an average pulling force of about 45 N throughout the event and a centripetal force of about 2.8 kN at release.

Because forces act in pairs (action-reaction), the hammer must exert a force on the thrower (inertial force) of 2.8 kN at release. You can quite easily visualize this motion-dependent effect of the hammer by imagining what would happen to the athlete if he forgot to let go of the hammer but instead held on past the release point. As soon as the centripetal force exerted by the athlete on the hammer declined below 2.8 kN, the inertial force of the hammer, due to its motion, would cause the athlete to be dragged out of the throwing circle.

Motion-dependent effects are significant in many of our everyday activities, such as running, kicking, and piano playing. In running, for example, the motion of the thigh can readily influence the motion of the shank (Piazza & Delp, 1996). Figure 2.11 shows the changes that occur in knee angle and resultant muscle torque (rotary force exerted by muscle) about the hip and knee joints during the swing phase of a running stride (Phillips & Roberts, 1980). The net muscle torque about the hip joint exhibits a biphasic profile: It has a flexor direction (shown as a positive torque) for the first half of the swing phase and then acts in an extensor direction for the second half of the swing phase. The flexor hip torque accelerates the thigh in a forward direction whereas the extensor torque accelerates the thigh in a backward direction. Similarly, the resultant muscle torque about the knee joint (dashed line in figure 2.11) appears biphasic, with an intermediate period of a zero torque. However, whereas the hip torque has a flexor-extensor sequence, the resultant muscle torque about the knee has an extensor-flexor sequence. Thus a net hip flexor torque is associated with a net knee extensor torque, and the latter part of the swing phase comprises a sequence of net hip extensor and knee flexor torques.

The role of the resultant muscle torque about the knee joint becomes clearer when we consider how knee angle changed during the swing phase. We can do this by comparing the two graphs in figure 2.11. Knee angle first decreased and then increased during the swing phase (knee angle graph in figure 2.11). When the knee joint flexed (first half of the swing phase), this was controlled by a resultant extensor torque about the knee joint; that is, flexion was controlled by the extensor muscles. This condition represents a lengthening of the active muscles (extensors, quadriceps femoris) and is referred to as a lengthening contraction. Similarly, during the phase of knee joint extension, there was a resultant flexor (hamstrings) torque about the knee joint. This means that extension of the knee joint was controlled by a lengthening contraction of the knee flexor muscles; knee flexion was controlled, not by the knee flexors, but by the knee extensors, and knee extension was controlled by the knee flexors.

Most movements that we perform involve lengthening contractions. The function of lengthening contractions is to control the effect of a load on the body. In this instance, the load was the inertial force exerted by the thigh on the shank. By using the muscles that cross the hip joints, subjects are able to cause the knee joint to flex and to extend. The muscles that cross the knee joint are used to control the effect of the thigh inertia force on the shank; the lengthening contractions brake the forward and backward rotation of the shank. This example illustrates that the inertial force exerted by one body segment on another is an important consideration in the analysis of human movement (Bizzi & Abend, 1983; Ganley & Powers, 2006; J. P. Hunter et al., 2004; Phillips et al., 1983; Putnam, 1991). We will return to this concept in chapter 3 when we discuss interaction torques.

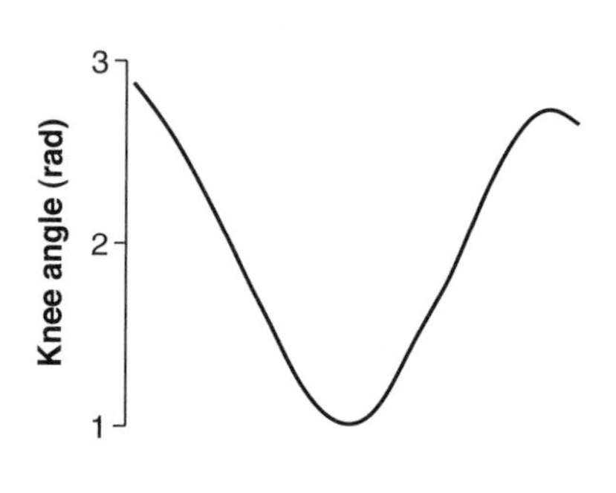

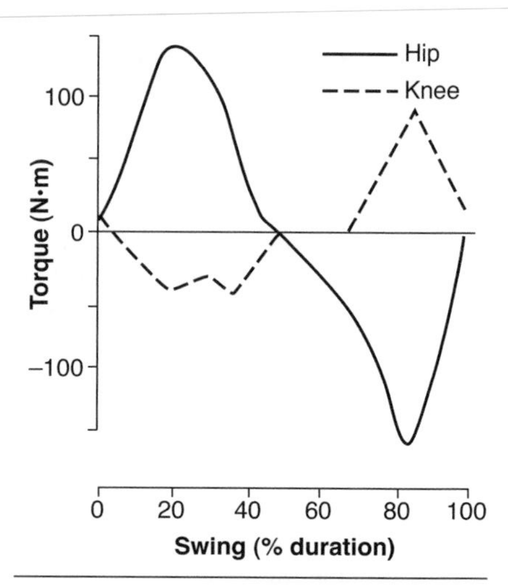

Figure 2.11 Knee angle and resultant muscle torques about the hip and knee joints during the swing phase of a running stride. The subject ran at 5.1 m/s. Knee angle was measured as the relative angle between the thigh and the shank. Positive torque indicates a net muscle force in the direction of flexion.

Data from S.J. Phillips and E.M. Roberts, 1980, "Muscular and non-muscular moments of force in the swing limb of masters runners." In *Proceedings of the Biomechanics Symposium*, edited by J.M. Cooper and B. Haven (Bloomington, IN: Indiana State Board of Health), 256-274.

FORCES DUE TO THE SURROUNDINGS

The other major class of forces that we must consider in the analysis of human movement involves those that are imposed on the body by the surroundings. These include the reaction force provided by the ground and other support surfaces, as well as the interaction between the body and the fluid (air, water) in which the movement occurs.

Ground Reaction Force

The **ground reaction force** describes the reaction force provided by the support surface on which the movement is performed. It is derived from Newton's law of action-reaction to represent the reaction of the ground to the accelerations of all the body segments. The ground reaction force can be measured with an instrument known as a force platform, which essentially operates like a scale for measuring weight. Researchers began using this technique in the 1930s (Elftman, 1938, 1939; Fenn, 1930; Manter, 1938), although the idea had been proposed some time earlier (Amar, 1920; Marey, 1879).

One important difference between a force platform and a weight scale is that the force platform can measure the ground reaction force in three dimensions, and it can do so with greater temporal resolution. The resultant ground reaction force can be resolved into three components whose directions are functionally defined as vertical (up-down), forward-backward, and side to side. These components represent the reaction of the ground to the actions of the person that are transmitted through the feet to the ground and that correspond to the acceleration of the body in these respective directions. The extent to which any body segment influences the ground reaction force depends on its mass and the acceleration of its CM. For example, Miller (1990) estimates that the trunk and head account for about 50% of a runner's acceleration whereas each leg contributes about 17% and the arms about 5% each.

An instructive example of the association between the ground reaction force and the associated movement kinematics is shown in figure 2.12. The movement is the vertical jump. The height of a vertical jump depends on the magnitude of the vertical velocity at takeoff, which in turn is determined by the vertical component of the ground reaction force. For the performance shown in figure 2.12, the subject began from an upright position and then lowered her CM by approximately 0.2 m before changing direction (velocity crosses zero and goes from negative to positive) and moving upward to the takeoff position. Takeoff occurred when the vertical component of the ground reaction force fell to zero (time = 0.53 s). The subject was in the air for about 0.41 s, during which time her CM was raised 0.49 m from the starting position. Recall that figure 2.12 indicates the net effect of the interaction between the subject and the ground; accordingly, the kinematics describe the motion of her total-body CM.

In this vertical jump, the peak downward velocity (acceleration = 0) occurred about midway during the downward movement (time = 0.19 s), and the peak upward velocity (positive values; acceleration = 0) occurred just prior to takeoff. The ground reaction force was zero and the vertical component of acceleration had a value of –9.81 m/s² (i.e., the effect of gravity at the jumper's CM) during the flight phase. As expected from our discussion of kinematics in chapter 1, changes in acceleration preceded changes in position. As noted previously in the context of Newton's law of acceleration, the ground reaction force and the acceleration graphs parallel each other. Initially, the acceleration of the system (total-body CM) was zero, and the ground reaction force was equal to body weight. As the body segments began to accelerate, the system acceleration changed, and the ground reaction force changed in parallel but oscillated about the body weight line. The ground reaction force record consisted of four phases: (a) an initial phase in which the ground reaction force was less than body weight (negative acceleration); (b) a phase in

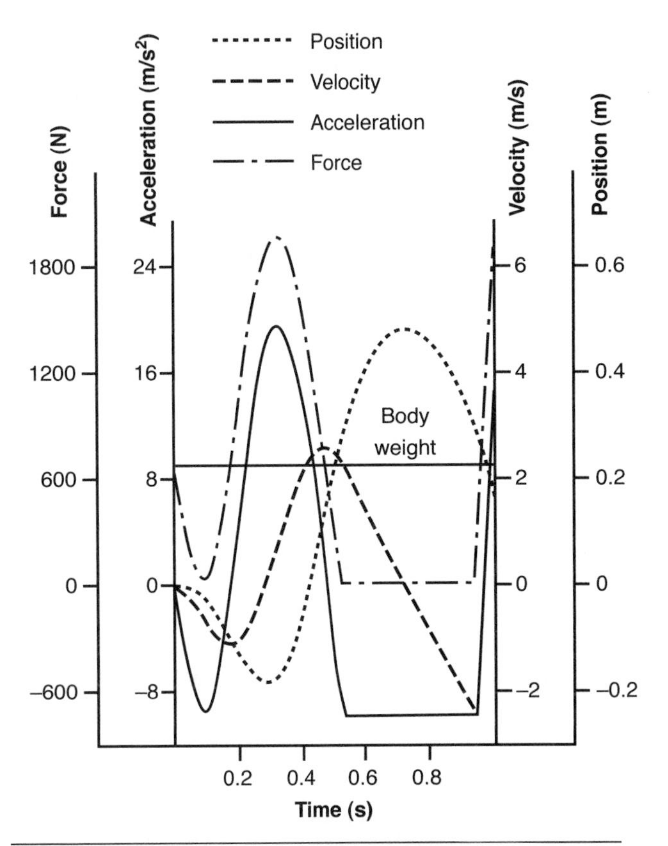

Figure 2.12 The vertical components of the kinematics and kinetics associated with a vertical jump. The position, velocity, and acceleration data are for the CM of the jumper. The force record corresponds to the vertical component of the ground reaction force ($F_{g,y}$).

Data from Miller 1976.

which the ground reaction force was greater than body weight (positive acceleration); (c) the flight phase, in which the ground reaction force was zero (acceleration = –9.81 m/s^2); and (d) the impact phase, in which the jumper returned to the ground.

Figure 2.13 illustrates the vertical component ($F_{g,y}$) of the ground reaction force associated with walking and with running. These data indicate the manner in which $F_{g,y}$ changed from the instant of foot contact with the ground (time zero on the abscissa) until the instant that the same foot left the ground (the time at which $F_{g,y}$ returns to zero). This interval is known as the stance or support phase. $F_{g,y}$ is nonzero only when the foot is in contact with the ground and changes continuously throughout this period of support. Recall from the discussion of weight that the magnitude of the weight vector is equivalent to $F_{g,y}$ when the system (body) is not accelerating in the vertical direction. Accordingly, when $F_{g,y}$ differs from body weight, the system experiences a vertical acceleration; when $F_{g,y}$ is greater than body weight, the vertical acceleration of the CM is upward; and when $F_{g,y}$ is less than body weight, the vertical acceleration of the CM is downward.

Because the ground reaction force represents the reaction of the ground to the action (acceleration of the CM) of the runner, the movement of the runner while the foot is on the ground is reflected in the ground reaction force (figure 2.14) (De Wit et al., 2000; Dixon et al., 2000; Gottschall & Kram, 2005; Wright & Weyand, 2001). A runner in the nonsupport phase of a stride experiences a downward acceleration (due to gravity) of 9.81 m/s^2, which means that when the foot contacts the ground there is an upward-directed vertical component ($F_{g,y}$) to counteract the downward motion of the runner. Furthermore, when the foot contacts the ground it is initially in front of the runner's CM, which causes the ground to respond with a backward-directed (braking) horizontal component ($F_{g,x}$). As the runner's CM passes over the support foot, the horizontal component changes direction so that it acts forward (propulsion). The side-to-side component ($F_{g,z}$) is more difficult to explain, but it has a lesser magnitude and is more variable than $F_{g,x}$ and $F_{g,y}$. However, $F_{g,z}$ is correlated (r = 0.71) with the position of the foot relative to the midline between the feet during contact with the ground (Williams, 1985).

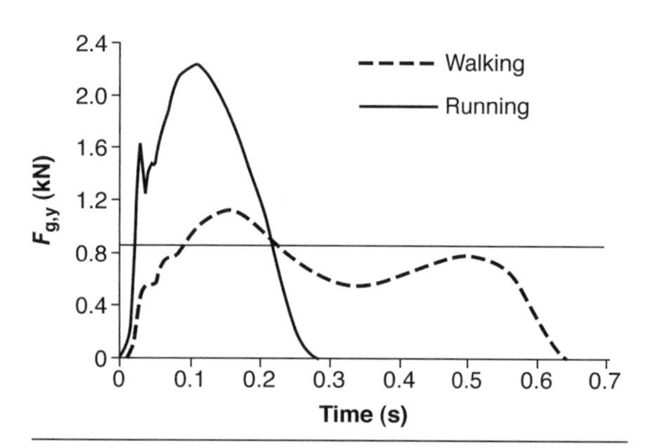

Figure 2.13 Vertical component of the ground reaction force ($F_{g,y}$) during the period of support in walking (dashed line) and running (solid line). The foot was placed on the ground at time zero and left the ground when $F_{g,y}$ returned to zero.

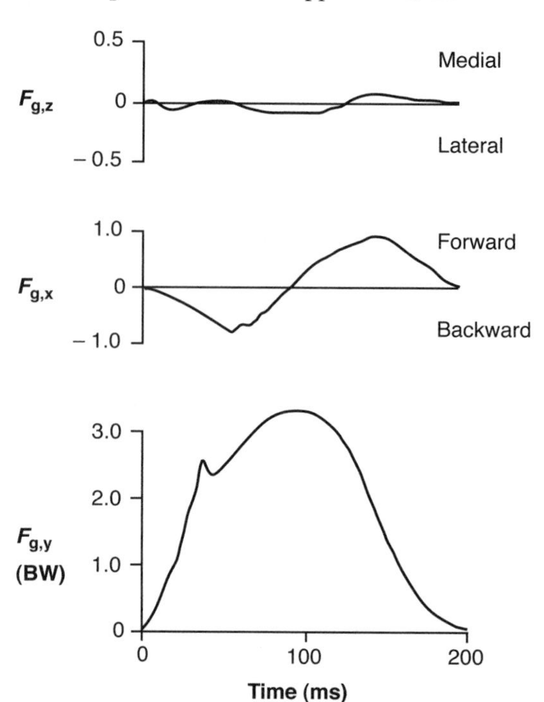

Figure 2.14 Generalized force–time curves for the three components of the ground reaction force during the support phase of a running stride. The forces are expressed relative to body weight (BW).

EXAMPLE 2.4

Calculating the Resultant Ground Reaction Force

Suppose the runner depicted in figure 2.1*c* experienced the following ground reaction force components at one point in time during the stance phase:

Forward-backward ($F_{g,x}$)	–286 N (positive = forward)
Vertical ($F_{g,y}$)	812 N (positive = upward)
Side to side ($F_{g,z}$)	61 N (positive = medial)

A. Draw a diagram of the two components and the resultant force in each of the *sagittal* (divides body into left and right), *frontal* (divides body into front and back), and *transverse* (divides body into upper and lower) planes (figure 2.15).

B. Calculate the magnitude and direction for each of these resultants.

Sagittal plane:

$$\text{Magnitude} = \sqrt{F_{g,x}^{\ 2} + F_{g,y}^{\ 2}}$$
$$= \sqrt{(-286)^2 + 812^2}$$
$$= 861 \text{ N}$$
$$\text{Direction} = \tan^{-1}\frac{286}{812}$$
$$= 0.34 \text{ rad relative to the vertical}$$

Frontal plane:

$$\text{Magnitude} = \sqrt{F_{g,y}^{\ 2} + F_{g,z}^{\ 2}}$$
$$= \sqrt{812^2 + 61^2}$$
$$= 814 \text{ N}$$
$$\text{Direction} = \tan^{-1}\frac{61}{812}$$
$$= 0.07 \text{ rad relative to the vertical}$$

Transverse plane:

$$\text{Magnitude} = \sqrt{F_{g,x}^{\ 2} + F_{g,z}^{\ 2}}$$
$$= \sqrt{(-286)^2 + 61^2}$$
$$= 292 \text{ N}$$
$$\text{Direction} = \tan^{-1}\frac{61}{286}$$
$$= 0.21 \text{ rad relative to the forward horizontal}$$

C. Use the Pythagorean relation to determine the magnitude of the resultant (F_g) ground reaction force.

$$F_g = \sqrt{F_{g,z}^{\ 2} + F_{g,y}^{\ 2} + F_{g,x}^{\ 2}}$$
$$= \sqrt{(-286)^2 + 812^2 + 61^2}$$
$$F_g = 863 \text{ N}$$

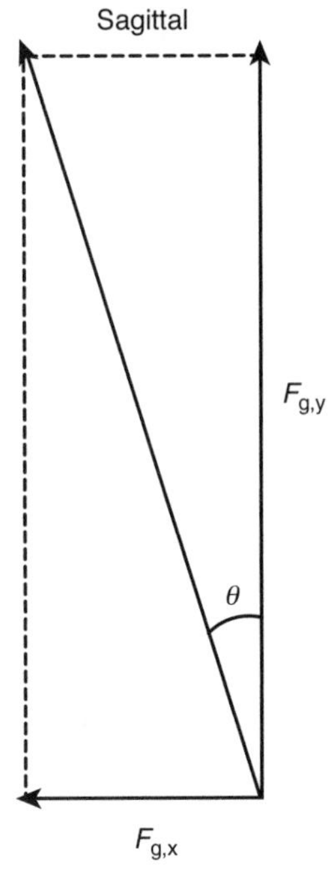

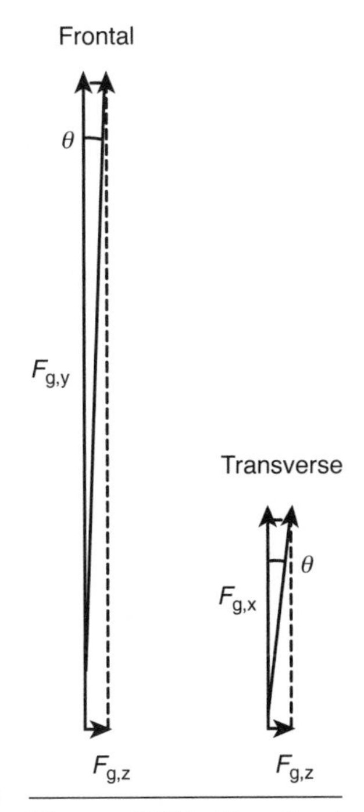

Figure 2.15 Resultant ground reaction forces in the three planes.

Center of Pressure

Figure 2.14 shows the three components of the resultant ground reaction force vector that acts on the foot during the stance phase of running. This force, however, does not act at a single point but rather is distributed over part of the foot. This is obvious when you focus on how your foot touches the ground as you walk and run. The distribution of force over an area is measured as pressure, which has the unit of measurement of pascal (Pa; 1 Pa = 1 N/m^2). One way to measure the pressure distribution under the foot during the stance phase is to place a number of small pressure transducers in a shoe that a subject can wear while walking or running (Cavanagh & Ae, 1980; Hennig et al., 1982, 1996). Figure 2.16 shows such measurements. In this study, eight pressure transducers were placed in the shoe: two under the heel, two under the midfoot, three under the forefoot, and one under the big toe. The sequence of diagrams in figure 2.16 shows the distribution of pressure at these locations for various times (10 to 250 ms) after the foot first contacted the ground. For the first 50 ms of foot contact, the pressure was greatest under the heel and reached a peak value of ~1000 kPa at 30 ms. In the two middle records (70 and 90 ms), the pressure was more or less evenly distributed over much of the foot. Thereafter, the pressure was greatest under the forefoot.

The peak pressure experienced at different locations on the foot during walking and running varies across individuals. The factors that influence the variations in pressure include both structural features of the foot and details of the motion. Morag and Cavanagh (1999) examined the correlations between peak plantar pressure and eight sets of these structural and functional measures during walking. It was possible to explain approximately 50% of the variance in peak pressures under the rearfoot, the midfoot, the head of the first metatarsal, and the hallux with multiple regression equations that used various subsets of the structural and functional measures. The structural factors were the dominant predictors for peak pressure under the midfoot and the head of the first metatarsal, whereas both structural and functional factors were important at the heel and hallux.

When we measure the ground reaction force with a force platform, the magnitude of the force represents the sum of the pressure distributed under the foot (Forner Cordero et al., 2004). The location (point of application) of the ground reaction force under the foot corresponds to the **center of pressure.** It is simply the central

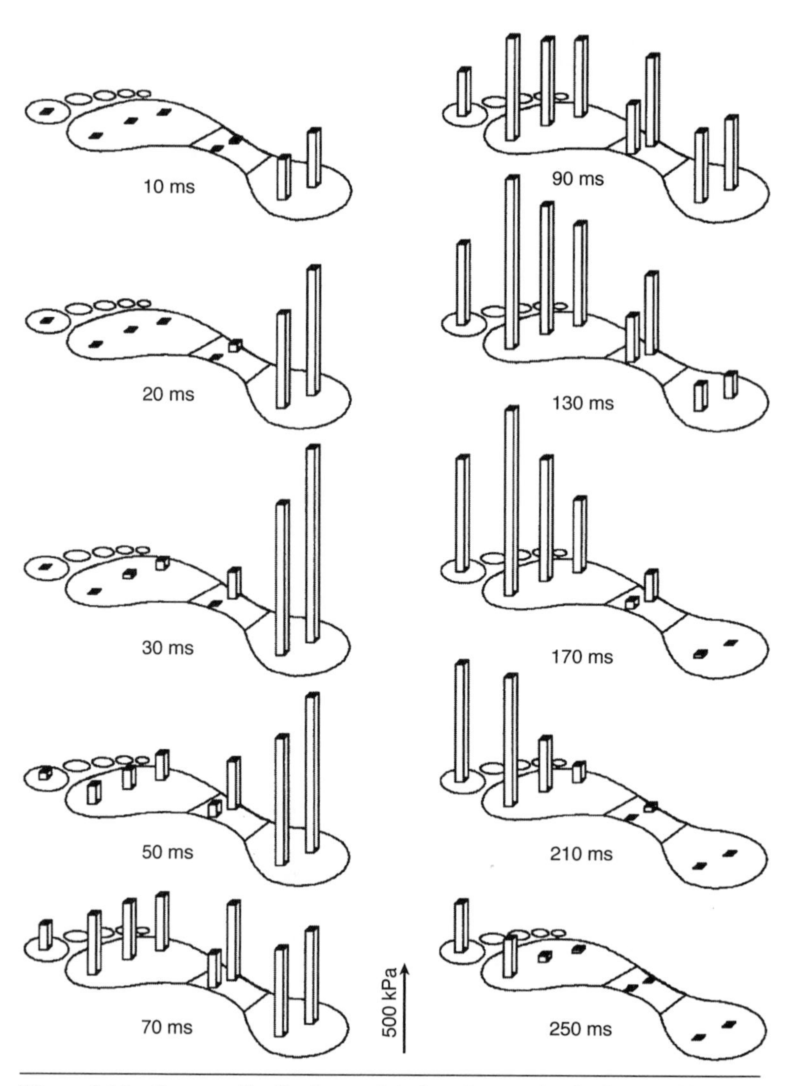

Figure 2.16 Pressure distribution at eight locations under the foot at 10 instants during the stance phase for a subject running at 3.3 m/s.

Reprinted, by permission, from E.M. Hennig and T.L. Milani, 1995, "In-shoe pressure distribution for running in various types of footwear," *Journal of Applied Biomechanics* 11: 303.

point of the pressure exerted on the foot. Cavanagh and Lafortune (1980) characterized runners as either midfoot or rearfoot strikers; this distinction referred to the initial location of the center of pressure on the foot when the foot first contacted the ground (figure 2.17). The initial location of the center of pressure was on the lateral border in the middle of the foot for midfoot strikers and on the rear of the foot for rearfoot strikers. For both types of runners, however, the center of pressure soon shifted to the central part of the foot and ended up under the big toe as the foot left the ground. The distribution of pressure and the trajectory of the center of pressure, however, can be modified by the characteristics of the shoe, the terrain over which the individual walks or runs, and the age of the person (Goske et al., 2006; Menz & Morris, 2006; Mohamed et al., 2005; Scott et al., 2007; Xu et al., 1999).

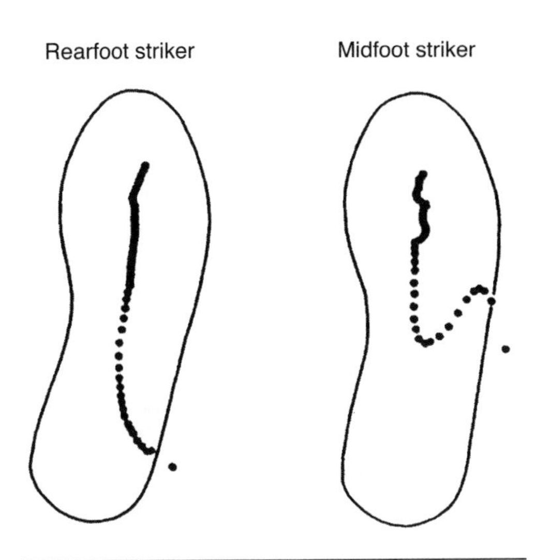

Figure 2.17 The displacement of the center of pressure (point of application of the ground reaction force vector) during the stance phase of a running stride (4.5 m/s).

Reprinted from *Journal of Biomechanics,* Vol. 13, P.R. Cavanagh and M.A. Lafortune, "Ground reaction forces in distance running," pp. 397-406. Copyright 1980, with permission from Elsevier.

Friction

The resultant of the two horizontal components of the ground reaction force ($F_{g,x}$ and $F_{g,z}$) represents the friction, or **shear force** (F_s), between the shoe and the ground. It is the reaction of the ground to the forces exerted in the horizontal plane by the person or object. Friction is important in locomotion because it provides the basis for the horizontal progression of the CM, and an inadequate friction force can contribute to falls (Burnfield & Powers, 2006; Redfern et al., 2001). Recall that the three components of the ground reaction force represent the acceleration of the total-body CM, which is the sum of the individual body segment CM accelerations in all three directions. F_s corresponds to the acceleration of the total-body CM in the two directions located in the horizontal plane (i.e., the forward-backward and side-to-side directions).

In general, the maximal friction force is determined by the magnitude of the force that is normal (perpendicular) to the surface, which for the ground reaction force is $F_{g,y}$, and a coefficient (μ) that characterizes the contact (rough-smooth, dry-lubricated, static-dynamic) between the two objects (figure 2.18):

$$F_{s,max} = \mu F_{g,y} \tag{2.10}$$

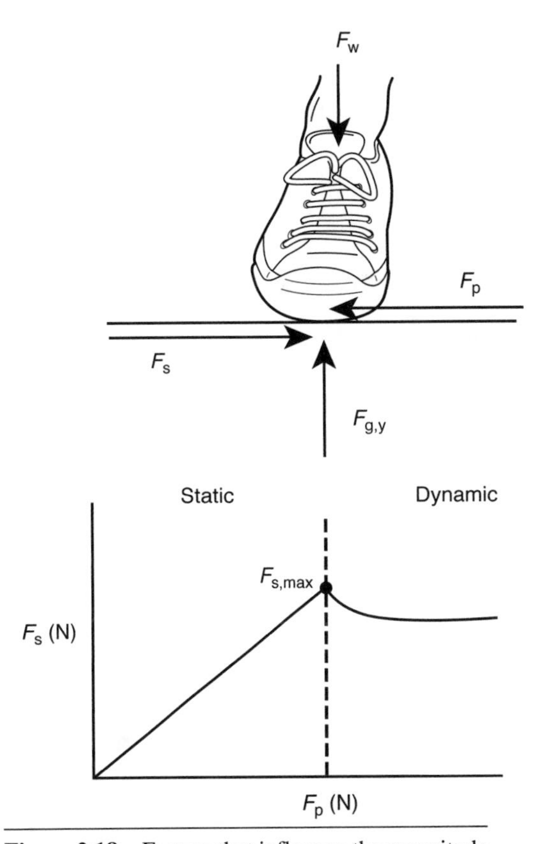

Figure 2.18 Factors that influence the magnitude of the friction force (F_s, shear force) when the shoe contacts the ground.

Data from Miller and Nelson 1973.

For a given shoe-ground contact, the friction coefficient will differ depending on whether the shoe is stationary relative to the ground (static, μ_s) or is moving (dynamic, μ_d). Because μ_s is greater than μ_d, F_s reaches greater values when the shoe does not move or slide on the ground; hence it is more difficult to make a rapid turn once the shoe begins to slip on the ground. We determine the magnitude of μ_s between shoes and surfaces experimentally by calculating the ratio of F_s to $F_{g,y}$ at the point in time just before the shoe moves relative to the ground. With this approach, μ_s has been reported to range from 0.3 to 2.0, with 0.6 for a cinder track and 1.5 for grass (Gao et al., 2004; Nigg, 1986; Stucke et al., 1984); and μ_d ranges from 0.003 to 0.007 during racing on ice skates (De Koning et al., 1992). Friction is simply equal to the resultant force in the horizontal plane when F_s is less than $F_{s,max}$.

When a shoe contacts the ground, the friction force (F_s) depends on the magnitude of the pushing force (F_p) exerted by the shoe on the ground (figure 2.18): (a) static—when the shoe does not slip, $F_s = F_p = \sqrt{F_{g,x}^2 + F_{g,z}^2}$; (b) maximum—the peak friction force ($F_{s,max}$) occurs just prior to the moment of the shoe slipping on the ground, when $F_s = \mu_s F_{g,y}$; (c) dynamic—once the shoe slips on the ground, F_s varies as a function of $\mu_d F_{g,y}$. These characteristics indicate that, for a given shoe-ground condition as represented by μ, friction increases with $F_{g,y}$; and because $F_{g,y}$ is predominantly influenced by body weight (F_w), the amount of friction is often greater for heavier individuals. Furthermore, as long as the shoe does not slip on the ground, the friction force is smaller than the maximum possible ($F_{s,max}$).

EXAMPLE 2.5

Friction on a Sled

Determine the acceleration of a person sledding down a snow-covered slope ($\mu_d = 0.085$) that was inclined (θ) above the horizontal at 0.6 rad (figure 2.19).

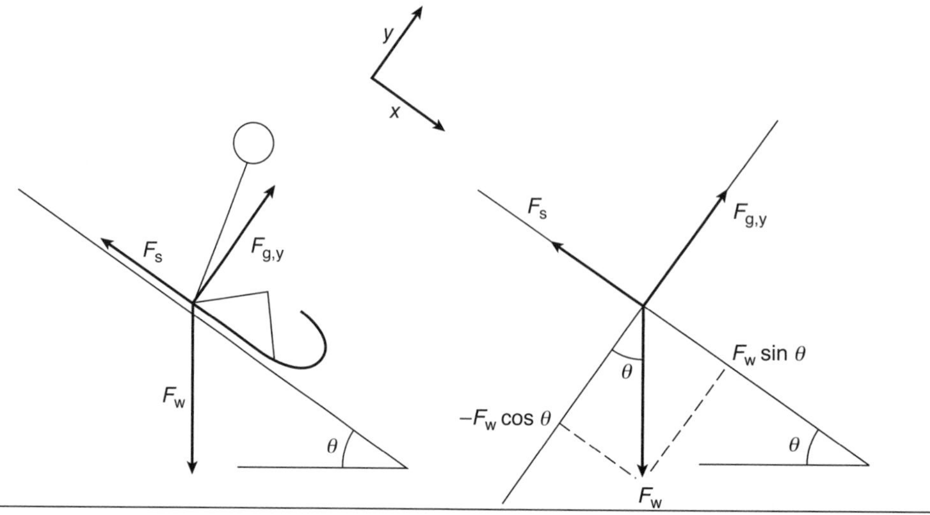

Figure 2.19 Free body diagram of a person sledding down a slope.

As shown in the figure 2.19, the system (person + sled) experienced two forces: gravity (weight) and the ground reaction force. When we consider the effects of friction, the ground reaction force is usually resolved into two components, one acting along the surface (F_s) and one perpendicular or normal to the surface ($F_{g,y}$). To determine the acceleration of the system, we can use Newton's law of acceleration:

$$\boldsymbol{F}_w + \boldsymbol{F}_{g,y} + \boldsymbol{F}_s = m\boldsymbol{a}$$

As we will discuss in more detail later, one way to solve such problems is to consider each direction (*x* and *y*) separately. According to the coordinate system in the figure, the forces acting in the *x* direction include F_s ($\mu_d F_{g,y}$) and a component of the system weight ($F_{w,x} = mg \sin \theta$, where g = the acceleration due to gravity).

$$mg \sin \theta - \mu_d F_{g,y} = ma$$

In the y direction, the forces are $F_{g,y}$ and a component of the system weight ($F_{w,y} = mg \cos \theta$).

$$F_{g,y} - mg \cos \theta = 0$$
$$F_{g,y} = mg \cos \theta$$

The identity for $F_{g,y}$ can be inserted into the equation for the x direction; then we can divide through by the mass of the system and derive an expression for a:

$$mg \sin \theta - \mu_d mg \cos \theta = ma$$
$$a = g \sin \theta - \mu_d g \cos \theta$$
$$a = g(\sin \theta - \mu_d \cos \theta)$$
$$a = 9.81[\sin 0.6 - (0.085 \times \cos 0.6)]$$
$$a = 9.81[0.5646 - (0.085 \times 0.8253)]$$
$$a = 4.85 \text{ m/s}^2$$

Fluid Resistance

Both human motion (e.g., ski jumping, cycling, swimming, skydiving) and projectile motion (e.g., the flight of a discus or golf ball) can be profoundly influenced by the fluid (gaseous or liquid) medium in which they occur (Hubbard, 2000). This phenomenon, known as **fluid resistance,** occurs because the fluid opposes the motion of the object. This interaction causes the object to experience two main effects. One effect, a **drag force,** opposes the forward motion of the object. The other effect, a **lift force,** usually exerts an upward force on the object. The drag force is drawn on a free body diagram with a direction parallel to the direction of the fluid flow around the object; the lift force is perpendicular to the direction of the fluid flow (figure 2.20). The net fluid resistance is the resultant of the drag and lift forces.

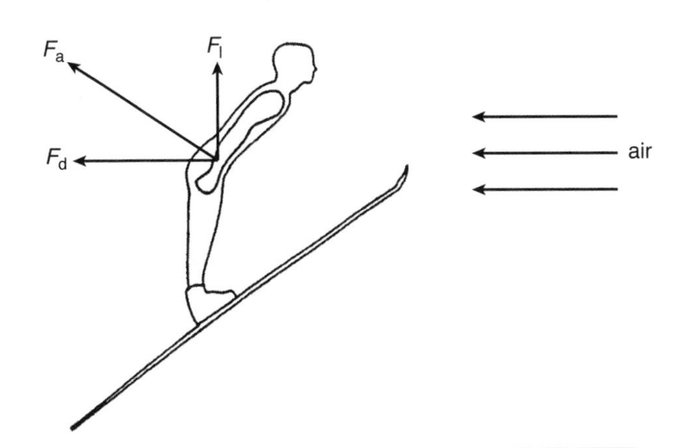

Figure 2.20 The air resistance encountered by a ski jumper. The air is shown moving relative to the jumper. The resultant air resistance vector (F_a) is resolved into two components, drag (F_d) and lift (F_l).

The interaction between the fluid and the object is typically schematized as in figure 2.21, where the fluid is shown flowing around a stationary object. The schematized lines of fluid flow are referred to as **streamlines;** they conceptually represent consecutive layers of particles in the fluid. One streamline, or layer, is adjacent to the object; another streamline lies on top of this, another on top of that, and so forth. The streamline closest to the object is referred to as the **boundary layer.** By studying the motion of the streamlines, it is possible to identify two main factors that contribute to the drag force experienced by the object: One is related to the different velocities of the streamlines, and the other depends on the extent to which the relative motion of the streamlines is disturbed.

First, when a dye is placed in the fluid, it is apparent that the streamlines can move at different velocities. Typically, the streamline touching the object (boundary layer) has the lowest velocity because it is slowed down by the effect of friction between the layer of particles and the object; this effect is known as **friction drag** or **surface drag.** The amount of surface drag depends on the smoothness of the object's surface. The rougher the surface of the object, the greater the friction and thus the greater the surface drag. Concern for surface drag is evident in many sports, including swimming, cycling, and rowing. This is the reason many swimmers shave body hair, wear caps, and don full body suits and the reason cyclists experiment with exotic materials and designs for clothing.

Second, when the streamlines move around an object, they may remain uniform or become nonuniform: Uniform flow is referred to as **laminar flow** (figure 2.21*a*), and nonuniform flow is called **turbulent flow** (figure 2.21*b*). When the flow is turbulent, there is a pressure differential from the front to the back of the object; this exerts a force on the object that is known as **pressure drag** or **form drag.** The pressure is greater on the leading side of the object, and the front-to-back difference increases as the flow becomes more turbulent. This is why designers often streamline an object in order to minimize the tendency for the fluid flow to become turbulent. Since the 1984 Los Angeles Olympics, cyclists' helmets have become more pointed in the rear because such a design reduces airflow turbulence. Furthermore, the surface texture of an object, such as the dimples in a golf ball, can influence the width of the turbulent wake and hence decrease the pressure drag experienced by the object. The golf ball dimples decrease the width of the wake and pressure drag by delaying the point along the object at which the streamlines become turbulent. The net effect is that for a given golf swing, a dimpled golf ball can be driven 239 m, whereas a smooth golf ball would travel only 46 m (Townend, 1984).

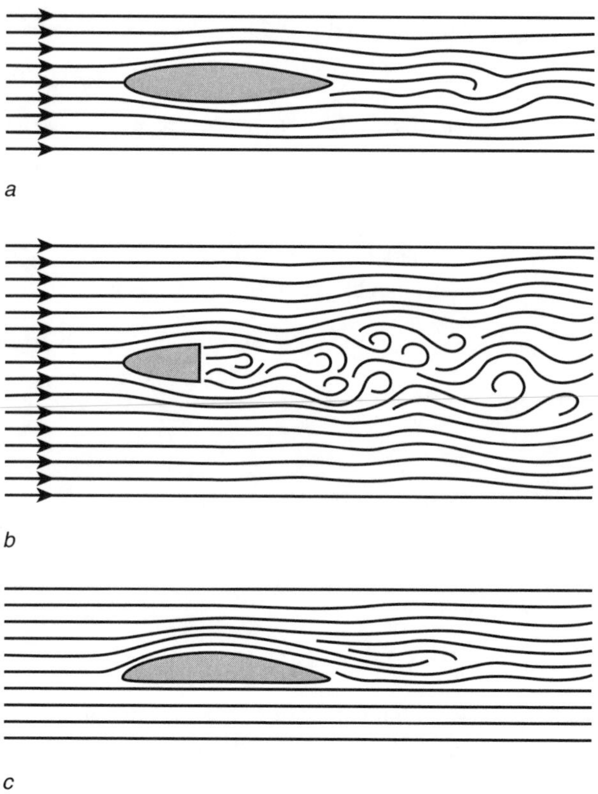

Figure 2.21 Tracings of smoke streamlines show the air traveling *(a)* around a streamlined object, *(b)* around the same object when the object was cut in half, and *(c)* around an airfoil.

In addition to pressure and surface drag, a third factor that influences the drag component of fluid resistance is **wave resistance.** The presence of waves decreases the average proportion of the object's body that is out of the water and therefore increases fluid density. Wave resistance sets the upper limit on the speed of surface ships and may be important at the higher velocities associated with competitive swimming. The resistive effects of waves are due to the uneven density of the fluid (water vs. air) that the swimmer encounters. The current practice in pool design is to minimize wave turbulence with specially designed gutters and lane markers.

When an object moves through a fluid, it always experiences a drag force, and the pressure drag is usually much greater than the surface drag. In addition, an object can also experience a lift force. This occurs when the object has an asymmetrical shape (e.g., an airfoil), when it is inclined at an angle relative to the fluid flow (angle of projection), or when it is spinning. In the case of an airfoil (or nonzero angle of projection), the path that the particles must travel over the topside of the airfoil is greater than that traveled by the particles on the underside (figure 2.21*c*). As a result, the streamline traveling the greater distance around the object (top side) travels at a greater velocity, and the pressure on that side of the object is less than on the other side. This creates a pressure differential in the up-down direction across the object. This effect is known as **Bernoulli's principle,** which states that fluid pressure is inversely related to fluid velocity. When the pressure on the top side of the airfoil is less because the velocity of the streamlines is greater, there is a net force that pushes the airfoil upward. This effect is known as lift (F_l) and represents the fluid resistance force that acts perpendicular to the direction of fluid flow (figure 2.20).

Similarly, a projectile that spins about an axis not in the same direction as the fluid flow will create a pressure gradient across itself by influencing the velocity of the streamlines. For example, if a baseball has a counterclockwise spin as it passes through the air, the streamlines flowing in the counterclockwise direction will have a greater velocity (due to reduced surface drag) than those flowing in the clockwise direction, and thus the pressure on the left side of the ball is lower. This creates a pressure gradient across the baseball from right to left and causes the baseball to curve to the left (figure 2.22). These effects on the trajectory of a projectile are due to the lift component of air resistance. This effect, sometimes referred to as the Magnus force, causes baseballs to curve, golf balls to slice, and tennis balls to drop. For most projectile motion, however, the magnitude of the Magnus force is much smaller than the lift forces caused by shape asymmetries and a nonzero angle of projection.

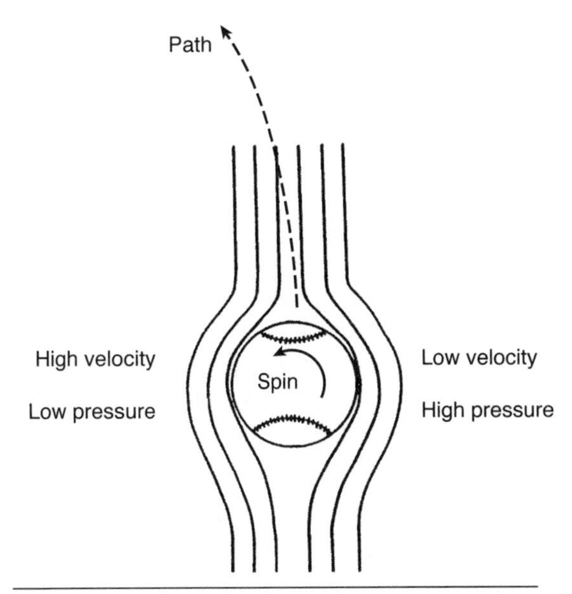

Figure 2.22 The effect of the lift component of air resistance on the trajectory of a baseball. The baseball has a spin in the counterclockwise direction, which will cause it to curve to the left.

Magnitude of Fluid Resistance

The magnitude of the fluid resistance vector due to pressure drag can be determined from

$$F_f = kAv^2 \qquad (2.11)$$

where k is a constant, A represents the projected area of the object, and v refers to the velocity of the fluid relative to the object. The projected area is a silhouette view of the frontal area of the object as it moves through the fluid. The constant k is an abbreviation for the term $0.5\rho C_D$ for drag and $0.5\rho C_L$ for lift; ρ accounts for fluid density; C_D distinguishes the effects of laminar and turbulent flow; and C_L is proportional to the angle between fluid flow and the orientation of the object. We can determine the net drag force by substituting values for the term $0.5\rho C_D$; C_D ranges from 0.4 for a smooth sphere to ~1.1 for an adult male running in an upright posture. Similarly, the lift force can be determined with the expression $0.5\rho C_L$; C_L for a discus ranges from 0.1 for a 0.03 rad (2°) angle of projection to 1.2 for a 0.45 rad (26°) angle of projection.

Consider the action of the air resistance vector on the runner (Hill, 1928) in figure 2.1*c*. The constant k is about 0.72 kg/m³ for a runner moving through air ($0.\rho C_D = 0.5 \times 1.2\ \text{kg/m}^3 \times 1.2$). The projected area of the runner is approximately 0.45 m² (Shanebrook & Jaszczak, 1976), which compares with estimated values of 0.27 m² for a skier in a semisquat position and 0.65 m² for a skier in an upright position (Spring et al., 1988). If the runner had a speed of 6.5 m/s and experienced a tail wind of 0.5 m/s, then the relative velocity (v) would be 6.0 m/s. Thus

$$F_f = 0.72 \times 0.45 \times 6.0^2$$
$$= 11.7\ \text{N}$$

Calculations by Shanebrook and Jaszczak (1976) have indicated that, at middle-distance running speeds (approximately 6 m/s), up to 8% of the energy expended by the runner is used in overcoming air resistance, whereas at sprinting speeds this value can be up to 16% of the total energy expenditure.

Ward-Smith (1985) has calculated the effects of head and tail winds on performance times in a 100 m sprint (see table 2.8). With a 3 m/s head wind, 100 m time would be increased by 0.26 s, whereas a 3 m/s tail wind would decrease the time by 0.34 s. The effects of the head and tail winds at a similar speed are asymmetrical because of the differences in relative velocity; recall from equation 2.11 that a doubling of the relative velocity will cause a fourfold increase in the fluid resistance.

Table 2.8 Influence of Wind Speed on the Time to Complete a 100 m Run

Wind speed (m/s)	Head wind (s)	Tail wind (s)
1	+0.09	−0.10
3	+0.26	−0.34
5	+0.38	−0.62

The drag acting on a swimmer has been estimated via measurement of the propulsive force exerted by a subject swimming at a constant velocity (Vorontsov & Rumyantsev, 2000). Under these conditions, the acceleration of the swimmer is zero, and thus the forces acting on the swimmer are balanced. In the horizontal direction, the propulsive force is equal to the resistive force (drag). With this method, the average drag experienced by a swimmer during the front crawl has a value of 53 N at a constant speed of 1.48 m/s (van der Vaart et al., 1987).

In cycling, air resistance is one of five opposing forces that a cyclist can experience. The other four are rolling resistance of the tires on the ground, the frictional resistance of the drive train, the force due to gravity during riding on a slope, and the resistance to acceleration due to the inertia of the system (di Prampero, 2000). During riding on a flat surface at a constant velocity, the two main forces are air resistance and rolling resistance, which collectively are known as **tractive resistance.** By measuring the power produced by a cyclist with a device mounted on the rear hub, one can determine the force produced by the cyclist, which equals the tractive resistance when cycling at a constant speed. When such measurements are made over a range of speeds and the tractive resistance is graphed as a function of velocity squared, the intercept corresponds to the rolling resistance, and the slope indicates air resistance independent of velocity. Edwards and Byrnes (2007) used this approach on 13 elite cyclists riding at speeds that ranged from 25 to 45 km/h and determined average values of 0.422 m^2 for projected frontal area, 0.771 for the drag coefficient (C_D), and 6.20 N for the rolling resistance. They also found that C_D declined by 42.4% and power production decreased by 33.3% during drafting behind another rider, and that the magnitude of the decrease in air resistance was directly related to the drag area ($C_D \times$ projected frontal area) of the leading cyclist.

Terminal Velocity

Because the relative velocity in equation 2.11 is a squared term, it is the single most important factor influencing air resistance. A sky diver represents an interesting example of the interaction among the variables in equation 2.11. After the sky diver has jumped from the airplane, his speed will increase up to some terminal value. To determine the value of terminal velocity, let's begin with a diagram (figure 2.23).

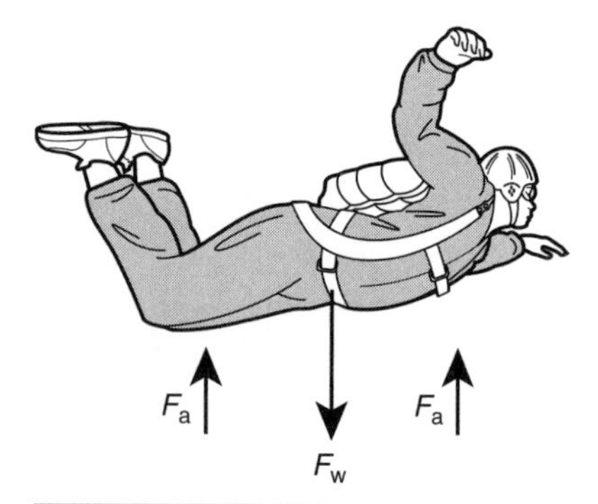

Figure 2.23 The forces experienced by a sky diver during free fall.

When the sky diver jumps out of the airplane, the system (sky diver, parachute, and associated equipment) is accelerated, due to gravity (F_w), toward the ground. But the downward acceleration of the system is only briefly equal to the value for gravity (–9.81 m/s^2), because as system speed increases, the opposing effect of air resistance increases. We know from equation 2.11 that the magnitude of air resistance increases as the square of relative velocity: Thus, as v increases, so does the force due to air resistance (F_a). The speed of the system will continue to increase (i.e., acceleration will be nonzero) until the force due to air resistance is equal to the weight of the system (i.e., the acceleration due to gravity). After the two forces become equal, speed will remain constant, and the system will have reached a terminal velocity. Because velocity remains constant (i.e., zero acceleration) and the system is in equilibrium at terminal velocity, the forces must be balanced. Thus

$$F_w = F_a$$

Further, because we have defined F_a (equation 2.11),

$$F_w = kAv^2$$

we can rearrange this relation to determine the terminal velocity:

$$v = \sqrt{\frac{F_w}{kA}}$$

If $k = 0.55$ kg/m^3, $A = 0.36$ m^2, and $F_w = 750$ N, then

$$v = \sqrt{\frac{750}{0.55 \times 0.36}}$$

Terminal velocity = 61.6 m/s (138 mph)

Despite being in free-fall conditions, experienced sky divers are able to perform somersaults, cartwheels, and other sorts of movements. How can they do this? Sky divers experience two forces as they fall: weight and air resistance (figure 2.23). Both of these forces are distributed forces; that is, they are distributed over the entire system but are drawn as acting at one or two points. The weight vector is always drawn as acting at the CM. Similarly, the force due to air resistance has a central balance point: the center of pressure for air resistance. By moving her limbs, a sky diver can alter her projected area and shift the center of pressure for air resistance so that the vector does not pass through the CM. If the line of action of the air resistance vector is not collinear with the weight vector, then the air resistance will exert a rotary force (torque) about the CM and cause the sky diver to experience angular motion. The sky diver can stop the angular (but not linear) motion by aligning the lines of action of the two vectors.

EXAMPLE 2.6

Air Resistance During Skydiving

A sky diver experiences at least four phases during a jump: (a) an initial phase of acceleration from the moment of leaving the plane until reaching terminal velocity; (b) a phase of terminal velocity; (c) a phase of decreased vertical velocity immediately after the parachute opens; and (d) a final phase of free fall (constant velocity) with the parachute opened and the sky diver preparing to land (figure 2.24). Draw a velocity–time graph that represents these four phases. Graph the air resistance as a function of time to show how the air resistance changes to produce the velocity–time graph.

From these graphs, we can answer a number of questions about the experiences of the sky diver.

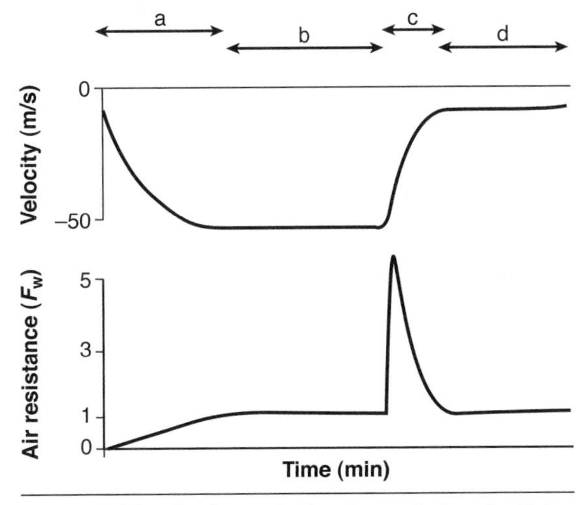

Figure 2.24 The four velocity phases during skydiving are caused by a change in air resistance.

A. Why does velocity increase during phase a? This happens because the forces acting on the sky diver (F_W and F_a) are not balanced and there is a net force in the downward direction.

B. Why is velocity constant in phase b? The reason is that F_w and F_a are equal in magnitude but opposite in direction, so there is no net force acting on the sky diver. In the absence of a net force, the motion of a system remains constant (Newton's law of inertia).

C. Why does velocity decrease in phase c? It decreases because of the marked increase in F_a. This occurs because the sky diver opens her parachute, which increases the projected area (equation 2.11) and therefore F_a.

D. Why is the constant velocity in phase d less than that in phase b? The reason is that the magnitude of F_a required to oppose F_w is achieved at a lower relative velocity, due to the larger projected area of the sky diver and parachute.

Buoyancy

When submerged in a fluid, an object will experience an upward-directed force known as the **buoyancy force.** As described by Archimedes, the magnitude of the buoyancy force (F_b) is equal to the weight of the fluid displaced by the object. We can determine this by calculating the product of the volume of the object (V_o) and the specific weight of the fluid (γ); specific weight refers to the weight for a standard volume of fluid.

$$F_b = V_o \gamma \tag{2.12}$$

For example, a person floating in a pool and displacing 0.064 m^3 of water (20° C; $\gamma = 9810$ N/m^3) will experience a buoyancy force of 628 N. When stationary, this person will assume a position in the water where the magnitude of the buoyancy force is equal to body weight; that is, buoyancy force and weight will have equal magnitudes but opposite directions. The line of action for the buoyancy force acts vertically upward through the center of the volume for the part of the object that is submerged. When a person is in a floating position (i.e., the forces are balanced), the lines of action for the buoyancy force and the weight vector will be coincident. When the person becomes active, as in swimming, the significant forces will include weight, buoyancy force, and fluid resistance (McLean & Hinrichs, 1998). These forces must be coincident when swimming at a constant speed (acceleration = 0).

MOMENTUM

One of the fundamental principles of motion is Newton's law of acceleration, which essentially says that the motion of an object remains constant unless it acted upon by a force. In the formulation of this law, the motion of an object was quantified as momentum. If you think of two people running, what might you measure to determine who has more motion? Is it simply the faster person who has more motion? No, this also depends on the size of each person. It is the combination of speed and size that we use to express the amount of motion that a person has. We call this combination momentum, and calculate it as the product of mass and velocity. Momentum is a vector quantity whose direction is the same as velocity. As with several other physical quantities we have considered, we distinguish between **linear momentum** (**G**) and **angular momentum** (**H**). By definition, the unit of measurement for linear momentum is kg•m/s (mass × velocity), and the unit of measurement for angular momentum is kg•m^2/s (moment of inertia × angular velocity).

Impulse

Newton's law of acceleration states that when a force acts on an object, it changes the momentum of the object. In the analysis of human movement, however, it is necessary to extend this idea. The reason is that forces are not applied instantaneously but rather over an interval of time. For example, figure 2.25 shows the vertical component of the ground reaction force ($F_{g,y}$) that is applied to the foot of a runner during the stance phase of running. An impulse is defined graphically as the area under a force–time curve, numerically as the product of the average force (N) and time (s), and mathematically as the integral of force with respect to time (equation 2.13).

$$\text{Impulse} = \int_{t_1}^{t_2} \boldsymbol{F} dt \tag{2.13}$$

where t_1 and t_2 define the beginning and end of the force application. For example, if the average force ($F_{g,y}$) in figure 2.25 was 1.3 kN and the time of application was 0.29 s, the impulse would be 377 N•s. Thus, the magnitude of an impulse can be altered through variation in either the average force or its time of application.

When Newton's law of acceleration is interpreted to focus on intervals of time rather than instants of time, the law indicates that the application of an impulse will result in a change in the momentum of the system. This is the basis of the impulse–momentum approach to the analysis of motion, which can be stated as

$$\int_{t_1}^{t_2} \mathbf{F} dt = \Delta\mathbf{G} \tag{2.14}$$

Furthermore, the derivation of the impulse–momentum relation from Newton's law of acceleration confirms this interpretation:

$$\sum \mathbf{F} = m\mathbf{a}$$

$$\sum \mathbf{F} = m\frac{(\mathbf{v}_f - \mathbf{v}_i)}{t}$$

$$\sum \mathbf{F}t = m(\mathbf{v}_f - \mathbf{v}_i)$$

$$\sum \mathbf{F}t = \Delta m\mathbf{v} \text{ or } \bar{F} \bullet t = \Delta mv \tag{2.15}$$

where the term $\Sigma\mathbf{F}t$ represents the area under a force–time curve (figure 2.25) and is equivalent to the product of the average force and its time of application ($\bar{F} \bullet t$). Equation 2.15 suggests that if the magnitude of the impulse is known, then its effect on the momentum of the system can be calculated. Conversely, if the change in momentum can be measured, it is possible to determine the applied impulse.

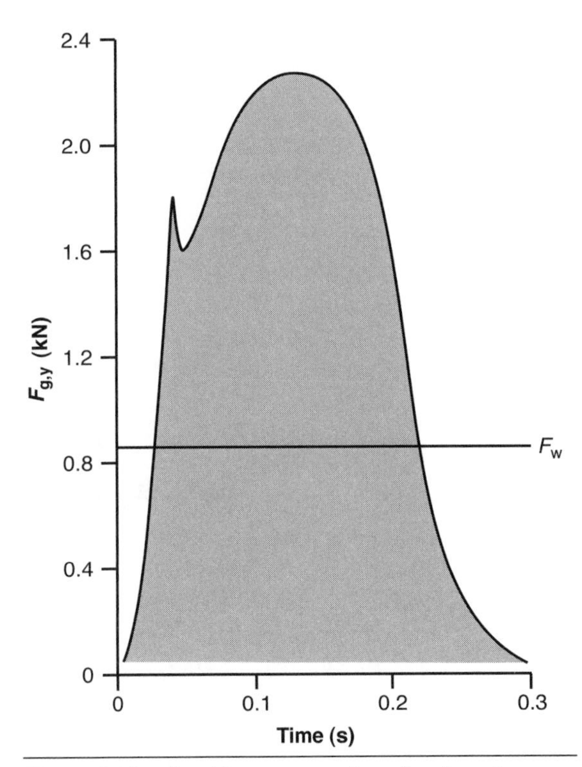

Figure 2.25 Graphic representation of an impulse as an area (shaded) under a force–time curve.

EXAMPLE 2.7

Spiking a Volleyball

By videotaping a person spiking a volleyball and measuring the mass (m) of the ball, we can determine the impulse applied to the ball. From the videotape it would be necessary to measure both the velocity of the ball before (v_b) and after (v_a) contact and the total time the hand is in contact with the ball (t_c). For one performance of this skill, we obtained

$$v_b = 3.6 \text{ m/s}$$

$$v_a = 25.2 \text{ m/s}$$

$$m = 0.27 \text{ kg}$$

$$t_c = 18 \text{ ms}$$

Equation 2.15 is applied as follows:

$$\begin{aligned}\bar{F} \bullet t &= \Delta mv \\ &= m\Delta v \\ &= m(v_b - v_a) \\ &= 0.27(25.2 - 3.6) \\ \bar{F} \bullet t &= 5.83 \text{ N•s}\end{aligned}$$

Because we know the contact time (t_c), we can determine the average force ($\bar{F}$) exerted by the spiker during the contact:

$$\bar{F} \cdot t = 5.83 \text{ N} \cdot \text{s}$$
$$= \frac{5.83}{t_c}$$
$$= \frac{5.83}{0.018}$$
$$\bar{F} = 324 \text{ N}$$

Thus, although the impulse appeared to be small (5.83 N•s), the brief duration of the force application resulted in forces that were quite substantial ($\bar{F}$ = 324 N). Incidentally, the time of contact with the volleyball in this example is quite similar to those (10 to 16 ms) recorded during the kicking of a ball (Asami & Nolte, 1983).

EXAMPLE 2.8

Momentum Depends on Average Force and Contact Time

In most contact events, such as spiking a volleyball, the momentum of an object is altered by the application of relatively high forces for brief periods of time. There are instances, however, in which the change in momentum is accomplished by the application of smaller forces for longer periods. As an example (Brancazio, 1984) of these two strategies, consider the distinction between the consequences of one person (mass = 71 kg) jumping off a 15 m building onto the pavement and another person (mass = 71 kg) diving off a 15 m cliff into the ocean. In each instance the individual will have a speed of about 17.3 m/s just prior to contact and a linear momentum of 1228 kg•m/s. Eventually, however, the speed (and thus momentum) of each person will reach zero. The jumper will experience large forces (perhaps fatal) for a brief interval upon contact with the pavement. The diver, however, will encounter smaller forces, due to contact with the water, over a longer period of time. Nonetheless, the change in momentum for each individual will be the same (ΔG = 1228 kg•m/s). The impulse (area under the force–time curve) provided by the landing surface will be identical for the two landings, although the shapes of the two force–time curves will differ.

Figure 2.26 shows the reaction force provided by the landing surface for the two jumps. For the landing onto the pavement, the upward-directed vertical component of the ground reaction force is a consequence of the structural properties of the pavement. If this ground reaction force was applied for 25 ms, then the average force ($\bar{F}$) required to change the linear momentum from 1228 kg•m/s to zero would be 49,120 N. For the water landing, the reaction force is due to a fluid resistance drag force (equation 2.11) and a buoyancy force (equation 2.12). The drag force is influenced by the density of water (998 kg/m^3), a drag coefficient (~0.1), the projected

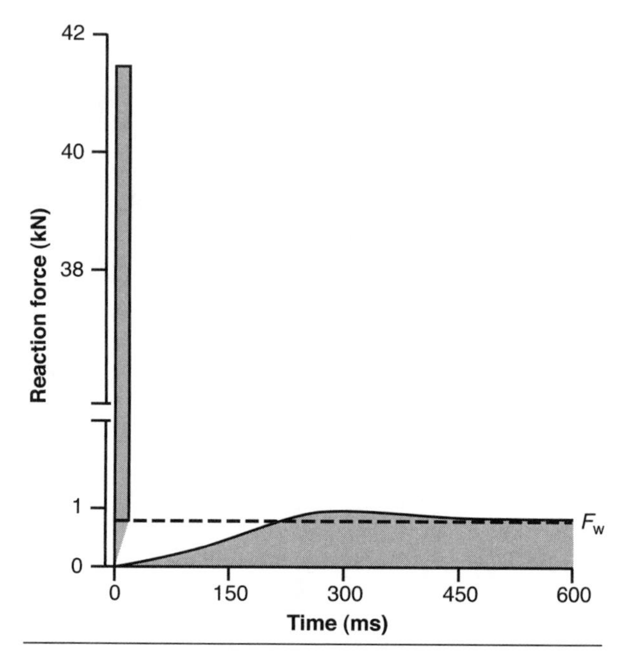

Figure 2.26 The reaction forces provided by pavement (briefer impulse) and water during landing from a 15 m jump.

area (~0.07 m^2), and the relative velocity (17.3 m/s at first contact). For this example, the drag force would reach a peak value of about 1000 N soon after contact with the water but then would decline to zero as the relative velocity decreased. The buoyancy force depends on the volume of water that is displaced; when a person is floating in water, the magnitude of the upward-directed buoyancy force is equal to body weight. So during the landing, the buoyancy force would increase from zero, at first contact with the water, up to body weight when the person was floating in the water at the end of landing. The combined effect of the drag force and the buoyancy force would be something like that shown in figure 2.26. The key feature of this example, however, is that each individual will experience the same net vertical impulse to change the same quantity of linear momentum; therefore, the area under each curve in figure 2.26 should be equal.

EXAMPLE 2.9

Forward Momentum in Running

Whenever an impulse is applied to a system, the momentum of the system will change in proportion to the net impulse. Furthermore, the effect on momentum will be confined to the direction of the impulse. Consider the forward-backward component ($F_{g,x}$) of the ground reaction force during the support phase of running (figure 2.14). The graph illustrates that the runner experiences two forward-backward impulses during support. Initially $F_{g,x}$ is directed backward, creating a retarding or braking impulse; then $F_{g,x}$ changes direction, eliciting a propulsive impulse. Because these impulses act in opposite directions, the change in momentum that the runner experiences in the x direction depends on the difference between the braking and propulsion impulses. When the individual is running at a constant speed (no change in momentum), the two impulses are equal. For a runner to increase speed, however, the propulsive impulse must exceed the braking impulse; to decrease speed, the braking impulse must be greater than the propulsion impulse. These relations are shown in figure 2.27. The change in momentum due to the ground reaction force is equal to the difference between the initial (before stance) and final (after stance) values for momentum.

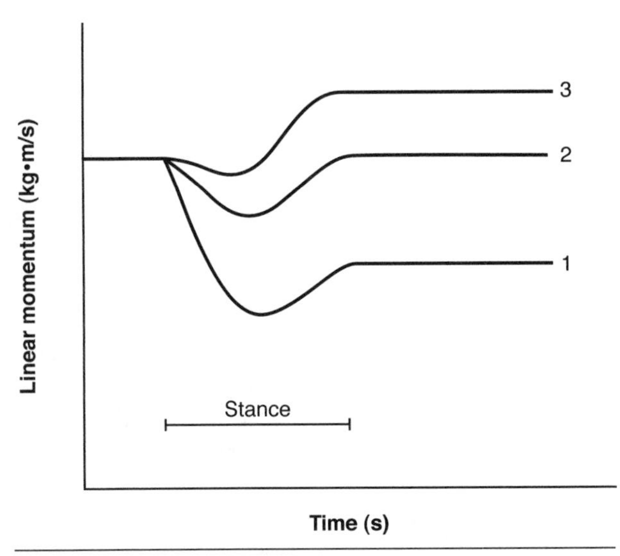

Figure 2.27 Change in the forward-backward linear momentum of a runner in response to (1) a braking impulse that is larger than the propulsive impulse; (2) equivalent braking and propulsive impulses; and (3) a propulsive impulse that is larger than the braking impulse.

Because impulses in locomotion and other such activities are influenced by the weight of the individual, it is common practice to normalize the impulses and to express them as a proportion of the body weight impulse. For example, if a runner weighed 630 N and experienced a braking impulse of 15.8 N•s that lasted for 0.1 s, then the braking impulse was 0.25 times the body weight impulse (15.8 / [630 × 0.1] = 0.25). Munro and colleagues (1987) reported braking and propulsive impulses of 0.15 to 0.25 times body weight impulse with runners traveling at 3.0 to 5.0 m/s. These impulses increased with running speed.

Collisions

A **collision** is a brief contact between objects in which the force associated with the contact is much larger than other forces acting on the objects. When a tennis ball collides with a racket, for example, the contact force is much larger than the gravitational and air resistance forces that the ball and racket experience (Hennig, 2007). Many of our physical activities involve collisions, such as those between athletes (e.g., rugby, martial arts), between a participant and an inanimate object (e.g., handball, soccer), and between inanimate objects (e.g., badminton, golf, hockey).

When two objects collide, they exert equal and opposite forces on each other. When two objects are involved in the collision, they both experience the same force. And because the contact time associated with the collision is the same for the two objects, the impulse applied to each object as well as the change in momentum experienced by each object is the same. So in a collision there is a conservation of linear momentum. This means that the sum of the linear momenta of the two objects (*A* and *B*) before the collision is the same as the sum of the linear momenta after the collision:

$$(m_A v_A)_{\text{before}} + (m_B v_B)_{\text{before}} = (m_A v_A)_{\text{after}} + (m_B v_B)_{\text{after}} \tag{2.16}$$

The left-hand side of equation 2.16 represents the momentum of the system before the collision, and the right-hand side indicates the momentum after the collision. Because mass does not usually change in human movement, this equation can be rearranged to show that the change in momentum of object *A* is equal to the change in momentum of object *B* (equation 2.17), and that the change in velocity experienced by each object is inversely proportional to its mass (equation 2.18).

$$\begin{aligned} (m_A v_A)_{\text{before}} - (m_A v_A)_{\text{after}} &= (m_B v_B)_{\text{after}} - (m_B v_B)_{\text{before}} \\ m_A (v_{A,\text{before}} - v_{A,\text{after}}) &= m_B (v_{B,\text{after}} - v_{B,\text{before}}) \\ m_A \Delta v_A &= m_B \Delta v_B \end{aligned} \tag{2.17}$$

$$\frac{\Delta v_A}{\Delta v_B} = \frac{m_B}{m_A} \tag{2.18}$$

For example, consider hitting a tennis ball with a racket. After contact between the ball and the racket (the collision), the velocity of the ball is usually much greater than the velocity of the racket. The difference in the velocity of the ball and the racket is determined by the ratio of their masses. Hence the advantage of size (mass) in contact sports such as football and wrestling.

One important feature of collisions is whether or not they are elastic; that is, do the objects bounce off one another or do they remain joined after the collision? If the collision is elastic, then each object preserves most of its kinetic energy (velocity after the collision is nonzero). The elasticity of a collision is indicated by the **coefficient of restitution** (*e*):

$$\text{Coefficient of restitution} = \frac{\text{Speed after collision}}{\text{Speed before collision}}$$

A perfectly elastic collision has a coefficient of restitution equal to 1, which indicates that the speed (velocity) after the collision is identical to that before the collision. However, the coefficient of restitution is less than 1 for most collisions in human movement. One can measure the coefficient of restitution of a ball by dropping it from a known height onto the ground and measuring how high it rebounds. If a ball dropped from a height of 1.0 m rebounds to 0.5 m, then the coefficient of restitution is 0.5. Selected coefficients of restitution for various balls at an impact speed of 24.6 m/s (55 mph) include softball, 0.40; tennis ball, 0.55; golf ball, 0.58; basketball, 0.64; soccer ball, 0.65; and Superball, 0.85. The coefficient of restitution tends to decrease as the speed of the collision increases.

The coefficient of restitution quantifies the extent to which a perfect collision is modified by the material properties of the objects involved in the collision. After a bat contacts a ball, the velocities of the ball and the bat depend not only on the mass of each but also on the coefficient of restitution. The coefficient of restitution represents the constant of proportionality between the speed before the collision and the speed after the collision. For example, consider the contact of a baseball and a bat, which is described by the equations that define the coefficient of restitution and the conservation of momentum:

$$\text{Speed after collision} = -e\ (\text{Speed before collision})$$

$$v_{B,a} - v_{b,a} = -e\ (v_{B,b} - v_{b,b}) \tag{2.19}$$

where B represents the bat and b indicates the ball. If we assume that the velocity vectors of the ball and the bat are collinear (lie in the same line as shown in figure 2.28), rearrangement of equation 2.19 allows us to specify the velocity of the ball ($v_{b,a}$) and the bat ($v_{B,a}$) after contact:

$$v_{b,a} = v_{B,a} + e\ (v_{B,b} - v_{b,b}) \tag{2.20}$$

$$v_{B,a} = v_{b,a} - e\ (v_{B,b} - v_{b,b}) \tag{2.21}$$

We can determine the velocity of the bat after the collision ($v_{B,a}$) by substituting equation 2.20 into the expression for the conservation of linear momentum (equation 2.16) and rearranging the expression:

$$v_{B,a} = \frac{m_b v_{b,b}(1+e) + v_{B,b}(m_B - m_b e}{m_b + m_B} \tag{2.22}$$

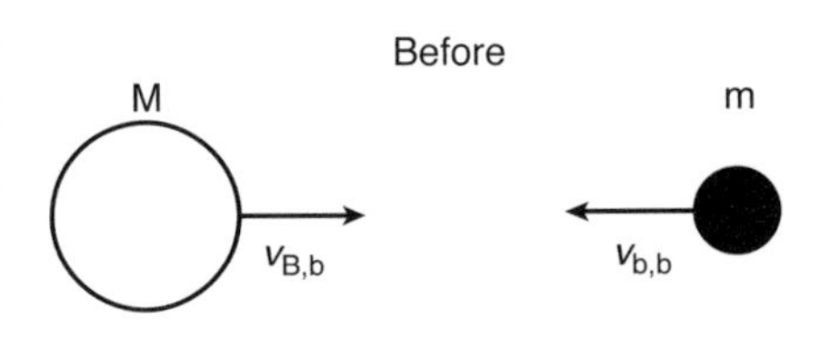

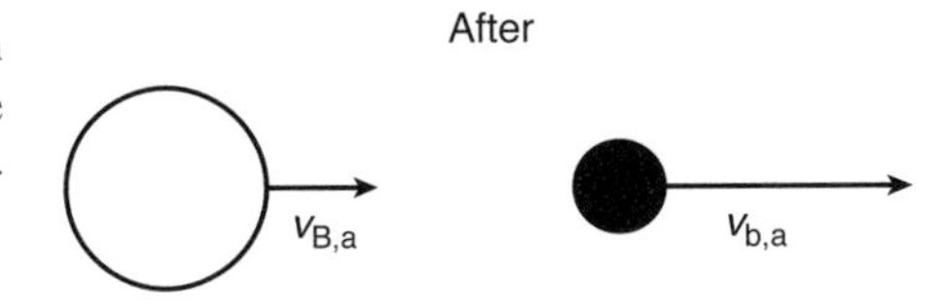

Figure 2.28 Mass and velocity of the bat (B) and ball (b) before ($v_{B,b}$, $v_{b,b}$) and after ($v_{B,a}$, $v_{b,a}$) a collision.

Similarly, we can derive an expression for $v_{b,a}$ by substituting the expression for $v_{B,a}$ with equation 2.20:

$$v_{b,a} = \frac{m_B v_{B,b}(1+e) + v_{b,b}(m_b - m_B e)}{m_B + m_b} \tag{2.23}$$

Equations 2.22 and 2.23 represent the general case for the velocity of a bat ($v_{B,a}$) and ball ($v_{b,a}$) after a collision. Typically, however, ball velocities are rarely collinear before ($v_{b,b}$) and after ($v_{b,a}$) impact, and the directions of $v_{b,b}$ and $v_{b,a}$ are usually opposite (i.e., toward and away from the batter). We accommodate the noncollinearity of the ball and bat velocities by inserting a cosine term next to $v_{b,b}$ in equations 2.22 and 2.23 (see Townend, 1984, p. 142, for an example).

EXAMPLE 2.10

A Ball-Bat Collision

Let us apply these ball-bat equations to determine the velocity of the baseball and the bat after a hit:

Mass of bat (m_B) = 0.93 kg (33 oz)

Mass of ball (m_b) = 0.16 kg (5.6 oz)

Velocity of ball before impact ($v_{b,b}$) = –38 m/s (85 mph)

Velocity of bat before impact ($v_{B,b}$) = 31 m/s (70 mph)

Coefficient of restitution (specified in rules) = 0.55

A. Calculate the velocity of the baseball and bat after the hit.

$$v_{B,a} = \frac{m_b v_{bb}(1+e) + v_{Bb}(m_B - m_b e)}{m_b + m_B}$$

$$v_{B,a} = \frac{-6.08(1+0.55) + 31(0.93 - 0.088)}{0.16 + 0.93}$$

$$v_{B,a} = 15.3 \text{ m/s}$$

and

$$v_{b,a} = \frac{m_B v_{Bb}(1+e) + v_{bb}(m_b - m_B e)}{m_B + m_b}$$

$$v_{b,a} = \frac{28.8(1+0.55) + (-38)(0.16 - 0.51)}{0.16 + 0.93}$$

$$v_{b,a} = 53.2 \text{ m/s}$$

B. Would such a hit produce a home run in baseball? Assume that the angle of the ball's trajectory after it left the bat was 0.785 rad, and ignore the effects of air resistance. To find the horizontal distance that the ball will travel, we need to use the equations of motion that were introduced in chapter 1. First, we find out how long the ball would be in the air by using the y direction information.

$$\bar{a} = \frac{\Delta v}{\Delta t}$$

$$t_{up} = \frac{v_f - v_i}{\bar{a}}$$

$$t_{up} = \frac{0 - 53.3 \sin 0.78}{-9.81}$$

$$= 3.82 \text{ s}$$

$$t = 2 \times t_{up}$$

$$t = 7.64 \text{ s}$$

Next, we can find the horizontal distance the ball will travel. Because we are ignoring the effect of air resistance, the horizontal velocity remains constant.

$$\bar{v} = \frac{\Delta r}{\Delta t}$$

$$\Delta r = 53.3 \cos 0.78 \times 7.64$$

$$= 289 \text{ m } (940 \text{ ft})$$

C. It is highly unlikely that the ball would be hit a distance of 289 m. Because the numbers used in the calculations are within the range of those observed in baseball, the calculated distance suggests that we are missing an important factor in the analysis. Perhaps we need to consider the effect of air resistance, which does influence the actual trajectory of a baseball. How can we do this? We could use equation 2.11, which provides a direct method for calculating the magnitude of the air resistance. However, there is a simpler way to estimate the effect of air resistance. We could assume that the effect of the air resistance is to cause a linear decrease in the horizontal velocity of the ball throughout its trajectory; thus, $v_i = 37.9$ m/s, $v_f = 0$ m/s.

$$\bar{v} = \frac{\Delta r}{\Delta t}$$

$$\Delta r = \frac{37.9 + 0}{2} \times 7.64$$

$$= 145 \text{ m } (475 \text{ ft})$$

The distance for the hit is more reasonable, but it would still be a home run.

Another aspect of ball-bat collisions is the notion of a sweet spot. Contact in a ball-bat collision is said to occur at the sweet spot when no reaction force is felt at the hands. Three theories seek to account for the sweet spot: the center of percussion, the natural frequency node, and the coefficient of restitution (Brancazio, 1984). Of these, the explanation provided by the center of percussion is most popular. When a ball contacts a bat, the location along the bat where no reaction force ($\mathbf{F}_r$) is felt at the hands is referred to as the center of percussion. When a force ($\mathbf{F}_c$) is applied at the CM of the bat, the bat experiences only linear motion (figure 2.29*a*). However, when the force is not applied at the CM, the bat experiences both linear and angular motion (figure 2.29*b*). When this happens, there is a point where the linear translation and the angular rotation cancel and the bat does not move. The contact point that produces a stationary pivot point is the center of percussion (figure 2.29*c*). It is located distal on the bat to the CM. The location of the ball's contact on the bat relative to the centers of mass and percussion determines the nature of the reaction force at the hands (figure 2.29*c*).

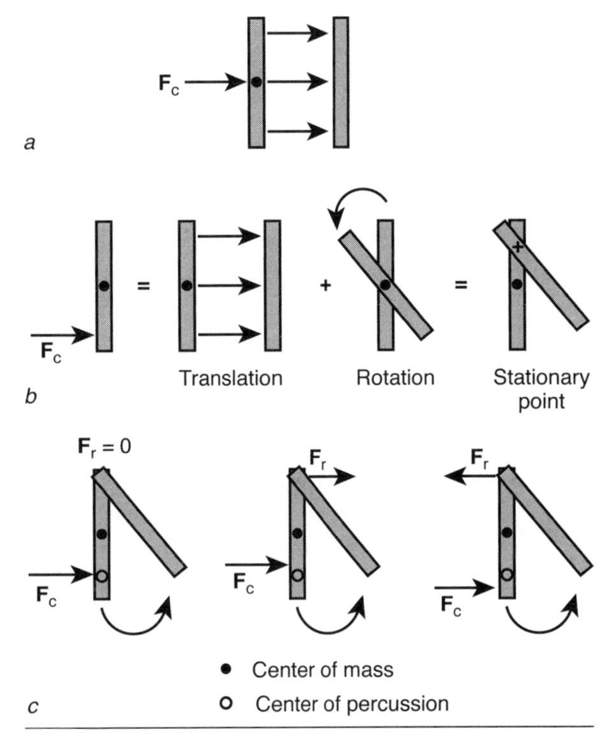

Figure 2.29 The effect of a force applied to a bat on the motion of the bat: *(a)* a contact force ($\mathbf{F}_c$) applied at the center of mass; *(b)* $\mathbf{F}_c$ applied at a place other than the center of mass; *(c)* $\mathbf{F}_c$ applied at the center of percussion.

Adapted from P. J. Brancazio, 1984, *Sport Science: Physical laws and optimum performance* (New York: Simon & Schuster). By permission of Peter J. Brancazio.

The three sweet spots (center of percussion, the natural frequency node, and the coefficient of restitution) do not coincide on a tennis racket (Hennig, 2007). When the ball contacts the racket where the coefficient of restitution is maximum, it leaves the racket with the maximal velocity. When it contacts the racket at its natural frequency node, the oscillations at the grip are minimal and have a low frequency. Players typically contact the ball at a location between the three sweet spots. The location of the sweet spots and the result of the ball-racket collision depend on the characteristics of the racket and grip force applied by the player. For example, stiff rackets produce higher rebound velocities and transfer lower vibration amplitudes to the player. Furthermore, although grip force does not influence the rebound speed of the ball, a tighter grip will result in a greater transfer of vibration to the player (Hatze, 1976).

Angular Momentum

When we analyze angular motion, we do so in terms of the torques acting on the system. Torque (moment of force) represents the rotary effect of a force and is calculated as the vector product of a moment arm (**r**) and force (**F**). In a similar vein, angular momentum (**H**) is derived as the moment of linear momentum ($\mathbf{r} \times \mathbf{G}$). Angular momentum describes the quantity of angular motion and is calculated as the product of moment of inertia (I) and angular velocity ($\boldsymbol{\omega}$). For example, the angular momentum of the bat in figure 2.29*b* after the application of $\mathbf{F}_c$ would be 0.258 kg•m²/s, calculated as the product of the bat's moment of inertia (0.05 kg•m²) and its subsequent angular velocity (5.2 rad/s). Because the direction of **H** is the same as that for $\boldsymbol{\omega}$, $\mathbf{H}_{bat}$ would be out of the page in figure 2.29*b*.

The impulse–momentum relation also applies for angular motion. In this context, however, the impulse is the area under the torque–time curve (left-hand side of equation 2.24).

$$\int_{t_1}^{t_2} (\boldsymbol{r} \times \boldsymbol{F})\, dt = \Delta \boldsymbol{H} \quad (2.24)$$

Again considering the bat in figure 2.29*b*, which experienced a $\Delta\mathbf{H}$ of 0.258 kg•m²/s, we know that the angular impulse must have been 218 N•m•s. The angular impulse applied to the bat was due to the torque ($\mathbf{r} \times \mathbf{F}_c$) about the axis of rotation (CM of the bat) applied over an interval of time. The moment arm ($\mathbf{r}$) was the perpendicular distance from the line of action of $\mathbf{F}_c$ to the axis of rotation.

EXAMPLE 2.11

Arm Action During Running

When a person runs, the upper body and lower body rotate in opposite directions about the twist axis during a single stride. This motion can be quantified as the angular momentum about the twist axis. When the left foot is on the ground, the upper body has an angular momentum in the downward direction (same direction as angular velocity), and the lower body has an angular momentum in the upward direction (figure 2.30). The net result is a relatively small angular momentum in the upward direction. The angular momentum experienced by the lower body is a result of the angular impulse (left-hand side of equation 2.24) due to the ground reaction force. Because the foot is not placed beneath the twist axis when we run, the resultant of the forward-backward and side-to-side components (**F**) exerts a torque about the twist axis. The perpendicular distance from the line of action of this resultant and the twist axis is **r** in equation 2.24. To prevent whole-body angular motion (nonzero angular momentum) about the twist axis, the runner responds to the angular impulse provided by

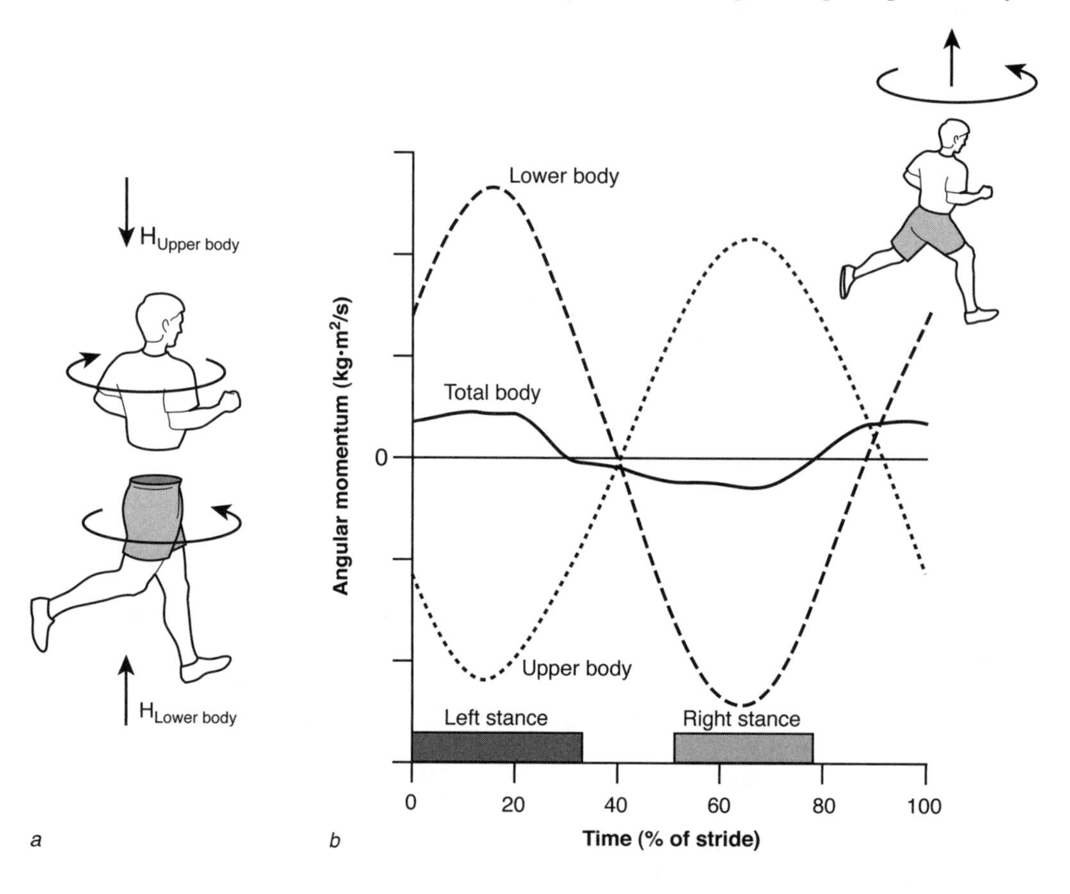

Figure 2.30 Angular momentum about the twist axis during a single running stride: *(a)* two-segment model of a runner; *(b)* angular momentum for the upper body, lower body, and total body.

Reprinted from *International Journal of Sport Biomechanics*, Vol. 3, R.N. Hinrichs, "Upper extremity function in running. II: Angular momentum considerations," pp. 242-263. Copyright 1987, with permission from Elsevier Science.

the ground reaction force to the lower body by contracting the trunk muscles and generating an opposing angular momentum for the upper body; that is, the arms and trunk rotate in the opposite direction to the legs. This interaction is shown in figure 2.30*b:* For the first 40% of the stride, the left foot is in contact with the ground, producing a positive angular momentum in the vertical direction (H_y) for the lower body, while concurrently the upper body rotates with a negative H_y. In the second part of the stride, the converse occurs while the right foot is in contact with the ground. The primary function of arm motion during running, therefore, is to counteract the angular momentum of the legs about the twist axis (Hinrichs, 1987; Hinrichs et al., 1987; McDonald & Dapena, 1991).

Because the human body does not behave as a single rigid body, such as the bat, it is necessary to consider the motion of each body segment when calculating angular momentum. This requires a **linked-system** analysis, in which we calculate the angular momentum of each segment about its CM (local angular momentum) and then determine the angular momentum of the CM for each segment about the system (whole-body) CM (remote angular momentum). This relation has the following form:

$$\begin{aligned} \boldsymbol{H}^{S/CS} &= \boldsymbol{H}^{B1/C1} + \boldsymbol{H}^{B2/C2} + \boldsymbol{H}^{B3/C3} + \ldots \text{ (local terms)} \\ &\quad \boldsymbol{H}^{C1/CS} + \boldsymbol{H}^{C2/CS} + \boldsymbol{H}^{C3/CS} + \ldots \text{ (remote terms)} \end{aligned} \tag{2.25}$$

where

$\boldsymbol{H}$ = angular momentum
B1 = body segment 1
C1 = center of mass for segment 1
CS = center of mass for the system
S = system
/ = with respect to

Local terms (e.g., $\boldsymbol{H}^{B1/C1}$) have the form $I_g\omega$ whereas remote terms (e.g., $\boldsymbol{H}^{C1/CS}$) consist of the vector (cross) product $\mathbf{r} \times m\mathbf{v}$, which represents the moment of linear momentum relative to the system CM.

Figure 2.31 shows an example of the time course of angular momentum of the whole body as determined by a linked-system analysis (Hay et al., 1977). In this example, a gymnast performed a vault that included periods in which he contacted either the ground or the vaulting horse and periods when he was not in contact with either. As we would expect from the impulse–momentum relation (equation 2.24), the angular momentum of the gymnast changed when he contacted his surroundings and remained constant during the flight phases. The change in angular momentum was caused by an angular impulse due to the ground

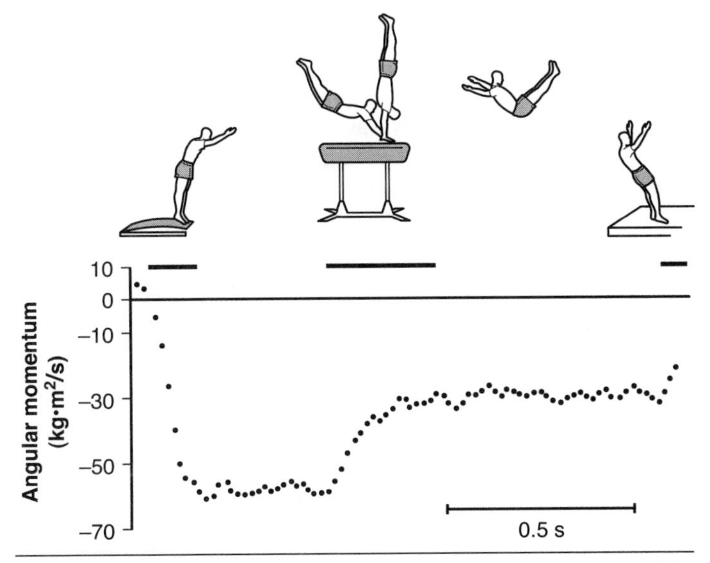

Figure 2.31 Angular momentum of a gymnast during a vault. The lines at the top of the graph indicate when the gymnast was in contact with the ground or the vaulting horse. Angular momentum was constant during the flight phases of the event but changed when the surroundings applied an angular impulse during the contact phases. Negative = forward angular momentum.

Reprinted from *Journal of Biomechanics,* 10, Hay et al., "A Computational Technique to Determine the Angular Momentum of the Human Body," pp. 269-277. Copyright 1997, with permission from Elsevier Science.

reaction force. As we discussed previously, the angular impulse represents the area under a torque–time curve, with the torque being the rotary effect of the ground reaction force about the CM of the gymnast. Because he rotated forward, the line of action of the ground reaction force had to pass behind the hips (CM) and produce an angular momentum vector directed into the page.

Conservation of Momentum

When the left-hand side of equation 2.24 is zero, so too is the right-hand side. If an object's momentum does not change ($\Delta \mathbf{H} = 0$), then it must remain constant and is described as being conserved. Momentum can be conserved in either the linear or angular direction, and not necessarily at the same time. This is evident in such activities as gymnastics (e.g., Dainis, 1981; Nissinen et al., 1985) and diving (e.g., Bartee & Dowell, 1982; Frohlich, 1980; Stroup & Bushnell, 1970; Wilson, 1977) and is classically demonstrated as the air-righting reaction whereby cats (and other animals) always land on their feet when they are dropped from a low height (Kane & Scher, 1969; Laouris et al., 1990; Magnus, 1922; Marey, 1894).

EXAMPLE 2.12

Angular Momentum Is Conserved During a Dive

Once a springboard diver has left the board, she will experience two forces, weight and air resistance (figure 2.32). Let us consider the effect of these forces on her linear (**G**) and angular (**H**) momentum. From the impulse–momentum relation we have

Impulse = Δ linear momentum

$$\bar{F} \bullet t = \Delta G$$

$$\bar{F} \bullet t = \Delta mv$$

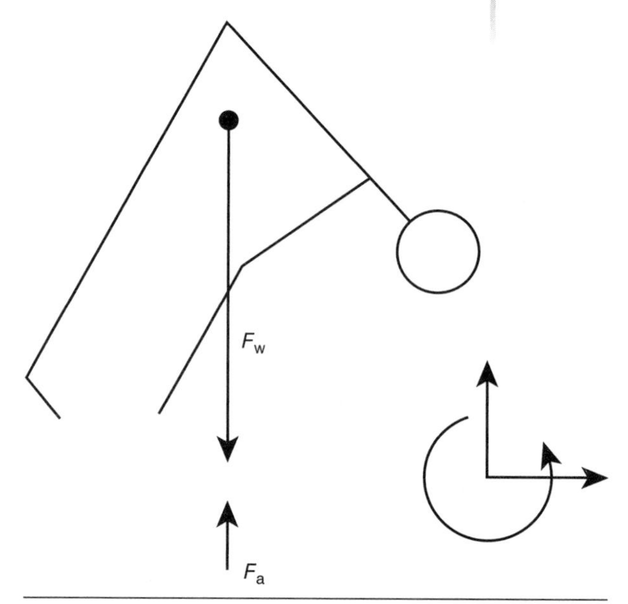

Figure 2.32 Forces experienced by a diver during free fall. F_a = air resistance; F_w = weight.

Because the average force includes the effect of weight (F_w) and air resistance (F_a), $\bar{F}$ is not zero and there is a change in linear momentum; that is, $\Delta mv \neq 0$. Therefore, linear momentum does not remain constant. In contrast, consider the effect on angular momentum:

Angular impulse = Δ angular momentum

$$(\bar{F} \times r) \bullet t = \Delta H$$

$$(\bar{F} \times r) \bullet t = \Delta I\omega$$

The **angular impulse** is equal to the product of the average torque ($\bar{F} \times r$) and its time of application, where r is the moment arm from the line of action of each force to the axis of rotation (the CM). Because both F_w and F_a act through the CM in figure 2.32, the moment arm for each is equal to zero and thus there is no torque about the CM during the dive. The absence of a torque means that the right-hand side of the equation ($\Delta I\omega$) is equal to zero and therefore momentum does not change but remains constant. This represents an example of the conservation of momentum.

Because angular momentum is equal to the product of moment of inertia and angular velocity, any change in one parameter (i.e., moment of inertia or angular velocity) will be accompanied by a complementary change in the other parameter when angular momentum is constant. For example, suppose a diver performed a multisomersault event in the pike position. If, during the dive, it became apparent that he would not make the appropriate number of revolutions, then one alternative would be to assume a tuck position, which would (because **H** remains constant) be accompanied by an increase in the speed of rotation. His moment of inertia (I_g) about the somersault axis passing through the CM is about 7.5 kg•m² in the pike position, as opposed to 4.5 kg•m² in the tuck position. If, in the pike position, he had an angular velocity (ω) of 6 rad/s, then on changing to a tuck his speed would increase to 10 rad/s such that the product of the two parameters ($H = I_g\omega$) would remain constant ($H = 45$ kg•m²/s). Specifically,

Pike $\quad H = 7.5\ \text{kg}\cdot\text{m}^2 \times 6\ \text{rad/s}$

$\quad H = 45\ \text{kg}\cdot\text{m}^2/\text{s}$

Tuck $\quad H = 4.5\ \text{kg}\cdot\text{m}^2 \times 10\ \text{rad/s}$

$\quad H = 45\ \text{kg}\cdot\text{m}^2/\text{s}$

and thus angular momentum (H) remains constant. He could also slow the speed of rotation by increasing his moment of inertia, in this case by assuming a greater layout position. This exchange between angular velocity and moment of inertia in order to conserve angular momentum during a dive is illustrated in figure 2.33.

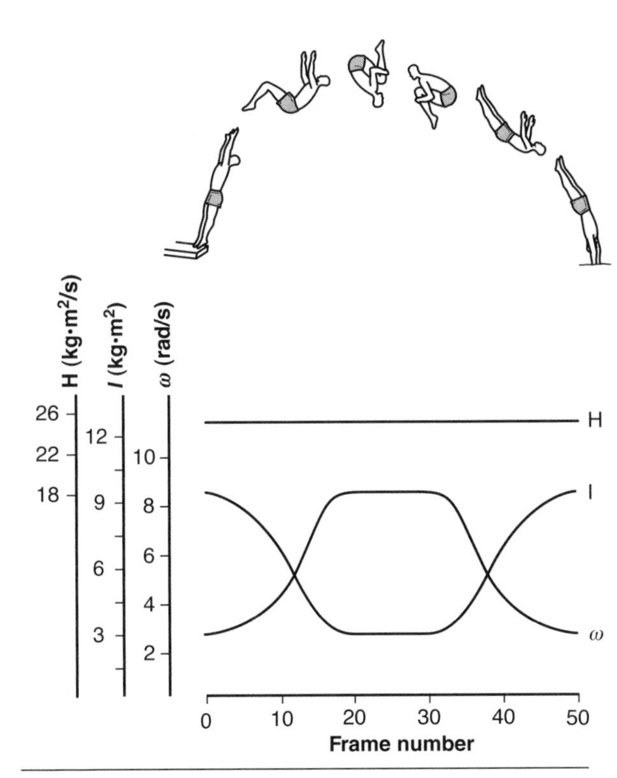

Figure 2.33 The conservation of angular momentum is accompanied by reciprocal changes in moment of inertia and angular velocity during a backward one-and-a-half somersault dive.

EXAMPLE 2.13

Initiating a Twist in Diving

The impulse–momentum relation can also be used to explain how a diver is able to initiate a twist even though no force is available to assist in the maneuver (Yeadon, 1993*a, b, c, d*). Figure 2.34 shows a diver performing a somersault in a layout position; his angular momentum ($\mathbf{H}_g$) is acting at the CM. According to the right-hand rule, he is preparing to perform a forward somersault. To initiate the twist, he rotates his arms about a cartwheel axis passing through his chest so that the right arm goes above his head and the left arm crosses his trunk (figure 2.34*b*). Because his arms rotate about a cartwheel axis, this does not alter $\mathbf{H}_g$ about the somersault axis. However, rotation of the arms generates an angular momentum in one direction about a cartwheel axis ($\mathbf{H}_{g,arms}$) that is counteracted (to keep $\mathbf{H}_g$ about the cartwheel axis zero) by an equivalent angular momentum of the trunk in the opposite direction about the cartwheel axis ($\mathbf{H}_{g,trunk}$) ($\mathbf{H}_{g,arms} + \mathbf{H}_{g,trunk} = 0$). Following this action-reaction maneuver, his arms are displaced as shown in figure 2.34*b* and his trunk is inclined to the left of vertical by an angle θ. Because he is in free fall, the angular momentum remains constant at the value that he possessed when he left the board. Now, however, because his orientation has changed, $\mathbf{H}_g$ has components about both the somersault and twist axes relative to his body; the direction of $\mathbf{H}_g$ remains constant, but the axes move with him (figure 2.34*c*). The component of $\mathbf{H}_g$ about the twist axis ($H_{g,t}$) is equal to $H_g \sin \theta$, and the component about the somersault axis ($H_{g,s}$) can be determined as $H_g \cos \theta$.

For example, consider the tilt angle (θ = angle that the trunk is inclined to the left of vertical) necessary to execute a forward layout dive with a full twist from a 3 m diving board. Assume that no twist is initiated from the board. In this example, the diver has 30 kg•m²/s about the somersault axis at the moment of takeoff; $I_{g,s}$ represents the moment of inertia of the diver about the somersault axis passing through the CM; $I_{g,t}$ indicates the moment of inertia of the diver about the twist axis passing through the CM; and t is the time taken to perform the dive. The following initial conditions are given: $\mathbf{H}_g$ = 30 kg•m²/s; $I_{g,s}$ = 14 kg•m²; $I_{g,t}$ = 1 kg•m²; and t = 1.5 s.

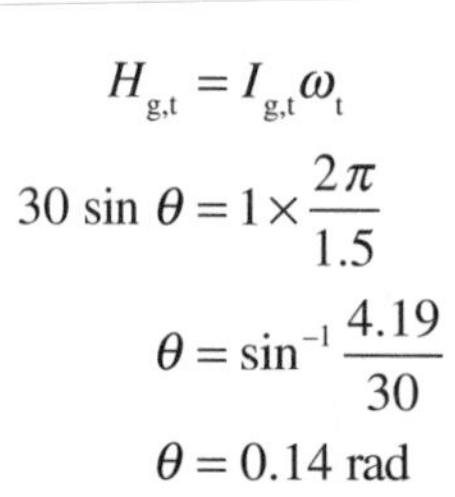

$$H_{g,t} = I_{g,t}\omega_t$$

$$30 \sin \theta = 1 \times \frac{2\pi}{1.5}$$

$$\theta = \sin^{-1} \frac{4.19}{30}$$

$$\theta = 0.14 \text{ rad}$$

Given these initial conditions, the diver will need to tilt by 0.14 rad to complete a full twist in the dive. How much rotation about the somersault axis would accompany a tilt of 0.14 rad?

$$H_{g,s} = I_{g,s}\omega_s$$

$$30 \cos 0.14 = 14\ \omega_s$$

$$29.7 = 14\ \omega_s$$

$$\omega_s = 2.12 \text{ rad/s}$$

$$\frac{\Delta\theta}{t} = 2.12 \text{ rad/s}$$

$$\Delta\theta = 3.18 \text{ rad}$$

Thus, a tilt angle (θ) of 0.14 rad will result in a full-twisting, one-half (3.18 rad) somersault dive.

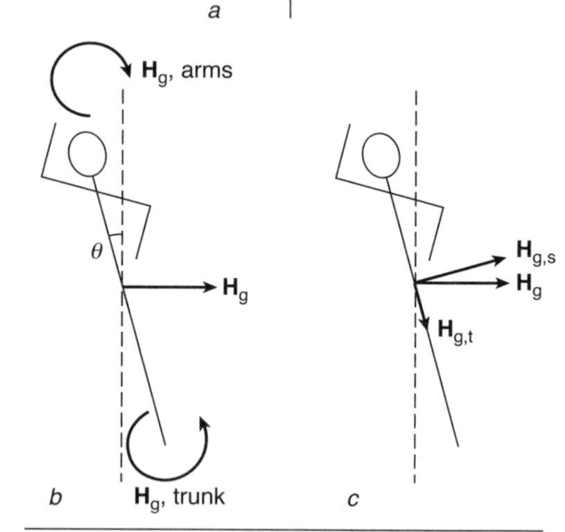

Figure 2.34 Initiation of a twist during a dive. *(a)* Orientation of the diver and $\mathbf{H}_g$ at takeoff. *(b)* Rotation of the arms causes the body to tilt relative to the vertical. *(c)* The tilt causes H_g to have components about the twist and cartwheel axes.

Yeadon (1993*e*) found that seven out of eight divers performing the reverse 1.5 somersault dive with 2.5 twists obtained most of the twist from asymmetrical movements of the arms and hips during the aerial phase of the dive. However, about one-third of the tilt was derived from actions performed while the diver was in contact with the diving board.

EXAMPLE 2.14

How Do Cats Land on Their Feet?

The impulse–momentum relation can also be used to aid understanding of the controversial air-righting reaction, the so-called cat twist—the ability of cats (and other animals) to land on their feet when they are dropped from a low height in an upside-down position. This is an interesting issue because it raises the question of how a cat can "acquire" angular momentum and perform the twist if there is no initial angular momentum and no angular impulse to cause a change in angular momentum.

One explanation (figure 2.35) of this feat is based on a muscle contraction–induced conservation of angular momentum for the upper and lower body (Hopper, 1973). As shown in figure 2.35*a*, the cat is modeled as a two-segment system comprising an upper (G_1) and a lower (G_2) body segment that are linked by a set of muscles (*PQ*). In this scheme, the twist is initiated by a contraction of the muscles linking the upper and lower body (*PQ*), which causes the two segments (G_1 and G_2) to rotate (as indicated by the arrows) about their respective longitudinal axes. These rotations can be represented by angular momentum vectors, one for each segment ($\mathbf{H}_1$ and $\mathbf{H}_2$), as shown in figure 2.35*b*. By the right-hand rule, $\mathbf{H}_1$ is directed diagonally downward, while $\mathbf{H}_2$ is directed diagonally upward. Hence $\mathbf{H}_1$ and $\mathbf{H}_2$ have components in both the horizontal and vertical directions. The angular momentum components in the vertical direction cancel; but those in the horizontal direction sum to nonzero and indicate a positive angular momentum to the right, which reveals a rotation that will enable the cat to twist and to land on its feet.

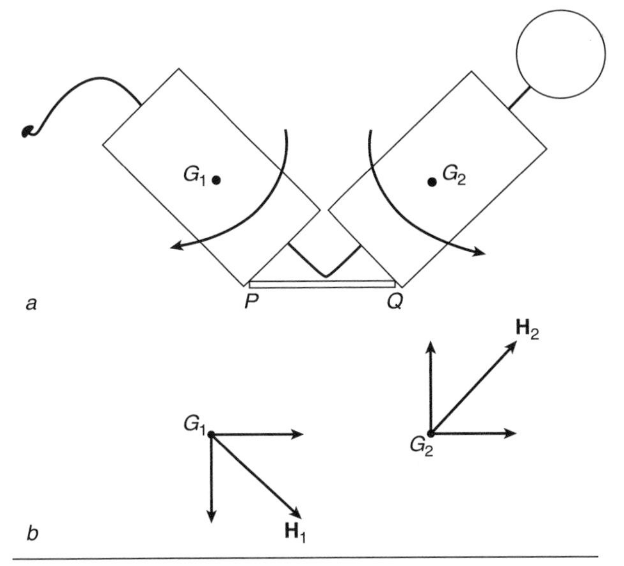

Figure 2.35 Model of a cat performing the air-righting reaction. *(a)* The two-segment model; *(b)* the angular momentum experienced by each segment.

Data from Hooper 1973.

WORK

A force applied to an object not only changes the momentum of the object but, if it displaces the object, does work on the object. **Work** (*U*) is a scalar quantity that is calculated as the product of the displacement experienced by the object and the component of the force acting in the direction of the displacement (equation 1.6). The unit of measurement for work is the joule (J; 1 J = 1 N•m). Work can be represented graphically as the area under a force–position (displacement) curve. This is a useful way to analyze movement when the force varies as a function of position. For example, when a person pulls on a linear spring and stretches it, the work done on the spring can be displayed as a force–length graph (figure 2.36*a*). The force needed to stretch the spring increases as a linear function of the amount of stretch. Because of this relation, the

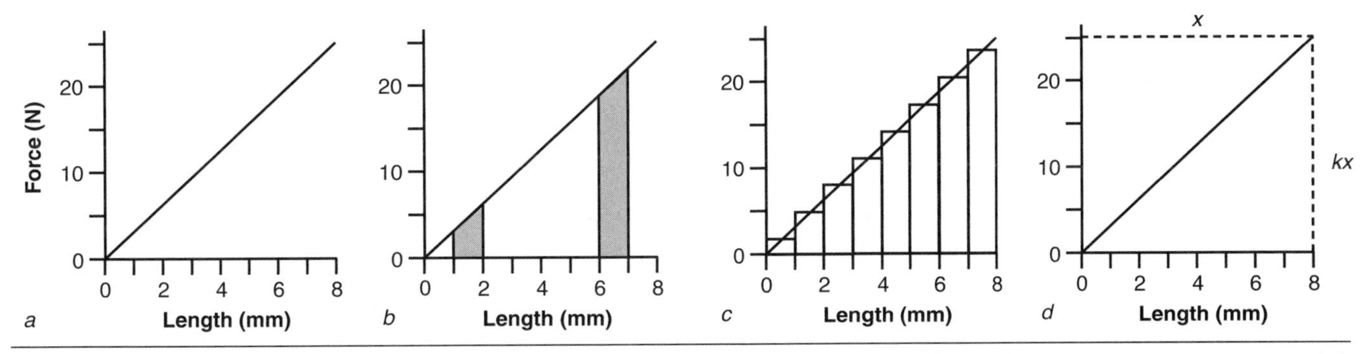

Figure 2.36 *(a)* The force applied to a spring increases linearly as the spring is stretched, *(b)* which means that the work done to stretch it by 1 mm is greater at longer lengths. *(c)* One can determine the area under the curve by calculating the amount of work represented by each of the eight rectangles and then summing the values for all the rectangles or *(d)* by taking one-half of the rectangle that has the sides of *kx* (force) and *x* (length).

Data from Hooper.

amount of work that must be done on the spring to stretch it by 1 mm is much greater at longer lengths; this is evident by the larger shaded area at the longer length (figure 2.36*b*).

Work can be positive or negative. When the force acts in the same direction as displacement, the work it does is positive. Conversely, when the force acts in the opposite direction from displacement, the work it does is negative. For example, when a person lifts a barbell by using the elbow flexor muscles, the muscles perform **positive work** by pulling on the forearm to lift the load so that the force and displacement occur in the same direction. When the person lowers the barbell, however, the muscles perform **negative work** because the force exerted by the elbow flexor muscles still pulls on the forearm while the displacement occurs in the opposite direction. When a muscle does positive work, it performs a shortening contraction. Conversely, when a muscle does negative work it performs a lengthening contraction. Due to a difference in muscle efficiency, the metabolic energy required to perform negative work is much less than that required for an equivalent amount of positive work.

We can calculate the work performed by the force by measuring the area under the force–position curve. Mathematically, this involves taking the integral of force with respect to position.

$$U = \int F\,dr \tag{2.26}$$

Numerically, however, we can determine the area by dividing it into many small rectangles (figure 2.31*c*), calculating the amount of work represented by each rectangle, and then summing the rectangles.

$$U = \sum_{i=1}^{N} F_i\,\Delta x \tag{2.27}$$

where N = the number of rectangles and Δx = the width of the rectangles. The procedure shown in figure 2.36*c* involves determining the height (F) of each rectangle, multiplying the height by the width (Δx), and summing the values for all eight rectangles. It should be obvious from figure 2.36*c* that the thinner the rectangles, the closer the estimated area will be to the actual area.

When the displacement experienced by an object is not linear but rather follows a curved path, which often occurs in human movement, the work done by the force is determined slightly differently. The procedure is to divide the path into many small segments, over which both the force and the displacement can be considered to be linear; calculating the work for each segment; and then summing the segments. This is represented as the line integral (equation 2.28) and involves taking the scalar product of the force vector with a particular path ($\Delta\mathbf{r}$), where the number of $\Delta\mathbf{r}$ segments is sufficient to cover the entire path of the object.

$$U = \int \mathbf{F} \cdot \mathbf{dr} \tag{2.28}$$

Kinetic Energy

In the time domain, we can determine the amount of motion an object has by calculating its momentum. In the length (position) domain, however, an object's motion is quantified as **kinetic energy** (E_k). Kinetic energy is a scalar quantity that depends on the mass of the object and its velocity (equation 2.29). As with many other quantities, it has both linear ($\frac{1}{2}mv^2$) and angular ($\frac{1}{2}I\omega^2$) terms.

$$E_k = \frac{1}{2}mv^2 + \frac{1}{2}I\omega^2 \tag{2.29}$$

As with work, the unit of measurement for energy is the joule (J; 1 J = 1 N•m = 1 kg•m^2/s^2). The kinetic energy of a volleyball after being served by a player, for example, could include both linear and angular terms or just linear, depending on whether or not the ball had any spin. A volleyball (mass = 0.27 kg; moment of inertia = 0.002 kg•m^2) traveling at 73 mph (32.6 m/s) and rotating at 48 rpm (5.02 rad/s) has the following amount of kinetic energy:

$$E_k = \frac{1}{2}mv^2 + \frac{1}{2}I\omega^2$$
$$= (0.5 \times 0.27 \times 32.6^2) + (0.5 \times 0.002 \times 5.02^2)$$
$$= 143 + 0.025 \text{ J}$$
$$E_k = 143 \text{ J}$$

As in this example, the magnitude of the linear component is often much larger than that of the angular component in most activities we perform. In a more complex activity, as when a person runs or cycles, the kinetic energy of the body depends on the linear and angular terms for all the body segments.

In parallel with the impulse–momentum relation (equation 2.15) in which an impulse changes the momentum of an object, the work–energy theorem states that the change in kinetic energy of an object is equal to the amount of work done on the object.

$$\Delta E_k = U \tag{2.30}$$

Because of this relation, energy is sometimes defined as the capacity to do work.

EXAMPLE 2.15

Work by the Knee Extensor Muscles

When a person performs a knee extension exercise on an isokinetic dynamometer and the torque exerted with the knee extensor muscles displaces the lever arm of the dynamometer by 1.57 rad, the work done by the muscles is apparent as the area under the torque–angle graph (figure 2.37) (1 N•m = 1 J). The exercise involves contracting the muscles to rotate the knee joint from a right angle (1.57 rad) until it is fully extended (3.14 rad).

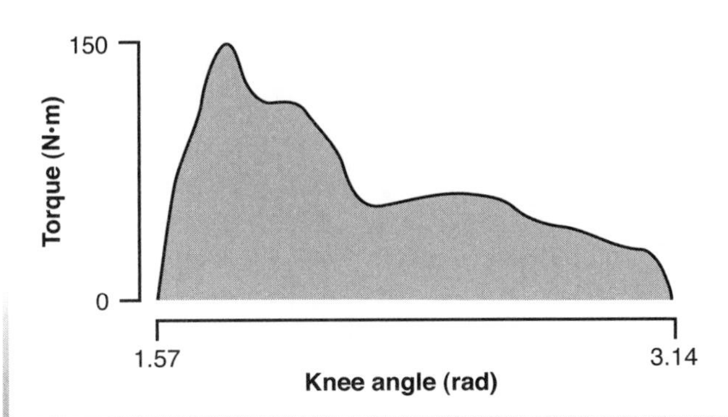

Figure 2.37 A torque–angle graph for one repetition of a knee extension exercise performed on an isokinetic dynamometer.

Potential Energy

When work is done on an object, its kinetic energy is changed and it is displaced. For most movements that we perform, the displacement often occurs against an opposing force, such as gravity or friction. If the direction of the opposing force never changes, as is the case with gravity, then displacement against the force provides the object with energy that can be used when the displacement-producing force is removed. This energy is known as **potential energy.** The two main forms that occur in human movement are gravitational ($E_{p,g}$) and strain or elastic ($E_{p,s}$) potential energy. Gravitational potential energy represents the energy that an object has due to its mass and its location in a gravitational field.

$$E_{p,g} = mgh \tag{2.31}$$

where g is the acceleration due to gravity and h is the vertical height above a baseline location, such as the ground. The $E_{p,g}$ of an object comes from the negative work done by gravity when the object is displaced to its location (h); the work is negative because gravity acts downward while the displacement is upward. Because h is the vertical location, the actual path (linear or not) taken to achieve this location has no effect on either the $E_{p,g}$ of the object or the negative work done by gravity.

The other form of potential energy that concerns us is **strain** or **elastic energy.** One example of this is seen in a spring. As described in figure 2.36, the force exerted by an ideal spring is a linear function of the amount of stretch that it experiences; the greater the stretch, the greater the force exerted by the spring. The slope of the line indicates the stiffness (k) of the spring. The more difficult it is to stretch a spring, the greater is its stiffness. The stiffness of the spring in figure 2.36 is 3.13 N/mm (25 N divided by 8 mm). The stiffness of a spring (or such tissues as tendon and ligament) depends on its material properties, which determine the resistance that it offers to any increases in its length. When a spring is stretched, the work is done on the spring by the stretching force and can be visualized as the area under the force–length graph (figure 2.36*d*). This area is one-half of the rectangle that has the amount of stretch (x) as the length of one side and the force (kx) as the length of the other side. This area represents the potential energy that the stretched spring has to do work.

$$E_{p,s} = \frac{1}{2}kx^2 \qquad (2.32)$$

EXAMPLE 2.16

Work Done on Elastic Bands

A physical therapist directing the rehabilitation of a patient recovering from knee surgery prescribes some knee extension exercises that involve a therapeutic elastic band. The stiffness of the band is 22 N/cm.

A. Calculate the force exerted by the band when it is stretched by 5 cm and 13 cm.

$$\begin{aligned} F_5 &= kx \\ &= 22 \text{ N/cm} \times 5 \text{ cm} \\ &= 110 \text{ N} \\ F_{13} &= kx \\ &= 22 \text{ N/cm} \times 13 \text{ cm} \\ &= 286 \text{ N} \end{aligned}$$

B. Draw a force–deformation graph that shows the elastic force provided by the therapeutic band.

C. On the same graph, draw a line to show the force and deformation when two bands are used (see figure 2.38).

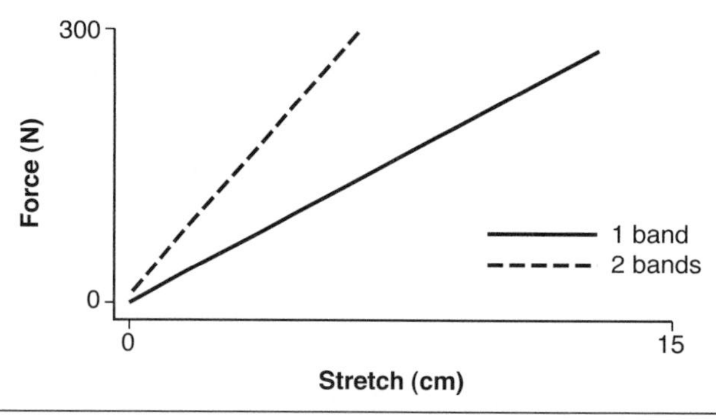

Figure 2.38 The force required to stretch one or more elastic bands increases with the amount of stretch.

D. Calculate the elastic energy stored in the two bands when they are stretched by 5 cm and that stored in one band when it is stretched by 13 cm.

$$E_{p,s} = 0.5kx^2$$
$$= 0.5 \times (2 \times 22 \text{ N/cm}) \times 5^2 \text{ cm}$$
$$= 550 \text{ J}$$
$$E_{p,s} = 0.5kx^2$$
$$= 0.5 \times 22 \text{ N/cm} \times 13^2 \text{ cm}$$
$$= 1859 \text{ J}$$

Conservation of Mechanical Energy

Equation 2.30 indicates that the work done on an object changes its kinetic energy. From the discussion of potential energy, we know that the work done on an object overcomes two types of forces, those that exert a constant opposing effect (e.g., gravity) and those that are not constant (e.g., friction, air resistance). Hence, we can rewrite equation 2.30 by representing the work done by these two types of forces:

$$\Delta E_k = U_c + U_{nc}$$

where U_c indicates the work done by constant opposing forces and U_{nc} represents the work done by the nonconstant opposing forces. Because U_c corresponds to potential energy acquired by the object, this equation becomes

$$\Delta E_k = \Delta E_p + U_{nc}$$
$$\Delta E_k + \Delta E_p = U_{nc} \quad (2.33)$$

When there are no nonconstant opposing forces acting on an object, the right-hand side of equation 2.33 becomes zero and we have a condition in which the sum of the changes in kinetic and potential energy is zero. That is, the sum of kinetic and potential energy is constant. This represents the law of **conservation of mechanical energy** (equation 2.34).

$$E_k + E_p = \text{Constant} \quad (2.34)$$

EXAMPLE 2.17

Exchange of Potential and Kinetic Energy

Consider a 60 kg (mass) acrobat about to leap off a 10 m tower. Assume that U_{nc} is zero. The acrobat has a potential energy of 5.88 kJ. The instant she leaves the tower, there will be no supporting surface to provide a ground reaction force; and because of gravity, she will fall and there will be a conversion of energy from one form to another. Her fall can be described as a change of energy from potential ($E_{p,g}$) to kinetic ($E_{k,t}$). Because the acceleration due to gravity is constant, the speed of falling increases with the distance covered in the fall. This correlates with an increasing conversion from potential to kinetic energy, in accordance with equation 2.34.

Consider her energy after she has fallen 4 m. At this position, her potential energy is 3.53 kJ, a decrease of 2.35 kJ. Her kinetic energy ($E_{k,t}$) can be determined by

$$E_{k,t} = \frac{1}{2}mv^2$$

To calculate her kinetic energy after falling 4 m, we need to determine her velocity (v) at that point. We can calculate velocity by using equation 1.4 to determine the time (0.9045 s) it took for her to fall 4 m and then using the definition of acceleration:

$$a = \frac{\Delta v}{\Delta t}$$

$$a = \frac{v_f - v_i}{t}$$

$$-9.81 = \frac{v_f - 0}{0.9045}$$

$$v_f = -8.86 \text{ m/s}$$

where v_i and v_f refer to the initial (at the beginning of the fall) and the final (at 4 m) velocities, respectively. The final velocity value (–8.86 m/s) can then be used to determine her kinetic energy at 4 m.

Alternatively, because $E_{k,t} = \frac{1}{2}mv^2$ and we know how much $E_{p,g}$ has been reduced (2.35 kJ) due to the law of conservation of mechanical energy, we can determine v by rearranging this relation:

$$E_{k,t} = \frac{1}{2}mv^2$$

$$v = \sqrt{\frac{2 \times E_{k,t}}{m}}$$

$$v = \sqrt{\frac{2 \times 2350}{60}}$$

$$v = \sqrt{78.33 \text{ m}^2/\text{s}^2}$$

$$v = 8.85 \text{ m/s}$$

Using either method, we obtain a magnitude for the final velocity of about 8.85 m/s. Thus her kinetic energy changed from zero at the beginning of the fall to a value of 2.35 kJ after 4 m, an increase that matched the decrease in potential energy.

EXAMPLE 2.18

Jumping on a Trampoline

A child, jumping on a trampoline, depresses the trampoline bed by 0.72 m. If the child weighs 391 N and the trampoline bed has a stiffness of 18 N/cm, how high will the child jump?

$$E_k + E_{p,g} + E_{p,s} = \text{Constant}$$

This is an energy transformation problem in which the potential strain energy is transformed to potential gravitational energy. Because E_k is zero at both the bottom of the bounce and the peak height reached, the equation becomes

$$E_{p,g} = E_{p,s}$$

$$mgh = \frac{1}{2}kx^2$$

$$h = \frac{kx^2}{2mg}$$

$$h = \frac{2 \times 1800 \times 0.72^2}{2 \times 391}$$

$$h = 1.19 \text{ m}$$

Power

In many short-duration athletic events (e.g., sprinting, Olympic weightlifting, arm wrestling, vertical jump) and even in some activities of daily living (e.g., recovering from a stumble), the rate at which muscles can produce work, referred to as power production, is the critical performance variable. Power production is measured as the amount of work done per unit time. **Power** ($\bar{P}$) can be determined as the work done (U) divided by the amount of time (Δt) it took to perform the work, or as the product of average force ($\bar{F}$) and velocity ($\bar{v}$):

$$\bar{P} = \frac{U}{\Delta t}$$

$$\bar{P} = \frac{\bar{F} \bullet \Delta r}{\Delta t}$$

$$\bar{P} = \bar{F} \bullet \bar{v}$$

As a measure of the rate of work performance, power is a scalar quantity that is measured in watts (W); 1 kW (1.36 horsepower) is the metabolic power corresponding to an oxygen consumption of about 48 ml/s. Because work represents a change in the energy of the system, power can also be written as the rate of change in energy (Ingen Schenau & Cavanagh, 1990).

$$\bar{P} = \frac{U}{\Delta t} = \frac{\Delta E}{\Delta t} = \bar{F} \bullet \bar{v}$$

The duration over which an activity can be sustained is inversely related to the power requirements of the activity (Lakomy, 1987).

Because power is defined as the rate of performing work, we can calculate it either by using the force and displacement associated with the work done or by using the change in mechanical energy (potential and kinetic) that enables the work to be done. Furthermore, these calculations can be applied to the whole body, applied to one body segment, or summed from the multiple segments in a limb (Arampatzis et al., 2000; Hatze, 1998). There is some debate among biomechanists, however, on the physical meaning of the power calculations applied to the whole-body CM (Ingen Schenau & Cavanagh, 1990). The more common application in human movement is to evaluate the instantaneous power about selected joints during a movement (DeVita & Hortobágyi, 2000; DeVita et al., 1996; Graf et al., 2005). This calculation, as presented in chapter 3, is typically performed with the angular equivalents of force and velocity. For example, the joint power during a movement is often estimated as the product of torque (τ) and angular velocity (ω).

EXAMPLE 2.19

Muscle Power for an Elbow Movement

Because power can be determined as the product of force and velocity and because acceleration and force are proportional to each other, we can obtain the power–time profile associated with the movement by multiplying the velocity and acceleration curves. When velocity and acceleration have the same sign (positive or negative), power is positive and represents an energy flow from the muscles to the arm. Conversely, when power is negative (i.e., velocity and acceleration have opposite signs), energy flows from the arm to the muscles. These conditions are known as **power production** and **power absorption,** respectively, indicating energy flow from and to the muscles.

With regard to the elbow extension-flexion movement in chapter 1 (figure 1.19), positive velocity (greater than zero) indicates elbow extension, positive acceleration represents an extension-directed force, and positive power (production) represents energy flowing from the muscles to the system (forearm-hand). We determine the power–time curve by multiplying the velocity and acceleration (force) graphs. When the velocity or acceleration graph crosses zero (i.e., changes sign), so does the power curve. The resulting power–time graph for the elbow extension-flexion movement is depicted as a four-epoch event (figure 2.39). The first two epochs, power production and absorption, respectively, occur during elbow extension (see velocity–time graph) and represent periods of positive and negative work. During positive work, the muscles do work on the system; during negative work, the system (due to its inertia) does work on the muscles. A similar sequence (power production, then absorption) occurs during the flexion phase of the movement.

This example emphasizes the correlation between the concepts of positive and negative work, power production and absorption, and shortening and lengthening muscle contractions. As shown in the lower panel of figure 2.39, the elbow extension-flexion movement is associated with a four-phase power–time profile. The epochs of power production correspond to periods of positive work and, therefore, a shortening muscle contraction. Recall that in chapter 1 we determined that this movement involved the following sequence of contractions: shortening extensor, lengthening flexor, shortening flexor, and lengthening extensor. According to this scheme, therefore, power absorption is related to negative work and lengthening muscle contractions.

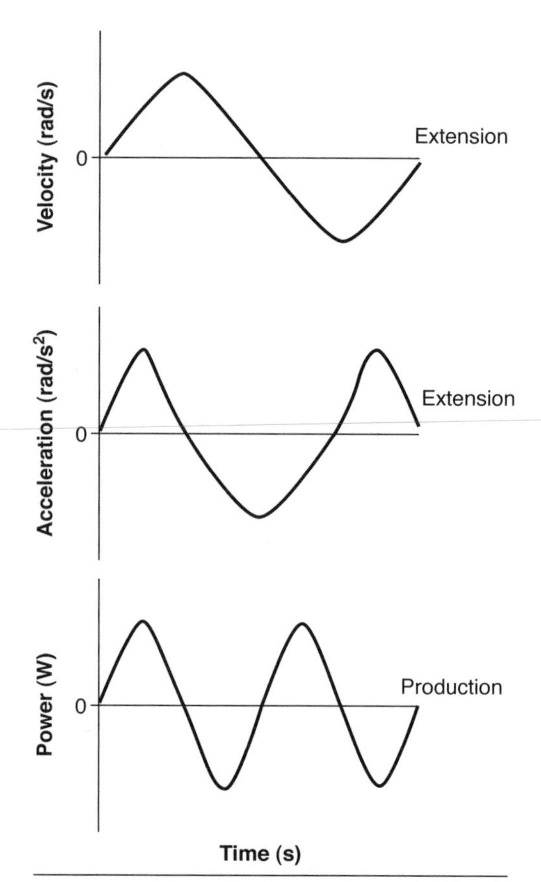

Figure 2.39 Power–time profile associated with the muscles across the elbow joint during the elbow extension-flexion movement shown in figure 1.19. Positive velocity represents an extension movement; positive acceleration indicates acceleration in the direction of extension; and positive power (production) refers to the power produced by the system.

EXAMPLE 2.20

Power in Jumping

For some athletic endeavors, the ability of an individual to produce power is a significant criterion for success. This capability can be assessed by measurement of the performance of an individual during a vertical jump for maximal height (Sayers et al., 1999), although there are some limitations to the accuracy of this procedure (Hatze, 1998). By simply measuring either the height that the CM is raised or the vertical impulse from the ground reaction force during the takeoff phase, we can estimate the average power produced by the individual during the jump. An assessment based on kinematics uses the equations of motion and the principles of projectile motion that were discussed in chapter 1. Consider the trajectory of an individual (mass = m) from the peak of a vertical jump to the landing. The height of the jump (r = the distance the CM was raised) can be determined with equation 1.4:

$$r = v_i t + 0.5at^2$$

$$r = 0.5 \times 9.81 \times t^2$$

$$t = \sqrt{\frac{r}{4.9}}$$

where t = one-half of the flight time. From equation 1.1, the average velocity is defined as

$$\bar{v} = \frac{\Delta r}{\Delta t}$$

and substituting for t,

$$\bar{v} = \frac{r}{\sqrt{r / 4.9}}$$

$$\bar{v} = \sqrt{r \cdot 4.9}$$

From the definition of power ($F \times v$), the average power ($\bar{P}$) during the descent phase for the individual (which is the same as that during the ascent phase and comes from the actions of the jumper during the takeoff phase) can be determined as

$$\bar{P} = (9.81 \cdot m) \cdot \sqrt{r \cdot 4.9} \tag{2.35}$$

Another way to determine average power is based on the kinetics of the vertical jump. This involves measuring the vertical impulse from the ground reaction force and using the impulse–momentum relation (equation 2.15).

$$\text{Impulse} = \Delta \text{ momentum}$$

$$\int F_{\text{net}}\, dt = m \cdot \Delta v$$

$$\Delta v = \frac{\int F_{\text{net}}\, dt}{m}$$

This expression indicates that the takeoff velocity ($\Delta v = v_f$) is equal to the net vertical impulse divided by the mass of the individual. To determine the net vertical impulse, we measure the area under the force–time curve for the ground reaction force and then subtract the body weight impulse (body weight times the duration of the takeoff phase). Because initial velocity (v_i) is zero, average velocity ($\bar{v}$) can be calculated as

$$\bar{v} = \frac{v_i + v_f}{2}$$

$$\bar{v} = \frac{v_f}{2}$$

$$\bar{v} = \frac{\int F_{\text{net}}\, dt}{2m}$$

The average force ($\bar{F}$) exerted during the takeoff phase of the jump is calculated by dividing the absolute impulse ($\int F_{g,y}\, dt$) by the duration of the takeoff phase:

$$\bar{F} = \frac{\int F_{g,y}\, dt}{t}$$

Average power produced by the jumper can be determined from

$$\bar{P} = \bar{F} \cdot \bar{v}$$

$$\bar{P} = \frac{\int F_{g,y}\, dt}{t} \cdot \frac{\int F_{\text{net}}\, dt}{2m} \tag{2.36}$$

The advantage of equation 2.36 is that the average power can be estimated from the measurement of the vertical component of the ground reaction force and the mass of the individual. However, a more accurate measurement of average power can be achieved if we consider the change in mechanical energy during the takeoff phase for the jump (Hatze, 1998).

$$U = \frac{1}{2}mv^2 + mgh$$

where v = the velocity of the CM at takeoff and h = the vertical displacement of the CM from the start of the jump (crouch position) to the takeoff. Because the takeoff velocity can be obtained from jump height ($v = \sqrt{2gr}$), the equation can be rewritten as

$$U = \frac{1}{2}mv^2 + mgh$$

$$U = mgr + mgh$$

where m = jumper's mass, g = acceleration due to gravity, r = height that the CM was raised, and h = CM displacement during the takeoff phase. Average power can then be calculated as the work done divided by the duration (t) of the takeoff phase:

$$\bar{P} = \frac{mg(r+h)}{t} \tag{2.37}$$

These measurements, although more involved than those for equation 2.36, can also be made from measurement of the vertical component of the ground reaction force (Hatze, 1998).

SUMMARY

This chapter is the first of two to address the forces associated with human movement. Our approach is based on Newton's laws of motion and an analysis technique known as the free body diagram. Because human movement involves the rotation of body segments about one another, the concept of a rotary force (torque) and the relevant calculations are introduced. After these introductory sections, the chapter introduces the forces experienced by the human body during movement. These are categorized into forces due to body mass (gravity and inertial force) and forces due to the surroundings (ground reaction force, fluid resistance). The last part of the chapter describes the effects of these forces on the mechanical actions of the body, which are quantified as momentum and work. With these techniques, the motion of an object is characterized in terms of its momentum or its mechanical energy.

SUGGESTED READINGS

Hamill, J., & Knutzen, K.M. (2003). *Biomechanical Basis of Human Movement.* Baltimore: Williams & Wilkins, chapters 10 and 11.

McGinnis, P.M. (1999). *Biomechanics of Sport and Exercise.* Champaign, IL: Human Kinetics, chapters 3, 4, and 6-10.

Medved, V. (2001). *Measurement of Human Locomotion.* Boca Raton, FL: CRC Press LLC, chapter 5.

Nigg, B.M., MacIntosh, B.R., & Mester, J. (2000). *Biomechanics of Biology and Movement.* Champaign, IL: Human Kinetics, chapter 14.

Watkins, J. (1999). *Structure and Function of the Musculoskeletal System.* Champaign, IL: Human Kinetics, chapter 1.

Winter, D.A. (1990). *Biomechanics and Motor Control of Human Movement.* New York: Wiley, chapters 3, 4, and 5.

chapter 3

Forces Within the Body

The motion of an object, which is described in terms of position, velocity, and acceleration, is characterized as the consequence of the interaction between the object and its surroundings. This interaction is commonly represented as a force. In chapter 2 we considered the forces due to body mass and those due to the surroundings that can be imposed on the human body to change its momentum or its mechanical energy. The forces involved in human movement, however, include not only these interactions but also the mechanical interactions within the musculoskeletal system. To examine these musculoskeletal forces, we draw free body diagrams and apply Newton's laws of motion to each body segment that participates in the movement. In chapter 2, this approach was described as a segmental analysis. The purpose of chapter 3 is to describe the properties of the musculoskeletal forces that are encountered in such an analysis and to present the procedures used to estimate the magnitudes and directions of these forces.

MUSCULOSKELETAL FORCES

Through the application of Newton's law of acceleration to a segmental analysis of the human body, it is possible to determine the magnitude and direction of forces inside the human body. We call these interactions **musculoskeletal forces.** To accomplish this we draw a free body diagram that ends at a joint; this allows us to include these forces as external effects acting on a system (e.g., figures 2.2, 2.3, 2.10). Whenever we draw a free body diagram that includes part of the body and ends at a joint, we need to include a joint reaction force and a muscle force in the diagram. The joint reaction force is typically drawn as a force acting into the joint, which represents the reaction of the missing body segment to the compressive force in the joint. The muscle force is usually shown acting back across the joint to represent the net pulling action of the muscles that cross the joint. In addition, when the free body diagram ends anywhere in the trunk, we need to include a force due to the pressure inside the abdomen or the thorax.

Joint Reaction Force

When a system for a free body diagram is defined so that it ends at a joint, the concept of a **joint reaction force** (F_j) is invoked to represent the reaction of the adjacent body segment to the forces exerted by the isolated system. This is a three-dimensional force that has one component normal to the joint surface and, like the ground reaction force, two components that are tangential to the surface. The normal component is typically directed into the joint surface and represents a compressive force. The two tangential components comprise the shear force that acts along the joint surface.

The joint reaction force can be influenced by any effect included on the free body diagram. Examples are forces that are transmitted from one end of the segment to the other end (e.g., ground reaction force), forces due to joint-related soft tissue structures (e.g., ligaments, joint capsule), and the forces exerted by the muscles. The magnitude of these forces can be large. Although the magnitude of the forces transmitted by soft tissues, especially the ligaments, has been controversial, it now appears that these forces can be significant and that they vary over the range of motion of the joint (Mommersteeg et al., 1997; Shelburne & Pandy, 1997).

The most significant and consistent contributor to the joint reaction force is the force due to muscle activity (Duda et al., 1997; Komistek et al., 2005; Lu et al., 1997; Taylor et al., 2004). When a muscle contracts, one component of the muscle force vector is transmitted into the joint as a compressive force. Given that muscles have a shallow angle of pull, most of the muscle force is directed into the joint. For example, Lu and colleagues (1997) found that when subjects stood in an upright position and performed isometric contractions with the hip flexor, extensor, abductor, and adductor muscles, the compressive force along the femur was ~20 times greater than the shear force measured at the ankle. For these contractions, the hip muscles exerted average forces of ~2000 N.

It is difficult to measure F_j experimentally, and such measurement usually involves either invasive procedures or mathematical modeling (Komistek et al., 2005). For example, Bergmann and colleagues (1993) implanted force transducers to measure the hip joint reaction force in two patients who had been fitted with hip joint prostheses. When these patients walked on a treadmill at 1.1 m/s, the joint reaction forces at the hip varied during a stride as shown in figure 3.1. The vertical component was directed downward (compression) with a peak magnitude of three times body weight; the medial-lateral component was directed medially with a peak magnitude of about body weight; and the front-to-back component was directed first forward and then backward with a peak magnitude of about 0.5 times body weight. Similar measurements have been made on a patient during performance of activities of daily living and exercises in the first year after total knee arthroplasty (D'Lima et al., 2005).

In most studies, however, researchers estimate the magnitude of F_j by determining all the other forces on a free body diagram and assuming that the remaining effect is due to F_j (Shelburne et al., 2005; Thambyah et al., 2005); this procedure is known as a residual analysis. We can use a residual analysis, for example, if the system is in equilibrium, which means that all the forces acting on the system must be balanced. Alternatively, it is possible to use various mathematical procedures, such as minimizing muscle stress, to estimate the magnitude of F_j. An and colleagues (1984) used such an approach on the elbow (humeroulnar) joint when a load was applied at the wrist perpendicular to the forearm. When the joint was rotated over a range of motion from complete extension to a right angle, F_j at the elbow joint was 6 to 16 times greater than the load at the wrist. When the loads encountered in normal daily activities were considered, this meant that values for F_j of 0.3 to 0.5 times body weight were commonly encountered at the elbow joint.

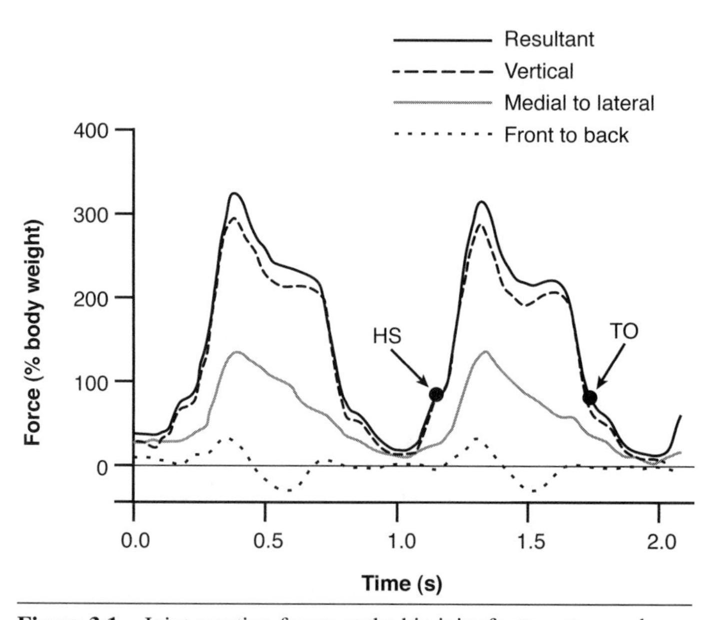

Figure 3.1 Joint reaction forces at the hip joint for two stance phases (HS = heel strike; TO = toe-off) during walking on a treadmill.

Data from Bergmann et al. 1993.

Values for F_j have been estimated by residual analysis for such activities as standing, moving from sitting to standing, walking, running, weightlifting, and landing from a drop (summarized in Harrison et al., 1986). Even the common task of going from an erect posture to a squat position and then rising again is associated with large joint reaction forces. For this task, the maximal compression component of the tibiofemoral joint reaction force ranged from 4.7 to 5.6 times body weight, whereas the shear component ranged from 3.0 to 3.9 times body weight (Dahlkvist et al., 1982). Powerlifters experience maximal compressive forces of 17 times body weight and maximal shear forces of 2.3 times body weight at the L4-L5 joint during the performance of the deadlift (Cholewicki et al., 1991). Results such as these emphasize that the joint reaction force varies over the range of motion and that it can have a substantial magnitude, especially in comparison with the loads that are encountered by the limbs in daily activities.

An alternative approach to residual analysis for estimating F_j is to measure acceleration and use inverse dynamics to determine joint reaction forces and torques (Bogert et al., 1996). Attaching four accelerometers at known locations to the trunks of individuals made it possible to estimate the joint reaction force at the hip during walking, running, and skiing (Bogert et al., 1999). The peak forces, in terms of body weight (F_w), averaged about 2 times F_w for walking at 1.5 m/s, 5 times F_w for running at 3.5 m/s, 4 to 7 times F_w for alpine skiing, 4 times F_w for cross-country skiing, and 7 to 13 times F_w for skiing moguls. Such information indicates the range of forces that a hip prosthesis must be able to withstand in an active individual (Heller et al., 2005).

EXAMPLE 3.1

Absolute Magnitude of the Joint Reaction Force

The joint reaction force is usually determined from the *net* forces between adjacent body segments. To determine the absolute magnitude of the joint reaction force, however, it is necessary to consider each force rather than calculating a net effect. An interesting example of the difference between the net and absolute effects was provided by Galea and Norman (1985) in a study of the muscle actions across the ankle joint during a rapid ballet movement, a spring from flat feet onto the toes. In this analysis, the model used to perform these calculations (figure 3.2*a*) took into account the major muscles crossing the ankle joint and therefore those that contributed to the muscle-dependent component of the joint reaction force. The muscles included extensor hallucis longus, tibialis anterior, flexor hallucis longus, peroneus longus, and gastrocnemius/soleus. For the calculations, the force exerted by each of these muscles was estimated based on its electromyogram (EMG), length, and rate of change in length (Hof, 1984; Hof & van den Berg, 1981*a, b, c,* and *d*).

Once Galea and Norman (1985) had estimated a force for each muscle, the forces were grouped (figure 3.2*b*) into those that contributed to a plantarflexor torque ($F_{m,pf}$) and those that contributed to a dorsiflexor torque ($F_{m,df}$). The absolute force was calculated as the compressive component of the $F_{m,pf}$ vector plus the compressive component of the $F_{m,df}$ vector. The net joint reaction force (figure 3.2*c*), however, was calculated as the compressive component of the resultant muscle force (F_m), where F_m is the difference between $F_{m,df}$ and $F_{m,df}$. For

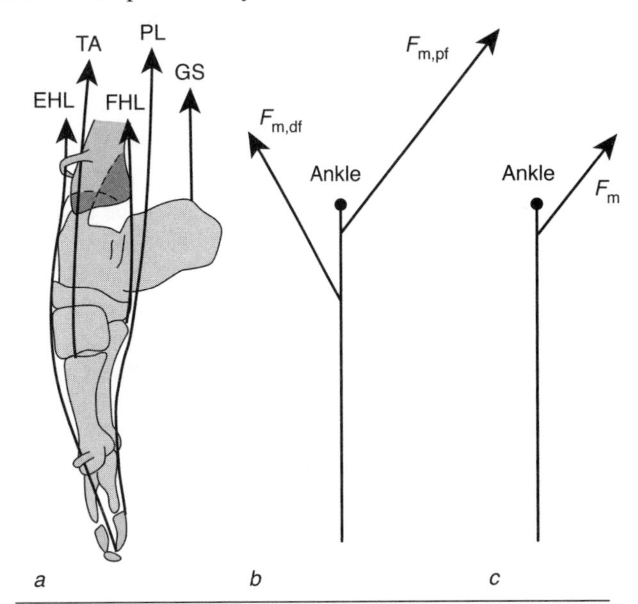

Figure 3.2 Muscles that are active across the ankle joint during a full pointe: *(a)* lines of action of the involved muscles; *(b)* separation of the muscular effects (F_m) into those that exert dorsiflexor (df) and plantarflexor (pf) torque; *(c)* resultant muscle force. EHL = extensor hallucis longus; FHL = flexor hallucis longus; GS = gastrocnemius/soleus; PL = peroneus longus; TA = tibialis anterior.

Adapted, by permission, from V. Galea and R.W. Norman, 1985. Bone-on-bone forces at the ankle joint during a rapid dynamic movement. In *Biomechanics IX-A*, edited by D.A. Winter et al. (Champaign, IL: Human Kinetics), 72.

one subject, Galea and Norman calculated a net joint reaction force due to muscle activity of 732 N during the movement and an absolute joint reaction force of 6068 N, which is a huge difference.

For movements or parts of movements that do not involve concurrent activation of opposing muscles, it is possible to estimate the bone-on-bone force from the calculated net forces. Simonsen and colleagues (1995) performed such an analysis at the ankle, knee, and hip joints as individuals walked with and without a load (10 and 20 kg). They recorded EMG activity over opposing muscles at the three joints to confirm the absence of coactivation. The peak bone-on-bone forces ranged from 3.32 kN at the ankle joint without a load up to 6.4 kN at the hip when people were carrying a load of 20 kg. The bone-on-bone forces increased at both the ankle and hip joints when the individuals carried the loads, whereas there was no difference at the knee joint. There were, however, considerable differences between individuals.

Muscle Force

In the segmental analysis of human movement, **muscle force** is often the most significant component that is included on a free body diagram. When the system defined in a free body diagram ends at a joint, the muscle force vector is shown as acting back across the joint to represent the pulling effect of the muscle on the segment. This vector usually represents the net muscle activity about a joint.

Because muscles can exert only a pulling force on a segment, opposing sets of muscles (agonists and antagonists) control movement about a joint. An extension-flexion movement about the elbow joint, for example, is controlled by one muscle group that causes acceleration in the extension direction (elbow extensors) and another group that causes acceleration in the flexion direction (elbow flexors).

The muscle force vector can be represented as an arrow and described in terms of its magnitude and direction. In a segmental analysis, the force is represented by a line of action that extends between its proximal and distal attachments. Figure 3.3 shows an example for 24 muscles in the human leg; this model was used to estimate the joint reaction force at the hip (Duda et al., 1997). In this scheme, the application of force is represented as acting at a point, which is reasonable for most muscles (Brand et al., 1982). If the attachment site is substantial (e.g., trapezius, pectoralis major), a muscle may be represented by several lines of action (Davis & Mirka, 2000; van der Helm & Veenbaas, 1991). Furthermore, the lines of action of some muscles (e.g., neck and trunk muscles) are most appropriately represented by curved paths (Arjmand et al., 2006; Kruidhof & Pandy, 2006). Obviously, care must be taken to determine the lines of action as accurately as possible (Nussbaum et al., 1995).

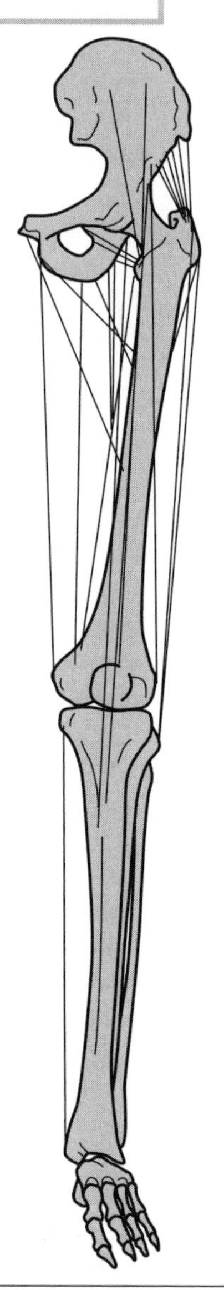

Figure 3.3 Lines of action of 24 muscles in the human leg.

Reprinted from *Journal of Biomechanics*, 30, Duda et al., "Internal forces and moments in the femur during walking," pp. 933-941. Copyright in 1997, with permission from Elsevier Science.

Magnitude of Muscle Force

Both the magnitude and direction of the muscle force vector are difficult to measure. To measure the magnitude of muscle force directly, we must measure the force transmitted by the tendon (Fukashiro et al., 1995; Komi, 1990). In an isolated-muscle experiment, this measurement involves connecting the tendon to a force transducer (Ralston et al., 1947). In human experiments, in which the tendon is not detached from the bone, we can measure muscle force by using a surgical procedure either to thread a tendon through an E-shaped buckle or to insert a fiber optic cable (0.5 mm diameter) through the tendon. When a buckle transducer was placed on the Achilles tendon during cycling, Gregor and colleagues (1987) measured peak tendon forces of about 700 N in the right Achilles tendon for subjects pedaling at a rate of 90 rpm while producing 265 W of power. When a fiber optic cable was inserted through the Achilles tendon, Finni and colleagues (1998) recorded peak forces of 1430 N during the stance phase of walking. The peak force was similar across walking speeds (1 to 2 m/s), but the rate of force development increased by 32% when individuals went from a slow to a fast walk.

Most human subjects, however, are not willing to volunteer for such invasive procedures, and it is necessary to use more indirect techniques to assess muscle force. Much of the information available on the magnitude and direction of muscle forces has been derived from indirect estimates. One common approach has been to determine the **cross-sectional area** of muscle from sections that are made perpendicular to the orientation of the muscle fibers and to use this information to estimate maximal muscle force (Fick, 1904). Cross-sectional area, a measurement of the end-on view of the area at the level at which the section (cut) has been made, indicates the number of force-generating units (myofibrils) that are lying in parallel in the muscle. These measurements can be obtained through dissection of cadavers or through the use of imaging procedures on volunteers (ultrasound, computed tomography, magnetic resonance imaging) (Bamman et al., 2000; Kawakami et al., 1994; Klein et al., 2001; Morse et al., 2007; Tate et al., 2006).

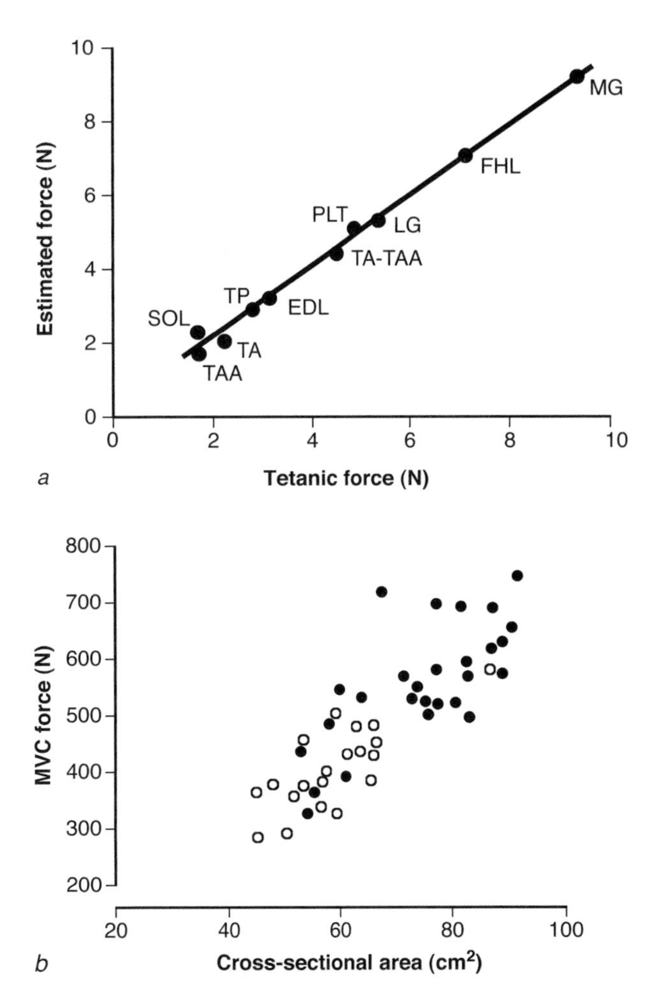

Figure 3.4 Relation between the cross-sectional area of muscle and its maximal force. *(a)* The measured tetanic force was linearly related to the maximal force estimated from the product of cross-sectional area and specific tension for selected muscles in the hindlimb of the guinea pig. *(b)* The peak force exerted by the quadriceps femoris muscle during an isometric maximal voluntary contraction was linearly related to cross-sectional area as measured with computed tomography in men (filled circles) and women (open circles). EDL = extensor digitorum longus; FHL = flexor hallucis longus; LG = lateral gastrocnemius; MG = medial gastrocnemius; PLT = plantaris; SOL = soleus; TA = tibialis anterior; TAA = tibialis anterior accessorius.

Data from Powell et al. 1984; Jones et al. 1989.

The relation between the cross-sectional area of a muscle and its maximal force is shown in figure 3.4*a*. There is a linear relation between the maximal tetanic force evoked by electrical stimulation and the estimated peak force based on cross-sectional area. The slope of the regression line is referred to as **specific tension** and has an average value of 22.5 N/cm^2 for these data. The soleus muscle, however, has a significantly lower specific tension of 15.7 N/cm^2, presumably due to the high proportion of type I muscle fibers.

Despite this association, the cross-sectional area of muscle accounts for only about 50% of the variance in strength across individuals during a voluntary contraction (figure 3.4*b*). In studies on humans, normalized force varies from 16 to 60 N/cm^2 across studies, with a nominal value of 30 N/cm^2 (Edgerton et al., 1990; Maganaris et al., 2001; McDonagh & Davies, 1984; Narici et al., 1992). Five factors that often contribute to this variation are (a) the use of a single measurement of cross-sectional area when, for most muscles, cross-sectional area varies along the length of the muscle (Kawakami et al., 1995; Narici et al., 1988); (b) the need to identify all of the muscles that contribute to the force; (c) the difficulty associated with keeping the antagonist muscles silent during performance of a maximal contraction with the agonist muscles;

(d) the assumption that the entire muscle mass can be activated (Kandarian & Williams, 1993; Kandarian & White, 1990); and (e) the variation in muscle architecture (the way muscle fibers are organized), which appears to modulate the estimated specific tension (Fukunaga et al., 1996). Because of these limitations, the measurement of muscle force relative to cross-sectional area at the whole-muscle level should be referred to as **normalized muscle force** rather than specific tension. The term *specific tension* should be used to denote the intrinsic force capacity of single muscle fibers and in situ measurements on isolated muscles.

Given an accurate estimate of specific tension, the maximal force a muscle (F_m) can exert is estimated from

$$F_m = \text{specific tension} \times \text{cross-sectional area} \quad (3.1)$$

If biceps brachii has a cross-sectional area of 5.8 cm^2 (table 3.1), then on the basis of a specific tension of 30 N/cm^2 we estimate that it can develop a maximum force of 174 N. Together, the maximal force that could be exerted for the elbow flexor muscles would be 657 N based on the cross-sectional data listed in table 3.1. However, Edgerton and colleagues (1990) report that the largest cadaver in their sample of four had a cross-sectional area of 34.7 cm^2 and therefore that the elbow flexor muscles could have exerted a maximal force of 1297 N. In addition to the data available on the muscles that cross the elbow joint (table 3.1), there are data for most other muscle systems, such as the hip, leg, and wrist (Brand et al., 1986; Clark & Haynor, 1987; Häkkinen & Keskinen, 1989; Lieber et al., 1990; Murray et al., 2000).

One issue of interest to muscle physiologists is whether specific tension can vary. Is it possible, for example, for the cross-sectional area of a muscle in two individuals to be the same but for the maximal force to differ? Specific tension is a functional measure (the intrinsic force capacity) of the number of myofibrils per unit of cross-sectional area. Theoretically, therefore, specific tension would vary when the density of the myofibrils changes, as might occur when a muscle fiber swells or when the myofibrils are packed more closely. There is some evidence, based on single-fiber measurements, that specific tension can vary with physical activity (Frontera et al., 2003; Kawakami et al., 1995; Riley et al., 2000) and that it differs across fiber types. For example, Larsson and colleagues (1996) found that the specific tension of muscle fiber segments taken from the vastus lateralis muscle of volunteers decreased by 40% after six weeks of bed rest. Similarly, the specific tension of type II muscle fibers appears to decline with age (Larsson et al., 1997). Furthermore, specific tension can vary across muscle fiber types and can be different for the same fiber type in different muscles. Harridge and colleagues (1996) found the specific tension of type IIa muscle fibers to be 22.3 N/cm^2 in soleus and vastus lateralis and 38.6 N/cm^2 in triceps brachii, whereas it was 22.6 N/cm^2 for type I fibers in triceps brachii. Similarly, at the whole-muscle level, the maximal force capacity normalized to cross-sectional area differs among muscles. The ankle plantarflexor muscles, for example,

Table 3.1 Summary Data on Cross-Sectional Areas (CSA) and Moment Arms for the Elbow Flexor and Extensor Muscles

	CSA	Predicted force	Moment arm	Torque	
	(cm^2)	(N)	(cm)	(N·m)	(% maximum)
Biceps brachii	5.8	174	3.8	6.6	32
Brachialis	7.4	222	2.9	6.4	31
Brachioradialis	2.0	60	6.1	3.7	18
Pronator teres	3.6	108	1.6	1.2	6
Extensor carpi radialis longus	3.1	93	3.0	2.8	14
Triceps brachii	23.8	714	—	—	—

Note. Predicted force was estimated by multiplying the CSA values by a specific tension of 30 N/cm^2. Torque was determined as the product of predicted force and moment arm. The % maximum data indicate the contributions of the respective elbow flexor muscles to the total elbow flexor torque.

Data are from Edgerton et al. 1990.

exhibit a maximal normalized force that is twice as large as that of the dorsiflexor muscles (Fukunaga et al., 1996).

In contrast to estimating the maximal force capacity of muscle based on its cross-sectional area, we can use the amplitude of the EMG to estimate the actual muscle force exerted during a contraction. Because the EMG measures action potentials as they travel along many of the activated muscle fibers, EMG amplitude changes with muscle activation. Under isometric conditions, the EMG amplitude can be highly correlated with muscle force (Doorenbosch et al., 2005; Staudenmann et al., 2006). Although this association is less direct for anisometric conditions (Calvert & Chapman, 1977; Milner-Brown & Stein, 1975), it is possible to estimate the magnitude of the muscle force from the EMG (Davis & Mirka, 2000; Lee et al., 2003; Lloyd & Besier, 2003; Wang & Buchanan, 2002). The most typical strategy in using EMG to estimate muscle force is to measure the EMG signal during a maximal voluntary contraction and then to normalize subsequent EMG recordings to this maximum (Burden et al., 2003; Keenan et al., 2005). With this approach we can describe the quantity of EMG during a particular movement as a percentage of the value recorded during a maximal voluntary contraction performed in the same testing session.

EXAMPLE 3.2

Muscle Force in Humans

The direct measurement of muscle force in humans can be accomplished only with invasive techniques, which were pioneered by Komi and colleagues (Komi, 1990, 1992). These procedures involve either the attachment of a force transducer to a tendon or the insertion of an optic fiber through a tendon (Finni et al., 1998; Fukashiro et al., 1995; Gregor et al., 1991; Komi et al., 1996). With such procedures, it is possible to estimate the tendon force during various activities. In addition to the force measurement, subjects are videotaped while they walk, run, hop, jump, or ride a bicycle so that changes in muscle length can be estimated. The results of one such experiment, in which tendon forces were measured with the optic fiber technique, are shown in figure 3.5. Changes in muscle length were estimated as a function of joint angles (Hawkins & Hull, 1990). The data represent Achilles tendon and patellar tendon forces and the length changes of the triceps surae and quadriceps femoris muscles; negative velocity indicates lengthening of the tendon. When the foot contacted the ground, the force increased to a peak value of about 1.7 kN in the Achilles tendon and 2.2 kN in the patellar tendon. The peak forces occurred at the transition from the lengthening to the shortening contractions.

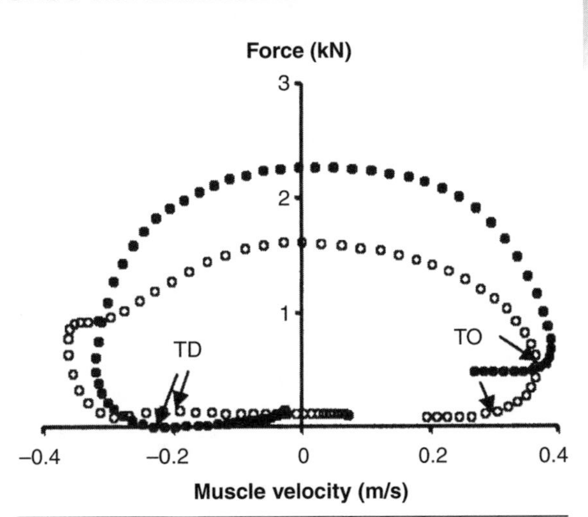

Figure 3.5 Forces transmitted by the Achilles tendon (open circles) and patellar tendon (filled circles) relative to the velocities of the triceps surae and quadriceps femoris muscles during two-legged hopping. Touchdown (TD) occurred at the left endpoint and toe-off (TO) at the right endpoint of the graph. The time between each data point was 5 ms.

Data from Finni et al., 2000.

Direction of Muscle Force

When a muscle force is included on a free body diagram, the force vector should be drawn so that it is directed back across the joint. We can think of the muscle as pulling on the body segment so that it rotates the segment about the joint. Most joint systems in the human body are designed as third-class levers in which the muscle force applied to the skeleton and the load experienced by the lever lie on the same side of the joint, with the point of application of the muscle force being proximal to the load. This type of arrangement maximizes the linear velocity of the endpoint of the lever ($v = r\omega$) but requires large muscle forces to control the motion of the lever about the fulcrum.

The angle between the muscle force vector and the skeleton is referred to as the **angle of pull.** Although the angle of pull of a muscle is generally shallow when a joint is in its anatomical resting position, the angle does change over the range of motion and can become substantial. When the elbow joint is at a right angle (1.57 rad), for example, the angles of pull of the muscles that cross the elbow joint were shown to be 1.4 rad for biceps brachii, 1.2 rad for brachialis, 0.4 rad for brachioradialis, 0.2 rad for pronator teres, and 0.05 rad for triceps brachii (An et al., 1984).

EXAMPLE 3.3

Muscle and Load Torques During Knee Extension

Consider a person recovering from knee surgery who does seated knee extension exercises with a weighted boot. The exercise involves sitting at the end of an exercise bench and raising the lower leg (shank + foot) from a vertical (knee angle = 1.57 rad) to a horizontal position and then lowering the leg again. What torques about the knee joint (K) would be involved in this exercise? Figure 3.6 depicts the appropriate system, from the knee joint down to the toes, and the four forces—joint reaction force (F_j), resultant muscle force (F_m),

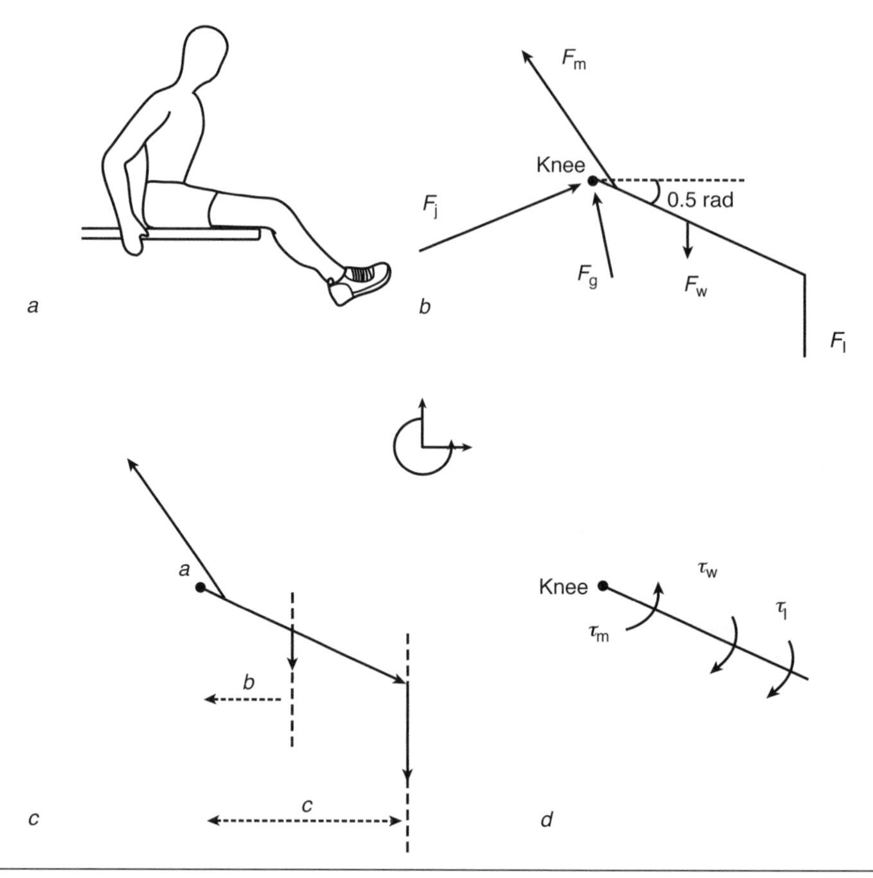

Figure 3.6 Free body diagram of the leg of a person performing leg extension exercises: *(a)* whole-body diagram; *(b)* free body diagram of the lower leg; *(c)* the forces that exert a torque about the knee joint; *(d)* the three torques produced by these forces.

limb weight ($F_{w,l}$), and boot weight ($F_{w,b}$)—that represent the interaction of this system with its surroundings. The first step in determining the torques is to draw in the moment arms (*a*, *b*, *c*). This involves extending the line of action for each force and then drawing a perpendicular line from the line of action to the axis of rotation (figure 3.6*c*). Because the line of action for the joint reaction force passes through the axis of rotation, its moment arm, and therefore its torque about the knee joint, is equal to zero. Thus, for the system indicated in figure 3.6, there are three forces that produce a torque about the knee joint during this exercise. Figure 3.6*d* shows the resultant muscle torque (τ_m), the torque due to the weight of the boot (τ_l), and the torque due to the weight of the leg (τ_w). The torque due to the total load is determined as the sum of τ_l and τ_w. The direction of the curved torque arrow is the same as the rotation that the torque causes.

Now suppose that the person performing this exercise has an 80 N weight attached to the ankle, a body weight of 700 N, and a resultant muscle force with a magnitude of 1000 N and an angle of pull of 0.25 rad; calculate the three torques about the knee joint. We also know that the distance along the shank from the muscle force vector to the knee is 5 cm and that the length of the shank (knee to ankle) is 36 cm. The simplest approach for now is to calculate each of these torques separately. First, let us determine the torque produced by the resultant muscle force about the knee joint and draw a diagram (figure 3.7) that shows F_m and its moment arm (*a*). As we discussed in chapter 2, torque is calculated as the product of a force and its moment arm, which is the shortest distance from the line of action of a force vector to the axis of rotation.

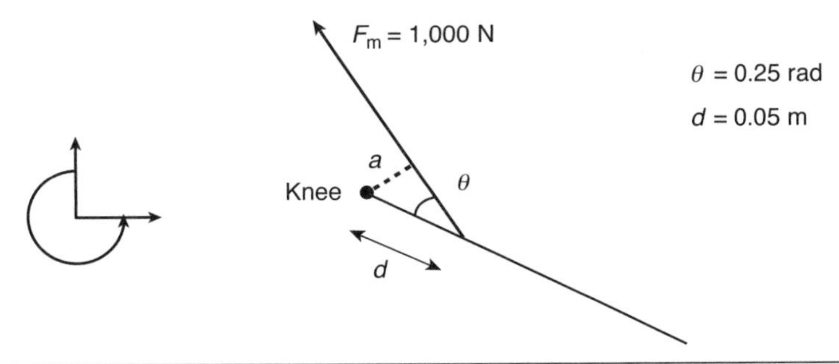

Figure 3.7 The magnitude and direction of muscle force (F_m) and the associated moment arm (*a*) to the knee joint.

$$\tau_m = F_m \times a$$
$$= 1000 \times a$$

$$\sin 0.25 = \frac{a}{0.05}$$
$$a = 0.05 \sin 0.25$$
$$= 0.0124 \text{ m}$$

$$\tau_m = 1000 \times 0.0124$$
$$\tau_m = 12.4 \text{ N•m}$$

The torque due to the weight of the limb can be determined in a similar manner. We can also take the opportunity to review the methods we learned in chapter 2 to estimate the weight and center-of-mass location of body segments. The weight of the shank and foot can be estimated from the combined shank and foot regression equations in table 2.1. Similarly, from table 2.1 we can estimate the center-of-mass location (i.e., point of application for the weight vector) for the lower leg (shank + foot) as being at a distance (*d*) of 43.4% of shank length, as measured from the knee joint. We begin by drawing a diagram (figure 3.8) that shows F_w and its moment arm (*b*). Once these two variables have been determined, we can calculate the torque due to the weight of the system (τ_w) (figure 3.9).

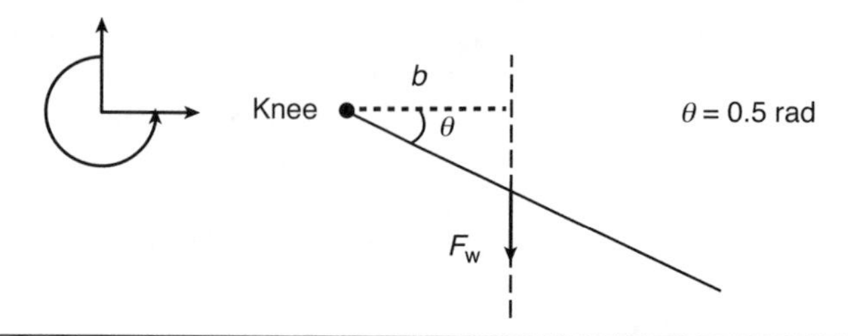

Figure 3.8 The magnitude and direction of the weight vector (F_w) and the associated moment arm (b) to the knee joint.

$$d = 0.36 \times 0.434$$
$$= 0.156 \text{ m}$$
$$F_w = (0.044 \times 700 - 1.75) + (0.009 \times 700 + 2.48)$$
$$= 37.8 \text{ N}$$

$$\tau_w = F_w \times b$$
$$= 37.8 \times b$$

$$\cos 0.5 = \frac{b}{d}$$
$$b = 0.156 \cos 0.5$$
$$b = 0.14 \text{ m}$$

$$\tau_w = 37.8 \times 0.14$$
$$\tau_w = 5.2 \text{ N}\bullet\text{m}$$

The torque about the knee joint due to the ankle weight (τ_l) can be determined in a similar manner.

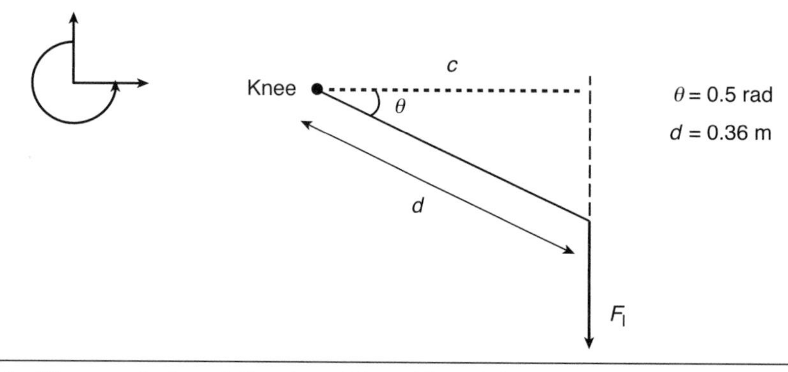

Figure 3.9 The magnitude and direction of the ankle weight vector (F_l) and the associated moment arm (c) to the knee joint.

$$\tau_l = F_l \times c$$
$$= 80 \times c$$

$$\cos 0.5 = \frac{c}{0.36}$$
$$c = 0.36 \cos 0.5$$
$$c = 0.32 \text{ m}$$

$$\tau_l = 80 \times 0.32$$
$$\tau_l = 25.6\ \text{N}\cdot\text{m}$$

We can determine the net torque about the knee joint in this example by summing the magnitudes and taking into account the direction of each torque (i.e., counterclockwise indicated as positive). As discussed previously, we can determine the direction of each torque vector with the right-hand-thumb rule. In the present example, the forces F_w and F_l produce clockwise rotation and thus are identified as negative torques.

To determine the net effect of these forces on the system, we sum (Σ) the moments of force (torques) about the knee joint (τ_K):

$$\sum \tau_k = (F_m \times a) - (F_w \times b) - (F_l \times c)$$
$$= 12.4 - 5.2 - 25.6$$
$$\sum \tau_k = -18.4\ \text{N}\cdot\text{m}$$

The net torque, therefore, has a magnitude of 18.4 N•m and acts in a clockwise direction. Does this mean that the leg is being lowered? No, the clockwise direction of the torque vector says nothing about the direction of the limb displacement. Recall from chapter 1 that a ball tossed into the air first goes up and then down while experiencing a downward acceleration throughout the entire trajectory. Similarly, a negative torque (acceleration) provides no information on the displacement about the knee joint. The leg could be going up or down while experiencing a net torque about the knee joint that acts in a clockwise direction. We need to have position, velocity, and acceleration (torque) information for a complete description of the motion.

Moment Arm

Because torque is equal to the product of force and moment arm (figure 2.4), the rotary effect of a force can be altered by either factor. As we just discussed, muscles generally have a shallow angle of pull and are located close to the joint about which they exert a torque. Because the angle of pull is shallow, anatomical moment arms are typically short. But moment arms change throughout the range of motion, as shown for the knee extensors in figure 3.10. Three techniques can be used to estimate the moment arm: the geometric imaging method, the tendon excursion method, and the direct load measurement method (Tsaopoulos et al., 2006). These methods can yield different estimates for a moment arm (figure 3.10). Nonetheless, the moment arm at many joints is maximal at intermediate joint angles. Because strength is a measure of the capacity of muscle to exert torque, it is influenced not only by the size of the muscle (muscle force) but also by differences in the moment arm from the muscle force vector to the joint. Based on the moment arm values in figure 3.10, if the force exerted by the knee extensors is constant over the indicated range of motion, then strength will be greatest at intermediate knee angles when the moment arm is greatest.

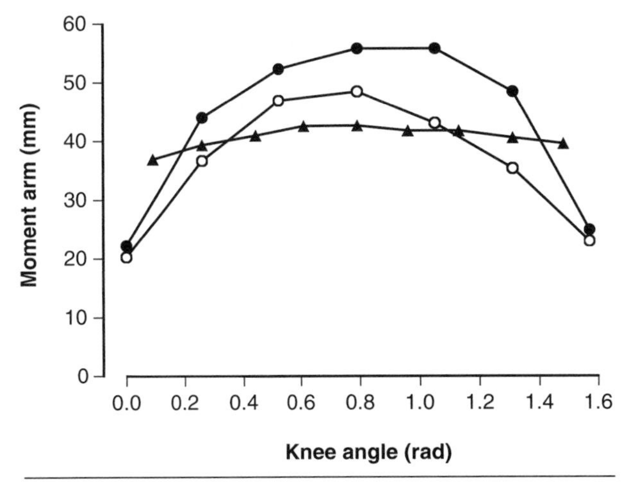

Figure 3.10 Values from two studies for the moment arm of the patellar tendon relative to the flexion-extension axis of rotation for the knee joint. In one study, the measurements were performed on six cadavers using a three-dimensional analysis with the geometric imaging method. The moment arms of the three women (open circles) were less than those of the three men (filled circles). In another study, the measurements were performed on 10 young men (filled triangles) using a geometric imaging method based on estimating the location of the point of application of the joint reaction forces between the tibia and femur.

Data from Krevolin et al. 2004; Kellis and Baltzopoulos 1999.

Data from An and colleagues (1981) indicate that the moment arms for the major elbow flexor muscles double as the elbow goes from a fully extended position to 1.75 rad (100°) of flexion. In contrast, the moment arm for triceps brachii (elbow extensor) decreases by about one-third over the same range of motion (table 3.2). Similar observations have been made for other arm muscles (Bremer et al., 2006; Loren et al., 1996; Murray et al., 2002; Pigeon et al., 1996) and those around the shoulder joint (Kuechle et al., 2000), for muscles that cross the hip joint (Németh & Ohlsén, 1985; Visser et al., 1990), and for muscles that cross the ankle and knee joints (Herzog & Read, 1993; Maganaris, 2004; Rugg et al., 1990; Spoor et al., 1990). Furthermore, the length of a moment arm can change as the line of action of the muscle force varies during a contraction (Ito et al., 2000; Maganaris et al., 1999).

Table 3.2 Moment Arms (cm) Associated With the Elbow Flexor and Extensor Muscles

	Elbow extended		Elbow flexed	
Muscle	**Neutral**	**Supinated**	**Neutral**	**Supinated**
FLEXOR				
Biceps brachii	1.47	1.96	3.43	3.20
Brachialis	0.59	0.87	2.05	1.98
Brachioradialis	2.47	2.57	4.16	5.19
EXTENSOR				
Triceps brachii	2.81	2.56	2.04	1.87

Note. The moment arms are listed with the elbow in full extension and in 1.75 rad of flexion and with the hand in neutral and supinated positions.

Data from An et al. 1981.

EXAMPLE 3.4

Moment Arm Changes Influence the Point of Failure During Push-Ups

Let us examine the significance of the variation in a moment arm over a range of motion. Consider an individual who performs push-ups to exhaustion. The prime mover for push-ups is the elbow extensor muscle, triceps brachii. Suppose this muscle is maximally active as the individual approaches failure and, on the basis of the cross-sectional data presented previously, is exerting a force of 714 N. According to the data in table 3.2, the moment arm for the triceps brachii is about 2.81 cm with elbow extended (figure 3.11*a*) and about 2.04

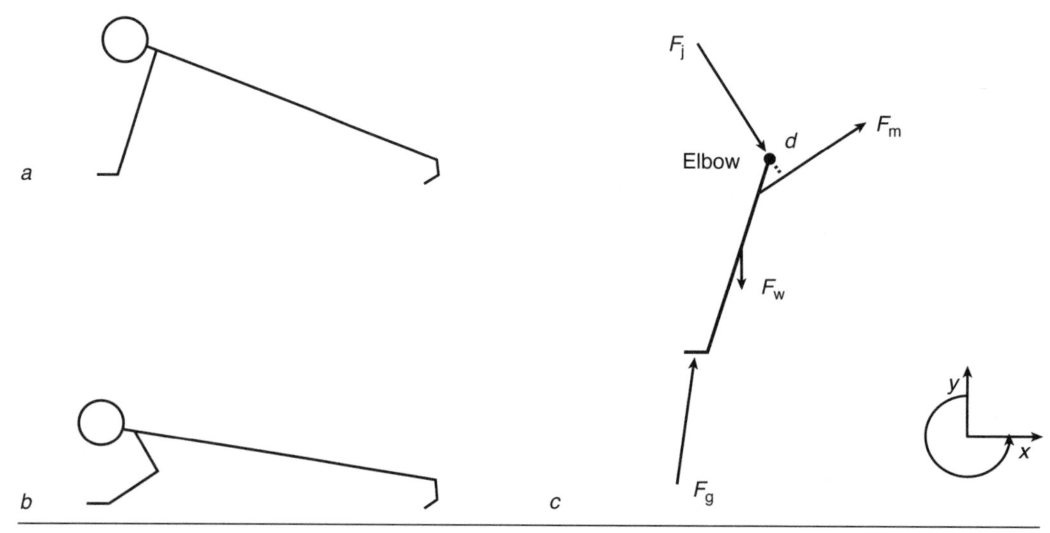

Figure 3.11 A person performing a push-up exercise: *(a)* straight-arm position; *(b)* bent-arm position; *(c)* free body diagram that isolates the resultant muscle force (F_m) exerted by the extensor muscles about the elbow joint. The net torque (τ_m) about the elbow joint is equal to the product of F_m and the moment arm, *d*.

cm with the elbow flexed to 1.75 rad (figure 3.11*b*). This variation in moment arm length is indicated as a change in length of *d* in figure 3.11*c*. Because of this variation, the maximal torque would be approximately 20.1 N•m for the extended position and 14.6 N•m for the flexed position. That is, in the flexed position the torque due to the triceps brachii force, the prime mover for the exercise, is less than in the extended position. Donkers and colleagues (1993) found that the peak elbow extensor torque during a normal push-up was 56% of the maximal isometric torque that could be exerted by the elbow extensor muscles and that this value declined to 29% when the hands were placed farther apart and increased to 71% when the hands were placed together.

Consequently, failure to perform any more push-ups is more likely to occur in the flexed position, where the maximal torque is least. In this case, failure occurs because of an inability to raise body weight, a constant load, up and down. A similar rationale applies to the point of failure during pull-ups. The moment arms for the elbow flexor muscles are minimal with complete elbow extension, and if the muscle force is reasonably constant throughout the range of motion, that is the point at which failure occurs.

Tendon and Aponeurosis

The movements we perform are a consequence of the forces applied to the skeleton by our muscles. The work done by muscles, however, depends not only on the activity of the cross-bridges within the sarcomeres but also on the elasticity of the tissues that connect the force generators to the skeleton. When a muscle is activated by the nervous system, the force it exerts is transmitted to the skeleton by the aponeurosis and tendon and is modified by the material properties of these tissues (Bojsen-Møller et al., 2004; Fukunaga et al., 2001; Hansen et al., 2006; Kubo et al., 2001*a*).

The mechanical properties of these tissues can be characterized through measurement of the stretch that occurs when a force is applied. When the magnitude of the stretch is relatively small (within the range of A to B in figure 3.12*a*), the resisting force provided by the tissue can be represented by the relation for an ideal spring (Hooke's law):

$$F_e = kx \tag{3.2}$$

where F_e = elastic force, k = spring stiffness, and x = amount of stretch. Equation 3.2 is an equation for a straight line, which has an intercept of zero and a slope of k. The slope corresponds

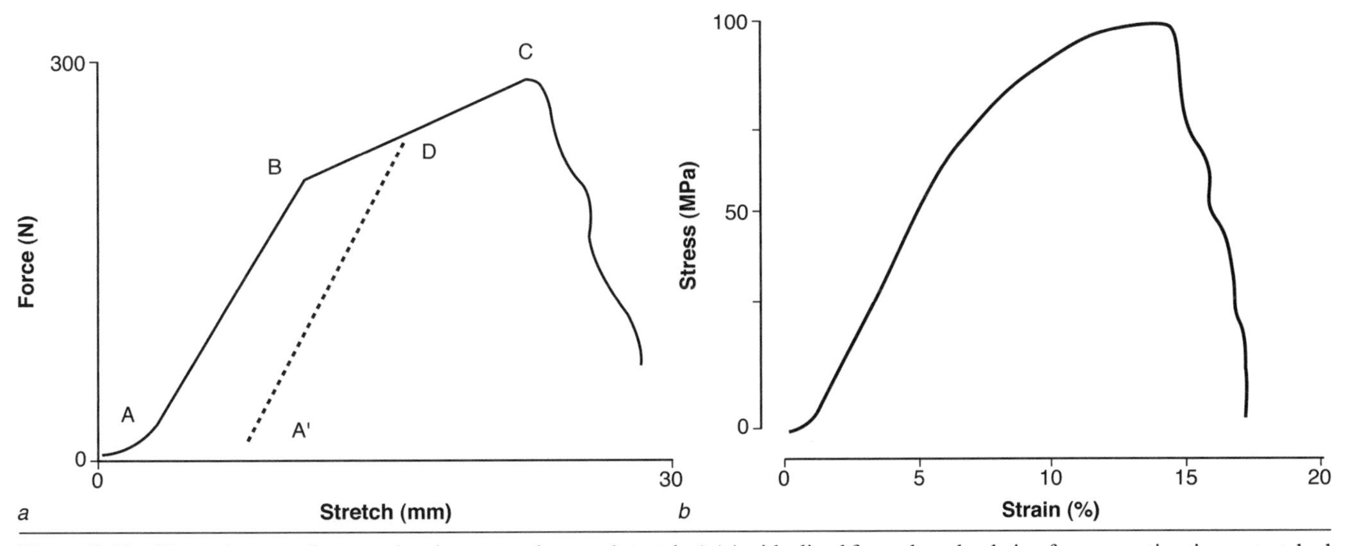

Figure 3.12 The resistance of connective tissue to an imposed stretch. *(a)* An idealized force–length relation for connective tissue stretched beyond its elastic region into its plastic region. *(b)* Stress–strain relation for an extensor digitorum longus tendon. Strain indicates the change in length of the tendon relative to its initial length [(Δ*l* / *l*) × 100], whereas stress represents the force per unit area (N/m² = Pa).

Data for part *b* from Schechtman and Bader 1997.

to the stiffness of the tissue. According to equation 3.2, the elastic force exerted by the tissue is proportional to the amount of stretch. The proportionality constant (k) of a tissue depends on both the composition and the organization of its constituents, which can change with physical activity and with aging (Karamanidis & Arampatzis, 2005; Kubo et al., 2003; Magnusson et al., 2003*b;* Reeves et al., 2006).

When young adults exerted a near-maximum force with the triceps surae muscles, the stiffness of the Achilles tendon averaged 306 N•m/rad (Hof, 1998). Under these conditions, the muscle-tendon length shortened by about 28 mm (8% of tendon + aponeurosis length) during a maximal isometric contraction. For high-force lengthening contractions of the triceps surae muscles, the Achilles tendon might be stretched by as much as 10% of its resting length (Hof, 1998). Such findings underscore two features of muscle function. First, some of the work performed by the cross-bridges is used to stretch the tendon and aponeurosis, which then function as a more rigid connector between the muscle and the skeleton. For this reason, larger muscles have thicker (cross-sectional area) tendons (Loren & Lieber, 1995), and the elastic properties of tendon are correlated with muscle strength (Muraoka et al., 2005). Second, the elasticity of these tissues enables them to store and release energy, which can significantly reduce the metabolic cost of performing some movements (Alexander, 1997; Biewener & Roberts, 2000; Gabaldon et al., 2004; Prilutsky et al., 1996*a*).

There is a limit to how far tendon and aponeurosis can be stretched until they break (point C in figure 3.12*a*). Based on in vitro measurements, rupture occurs in tendon (extensor digitorum longus) when it is stretched by about 15% of its initial length (Schechtman & Bader, 1997) and in ligament (medial collateral) after a stretch of about 20% (Liao & Belkoff, 1999). With less substantial stretches, the tissues either behave like a spring (equation 3.2) as they operate in the **elastic region** (between points A and B in figure 3.12*a*) or the stretch extends beyond the yield point (point B in figure 3.12*a*) and into the **plastic region** (between points B and C in figure 3.12*a*) where the structure of the tissue is altered and the slope of the force–length relation changes. If the tissue is stretched to point D (figure 3.12*a*) and then released, it will assume a new resting length (point A′) that is longer than the initial length (point A) due to plastic changes in its structure. The controlled lengthening of connective tissue into its plastic region will increase its resting length, which expands the range of motion about a joint.

The force–length relation of tendon and aponeurosis varies, both between tissue types and among different structures of the same tissue type. Many of these differences are due to variation in cross-sectional area and length. For example, the stiffness of two tendons with the same length varies according to the difference in cross-sectional area; a tendon with twice the cross-sectional area of another has twice the stiffness. Similarly, the stiffness of two tendons with the same cross-sectional area varies on the basis of differences in length; a tendon that is twice as long as another will have one-half the stiffness. For this reason, the comparison of connective tissue properties across conditions and subjects is based on normalized values, which are expressed as a stress–strain relation (figure 3.12*b*). Stress (Pa) represents the force applied per unit area of the tissue, where area is measured in the plane that is perpendicular to the force vector (cross-sectional area). **Strain** (%) indicates the change in length of the tissue relative to its initial length, which can vary along the length of a tendon and aponeurosis (Magnusson et al., 2001; Stafilidis et al., 2005). Stress and strain characterize the intrinsic force capacity and extensibility of connective tissue. The slope of the elastic region of a stress–strain relation is quantified as the **modulus of elasticity** (E), which is defined as the ratio of stress (σ) to strain (ε) and represents the normalized stiffness of the tissue.

$$E = \frac{\sigma}{\varepsilon} \tag{3.3}$$

Normalization of the force–length relation as a stress–strain curve, however, does not account for all the differences between various structures. The modulus of elasticity for mammalian tendon, for example, varies from about 0.8 to 2 GPa, with an average value of 1.5 GPa (Bennett et al., 1986; Magnusson et al., 2003*a*). Some of this variability can probably be explained by the location on the stress–strain curve where the slope (modulus) is measured. For example,

EXAMPLE 3.5

Tendon Properties

As indicated by the stress–strain normalization procedure, tendons vary in thickness (cross-sectional area) and length. The major determinant of these differences appears to be the magnitude of the physiological loads experienced by the tendon. This is evident, for example, if we compare the stress–strain relation of two tendons that differ in size (table 3.3).

Table 3.3 Mechanical Properties of Two Tendons

	Extensor carpi radialis brevis tendon	Achilles tendon
Maximal muscle force (N)	58	5000
Tendon length (mm)	204	350
Tendon thickness (mm^2)	14.6	65
Elastic modulus (MPa)	726	1500
Stress (MPa)	4.06	76.9
Strain (%)	2.7	5
Stiffness (N/cm)	105	2857

These two tendons have similar values for strain and somewhat similar values for stress at the maximal muscle force, but the Achilles tendon is much stiffer than the tendon of the wrist extensor muscle.

Loren and Lieber (1995) found no difference in the modulus of elasticity among the muscles that cross the wrist when the slope was measured at the maximal force the muscle could exert. These values ranged from 0.438 GPa for extensor carpi radialis longus to 0.726 GPa for extensor carpi radialis brevis. However, when the modulus was measured at forces less than maximum, there were significant differences among tendons, which suggested differences in material properties between tendons at low forces.

Intra-Abdominal Pressure

The pressure inside the abdominal cavity varies in the course of our daily activities. It can range from low values that cause air to flow into the lungs to high values that make the trunk rigid. The magnitude of the intra-abdominal pressure and trunk rigidity is controlled by the activity of the trunk muscles. The principal muscles are those that surround the abdominal cavity, including the abdominal muscles anteriorly and laterally (rectus abdominis, external and internal obliques, and transversus abdominis), the diaphragm above, and the muscles of the pelvic floor below (Hodges & Gandevia, 2000).

Intra-abdominal pressure is increased through closing of the epiglottis and activation of these muscles; fluctuations in the intra-abdominal pressure tend to parallel most closely changes in the EMG activity of the transversus abdominis muscle (Cresswell et al., 1992). Voluntarily pressurizing the abdominal cavity, which we do when lifting heavy loads or anticipating high-impact forces, is referred to as the **Valsalva maneuver.** In most activities we perform, however, the diaphragm and abdominal muscles are activated automatically and the **intra-abdominal pressure** is altered without a need for voluntary (conscious) intervention (Hodges et al., 2004). When the trunk muscles are activated, both the **intrathoracic pressure** and the intra-abdominal pressure increase (Hodges et al., 2005). The pressures in these two cavities tend to change in parallel during many activities, with the intra-abdominal pressure usually being greater (Harman et al., 1988). The force generated by intra-abdominal pressure is estimated as

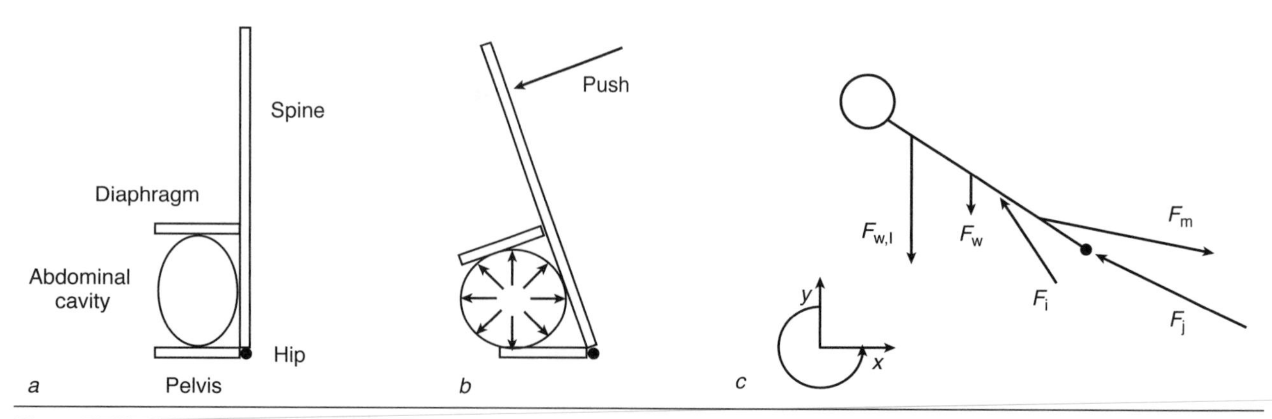

Figure 3.13 The effect of intra-abdominal pressure on the trunk: *(a)* a three-segment system with the intra-abdominal cavity represented as a balloon; *(b)* increase in pressure inside the balloon when the trunk is pushed forward; *(c)* a free body diagram that includes the force due to the intra-abdominal pressure.

the product of the pressure and the cross-sectional area of the cavity (smallest transverse section), with the force acting at the center of pressure (Daggfeldt & Thorstensson, 1997).

Intra-abdominal pressure has been proposed as a mechanism to reduce the load on back muscles during lifting tasks (Cholewicki et al., 1999; Daggfeldt & Thorstensson, 1997; Morris et al., 1961). Pressurization of the abdominal cavity provides a force that causes the trunk to extend about the hip joint. This effect is shown in figure 3.13. Imagine a three-segment system that comprises a base (pelvis), an upright (trunk) component, and an upper support (diaphragm). Between these elements is an inflated balloon (intra-abdominal cavity). When the system is pushed forward about the hip, the balloon will be compressed and the pressure inside the balloon will increase (figure 3.13*b*). When the pushing force is released, the pressure inside the balloon pushes the system back to an upright position (Hagins et al., 2006; Hodges et al., 2001). Similarly, the intra-abdominal pressure provides a force (F_i) that acts through the diaphragm to oppose hip flexion loads (figure 3.13*c*). For example, if the person shown in figure 3.13*c* lifted a load of 91 kg ($F_{w,l}$) without increasing the intra-abdominal pressure, the back and hip muscles (F_m) would have to exert a force of 8223 N, and the joint reaction force (F_j) would be 9216 N just to support the load. However, if this lifter increased his intra-abdominal pressure to 19.7 kPa (force of 810 N), then F_m would be reduced to 6403 N and F_j would decrease to 6599 N (Morris et al., 1961).

Intra-abdominal pressure can be measured with a pressure transducer that is attached to a catheter and inserted into the abdominal cavity through the nasal cavity. Figure 3.14 provides an example of such a recording during a weightlifting exercise (Lander et al., 1986). In this lift, the individual began from an

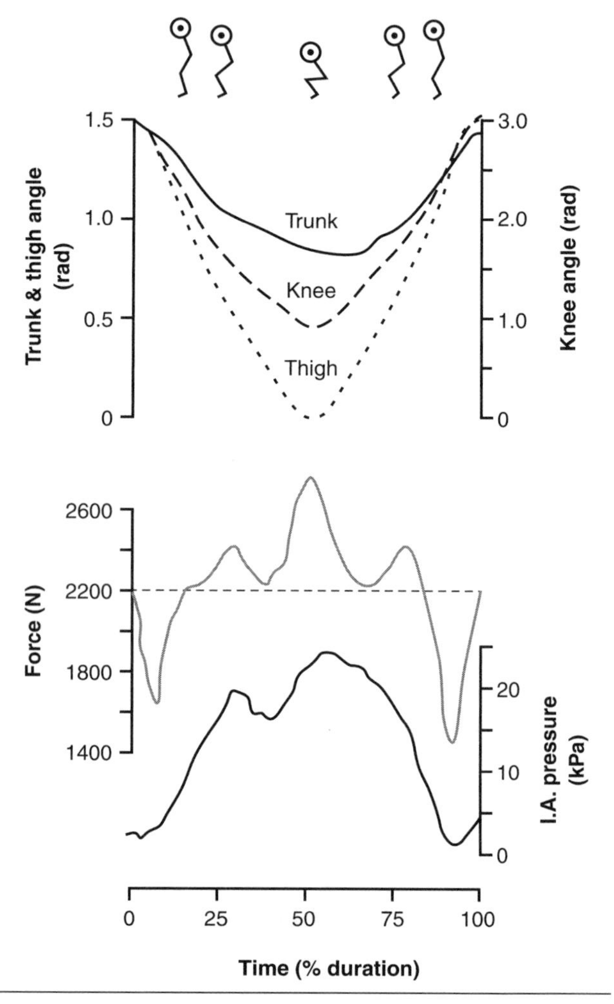

Figure 3.14 Changes in joint angles, vertical component of the ground reaction force, and intra-abdominal pressure during a squat lift by a weightlifter. Trunk and thigh angles are indicated as absolute angles relative to the horizontal such that the angles are 1.57 rad when the individual was vertical. The knee angle is a relative angle between the thigh and shank.

Adapted, by permission, J.E. Lander, B.T. Bates, and P. DeVita, 1986, "Biomechanics of the squat exercise using a modified center of mass bar," *Medicine and Science in Sports and Exercise* 18: 473.

erect position and lowered the load until the thighs were parallel to the ground (the trunk, knee, and thigh angles reached minimal values) and then returned to an erect posture. Because the load (weight of the lifter and the barbell) was 2.2 kN, the vertical component of the ground reaction force ($F_{g,y}$) fluctuated about this value. $F_{g,y}$ values less than 2.2 kN indicate that the system was accelerating downward; conversely, values greater than 2.2 kN depict an upward acceleration of the system. The increase in intra-abdominal pressure tends to coincide with values greater than 2.2 kN when the load (system acceleration) on the trunk muscles is greatest.

As a pressure in a confined volume, the intra-abdominal pressure exerts a force over the surface area of the abdominal cavity. The force that the intra-abdominal pressure exerts on the trunk is usually estimated as the product of intra-abdominal pressure and the surface area of the diaphragm, which Morris and colleagues (1961) estimated to be about 0.0465 m^2 for an adult. If this estimate is combined with the peak intra-abdominal pressure of 25 kPa shown in figure 3.14, then the force acting on the diaphragm due to the intra-abdominal pressure would have been approximately 1163 N during the squat lift. The magnitude of the intra-abdominal pressure can be influenced by the use of a weightlifting belt (Harman et al., 1989; Ivancic et al., 2002; Lander et al., 1992; McGill et al., 1990). Clearly, this is a significant force in human movement.

Despite the correlation between various movements and changes in intra-abdominal pressure, there is some controversy over the functional role of this mechanical effect (Arjmand & Shirazi-Adl, 2006; Marras & Mirka, 1992). It has been proposed, for example, that one effect of intra-abdominal pressure is to reduce the compressive forces that act on the intervertebral disks. However, Nachemson and colleagues (1986) showed that although the Valsalva maneuver does increase the intra-abdominal pressure, it can also increase the pressure on the nucleus of the L3 disk for some moderate tasks. Nonetheless, for the most strenuous task, in which the subjects leaned forward 0.53 rad while holding an 8 kg load in outstretched arms, a Valsalva maneuver increased the intra-abdominal pressure from 4.35 kPa to 8.25 kPa and reduced the **intradiscal pressure** from 1625 kPa to 1488 kPa. Such a mechanism would likely have a significant cumulative effect on the stress experienced when performing manual labor (Essendrop et al., 2002).

STATIC ANALYSIS

Now that we have discussed musculoskeletal forces, we have considered all the forces that we might need to include on a free body diagram. The three categories of forces are those due to body mass (gravity and inertial force), those due to the surroundings (ground reaction force and air resistance), and musculoskeletal forces (joint reaction force, muscle force, and the force due to intra-abdominal pressure). Our next task is to consider the formal procedures that we use to determine unknown forces after we have drawn a free body diagram.

When we apply Newton's law of acceleration ($\Sigma F = ma$) to study the motion of an object, we commonly distinguish among movements in which acceleration is zero and those in which it is not zero. When acceleration is zero, the right-hand side of the equation is zero, which means that the sum of the forces acting on the object is equal to zero. This is referred to as a static condition because acceleration is zero and all the forces are balanced. When this occurs, the object has zero acceleration, which means that it has a constant velocity; that is, it is either stationary or moving at a constant speed.

In a static analysis, the sum of the forces in any given direction is zero ($\Sigma F = 0$) and the sum of the torques is also equal to zero ($\Sigma \tau_o = 0$). At most, there are three independent scalar equations available to solve statics problems in which the movement is confined to a single plane:

$$\sum F_x = 0 \tag{3.4}$$

$$\sum F_y = 0 \tag{3.5}$$

$$\sum \tau_o = 0 \tag{3.6}$$

Equations 3.4 and 3.5 refer to the sum (Σ) of the forces in two linear directions (x and y) that are perpendicular to one another. Equation 3.6 represents the sum of the torques about point o, which may or may not be the center of mass of the system.

When performing a static analysis, we follow a number of steps. These include drawing a free body diagram, writing the equation to be used, expanding the equation to include all the forces shown on the free body diagram, and then solving for the unknown term. The following examples demonstrate these procedures.

EXAMPLE 3.6

Finding the Magnitude and Direction of an Unknown Force

Consider the rigid body in figure 3.15*a,* which is in equilibrium ($a = 0$) and upon which are acting three known forces and one unknown force. What is the magnitude of the unknown force? The first step is to draw a free body diagram. Although figure 3.15*a* indicates all the forces acting on the rigid body and does comprise a free body diagram, the most convenient form is shown in figure 3.15*b.* A free body diagram must include a coordinate system, which indicates the positive linear and angular directions. For most of our problems this will include the *x-y* directions and a rotation. In figure 3.15*b,* the positive *x* direction is to the right, the positive *y* direction is up, and the positive rotation is counterclockwise. This is the most common convention; any horizontal force directed to the right is regarded as positive, and any to the left is negative. The declaration of the horizontal-vertical directions is also an indication that only forces in these directions will be used in the analysis and thus any force (e.g., **R**) not in either direction must be *resolved* into such directions (figure 3.15*b*).

The second step in the analysis is to write the equation of motion, which will be one of the versions of Newton's law of acceleration (equations 3.4 through 3.6). The actual equation needed depends on the question. For example, if we want to determine the magnitude of a force in the *x* direction, then we should used equation 3.4 as it involves this direction. Once

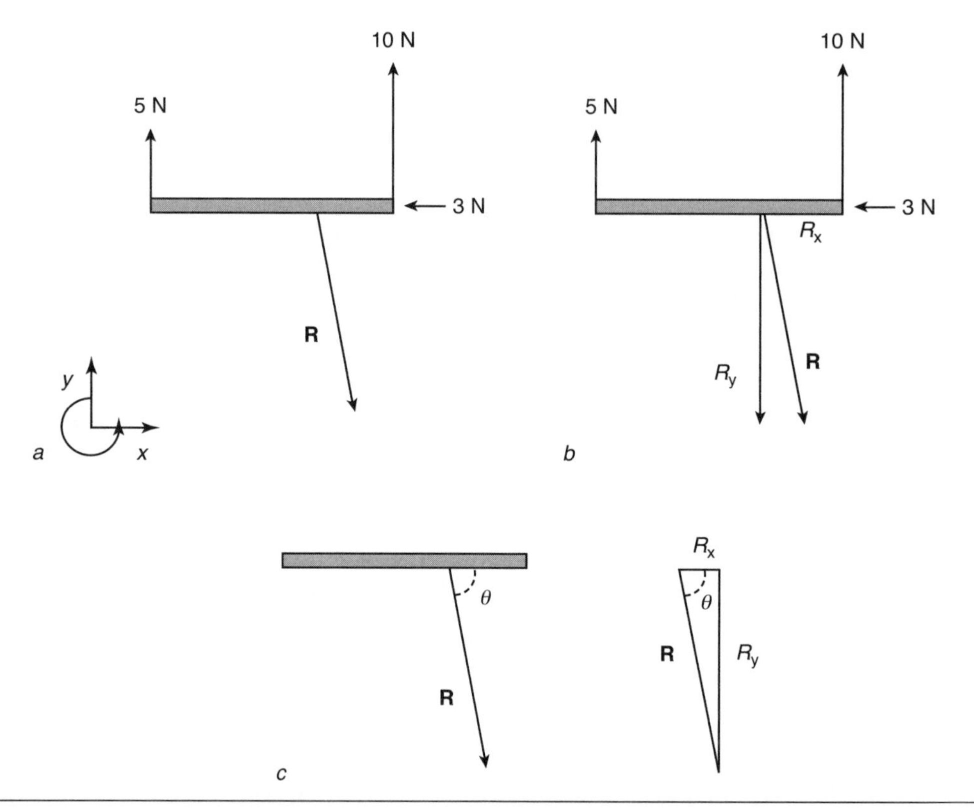

Figure 3.15 Distribution of forces acting on a rigid body: *(a)* free body diagram; *(b)* resolution of **R** into *x* and *y* components; *(c)* calculation of the direction and magnitude of **R**.

the equation has been written, the third step is to expand the equation based on the forces identified in the free body diagram. This step, which represents the second line of the solution, should identify the magnitude and direction of the forces acting in the chosen direction (x, y, or a rotation). So to determine the magnitude of R_x, we should proceed as follows:

$$\sum F_x = 0$$
$$R_x - 3 = 0$$
$$R_x = 3 \text{ N}$$

Note that in this example, the direction of the unknown force (R_x) is indicated as positive on the second line of the solution. This direction came from the free body diagram (figure 3.15*b*). There are two forces acting on the rigid body in the x direction: one in the negative x direction with a magnitude of 3 N and the other in the positive x direction with an unknown magnitude. Because the system is in equilibrium, the calculation shows that the magnitudes of the two forces are equal but that the directions are opposite.

Now we do the same procedure in the y direction. Again, we begin by writing the necessary equation; then we expand the equation from the free body diagram, and finally we solve for the unknown variable.

$$\sum F_y = 0$$
$$5 + 10 - R_y = 0$$
$$R_y = 15 \text{ N}$$

According to figure 3.15*b*, there are three forces acting in the y direction. The direction of all three forces is known. Because the system is in equilibrium and the magnitude of two forces is known, we are able to determine the magnitude of the unknown force (R_y).

Once we have determined R_x and R_y, we can find the magnitude and direction of **R** (figure 3.15*c*). The magnitude of the resultant (R) can be determined by the Pythagorean relation:

$$R = \sqrt{R_x^2 + R_y^2}$$
$$= \sqrt{3^2 + 15^2}$$
$$R = 15.3 \text{ N}$$

This result, however, specifies only the magnitude of the resultant and not its direction. We can actually indicate the direction of **R** relative to several references; for example, we could determine the angle relative to a horizontal (R_x) or a vertical (R_y) reference. Let us calculate the direction of **R** relative to the system (figure 3.15*c*). By the parallelogram rule, R is the diagonal of the rectangle, which has R_x and R_y as its sides. Thus R_x, R_y, and R represent the sides of a triangle and we can determine θ as

$$\cos\theta = \frac{R_x}{R} \qquad \sin\theta = \frac{R_y}{R} \qquad \tan\theta = \frac{R_y}{R_x}$$

Accordingly,

$$\cos\theta = \frac{R_x}{R}$$
$$\theta = \cos^{-1}\frac{R_x}{R}$$
$$\theta = \cos^{-1}\frac{3}{15.3}$$
$$\theta = \cos^{-1} 0.1961$$
$$\theta = 1.37 \text{ rad}$$

The answer to the question is that **R** has a magnitude of 15.3 N and is directed to the right at an angle of 1.37 rad below the horizontal.

EXAMPLE 3.7

Calculating a Net Muscle Torque

A person sitting on an exercise bench has a light load attached to his ankle while performing a knee extension exercise. Suppose we want to estimate the magnitude of the force exerted by his knee extensor muscles (quadriceps femoris) when he holds the load stationary in the middle of the range of motion. We can do this by performing a static analysis because the forces would have to be balanced to hold the load stationary.

To estimate the muscle force about the knee joint during the knee extension exercise, the free body diagram must include the knee joint at one end (figure 3.16). One way to draw the free body diagram for this analysis is to have the foot and shank comprise the system and then to show how the surroundings interact with this system. In this type of analysis, the muscle force (F_m) and the joint reaction force (F_j) indicate separate effects of the surroundings (the rest of the body) on the system. The object of the static analysis is to determine the magnitude and direction of the resultant muscle torque about the knee joint (K), required to hold the leg in the specified position. Figure 3.16 shows two moment arms (b = 0.11 m and c = 0.32 m), the joint reaction force (F_j), the system (shank + foot) weight (F_w = 41 N), and the weight of the load (F_l = 80 N). Suppose the individual had to hold the limb at an angle of 0.5 rad below the horizontal: What resultant muscle torque (τ_m) must the person generate to accomplish this task?

The first step is to construct the free body diagram (figure 3.16). This involves defining the system, drawing the system as a simplified diagram such as a stick figure, and showing how the system interacts with its surroundings. These interactions can include forces due to body mass, forces due to the surroundings, and forces due to the musculoskeletal system. The second step is to select the appropriate equation (equations 3.4 through 3.6). To determine angular effects, as in this example, we should choose equation 3.6 to sum the torque about the somersault axis through the knee joint.

$$\sum \tau_K = 0$$

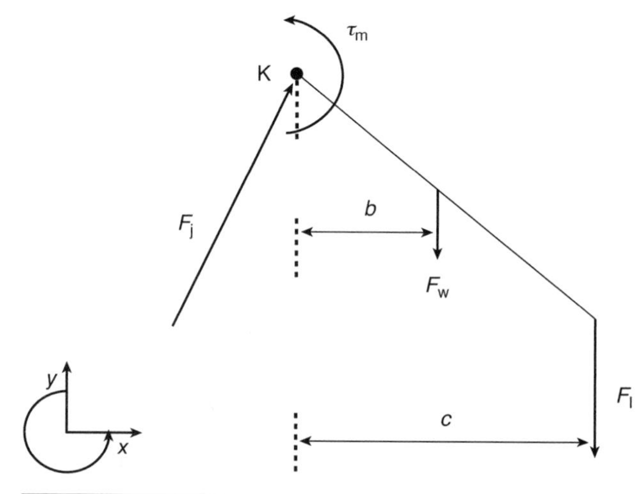

Figure 3.16 Free body diagram of the shank and foot of an individual performing a knee extension exercise.

The third step is to expand the equation, which is the second line of the solution. In this example, there are three forces (F_w, F_l, and F_j) and one torque (τ_m) acting on the system as indicated by the vectors in figure 3.16. By choosing point K (the knee joint) as the point about which to sum the torques, we can ignore the joint reaction force because its line of action passes through this point and its moment arm is therefore zero. Consequently,

$$\sum \tau_K = 0$$

$$\tau_m - (F_w \times b) - (F_l \times c) = 0$$

$$\tau_m = (F_w \times b) + (F_l \times c)$$

$$= (41 \times 0.11) + (80.0 \times 0.32)$$

$$= 4.51 + 25.6$$

$$\tau_m = 30.1 \text{ N}\bullet\text{m}$$

This analysis indicates that when the shank-foot is held in the middle of the range of motion for the exercise, the weight of the system (F_w) and the load attached to the ankle (F_l) exert a torque of 30.1 N•m about the knee joint. To hold the shank-foot stationary, the net muscle torque must equal the magnitude of this load torque.

EXAMPLE 3.8

Solving for Two Unknown Forces

A student sits at the end of an exercise bench and uses a rope-pulley apparatus to strengthen her quadriceps femoris muscle group with an isometric exercise. Let's determine the musculoskeletal forces at the knee joint. Again, the first step is to draw the free body diagram. If we are to calculate the musculoskeletal forces at the knee joint, one end of the free body diagram must be the knee joint. The simplest system comprises the shank and the foot, from the knee (*K*) joint down to the toes (figure 3.17). The forces that must be included on the free body diagram are the weight of the system (F_w), the musculoskeletal forces (F_m and

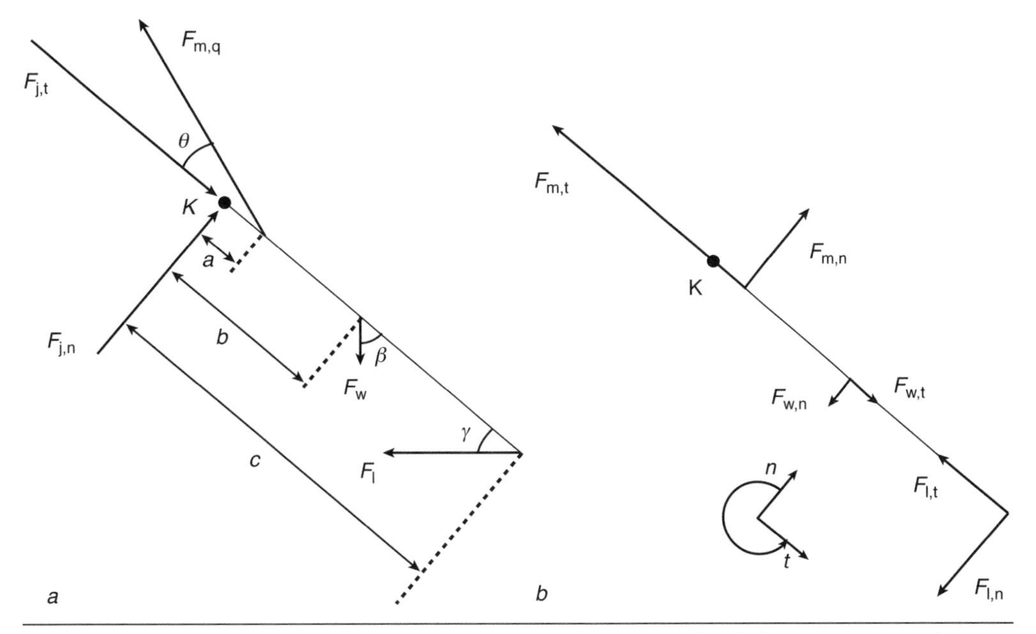

Figure 3.17 *(a)* Free body diagram of the lower leg (shank + foot) of an individual as she performs an isometric exercise to strengthen her quadriceps femoris muscle group. *(b)* The forces can also be resolved into normal (n) and tangential (t) components.

F_j), and the load due to the apparatus (F_l). The student's leg weighs 30 N, and the center of mass of the system is 20 cm from the knee joint (*b*). She applies a force of 100 N to the rope-pulley apparatus, which is attached to an ankle cuff at a distance of 45 cm from her knee joint (*c*). The muscle force (quadriceps femoris) vector acts back across the knee joint with an angle of pull (θ) of 0.25 rad and a point of application that is 7 cm from the knee joint (*a*). The joint reaction force (F_j) is shown as two components, one in the normal ($F_{j,n}$) direction and the other in the tangential ($F_{j,t}$) direction. The free body diagram shows the system at an angle of 0.7 rad below the horizontal, which means that the angle of F_w (β) is 0.87 rad and the angle of F_l (γ) is 0.7 rad.

The next step is to write an equation for the calculation. Because this problem has two unknowns (F_m and F_j), we must choose an equation that does not include both terms. Thus we choose equation 3.6. If we sum the torques about point K, we can momentarily ignore

F_j because the moment arm from its line of action to K is zero. This involves summing the torques about point K due to the weight of the system (F_w) and the load exerted by the rope-pulley apparatus (F_l) and setting these equal to the torque associated with the quadriceps femoris muscle activity ($F_{m,q}$). This equality exists because the person is performing an isometric contraction, which means that the forces are balanced and the system is in equilibrium. So we perform the calculation by writing the equation (first line) and then expanding the equation to include all the torques acting on the system about point K (second line). The magnitude of each torque is calculated as the product of the force and its moment arm. The calculation proceeds as follows:

$$\sum \tau_K = 0$$

$$(F_{m,q} \times a \sin\theta) - (F_{w,l} \times b \sin\beta) - (F_l \times c \sin\gamma) = 0$$

$$(F_{m,q} \times 1.7) - (30 \times 15.3) - (100 \times 29.0) = 0$$

$$(F_{m,q} \times 1.7) = (30 \times 15.3) + (100 \times 29.0)$$

$$(F_{m,q} \times 1.7) = 459 + 2900$$

$$F_{m,q} = \frac{3359}{1.7}$$

$$F_{m,q} = 1976 \text{ N}$$

To determine the magnitude of $F_{j,n}$ and $F_{j,t}$, we need to resolve each force into its normal and tangential components (figure 3.17*b*) and then sum the forces in each direction. The components are the following:

$$F_{m,n} = F_{m,q} \sin 0.25$$
$$= 1975 \sin 0.25$$
$$= 489 \text{ N}$$

$$F_{w,n} = F_{w,l} \sin 0.87$$
$$= 30 \sin 0.87$$
$$= 23 \text{ N}$$

$$F_{l,n} = F_l \sin 0.70$$
$$= 100 \sin 0.70$$
$$= 64 \text{ N}$$

$$F_{m,t} = F_{m,q} \cos 0.25$$
$$= 1975 \cos 0.25$$
$$= 1914 \text{ N}$$

$$F_{w,t} = F_{w,l} \cos 0.87$$
$$= 30 \cos 0.87$$
$$= 19 \text{ N}$$

$$F_{l,t} = F_l \cos 0.70$$
$$= 100 \cos 0.70$$
$$= 76 \text{ N}$$

These values represent the magnitudes of the normal and tangential components as they are shown in figure 3.17*b*. The magnitude of the *tangential* component of the joint reaction force is determined as

$$\sum F_t = 0$$

$$F_{j,t} - F_{m,t} + F_{w,t} - F_{l,t} = 0$$

$$F_{j,t} = F_{m,t} - F_{w,t} + F_{l,t}$$

$$= 1914 - 19 + 76$$

$$F_{j,t} = 1971 \text{ N}$$

The magnitude of the *normal* component can be found in a similar manner:

$$\sum F_n = 0$$

$$F_{j,n} + F_{m,n} - F_{w,n} - F_{l,n} = 0$$

$$F_{j,n} = - F_{m,n} + F_{w,n} + F_{l,n}$$

$$= - 489 + 23 + 64$$

$$F_{j,n} = -402 \text{ N}$$

The magnitude of $F_{j,n}$ is determined as negative 402 N. This does not mean that it is a negative force but rather that its direction is incorrect on the free body diagram (figure 3.17*a*). When the free body diagram was drawn, we were told that the system was in equilibrium. This means that the forces in each direction must add to zero. For the tangential direction, we assumed that $F_{j,t}$ was acting in the positive direction; this turned out to be correct. Also, we assumed that $F_{j,n}$ would have to act in the positive direction in order for the system to be in equilibrium. But this was incorrect, as indicated by the negative value that we calculated for $F_{j,n}$. The calculation indicates that $F_{j,n}$ acts in the negative direction in order for the system to be in equilibrium. Now that we know the magnitude of $F_{j,n}$ and $F_{j,t}$, we can determine the magnitude of the resultant joint reaction force (F_j) with the Pythagorean relation:

$$F_j = \sqrt{F_{j,n}^2 + F_{j,t}^2}$$

$$= \sqrt{402^2 + 1972^2}$$

$$F_j = 2013 \text{ N}$$

Thus, when a 100 N load is applied at the ankle, the knee joint experiences a joint reaction force that is almost 20 times larger than the load.

Finally, we can calculate the direction of the resultant joint reaction force with respect to the axis of the shank as

$$\tan\theta = \frac{F_{j,n}}{F_{j,t}}$$

$$\theta = \tan^{-1}\frac{F_{j,n}}{F_{j,t}}$$

$$\theta = \tan^{-1}\frac{402}{1972}$$

$$\theta = 0.2 \text{ rad}$$

The resultant joint reaction force, therefore, has a magnitude of 2013 N and is directed 0.2 rad below the longitudinal axis of the shank.

EXAMPLE 3.9

Locating the Balance Point

A rigid body of uniform density, which weighs 20 N and has a length of 22 cm, has a load suspended from each end (figure 3.18*a*). At one end the load is 30 N, and at the other end the load is 60 N. To balance the rigid body on an extended finger, what is the magnitude of the force that we must exert with the finger, and where should the finger be placed? To answer this question, the first step is to draw a free body diagram. The rigid body has a uniform density, which means that the center of mass is located in the middle of the object. So we know the magnitude and direction of three forces acting on the rigid body. For the rigid body to balance on a finger, the finger must exert a force equal in magnitude but opposite in direction so that the forces are all balanced. With this information, we can draw the free body diagram (figure 3.18*b*). Once we have the free body diagram, we can choose an equation, expand it, and solve for the finger force:

$$\sum F_y = 0$$
$$-30 - 20 - 60 + \text{finger force} = 0$$
$$\text{finger force} = 110 \text{ N}$$

Because the loads at the two ends of the system are not equal, the balance point (finger location) will not be in the middle of the rigid body. The balance point is essentially a fulcrum: The loads on one side pull the rigid body in one direction, and those on the other side pull it in the other direction. The balance point, therefore, is the location about which the torques in each direction are equal and the rigid body does not rotate. This is the definition of the center of mass (CM); it is the point (a location) about which the mass of the system is evenly distributed. Because the balance point (CM) represents the location about which the system is in equilibrium, we can find it by performing a static analysis. To find the balance point, we can sum the torques about the origin of the scale in figure 3.18 and find where the 110 N force must be located for the rigid body to be balanced.

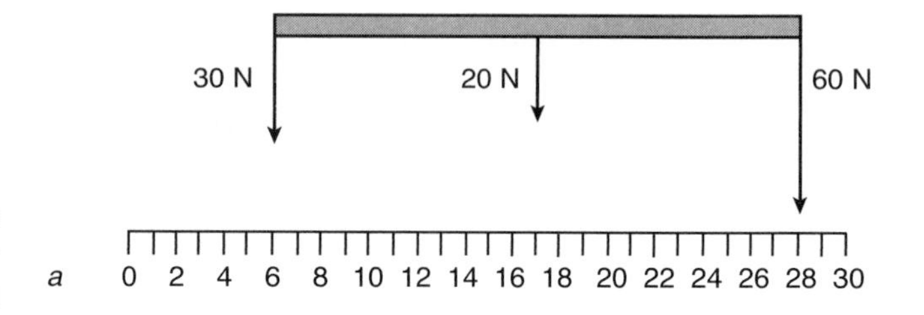

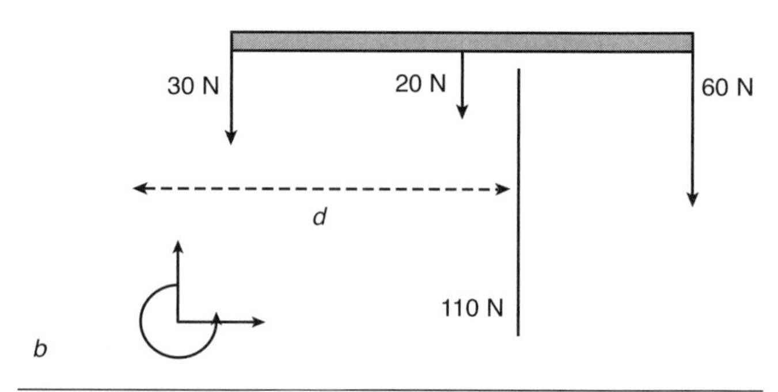

Figure 3.18 The location and magnitude of an opposing force that will balance a loaded rigid body: *(a)* the system; *(b)* the free body diagram.

$$\sum \tau_o = 0$$
$$-(30 \times 6) - (20 \times 17) + (110 \times d) - (60 \times 28) = 0$$
$$d = \frac{(30 \times 6) + (20 \times 17) + (60 \times 28)}{110}$$
$$d = 20 \text{ cm}$$

So, to balance the rigid body shown in figure 3.18*a,* it is necessary to exert an upward force of 110 N that is applied at 20 cm along the scale.

EXAMPLE 3.10

Center of Mass for the Human Body

The same approach as described in example 3.9 can be used to find the location of the CM of a system that comprises several different parts, such as the human body. The approach involves using equation 3.6 to find the point about which the mass of the system is evenly distributed—that is, the balance point. The location of the balance point depends on the relative position of the different parts in the system. For the human body, this means the location of the individual body segments.

To demonstrate this technique, let us determine the location of the total-body CM of a gymnast about to perform a backward handspring (figure 3.19). The necessary steps include the following:

1. Identify the appropriate body segments. The human body is divided into multiple parts for a segmental analysis (figures 2.7 and 2.9). The number of segments to be used in an analysis is determined by the number of joints that experience an angular displacement during the movement. If there is no rotation about the elbow joint, for example, then the arm (upper arm + forearm) can be represented as a single segment. For this example, we will use 14 segments: head, trunk, upper arms, forearms, hands, thighs, shanks, and feet. These segments are indicated in figure 3.19*a* by the marks over the joint centers that represent the proximal and distal anatomical landmarks for each segment.

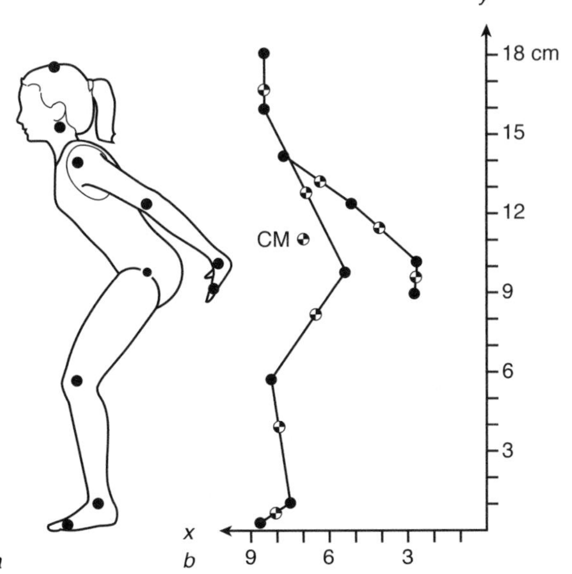

Figure 3.19 Location of whole-body center of mass (CM) as a function of the position of the body segments: *(a)* limits of the respective body segments; *(b)* location of the segmental CMs as a percentage of segment length.

2. Connect the joint-center markers to construct a stick figure (figure 3.17*b*).
3. From table 2.1, determine the location of the CM for each segment as a percentage of segment length. These lengths are measured from the proximal end of each segment (table 3.4).
4. Estimate segmental weights as a function of body weight (F_w) (table 3.5). The gymnast weighs 450 N.
5. Measure the location of segmental CMs relative to an *x*-*y* axis (table 3.6). The location of this axis is arbitrary and does not influence the location (with respect to the body) of the total-body CM; you can convince yourself of this by doing the calculation twice with the *x*-*y* axis in a different location each time.
6. With the segmental weight ($F_{w,s}$) and location (*x*, *y*) data, sum the segmental torques about the *y*-axis ($\Sigma\tau_y = F_{w,s} \times x$) and the *x*-axis ($\Sigma\tau_x = F_{w,s} \times y$). Double the limb segmental weights to account for both limbs (table 3.7).
7. Find the location of the balance point. This is the point that produces the same torque for total-body weight about the *x*- and *y*-axes as that due to the sum of the segmental effects. This procedure is the same as the one used in example 3.9. This similarity is shown in figure 3.20, which illustrates an end-on view of the gymnast represented as a rectangular object. The arrows indicate the magnitudes of the segmental weight vectors (1 = head, 2 = trunk, 3 = upper arm, 4 = forearm, 5 = hand, 6 = thigh, 7 = shank, 8 = foot) and the total-body weight vector (F_w) with the locations representing the *x*-coordinate for the CMs.

Table 3.4 Center of Mass (CM) Locations for Figure 3.19

Segment	CM location (%)	Proximal end
Head	66.3	Top of the head
Trunk	52.2	Top of the neck
Upper arm	50.7	Shoulder
Forearm	41.7	Elbow
Hand	51.5	Wrist
Thigh	39.8	Hip
Shank	41.3	Knee
Foot	40.0	Ankle

Table 3.5 Segmental Weights for Figure 3.19

Segment	Equation	Weight (N)
Head	$0.032 \times F_w + 18.70 =$	33.10
Trunk	$0.532 \times F_w - 6.93 =$	232.47
Upper arm	$0.022 \times F_w + 4.76 =$	14.66
Forearm	$0.013 \times F_w + 2.41 =$	8.26
Hand	$0.005 \times F_w + 0.75 =$	3.00
Thigh	$0.127 \times F_w - 14.82 =$	42.33
Shank	$0.044 \times F_w - 1.75 =$	18.05
Foot	$0.009 \times F_w + 2.48 =$	6.53

Table 3.6 *x-y* coordinates of the Centers of Mass for Figure 3.19

Segment	*x*-coordinate (cm)	*y*-coordinate (cm)
Head	8.6	16.8
Trunk	6.9	12.8
Upper arm	6.5	13.3
Forearm	4.2	11.5
Hand	2.8	9.6
Thigh	6.6	8.2
Shank	8.0	3.9
Foot	8.0	0.8

Table 3.7 Resultant Weight Torques for Figure 3.19

Segment	x (cm)	y (cm)	$F_{w,s}$ (N)	$\Sigma\tau_y$ (N•cm)	$\Sigma\tau_x$ (N•cm)
Head	8.6	16.8	33.10	284.7	556.1
Trunk	6.9	12.8	232.47	1604.0	2975.6
Upper arm	6.5	13.3	29.32	190.6	390.0
Forearm	4.2	11.5	16.52	69.4	190.0
Hand	2.8	9.6	6.00	16.8	57.6
Thigh	6.6	8.2	84.66	558.8	694.2
Shank	8.0	3.9	36.10	288.8	140.8
Foot	8.0	0.8	13.06	104.5	10.5
				3117.6	5014.7

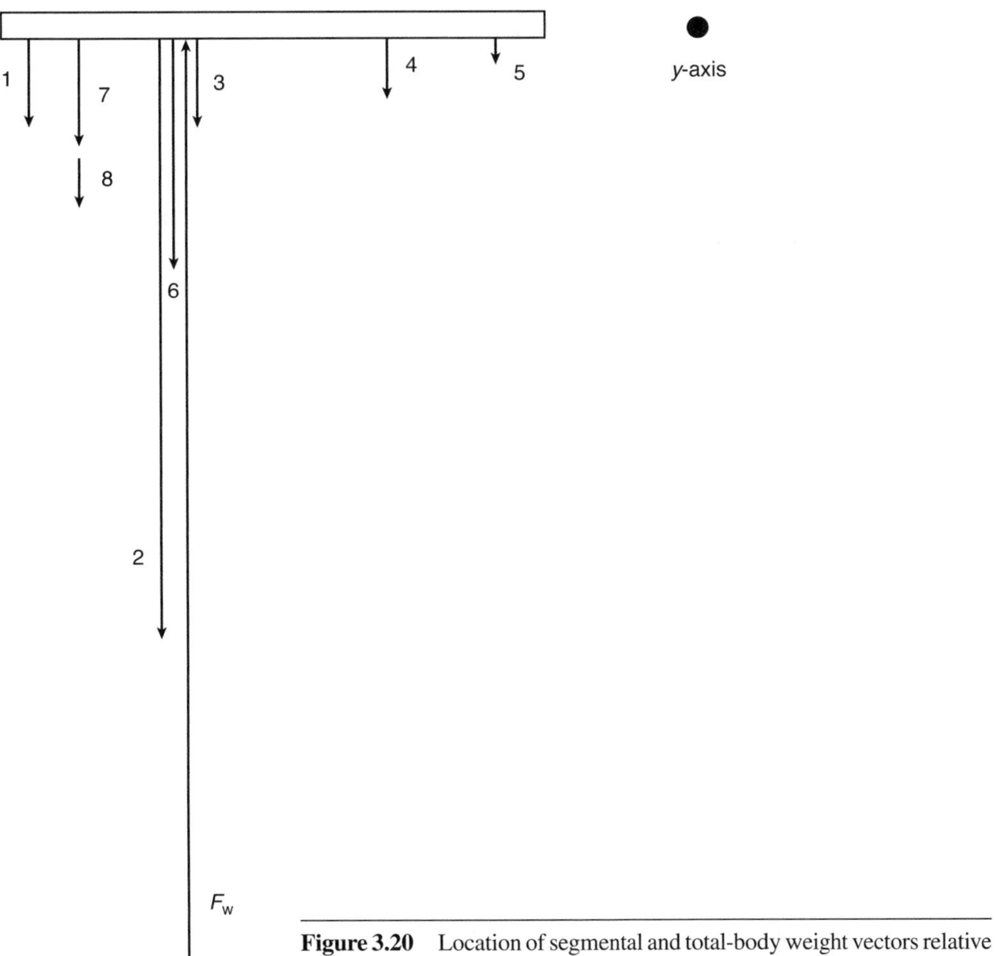

Figure 3.20 Location of segmental and total-body weight vectors relative to the *y*-axis for the gymnast in figure 3.19.

This final calculation uses equation 3.6 to determine the *x*- and *y*-coordinates for F_w. The net torque due to segmental weights ($F_{w,s}$) about the *x*- and *y*-axes (3018.0 N•cm and 5014.7 N•cm, respectively) is the same as the net torque due to total body weight. Thus

$$F_w \times \text{moment arm} = \sum_{i=1}^{8} (F_{w,s} \times \text{moment arm})_i$$

$$\text{moment arm} = \frac{\sum (F_{w,s} \times \text{moment arm})}{F_w}$$

To find the *x*- and *y*-coordinates for the total-body CM,

$$x = \frac{3117.6 \text{ N•cm}}{450 \text{ N}}$$
$$= 6.93 \text{ cm}$$
$$y = \frac{5014.7 \text{ N•cm}}{450 \text{ N}}$$
$$= 11.1 \text{ cm}$$

With these coordinates, the location of the total-body CM is indicated in figure 3.19*b* relative to the *x*- and *y*-axes established at the beginning of the example. These coordinates (6.93, 11.1) represent the point about which the mass of the system is evenly distributed (balanced).

DYNAMIC ANALYSIS

A static analysis is the most elementary approach to the kinetic analysis of human movement. In contrast, when the system is subjected to unbalanced forces it will be accelerated, and this requires a more complicated type of analysis, known as a **dynamic analysis.** The general form of the three independent planar equations (equations 3.4 through 3.6) used in the static approach also applies to the dynamic condition, with the exception that the right-hand side of the equations is now equal to the nonzero product of mass and acceleration. Hence, the scalar components and the angular term may be written as

$$\sum F_x = ma_x \quad (3.7)$$

$$\sum F_y = ma_y \quad (3.8)$$

$$\sum \tau_o = I_o\alpha + mad \quad (3.9)$$

where $a = \sqrt{a_x^2 + a_y^2}$ and $\Sigma F = \sqrt{(\Sigma F_x)^2 + (\Sigma F_y)^2}$ in the *x-y* plane. As in the static case, these equations are independent and represent two linear directions (*x* and *y*) and one angular direction. In expanded form, equation 3.7 states that the sum (Σ) of the forces (F) in the *x* direction is equal to the product of the mass (m) of the system and the acceleration of the system's CM in the *x* direction (a_x). Equation 3.8 similarly addresses forces and accelerations in the *y* direction. Equation 3.9 states that the sum of the torques about point *o* is equal to two effects, one related to the angular kinematics of the system and the other related to the linear kinematics of the system. The term for angular kinematics includes the product of the relevant moment of inertia (I_o) and the angular acceleration (α) of the system about the axis of rotation (o). The linear kinematics term includes the product of the mass of the system, the linear acceleration of the system CM (a), and the distance (d) between point *o* and the system CM.

Point *o* can be any point about which the moments are summed. Equations 3.7 through 3.9 are modified if point *o* is the CM or is a fixed point. If point *o* is the CM, then $d = 0$ and equation 3.9 is reduced to $\Sigma\tau_g = I_g\alpha$. If point *o* is a fixed point, then $a = 0$ and the resultant of the applied forces is equal to $I_o\alpha$. If the system comprises a single rigid body, such as one body segment, and the moments are summed about the CM, then the resultant effect of the forces acting on the system with regard to a normal-tangential (n-t) reference frame can be calculated as follows (Meriam & Kraige, 1987):

$$\sum F_n = mr\omega^2 \quad (3.10)$$

$$\sum F_t = mr\alpha \quad (3.11)$$

$$\sum \tau_g = I_g\alpha \quad (3.12)$$

Kinetic Diagram

Because the right-hand side of equations 3.7 through 3.9 is nonzero, the free body diagram of the system can be equated to a **kinetic diagram.** That is, by Newton's law of acceleration (***F*** **=** ***ma***), force (free body diagram) equals mass times acceleration (kinetic diagram). In this context, the free body diagram represents the left-hand side of the equation and the kinetic diagram the right-hand side. In this sense, the free body diagram defines the system and how it interacts (forces shown with arrows) with its surroundings. The kinetic diagram shows the effects of these interactions on the system—that is, how the interactions alter the motion of the system.

EXAMPLE 3.11

Finding the Resultant Muscle Force

A volleyball player serves the ball using the overhand technique. This involves tossing the ball up in the air above the head and then striking the ball with a hand so that it travels over the net and into the opponents' court. Suppose we want to calculate the resultant muscle force about the CM at one point in time during the serve when the ball is in contact with the hand. As with a static analysis, the first step is to draw an appropriate diagram. For a dynamic analysis, this means both a free body diagram and a kinetic diagram. For this analysis we can define the system as the forearm and hand of the volleyball player so that the muscles crossing her elbow joint (F_m) appear as an external force (figure 3.21*b*). The forces that must be included on the free body diagram include the weight of the system (F_w), the contact force between the ball and the hand (F_b), and the musculoskeletal forces (muscle force [F_m] and joint reaction force [F_j]). The kinetic diagram shows the effects of these forces on the system; this includes vectors for the three terms in equations 3.7 through 3.9 (ma_x, ma_y, and $I_o\alpha$).

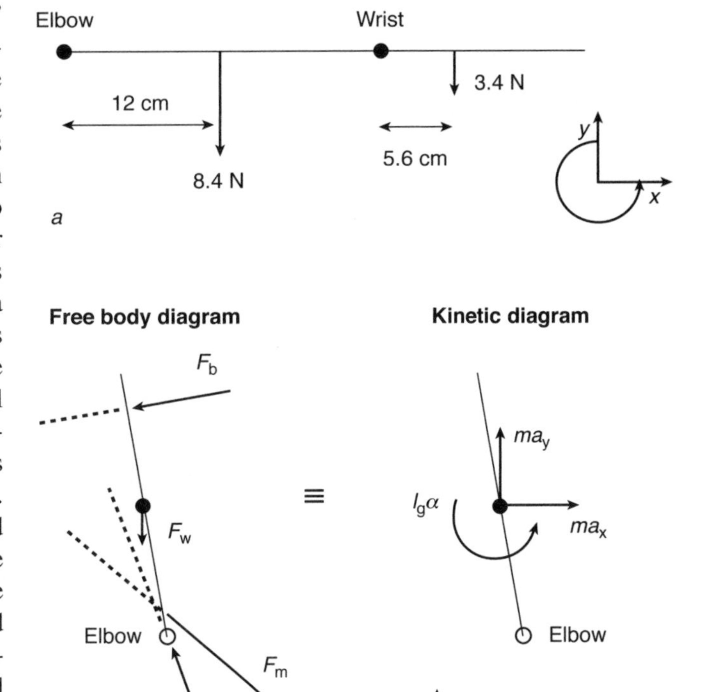

Figure 3.21 A dynamic analysis to determine the resultant muscle torque about the elbow joint during a volleyball serve: *(a)* the weight and CM location of the forearm and hand segments; *(b)* the free body and kinetic diagrams for the forearm plus hand.

The next step in the procedure is to identify the equation we will use. Because we are attempting to determine the resultant muscle torque about the CM, we should choose equation 3.9 so that we can calculate angular effects. Before we can use this equation, however, we need to know more details about both the geometry of the system and the movement. This includes the location of the system CM and the moment of inertia about the system CM. Because we know that the volleyball player weighs 608 N, we can estimate the weight of her forearm as 8.4 N and the weight of her hand as 3.4 N from the anthropometric data in table 2.5. Also, because we know the length of her forearm (26 cm) and the length of her hand (7.5 cm), we can use table 2.5 to estimate that the forearm CM is located 12 cm from the elbow joint and that the hand CM is 5.6 cm from the wrist (figure 3.21*a*). To calculate the location of the system (forearm + hand) CM, we sum the torques about the elbow joint and perform a static analysis, as we did in figure 3.19. The system weight is 11.8 N, and the distance to the system CM from the elbow joint (d) can be determined as

$$\sum \tau_{\text{elbow}} = 0$$

$$-(8.4 \times 12) + (11.8 \times d) - [3.4 \times (26 + 5.6)] = 0$$

$$d = \frac{(8.4 \times 12) + (3.4 \times 31.6)}{11.8}$$

$$d = 17.6 \text{ cm}$$

Next, we need to determine the moment of inertia of the system about a somersault axis through its CM. From table 2.6, we can estimate the moment of inertia about the somersault axis for the forearm as 0.0039 kg•m² and for the hand as 0.0004 kg•m². To determine the moment of inertia for the system, we use the parallel axis theorem (equation 2.8) to transfer the known value for each segment (forearm and hand) to the system CM. Because the system CM is located at 17.6 cm from the elbow, the transfer distance from the forearm CM to the system CM is 5.6 cm (17.6 – 12), and the transfer distance for the hand is 14 cm (31.6 – 17.6). Given that we estimated the mass of the forearm as 0.86 kg (8.4/9.81) and the mass of the hand as 0.35 kg (3.4/9.81), the moment of inertia of the system about a somersault axis passing through its CM (I_g) can be calculated as

$$I_g = [0.0039 + (0.86 \times 0.056^2)] + [0.0004 + (0.35 \times 0.14^2)]$$
$$= (0.0039 + 0.0027) + (0.0004 + 0.0069)$$
$$= 0.0066 + 0.0073$$
$$I_g = 0.0139 \text{ kg} \cdot \text{m}^2$$

We also need to know some kinematic details of the performance, which can be obtained with a motion analysis system. Suppose the video record indicates that the system (forearm + hand) was rotated 0.26 rad to the left of vertical and a numerical analysis of the position–time data (e.g., table 1.2) yielded an angular acceleration for the system (α) of 489 rad/s^2. From the video record we can also determine that the moment arm for F_m (a) to the axis of rotation (g) was 8.2 cm, for F_b (b) it was 14.0 cm, and for F_j (c) it was 3.5 cm. Finally, other measurements indicated that, at the moment of interest, the contact force between the ball and the hand (F_b) was 290 N and that the joint reaction force (F_j) was 1821 N.

Given all this information, we can use figure 3.21*b* to write the equation of motion for a dynamic analysis (equation 3.9), expand the equation (second line), and solve for F_m:

$$\sum \tau_g = I_g \alpha$$
$$(F_m \times a) + (F_b \times b) - (F_j \times c) = I_g \alpha$$
$$(F_m \times a)\ I = I_g \alpha - (F_b \times b) + (F_j \times c)$$
$$F_m = \frac{I_g \alpha - (F_b \times b) + (F_j \times c)}{a}$$
$$F_m = \frac{(0.0139 \times 489) - (290 \times 0.14) + (1821 \times 0.035)}{0.082}$$
$$F_m = \frac{6.8 - 40.6 + 63.7}{0.082}$$
$$F_m = 365 \text{ N}$$

Although this example may seem relatively simple, it is often impossible to determine the magnitude of the resultant muscle force this way because we are required to know the direction of the resultant muscle force and the magnitude and direction of the resultant joint force. The more common approach is to sum the torques about the joint, as described in the next example.

EXAMPLE 3.12

Finding the Resultant Muscle Torque

In this example, a weightlifter raises a barbell to his chest. Suppose that we were interested in determining the torque developed by the back and the hip extensor muscles when the barbell is about knee height. The first step is to draw the free body and kinetic diagrams. Because we want to determine a musculoskeletal force at the lumbosacral joint, one end of the system must include this joint. An appropriate system, as identified previously, would include the upper body from the lumbosacral (LS) joint to the head (figure 3.22). We can identify five forces that act on this system and that we should include on the free body diagram: the resultant

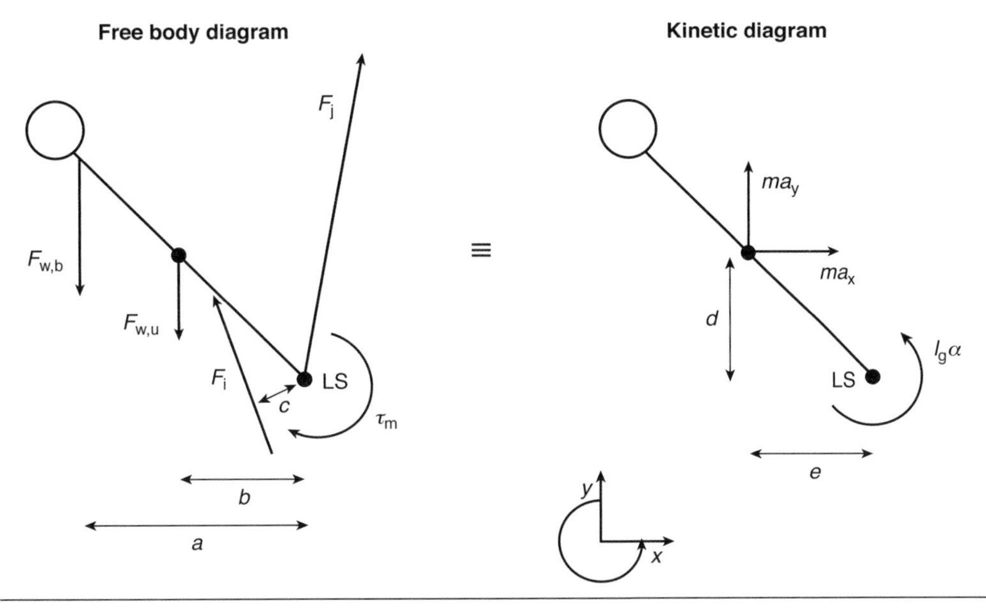

Figure 3.22 A dynamic analysis (free body diagram = kinetic diagram) of a weightlifter performing the clean lift.

muscle torque (τ_m) about the lumbosacral joint, the joint reaction force (F_j), the load due to the barbell ($F_{w,b}$), the weight of the system ($F_{w,u}$), and a force due to the intra-abdominal pressure (F_i). The kinetic diagram should include the three terms on the right-hand side of equations 3.7 through 3.9. The typical approach is to draw these in the positive direction on the kinetic diagram and to derive the equation based on these directions.

The next steps are to write the equation of motion, expand it, and solve for τ_m. As in the previous example, however, we need information about the performance before we can complete the calculation, specifically the following:

1. Estimates of the magnitudes of the three forces that produce moments about LS. The magnitude of $F_{w,b}$, which can be determined with a weight scale, is measured as 1003 N. The magnitude of $F_{w,u}$, which can be estimated from anthropometric tables (e.g., tables 2.1 and 2.5), is set at 525 N. The magnitude of F_i, which can be estimated from values published in the literature or can be measured with an intra-abdominal catheter and an estimate of diaphragm area, is assigned a value of 1250 N.
2. The mass of the system and its distribution. The mass of the system (upper body of the lifter) is estimated as 54 kg. The location of the system CM can be determined by the procedures outlined in example 3.10 and is found to be 47 cm from LS. The moment of inertia of the system about its CM (I_g), which can be determined from known segmental values (e.g., table 2.6) and the parallel axis theorem, is estimated as 7.43 kg•m^2.
3. Kinematics of the movement (figure 1.17), as recorded from a video analysis. These measurements include the moment arms for the forces and the accelerations of the system. The moment arms are as follows: $F_{w,b}$ (a = 38 cm), $F_{w,u}$ (b = 24 cm), F_i (c = 9 cm), ma_x (d = 40 cm), and ma_y (e = 24 cm). The angular acceleration (α) of the system was 8.7 rad/s^2, the horizontal acceleration (a_x) of the system CM was 0.2 m/s^2, and the vertical acceleration (a_y) was –0.1 m/s^2.

With this information, we can calculate the magnitude of the resultant muscle torque (τ_m) about the lumbosacral (LS) joint by using equation 3.9,

$$\sum \tau_{LS} = I_g\alpha + mad$$

and replacing d with the distances d and e in figure 3.22:

$$(F_{w,b} \times a) + (F_{w,u} \times b) - (F_i \times c) - \tau_m = I_g\alpha - ma_x d - ma_y e$$

$$\tau_m = (F_{w,b} \times a) + (F_{w,u} \times b) - (F_i \times c) - I_g\alpha + ma_x d + ma_y e$$

$$\tau_m = (1003 \times 0.38) + (525 \times 0.24) - (1250 \times 0.09) - (7.43 \times 8.7) + (54 \times 0.2 \times 0.40) + (54 \times -0.1 \times 0.24)$$

$$\tau_m = 381 + 126 - 113 - 65 + 4 - 1$$

$$\tau_m = 332\ \text{N}\cdot\text{m}$$

The weightlifting event discussed in this example takes an experienced athlete about 0.4 s to complete. If we record the event at 100 frames per second, we will have 40 frames of video that contain relevant information on the event. To completely describe the time course of τ_m, it is necessary to perform the calculation shown for each frame of data. The result is a set of instantaneous torques that can be plotted as a torque–time curve for the movement. An example for the knee joint is shown in figure 3.23; the data represent the mean muscle torque about the knee joint for a group of 15 experienced weightlifters as they lifted a barbell (1141 N, which was 1.5 times body weight) to chest height. The graph indicates that the resultant muscle torque about the knee joint reached maximal values of about 100 and 50 N•m in the extensor and flexor directions, respectively, and that the direction fluctuated between extensor and flexor during the movement. This torque–time graph corresponds to the angle-angle diagram shown in figure 1.17.

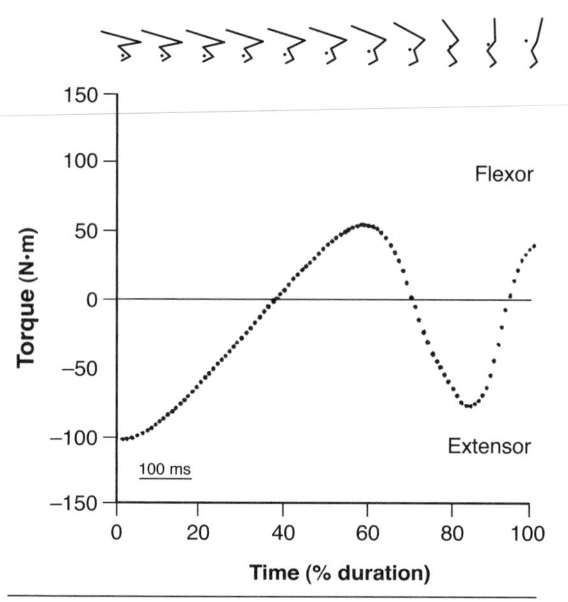

Figure 3.23 Resultant muscle torque about the knee joint during the clean lift in weightlifting.

Data from Enoka, 1983.

EXAMPLE 3.13

When Is a Movement Fast?

An issue that often arises in the study of human movement is whether a dynamic analysis is necessary or whether the movement is slow enough that it can be assumed to be **quasi-static**—in other words, that we can analyze it using static techniques. Rogers and Pai (1990) examined this issue in a simple movement that involved a human subject moving from double- to single-leg support, as occurs when an individual is about to start walking. In the initiation of a step, body weight is shifted from two-legged support to single-leg support while the other leg flexes at the knee. This is accomplished by an increase in the activity of the lateral hip muscles of the flexing leg, an increase in the vertical component of the ground reaction force under the flexing leg, and a shift of the center of pressure toward the single-support leg (Rogers & Pai, 1990). Because the ground reaction force, as we discussed in chapter 2, represents the acceleration of the individual's CM, Rogers and Pai sought to determine how well the change in the vertical component of the ground reaction force under the flexing leg could be determined using a static analysis.

To answer this question, we first draw a free body diagram for the analysis. We do not need to draw a kinetic diagram because we are going to perform a static analysis to see how

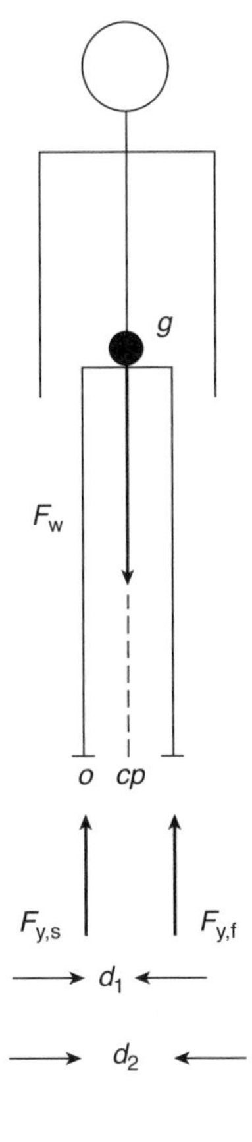

Figure 3.24 Quasistatic model used to predict the vertical component of the ground reaction force during step initiation. cp = center of pressure; d_1 = distance between cp and $F_{y,s}$; d_2 = distance between o and $F_{y,f}$; $F_{y,f}$ = vertical component of the ground reaction force beneath the flexing leg; $F_{y,s}$ = vertical component of the ground reaction force beneath the single-support leg; F_w = body weight; g = center of mass; and o = point of application of $F_{y,s}$.

Adapted, by permission, from M.W. Rogers and C.-Y. Pai, 1990, "Dynamic transitions in stance support accompanying leg flexion movements in man," *Experimental Brain Research* 81: 399. Copyright 1990 by Springer-Verlag.

well we can predict the measured ground reaction force. The free body diagram involves the front view of a person with a ground reaction force acting on each foot and the weight of the system (figure 3.24). Next we must write an equation and then expand it using the information on the free body diagram. To find out whether a static analysis is sufficient to estimate the ground reaction force acting on the flexing leg, we can sum the torques about point o and estimate the magnitude of $F_{y,f}$.

$$\sum \tau_o = 0$$

$$(F_{y,f} \times d_2) - (F_w \times d_1) = 0$$

$$F_{y,f} = \frac{F_w \times d_1}{d_2}$$

Because F_w is constant, $F_{y,f}$ can be predicted as the ratio of changes in d_1 and d_2. To obtain the data necessary to test this prediction, Rogers and Pai (1990) had subjects stand on two force platforms, one foot on each, in order to measure the vertical component of the ground reaction force and its point of application (center of pressure) on each foot. In addition, they used a motion analysis system and a segmental analysis to determine the location of the total-body CM; d_1 represents the distance between the vertical projection of F_w and the point of application of $F_{y,s}$. The results are shown in figure 3.25. Focus on the graphs that show the change in $F_{y,f}$ with time: The solid line represents the actually measured $F_{y,f}$; the dashed line indicates the estimated $F_{y,f}$ based on the quasistatic analysis; and the shaded area represents the difference between the two. In figure 3.25, the graphs on the left are for a rapid step initiation (leg flexion), whereas those on the right are for a slow step initiation. The results indicate that a quasistatic analysis is appropriate for this task when it is performed slowly but not when it is performed at normal or fast speeds. Similarly, slow lifts can be analyzed with a quasistatic analysis (Toussaint et al., 1992), but a quasistatic analysis is not appropriate for determining when a person will take a step after an unexpected disturbance (Pai et al., 1998).

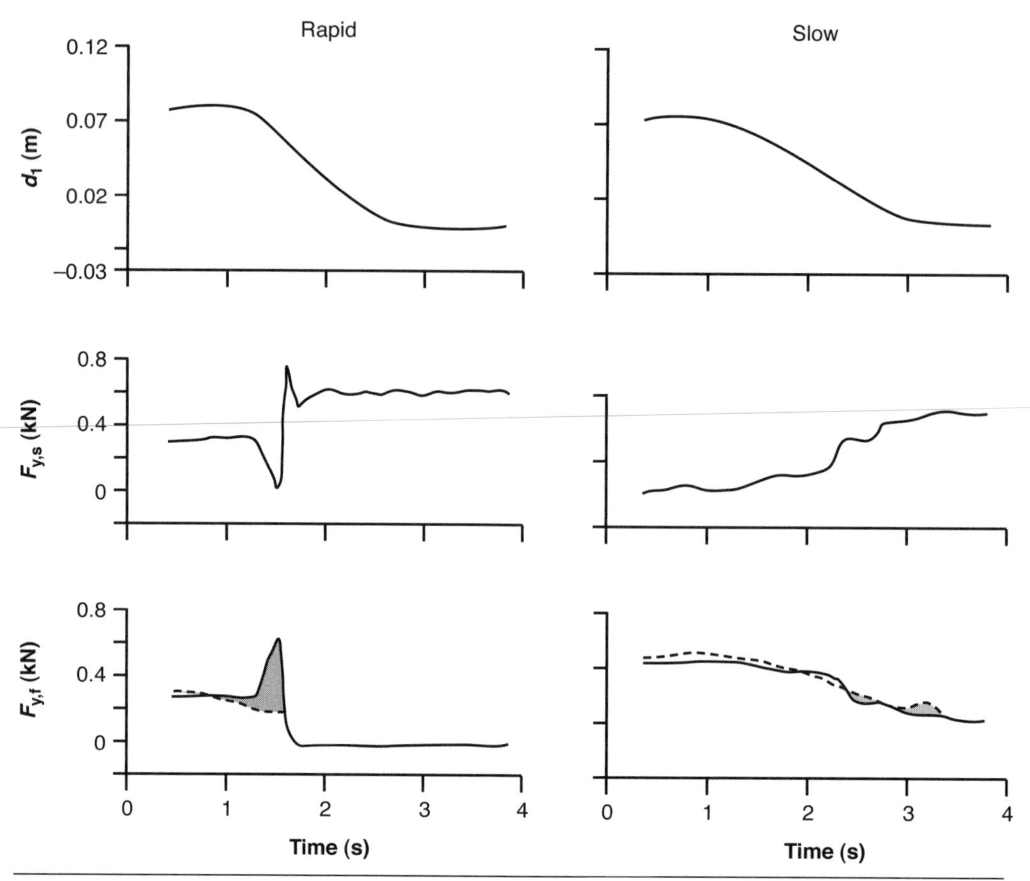

Figure 3.25 Changes in d_1, $F_{y,s}$, and $F_{y,f}$ with time during the step-initiation task when it is performed rapidly (left) and slowly (right).

Reprinted, by permission, from M.W. Rogers and C.-Y. Pai, 1990. "Dynamic transitions in stance support accompanying leg flexion movements in man," *Experimental Brain Research* 81: 400. Copyright 1990 by Springer-Verlag.

Forward and Inverse Dynamics

A dynamic analysis can, in general, proceed in either of two directions. Given information on the forces and torques, we can determine the associated kinematics; or given the kinematics, we can estimate the underlying forces and torques. These two approaches are referred to as **forward dynamics** and **inverse dynamics,** respectively (Otten, 2003). The inverse dynamics approach involves obtaining the derivative of position–time data to yield velocity– and acceleration–time data (Forner Cordero et al., 2006; Silva & Ambrosio, 2004)—for example, measuring position and calculating joint forces. The principal disadvantage of this method is that errors embedded in the position–time data are greatly magnified by the time the data have been processed to yield acceleration (Cahouet et al., 2002; Hatze, 2002; Valero-Cuevas et al., 2003). In contrast, the forward dynamics approach involves the integration (in the calculus sense) of forces and torques (or accelerations) to produce the related kinematic information—for example, measuring acceleration and calculating velocity and position (Thelen et al., 2005). The main difficulty associated with this technique is the need to accurately specify initial conditions. An alternative technique involves the use of features from both the inverse and forward approaches. In this method, measurements of position, linear acceleration, and angular velocity are combined to provide reliable estimates of link kinematics and joint loads (Ladin & Wu, 1991; Seth & Pandy, 2007; Wu & Ladin, 1993).

Calculation of the net torque exerted about the hip by the weightlifter in example 3.12 used the inverse dynamics approach; in that example we proceeded from the kinematics of the movement, along with some force information, to determine an unknown torque. It has been suggested that the nervous system uses inverse dynamics when it plans movements (Hollerbach & Flash, 1982). In this scheme, the nervous system determines the desired kinematics for a movement and then uses inverse dynamics to calculate the muscle torques needed to produce these kinematics. The complexity of these calculations, however, makes it unlikely that the nervous system organizes movements in this manner (Hasan, 1991).

Bobbert and colleagues (1991) used inverse dynamics with reasonable accuracy to estimate the magnitude of the vertical component of the ground reaction force during the support phase of running. They used four video cameras to obtain the kinematic information needed to determine the vertical acceleration of the CM of each body segment and combined this with body segment mass data (Clauser et al., 1969) to calculate the vertical inertia force (mass × acceleration) for each segment. The inertia forces for each segment were summed to calculate $F_{g,y}$, which was then compared with the measured $F_{g,y}$. With this technique, the magnitude of the initial peak in the $F_{g,y}$–time graph was estimated with less than 10% error and its time of occurrence within 5 ms. To achieve such accuracy in the estimation of forces, however, it is necessary to acquire extremely accurate kinematic data (Bobbert et al., 1991; Hatze, 2002; Seth & Pandy, 2007).

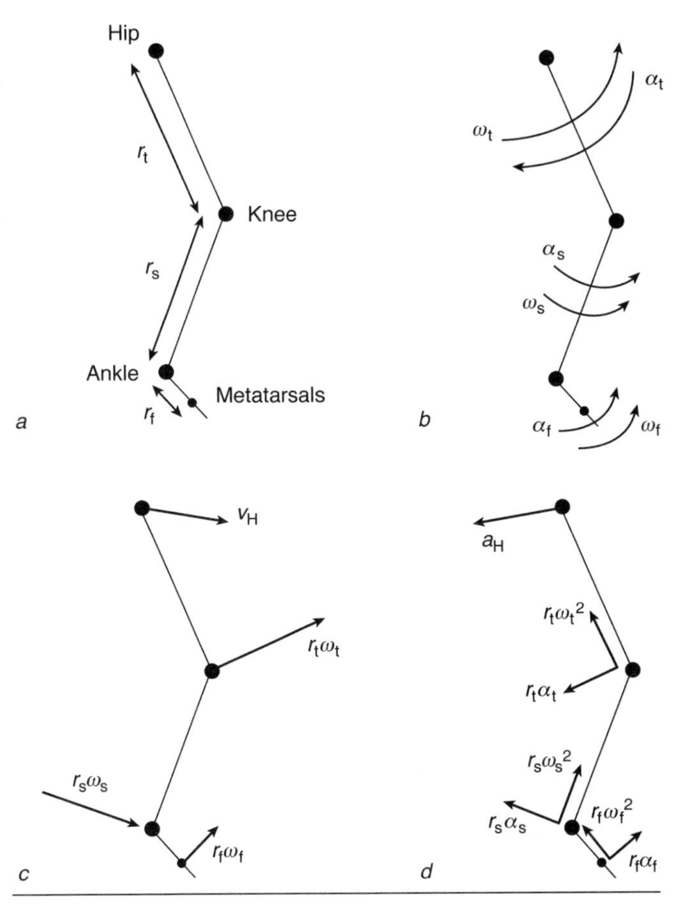

Figure 3.26 Kinematics of the leg during a kick. *(a)* Geometry of the leg. *(b)* The foot (f), shank (s), and thigh (t) each experience an angular velocity (ω) and acceleration (α) as the extremity moves to contact the ball at the metatarsals. The foot rotates about the ankle (A), the shank about the knee (K), and the thigh about the hip (H) joint. *(c)* The kinematic factors that influence the linear velocity of the metatarsals. *(d)* The kinematic variables that influence the linear acceleration of the metatarsals.

Interaction Torques

Because human movement typically involves the motion of more than one body segment, the cumulative motion of all the involved body segments determines the kinematics experienced by an anatomical landmark. For example, consider an individual about to punt a football for distance. This task clearly involves the coordinated motion of the thigh, shank, and foot segments: The motion of the contact point on the foot (*m*) depends on the relative motion of the foot, shank, and thigh segments (figure 3.26*a*). The displacement of point *m* (s_M) is given by

$$s_M = s_{M/A} + s_{A/K} + s_{K/H} + s_H \tag{3.13}$$

where the expression $s_{M/A}$ refers to the displacement of the metatarsals relative to the ankle joint and the subscripts A, K, and H represent the ankle, knee, and hip, respectively. According to this equation, the displacement of the metatarsals depends on four displacement terms: the displacement relative to the ankle joint, the displacement of the ankle joint relative to the knee joint, the displacement of the knee joint relative to the hip joint, and the absolute displacement of the hip joint.

Similarly, if each segment also has an angular velocity (ω) and acceleration (α), then the magnitude of the linear velocity (v) and the acceleration (a) of the metatarsals depend on the relative kinematics of each segment in combination with the absolute kinematics of the hip joint. The angular velocity and acceleration of each segment are shown in figure 3.26*b*. An equation parallel to the one for s_M can be developed for v_M:

$$v_M = v_{M/A} + v_{A/K} + v_{K/H} + v_H$$

Because linear and angular velocity are related by the equation $v = r\,\omega$,

$$v_M = r_f\,\omega_f + r_s\,\omega_s + r_t\,\omega_t + v_H \quad (3.14)$$

where r_f corresponds to the distance from the ankle to the metatarsals, r_s indicates the length of the shank, and r_t represents the length of the thigh (figure 3.26*a*). The direction of each term is tangent to the path traced by the relevant anatomical landmark (figure 3.26*c*): $r_f\omega_f$ is tangent to the path of the metatarsals; $r_s\omega_s$ is tangent to the path of the ankle (the endpoint of the shank); and $r_t\omega_t$ is tangent to the path of the knee (the endpoint of the thigh).

The acceleration of the metatarsals (a_M) also depends on the acceleration of the involved segments (figure 3.26*d*) and, based on the relation between linear and angular acceleration, can be written as

$$a_M = a_{M/A} + a_{A/K} + a_{K/H} + a_H$$

$$a_M = \sqrt{(r_f\omega_f^{\,2})^2 + (r_f\alpha_f)^2} + \sqrt{(r_s\omega_s^{\,2})^2 + (r_s\alpha_s)^2} + \sqrt{(r_t\omega_t^{\,2})^2 + (r_t\alpha_t)^2} + a_H \quad (3.15)$$

where $r\omega^2$ represents the change in direction of **v** for a segment and $r\alpha$ indicates the change in magnitude of **v**. An important feature of this relation is that the acceleration of each segment in this system is influenced by the acceleration of all the other segments. For example, rearrangement of equation 3.15 indicates that the acceleration of the shank during the football punt is dependent on the angular acceleration of the foot and thigh and the linear acceleration of the hip and a_M:

$$\sqrt{(r_s\omega_s^{\,2})^2 + (r_s\alpha_s)^2} = a_M - \sqrt{(r_f\omega_f^{\,2})^2 + (r_f\alpha_f)^2} - \sqrt{(r_t\omega_t^{\,2})^2 + (r_t\alpha_t)^2} - a_H$$

This equation can be used to determine the linear acceleration of any point along the shank, where r represents the distance from the knee to the point of interest (e.g., CM and ankle joint).

This kinematic coupling between segments occurs because of the dynamic interactions between segments. On the basis of what we know about the relation between force and acceleration (***F*** = ***ma***), these interactions between the accelerations of the body segments indicate that there must be interactive forces between body segments during human movement. The study of **interaction torques** examines these motion-dependent interactions between segments (Dounskaia, 2005; Hore et al., 2005; J.P. Hunter et al., 2004).

As an example of these effects, consider the interactions between the thigh and shank during a kick in which the motion of the two segments is confined to a single plane. Figure 3.27*a* shows the two-segment system and its orientation, which can be defined by four coordinates: the *x*- and *y*-coordinates of the hip (h) joint and the angles of the thigh (θ_t) and shank (θ_s). Each segment has a weight vector ($\boldsymbol{F}_{w,s}$ and $\boldsymbol{F}_{w,t}$), and there are resultant joint forces ($\boldsymbol{F}_{j,k}$ and $\boldsymbol{F}_{j,h}$) and resultant muscle torques ($\boldsymbol{\tau}_{m,k}$ and $\boldsymbol{\tau}_{m,h}$) acting about the knee and hip joint, respectively. To determine the interactive effects, we need to express the resultant joint forces in terms of kinematic variables (position, velocity, and acceleration) and derive an expression for the resultant muscle torques. We begin by writing the dynamic equation of motion for the shank.

$$\sum \mathbf{F} = m\boldsymbol{a}$$

$$\boldsymbol{F}_{j,k} + \boldsymbol{F}_{w,s} = m_s \cdot \boldsymbol{a}_s \quad (3.16)$$

where m_s is the mass of the shank and $\boldsymbol{a}_s$ is the linear acceleration of the shank CM. Because we are dealing with a linked, two-segment system, $\boldsymbol{a}_s$ can be expressed in the form of equation 3.15:

$$\boldsymbol{a}_s = \boldsymbol{a}_h + (\boldsymbol{\alpha}_t \times \boldsymbol{r}_{k/h}) + (\boldsymbol{\omega}_t \times \boldsymbol{\omega}_t \times \boldsymbol{r}_{k/h}) + (\boldsymbol{\alpha}_s \times \boldsymbol{r}_{s/k}) + (\boldsymbol{\omega}_s \times \boldsymbol{\omega}_s \times \boldsymbol{r}_{s/k}) \quad (3.17)$$

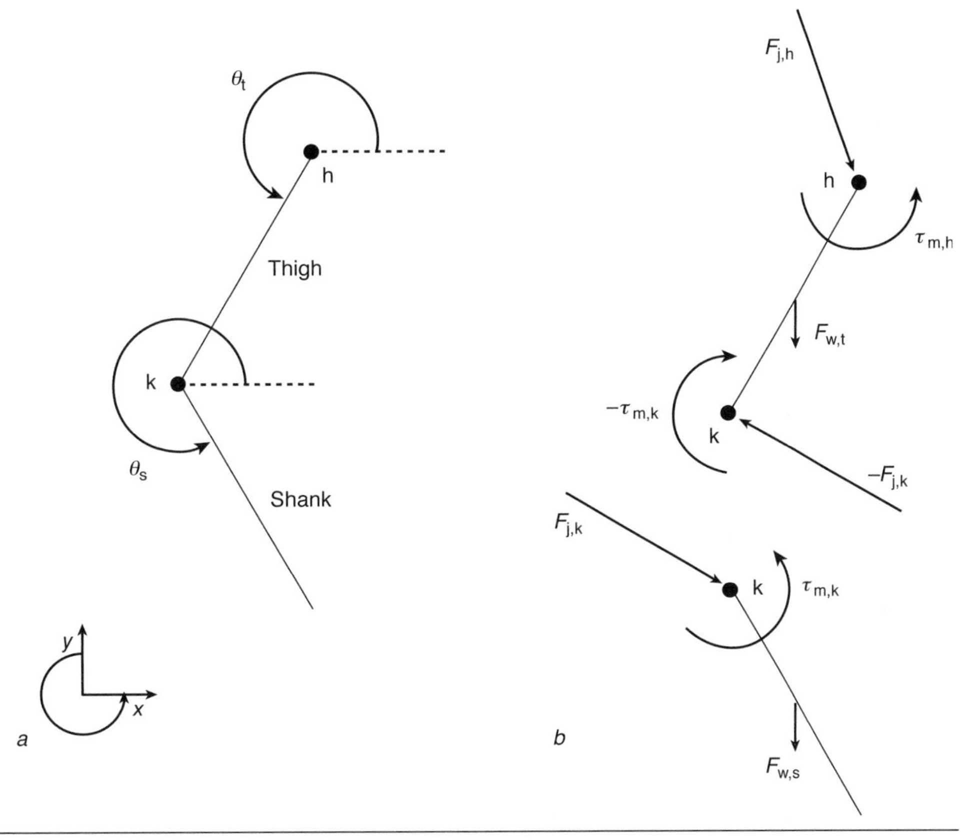

Figure 3.27 The *(a)* position and *(b)* free body diagram of a two-segment system (thigh + shank).

And this expression (equation 3.17) can be inserted into equation 3.16 and rearranged to solve for $\boldsymbol{F}_{\mathrm{j,k}}$.

$$\begin{aligned}\boldsymbol{F}_{\mathrm{j,k}} &= m_{\mathrm{s}}\boldsymbol{a}_{\mathrm{h}} + m_{\mathrm{s}}(\boldsymbol{\alpha}_{\mathrm{t}} \times \boldsymbol{r}_{\mathrm{k/h}}) + m_{\mathrm{s}}(\boldsymbol{\omega}_{\mathrm{t}} \times \boldsymbol{\omega}_{\mathrm{t}} \times \boldsymbol{r}_{\mathrm{k/h}}) + m_{\mathrm{s}}(\boldsymbol{\alpha}_{\mathrm{s}} \times \boldsymbol{r}_{\mathrm{s/k}}) \\ &+ m_{\mathrm{s}}(\boldsymbol{\omega}_{\mathrm{s}} \times \boldsymbol{\omega}_{\mathrm{s}} \times \boldsymbol{r}_{\mathrm{s/k}}) - \boldsymbol{F}_{\mathrm{w,s}}\end{aligned} \tag{3.18}$$

where h = hip, s = shank, t = thigh, $\boldsymbol{r}_{\mathrm{k/h}}$ = the distance from the knee to the hip, and $\boldsymbol{r}_{\mathrm{s/k}}$ = the distance from the shank CM to the knee.

To derive the expression for the resultant muscle torque about the knee ($\boldsymbol{\tau}_{\mathrm{m,k}}$), we write the moment-of-force equation for the shank about its CM.

$$\sum \boldsymbol{\tau}_{\mathrm{g,s}} = I_{\mathrm{g,s}}\,\boldsymbol{\alpha}_{\mathrm{s}}$$

$$\boldsymbol{\tau}_{\mathrm{m,k}} + (\boldsymbol{r}_{\mathrm{k/s}} \times \boldsymbol{F}_{\mathrm{j,k}}) = I_{\mathrm{g,s}}\,\boldsymbol{\alpha}_{\mathrm{s}} \tag{3.19}$$

In the final step, equation 3.19 is rearranged to solve for $\boldsymbol{\tau}_{\mathrm{m,k}}$; equation 3.18 is substituted for the joint reaction force ($\boldsymbol{F}_{\mathrm{j,k}}$) in equation 3.19, and the expression is changed from vector to scalar variables (Dounskaia, 2005; Putnam, 1991).

$$\tau_{\mathrm{m,k}} = I_{\mathrm{s}}\alpha_{\mathrm{k}} + (I_{\mathrm{s}} + r_{\mathrm{s}}l_{\mathrm{t}}m_{\mathrm{s}}\cos\phi)\alpha_{\mathrm{t}} + (r_{\mathrm{s}}l_{\mathrm{t}}m_{\mathrm{s}}\sin\phi\,\omega_{\mathrm{k}}^{2})$$

where ϕ = knee angle ($\theta_{\mathrm{t}} - \theta_{\mathrm{s}}$) and ω_{k} and α_{k} represent its angular velocity and acceleration; l_{t} = length of the thigh; r_{s} = distance from the knee joint to the shank CM; and I_{s} = the moment of inertia of the shank about its proximal end.

Similar procedures can be used to identify the interactive effects for the thigh. Again, we begin by writing the moment-of-force equation for the thigh relative to its CM.

$$\sum \boldsymbol{\tau}_{\mathrm{g,t}} = I_{\mathrm{g,t}}\,\boldsymbol{\alpha}_{\mathrm{t}}$$

$$\boldsymbol{\tau}_{\mathrm{m,h}} - \boldsymbol{\tau}_{\mathrm{m,k}} + (\boldsymbol{r}_{\mathrm{h/t}} \times \boldsymbol{F}_{\mathrm{j,h}}) - (\boldsymbol{r}_{\mathrm{k/s}} \times \boldsymbol{F}_{\mathrm{j,k}}) = I_{\mathrm{g,t}}\,\boldsymbol{\alpha}_{\mathrm{t}} \tag{3.20}$$

As for the shank, we derive expressions for $\boldsymbol{F}_{j,h}$ and $\boldsymbol{F}_{j,k}$ in terms of kinematic variables (e.g., equation 3.17) and substitute these into equation 3.20. The expressions for $\boldsymbol{F}_{j,h}$ and $\boldsymbol{F}_{j,k}$ will contain effects due to the linear acceleration of the segment endpoint (hip), the change in direction ($r\omega$) and magnitude ($r\alpha$) of the linear velocity of the thigh and shank, and segment weight.

$$\tau_{m,h} = I_t\alpha_t + [I_s + m_s(l_t^2 + 2r_s l_t \cos\phi)]\alpha_t + (I_s + r_s l_t m_s \cos\phi)\alpha_k$$
$$-r_s l_t m_s \sin\phi\ \omega_k - 2\, r_s\, l_t\, m_s \sin\phi\ \omega_k\ \omega_t$$

where I_t = the moment of inertia of the thigh about its proximal end.

Figure 3.28*a* shows the time course of the resultant muscle torque about the knee joint and the interactive torque exerted by the thigh on the shank during the kick. This graph shows that the effect of thigh motion on the shank is as large as that due to the muscles that cross the knee joint. The interactive torque exerted by the thigh on the shank is negative for about the first 60 ms of the kick; the negative direction means that the thigh accelerated the shank in a backward-rotation direction. For most of the kick, the thigh motion accelerated the shank in a forward direction.

By a similar analysis it is possible to identify the effect of shank motion on the thigh. As described by Putnam (1991), the effect of shank motion on the thigh is not simply a mirror image of the effect of thigh motion on the shank. The shank is the distal segment in this two-segment system, whereas the thigh is the proximal segment with the shank on one end and the hip joint at the other. As with the thigh-on-shank interaction, we are interested in the magnitude of the shank-on-thigh interactive torque. Figure 3.28*b* shows the net torque acting on the thigh due to the motion of the shank. Once again, the magnitude of the interactive torque is large compared

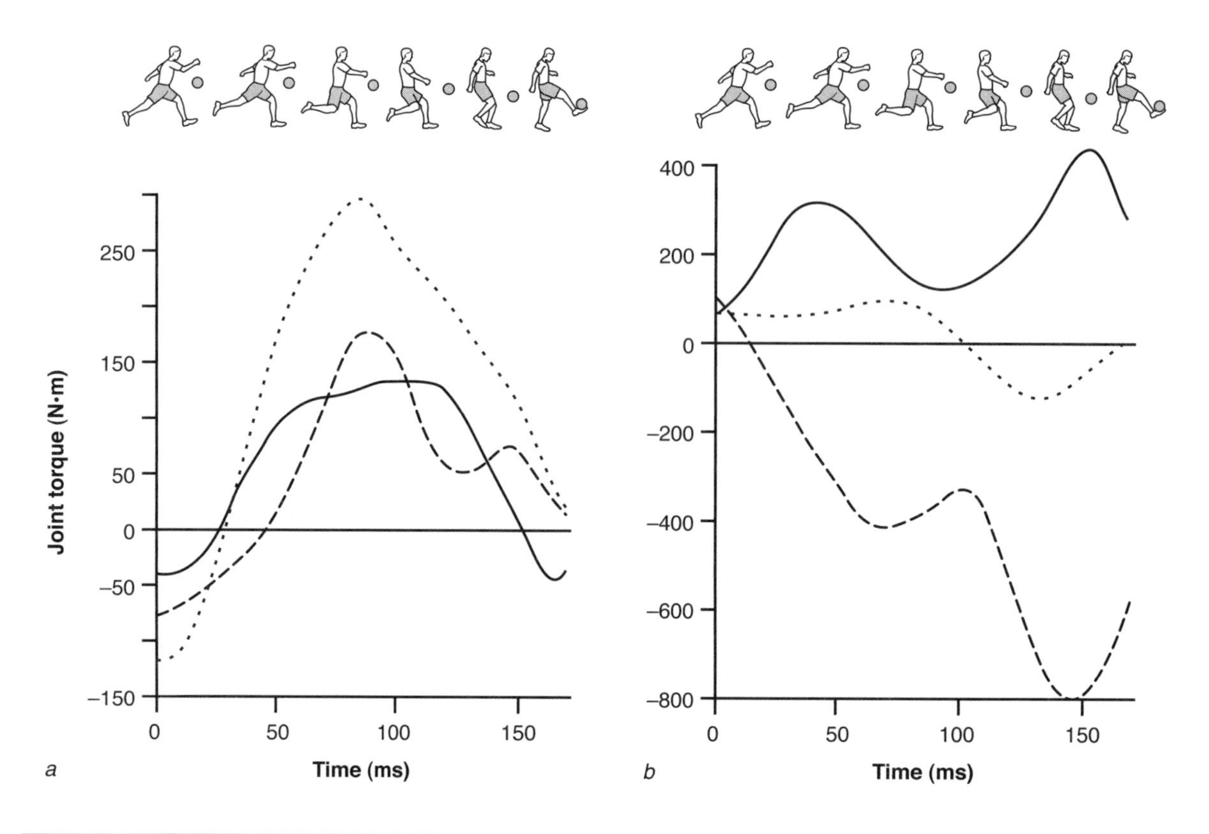

Figure 3.28 Interactive effects during a kick: *(a)* the resultant muscle torque about the knee joint (solid line), the interactive torque exerted by the thigh on the shank (dashed line), and the net effect (dotted line) of all torques acting on the shank; *(b)* the resultant muscle torque about the hip joint (solid line), the interactive torque exerted by the shank on the thigh (dashed line), and the net effect (dotted line) of all the torques acting on the thigh.

Adapted, by permission, from C.A. Putnam, 1983, "Interaction between segments during a kicking motion." In *Biomechanics VIII-B*, edited by H. Matsui and K. Kobayashi (Champaign, IL: Human Kinetics), 691, 692.

with the resultant muscle torque about the hip joint. Even though the resultant muscle torque about the hip was in the direction of flexion for the entire movement, which is consistent with the EMG activity (Dörge et al., 1999), the net torque indicated that the thigh accelerated in the direction of extension toward the end of the kick. The net effect has a smaller magnitude than the absolute torque associated with the muscle activity and the interaction. To achieve this net effect, therefore, the primary function of the muscles about the hip joint during a kick is to counteract the interactive effects between the shank and thigh.

These effects are substantial for whole-limb, rapid movements such as kicking and throwing, but are also significant even in finger movements (Darling & Cole, 1990; Hirashima et al., 2003). Consequently, the control of movement by the nervous system must accommodate interactive torques (Debicki & Gribble, 2005; Koshland et al., 2000; Sainburg et al., 1999; Smith & Zernicke, 1987). Skilled performers, however, can learn to exploit the interactive torques to enhance a particular action, such as throwing (Hirashima et al., 2007*b;* Hore et al., 2005; Kadota et al., 2004). The study of these effects is not trivial, as the calculation of interactive torques during three-dimensional movements requires a method that can account for differences in the joint axes and the principal axes of inertia (Hirashima et al., 2007*a*).

JOINT FORCES, TORQUES, AND POWER

On the basis of these descriptions of the characteristics of the musculoskeletal forces and the techniques that can be used to determine the magnitude and direction of each force, we can compare the relative contributions of different muscle groups to the performance of a movement. It is possible, for example, to determine the net joint reaction force and the net muscle torque acting at various joints in a subject at a particular point in the performance of a movement (example 3.14). By extension, we can perform these calculations for the duration of a movement and compare the performance between groups of individuals and after various interventions (example 3.15). It is even possible to estimate the distribution of the musculoskeletal forces among the structures that compose a joint if we determine the relative activity of the involved muscles (examples 3.16 and 3.17).

EXAMPLE 3.14

Segmental Analysis of Joint Forces and Torques

Because human movement is the result of the muscle-controlled rotation of body segments about one another, we are frequently interested in determining the quantity of muscle activity that contributes to the rotation. We can accomplish this by treating the human body as a series of connected rigid links (body segments) and performing a dynamic analysis on each segment. Three sets of information are required:

1. The kinematics (position, velocity, and acceleration) of the body segments
2. Estimates of the mass and the mass distribution (CM location and moment of inertia) of the segments
3. A known boundary constraint, such as the ground reaction force or a load acting on a segment

We begin the analysis with the body segment that includes the boundary constraint (the foot, in the case of the ground reaction force) and then back-calculate through the body, one segment at a time, to determine the net forces and torques acting about each joint. With this segmental analysis, we can determine the musculoskeletal forces and torques at the ankle, knee, and hip joints at one point in a weightlifting movement (Enoka, 1983). The analysis involves four steps. First, we define the system and its orientation. Second, we separate the human body into an appropriate number of rigid segments. Third, we draw the free body diagram and the kinetic diagram for each body segment. Fourth, we derive and solve the

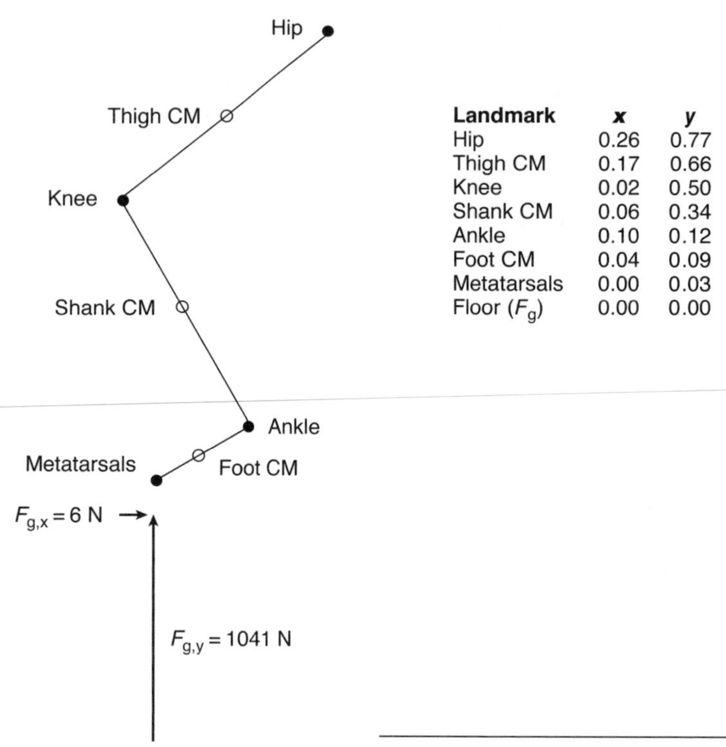

Landmark	x	y
Hip	0.26	0.77
Thigh CM	0.17	0.66
Knee	0.02	0.50
Shank CM	0.06	0.34
Ankle	0.10	0.12
Foot CM	0.04	0.09
Metatarsals	0.00	0.03
Floor (F_g)	0.00	0.00

Figure 3.29 The *x-y* coordinates (m) of selected landmarks in the middle of the weightlifting movement.

equations of motion (equations 3.7, 3.8, and 3.9) for each segment. For this weightlifting example, the system includes the legs in the configuration shown in figure 3.29.

Because the purpose of the analysis is to determine the musculoskeletal forces at the ankle, knee, and hip, the leg should be separated into the foot, shank, and thigh for the analysis. We can obtain estimates of the mass distribution for these segments from the tables included in chapter 2. Furthermore, assume that a video-based motion analysis has provided estimates of the angular and linear accelerations of each segment at the instant of interest in the movement.

Because the I_g values obtained from chapter 2 are relative to the CM of each segment and the analysis is performed about the proximal end of each segment, the I_g values must be transferred to the proximal end of each segment with the parallel axis theorem (equation 2.8). The transferred segmental moments of inertia are indicated in the bottom row of table 3.8.

The next step is to draw the free body and kinetic diagrams for each segment. We should begin with the foot (figure 3.30) because that is the segment for which we have some boundary information (ground reaction force). From the free body diagram (figure 3.30), we can derive the three equations of motion and solve for the unknown terms, which are the *x* and *y* components of the net joint reaction force at the ankle and the net muscle torque about the ankle:

$$\sum F_x = ma_x$$

$$-F_{a,x} + F_{g,x} = ma_x$$

$$-F_{a,x} = -F_{g,x} + ma_x$$

$$= -6 + (1 \times -0.36)$$

$$F_{a,x} = 6.36 \text{ N}$$

Table 3.8 Segmental Data for Figure 3.29

Parameter	Thigh	Shank	Foot
I_g (kg•m²)	0.1995	0.0369	0.0040
Mass (kg)	8	3	1
α (rad/s²)	0.32	-9.39	-3.41
a_x (m/s²)	2.34	1.56	-0.36
a_y (m/s²)	-1.96	-1.64	-0.56
$I_g + md^2$ (kg•m²)	0.3611	0.1185	0.0065

$$\sum F_y = ma_y$$

$$F_{g,y} - F_{a,y} - F_{w,f} = ma_y$$

$$F_{a,y} = F_{g,y} - F_{w,f} - ma_y$$

$$= 1041 - (1 \times 9.81) - (1 \times -0.56)$$

$$F_{a,y} = 1032 \text{ N}$$

$$\sum \tau_A = I_A\alpha + (ma_x \times d) - (ma_y \times a)$$

$$\tau_{m,a} + (F_{w,f} \times a) - (F_{g,y} \times b) + (F_{g,x} \times c) = I_A\alpha + (ma_x \times d) - (ma_y \times a)$$

$$\tau_{m,a} = (F_{g,y} \times b) - (F_{g,x} \times c) - (F_{w,f} \times a) + I_A\alpha + (ma_x \times d) - (ma_y \times a)$$

$$= (1041 \times 0.10) - (6 \times 0.12) - (9.81 \times 0.04) + (0.0065 \times -3.41) + (1 \times -0.36 \times 0.03) - (1 \times -0.56 \times 0.04)$$

$$= 104.1 - 0.72 - 0.392 - 0.022 - 0.011 + 0.022$$

$$\tau_{m,a} = 103 \text{ N}\cdot\text{m}$$

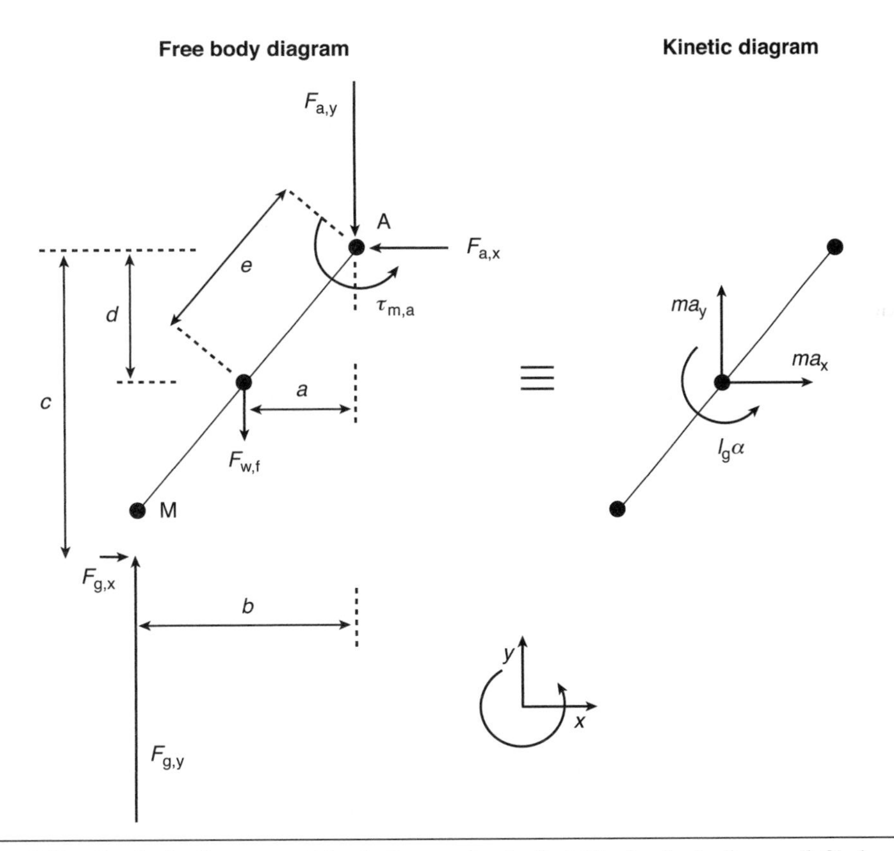

Figure 3.30 Free body diagram and kinetic diagram for the foot. The free body diagram (left) shows the forces (F) acting on the foot at the metatarsals (M), the distances from the line of action of each force to the ankle joint (A), and the net muscle torque ($\tau_{m,a}$) about the ankle joint. The forces include the ground reaction force ($F_{g,x}$, $F_{g,y}$), the weight of the foot ($F_{w,f}$), and the joint reaction force ($F_{a,x}$, $F_{a,y}$). The kinetic diagram shows the horizontal inertia force (ma_x), the vertical inertia force (ma_y), and the inertia torque ($I_A\alpha$) about the proximal end of the segment (A).

Once the musculoskeletal forces at the ankle are known, we proceed to the next segment in the system, the shank (figure 3.31). We can derive the three equations of motion and solve for the x and y components of the net joint reaction force at the knee and the net muscle torque about the knee:

$$\begin{aligned}\sum F_x &= ma_x \\ F_{a,x} - F_{k,x} &= ma_x \\ F_{k,x} &= F_{a,x} - ma_x \\ &= 6.36 - (3 \times 1.56) \\ F_{k,x} &= 1.68 \text{ N}\end{aligned}$$

$$\begin{aligned}\sum F_y &= ma_y \\ F_{a,y} - F_{k,y} - F_{w,s} &= ma_y \\ F_{k,y} &= F_{a,y} - F_{w,s} - ma_y \\ &= 1032 - (3 \times 9.81) - (3 \times -1.64) \\ F_{k,y} &= 1007 \text{ N}\end{aligned}$$

$$\begin{aligned}\sum \tau_K &= I_K\alpha + (ma_x \times d) + (ma_y \times a) \\ \tau_{m,k} - \tau_{m,a} - (F_{w,s} \times a) + (F_{a,y} \times b) + (F_{a,x} \times c) &= I_K\alpha + (ma_x \times d) + (ma_y \times a) \\ \tau_{m,k} &= \tau_{m,a} + (F_{w,s} \times a) - (F_{a,y} \times b) - (F_{a,x} \times c) \\ &\quad + I_K\alpha + (ma_x \times d) + (ma_y \times a) \\ &= 103 + (29.4 \times 0.04) - (1031 \times 0.08) \\ &\quad - (6.36 \times 0.38) + (0.1185 \times -9.39) \\ &\quad + (3 \times 1.56 \times 0.16) + (3 \times -1.64 \times 0.04) \\ &= 103 + 1.18 - 82.5 - 2.42 - 1.11 \\ &\quad + 0.749 - 0.197 \\ \tau_{m,k} &= 18.7 \text{ N}\bullet\text{m}\end{aligned}$$

Once the musculoskeletal forces at the knee are known, we proceed to the next segment in the system, the thigh (figure 3.32). We can derive the three equations of motion and solve for the x and y components of the net joint reaction force at the hip and the net muscle torque about the hip:

$$\begin{aligned}\sum F_x &= ma_x \\ F_{k,x} - F_{h,x} &= ma_x \\ F_{h,x} &= F_{k,x} - ma_x \\ &= 1.68 - (8 \times 2.34) \\ F_{h,x} &= -17 \text{ N}\end{aligned}$$

The value of –17 means that $F_{h,x}$ has a magnitude of 17 N but that it is actually acting in the direction opposite to that drawn on the free body diagram.

$$\begin{aligned}\sum F_y &= ma_y \\ F_{k,y} - F_{h,y} - F_{w,t} &= ma_y \\ F_{h,y} &= F_{k,y} - F_{w,t} - ma_y \\ &= 1007 - (8 \times 9.81) - (8 \times -1.96) \\ F_{h,y} &= 944 \text{ N}\end{aligned}$$

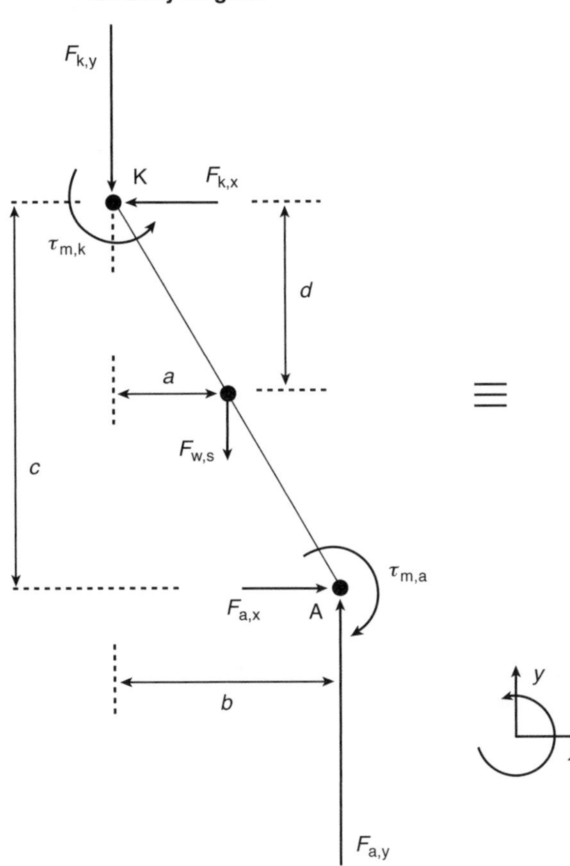

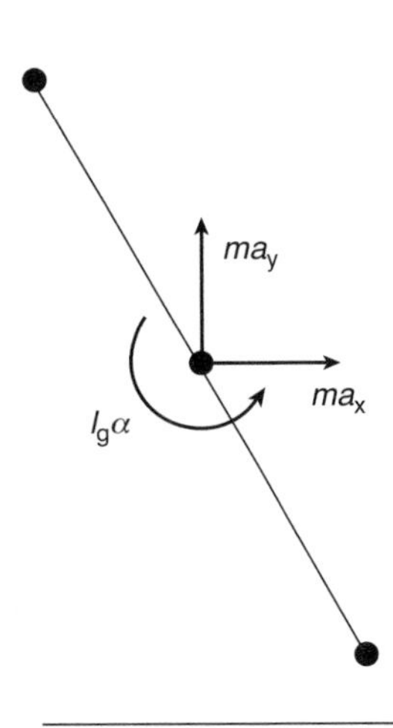

Figure 3.31 Free body diagram and kinetic diagram for the shank. The free body diagram (left) shows the forces (F) acting on the shank, the distances from the line of action of each force to the knee joint (K), and the net muscle torque ($\tau_{m,k}$) about the knee joint. The forces include the weight of the shank ($F_{w,s}$) and the joint reaction forces at the ankle ($F_{a,x}$, $F_{a,y}$) and the knee ($F_{k,x}$, $F_{k,y}$). The kinetic diagram shows the horizontal inertia force (ma_x), the vertical inertia force (ma_y), and the inertia torque ($I_K\alpha$) about the proximal end of the segment (K).

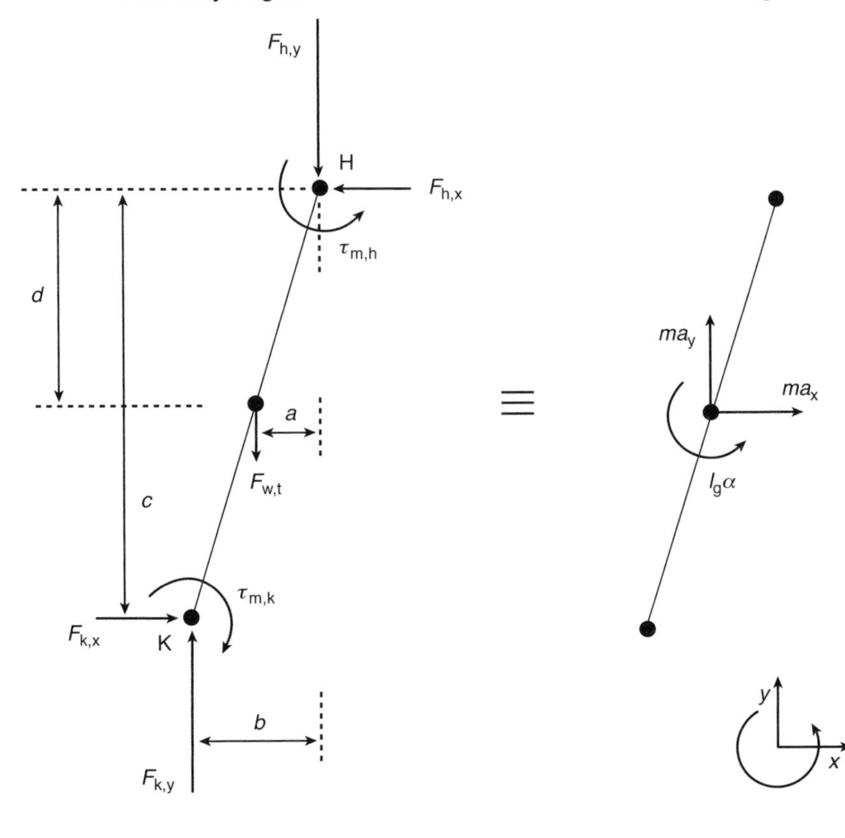

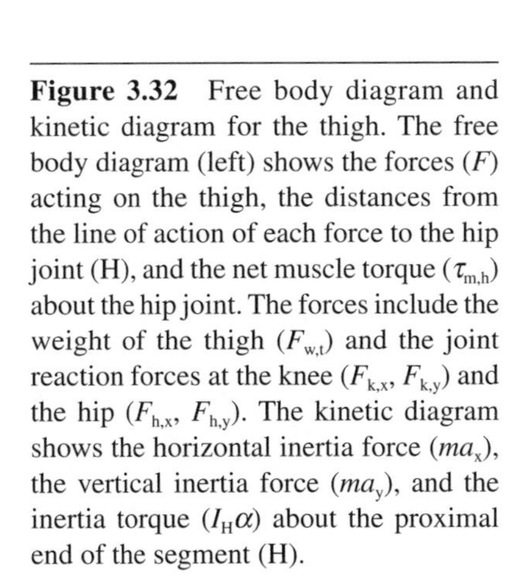

Figure 3.32 Free body diagram and kinetic diagram for the thigh. The free body diagram (left) shows the forces (F) acting on the thigh, the distances from the line of action of each force to the hip joint (H), and the net muscle torque ($\tau_{m,h}$) about the hip joint. The forces include the weight of the thigh ($F_{w,t}$) and the joint reaction forces at the knee ($F_{k,x}$, $F_{k,y}$) and the hip ($F_{h,x}$, $F_{h,y}$). The kinetic diagram shows the horizontal inertia force (ma_x), the vertical inertia force (ma_y), and the inertia torque ($I_H\alpha$) about the proximal end of the segment (H).

$$\sum \tau_{H} = I_{H}\alpha + (ma_{x} \times d) - (ma_{y} \times a)$$

$$\tau_{m,h} - \tau_{m,k} + (F_{w,t} \times a) - (F_{k,y} \times b) + (F_{k,x} \times c) = I_{H}\alpha + (ma_{x} \times d) - (ma_{y} \times a)$$

$$\begin{aligned} \Sigma\tau_{m,h} &= \tau_{m,k} - (F_{w,t} \times a) + (F_{k,y} \times b) - (F_{k,x} \times c) \\ &\quad + I_{H}\alpha + (ma_{x} \times d) - (ma_{y} \times a) \\ &= 18.7 - (78.5 \times 0.09) + (1006 \times 0.24) \\ &\quad - (1.68 \times 0.27) + (0.3611 \times 0.32) \\ &\quad + (8 \times 2.34 \times 0.11) - (8 \times -1.96 \times 0.09) \\ &= 18.7 - 7.07 + 241 - 0.454 + 0.116 \\ &\quad + 2.06 + 1.41 \end{aligned}$$

$$\tau_{m,h} = 256 \text{ N}\cdot\text{m}$$

The major point of this example is to demonstrate how to determine musculoskeletal forces in the human body at one instant in the course of a movement. To perform such an analysis, we need to know the forces due to body mass, the forces due to the surroundings, and the kinematics of the system. For this specific movement, the magnitude of the joint reaction force in the x direction was much smaller than that in the y direction, and the magnitude of the net muscle torque was quite different across the ankle, knee, and hip, with the greatest value at the hip. Be careful not to interpret these data as representing the absolute magnitude of the muscle action about each joint; these data correspond to the net effect. To determine the relative activity among the muscles about a joint, it is necessary to make additional measurements, such as EMG recordings of muscle activity (Buchanan et al., 2005).

EXAMPLE 3.15

Joint Torque and Power During Walking

With the procedure described in the preceding example, it is possible to calculate the musculoskeletal forces at one point in time during a movement. To obtain a complete description of the movement, it is necessary to repeat these calculations approximately 100 times over the course of the movement. DeVita and colleagues (1998) used such an approach when evaluating the effectiveness of an aggressive rehabilitation protocol for patients recovering from surgical repair of the anterior cruciate ligament (ACL).

The analysis focused on a comparison of the torque and power at the ankle, knee, and hip joints during walking. DeVita and colleagues (1998) compared the performance of patients after three weeks and six months of rehabilitation with that of healthy control subjects. The data required for such a comparison included estimates of body mass distribution, the measurement of a boundary condition, and a description of the kinematics of the leg. The mass, CM location, and moment of inertia for the foot, shank, and thigh were estimated using some of the cadaver data and mathematical models described in chapter 2. The kinematic data were obtained with a motion analysis system that was used to determine the position of the lateral malleolus, lateral femoral condyle, greater trochanter, and shoulder. These landmarks were used to determine the relative angles between the foot and shank (ankle joint), the shank and the thigh (knee joint), and the thigh and the trunk (hip joint). The position data were smoothed with a second-order, Butterworth digital filter, and then the angles at the ankle, knee, and hip joints were determined. Angular velocity was calculated by the finite-differences method (chapter 1).

After six months of rehabilitation, DeVita and colleagues (1998) found that the angular kinematics of the patients were almost identical to those of the healthy control subjects. At the three-week point, however, the patients presented with greater flexion and a lesser range of motion than normal at the ankle, knee, and hip joints. These adaptations were also evident in the profiles of the net torque about the three joints (figure 3.33). After three weeks of rehabilitation, the net torques during the stance phase were greater at the ankle and hip

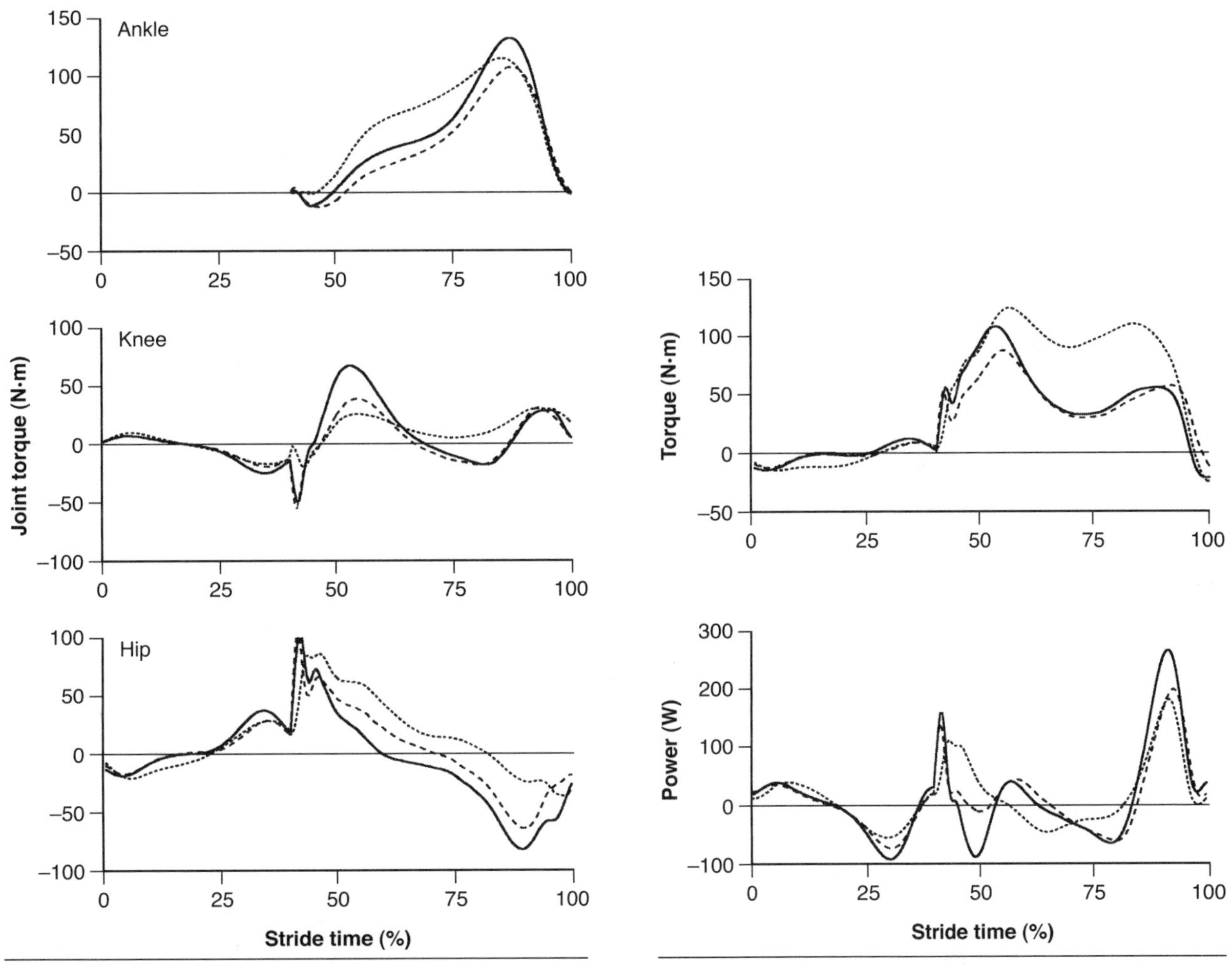

Figure 3.33 Net muscle torque about the ankle, knee, and hip joints during walking for patients after three weeks (dotted line) and six months of rehabilitation (dashed line) and for healthy control subjects (solid line). Positive torque indicates an extensor (and plantarflexor) torque. The foot was on the ground (stance phase) from 40% to 100% of stride time.

Figure 3.34 Summed torque and power about the ankle, knee, and hip joints during a single stride in walking. The data are for patients after three weeks (dotted line) and six months of rehabilitation (dashed line) and for healthy control subjects (solid line). Positive torque indicates an extensor (and plantarflexor) torque; positive power represents power production and negative power corresponds to power absorption. The foot was on the ground (stance phase) from 40% to 100% of stride time.

joints and less at the knee joint compared with the values at six months, which were similar to the data for the healthy control subjects. These results suggest that the patients initially compensated for the intervention by diverting activity away from the knee joint to the ankle and hip joints. Even after six months of rehabilitation, the patients exhibited a lesser net torque about the knee joint at about 50% to 60% of stride time and a greater net torque about the hip joint over the same interval. Despite these differences, the summed torques about the three joints, which corresponds to the net support torque (Winter, 1983), did not differ between the patients at six months and the healthy subjects (figure 3.34).

As with the torque profiles (figure 3.33), the joint power at the ankle, knee, and hip joints was significantly different in the patients after three weeks of rehabilitation but after six months of rehabilitation became more similar to that of healthy control subjects (figure 3.35). For example, the positive work (area under the power–time curve) done at the hip from 40% to 80% of stride time by the patients at three weeks was decreased by 44% after six months of rehabilitation. Similarly, the amount of negative work done at the knee joint (power absorption due to a lengthening contraction) increased with the duration of rehabilitation. Neither the positive work done at the hip nor the negative work done at the knee fully recovered even after six months of rehabilitation. The scale on the vertical axes in figure 3.35 indicates that the magnitude of power was greatest at the ankle joint and least at the hip joint. As a result, the summed power (figure 3.34) was dominated by the power about the knee and ankle joints. The summed power comprised alternating phases of power production and absorption, with the peak power production occurring just prior to toe-off. After six months of rehabilitation, the summed power for the patients was essentially similar to that for the healthy subjects with the exception at around 50% of stride time.

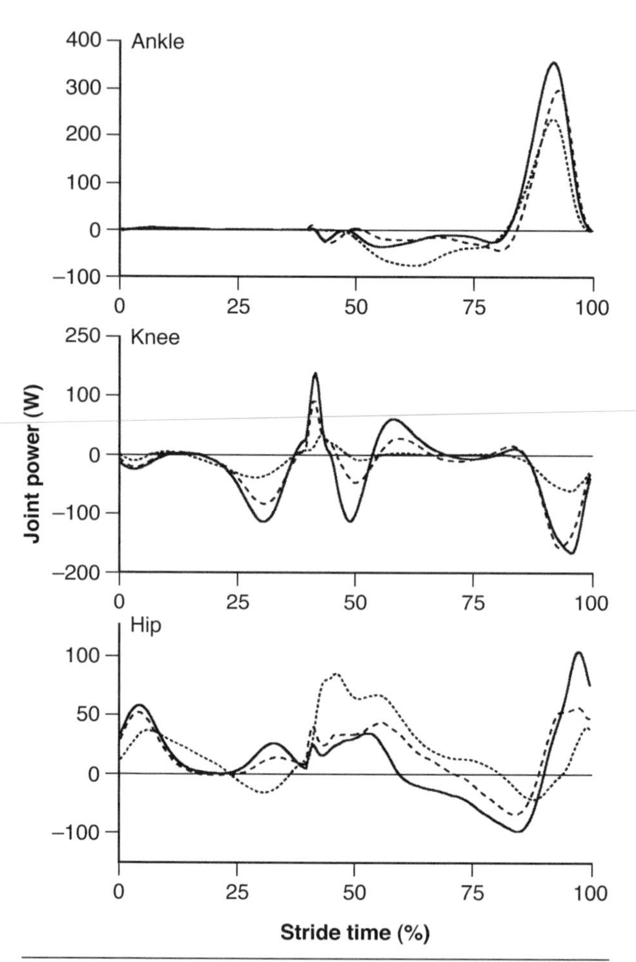

Figure 3.35 Net power about the ankle, knee, and hip joints during walking for patients after three weeks (dotted line) and six months of rehabilitation (dashed line) and for healthy control subjects (solid line). Joint power was calculated as the product of angular velocity and joint torque; positive power indicates production of power by muscles that cross the joint, whereas negative power represents the absorption of power. The foot was on the ground (stance phase) from 40% to 100% of stride time.

The interesting feature of this analysis, however, is that although the summed joint torques and powers were similar between healthy control subjects and patients after six months of rehabilitation, there remained significant differences between the two groups at the level of the individual joints. This result suggests that the patients developed a coping strategy to deal with an impaired knee joint mechanism (Alkjaer et al., 2003; Barrance et al., 2006, 2007). Similarly, the use of a functional knee brace by individuals recovering from ACL injury changes the joint kinetics during both walking and running (DeVita et al., 1998, 1996; Lu et al., 2006).

EXAMPLE 3.16

Joint Reaction Force in the Knee

Whereas the previous two examples dealt with the net mechanical action at various joints, the third example shows how it is possible to estimate absolute effects. In this example we compare the joint reaction force at the knee joint during a squat lift and a knee extension exercise (Escamilla et al., 1998*b*). In the rehabilitation literature, these activities are described as closed and open kinetic chain exercises, respectively (Blackard et al., 1999; Stensdotter et al., 2003). The major difference between such exercises is the point of application and

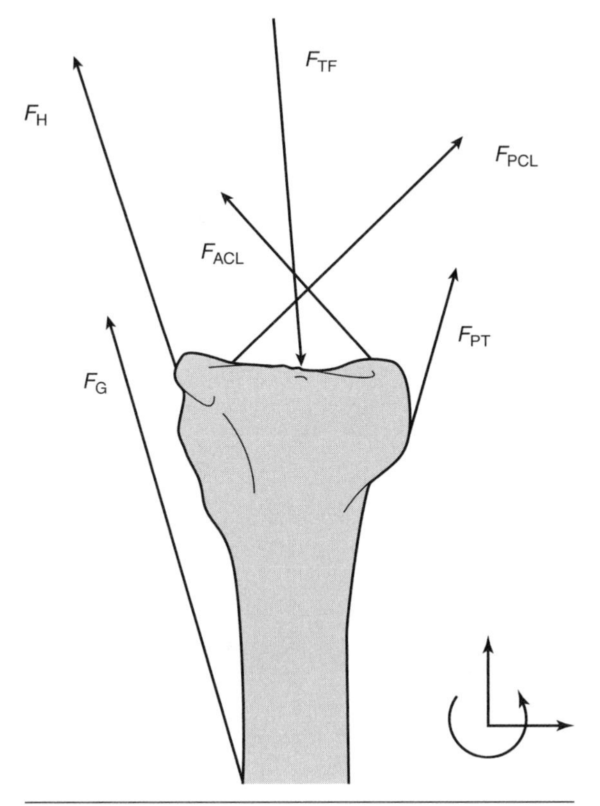

Figure 3.36 Forces acting on the tibia. F_{ACL} = tensile force of the anterior cruciate ligament; F_G = tensile force exerted by the gastrocnemius muscle; F_H = tensile force exerted by the hamstring muscles; F_{PT} = tensile force of the patella tendon; F_{PCL} = tensile force of the posterior cruciate ligament; and F_{TF} = compressive force due to tibiofemoral contact.

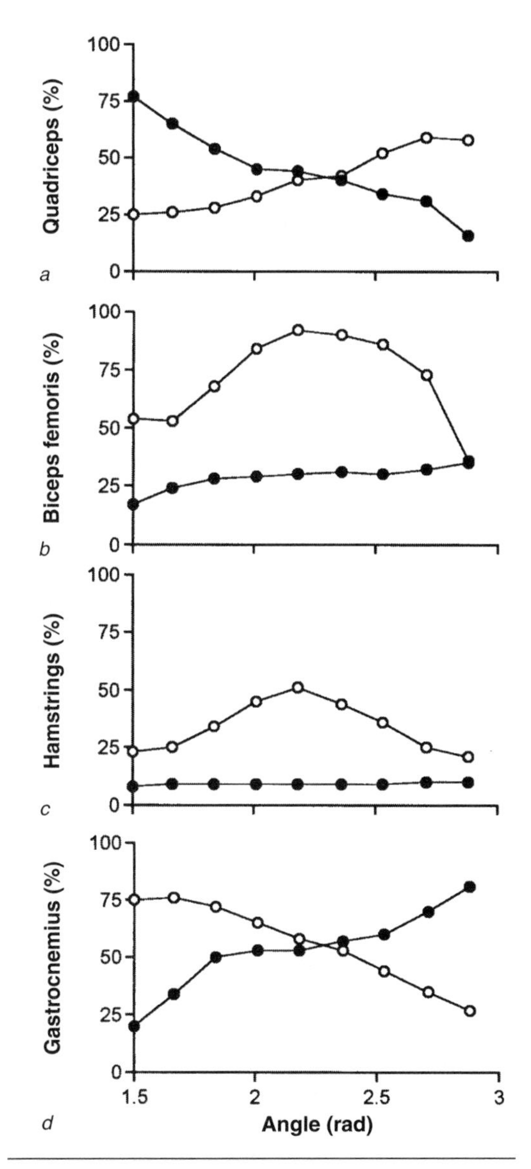

Figure 3.37 Average EMG of the *(a)* quadriceps femoris, *(b)* biceps femoris, *(c)* medial hamstrings, and *(d)* gastrocnemius muscles during the knee extension phase of the squat lift (open circles) and the knee extension exercise (filled circles). The EMG data are normalized to the values recorded during a maximal voluntary contraction.

Data from Escamilla et al., 1998*b*.

direction of the force exerted by the surroundings: In this example, these correspond to the ground reaction force for the squat lift and the contact force with the machine for the knee extension exercise.

As in examples 3.14 and 3.15, a motion analysis system was used to measure the kinematics of each movement in the sagittal plane, and the contact force was measured with a force transducer placed between the subject and the surroundings (i.e., ground or machine). These data were then used to perform a dynamic analysis and to determine the musculoskeletal forces at the knee joint. In this study, however, these forces included the sum of the individual muscle forces, the net ligament force, and the contact force between the femur and the tibia (Zheng et al., 1998). To achieve this level of detail, it was necessary to estimate the force exerted by each of the major muscles that cross the knee joint and to resolve the joint reaction force into a normal compressive component and a ligament force (figure 3.36).

Estimates of the force exerted by each muscle during the two tasks were based on the EMG activity of the muscle (figure 3.37). The muscles were quadriceps femoris (rectus femoris and the three vasti), medial hamstrings (semimembranosus and semitendinosus), and gastrocnemius. The force of each muscle was estimated from the following relation:

$$F_m = ckA\ \sigma\,\text{EMG}$$

where c = weighting coefficient, k = a factor that accounted for the effect of a change in muscle length on the force exerted by the muscle, A = cross-sectional area, σ = specific tension, and EMG = the amount of muscle activity normalized to the value during a maximal voluntary contraction. Once the summed muscle force was determined (sum of agonist and antagonist muscles), the difference between this value and the resultant force was set equal to the sum of the tibiofemoral contact force and the ligament force. The tibiofemoral contact force was assumed to act perpendicular to the articulating surface of the tibia. The ligament force, however, was derived from the shear component of the joint reaction force and the line of action of the ligament force. To calculate the ligament force, the shear component of the joint reaction force was multiplied by the cosine of the angle for the line of action for the ligament based on functions reported by Herzog and Read (1993).

For the knee extension exercise, the subject pushed against a pad that contacted the lower shank. The range of motion was about 1.57 rad, beginning with the knee joint flexed at a right angle. While the range of motion was also about 1.57 rad at the knee joint for the squat lift, the subject began from an erect position with the knee joint extended (3.14 rad). Because of these different starting positions, knee angle changed from left to right for the knee extension exercise and from right to left for the squat lift in figure 3.37. As a result, figure 3.37*a* shows that the EMG for quadriceps femoris was low at the beginning of each movement and increased as the knee was extended, that is, as the load was raised. A similar EMG pattern was recorded for gastrocnemius (figure 3.37*d*) but not for the other two antagonist muscles (figure 3.37, *b* and *c*). There was greater coactivation of the biceps femoris and hamstrings during the squat lift.

The net extensor torque about the knee joint changed with a parabolic-like shape over the range of motion for the knee extension exercise (figure 3.38*a*). For the squat lift, however, the extensor torque increased almost linearly throughout the lift. The tibiofemoral compressive force was lowest at the beginning of each exercise but increased to similar levels (~3000 N) by about the middle of the range of motion (figure 3.38*b*). Because the tibiofemoral force was based on the absolute activity of the agonist and antagonist muscles, it was about three times greater than the compressive component of the net joint reaction force. The compressive force was greatest when the knee joint approached complete extension during the knee extension exercise, but the peak values occurred in the middle of the range of motion during the squat lift. In contrast, the ligament force was generally greatest when the knee joint was flexed. The ligament force was provided by the posterior cruciate ligament for the entire

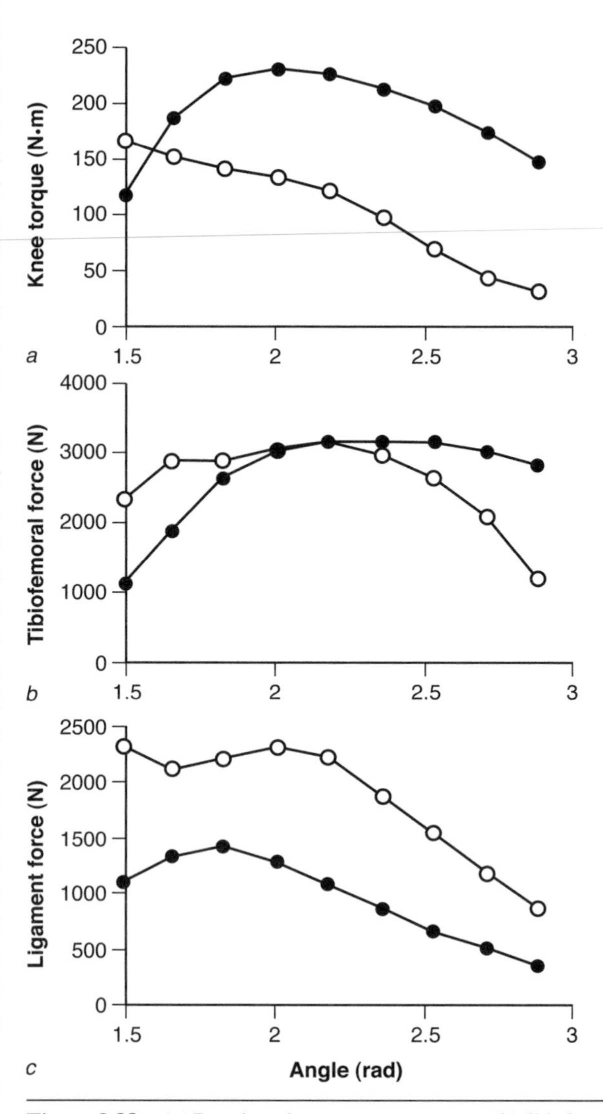

Figure 3.38 *(a)* Resultant knee extensor torque, *(b)* tibiofemoral compressive force, *(c)* and ligament force during the knee extension phase of the squat lift (open circles) and the knee extension exercise (filled circles). A positive ligament force indicates tension in the posterior cruciate ligament, whereas a negative ligament force indicates tension in the anterior cruciate ligament.

Data from Escamilla et al. 1998*b*.

squat lift and for most of the knee extension exercise (figure 3.38*c*). The peak ligament force was twice as great during the squat lift, and remained greater than the peak value for the knee extension exercise over most of the range of motion. Such comparisons indicate the effects of different exercises on the musculoskeletal forces.

EXAMPLE 3.17

Modeling of Knee Forces

One can obtain more detail about the distribution of forces within a joint by performing modeling studies (Lloyd et al., 2005; Shelburne et al., 2005). For example, Pandy and Shelburne (1997) developed a sagittal plane model of the human knee to examine load sharing between the muscles, ligaments, and bones during isometric exercises. The key features of the model were as follows (Shelburne & Pandy, 1997):

- The geometry of the distal femur was based on cadaver data.
- The tibial plateau and patellar facet were represented as flat surfaces.
- The ligaments and capsule of the knee joint were modeled as 11 elastic elements.
- Eleven muscles crossed the knee joint, with each muscle represented by a Hill-type model (chapter 6) and attached to an elastic tendon.

With such a model, it was possible to estimate the forces experienced by the different structures in the joint over its entire range of motion in response to varying levels of muscle activation. Figure 3.39 shows an example of such forces experienced by the ACL. The results indicate that the ACL force increased with the level of activation of the quadriceps femoris but that the range of motion over which the force acted remained the same for the different levels of activation. The model also demonstrated that coactivation of the hamstring muscles decreased the ACL force and the range of motion over which the ACL was loaded. However, coactivation of the hamstrings increased the anterior force due to tibiofemoral contact and hence the anterior force applied by the patellar tendon.

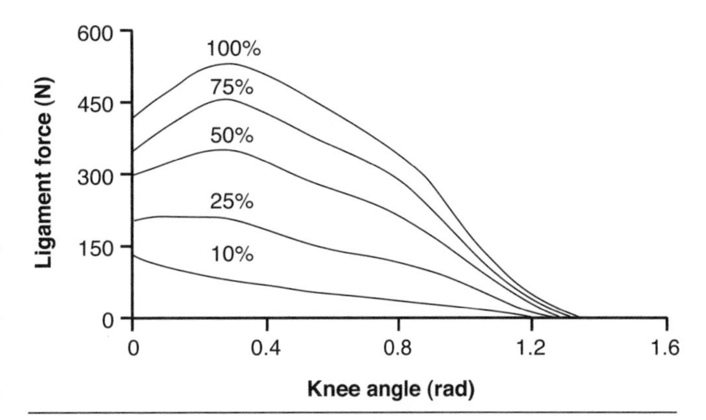

Figure 3.39 Anterior cruciate ligament force associated with isolated contractions of the quadriceps femoris. The knee joint is completely extended at 0 rad.

Data from Pandy and Shelburne, 1997.

Shelburne and colleagues (2005) used a similar approach to estimate muscle, ligament, and joint contact forces at the knee during walking. They estimated the distribution of forces in model knees with and without an ACL. The peak ACL force occurred early in the stance phase and was caused by the anterior pull of the patellar tendon on the tibia, which was opposed by the medial collateral ligament. The anterior tibial translation in the ACL-deficient knee could be counteracted by an increase in hamstring activity, but not by a reduction in quadriceps activity. With a different approach, Lloyd and colleagues (2005) used EMG-driven models to identify the patterns of muscle activation that can improve knee stability in a range of movements, including those actions that can minimize the ligament injuries.

SUMMARY

This chapter focuses on the mechanical interactions that occur inside the human body when we perform movements. The purpose of the chapter is to characterize these interactions, introduce the techniques used to analyze them, and provide detailed examples of how these within-body interactions can be estimated. We refer to the interactions as musculoskeletal forces, and three main ones are identified: joint reaction force, muscle force, and the force due to intra-abdominal pressure. These forces occur among body segments and are associated with the rotation of one body segment relative to its neighbors. The first part of the chapter provides a description of the three musculoskeletal forces. The second part introduces the formal analysis techniques employed in biomechanics. Known as static and dynamic analyses, these are used to determine the quantitative details of a movement, including the magnitude and direction of unknown musculoskeletal forces. The final part of the chapter presents several detailed examples of how these techniques are used to determine the torque, work, and power at selected joints in the human body during movement.

SUGGESTED READINGS

Hamill, J., & Knutzen, K.M. (2003). *Biomechanical Basis of Human Movement.* Baltimore: Williams & Wilkins, chapters 10 and 11.

Winter, D.A. (1990). *Biomechanics and Motor Control of Human Movement.* New York: Wiley, chapters 5 and 8.

chapter 4

Running, Jumping, and Throwing

Human movement can take a variety of forms, ranging from transporting the center of mass to the expression of emotions. Although there is some literature on the use of biomechanics to develop a taxonomy of postures and movements employed in various activities, such as dance, the most common application of biomechanics has been to study the fundamental features of such activities as running, jumping, and throwing. The purpose of this chapter is to describe some of the biomechanical characteristics of these movements, which provides an opportunity to review and apply the concepts presented in chapters 1 through 3.

WALKING AND RUNNING

Human gait involves alternating sequences in which the body is supported first by one limb, which contacts the ground, and then by the other limb. Human gait has two modes, walking and running. One distinction between these two modes lies in the percentage of each cycle during which the body is supported by foot contact with the ground. When we walk, there is always at least one foot on the ground; and for a brief period of each cycle, both feet are on the ground. Accordingly, walking can be characterized as an alternating sequence of single and double support. In contrast, running involves alternating sequences of support and nonsupport, with the proportion of the cycle spent in support varying with speed. For both walking and running, however, each limb experiences a sequence of support and nonsupport during a single cycle. The period of support is referred to as the **stance phase,** and nonsupport is known as the **swing phase.** The stance phase begins when the foot contacts the ground (footstrike) and ends when the foot leaves the ground (toe-off). Conversely, the swing phase extends from toe-off to footstrike. Gait cycles are usually defined relative to these events. For example, one complete cycle, such as from left foot toe-off to left foot toe-off, is defined as a **stride** (figure 4.1).

The stride contains two steps (figure 4.1). A **step** is defined as the part of the cycle from the toe-off (or footstrike) of one foot to the toe-off (or footstrike) of the other foot. Within a stride, there are four occurrences of footstrike and toe-off, two events for each limb. These are right footstrike, right toe-off, left footstrike, and left toe-off. Figure 4.1 shows the stance (shaded areas) and swing (open rectangles) for the right (R) and left (L) legs during walking, racewalking, running, and sprinting.

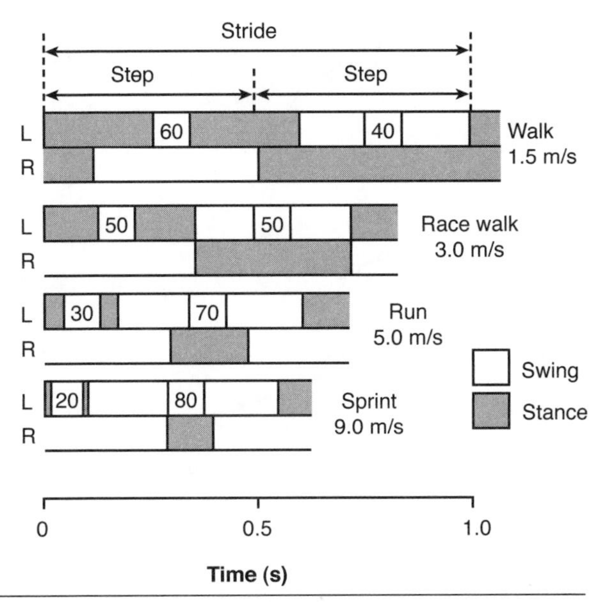

Figure 4.1 The events and phases characterizing walking and running gaits (R = right, L = left). The numbers in various rectangles indicate the relative duration (% stride) of that stance or swing phase.

Data from Vaughan 1984.

Each foot is on the ground (stance phase) during walking for about 60% of the stride and off the ground (swing phase) for 40%. The duration of the stance phase decreases to 50% for racewalking, 30% for running, and 20% for sprinting.

The absolute duration of the stride decreases with speed for both walking and running (figure 4.2*a*). The reduction in stride time is attributable mainly to a decrease in stance duration, as the duration of the swing phase does not change much with speed (figure 4.2*b*). Although the speed-related decline in the absolute duration of the stance phase is greater with walking, the decrease in the relative duration (% stride time) is greater for running (figure 4.2*c*).

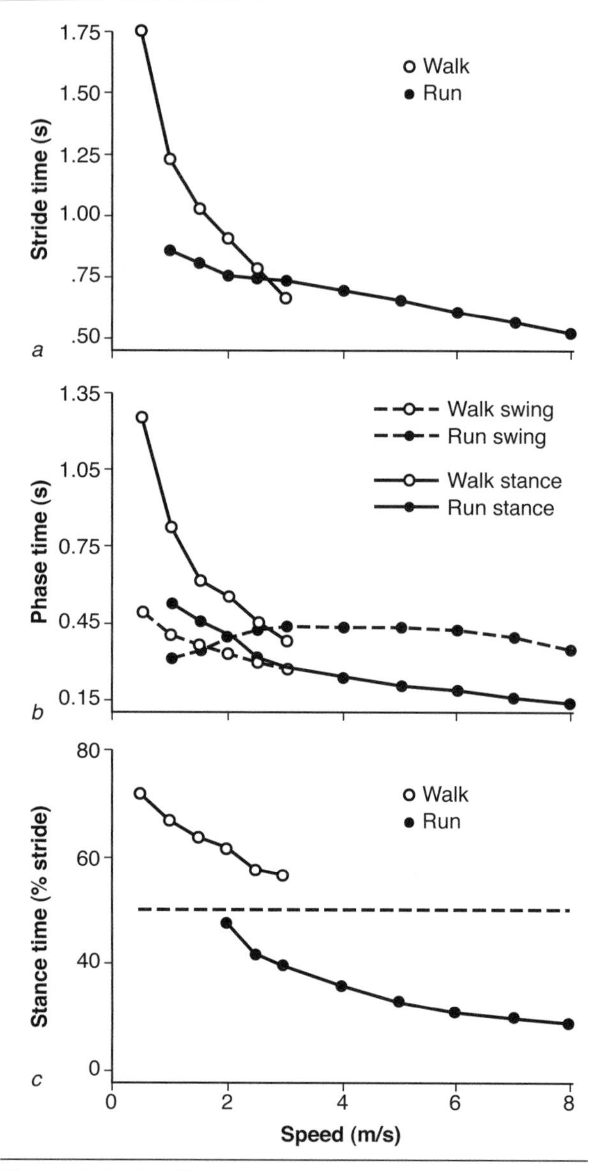

Figure 4.2 (*a*) Changes in stride time and (*b*) the absolute and (*c*) relative durations of the stance and swing phases with an increase in the speed of walking and running.

Adapted, by permission, from J. Nilsson and A. Thorstensson, 1987, "Adaptability in frequency and amplitude of leg movements during human locomotion at different speeds," *Acta Physiologica Scandinavica* 129: 109.

Stride Length and Rate

Running speed depends on two variables, stride length and stride rate (Vaughan, 1984; Weyand et al., 2000). For example, running speed increases if stride length remains constant and stride time decreases (i.e., stride rate increases). Similarly, speed increases if stride rate remains constant and stride length increases. Within certain limits, a number of length–rate combinations will produce a desired speed. The average combinations are shown in figure 4.3. For example, an individual running at a speed of 8 m/s will use a stride rate of about 1.75 Hz and a stride length of about 4.6 m. Figure 4.3 illustrates that, on average, a runner increases speed over the range from 4 to 9 m/s by increasing stride rate continually, although more slowly (the slope is not that steep) at lower velocities, but increases stride length only up to about 8 m/s. Notice that the contributions of changes in stride length and stride rate to running speed differ at low and high speeds; this is apparent by the differences in the slope of each curve (stride length and rate) at different speeds. Nonetheless, the top speeds that runners can achieve depend on the magnitude of the ground reaction force and not stride rate (Weyand et al., 2000).

Based on the average data shown in figure 4.3, it appears that initial increases in running speed depend more on increases in stride length than stride rate. The relative contributions of stride length and stride rate are more obvious when the data are plotted in an *x-y* graph (stride rate vs. stride length), such as figure 4.4. For this graph, four subjects were studied while running at several speeds that ranged from 4 to 9 m/s. Each subject ran at 5

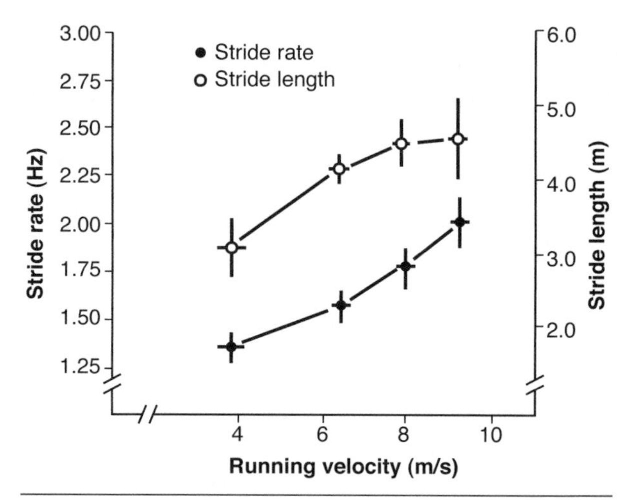

Figure 4.3 Average values of stride length and stride rate at four running velocities.

Data from Luhtanen and Komi 1978.

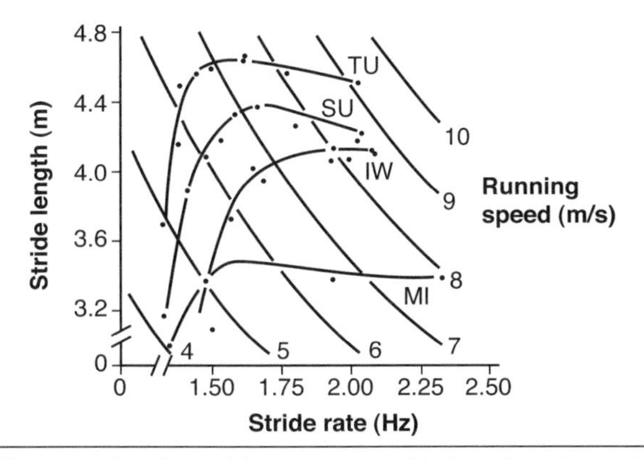

Figure 4.4 The stride rates and stride lengths selected by four subjects (TU, SU, IW, and MI) to achieve a range of running speeds.

Data from Saito et al. 1974.

to 12 constant speeds using self-selected combinations of stride length and stride rate. Stride length was measured from footprints, and stride rate from foot switches that indicated the stance phase. The combination chosen by each subject for a given speed is plotted as a small dot on figure 4.4.

Subject SU ran at speeds that ranged from 4.3 to 8.5 m/s. His initial changes in speed (4.3 to 7.0 m/s) were due to a combined increase in stride length (3.2 to 4.4 m) and stride rate (1.4 to 1.7 Hz). Based on the slope of the length–rate relation over this range, changes in stride length were more significant for these initial increases in speed. Subsequently, SU increased speed (7.0 to 8.5 m/s) by slightly decreasing stride length (about 20 cm) and substantially increasing stride rate (1.7 to 2.0 Hz). In general, the trained runners (Subjects TU, SU, and IW) increased stride length up to 7.0 m/s, whereas the untrained runner (MI) did so only up to about 5.5 m/s. All four runners, however, achieved initial increases in speed (up to ~6.0 m/s) mainly by increasing stride length, which requires less energy (Cavagna et al., 1997; Martin et al., 2000). It seems that anthropometric variables (e.g., stature, leg length, limb segment mass) are not the primary determinants of preferred stride frequency and length (Cavanagh & Kram, 1989). However, the performance of elite runners (100 m to 10 km) is related to mass, stature, and the ground reaction force (Weyand & Davis, 2005).

Angle-Angle Diagrams

Increases in stride length are accomplished by alterations of the kinematics of the limbs. The changes needed include both the range of motion about a joint (quantity) and the pattern of displacement (quality). For example, figure 4.5 shows that angular displacement about the knee joint increases as the runner goes from a walk to a run

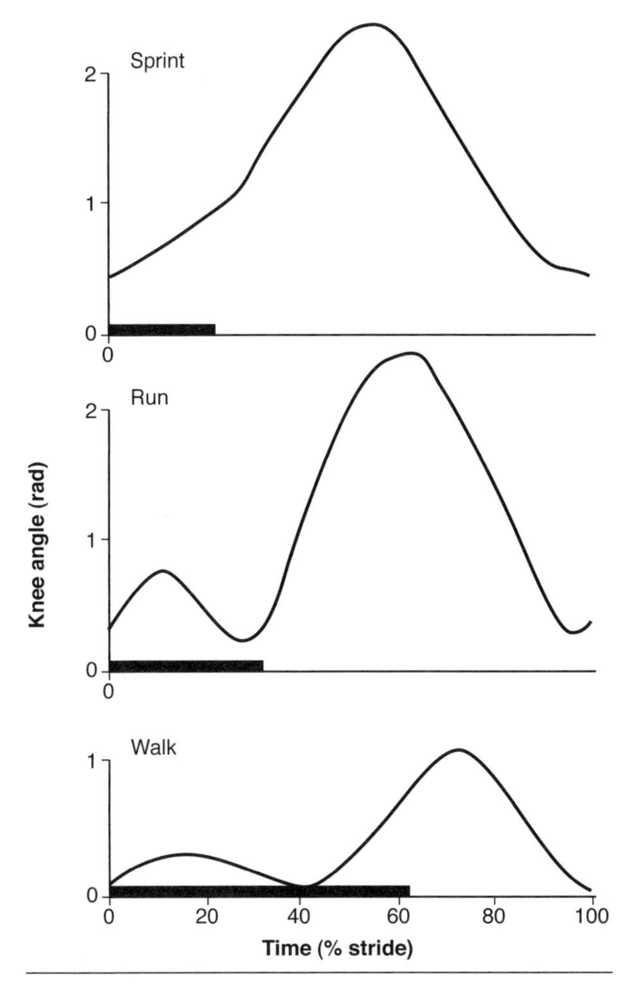

Figure 4.5 Knee angle during a stride for a sprint, run, and walk. A knee angle of 0 rad indicates complete extension. The shaded bar indicates the stance phase, from footstrike (left end) to toe-off (right end).

Data from C.L. Vaughan, 1984, "Biomechanics of running gait," *CRC Critical Reviews in Biomedical Engineering* 12: 11.

and that the stance phase (indicated by the shaded horizontal bar) includes only knee flexion during a sprint but both flexion and extension during walking and running. Similarly, there is an increase in arm motion as running speed increases, which includes an increase in the range of motion about both the shoulder and elbow joints.

These changes in the quantity and quality of displacement with speed can be characterized with angle-angle diagrams (Hershler & Milner, 1980*a, b;* Marey, 1879; Miller, 1978). When you examine angle-angle diagrams, the two features to focus on are the shape of the diagram and its location relative to reference angles. Figure 4.6 shows a typical thigh-knee angle-angle diagram for a person running at a moderate speed. The angles included on the diagram are indicated by the stick figure in the lower right portion of the figure. As described in chapter 1, these comprise an absolute angle (thigh angle with

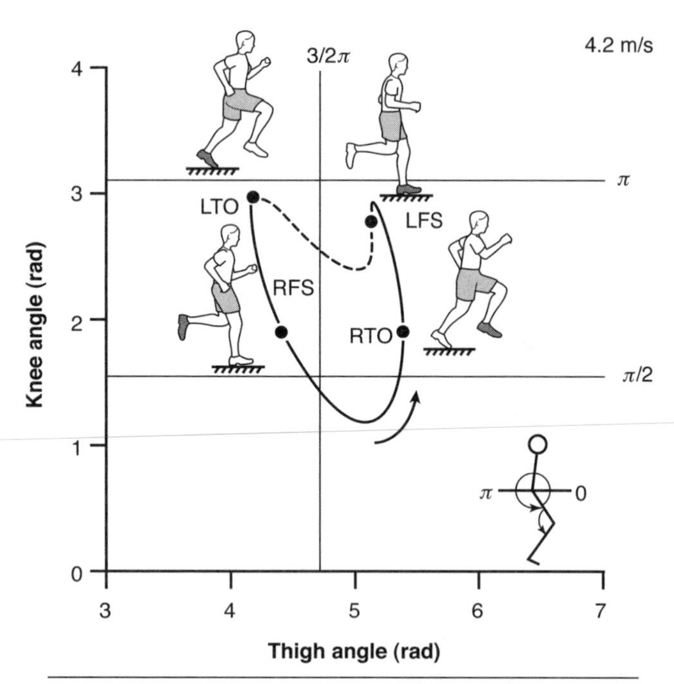

Figure 4.6 Thigh-knee diagram of the left leg of a person running at 4.2 m/s. The left foot has the gray shoe. LFS = left footstrike; LTO = left toe-off; RFS = right footstrike; RTO, right toe-off.

Reprinted, by permission, from R.M. Enoka, D.I. Miller, E.M. Burgess, 1982, "Below-knee amputee running gait," *American Journal of Physical Medicine* 61: 70.

respect to the forward horizontal) and a relative angle (knee angle as the angle between the thigh and the shank). Three reference angles are included on the diagram, one for the thigh and two for the knee. The reference angle for the thigh is $3/2\pi$ rad, which indicates when the thigh was vertical. The reference angles for the knee indicate an extended knee joint (π rad) and the knee joint flexed to a right angle ($\pi/2$ rad). According to the location of the thigh-knee diagram in figure 4.6, the thigh rotates in front of and behind the thigh-vertical position, while the knee is never completely extended but is flexed to more than a right angle.

The shape of an angle-angle diagram indicates the relation between the included angles throughout the movement. For cyclic activities, the diagram indicates the angles during a single cycle, which corresponds to one stride for walking and running. These angles change in a counterclockwise direction on the thigh-knee diagram for walking and running. The interpretation of the shape of the diagram is aided by the possibility of including the location of specific events, such as footstrike and toe-off, on the diagram. Figure 4.6 includes the location of footstrike and toe-off for both the right and left legs. Because figure 4.6 shows the thigh and knee angles for the left leg, the stance phase includes the region from left footstrike (LFS) to left toe-off (LTO). The inclusion of right footstrike (RFS) and right toe-off (RTO) indicates the thigh and knee angles of the left leg when these right-foot events occurred.

The shape of the thigh-knee diagram from left leg footstrike to toe-off (stance phase—dashed line) comprises two parts. First, the knee angle decreases (flexion) while the thigh angle does not change. Second, the knee angle increases (extension) while the thigh angle decreases (backward rotation). From left leg toe-off to footstrike (swing phase—solid line), the angular displacement again comprises two distinct parts. First, from LTO to the minimal knee angle, the thigh angle increases (forward rotation) and the knee angle decreases (flexion). Second, from the minimal knee angle through to LFS, the knee angle increases (extension) while the thigh rotates slightly forward and then backward.

Although it is sometimes useful to dissect the angular displacement of an angle-angle diagram, the most useful feature of this type of diagram is as a template for comparing the coordination of a movement across conditions and between subjects. In contrast to this typical thigh-knee diagram, the three graphs for subjects with below-knee amputations depicted in figure 4.7 indicate a substantial difference during the stance phase (i.e., the region from LFS to LTO). Specifically, the amputee thigh-knee diagrams reveal a knee joint pattern of a constant angle followed by flexion rather than the normal flexion-extension sequence evident in figure 4.6. Because figure 4.7 shows the thigh-knee diagrams for the prosthetic limbs of the below-knee amputees, the pattern of the graphs is perhaps not surprising. This failure to flex the knee during stance, shown in figure 4.7 by the lack of a decrease in the knee angle immediately after footstrike, means that the amputees just used their limbs as a rigid strut about which to rotate while the prosthetic foot was on the ground. This type of graphic display could be used in a clinical setting to monitor a rehabilitation program aimed at modifying this strategy so that the gait would appear more symmetrical. Similarly, angle-angle diagrams can be used to indicate the reduced coordination of various patient populations, such as those with cerebellar ataxia (Palliyath et al., 1998).

Angle-angle diagrams have also been used to represent the kinematics of the arms during running. Because the motion of the arms is frequently not confined to the sagittal plane during running, imaging techniques that can capture three-dimensional motion are necessary. From such measurements, figure 4.8 shows the displacement of the upper arm about the shoulder and the relative angle between the upper and lower arms (elbow angle) for an individual running at 11.4 m/s. A positive shoulder angle indicates flexion (forward of vertical) and a negative one represents extension (backward of vertical), whereas an elbow angle of 0 rad corresponds to complete extension.

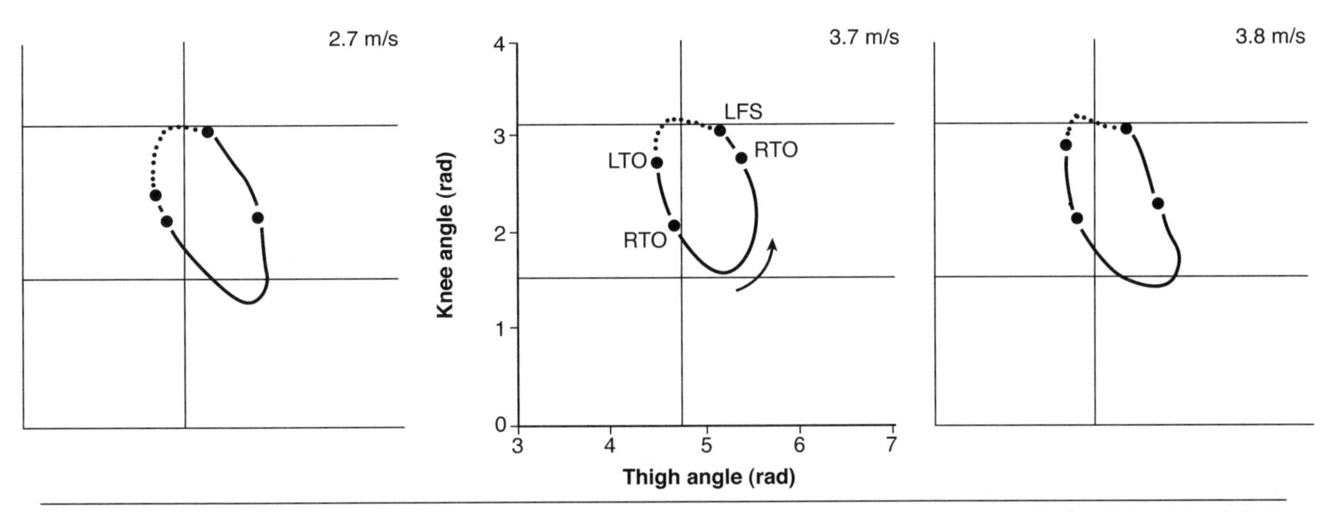

Figure 4.7 Thigh-knee diagrams for the prosthetic limb of three below-knee amputees running at speeds from 2.7 to 3.8 m/s. LFS = left footstrike; LTO = left toe-off; RFS = right footstrike; RTO = right toe-off.

Reprinted, by permission, from R.M. Enoka, D.I. Miller, E.M. Burgess, 1982, "Below-knee amputee running gait," *American Journal of Physical Medicine* 61: 78.

The amplitude and timing of the displacement vary as a function of running speed (Lusby & Atwater, 1983).

The relative changes in the shoulder and elbow angles can be represented in an angle-angle diagram (figure 4.9). Essentially, the pattern of displacement about the shoulder and elbow joints is confined to the upper right and lower left quadrants of the angle-angle diagram. The upper right quadrant represents concurrent shoulder and elbow flexion, whereas the lower quadrant indicates concurrent shoulder and elbow extension. The stance phase of the right leg (RFS to RTO) is mainly accompanied by concurrent flexion at the two joints. But the pattern is not one of a tight coupling of the two actions (i.e., flexion or extension), because there are instances when opposing motion occurs at the two joints. Can you see where this occurs in figure 4.9? One example is in the phase from LFS to LTO when the shoulder extends and the elbow flexes. The shoulder-elbow angle-angle diagrams, like

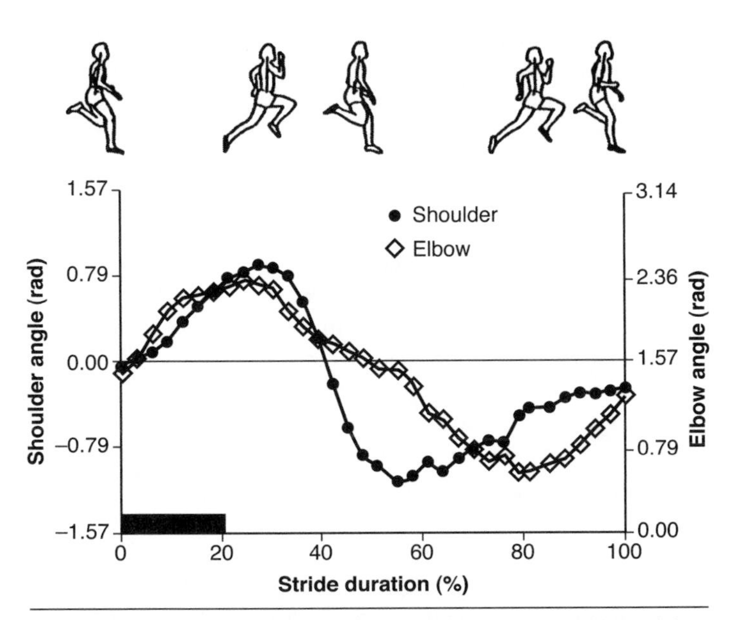

Figure 4.8 Displacement about the right shoulder and elbow joints during a sprinting stride. A shoulder angle of 0 rad indicates a vertical position of the upper arm; positive angles correspond to forward rotation (flexion). An elbow angle of 0 rad represents complete extension.

Reprinted, by permission, from C. Li and A.E. Atwater, 1984, "Temporal and kinematic analysis of arm motion in sprinters." Presented at the Olympic Scientific Congress, Eugene, OR.

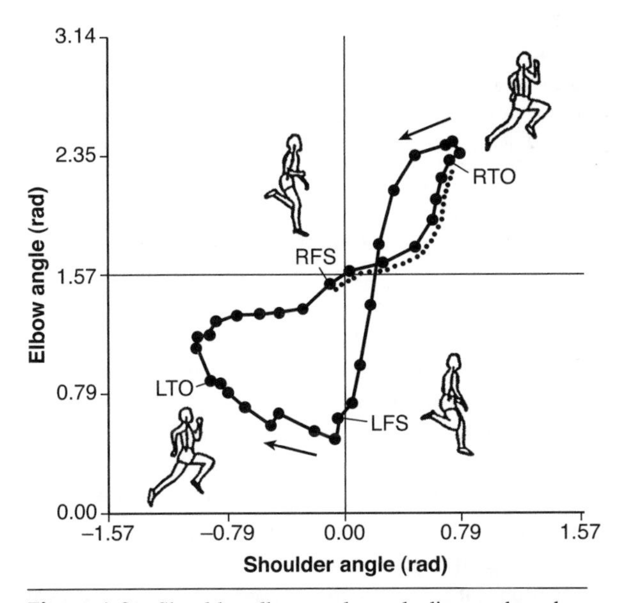

Figure 4.9 Shoulder-elbow angle-angle diagram based on the data shown in figure 4.8. LFS = left footstrike; LTO = left toe-off; RFS = right footstrike; RTO = right toe-off.

Reprinted, by permission, from C. Li and A.E. Atwater, 1984, "Temporal and kinematic analysis of arm motion in sprinters." Presented at the Olympic Scientific Congress, Eugene, OR.

those for the leg, provide a qualitative means to evaluate the coordination of the movement.

Another useful feature of these cyclic angle-angle diagrams is that, in addition to shape comparisons, the size of the diagram indicates the range of motion experienced at each joint during the event. For example, we would expect increases in stride length as a runner increases speed to be due to changes in the range of motion (amount of motion) at various lower extremity joints. Figure 4.10 confirms this expectation by showing that as speed increases (3.9 and 7.6 m/s), the amount of rotation of the thigh and about the knee joint increases; the larger angle-angle diagram represents the faster speed.

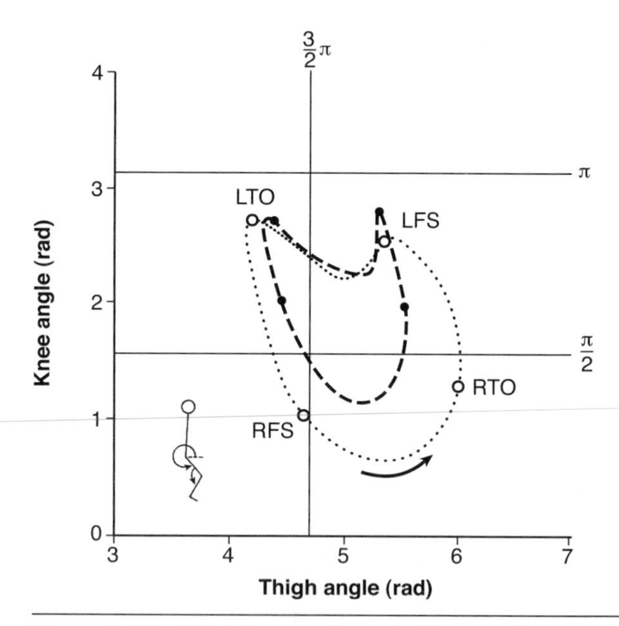

Figure 4.10 Thigh-knee angle-angle diagrams for a person running at 3.9 m/s (dashed line) and at 7.6 m/s (dotted line).

Data from Miller et al. 1979.

Ground Reaction Force

The ground reaction force represents the response of the ground (support surface) to the actions of the body segments. It represents the resultant effect of these actions and so corresponds to the acceleration experienced by the center of mass of the body.

As shown in figure 2.13, the $F_{g,y}$ curve has two peaks for walking but usually a single peak for running (Alexander, 1984). The reason for this difference is the action of the leg during the stance phase; it functions as a spring during running and as a rigid strut during walking (Farley & Ferris, 1998; Geyer et al., 2006; Saibene & Minetti, 2003; Segers et al., 2007). Although the center of mass is lowered and then elevated during the stance phase of running, it is accelerated in the direction of extension throughout the stance phase due to the continuous activation of the knee extensor muscles (quadriceps femoris) and the plantarflexor muscles (soleus and gastrocnemius) (Dietz et al., 1979; Mero & Komi, 1987). The sustained acceleration in the direction of extension results in a single-peaked curve for the ground reaction force.

In contrast to the knee flexion that occurs during the stance phase of running (figure 4.5), the knee joint flexes only slightly (figure 4.5) during the stance phase of walking. For simplicity, we can assume that the knee remains extended, which means that the leg is used as a rigid strut during the stance phase; this is known as the **inverted-pendulum model** of walking. As a consequence, the center of mass (approximately the hip joint) follows the arc of a circle during the stance phase as the leg rotates about the ankle joint, and it covers a greater range of motion compared with running (Alexander, 1984; Cavagna et al., 2002; Lee & Farley, 1998). When the hip joint is over the foot, therefore, the center of mass will be at its maximal height and the forward horizontal velocity will be at a minimum, but not zero (figure 4.11). This produces a minimum in the $F_{g,y}$ curve in the middle of the stance phase. Whereas the center of mass reaches a minimum vertical position near the middle of the stance phase in running, it reaches a maximum at about the same point during walking (Cavagna et al., 1976). This is another way to distinguish between the two gaits (Lee & Farley, 1998; McMahon et al., 1987).

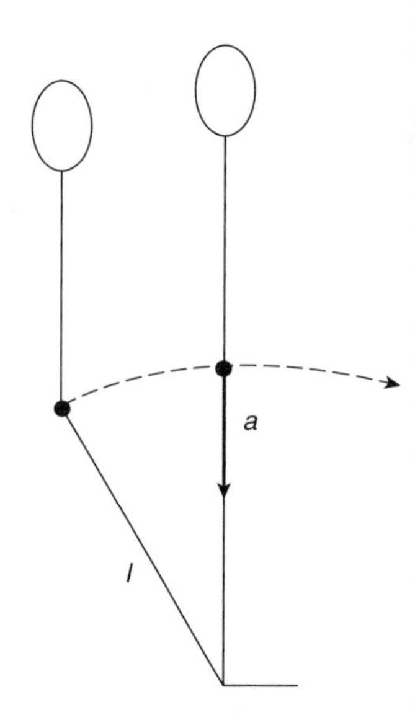

Figure 4.11 Path of the hip joint during the stance phase of walking, based on the simplification that the leg (l) remains straight.

Data from Alexander 1984.

Munro and colleagues (1987) characterized some of these changes in running gait as the speed of 20 men increased from 3.0 to 5.0 m/s. They found that, on average, the stance time decreased from 270 to 199 ms, the peak $F_{g,y}$ increased from 2.51 to 2.83 times body weight, and the average force during the stance phase increased from 1.4 to 1.7 times body weight. These results indicated that the increase in running speed from 3.0 to 5.0 m/s involved a greater $F_{g,y}$ applied over a shorter duration (figure 4.12), which is consistent with the finding that faster runners are capable of producing greater ground reaction forces (Weyand et al., 2000). The average force for both the braking and propulsion directions of $F_{g,x}$ also increased with running speed, although the *net* forward-backward impulse was zero because the running speeds were constant.

Given the association between stance duration and the peak value of $F_{g,y}$ (figure 4.12), Breit and Whalen (1997)

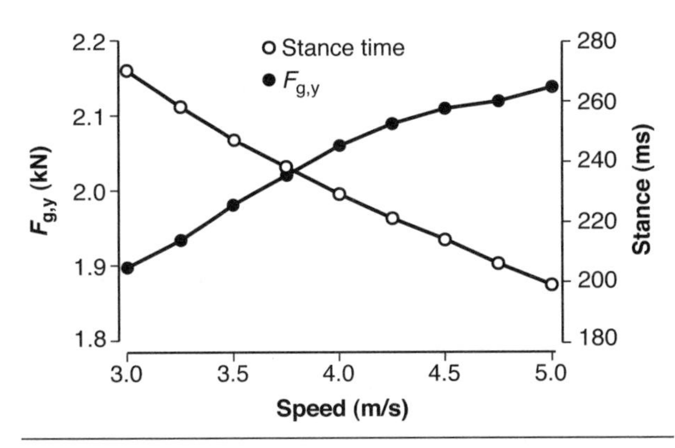

Figure 4.12 Change in the maximum $F_{g,y}$ and stance time with running speed.

Data from Munro et al. 1987.

examined the relation between temporal measures of foot-ground contact and the ground reaction force. For 218 walking steps and 199 running steps, they found that the magnitude of the ground reaction force components increased with gait speed. There was a strong association between the magnitude of the peak forces (vertical and forward-backward components) and the reciprocal of stance duration for both walking (0.91-2.34 m/s) and running (1.78-5.9 m/s). The best predictor of the peak vertical force ($F_{g,y}$), expressed relative to body weight (BW), was the ratio of stride duration (T) to stance duration (t_c). The regression equation for running was

$$\text{Peak } F_{g,y}(\text{BW}) = 0.856\frac{T}{t_c} + 0.089 \tag{4.1}$$

This relation can be used to estimate the daily loading history of the musculoskeletal system via measurement of durations rather than the ground reaction force.

EXAMPLE 4.1

Maximal Walking Speed

The inverted-pendulum model of leg function (figure 4.11) during walking provides a reasonable estimate of the maximal walking speed that a human can achieve (Alexander, 1984). Suppose that the center of mass has a constant linear velocity v and the length of the leg is denoted by l; the angular velocity of the leg can be obtained from equation 1.9:

$$\omega = \frac{v}{l}$$

Furthermore, the acceleration experienced by the hip joint (center of mass) during the stance phase is equal to the effect that causes the hip to move along the arc of a circle. This is the component of acceleration that accounts for the change in direction of the linear velocity vector, which produces equation 1.12:

$$a = \frac{v^2}{l}$$

From the peak vertical position of the hip joint (hip over the foot) through to toe-off, the maximal vertical acceleration experienced by the center of mass is that due to gravity (g). Thus,

$$g = \frac{v^2}{l}$$

$$v \leq \sqrt{gl}$$

For an adult with a leg length (hip to ground) of 0.85 m (McMahon et al., 1987), this equation suggests that the maximum walking speed is

$$v = \sqrt{9.81 \times 0.85}$$

$$v = 2.9 \text{ m/s}$$

This value is slightly faster than the acknowledged speed (2.0 m/s) when adults switch from a walk to a run. Furthermore, the model also explains why people with shorter legs, such as children, cannot walk as fast as adults.

Muscle Activity

The angular displacements experienced by the limbs during walking and running are controlled by the actions of muscles. The muscles that are activated during locomotion include those that control the limbs, those that stabilize the trunk, and those that maintain the orientation of the head (Davis & Vaughan, 1993; Patla, 1985; Winter & Yack, 1987). The amount of electromyographic (EMG) activity differs for walking and running and varies with gait speed. Figure 4.13 shows EMG recordings for 32 muscles at a range of walking (0.28-2.5 m/s) and running (1.39-3.33 m/s) speeds (Cappellini et al., 2006; Ivanenko et al., 2006*a, b*). The EMG traces were obtained from one side of the body and were averaged across eight subjects as they walked and ran on a treadmill.

Despite the involvement of many muscles in walking, statistical analysis indicates that the EMG activity of the 32 limb, trunk, and shoulder muscles on one side of the body can be explained by five basic components that are aligned with the following events during a stride (Ivanenko et al., 2006*a, b*): (1) braking impulse, (2) propulsion impulse, (3) trunk stabilization during double support, (4) toe-off, and (5) footstrike. Each muscle contributes to one or more of these components. When a similar analysis was performed on the EMG recordings of 32 muscles as subjects ran, the muscle activity again comprised

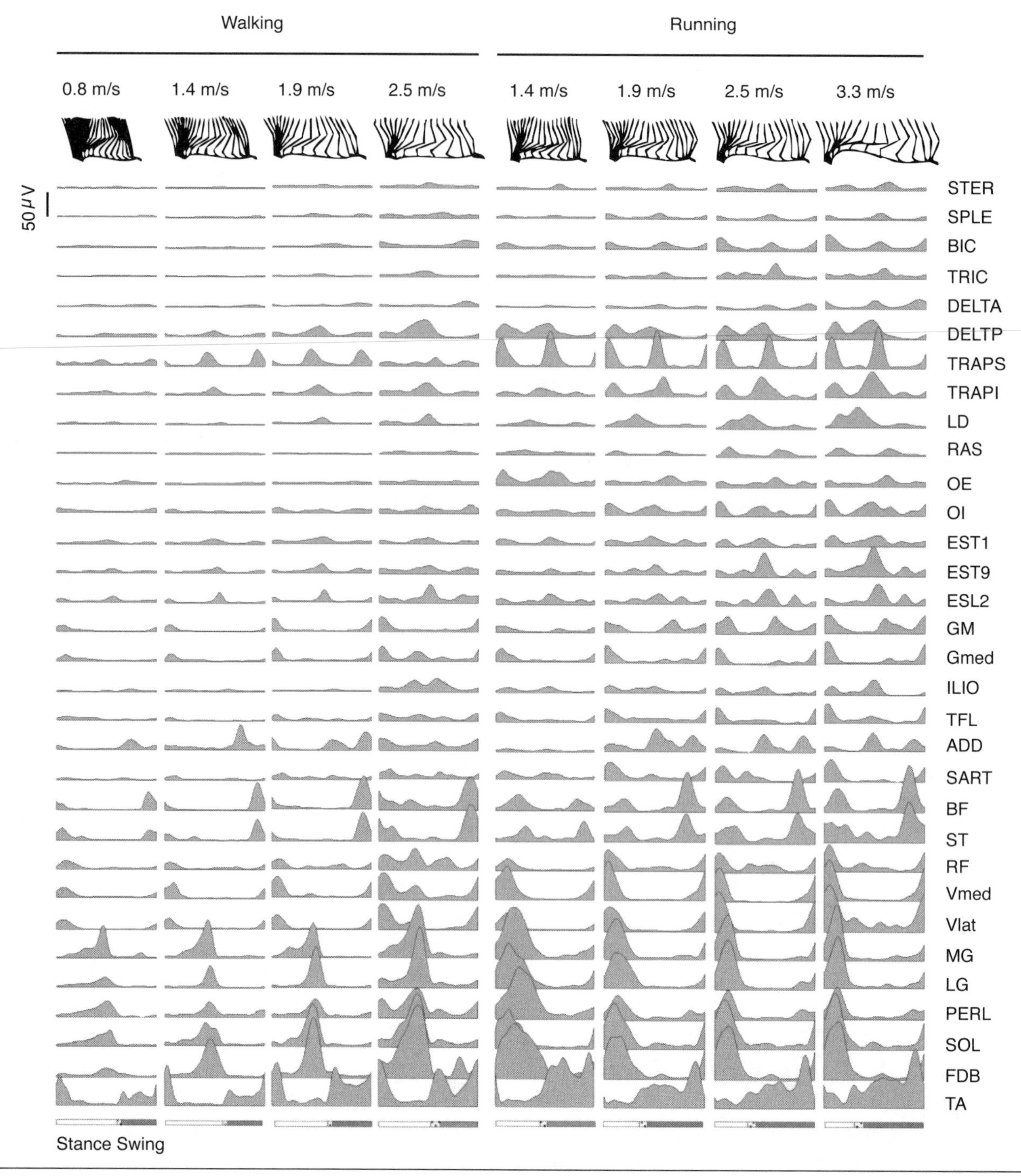

Figure 4.13 Average EMG activity in 32 muscles during walking and running at a range of speeds. Each trace begins with the stance phase and ends with the swing phase. Step duration is normalized across the columns to show the change in the duration of the stance and swing phases. STER = sternocleidomastoideus; SPLE = splenius; BIC = biceps brachii; TRIC = triceps brachii; DELTA = anterior deltoid; DELTP = posterior deltoid; TRAPS = superior trapezius; TRAPI = inferior trapezius; LD = latissimus dorsi; RAS = superior rectus abdominis; OE = external oblique; OI = internal oblique; EST1 = erector spinae at T1; EST9 = erector spinae at T9; ESL2 = erector spinae at L2; GM = gluteus maximus; Gmed = gluteus medius; ILIO = iliopsoas; TFL = tensor fascia latae; ADD = adductor longus; SART = sartorius; BF = biceps femoris; ST = semitendinosus; RF = rectus femoris; Vmed = vastus medialis; Vlat = vastus lateralis; MG = medial gastrocnemius; LG = lateral gastrocnemius; PERL = peroneus longus; SOL = soleus; FDB = flexor digitorum brevis; TA = tibialis anterior.

Reprinted, by permission, from G. Cappellini, Y.P. Ivanenko, R.E. Poppele, and F. Lacquaniti, 2006, "Motor patterns in human walking and running," *Journal of Neurophysiology* 95: 3431. Used with permission.

five basic components (Cappellini et al., 2006). The five components were similar to those observed during walking, except that the second component, which is aligned with the propulsion impulse, occurred earlier in the stance phase. For example, the EMG amplitude of muscles in the lower leg increased with walking speed and then occurred earlier in the gait cycle as the subjects changed from a walk to a run.

Furthermore, the five components were also evident in EMG recordings obtained with varying levels of body weight support (Ivanenko et al., 2004). However, the performance of stepping actions similar to walking while in a recumbent position involve some similarities but also some differences in EMG activity compared with walking (Stoloff et al., 2007). These results suggest that the motor program responsible for walking and running is linked to specific kinematic and kinetic events in the gait cycle, which ensures the appropriate timing of muscle activation.

Walk-Run Transition

On the basis of the inverted-pendulum model of walking (Alexander, 1984; Cavagna et al., 1977), the major force acting on the body during walking is gravity, which causes the center of mass to be accelerated downward (a in figure 4.11) throughout the stance phase (Chang et al., 2000). This provides a centripetal force (equation 2.1) that enables the center of mass to follow a circular path during the stance phase. This centripetal force is equal to body mass (m) times acceleration,

$$\text{Centripetal force} = \frac{mv^2}{l} \quad (4.2)$$

as described in Example 4.1. Thus, the maximal speed at which we can walk (v) and still maintain the circular path for the center of mass is limited by gravity:

$$\frac{mv^2}{l} \leq mg$$

The ratio of these two forces is defined as the **Froude number,** which is described as dimensionless speed (Kram et al., 1997).

$$\text{Froude number} = \frac{mv^2}{l} \bullet \frac{1}{mg} = \frac{v^2}{gl} \quad (4.3)$$

We cannot walk at Froude numbers greater than 1.0, because that would mean that the centripetal force exceeded the gravitational force. Interestingly, many bipeds, including humans and birds, prefer to switch from a walk to a run at a Froude number of ~0.5 (Gatesy & Biewener, 1991). This indicates that the different speeds (v) at which bipeds change from a walk to a run depend mainly on differences in leg length (l).

To test this simple model, Kram and colleagues (1997) determined the speed of the walk-run transition when they simulated a reduction in the acceleration due to gravity by partially supporting body weight with a suspension harness. Based on the definition of a Froude number (v^2/gl), the speed of the walk-run transition (v) should decrease if the acceleration due to gravity declines so that the Froude number remains constant at ~0.5. The results of the experiment indicated that the transition speed did decline with a reduction in the acceleration due to gravity (g), but the decrease was significant only down to g values of 0.4 (table 4.1). Kram and colleagues (1997) found that at the lowest g values, the acceleration due to the swinging arms and legs contributed significantly to the task of walking, so gravity was not the only force acting on an individual (Donelan & Kram, 2000). As a consequence, it appears that the inverted-pendulum model provides a reasonable explanation of the walk-run transition, although many other factors can also influence the transition (Biewener et al., 2004; Prilutsky & Gregor, 2001; Raynor et al., 2002).

For example, Neptune and Sasaki (2005) suggest that the transition from a walk to a run occurs when the contribution of the plantarflexor muscles, the major contributor to the ground reaction force, decreases below a critical value during the propulsion phase of stance. Although muscle activity increases with walking speed (Cappellini et al., 2006), the estimated gastrocnemius muscle force decreases, which causes the propulsion impulse to decline below a value necessary to sustain the speed of walking; hence the gait changes from a walk to a run. The result is a shift in the second component of EMG to earlier in the stance phase (figure 4.13). The decrease in the gastrocnemius muscle force is attributed to the decline in force that accompanies an increase in contraction speed, which

Table 4.1 Speed of the Walk-Run Transition for Variations in Acceleration Due to Gravity

Gravity (g)	Measured speed (m/s)	Froude number	Predicted speed (m/s)
1.0	1.98	0.45	1.98
0.8	1.84	0.49	1.77
0.6	1.65	0.53	1.54
0.5	1.55	0.56	1.40
0.4	1.39	0.56	1.25
0.2	1.18	0.83	0.89
0.1	0.97	1.13	0.63

Note. The predicted transition speed was based on a Froude number of 0.45 and a leg length (l) of 0.89 m.

Data from Kram et al. 1997.

suggests that the walk-run transition depends mostly on intrinsic muscle properties.

Leg Spring Stiffness

Running and hopping have been characterized as actions that enable an individual to move along the ground like a bouncing ball (Cavagna et al., 1964; Farley et al., 1998). To accomplish this action, the legs behave as springs, and this enables us to model the human body as a **spring-mass system** (Arampatzis et al., 1999; McMahon & Cheng, 1990; Seyfarth et al., 2002). The spring is compressed during the first half of the stance phase and rebounds during the second half (figure 4.14). Although the leg spring model gets shorter during the stance phase, the actual anatomical springs (i.e., tendons) in fact get longer. The mechanical characteristics of the spring are expressed as leg stiffness (example 4.2). The magnitude of leg stiffness influences the duration of the stance phase and the vertical displacement of the center of mass during the stance phase. This capability enables a person to use a range of stride rates and stride lengths to run at a given speed (Farley & Gonzalez, 1996). Furthermore, subjects tend to change leg stiffness when running on surfaces of different stiffness (e.g., concrete floor vs. rubber mats) so that the total vertical stiffness (leg stiffness + surface stiffness) remains constant (Moritz & Farley, 2006; Moritz et al., 2004). Regulation of the total vertical stiffness enables a person to use similar running mechanics on different surfaces, such as constant stance phase duration, stride rate, and vertical displacement of the center of mass (Ferris et al., 1998, 1999; Kerdok et al., 2002; Kuitunen et al., 2002). Furthermore, the application of an elastic ankle-foot orthosis that supplements ankle stiffness enables an individual to reduce the leg stiffness that must be provided by muscle activity (Ferris et al., 2006).

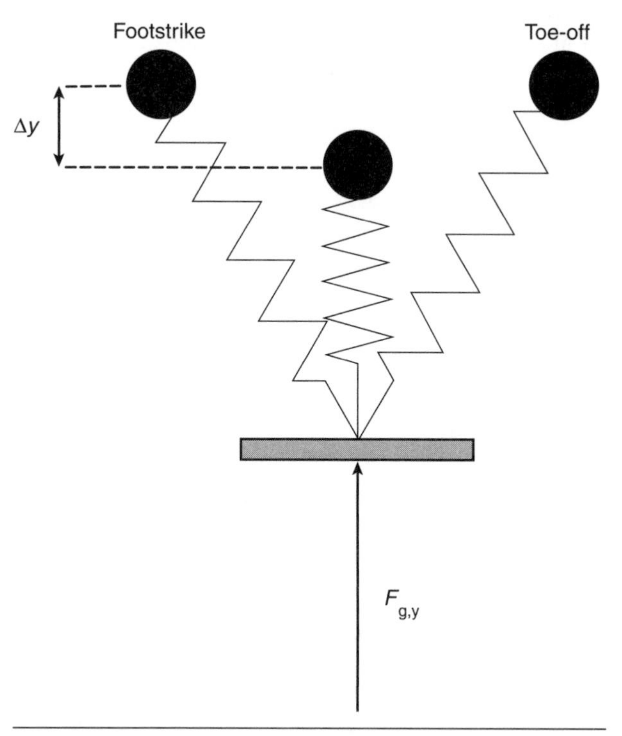

Figure 4.14 A spring-mass model of the human body at three points during the stance phase of running. The large filled circle represents body mass; the spring indicates the leg. Compression of the leg spring is indicated by the vertical displacement (Δy) of body mass.

Adapted from *Journal of Biomechanics*, D.P. Ferris, L. Kailine, and C.T. Farley, "Runners adjust leg stiffness for their first step on a new running surface," pg. 790, copyright 1999, with permission from Elsevier.

Leg stiffness in running depends on the level of muscle activity and the geometry of the leg when the foot contacts the ground. The effect of varying muscle activity is to alter the stiffness at a joint. Previously, we have discussed linear springs, such as tendons, ligaments, and therapeutic bands. There are also angular springs, which are known as **torsional springs.** A common example of a torsional spring is a snapping mousetrap. These springs resist angular displacement because of the property of angular stiffness (κ). For example, when muscles perform lengthening contractions, they resist the angular displacement caused by a load and therefore act as torsional springs. This occurs during the first half of the stance phase of running when the leg spring is compressed and the leg extensor muscles (ankle, knee, and hip) perform lengthening contractions. Because of this relation, leg stiffness depends on the quantity of muscle activity about each joint, which we can determine by performing a dynamic analysis (chapter 3) to calculate the resultant muscle torque (Arampatzis et al., 1999; Kuitunen et al., 2002).

The other factor that influences leg stiffness in running is the orientation of the leg at footstrike. By varying leg geometry, this alters the perpendicular distance (moment arm) from the line of action of the ground reaction force vector to each of the major joints (McMahon et al., 1987). The product of the ground reaction force and moment arm corresponds to the load that the muscles must counteract (Glitsch & Baumann, 1997). Figure 4.15 shows that increasing knee flexion by 0.35 rad (20°) at footstrike results in a markedly different load torque about the knee and hip. For a given amount of muscle activity, for example, a greater load torque (ground reaction force × moment arm) will produce a greater angular displacement. This occurs during Groucho running, which involves an exaggerated knee flexion posture (Grasso et al., 2000; McMahon et al., 1987).

To examine the relative contributions of muscle activity and leg geometry to variation in leg stiffness, Farley and colleagues (1998) studied humans hopping on surfaces of different stiffness. They characterized the behavior of the leg spring by plotting the vertical compo-

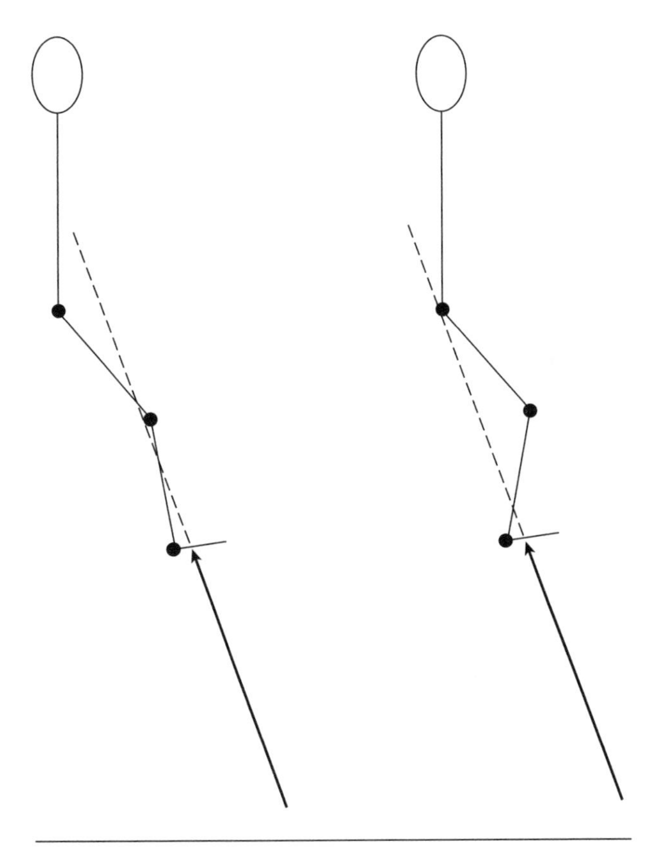

Figure 4.15 Changes in leg geometry at footstrike alter the load torque about each joint.

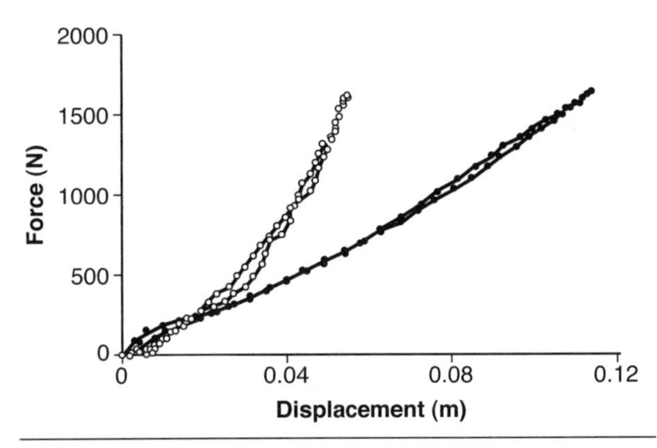

Figure 4.16 The change in leg length and the vertical component of the ground reaction force for a subject hopping in place on two surfaces with different stiffnesses (filled circles = stiffest surface; open circles = least stiff surface).

Data from Farley et al. 1998.

nent of the ground reaction force ($F_{g,y}$) against the change in leg length (Δl; figure 4.16). The bottom left corner of the graph (figure 4.16) represents the moment the foot contacted the ground, after which the length of the leg spring decreased and $F_{g,y}$ increased. Peak $F_{g,y}$ occurred when the leg spring was maximally compressed. The slope of this relation (Δl and $F_{g,y}$) indicates the stiffness of the leg spring (N/m). The stiffness of the leg spring was least when subjects hopped on the stiffest surface. For the subject shown in figure 4.16, leg stiffness went from 14.3 kN/m on the stiffest surface to 29.4 kN/m on the least stiff surface.

To determine how subjects changed leg stiffness, Farley and colleagues (1998) calculated joint stiffness when the subjects hopped on the various surfaces. **Joint stiffness** was determined as the ratio of the change in resultant muscle torque to the angular displacement (N•m/rad):

$$\kappa = \frac{\Delta\tau_m}{\Delta\theta} \tag{4.4}$$

On the basis of experimental measurements and a computational model, Farley and colleagues (1998) found that the stiffness of the ankle joint had the greatest effect on leg stiffness for this task. Ankle stiffness changed from 396 N•m/rad on the stiffest surface to 687 N•m/rad on the least stiff surface, accounting for 75% of the change in leg stiffness. Farley and colleagues also found that subjects flexed the knee by 0.16 rad when going from the least to the most stiff surface; this changed the moment arm for $F_{g,y}$ relative to the knee joint from 0.001 m to 0.054 m. This change in leg geometry caused the leg stiffness to change from 17.1 to 22.2 kN/m. Of these two factors, changes in the stiffness of the ankle joint had the greater effect on the variation in leg stiffness, at least for hopping in place. The role of leg geometry may be more significant in running, where it appears that stiffness about the knee joint has the greatest effect on changes in leg stiffness (Arampatzis et al., 1999).

EXAMPLE 4.2

Spring Stiffness of the Leg

The calculation of leg stiffness for such studies as those shown in figure 4.16 involves the application of rules from geometry, trigonometry, and calculus to a few experimental measurements (Ferris et al., 1998). The required measurements include leg length (l; distance from the greater trochanter to the point of foot contact with the ground), the vertical component of the ground reaction force ($F_{g,y}$), and the speed of the runner (v). From these data, it is possible to determine leg stiffness (k_{leg}) from equation 3.2:

$$k_{leg} = \frac{\text{Peak } F_{g,y}}{\Delta l} \tag{4.5}$$

where peak $F_{g,y}$ refers to the maximal amplitude of the vertical component of the ground reaction force and Δl indicates the change in leg length from footstrike to the middle of the stance phase, which corresponds to the extent to which the spring is compressed

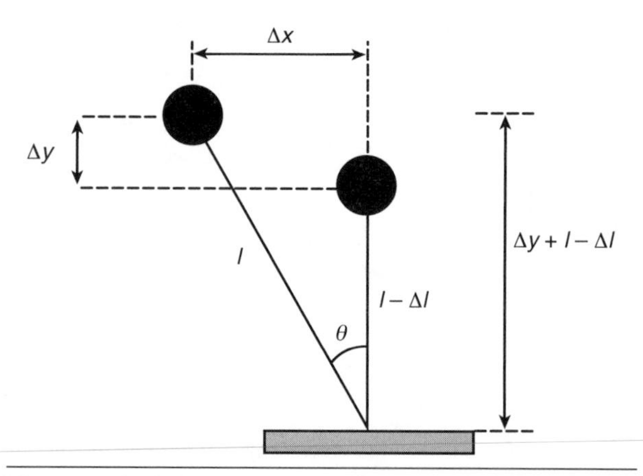

Figure 4.17 Model of leg spring compression during the stance phase of running.

(figure 4.17). Calculation of k_{leg} requires that we know l, Δl, Θ, Δy, and Δx.

The length of the leg in the midstance position is equal to leg length (l) minus the length the leg has shortened due to flexion at the ankle, knee, and hip joints (Δl). The angle (θ) represents the angular displacement of the leg about the ankle joint from footstrike through to midstance, and Δy indicates the vertical displacement of the center of mass. From trigonometry we have

$$\cos\theta = \frac{\Delta y + l - \Delta l}{l}$$

$$\Delta l = \Delta y + l - l\cos\theta$$

This expression is for the simplest case, when an individual runs on a rigid surface; if the surface is not rigid, then an additional Δy term must be added to account for the compliance of the surface (Ferris et al., 1998). The vertical displacement of the center of mass (Δy) can be obtained by double integration (calculus) of the vertical acceleration of the center of mass, which was obtained from $F_{g,y}$ after subtraction of the subject's weight and division by body mass.

θ can be determined from

$$\theta = \sin^{-1}\frac{\Delta x}{l}$$

The horizontal displacement of the center of mass (Δx) can be calculated with equation 1.1. This involves multiplying the average horizontal velocity (v) by one-half of the stance phase duration, that is, the time it took for the leg to rotate from footstrike through to midstance. The duration of the stance phase can be measured from the ground reaction force record. With these procedures, it is possible to determine the average spring stiffness of the leg during the stance phase of running.

Energy Fluctuations

When most of us walk or run, we pay little attention to either the length or the rate of the strides that we use to achieve a particular speed. If we do experiments with different combinations of stride length and rate, it becomes obvious that there is one combination that seems to require the least effort to achieve the desired speed. To assess the validity of this perception, biomechanists determine the energy costs needed to perform the work of walking and running at different speeds.

Most of the work done during walking and running is used to displace the center of mass in the vertical (U_v) and forward (U_f) directions (Cavagna et al., 1976). These two components are sometimes referred to as **external work,** to distinguish them from the work (internal) done to move the limbs (Donelan et al., 2002; Fenn, 1930; Saibene & Minetti, 2003). The mechanical energy used to perform external work is derived from the kinetic (E_k) and potential ($E_{p,g}$) energy of the center of mass. In the vertical direction, the amount of work done (U_v) depends on the change in both $E_{k,v}$ and $E_{p,g}$. In the forward direction, however, the amount of work done (U_f) depends only on the change in $E_{k,f}$. Because the fluctuations in $E_{k,v}$ are so small compared with those of the other two energy terms, it is often neglected. Thus the total mechanical energy of the center of mass (E_{cm}) is mainly due to the sum of $E_{p,g}$ and $E_{k,f}$ (figure 4.18).

One of the features that distinguishes walking from running is the trajectory of the center of mass during the stance phase. The vertical position of the center of mass reaches a maximum in midstance during walking but a minimum at the same point during running. This distinction influences the contributions of the energy fluctuations to the external work done on the center of mass. When the knee is kept extended and the leg used as a strut during walking, $E_{p,g}$ reaches a maximum and $E_{k,f}$ achieves a minimum at midstance (Cavagna & Franzetti, 1986). The fluctuations in $E_{p,g}$ and $E_{k,f}$ are inversely related (figure 4.18), and, as with the exchange of energy in example 2.17, some of the change in $E_{p,g}$ results from the forward motion of the body ($E_{k,f}$), and some of the change in $E_{k,f}$ is produced by changes in the vertical position of the center of mass ($E_{p,g}$). Because of this interaction, human walking has been described as an inverted pendulum in which the center of mass rises to a peak during the stance phase and then falls forward (Cavagna et al., 1976, 2000).

The amount of energy that is exchanged between potential and kinetic can be quantified as the percent of energy that is recovered (% recovery). To calculate this we subtract the external work (U_e) from the work done in the vertical and forward directions (Cavagna et al., 1976):

$$\%\ \text{recovery} = \frac{U_v + U_f - U_e}{U_v + U_f} \bullet 100 \qquad (4.6)$$

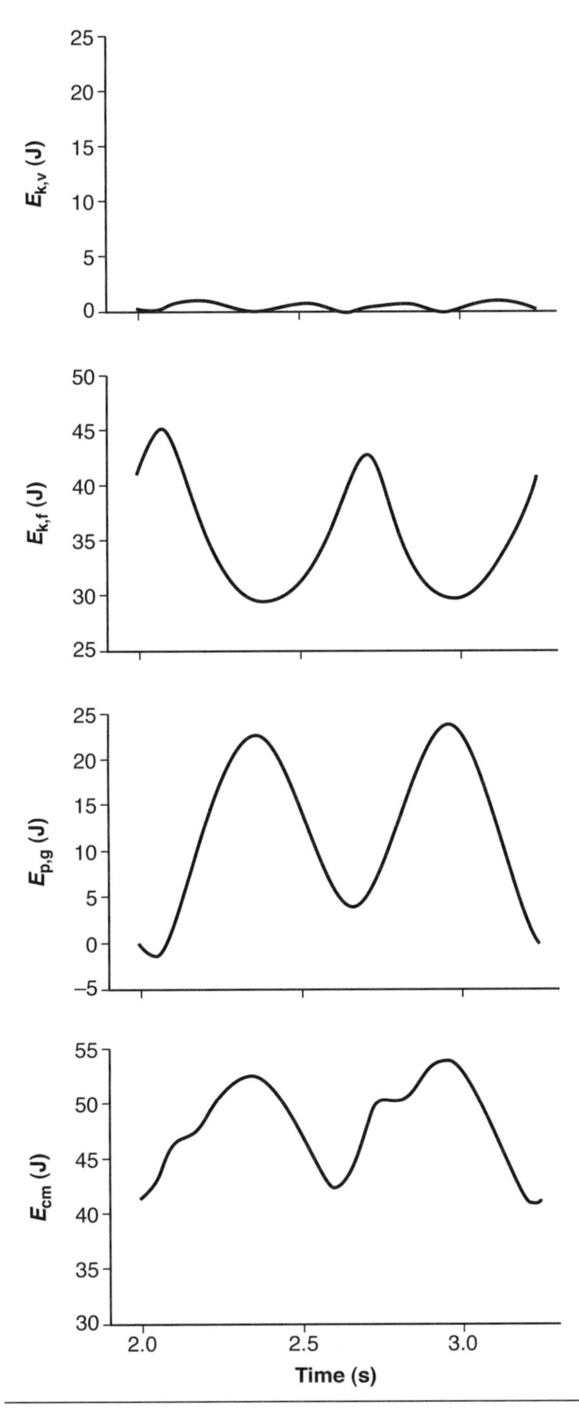

Figure 4.18 Fluctuations in mechanical energy of the center of mass within one stride (two support phases) during walking. Note that the gravitational potential energy ($E_{p,g}$) reaches a minimum when the forward kinetic energy ($E_{k,f}$) is at a maximum, resulting in a sum (E_{cm}) that fluctuates less than the individual components. Fluctuations in E_{cm} indicate the external work done on the center of mass. $E_{k,v}$ denotes the kinetic energy in the vertical direction.

Data from Griffin et al. 1999.

The amount of mechanical energy recovered during walking at intermediate speeds is about 65%, compared with <5% in running (Cavagna et al., 1976). The % recovery is greatest when U_v and U_f are similar, which occurs at intermediate walking speeds. At lower ($U_v > U_f$) and higher ($U_v < U_f$) speeds, the external work increases and the % recovery declines. Similarly, when reductions in gravity are simulated by means of partial support of body weight, the % recovery decreases as the amount of body weight support increases, and the speed at which maximal recovery occurs declines with an increase in support (Bastien et al., 2005; Griffin et al., 1999).

EXAMPLE 4.3

Ground Reaction Force and Energy Fluctuations

The energy fluctuations during walking and running can be calculated based on the measurement of the ground reaction force during the stance phase (Cavagna, 1975). We know from chapter 2 that the necessary equations are

$$E_{p,g} = mgh$$

$$E_{k,f} = \frac{1}{2}mv_f^{\,2}$$

For the fluctuations in gravitational potential energy ($E_{p,g}$), we need to obtain the mass of the subject (m), the acceleration due to gravity (g), and the position of the subject's center of mass (h) throughout the stance phase. For the fluctuations in forward kinetic energy ($E_{k,f}$), we need m and the forward velocity (v_f) of the center of mass during the stance phase.

From Newton's law of acceleration ($F = ma$), we know that the acceleration of the center of mass depends on the mass of the person and the forces acting on the body; that is,

$$a = \frac{F}{m}$$

In the forward-backward direction during walking and running, the only significant force acting in this direction is the ground reaction force ($F_{g,x}$). As a result, we can obtain the forward velocity of the center of mass in this direction (v_x) by integrating the forward-backward acceleration:

$$v_x(t) = \int \frac{F_{g,x}}{m}\,dt + c \qquad (4.7)$$

where c represents the integration constant, which is the average velocity of the person's center of mass, and the limits of integration are the beginning and end of the stance phase. Once the velocity of the center of mass has been determined in the forward-backward direction, the kinetic energy of the center of mass in this direction can be calculated as a function of time:

$$E_{k,f}(t) = \frac{m}{2}[v_x(t)]^2 \quad (4.8)$$

Similarly, the vertical position of the center of mass (h) can be obtained from the double integration of acceleration in the vertical direction. Because there are two significant forces in the vertical direction, the term for acceleration is a little more involved:

$$v_y(t) = \int \frac{F_{g,y} - F_w}{m} dt + c_1 \quad (4.9)$$

In this instance, c_1 is set so that the average vertical velocity is zero over one stride. Variations in vertical height (h) are obtained from the integration of $v_y(t)$.

$$h(t) = \int v_y \, dt + c_2 \quad (4.10)$$

and c_2 is set so that the average displacement is zero. Fluctuations in the potential energy due to gravity ($E_{p,g}$) of the center of mass are calculated from

$$E_{p,g}(t) = mgh(t) \quad (4.11)$$

These procedures are shown graphically in figure 4.19.

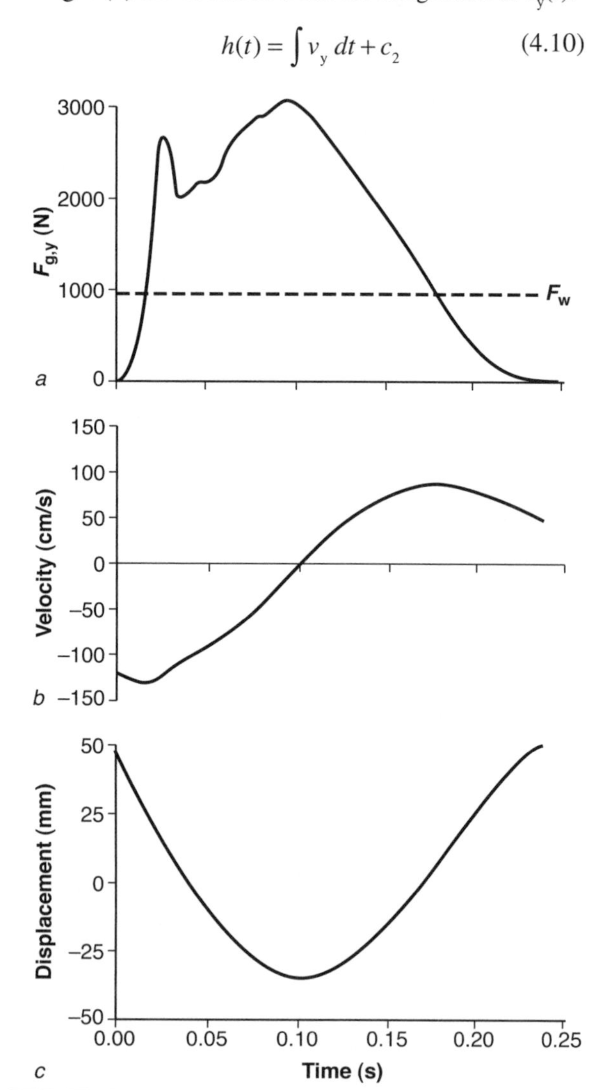

Figure 4.19 Double integration of the forces ($F_{g,y}$ and F_w) acting in the vertical direction during the *(a)* stance phase of running to yield *(b)* the vertical velocity of the center of mass and then *(c)* the vertical position of the center of mass.

Energy Cost

If we run a short distance several times at various speeds, it is evident that the rate at which we consume energy increases as we run faster. This is obvious by the increases in heart rate, ventilation rate, and the rate of oxygen consumption when we run faster. The increase in the rate of energy expenditure appears to be mainly due to the increase in the intensity of muscle activity that is needed to run faster (Griffin et al., 2003; Wright & Weyand, 2001). In general, the muscle activity during walking and running serves two major functions: to support body weight and to generate a propulsive impulse (Kram, 2000). Because the magnitude of the vertical component of the ground reaction force is much greater than the forward-backward component (figure 2.14), most of the metabolic energy we expend during running is used to support body weight.

Three types of experiments have been performed to determine the contributions of body weight support to the metabolic energy needs during locomotion. These experiments have included measuring the metabolic cost of locomotion (rate of oxygen consumption) when extra mass is added, when gravity is reduced, and when species of different size locomote at various speeds (Farley & McMahon, 1992; Kram & Taylor, 1990; Taylor et al., 1980). The general conclusion from these studies is that the magnitude of the vertical component of the ground reaction is a major determinant of the metabolic cost during running at a constant speed. For example, when Farley and McMahon (1992) simulated a reduction in gravity by providing partial support of body weight, they found a linear relation between the metabolic cost of running ($\leq$3 m/s) and the average amplitude of the vertical component of the ground reaction force.

Similarly, the amount of time that the leg muscles have to generate the necessary force seems to be directly related to the metabolic cost of locomotion. Kram and Taylor (1990) measured the rate of oxygen consumption and stance time when various species ran at a range of steady-state speeds. The animals ranged from kangaroo rats (32 g) to ponies (141 kg). The investigators found, as expected, that the rate of oxygen consumption increased with running speed. Furthermore, as we noted previously with humans (figure 4.12), stance time decreased with running speed. The ratio of these two measures, which represents a **cost coefficient,** was nearly constant across speed. This meant that much of the increase in metabolic cost was explained by the decrease in stance time—that is, the intensity of muscle activity.

The energy cost of running varies with conditions in which the movement is performed, such as the presence of tail and head winds, the slope of the terrain, and the stiffness of the support surface. Chang and Kram (1999) examined the influence of tail and head winds by applying horizontal forces to treadmill runners that aided or impeded the task of running at a constant speed (3.3 m/s). The effect of this intervention was to modify the relative magnitudes of the braking and propulsive impulses in the forward-backward direction. The force that impeded running was similar to a head wind, which caused an increase in the propulsive impulse (figure 4.20). Conversely, a force that aided the runner was similar to a tail wind and produced an increase in the braking impulse. By comparing the rate of oxygen consumption with changes in the propulsive impulse, Chang and Kram (1999) concluded that the muscle activity associated with generating the propulsive impulse accounts for about 30% of the metabolic energy we use when running at slow speeds.

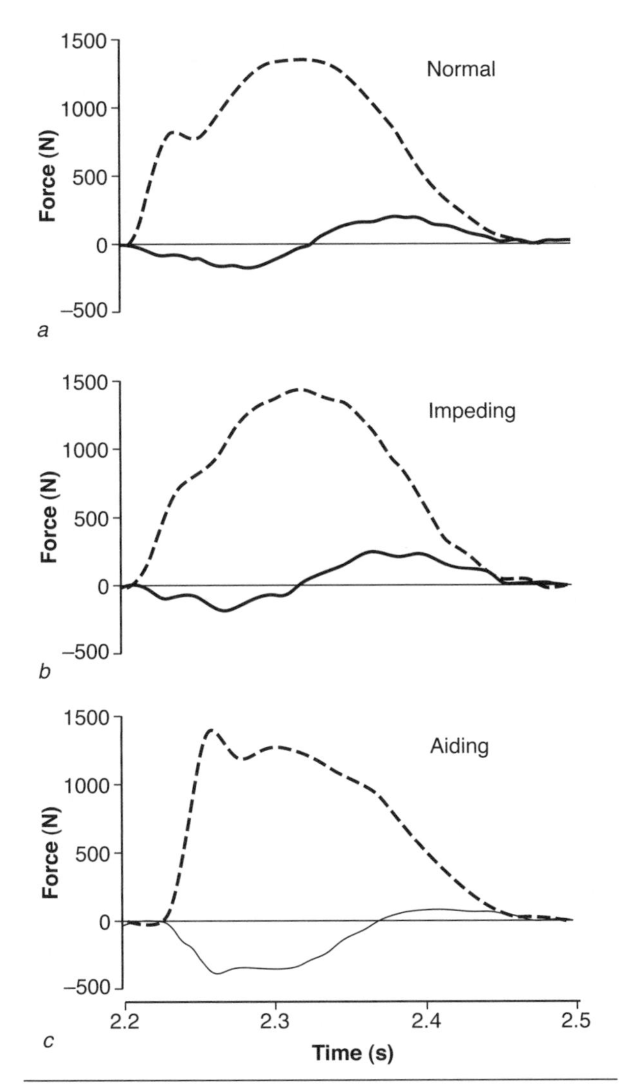

Figure 4.20 Vertical (dotted line) and forward-backward (solid line) components of the ground reaction force during running at 3.3 m/s *(a)* under normal conditions, *(b)* with an impeding horizontal force, and *(c)* with an aiding horizontal force. The magnitudes of the braking (negative) and propulsion (positive) impulses change with the presence of applied horizontal forces.

Data from Chang and Kram 1999.

The slope and stiffness of the terrain also influence the energy cost of running. Running on a slope requires changes in both the vertical and forward-backward components of the ground reaction force (Gottschall & Kram, 2005). During running downhill, the peak vertical component occurs earlier in the stance phase and the forward-backward component is dominated by the braking impulse; during running uphill, the converse occurs, with a later peak in the vertical component and an increase in the propulsion impulse. As a result of these adjustments, the most economical gradient is about –10% during running at 3 to 4 m/s, where the energy cost is 3.1 $J/kg \cdot m^{-1}$ (Saibene & Minetti, 2003). The increased metabolic cost of running uphill is related to the cost of elevating the center of mass; about two-thirds of the increase in energy expenditure is due to the greater requirement by the muscles that are active during the stance phase, and the other one-third is used by the muscles that control the swing phase (Gabaldon et al., 2004; Rubenson et al., 2006). Similarly, a decrease in surface stiffness from 946 to 75 kN/m was accompanied by a 12% decrease in the metabolic cost of running at 3.7 m/s (Kerdok et al., 2002). The runners increased leg stiffness by 29% on the stiffest surface, which presumably enabled them to substitute some rebound energy from the compliant surface to the metabolic cost of sustaining the running speed.

Due to differences in leg configuration relative to the ground reaction force vector during the stance phase, the metabolic cost of walking is less than that of running (Biewener et al., 2004; Ortega & Farley, 2005). In addition to the need for energy to support body weight and to swing the limbs (Ellerby et al., 2005), there is a cost in walking associated with the transition between steps. According to the inverted-pendulum model of walking, the center of mass rotates up and over the foot during the stance phase with an arc-like trajectory. Consecutive steps, therefore, require that the trajectory be changed from downward to upward at the transition between steps, which incurs a significant metabolic cost (Donelan et al., 2002; Grabowski et al., 2005). Furthermore, the side-to-side placement of the feet during walking provides lateral stability that involves a metabolic cost (Donelan et al., 2004). The dominant energetic cost during walking is related to the work performed on the center of mass.

Gait Disorders

Because orthopedic and neurological impairments are often expressed as movement disturbances before they can be detected by the physical signs and symptoms obtained in a clinical examination, many institutions have established movement disorders clinics to assist with the management of patients (Hallett et al., 1994). A central feature of such clinics is the biomechanical assessment of gait, often including the measurement of whole-body kinematics, the recording of EMG activity from leg muscles, and the evaluation of contact forces between the feet and the ground. The performance of patients can be compared with that of healthy persons, and inferences can be made about the locus of the impairment in the motor system (Dietz, 1997).

Much is known about the control of locomotion by the nervous system (Ivanenko et al., 2006*a;* Rossignol et al., 2006), and this facilitates the identification of sites that might be responsible for a movement disorder. Neuronal circuits located in the spinal cord generate the essential rhythm of locomotion. The rhythm comprises alternating activation of the flexor and extensor muscles of the leg. The output generated by these circuits specifies the timing and intensity of the muscle activity required for the movement. These stereotypic patterns of muscle activity are modulated by information that is sent back to the spinal cord from peripheral sensory receptors to ensure that the activation pattern can accommodate variations in the surroundings. Furthermore, descending commands from more rostral parts of the nervous system (e.g., brain stem, cerebellum, and motor cortex) can modify both the stereotypic patterns and the sensory feedback.

Studies on patients with Parkinson's disease, spasticity, and paraplegia illustrate the utility of examining their ability to walk (Dietz, 1997). Parkinson's disease comprises an impairment of the basal ganglia, a group of nuclei involved in planning movements that project to locomotor centers in the brain stem. The function of the basal ganglia is impaired in Parkinson's disease because of the depletion of a neurotransmitter (dopamine). One of the earliest signs of the disease is a decrease in the speed of locomotion and a reduction in the range of motion about the joints of the legs. The result is a rigid and poorly modulated gait that comprises small, shuffling steps. From these signs, it is evident that the dysfunction involves a problem with the regulation by descending commands and a reduced modulation by sensory feedback that can be manifested as an inadequate activation of the muscles required for locomotion (Albani et al., 2003; Mitoma et al., 2000). Not only can gait studies be used as a diagnostic tool; they can also be used to monitor the effectiveness of various therapeutic interventions. For patients with Parkinson's disease, gait studies can provide an evaluation of L-dopa medication (Blin et al., 1991), which facilitates the replacement of the depleted neurotransmitter.

Spasticity is a disorder that results from a lesion in the brain or spinal cord. The symptoms of spasticity include exaggerated reflexes, clonus, and muscle hypertonia. Clonus involves the repeated rapid contraction and relaxation of a passively stretched muscle, whereas muscle hypertonia refers to the reflex resistance of a muscle to stretch, which varies with the speed of the stretch. The physical signs of spasticity are largely unrelated to a patient's disability, which is a movement disorder (Dietz, 1997; Perry, 1993). For example, studies on functional limb movements have shown no relation between the exaggerated reflexes and the movement disorder (Berger et al., 1984; Powers et al., 1989). In walking, patients with spasticity exhibit a lower amplitude and reduced modulation of activation in the calf muscles. The stereotypic pattern of muscle activation required for locomotion appears to be intact in these patients, whereas the modulation by sensory feedback is impaired (Damiano et al., 2006; Morita et al., 2001). Therapy should focus on training the residual motor functions and preventing secondary complications, such as muscle spasms (Dietz, 1997; Young, 1994). Furthermore, surgical interventions should be based on the outcome of gait analyses rather than clinical exams (Granata et al., 2000; Perry, 1993; Remy-Neris et al., 2003).

A final example involves locomotor training of patients with paraplegia due to partial or complete spinal cord transections. In these patients, the connections between the rostral centers of the nervous system and the generator circuits in the spinal cord are disrupted. Nonetheless, with appropriate training, patients with paraplegia are often able to improve muscle activation during assisted locomotion (Behrman et al., 2006; Dietz & Harkema, 2004; Hicks et al., 2005; Wirz et al., 2005). The typical strategy involves suspending the patient in a harness over a treadmill so that the legs do not have to support the entire weight of the body. Over the course of several months, patients with paraplegia can learn to modulate the timing of activity in leg muscles to resemble that of healthy individuals, although the amplitude remains reduced and the motor strategy tends to differ from that used by individuals without disabilities (Grasso et al., 2004). Patients with an incomplete paraplegia can learn to perform unsupported stepping movements on the ground (Field-Fote et al., 2005). The benefits experienced by patients with complete paraplegia include an enhancement of cardiovascular function and reduced symptoms of spasticity. Such training appears to be a necessary adjunct to developing therapies, including pharmacologic interventions and tissue implants (Dietz & Muller, 2004; Edgerton et al., 2004).

JUMPING

A jump is a movement that causes the center of mass to be projected upward and the feet of the performer to leave the ground. Jumps can be organized to achieve several different goals, such as the maximal height that the hands can reach, the maximal horizontal distance that can be covered, the maximal height to which the center of mass can be raised, and the maximal time that can be spent off the ground. These goals correspond to such actions as rebounding in basketball; the long jump in athletics and the standing broad jump; the high jump in athletics; and performances in gymnastics, diving, and dance. In terms of biomechanics, these movements can be categorized into those that seek to maximize the vertical velocity of the center of mass at takeoff, those designed to maximize the horizontal distance, and those that involve somersaults and twists during the flight phase. As examples of these movements, we will examine the vertical jump, the long jump, and springboard diving.

The Vertical Jump

When a person is asked to jump and reach as high as possible with the hands, the typical strategy involves a technique called the **countermovement jump.** The person begins from an upright erect position and then performs a small-amplitude downward movement that involves flexion at the hip, knee, and ankle; this is followed by a rapid extension of the legs and a forward and upward rotation (flexion) of the arms about the shoulders. This strategy is called a countermovement jump because it begins with an initial movement in the opposite direction; that is, the jumper first moves downward even though the goal is to maximize the upward vertical velocity at takeoff.

The kinematics of the jumper's center of mass during a countermovement jump are shown in the column on the left in figure 4.21. The upward and downward

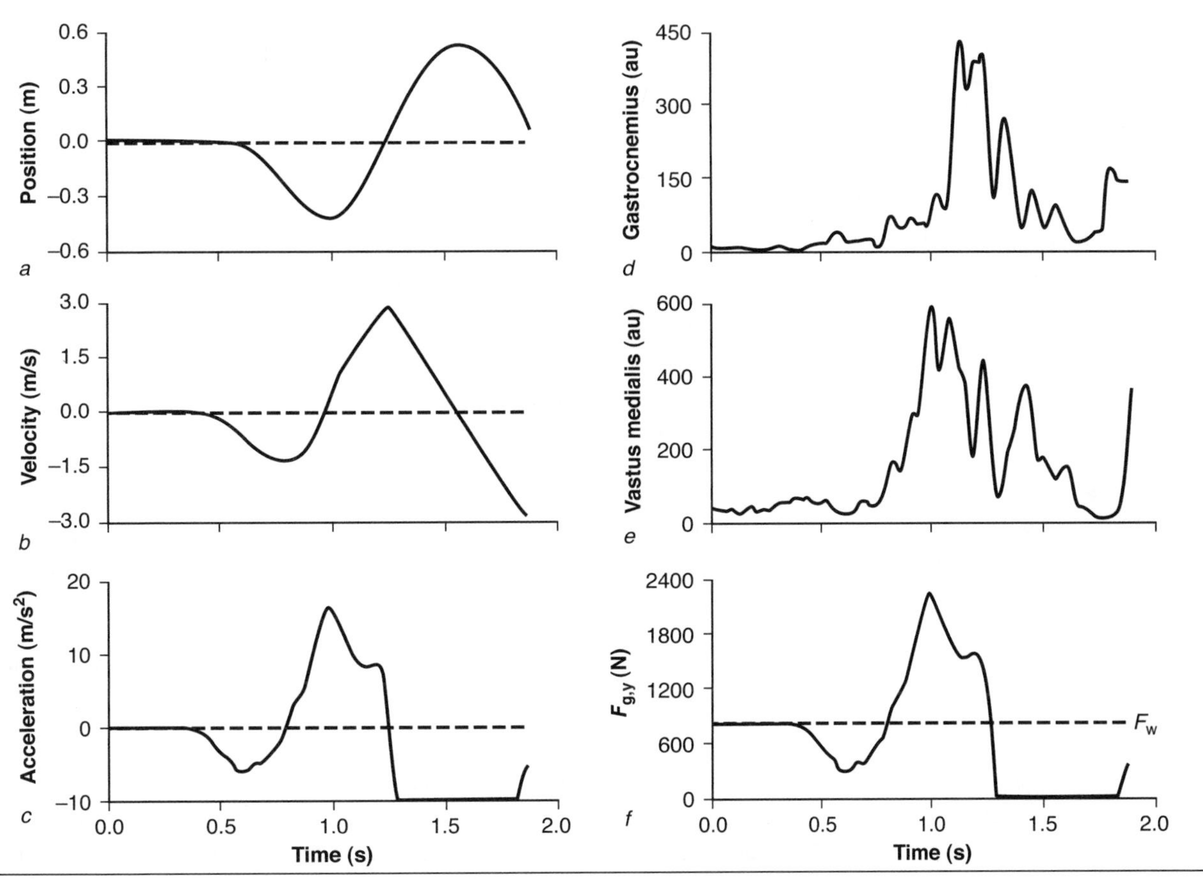

Figure 4.21 Kinematics, EMG, and ground reaction force for the vertical jump. The three panels on the left indicate the *(a)* position, *(b)* velocity, and *(c)* acceleration of the center of mass during a countermovement jump performed by an elite athlete. The top two panels on the right show the EMG of the *(d)* gastrocnemius and *(e)* vastus medialis muscles, which was measured in arbitrary units (au). *(f)* The bottom panel on the right represents the vertical component of the ground reaction force.

Data provided by Marten F. Bobbert, Ph.D.

displacements of the center of mass are obvious in the velocity–time graph (figure 4.21*b*). As we discussed in chapter 1, a negative velocity indicates a downward displacement of the center of mass whereas a positive velocity represents an upward displacement. Accordingly, the velocity–time graph shows that the center of mass experiences a downward-upward-downward pattern of displacement during the jump. The peak positive velocity occurred just prior to the jumper leaving the ground.

Similarly, the jumper's center of mass first accelerates downward, then upward, and finally downward (figure 4.21*c*). However, this pattern includes two phases when the acceleration is constant. First, acceleration had a value of zero at the beginning of the jump, which means that the two vertical forces (body weight and ground reaction force) acting on the jumper were balanced. Second, acceleration was constant at about –10 m/s^2 during the flight phase of the jump, which indicates the effect of gravity on the center of mass of the jumper. When acceleration is not constant, the ratio of the vertical component of the ground reaction force to the force due to gravity (body weight) changes. Negative accelerations indicate intervals in which the vertical component of the ground reaction force is less than body weight. Conversely, positive accelerations represent intervals in which the vertical component of the ground reaction force is greater than body weight.

This association between the vertical acceleration of the center of mass and the vertical component of the ground reaction force can be observed in figure 4.21*c* and *f*. The shape of the two graphs is the same; they differ only in the values on the *y*-axis. If body weight (F_w) is subtracted from $F_{g,y}$, then $F_{g,y}$ fluctuates about F_w just as the acceleration varies about zero. This comparison suggests that when we interpret graphs of the vertical component of the ground reaction force, variation in $F_{g,y}$ about F_w provides information about the direction of the vertical acceleration experienced by the center of mass.

The EMGs of two muscles that contribute significantly to the countermovement jump are shown in figure 4.21*d* and *e*. These two muscles produce an extensor (plantarflexor) torque about the ankle joint (gastrocnemius) and an extensor torque about the knee joint (vastus medialis). These muscles, and their synergists, control both the downward and the upward displacement of the center of mass during the takeoff phase. The initial downward displacement is achieved by a reduction in the level of muscle activation so that the extensor torque is less than the effect of gravity. This involves a lengthening contraction by these muscles. To accelerate the center of mass upward, however, the effect of the muscle activity must be greater than gravity. This is apparent in figure 4.21*e* as the increase in the EMG of vastus medialis at about the time $F_{g,y}$ begins to exceed F_w. Also note that muscle activation proceeds in a proximal-to-distal sequence (vastus medialis before gastrocnemius), as has been reported for the EMGs of several leg muscles during the countermovement jump (Bobbert & Ingen Schenau, 1988).

To identify the critical variables that determine the height an individual can jump, biomechanists compare performances in the countermovement and squat jumps. In contrast to the countermovement jump, the squat jump has no initial downward displacement, and the jump begins from the crouched position (Hasson et al., 2004). When experienced volleyball players perform these two jumps, they are able to jump 3 to 11 cm higher with the countermovement jump (Bobbert et al., 1996; Ravn et al., 1999). A key difference between these two jumps lies in the types of contractions performed by the knee and ankle extensor muscles. These muscles perform a lengthening and then a shortening contraction in the countermovement jump, but an isometric followed by a shortening contraction in the squat jump. A typical explanation for the superior performance of the countermovement jump is that the initial lengthening contraction maximizes muscle force at the beginning of the push-off phase, that is, the interval involving upward displacement of the center of mass (Bobbert et al., 1996). This effect is most evident at the hip joint, where the resultant muscle torque reaches a peak value of about 313 N•m during the countermovement jump compared with 183 N•m during the squat jump (Fukashiro & Komi, 1987).

The maximal height that a person can reach in the vertical jump depends primarily on the amplitude and timing of the muscle activity in the legs (Ravn et al., 1999; Rodacki et al., 2002). One way to compare performances in the countermovement and squat jumps is to perform a dynamic analysis (chapter 3) and determine the resultant muscle torque about the hip, knee, and ankle joints. Such a data set is shown in figure 4.22, where the resultant muscle torque is graphed as a function of angular velocity at each joint. A negative angular velocity indicates flexion at the joint, and a positive torque represents an extensor muscle torque. Because the squat jump does not include an initial downward movement, there was no flexion at the three joints. For the countermovement jump, however, there was an initial flexion at each joint that involved lengthening contractions, that is, flexion combined with extensor muscle activity. At all three joints, the peak torque occurred at the transition from the lengthening to the shortening contraction—consistent with the concept that the initial lengthening contraction maximizes the muscle force at the onset of the push-off phase. The greatest difference between the two jumps occurred at the hip joint, where the peak torque is much greater during the countermovement jump.

Despite the obvious superiority of the countermovement jump over the squat jump in the maximal vertical

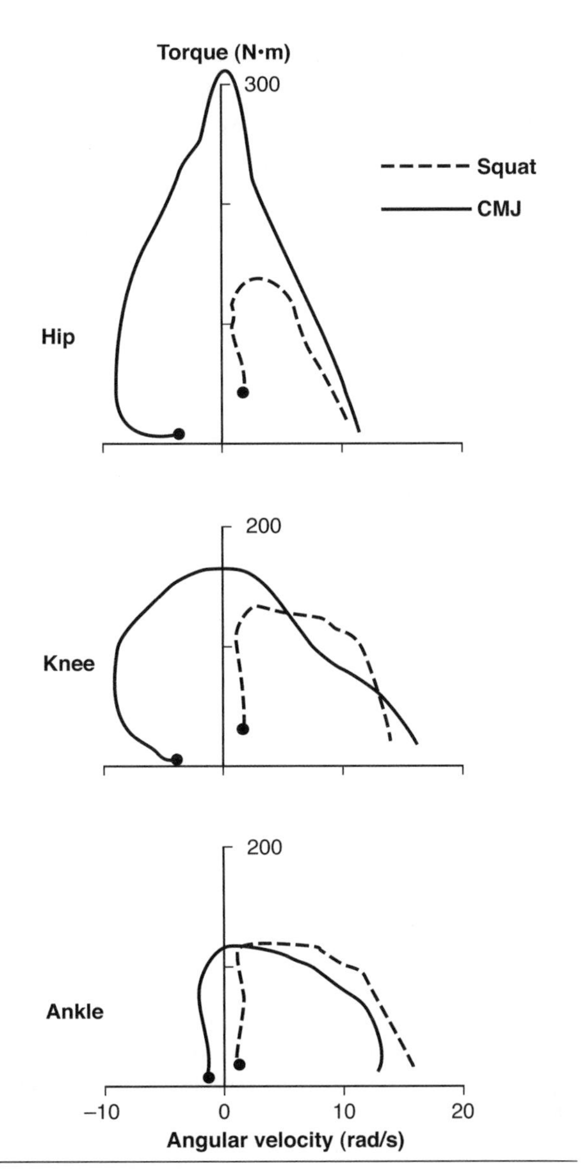

Figure 4.22 Resultant muscle torque and angular velocity at the hip, knee, and ankle joints during the countermovement (CMJ) and squat jumps performed by a single subject. Positive torque indicates extensor activity. Each trial began from the filled circle.

Data from Fukashiro and Komi 1987.

height that can be achieved, there is no consensus among biomechanists and muscle physiologists on the reason for this difference (Bobbert & Casius, 2005; Ingen Schenau et al., 1997).

The Long Jump

The purpose of the long jump is to maximize the horizontal distance between the takeoff and the landing positions. The two principal factors that contribute to this distance are the displacement of the center of mass and the lean of the body at takeoff and at landing. Leaning forward at takeoff and backward at landing adds distance to the displacement of the center of mass and increases the jump distance. The primary determinant, however, is the horizontal distance that the center of mass can be displaced. When experienced athletes perform the long jump, for example, 90% of the distance achieved is due to the displacement of the center of mass and 5% is due to the lean at each of the takeoff and landing positions (Hay et al., 1986).

According to the laws of projectile motion, the horizontal displacement of the center of mass is greatest when the takeoff angle is 0.785 rad (45°), which requires that the horizontal and vertical velocity of the center of mass be similar at takeoff. However, the takeoff angle is less than this optimum value whether the long jump is performed with a running approach or from a stationary position. The takeoff angle is about 0.35 rad (20°) with the running approach and 0.51 rad (29°) from a standing position (Hay et al., 1986; Horita et al., 1991; Kakihana & Suzuki, 2001). The reason for this discrepancy is that individuals can achieve a much greater horizontal velocity than vertical velocity (Brancazio, 1984). The horizontal and vertical velocities are around 9.0 and 3.2 m/s for the running long jump and 3.27 and 1.83 m/s for the standing long jump.

The horizontal distance achieved in a long jump depends on the time during which the jumper is in the air, which is determined by the vertical velocity at takeoff. This does not involve converting horizontal velocity to vertical velocity but rather performing actions that accelerate the jumper's center of mass in the upward direction during the takeoff phase. One concise way to characterize these actions is with angle-angle diagrams. Figure 4.23 shows the relative angular displacements of the thigh, shank, and foot segments of an athlete during the last stance phase (takeoff) of a running long jump (Kakihana & Suzuki, 2001). The thigh-knee angle-angle diagram (figure 4.23*a*) indicates a displacement profile similar to those we observe during running (figure 4.6). After the foot contacted the ground (TD; touchdown), the stance phase comprised two parts. First, the knee flexed while thigh angle did not change. Second, the thigh rotated backward while the knee extended. In contrast, the shank and ankle (figure 4.23*b*) first experienced backward rotation of the shank and flexion (dorsiflexion) of the ankle, and then extension (plantarflexion) of the ankle while the shank angle did not change much.

The timing of the angular displacements for the leg segments can also be examined in a velocity–time graph. Figure 4.23*c* shows the angular velocity for the thigh (an absolute angle) and the knee and ankle (relative angles) during the takeoff phase of a running long jump. A negative angular velocity indicates that the thigh segment was rotating backward and that the knee and ankle joints were

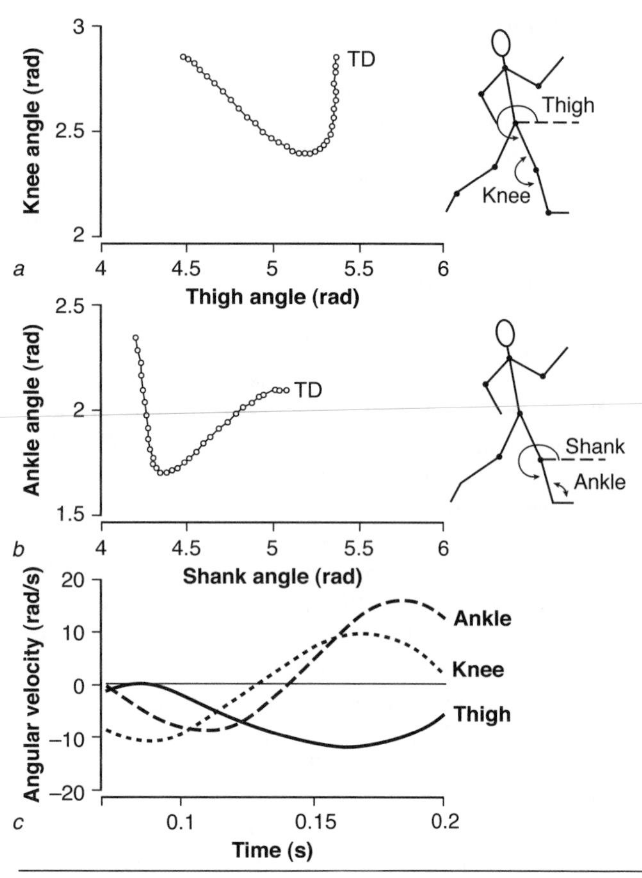

Figure 4.23 Angular kinematics of the support leg during the last stance phase of a running long jump: *(a)* thigh-knee angle-angle diagram; *(b)* shank-ankle angle-angle diagram; *(c)* angular velocity of the thigh, knee, and ankle angles. The data are from a single subject who was an Olympic-caliber sprinter. The thigh and shank angles are absolute segment angles; the knee and ankle represent the joint angles between the adjacent segments. TD = touchdown.

Data from Kakihana & Suzuki 2001.

flexing. According to figure 4.23*c*, the thigh segment of the athlete rotated backward for the entire duration of the stance phase while the knee and ankle joints flexed for about the first half of takeoff and then extended. Figure 4.23*c* also indicates the timing of the change in direction of angular displacement from flexion to extension; the change in direction at the knee joint preceded that at the ankle joint.

To characterize the actions necessary to maximize long-jump distance, Kakihana and Suzuki (2001) compared the performances of two athletes when they used three-, five-, and nine-step approaches for a running long jump. Some of the details of the performances by the more accomplished athlete are listed in table 4.2. As you might expect, he jumped farther as the approach distance increased, due to greater horizontal and vertical velocity of his center of mass at takeoff. Although the kinematics of the support leg did not differ markedly, the magnitude

Table 4.2 Takeoff Characteristics for Long Jumps Performed With Three-, Five-, and Nine-Step Approaches by an Accomplished Athlete

	3 step	5 step	9 step
Distance jumped (m)	2.63	2.80	4.22
Takeoff duration (ms)	153	131	124
Touchdown velocity (m/s)			
Horizontal	4.82	6.19	7.37
Vertical	−0.31	−0.29	−0.10
Takeoff velocity (m/s)			
Horizontal	4.46	5.71	6.85
Vertical	2.26	2.54	3.01
$F_{g,y}$ impulse (N•s)	291	293	319
$F_{g,x}$ impulse (N•s)			
Backward (braking)	46	51	88
Forward (propulsion)	7	8	8

of the ground reaction force and the amplitude of the muscle activity increased with the distance jumped.

With the three-step approach, the shape of the vertical component of the ground reaction force ($F_{g,y}$) was similar to that during running (figure 4.24). For the five- and nine-step approaches, however, the magnitude of the first peak (the "impact" maximum) increased to values that were about 10 times greater than body weight (F_w) (Hatze, 1981*a;* Gottschall & Kram, 2005). This peak occurred about 10 to 15 ms after the foot contacted the ground. In contrast, the forward-backward component of the ground reaction force ($F_{g,x}$) was not similar to those measured during running, due to the pronounced braking impulse. During running at a constant speed, for example, the braking and propulsion impulses have similar magnitudes so that the net horizontal impulse is close to zero. As indicated in table 4.2, however, the braking impulse during the long jump increased from 46 to 88 N•s with approach distance, while the propulsion impulse did not change much. The greatest increase in the braking impulse appears to have been due to a large peak in the backward component, which coincided with the peak in the $F_{g,y}$ record (figure 4.24).

Because the ground reaction force represents the reaction of the ground to the actions of the athlete while his foot was on the ground, the variation in $F_{g,y}$ and $F_{g,x}$ must be accompanied by changes in the amount of muscle activity. This is indicated by variation in the rectified and averaged EMGs for several muscles of

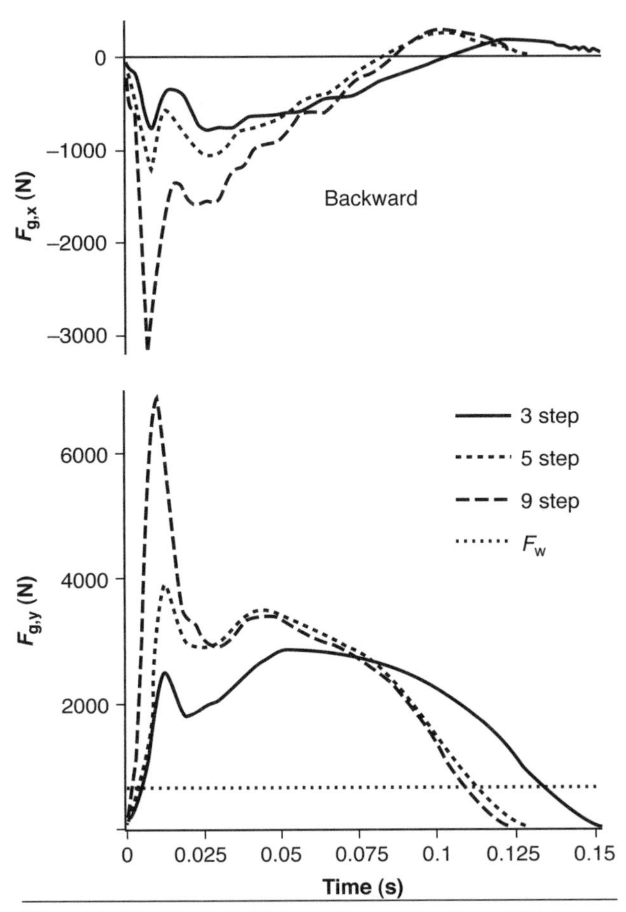

Figure 4.24 Vertical ($F_{g,y}$) and forward-backward ($F_{g,x}$) components of the ground reaction force during the takeoff phase for the long jump. Three trials of data are shown: one each for a three-, five-, and nine-step approach.

Data from Kakihana and Suzuki 2001.

the support leg for the three approach distances (figure 4.25). In these records, ▼ indicates touchdown of the foot and ▲ represents the moment when the foot left the ground. The muscles that experienced the greatest increase in activity across the three approach distances were vastus medialis, tibialis anterior, and lateral gastrocnemius. In addition to the differences in EMG amplitude, the timing of the muscle activity also changed as more of the EMG occurred prior to foot touchdown with the nine-step approach. Such information can often help explain why one person can jump farther than another (Kakihana & Suzuki, 2001).

Because the line of action of the ground reaction force vector does not pass through the jumper's center of mass for most of the takeoff phase, the jumper has some angular momentum during the flight phase of the jump. It is the magnitude and duration of the backward component of $F_{g,x}$ during the takeoff phase that cause the line of action of the ground reaction force to pass behind the center of mass for most of the takeoff phase (Kakihana & Suzuki, 2001). This generates angular momentum in the direction of a forward somersault; that is, the jumper experiences an angular impulse about his center of mass. This effect can be visualized as the ground reaction force times the moment arm from its line of action to the center of mass, which acts over the duration of the takeoff phase. The jumper controls the forward angular momentum by rotating his legs and arms during the flight phase. These actions must be sufficient to prevent the athlete from performing a forward somersault while placing the body in a pike-like position for landing.

In athletic competition, the length of the approach must be adequate for the athlete to obtain a horizontal velocity of about 10 m/s at the beginning of the takeoff phase. Once this speed has been achieved, the challenge is to obtain as much vertical velocity at takeoff as possible while minimizing the decline in horizontal velocity. This constraint leaves the athlete with approximately 100 ms to increase the vertical velocity of the center of mass, which limits the distance of the jump (Graham-Smith & Lees, 2005).

Springboard Diving

In addition to the jumps designed to maximize vertical and horizontal displacements of the center of mass, there are jumps that enable individuals to perform rotations while in the air. These jumps, which are included in such activities as dancing, diving, freestyle skiing, gymnastics, skating, and trampolining, require the performer to generate sufficient angular momentum during the takeoff phase to accomplish the desired rotation. Many of the principles associated with these types of movements can be illustrated by examining springboard diving (Miller, 2000).

There are four main types of springboard dives: forward, backward, reverse, and inward (figure 4.26). These dives differ in the number of steps taken prior to the takeoff phase, the direction the diver faces relative to the board, and the direction of the rotations about the somersault axis. The *forward* dive involves several steps in the approach, with the diver facing the water and the somersaults occurring in the forward direction. The *backward* dive is performed from a standing position, with the diver facing the board, and the somersaults are done in the backward direction. The *reverse* dive also involves several steps in the approach and the diver facing the water, but the somersaults occur in the backward direction. The *inward* dive too is performed from a standing position with the diver facing the board, but the somersaults occur in the forward direction.

Each dive can comprise up to five phases: approach, hurdle, takeoff, flight, and entry. The approach occurs only in the forward and reverse dives and corresponds to

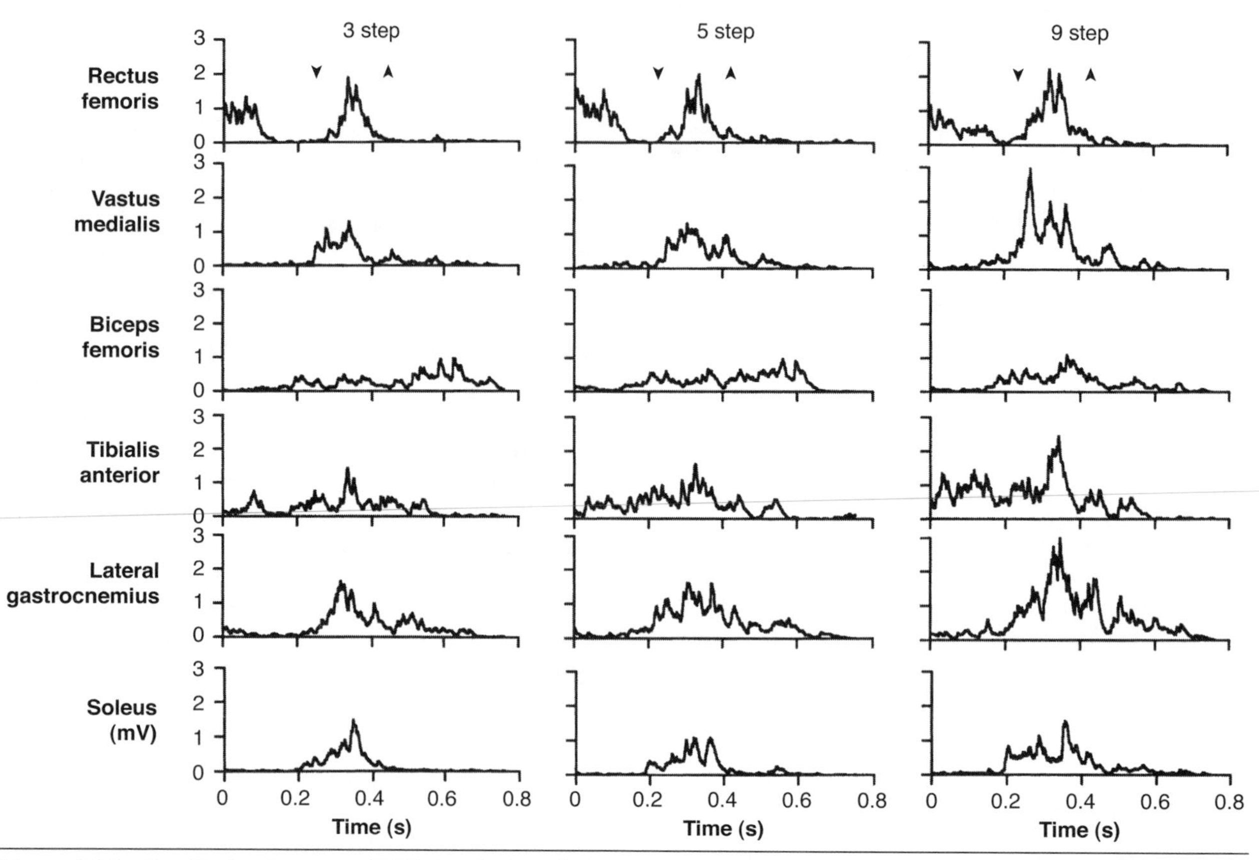

Figure 4.25 Rectified and averaged EMG signals from six leg muscles during the takeoff phase for long jumps with three-, five-, and nine-step approaches. The takeoff phase occurred between ▲ and ▼.

Data from Kakihana and Suzuki 2001.

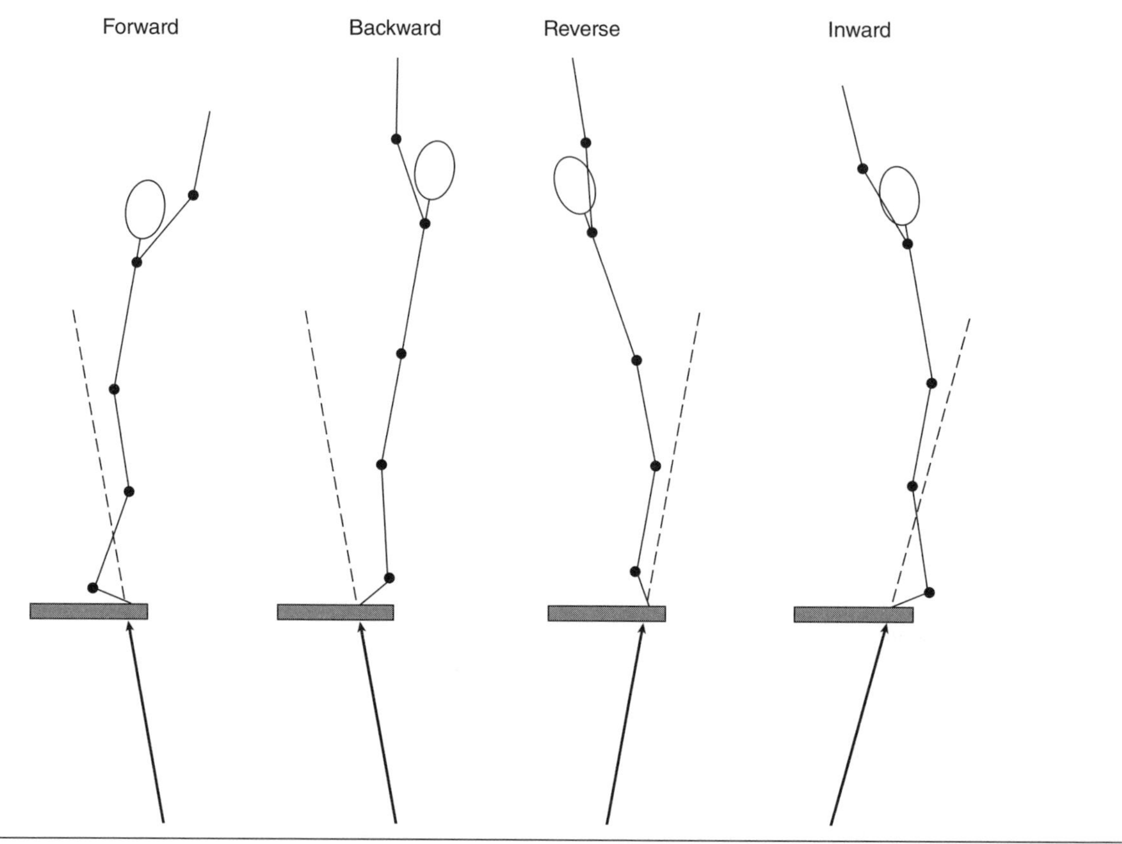

Figure 4.26 Position of the diver relative to the board reaction force for four types of dives.

Data from Miller 1981.

the initial steps taken by the diver. The hurdle refers to the movements performed by the diver to raise and lower the whole-body center of mass prior to the takeoff phase. For the dives with an approach phase (forward and reverse dives), the diver rotates the thigh of one leg forward as if performing a high step up, and the diver's feet leave the board. For the dives performed from a standing position (backward and inward dives), the diver's feet remain in contact with the board during the hurdle while the arms are raised and the ankles are extended, then flexed, to raise and lower the center of mass. The takeoff phase involves the final depression and recoil of the springboard, which projects the diver into the flight phase and then the entry phase into the water.

Reaction Force

Probably the most critical component of the dive is the takeoff phase, in which the board provides a reaction force on the feet of the diver (Cheng & Hubbard, 2005; Yeadon et al., 2006). The takeoff phase involves two actions by the diver: pressing down on the board and lifting up off the board. These actions cause the board first to be depressed and then to recoil from the imposed load. As the board is depressed, it stores elastic energy, which it returns to the diver when it recoils. When the board is depressed, the center of mass of the diver is displaced downward and reaches a minimum at the transition from the depression to the recoil (figure 4.27). Accordingly, the center of mass has a downward (negative) velocity during depression of the board and an upward velocity during the recoil. The diver's upward velocity peaks just prior to the moment the toes leave the board. For most of the takeoff phase, however, the center of mass of the diver accelerates in the upward (positive) direction due to the reaction force provided by the board. In addition to these vertical kinematics, the reaction force must include a horizontal component that enables the diver to clear the end of the board. The horizontal velocity of the diver at the end of the takeoff phase is about 0.5 to 1.0 m/s (Miller, 1981).

As the board recoils, the position of the diver relative to the reaction force provided by the board determines the direction of the diver's angular momentum vector during the flight phase. The effect of the reaction force is to exert a torque about the diver's center of mass, which can be visualized as the product of the force and its moment arm relative to the center of mass. Figure 4.26 shows the different body positions that are required to perform the four types of dives. For the forward dive, for example, the reaction force exerts a torque about the center of mass in the direction of forward rotation. Because the magnitude and direction of the reaction force change throughout the takeoff phase, the effect of the reaction force is graphed as the torque relative to the center of mass over the course of the takeoff phase. Such a torque–time graph is shown for a reverse 2.5-somersault dive (figure 4.28); the positive torque indicates a backward rotation during the flight phase. The middle stick figure at the top of figure 4.28 indicates maximal depression of the board. For both the depression and recoil parts of this dive, the diver experienced

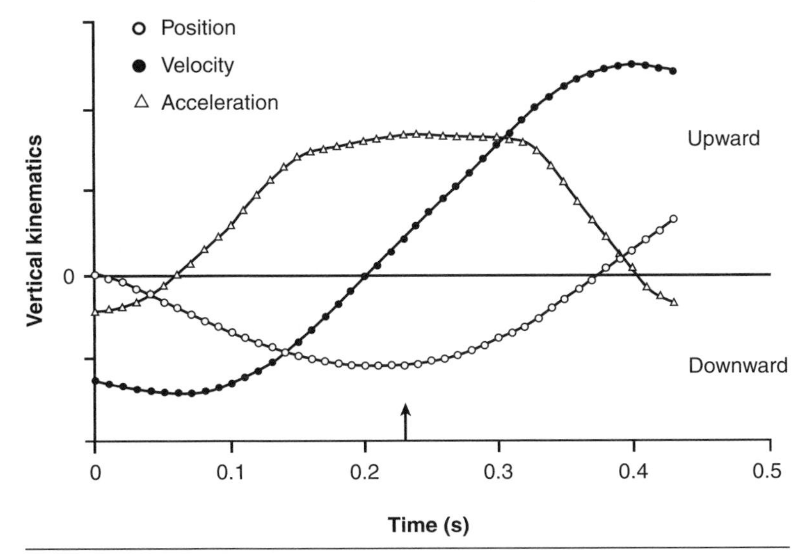

Figure 4.27 Vertical kinematics of a diver's center of mass relative to the water during the takeoff phase of an inward 2.5-somersault dive in the tucked position. The up arrow on the x-axis indicates the transition of the board from the depression to the recoil.

Data from Miller 1981.

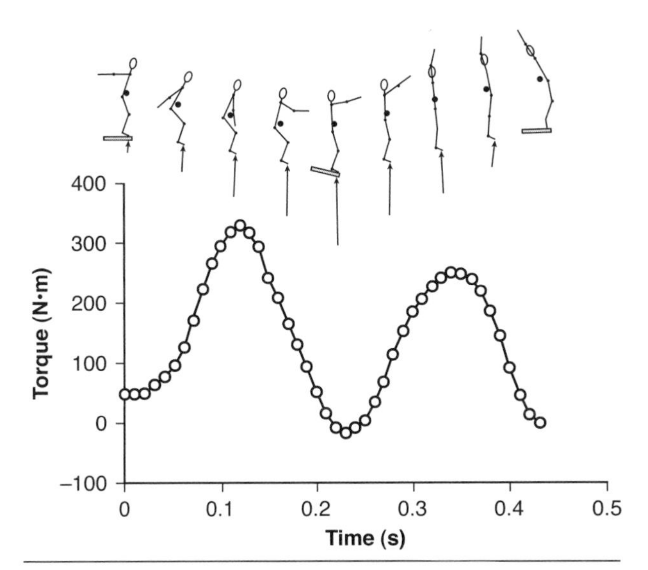

Figure 4.28 Torque about a diver's center of mass due to the reaction force during the takeoff phase for a reverse 2.5-somersault dive. The largest dot on each stick figure indicates the position of the center of mass. The arrow at the diver's feet represents the reaction force exerted by the board.

Data from Miller 1981.

a torque that produced a backward rotation; that is, the line of action of the reaction force vector passed in front of the diver's center of mass.

Angular Momentum

Based on the impulse–momentum relation (chapter 2), the effect of the diver's actions during the takeoff phase can be quantified as the area under the torque–time curve, which represents the impulse applied to the diver by the board. The impulse during the recoil of the board determines the magnitude and direction of the angular momentum vector during the flight phase. To perform forward somersaults during the flight phase, for example, the diver needs to experience an impulse that will produce such rotation, which is characterized by an angular momentum vector pointing to the diver's left. Accordingly, the direction of the angular impulse is different for dives that involve forward (forward and inward) and backward (backward and reverse) somersaults (figure 4.29).

Once a diver leaves the board, the only significant force acting on the diver is gravity. This means that the linear momentum of the diver changes during the flight phase but the angular momentum of the diver remains constant (example 2.12). As a consequence, the magnitude and direction of the angular momentum obtained by the diver during the takeoff phase do not change during the flight phase. This whole-body angular momentum could have components in the somersault, twist, and cartwheel directions (Yeadon, 1993*b*). The somersault component, for example, would act along the somersault axis passing through the center of mass, with the direction to the left for forward somersaults and to the right for backward somersaults. Recall that the direction of the angular momentum vector is perpendicular to the plane in which the rotation occurs.

Although angular momentum remains constant during the flight phase, this does not mean that the speed of the angular rotations cannot change. A constant angular momentum means that the product of the moment of inertia (I) and angular velocity (ω) for the whole body is constant. Consequently, when I changes, ω changes also. For example, when a diver goes from a tuck to a layout position, I decreases about the somersault axis and thus the speed of the somersault increases. Divers use this interaction to control the orientation of the body for the entry phase of the dive.

While the angular momentum for the whole body remains constant during the flight phase, the angular momentum for individual body segments can change. As stated in equation 2.25, the angular momentum of each body segment comprises two terms: local and remote angular momentum. The local angular momentum ($I\omega$) refers to the angular momentum of the segment relative to its own center of mass. The remote angular momentum ($\mathbf{r} \times m\mathbf{v}$) corresponds to the effect of segment center of mass about the whole-body center of mass. If the angular momentum of a body segment changes during a dive, then the angular momentum of another body segment must change by an equal magnitude but in the opposite direction so that the net angular momentum about the whole-body center of mass remains constant. For example, when a diver goes from a pike position to a layout position, the upper body rotates backward while the lower body rotates in the forward direction. Similarly, when a diver rotates the arms about a cartwheel axis through the chest to initiate a twist (example 2.13), the trunk and legs rotate in the opposite direction so that the two angular momentum vectors cancel (Yeadon, 1993*c*).

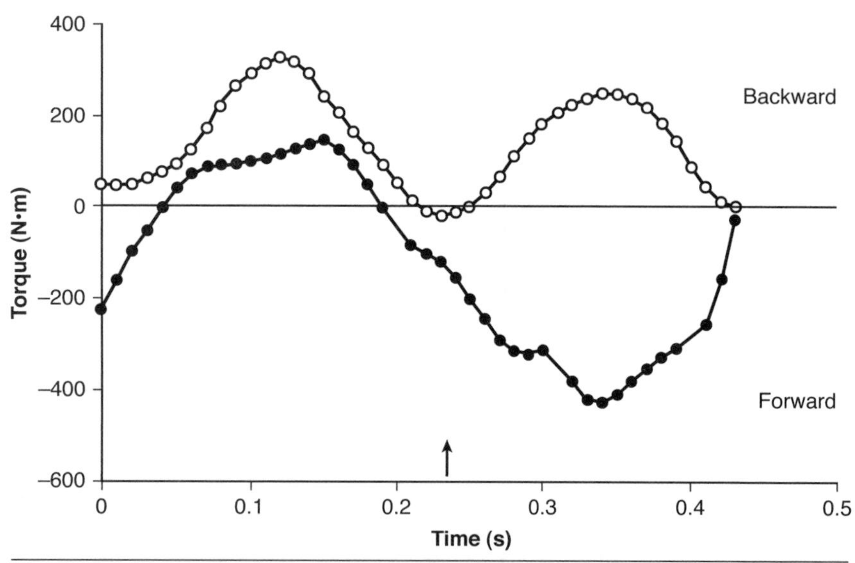

Figure 4.29 Torque about a diver's center of mass due to the reaction force during the takeoff phase for a reverse 2.5-somersault dive (open circles) and a forward 3.5-somersault dive (filled circles). The up arrow indicates the transition between the depression and recoil of the board. Positive torques produced a backward somersault.

Data from Miller 1981.

The most difficult dives involve various combinations of somersaults and twists. While the angular momentum needed for the somersaults must be obtained from the springboard, the angular momentum for the twists can come from either the board or from asymmetrical movements of the arms, chest, or hips about the sagittal plane during the flight phase (Yeadon, 1993*a*). However, both computer simulations (Yeadon, 1993*d*) and performance measurements (Yeadon, 1993*e*) indicate that most of the twist is produced by the movements performed during the flight phase. Elite performances in springboard diving, therefore, seem to require the generation and control of angular momentum during both the takeoff and flight phases as well as adherence to strict requirements on the orientation of the body.

THROWING AND KICKING

The throw and the kick are two of the basic elements of human movement. Although the purpose of both is to project an object so that it has a flight phase, the distinction between the two is the manner in which the body imparts the flight phase to the object. In a throw, the object is supported by a limb, usually the hand, and displaced through a range of motion while the limb increases the momentum of the object. Typically several body segments, in a proximal-to-distal sequence, contribute to the momentum of the object (Bobbert & van Soest, 2001; Hirashima et al., 2007*b;* Hore et al., 2005). Although the kick is also characterized by a proximal-to-distal involvement of body segments, it differs from a throw in that it is a striking event in which the momentum of the object is increased by a brief impact between a limb and the object (Elliott, 2000).

Throwing Motion

While the throw can be characterized by the progressive contribution of the body segments to the momentum of the object to be projected (with a constant mass, the change in momentum corresponds to a change in velocity), the task can be accomplished with a variety of motions. These different forms include the overarm throw (e.g., baseball, cricket, javelin, darts), the underarm throw (e.g., bowling, softball pitch), the push throw (e.g., shot put), and the pull throw (e.g., discus, hammer). The kinematics of the throwing motion are typically three-dimensional, especially when the throw is for maximal distance or speed (Escamilla et al., 1998*a;* Feltner, 1989; Feltner & Dapena, 1989; Feltner & Taylor, 1997). For example, the contributions of the body segments to the overarm throw (figure 4.30) include displacements in the vertical, side-to-side, and forward-backward directions (figure 4.31). In contrast, when the task requires accuracy, as in throwing a dart or shooting a free throw in basketball, the throwing motion is generally planar; and the strategy, especially for beginners, is to minimize the number of body segments involved in the movement.

The sequence of a typical baseball pitch is shown in figure 4.30. A qualitative inspection of this sequence indicates that the movement involves the progressive contribution of the body segments, beginning from the base of support and progressing through to the hand. The baseball pitch consists of two phases (figure 4.30): (a) positions *a* through *k*—the velocity of the ball is increased mainly by the action of the legs; and (b) positions *l* through *u*—the velocity of the ball is increased by the action of the trunk and arms. The second phase, which produces the greater increase in the velocity of the ball, involves the progressive increase in the angular velocity of the body segments in the following order: pelvis, upper trunk and upper arm, forearm, and hand (Atwater, 1979; Hirashima et al., 2007*b*). This means that the peak angular velocity of the pelvis occurs before that of the upper trunk and upper arm, the peak angular velocity of the upper trunk and upper arm occurs before that of the forearm, and so on (Wight et al., 2004). In this progression of segmental activity, which is also seen in striking movements such as the kick, proximal segments begin to rotate before the distal segments, and the proximal segments begin to slow down before the distal segments have reached peak angular velocity. The result of this proximal-to-distal progression is that the velocity of the ball does not increase substantially until the last 100 ms of the movement (positions *r* through *t* in figure 4.30). Nonetheless, Hirashima and colleagues (2007*b*) found that skilled baseball players increased the speed of a pitch mainly by varying trunk rotation, internal rotation at the shoulder, elbow extension, and wrist flexion. Others have similarly found that changes in the

Figure 4.30 Sequence of actions for a baseball pitch.

Reprinted, by permission, from M. Feltner and J. Dapena, 1986, "Dynamics of the shoulder and elbow joints of the throwing arm during a baseball pitch," *International Journal of Sport Biomechanics* 2: 236.

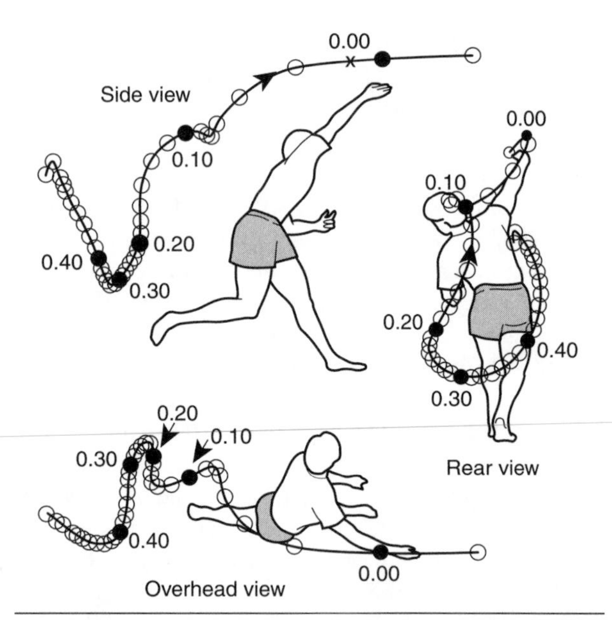

Figure 4.31 Three views of the path of the ball during an overarm throw. The ball was released at 37.3 m/s (83.5 mph), and the instant of release is indicated as the final shaded-ball position (0.00). Prior to release, the shaded-ball positions indicate intervals of 100 ms. These measurements were obtained from a film (64 frames/s) of the movement.

Adapted, by permission from A.E. Atwater, 1977, *Biomechanics of throwing: Correction of common misconceptions.* Paper presented at the Joint Meeting of the National College Physical Education Association for Men and the National Association for Physical Education of College Women, Orlando, FL.

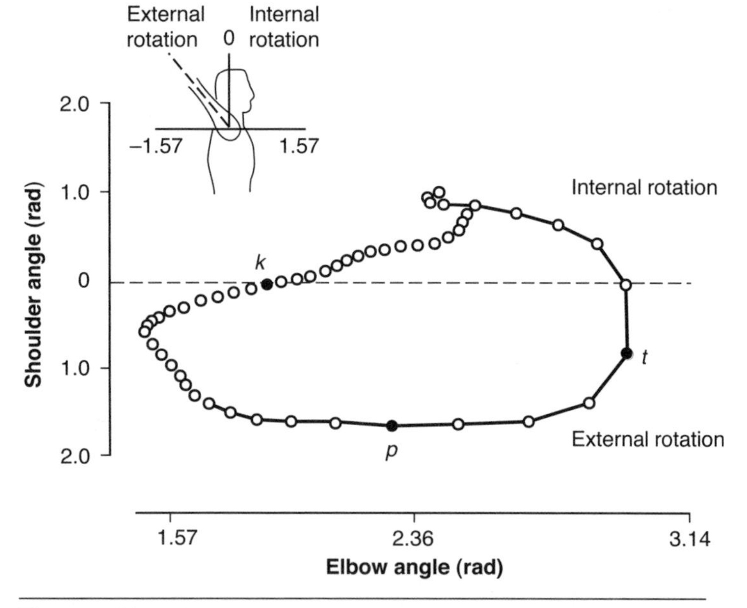

Figure 4.32 Elbow-shoulder angle-angle diagram for a baseball pitch. The interval between data points is 5 ms; the farther apart the data points, the faster the movement. The data points at *k, p,* and *t* correspond to positions identified in figure 4.30.

Reprinted, by permission, from M. Feltner and J. Dapena, 1986, "Dynamics of the shoulder and elbow joints of the throwing arm during a baseball pitch," *International Journal of Sport Biomechanics* 2: 249.

forces and torques at the shoulder and elbow are largely responsible for variation in the speed of the ball across types of pitches and successive pitches of the same type (Fleisig et al., 2006; Stodden et al., 2005).

The overarm throw involves an extensive range of motion for the arm (Escamilla et al., 1998*a;* Stodden et al., 2005). These displacements occur at the shoulder, elbow, and wrist joints but are most extensive about the shoulder joint. Because of its geometry, the shoulder joint is capable of displacement about three separate axes of rotation: rotation about an anterior-posterior (cartwheel) axis, which is referred to as abduction-adduction; rotation about a side-to-side (somersault) axis, which is known as flexion-extension; and rotation about a longitudinal (twist) axis, which is described as external-internal rotation. Figure 4.32 shows that during a baseball pitch there are about 1.57 rad of elbow flexion-extension, 1.57 rad of external rotation, and 1.0 rad of internal rotation. Position *p* in figure 4.30 shows the arm in the position of maximum external rotation. This extreme position probably causes many of the arm injuries that are sustained by baseball pitchers (Ogawa & Yoshida, 1998; Sabick et al., 2004, 2005).

Because throws and kicks involve multiple body segments, the actions are influenced by interaction torques between segments. As described in chapter 3, an interaction torque is an effect that one segment exerts on its neighbors due to its motion. The magnitude of an interaction torque can be much greater than that of a net muscle torque (figure 3.28). Skilled baseball players are able to exploit the interaction torques to increase the speed at which a ball can be thrown (Hirashima et al., 2007*a;* Hore et al., 2005; Kadota et al., 2004). According to Hirashima and colleagues (2007*a*), these players achieve this effect by increasing the torque exerted by shoulder and trunk muscles, but not the muscles that cross the elbow or wrist. Increases in ball speed, therefore, are critically dependent on the actions of the proximal muscles and the orientation of the arm as the throw progresses.

Kicking Motion

The kick can be described as a striking skill in which a flight phase is imparted to an object as the result of a brief impact between a limb (or implement) and the object. By this criterion, we include as striking skills such activities as kicking a ball, hitting a volleyball, and striking a projectile in racket and bat sports. As with the throw, the motion underlying a striking skill can vary depending on the objectives of the activity; these goals can be related to horizontal distance, time in the air, accuracy, or the speed of the movement. For many striking skills (e.g., soccer kick, punt, volleyball serve, tennis serve), the motion is similar to that for the overarm throw and

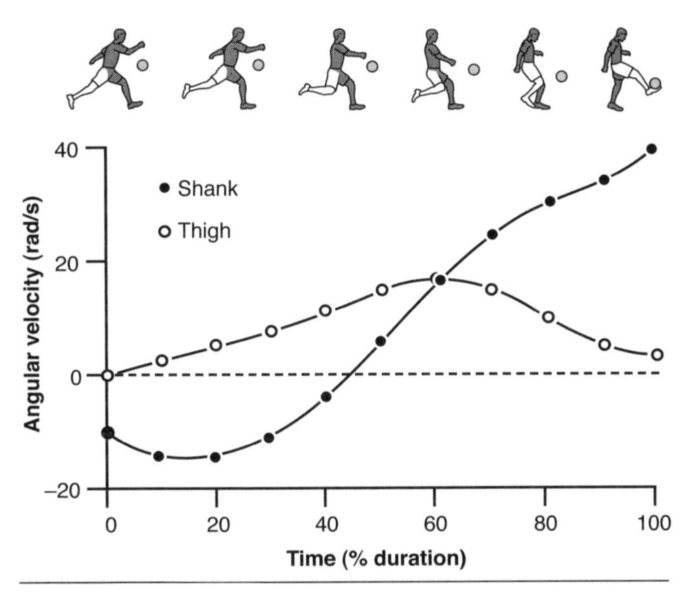

Figure 4.33 Angular velocity of the thigh and shank of the kicking leg during the final step of a kick.

Adapted, by permission, from C.A. Putnam, 1991, "A segment interaction analysis of proximal-to-distal sequential segment motion patterns," *Medicine and Science in Sports and Exercise* 23: 134.

involves a proximal-to-distal sequence of body segment contributions to the velocity of the endpoint (hand, foot, implement) that will strike the projectile (Dörge et al., 2002; Nunome et al., 2006). For example, when kicking a ball the thigh reaches a peak positive (forward) angular velocity before the shank and decreases its angular velocity while that for the shank continues to increase through to contact with the ball (figure 4.33).

Because striking skills alter the momentum of the projectile through an impact, an important difference between a throw and a kick is the rigidity of the limb during contact with the ball (Naunheim et al., 2003; Nicholls et al., 2006). When performing striking skills, athletes frequently manipulate the rigidity of the limb to influence the impact with the object (Sterzing & Hennig, 2008). For example, the change in ball velocity during a kick is greater when the lower leg is more rigid at impact. Athletes accomplish this by contracting the muscles in the foot and those that cross the ankle joint so that the foot is more rigidly attached to the shank. Another example of variation in rigidity is the underhand pass (bump, dig) in volleyball. This skill requires that the individual allow the volleyball to bounce off the ventral surface of the forearms with a prescribed trajectory. Skillful players accomplish this task by varying the rigidity of the arms and thereby determining the extent to which the ball will bounce off the forearms. Similarly, the strings on rackets can be strung to various levels of tightness, which represents one factor that contributes to rigidity in racket sports (Hennig, 2007).

SUMMARY

This chapter applies the principles and concepts described in chapters 1 through 3 to the basic forms of human movement: walking, running, jumping, throwing, and kicking. Human locomotion comprises alternating phases of support (stance) and nonsupport (swing), whose durations change with speed. The displacement of the limbs during walking and running is most succinctly characterized with angle-angle diagrams, and the interaction of the foot with the ground is quantified by measurement of the ground reaction force. Because of differences in the activity of leg muscles, the path followed by the center of mass during the stance phase differs between walking and running. In the stance phase of walking, there is little knee flexion, which provides an explanation for the speed at which humans prefer to switch from a walk to a run and for the fluctuations in mechanical energy. In the stance phase of running, there is greater modulation of muscle activity, which varies the stiffness of the leg during the stance phase and the energy cost of locomotion. These characteristics provide the foundation for the clinical evaluation of gait.

Similar analyses are applied to jumping and throwing, which have different performance goals. In jumping, the object is to project the center of mass upward so that the feet leave the ground. The different trajectories experienced by the center of mass for the different types of jumps are a consequence of variations in the ground reaction force. In throwing and kicking, the goal is to apply an impulse to an object so that it has a flight phase. Throws tend to involve actions in which forces are applied by a limb over a relatively long duration. The impulses applied in kicks involve much briefer contact times. For both throws and kicks, however, the limb motion involves a coordinated sequence of proximal-to-distal muscle activity. The examples presented in this chapter indicate the strategies that can be used to describe and understand the biomechanical details of most movements.

SUGGESTED READINGS

Alexander, R.M. (2003). *Principles of Animal Locomotion.* Princeton, NJ: Princeton University Press, chapters 3, 7, and 8.

Biewener, A.A. (2003). *Animal Locomotion.* Oxford UK: Oxford University Press, chapter 3.

Kram, R. (2000). Muscular force or work: What determines the metabolic energy cost of running? *Exercise and Sport Sciences Reviews, 28,* 138-142.

Winter, D.A. (1990). *Biomechanics and Motor Control of Human Movement.* New York: Wiley, chapter 5.

Part I Summary

At the beginning of part I, a number of specific objectives were listed to help us achieve the goal of defining the mechanical bases of movement. Completing part I should have enabled you to do the following:

- Understand the definitions of and relations (numeric and graphic) among position, velocity, and acceleration, which comprise the kinematic variables used to describe movement
- Know how to read a graph carefully and how to interpret the relation between the two or more variables shown on the graph
- Appreciate the relations between linear and angular motion
- Realize that many of the details of projectile motion can be determined from the definitions of position, velocity, and acceleration
- Consider force as a concept used to describe an interaction between two objects, and understand that the magnitude of the interaction can be determined using Newton's laws, particularly the law of acceleration
- Use a free body diagram to define the conditions of an analysis and use of free body and mass acceleration diagrams as graphic versions of Newton's law of acceleration
- Conceive of torque as the rotary effect of a force for which torque is defined as the product of force and moment arm
- Identify the ways in which the human body interacts with its surroundings to influence movement
- Recognize that force acting over time (impulse) causes a change in the momentum (quantity of motion) of a system
- Acknowledge that the performance of work (force × distance) requires the expenditure of energy and that the work can be done either by the system (positive) or on the system (negative)
- Perceive of power as a measure of the rate of doing work or the rate of using energy
- Comprehend the concept of musculoskeletal forces that occur within the body as body segments rotate about one another
- Perform static and dynamic analyses to estimate the magnitude and direction of musculoskeletal forces
- Distinguish the kinematic and kinetic descriptions of running, throwing, and kicking
- Differentiate the biomechanical characteristics of walking and running
- Note the mechanical energy fluctuations and the energy cost of human locomotion
- Appreciate the role of biomechanics in the clinical evaluation of gait
- Grasp the significance of variations in the ground reaction force in determining the details of a movement, such as the height, distance, and rotations that occur during a jump
- Be able to distinguish between throws and kicks
- Understand that the sequencing of muscle activity is critical in throws and kicks

part II

The Motor System

This text is about the neural control of movement, recognizing that the activation signals generated by the nervous system and the force exerted by muscle to produce a movement must accommodate the laws of physics. Part I of the text reminded us about the laws of motion and their application to the study of human movement. Part II presents the **motor system**, which corresponds to those parts of the nervous system and muscle that are responsible for movement. We use a bottom-up approach that begins from the principles of electricity responsible for the excitability of cell membranes and extends up to connections within the cerebral cortex that contribute to the motor command that is dispatched from the motor cortex to the spinal cord.

Part II consists of three chapters. Chapter 5 ("Excitable Membranes") describes the electrical basis of the resting membrane potential, the electrical properties of neurons, the mechanisms involved in synaptic transmission, the connection between the activation signal and a muscle contraction, and a technique (electromyography) that is used to measure the activation signal. Chapter 6 ("Muscle and Motor Units") details the connection between the nervous system and muscle, the mechanical properties of muscle, and the organization and activation patterns of muscles. Chapter 7 ("Voluntary Movement") addresses three types of movements generated by the nervous system (spinal reflexes, automatic responses, and voluntary actions) by focusing on the neural circuits and signals responsible for each type of movement.

OBJECTIVES

The goal of this text is to provide a foundation for the study of how the nervous system controls the actions of muscles to exert forces on the surroundings and thereby produce movement. In part I, we examined the biomechanics of movement, focusing on the relation between force and motion. The aim of part II is to describe the function of the motor system. Specific objectives include the following:

- To describe the essentials of electricity responsible for the excitability of cell membranes
- To explain the characteristics of an excitable membrane underlying its resting potential
- To list the electrical properties of neurons
- To outline the currents that flow across excitable membranes
- To define the transmission of electrical signals from one cell to another cell
- To characterize the connection between the activation signal generated in the nervous system and the actions of the contractile proteins in muscle
- To examine a technique that can be used to measure the activation of muscle by the nervous system
- To describe the structure of muscle and the associated connective tissues
- To explain how the contractile proteins are activated by the nervous system
- To outline the interaction between the contractile proteins that generate muscle force
- To introduce an imaging technique that can be used to measure muscle activity
- To discuss the functional relations between spinal neurons and muscle fibers
- To characterize the mechanical properties of muscle
- To indicate the behavior of muscle during human movement
- To document the properties of the sensory receptors that provide afferent feedback for the control of movement
- To describe spinal reflex pathways
- To outline the connections between neurons in the spinal cord
- To examine the activity in spinal pathways during movement
- To characterize the postural contractions that enable movement
- To discuss the properties and function of neuronal circuits that produce coordinated motor patterns
- To explain the organization of the motor system
- To describe the control of reaching and pointing movements

chapter 5

Excitable Membranes

To understand the activation of muscle by the nervous system, we begin by reviewing selected concepts from the physics of electricity and then examine how the flow of current along excitable membranes can enable the interaction of the contractile proteins in muscle. This chapter also includes a description of the technique known as electromyography, which can be used to measure the activation of muscle by the nervous system.

ESSENTIALS OF ELECTRICITY

The activation signal that is critical for the operation of the motor system is based on four electrical concepts: potential difference, current, conductance, and capacitance. Once these concepts are understood, it is possible to describe the flow of current in electrical circuits and to appreciate the events responsible for the activation of muscle by the nervous system.

Basic Concepts

The membranes of nerve and muscle cells have the capacity to control the location and flow of charged particles into and out of the cell. This capacity is critical to the function of the motor system and depends on four electrical properties.

Potential Difference

Because electrical charges exert an electrostatic force on other charges (i.e., like charges repel and opposite charges attract), work must be done to bring together two charges that are initially separated. Opposite charges are brought together by negative work, whereas like charges require positive work. A **potential difference** is a measure of the potential energy that must be used to move a positive charge from one location to another. The amount of work that must be done depends on the size of the charge and the distance the charge must be moved. Potential difference is measured in volts (V), which is defined as the work (J) done to move one coulomb (C) of charge (Q) between two points (equation 5.1).

$$V = \frac{J}{C} \tag{5.1}$$

1 V is the energy required to move 1 C a distance of 1 m against a force of 1 N. Although it is common to use the term voltage when referring to a potential difference, it is important to appreciate that both terms indicate a difference between two locations.

A **battery** (E) is a common source of a potential difference in that it stores potential energy as a result of a chemical reaction. A battery has two terminals, positive and negative, for connection to other circuit elements. Energy conversion processes within the battery move an excess of positive charge to one terminal and negative charge to the other terminal. Thus, there is a potential difference between the two terminals. The circuit symbol for a battery is shown in figure 5.1. In an ideal battery, the amount of stored chemical energy is considered to be unlimited; hence the potential difference should not be influenced by the rate at which energy is released when the battery is placed in a circuit. Under these conditions, the battery is regarded as a source of constant potential energy. Real batteries, however, have a limited capacity to sustain the chemical reactions that provide the potential difference. For example, both a type D battery and a calculator battery have a potential difference of 1.5 V, but the type D battery can sustain the chemical reactions longer than the calculator battery. This capacity can be represented as the internal resistance of a battery. As a consequence, the effectiveness of a battery depends on the size of its internal resistance relative to the resistance in the circuit in which it operates.

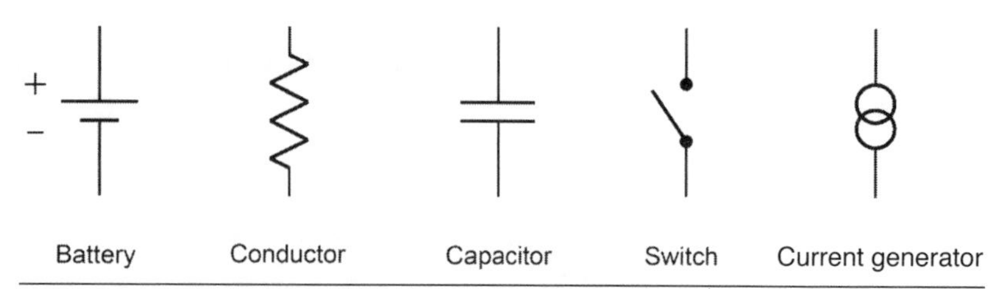

Figure 5.1 Common symbols in an electrical circuit.

Current

Charged particles move between locations that comprise a potential difference. Positive charges are attracted to the region with a more negative potential, and negative charges go to the regions of positive potential. Current (I) is defined as the rate at which positive charges move between the two locations that represent a potential difference (equation 5.2). The unit of measurement for current is the ampere (A). 1 A corresponds to 1 Q (6.24×10^{18} electron charges) passing a given point each second. The currents that cross excitable membranes are small and are expressed in mA, µA, nA, and pA (consult table 1.1 for definitions of these prefixes). Currents travel only in complete circuits, that is, a path between the two locations of the potential difference.

$$I = \frac{Q}{t} \tag{5.2}$$

In metallic conductors, current is carried by electrons, with the flow of current indicated as the direction opposite to the movement of the electrons. For example, for the circuit shown in figure 5.2, current flows from the positive to the negative terminal of the battery. In nerve and muscle cells, current is carried by positive and negative ions in solution.

Resistance and Conductance

A conductor is an object through which a current can flow. The conductivity (σ) of the object depends on its molecular structure: Metallic conductors have high conductivities, ionized salt solutions have lower conductivities, and lipids have very low conductivities. The term **conductance** (g) denotes the capacity of the object to transmit a current. It is measured in siemens (S). Conductance is equal to the product of conductivity and the ratio of the cross-sectional area (A) and length (L) of the object (equation 5.3).

$$g = \sigma \frac{A}{L} \tag{5.3}$$

The current that flows through a conductor is proportional to the potential difference that exists across it, with the proportionality constant indicated by the conductance of the object. This relation is known as **Ohm's law** (equation 5.4):

$$I = gV \tag{5.4}$$

where V indicates the potential difference. Thus, the amount of current that flows through an object varies directly with its conductance. Sometimes the conductance of a tissue, such as a membrane, is described in terms of the amount of current that can flow through a given area of tissue. **Specific membrane conductance** (S/cm^2) is a measure of the current that can cross 1 cm^2 of membrane. The specific membrane conductance of a nerve cell can range from 50 to 500 µS/cm^2.

More typically, the ability of an object to transmit current is characterized by the **resistance** it offers to the flow, which is the reciprocal of conductance. In membrane physiology, however, it is more helpful to think in terms of the ease with which current can flow. When written in terms of resistance, Ohm's law states that current is equal to the ratio of the potential difference (V) to the resistance (R). The unit of measurement for resistance is the ohm (Ω). Because conductance and resistance are the same physical entities, differing only in the units of measurement, they are both represented by the symbol for a conductor in an electrical circuit (figure 5.1).

Capacitance

A capacitor consists of two conducting plates that are separated by an insulating layer. The conducting plates can store charges of opposite sign. The symbol for a capacitor in an electrical circuit is shown in figure 5.1. Capacitors are used for short-term energy storage when it is necessary to store and release energy quickly. The net excess of positive charges on one plate and negative charges on the other plate creates a potential difference across the capacitor. The greater the density of charges on the plates of the capacitor, the greater the potential

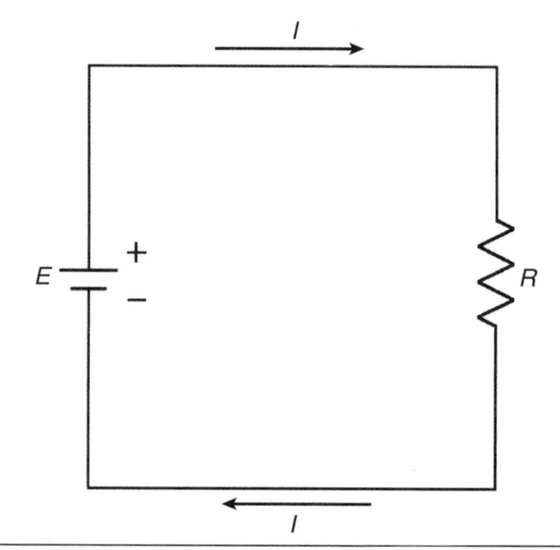

Figure 5.2 A battery (E) passes a current (I) around a circuit that includes a resistor (R).

difference between the plates. **Capacitance** (C) is defined as the ratio of the amount of charge stored on the plates (Q) and the potential difference (V) across the plates (equation 5.5).

$$C = \frac{Q}{V} \tag{5.5}$$

The unit of measurement for capacitance is the farad (F). **Specific membrane capacitance** is a measure (F/cm^2) of the capacitance of 1 cm^2 of membrane. The specific membrane capacitance of a nerve cell is about 1 μF/cm^2.

Current Flow in an Electrical Circuit

We can describe the essential features of current flow across an excitable membrane by considering the flow of current in circuits that contain conductors (resistors) in series, a conductor and capacitor in series, and a conductor and capacitor in parallel. Recall that an in-series arrangement refers to an end-to-end arrangement of the components, whereas in parallel describes a side-by-side arrangement.

Conductors in Series

Figure 5.3*a* shows the flow of current from the positive terminal of an ideal battery through two conductors in series and back to the negative terminal of the battery. This is known as a **voltage-divider circuit.** The current that flows through the two conductors is the same. The potential difference across the two conductors (R_1 and R_2) can be measured as the potential differences between locations *a* and *b*, and *b* and *c*, respectively. From Ohm's law, it is possible to determine the potential difference across the first conductor (V_{ab}):

$$I = \frac{E}{R}$$

$$= \frac{E}{R_1 + R_2}$$

Substituting $V_{ab} = I\,R_1$,

$$V_{ab} = E\frac{R_1}{R_1 + R_2} \tag{5.6}$$

Equation 5.6 indicates that the potential difference across the first conductor (V_{ab}) is a fraction of the battery voltage, and that the fraction depends on the ratio of R_1 to the total resistance ($R_1 + R_2$). When E and R_2 are constant, then V_{ab} depends on R_1. When R_1 is large, then V_{ab} is essentially the same as E. This example illustrates that to measure the voltage of a battery, including those that exist in cell membranes, the input resistance of the measuring device (R_1) must be large compared with the resistance (R_2) of the voltage source.

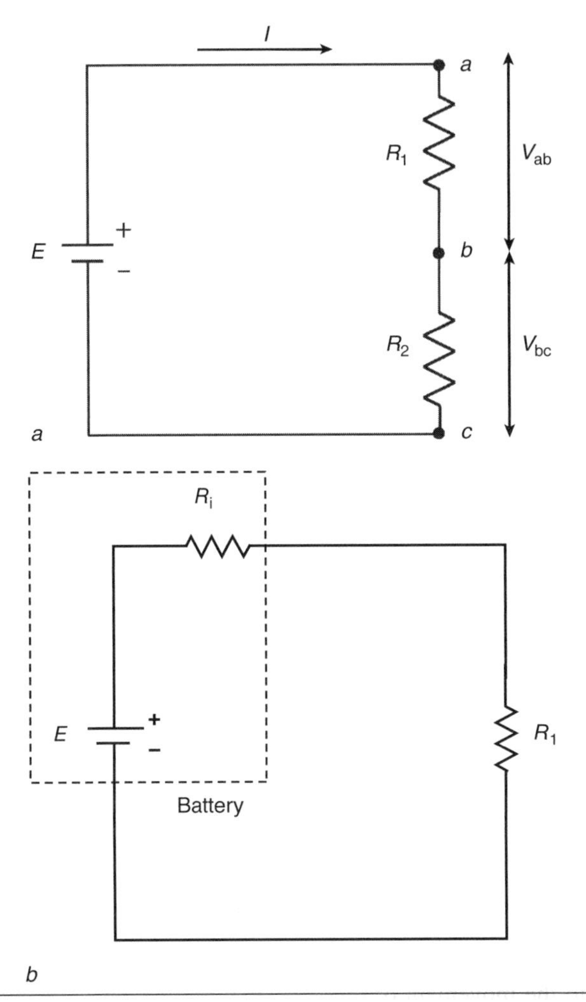

Figure 5.3 Conductors in series. *(a)* A battery (E) passes a current (I) around a circuit that includes two conductors (R_1 and R_2) in series. The potential difference between points *a* and *b* (V_{ab}) represents the potential difference across R_1, whereas V_{bc} denotes the potential difference across R_2. *(b)* Rearrangement of the circuit in part *a* so that the second conductor (R_2) now represents the internal resistance (R_i) of a real battery.

The voltage-divider circuit helps us to understand the function of a real battery. This is accomplished by rearrangement of the circuit so that R_2 now represents the internal resistance of the battery (R_i) (figure 5.3*b*). When R_1 is negligible, the current that the battery can generate is equal to E/R_i. As the chemical potential energy of a battery declines, E decreases and there is a corresponding reduction in the current that the battery can provide. Effective batteries, therefore, have a low internal resistance.

Conductor and Capacitor in Series

As a device that can store charged particles for brief periods of time, a capacitor requires a voltage source to provide a current that delivers the charged particles to be stored. The circuit shown in figure 5.4*a* can accomplish

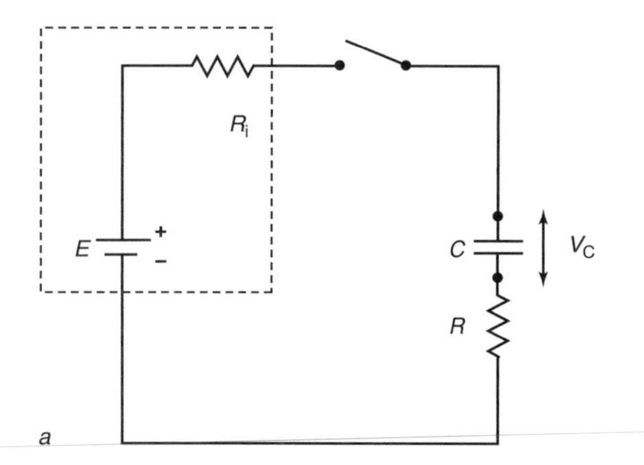

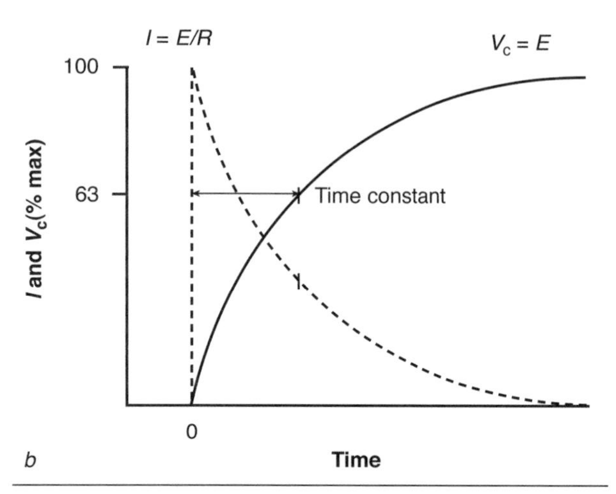

Figure 5.4 Charging a capacitor. *(a)* When the switch is closed, the battery charges the capacitor (C) until the potential difference across the capacitor (V_C) is equal to the voltage of the battery (E). *(b)* The exponential decrease in I (dashed line) and increase in V_C (solid line) are characterized by the time constant of the circuit, which depends on the magnitudes of R and C. The switch was closed at time 0. $V_C = E$ when the internal resistance of the battery (R_i) is negligible.

this task. When the switch is closed to activate the circuit, current (positively charged particles) flows from the positive terminal of the battery and accumulates on one of the plates of the capacitor. Positive charges are attracted from the other plate to the negative terminal of the battery. Thus, one plate accumulates positive charges and the other plate loses them, which results in a separation of charged particles and produces a potential difference between the two plates of the capacitor (V_C). Note that although the current does not actually cross the insulating layer between the two plates of the capacitor, it is possible to measure a current flowing into and out of the capacitor. This is called a capacitive current (I_C).

As positive charges accumulate on the plate closest to the positive terminal of the battery, the positive charges begin to repel the like charges that are flowing from the positive terminal of the battery. The amount of repulsion will increase as more charges accumulate on the plate of the capacitor. At some point, the plate of the capacitor will contain enough positive charges to oppose the electromotive force that propels the current from the battery. When this occurs, no more current will flow around the circuit, and the potential difference across the capacitor (V_C) will equal the voltage of the battery.

The rates at which both the current decreases and V_C increases are exponential (figure 5.4*b*). The exponential curves are characterized by a parameter known as the **time constant,** which is the time that the curve takes to reach 63% $(1 - 1/e)$ of its final value. The time constant for this circuit can be calculated as the product of R and C and has a magnitude of milliseconds for cell membranes.

The flow of current around this circuit depends on both the capacity of the battery and the difference between the voltage of the battery and the potential difference across the capacitor. For this circuit, therefore, Ohm's law can be rewritten as

$$I = \frac{E - V_C}{R} = g(E - V_C) \qquad (5.7)$$

The amount of current that flows around this circuit is directly proportional to the difference between the two potential differences, with the result that the term $E - V_C$ is known as the **driving force** of the circuit. When $E = V_C$, the driving force is zero and no current flows around the circuit.

Conductor and Capacitor in Parallel

Excitable membranes contain conductors and capacitors arranged in parallel. When current is provided to such a circuit, the flow of current through the two components will vary in proportion to the potential difference that develops across the capacitor. To describe the function of this type of circuit, we represent the response to a constant current delivered by a current generator (figure 5.5*a*), as occurs with synaptic input, rather than a constant-voltage source (battery). When a circuit contains components that are arranged in parallel, the current will follow the path of greatest conductance (least resistance). Accordingly, all the current provided by the generator (I_G) will initially flow through the capacitor and V_C will increase from its initial value of zero. As V_C increases, the flow of current across the capacitor will decrease and more current will travel through the conductor. Because the current traveling around the circuit is constant (I_G), the exponential changes in the current passing through the capacitor (I_C) and the conductor (I_R) must sum to a constant value (figure 5.5*b*).

As defined by Ohm's law, the potential difference across the conductor (V_R) is equal to the product of I_R and R. Because R is constant, the increase in V_R parallels the

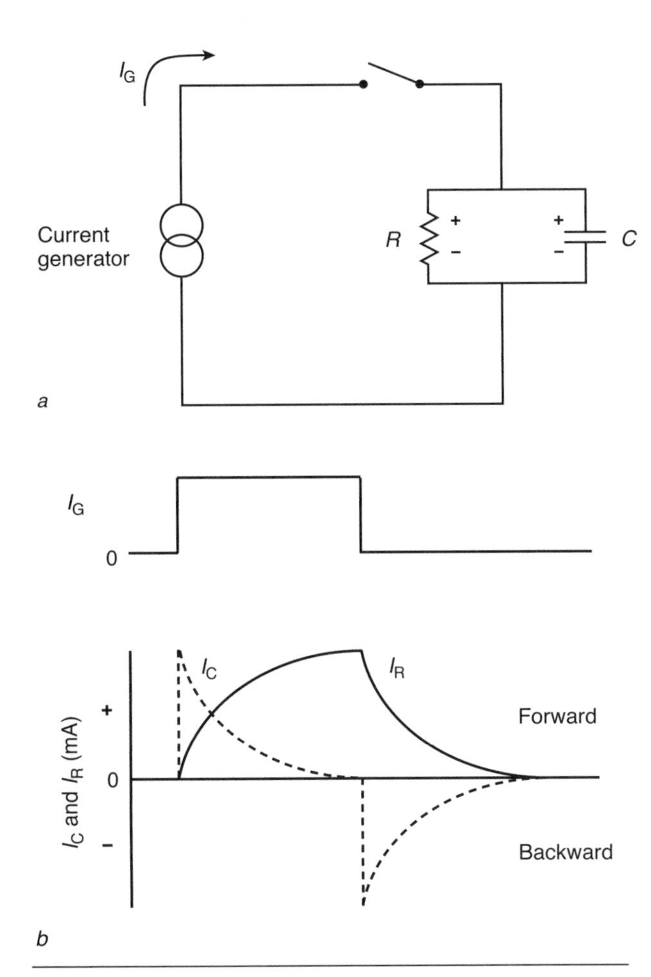

Figure 5.5 Conductor and capacitor in parallel. *(a)* When the switch is closed, current will flow from the current generator through both the conductor (*R*) and the capacitor (*C*). The proportion of the current that flows through each component depends on the magnitude of V_C. *(b)* Once the switch is closed (time 0), the current produced by the generator (I_G) initially all flows through the capacitor (I_C, dashed line) and then finally all travels through the conductor (I_R, solid line). When the switch is opened, the circuit receives no current from the generator ($I_G = 0$), and the potential difference that has accumulated on the capacitor is disbursed. This is accomplished as a current flow from the positive plate of the capacitor back through the conductor and onto its negative plate.

change in I_R. Furthermore, due to the rule that potential differences across branches of a parallel circuit are equal, the change in V_C matches the change in V_R, and the maximal value for V_C is equal to the product of I_R and *R*.

When the switch is opened so that no current is provided to the circuit by the generator, the capacitor will be discharged. This involves positive charges (I_G) leaving the plate that has an excess of them and traveling backward through the conductor (I_R) and onto the other plate of the capacitor. This action reduces V_C to zero. Because the positive charges leave the capacitor in the direction opposite to that in which they were accumulated, I_C is assigned a negative value as the capacitor discharges (figure 5.5*b*). In contrast, I_R occurs in the same direction and remains positive; hence, the sum of the two currents is zero.

Properties of Excitable Membranes

One can demonstrate the significance of these concepts by passing a current across the excitable membrane of a neuron and recording the accompanying change in the potential difference across the membrane. The experimental setup is shown in figure 5.6*a*. A current is passed across the membrane (I_m) as one probe, connected to a

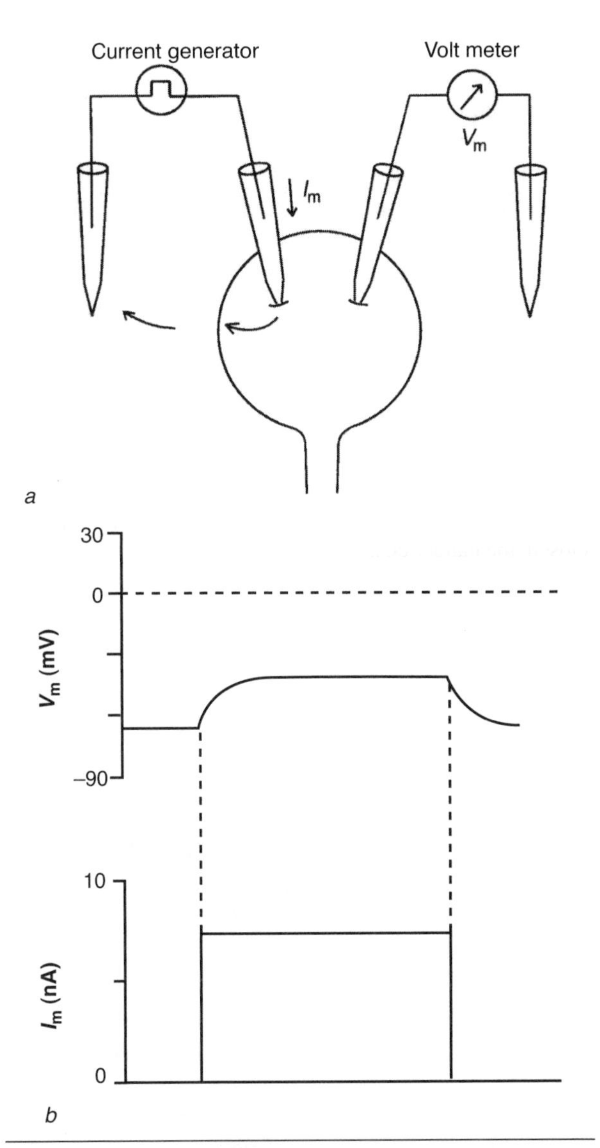

Figure 5.6 Recording the potential difference across the membrane (V_m) of a neuron. *(a)* Probes connected to a current generator and a voltmeter are placed inside and outside the neuron. *(b)* The generator delivers a current that crosses the membrane, and the accompanying change in membrane potential is recorded.

current generator, is inserted into the neuron and another probe is placed just outside the cell. The potential difference across the membrane (V_m) is recorded with two other probes that are connected to a voltmeter and similarly placed inside and outside the cell. When the generator is activated, a current is passed from the probe inside the cell to the one outside, and the effect on the membrane potential is recorded (figure 5.6*b*).

The membrane potential recording provides information about two important properties of the membrane. The first property is evident by the change in V_m when the current is turned on and then when it is turned off. The generator provided a step change in current, from zero to some preset value and back to zero again. The change in V_m, however, did not parallel the change in current. Rather, V_m changed gradually both when the current was turned on and when it was turned off. This behavior is consistent with an electrical circuit that contains a conductor and a capacitor arranged in parallel. Figure 5.5*b* indicates that when a step current was provided by a generator, I_R increased exponentially (as do the potential differences across the resistor [V_R] and capacitor [V_C]) when the switch was closed and decreased exponentially when the current was turned off. The corresponding circuit that would produce these measurements in the neuron is shown in figure 5.7*a*.

The second property evident in the V_m recording is its magnitude in the absence of a current from the generator. V_m is not zero, but has a value of about –65 mV. This is known as the resting membrane potential. This means that the membrane must include a voltage source capable of providing the resting membrane potential. This effect can be represented in an electrical circuit by inclusion of a battery. The revised circuit that accommodates both the exponential changes in V_m and the nonzero resting values is shown in figure 5.7*b*. These properties form the basis for understanding how nerve and muscle cells can generate rapid activation signals.

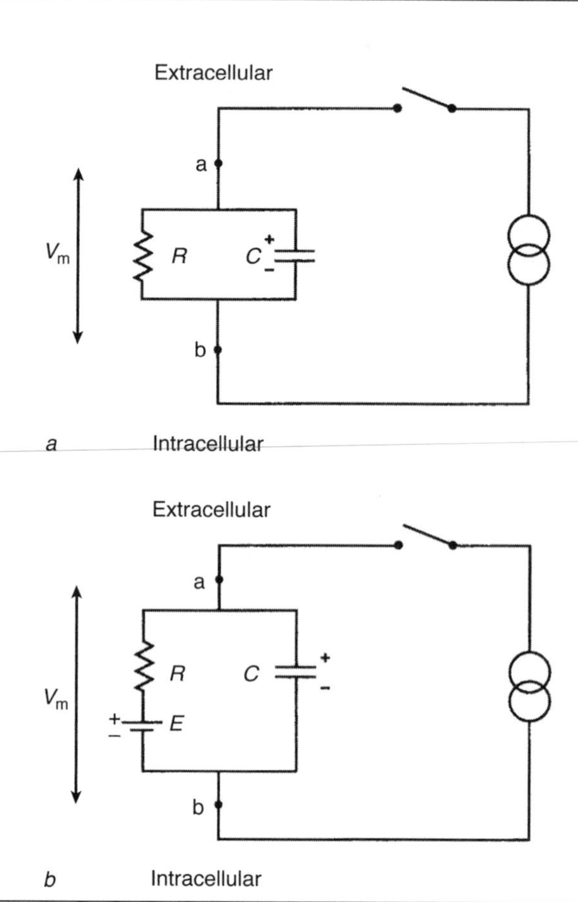

Figure 5.7 An electrical circuit for an excitable membrane. *(a)* The exponential changes in V_m can be represented by a circuit that contains a conductor and a capacitor arranged in parallel. *(b)* The nonzero value of V_m in the absence of an external current source requires the addition of a battery to the circuit. When the switch is closed, current flows around each circuit in a counterclockwise direction.

RESTING MEMBRANE POTENTIAL

The rapid transmission of an activation signal from the command centers in the central nervous system to the contractile proteins depends on two properties of the membranes of nerve and muscle cells: (1) the existence of a potential difference across the membrane when the cell is at rest and (2) the capacity to produce transient changes in the membrane potential. As described in this chapter, both processes involve the controlled movement of ions across the membrane. Because ions are charged particles, an activation signal involves the flow of current and the development of potential differences across the membrane.

Gating of Ion Channels

The structure of a cell membrane enables it to generate a potential difference, to pass current, and to store charged particles. The membrane consists of lipids and proteins. The foundation is a double layer of phospholipids that forms the surface of the membrane. Phospholipids have a hydrophilic head and a hydrophobic tail, and they join together so that the heads face the aqueous solutions of the extracellular fluid and the cytoplasm. The bilayer of phospholipids excludes water and ions, which is significant because the net charge of an ion in solution attracts water molecules. Although the net charge on a water molecule is zero, the oxygen atom has a slight negative charge and the hydrogen atoms contain a small positive charge. The lipid bilayer structure of the membrane enables it to act as a capacitor.

The phospholipid bilayer is almost impermeable to ions. However, ions can cross the membrane through specialized proteins that are part of the membrane; these are known as **ion channels.** Ion channels function as conductors and enable nerve and muscle cells to transmit changes in membrane potential rapidly over long distances. An ion channel consists of a large protein that spans the membrane with carbohydrate groups attached to its surface. The protein includes a central pore region of two or more subunits, which may be identical or different. Some channels also have auxiliary subunits that can modify the functional properties of the channel. Ion channels have three basic properties: (1) They conduct ions rapidly; (2) they recognize and select specific ions; and (3) they open and close in response to specific electrical, mechanical, or chemical stimuli.

The flux of ions through a channel is a passive event that depends on the electrostatic and concentration differences across the membrane. Most channels allow only one type of ion to pass through the membrane. The flux of an ion through a channel corresponds to a current. According to Ohm's law (equation 5.4), current depends on the conductance of the membrane to that ion and the potential difference across the membrane. Figure 5.8 demonstrates the change in current (magnitude and direction) through an ion channel with changes in the potential difference across the membrane (V_m). The ion moved into the cell (negative current) when V_m was negative, which corresponds to the inside of the membrane being negative relative to the outside, and it moved out of the cell (positive current) when V_m was positive. Some channels, however, transmit current more readily in one direction than the other, which results in a graph that has a different slope for the relation between I and V_m in the two quadrants.

Each ion channel has at least one open state and one or two closed states. Switching between these states is known as **gating.** The transition from a closed state to an open state is triggered by the occurrence of specific stimuli: (1) ligand-gated channel—the binding of a chemical transmitter; (2) phosphorylation-gated channel—phosphorylation of the protein; (3) voltage-gated channel—a change in V_m; (4) stretch- or pressure-gated channel—the presence of a mechanical stimulus such as

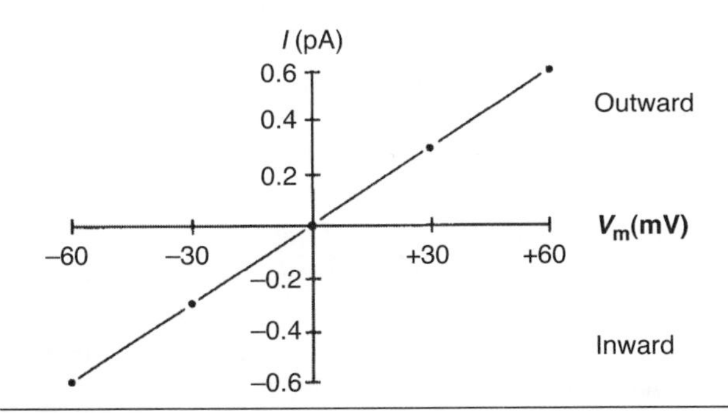

Figure 5.8 Flow of current through a single ion channel in response to variation in the potential difference across a membrane (V_m).

Table 5.1 Resting Distribution of the Major Ions Across the Membrane of a Squid Nerve Cell

Ion	Cytoplasm [mM]	Extracellular [mM]	Equilibrium potential (mV)
K^+	400	20	−75
Na^+	50	440	+55
Cl^-	52	560	−60
Ca^{2+}	0.0001	10	+145
A^-	385	—	—

stretch or pressure. Once a channel opens, it generally remains open for only a few milliseconds and then closes again. Some voltage-gated channels are more difficult to open for several milliseconds after closing due to inactivation of the channel by one of several processes.

Resting Potentials and Currents

The potential difference that exists across a cell membrane in the absence of stimulus-gated activation of ion channels is known as the **resting membrane potential.** As with all potential differences, the resting membrane potential is due to a separation of charged particles by the capacitance of the membrane. Resting nerve and muscle cells have an excess of negative charges on the inside of the membrane and positive charges on the outside, with the result that the inside is negative with respect to the outside. In many neurons, the resting membrane potential is around −65 mV. The resting membrane potential is caused by the uneven distribution of ions across the membrane. Table 5.1 lists the concentrations (mM) of four key ions on the inside and outside of a commonly studied nerve cell. Potassium (K^+) and organic anions (A^-) are more concentrated in the cytoplasm, whereas sodium (Na^+) and chloride (Cl^-) are more concentrated in the extracellular fluid. A difference in the concentration in the two spaces causes a **concentration gradient** from the high to the low concentration that functions as a battery (i.e., a voltage source). Thus, the K^+ and A^- ions experience an outward concentration gradient, whereas the Na^+ and Cl^- ions are subjected to an inward concentration gradient. Although the absolute concentrations listed in table 5.1 vary across the cell membranes of different vertebrates, the concentration gradients are similar.

A key factor that contributes to the uneven distribution of ions across the resting membrane is the ion channels that are open when the cell is at rest (figure 5.9). These are known as **resting channels.** Nerve cells contain resting channels for K^+, Na^+, and Cl^-. The diffusion of an ion

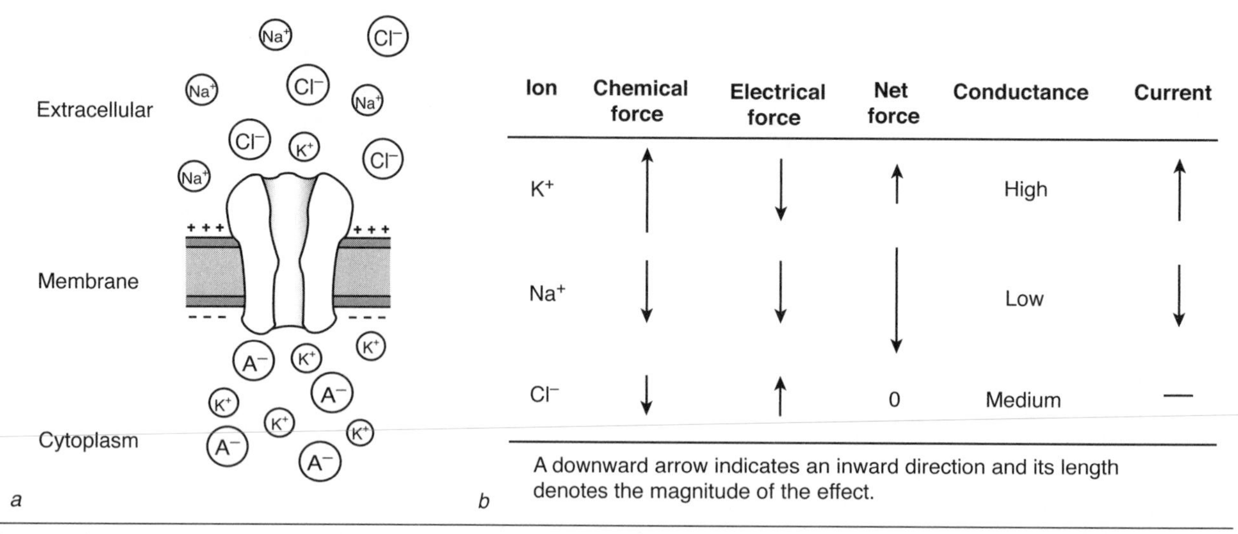

Figure 5.9 Ions involved in establishing the resting membrane potential. *(a)* A resting K^+ channel and the uneven distribution of ions across the membrane of a nerve cell. *(b)* The opposing currents carried by K^+ and Na^+ are due to different combinations of forces and conductances that they experience.

across the membrane depends on the balance of the two forces that it experiences: (1) *chemical force*—the force produced by the concentration gradient; and (2) *electrical force*—the electrostatic force associated with the potential difference across the membrane. For example, the flux of K^+ ions across the membrane depends on the balance between the outward chemical force and the inward electrical force that it experiences and the conductance of the membrane to the ion. The chemical force acting on the K^+ ions is due to its concentration gradient, whereas the electrical force is due to the repulsion that the positively charged ion experiences from the positive charges on the extracellular side of the membrane. When diffusion of K^+ has proceeded to the point where the chemical and electrical forces it experiences are balanced, the membrane potential is known as the **equilibrium potential** for K^+ (E_K).

The equilibrium potential for an ion (X) can be calculated from the **Nernst equation:**

$$E_X = \frac{58\text{ mV}}{z}\log\frac{[X]_o}{[X]_i} \quad (5.8)$$

where z is the valence of the ion and $[X]_o$ and $[X]_i$ are the concentrations of the ion outside and inside the cell. The first term in the Nernst equation (58 mV/z) corresponds to the electrical force, and the second term ($[X]_o/[X]_i$) corresponds to the chemical force. The equilibrium potential for K^+, which has a valence of +1 and the concentrations listed in table 5.1 is as follows:

$$E_K = \frac{58\text{ mV}}{1}\log\frac{[20]}{[400]} = -75\text{ mV}$$

The equilibrium potentials for Na^+, Cl^-, and Ca^{2+} are listed in table 5.1. Because E_{Cl} is so close to the resting membrane potential, there is little movement of Cl^- ions across the resting membrane.

Because the resting membrane potential of a typical nerve cell is about −65 mV and not −75 mV, this means that other ions besides K^+ must be involved in establishing the potential difference across a resting membrane. The other key ion is Na^+. Both the chemical force and the electrical force acting on Na^+ ions when the cell is at rest are directed inward, which means that Na^+ ions experience a large net force driving Na^+ into the cell (figure 5.9). The inward current carried by Na^+ ions deposits positive charges on the inside of the membrane, which causes the resting membrane potential to be less negative than the equilibrium potential for K^+. The average value of the resting membrane potential is relatively constant due to a balance between the inward Na^+ current and the outward K^+ current (figure 5.9).

As described by equation 5.7, the current carried by an ion depends on the product of the membrane conductance for the ion and the driving force ($E - V_C$) that it experiences. The driving force, as we have just discussed, is the result of the magnitude and direction of the chemical and electrical forces. The conductance, however, depends on the number of ions available to provide the current (i.e., concentration gradient) and the number of resting channels that are open. Because there are many more resting channels for K^+ than for Na^+, membrane conductance is much greater for K^+. Thus, although the net driving force acting on the K^+ ions is smaller than that for the Na^+ ions, the difference in the number of resting channels results in currents that have similar magnitudes (figure 5.9*b*). Because the efflux of K^+ and the influx of Na^+ involve the passive movement of ions, their diffusion is sometimes referred to as *leak currents.*

Note in figure 5.9*b* that to assess the effect of a current on the membrane potential, it is necessary to account for both the magnitude and direction of the current. Identifying the direction involves specifying whether the ion moves into or out of the cell and assigning it a sign (i.e., negative or positive). Because a current is defined as the flow of positive charges, both the influx of Na^+ into a cell and the efflux of Cl^- out of the cell are described as inward currents. When currents are graphed, the convention is to indicate an inward current as a negative current (e.g., figure 5.8).

Due to the role of the concentration gradients in generating the resting membrane potential, a cell cannot sustain the passive diffusion of K^+ and Na^+ ions indefinitely without a mechanism to maintain the gradients. The primary mechanism is a protein embedded in the membrane that can return K^+ and Na^+ ions to their regions of high concentration. Because the two ions have to be moved against their electrochemical gradients, the action of the protein requires it to use energy and it is therefore referred to as a pump. The **Na^+-K^+ pump** is a large protein that spans the membrane and has binding sites for K^+ on the extracellular surface and for Na^+ and adenosine triphosphate (ATP) on the cytoplasmic side. One ATP molecule is hydrolyzed during each cycle of the pump, and the energy that is released is used to remove three Na^+ ions from the cytoplasm and to return two K^+ ions to the cytoplasm. The unequal transfer of the two ions results in a net outward current that maintains the negative charge on the inside of the membrane. This is known as the *electrogenic* action of the pump. The electrogenic action of the Na^+-K^+ pump contributes about –5.6 mV to the resting membrane potential.

EXAMPLE 5.1

Na^+-K^+ Pump Action

The Na^+-K^+ pump is a transmembrane molecule that moves three Na^+ ions and two K^+ ions across the membrane for each cycle of action. The protein uses energy from ATP to perform this function. It has been estimated that each pump can split a maximum of 8000 molecules of ATP per minute (Clausen et al., 1998). In the skeletal muscles of mammals, the concentration of Na^+-K^+ pumps is about 0.2 to 0.8 nmol/g wet weight, but this increases with strength training and endurance training (Green et al., 1999*a*) and decreases under hypoxia (Green et al., 1999*b*). Under resting conditions, only about 5% of these pumps are active in exchanging Na^+ and K^+. When a muscle is activated, however, either by voluntary activation or by electrical stimulation, most of the Na^+-K^+ pumps become active (Hicks & McComas, 1989; Nielsen & Harrison, 1998). Nonetheless, the pumps seem unable to deal with the huge flux of ions that occurs with sustained excitation, which probably contributes to development of muscle fatigue under some conditions (Clausen et al., 1998; Nielsen & Harrison, 1998, Sjøgaard, 1996; Verburg et al., 1999). For example, the duration over which subjects could cycle on an ergometer at a high workload was prolonged by the administration of an antioxidant compound that reduced the depressive effect of reactive oxygen species on Na^+-K^+ pump activity (McKenna et al., 2006).

An Equivalent Electrical Circuit

The resting membrane potential is a product of four properties: the capacity of the membrane to separate charged particles, the existence of concentration gradients across the membrane, the ability of ion channels to allow selected ions to cross the membrane, and the restorative activity of the Na^+-K^+ pump. The function of these four components can be represented in an electrical circuit. Let us begin developing the electrical circuit by considering only the resting K^+ channels, the concentration gradient for K^+, and the capacity of the membrane to separate charged particles. These three properties can be represented as a conductor, a battery, and a capacitor, respectively. The corresponding circuit is shown in figure 5.10*a*. Because the current flows from the positive terminal of a battery, the orientation of the K^+ battery (E_K) indicates the direction of the current (I_K) through the resting K^+ channels (g_K) due to the concentration gradient for K^+. I_K deposits positive charges on the outside of the membrane, and the negative terminal of the battery draws positive charges from the inside of the membrane; thus, the membrane functions as a capacitor (C_m) and is responsible for the potential difference across the membrane (V_m).

Now we need to add to the circuit the resting Na^+ channels, the concentration gradient for Na^+, and the Na^+-K^+ pump. This involves adding a conductor for Na^+ ions (g_{Na}), a battery for the equilibrium potential (E_{Na}), and current generators for the Na^+-K^+ pump. The positive terminal of the battery for Na^+ faces into the cell to represent the direction of its equilibrium potential. Because the Na^+-K^+ pump uses energy to move ions against an electrochemical gradient, its actions are represented as two current generators: one for K^+ (I'_K) and another for Na^+ (I'_{Na}). Note the difference in the length of the arrows for the Na^+-K^+ pump currents to represent the influx of two K^+ ions and the efflux of three Na^+ ions. The revised circuit is shown in figure 5.10*b*.

NEURONS

Now that we understand how the membranes of nerve and muscle cells can establish a resting membrane potential, we next consider the means by which these cells can transmit

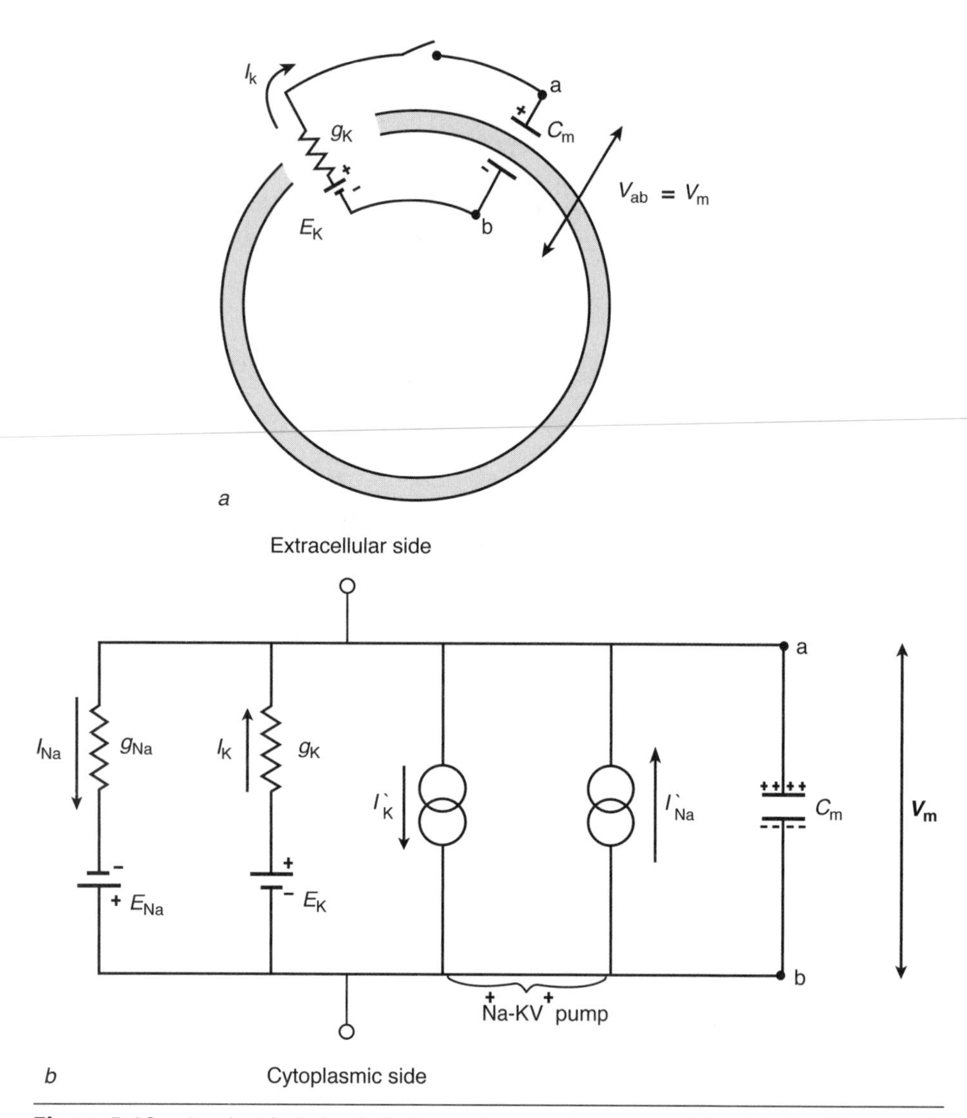

Figure 5.10 An electrical circuit for the resting membrane potential. *(a)* A simplified circuit that shows the contribution of the K^+ current (I_K) to the separation of charges across the membrane (C_m) and the potential difference (V_m) that develops across the capacitor. *(b)* A more complete circuit that includes the contributions of the Na^+ ions and the Na^+-K^+ pump to V_m.

a rapid activation signal from the central nervous system to the contractile proteins in muscle. The basic unit of the activation signal is an *action potential,* which corresponds to a transient reversal of the potential difference across the membrane of the cell. This section describes the properties of nerve cells (neurons) that enable them to generate action potentials, a technique that can be used to measure the associated membrane currents, and the characteristics of an action potential.

Properties of Neurons

Each neuron receives thousands of contacts, which are known as *synapses,* from other neurons (figure 5.11). In terms of the electrical circuits that we have discussed in this chapter, a synapse acts as a current source for a neuron. Even when a neuron is at rest, some synapses are activated occasionally and provide current inputs to the neuron. Because the delivery of a current to a neuron can alter the potential difference across the membrane, the continual current inputs received by a neuron cause its resting membrane potential to fluctuate about a steady-state value. When the input is sufficient to cause the membrane potential to exceed a threshold value that is slightly above the resting membrane potential, the neuron will discharge an action potential. The change in membrane potential in response to an applied current differs among neurons because it depends on the input conductance and the membrane capacitance of the neuron, that is, the time constant of the neuron. Furthermore, the capacity of the neuron to conduct changes in membrane potential, such as action potentials, is influenced by its axial conductance and the associated length constant.

Input Conductance

Figure 5.6 shows the effect of an applied current on the membrane potential of a neuron. The resting membrane potential of the neuron is about –65 mV. When a current is passed from an electrode inside the neuron to one that is outside, the membrane potential becomes less negative. This change in membrane potential, which is known as **depolarization** of the membrane, is caused by the accumulation of some of the positively charged ions carrying the current on the inside of the membrane so that the potential difference across the membrane (capacitor) is reduced. The converse effect occurs when the current travels in the opposite direction, from the electrode outside the cell to one inside the cell. In this instance, positive ions on the inside of the membrane are attracted toward the intracellular electrode, which increases the potential difference across the membrane. This change in membrane potential is known as **hyperpolarization** of the membrane. The magnitude of the change in membrane potential experienced by a neuron in response to an applied current, whether the change involves depolarization or hyperpolarization of the

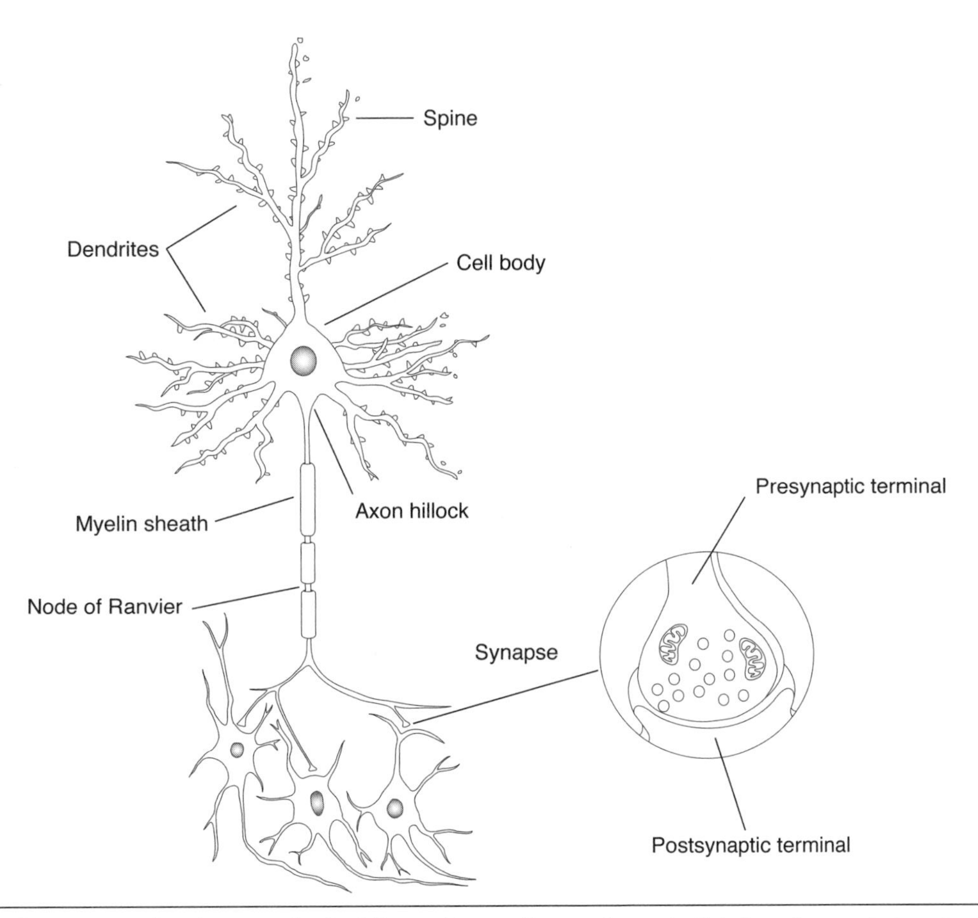

Figure 5.11 A neuron receives inputs on its dendrites and transmits an action potential along its axon to target cells. Most of the synapses that deliver input to the neuron occur on the dendrites, many of them on dendritic spines. Each synapse includes a presynaptic neuron that delivers the input and the postsynaptic neuron that receives it.

membrane potential, depends on the input conductance of the neuron.

The **input conductance** (g_{in}) of a neuron is a measure of its ability to conduct an applied current. It is determined by means of the experiment shown in figure 5.6*a*, which involves passing currents of different magnitude across the membrane in both directions and measuring the resulting change in membrane potential. When the data are graphed as in figure 5.8, we know from Ohm's law that the slope of the current–voltage relation represents g_{in} of the neuron (equation 5.9), which is the reciprocal of its **input resistance** (R_{in}). The functional significance of a neuron's g_{in} is that it indicates the change in membrane potential (ΔV_m) that the neuron will experience in response to an applied current.

$$\Delta V_m = \frac{I_m}{g_{in}} \tag{5.9}$$

For an ideal spherical neuron with no processes, g_{in} depends on the density of the resting ion channels and the size of the neuron. If the conductance per unit area of membrane, which is known as specific membrane conductance, is the same for all neurons, g_{in} is least in the smallest neuron. Consequently, the ΔV_m in response to a given I_m will be greatest in the smallest neuron (equation 5.9). Because the voltage threshold is similar in neurons of all sizes, the smallest neuron will achieve the required ΔV_m to reach threshold with the smallest I_m. This means that when a current applied to a group of neurons increases gradually, the smallest neuron will be the first to reach threshold and to discharge an action potential. Although specific membrane conductance varies among neurons and the size of the dendrites that extend from the cell body also influences g_{in}, there is a strong association between the size of a neuron and its response to applied currents.

Membrane Capacitance

The electrical-circuit model of an excitable membrane comprises a conductor and a capacitor in parallel (figure 5.7). When a current is applied to an excitable membrane, therefore, there are two paths along which it can travel. The current that charges the capacitor is known as the **capacitive current,** and the current that travels through the conductor is called the **ionic current.** The capacitive current corresponds to the accumulation of positive charges on one side of the membrane and the loss of positive charges on the other side. The ionic current represents the movement of ions through the ion channels in the membrane.

When a source delivers a current to such a circuit, all the current initially appears as a capacitive current that declines over time and finally exists as just an ionic current (figure 5.5*b*). Although it is the capacitive current that produces the potential difference across a membrane, the change in the membrane potential parallels the change in the ionic current (figure 5.5*b*). The shift from a capacitive current to an ionic current depends on the size of the capacitor (surface area of the neuron) that has to be charged. Because large neurons require a greater amount of capacitive current to charge the capacitor, it takes them longer to realize a significant ionic current and the corresponding change in membrane potential. Thus, the voltage response of a neuron to an applied current is altered by its membrane capacitance.

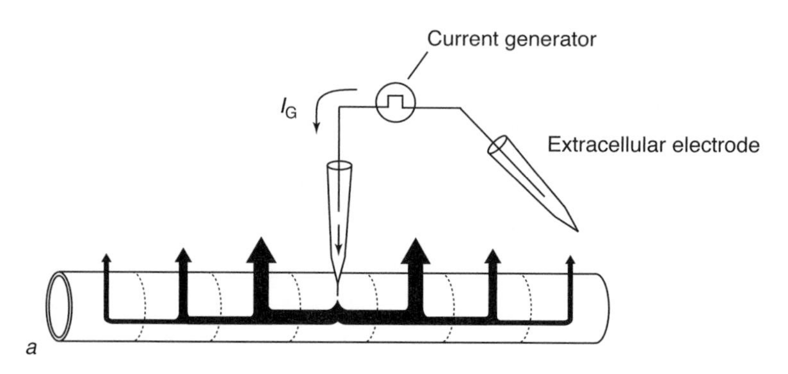

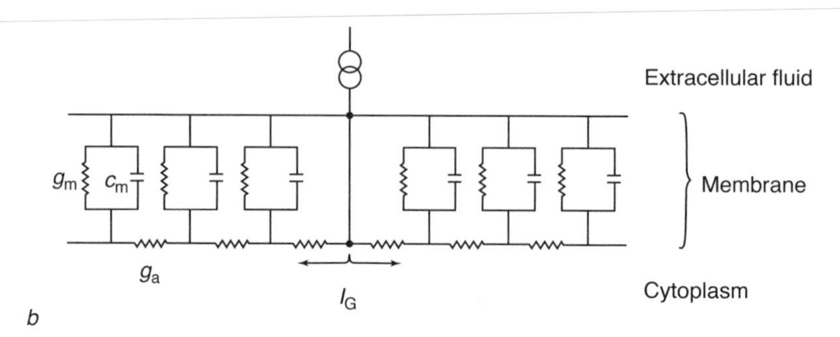

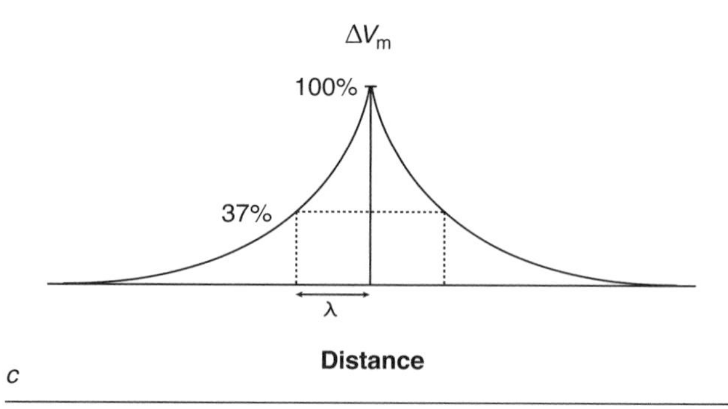

Figure 5.12 The change in membrane potential (ΔV_m) of a dendrite produced by an applied current. *(a)* A section of a dendrite is represented as a series of seven length segments. A current generator delivers a current (I_G) through an intracellular electrode. The current flows along the dendrite in both directions and across the membrane. The amount of current that crosses the membrane declines with an increase in the distance from the current source. *(b)* The section of dendrite shown in part *a* is represented by an electrical circuit that comprises a membrane conductor (g_m) and capacitor (c_m) in parallel for each length segment, with each length segment connected by an axial conductance (g_a). *(c)* The ΔV_m decays exponentially with an increase in the distance from the current source. The decrease is characterized by the length constant (λ), which indicates the distance from the current source over which ΔV_m has declined to 37% of its original value.

Axial Conductance

Once the potential difference across a local region of the membrane has been changed by a current source, the change in membrane potential is transmitted along the neuron. There are active and passive processes that can transmit the change in membrane potential. The active process involves the generation of an action potential, which will be discussed in the next section. The passive process concerns the extent to which the change in membrane potential will spread simply due to flow of current in an electrical circuit.

Neurons comprise three distinct regions: dendrites, cell body, and axon (figure 5.11). Much of the input arrives through synapses on the dendrites and is transmitted passively toward the cell body and the axon hillock. One can visualize the passive spread of a change in membrane potential along a dendrite, which is known as **electrotonic conduction,** by representing a dendrite as a cable that comprises a series of membrane circuits connected by conductors (figure 5.12). The conductors that connect adjacent sections of the dendrite represent the **axial conductance** (g_a) of the dendrite, which corresponds to the capacity of the dendrite to transmit a current along its length. Axial conductance depends on the length and diameter of the dendrite; it decreases with length and increases with diameter. Short, large-diameter dendrites have a higher axial conductance and will passively transmit the change in membrane potential a greater distance.

The spread of the change in membrane potential depends on the extent to which the current travels along the dendrite. When a current is delivered to a dendrite, it will follow the path of greatest conductance. Because conductance decreases with distance along the dendrite due to an increase in the number of g_a units (figure 5.12*b*), most of the current crosses the membrane close to the current source. Hence, the change in membrane potential (ΔV_m) is greatest for those sections of the membrane that are closest to the current source and decreases exponentially with distance from the source (figure 5.12*c*). The distance the current will spread passively to change the membrane potential to 37% of the ΔV_m occurring at the current source is known as the **length constant** (λ). The length constant is proportional to the square root of the ratio of g_a and g_m (figure 5.12*c*), which means that ΔV_m decays less as g_a increases or as g_m decreases.

Membrane Currents

Several different ions can contribute to the current that crosses the membrane. The role of different ions has been determined with a technique known as the **voltage clamp.** In this chapter, we have learned that the flow of current contributes to a potential difference between two locations, such as those across a membrane. The voltage-clamp technique works in the opposite way by imposing a potential difference and then measuring the currents that flow to maintain the potential difference. The voltage clamp is a current generator that is connected to a pair of electrodes placed on each side of a membrane. The current provided by the generator drives the potential difference across the membrane to a value set by an experimenter, and the voltage-clamp device measures the requisite currents.

The voltage-clamp technique has been used to identify the currents responsible for the action potential. When the voltage clamp is set to impose 10 mV of depolarization (figure 5.13*a*), which involves reducing the potential difference across the membrane by 10 mV, the current across the membrane (I_m) includes both a capacitive current (I_C) and a leak current (I_l). Recall that I_C represents the current associated with charging and discharging the membrane capacitance (figure 5.5) and that I_l corresponds to the ionic current associated with the resting channels. When the voltage clamp imposes 60 mV of depolarization, I_m is larger and more complex (figure 5.13*b*). In addition to both I_C and I_l, I_m now includes a large inward current followed by a large outward current. The large depolarization exceeded the threshold of the cell and activated voltage-gated ion channels that were not required for the 10 mV of depolarization. The ions responsible for the inward and outward currents in figure 5.13*b* are identified when the voltage-clamp experiment is repeated on preparations in which selective drugs and toxins are used to block different ion channels. The results reveal that the inward current is carried by Na^+ ions and that the outward current involves K^+ ions.

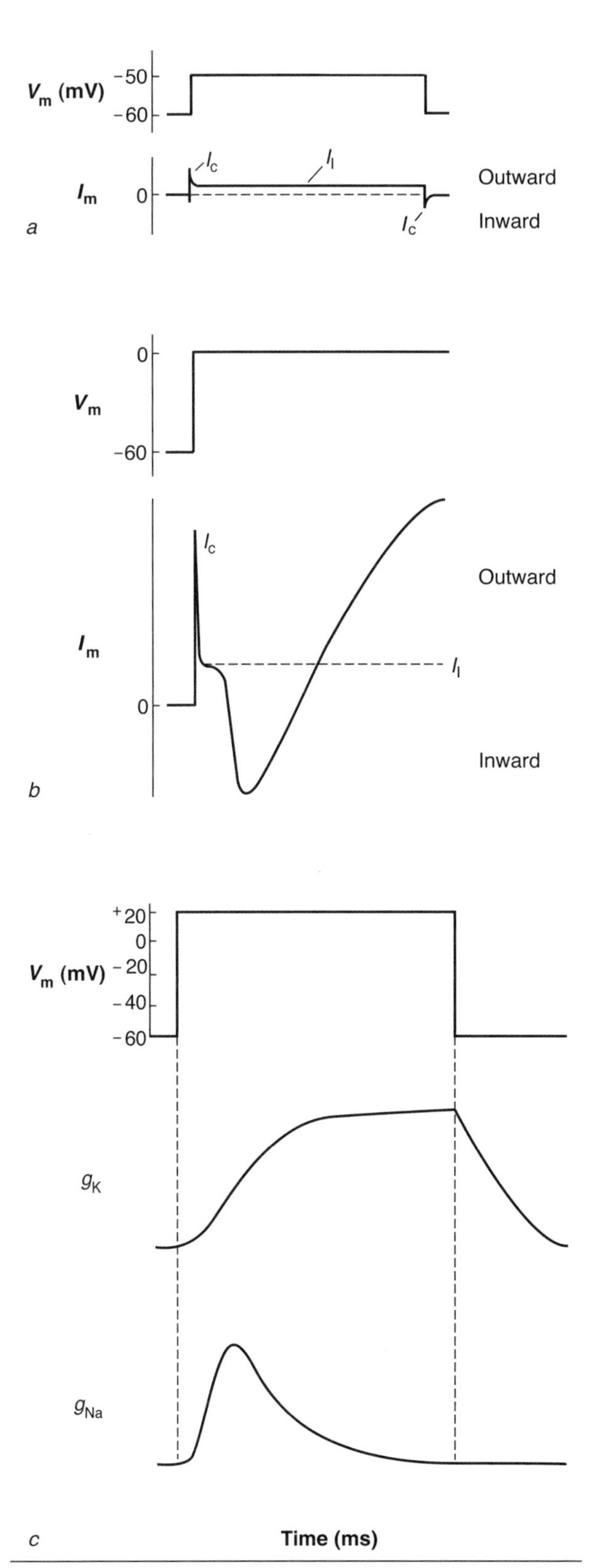

Figure 5.13 Voltage-clamp measurements of membrane currents (I_m) and conductances. *(a)* A small depolarization is accompanied by capacitive (I_C) and leak (I_l) currents. *(b)* A large depolarization activates inward and outward currents in addition to I_C and I_l. *(c)* Calculated changes in g_K and g_{Na} of a neuron in response to a large depolarization.

The currents measured with the voltage-clamp technique can be used to calculate the voltage-gated changes in sodium and potassium conductances, which are the basis of the action potential. The calculation relies on Ohm's law, which can be rearranged as expressions for the K^+ and Na^+ conductances:

$$g_K = \frac{I_K}{(V_m - E_K)} \text{ and } g_{Na} = \frac{I_{Na}}{(V_m - E_{Na})} \qquad (5.10)$$

The calculation of g_K and g_{Na} requires knowledge of V_m, E_K, E_{Na}, I_K, and I_{Na}. The experimenter sets the potential difference across the membrane (V_m), and the two currents (I_K and I_{Na}) are determined from the voltage-clamp experiments. The denominator indicates the driving force experienced by the ion. The equilibrium potentials (E_K and E_{Na}) are constant values that correspond to the V_m when the ion experiences a balance between its chemical and electrical forces. Experimenters can measure the equilibrium potentials by using a microelectrode to record the current that flows through a single ion channel during a voltage-clamp experiment and determining the value of V_m at which there is a change in the direction of the current. The voltage at which the current changes direction is known as the **reversal potential.** With these data, the two conductances can be calculated to reveal that when the depolarization of the neuron exceeds threshold, there is a rapid increase and subsequent decline in g_{Na} and a more gradual sustained increase in g_K (figure 5.13*c*).

Action Potential

The resting membrane potential of a neuron is not constant but rather fluctuates about a steady-state value due to the actions of the resting channels, the Na^+-K^+ pump, and the activation of a few synapses. When a number of synapses are activated concurrently and the current input depolarizes the membrane potential by more than about 15 mV, the voltage-gated Na^+ and K^+ conductances are activated to generate an action potential. An **action potential** is a transient reversal in the potential difference across the membrane that is transmitted rapidly along an excitable membrane.

An action potential is typically an all-or-none event, which means that once the voltage threshold has been crossed, the reversal of the membrane potential will occur within a few milliseconds. The ion channels through which Na^+ can pass are typically closed in a resting state, but they open rapidly when the membrane is depolarized above threshold; they are voltage-gated channels. The opening of the Na^+ channels corresponds to an increase in g_{Na}. With an increase in g_{Na}, Na^+ ions are driven into the cell and represent an inward current that deposits positive charges on the inside of the membrane. The accumulation of positive charges on the inside of the membrane further reduces the potential difference across the membrane, which opens more voltage-gated Na^+ channels; and the influx of Na^+ further depolarizes the membrane. Because both the chemical and electrical forces acting on Na^+ drive the ions into the cell (figure 5.9), the inward Na^+ current due to the increase in g_{Na} causes the membrane potential to approach E_{Na}. The influx of Na^+ ions produces the depolarization phase of the action potential (figure 5.14*a*), which produces a brief **overshoot** phase when the polarity of the membrane potential is reversed so that the inside is positive with respect to the outside.

The increase in g_{Na} is transient (figure 5.14*b*), however, as Na^+ channels can remain open for only a few milliseconds before they are inactivated. Concurrently, there is a gradual increase in g_K due to the opening of voltage-gated K^+ channels. Because the influx of Na^+ ions reverses the direction of the electrical force acting on the K^+ ions (figure 5.9), the increase in g_K results in an outward K^+

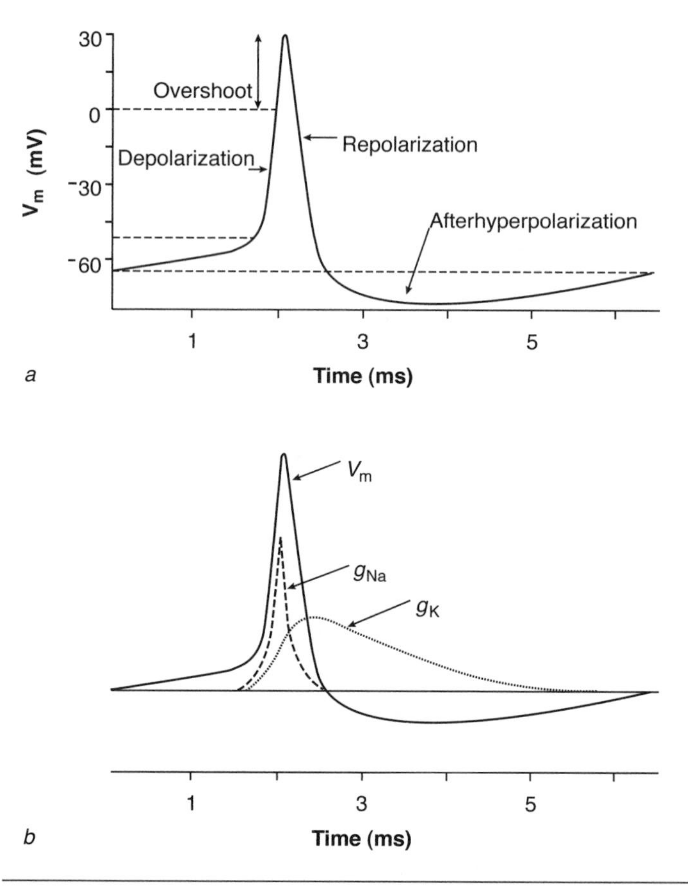

Figure 5.14 (*a*) Changes in the membrane potential (V_m) and (*b*) the responsible Na^+ (g_{Na}) and K^+ (g_K) conductances comprising an action potential.

current that reduces the net positive charge on the inside of the membrane and returns the membrane potential to its initial state. The efflux of K^+ ions, therefore, produces the **repolarization** phase of the action potential (figure 5.14*a*). The increase in g_K lasts for so long (figure 5.14*b*) that the sustained outward K^+ current hyperpolarizes the membrane and produces the **afterhyperpolarization** phase of the action potential (figure 5.14*a*). For a few milliseconds after the action potential, the membrane is **refractory** and it is difficult to generate another action potential. The duration of the refractory period depends on the time it takes for the Na^+ channels to be released from the state of inactivation and the time needed for the voltage-gated K^+ channels to close. Although the basic characteristics of the action potential can be explained by the actions of the Na^+ and K^+ channels, there are at least 13 conductances that can influence the profile of the action potential (Nordstrom et al., 2007).

Now that we know how an action potential is generated, we next consider how it is transmitted. Recall that we previously distinguished between active and passive transmission of a local change in membrane potential. Passive transmission is achieved by electrotonic conduction, whereas active transmission involves the **propagation** of an action potential. This requires the displacement of the transient reversal in the membrane potential along an excitable membrane. In a neuron, the action potential is generated at the axon hillock (figure 5.11) and is propagated along the axon to the target cells. We can explain the processes involved in propagating an action potential by considering how an action potential propagates along a myelinated axon (figure 5.11) by jumping from one node of Ranvier to the next.

Figure 5.15 shows part of a myelinated axon and indicates the membrane potential across three adjacent nodes of Ranvier. An action potential, which is indicated as a reversal of the membrane potential, is located at node

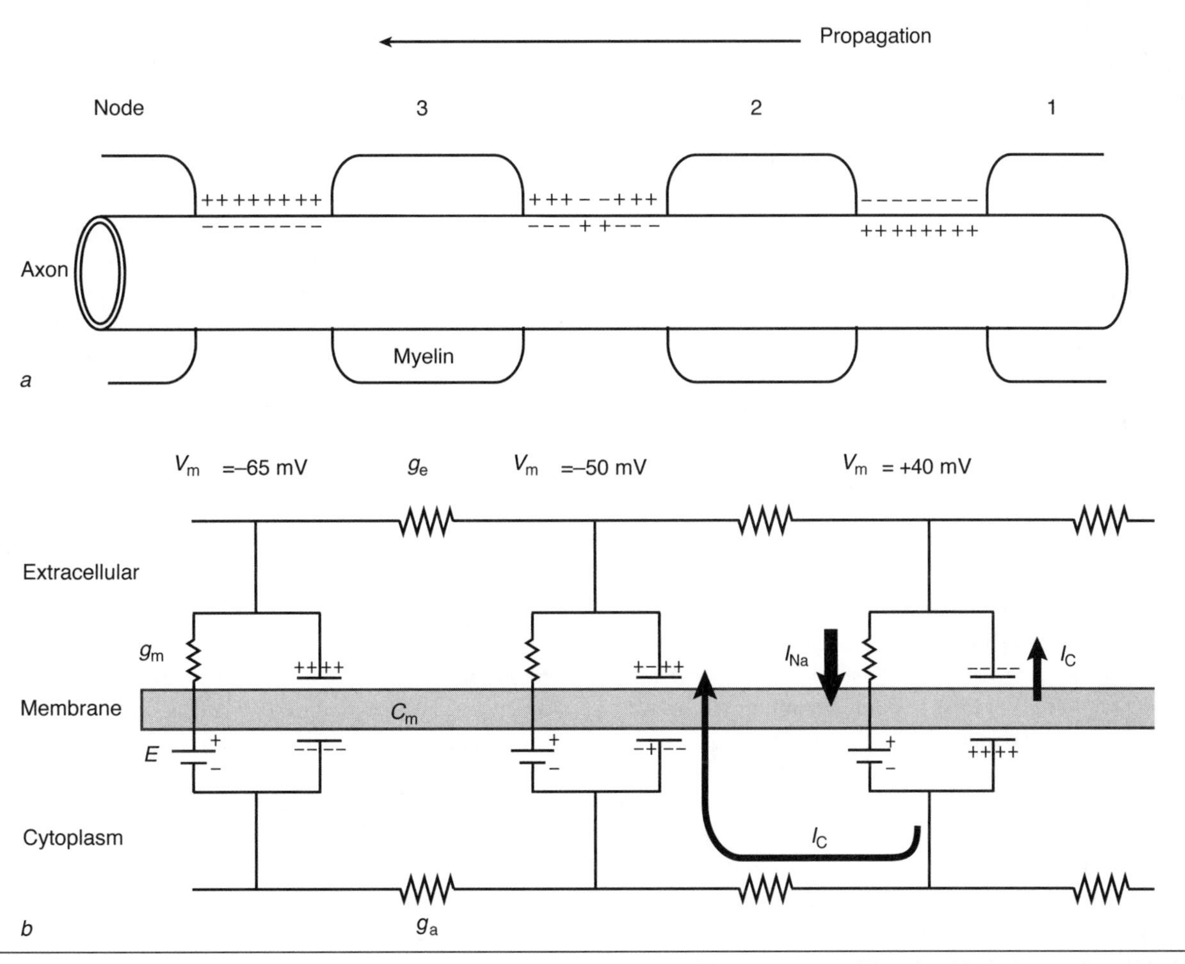

Figure 5.15 Action potential propagation. *(a)* Part of a myelinated axon with three nodes of Ranvier: Node 1 contains an action potential with the membrane potential reversed (V_m = +40 mV); node 3 is at resting state, with the inside negative with respect to the outside (V_m = –65 mV); and node 2 is slightly depolarized, with a reduced membrane potential (V_m = –50 mV). *(b)* Equivalent circuits for each node and conductors that connect the nodes by an axial conductance (g_a) and an extracellular conductance (g_e). The inward Na^+ current (I_{Na}) produces an outward capacitive current (I_C) at both nodes 1 and 2.

1 and is about to be propagated to nodes 2 and 3 (figure 5.15*a*). In terms of our electrical circuits, the inward Na^+ current (I_{Na}) that depolarizes the membrane represents a discharging of the membrane capacitance and is indicated as an outward capacitive current (I_C) at node 1 (figure 5.15*b*). However, not all of I_{Na} flows as an outward I_C at node 1, as some travels passively by electrotonic conduction to node 2 and discharges the capacitance of the membrane at this location. This means that some positive charges (current) accumulate on the inside of the membrane at node 2, which depolarizes the membrane there. When the amount of depolarization exceeds the voltage threshold (~15 mV), the voltage-gated Na^+ and K^+ channels at node 2 are activated and an action potential is generated. By the same processes, the action potential will propagate to node 3, but it will not travel back to node 1 because of the refractoriness of that membrane. Action potentials propagate along excitable membranes that are not myelinated, such as muscle fibers, by the same mechanisms.

Two factors cause the speed of action potential propagation to depend on the diameter of the axon. The first factor is the speed of electrotonic conduction from one node to the next. This depends on two passive properties: axial conductance (g_a) and membrane capacitance (C_m). The speed at which the action potential is propagated from node 1 to node 2 (figure 5.15*b*) depends on how quickly the I_{Na} at node 1 can cause an I_C at node 2 and depolarize the membrane enough to activate the voltage-gated channels. This involves electrotonic conduction from node 1 to node 2 through the axial conductor and then depolarization of the membrane at node 2. The speed at which this occurs depends on g_a/C_m. Of these two passive properties, g_a has the greater influence because it varies as the square of axon diameter, whereas C_m increases in direct proportion to axon diameter. Thus, the passive properties cause propagation velocity to increase with axon diameter.

The second factor that influences propagation velocity is myelination. Myelination involves glial cells (oligodendrocytes in the central nervous system and Schwann cells in the peripheral nervous system) wrapping around the axon of a neuron (figure 5.11). Because the capacitance of a parallel-plate capacitor, such as a membrane, is inversely related to the distance between the plates, myelination reduces C_m of the myelinated region of the axon and thereby prevents much of the surface area of the axon from participating in the propagation of the action potential. Furthermore, the membrane at each node (~2 µm in length) has a high density of voltage-gated Na^+ channels, which enables it to generate a significant inward Na^+ current. As a consequence, myelination increases the propagation velocity of action potentials along axons.

EXAMPLE 5.2

Demyelinating Diseases

The presence of myelin is somewhat fragile, and there are a number of neurological diseases that cause the focal or patchy destruction of myelin sheaths, often as a result of an inflammatory response. These demyelinating diseases can influence myelin in the central nervous system (e.g., multiple sclerosis, encephalomyelitis, myelopathy) and the peripheral nervous system (e.g., Guillian-Barré syndrome, some types of neuropathies) and can impair the propagation of action potentials.

SYNAPTIC TRANSMISSION

Before an action potential can be initiated and propagated, the nerve or muscle cell has to receive current inputs that cause the membrane potential to exceed the voltage threshold of the cell. The current inputs occur at synapses (figure 5.11). A synapse is a specialized connection that enables a change in membrane potential in one cell to be transmitted to another cell. The cell that delivers the change in membrane potential is known as the *presynaptic cell,* and the one that receives it is the *postsynaptic cell.* This section describes two types of synapses, the properties of the synapse that connects a nerve cell to a muscle cell, the integration of multiple synaptic inputs, and signaling pathways inside a postsynaptic cell that can be activated by synaptic input.

Electrical and Chemical Synapses

There are two ways in which the change in membrane potential in the presynaptic cell can be transmitted to the postsynaptic cell. One way, which is known as an **electrical synapse,** involves the flow of current directly between the two cells. The other way, which is known as a **chemical synapse,** involves the release of a chemical agent by the presynaptic cell and its attachment to the postsynaptic cell.

At an electrical synapse, the membranes of the pre- and postsynaptic cells are separated by about 3 nm, and the current flows through specialized channels known as **gap junctions** (figure 5.16). Gap junctions comprise a pair of channels in the membranes of the pre- and postsynaptic cells. Gap junctions can be up to 2 nm in diameter, which is large enough to allow the passage of all the ions that contribute to the membrane potential of a cell as well as many small organic molecules that may function as metabolic signals. Gap junctions are relatively rare between neurons in the adult mammal; but

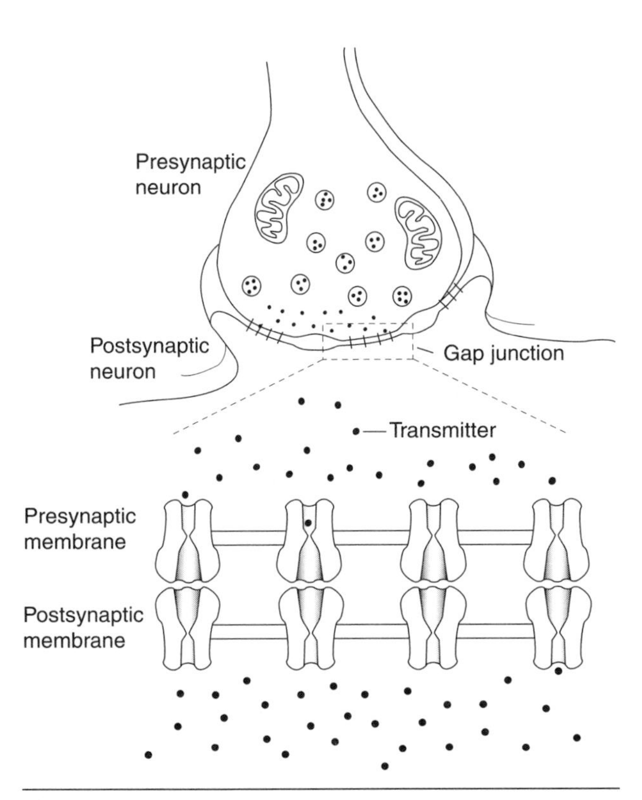

Figure 5.16 Transmission at an electrical synapse. Ions flow through gap junctions that connect pre- and postsynaptic cells.

they are common between such cells as glia, epithelial cells, smooth and cardiac muscle cells, liver cells, and some glandular cells.

The properties of electrical synapses include the high conductance of the pathway, the capacity for the current to flow in either direction, the speed of the transmission between cells, and the ability to close. Voltage-gated ion channels in the presynaptic cell generate the current that passes through gap junctions, which requires the existence of an action potential in the presynaptic cell. To evoke an action potential in the postsynaptic cell, the synapse must contain enough gap junctions that the flow of current is sufficient to depolarize the postsynaptic membrane above its voltage threshold. This process can be facilitated by the smaller size of the postsynaptic cell, whereby it has a lower input conductance (higher input resistance) and therefore experiences a greater change in membrane voltage in response to a given presynaptic current (Ohm's law). Often, however, many cells are connected by electrical synapses and are electrically equivalent to a larger cell—one that is less responsive to the flow of current and in which it is relatively difficult to initiate an action potential. When the voltage threshold is reached, however, the cells will generate action potentials synchronously and produce an enhanced response. This mechanism is used to produce defensive responses in several species and to facilitate a burst of hormone secretion by neurons in the hypothalamus.

In addition to being bidirectional, most electrical synapses can pass both depolarizing and hyperpolarizing currents. A depolarizing current causes positive ions to be deposited on the cytoplasmic side of the postsynaptic membrane, whereas a hyperpolarizing current corresponds to the transfer of negative ions. Some gap junctions, however, are voltage gated, so that they are restricted to passing depolarizing currents only. When the change in membrane potential in the postsynaptic cell does not evoke an action potential, the change in membrane potential will spread along the membrane by electrotonic conduction (figure 5.12).

Because electrical synapses effectively connect the cytoplasmic content of cells, gap junctions have the capacity to close and isolate a cell. Such actions can be triggered by a decrease in cytoplasmic pH or an increase in cytoplasmic Ca^{2+}, which are common markers of cell damage. The closing of a gap junction may involve a conformational change in the orientation of the six subunits, known as **connexins,** in the channel located in the postsynaptic membrane.

Compared with what occurs with an electrical synapse, transmission of the change in membrane potential at a chemical synapse is relatively slow, although it may take only about 0.5 ms. The process involves the release of a chemical agent by the presynaptic cell and its attachment to ligand-gated channels in the membrane of the postsynaptic cell (figure 5.17). The chemical agent is known as a **neurotransmitter.** There are two main classes of neurotransmitters: small-molecule transmitters (e.g., acetylcholine, glutamine, glycine) and neuroactive peptides that are short polymers of amino acids. The release of neurotransmitter occurs from **presynaptic terminals,** which are specialized swellings of the axon. Neurotransmitter is stored in **synaptic vesicles** in the presynaptic terminals. Each vesicle contains ~5000 molecules of a specific neurotransmitter.

The release of neurotransmitter by a presynaptic cell requires the arrival of an action potential at the presynaptic terminal. The change in membrane potential associated with the action potential activates voltage-gated Ca^{2+} channels and results in the flux of Ca^{2+} down its concentration gradient and into the presynaptic terminal (figure 5.17*a*). The increase in intracellular Ca^{2+} concentration enables vesicles to be mobilized from the cytoskeleton and, by a process known as **exocytosis,** to fuse with the presynaptic membrane and release the neurotransmitter (figure 5.17*b*). The neurotransmitter diffuses across the 20 to 40 nm space, known as the **synaptic cleft,** between the pre- and postsynaptic membranes and attaches to specific receptors in the postsynaptic membrane. Activation of

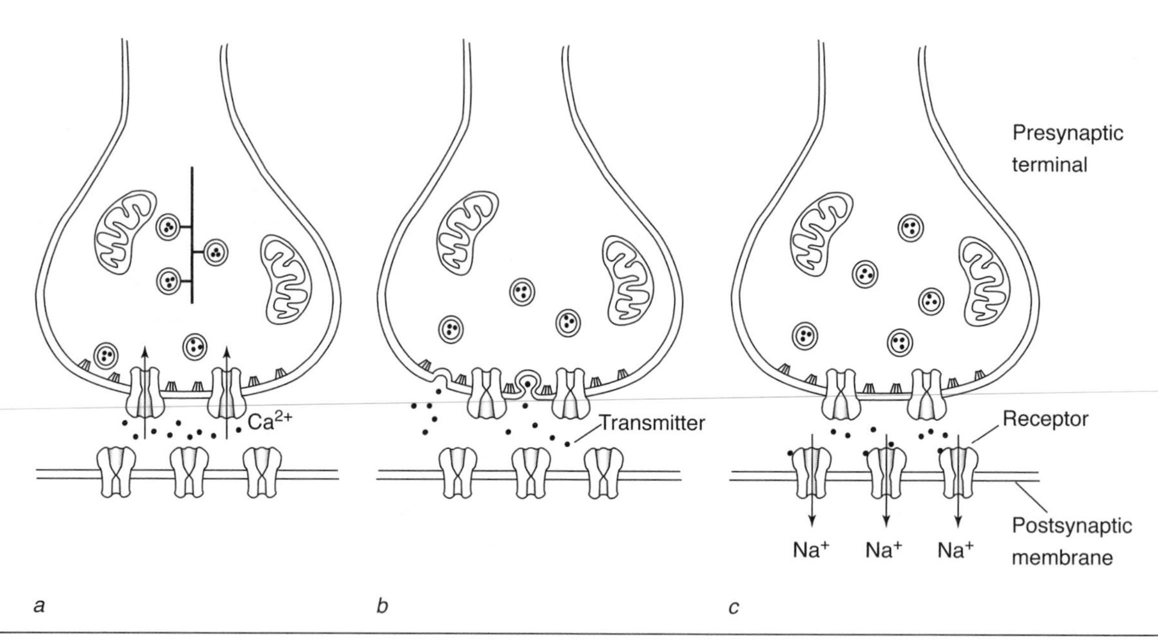

Figure 5.17 Transmission at a chemical synapse. The arrival of an action potential at the presynaptic terminal results in *(a)* the influx of Ca^{2+}, *(b)* the fusion of vesicles to the presynaptic membrane at the active zone and subsequent release of neurotransmitter, and *(c)* binding of the neurotransmitter to a receptor and the flow of current across the postsynaptic membrane.

the ligand-gated receptors opens ion channels, and the resulting current changes the membrane potential of the postsynaptic cell (figure 5.17*c*). Whether a neurotransmitter evokes an excitatory or inhibitory postsynaptic potential depends on the type of receptor to which the neurotransmitter binds and not on the type of neurotransmitter.

The fusion of a vesicle to a presynaptic terminal occurs only at **active zones,** which are specialized release sites that enable the release of neurotransmitter to be focused toward a specific postsynaptic membrane. The number of active zones varies with the size of the synapse, ranging from one at the synapse between a sensory neuron and a neuron in the spinal cord to up to 300 at a nerve–muscle synapse. The Ca^{2+} current that activates exocytosis is greatest around the active zones. The influx of Ca^{2+} appears to align preexisting half-channels in the membranes of the vesicle and the presynaptic cell that then form gap junctions through which the neurotransmitter passes. The amount of neurotransmitter released by a vesicle is called a **quantum.** It produces a postsynaptic potential of fixed size. At most synapses in the central nervous system, each action potential in a presynaptic terminal results in the release of 1 to 10 quanta, which is fewer than the 150 quanta released at the nerve–muscle synapse.

Nerve–Muscle Synapse

Receptors in the postsynaptic membrane at a synapse can act either directly or indirectly to open ion channels in the postsynaptic cell. The nerve-muscle synapse involves a direct action as attachment of the neurotransmitter to the receptor activates the associated ion channel. Membrane proteins that provide both attachment sites for the neurotransmitter and a channel through which ions can flow are known as **ionotropic receptors.**

Because the nerve-muscle synapse corresponds to the junction between a motor neuron and a skeletal muscle fiber, it is often referred to as the **neuromuscular junction.** The synapse is relatively large (2000-6000 μm^2) and involves multiple contacts between the axon of the motor neuron and the specialized region of the muscle cell membrane known as the **end plate.** The axon forms several fine branches that each contain a few swellings known as **synaptic boutons.** The neurotransmitter, which at the nerve-muscle synapse is **acetylcholine** (ACh), is released from the boutons. Beneath each bouton in the end plate are several junctional folds that contain a high density of ACh receptors (10,000 receptors/μm^2). Once released into the synaptic cleft, ACh disappears rapidly due to enzymatic hydrolysis by acetylcholinesterase and diffusion across the 100 nm to the end-plate receptors. Each receptor contains five subunits that include an extracellular domain to which ACh binds and a membrane-spanning domain that forms the ion channel.

Activation of the ACh-gated receptor results in depolarization of the end plate. Voltage-clamp experiments have demonstrated that the reversal potential for a change in the direction of current flow is 0 mV, which indicates that the end-plate potential is produced by a combined influx of Na^+ and efflux of K^+. Because the membrane is depolarized, however, the influx of Na^+ must dominate at the resting membrane potential to produce a net inward current. At the resting membrane potential of –90 mV, a current of 2.7 pA flows through a single channel, which corresponds to the flux of about 17 million ions per second. By varying the membrane potential and measuring the current that flows through a single channel, it is possible to graph the relation between voltage and current (figure 5.8) and to determine that the conductance of a single channel is 30 pS. Stimulation of a motor nerve evokes an inward end-plate current of about 500 nA, which would require the simultaneous activation of about 200,000 ACh-gated channels.

The ACh-gated channels that generate the end-plate potential differ from the voltage-gated channels that produce the action potential. An important difference between the two types of channels is that the flow of Na^+ through the voltage-gated channels is regenerative, as the depolarization caused by the influx of Na^+ opens more voltage-gated Na^+ channels and produces the all-or-none property of the action potential. In contrast, the amplitude of the end-plate potential depends on the amount of ACh that is available. Because the two types of channels coexist in the end-plate membrane, however, the depolarization associated with the end-plate potential provides the trigger for the voltage-gated Na^+ and K^+ channels to generate an action potential that propagates along the excitable membrane of the muscle fiber. Due to the amplitude of the end-plate potential, there is a high safety factor and the end-plate potential invariably leads to an action potential.

EXAMPLE 5.3

Impairment of Action Potential Generation

The processes involved in the transformation of an axonal action potential into a muscle fiber action potential are collectively referred to as **neuromuscular propagation.** One or more of these processes can be impaired under various conditions, as with pharmacological agents, diseases that affect the neuromuscular junction, or prolonged exercise. For example, drugs that bind to the ACh receptor (suxamethonium, tubocurarine) or prevent the breakdown of ACh (anticholinesterase drugs) can be used to paralyze muscle during surgical procedures. Similarly, the snake venom alpha-bungurotoxin reduces the ability of the nerve terminal to release ACh.

The diseases that influence the neuromuscular junction are transmitted genetically and produce defects that range from faulty packaging of ACh to an impairment of the receptors and enzymes that manage ACh. *Myasthenia gravis,* for example, is an immunological disorder in which the ACh receptors are targeted by the immune system. The result is a reduction in the number of ACh receptors, which produces muscle weakness. The disorder usually occurs first in the ocular muscles but can progress to involve swallowing, talking, and chewing. One unique feature of the disorder is that the muscle weakness can vary over the course of a day and from day to day. The disorder can be detected by the presence of antibodies to the ACh receptor in the plasma and by electrophysiological tests. In a clinical exam, the disturbance of neuromuscular propagation can be detected through comparison of the timing of action potentials in two muscle fibers belonging to the same motor unit. In healthy individuals, there is a slight variation (<20 μs)

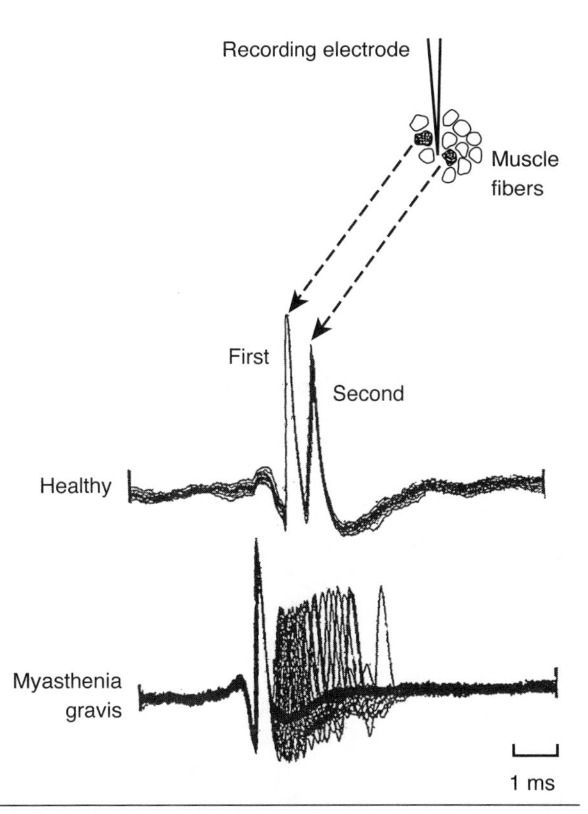

Figure 5.18 Jitter of action potentials in two muscle fibers belonging to the same motor unit. Superimposed records in a healthy subject indicate a slight variation in the timing of the second action potential relative to the first. Similar recordings in a patient with myasthenia gravis show a pronounced increase in the jitter that appears as a spread in the timing of the second action potential.

Adapted, by permission, from A.J. McComas, 2005, *Skeletal muscle: Form and function,* 2nd ed. (Champaign, IL: Human Kinetics), 146.

in the arrival time of action potentials in two muscle fibers; this variability is referred to as **jitter.** Patients with myasthenia gravis, however, experience a substantial increase in jitter due to a decrease in the number of ACh receptors (figure 5.18).

The *Lambert-Eaton myasthenic syndrome* is another immune disorder that affects the neuromuscular junction. This condition results in the destruction of the voltage-gated Ca^{2+} channels in the presynaptic motor terminal. Patients with this disorder experience myasthenic-like weakness, but electrical stimulation of the nerve to the muscle evokes responses that increase in amplitude, in contrast to the decrease that occurs in patients with myasthenia gravis. Such results suggest the impairment of presynaptic processes.

Synaptic Integration

In comparison with the nerve-muscle synapse, synaptic transmission between central neurons can be influenced by many more factors. These include the number of connections received by each neuron; the type of neurotransmitter, receptor, and current involved in the transmission; and the size of the response in the postsynaptic cell. As a consequence, an action potential in a presynaptic neuron rarely elicits an action potential in the postsynaptic neuron as it does at the nerve-muscle junction. Rather, the responsiveness of a postsynaptic neuron is based on the integration of all the inputs that it receives.

A neuron can receive currents that depolarize or hyperpolarize its membrane potential. The response produced by a current that depolarizes the membrane is known as an **excitatory postsynaptic potential** (EPSP), whereas one that hyperpolarizes the membrane is referred to as an **inhibitory postsynaptic potential** (IPSP). Although the type of current produced in the postsynaptic neuron depends on the type of ion channel gated by the neurotransmitter and not on the neurotransmitter, many neurotransmitters attach to specific receptors and consistently produce either excitatory or inhibitory postsynaptic potentials. For example, the neurotransmitter glutamate generally produces excitation in the vertebrate brain and spinal cord, whereas gamma-aminobutyric acid (GABA) and glycine generally produce inhibition. Glutamate-gated channels are permeable to both Na^+ and K^+, similar to the ACh-gated channels of the nerve-muscle synapse; GABA- and glycine-gated channels usually conduct Cl^-.

These postsynaptic actions, however, are complicated by the existence of two types of receptors in neurons. In addition to ionotropic receptors that directly gate a current, there are **metabotropic receptors** that can indirectly gate ion channels through second messengers. When a neurotransmitter, such as glutamate or GABA, binds to a metabotropic receptor, it evokes a series of intracellular actions in the postsynaptic neuron that alter the biochemical state of the cell. These intracellular actions can involve the gating of ion channels, but this effect is slower (seconds to minutes) than that produced directly by ionotropic receptors (milliseconds).

Many types of neurons can release more than one neurotransmitter. Neurotransmitters that are released from the same presynaptic terminal are called **cotransmitters.** Different cotransmitters are contained in separate vesicles and are not necessarily released at the same time. Typical cotransmitters include one of the small-molecule neurotransmitters and a peptide. In addition to ACh, for example, the presynaptic terminals of spinal motor neurons can contain calcitonin gene-related peptide (CGRP) that acts through metabotropic receptors to influence the phosphorylation state of the muscle cell.

A presynaptic neuron can synapse with a postsynaptic neuron at any of its three regions: dendrites, cell body, or axon (figure 5.11). The influence of a specific synapse on the likelihood that the postsynaptic neuron will discharge an action potential depends on the amplitude and sign (excitatory or inhibitory) of the postsynaptic potential and the location of the synapse relative to the axon hillock of the postsynaptic neuron. Because postsynaptic potentials, both excitatory and inhibitory, are conducted electrotonically along the cell membrane, the amplitude decays with distance, and its contribution to the change in membrane potential at the axon hillock decreases with an increase in distance (figure 5.19). The rate at which it decays depends on the length constant of the postsynaptic neuron (figure 5.12).

Because synapses on the cell body of a postsynaptic neuron often produce inhibitory postsynaptic potentials, these inputs tend to dominate the membrane potential of the axon hillock and drive it away from its voltage threshold. This effect, however, can be overcome by the summation of multiple excitatory postsynaptic potentials. The amplitude of a net excitatory postsynaptic potential corresponds to the sum of inputs from multiple synapses **(spatial summation)** and the sum of multiple inputs at a single synapse **(temporal summation)**. Spatial summation depends on the distance between the synapses due to the influence of the length constant on the rate at which the amplitudes decay before they are summed. The amplitude of the excitatory postsynaptic potential produced at synapse B in figure 5.19 was insufficient to exceed the voltage threshold at the axon hillock, but if it overlapped with the potential from synapse A, then an action potential would be generated. Temporal summation refers to the overlap of postsynaptic potentials produced by consecu-

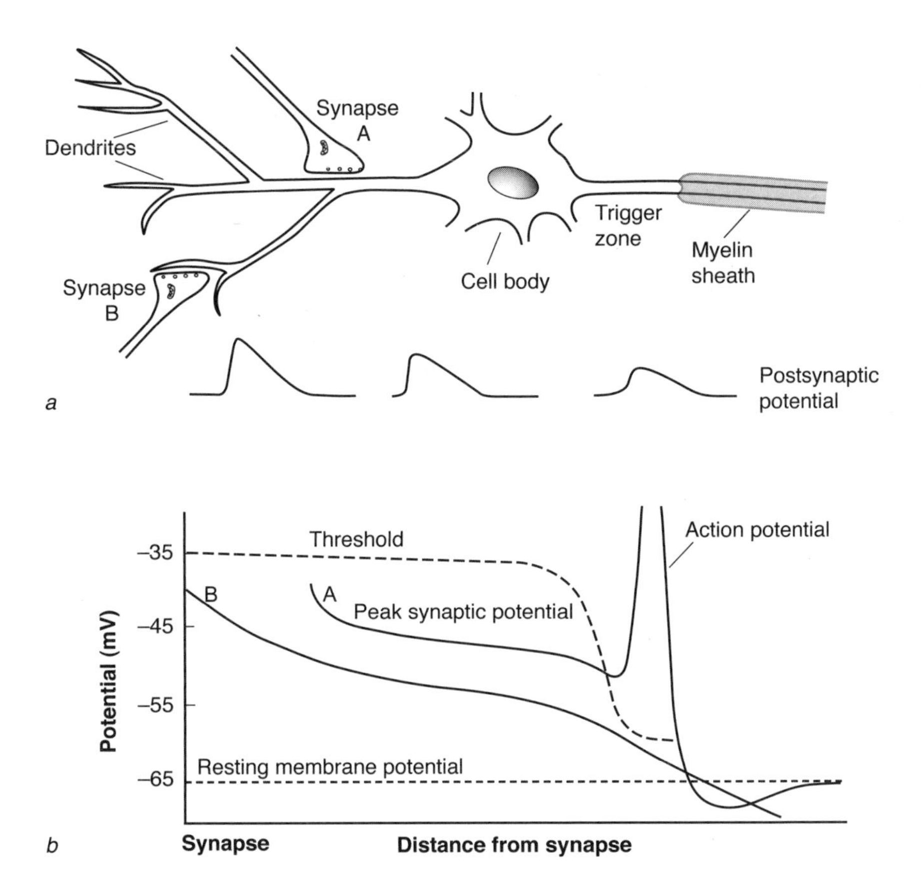

Figure 5.19 Electrotonic conduction of an excitatory postsynaptic potential from two synapses (A and B) on the dendrites to the cell body of the postsynaptic neuron. *(a)* The amplitudes of excitatory postsynaptic potentials decrease as they are conducted along the neuron. *(b)* Comparison of the voltage threshold for generating an action potential (long-dashed line) and the amplitudes of the excitatory postsynaptic potentials evoked at synapses A and B as they travel along the dendrites. When the postsynaptic potential from synapse B reaches the trigger zone, its amplitude is less than the voltage threshold and it does not generate an action potential. In contrast, the amplitude of the postsynaptic potential associated with synapse A exceeds the voltage threshold in the trigger zone and elicits an action potential.

tive inputs from the presynaptic neuron. The extent to which the consecutive potentials overlap depends on the time course of the postsynaptic potential, which is a function of the time constant of the membrane (figure 5.4). A long time constant prolongs the postsynaptic potential in response to a synaptic current and enables multiple potentials to sum together and increase the amplitude of the net postsynaptic potential.

Second Messengers

Metabotropic receptors increase the types of effects that can be evoked by synaptic transmission, with the range extending from the gating of ion channels to altering the gene expression of a cell. Metabotropic receptors can open and close ion channels, but this is accomplished indirectly through one or more metabolic steps. The capacity to close an ion channel, however, means that metabotropic receptors can reduce the conductance of the membrane to a specific ion. Metabotropic receptors comprise an extracellular domain that contains the neurotransmitter binding site and an intracellular domain that activates intracellular messengers to change the biochemical state of the cell. Either these second messengers activate specific protein kinases that phosphorylate various proteins or they mobilize Ca^{2+} ions from intracellular stores. In some instances, the second messengers can act directly on an ion channel.

There are two categories of second messengers, nongaseous and gaseous. An example of a nongaseous second messenger is cyclic adenosine monophosphate (cAmp). When a neurotransmitter binds to a receptor that is linked to a cAMP cascade, it first activates an intermediate molecule called a G protein (guanine nucleotide-binding protein). The activated G protein then stimulates adenylyl cyclase, which is an integral membrane protein, to catalyze the conversion of ATP to cAMP. The major target of action for cAMP is the cAMP-dependent protein kinase. When activated by cAMP, the protein kinase can phosphorylate specific substrate proteins that regulate a cellular response. One significant feature of this system is that each metabotropic receptor can activate many G proteins and thereby greatly amplify the postsynaptic effect of a small presynaptic input. Other prominent nongaseous second messengers are the inositol phosphates, diacylglycerol, and arachidonic acid, which are produced by hydrolysis of the phospholipids on the cytoplasmic side of the membrane.

The gaseous second messengers include nitric oxide (NO) and carbon monoxide (CO). Nitric oxide is produced by the enzyme NO synthase, which is activated by the influx of Ca^{2+} ions through a metabotropic receptor that binds glutamate. Carbon monoxide is produced by the enzyme heme oxygenase. These second messengers can operate in other cells besides neurons. For example, NO is released by endothelial cells of blood vessels and causes the smooth muscle cells to relax and the blood vessel to dilate. The gaseous second messengers pass through membranes readily and do not act through a receptor, but the effect is brief. Both gaseous second messengers stimulate the synthesis of cyclic guanine monophosphate (cGMP), which produces a cascade of effects similar to that with cAMP. In the Purkinje cells of the cerebellum, the cGMP cascade contributes to the long-term depression of synaptic transmission that contributes to some forms of motor learning.

Because metabotropic receptors recruit diffusible second messengers, their effects can be distributed widely throughout the neuron. As a consequence of this capacity, the actions evoked by metabotropic receptors can influence a number of neuronal properties, including resting membrane potential, input conductance, length and time constants, voltage threshold, action potential duration, and discharge characteristics. Thus, metabotropic receptors are described as producing **modulatory synaptic actions.** There are three classes of modulation: the amount of transmitter released from the presynaptic terminal, the responsiveness of ionotropic receptors in the postsynaptic neuron, and the function of resting and voltage-gated channels of the cell body.

Two examples serve to emphasize the significance of these actions. The first example involves an action known as **presynaptic inhibition.** As shown in figure 5.20, presynaptic inhibition occurs at an axoaxonic synapse when input by an interneuron reduces the amount of transmitter released by a presynaptic cell (sensory neuron) in response to an action potential generated in its axon hillock. Presynaptic inhibition can be produced by three different mechanisms: Two involve metabotropic receptors, and one utilizes an ionotropic receptor. The mechanism used by one metabotropic receptor is to close Ca^{2+} channels and thereby reduce the influx of Ca^{2+} into the presynaptic terminal that is necessary for exocytosis. The other metabotropic receptor directly inhibits the processes involved in the release of the neurotransmitter. The ionotropic receptor that can produce presynaptic inhibition is a GABA-gated Cl^- channel. Activation of this channel increases Cl^- conductance, which depolarizes the presynaptic membrane that reduces the amplitude of the incoming action potential and thereby the influx of Ca^{2+}. Presynaptic inhibition is frequently used in the nervous system to modulate incoming sensory information (Rudomin & Schmidt, 1999; Stein, 1995). The converse effect, **presynaptic facilitation,** is produced by mechanisms that enhance the influx of Ca^{2+} into the presynaptic terminal and prolong the postsynaptic potential.

The other example of modulation involves an effect on the resting and voltage-gated channels of the neuron. The dendrites of motor neurons contain metabotropic receptors that bind such neurotransmitters as serotonin and norepinephrine. The presynaptic neurons that provide this input are located in the brain stem; other neurotransmitters that produce similar effects arise in the spinal cord (Heckman et al., 2003). Once activated, these metabotropic receptors can mobilize second messengers to open voltage-gated ion channels for Ca^{2+} and Na^+, which produce long-lasting inward currents known as **persistent inward currents** (Heckman, 2003). When ionotropic channels in the dendrites are activated concurrently with the release of the monoamines (serotonin and norepinephrine), the inward currents generated by the metabotropic receptors depolarize the membrane and amplify the ionotropic input (figure 5.21). Furthermore, when the ionotropic input stops, the persistent

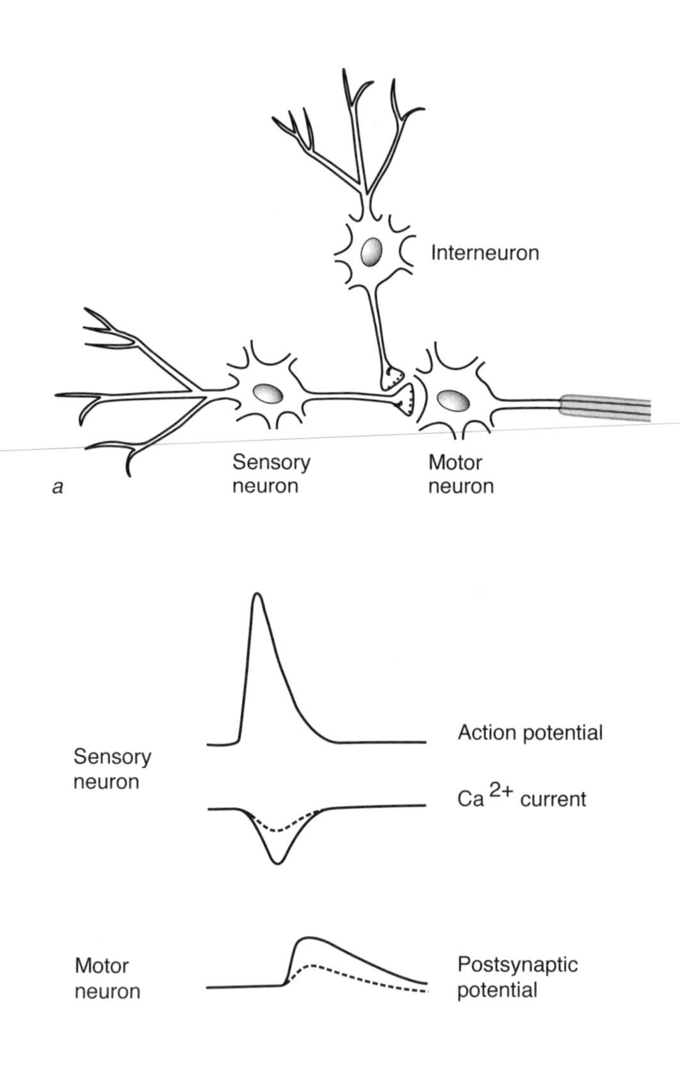

Figure 5.20 Presynaptic inhibition. *(a)* An interneuron reduces the amount of neurotransmitter released by a sensory neuron by reducing the Ca^{2+} current in its presynaptic terminal. *(b)* The result is a smaller postsynaptic potential in the motor neuron.

inward current continues to produce a sustained current that depolarizes the membrane (**plateau potential**) and can be sufficient to cause the motor neuron to continue discharging action potentials—a phenomenon known as **self-sustained firing** (figure 5.21). Consequently, the responsiveness of motor neurons to synaptic (ionotropic) input is not constant but can be modulated by descending input from higher centers. Such a mechanism might be important for enhancing motor output when the level of arousal is high or, conversely, depressing motor output when necessary, for example during sleep.

Axonal Transport

Neurons have developed an effective mechanism to control the synthesis, packaging, release, and degradation of neurotransmitter at synapses. The preparation of neurotransmitters involves the transport of enzymes and precursors from the cell body to the presynaptic terminal where they are synthesized and packaged into vesicles. The enzymes required for the synthesis of small-molecule neurotransmitters are produced in the cell body and transported out to the terminal by a mechanism called **slow axonal transport** (0.5-5 mm/day). The peptides that can function as neurotransmitters, as well as the enzymes that modify their precursors, are synthesized in the cell body and transported out to the terminal by **fast axonal transport** (up to 400 mm/day).

The transport system involves the fibrillar elements that compose the cytoskeleton of the neuron. These are microtubules, neurofilaments, and microfilaments, shown as long cylinders in the axon of figure 5.22. Microtubules, which can be up to 0.1 mm in length and about 26 nm in diameter, are arranged longitudinally in the axon with polarity in the same direction. They comprise long polymers of tubulin dimers. Specific proteins regulate the stability and orientation of the microtubules. Neurofilaments are comparable to the intermediate filaments in other cells, such as the muscle fiber, and include such proteins as vimentin, desmin, keratin, and glial fibrillary acidic protein. They have a diameter of about 10 nm and are 3 to 10 times more abundant than microtubules. Microfilaments are similar to the thin filaments in muscle, comprising a two-stranded helix of actin polymers. Microfilaments have a diameter of about 4 nm.

Axonal transport along the cytoskeleton occurs in both directions and at two different speeds, fast and slow. Transportation away from the cell body is known as **orthograde transport** (or anterograde transport), whereas that toward the cell body is known as **retrograde transport.** Fast orthograde transport occurs along the microtubules in a saltatory manner—that is, stop and

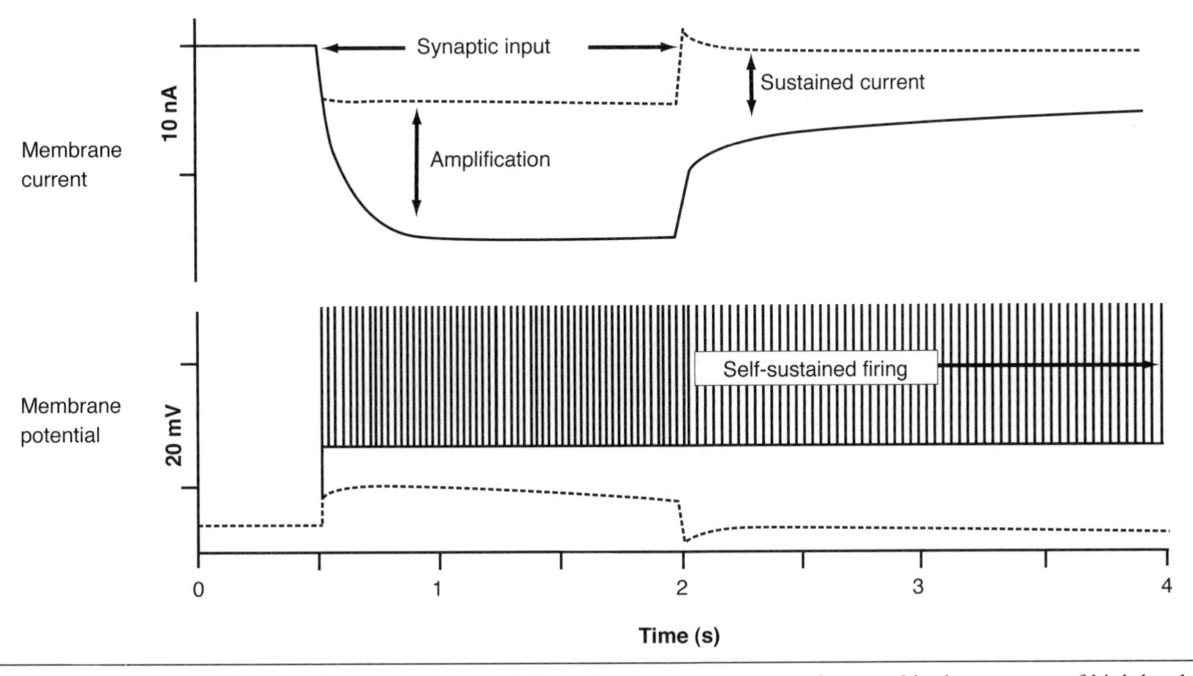

Figure 5.21 The consequences of an inward current delivered to a motor neuron are increased in the presence of high levels of monoaminergic neurotransmitters. When the level of the monoamines was low (dashed line), the inward current (synaptic input) did not depolarize the membrane enough to cause it to reach its voltage threshold. When the level of the monoamines was high (solid line), the inward current was greater due to the combined contributions from the ionotropic and metabotropic receptors. When the synaptic input ceased, the sustained current continued to depolarize the membrane above its voltage threshold, and the neuron continued to discharge action potentials (self-sustained firing).

Adapted from Heckman and Enoka 2004.

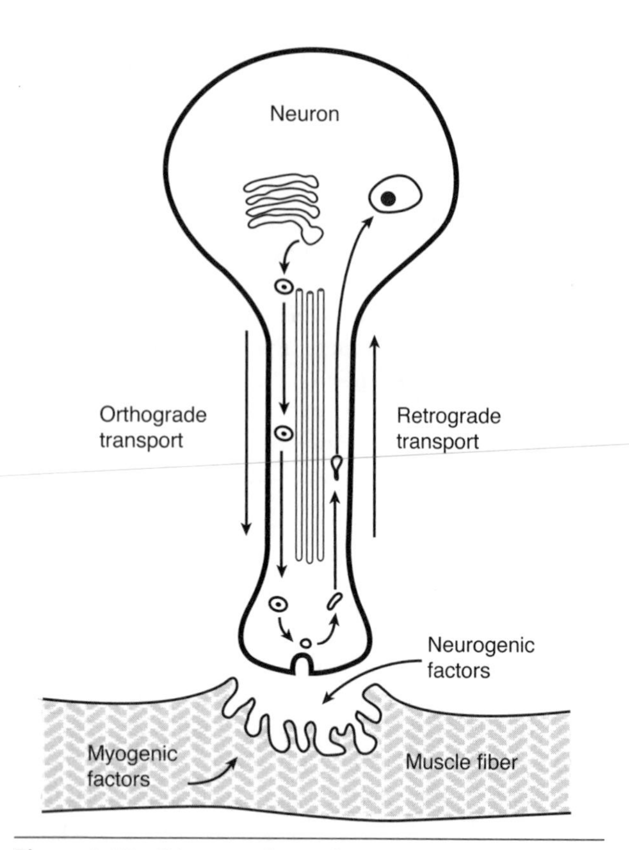

Figure 5.22 Diagram of axonal transport systems.

go. The displacement is driven by molecular motors, which include the enzymes kinesin and cytoplasmic dynein. These motors, attached to the microtubules, function similarly to the cross-bridge in muscle. Each motor translocates material in only one direction. Kinesin drives orthograde transport, whereas cytoplasmic dynein supports retrograde transport. On the basis of measurements made with a laser beam, it has been estimated that a single kinesin molecule can exert a force of about 2 pN and can move material at about 0.5 to 2.0 μm/s. These motors can move material at speeds up to 400 mm/day in the orthograde direction and 200 mm/day in the retrograde direction. In contrast, slow axonal transport displaces material in the orthograde direction only, and at much slower rates.

In addition to providing a supply of neurotransmitter, axonal transport also modulates the properties of nerve and muscle cells. These effects can occur in both directions at a nerve-muscle synapse (figure 5.22) and are known as nerve-muscle (neurogenic) and muscle-nerve (myogenic) trophism (Hyatt et al., 2006). For example, when a muscle is denervated (its nerve is cut), the changes involve the following: a loss of muscle mass (atrophy); a degeneration of muscle fibers as nuclei migrate to the center of the fiber and the mitochondria disintegrate; decreases in the force output and the time course of the twitch; depolarization of the resting membrane potential within 2 h after denervation; an increase in the resting membrane resistance, reflecting a decrease in membrane permeability; alterations in the sodium channel structure; the spontaneous generation of muscle fiber action potentials; the synthesis of extrajunctional ACh receptors; a reduction in the enzyme acetylcholinesterase; and the development of sprouts by intact motor axons. Although denervation causes the removal of both action potentials and axonal transport, the observation that the onset of these changes depends on how far the nerve is cut from the muscle strongly suggests a prominent role for neurogenic factors in regulating the normal properties of the nerve-muscle system. Furthermore, **reinnervation** (reconnecting the nerve to the muscle) results in a reversal of these changes.

Similar effects occur in the opposite direction, whereby the properties of the motor neuron change when its connection to the muscle is disturbed (Lowrie & Vrbová, 1992). This can be shown by changes in the afterhyperpolarization phase of the motor neuron action potential (Czéh et al., 1978). When a drug (tetrodotoxin) that blocks propagation of an action potential but does not impair axonal transport is applied to an axon, the afterhyperpolarization phase changes if the axon is stimulated to generate action potentials on one side of the block but not the other. When the axon is artificially stimulated for 14 weeks on the motor neuron side of the block so that no action potential reaches the muscle, the afterhyperpolarization phase decreases in duration. In contrast, when the stimulation is applied on the muscle side of the block and the muscle contracts, the afterhyperpolarization phase does not change. This finding suggests that activation of the muscle, albeit with artificial stimulation, maintains the health of the motor neuron as displayed in the afterhyperpolarization phase of the action potential. Other electrophysiological properties of the motor neurons (e.g., action potential overshoot, resting membrane potential, axonal conduction velocity) also change after the axon is cut (Huizar et al., 1978), but these effects appear to differ for motor neurons that innervate slow and fast muscle fibers (Cormery et al., 2000). Furthermore, the development of a functional contact between the axon terminals of a motor neuron and muscle seems critical for the maturation of the neuron (Greensmith & Vrbová, 1996).

As a physiological process, axonal transport can adapt to altered patterns of usage. Jasmin and colleagues (1987) examined the transport of the enzyme that degrades ACh (acetylcholinesterase) from the cell body to the neuromuscular junction in the motor neurons of rats following eight weeks of participation in either a swimming or a running program. After training, the axonal transport of the enzyme increased in the runners but not in the swimmers. This finding indicates that axonal transport can adapt to habitual changes in physical activity but that the adaptations are specific to the type of activity; motor neuron activity increased in both the swimmers and runners during training, yet only the runners experienced an increase in the axonal transport of the enzyme.

The rate of axonal transport can also decrease, as occurs with aging (Frolkis et al., 1985). Axonal transport occurs at 200 mm/day in older rats compared with 380 mm/day in younger rats. Furthermore, when axonal transport is halted, there is less of a change in the resting membrane potential and excitability of single muscle fibers in older rats. This means that axonal transport plays less of a role in specifying some properties of muscle fibers in older rats. The slowing of axonal transport with age is a primary factor involved in neuronal aging.

Axonal transport also appears to be an important mechanism by which disease can invade the nervous system. Viruses and bacteria can be taken up by vesicles during **pinocytosis** (the closure and release of the vesicle from the membrane after exocytosis) and become internalized inside the axon. Retrograde transport has been implicated in the movement of viruses (poliomyelitis and herpes) and the tetanus toxin (due to bacterial infection in the skin) from the periphery to the cell body.

ELECTROMYOGRAPHY

When an end-plate potential is generated at a nerve-muscle synapse, it usually results in a muscle fiber action potential that propagates from the synapse to the ends of the muscle fiber. The currents associated with the muscle fiber potential can be measured with electrodes. Such a recording is known as an **electromyogram** (EMG; *myo* = muscle) (Merletti & Parker, 2004). Electromyogram measurements are used by clinicians to diagnose problems in the neuromuscular junction, by ergonomists to determine the requirements of job-related tasks, by physiologists to identify the mechanisms involved in various adaptations within the neuromuscular system, and by biomechanists to estimate muscle force.

Recording and Measurement

The action potential recordings shown in figures 5.14*a* and 5.21 were obtained through measurement of the difference between one electrode inside the cell and one outside the cell. These types of measurements are often referred to as intracellular recordings. In contrast, EMG recordings are extracellular recordings because they involve measuring the difference between two electrodes placed outside the cell. Extracellular recordings detect the current fields associated with the propagating action potentials and not the transmembrane currents (Moritani et al., 2004). An example of an extracellular recording of action potential propagating along a single muscle fiber is shown in figure 5.23. Such measurements are typically made with two electrodes,

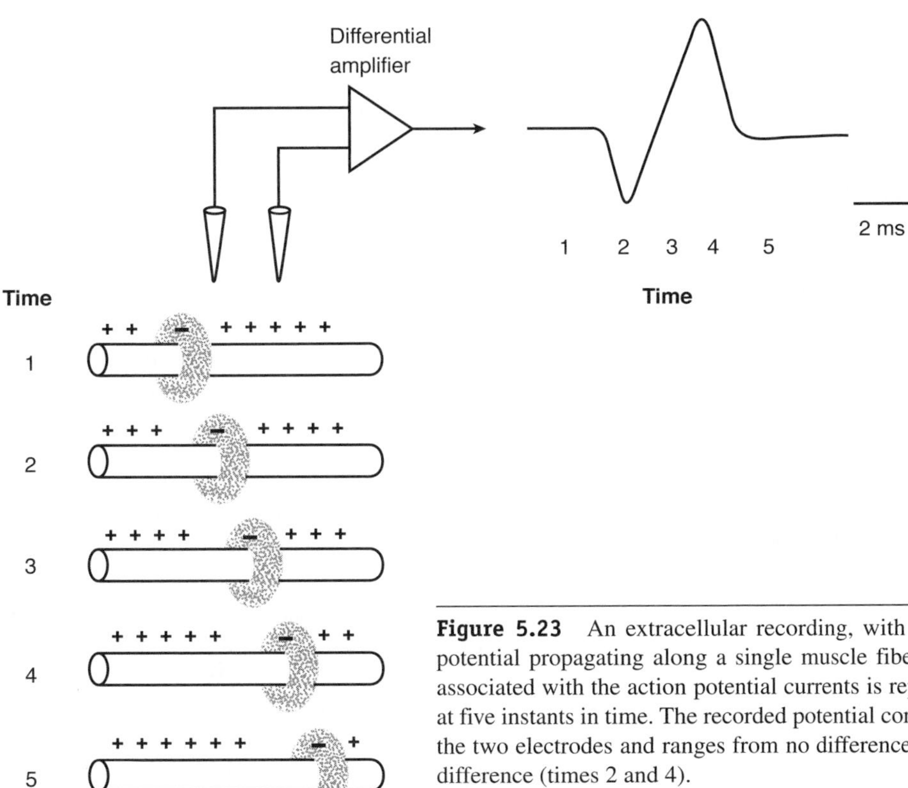

Figure 5.23 An extracellular recording, with a pair of electrodes, of an action potential propagating along a single muscle fiber. The extracellular field potential associated with the action potential currents is represented as a donut-shaped object at five instants in time. The recorded potential corresponds to the difference between the two electrodes and ranges from no difference (times 1, 3, and 5) to a detectable difference (times 2 and 4).

Data from Camhi 1984.

so-called **bipolar recordings,** and the resulting signal represents the difference between the two electrodes to yield a differential recording.

Electrodes can vary in size and material—for example, large (30 cm^2) rubber–carbon pads, small (4 mm diameter) silver–silver chloride disks, and fine wires (25 μm diameter). For recording an EMG signal, the electrodes can be placed on the skin over a muscle (surface EMG), under the skin but over the muscle (subcutaneous EMG), or in the muscle between the fibers (intramuscular EMG). The size and location of the electrode determine the composition of the recording, which can range from the action potential of a single motor unit to many superimposed action potentials (figure 5.24). Electromyogram recordings of many overlapping action potentials are known as an **interference EMG** (Adrian, 1925; Fuglsang-Frederiksen, 2000; Sanders et al., 1996). Electrodes placed on the skin provide a global measure of action potential activity in the underlying muscle, whereas fine-wire electrodes placed in the muscle are able to record single action potentials in a few adjacent muscle fibers.

Clinicians and experimentalists often use EMG recordings to provide information about the timing and intensity of muscle activation by the nervous system. However, it is necessary to appreciate that the EMG signal, especially its amplitude, is profoundly influenced by the measurement technique and the physical and physiological properties of the neuromuscular system (table 5.2). Let us consider four examples of the problems introduced by these factors and the strategies that are used to minimize their effects:

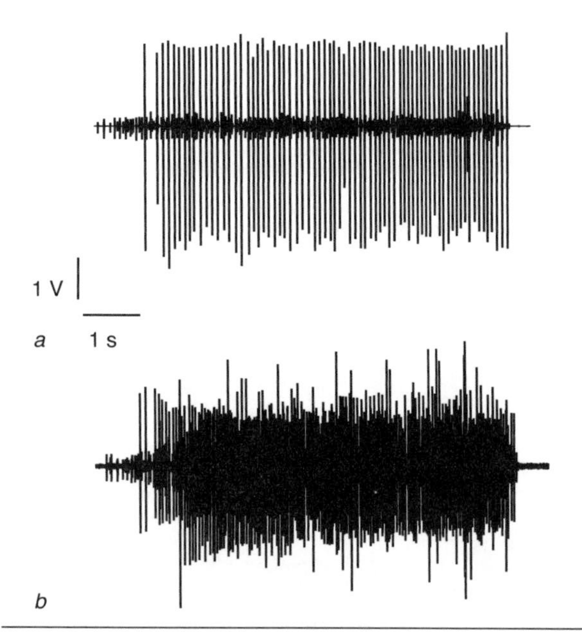

Figure 5.24 Examples of EMG recordings: *(a)* a train of action potentials for muscle fibers belonging to one motor unit; *(b)* multiple, overlapping action potentials due to the concurrent activation of many motor units.

Table 5.2 Nonphysiological Factors and Physiological Properties That Influence Recordings of the Surface EMG

Category	Factors
NONPHYSIOLOGICAL FACTORS	
Anatomical	Shape of the neuromuscular system
	Thickness of the subcutaneous tissue
	Size and distribution of motor unit territories
	Number and distribution of muscle fibers innervated by each motor neuron
	Muscle fiber length
	Spread of the end plates and tendon junction within and among motor units
	Range of pennation angles
Measurement system	Electrode size, shape, and placement
	Skin-electrode contact (impedance, noise)
Geometrical	Muscle fiber shortening
	Movement of the muscle relative to the electrodes
Physical	Conductivity of the tissues
	Cross-talk from nearby muscles
PHYSIOLOGICAL PROPERTIES	
Fiber membrane	Average conduction velocity
	Distribution of conduction velocities
	Shape of intracellular action potentials
Motor unit	Number of activated motor units
	Distribution of discharge rates
	Synchronization

Note. A motor unit corresponds to a motor neuron and the muscle fibers it innervates.

Adapted from Farina et al. 2004*a*.

amplitude cancellation, influence of the innervation zone, unwanted signal content, and cross-talk from neighboring muscles.

Because a muscle fiber action potential corresponds to an extracellular recording of the field potential, it comprises positive and negative phases that vary about a zero potential (figure 5.23). When the positive and negative phases of concurrent action potentials overlap, the amplitude of the resulting sum is reduced (figure 5.25). This effect is known as **amplitude cancellation** (Day & Hulliger, 2001; Keenan et al., 2005). Due to amplitude cancellation, an EMG recording underestimates the output from the spinal cord that is sent to muscle. Expression of EMG amplitude relative to the peak values that are recorded during a maximal voluntary contraction (MVC), however, can reduce this limitation. Compared with the EMG amplitude for the no-cancellation condition (figure 5.25), the normalized EMG (% MVC) provides a reasonable measure of muscle activation; the only discrepancy between the two measures is at intermediate levels of muscle activation (Keenan et al., 2005).

The second example of factors that influence an EMG recording is the effect of electrode placement. This involves both the distance between the electrodes and the location of the electrodes relative to the neuromuscular junctions (Farina et al., 2002*c;* Merletti et al., 2001). To illustrate these effects, figure 5.26 shows the EMG recorded by eight different pairs of electrodes placed over a set of muscle fibers. The neuromuscular junctions are located in the middle of the fibers and span an area known as the **innervation zone.** Each fiber is activated at the neuromuscular junction, which results in the propagation of action potentials in each direction toward the ends of the fiber. The scheme shown in figure 5.26*a* can be used to interpret the actual EMG records made from biceps brachii when it was contracting at 70% of maximum (Merletti et al., 2001). These data indicate the following points:

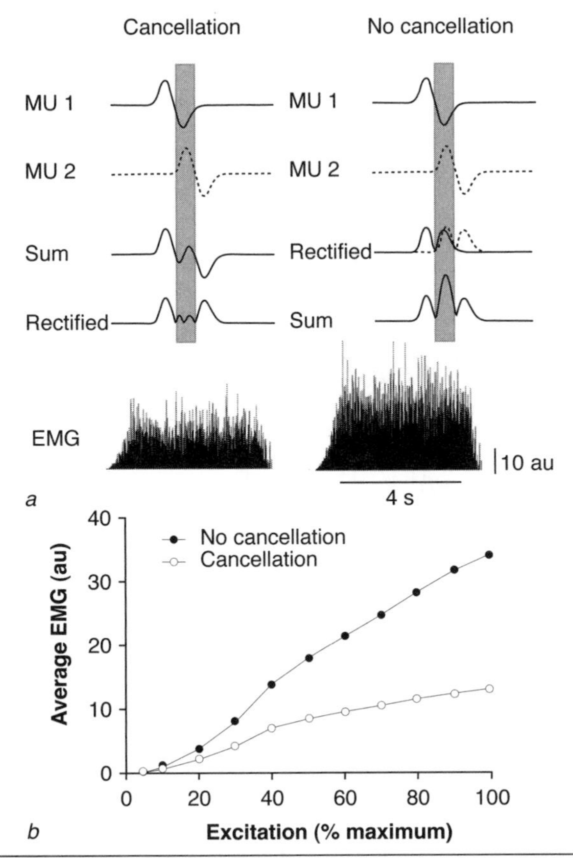

Figure 5.25 The influence of cancellation on the amplitude of the summed activity of muscle fiber action potentials. *(a)* An overlap in the negative and positive phases of two motor unit (MU) potentials reduces the sum of the two potentials (left column: cancellation condition). When the motor unit potentials are rectified (converted to absolute values) before being summed, there are no negative phases and there is no cancellation due to overlapping phases (right column: no-cancellation condition). The difference in the amplitude of the EMG signals at the bottom of the two columns is attributable to cancellation. In most EMG recordings, it is not possible to rectify the individual potentials before they are summed, which means that EMG amplitude underestimates the actual level of total activity. *(b)* The magnitude of amplitude cancellation, which is indicated by the difference between the two lines, increases with the intensity of the muscle contraction and can be as much as 80% at maximal levels of activation.

Data from Keenan et al. 2005.

- The amplitude of the recorded signal is low for small interelectrode distances because the potential detected by each electrode in the pair is relatively similar—compare traces 1 and 3 with trace 8.
- When the electrodes span the innervation zone (trace 2), the amplitude is low because the electrodes record symmetrical potentials propagating in each direction.
- The EMG amplitude for large interelectrode distances (traces 6 and 7) is less than that for smaller distances (traces 3 and 4) if the electrodes span the innervation zone.
- If one electrode pair is placed over the innervation zone (trace 4), the EMG amplitude can be similar for a signal recorded with a smaller interelectrode distance (trace 3) because the innervation zone contributes minimally to the differential recording.
- The greatest amplitude is recorded when the interelectrode distance is relatively large and the electrodes are placed on the same side of the innervation zone (trace 8). The optimum interelectrode distance depends on the length of the muscle fibers.

On the basis of these effects, there are recommendations regarding interelectrode distance and electrode placement for various muscles (Freriks et al., 1999; Rainoldi et al., 2004).

The third example involves a process that can be used to limit the frequency content of an EMG signal, which is necessary when the recording is contaminated by noise and movement artifact. The process, called **filtering,** can be implemented either with hardware or numerically. There are basically four types of filters (figure 5.27): (1) low-pass filter—retains only the low-frequency data; (2) high-pass filter—keeps the high-frequency data; (3) band-pass filter—eliminates frequencies below and above specified values so that the data comprise a desired band

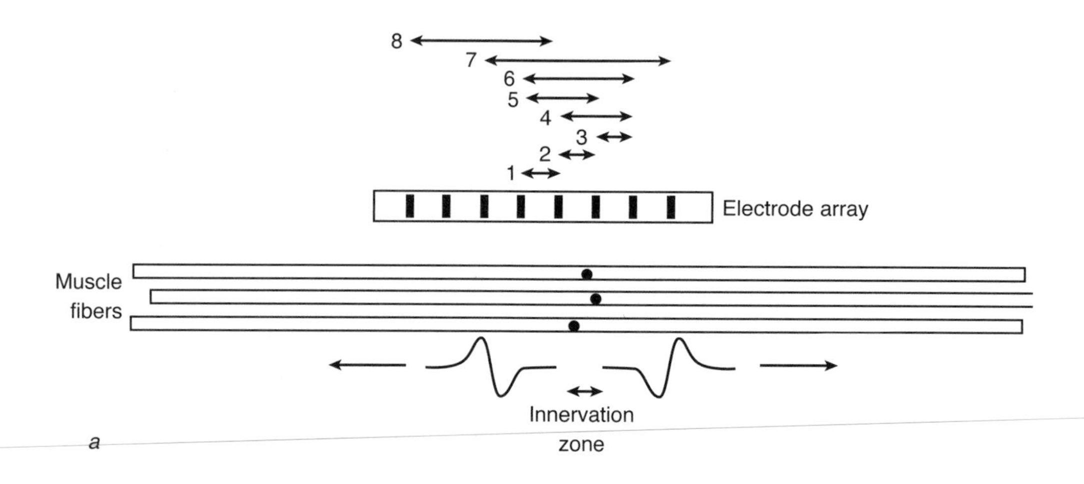

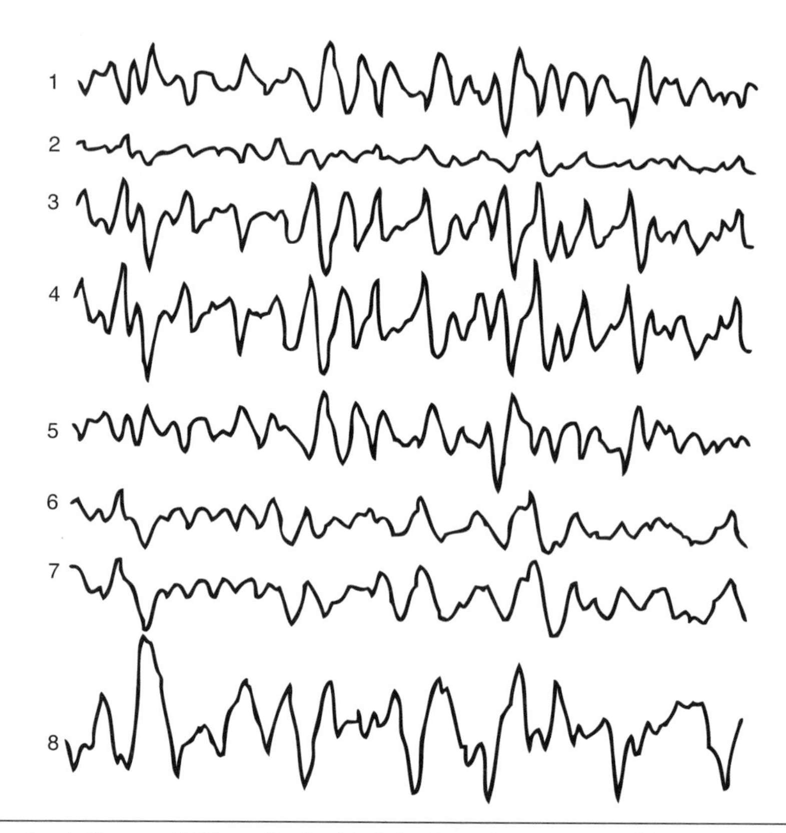

Figure 5.26 Electrode location influences EMG amplitude. *(a)* Eight electrodes are arranged in an array placed over some muscle fibers. The distance between each electrode is 10 mm. The lines above the electrode array indicate eight different combinations of electrodes that were used to make bipolar recordings of the EMG signal. The interelectrode distances are 10 mm for pairs 1, 2, and 3; 20 mm for pairs 4 and 5; 30 mm for pair 6; 40 mm for pair 8; and 50 mm for pair 7. *(b)* EMG recordings with the eight pairs of electrodes. Traces 1 to 8 correspond to the electrode pairs shown in *a*.

Adapted, by permission, from R. Merletti, A. Rainoldi, and D. Farina, 2001, "Surface EMG for the noninvasive characterization of muscle," *Exercise and Sport Sciences Reviews* 29: 20-25.

of frequencies; and (4) band-stop filter—reduces the presence of a specified frequency or range of frequencies in the signal (e.g., 60 Hz). As discussed in chapter 1 (figure 1.26), a critical characteristic of a filter is its cutoff frequency, which corresponds to the frequency above or below which the other frequencies will be modified. For example, a cutoff frequency of 250 Hz on a low-pass filter will cause the filter to reduce the amplitude of the frequencies above this value. Conversely, a high-pass filter of 400 Hz will eliminate 90% of the power in surface EMG signal and keep only frequencies above this value. Similarly, cutoff frequencies of 10 and 50 Hz for a band-

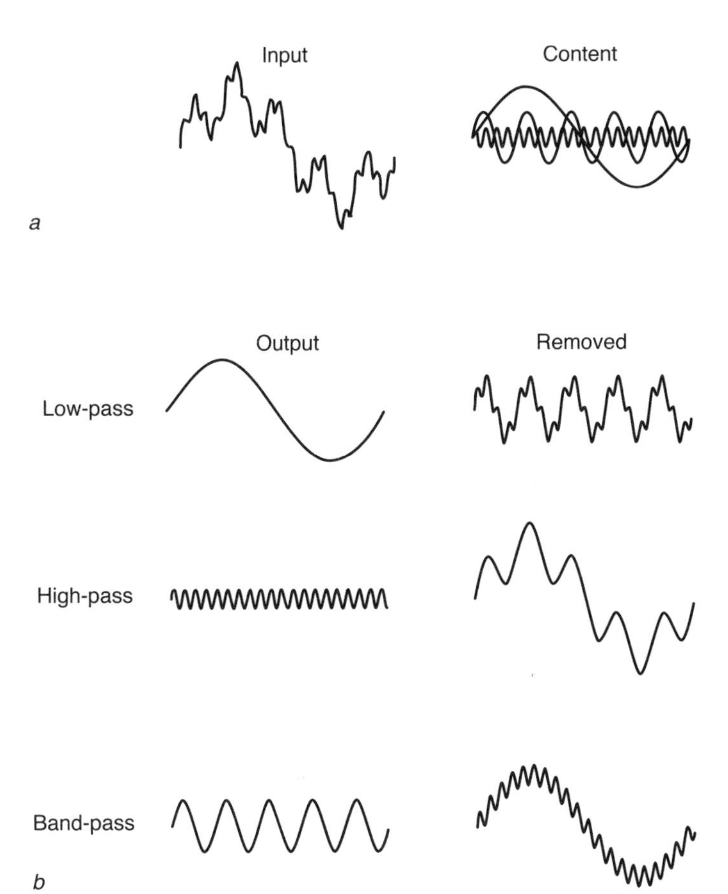

Figure 5.27 Filters can alter the frequency content of a signal: *(a)* an input signal that comprises three sine waves (content—1, 5, and 20 Hz); *(b)* the output of three filters and the frequencies that were removed by each filter. The low-pass filter removed the 5 and 20 Hz waves; the high-pass filter eliminated the 1 and 5 Hz waves; and the band-pass filter excluded the 1 and 20 Hz waves.

pass filter will result in an attenuation of the frequencies below and above these values (figure 5.28*a*).

The fourth example of factors that influence an EMG recording is **cross-talk,** which refers to a signal that is recorded by electrodes placed over one muscle but that originates from a nearby muscle (Farina et al., 2004*a*). When an action potential is generated at a neuromuscular junction, it propagates along the muscle fiber and is extinguished at the ends of the fiber. Cross-talk is due to the extinction phase of muscle fiber potentials (Farina et al., 2002*d*, 2004*b*). An EMG recording, therefore, comprises potentials due to both the propagation and extinction components of the muscle fiber potentials. The relative contribution of the extinction component increases with fat thickness and when the electrodes are placed over the tendon compared with the belly of the muscle, but it is not influenced by interelectrode distance.

It is difficult to quantify and remove cross-talk from an EMG recording. One approach that is often used is to determine the cross-correlation coefficient between the EMG signals detected from two muscles and to interpret significant values as indicating the presence of cross-talk. This is not a valid approach, however, as the propagating and extinction components contribute differently to the EMG recording at small and large distances (Farina et al., 2002*d;* Lowery et al., 2003). Furthermore, because the frequency content of the cross-talk signals, which can be determined by spectral analysis (figure 1.23), overlaps with that of the muscle of interest, it is also not possible to separate the two components based on filtering of the data (Farina et al., 2002*d*). One strategy that can be useful is to compare a differential recording with a double-differential recording. This involves measuring the EMG with two pairs of electrodes and passing the differential recording of each pair through a second differential amplifier. In the absence of cross-talk, the single- and double-differential signals will have similar amplitudes. Conversely, the presence of cross-talk will result in a lower amplitude signal for the double-differential recording due to the attenuation of the cross-talk by this process (Farina et al., 2002*d*).

Analysis and Interpretation

Because an EMG signal provides information about the activation of muscle by the nervous system, it is often used to estimate the timing and amplitude of a muscle contraction. Although these parameters can be estimated from the interference EMG, the typical approach is to rectify and smooth the signal before beginning an analysis. As shown in figure 5.28, **rectification** consists of taking the absolute value of an EMG signal; an electronic module or software can be used to remove or flip over the negative phases (below the isoelectric line) of the interference EMG. The rectified EMG is then typically filtered (low-pass or band-pass filter) to produce a smoothed signal, with the amount of smoothing varying according to the purpose of the analysis. Figure 5.28 shows two examples of such processing: (1) the parallel increase in the rectified and smoothed EMG with the force exerted by a hand muscle as the contraction intensity gradually increased from zero to maximum, and (2) the bursts of EMG generated by agonist and antagonist muscles during rapid contractions.

To estimate the amplitude of an EMG signal, the rectified EMG, either smoothed or not, can be averaged over an interval of time. During a maximal voluntary contraction, for example, the peak EMG might be averaged over a 0.5 s interval to represent the maximal value achieved by the given individual under those recording conditions. Similarly, the rectified EMG might be averaged over 100 ms intervals during a submaximal fatiguing contraction to indicate the change in contraction intensity during the task. An alternative measure of EMG amplitude is

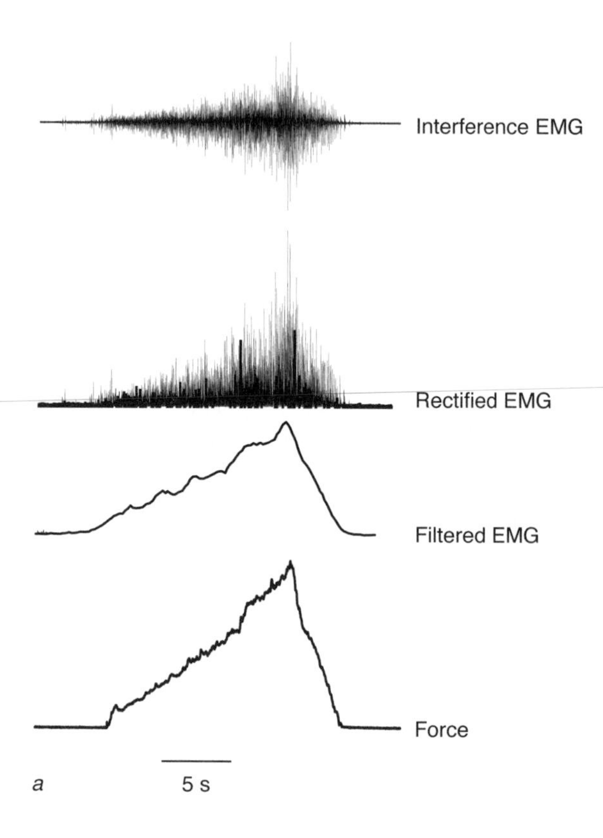

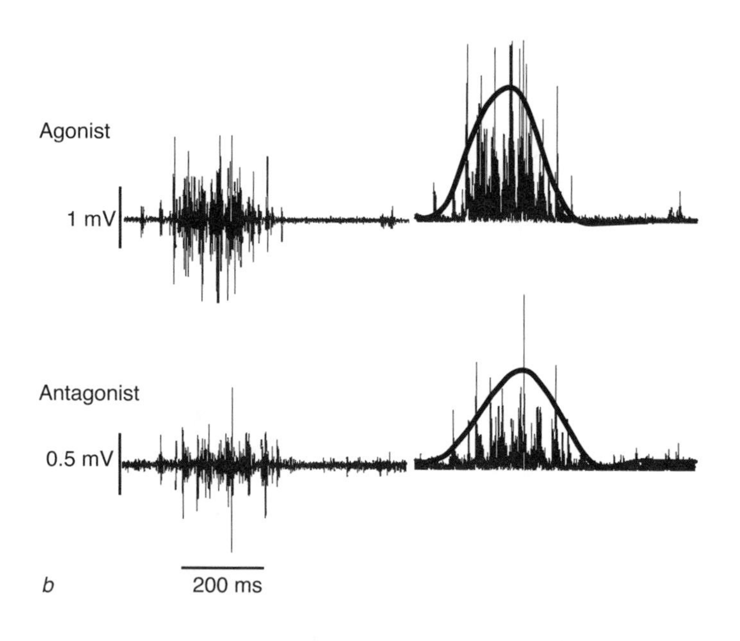

Figure 5.28 Processing of an interference EMG signal. *(a)* Rectified and filtered (band-pass filter set at 10-50 Hz) EMG for a hand muscle and the force it exerted during a maximal contraction. *(b)* Rectified and filtered (low-pass filter set at 6 Hz) EMGs for agonist and antagonist muscles during a rapid contraction.

the **root mean square** *(rms)* value of the interference EMG. The *rms* value is calculated with the following equation:

$$EMG_{rms} = \sqrt{\frac{1}{N}\sum_{i=1}^{N} x_i^2} \tag{5.11}$$

where x_i is the *i*th sample of the interference EMG and N is the number of samples in the interval of interest. The EMG_{rms} reflects the mean power of the signal, whereas the average value provides information about the area under the signal. For a given EMG recording, the absolute values of these amplitude estimates will differ, but the variations during a contraction will be highly correlated (Keenan et al., 2005).

When the rectified and filtered EMG is compared with the force exerted by muscle, there is a close association between the two (Bigland-Ritchie, 1981; Fuglevand et al., 1993). This association, however, is limited to isometric conditions in which the muscle contracts without changing its overall length and the force is relatively constant. Under these conditions, there is a linear or a curvilinear relation between the rectified and filtered EMG and force (Lawrence & De Luca, 1983). One can improve the estimation of muscle force from an EMG signal by recording the signal with an array of electrodes (Staudenmann et al., 2006) and by determining the amplitude of the high-pass–filtered (400 Hz) signal (Potvin & Brown, 2004). With these procedures, the magnitude of the EMG provides a reasonable index of

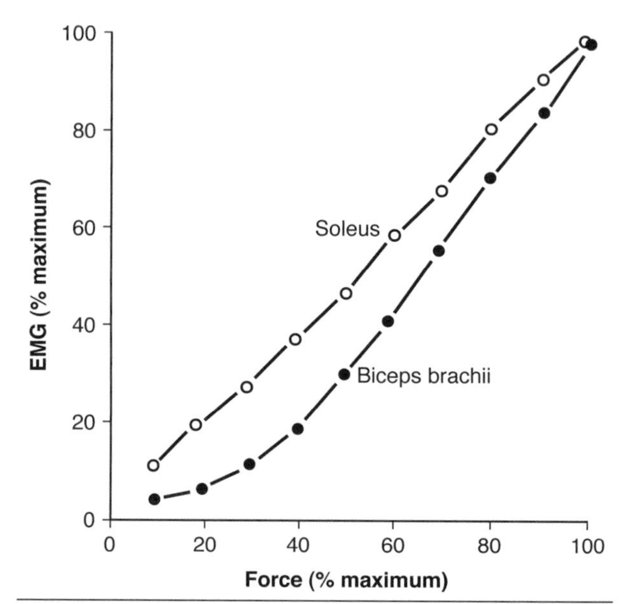

Figure 5.29 Relation between rectified and filtered EMG and force for the soleus and biceps brachii muscles during an isometric contraction. Both EMG and force have been normalized to their respective maximal values.

Data from Bigland-Ritchie 1981.

the force exerted by muscle (figure 5.29) and enables the EMG to be a useful signal to drive prosthetic devices (Parker et al., 2004).

When muscle length and force change (so-called dynamic contractions), however, three factors complicate the interpretation of the EMG: stationarity, shift in the relative position of the electrodes, and changes in tissue conductivities (Farina, 2006). Classic analysis techniques require a stationary signal in which the mean and correlation between samples do not change with time. The rapid recruitment of motor units and changes in muscle length usually cause the EMG to be nonstationary. Furthermore, changes in joint angle cause both the recording electrodes to shift relative to the active muscle fibers (e.g., figure 5.26) and the change in muscle fiber direction to alter tissue conductivity. Although these effects have only a minor influence on estimates of the on and off times of a muscle, provided there is minimal cross-talk, the use of EMG amplitude as an estimate of contraction intensity is confounded by changes that are unrelated to the activation of the muscle by the nervous system (Gerilovsky et al., 1989; Heckathorne & Childress, 1981; Inman et al., 1952).

Provided that an EMG signal is stationary, a spectral analysis can be used to determine its frequency content (figure 1.23). The spectral frequencies can be computed with several different techniques, such as the classic periodogram, autoregressive-based approaches, Cohen's class time–frequency distributions, and wavelet analysis (Farina, 2006; Farina et al., 2004*a*). Three features of an EMG signal can influence its frequency content: the shape of the action potentials, the discharge rate of the motor units, and the relative timing of the action potentials discharged by the different motor units (Lago & Jones, 1977; Weytjens & van Steenberghe, 1984).

The shape of the action potentials is the major factor that influences the EMG power spectrum. The shape of an action potential depends on the average conduction velocity of the muscle fibers and on the length and location of the fibers relative to the recording electrodes. Although there is a linear relation between changes in the spectral frequencies and conduction velocity during sustained isometric contractions at high-to-moderate intensities (Arendt-Nielsen et al., 1989; Merletti et al., 1990), it is inappropriate to use this relation when the number of activated motor units changes significantly during a contraction (Farina, 2006; Farina et al., 2003). Furthermore, spectral analysis of the surface EMG cannot be used to estimate the activation of type I and II muscle fibers and the recruitment of motor units during a contraction (Farina et al., 2004*a*). For example, the central frequency (mean or median) in the EMG power spectrum has been reported to increase (Gerdle et al., 1990; Moritani & Muro, 1987), not change (Petrofsky & Lind, 1980; Viitasalo & Komi, 1978), or decrease (Rainoldi et al., 1999; Westbury & Shaughnessy, 1987) with an increase in the force of the muscle contraction.

The rate at which action potentials are discharged has only a minor effect on the power spectrum of the interference EMG (Lago & Jones, 1977) and does not produce reliable peaks in the spectra derived from either the interference or rectified signals (Farina et al., 2004*a*). However, the correlated discharge of action potentials by different motor units can influence the mean frequency of an EMG spectrum (Kleine et al., 2001; Yao et al., 2000), but not independently enough to be used as an index of the amount of correlated activity (Farina et al., 2002*b;* Fattorini et al., 2005).

Surface Mechanomyogram

When a muscle contracts, the actions of the contractile proteins cause the muscle to be displaced in a transverse direction relative to the skin. This displacement appears to be caused by an increase in the transverse width of the muscle as it shortens and the lateral movement of activated muscle fibers (Orizio & Gobbo, 2006; Shinohara & Søgaard, 2006). Motion in the transverse direction can be detected with accelerometers, laser distance sensors, microphones, and piezoelectric contact sensors. Such recordings are collectively known as the **surface mechanomyogram** (MMG).

To compare the information conveyed by EMG and MMG (acceleration) signals, Cescon and colleagues (2006) measured the contributions of single motor units in biceps brachii to the two signals as subjects performed contractions at 20%, 50%, and 80% of maximum. There was no correlation between either the amplitudes or the mean spectral frequencies of the two signals. Whereas the amplitude of the EMG for the motor units increased with target force, the MMG amplitude increased from 20% to 50% but did not increase further at 80%. The likely explanation for the trend in MMG amplitude was an increase in the longitudinal stiffness of the muscle with contraction intensity that attenuated the transverse actions. These results suggest that the MMG signal is influenced by other factors besides just motor unit activity.

SUMMARY

This chapter is the first of three on the properties of the motor system. It focuses on the electrical properties of excitable membranes, which enable nerve and muscle cells to transmit activation signals rapidly, and the transfer of the activation signal between cells. Electrical signaling in nerve and muscle cells involves a transient change in the potential difference across the cell membrane that can be spread passively or actively. The active propagation of the change in membrane potential corresponds to the action potential, which is the elementary unit for rapid signaling between cells. The transfer between cells occurs at synapses; nerve cells have thousands of synapses, whereas muscle cells have a single synapse. The generation of an action potential in a nerve cell depends on the net effect of concurrent input at many synapses. In contrast, transmission at a nerve-muscle synapse usually evokes a muscle fiber action potential. The nerve-muscle synapse represents the pathway by which neurons in the spinal cord and brain stem elicit action potentials in muscle fibers and cause a muscle to contract. Muscle fiber action potentials can be recorded with a technique known as electromyography, which is used to indicate the activation of muscle by the nervous system.

SUGGESTED READINGS

Kernell, D. (2006). *The Motoneurone and Its Muscle Fibres*. Oxford, Great Britain: Oxford University Press, chapters, 1, 5, and 6.

MacIntosh, B.R., Gardiner, P.F., & McComas, A.J. (2006). *Skeletal Muscle: Form and Function.* Champaign, IL: Human Kinetics, chapters 7-10.

Medved, V. (2001). *Measurement of Human Locomotion.* Boca Raton, FL: CRC Press LLC, chapter 6.

Merletti, R., & Parker, P.A. (2004). *Electromyography. Physiology, Engineering, and Noninvasive Applications.* Hoboken, NJ: Wiley.

chapter 6

Muscle and Motor Units

To understand the operation of the motor system, we continue by discussing the mechanical consequences of muscle fiber action potentials in activating the contractile proteins and enabling a muscle contraction. This chapter examines the structure of muscle, the activation of the contractile proteins, the functional relations between spinal neurons and muscle fibers, the mechanical properties of muscle, and the net effects of muscle activity about joints in the human body.

MUSCLE

Muscles are molecular structures that convert chemical energy, initially derived from food, into force. The properties of muscle include (a) irritability—the ability to respond to a stimulus; (b) conductivity—the capacity to propagate a wave of excitation; (c) contractility—the ability to modify its length; and (d) adaptability—a limited growth and regenerative capacity. Histology identifies three types of vertebrate muscle: cardiac, smooth, and skeletal. We will focus on skeletal muscle, which comprises fused cells with well-defined striations. With the exception of some facial muscles, skeletal muscles act across joints to rotate body segments and thus produce movement. Muscle properties, therefore, have a major influence on the movement capabilities of humans.

Muscle contains many identifiable elements. Our focus, however, is on the function of muscle and the processes by which that function is achieved. Two critical processes of muscle function include (a) the interaction between the sarcolemma and the sarcoplasmic reticulum and (b) the capabilities of the sarcomere. These processes contribute significantly to the connection between the nervous system and muscle (sarcolemma–sarcoplasmic reticulum) and the force that muscle can exert (sarcomere).

Gross Structure

Muscle fibers are linked together by a three-level network of collagenous connective tissue. **Endomysium** surrounds individual muscle fibers; **perimysium** collects bundles of fibers into fascicles; and **epimysium** ensheathes the entire muscle (figure 6.1, *a* and *b*). This connective tissue matrix, which exists throughout the entire muscle and not just at its ends, connects muscle fibers to tendon and hence to the skeleton (Huijing, 2003). Because of this relation, muscle fibers and connective tissue (including tendon) operate as a single functional unit, and the term *muscle* is used to denote the entire structure.

Muscle fibers vary from 1 to 400 mm in length and from 10 to 60 μm in diameter. The cell membrane encircling each set of myofilaments that composes a muscle fiber is known as the **sarcolemma.** The sarcolemma is an excitable membrane with the properties described in chapter 5. The sarcolemma is about 7.5 nm thick. The fluid enclosed within the fiber by the sarcolemma is known as **sarcoplasm.** The sarcoplasm contains fuel sources (e.g., lipid droplets, glycogen granules), organelles (e.g., nuclei, mitochondria, lysosomes), enzymes (e.g., myosin adenosine triphosphatase [ATPase], phosphorylase), and the contractile apparatus (bundles of myofilaments arranged into myofibrils).

The sarcoplasm also contains an extensive, hollow, membranous system that is functionally linked to the surface sarcolemma and that assists the muscle in conducting signals from the nervous system. This membranous system includes the sarcoplasmic reticulum, lateral sacs (terminal cisternae), and transverse (T) tubules (figure 6.2). The **sarcoplasmic reticulum** runs longitudinally along the fiber, which places it parallel to and surrounding the myofibrils. At specific locations along the myofibril, the sarcoplasmic reticulum bulges into **lateral sacs.** Perpendicular to the sarcoplasmic reticulum and

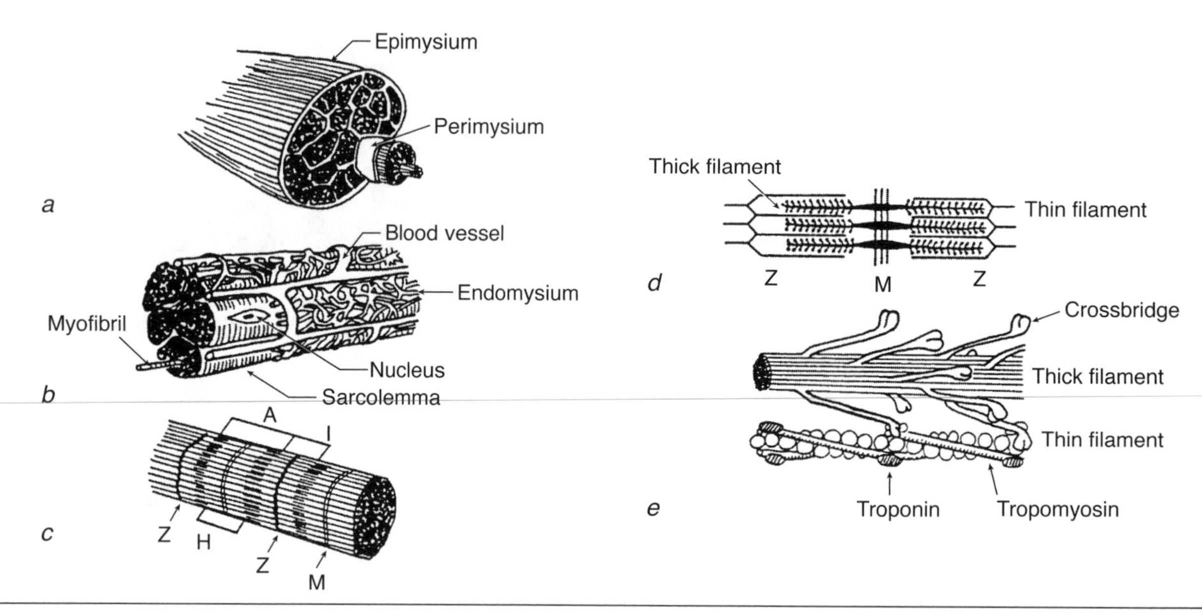

Figure 6.1 Organization of skeletal muscle from the gross to the molecular level. Note the levels of connective tissue within muscle, the bands and zones that comprise the sarcomere, and the molecular components of the thick and thin filaments: *(a)* whole muscle; *(b)* a group of muscle fibers; *(c)* one myofibril; *(d)* a sarcomere; *(e)* one thick and one thin filament.

Reprinted, by permission, from M.I. Pittman and L. Peterson, 1989, Biomechanics of skeletal muscle. In *Basic biomechanics of the musculoskeletal system,* edited by M. Nordin and V.H. Frankel (Philadelphia: Lea & Febiger), 90.

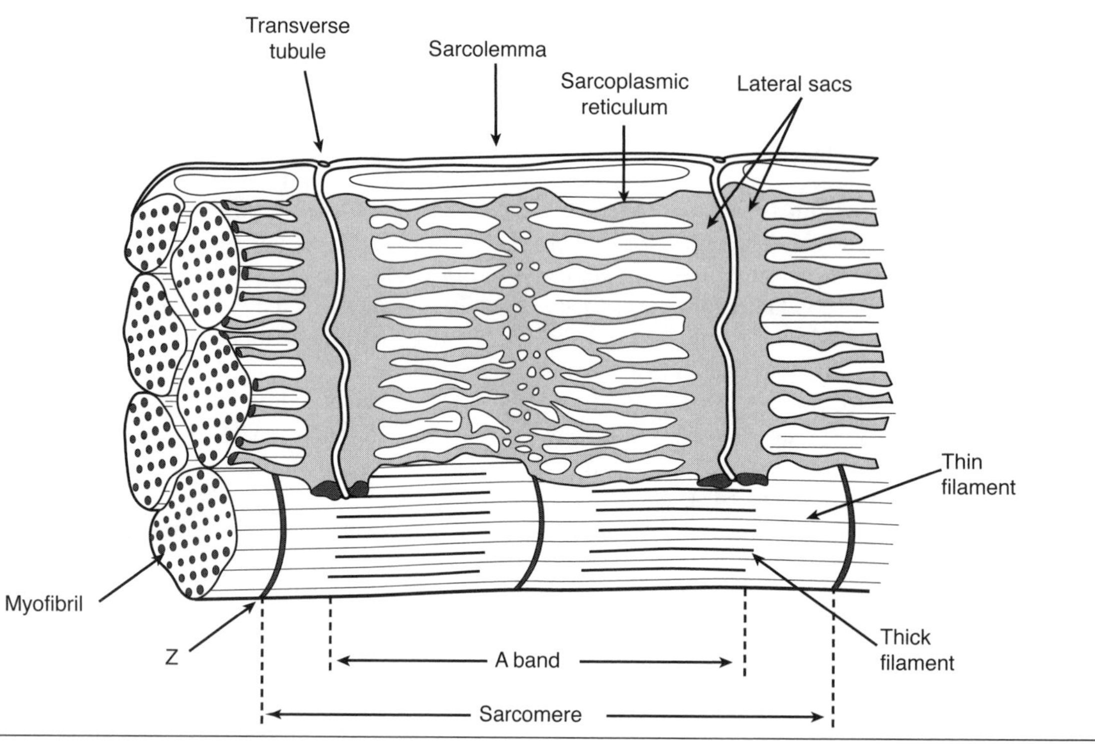

Figure 6.2 Alignment of the transverse tubules and sarcoplasmic reticulum with respect to the myofibrils. The figure shows six of the myofibrils that are part of a single muscle fiber.

Adapted from *The Journal of Cell Biology*, 1969, 42: 46-67. Copyright 1969, The Rockefeller University Press.

associated with the lateral sacs are **transverse tubules,** which are branched invaginations of the sarcolemma. The rapid transmission of the activation signal from the sarcolemma to the contractile apparatus is facilitated by the connection between the sarcoplasmic reticulum and the transverse tubules.

Sarcomere

Skeletal muscle comprises a series of repeating units, each of which contains the same characteristic banded structure. This unit, the sarcomere, includes the zone of a myofibril from one Z band to the next Z band (figures 6.1*d* and 6.2). The sarcomere is the basic contractile unit of muscle and comprises an interdigitating set of thick and thin contractile proteins (figure 6.1*d*). Thick filaments have a diameter of 12 nm compared with 7 nm for thin filaments. Thick filaments are also slightly longer (1.6 µm) than thin filaments (1.27 µm). A **myofibril** is a series of sarcomeres that are connected end to end. Each myofibril is composed of bundles of myofilaments (thick and thin contractile proteins) and has a diameter of about 1 µm. Because a sarcomere has a length of about 2.5 µm in resting muscle, a 10 mm myofibril represents 4000 sarcomeres in series.

The obvious striations of skeletal muscle are due to the differential refraction of light as it passes through the sarcomere. The thick-filament zone (figures 6.1*c* and 6.2), which includes some interdigitating thin filaments, is doubly refractive (i.e., forms two refracted rays of light from a single incoming ray) and comprises the dark band, called the **A band** (anisotropic). Within the A band is a zone that contains only thick filaments (figure 6.1*c*). Because this zone is clear of thin filaments, it is known as the **H band** (Hellerscheibe, or clear disk). The area between the A bands contains predominantly thin filaments and, because it is singly refractive, is called the **I band** (isotropic).

Each set of thick and thin filaments is attached to a central transverse band: The thick filaments attach to the **M band** (Mittelscheibe, or middle disk—the band located in the middle of the A band), and the thin filaments connect to the **Z band** (Zwischenscheibe, or between disk). About 3000 to 6000 thin filaments connect to each Z band. A cross section through the A band shows that six thin filaments surround each thick filament, whereas a single thin filament can interact with only three thick filaments.

Myofilaments

The myofibril contains two myofilaments, known as thick and thin filaments. Each myofilament is composed of several proteins (figure 6.3). The structure of the thin filaments is dominated by actin but also includes the proteins tropomyosin and troponin, which regulate the interaction between actin and myosin. Each thin filament is composed of two helical strands of fibrous actin **(F actin)** (figure 6.3*a*). Each F-actin strand is a polymer (i.e., chemical union of two or more molecules) of some 200 globular actin **(G actin)** molecules (figure 6.3*b*). A G-actin molecule is a protein containing about 374 amino acids. Located in the groove of the F-actin helix are two coiled strands of tropomyosin (figure 6.3, *a* and *b*). The structure of **tropomyosin** is referred to as a two-chain coiled coil; each of these chains contains approximately 284 amino acids. The **troponin** (TN) complex has a globular structure that includes three subunits (figure 6.3*b*): The **TN-T** unit binds troponin to tropomyosin; **TN-I** inhibits four to seven G-actin molecules from binding to myosin when tropomyosin is present; and **TN-C** can reversibly bind Ca^{2+} ions as a function of calcium concentration. Troponin C has four binding sites, two for

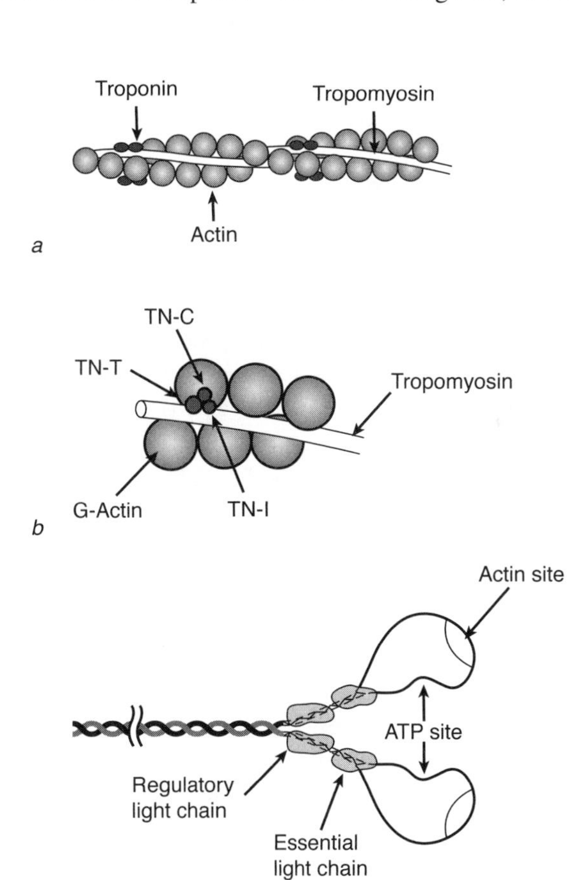

Figure 6.3 Organization of the myofilaments: *(a)* the thin filament; *(b)* troponin (TN) elements; *(c)* the myosin molecule. ATP = adenosine triphosphate; LMM = light meromyosin; S1, S2 = subfragments 1 and 2.

Adapted, by permission, from R.G. Whalen, 1985, "Myosin isoenzymes as molecular markers for muscle physiology," *Journal of Experimental Biology* 115: 46. Copyright 1985 by The Company of Biologists Limited.

Ca^{2+} and two for Ca^{2+} or Mg^{2+}. Thus the thin filament consists of two strands of actin molecules (F actin) that have superimposed upon them (wrapped around or attached to them) two-stranded (tropomyosin) and globular (troponin) proteins. Tropomyosin and troponin are referred to as regulatory proteins because they permit the interaction of actin and myosin and hence muscle contraction.

A set of thin filaments that projects longitudinally into one sarcomere connects the Z band region to another set, which projects in the opposite direction into the adjacent sarcomere. At this connection, each thin filament appears to be linked to its four closest neighbors. A region of considerable flexibility, the Z band changes its shape as contractile conditions vary. Z band width can vary from one muscle fiber to another (e.g., in different muscle fiber types), and it probably also varies as a consequence of training (Schroeter et al., 1996; Sjöström et al., 1982; Thornell et al., 1987).

The thick filaments contain myosin and several myosin-binding proteins: C protein, H protein, M protein, and myomesin (Schiaffino & Reggiani, 1996). The myosin molecule is a long, two-chain, helical structure that terminates in two large globular heads (figure 6.3*c*). The myosin molecule is composed of six proteins: two myosin heavy chains (~200 kilodaltons [kDa]) with one essential myosin light chain and one regulatory myosin light chain on each heavy chain. Each globular head, which is ~17 nm long, has an adenosine triphosphate (ATP)- and an actin-binding site and a site for hydrolyzing ATP. With the aid of an enzyme (protease), the myosin molecules can be decomposed into **light meromyosin** (LMM) and **heavy meromyosin** (HMM) fragments. The HMM fragment can be further subdivided into **subfragments 1** and **2** (S1 and S2). Light meromyosin has a relatively low molecular weight (135 kDa), whereas HMM is heavier (335 kDa). The globular heads of the myosin molecule, which contain the ATP- and actin-binding sites, are known as S1. The two binding sites are connected by a cleft. The remaining portion of HMM is called S2.

Because LMM binds strongly to itself under physiological conditions, approximately 300 myosin molecules aggregate to form the dominant element of the thick filament (Pepe & Drucker, 1979). The union is not random, but structured. The molecules are aligned in pairs, and the S1 element of each molecule is oriented to its partner at 3.14 rad (180°). The next pair is displaced by about 0.0143 μm and 2.1 rad (120°). The result is an ordered alignment of myosin molecules in which the HMM projections (cross-bridges) encircle the thick filament (figure 6.1*e*). Each sarcomere actually contains two such sets of myosin molecules; however, because the S1 elements of the two sets point in opposite directions, the LMM fragments unite in the M band (figure 6.1*c*) to form a single filament.

The myosin molecule contains two zones of greater flexibility. These occur at the LMM-HMM and S1-S2 junctions. In the resulting alignment, the HMM fragment can extend from the thick filament to within close proximity of the thin filament (figure 6.1*e*). Due to the ability of S1 to interact with actin, the HMM extension is called the **cross-bridge.** The S1 region of HMM contains the sites that hydrolyze ATP and a motor domain that converts chemical energy to mechanical work (Lutz & Lieber, 1999). Each thick filament is surrounded by and can interact with six thin filaments because the cross-bridges encircle the thick filament. There are about 1600 thick filaments per square micrometer in human quadriceps femoris muscle (Claasen et al., 1989).

The proteins that compose the contractile apparatus can be distinguished as the products of different genes. Eight multigene families contribute the major components of the sarcomere: myosin heavy chain, alkali light chain, DTNB (dithionitrobenzoic acid) light chain, actin, tropomyosin, troponin C, troponin I, and troponin T (Gunning & Hardeman, 1991; Tsika et al., 1987). The first three of these components form the myosin molecule. Because proteins comprise sequences (chains) of amino acids, a protein with a high molecular weight (200 kDa) is referred to as a **heavy chain**, and a protein with a low molecular weight (<30 kDa) is identified as a **light chain.** And a given protein that is synthesized with a slightly different amino acid composition is known as an **isoform;** that is, different isoforms of the same protein can exist that may or may not be the product of different genes and have different effects on the cross-bridge cycle (Andruchov et al., 2006; Babij & Booth, 1988; Caiozzo & Rourke, 2006). There can be different isoforms of the myosin heavy chain and the light chain (table 6.1). Although the different heavy chain isoforms appear to have physiological significance, less is known about the functional consequences of differences in other contractile protein isoforms. However, isoforms of the contractile proteins do change during development and as a consequence of alterations in habitual physical activity.

The myosin molecule consists of two coiled heavy chains with light chains attached to the myosin heads (figure 6.3*c*). The isoforms differ in cardiac, smooth, and skeletal muscle. The heavy chains in skeletal muscle have a molecular weight of 220 kDa. There is a strong relation between the maximal velocity at which muscle fibers can shorten and the ATPase activity of the heavy chain isoform contained in the fiber (Schiaffino & Reggiani, 1996). There appear to be one slow and five fast heavy chain isoforms (table 6.1). Four light chains are attached to the globular heads, and these are distinguishable by molecular weight (~16-30 kDa), by whether or not they can be phosphorylated, and by the experimental agent (alkali or DTNB) that separates them from the heavy

Table 6.1 Isoforms of the Contractile Proteins in Various Adult Skeletal Muscles

	SKELETAL MUSCLE	
Gene family	**Slow**	**Fast**
Myosin heavy chain	S	F_{2A}, F_{2B}, F_{2X}, F_{EO}, F_{SF}
Alkali light chain	1_{Sa}, 1_{Sb}	1_F, 3_F
DTNB light chain	2_S, $2_{S'}$	2_F
Actin	α_{sk}	α_{sk}
α-Actinin	S	F
Myosin-binding protein C	S	F
Tropomyosin	β, α_S	β, α_F
Troponin C	S	F
Troponin I	S	F
Troponin T	S	F

Note. S = slow, F = fast; F_{2A}, F_{2B} correspond to the two fast-twitch fibers defined by histochemistry; F_{2X} is defined by antibody staining and protein analysis; F_{EO} is found in adult extraocular muscle; F_{SF} are superfast contractile proteins of jaw muscle; DTNB = dithionitrobenzoic acid.

Data from Caiozzo and Rourke, 2006; Gunning and Hardeman, 1991; Schiaffino and Reggiani, 1996.

chain. As with heavy chains, there appear to be different isoforms of the light chains for fast- and slow-twitch muscle fibers (table 6.1). Although the specific function of the light chains is unknown, they are necessary for contractile function and probably modulate the interaction between actin and myosin (Hernandez et al., 2007; Lowey et al., 1993).

Cytoskeleton

Since the initial proposal of the sliding filament theory of muscle contraction, it has become obvious that there must exist additional structures that are necessary for a muscle contraction (Huijing, 2003; Kovanen, 2002; Monti et al., 1999; Sheard, 2000; Trotter, 1993). These structures, which are termed the **cytoskeleton,** are involved in the alignment of the thick and thin filaments and in the transmission of force from the sarcomeres to the skeleton. The cytoskeleton has been described as comprising two lattices: The **endosarcomeric** cytoskeleton maintains the orientation of the thick and thin filaments within the sarcomere, and the **exosarcomeric** cytoskeleton maintains the side-by-side alignment of the myofibrils (Clark et al., 2002; Patel & Lieber, 1997; Waterman-Storer, 1991).

The major components of the endosarcomeric cytoskeleton are the proteins **titin** and **nebulin.** Titin appears to connect the thick filaments and the Z band (figure 6.4*a*) to keep the myofilaments aligned and contributes to the banding structure of skeletal muscle. Because the titin connection provides a continuous link along the sarcomere, it probably contributes significantly to the passive tension of muscle (Granzier & Labeit, 2006; Prado et al., 2005; Wang et al., 1993). Nebulin appears to regulate the length of the thin filament and to influence the interaction between actin and myosin. Because the thin-filament proteins (actin, tropomyosin, troponin) can be of variable lengths, it has been proposed that nebulin sets the number of elements that should be connected (polymerized) and therefore determines the length of the thin filament. Nebulin may also, like tropomyosin and troponin, play a regulatory role in muscle contraction (Patel & Lieber, 1997).

The exosarcomeric cytoskeleton provides connections that transmit the force generated by actin and myosin to intramuscular connective tissues and the skeleton. The exosarcomeric proteins include the intermediate filaments and focal adhesions. The **intermediate fibers** are arranged longitudinally along and transversely across sarcomeres (figure 6.4*b*), between the myofibrils within a muscle fiber (figure 6.4*c*), and between muscle fibers (figure 6.4*d*). The intermediate fibers—which consist of such proteins as desmin, vimentin, and skelemin—are localized at the Z and M bands and connect each myofibril to its neighbor. The intermediate fibers are probably responsible for the alignment of adjacent sarcomeres and undoubtedly provide a pathway for the longitudinal and lateral transmission of force between sarcomeres, myofibrils, and muscle fibers. Much of the force generated by the contractile proteins is transmitted laterally (Monti et al., 1999; Street, 1983). Additionally, when myofibrils are added to a muscle fiber, as occurs with muscle hypertrophy, the intermediate fibers can align the new contractile proteins.

Focal adhesions connect myofibrils to the sarcolemma (figure 6.4*c*) and muscle fibers to the endomysium and the muscle-tendon junction (figure 6.4*d*). The major role of focal adhesions is to connect intracellular proteins to extracellular space; in muscle, focal adhesions are also known as costameres (Bloch & Gonzalez-Serratos, 2003; Caiozzo & Rourke, 2006). Focal adhesions include a number of cytoskeletal anchor proteins: ankyrin, *α*-dystrobrevin, *α*/*β*-dystroglycan, dystrophin, α-fodrin, integrins, *α*-sarcoglycan, *α*/*β*-spectrin, syntrophins, talin, and vinculin. One preliminary model of the focal adhesion (costamere) proposes that force is transmitted from *α*-actinin in the Z band to the talin-vinculin complex and then to the transmembrane integrins (Patel & Lieber, 1997). Dystrophin appears to be necessary to prevent mechanical damage (Petrof et al., 1993). The properties of the cytoskeleton, therefore, have a major influence on the capacity of muscle to move the skeleton (Clark et al., 2002).

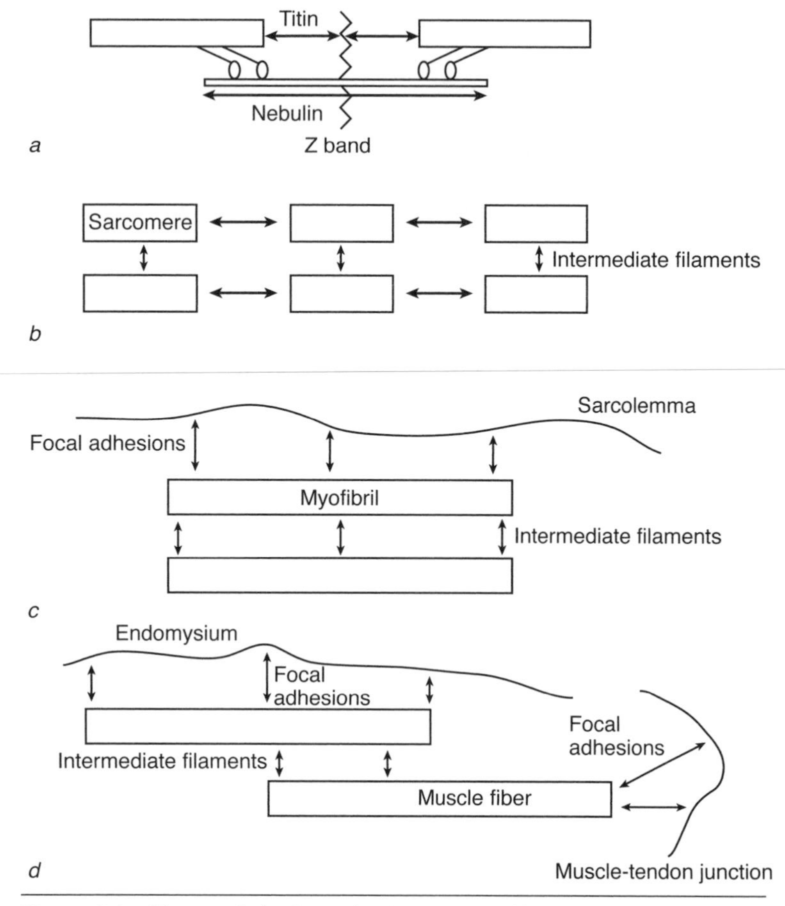

Figure 6.4 The cytoskeletal proteins (arrows) provide connections *(a)* between the myofilaments, *(b)* between sarcomeres within a myofibril, *(c)* between myofibrils and the sarcolemma, and *(d)* between muscle fibers and associated connective tissues.

Adapted, by permission, from T.J. Patel and R.L. Lieber, 1997, "Force transmission in skeletal muscle: From actomyosin to external tendons," *Exercise and Sport Sciences Reviews* 25: 322.

EXCITATION–CONTRACTION COUPLING

At the neuromuscular junction, the neurotransmitter acetylcholine (ACh) takes less than 100 μs to diffuse across the synaptic cleft and attach to receptors on the postsynaptic (muscle fiber) membrane. The attachment of ACh to the receptors results in the opening of the transmitter-gated Na^+-K^+ channels and in the influx of Na^+ into and efflux of K^+ from the muscle fiber. The movement of Na^+ and K^+ across the muscle fiber membrane results in the development of the end-plate potential that can trigger the generation of a muscle fiber action potential. The processes involved in the conversion of an axonal action potential into a muscle fiber potential are referred to as neuromuscular propagation.

Once a muscle fiber action potential has been generated, several processes contribute to the generation of a muscle fiber force. These processes are known as **excitation–contraction coupling** (Payne & Delbono, 2004; Rossi & Dirksen, 2006). The steps involved in excitation–contraction coupling are indicated in figure 6.5: (1) propagation of the action potential along the muscle fiber; (2) propagation of the action potential down the transverse tubule; (3) coupling of the action potential to the change in Ca^{2+} conductance of the sarcoplasmic reticulum; (4) release of Ca^{2+} from the sarcoplasmic reticulum; (5) reuptake of Ca^{2+} into the sarcoplasmic reticulum; (6) Ca^{2+} binding to troponin; and (7) interaction of the contractile proteins. Most of these steps involve events that permit the interaction of actin and myosin (Ca^{2+} disinhibition), whereas only step 7 corresponds to the cross-bridge cycle.

Ca^{2+} Disinhibition

Under resting conditions, the thick and thin filaments are prevented from interacting by the regulatory action of troponin and tropomyosin, and Ca^{2+} is stored largely in the sarcoplasmic reticulum. Steps 1 through 6 in figure 6.5 identify the events involved in the removal of this inhibition. Essentially, these events enable the muscle fiber action potential to trigger the release of Ca^{2+} from the sarcoplasmic reticulum and the subsequent inhibition of the regulatory action of troponin and tropomyosin. Because the net effect of this series of events is the removal, or inhibition, of inhibition, this process is referred to as **disinhibition.**

Ca^{2+} disinhibition begins with the muscle fiber action potential as it is propagated along the sarcolemma at speeds up to 6 m/s. The muscle fiber action potential is propagated down the T tubule and into the interior of the muscle fiber (Bastian & Nakajima, 1974); the inward spread of activation has been estimated to occur at a speed of about 70 μm/ms. The T tubule action potential activates the dihydropyridine (DHP) receptors that are distributed throughout the T tubules (figure 6.5). These receptors act as a voltage sensor that transmits a signal to the ryanodine receptors in the sarcoplasmic reticulum, through which Ca^{2+} is released from the sarcoplasmic reticulum. The mechanism by which the DHP receptors span the 15 nm between the T tubule and the sarcoplasmic reticulum and contact the ryanodine receptors is unknown; it may involve a chemical connection, a mechanical signal, or an effect due to changes in Ca^{2+} concentration. Some evidence favors the mechanical connection between con-

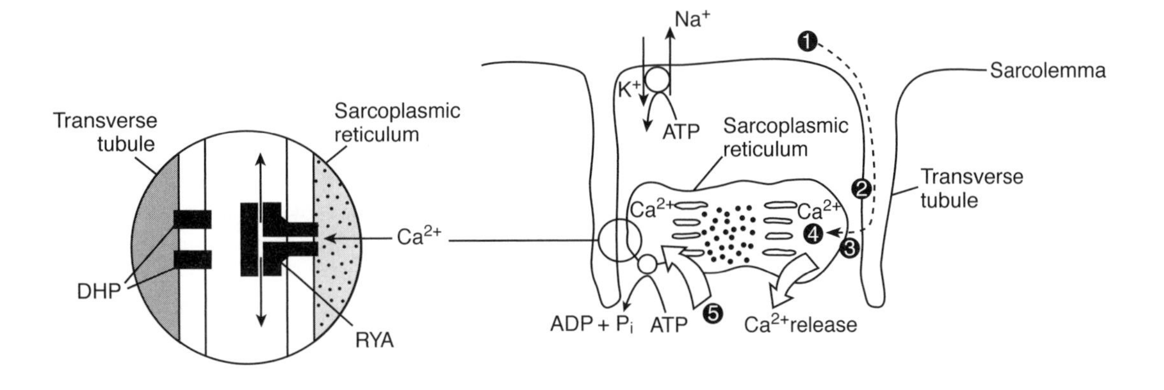

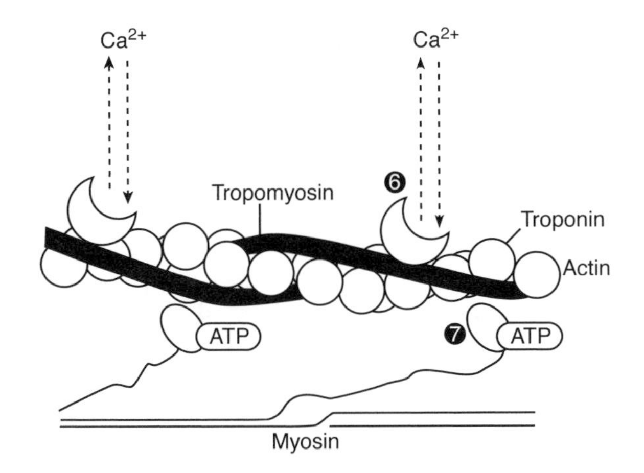

Figure 6.5 The seven steps involved in excitation–contraction coupling. The inset shows the structures involved in the release of Ca^{2+} from the sarcoplasmic reticulum. The dihydropyridine receptors (DHP) are located in the transverse (T) tubule and act as a voltage sensor. In the presence of an action potential, the DHP receptors activate the ryanodine receptors (RYA) in the sarcoplasmic reticulum and efflux of Ca^{2+} into the sarcoplasm of the muscle fiber.

Adapted, by permission, from R.H. Fitts and J.M. Metzger, 1993, Mechanisms of muscular fatigue. In *Principles of exercise biochemistry*, edited by J.R. Poortmans (Basel: Karger), 214.

formational (structural) changes in DHP channels and the ryanodine receptors (Caiozzo & Rourke, 2006; Payne & Delbono, 2004; Ríos et al., 1991; Ríos & Pizarró, 1988; Westerblad et al., 2000).

Figure 6.6 shows the role of Ca^{2+} in excitation–contraction coupling. In the resting state, the potential across the sarcolemma is negative on the inside with respect to the outside; and most of the Ca^{2+} is stored in the terminal cisternae, which are the enlargements of the sarcoplasmic reticulum near the T tubules (figure 6.6*a*). The action potential is propagated down the T tubule and triggers an increase in the Ca^{2+} conductance (g_{Ca}) of the terminal cisternae, which corresponds to an opening of the ryanodine receptors. In the absence of an action potential, g_{Ca} is normally low so that Ca^{2+} has difficulty crossing the membrane of the sarcoplasmic reticulum. Once g_{Ca} has been increased, Ca^{2+} moves down its concentration gradient from the terminal cisternae through the ryanodine release channels and into the sarcoplasm (figure 6.6*b*). The quantity of ryanodine receptors is two- to threefold greater in fast-twitch fibers than in slow-twitch fibers, enabling a greater amount of Ca^{2+} to be released by each action potential (Damiani & Margreth, 1994).

When the Ca^{2+} concentration in the sarcoplasm is above a threshold level (10^{-7} M), Ca^{2+} binds to the regulatory protein troponin that is attached to the actin filament. The binding of Ca^{2+} to troponin probably causes a structural change in the thin filament such that the myosin-binding site on actin is uncovered, and the two proteins (actin and myosin) are then able to interact. The uncovering of the binding site may involve a transient rotation of the regulatory complex (troponin-tropomyosin).

The change in g_{Ca} is transient, and once the action potential has passed (figure 6.6*c*), g_{Ca} returns to a resting level and Ca^{2+} is returned to the sarcoplasmic reticulum by Ca^{2+} pumps (Ca^{2+} ATPase) attached to the longitudinal regions of the sarcoplasmic reticulum. The enzyme Ca^{2+} ATPase pumps Ca^{2+} into the sarcoplasmic reticulum against a concentration gradient requiring the hydrolysis of one molecule of ATP for the translocation of two Ca^{2+} ions. The rate at which Ca^{2+} is returned to the sarcoplasmic reticulum determines the rate of decline in force after the cessation of an action potential. Fatigued muscle, for example, exhibits a reduction in the rate of reuptake of Ca^{2+} (due to a decline in the activity of the Ca^{2+} pumps), which produces a decline in the rate of relaxation. This reuptake of Ca^{2+} lowers the concentration of Ca^{2+} in the sarcoplasm, which inhibits the activity of the enzyme (actomyosin ATPase) that regulates the interaction of actin and myosin.

The relative duration of the events is shown in figure 6.7. These data, which were measured on the large muscle

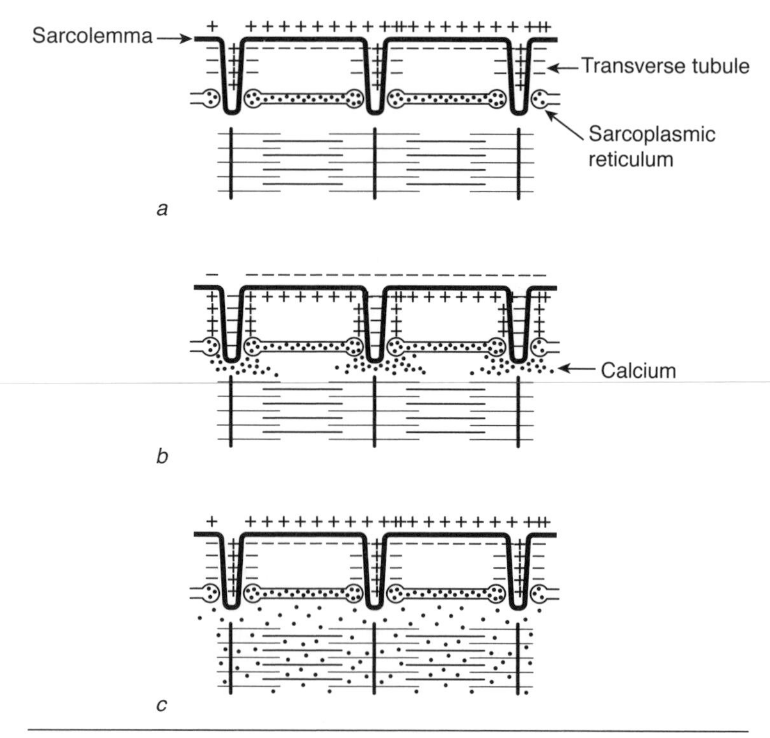

Figure 6.6 Role of Ca^{2+} in excitation–contraction coupling: *(a)* Ca^{2+} is stored in the sarcoplasmic reticulum at rest, *(b)* is released into the sarcoplasm in the presence of an action potential, and *(c)* remains in the sarcoplasm after the action potential has passed until it is returned to the sarcoplasmic reticulum by the pumps.

Reprinted, by permission, from J.C Rüegg, 1983, Muscle. In *Human physiology,* edited by R.F. Schmidt and G. Thews (New York: Springer-Verlag), 37. Copyright 1983 by Springer-Verlag.

fibers of the barnacle, comprised the potential across the excitable membrane, intracellular Ca^{2+} concentration, and muscle force (Ashley & Ridgway, 1968, 1970). Ca^{2+} concentration was visualized through use of aequorin, which is a protein that luminesces when Ca^{2+} is present. Such measurements indicate that the membrane potential changes first, then Ca^{2+} concentration, and finally muscle force. The duration of a typical action potential is much briefer than that shown in figure 6.8, which means that the duration of these events is briefest for the action potential and longest for the muscle force. Intracellular Ca^{2+} concentration increases with the rate of muscle activation (Blinks et al., 1978).

Cross-Bridge Cycle

The interaction between the contractile proteins that occurs as a result of Ca^{2+} disinhibition involves several biochemical events that produce transient structural changes in the proteins. Because these events involve the globular heads of myosin attaching to actin, the sequence is known as the cross-bridge cycle. The process involves the use of chemical energy contained in ATP by the globular heads of myosin to perform work. The cross-bridge cycle comprises several steps that can be distinguished based on the absence (nonactivated states) or presence (activated states) of Ca^{2+} bound to troponin (Gordon et al., 2001; Rayment et al., 1993; Warshaw, 1996).

For each globular head, the cycle involves a detachment phase, an activation phase, and an attachment phase. The cycle begins at the top of figure 6.8 with the binding of ATP to the myosin head, as a result of which the myosin head detaches from actin, ATP is hydrolyzed to adenosine diphosphate (ADP) and inorganic phosphate (P_i), and the myosin head is weakly bound to actin. The rate of ATP hydrolysis varies about fourfold with muscle fiber type, being greatest for the fastest-contracting fibers (Steinen et al., 1996). The binding of Ca^{2+} to troponin, which occurs at the bottom of figure 6.8, enables the two myosin heads to close, resulting in the release of P_i and the power stroke of the cross-bridge cycle (Volkmann et al., 2000). Each cross-bridge exerts a force of about 2 pN for several hundreds of milliseconds during the strong binding phase that concludes with the release of ADP. The work performed by the myosin heads causes the thick and thin filaments to be displaced by 5 to 10 nm in one cross-bridge cycle (Kitamura et al., 1999). The cycling of cross-bridges continues as long as there are sufficient amounts of Ca^{2+} and ATP in the muscle cell.

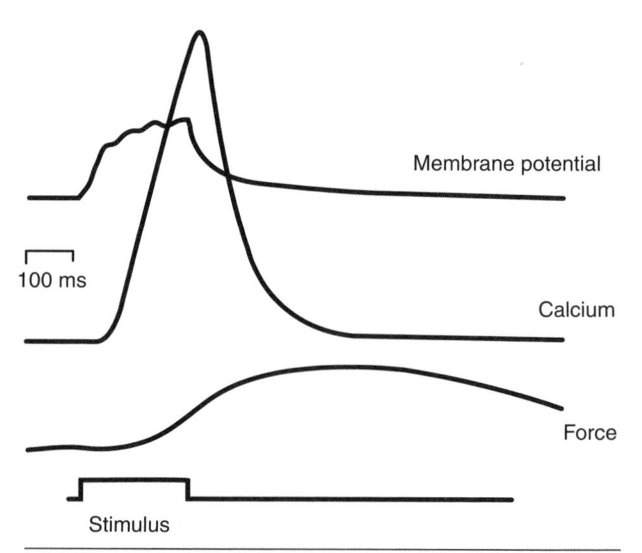

Figure 6.7 Depolarization of a muscle fiber membrane causes an increase in intracellular Ca^{2+} and enables the muscle fiber to generate force.

Adapted, by permission, from C.C. Ashley and E.B. Ridgway, 1970, "On the relationships between membrane potential, calcium transient and tension in single barnacle muscle fibres," *Journal of Physiology* 209: 111.

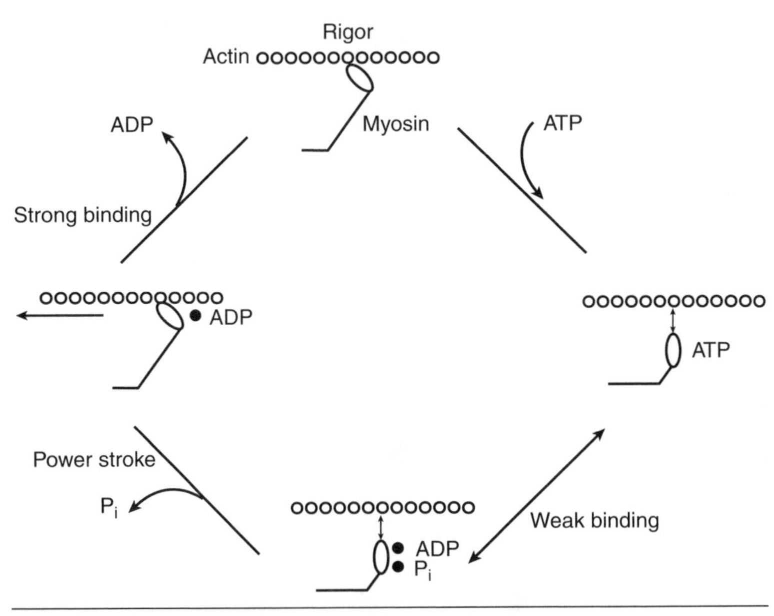

Figure 6.8 The cross-bridge cycle. ATP = adenosine triphosphate; ADP = adenosine diphosphate; P_i = inorganic phosphate.

Because the displacement of the myosin head occurs while actin and myosin are connected, the thick and thin filaments slide past one another and exert a force on the cytoskeleton. The sliding of the filaments relative to one another has given rise to the **sliding filament theory** of muscle contraction (Huxley & Niedergerke, 1954; Huxley & Hanson, 1954). The force exerted by muscle is usually explained as a consequence of the concurrent, but not synchronous, cycling of many cross-bridges following Ca^{2+} disinhibition; this is known as the *cross-bridge theory of muscle contraction* (Huxley, 2000, 1957; Huxley & Simmons, 1971). The term **contraction** refers to the state of muscle activation in which cross-bridges are cycling in response to a muscle fiber action potential. When the action potential has passed, g_{Ca} of the sarcoplasmic reticulum returns to its normal low level, Ca^{2+} is actively removed from the sarcoplasm, and the inhibitory actions of troponin and tropomyosin are resumed.

Imaging a Muscle Contraction

Similar to the use of electromyography (EMG) to measure muscle fiber action potentials, magnetic resonance imaging (MRI) has been used to detect the contractile activity of muscle. As a technique that can localize atomic nuclei, MRI can be used to study the structure and function of muscle. Conceptually, the technique involves three steps: (1) exposing an object to a magnetic field that aligns chemical elements with odd atomic weights; (2) perturbing the alignment of these elements; and (3) measuring the rate, called the relaxation rate, at which atomic nuclei return to the initial alignment. The principal relaxation rates are known as the longitudinal (T_1) and transverse (T_2) relaxation times. Increases in both T_1 and T_2 **relaxation times** have been correlated with increases in the water content of muscle and, as a result, have proved useful for the study of muscle activation, muscle soreness, and musculotendinous strains (Fleckenstein & Shellock, 1991; Meyer & Prior, 2000; Patten et al., 2003).

In a typical experiment, measurements are made of a body part before and after the performance of a prescribed task so that changes in T_1 and T_2 can be determined. By varying the details of the performance in repeat measurements, it is possible to determine associations between the relaxation times and such performance criteria as the work done and the magnitude of the force exerted by the muscle. With this protocol, the MRI data can provide information on the cumulative effects of the exercise but do not indicate the moment-to-moment activity during the performance. Changes in T_2 relaxation time, for example, appear to be related to an exercise-induced change in the intracellular water content of muscle fibers following a muscle contraction (Fleckenstein et al., 1988; Fisher et al., 1990). Although T_2 relaxation time can be influenced by extracellular water, intracellular pH, level of oxygenation in the blood, fluid viscosity, temperature, and intramuscular fat, exercise-induced changes seem to be caused by alterations in H_2O exchange or binding with proteins rather than either a relative increase in the extracellular fluid volume or a decrease in the paramagnetic solutes. For this reason, the measurement of changes in T_2 with physical activity is a technique that offers considerable potential for expanding our knowledge of muscle function (Patten et al., 2003).

An example of the application of this technology is shown in figure 6.9. These data indicate the relative use of the thigh muscles to perform brief bouts of high-intensity exercise on a cycle ergometer compared with a knee extension device (Richardson et al., 1998). Muscle usage for these two exercises can be evaluated qualitatively through comparison of the intensities of the images after exercise. Lighter regions after exercise indicate a greater use of those muscles. From these images, we can determine that the knee extensor muscles were used most during the knee extension exercise and that substantial activation of both the anterior and posterior compartments occurred during the cycling exercise. The magnitude of the change in intensity can be estimated from the calculated T_2 values. Richardson and colleagues found that there was a significant increase in T_2 in the four knee extensor muscles after the knee extension exercise and

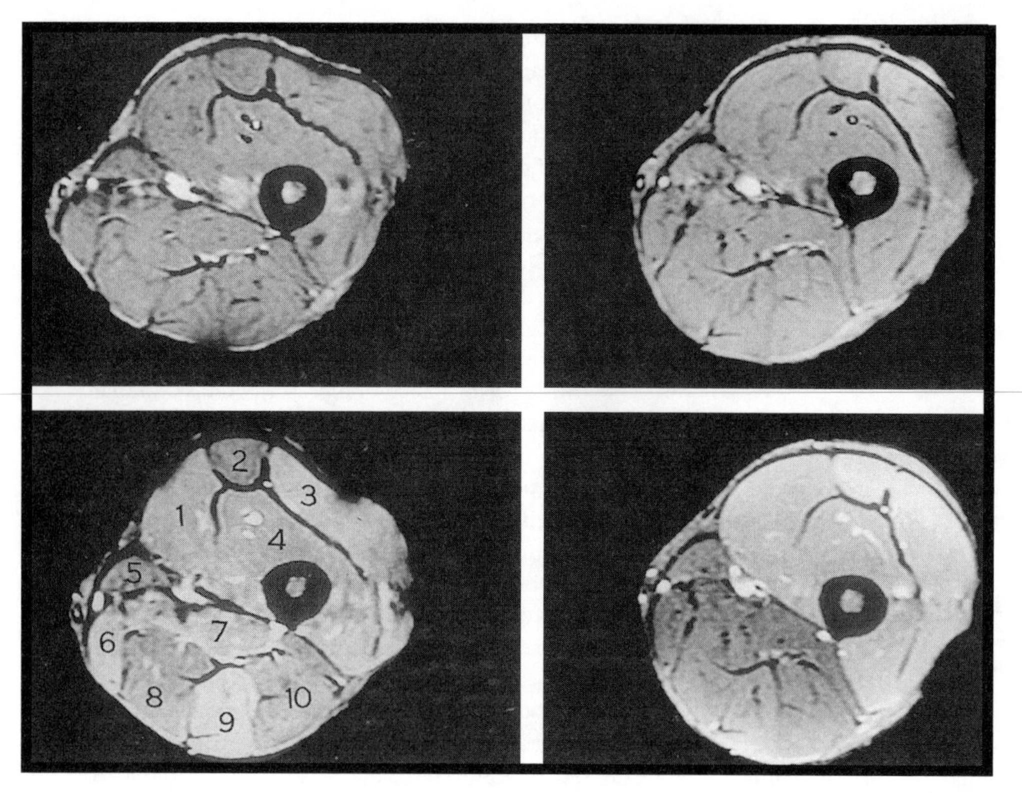

Figure 6.9 Magnetic resonance images of the thigh muscles in the right leg at rest (upper row) and after exercise (bottom row). The images in the left column are from before and after exercise on a cycle ergometer. The images in the right column are from before and after exercise on a knee extension machine. The lighter image after exercise indicates an increase in signal intensity, which is used to determine T_2 values. The muscles labeled on the bottom left image correspond to (1) vastus medialis, (2) rectus femoris, (3) vastus lateralis, (4) vastus intermedius, (5) sartorius, (6) gracilis, (7) adductor longus, (8) semimembranosus, (9) semitendinosus, and (10) biceps femoris.

Photograph provided by Russell S. Richardson, Ph.D.

Table 6.2 T_2 Values (ms) in Thigh Muscles Before and After Cycling Exercise and Knee Extension

	CYCLING EXERCISE		KNEE EXTENSION EXERCISE	
Muscle	**Before (ms)**	**After (ms)**	**Before (ms)**	**After (ms)**
Vastus medialis	27.7	31.4*	27.7	32.6*
Rectus femoris	26.3	29.1*	27.3	37.4*
Vastus lateralis	27.5	33.0*	28.3	34.2*
Vastus intermedius	28.8	32.0*	28.8	32.0*
Sartorius	26.8	31.4*	30.2	29.4
Adductor magnus	27.9	33.5*	29.0	28.6
Semimembranosus	21.6	25.2*	24.5	25.5
Semitendinosus	24.7	29.3*	26.1	27.0
Biceps femoris	28.1	30.5*	28.0	27.1

Note. *Significant increase in T_2 after exercise.

Data from Richardson et al. 1998.

a significant increase in T_2 for all 10 muscles after the cycling exercise (table 6.2).

Two observations underscore the usefulness of MRI for studying muscle function. First, there is a positive linear relation between the intensity of the T_2 signal and the intensity of a muscle contraction. This relation has been reported for shortening (concentric) and lengthening (eccentric) contractions and for several different muscle groups. When a muscle group lifts a load with a shortening contraction, for example, the oxygen consumption is greater and more motor units are recruited than when the muscles lower the same load with a lengthening contraction. As a consequence, the intensities of both the EMG and the T_2 signals are less during the lengthening contraction (Adams et al., 1992; Fisher et al., 1990; Jenner et al., 1994; Shellock et al., 1991). Second, the MRI measurement has enough spatial resolution to identify the subvolume of finger flexor muscles (flexor digitorum superficialis and flexor digitorum profundus) that control the activity of individual fingers (Fleckenstein et al., 1992). This capability enables researchers to determine the prevalence of functional compartments within a single muscle and within a group of synergist muscles (Enocson et al., 2005; Kinugasa et al., 2005). It also makes it possible to determine whether a task involves coactivation of an agonist–antagonist set of muscles (figure 6.9). Although the measurement is not reliable enough to allow the development of muscle activation maps based on pixel-by-pixel estimates of T_2 (Prior et al., 1999), the distribution of changes in signal intensity after exercise can provide insight into the in vivo function of muscle as it yields information complementary to EMG recordings (Price et al., 2003).

MOTOR UNIT

The motor unit is the basic functional unit of the nervous system and muscle that produces movement. A **motor unit** consists of a motor neuron in the ventral horn of the spinal cord or brain stem, its axon, and the muscle fibers that it innervates (Liddell & Sherrington, 1925; Sherrington, 1925). The central nervous system controls muscle force by varying the activity of motor units that compose the muscle.

Most skeletal muscles comprise a few hundred motor units, ranging from about 10 for small muscles to 1500 for large muscles (table 6.3). The group of motor neurons that innervate a single muscle is referred to as a motor pool or motor nucleus. The motor neurons in a motor pool vary widely in their electrical properties and in the relative amplitudes of the synaptic inputs that they receive from different sources. The force that a muscle exerts during a contraction depends on the number of motor neurons that are activated and the rates at which they discharge action potentials (Adrian & Bronk, 1929). These two features of motor unit activity are known as recruitment and rate coding, respectively.

Table 6.3 Estimated Numbers of Motor Neurons in the Motor Pools of Selected Upper Limb Muscles of a Primate

Muscle	Motor neurons
Abductor and flexor pollicis brevis	115
Adductor pollicis	370
Abductor pollicis longus	126
Biceps brachii	1051
Extensor carpi radialis	890
Extensor carpi ulnaris	216
Extensor digiti secundi proprius	87
Extensor digitorum communis	273
Extensor pollicis longus	14
First dorsal interosseus	172
Flexor carpi radialis	235
Flexor carpi ulnaris	314
Flexor digitorum profundus	475
Flexor digitorum superficialis	306
Lumbricals (lateral)	57
Triceps brachii	1271

Note. The motor neurons were visualized for counting by retrograde transport of horseradish peroxidase injected into each muscle.

Data from Jenny and Inukai 1983.

Motor Neuron

The neural component of the motor unit, which consists of the motor neuron and its dendrites (figure 5.11), represents the final common pathway by which the commands from the nervous system are sent to muscle (Liddell & Sherrington, 1925). Motor neurons are located in the ventral horn of the spinal cord and brain stem, and each motor neuron sends its axon via a peripheral nerve to the muscle fibers that it innervates. Figure 6.10 shows two populations of motor neurons (motor pools), one on the

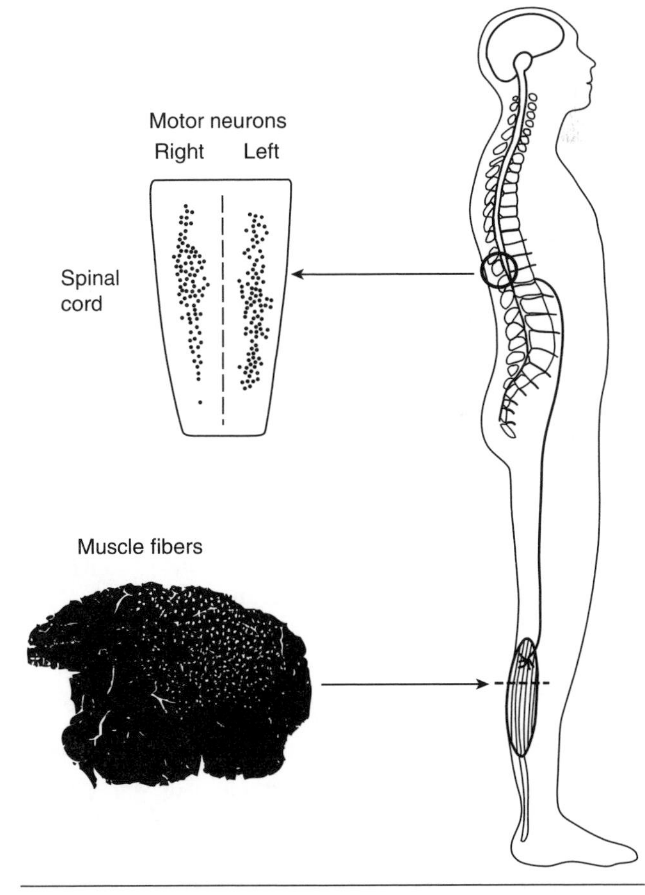

Figure 6.10 Motor neurons that innervate muscles on the right side of the body are located on the right side of the spinal cord, and correspondingly for muscles on the left side. The motor neurons that compose a motor pool are distributed across several spinal segments and are shown as dots in the frontal section of the spinal cord. Each motor neuron innervates, on average, a few hundred muscle fibers, and the distribution of fibers innervated by one motor neuron is indicated as white dots in the muscle cross section.

right and one on the left side of the spinal cord, which are distributed across several segments in the spinal cord. Each motor pool innervates a muscle on the corresponding side of the body.

The features of the motor neuron that vary across the motor pool include its morphology, excitability, and distribution of inputs (Heckman & Enoka, 2004; Kernell, 2006). The morphological feature that has received the greatest attention is motor neuron size, due to its association with the activation order of motor neurons. The diameter, surface area, number of dendrites, and capacitance of the cell body are used as indexes of motor neuron size (Binder et al., 1996). As initially reported by Henneman (1957), there is a strong correlation between the size of a motor neuron and the activation order within a motor pool. When Henneman evoked a stretch reflex in a muscle of a decerebrate cat, he noted that the order in which motor units were activated varied with action potential amplitude. Because action potential amplitude depends on axon diameter and thus cell size, he deduced that motor units were activated in order of increasing size. This observation became known as the **size principle** (Henneman, 1979).

Subsequent studies have continued to characterize motor neurons by placing a microelectrode inside the cell and measuring its electrical properties. As summarized in table 6.4 for cat motor neurons, these properties include input resistance (MΩ), rheobase current (nA), voltage threshold (mV), afterhyperpolarization duration (ms), and axonal conduction velocity (m/s). Input resistance represents the electrical resistance exhibited by a cell as the change in membrane potential it experiences in response to current injected by an intracellular microelectrode (Ohm's law; equation 5.9). Because input resistance depends on the density of the resting ion channels and the size of the neuron, small motor neurons tend to have a high input resistance, which results in a larger postsynaptic potential in response to a given synaptic current. **Rheobase current** is a measure of excitability as indicated by the amount of current that has to be injected into the motor neuron for it to generate an action potential. The rheobase current depends on the resting membrane potential, voltage threshold, and input resistance. It is much less for small motor neurons than for larger motor neurons. **Voltage threshold** denotes the amount of depolarization that is necessary to cause a neuron to discharge an action potential. The voltage threshold, which is primarily determined by the voltage sensitivity of the Na^+ channels, is least for small motor neurons. Afterhyperpolarization duration is the length of time the membrane potential during the trailing part of the action potential is more hyperpolarized than the normal resting membrane potential. The afterhyperpolarization is caused by a g_K that is activated by Ca^{2+}; hence it is known as a calcium-activated potassium conductance. Afterhyperpolarization duration is longest in small motor neurons. **Conduction velocity** refers to the speed at which action potentials are propagated along the axon, which varies with axon diameter. These associations indicate that small motor neurons are more excitable but that they generate and propagate action potentials at a slower rate than larger motor neurons.

To understand the function of the motor unit, it is necessary to combine the information on the excitability of motor neurons with details of the inputs they receive from the cortex, brain stem, and peripheral sensory receptors (Heckman & Enoka, 2004). Motor neurons receive up to 50,000 synapses, and about 95% of these occur on the dendrites (Cullheim et al., 1987; Moore & Appenteng, 1991; Rose et al., 1985; Ulfhake & Cullheim, 1988; Ulfhake & Kellerth, 1983). It appears, however, that the input from different sources can have a variable effect on the generation of an action potential by the motor neuron.

Table 6.4 Distribution of Motor Neuron Properties Based on Differences in Size

	Large	Intermediate	Small
Morphology			
Cell diameter (μm)	53	53	49
Cell surface area (μm²)	369	323	249
Number of main dendrites	10	11	11
Axon diameter (μm)	7	7	6
Membrane properties			
Axon conduction velocity (m/s)	101	103	89
Input resistance (MΩ)	0.6	0.9	1.7
Rheobase current (nA)	20	13	5
Voltage threshold (mV)	20	19	14
Afterhyperpolarization duration (ms)	65	78	161
Synaptic input			
Group Ia connectivity (%)	87	97	94
Group Ia EPSP amplitude (μV)	71	118	179
Recurrent inhibition amplitude (μV)	280	679	1173

Note. EPSP = excitatory postsynaptic potential.

Data from Burke et al. 1982; Cullheim & Kellerth 1978; Fleshman et al. 1981*a*, *b*; Friedman et al. 1981; Gustafsson and Pinter 1984; Ulfhake and Kellerth 1982; Zengel et al. 1984.

This difference can be a consequence of the number and location of synapses associated with each input system. The effect of different inputs can be assessed through use of a microelectrode to measure the **effective synaptic current** (nA) generated in the motor neuron in response to a given input (Binder et al., 1996, 2002; Heckman & Binder, 1991). This measurement indicates the net effect of activating an input system and represents the signal that will be conducted to the axon hillock where the action potential will be generated. Synaptic inputs to populations of motor neurons exhibit three patterns of distribution (Heckman & Binder, 1990): (1) least input (smallest effective synaptic current) to the largest motor neurons—input from the group Ia afferent of the muscle spindle; (2) uniform input to all motor neurons—inhibitory input from a muscle spindle located in an antagonist muscle and from an identified interneuron (recurrent inhibition from the Renshaw cell); and (3) greatest input to the largest motor neurons—input from a brain stem nucleus (red nucleus) and from a nerve (sural) containing feedback from cutaneous receptors. These observations suggest that the activation of a motor neuron depends both on its intrinsic excitability and on the distribution of input it receives (Kernell & Hultborn, 1990).

Muscle Unit

The muscle fibers that are innervated by a single motor neuron are sometimes referred to as a **muscle unit** (Burke & Tsairis, 1973). The number of muscle fibers in a muscle unit, which is known as the **innervation number,** ranges from around 10 to several thousand (figure 6.11). Based on autopsy specimens, the average innervation number appears to vary with muscle size (table 6.5). For example, first dorsal interosseus (a hand muscle) has about 120 motor units and an average innervation number of 340, compared with 580 motor units and an average innervation number of 1934 for medial gastrocnemius (Feinstein et al., 1955). The peak force that a muscle unit can produce largely depends on its innervation number. Although table 6.5 indicates that the average innervation number is known for some muscles, the range of innervation numbers in muscles is unknown (Enoka & Fuglevand, 2001).

The fibers belonging to a muscle unit tend to be confined to a relatively restricted volume of muscle. Figure 6.12 shows the location of the fibers belonging to a single muscle unit in the medial gastrocnemius muscle of a cat. The muscle unit had 500 muscle fibers. Such measurements have indicated that the territory of a single muscle unit can extend up to about 15% of the volume in this muscle, with a density of 2 to 5 muscle fibers per 100 belonging to the same muscle unit (Burke, 1981). Consequently, a given volume of muscle contains 20 to 50

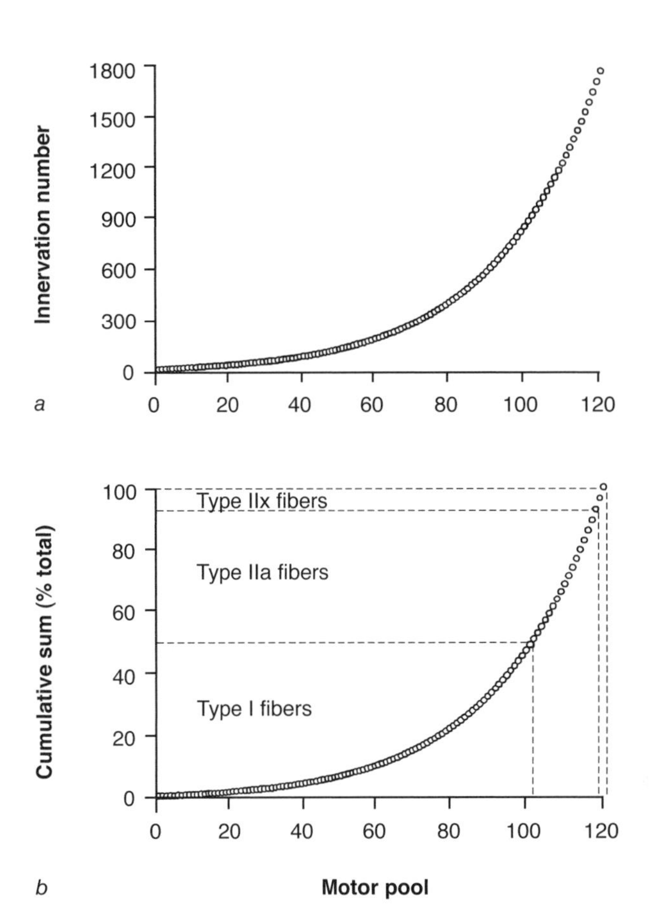

Figure 6.11 Distribution of *(a)* innervation number and *(b)* muscle fiber types across a motor pool that comprises 120 motor units. The cumulative sum in *b* denotes the progressive sum of the number of muscle fibers in successive motor units. In this example, the muscle contained ~48,000 muscle fibers. The exponential relation is based on the distribution of motor unit forces within a motor pool. A similar relation is assumed to exist for most muscles.

Adapted from R.M. Enoka and A.J. Fuglevand, 2001, "Motor unit physiology: Some unresolved issues," *Muscle & Nerve* 24(1): 7. By permission of John Wiley & Sons, Inc.

different muscle units. The territory of a muscle unit in the cat tibialis anterior muscle can range from 8% to 22% of cross-sectional area, but in the soleus muscle the territory ranges from 41% to 76% (Bodine et al., 1988).

Not only is the territory of muscle units constrained to specific volumes within a muscle, but also different parts of a muscle can contain distinct populations of motor units. This observation has given rise to the concept of a **neuromuscular compartment** (Peters, 1989; Windhorst et al., 1989), which corresponds to the volume of muscle supplied by a primary branch of the muscle nerve (figure 6.13). The muscle fibers belonging to a muscle unit are confined to a single neuromuscular compartment. Neuromuscular compartments have been found in some, but not all, muscles. Because compartments can be activated

Table 6.5 Anatomical Estimates of the Number of α Motor Axons and Muscle Fibers in Human Skeletal Muscles

	Specimen	α Motor axons	Number of muscle fibers	Innervation number
Abductor digiti mini[h]	Ten adults	380	72, 300	190
Abductor pollicis brevis[i]	Ten adults	171	15,400	90
Adductor pollicis[i]	Ten adults	128	13,600	106
Biceps brachii[b,d]	Stillborn infants	774	580,000	750
Brachioradialis[g]	Man 40 years	315	>129,000	>410
		350		
Cricothyroid[b,f]	Four adults	112	18,550	155
First dorsal interosseus[g]	Man 22 years	119	40,500	340
First lumbrical[g]	Man 54 years	93	10,038	108
	Woman 29 years	98	10,500	107
Flexor pollicis brevis[i]	Ten adults	172	15,300	89
Masseter[c]	Man 54 years	1452	929,000	640
Opponens pollicis[b,d,i]	Stillborn infants	133	79,000	595
	Ten adults	172	15,300	89
Platysma[g]	Woman 22 years	1096	27,100	25
Medial gastrocnemius[g]	Man 28 years	579	1,120,000	1934
	Man 22 years		964,000	1634
Plantaris[e]	Ten cadavers	204	64,300	372
Posterior cricoarytenoid[b,f]	Four adults	140	16,200	116
Rectus lateralis[j]	Two cadavers	4150	22,000	5
Stapedius[a]	Twenty cadavers	256	1081	7
Temporalis[c]	Man 54 years	1331	1,247,000	936
Tensor tympani[j]	Two cadavers	146	1100	8
Tibialis anterior[g]	Man 40 years	445	250,000	562
	Man 22 years		295,500	657
Transverse arytenoids[b,f]	Four adults	139	34,470	247

a. Blevins 1967; b. Buchthal 1961; c. Carlsöö 1958; d. Christensen 1959; e. de Carvalho 1976; f. Faaborg-Andersen 1957; g. Feinstein et al. 1955; h. Santo Neto et al. 1985; I. Santo Neto et al. 2004; j. Torre 1953.

independently, a single muscle can consist of several distinct regions each of which serves a different physiological function (English, 1984; English & Ledbetter, 1982; Fleckenstein et al., 1992; Keen & Fuglevand, 2004; Kilbreath et al., 2002). The existence of neuromuscular compartments makes it misleading to infer the function of a muscle based solely on the location of its attachments to the skeleton without considering both the architecture and innervation pattern of the muscle.

The muscle biceps brachii is often cited as an example of a muscle with distinct neuromuscular compartments. It is innervated by three to five primary branches of the musculocutaneous nerve, and the motor units appear to function in two distinct populations. One population, which is located in the lateral aspect of the long head, is active when a flexion torque is exerted about the elbow joint. The other population is active when the torque about the elbow joint includes both flexion and supination

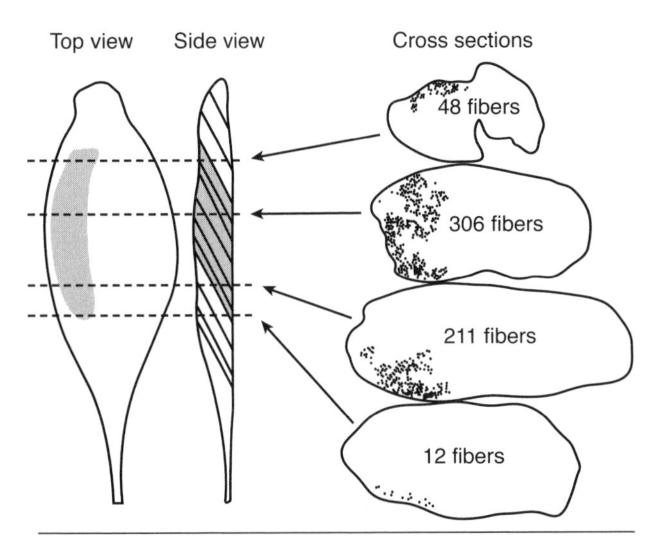

Figure 6.12 The distribution of fibers belonging to a single muscle unit in the medial gastrocnemius is shown in longitudinal sections on the left and in cross sections on the right. Each dot in the cross sections represents a single muscle fiber.

Adapted, by permission, from R.E. Burke and P. Tsairis, 1973, "Anatomy and innervation ratios in motor units of cat gastrocnemious," *Journal of Physiology* 234: 755.

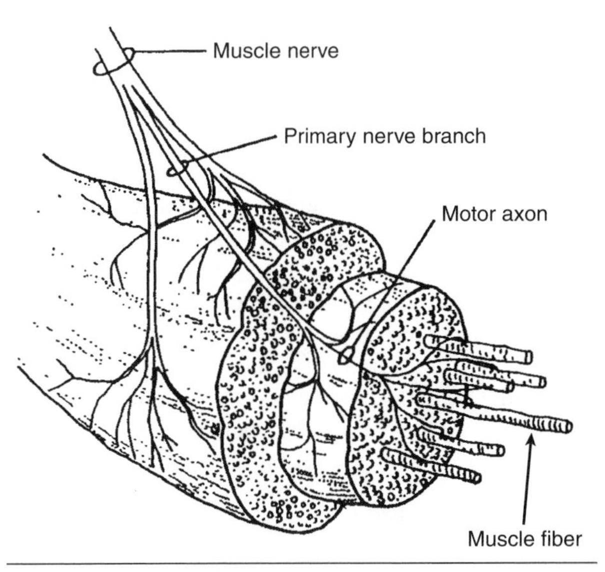

Figure 6.13 Subdivision of a muscle nerve down to the level of single axons that innervate single muscle fibers. The primary nerve branches can produce discrete activation of different parts of a muscle.

Adapted, by permission, from S.E. Peters, 1989, "Structure and function in vertebrate skeletal muscle," *American Zoologist* 29: 222. Copyright 1989 by The Society for Integrative and Comparative Biology.

components (ter Haar Romeny et al., 1982; van Zuylen et al., 1988).

EXAMPLE 6.1

Number of Motor Units in a Muscle

Changes in the number of functioning motor units in a muscle can be estimated with electrophysiological techniques (McComas, 1995; Shefner, 2004). One common procedure involves applying electric shocks of varying intensity to a peripheral nerve and measuring the evoked responses in the muscle (figure 6.14*a*); this is known as the Motor Unit Number Estimation (MUNE) technique. The protocol begins with the application of a weak electric shock that will generate an action potential in a single motor axon, which is recorded as a motor unit action potential. Then the intensity of the stimulus is increased slightly to activate another motor axon. By gradually increasing the stimulus intensity, the investigator is able to measure the evoked potentials from 11 to 20 motor axons (figure 6.14*b*). From this set, the average amplitude of the evoked response for a single motor unit is determined and compared with the response evoked by a maximal stimulus (figure 6.14*c*). To determine the estimated number of functioning motor units, the amplitude of the maximal response is divided by the average amplitude for the single motor units. Although the reliability of this technique has been questioned (Stein & Yang, 1990), it does appear to provide valuable information about changes in motor unit number that occur in patient populations (Boe et al., 2006; Kwon & Lee, 2004; McNeil et al., 2005).

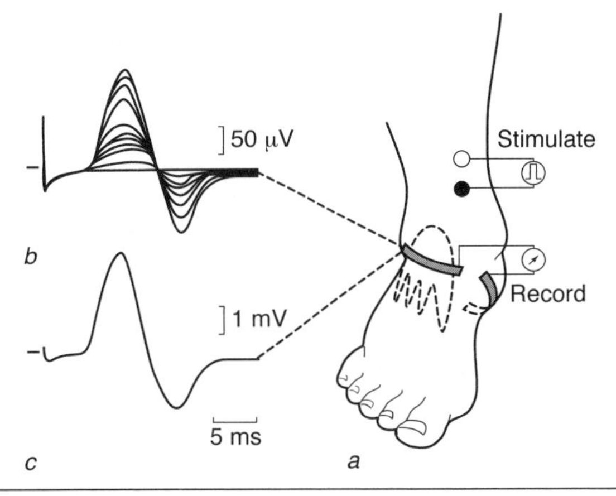

Figure 6.14 Electrophysiological technique for estimating the number of functioning motor units in a muscle: *(a)* stimulating electrodes placed over the nerve to evoke responses in the extensor digitorum brevis muscle; *(b)* set of responses evoked in single motor axons; and *(c)* the maximal evoked response.

Reprinted, by permission, from B. MacIntosh, P. Gardiner, and A. McComas, 2005, *Skeletal muscle: Form and function*, 2nd ed. (Champaign, IL: Human Kinetics), 193.

Contractile Properties

Motor units can be compared with one another based on a number of physiological properties, including the discharge characteristics of the motor neuron, the speed of contraction, the magnitude of force, and the resistance to fatigue. Most comparisons are based on the contractile properties of the motor unit. Two methodologies are commonly used for evaluating these parameters, one direct and the other indirect. A direct evaluation refers to the physiological measurement of the motor unit properties. Alternatively, these parameters can be estimated indirectly from histochemical, biochemical, and molecular measures that are related to contractile function.

Contractile Speed

The basic contractile property of a motor unit is a **twitch.** This is the force–time response to a single action potential. A twitch is characterized by three measurements: the time to peak force, the amplitude of the peak force, and the time it takes for the force to decline to one-half of its peak value (half-relaxation time) (figure 6.15). The time to peak force is often used as an index of contraction speed, although it mainly depends on the rate at which Ca^{2+} is released from the sarcoplasmic reticulum. A long time to peak force corresponds to a slow-twitch motor unit, whereas a brief time indicates a fast-twitch motor unit.

Motor Unit Force

Motor units are rarely activated to produce individual twitches. Rather, muscle units receive a number of action potentials that result in overlapping twitches, which summate to produce a force profile known as a **tetanus.** The degree of overlap between successive twitches depends on the rate at which the action potentials are generated and the time course of the twitch response (figure 6.15). The tetani produced during most voluntary contractions involve the partial summation of twitches, such as the tetanus that resulted from activation of the fast-twitch unit at 56 pps or the slow-twitch unit at 21 pps (figure 6.15).

The effect of activation rate on the summation of twitches for a unit can be determined in an experiment in which action potentials are generated artificially (electric shocks) and the evoked force is measured. The results of such an experiment are shown in figure 6.16. A tungsten microelectrode was inserted into a nerve to stimulate the axons of single motor units at various frequencies (2-100 pps), and the force exerted by the motor units was measured (Fuglevand et al., 1999; McNulty & Macefield, 2005; Thomas et al., 1991). The results indicate that human motor units need to be activated at 80 to 100 pps to reach maximal force, that the greatest changes in force

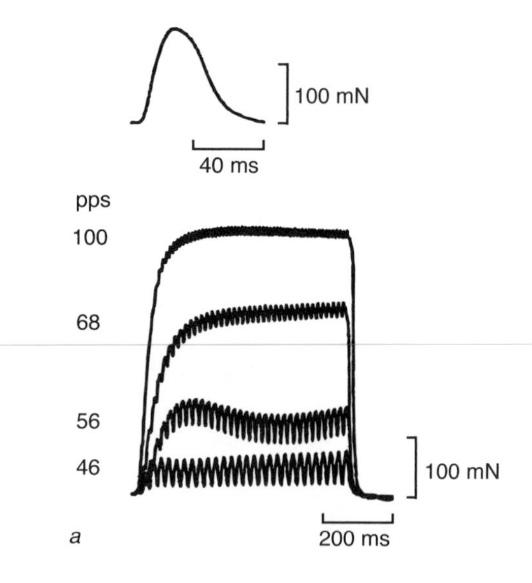

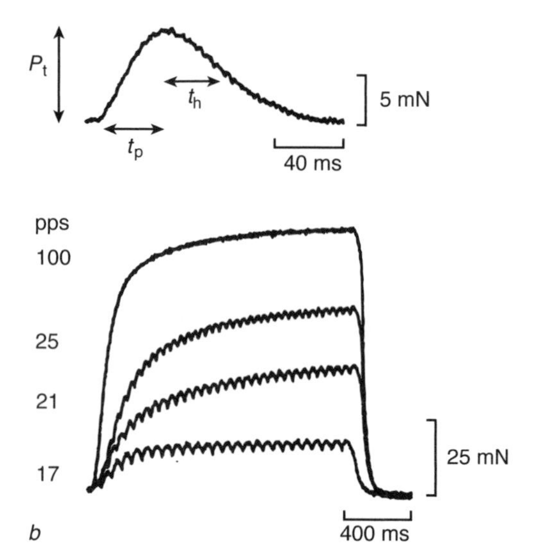

Figure 6.15 Twitch and tetanic forces for *(a)* a fast-twitch and *(b)* a slow-twitch motor unit. The twitches are shown as the upper traces in each panel. The twitch is characterized by the time to peak force (t_p), the magnitude of peak force (P_t), and half-relaxation time (t_h). Four tetani are shown for each motor unit. The tetani show the degree of summation of the twitches for a range of rates at which action potentials were evoked (pulses per second, pps). Due to the longer time course for the twitch of the slow-twitch unit, summation of twitches was greater at lower activation rates. Also, note the shape of the second tetanus for each unit (56 and 21 pps, respectively), where the force sags for the fast-twitch unit but continues to increase for the slow-twitch unit. Note the different scales for the different twitch and tetanic forces.

Adapted, by permission, from B.R. Botterman, G.A. Iwamoto, and W.J. Gonyea, 1986, "Gradation of isometric tension by different activation rates in motor units of cat flexor carpi radialis muscle," *Journal of Neurophysiology* 56: 497. Used with permission.

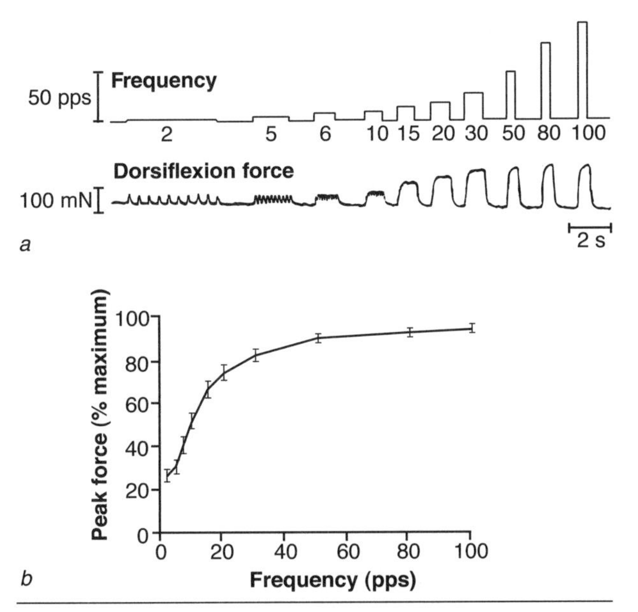

Figure 6.16 The relation between activation rate (frequency) and the force exerted by single motor units in human foot muscles. *(a)* Motor units activated by intraneural stimulation (upper trace) evoked a force in the dorsiflexor muscles (lower trace); *(b)* force was normalized to the maximal value for each of the 13 motor units and graphed to show the relation between activation rate and force. For these motor units, the half-maximal force was produced with an activation rate of about 10 pps.

Data from Macefield et al. 1996.

occur at activation rates of 8 to 20 pps, and that twitches begin to summate at 5 to 8 pps.

As indicated in figure 6.15, the shape of a tetanus, in addition to its amplitude, can vary with the activation rate. When the time between successive action potentials is equal to 1.25 times the time to peak force for the motor unit, the tetanus may reach an initial peak and subsequently decline briefly before reaching a peak again. This property is known as **sag** and is illustrated by the tetanus for the fast-twitch unit when it was activated at 56 pps (figure 6.15*a*). In many mammalian muscles, fast-twitch motor units exhibit sag whereas slow-twitch units do not. Sag appears to be caused by differences in the release and reuptake of Ca^{2+} by muscle fibers after each action potential (Carp et al., 1999).

The maximal force that a motor unit can exert (P_o) varies from 1.5 to 10 times greater than its twitch force (P_t); this value is known as the twitch-tetanus ratio. Peak tetanic force for human motor units does not vary as a function of the time to peak twitch force (figure 6.17*a*) as it does in most other mammalian muscles. Rather, P_o can vary over a substantial range for the same time to peak twitch force, and conversely the time to peak force can vary among motor units that have the same P_o (Bigland-Ritchie et al., 1998; Van Cutsem et al., 1997). Thus, some fast-twitch units are relatively weak.

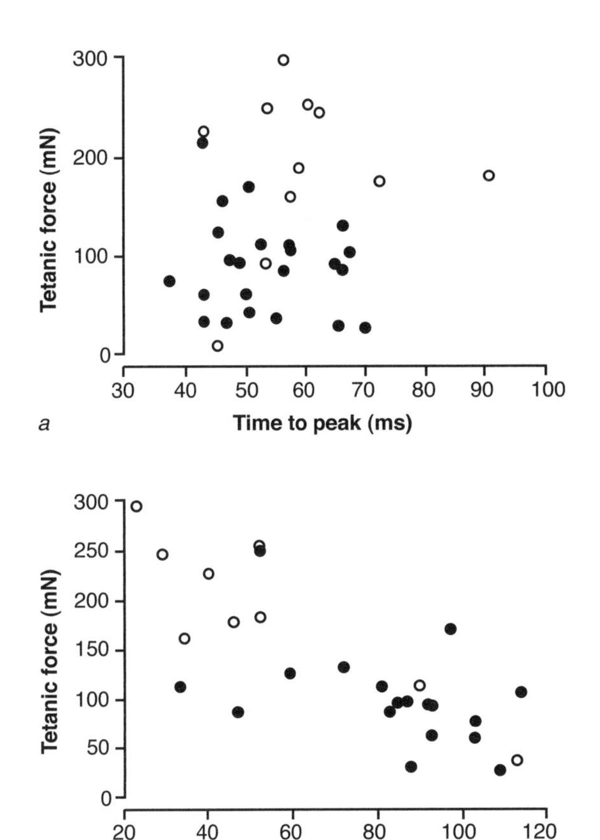

Figure 6.17 Relations between *(a)* maximal tetanic force and time to peak twitch force and between *(b)* maximal tetanic force and the decline in force after a fatiguing contraction for motor units of human hand and arm muscles. Each data point represents the values for one motor unit.

Data provided by Christine K. Thomas, PhD (○) and by Andrew J. Fuglevand, PhD (●).

Fatigability

The force produced in a single tetanus decreases when the motor unit is activated to produce a series of tetani. Motor units that do not exhibit this decline in force are described as resistant to fatigue. The rate at which the peak force decreases across successive tetani can be assessed with various fatigue tests. A standard fatigue test (Burke et al., 1973) designed for motor units in the cat hindlimb involves eliciting tetani for 2 to 6 min at a rate of one tetanus each second—each tetanus lasts for 330 ms and includes 13 stimuli. The ratio of the peak tetanic force exerted after 2 min of this stimulus protocol to the force in the initial tetanus is used as an index of fatigability.

This test distinguishes motor units that are resistant to fatigue from those that are not. Fatigue-resistant motor units exhibit ratios greater than 0.75, which means that after 2 min of this repetitive activation they can still produce 75% of the initial force. In contrast, motor units that are not fatigue resistant are characterized by ratios that are less than 0.25. This fatigue test appears to stress excitation-contraction coupling, which means that the processes associated with transforming a muscle fiber action potential into activation of the cross-bridge cycle are less susceptible to impairment during prolonged activity in fatigue-resistant motor units (Enoka & Stuart, 1992; Jami et al., 1983).

Motor Unit Types

Based on the distributions of these contractile properties, it is often possible to identify different groups of motor units in a muscle (Kernell, 2006). For example, motor units in the hindlimb muscles of the cat can be classified into three groups based on the presence or absence of sag in the tetanus and the resistance to fatigue (Burke, 1981): slow contracting, fatigue resistant **(type S)**; fast contracting, fatigue resistant **(type FR)**; and fast contracting, fast to fatigue **(type FF)**. Type S motor units do not exhibit sag, whereas sag is present in submaximal tetani of type FR and FF motor units (figure 6.15). Type S motor units are also the weakest, due to a combination of lower innervation numbers and smaller fibers (table 6.6).

Human motor units do not appear to fit this scheme (Bigland-Ritchie et al., 1998; McComas, 1998; Van Cutsem et al., 1997). For example, there is no relation between peak tetanic force and time to peak force in a twitch (figure 6.17a), whereas the most fatigue-resistant motor units in human thenar muscles tend to produce a lesser tetanic force (figure 6.17*b*) (Fuglevand et al., 1999). As a consequence of these relations, the first motor units activated during a voluntary contraction performed by a human are weak and fatigue resistant, but they may have slow or fast times to peak twitch force. One feature of contraction speed that may distinguish among human motor units is the activation rate required to achieve half the maximal force. Fuglevand and colleagues (1999) identified two groups of motor units when assessing the force–frequency relation (figure 6.16) of motor units in the long finger flexors and hand muscles. One group of motor units required an activation rate of 9 pps whereas the other group needed 16 pps to achieve half-maximal force. The average times to peak twitch force for the two groups of motor units were 66 ms and 46 ms, respectively. However, there was no difference in the fatigability of these two groups of motor units.

Muscle Fiber Types

In contrast to the use of direct physiological measurements to distinguish motor unit types, some classification schemes are based on histochemical, biochemical, and molecular properties of the muscle fibers (Caiozzo & Rourke, 2006; Kernell, 2006; Pette et al., 1999). The histochemical and biochemical measures involve determining the enzyme content of the muscle fibers. Because enzymes are the catalysts for chemical reactions, measuring the amount of enzyme provides an index of the rate at which a reaction can occur. Similarly, molecular measures can be used to determine the distribution of different isoforms of key molecules involved in a contraction. Once a correlation can be determined between a biochemical reaction or the abundance of a molecule and a physiological response, the quantity of the enzyme or molecule can be interpreted as a correlate of the physiological property.

One commonly used approach is to assay for the enzyme myosin ATPase, which is an index of the maximal velocity of shortening during a contraction (Bárány, 1967; Edman et al., 1988). One scheme is based on a histochemical analysis and identifies three types of fibers in human skeletal muscle: type I, type IIa, and type IIx. The distinction between type I and type II muscle fibers is based on the amount of ATPase activity remaining in the muscle fibers after preincubation in a solution with a pH

Table 6.6 Motor Unit and Muscle Fiber Characteristics for Three Cat Hindlimb Muscles

	MG	FDL	TA
Number of muscle fibers	170,000	26,000	—
Number of motor units	270	130	—
Innervation number			
S	611	180	93
FR	553	132	197
FF	674	328	255
Mean tetanic force (mN)			
S	76	11	40
FR	287	53	101
FF	714	300	208
Mean fiber area (μm²)			
I	1980	1023	2484
IIa	2370	1403	2430
IIb	4503	2628	3293
Muscle fiber types (%)			
I	25	10	11
IIa	20	37	50
IIb	55	53	39

Note. MG = medial gastrocnemius; FDL = flexor digitorum longus; TA = tibialis anterior.

Data from Bodine et al. 1987; Burke 1981; Dum & Kennedy 1980.

of 9.4. Type I fibers are often referred to as slow-twitch fibers, and type II as fast-twitch muscle fibers. Type II fibers can be further separated into two groups (IIa and IIx) after preincubation in solutions with pHs of 4.3 (IIa) and 4.6 (IIx) (Brooke & Kaiser, 1974). The distinction between the muscle fiber types is shown in figure 6.18 with a myosin ATPase stain of a thin cross section of a human vastus lateralis muscle.

Another commonly used classification scheme assesses the distribution of the genetically defined isoforms of the myosin heavy chain (MHC) (Bottinelli & Reggiani, 2000; Sant'ana et al., 1994; Staron & Pette, 1986). With this technique, the molecular components of a muscle fiber specimen can be separated by gel electrophoresis and the quantity of each element measured by densitometry. Three types of muscle fibers have been identified with this technique in human skeletal muscle: MHC-I, MHC-IIa, and MHC-IIx. There is a high correspondence between the I, IIa, and IIx fiber types identified by histochemistry and the MHC-I, MHC-IIa, and MHC-IIx types that emerge from the molecular analysis. However, MHC analyses have demonstrated that muscle fibers in adults can express from one to three of the MHC isoforms (Caiozzo et al., 2003; Parry, 2001; Talmadge et al., 1999).

Despite the diversity of fiber types that have been identified (Caiozzo & Rourke, 2006), there does appear to be some association between the basic muscle fiber types and muscle function (Aagaard & Bangsbo, 2006).

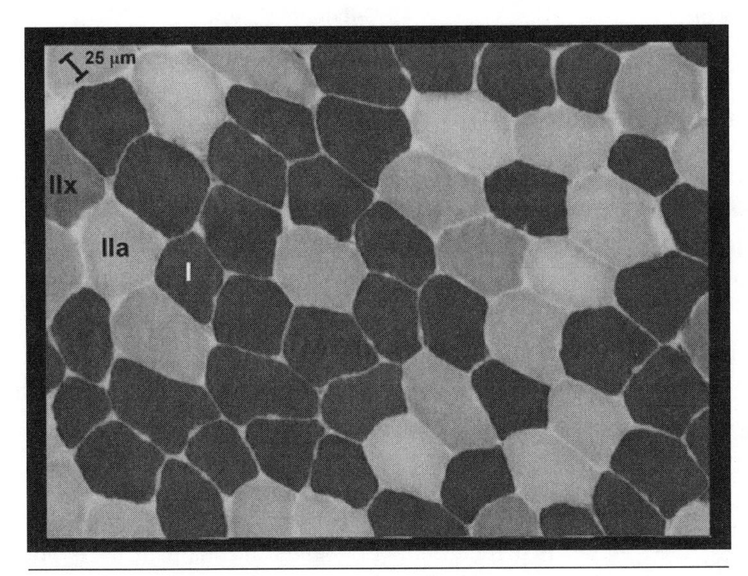

Figure 6.18 Photomicrograph of muscle biopsy cross section stained for myofibrillar ATPase after preincubation at pH 4.6. With this preparation, type I fibers stained dark, type IIa fibers stained light, and type IIx fibers stained intermediate.

Reprinted, by permission, from P. Aagaard et al., 2001, "A mechanism for increased contractile strength of human pennate muscle in response to strength training: Changes in muscle architecture," *Journal of Physiology* 534: 615.

When Monster and colleagues (1978) measured the usage of 15 muscles during an 8 h working day, they found that muscles with a higher proportion of type I fibers were used more frequently (table 6.7). Harridge and colleagues (1996) further explored this association by obtaining biopsy samples from three different muscles of human volunteers and measuring the fiber type distributions and the physiological properties of the muscles. The muscles examined in this study were soleus (MHC-I = 70%), vastus lateralis (MHC-I = 47%), and triceps brachii (MHC-I = 33%). Across these three muscles, whole-muscle measurements of time to peak twitch force (figure 6.19*a*), the rate of increase in tetanic force when the muscle was activated by electrical stimulation (figure 6.19*b*), and the amount of fatigue exhibited after 2 min intermittent electrical stimulation (figure 6.19*c*) were all associated with the proportion of the MHC-II fibers in the muscles. An important feature of these data, however, is the absence of significant associations within each muscle. These results suggest that when we compare the fiber type proportions for a specific muscle across subjects, variation in fiber type composition of a muscle can only partially explain the physiological differences.

Although many muscles in humans contain similar proportions of type I and type II fibers (table 6.7), the exponential distribution of innervation number across the motor pool means that most motor neurons innervate type I muscle fibers. This association is illustrated in figure 6.11*b* for a muscle that comprises 50% type I fibers, 45% type IIa fibers, and 5% type IIx fibers. In this example, 101 of the 120 motor neurons in the motor pool innervate type I fibers, 17 motor neurons innervate type IIa fibers, and only 2 motor neurons innervate type IIx fibers (Enoka & Fuglevand, 2001). Based on this scheme, 84% of the motor pool innervates 50% of the muscle fibers.

Motor Unit Activation

The force a muscle exerts depends on the number of motor units that are activated (motor unit recruitment) and the rate at which the motor neurons discharge action potentials (rate coding) (Kernell, 1992, 2006).

Motor Unit Recruitment

A voluntary contraction involves the activation of many motor units that are recruited in order of increasing size, as described by the size principle. The force at which a motor unit is recruited is known as its **recruitment threshold.** Figure 6.20 illustrates the recruitment of two motor units during a gradual increase in muscle force and indicates that the first recruited unit is weaker than the one

Table 6.7 Percentage of Type I Fibers in Human Skeletal Muscles

Muscle	Percentage
Obicularis oculi	15
Biceps brachii	38-42
Triceps brachii	33-50
Extensor digitorum brevis	45
Vastus lateralis	46
Lateral gastrocnemius	49
Diaphragm	50
Quadriceps femoris	52
First dorsal interosseus	57
Abductor pollicis brevis	63
Masseter	60-70
Tibialis anterior	73
Adductor pollicis	80
Soleus	80

recruited later. These two motor units are characterized as having low and high recruitment thresholds, respectively. A motor pool contains many weak motor units and only a few strong motor units (figure 6.11*a*).

When the same contraction is performed several times, motor units are activated in a relatively fixed order. This is known as **orderly recruitment** (Denny-Brown & Pennybacker, 1938). The typical way to study orderly recruitment is to determine the recruitment order of pairs of motor units during repeat performances of a task. Under normal conditions, the recruitment order varies in only about 5% of the trials, and then only when the two units have relatively similar recruitment thresholds (Feiereisen et al., 1997; Gordon et al., 2004; Thomas et al., 1991). Recruitment order does not vary with contraction speed, although the recruitment thresholds of motor units decrease for fast contractions (figure 6.21*a*) (Desmedt & Godaux, 1977). However, the recruitment order of motor units can vary when the synaptic input that the motor pool receives is altered, as occurs when electrical stimulation of the skin excites cutaneous receptors (Garnett & Stephens, 1981; Kanda et al., 1977). Motor units are also derecruited in a fixed order such that the motor unit recruited last when the force is increasing is the first unit derecruited when the force is decreasing. The force of the **derecruitment** threshold is similar to that for the recruitment threshold.

One advantage of orderly recruitment is that when a muscle is commanded to exert a force, the sequence of

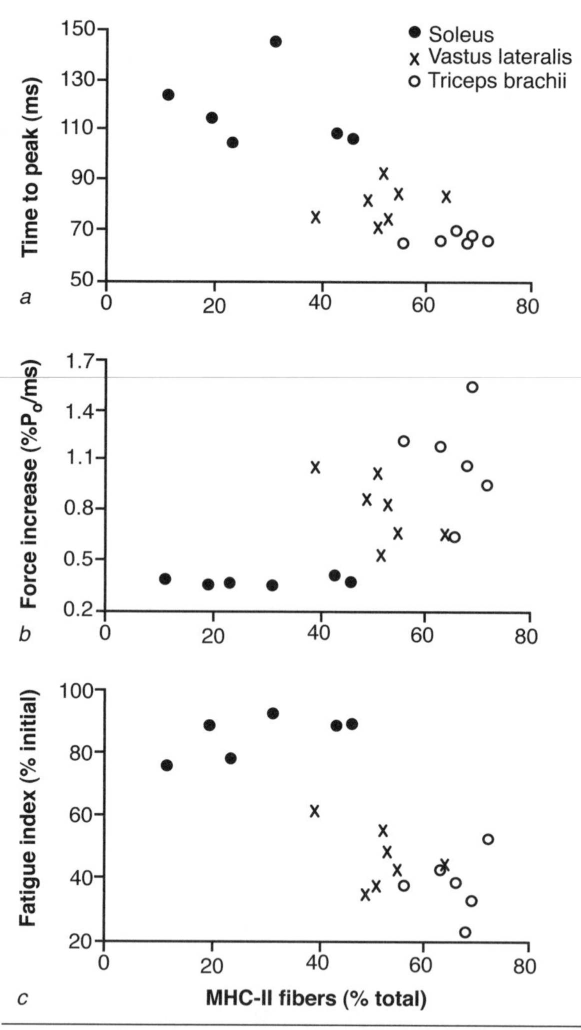

Figure 6.19 Associations among the proportion of myosin heavy chain II (MHC-II) fibers and *(a)* time to peak twitch force, *(b)* rate of increase in tetanic force, and *(c)* fatigability for three muscles. Each data point represents the average fiber type proportion and physiological measurement for a muscle in one subject.

Data from Harridge et al. 1996.

motor unit recruitment is determined by spinal mechanisms and does not need to be specified by the brain. Therefore, the command generated by the brain does not have to include information on which motor units to activate; this relieves the brain of the need to be concerned with this level of detail for the performance of a movement (Henneman, 1979). However, because recruitment order is predetermined by spinal mechanisms, it is not possible to activate motor units selectively.

Rate Coding

In addition to the recruitment of motor units, the force that a muscle can exert depends on the rate at which the motor neurons discharge action potentials (Kernell, 2006).

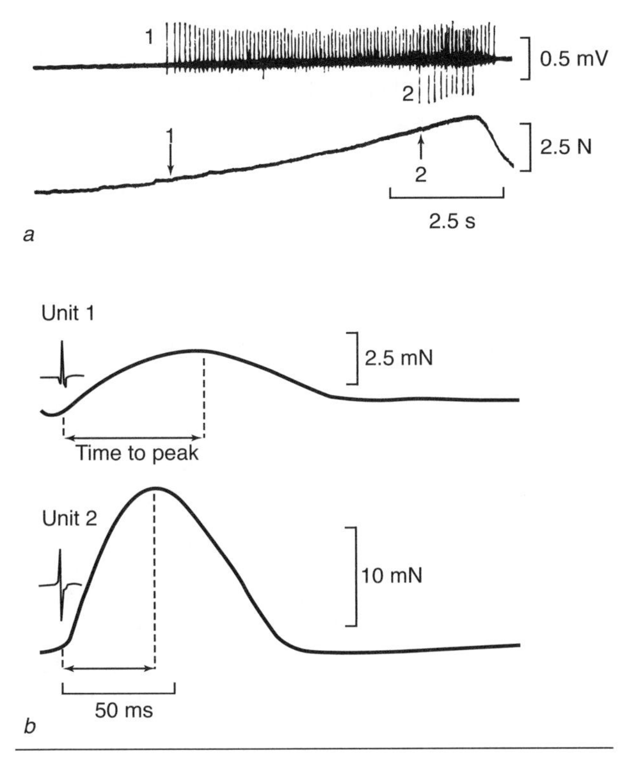

Figure 6.20 Recruitment of motor units during a voluntary contraction. *(a)* The recruitment of two motor units. Unit 1 has a lower recruitment threshold (arrow) than unit 2. *(b)* The average force response of each motor unit to its action potential. Unit 1 is weaker and has a longer time to peak force.

Adapted, by permission, from J.E. Desmedt and E. Godaux, 1977, "Fast motor units are not preferentially activated in rapid voluntary contractions in man," *Nature* 267: 717-719. Copyright 1977 by Macmillan Journals Ltd.

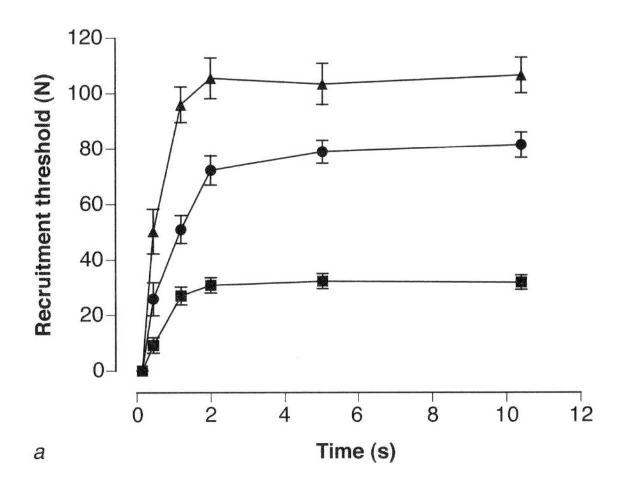

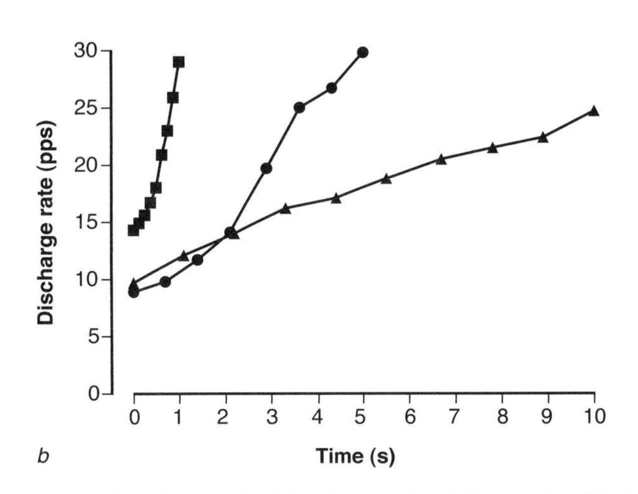

Figure 6.21 Influence of contraction speed on the activity of motor units in tibialis anterior. *(a)* The recruitment thresholds of three motor units (indicated by different symbols) when the force was increased linearly to 120 N in 0.4, 1.2, 2.3, 5.0, and 10.0 s. The recruitment thresholds were close to zero when the time to peak force was less than 0.15 s. *(b)* The increase in discharge rate for the same motor unit when the task involved a linear increase in contraction force up to 120 N in 1 s (squares), 5 s (circles), and 10 s (triangles).

Data from Desmedt and Godaux 1977.

The discharge rate of a motor neuron is proportional to the amount of synaptic input it receives (Kernell, 1965; Schwindt, 1973; Schwindt & Calvin, 1972), which may be modified by metabotropic inputs (Heckman & Enoka, 2004). As indicated in figure 6.16, there is a sigmoidal relation between mean discharge rate and motor unit force that results in the occurrence of the greatest changes in peak force when discharge rate varies within the range of 8 to 20 pps (Howells et al., 2006). Gradual increases in discharge rate produce linear increases in muscle force, with a strong relation between the rates of increase in discharge rate and force (figure 6.21*b*). Minimal discharge rates are around 5 to 8 pps for low-threshold motor units, but appear to be higher (10 to 15 pps) for high-threshold motor units (Gydikov & Kosarov, 1974; Moritz et al., 2005*a;* Van Cutsem et al., 1997). Maximal discharge rates extend across a greater range and depend on the task being performed; maximal rates range from 20 to 60 pps during gradual increases in force and can reach up to 100 pps during rapid, brief contractions (Hannerz, 1974; Monster & Chan, 1977; Van Cutsem et al., 1997, 1998). Maximal discharge rates also seem to differ across muscles, with the lowest at 11 pps for soleus (Bellemare et al., 1983).

Because peak discharge rates achieved during rapid contractions are much greater than those recorded during slow contractions (Bawa & Calancie, 1983; Desmedt & Godaux, 1977; Van Cutsem & Duchateau, 2005), mechanisms must exist that constrain the discharge rates that can be generated during gradual increases in muscle force. When a brief voluntary contraction is performed as fast as possible, the involved motor units will discharge only a few action potentials, and these will occur before muscle force has increased significantly. In contrast, motor units discharge action potentials for most of the duration of a slow contraction. Consequently, the lesser peak discharge rates during slow contractions are likely due to constraints imposed by inhibitory feedback pathways (Klass et al., 2007*a*). Furthermore, the decrease in recruitment thresholds with contraction speed (figure 6.21) means that a greater proportion of the force during a rapid contraction depends on rate coding.

Most contractions performed during activities of daily living involve the concurrent recruitment of motor units and variation in their discharge rates. This interaction is illustrated in figure 6.22*a* during the gradual increase and then decrease in force during an isometric contraction. The force reached a peak value of 35% of maximum. The contribution of four motor units (out of many) is shown with the thin lines; the left-most point of each line indicates the moment of recruitment, and the right-most point denotes derecruitment. The recruitment threshold of unit 1 was 10% of maximal force and its initial discharge rate was 9 pps. Units 2, 3, and 4 were recruited at higher forces and slightly different discharge rates. Also note that discharge rate reached greater values for the earlier-recruited motor units (De Luca et al., 1982).

The upper limit of motor unit recruitment during contractions that involve gradual increases in force ranges from 65% of maximal force for some hand muscles to 85% of maximal force for limb and trunk muscles (De Luca et al., 1982; Kukulka & Clamann, 1981; Thomas et al., 1991; Van Cutsem et al., 1997). The significance of this upper limit is that greater forces are produced solely by increases in discharge rate. For rapid, brief contractions, however, the involved motor units are recruited at the onset of the task (figure 6.21*b*), and discharge rate is relatively constant for the few action potentials that are discharged (Desmedt & Godaux, 1977; Van Cutsem et al., 1998). On the basis of these patterns, Petit and colleagues (2003) suggest that contraction speed is controlled largely by variation in rate coding, whereas the power produced during a contraction is more related to motor unit recruitment.

The relative contributions of recruitment and rate coding to the maximum muscle force can be estimated from a model of motor unit recruitment and rate coding (Barry et al., 2007; Fuglevand et al., 1993). The output of the model is shown in figure 6.22*b*. When the forces exerted by each of the 180 motor units at recruitment (minimal discharge rate) were summed, the value represented 25% of the peak force achieved when the motor unit pool was fully activated. This result suggests that rate coding accounts for about 75% of the force exerted by a muscle during a maximal contraction.

The rate at which a motor neuron discharges action potentials is never as regular as that suggested by the thin lines in figure 6.22*a* and varies across the motor unit pool (figure 6.22*b*). Rather, the time between consecutive action potentials, known as the **interspike interval,** varies depending on the mean discharge rate (figure 6.23). At long interspike intervals, which correspond to low discharge rates, the standard deviation of discharge is high, and it declines as the mean interval decreases (Person & Kudina, 1972). A similar relation is observed when the variability is normalized as the coefficient of variation (equation 6.1) (Moritz et al., 2005*a*).

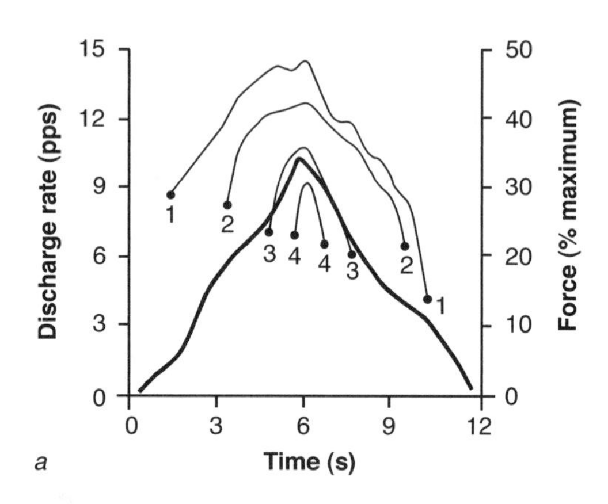

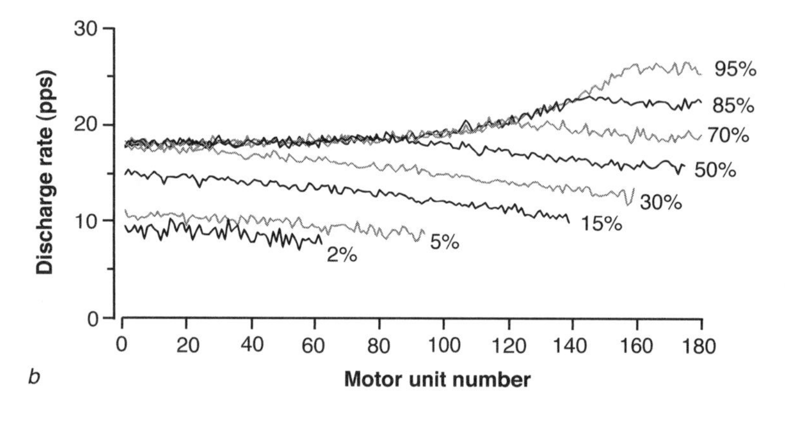

Figure 6.22 Patterns of motor unit recruitment and rate coding. *(a)* The recruitment and rate coding of four motor units (thin lines) that contributed to a gradual increase and decrease in the force (thick line) exerted by the knee extensor muscles during an isometric contraction. *(b)* Simulated discharge rates of motor units during isometric contractions that ranged from 2% to 95% of the maximal force. The right-most point on each line denotes the last motor unit recruited for the contraction. The upper limit of recruitment for the pool of 180 motor units was 60% of the maximal force.

Panel *a* modified from Person and Kudina 1972. Panel *b* from Moritz et al. 2005*a*.

$$\text{Coefficient of variation (\%)} = \left(\frac{\text{Standard deviation}}{\text{Mean}}\right) \times 100 \quad (6.1)$$

When the variability of discharge times is calculated with equation 6.2 (Stein et al., 2005), the standard deviation (SD) of discharge rate exhibits a minimum at about 10 to 12 pps (figure 6.23*c*).

$$\text{Standard deviation (pps)} = \sqrt{\frac{(\text{SD of interspike interval})^2}{(\text{Mean of interspike interval})^3}} \quad (6.2)$$

where the interspike interval is expressed in seconds.

Because voluntary contractions involve discharge rates that produce the partial summation of overlapping twitches, variation in this summation has an obvious influence on the force exerted by a muscle (figure 6.24). For example, the amount of discharge rate variability within the motor pool influences the steadiness of the force exerted by a muscle, especially during low-force contractions (Barry et al., 2007; Galganksi et al., 1993; Moritz et al., 2005*a*). Additionally, the occurrence of a brief interspike interval (~10 ms) at the beginning of a contraction can produce a substantial increase in force (Burke et al., 1970*b*; Macefield et al., 1996; Thomas et al., 1999; Van Cutsem et al., 1998). Such intervals are often referred to as **double discharges** (Garland & Griffin, 1999) and have been observed in some tasks performed by humans (Bawa & Calancie, 1983; Christie & Kamen, 2006; Griffin et al., 1998; Gydikov et al., 1987).

Correlated Discharges

The action potentials generated by motor neurons during a voluntary contraction can exhibit two forms of correlated activity: synchronization and frequency modulation. Motor unit discharge is described as synchronized when some of the action potentials discharged by different motor neurons occur at about the same time. **Motor unit synchronization** is quantified as the proportion of coincidental discharges of action potentials by pairs of

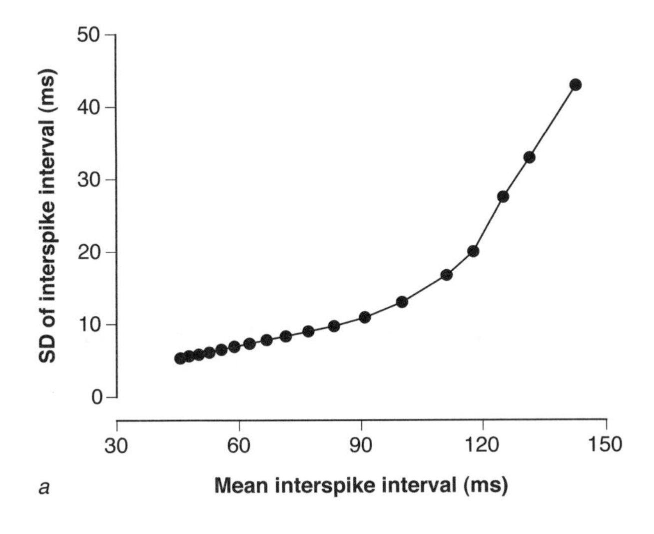

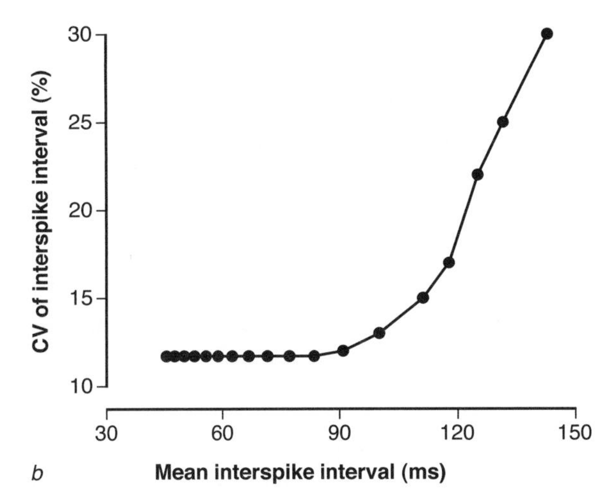

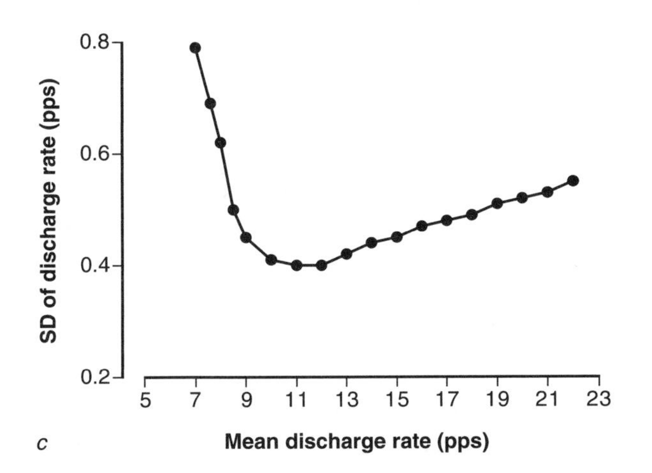

Figure 6.23 Association between variability and mean discharge rate. *(a)* The standard deviation (SD) of the time between consecutive action potentials (interspike interval) decreases as discharge rate becomes faster (reduction in interspike interval). *(b)* The coefficient of variation (CV) for interspike interval also declines with a decrease in interspike interval. *(c)* The standard deviation of discharge rate exhibits a minimum at about 10 to 12 pps.

Data provided by Benjamin K. Barry, Ph.D.

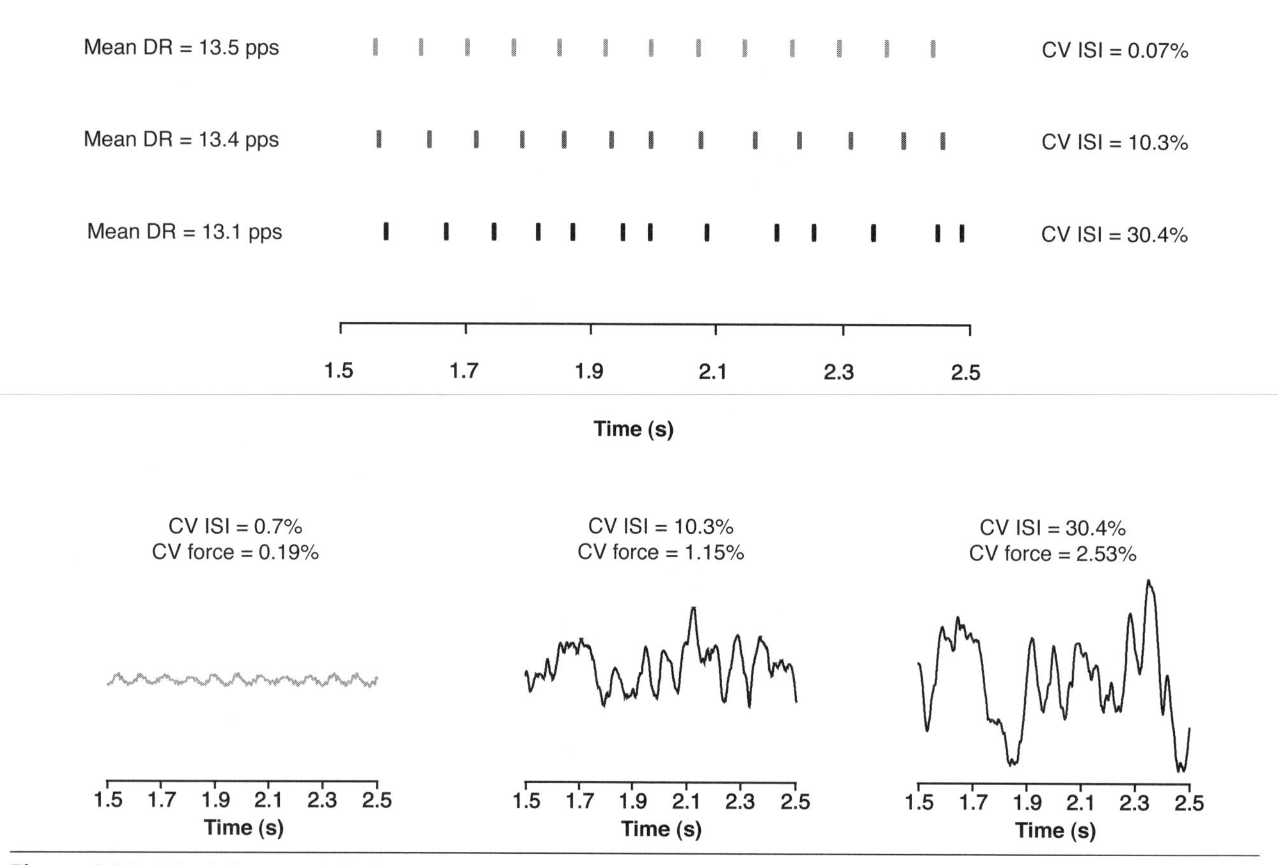

Figure 6.24 The influence of discharge rate variability on simulated force fluctuations. The mean discharge rate (DR) and coefficient of variation (CV) for interspike interval (ISI) for three trains of action potentials and the corresponding forces produced by a motor unit with a contraction time of 90 ms.

Figure provided by Mark Jesunathadas.

motor units and is attributed to the shared input delivered by branched axons of last-order neurons (figure 6.25) (Kirkwood, 1979; Nordstrom et al., 1992; Sears & Stagg, 1976). The analysis involves constructing a cross-correlation histogram from the discharges of a pair of concurrently active motor units and measuring the size of the peak in the histogram. The amount of synchronization between pairs of motor units is variable and is influenced by such factors as the task that is examined, the motor units and muscles involved in the task, and the habitual physical activity that is performed by the individual (Bremner et al., 1991*a, b, c;* Kamen & Roy, 2000; Santello & Fuglevand, 2004; Schmied et al., 1994, 2000; Semmler et al., 2002). For example, it is greater in the nondominant hand compared with the dominant hand of most individuals, greatest in strength-trained individuals, and least in skill-trained individuals (Semmler & Nordstrom, 1998).

The other form of correlation is modulation of the discharge rates for different motor neurons at common frequencies. Frequency modulation of motor unit discharge has been observed as peaks in the coherence spectrum at 1 to 2 Hz and 16 to 32 Hz. The modulation of average discharge rates at 1 to 2 Hz is known as **common drive** (De Luca & Erim, 1994, 2002). The magnitude of common drive appears to change across tasks (De Luca & Mambrito, 1987) and to be unrelated to motor unit synchronization (Semmler et al., 1997). In contrast, there is a strong association between modulation at 16 to 32 Hz and motor unit synchronization (Davey et al., 1993; Farmer et al., 1993*a;* Moritz et al., 2005*b*). The 16 to 32 Hz modulation arises from supraspinal centers (Baker et al., 1988; Farmer et al., 1993*b;* Marsden et al., 1999; Schmied et al., 1995, 2000). Accordingly, the two forms of correlated discharge can provide information about functional connections in the human central nervous system and can be used to test the integrity of the nervous system in patient populations (Baker et al., 1992; Datta et al., 1991; Farmer et al., 1998; Schmied et al., 1995, 1999).

MUSCLE MECHANICS

The force that a muscle exerts during a voluntary contraction depends on the rate at which motor units are activated, the properties of the activated motor units,

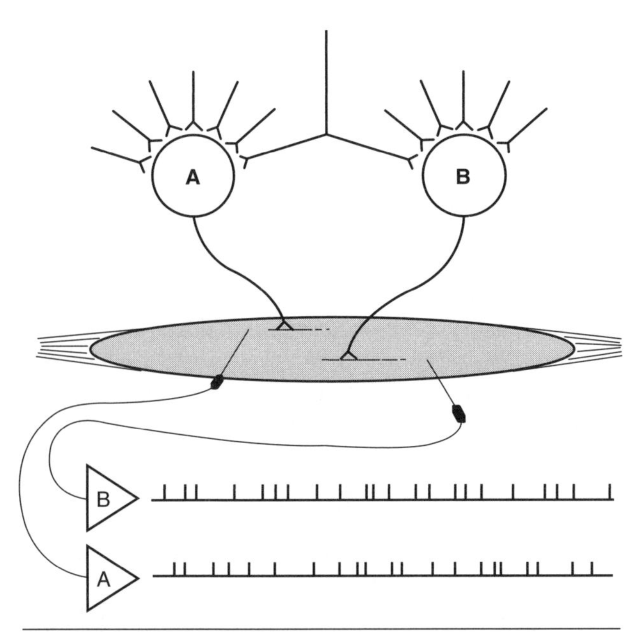

Figure 6.25 Timing of motor unit action potentials. The motor neurons of two motor units (A and B) are shown receiving many inputs, including one input that is common to both motor neurons. The action potentials discharged by the two motor neurons (shown below as tick marks) are recorded with wire electrodes inserted into the muscle. Although the action potentials discharged by the two motor neurons appear to be independent, there is a significant statistical correlation between the two sets of discharge times, caused by the common input received by the two motor neurons.

and the characteristics of the connective tissues to which the contractile proteins are attached. The capabilities of muscle are typically evaluated through quantification of its mechanical properties, which represent the foundation for muscle function during movement.

Mechanical Properties

For a given level of motor unit activity, the force that a muscle can exert depends on its length and on the rate of change in length (velocity). When these capabilities are measured experimentally, the muscle is placed at a certain length or its length is changed at a constant velocity, and the force is measured. The outcome is often a single data point that, along with other data points, can be graphed to represent the force–length and force–velocity relations of muscle. Such measurements characterize the mechanical properties of muscle, but do not describe the force exerted by muscle during movement.

The force–length and force–velocity relations of muscle depend on three factors: (1) the contractile properties of the muscle fibers, (2) the organization of the fibers in the muscle, and (3) the arrangement of the muscle around the joint. These architectural details have a much greater influence on the function of a muscle than do the proportions of the different fiber types in the muscle. This section describes the mechanical properties of muscles at each of these three levels.

Single-Fiber Level

The sliding filament hypothesis of muscle contraction states that the exertion of force by muscle is accompanied by the sliding of thick and thin filaments past one another. The most common explanation for this phenomenon, the cross-bridge theory of muscle contraction, suggests that cross-bridges (S1 extensions) extending from the thick filaments are able to attach to the thin filaments and then undergo a structural-chemical transition, thereby generating a tensile force (Huxley, 2000; Piazzesi et al., 1999; Rayment et al., 1993). According to this scheme, the development of force depends on the attach–detach cycles of the cross-bridges. For example, muscle force is proportional to the number of cross-bridges occurring at the same time. And because force is exerted only during the attachment phase, the thick and thin filaments must be close enough to each other for the attachment to occur and thus for a force to be generated.

As the length of the muscle changes and the thick and thin filaments slide past one another, the number of actin-binding sites available for the cross-bridges changes. As a consequence of this effect, the tension that a sarcomere can generate varies with the amount of overlap between the thick and thin filaments (Gordon et al., 1966). The tension is maximal at intermediate lengths and decreases at shorter and longer lengths (figure 6.26). Measurements made on a wrist muscle (extensor carpi radialis brevis) during a surgical procedure showed that sarcomere length went from 3.4 µm in a flexed-wrist position (muscle stretched) to 2.6 µm in an extended-wrist position (Lieber et al., 1994). On the basis of figure 6.26, this change in sarcomere length would be associated with a ~50% decline in the maximal force the sarcomere could exert.

By dissecting a fiber, placing it in a bath, and connecting it to a device that can apply a desired force, it is possible to determine the force–velocity relation for the fiber. Optical devices are used to measure sarcomere length and fiber length, both at rest and during the contraction. The fiber is activated with brief electrical shocks and then released so that the velocity of the contraction can be measured. The velocity of the muscle contraction depends on the ratio between the maximal isometric force of the fiber (P_0) and the magnitude of the force applied to the fiber. When the applied force is less than P_0, the fiber will shorten during the contraction. Conversely, the fiber will lengthen when the applied force exceeds P_0. For each applied force, a single velocity is measured, and a series of measurements will indicate the force–velocity relation of the fiber (figure 6.27).

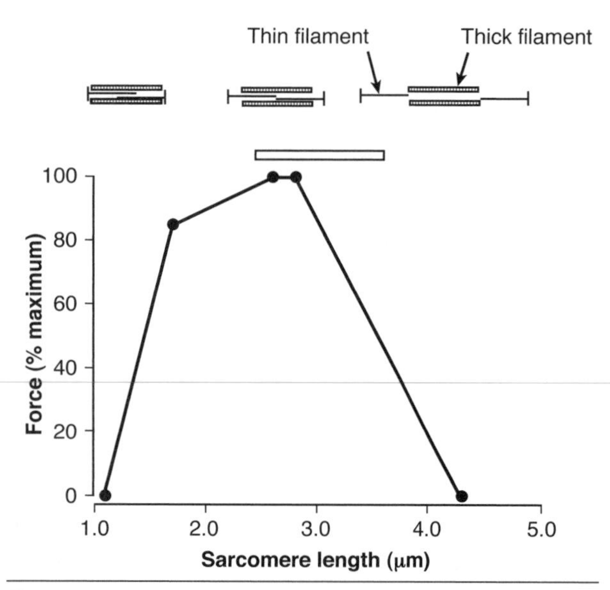

Figure 6.26 The force a muscle fiber can exert as a function of sarcomere length. The open bar at the top of the figure indicates how much sarcomere length changes over the physiological range of motion for the wrist extensor muscle.

Modified, by permission, from R.L. Lieber, F.J. Loren, and J. Fridén, 1994, "In vivo measurement of human wrist extensor muscle sarcomere length changes," *Journal of Neurophysiology* 71:880. Used with permission.

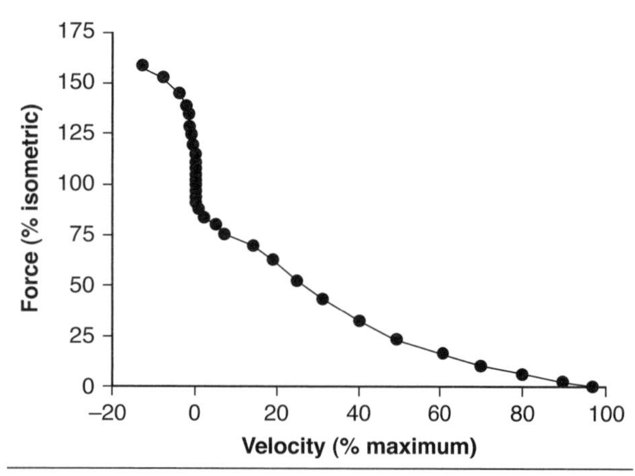

Figure 6.27 The relation between force and velocity for a single muscle fiber from a frog. The graph is based on 32 velocity measurements. Positive velocity indicates a shortening contraction, and negative velocity corresponds to a lengthening contraction.

Data from Edman 1988.

The force–velocity relation for a single fiber (figure 6.27) has three key features. First, the force that the fiber can exert decreases as the speed at which it shortens increases. Second, the force–velocity relation is reasonably flat around P_o, corresponding to a 2% change in muscle velocity for a 30% change in force. Third, the force exerted by the fiber is greatest during lengthening contractions (negative velocity in figure 6.27): Lengthening contractions occur only when the load is about 40% greater than P_o in this preparation.

The maximal rate at which a muscle fiber can shorten (V_{max}) is limited by the maximal cycling rate of the cross-bridges. When expressed in fiber lengths per second, V_{max} correlates strongly with the quantity of the enzyme myofibrillar ATPase (Bárány, 1967; Caiozzo & Rourke, 2006; Edman et al., 1988). Because this enzyme is responsible for controlling the splitting of ATP within the contractile system, increased amounts of the enzyme enhance the rate at which energy is made available for the cross-bridges. Furthermore, V_{max} remains constant over a range of sarcomere lengths and levels of activation (Edman, 1992).

The force that a muscle can exert at various velocities can be explained by the cross-bridge theory of muscle contraction. The amount of work done by muscle during a contraction depends on the number of cross-bridge attachments and the average work done during each cross-bridge cycle. The decrease in force with an increase in the speed of a shortening contraction can be explained by binding sites that are missed as the thick and thin filaments slide past one another more quickly. The more potential binding sites missed, the greater the reduction in force due to the decline in the number of cross-bridge attachments. The greater force that occurs during a lengthening contraction appears to be caused by the stretching of incompletely activated sarcomeres within each myofibril, an increase in the average force exerted during each cross-bridge cycle, and a more rapid reattachment phase (Lombardi & Piazzesi, 1990; Proske & Morgan, 2001).

EXAMPLE 6.2

The Hill Model of Muscle

The contractile properties of muscle are often characterized with a **rheological model,** such as the Hill model of muscle (figure 6.28). Rheology is the study of the deformation and flow of matter. Three elements are often included in these models: a linear spring to represent elasticity, a dashpot to denote viscosity, and a frictional element. Because the effect of internal friction is relatively small compared with those of the elastic and viscous properties of muscle and tendon, it is usually omitted from the model of muscle. The central component of the Hill model of muscle (figure 6.28) is the **contractile element** (CE), which can include a dashpot and is characterized by the force–length and force–velocity relations of muscle. The CE is surrounded by two elastic elements that represent the elasticity of the connective tissue. These are indicated as springs in the model. The **series**

elastic element (SE) and the **parallel elastic element** (PE) correspond to the passive effects of the connective tissue (Kawakami & Lieber, 2000), including the cytoskeleton, on the force exerted by the CE. The SE can be separated into active and passive components; the active component represents the elasticity in the cross-bridges and myofilaments, whereas the passive component denotes the elasticity due to the tendon and aponeuroses (Hof, 1998; Shorten, 1987). The two components of the series elasticity can respond differently to an intervention (Almeida-Silveira et al., 2000). The SE can increase the capacity of muscle to accelerate a mass (Roberts & Marsh, 2003).

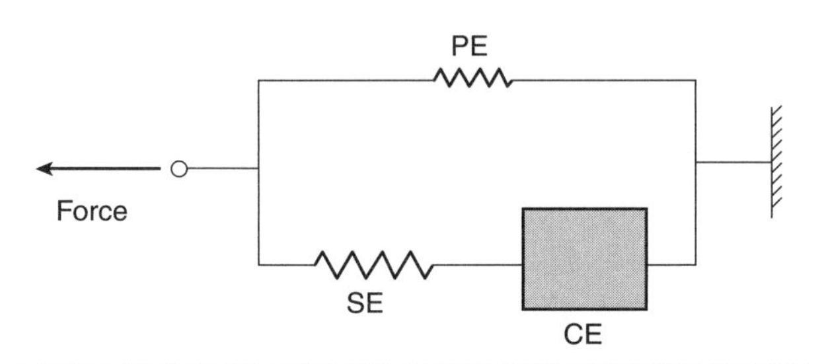

Figure 6.28 A Hill model of a single muscle fiber. CE = contractile element; PE = parallel elastic element; SE = series elastic element.

The Hill model can be used to explain why the peak force that a muscle exerts in response to a single action potential (twitch) is less than that exerted in response to multiple action potentials (tetanus). The difference in force is attributable to two factors: the quantity of calcium released by the sarcoplasmic reticulum and the mechanical behavior of muscle. First, the single action potential that evokes a twitch does not release sufficient Ca^{2+} to uncover enough binding sites for maximal force. With a series of action potentials, there is a progressive accumulation of intracellular Ca^{2+} that eventually maximizes cross-bridge activity, and the force reaches a maximum (Allen et al., 1989; Blinks et al., 1978). Second, the effect of Ca^{2+} on the muscle force is accompanied by a mechanical effect that can be explained by the Hill model. The state of activation of the contractile proteins reaches a maximal intensity within 4 ms after the action potential and is maintained at maximum for about 30 ms before it begins to decline (Ashley & Ridgway, 1968, 1970). In response to this activation, the CE generates force, but this force is registered externally only by being transmitted through the SE. Because the SE functions as a spring, the slack must be stretched out of the SE before it will transmit any force exerted by the CE (Kawakami & Lieber, 2000). The characteristics of the SE are such that the state of activation of the CE has begun to decline in a twitch before the SE has been fully stretched. With the delivery of multiple action potentials, the SE can be stretched completely, and the force generated by the CE can be transmitted through the SE more effectively.

Whole-Muscle Level

The basic contractile properties of muscle, as characterized by the force–length and force–velocity relations of the single fiber, are influenced by the way in which the fibers are organized to form a muscle (Aagaard & Bangsbo, 2006; Lieber & Fridén, 2000; Russell et al., 2000). This section describes the design features of muscle and then addresses their influence on the mechanical properties of muscle.

Muscle Design There are three main effects of muscle design: length, width, and angle. We can describe these effects by considering a muscle that consists of three muscle fibers (figure 6.29). The three fibers can be placed end to end (in series), side by side (in parallel), or at an angle to the line of pull of the muscle. When muscle fibers are arranged in series, this maximizes the range of motion (ΔL) and the maximal shortening velocity of the muscle. In contrast, an in-parallel arrangement maximizes the force (F) that the muscle can exert. However, when the fibers are aligned at an angle to the line of pull of the muscle, the force contributed by each fiber is less than its maximum (figure 6. 30).

To explain these effects of muscle design, we will consider what happens when each muscle is activated by an action potential. For the **in-series** muscle, each muscle fiber experiences a change in length (Δl), and the change in length of the muscle (ΔL) is equal to three times Δl. Consequently, the range of motion for a muscle depends on the number of fibers (n) that are arranged in series ($\Delta L = n\Delta l$). Similarly, the maximal velocity (V) of a contraction depends on the number of fibers arranged in series ($\Delta V = n\Delta v$). In contrast, the maximal force that a muscle can exert is maximized by the **in-parallel** arrangement of muscle fibers. When the fibers are arranged in parallel, muscle force (F) is equal to the sum of the forces exerted by each fiber (f); that is, $F = nf$. When the three fibers are arranged in series, muscle force is equal to the average force exerted by the fibers. We encountered the relation between the in-parallel arrangement of fibers and muscle force in chapter 3 when we used the cross-sectional area of muscle as an index of its maximal force (Aagaard et al., 2001; Lieber & Fridén, 2000; Roy & Edgerton, 1992).

The third feature of muscle design is the angle of **pennation** (β), which refers to the alignment of the muscle fibers at an angle to the line of pull of the muscle. When the angle of pennation is zero, the net force exerted by

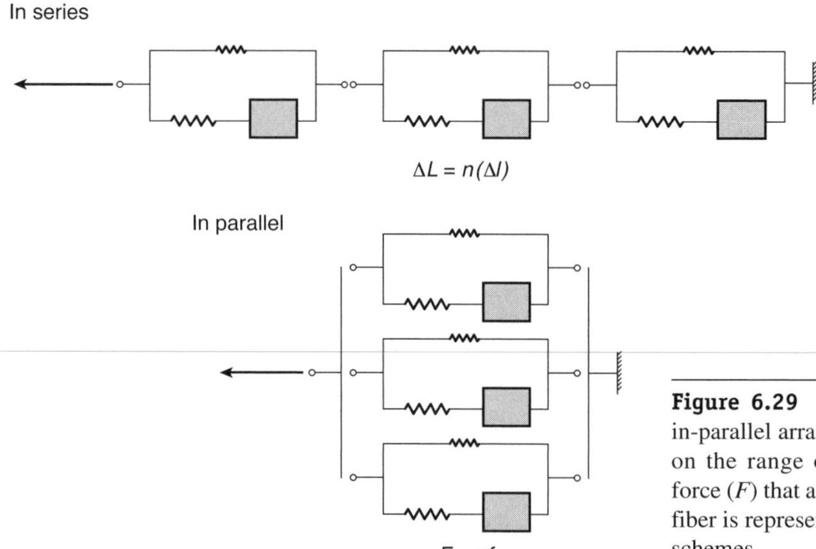

Figure 6.29 Influence of the in-series and in-parallel arrangement of three muscle fibers on the range of motion (ΔL) and maximal force (F) that a muscle can exert. Each muscle fiber is represented by a Hill model in the two schemes.

the fiber (f) acts in the direction of the whole-muscle force (F; figure 6.30). When the angle of pennation is not zero, however, the proportion of the force exerted by the fiber that acts in the direction of whole-muscle force varies as the cosine of the angle of pennation. Although much of the force exerted by a fiber is transmitted laterally through the cytoskeleton (figure 6.4), it is reasonable to conceptualize the whole-muscle force as being transmitted in a longitudinal direction.

One reason most muscles have a nonzero angle of pennation is that for a given volume of muscle, more fibers can be placed in parallel. This is shown in the bottom row of figure 6.30; there are 7 fibers in the volume when $\beta = 0$ and 13 fibers in the same volume when $\beta \neq 0$. As a result, the pennated arrangement can exert a greater maximal force. The number of fibers that can be contained in a given volume is an important consideration because the volume available for various organs is often

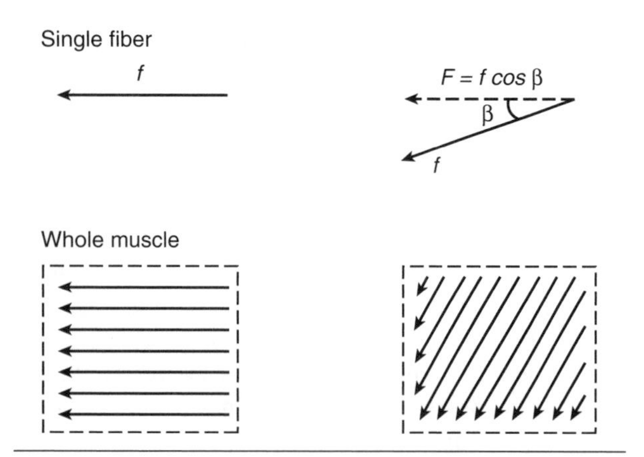

Figure 6.30 The angle of pennation (β) has an effect on the force exerted by both the single fiber and the whole muscle.

a limiting factor in biological design (Otten, 1988). For example, the muscles that control the fingers are located in either the hand or the forearm; those in the hand are small (restricted space), whereas those in the forearm are larger and stronger. If all the muscles that control the fingers were located in the hand, it would be bulky and more awkward for manipulation.

EXAMPLE 6.3

Estimating the Number of In-Parallel Fibers

The in-parallel effect on whole muscle is the same as that at the muscle fiber level; the force capacity of muscle varies with its cross-sectional area. When the pennation angle is not zero, it is difficult to determine the cross-sectional area of muscle. One cannot estimate it simply by making a section that is perpendicular to the long axis of the muscle; such a measurement is known as the anatomical cross-sectional area. Rather, it is necessary to include the effect of the pennation angle, which yields a measure referred to as the **functional cross-sectional area** (fCSA). The functional cross-sectional area provides a more accurate estimate of the force capacity of muscle than the anatomical cross-sectional area (Lieber & Fridén, 2000; Narici, 1999).

Functional cross-sectional area can be calculated from knowledge of muscle volume, average pennation angle, and average fiber length:

$$\text{fCSA} = \frac{\text{Volume (cm}^3) \times \cos\beta}{\text{Fiber length (cm)}} \quad (6.3)$$

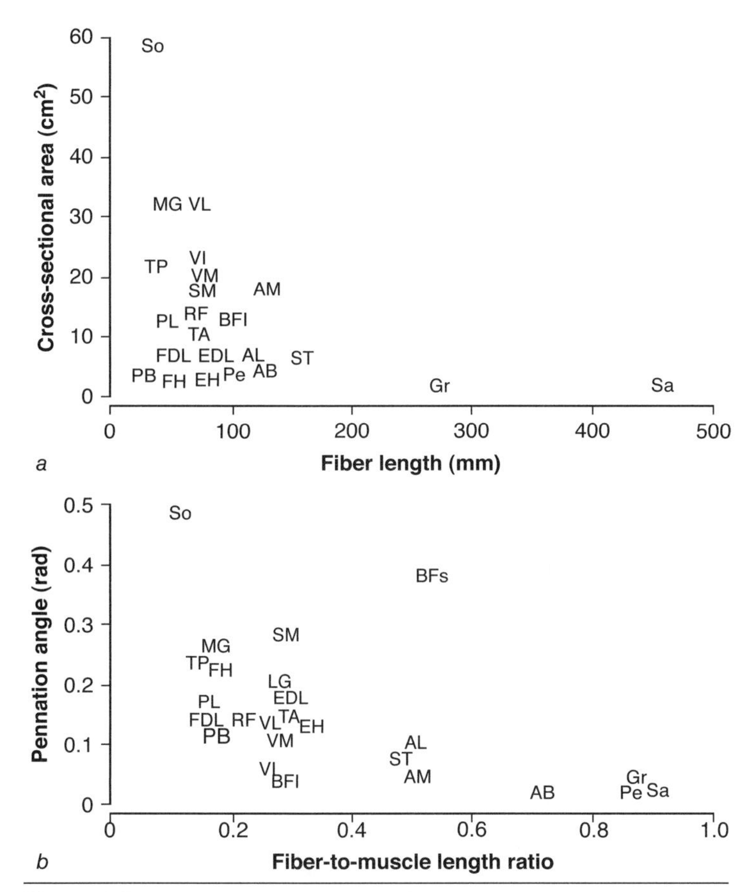

Figure 6.31 Design characteristics of human leg muscles: *(a)* association between anatomical cross-sectional area and fiber length; *(b)* relation between pennation angle and the ratio of fiber length to muscle length. AB = adductor brevis; AL = adductor longus; AM = adductor magnus; BFl = long head of biceps femoris; BFs = short head of biceps femoris; EDL = extensor digitorum longus; EH = extensor hallucis; FDL = flexor digitorum longus; FH = flexor hallucis; Gr = gracilis; LG = lateral gastrocnemius; MG = medial gastrocnemius; PB = peroneus brevis; Pe = pectineus; PL = peroneus longus; RF = rectus femoris; Sa = sartorius; SM = semimembranosus; So = soleus; ST = semitendinosus; TA = tibialis anterior; TP = tibialis posterior; VI = vastus intermedius; VL = vastus lateralis; VM = vastus medialis.

Data from Friederich and Brand 1990; Wickiewicz et al. 1983.

Muscle volume can be measured with an imaging technique, such as MRI or computerized tomography. Estimates of pennation angle and fiber length are usually taken from published data (Lieber & Fridén, 2000). For the leg muscles included in figure 6.31, fCSA ranged from 1.7 cm^2 (sartorius, gracilis, extensor hallucis) to 58 cm^2 (soleus).

Muscles in the human body comprise a mixture of the three design features. Some muscles are long, some are wide, and most have fibers arranged with a nonzero angle of pennation. This distinction is apparent in measurements made on 25 leg muscles in five cadavers (figure 6.31); consult Roy and Edgerton, 1992, and Lieber, 1992, for similar data on arm muscles and Yamaguchi and colleagues, 1990, for data on lower extremity, trunk, upper extremity, hand, and head-neck musculature. Although these measurements were made on preserved cadavers and probably differ from the values for living tissue, the data do allow comparisons between muscles. The magnitude of the difference between muscles in terms of in-series and in-parallel design can be substantial. In these specimens, for example, fiber length ranged from 25 mm in soleus to 448 mm in sartorius (figure 6.31*a*). If we assume an average sarcomere length of 2.2 μm, then the average number of in-series sarcomeres ranged from 11,364 for soleus to 203,636 for sartorius.

Similarly, there are marked differences among muscles in the number of muscle fibers arranged in parallel, which is indicated by the measurement of cross-sectional area (figure 6.31*a*). For example, muscles that support an upright posture (e.g., knee extensors, ankle plantarflexors) are generally twice as strong as their antagonists. This suggests that the cross-sectional areas of the antigravity muscles should be twice as large. Indeed, the cross-sectional area of quadriceps femoris is about double that for the hamstrings (87 vs. 38 cm^2), and the value for the plantarflexors is substantially larger than that for the dorsiflexors (139 vs. 17 cm^2). As shown in figure 6.31*a,* however, muscles that have large cross-sectional areas tend to have short muscle fibers. For example, the average muscle fiber in the hamstrings is longer than an average one in quadriceps femoris (43,000 vs. 31,200 sarcomeres in series). Consequently, although the hamstrings and dorsiflexors are weaker than their antagonists, they are capable of a greater change in length and shortening velocity.

Figure 6.31*b* indicates that muscle fibers are usually shorter than muscle length. The ratio of fiber length to muscle length ranges from 0.08 in soleus to 0.85 for some leg muscles (sartorius, pectineus, gracilis) and wrist muscles (extensor carpi radialis longus). Most muscles have ratios in the range of 0.2 to 0.5 (figure 6.31*b*). Deviation of the ratio from 1, where a value of 1 indicates equal fiber and muscle lengths, is due to two in-series effects: pennation and staggered fibers. Muscles with low ratios tend to have fibers arranged with greater pennation angles (figure 6.31*b*). Conversely, muscles with pennation angles close to zero have high ratios, which means that the fibers span most of the muscle length.

The other in-series effect that can account for a ratio of fiber-to-muscle length less than 1 is the serial attachment of short **staggered muscle fibers.** One can demonstrate this effect by plotting the longitudinal distribution of the muscle fibers that belong to a single motor unit (figure 6.32). Because of this arrangement, the force exerted by one motor unit is not transmitted from one end of the muscle to the other by its own fibers but rather is transmitted by the cytoskeleton and connective tissue (Huijing, 2003; Lieber & Fridén, 2000; Monti et al., 1999; Roy & Edgerton, 1992; Trotter, 1990). A way of demonstrating this organization is to record motor unit potentials as they propagate along the length of a muscle. Such an experiment was performed on sartorius, which is the longest muscle in the human body and can reach up to 600 mm in length (Harris et al., 2005). By placing five electrodes along the length of the muscle (figure 6.32*a*), it was possible to identify the location of the motor end plate for each motor unit and the propagation of its potentials toward the ends of the fibers. From recordings made on five individuals, it is apparent that the muscle had multiple end-plate zones and that 30% of the motor units had fibers that ran the entire length of the muscle, but that most motor units did not span the length of the muscle (figure 6.32*b*). Furthermore, the muscle fibers in a motor unit do not extend the length of its territory (Burke & Tsairis, 1973; Ounjian et al., 1991). Thus, the lateral transmission of force is critical in order for a muscle to function effectively.

Muscle Mechanics The major effect of these design features on the force–length relation of muscle is attributable to the connective tissue that combines single fibers into whole muscles. As a consequence, the force exerted by muscle is not solely dependent on the *active* process of cross-bridge cycling and filament overlap. In addition,

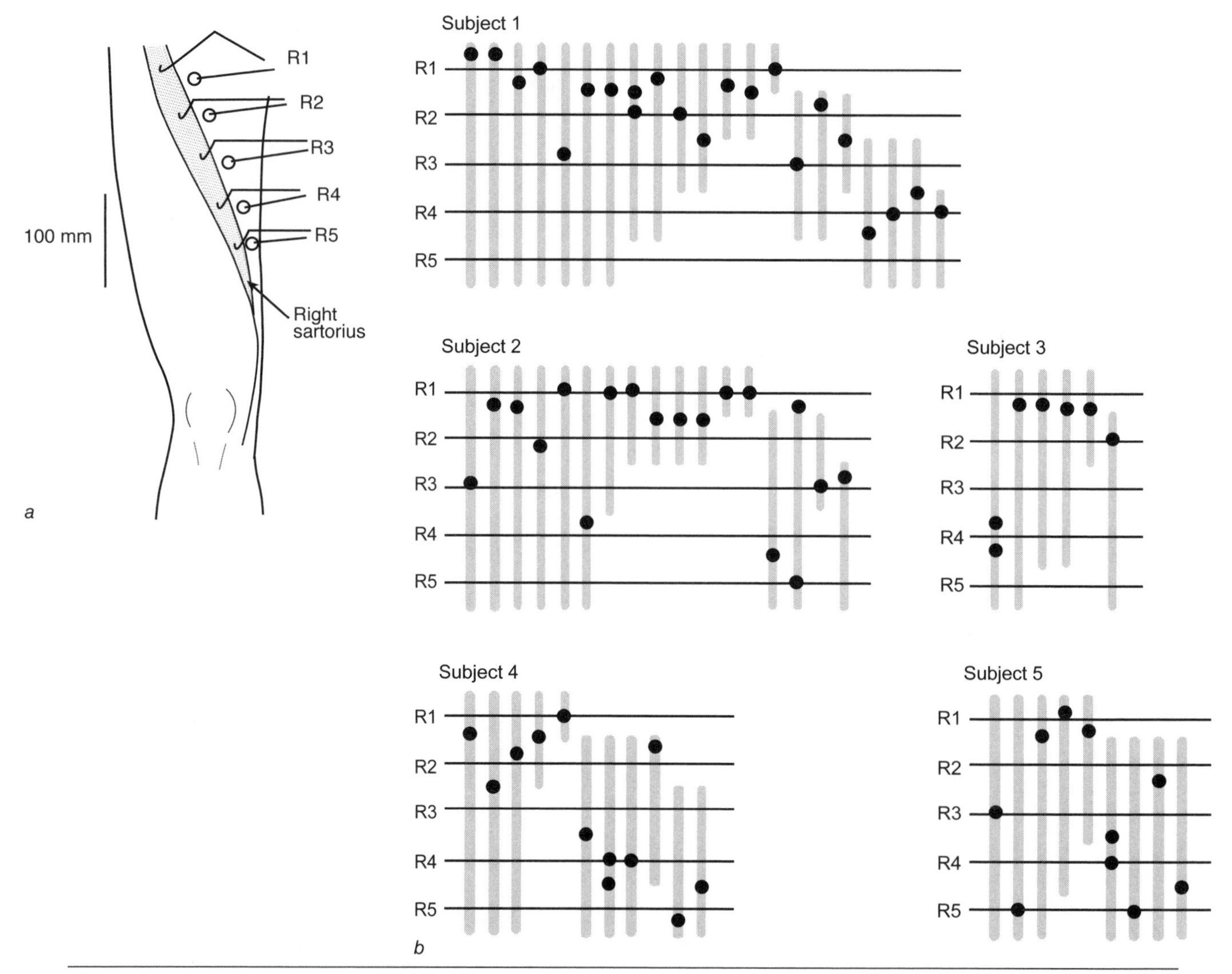

Figure 6.32 Motor unit territories in the sartorius muscle. *(a)* Arrangement of recording electrodes (R1 to R5) along the length of the sartorius muscle. *(b)* End-plate location (filled circles) and the extent of muscle length along which the potentials for single motor units were recorded. Each vertical bar represents a single motor unit. The number of motor units recorded in each subject ranged from 6 for Subject 3 through to 21 for Subject 1.

Adapted, by permission, from A.J. Harris et al., 2005, "Muscle fiber and motor unit behavior in the longest human skeletal muscle," *Journal of Neuroscience* 25: 8529-8532. Copyright 2005 by The Society for Neuroscience.

the connective tissue (e.g., endomysium, perimysium, epimysium, tendon) and cytoskeleton (figure 6.4) exert a *passive* force that combines with the cross-bridge activity. Because of this interaction, the force exerted by muscle depends on the properties of both the contractile (myofilaments) and structural (connective tissue and cytoskeleton) elements.

Figure 6.33 illustrates the contributions of the active and passive components to total muscle force as muscle length varied over the range from the minimal contraction length to the maximal stretched length. This graph shows the **force–length relation** of whole muscle. These data were obtained by the direct measurement of muscle force in patients who had a special type of below-elbow prosthesis. Two forces were measured at each muscle length: one when the subject was resting (passive = filled circles) and the other when the subject exerted a maximal voluntary force (total = open circles). When the subject remained relaxed, the passive force increased as the muscle length increased. When the subject performed a maximal voluntary contraction at each length, the total force was due to both the passive and active components (figure 6.33). The parabolic line, which was determined as the difference between the total and passive components, represents the change in force due to the active component as a function of muscle length. At shorter muscle lengths, all the force was due to the active component (cross-bridge activity), whereas most of the force at long lengths was due to the passive component. The actual shapes of these different components, however, vary among muscles (Baratta et al., 1993), as would be expected from the differences in muscle design (figure 6.31).

Changes in muscle length during voluntary contractions can be measured with ultrasound images (Bojsen-Møller et al., 2005; Kawakami & Fukunaga, 2006; Reeves et al., 2004). Ultrasound echoes can indicate the length and orientation of muscle fascicles, especially in pennated muscles (figure 6.34). Because muscle fibers span the entire length of the fascicles in a muscle, the assessment of fascicles in ultrasound images provides information about muscle fibers. With this approach, it has been found that fascicles can shorten by up to 30% of the initial length during a maximal isometric contraction (figure 6.34); this is accompanied by a corresponding elongation of the tendon and other connective tissue structures (Kawakami et al., 1998; Kawakami & Lieber, 2000). For the soleus and medial gastrocnemius muscles, sarcomere length shortens from resting values between 2.5 and 3.0 μm to about 2.0 μm during a maximal isometric contraction, which places the sarcomeres on the ascending part of the force–length relation (figure 6.26) (Kawakami et al., 2000).

As with the relation between force and length, the force–velocity relation for muscle (figure 6.35*a*) is influenced

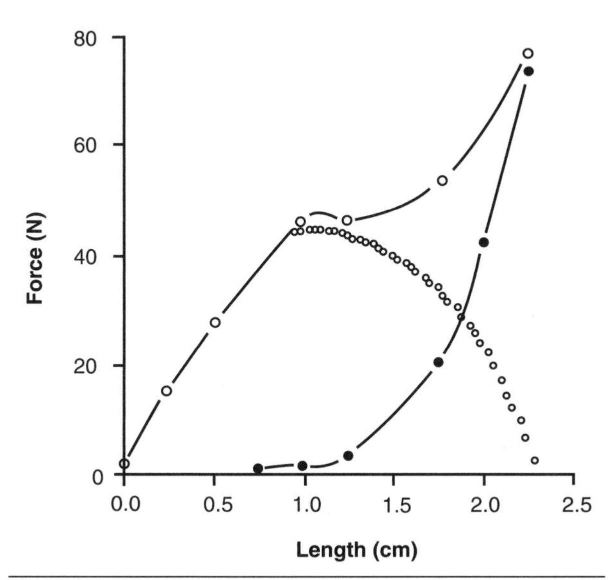

Figure 6.33 Contributions of the active (parabolic function) and passive (filled circles) elements to the total (open circles) force–length relation for whole muscle.

Data from Ralston et al., 1947.

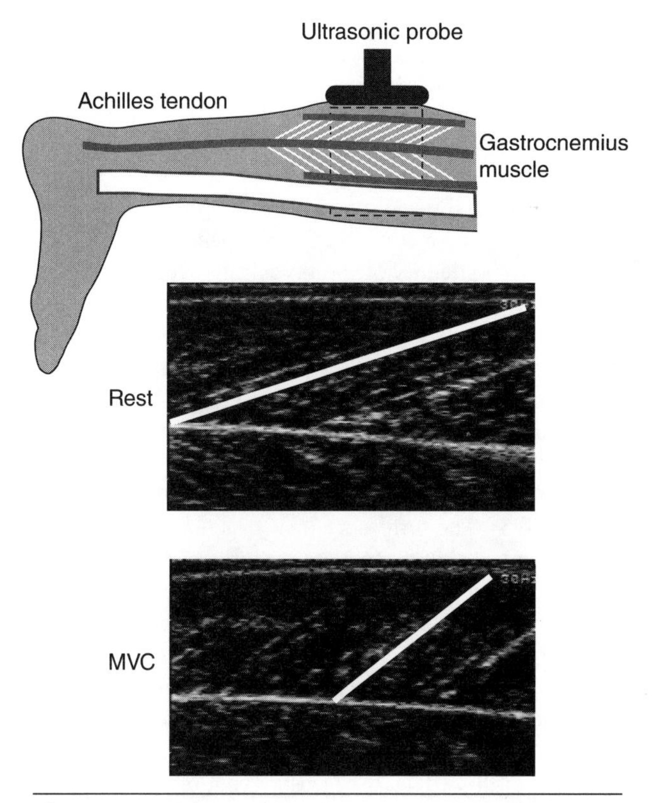

Figure 6.34 Ultrasound images of the gastrocnemius muscle at rest and during a maximal isometric contraction (MVC).

Images provided by Yasuo Kawakami, Ph.D.

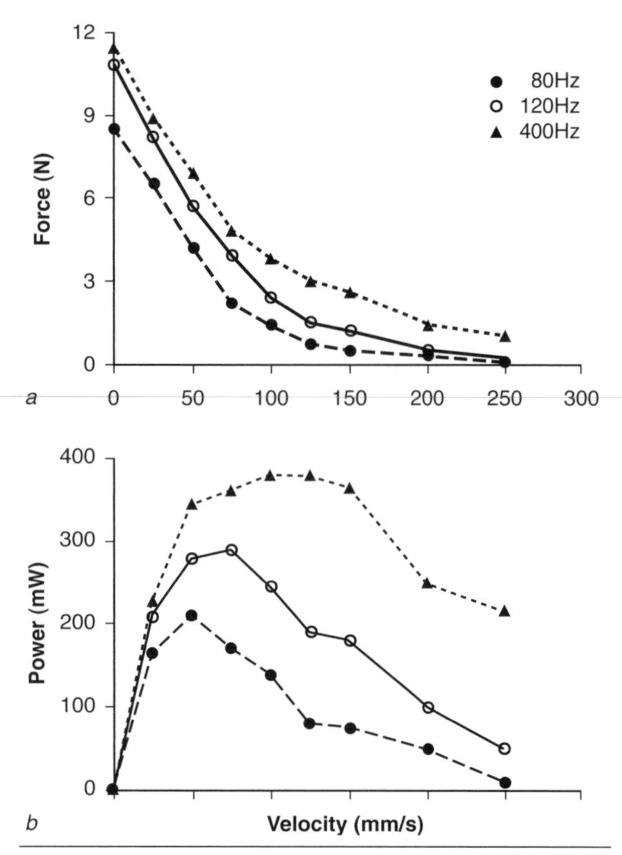

Figure 6.35 The *(a)* force–velocity and *(b)* power–velocity relations for the rat medial gastrocnemius muscle when stimulated at different frequencies (80, 120, and 400 Hz). The data are for shortening contractions.

Data from de Haan 1998.

by muscle design. Whereas the maximal shortening velocity (V_{max}) for a single fiber depends primarily on the fiber type details of the myosin molecule (Caiozzo & Rourke, 2006; Schiaffino & Reggiani, 1996), V_{max} for whole muscle depends on both its fiber type composition and two design features: fiber length and pennation angle. In the cat hindlimb, for example, the tibialis anterior and tibialis posterior muscles each have about 20% type I fibers, yet V_{max} for tibialis anterior is 28.4 cm/s compared with 4.2 cm/s for tibialis posterior (Baratta et al., 1995). The difference in V_{max} is due to the greater change in length (due to more sarcomeres in series) that can be achieved by tibialis anterior (5.83 cm) compared with tibialis posterior (2.33 cm). The shortening of a muscle during a contraction is also influenced by the rate of change in pennation angle. As indicated in figure 6.34, pennation angle increases during an isometric contraction, which contributes to the shortening velocity of the muscle. Because of this effect, the measurement of whole-muscle velocity overestimates the rate of change in fiber length and confounds the interpretation of a torque–angular velocity relation as an index of the force–velocity relation of muscle (Kawakami & Fukunaga, 2006).

The power that a muscle can produce during a shortening contraction is a product of the force it exerts and the speed at which it shortens. Because the product is zero during both an isometric contraction (velocity = 0) and at the fastest contraction speed (force = 0), power production reaches a peak value at intermediate shortening speeds (figure 6.35*b*). According to the classic hyperbolic function for the force–velocity relation (Hill, 1938), power production is maximum when a muscle shortens at about one-third of its maximal shortening velocity or when it acts against a load that is one-third of the maximal isometric force (Josephson, 1993). Measurements of single muscle fibers, for comparison, indicate that peak power occurs when the load is about 20% of the maximal isometric force (Trappe et al., 2000). The peak power that a muscle can produce and the shortening speed at which this occurs depend on the rate at which the muscle is activated (figure 6.35*b*).

EXAMPLE 6.4

Isotonic Contractions Underestimate Muscle Function

The classic way to measure the force–velocity relation for a muscle in situ is to disconnect one end from the bone and attach it to a load, then activate the muscle supramaximally by stimulating its nerve and measure the force and velocity during the evoked contraction. This procedure is known as isotonic loading: The load remains constant during the **isotonic contraction.** Different values for force and velocity are obtained when the load is varied between contractions; these measurements form the data set for the force–velocity and power–velocity curves (figure 6.35).

An alternative approach is to measure the muscle force and velocity without detaching the muscle from the bone. This requires devices that can measure force and velocity in vivo. Gregor and colleagues (1988) performed such measurements by attaching a force transducer to the tendon of the soleus muscle of a cat and videotaping the cat running on a treadmill; the change in muscle length was estimated from the videotape images. The force–velocity and power–velocity relations obtained with this approach were compared to those measured with isotonic loading (figure 6.36).

When the foot contacted the ground as the cat ran on the treadmill (filled dot) and for some time thereafter, soleus performed a lengthening contraction (negative velocity), which was associated with a rapid

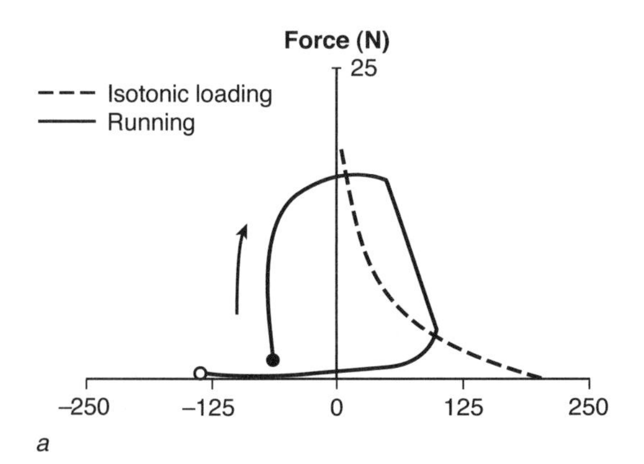

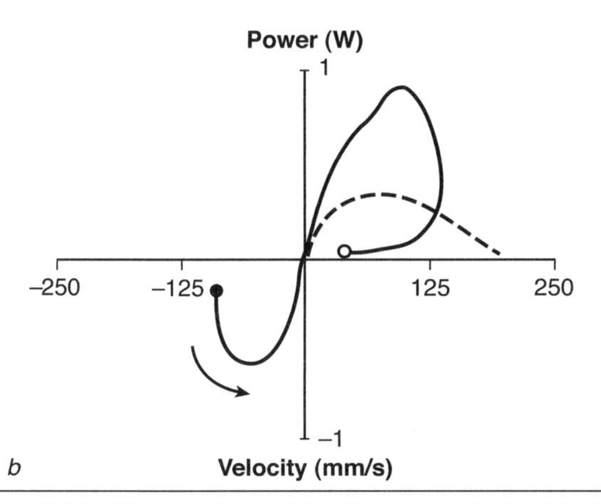

Figure 6.36 Force–velocity *(a)* and power–velocity *(b)* relations for the cat soleus muscle. The isotonic-loading curves were measured experimentally and are compared with those estimated during running. Footstrike during running is indicated by the filled circle, and toe-off by the open circle.

Data from Gregor et al. 1988.

increase in force to a peak value and the absorption of power (negative power). After the contraction reached zero velocity (isometric), soleus performed a shortening contraction (positive velocity) and produced power (positive power). Both the muscle force and power production during the movement (running at 2.2 m/s) were greater than the values measured during isotonic loading. These data indicate that during cyclic activity, especially different forms of locomotion, isotonic contractions underestimate the performance capabilities of muscle (Biewener & Roberts, 2000; Kawakami & Fukunaga, 2006; Josephson, 1993).

Joint Level

At the next level of analysis, we consider the force–length and force–velocity relations for muscles that are attached to the skeleton. In contrast to the situation at the single-fiber and whole-muscle levels, the significant mechanical action of muscle at this level is not the force it exerts but rather the torque produced around the joint (DeVita, 2005). The variation in net muscle torque throughout the range of motion about a joint depends on two principal factors: the location of the attachments on the skeleton and the contribution of multiple muscles to the net effect about a joint. The attachment locations influence the moment arm relative to the joint, the number of axes about which the muscle exerts a torque, and the number of joints spanned by a muscle.

Moment Arm The torque that a muscle can produce typically varies throughout its range of motion due to the force–length property of muscle (figures 6.26 and 6.33) and the variation that occurs in the moment arm. For most muscles, variation in the moment arm depends on where the muscle is attached to the skeleton relative to the joint (van Mameren & Drukker, 1979). For example, figure 6.37*a* shows two muscles that differ in the location of the distal attachment. The distance from the proximal attachment to the joint (*p*) is the same for the two muscles. The distance from the distal attachment to the joint for muscle *A* (d_a) is one-half of that for muscle *B* (d_b), and

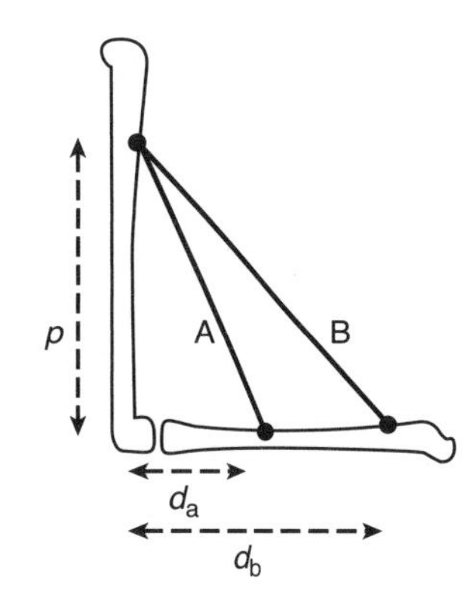

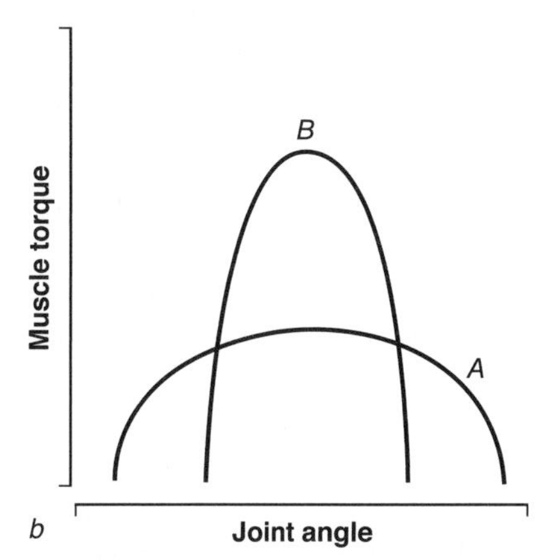

Figure 6.37 The location of the attachment sites of a muscle on the skeleton influences *(a)* its moment relative to the joint and *(b)* its torque–angle relation.

Data from Rassier et al. 1999.

the distances p and d_b are the same. As a result of this arrangement, the moment arm for muscle A is about one-half of that for muscle B (Rassier et al., 1999). As a consequence of this difference in the attachment sites, the muscle with the shorter moment arm (muscle A) exerts torque over a greater range of motion, but the peak torque is less than that for the identical muscle with the longer moment arm (muscle B) (figure 6.37).

On the basis of cadaveric arm specimens, Lieber and colleagues measured the moment arms and estimated muscle forces for a number of muscles that cross the wrist joint (Loren et al., 1996). The data for three wrist extensor muscles show that the moment arm and muscle force can change in different ways over the range of motion of a muscle and produce different torque–angle relations (figure 6.38). For many joints in the human body, the net muscle torque varies over the range of motion and has a peak value at intermediate joint angles. Although the relative contributions of variations in moment arm and muscle force to the shape of the torque–angle relation are generally unknown, the percentage change in moment arm over the range of motion is usually greater than that for muscle force (figure 6.38, *a* and *b*).

Lieber (1992) characterized these interactions by examining the ratio of muscle fiber length to moment arm length. The size of the ratio for a particular muscle indicates the extent to which the muscle contributes to the resultant muscle torque over the entire range of motion. When the ratio is high, the change in fiber length and thus sarcomere length is small relative to the angle-dependent change in moment arm. Recall from the force–length relation of sarcomeres (figure 6.26) that the force exerted by a sarcomere is greatest at intermediate lengths. When the ratio of muscle fiber length to moment arm length is high, therefore, the muscle fiber operates at an intermediate length and is capable of exerting maximal forces over a greater range of motion. Lieber (1992) reported that muscles with a high ratio include gluteus maximus (80), sartorius (11), and extensor carpi radialis longus (11); these ratios contrast with the low ratios of soleus (0.9), vasti (1.8), hamstrings (1.8), and dorsiflexors (3.1). Thus, the entire architecture of a muscle, from sarcomere arrangement to joint organization, has a major influence on the in vivo function of muscle throughout its range of motion (Lieber & Fridén, 2000).

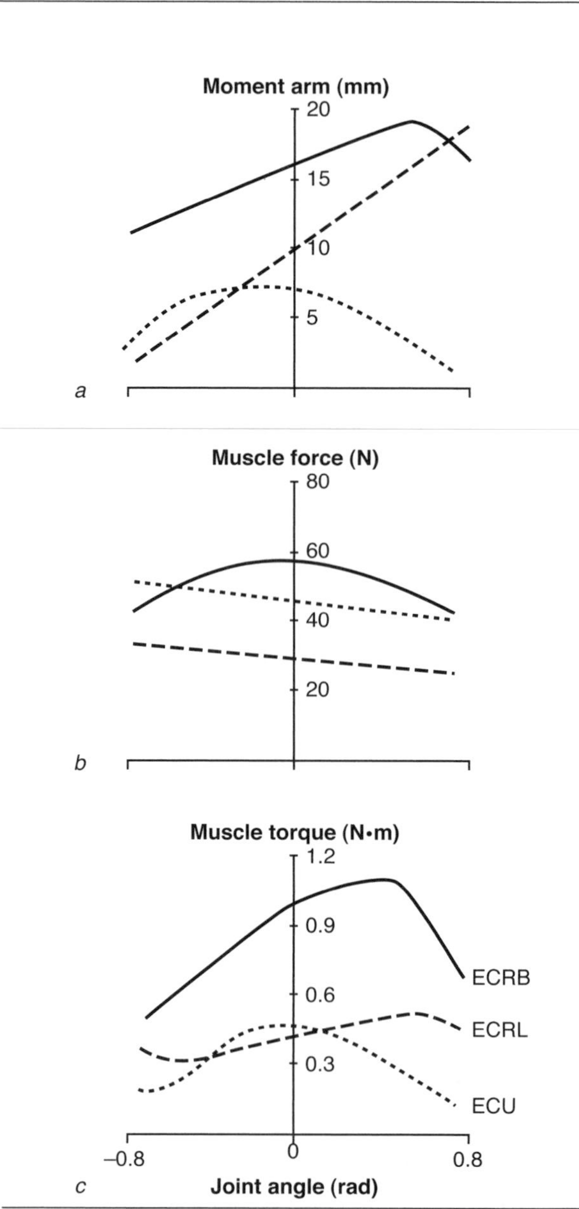

Figure 6.38 Changes in *(a)* moment arms, *(b)* muscle forces, and *(c)* muscle torques over the wrist joint range of motion for the muscles extensor carpi radialis brevis (ECRB), extensor carpi radialis longus (ECRL), and extensor carpi ulnaris (ECU). Negative joint angle corresponds to flexion relative to neutral position of the wrist (0 rad).

Data from Loren et al. 1996.

Off-Axis Attachments Rarely do muscles attach about a joint to produce an isolated function, such as extension or adduction. Rather, the attachments are slightly off axis so that the muscle can exert a torque about more than one axis (Lawrence et al., 1993). For example, the direction of the torque exerted by the muscles that cross the human elbow joint has components about more than one axis. Zhang and colleagues (1998) determined these components by applying electrical stimulation to each muscle at various joint angles throughout the range of motion. The evoked torques for each muscle were normalized relative to the maximal torque the muscle exerted in the flexion-extension direction, and the data were averaged across a number of subjects. The forearm was kept in a neutral position. The two heads of biceps brachii exerted torques in the flexion and supination direction, which changed as a function of joint angle (figure 6.39). The other muscles had lesser components in the direction of pronation and supination, and these also changed with joint angle. When the effects of off-axis attachments are combined with differences in muscle architecture (table 6.8), the individual elbow flexor muscles exhibit a range of mechanical capabilities (Buchanan et al., 1989; Ettema et al., 1998; Funk et al., 1987; Murray et al., 1995). This diversity exists at many joints throughout the body, such as the ankle, knee, hip, vertebral column, and shoulder (figure 6.31).

Given that the magnitude and direction of the force exerted by each muscle are unique, the relative contribution of synergist muscles to a net force can vary across tasks. The study of this issue, which is termed **force sharing,** involves determining how the force is distributed or shared among the involved muscles. The most direct way to determine force-sharing patterns is to measure muscle forces in the course of normal motor behavior by placing force transducers on the tendons of selected muscles and recording the forces and the kinematics of the behavior (Walmsley et al., 1978). Most of these experiments have been performed on the hindlimb of the cat, where the properties of the involved muscles differ substantially. Figure 6.40 shows a typical set of data for a cat locomoting at various speeds, ranging from a walk to a trot (Herzog, 1996). Each trace comprises a loop and corresponds to the forces exerted by two muscles during a single step cycle. The loops change with speed and differ for the two pairs of muscles. While the peak force exerted by soleus remained relatively constant across the three speeds, the peak forces exerted by both gastrocnemius and plantaris increased with speed (figure 6.40). These differences are consistent with the architecture and physiological properties of these muscles. On the basis of EMG measurements, it is likely that similar effects occur among synergist muscles in humans (Cappellini et al., 2006; Hof et al., 2002; Ivanenko et al., 2004).

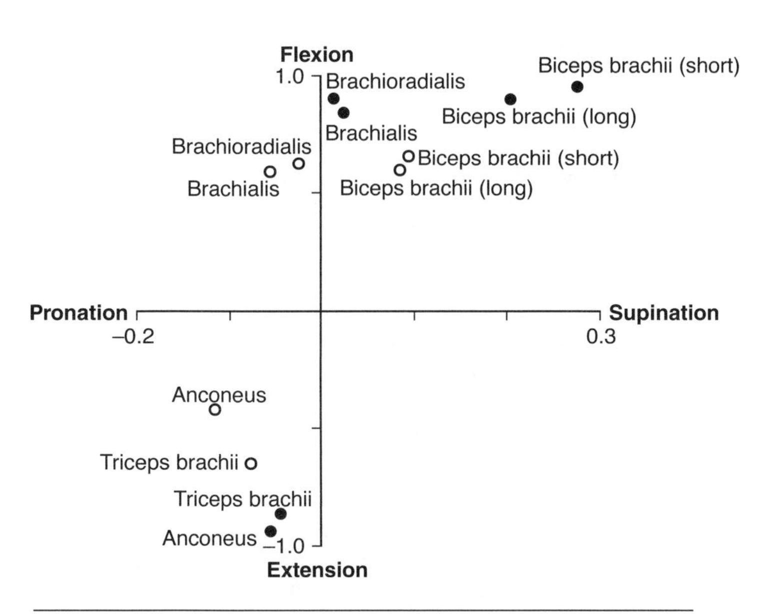

Figure 6.39 Normalized torques evoked by electrical stimulation in the muscles that cross the human elbow joint. The muscles studied were anconeus, the short and long heads of biceps brachii, brachialis, brachioradialis, and the lateral head of triceps brachii. The torques evoked in the two directions (flexion-extension and pronation-supination) were measured with the elbow joint at a right angle (filled circles) and with the elbow joint flexed by 0.5 rad from complete extension (open circles).

Data from Zhang et al. 1998.

Two-Joint Muscles One common variation in the points of attachment for a muscle is the number of joints that the muscle spans. For example, a significant number of muscles span two joints and are thus referred to as two-joint muscles. Two-joint muscles provide at least three advantages in the control of movement at the joint level. First, two-joint muscles couple the motion at the two joints that they cross (Arsenault & Chapman, 1974; Fujiwara & Basmajian, 1975; Ingen Schenau et al., 1990*a*). For example,

Table 6.8 Design Features of the Three Main Human Elbow Flexor Muscles

				DISTANCE TO JOINT		
	Muscle length (cm)	Fiber length (cm)	Muscle CSA (cm^2)	Proximal (cm)	Distal (cm)	Moment arm (cm)
Biceps brachii	24.0	17.8	5.8	30	4	4.3
Brachialis	20.1	14.4	7.4	10	3	2.3
Brachioradialis	23.7	14.7	2.0	6	26	—

Note. The distance to the joint indicates the location of the center of the attachment point (proximal and distal) to the axis of rotation of the elbow joint. CSA = cross-sectional area.

The distance data were provided by Scott L. Delp, PhD. The moment arms were measured with the elbow joint at a right angle by Klein et al., 2001.

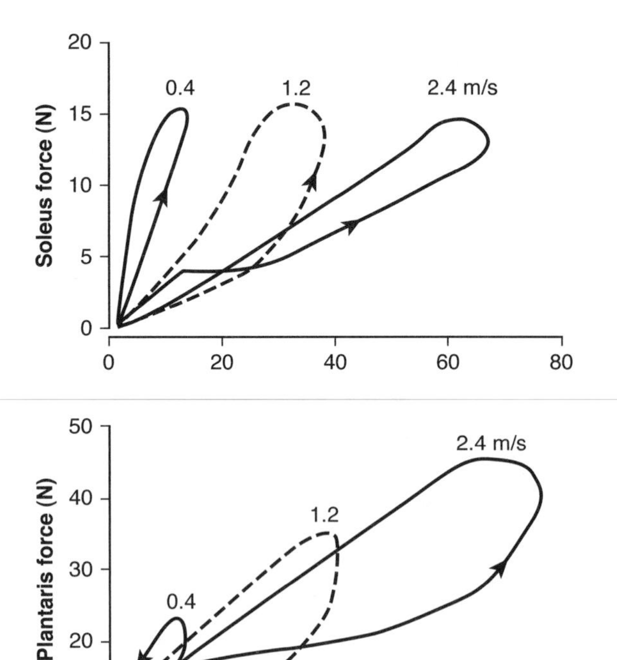

Figure 6.40 Force sharing between the gastrocnemius and soleus (top) and plantaris (bottom) muscles of the cat during the step cycle at various speeds of locomotion.

Data from Herzog 1996.

biceps brachii crosses both the elbow and shoulder joints and thus contributes to flexor torques about the two joints. Because these two actions occur concurrently in many daily activities, it is useful to have a muscle that contributes to both actions. This arrangement also provides the possibility of producing combined actions by reducing the EMG of the one-joint muscles and increasing the activity of a two-joint muscle. For example, Yamashita (1988) found a reduction in the EMG of gluteus maximus (a one-joint hip extensor) and an increase in the EMG of semimembranosus (a two-joint hip extensor and knee flexor) during concurrent hip extension and knee flexion. Furthermore, Zajac (1993) suggests that one-joint muscles produce the propulsive energy for a vertical jump, whereas the two-joint muscles refine the coordination.

Second, the shortening velocity of a two-joint muscle is less than that of its one-joint synergists (Ingen Schenau et al., 1990*a*). For example, the shortening velocity of rectus femoris (two joints) during concurrent hip and knee extension is less than the shortening velocity of the vasti (one joint). Similarly, the shortening velocity of gastrocnemius (two joints) is less than the shortening velocity of soleus (one joint) during concurrent knee extension and plantarflexion. The advantage of a lesser shortening velocity is that the two-joint muscle is higher on its force–velocity relation (figure 6.35*a*) compared with the one-joint muscle and hence is capable of exerting a force that is a greater proportion of its isometric maximum.

Third, two-joint muscles can redistribute muscle torque, joint power, and mechanical energy throughout a limb (Gielen et al., 1990; Ingen Schenau et al., 1990*a*; Prilutsky et al., 1996b; Prilutsky & Zatsiorsky, 1994; Toussaint et al., 1992). Figure 6.41 represents a model of the human leg comprising a pelvis, thigh, and shank with several one- and two-joint muscles. In this model, muscles 1 and 3 are one-joint hip and knee extensors; muscles 2 and 4 are one-joint hip and knee flexors; and muscles 5 and 6 are two-joint muscles. These muscles can be activated in various combinations to exert extensor torques about the hip and knee joints. One option is to activate muscles 1 and 3, which are the two one-joint muscles that produce the extensor torques. Alternatively, it would be possible to activate muscle 5 along with muscles 1 and 3. Because muscle 5 (two joints) exerts a flexor torque about the hip joint and an extensor torque about the knee joint, concurrent activation of muscles 1, 3, and 5 will reduce the net torque at the hip but increase the net torque at the knee. On the basis of this interaction, the two-joint muscle (muscle 5) is described as redistributing some of the muscle torque and joint power from the hip to the knee. Conversely, activation of muscle 6 along with muscles 1 and 3 will redistribute torque from the knee to the hip.

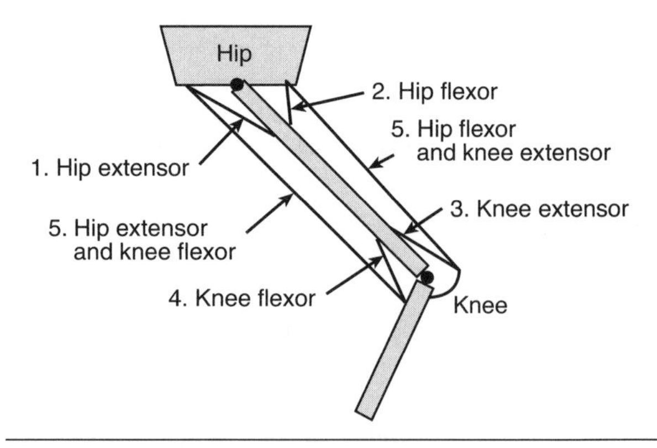

Figure 6.41 Model of the human leg with six muscles arranged around the hip and knee joints. Muscles 1 through 4 cross one joint, and muscles 5 and 6 cross two joints.

Multiple muscle systems: Biomechanics and movement organization, 1990, pg. 647, edited by J.M. Winters and S.L.Y. Woo, "The unique action of bi-articular muscles in leg extensions," G.J. van Ingen Schenau, M.F. Bobbert, and A.J. Van Soest. (New York: Springer-Verlag). With kind permission of Springer Science and Business Media.

EXAMPLE 6.5

Pedal Force Direction Requires Activation of Two-Joint Muscles

The downstroke in cycling, in which the foot pushes the pedal down, involves concurrent extension at the hip and knee joints. Although the downstroke could be accomplished through activation of the one-joint hip and knee extensor muscles (gluteus maximus and the vasti), EMG measurements have shown that several of the two-joint muscles also participate in this task (Li & Caldwell, 1998; Sarre et al., 2003; Suzuki et al., 1982). For example, rectus femoris is active at the beginning of the downstroke whereas the hamstrings are active later (Gregor et al., 1985). The explanation for this pattern of activity is based on the direction of the force that is applied to the pedal (Ingen Schenau, 1990). At the beginning of the downstroke, the pedal force is directed downward and forward (figure 6.42*a*), which requires net extensor torques about the hip and knee joints. The muscle torque about the knee, however, is much greater because of the longer moment arm (a) at the knee joint compared with the hip joint (b). Through coactivation of gluteus maximus and rectus femoris, the net muscle torque is kept small about the hip—which is necessary to keep the pedal force pointing forward—and an extensor torque is exerted about the knee joint by rectus femoris. The action of rectus femoris at the knee joint supplements that of the vasti, but can do so only because of the strong contraction by gluteus maximus.

Later in the downstroke, the direction of the pedal force changes from forward to backward (figure 6.42*b*). The line of action of this backward-directed pedal force passes in front of the knee joint as a result of a net extensor torque about the hip joint and a net flexor torque about the knee joint. Coactivation of the two-joint hamstring muscles contributes to both the extensor torque about the hip and the flexor torque about the knee. At the knee joint, however, both the hamstrings and vasti are active, but with the hamstrings exerting the dominant effect and producing the requisite flexor torque about the knee. Coactivation of the vasti and hamstrings increases the magnitude of the force exerted by the hamstrings, which is registered about both the hip and knee joints. By this mechanism, the net extensor torque about the hip is increased at this stage of the downstroke.

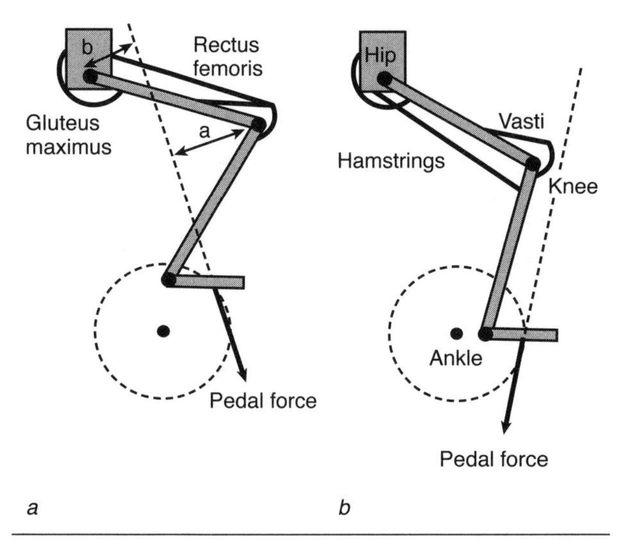

Figure 6.42 Model of the human leg during cycling at *(a)* the beginning and *(b)* the middle of the downstroke phase. The direction of the pedal force depends on the muscles that contribute to the movement.

Data from Ingen Schenau 1990*a*.

Torque–Velocity Relation The modifying effects of the moment arm relative to the joint, the off-axis attachments, the number of joints spanned, and the involvement of multiple muscles employed in an action complicate the influence of contraction velocity on muscle force. At the level of the joint, the force–velocity relation is expressed as the torque–velocity relation and is often measured with **isokinetic contractions,** which are analogous to the isotonic contractions used in single-fiber and whole-muscle studies. To perform an isokinetic contraction, one pushes against a lever arm that has its range of motion and angular velocity controlled by a motor. The motor maintains a constant angular velocity (isokinetic) throughout the range of motion; contraction velocity, however, is not constant. When the rotation about the joint occurs in the same direction as the displacement of the lever arm, the muscles perform a shortening (concentric) contraction. Conversely, the muscles perform a lengthening (eccentric) contraction when the individual resists the motion of the lever arm.

Figure 6.43*a* shows the torque exerted by the knee extensor muscles of one subject when he performed shortening contractions at six different angular velocities (Kawakami et al., 2002*b*). The data show that the peak torque in each contraction decreased as the angular velocity increased; the peak torque was greatest for the slowest speed (0.52 rad/s). The average effect of angular velocity on the peak isokinetic torque achieved by six men is shown in figure 6.43*b*. Furthermore, figure 6.43*a* indicates that the knee joint angle at which the peak torque occurred became more extended as the angular velocity increased. When humans perform lengthening isokinetic

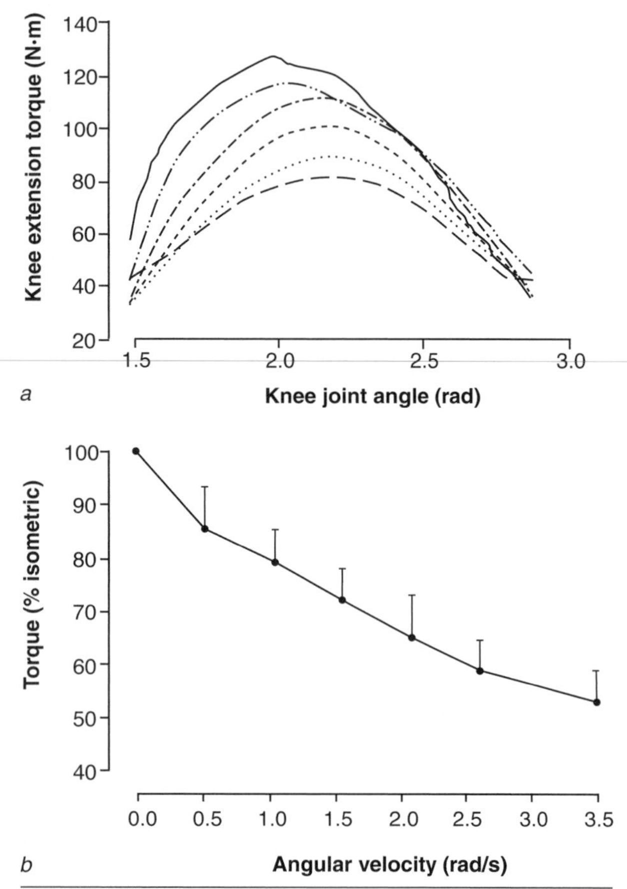

Figure 6.43 The torque exerted by the knee extensor muscles during isokinetic contractions. *(a)* Six contractions, each at a different speed, performed by one subject. The six speeds (top to bottom) were 0.52, 1.05, 1.57, 2.09, 2.62, and 3.49 rad/s. Knee joint angle corresponds to the angle between the thigh and the shank. *(b)* The torque–velocity relation during shortening contractions for six subjects. The data were normalized to the maximal isometric torque.

Data from Kawakami et al. 2002*b*.

contractions, the peak torque can be greater than that achieved during maximal isometric contractions and is relatively constant over a wide range of angular velocities (Aagaard et al., 1996; Linnamo et al., 2006; Webber & Kriellaars, 1997; Westing et al., 1990).

When the intensity of muscle activation is varied from low to maximal, both the magnitude of the muscle force and the shape of the torque–angle relation vary (figure 6.44). Marsh and colleagues (1981) used electrical stimulation of the tibialis anterior muscle at various frequencies to activate the muscle at different intensities. At the three lowest frequencies (1, 10, and 20 Hz), they found that the evoked torque increased as the ankle joint was moved into plantarflexion. At 30 and 40 Hz stimulation, however, they found a peak in the torque–angle relation at an intermediate joint angle. The shape of the 40 Hz curve was more similar to that achieved by maximal voluntary contraction of the dorsiflexor muscles. Similar effects have been observed in experiments on the length–force relation of isolated muscles (Rack & Westbury, 1969).

When isolated muscles are stimulated at different frequencies, there is a shift in the force–velocity relation. The effect differs for shortening and lengthening contractions. The force evoked and power produced at a given shortening velocity uniformly increase with stimulation frequency (figure 6.35). In contrast, submaximal stimulation alters the shape of the force–velocity relation for lengthening contractions. When the cat soleus muscle was stimulated at 35 Hz, the evoked force was relatively constant at five different velocities during lengthening contractions, and the force was greatest during these contractions (Joyce et al., 1969). Reduction of the stimulus rate to 7 Hz decreased the force evoked during the lengthening contractions so that it was less than the force exerted during an isometric contraction at 7 Hz. Thus the magnitude of the force exerted by muscle during a lengthening contraction relative to an isometric contraction depends on the intensity of muscle activation.

The relation between net muscle torque and angular velocity as determined with isokinetic contractions (figure 6.43) does not necessarily correspond to the force–velocity relation of isolated muscle (figure 6.27) due to the dependence of fascicle length on muscle force (Kawakami et al., 2002*b*). For this reason, the torque

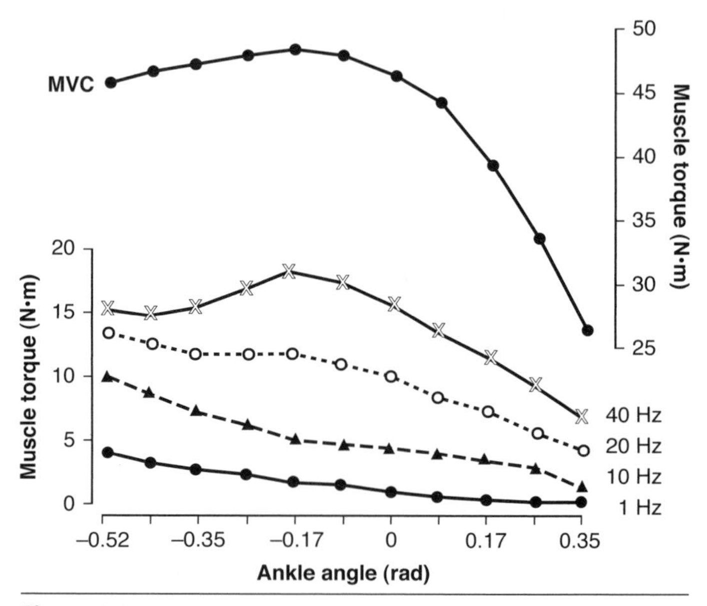

Figure 6.44 Torque exerted by the dorsiflexor muscles during submaximal evoked contractions and a maximal voluntary contraction (MVC). Negative angles designate plantarflexion. Each data point represents the peak torque achieved during a voluntary (MVC) or an evoked contraction. The evoked contractions involved electrical stimulation of the tibialis anterior muscle at a selected frequency (1-40 Hz).

Data from Marsh et al. 1981.

measurements extracted from isokinetic contractions should correspond to the peak value and not to the value at a specific joint angle (Kawakami & Fukunaga, 2006). Furthermore, the change in pennation angle during an isokinetic contraction results in a lower fascicle velocity than that estimated from the rate of change in muscle length derived from the angular velocity of the contraction (Chino et al., 2003). Because isokinetic contractions begin from a maximal isometric contraction, the relation between angular velocity and net muscle torque differs from that observed during most movements.

Muscles in Motion

The force–length and force–velocity relations of muscle, from the level of the single fiber to the joint, are often determined experimentally from isometric, isotonic, and isokinetic contractions. These conditions provide information about the force that a muscle can exert at a given muscle length or rate of change in length (velocity) when the muscle is activated maximally. In the performance of actual movements, however, muscle length changes at variable rates, activation is rarely maximal, and the load imposed on the muscle can change (Askew & Marsh, 2002; Marsh, 1999). As described in this section, the force exerted by muscle during movement often differs from that described by the standard measures of muscle function shown in figures 6.26, 6.27, 6.33, and 6.43 (Gillard et al., 2000; Herzog, 2004*b*).

Short-Range Stiffness

The force that a muscle can exert during a maximal, constant-velocity contraction is greatest when the muscle is being forcibly lengthened (figure 6.27). Even in less constrained movements, this capacity of muscle is evident (Gillard et al., 2000). For example, it is easier to lower a heavy load with the elbow flexor muscles, which requires a lengthening contraction, than it is to lift the same load with a shortening contraction. The force that a muscle exerts during a lengthening contraction, however, can be quite variable (Kirsch et al., 1994).

Muscle force varies nonlinearly when an active muscle performing an isometric contraction is forcibly lengthened. An example of this effect is shown in figure 6.45. In this experiment, the length of the extensor digitorum longus muscle was controlled while the nerve to the muscle was activated with electric stimuli and the muscle force was measured (McCully & Faulkner, 1985). For the isometric contraction, length did not change, whereas force rose to a plateau during the stimulation. When the muscle was allowed to shorten at a constant velocity during the stimulation, force declined, at first rapidly and then more slowly, from the initial isometric value. In contrast, constant-velocity lengthening of the muscle caused the force to increase, rapidly at first and then more slowly, from the initial isometric value. For a constant-velocity contraction, the slope of the force trace indicates the stiffness of the muscle. Stiffness during the lengthening contraction was initially high (i.e., steep slope) and then declined; this initial stiffness is referred to as the short-range stiffness (Gillard et al., 200; Rack & Westbury, 1974; Walmsley & Proske, 1981). It is greater for type I compared with type II muscle fibers (Malamud et al., 1996) and for type S compared with type FR and FF motor units (Petit et al., 1990).

This behavior can be explained by the function of the cross-bridges (Kirsch & Kearney, 1997). When a muscle is forcibly lengthened from an isometric contraction, the

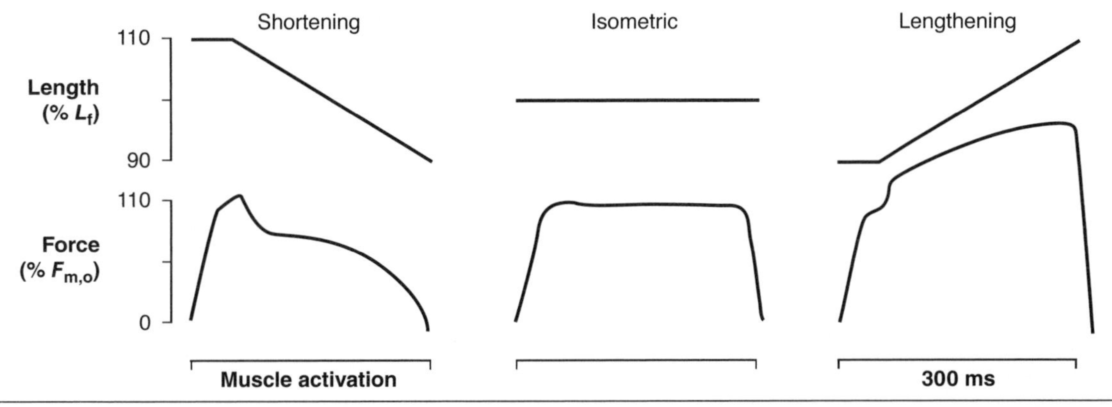

Figure 6.45 Change in muscle force during controlled changes in muscle length. Length is indicated as a percentage of muscle fiber length (% L_f) at the resting length of the muscle. Force is expressed as a percentage of the maximal isometric force (% $F_{m,o}$). For each contraction, the muscle was stimulated at 150 Hz for 300 ms. After 100 ms of stimulation, muscle length was decreased (shortening contraction), held constant (isometric), or increased (lengthening contraction). Compare the force profile during the shortening and lengthening contractions with that during the isometric contraction.

Modified, with permission from K.K. McCully and J.A. Faulkner, 1985, "Injury to skeletal muscle fibers of mice following lengthening contractions," *Journal of Applied Physiology* 59:120. Used with permission.

attached cross-bridges are stretched, and this increases the average force exerted by each cross-bridge. Fibers from frog muscles stretched this way exhibited first a rapid increase in force and then a steady-state force (Lombardi & Piazzesi, 1990). The steady-state force that was achieved during the lengthening was about twice the isometric force and occurred after the fiber had been stretched by ~20 nm per half sarcomere. The change in force at the beginning of the stretch depended on the velocity of the stretch; it increased continuously for slow stretches but increased and then decreased for fast stretches. By superimposing small stretches during the contraction, Lombardi and Piazzesi (1990) found that the stiffness of the fiber was 10% to 20% greater during the lengthening contraction than during the isometric contraction. They concluded that after the cross-bridges have been stretched by a certain amount, they detach and reattach rapidly; the reattachment rate is ~200 times faster than during an isometric contraction. Thus, the short-range stiffness that appears at the beginning of a lengthening contraction is due to an increase in the average force exerted by each cross-bridge. Furthermore, as the lengthening contraction continues, the cross-bridges reattach quickly, again generating a greater average force.

The stretch imposed on an active muscle fiber also has a prolonged effect that appears as an increase in the force that the fiber exerts at a given muscle length (Rassier & Herzog, 2004). This effect is known as **force enhancement.** The amount of force enhancement increases with stretch amplitude and fiber length, but is largely independent of the speed of stretch (Edman et al., 1978; Rassier et al., 2003). Force enhancement has both an active and a passive component (Herzog et al., 2006). The active component appears to involve a decrease in the detachment rates of cross-bridges after the stretch of the active muscle. The passive component may be produced by a change in the stiffness of a structural protein, such as titin, but appears only when the stretch begins from long fiber lengths. The force enhancement likely has a significant influence on subsequent muscle activity.

Stretch–Shorten Cycle

A common pattern of muscle activation during many movements is the stretch of an active muscle before it performs a shortening contraction. This is known as the **stretch–shorten cycle** (Ingen Schenau et al., 1997; Nicol et al., 2006). It occurs in most movements that we perform, involving, for example, the knee extensor and ankle plantarflexor muscles after footstrike in running (figures 4.5 and 4.6); the knee extensor muscles during kicking (figure 4.33); the trunk and arm muscles during throwing; and the hip, knee, and ankle extensor muscles during the countermovement jump (figures 4.21 and 4.22) and the long-jump takeoff (figure 4.23) (Prilutsky, 2000). Perhaps the only common physical activity not to include the stretch–shorten cycle is swimming.

The advantage of the stretch–shorten cycle is that a muscle can perform more positive work (Cavagna & Citterio, 1974; Fenn, 1924) and produce more power (Kawakami et al., 2002*a*) if it is actively stretched before being allowed to shorten. Four mechanisms have been proposed to explain the greater positive work a muscle can do with a stretch–shorten cycle: time to develop force, storage and release of elastic energy, force potentiation, and reflexes. The first mechanism, time to develop force, has to do with the increased time that the muscle has to become fully activated when there is an initial lengthening contraction. Because positive work corresponds to the area under the force–length curve, the area is greater with an initial lengthening contraction that increases the force at the start of the shortening contraction. The second mechanism, elastic energy, involves the storage of elastic energy in the SE (figure 6.28) during the stretch and the subsequent use of this energy during the shortening contraction. From the work–energy relation (chapter 2), we know that an increase in the available energy will increase the amount of work that can be done. The third mechanism, force potentiation, suggests that the force from individual cross-bridges is enhanced as a consequence of the preceding stretch. However, this effect appears only at relatively long muscle lengths (Edman & Tsuchiya, 1996). The fourth mechanism concerns the stretch reflexes that can be evoked by forced lengthening of the muscle at the beginning of the stretch-shorten cycle.

However, the role of these mechanisms in increasing the positive work done by muscle in the stretch–shorten cycle is controversial (Ingen Schenau et al., 1997). It appears that the relative contribution of each mechanism varies across movements (Ishikawa et al., 2007; Lichtwark et al., 2007). Some research on this topic has involved a comparison of the countermovement and squat vertical jumps (figures 4.21 and 4.22). The extra height that an individual can achieve with the countermovement jump can be completely explained by the first mechanism, the extra time the muscles have to generate force prior to the beginning of the shortening contraction (Ingen Schenau et al., 1997). Other movements, however, such as those that involve a rapid stretch–shorten cycle, appear to depend on the elastic energy mechanism (Henry et al., 2005; Ishikawa et al., 2006; Kawakami et al., 2002*a;* Kurokawa et al., 2003; Kubo et al., 2007*b*).

In contrast, force potentiation seems an unlikely contributor (Kawakami & Fukunaga, 2006), whereas the role of reflexes remains controversial. Although it is possible to elicit a stretch reflex in a stretch–shorten cycle (Avela et al., 2006; Nicol & Komi, 1998), the timing of the response is a problem. For example, it takes some time for the sensory receptors to be activated during the stretch, the afferent signal to be transmitted to the spinal cord, the reflex response to be transmitted back to the muscle, and the muscle to generate the force response. For this reason, it has been estimated by some (Ingen Schenau et al., 1997), but not others (Avela et al., 2006; Dietz et al., 1979), that a stretch-shorten cycle faster than 130 ms does not experience a contribution from the stretch reflex. Nonetheless, the rate of stretch experienced by muscles in many stretch-shorten cycles does seem to evoke a stretch reflex of sufficient magnitude to influence the force by muscle during the shortening contraction (Komi & Nicol, 2000; Nicol & Komi, 1998; Voigt et al., 1998*b*). Furthermore, four weeks of training with hopping improved the excitability of short-latency stretch reflexes in the soleus muscle (Voigt et al., 1998*a*).

Some of the controversy over the mechanisms underlying the improvement in performance with the stretch-shorten cycle is due to the difficulty in determining when the different parts of the muscle are stretched and shortened. It is important, for example, to know the separate length changes for the muscle fibers and tendon (Biewener & Roberts, 2000; Fellows & Rack, 1987; Fukunaga et al., 2001; Griffiths, 1991; Ishikawa et al., 2007; Marsh, 1999; Sousa et al., 2007). The change in muscle fiber length often does not parallel the change in joint angle during movements that involve a stretch-shorten cycle. Figure 6.46 compares these changes during a vertical jump that entails a countermovement with one that begins from a squat position. The countermovement jump (left column in figure 6.46) involved the subject beginning from a standing position and lowering himself down (ankle flexion) before jumping up. The squat jump, in contrast, simply involved a jump upward from an initial squatting position (note initial ankle angle of 1.40 rad). The squat jump entailed continuous extension about the ankle joint and shortening of the fascicles in the medial gastrocnemius muscle. The muscle fascicles also shortened when the ankle extended during the countermovement jump; but when the ankle flexed in the initial part of the jump, the fascicles first stretched and then remained at a constant length (isometric contraction) during the initial part of the jump for most of the downward phase. Thus, the countermovement jump involved a brief lengthening contraction, an isometric contraction, and then a shortening contraction of the medial gastrocnemius muscle.

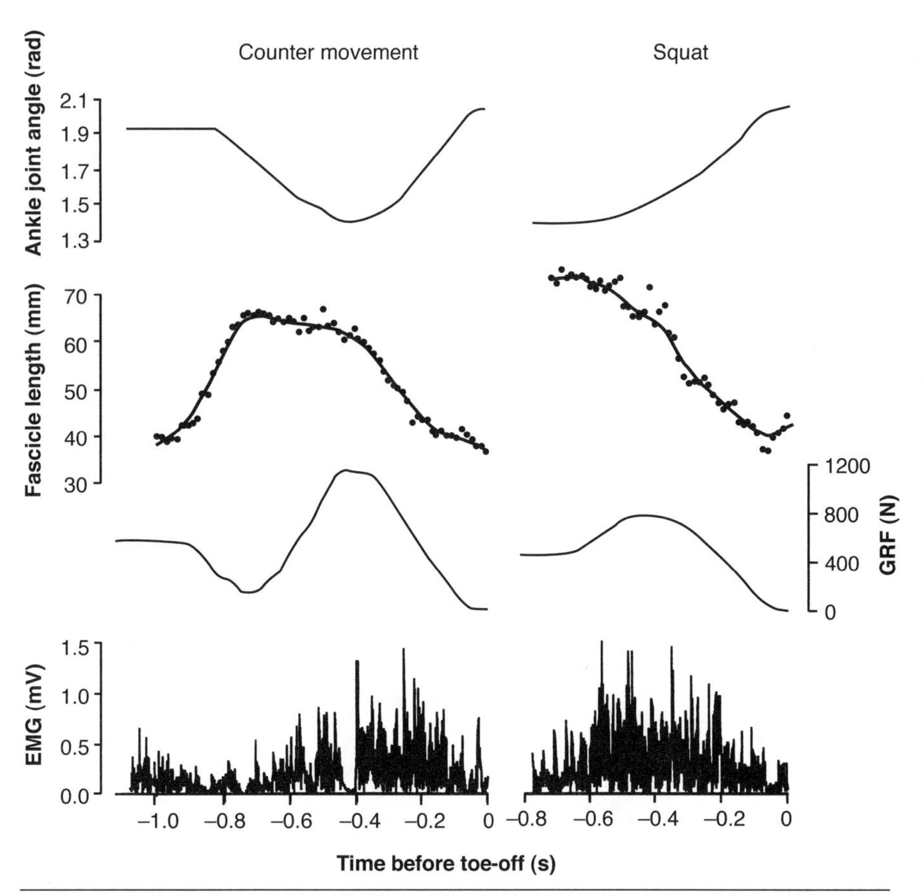

Figure 6.46 A countermovement jump (left column) and a squat jump (right column) performed by one subject. The traces (top to bottom) indicate ankle joint angle, the length of fascicles in the medial gastrocnemius muscle, vertical ground reaction force (GRF), and rectified EMG for medial gastrocnemius.

Because the stretch-shorten cycle increases positive work and power production, it also improves the economy of performance. This is most obvious for cyclic activities, such as walking, running, and hopping. The positive work that has to be done during the support phase of these activities can be accomplished with the use of two sources of energy: metabolic and elastic. When less metabolic energy is used to perform a given amount of work, as occurs with the stretch-shorten cycle, the performance is more economical. For example, Hof (1998) estimated that much of the 45 J of energy absorbed by the calf muscles of humans during the first half of the stance phase in a slow run can contribute to the 60 J of energy used to perform positive work in the second half of the stance phase. The capability of muscle and tendon to store elastic energy and then use it to perform positive work is greatest in muscles with long tendons (Hof, 2003). For example, Biewener (1998) reported that 92% to 97% of the work done by the plantaris and gastrocnemius muscles of the wallaby during hopping is recovered from elastic storage, compared with 60% for the turkey gastrocnemius muscle during trotting and 0% for the pigeon pectoralis muscle during flying. Similarly, the tendons of soleus, gastrocnemius, and plantaris in the cat contribute significantly to the work done by the muscles during walking and trotting (Prilutsky et al., 1996*a*).

There appear to be two reasons for the common occurrence of the stretch-shorten cycle in most movements. First, it can increase the positive work and power production by muscle during the shortening contraction. Second, it can lower the metabolic cost of performing a prescribed amount of positive work.

Shortening Deactivation

In contrast to the mechanisms that enhance muscle force in the stretch-shorten cycle, there are mechanisms that depress muscle force during a shortening contraction (de Ruiter et al., 1998; Edman, 1996; Herzog, 1998; Rousanoglou et al., 2006). This effect has been observed in both single muscle fibers and whole muscles. The depression is measured by comparing the force a muscle can exert during an isometric contraction with the force exerted when the muscle shortens to the same length and then performs an isometric contraction. Figure 6.47 compares the force exerted by a muscle fiber during an isometric contraction (trace a) with the force exerted after the fiber has shortened from a sarcomere length of 2.55 μm to 2.05 μm (trace b). The comparison indicates that the force exerted by the fiber was depressed immediately after the shortening but eventually reached the same value as the isometric contraction; it takes about 1 to 1.5 s to overcome the force depression due to the shortening. The force depression, which is called **shortening deactivation,** increases with the amount of shortening, the level of activation, and the magnitude of the force (Herzog, 1998; Leonard & Herzog, 2005). However, there is no evidence of force depression when a fiber is stimulated to produce a fused tetanus (Edman, 1996).

The mechanism that causes shortening deactivation after a shortening contraction appears to involve deactivation of the myofibrillar system. There is some disagreement, however, about whether the mechanism is chemical or mechanical. Edman (1996) suggests that the shortening contraction causes a temporary decrease in the affinity for calcium at the binding sites. In contrast, Rassier and Herzog (2004) note that the force depression lasts too long (>5 s) to be attributable to a transient chemical effect and instead argue in favor of a mechanical effect. The mechanism involves a stress-induced inhibition of cross-bridge attachment in the new overlap region of the thick and thin filaments after the shortening (Maréchal & Plaghki, 1979). According to this scheme, the thin filament is stretched during the shortening contraction, and this reduces the availability of attachment sites on actin in the new overlap region. The shortening-induced decrease in the proportion of attached cross-bridges causes a force depression that has a magnitude and duration sufficient to influence many movements.

Work-Loop Analysis

Measurement of the force exerted or the power produced by muscle during movement is often quite different from that during isotonic and isokinetic contractions (figure 6.36). One can assess the magnitude of this difference by directly measuring muscle force and displacement during a movement and performing a **work-loop** analysis (Biewener & Roberts, 2000; Caiozzo & Rourke, 2006; Josephson, 1985, 1999). The method quantifies the net work done by a muscle by subtracting the work done on the muscle during a stretch (lengthening contraction) from the work done by the muscle during the shortening contraction.

An example of the work-loop method is shown in figure 6.48. The change in length experienced by a muscle, along with the force it exerts, is measured during

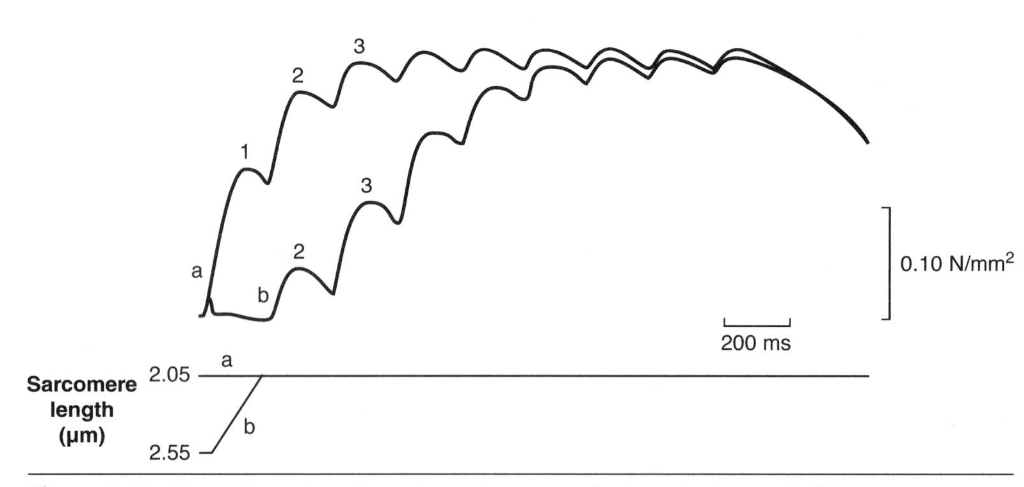

Figure 6.47 Force depression after a shortening contraction by a single muscle fiber.
Data from Edman 1996.

a repetitive movement; two cycles of the movement are shown in figure 6.48*a*. These measurements are then plotted on a force–length graph, which is interpreted in the counterclockwise direction. Beginning from the dot in figure 6.48*a,* the diagram has three phases: (1) Length increases with little change in force; (2) force increases while length decreases slightly; and (3) both length and force decrease. In this example, the first phase corresponds to the lengthening contraction, and the third phase indicates the shortening contraction. Because the work of a force is equal to the product of the force and the displacement, the area under the force–length diagram denotes the work associated with the movement. The work experienced by the muscle, however, comprises the work done on it during the lengthening contraction (work input) and the work it does during the shortening contraction (work output). The net work done by the muscle is the difference between the two, which is illustrated as the area inside the length–force diagram.

The shape of a work loop varies across movements and among muscles (Biewener & Roberts, 2000). For example, figure 6.49 contrasts the work performed by the medial gastrocnemius muscle of a man during the squat and countermovement jumps (Kawakami et al., 2002*b*). In the squat jump (figure 6.49*a*), the task began from a fascicle length of 75 mm that decreased to a final length of 35 mm at takeoff as the estimated Achilles tendon force first increased and then decreased. The squat jump did not involve an initial lengthening contraction, so the work done by the muscle corresponds to the area under the force–length graph during the shortening contraction. In the countermovement jump (figure 6.49*b*), the task began from a fascicle length of 35 mm, which increased to 70 mm during the lengthening contraction, remained constant during an isometric contraction, and then decreased to 30 mm during the shortening contraction. As indicated in figure 6.48, the net work done by the medial gastrocnemius is the difference between the work input (area during the lengthening contraction) and the work output (area during the shortening contraction). Although the work output, and hence the height of the jump, was greater for the countermovement jump, the net work was similar for the two tasks. The work-loop method has also demonstrated significant associations between the

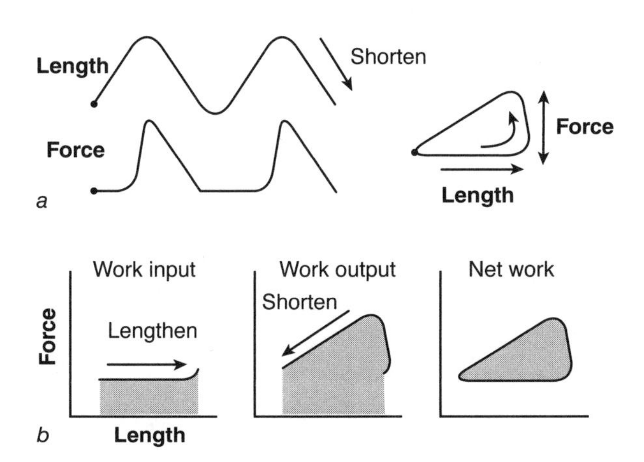

Figure 6.48 The work-loop method for determining the net work performed by muscle: *(a)* Length and force records are used to construct the work loop; *(b)* the work loop displays the differences between the work input (negative work) and the work output.

Data from Josephson, 1985.

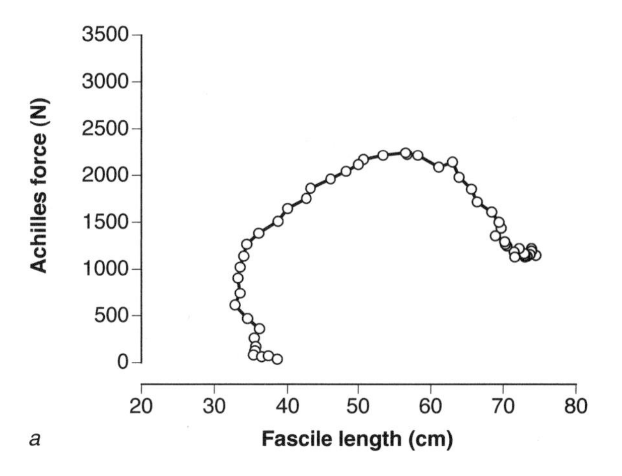

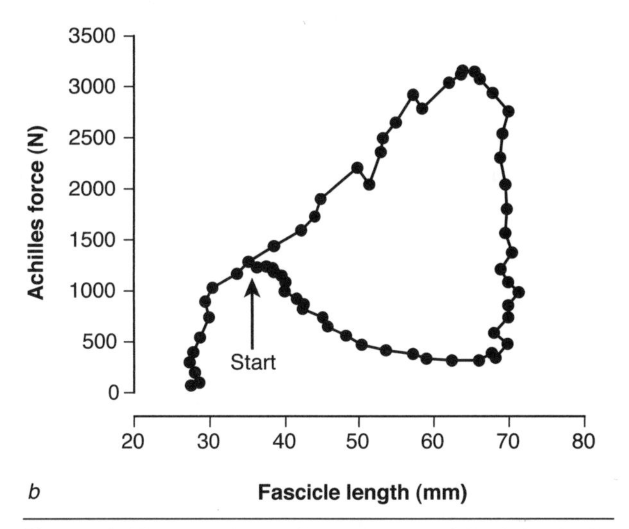

Figure 6.49 The change in the length of fascicles in the medial gastrocnemius muscle and the estimated Achilles tendon force of a man as he performed *(a)* a squat jump and *(b)* a countermovement jump. The countermovement jump begins from a fascicle length of 35 mm and proceeds in a counterclockwise direction around the graph. Short fascicle lengths correspond to plantarflexion of the ankle.

Adapted, by permission, from Y. Kawakami et al., 2002, "In vivo muscle fibre behaviour during counter-movement exercise in humans reveals a significant role for tendon elasticity," *Journal of Physiology* 540: 641.

structure of a muscle and the power it generates during a movement (Biewener & Roberts, 2000).

SUMMARY

This chapter is the second of three on the properties of the motor system. It focuses on the structure of muscle, the activation of the contractile proteins, the functional relations between spinal neurons and muscle fibers, the mechanical properties of muscle, and the net effects of muscle activity on human movement. The central theme of the chapter is the transformation of the activation signal discharged from the spinal cord into the output that is produced by a muscle during a contraction. The chapter begins by describing the organization of the contractile proteins in muscle and their activation by muscle fiber action potentials. The next major topic is how the output from the spinal cord is organized by the nervous system to vary the force that a muscle exerts during a voluntary contraction. This involves controlling the activity of single spinal neurons and the muscle fibers innervated by each neuron. The chapter concludes by examining the influence of the mechanical properties of muscle on the force exerted, the work performed, and the power produced by the activated muscle fibers.

SUGGESTED READINGS

Biewener, A.A. (2003). *Animal Locomotion.* Oxford, UK: Oxford University Press, chapters 1-3.

Enoka, R.M., & Pearson, K.G. (2008). The motor unit and muscle action. In E.R. Kandel, J.H. Schwartz, & T.M. Jessell (Eds.), *Principles of Neural Science* (5th ed.). New York: McGraw Hill, in press.

Heckman, C.J., & Enoka, R.M. (2004). Physiology of the motor neuron and the motor unit. In A. Eisen (Ed.), *Clinical Neurophysiology of Motor Neuron Diseases. Handbook of Clinical Neurophysiology* (Vol. 4, pp. 119-147). Amsterdam: Elsevier.

Jones, D.A., & Round, J.M. (1990). *Skeletal Muscle in Health and Disease.* Manchester: Manchester University Press, chapters 1-4.

Kernell, D. (2006). *The Motoneurone and Its Muscle Fibres.* Oxford, Great Britain: Oxford University Press, chapters 3 and 8.

Lieber, R.L. (1992). *Skeletal Muscle Structure and Function.* Baltimore: Williams and Wilkins, chapters 1-3.

MacIntosh, B.R., Gardiner, P.F., & McComas, A.J. (2006). *Skeletal Muscle: Form and Function.* Champaign, IL: Human Kinetics, chapters 1, 2, and 11-13.

Vogel, S. (2001). *Prime Mover.* New York: Horton, chapters 1-3 and 6.

chapter 7

Voluntary Movement

In the previous two chapters on the motor system, we discussed the properties of excitable membranes, the transfer of the activation signal between cells, and the mechanical consequences of activation of the contractile proteins by muscle fiber action potentials. To continue our examination of the motor system, we discuss the synaptic input provided to the motor neurons by the nervous system and the types of actions that this produces. The synaptic input received by motor neurons can arise from various sources, such as sensory receptors in the periphery and neurons in the brain, and it can involve only a single neuron or thousands of neurons. This chapter examines three types of actions that differ with respect to the way in which the nervous system organizes the activation of the motor neurons: spinal reflexes, automatic responses, and voluntary actions.

Spinal reflexes are fast responses that involve an afferent signal into the spinal cord and an efferent signal out to the muscle. The connection between the afferent and efferent axons may involve only a single synapse. **Automatic responses** can be triggered both by afferent input and by anticipated needs, but they involve neural circuits that are more complex than those responsible for spinal reflexes. These behaviors range from fight-or-flight responses associated with the preservation of life through to postural adjustments that precede the performance of a movement. **Voluntary actions** are contractions generated by the cerebral cortex in response to a perceived need. In contrast to spinal reflexes and automatic responses, voluntary contractions can be interrupted by the performer.

SPINAL REFLEXES

Three functional classes of neurons contribute to spinal reflexes: afferent, interneuron, and efferent neurons. Figure 7.1 shows the association between these neurons as information flows from the periphery into the central nervous system (CNS) and ends with a response that is transmitted to muscle. **Afferent** neurons convey sensory information into the CNS about stimuli that have been detected in the periphery. Sensory receptors convert energy from one form to another through a process known as **transduction** (Roatta & Passatore, 2006). Energy can exist in a variety of forms, such as light, pressure, temperature, and sound; but the common output of sensory receptors is electrochemical energy in the form of action potentials that are propagated along the associated afferent axons. The human body contains many different types of sensory receptors, and these can be distinguished on the basis of their location (exteroceptors, proprioceptors, interoceptors), function (mechanoreceptors, thermoreceptors, photoreceptors, chemoreceptors, nociceptors), and morphology (free nerve endings, encapsulated endings). In the study of human movement, however, the focus is often placed on somatosensory receptors and the information they provide about movement (Brooke & Zehr, 2006; Gandevia, 1996; Prochazka, 1996). With this input, the nervous system is able to organize a rapid response to a disturbance, to determine the location of the body, and to distinguish between self-generated and imposed movements.

Interneurons account for 99% of all neurons and represent the CNS component that modulates the interaction between input (afferent) and output (efferent). Interneurons can elicit excitatory and inhibitory responses in other neurons. The resting membrane potential of a neuron can be modulated either directly or indirectly by an interneuron. Direct modulation occurs when the interneuron forms part of the circuit between the afferent and efferent neurons. Indirect modulation is accomplished when the interneuron alters the excitability of the connection between the afferent and efferent neurons (figure 7.1). The excitability of an interneuron can itself be modulated by other neurons.

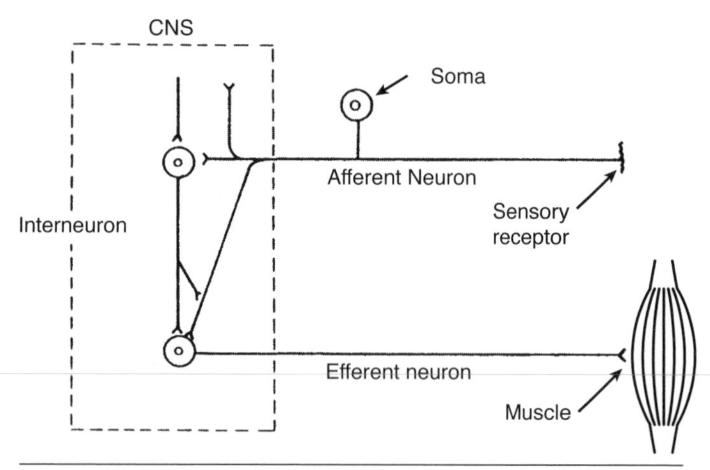

Figure 7.1 The three functional classes of neurons: the afferent neuron, interneuron, and efferent neuron. CNS = central nervous system.

Efferent neurons transmit action potentials from the CNS to the effector organ. Efferent neurons that innervate muscle are referred to as **motor neurons.** The somas of motor neurons are located in the brain stem and in the ventral horn of the gray matter of the spinal cord, and their axons exit the cord and are bundled together into peripheral nerves that course to the target muscles. Forty-three pairs of nerves (12 cranial and 31 spinal) in the human body leave the CNS and form the peripheral nervous system. The spinal cord is a segmented structure in which the segments correspond to the vertebrae. Between each pair of vertebrae, a set of axons exits and another set enters on each side (left and right) of the cord. The axons belonging to efferent neurons exit the spinal cord in the **ventral roots,** whereas the axons of the afferent neurons enter through the **dorsal roots** (figure 7.2). The ventral and dorsal roots are located on the front and back of the cord, respectively. Motor neurons have large-diameter, myelinated axons that traverse from the spinal cord directly to muscle.

Afferent Feedback

The ability of sensory receptors to initiate rapid responses to perturbations is based on the existence of short-latency connections between the input (afferent signal) and the output (motor response or efferent signal). Such input-output connections are termed reflexes (Pearson, 2000; Pierrot-Deseilligny & Burke, 2005; Prochazka et al., 2000). The simplest neural circuit underlying a reflex involves a sensory receptor and its afferent innervation, as well as a group of motor units that receive input from the afferent. This circuit, however, can be embedded in the neural elements controlling a single muscle, can be distributed among a group of synergists (muscles that contribute to a common mechanical action), can involve an interaction between an agonist-antagonist pair of muscles, or can require the coordination of muscles in contralateral limbs. The study of reflexes, therefore, provides information about the pathways over which sensory feedback is relayed during the performance of various tasks (Nielsen, 2004; Stein & Thompson, 2006).

We begin the discussion of reflexes by describing some of the sensory receptors that provide the afferent feedback to the CNS. The sensory information that is required to perform movements comes from **somatosensory receptors:** muscle spindles, tendon organs, joint receptors, cutaneous mechanoreceptors, nociceptors, and thermal sensors (Brooke & Zehr, 2006). The afferent axons that innervate some of these receptors, the movement stimulus for each receptor, and the velocity at which action potentials are propagated are listed in table 7.1. The afferents are categorized into four groups: I, II, III, and IV.

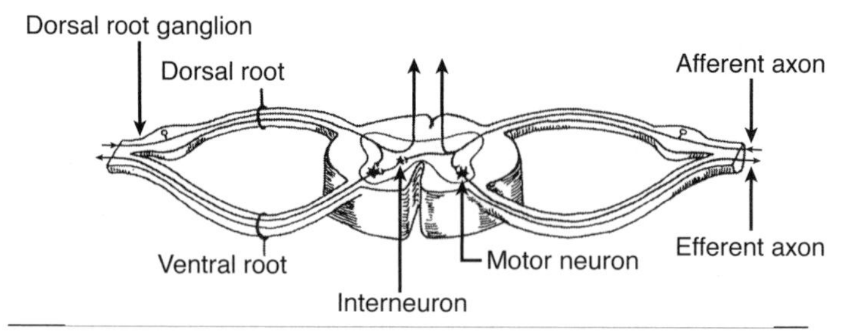

Figure 7.2 Segmental organization of the spinal cord. At the level of each vertebra in the spinal column, the spinal cord gives off a pair of dorsal and ventral roots to each side (left and right) of the body.

Table 7.1 Characteristics of Some Somatosensory Receptors

Somatosensory receptor	Movement stimulus	Afferent	Conduction velocity (m/s)
Muscle spindle primary	Rate of muscle stretch	Ia	40 to 90
Tendon organ	Muscle force	Ib	30 to 75
Muscle spindle secondary	Stretch of muscle	II	20 to 45
Joint receptors	Force around joint	II to III	4 to 45
Haptic receptors	Skin movement	I to III	4 to 80

Data from Brooke and Zehr 2006.

Muscle Spindle

Muscles that cross a joint and experience unexpected loads have a variable number of muscle spindles (6 to 1300) (Hasan & Stuart, 1984; Matthews, 1972). There are about 27,500 spindles in the human body, with 4000 in each arm and 7000 in each leg (Prochazka, 1996). Hand and neck muscles have the highest density of spindles. Spindles have a fusiform shape and lie in parallel with the skeletal muscle fibers (figure 7.3*e*). The ends of the spindles, which vary in length from 2 to 6 mm, attach to intramuscular connective tissue. The muscle spindle comprises 2 to 12 miniature skeletal muscle fibers that are enclosed in a connective tissue capsule (figure 7.4*b*). The muscle spindle fibers are referred to as **intrafusal fibers;** in contrast, the skeletal muscle fibers outside the spindle capsule are called **extrafusal fibers.** Intrafusal fibers do not contribute significantly to the force produced by a muscle. There are two types of intrafusal fibers based on differences in the arrangement of the nuclei, the motor innervation, and contraction speed. The nuclei of the **nuclear chain fiber** (~8 μm in diameter) are arranged end to end like the links in a chain, whereas those in the **nuclear bag fiber** (~17 μm in diameter and 8-10 mm long) cluster in a long group. There are two types of bag fibers: dynamic and static. Both types of intrafusal fiber are devoid of myofilaments in the central region (figure 7.4*a*).

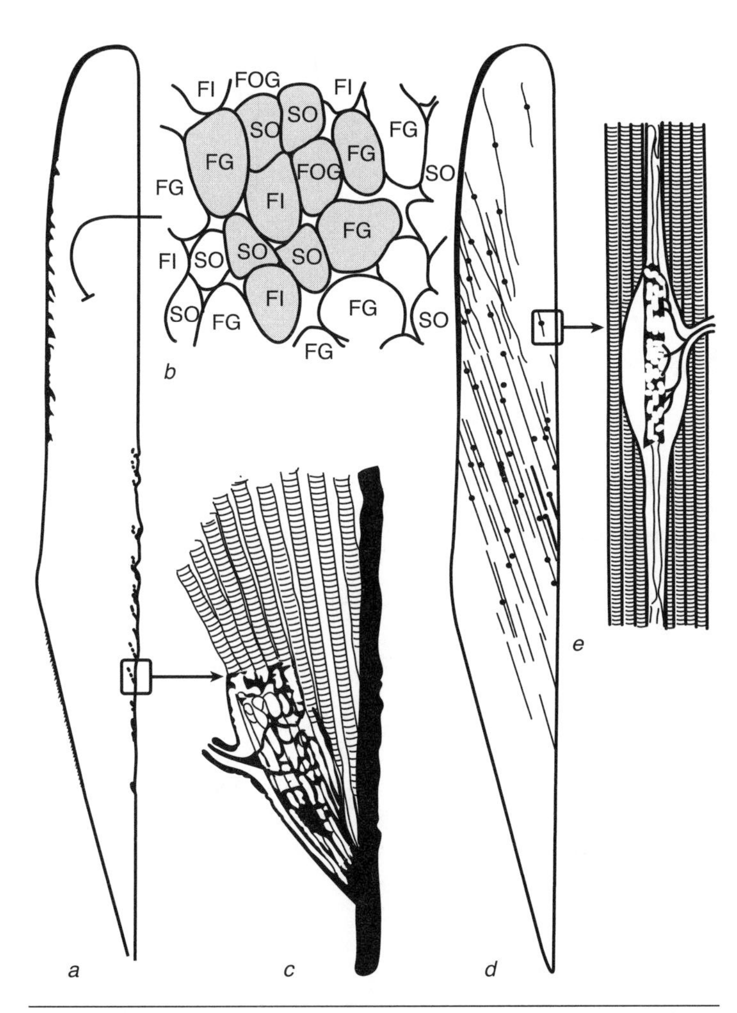

Figure 7.3 Distribution of the muscle spindle and tendon organ in the medial gastrocnemius of the cat: *(a)* a longitudinal section of the muscle with tendon organs stained; *(b)* a cross section through the muscle showing the mixture of muscle fiber types; *(c)* enlarged view of a single tendon organ in series with the skeletal muscle fibers as they attach to the aponeurosis; *(d)* a longitudinal section of the muscle showing the distribution of muscle spindles throughout the belly of the muscle; *(e)* enlarged view of a muscle spindle showing its location in parallel with the skeletal muscle fibers.

Adapted, by permission, from B.R. Botterman, M.D. Binder, and D.G. Stuart, 1978, "Functional anatomy of the association between motor units and muscle receptors," *American Zoologist* 118: 136. Copyright 1978 by The Society for Integrative and Comparative Biology.

The muscle spindle has an afferent supply along which action potentials are transmitted to the CNS. The capsular region of the spindle typically contains one group Ia afferent and one group II afferent (Boyd & Gladden, 1985). The Ia afferent has an ending that spirals around the central regions of both the nuclear chain and bag fibers (figure 7.4). The group II afferent has a nonspiral ending that connects principally to the chain fibers. Not all muscle spindles have group II afferents, but they all have Ia afferents. The somas of the Ia and II afferents are located in the dorsal root ganglion, close to the spinal cord (figure 7.1).

In addition to being innervated by afferent axons, intrafusal fibers receive efferent input. In general, skeletal muscle fibers are innervated by two main groups of motor neurons: alpha (α) and gamma (γ). **Alpha motor neurons** are the largest and innervate extrafusal fibers; **gamma motor neurons** are the smallest and connect exclusively to intrafusal muscle fibers. Each spindle is innervated by 10 to 12 gamma motor neurons. The axons of gamma motor neurons contact the intrafusal fibers in the polar regions, which is where the contractile proteins are located. The two types of nuclear bag fibers are innervated by different gamma motor neurons (figure 7.4*b*). When an action potential is initiated at the axon hillock of a gamma motor neuron and transmitted to the intrafusal fiber, the net effect is a contraction of the fiber in the polar region and a stretch of the central region. The central stretch of the intrafusal fiber that is induced by the gamma motor neurons determines the responsiveness of the muscle spindle to any change in length it detects. This means that a given change or rate of change in the length of a muscle does not always produce the

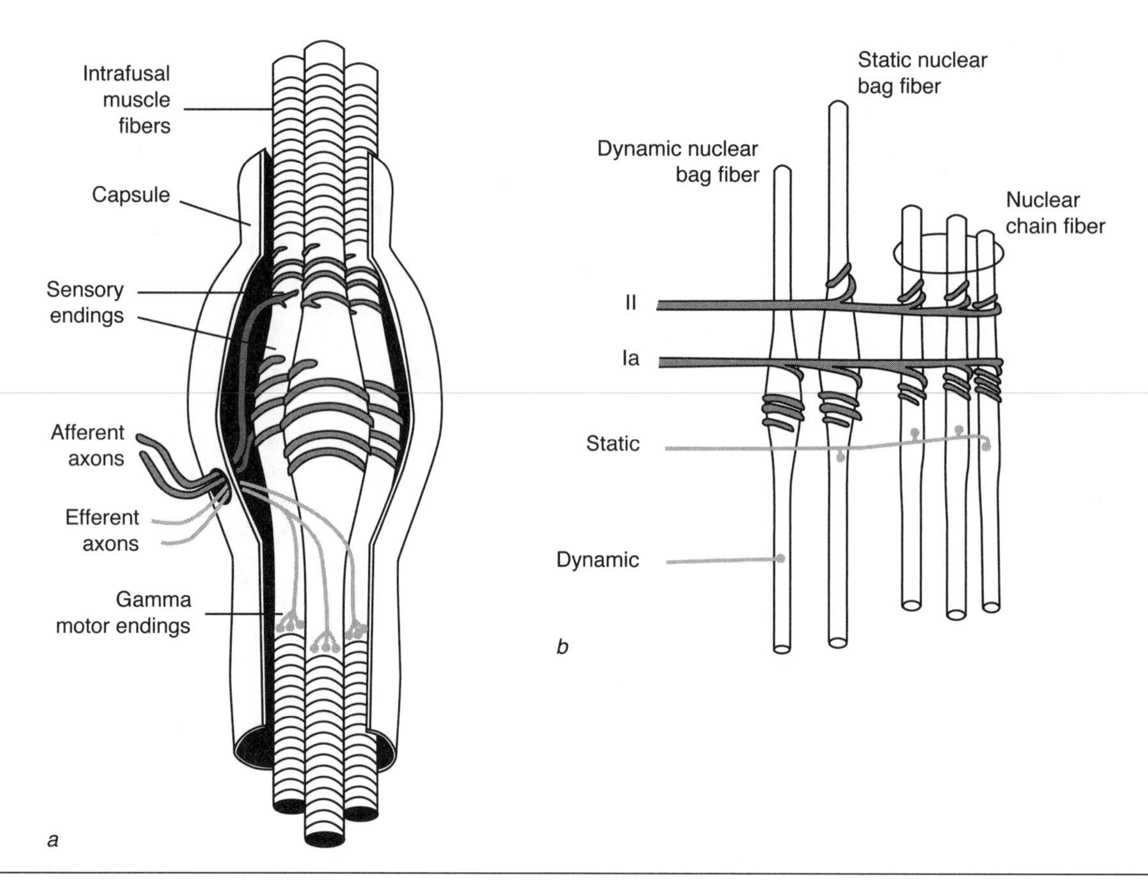

Figure 7.4 A schematized muscle spindle: *(a)* Intrafusal fibers are enclosed in a capsule and are innervated by both afferent (Ia and II) and efferent (gamma) axons; *(b)* distribution of innervation received by the nuclear chain and nuclear bag (dynamic and static) fibers.

Adapted, by permission, from K. Pearson and J. Gordon, 2000, Spinal reflexes. In *Principles of neural science,* 4th ed., edited by E.R. Kandel, J.H. Schwartz, and T.M. Jessel (New York: The McGraw-Hill Companies), 719. Adapted by permission of The McGraw-Hill Companies, Inc.

same amount of afferent feedback to the CNS (Pearson, 2000; Prochazka, 1996; Proske, 1997). Rather, the CNS prescribes the amount of afferent feedback that it needs during a particular movement, and the gamma motor neurons are activated to set the level of sensitivity of the muscle spindles (Kakuda et al., 1996; Prochazka, 1989; Taylor, 2002).

Figure 7.5 provides an example of how the gamma motor neurons can influence the feedback from a muscle spindle. The data show the effect of muscle stretch on the discharge rate of action potentials in a group Ia afferent from a muscle spindle (Pearson, 2000). In the baseline condition (figure 7.5*b*), discharge rate is initially low, increases during the muscle stretch (figure 7.5*a*), and then settles down at a new steady-state rate at the new muscle length. The change in discharge rate reflects two effects: one due to the velocity of the stretch and the other due to the change in muscle length (Cordo et al., 2002; Jones et al., 2001; Kakuda & Nagaoka, 1998). When a static gamma motor neuron was stimulated during the same amount of muscle stretch (figure 7.5*c*), the Ia discharge increased the amount of information about the change in muscle length. In contrast, stimulation of a dynamic gamma motor neuron increased the responsiveness of the Ia afferent to the velocity of the stretch (figure 7.5*d*).

By varying gamma motor neuron activity, the CNS can alter the amount and type of feedback it receives from muscle spindles. At one extreme, for example, an absence of gamma motor neuron activity might make it possible for a muscle to be stretched and for the CNS to receive no feedback from the muscle spindles about the change in length. The normal mode of operation, however, involves concurrent activation of the alpha and gamma motor neurons. This is known as **alpha-gamma coactivation;** the alpha motor neurons activate the extrafusal fibers to produce the force required for the task, and the gamma motor neurons activate the intrafusal fibers to set the desired level of feedback from the muscle spindles. Alpha-gamma coactivation enables the CNS to maintain the sensitivity of the muscle spindle even when the muscle (extrafusal fibers) changes length.

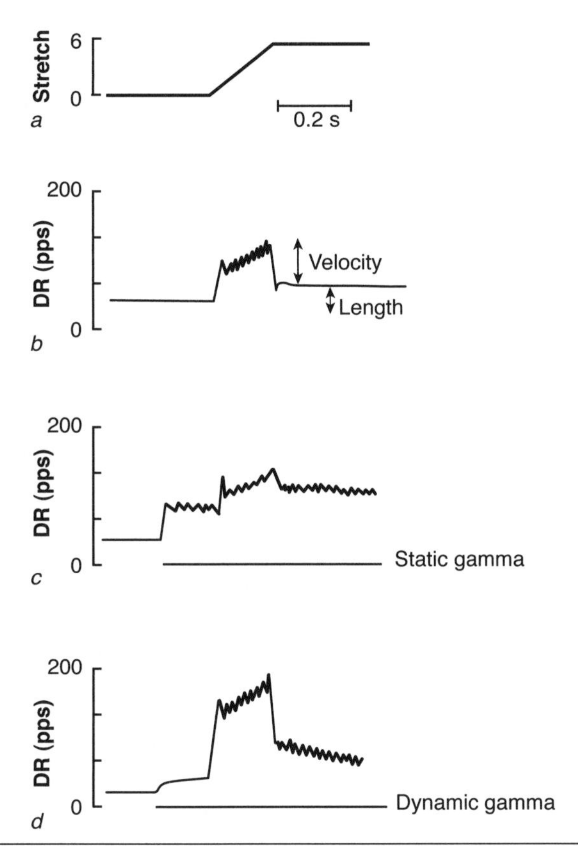

Figure 7.5 The response to muscle stretch of a group Ia afferent belonging to a single muscle spindle. *(a)* The magnitude of the stretch imposed on the muscle. *(b)* The increase in discharge rate (DR) was related to the magnitude and velocity of the stretch. *(c)* Stimulation of a static gamma motor neuron that innervated the muscle spindle increased the responsiveness of the Ia afferent to the change in muscle length. *(d)* Stimulation of a dynamic gamma motor neuron increased responsiveness to the velocity of the stretch.

Adapted, by permission, K. Pearson and J. Gordon, 2000, Spinal reflexes. In *Principles of neural science,* 4th ed., edited by E.R. Kandel, J.H. Schwartz, and T.M. Jessel (New York: The McGraw-Hill Companies), 719. Adapted by permission of The McGraw-Hill Companies, Inc.

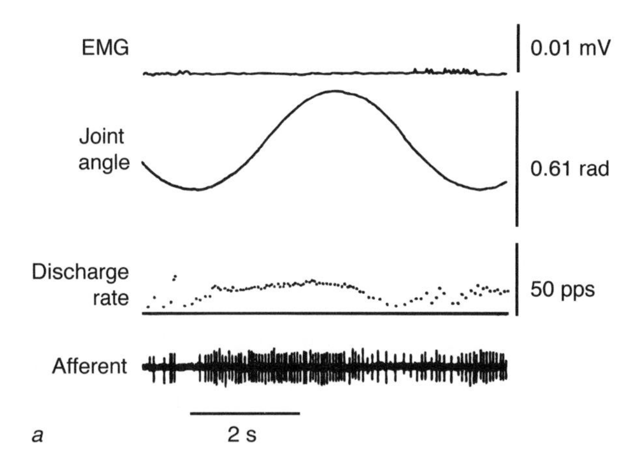

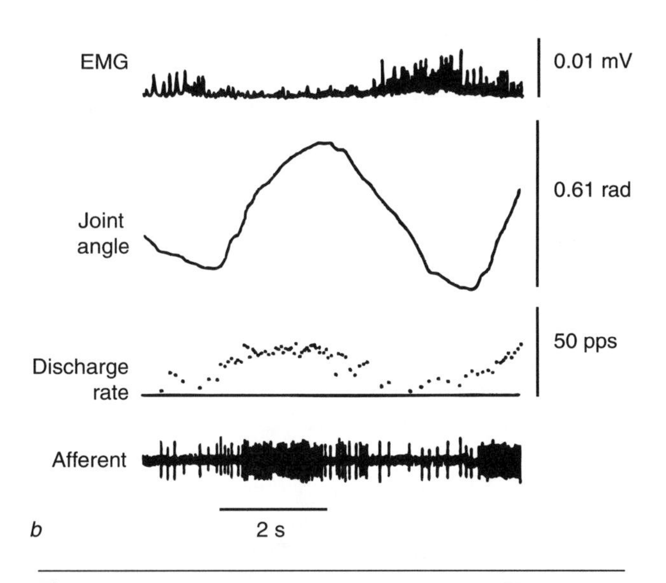

Figure 7.6 Discharge of a human muscle spindle during *(a)* a passive and *(b)* an active stretch (upward joint angle) of a forearm muscle.

Adapted, by permission, from N.A. Al-Falahe, M. Nagaoka, and A.B. Vallbo, 1990, "Response profiles of human muscle afferents during active finger movements," *Brain* 113: 339. Copyright 1990 by Oxford University Press.

EXAMPLE 7.1

Muscle Spindle Activity in Humans

Figure 7.6 shows examples of muscle spindle activity (Al-Falahe et al., 1990). The experiment measured the discharge of an afferent from a single muscle spindle in the extensor digitorum muscle in the forearm of a human subject while the finger was moved sinusoidally either by the experimenter (figure 7.6*a*) or by the subject (figure 7.6*b*). To record the afferent discharge, a probe (an electrode) was placed in the muscle nerve; this technique is known as **microneurography** (Pierrot-Deseilligny & Burke, 2005; Vallbo et al., 2004).

The bottom trace in each panel shows the action potentials discharged by the muscle spindle afferent during the two conditions. The second-to-bottom trace shows the discharge rate of the afferent, which reached peak values of 25 pps. The third trace from the bottom shows the angle of the metacarpophalangeal joint, with an upward deflection indicating flexion and stretch of the extensor digitorum muscle. The top trace shows the electrical activity (electromyogram; EMG) in the extensor digitorum muscle; the muscle was electrically silent when the movement was imposed by the investigator but was active when the subject performed the movement. These data show that for both types of movement, imposed and active, the afferent discharge increased when the extensor digitorum muscle was stretched during the flexion (upward) phase of the movement.

The axons of the group Ia and II afferents enter the spinal cord through the dorsal roots (figure 7.2) and make connections with many different classes of neurons. The branches of the afferent axons make connections at the same level as they enter the spinal cord and contribute to the ascending tracts in the white matter of the spinal cord. With the exception of the Ia afferent, all afferents that influence the motor neurons do so through pathways that involve one or more interneurons. There is substantial convergence of input from central and peripheral sources onto these interneurons; as a result, a given input produces a range of responses in a target motor neuron (Hultborn, 2001; Nielsen, 2004). Even for the Ia afferent, which makes direct connections with motor neurons, the transfer of information from the muscle spindle afferent to the motor neuron can be modulated by an interneuron that can exert a presynaptic effect on the synapse (figure 5.20).

The Ia afferent makes monosynaptic excitatory connections with motor neurons that innervate the same muscle from which the afferent originated and weaker monosynaptic excitatory connections with motor neurons that innervate synergist muscles (Eccles et al., 1957; Eccles & Lundberg, 1958; Hongo et al., 1984). The connections to the same muscle are known as **homonymous** connections, whereas those to synergists are referred to as **heteronymous** connections. The Ia afferents from one muscle can also make monosynaptic excitatory connections onto motor neurons that innervate muscles at other joints, which suggests that these inputs may contribute to the coordination of muscle activity within a limb. As with all other afferents, the Ia afferent synapses onto many different interneurons. One well-described pathway involves a synapse between a Ia afferent and an interneuron that connects to motor neurons innervating an antagonistic muscle (Tanaka, 1974). Because this pathway evokes inhibitory postsynaptic potentials in the motor neurons of the antagonist muscle, the interneuron is known as the **Ia inhibitory interneuron.**

The actions of the group II afferents on motor neurons are less well known (Pierrot-Deseilligny & Burke, 2005). There are monosynaptic connections from the group II afferents onto homonymous motor neurons, but these are less common than the connections through pathways that involve one or more interneurons. Group II afferents can evoke opposite effects in motor neurons that innervate flexor and extensor muscle: excitatory postsynaptic potentials in the flexor motor neurons and inhibitory postsynaptic potentials in the extensor motor neurons (Lundberg et al., 1977). Due to the convergence of afferent input onto interneurons, there is considerable overlap in the influence of group II afferents from the muscle spindle, joint receptors, and haptic receptors on motor neurons.

Tendon Organ

In contrast to the muscle spindle, the tendon organ is a relatively simple sensory receptor; it includes a single afferent and no efferent connections (Hasan & Stuart, 1984; Jami, 1992; Pearson, 2000). Tendon organs range in length from 0.2 to 1 mm. Few tendon organs are located in the tendon proper. Most are arranged around a few extrafusal muscle fibers as these connect with an aponeurosis of attachment (figure 7.3*a*). Because of this location, the tendon organ is described as being in series with skeletal muscle fibers. The sensory terminal of the afferent neuron is contained within a capsule, and it branches to encircle several strands of collagen in a myotendinous junction (figure 7.7). It is estimated that there are about 10 to 20 skeletal muscle fibers in a typical tendon organ capsule and that each of these muscle fibers is innervated by a different alpha motor neuron (figure 7.3*b*). Each motor unit can engage one to six tendon organs. The afferent axon associated with the tendon organ is referred to as the Ib afferent (table 7.1).

When a muscle and its connective tissues are stretched, either through pulling of the muscle (passive stretch) or through activation of the skeletal muscle fibers (active stretch), the strands of collagen pinch and excite the Ib afferent (inset in figure 7.7). Because the tendon organ is activated in this way, it provides information about muscle force. The level of force necessary to excite the tendon organ depends on the mode of activation. Passive stretch requires a muscle force of 2 N, whereas the activation of

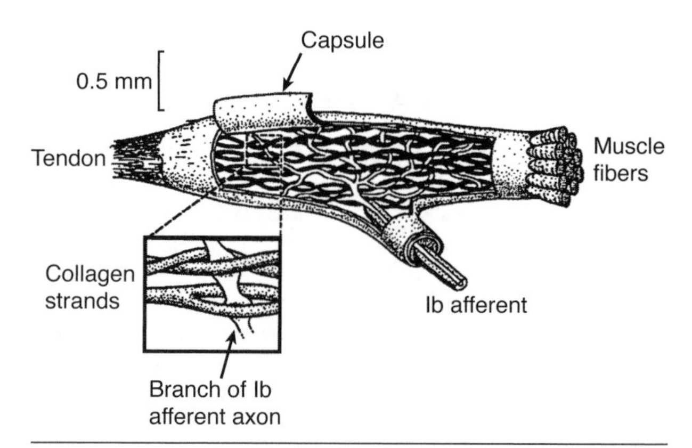

Figure 7.7 The tendon organ and its afferent axon.

Adapted, by permission, from A. Prochazka, 1996, Proprioceptive feedback and movement regulation. In *Handbook of physiology sect 12: Exercise,* edited by L.B. Rowell and J.T. Shepherd (Bethesda, MD: American Physiological Society), 94.

a single muscle fiber (30-90 μN) is sufficient during an active stretch (Binder et al., 1977).

EXAMPLE 7.2

Tendon Organ Activity in Humans

An example of the discharge of a tendon organ is shown in figure 7.8 (Al-Falahe et al., 1990). The microneurography technique was used to record the discharge of a single tendon organ located in the extensor digitorum muscle of a human volunteer. The electrode recorded action potentials as they propagated along the Ib afferent. The subject performed a finger movement against a zero load (figure 7.8*a*) and against a light load (figure 7.8*b*). In each panel, the top trace records the angle of the metacarpophalangeal joint, with flexion of the joint and stretch of the muscle shown as an upward deflection. The middle trace indicates the discharge of the tendon organ, which achieved peak rates of about 40 pps. The bottom trace shows the EMG (electrical activity) of the muscle performing the finger movement. Figure 7.8 shows how the discharge of a tendon organ parallels the EMG. Because there is a close association between muscle EMG and force under these conditions, these data indicate that tendon organ discharge monitors the force exerted by the muscle (Crago et al., 1982; Gregory et al., 2002; Petit et al., 1997; Prochazka & Gorassini, 1998; Taylor, 2002).

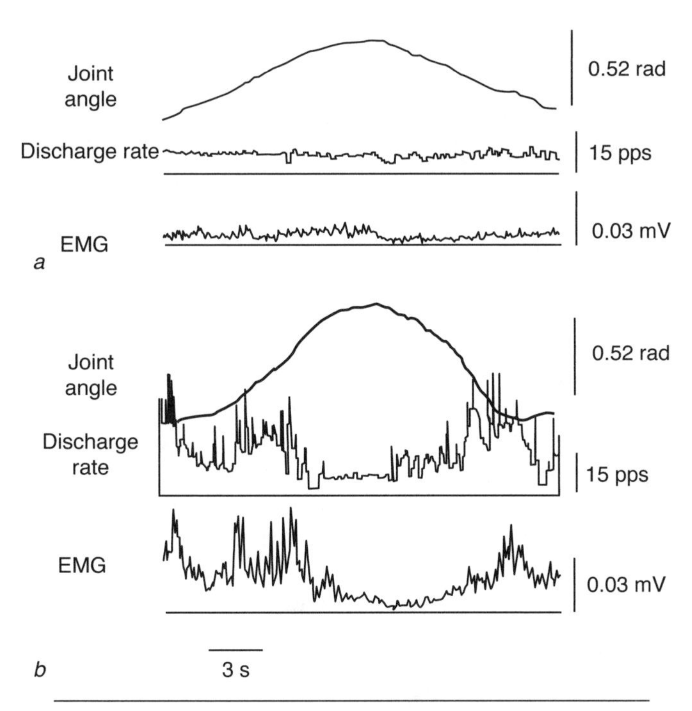

Figure 7.8 Discharge of a human tendon organ during finger movements against *(a)* a zero load and *(b)* a light load.

Adapted, by permission, from N.A. Al-Falahe, M. Nagaoka, and A.B. Vallbo, 1990, "Response profiles of human muscle afferents during active finger movements," *Brain* 113: 339. Copyright 1990 by Oxford University Press.

The Ib afferents enter the spinal cord through the dorsal roots and make connections onto motor neurons through pathways that involve one or two interneurons. The pathway that involves a single interneuron evokes inhibitory postsynaptic potentials in homonymous motor neurons. The pathway that involves two interneurons evokes excitatory postsynaptic potentials in motor neurons that innervate antagonist muscles (Eccles et al., 1957). The effects of the Ib afferent are more widespread than those of the Ia afferent and can be increased by descending input from the brain stem onto the single interneuron in the inhibitory pathway. Because the Ia afferents also synapse with this interneuron, the pathway transmits inhibitory input to homonymous motor neurons from both Ia and Ib afferents. The pathway is known as the **nonreciprocal group I inhibitory pathway,** with "nonreciprocal" referring to actions evoked in homonymous and heteronymous motor neurons and not antagonist muscles.

Joint Receptors

In contrast to the muscle spindle and the tendon organ, joint receptors are not well-defined entities. Nonetheless, joint receptors provide feedback that is critical for the control of movement (Burgess et al., 1987; Gandevia, 1996). Joint receptors vary in location (e.g., joint capsule, ligament, loose connective tissue) and type (e.g., Ruffini ending, Golgi ending, Pacinian corpuscle, free nerve ending). These receptors are innervated by smaller-diameter axons: II, III, and IV afferents.

The **Ruffini endings** typically consist of two to six thinly encapsulated, globular corpuscles with a single myelinated axon that has a diameter of 5 to 9 μm. These receptors may be categorized as static or dynamic mechanoreceptors and are capable of signaling joint position and displacement, angular velocity, and intra-articular pressure (Johansson et al., 1991). **Pacinian corpuscles** are thickly encapsulated, with an axon of 8 to 12 μm diameter. These receptors have low thresholds to mechanical stress and apparently detect acceleration of the joint (Bell et al., 1994). **Golgi endings** are thinly encapsulated, fusiform corpuscles that are similar to tendon organs. The afferent axon has a diameter of 13 to 17 μm. These receptors, which have high thresholds, monitor tension in ligaments, especially at the ends of the range of motion. **Free nerve endings** are widely distributed and constitute the joint nociceptive system. They have small-diameter axons (0.5-5 μm) and are active when a joint is subjected to an abnormal stress or to chemical agents.

These four types of joint receptors provide the CNS with information about position, displacement, velocity, and acceleration of movement as well as about noxious stimuli experienced by the joint (Dyhre-Poulson & Krogsgaard, 2000; Grigg, 2001; Hogervorst & Brand, 1998; Johansson et al., 1991; Sjölander et al., 2002). Although joint receptors do not provide unambiguous feedback about joint position (Gandevia, 1996), those located in distinct anatomical regions, as in the knee joint, are capable of providing unique afferent information (Edin, 2001; Zimny & Wink, 1991).

The significance of joint receptors for the control of movement has been convincingly demonstrated by the effect of joint pathology on muscle activation. For example, when normal subjects are given a large experimental effusion in the knee joint (fluid injected into the joint space), the ability to activate the quadriceps femoris muscle is greatly reduced, even in the absence of pain (Stokes & Young, 1984; Young et al., 1987). This inhibition of muscle activation depends on the volume of fluid added but can cause a 30% to 90% reduction in the maximal voluntary activation of quadriceps femoris. Conversely, the removal of fluid from the knee joint, for example after a meniscectomy, can markedly improve a patient's use of the muscle. In the absence of pain, chronic joint effusion can cause weakness and atrophy of muscle. Similarly, patients with old cruciate ligament tears typically exhibit a decrease in strength of both the quadriceps femoris and hamstrings (Grabiner et al., 1992; Johansson et al., 1991), and even reconstruction of a torn ligament can produce a prolonged deficit in function (DeVita et al., 1998; Holder-Powell et al., 2001).

The afferent axons arising from joint receptors distribute feedback extensively throughout the CNS, from the spinal cord up to the brain (Sjölander et al., 2002). Stimulation of afferents from an anterior cruciate ligament, for example, can evoke responses in the somatosensory cortex of a patient during a surgical procedure (Pitman et al., 1992). At the spinal level, joint receptors appear to project to both gamma and alpha motor neurons. One study showed that afferents arising from the knee joint could evoke short- and long-latency effects in gamma motor neurons that innervate leg muscles (Ellaway et al., 1996). The short-latency effects can be excitatory or inhibitory, whereas the long-latency effect is excitatory. The relatively long latencies suggested multiple synapses between the afferent input and the gamma motor neuron. By projecting to gamma motor neurons, however, joint receptors are able to influence the feedback from muscle spindles into the CNS.

The projections to alpha motor neurons can also produce mixed postsynaptic effects, which vary with the sensory receptor that is activated. The feedback from ligament afferents can have a potent effect on alpha motor neurons. For example, electrical stimulation of the posterior cruciate ligament inhibited ongoing muscle activity in both the quadriceps femoris and hamstring muscles of humans at a latency that suggested a more direct connection to alpha motor neurons (Fischer-Rasmussen et al., 2001). In general, however, feedback from joint receptors is spread to a number of ascending and reflex pathways where it mixes with feedback from the skin and muscles. Furthermore, it is likely that most individual joint receptors are capable of responding to several different stimuli and that the interpretation by the CNS is based on the feedback it receives from the population of activated sensory receptors (Sjölander et al., 2002).

Cutaneous Mechanoreceptors

Unlike the muscle spindle, tendon organ, and joint receptors, cutaneous mechanoreceptors provide information exclusively about external mechanical events. Cutaneous mechanoreceptors provide information about acceleration of the skin and deeper tissues; movement of hair; and displacement, stretch, and indentation of skin. Five types of receptors provide this feedback: Merkel disks, Meissner corpuscles, Ruffini endings, Pacinian corpuscles, and free nerve endings (Horch et al., 1977; Iggo & Andres, 1982). The **Merkel disk** is sensitive to local vertical pressure and does not respond to lateral stretch of the skin. This receptor is innervated by a large myelinated axon and responds with a rapid initial discharge of action potentials that is quickly reduced to a slow steady rate. Merkel disks are sensitive to edges and contribute significantly to tactile discrimination. **Meissner corpuscles** are innervated by two to six axons, and each axon may innervate more than one corpuscle. The Meissner corpuscle is sensitive to local, maintained pressure, but its response (discharge of action potentials) fades rapidly. Ruffini endings are innervated by a large myelinated axon and respond to stretch of the skin over a wide area. The endings are sensitive to the direction of the stretch; an ending will be excited by stretch in one direction and inhibited by stretch at a right angle to the preferred direction. The response of the Ruffini endings adapts slowly to a sustained stretch. Pacinian corpuscles, the largest receptors in the skin, are innervated by a single axon and are located deep in the dermis and the subcutaneous tissues. They have large receptive fields and detect rapid changes in pressure, responding mainly to the acceleration component of a stimulus. Free nerve endings provide feedback on movement of the skin and hair.

The cutaneous mechanoreceptors provide the CNS with information about the magnitude, speed, and direction of a mechanical stimulus. The afferents enter the spinal cord through the dorsal roots and connect to local interneurons and to an ascending tract known as the dorsal column. The dorsal column axons project principally to the contralateral thalamus and then to the cerebral cortex. Through this pathway, feedback from cutaneous mechanoreceptors can influence the excitability of the motor cortex (Maertens de Noordhout et al., 1992). At the spinal level, feedback by single afferents can alter the discharge of motor neurons, despite the presence of interneurons in the pathway from the cutaneous mechanoreceptors to the motor neurons (Fallon et al., 2005; McNulty & Macefield, 2001). Feedback from cutaneous mechanoreceptors, therefore, is distributed widely throughout the CNS and can contribute to a range of effects.

Reflex Pathways

The afferent feedback from sensory receptors to the CNS can evoke rapid responses that are known as reflexes (Pierrot-Deseilligny & Burke, 2005). Many of these pathways have been well defined and are used to provide information about the contribution of sensory feedback to the performance of a task (Brooke & Zehr, 2006; Hultborn, 2006; Nielsen, 2004; Stein & Thompson, 2006; Windhorst, 2007). This section describes some of the pathways that are studied for this purpose.

The Monosynaptic Reflex

The first reflex used to study spinal pathways was the monosynaptic projection of Ia afferents to homonymous motor neurons. The reflex can be evoked by application of a single electrical shock (stimulus) to a peripheral nerve, which evokes a twitch response in the muscle innervated by the nerve. The response can be measured as the EMG or the force associated with the twitch (figure 7.9). This electrically evoked monosynaptic response is known as the **Hoffmann reflex,** or H reflex (Hoffmann, 1918, 1922; Stein & Thompson, 2006). When a subject is at rest, the H reflex is most commonly studied in soleus, but it can be elicited in such other muscles as quadriceps femoris, hamstrings, tibialis anterior, and flexor carpi radialis. For example, one can evoke it in quadriceps femoris by laying a subject in a supine position, placing the stimulating electrode over the femoral nerve just below the inguinal ligament (figure 7.9*a*), and gradually increasing the stimulus intensity until a response is detected (figure 7.9*c*). The H reflex, however, can be recorded in most limb muscles with accessible peripheral nerves during a weak voluntary contraction. Because the H reflex requires the selective activation of the Ia afferents, the correct placement of the electrodes is critical and can be difficult to achieve in some subjects.

The pathway underlying the H reflex is shown in figure 7.10. It involves the activation of Ia afferents and the subsequent generation of action potentials in homonymous motor neurons. The electrical stimulus, however, likely activates both Ia and Ib afferents. The synchronous volley of action potentials generated by the stimulus is propagated centrally (2 in figure 7.10) to elicit postsynaptic potentials in the homonymous motor neurons. The pathway is monosynaptic (Magladery et al., 1951). The size of the excitatory postsynaptic potential evoked in the motor neurons is greatest in the smallest motor neurons and decreases with motor neuron size, which results in the activation of motor neurons according to the size principle (Buchthal & Schmalbruch, 1970; Trimble & Enoka, 1991). The probability that a motor neuron will discharge an action potential in response to Ia input depends on the amplitude of the excitatory postsynaptic potential and the difference between its voltage threshold and resting membrane potential. Because of this relation, the H reflex was originally used to assess the excitability of the motor neuron pool.

The capacity of the electrical stimulus to generate action potentials in the axons of a peripheral nerve depends on the distribution of the current and on the diameter of each axon. Largest-diameter axons are activated with the least amount of current. Because the diameter of Ia afferents is slightly larger than that of the

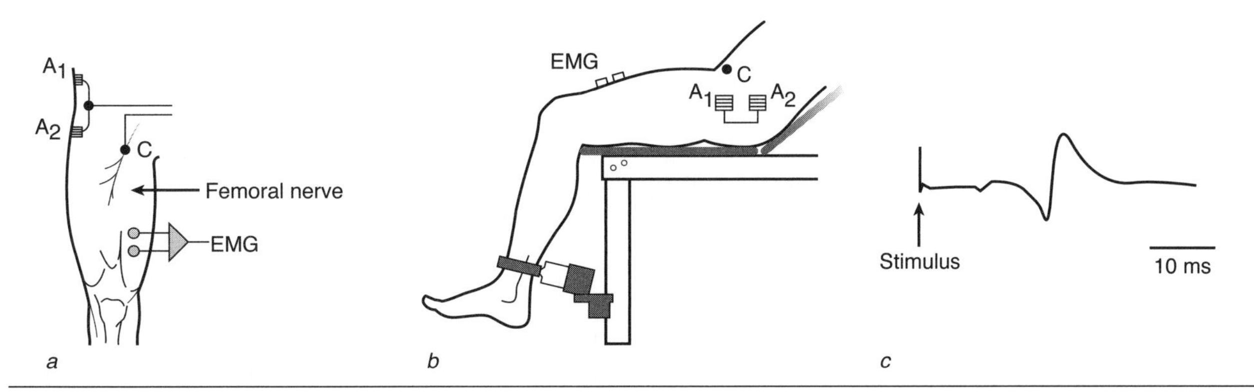

Figure 7.9 A Hoffmann (H) reflex elicited in quadriceps femoris: *(a)* location of the stimulating electrodes (C, cathode; A_1 and A_2, two anodes that are connected together); *(b)* position of the subject; *(c)* H reflex recorded as an EMG over vastus medialis.

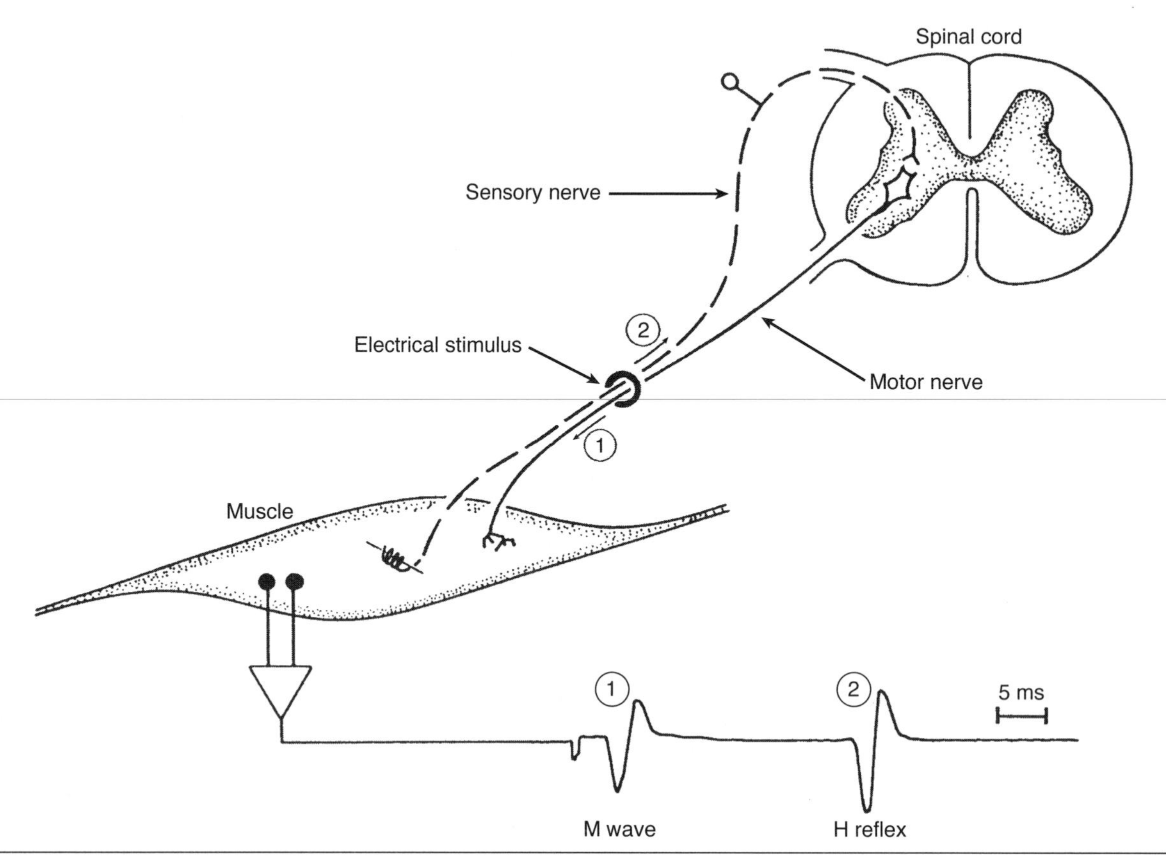

Figure 7.10 Pathway for the H reflex. The electrical stimulus is applied over the peripheral nerve, and the response is measured in the muscle innervated by the nerve.

axons belonging to alpha motor neurons, the Ia afferents can be activated selectively at low stimulus intensities. With an increase in the stimulus intensity, the action potentials generated in the efferent axons will evoke a short-latency (5 ms) response called the **M wave** (1 in figure 7.10). The M wave is elicited experimentally to probe the integrity of the pathway between the site of the stimulus (muscle nerve) and the recording site (muscle fiber action potentials); that is, the M wave tests the integrity of neuromuscular propagation (Bigland-Ritchie et al., 1982; Fuglevand et al., 1993).

The H reflex and M waves can interact with each other. When an action potential is generated in an efferent axon, the action potential is propagated both toward the neuromuscular junction and back toward the soma of the motor neuron. The action potential propagating in the **orthodromic** direction (toward the neuromuscular junction) will produce an M wave, whereas the action potential propagating in the **antidromic** direction (toward the motor neuron) will invade the motor neuron and reduce its responsiveness to incoming Ia afferent input (Pierrot-Deseilligny & Burke, 2005; Stein & Thompson, 2006). These interactions are typically characterized by measurement of the **recruitment curve** for H reflexes and M waves. This involves measuring the amplitude of the H reflex and M wave in response to a range of stimulus intensities. Figure 7.11 shows that the size of the H reflex is greatest when the M wave is small and that the size of the M wave increases with stimulus intensity as additional efferent axons are activated. Maximal stimulation provides an estimate of the response of the entire motor unit pool.

A frequently used method of examining the responsiveness of the motor neuron pool is to precede the Ia input with input onto the motor neurons either over another afferent pathway or over a descending pathway. This initial input is known as a **conditioning stimulus.** Conditioning stimuli can be applied to pathways that elicit either excitatory or inhibitory postsynaptic potentials in the test motor neurons. By varying the time between the conditioning and test stimuli and examining the size of the effect, it is possible to estimate the number of synapses in the pathway that deliver the conditioning stimulus. For example, a 4.8 ms interval between a conditioning stimulus (electrical shock) applied to the femoral nerve and the test stimulus (H reflex) delivered to the nerve innervating soleus suggested that a monosynaptic pathway was responsible for the observed heteronymous facilitation of the soleus H reflex from quadriceps femoris (Crone et al., 1990; Hultborn et al., 1987*a*).

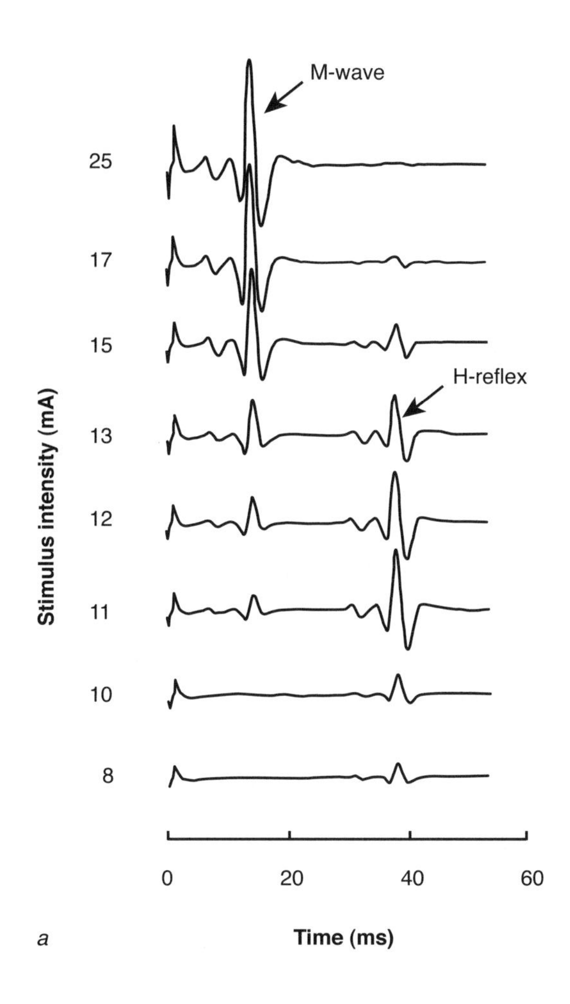

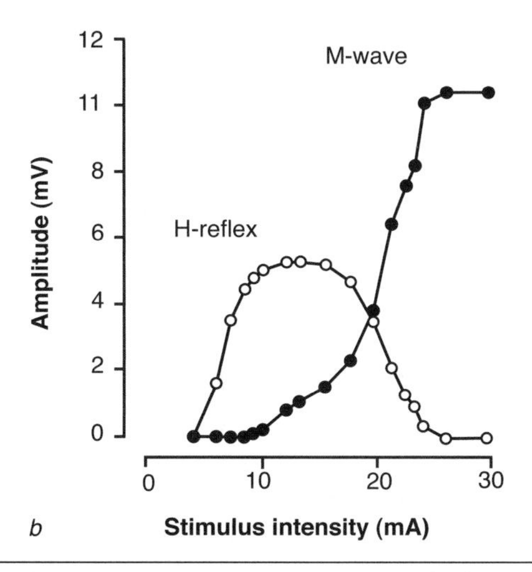

Figure 7.11 Recruitment curve of the H reflex and M wave in soleus. *(a)* Responses to eight stimulus intensities. *(b)* Variation in the amplitude of the H reflex and M wave with change in stimulus intensity.

Data provided by Jacques Duchateau, Ph.D.

The monosynaptic reflex can also be evoked with a mechanical stimulus in the form of a brief tap applied to the muscle tendon (Lloyd, 1943). The response is known as the **tendon tap reflex** (Stein & Thompson, 2006). This reflex can be elicited in a number of muscles: soleus, quadriceps femoris, semitendinosus, biceps brachii, triceps brachii, flexor carpi radialis, semitendinosus, extensor carpi radialis, and masseter. A major difference between the two monosynaptic reflexes is that the sensory stimulus for the tendon jerk is a stretch of the muscle spindle, whereas the sensory feedback for the H reflex is generated artificially with an electrical stimulus that bypasses the muscle spindle. Consequently, the amplitude of the tendon jerk reflex depends on both the responsiveness of the motor neuron pool and the sensitivity of the muscle spindles as established by the gamma motor neurons.

Since the original discovery of the monosynaptic reflex, it has become evident that the size of the response depends on additional factors besides the responsiveness of the motor neuron pool: modulation of the afferent volley and the distribution of the input to the motor neuron pool. The feedback provided by the afferent axons can change due to hyperpolarization of the active axons by presynaptic inhibition of the Ia terminals as they synapse onto motor neurons. Presynaptic inhibition reduces the release of neurotransmitter at the Ia synapse and thereby decreases the size of the H reflex (Pierrot-Deseilligny & Burke, 2005). Furthermore, protocols that involve a comparison of responses evoked by conditioning and test stimuli can be confounded by differences in the distribution of afferent input to the motor neuron pool over the two pathways. For example, more cutaneous input is distributed to larger motor neurons, which contrasts with the preferential distribution of Ia input to smaller motor neurons (Nielsen & Kagamihara, 1993). By measuring the influence of afferent input on the discharge of single motor units (example 7.3), however, it is possible to assess the magnitude of these effects (Stephens et al., 1976). Nonetheless, the H reflex remains a useful method to examine transmission in spinal pathways during motor performance (Zehr, 2006).

EXAMPLE 7.3

Poststimulus Time Histograms of Motor Unit Discharge

A **poststimulus time histogram** is derived from the discharge of a single motor unit during a voluntary contraction. The purpose of the histogram is to quantify the influence of an imposed stimulus on the time between successive action potentials (interspike

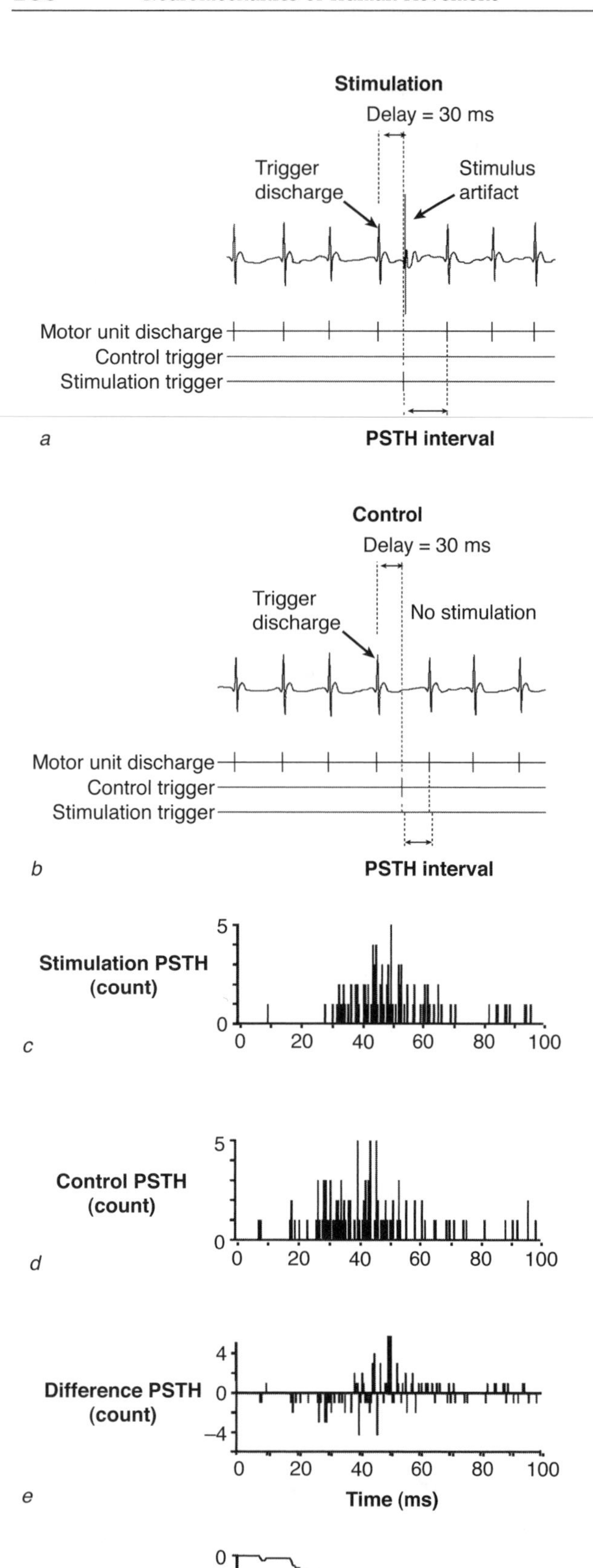

Figure 7.12 A poststimulus time histogram (PSTH) derived from the discharge of a motor unit in biceps brachii. The stimulus was an electric shock applied to the branch of the radial nerve that innervates brachioradialis. *(a)* The stimulus delayed the next appearance of the action potential (PSTH interval). *(b)* PSTH interval in the absence of stimulation (control). *(c)* PSTH obtained from ~100 stimuli. *(d)* PSTH derived from ~100 control triggers. *(e)* The difference between the stimulation PSTH and the control PSTH. *(f)* The stimulus-induced delay was evident as a depression in the Cumulative Sum derived from the Difference PSTH, which indicates that the afferent input evoked an inhibitory postsynaptic potential in the motor neuron.

Data provided by Benjamin K. Barry, Ph.D.

intervals); the technique is known as **spike-triggered stimulation.** The stimulus is delivered at some delay after an action potential discharged by an isolated motor unit, and the interspike intervals after the stimulus are plotted on the histogram. When the stimulus activates an afferent pathway that evokes an inhibitory postsynaptic potential in the motor neuron, the time to the next action potential is delayed because the membrane potential is moved away from the voltage threshold of the neuron (figure 7.12). Conversely, the generation of an excitatory postsynaptic potential will advance the time of the next action potential. The cumulative effect of multiple stimuli can be visualized as a depression (inhibition) or a facilitation (excitation) of the poststimulus time histogram when compared with a control histogram that involved no stimulation (Fournier et al., 1986).

Monosynaptic Ia Excitation

Ia afferents make widespread monosynaptic connections to both homonymous and heteronymous motor neurons (Eccles et al., 1957; Hultborn, 2006). Each Ia afferent synapses onto most of its homonymous motor neurons, but only to some motor neurons that innervate synergist muscles. For example, Ia afferents from the soleus muscle have monosynaptic projections to homonymous (soleus) motor neurons, to heteronymous motor neurons supplying a close synergist (medial gastrocnemius), and to motor neurons that innervate quadriceps femoris (Meunier et al., 1993). Tables 7.2 and 7.3 list the distribution of monosynaptic Ia excitation to heteronymous motor neurons in the human leg and arm (Pierrot-Deseilligny & Burke, 2005). These distributions indicate the presence of monosynaptic Ia excitation between muscles that span different joints in the human leg and widespread projections from distal to proximal muscles in the human arm.

Table 7.2 Distribution of Monosynaptic Heteronymous Ia Excitation From Six Nerves (Columns) to Seven Muscles in the Human Leg

	Soleus	Medial gastrocnemius	Superficial peroneal	Deep peroneal	Femoral	Tibial
Soleus	H	–	+	–	+	+
Medial gastrocnemius	+	H	+	+	+	+
Peroneus brevis	+	–	H	+	+	+
Tibialis anterior	–	–	–	H	+	+
Quadriceps femoris	+	+	–	+	H	+
Biceps femoris	+	+	+	–	–	+
Semitendinosus	+	+	+	+	–	+

H = homonymous monosynaptic Ia excitation; + = presence of heteronymous monosynaptic I excitation; – = absence of heteronymous monosynaptic I excitation.

Adapted from Pierrot-Deseilligny and Burke 2005.

Table 7.3 Distribution of Monosynaptic Heteronymous Ia Excitation From Six Nerves (Columns) to Ten Muscles in the Human Arm

	Musculocutaneous	Triceps brachii	Median	Radial	Ulnar	Median + ulnar
Deltoid	+	+	–	+	–	?
Biceps brachii	H	–	+	+	–	+
Triceps brachii	–	H	+	+	–	+
Flexor carpi radialis	–	–	H	–	+	+
Extensor carpi radialis	–	–	–	H	–	+
Flexor carpi ulnaris	–	–	+	?	H	+
Extensor carpi ulnaris	–	–	?	?	?	+
Flexor digitorum superficialis	–	?	?	?	?	+
Extensor digitorum	–	?	?	?	–	+
Hand	?	?	?	?	?	H

H = homonymous monosynaptic Ia excitation; + = presence of heteronymous monosynaptic I excitation; – = absence of heteronymous monosynaptic I excitation; ? = not known.

Adapted from Pierrot-Deseilligny and Burke, 2005.

When a muscle experiences a brief, unexpected increase in length (a stretch), the rapid response is known as the **stretch reflex** (Liddell & Sherrington, 1924; Sinkjær, 1997). An example of a stretch reflex is shown in figure 7.13. In this experiment, a human subject was grasping a handle that was unexpectedly displaced, resulting in a stretch of the elbow flexor muscles. The stretch reflex is indicated as the EMG elicited in the biceps brachii muscle. As figure 7.13 shows, the increase in EMG (response to stretch) begins soon after the onset of handle displacement (stimulus). The stretch reflex consists of at least two components (Matthews, 1991). One component is the short-latency response (M1) that is produced by homonymous Ia excitation. The second component (M2) has a longer latency and a more complicated origin that may involve the motor cortex in the brain (Lewis et al., 2006; Marsden et al., 1983*c;* Schieppati & Nardone, 1999). A third component (M3) is occasionally observed.

Figure 7.13 shows that the short- and medium-latency components of the stretch reflex preceded the earliest voluntary EMG (>120 ms), which underscores the ability

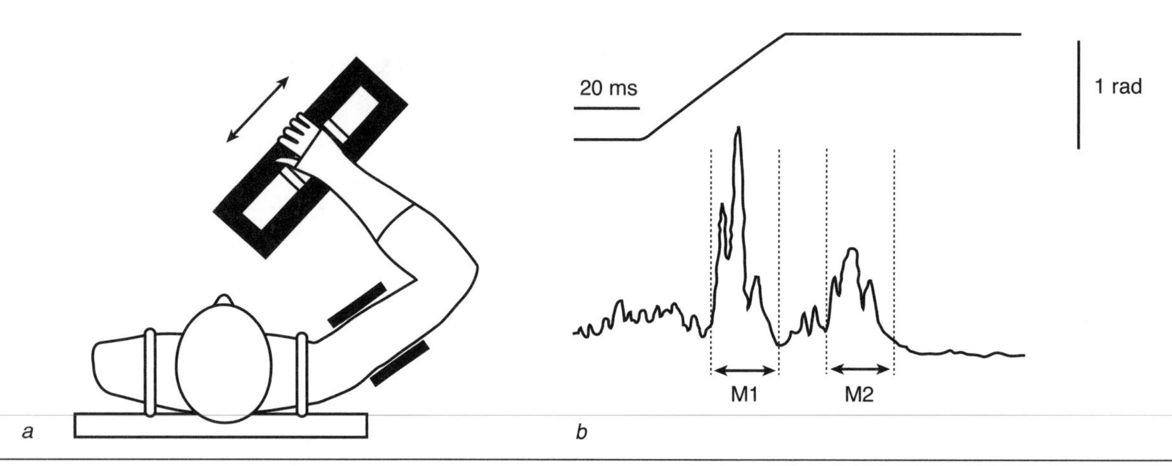

Figure 7.13 A stretch reflex that was elicited by an unexpected stretch of biceps brachii. *(a)* The subject was seated with the arm in a horizontal plane and the hand grasping a robotic actuator. *(b)* The subject maintained a weak contraction with the elbow flexor muscles before the muscle was unexpectedly stretched (~90°/s for 0.6 s). The bottom trace shows the short-latency (M1) and medium-latency (M2) responses in the biceps brachii.

of reflexes to provide rapid responses to perturbations. Although the latencies for the stretch reflex components can vary, the M1 component has a latency of about 30 ms and the M2 of around 50 to 60 ms, and the earliest voluntary EMG will begin at 170 ms. For the study shown in figure 7.13, the latency for M1 was 22 ms and that for M2 was 61 ms. When neurological disorders alter the stretch reflex (e.g., Parkinson's disease, hemiplegia, Huntington's disease, dystonia), the M2 component is usually affected (Hallet et al., 1994). Stretch reflexes have been observed in the leg muscles of humans during standing, walking, running, hopping, and landing (Avela et al., 1999; Diener et al., 1984; Dietz et al., 1979; Duncan & McDonagh, 2000; Grey et al., 2001; Voigt et al., 1998*b*), which indicates that feedback over the monosynaptic Ia pathway to homonymous motor neurons contributes to the control of these actions. Similarly, stretch reflexes in hand muscles are modulated during the performance of fine motor tasks (Xia et al., 2005).

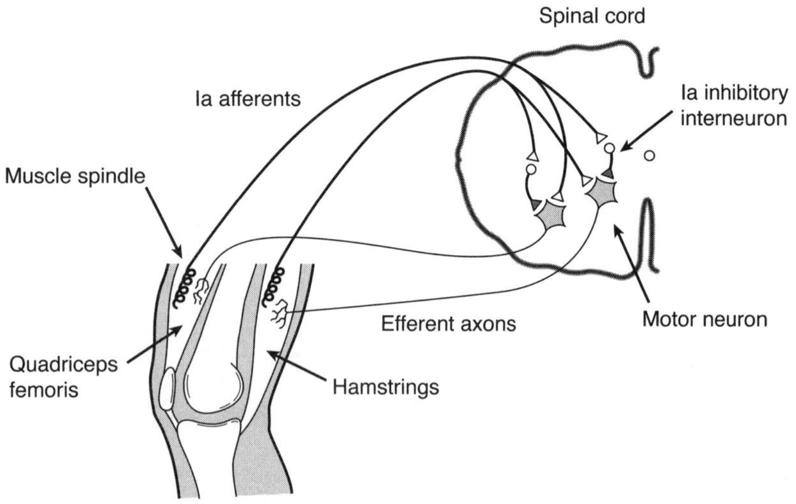

Because the reflexes evoked by monosynaptic Ia excitation are influenced by the excitability of the motor neuron pool and there is widespread distribution of this input, distant stimuli (e.g., loud and unexpected sounds) and remote muscle contractions can influence the amplitude of the response. For example, the H and stretch reflexes elicited in the soleus muscle increase in amplitude when the subject voluntarily activates other muscles, such as those involved in clenching the teeth (Zehr & Stein, 1999). The reflex enhancement involves central mechanisms (Burke et al., 1981). The use of remote muscle activity to increase the excitability of the motor neuron pool is referred to as the **Jendrassik maneuver** (Pereon et al., 1995). This effect has clinical applications. For example, a patient with weak leg muscles can be enabled to rise from a chair when a therapist provides resistance for the voluntary activation of arm and neck muscles by the patient, which can elicit a Jendrassik effect in the leg muscles.

Figure 7.14 Pathways responsible for the stretch and reciprocal Ia inhibition reflexes.

Reciprocal Ia Inhibition

The central connections of Ia afferents include an interneuron that evokes inhibitory postsynaptic potentials in the motor neurons that innervate antagonist muscles. This pathway is known as **reciprocal Ia inhibition** (Hultborn, 2006; Sherrington, 1897). The interneuron included in this disynaptic pathway is known as the Ia inhibitory interneuron (figure 7.14). Each Ia inhibitory interneuron projects to about 20% of the motor neurons in the pool and makes synaptic contacts on the soma and proximal dendrites (Burke et al., 1971). In general, all peripheral and descending inputs to motor neurons also project to Ia inhibitory interneurons (Jankowska & Lundberg, 1981). Furthermore, reciprocal Ia inhibition can be facilitated by feedback from cutaneous receptors and by input from descending systems. The function of the reciprocal Ia inhibition pathway is to link inhibition of an antagonist muscle to activation of an agonist muscle during movements that involve flexion and extension about a joint.

Reciprocal Ia inhibition is the most thoroughly studied spinal reflex pathway in humans. One can demonstrate it by determining the effect of an electric shock to the nerve innervating an antagonist muscle on the amplitude of an H reflex or on the discharge of single motor units in a test agonist muscle. These methods have indicated the presence of reciprocal Ia inhibition from the posterior tibial nerve to the tibialis anterior muscle, from the deep peroneal nerve to the soleus muscle, and between biceps and triceps brachii (Katz et al., 1991; Mao et al., 1984; Perez et al., 2003; Pierrot-Deseilligny et al., 1981). Disynaptic inhibition between the wrist flexors and extensors, however, does not seem to be mediated by reciprocal Ia inhibition (Aymard et al., 1995). Rather, this pathway appears to involve nonreciprocal group I inhibition.

Activation of the reciprocal Ia inhibition pathway during a voluntary contraction occurs in two phases. For example, about 50 ms prior to the onset of EMG in tibialis anterior, the amplitude of the H reflex in soleus is depressed. This suggests that the descending input to the motor neurons innervating the tibialis anterior muscle also activates the Ia inhibitory interneuron in the reciprocal Ia inhibition pathway from tibialis anterior to soleus. As the voluntary contraction of tibialis anterior progresses, the depression of the soleus H reflex increases, presumably due to afferent feedback from tibialis anterior that further activated the Ia inhibitory interneuron (Crone & Nielsen, 1989; Morin & Pierrot-Deseilligny, 1977). Reciprocal Ia inhibition declines during tasks that involve coactivation of agonist and antagonist muscles, increases during postural activity, and is modulated between ankle flexors and extensors during walking.

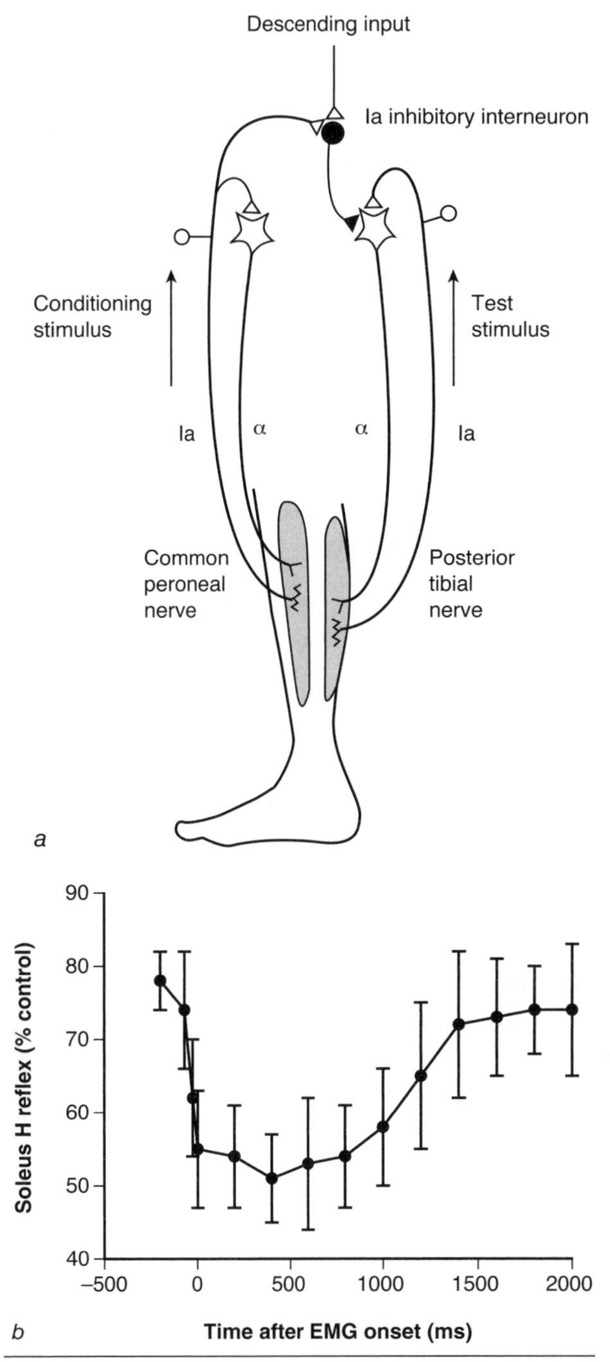

Figure 7.15 Changes in reciprocal Ia inhibition of soleus motor neurons during a voluntary contraction of the dorsiflexor muscles. *(a)* The conditioning stimulus was applied to the common peroneal nerve about 2 ms before an H reflex was evoked in soleus by a test stimulus applied to the posterior tibial nerve. *(b)* Depression of the soleus H reflex by reciprocal Ia inhibition during a ramp-and-hold contraction with the dorsiflexor muscles. The decline in the soleus H reflex began before the contraction and returned to resting values during the hold phase of the contraction. The first value for the soleus H reflex indicates the amount of depression due to reciprocal Ia inhibition in the resting muscle.

Part *b* adapted from Crone et al. 1987.

EXAMPLE 7.4

Reciprocal Ia Inhibition Is Modulated During a Contraction

By eliciting H reflexes in a pair of muscles during the course of an action, it is possible to monitor changes in the level of reciprocal Ia inhibition. Crone and colleagues (1987) used this strategy to track changes in reciprocal Ia inhibition from tibialis anterior to soleus during a voluntary contraction with tibialis anterior. The task was to increase the dorsiflexion torque from zero to a target of 6.4 N•m over a 600 ms interval and then maintain the target torque for 1.4 s. Prior to and during the voluntary contraction, a conditioning stimulus was applied to the common peroneal nerve about 2 ms before an H reflex was evoked in soleus by a test stimulus applied to the posterior tibial nerve (figure 7.15*a*). Stimulation of the Ia afferents in the common peroneal nerve has two effects: (1) It will activate the motor neurons that innervate tibialis anterior and evoke an H reflex, and (2) it will generate excitatory postsynaptic potentials in the Ia inhibitory interneurons that project to the soleus motor neurons. Consequently, the subsequent action potentials generated by the stimulus in the posterior tibial nerve will arrive at the soleus motor neurons while they are hyperpolarized by the inhibitory postsynaptic potentials evoked by the Ia inhibitory interneuron. The net effect of this interaction is a depression of the soleus H reflex due to reciprocal inhibition from tibialis anterior.

When the soleus muscle was at rest, the amplitude of the soleus H reflex was depressed by about 20% due to reciprocal inhibition (figure 7.15*b*). When the dorsiflexor muscles performed the ramp increase in torque, the amplitude of the soleus H reflex decreased, presumably due to an increase in the amount of reciprocal Ia inhibition from the common peroneal nerve. In combination with other inhibitory inputs to the soleus motor neurons, facilitation of reciprocal Ia inhibition reduces the possibility of evoking a stretch reflex in soleus during a voluntary contraction of the dorsiflexor muscles. As a consequence, the level of reciprocal inhibition differs between standing, walking, and running (Stein & Thompson, 2006).

Recurrent Inhibition

The flow of afferent feedback through the spinal cord is influenced by an interneuron known as a **Renshaw cell** (Renshaw, 1941). Renshaw cells are activated by branches of axons belonging to alpha motor neurons, by group II to IV afferents, and by descending pathways (Baldissera et al., 1981; Windhorst, 1996, 2007). Renshaw cells elicit inhibitory postsynaptic potentials in alpha motor neurons, Ia inhibitory interneurons, and gamma motor neurons. The connections to the alpha motor neuron and the Ia inhibitory interneuron are shown in figure 7.16. Because the branch of the alpha motor neuron axon that activates the Renshaw cell is known as a recurrent collateral, the inhibitory pathway to the Renshaw cell and back to the motor neuron is referred to as **recurrent inhibition.** Homonymous recurrent inhibition has been observed in many proximal arm and leg muscles of humans, but not in muscles of the foot and hand (Pierrot-Deseilligny & Burke, 2005). Heteronymous connections are more widespread in the human leg, between quadriceps femoris and all tested muscles at the ankle, for example, than in the human arm. Some muscles, such as tibialis anterior and medial gastrocnemius, share Ia excitation but not recurrent inhibition, which enables them to be coupled during various tasks. Recurrent inhibition of gamma motor neurons is much weaker than that for alpha motor neurons.

Each Renshaw cell is activated by many motor neurons, mainly from those innervating synergist muscles and not antagonist muscles. Although the strongest projections from Renshaw cells are to homonymous motor neurons and to the Ia inhibitory interneurons that receive the same Ia excitation as the alpha motor neurons, other motor neurons also receive significant inhibition (Eccles et al., 1961; Hultborn et al., 1971; Windhorst, 2007). Inhibition of the Ia inhibitory interneuron by the Renshaw cell is known as **recurrent facilitation** of the motor neurons supplying the antagonist muscle. Recurrent facilitation results in depression of reciprocal Ia inhibition (Baret et al., 2003), but descending input carried by the corticospinal tract can inhibit the Renshaw cell and thereby depress recurrent inhibition (Mazzocchio et al., 1994). Thus, recurrent inhibition can be modulated at several sites along its pathway.

Investigators test recurrent inhibition in humans by eliciting a pair of H reflexes, H1 and H′, with the same stimulating electrode. The two reflexes are elicited with the same submaximal stimulation, but they are separated by a supramaximal stimulus that evokes a maximal M wave. The H1 reflex represents a reference response, and the relative amplitude of H′ after the M wave denotes the amount of recurrent inhibition. Such comparisons have shown that recurrent inhibition is depressed during strong voluntary contractions, presumably due to inhibition of the Renshaw cell by descending input. Similarly, Renshaw cells are more inhibited at the same level of force during a dynamic contraction compared with a sustained contraction. In contrast, Renshaw cells are facilitated during weak voluntary contractions (Hultborn & Pierrot-Deseilligny, 1979; Iles & Pardoe, 1999). Renshaw cells are also facilitated during coactivation of antagonists, and both homonymous and heteronymous recurrent inhibition increase when subjects assume a standing posture (Barbeau et al., 2000; Nielsen & Pierrot-Deseilligny, 1996). The facilitation of recurrent inhibition during coactivation depresses reciprocal Ia inhibition and reduces the responsiveness of motor neurons to Ia afferent excitation.

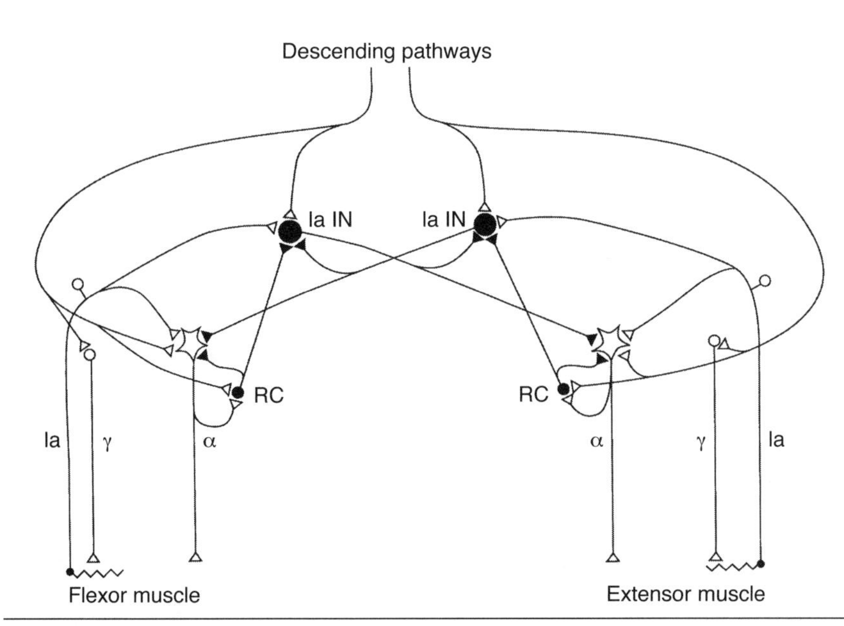

Figure 7.16 The pathways involving the Renshaw cell (RC) that produce recurrent inhibition of an alpha motor neuron (α) supplying an agonist muscle and recurrent facilitation of an alpha motor neuron that innervates an antagonist muscle. γ = gamma motor neuron; Ia = Ia afferent; Ia IN = Ia inhibitory interneuron.

Ib Pathways

Ib afferents originate from tendon organs (figure 7.7) and provide feedback on the force exerted by the muscle fibers in series with the tendon organ. Ib afferents project to interneurons, commonly identified as Ib interneurons, and then to motor neurons that innervate homonymous, synergist, and antagonist muscles. The input to homonymous and synergistic motor neurons is inhibitory, whereas that to antagonistic motor neurons is excitatory (Eccles et al., 1957). Ib afferents from many muscles often terminate on the same Ib interneurons and mutually facilitate one another (Jankowska, 1992).

Ib interneurons receive input from the periphery and from descending systems. The peripheral input includes direct connections from Ib and Ia afferents, with the Ib input being greater, and indirect connections through one or more interneurons from group II, cutaneous, and joint afferents. These inputs can both suppress (cutaneous afferents) and facilitate (cutaneous and joint afferents) transmission of Ib inhibition to motor neurons. The descending input comprises monosynaptic excitation from some pathways and monosynaptic inhibition from others. As a result of this extensive convergence of input onto the Ib interneuron, the inhibitory effects evoked in motor neurons during resting conditions can change with activity. For example, the Ib inhibition is reversed to Ib excitation during the stance phase of locomotion, which results in feedback from the tendon organ facilitating the muscle activity required to support body weight and to provide the propulsive force (Gossard et al., 1994; Hultborn, 2001; McCrea, 1998; Quevedo et al., 2000). Furthermore, the Ib input to the interneurons can be modulated by presynaptic inhibition that arises from the Ib afferents themselves and from a descending pathway. The presynaptic inhibition of Ib afferents by themselves contributes to the decline in inhibitory postsynaptic potentials in motor neurons during the course of a muscle contraction (Lafleur et al., 1993).

Because Ia and Ib afferents have similar diameters, they also have similar electrical thresholds. A stimulus applied to a peripheral nerve to elicit an H reflex, therefore, will also generate action potentials in Ib afferents. Consequently, motor neurons will experience first monosynaptic Ia excitation and then disynaptic Ib inhibition. One can quantify this effect by evoking a pair of H reflexes at various delays between the two stimuli; the first stimulus conditions the motor neuron pool, and the second stimulus tests its excitability. When a conditioning stimulus was applied to the inferior soleus nerve, subsequent test stimuli indicated that the amplitude of H reflexes in soleus and quadriceps femoris first increased, then decreased, and finally returned to control values. The depression of H-reflex amplitude, which was attributable to Ib inhibition, occurred at delays of 3 to 7 ms between the conditioning and test stimuli (Pierrot-Deseilligny et al., 1981). Similar effects have been observed from the medial gastrocnemius nerve to quadriceps femoris and biceps femoris, from the femoral nerve to soleus, and from the median nerve to biceps brachii (Pierrot-Deseilligny & Burke, 2005). The Ib inhibition of the test monosynaptic reflex lasts less than 10 ms.

Although tendon organs respond readily during isometric and shortening contractions (Burke et al., 1978), Ib depression of the test reflex during resting conditions is reduced during voluntary contractions, and the reduction increases as the contraction force increases (Fournier et al., 1983). The suppression of Ib inhibition to the active motor neurons is attributable to a decrease in transmission along Ib pathways due to modulation of interneurons by descending systems. In contrast, there is increased Ib inhibition to motor neurons that supply antagonist muscles, which would tend to minimize coactivation. However, Ib interneurons can be facilitated by cutaneous and joint afferents, and it has been proposed that such facilitation may help terminate some types of movements (Lundberg et al., 1977).

Group II Pathways

Most group II afferents in muscle nerves arise from muscle spindles, and they are more numerous than Ia afferents. The electrical threshold of group II afferents is about twice that of Ia afferents. Group II afferents from one muscle can reach multiple motor neuron pools, and each motor neuron receives group II input from several muscles. The major effects of group II afferents onto motor neurons are mediated through interneurons (Lundberg et al., 1977). Group II interneurons receive input from group II afferents (ipsilateral and contralateral), Ia and Ib afferents, cutaneous and joint afferents, and descending tracts. Group II interneurons commonly produce excitation of alpha motor neurons that innervate flexor muscles and inhibition of those supplying extensor muscles. These interneurons also provide strong excitation to gamma motor neurons, with most gamma motor neurons receiving input from several muscles (Gladden et al., 1998).

The group II afferents can be activated by stretch of the homonymous muscle or by electrical stimulation of a muscle nerve at a high intensity (Schieppati et al., 1995; Simonetta-Moreau et al., 1999). Because the conduction velocities of group II afferents are 40 to 50 m/s compared with 60 to 70 m/s for Ia afferents and because the group II pathway includes at least one interneuron, the influence of group II afferents on the activity of motor neurons occurs at a longer latency than that due to Ia input. Based on this difference, it has been shown that group II afferents can modulate H-reflex amplitude, the poststimulus time histogram (example 7.3), and ongoing EMG activity. For example, figure 7.17 shows that the medium-latency response to muscle stretch, which is mediated by group II afferents, depended on the degree to which the subject relied on the muscle to counteract a postural disturbance (Schieppati & Nardone, 1991). When the subject was able to hold on to a support, the medium-latency response was reduced because it was not so necessary to rely on the stretched muscle (soleus) to resist the disturbance and remain standing. Similarly, group II afferents have been shown to support homonymous muscle activation during walking and to mediate some of the homonymous and heteronymous reactions to disturbances that occur during walking (Dietz et al., 1987; Grey et al., 2001; Marchand-Pauvert & Nielsen, 2002; Sinkjær et al., 2000).

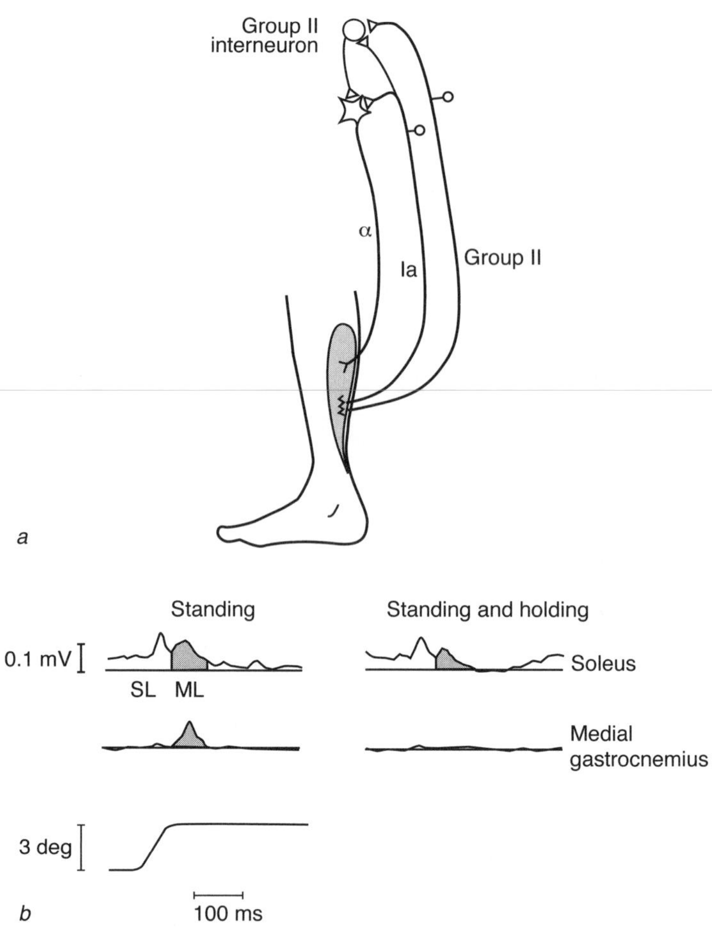

Figure 7.17 Change in group II excitation during postural tasks. *(a)* Pathways of Ia and group II afferents to soleus motor neurons. Both afferents connect to group II interneurons. *(b)* Toes-up rotation of the support surface by 3° evoked a stretch response in soleus. The response comprised short- and medium-latency components (SL and ML). The size of the medium-latency response (hatched area) was reduced in soleus and abolished in medial gastrocnemius when the subject held on to a support during the perturbation (right). The short-latency response in soleus did not differ for the two conditions.

M. Schieppati et al., "Free and supported stance in Parkinson's Disease. The effect of posture and postural set on leg muscle responses to perturbation, and its relation to the severity of the disease," *Brain,* 1991, Vol. 114, pg. 1236, by permission of Oxford University Press.

Presynaptic Inhibition of Ia Terminals

Presynaptic inhibition of Ia terminals involves multiple pathways that can modulate the release of neurotransmitter at the synapse of the Ia afferent onto the alpha motor neuron (figure 7.18). The interneurons that cause presynaptic inhibition enable the efflux of Cl^- from the Ia terminals, which depolarizes the membrane potential and reduces the influx of Ca^{2+} in response to the arrival of an action potential along the Ia afferent (Rudomin & Schmidt, 1999). The shortest pathway that can mediate presynaptic inhibition involves two interneurons (denoted as i and ii in figure 7.18). These interneurons receive input from Ia and Ib afferents, cutaneous and joint afferents, and descending tracts. The Ia and Ib afferents arising from flexor muscles produce presynaptic

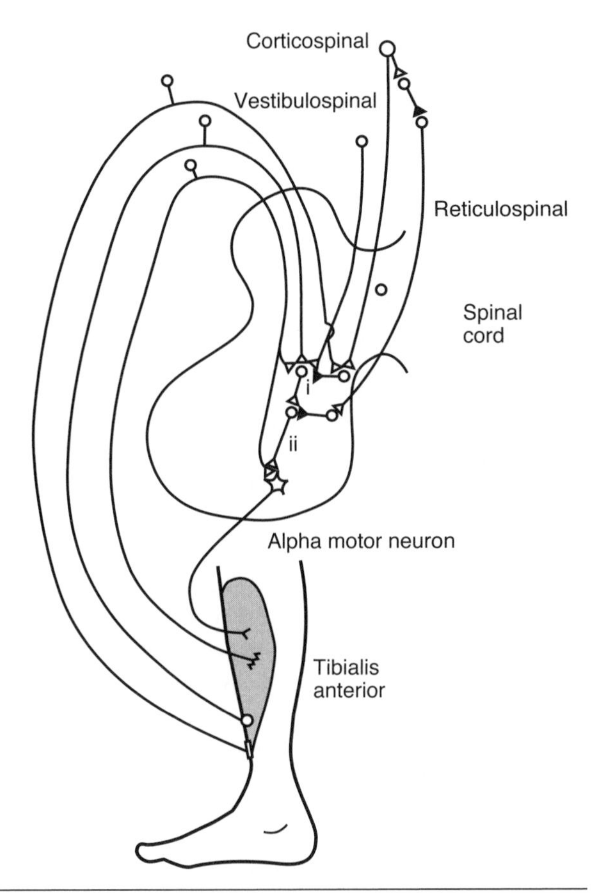

Figure 7.18 Connections from peripheral afferents and descending tracts onto the interneurons that mediate presynaptic inhibition. The descending pathways include the vestibulospinal, corticospinal, and reticulospinal tracts. Note that in this and other figures, the final interneuron connecting to the Ia terminal is shown as an inhibitory synapse even though the postsynaptic effect is to depolarize the membrane of the Ia terminal; functionally, however, the depolarization inhibits the release of neurotransmitter.

From E. Pierrot-Deseilligny and D. Burke, 2005, Presynaptic inhibition of Ia terminals. In *The circuitry of the human spinal cord. Its role in motor control and movement disorders* (New York: Cambridge University Press), 228. Adapted by the permission of Cambridge University Press.

inhibition in all ipsilateral muscles, whereas those arising from extensor muscles tend to contribute to presynaptic inhibition of the Ia afferents from extensor muscles only. The cutaneous and joint afferents connect to an inhibitory interneuron that depresses presynaptic inhibition. The descending tracts can facilitate and depress presynaptic inhibition through several different pathways.

The multiple pathways that can influence the interneurons involved in presynaptic inhibition provide multiple opportunities to modulate transmission from the Ia terminal (Brooke & Zehr, 2006; Hultborn, 2006). For example, the facilitation of the soleus H reflex evoked by electrical stimulation of the femoral nerve is similar at rest and during a brief sustained contraction of soleus, but it increases substantially at the onset of a voluntary contraction. The increase in facilitation is presumably due to a reduction in presynaptic inhibition of Ia afferents from quadriceps femoris projecting to soleus motor neurons (Hultborn et al., 1987*b*). Presynaptic inhibition returns to resting levels about halfway through the ramp increase in force (Meunier & Pierrot-Deseilligny, 1989). Conversely, synergist muscles not involved in the action experience an increase in presynaptic inhibition, as do some antagonist muscles. The gating of presynaptic inhibition begins about 50 ms before the onset of the contraction, which suggests that it is mediated by descending pathways.

Presynaptic inhibition of Ia terminals projecting to motor neurons that innervate soleus and quadriceps femoris varies during such activities as standing, walking, and running (Capaday & Stein, 1986; Faist et al., 1996; Katz et al., 1988; Zehr, 2006). The level of presynaptic inhibition is even modulated with the difficulty of the task. For example, Perez and colleagues (2005) found that facilitation of the soleus H reflex by electrical stimulation of the femoral nerve (quadriceps) was decreased after subjects practiced a difficult visuomotor task, but not when they performed voluntary contractions with the muscles that span the ankle joint. The function of presynaptic inhibition, therefore, is to gate the monosynaptic Ia input to motor neurons, and it is modulated according to the needs of the task (Nielsen & Sinkjær, 2002; Seki et al., 2003).

EXAMPLE 7.5

Depression of Presynaptic Inhibition in Patients With Spinal Cord Lesions

Presynaptic inhibition of Ia terminals onto motor neurons that innervate leg muscles is depressed in patients with spinal cord lesions (traumatic, amyotrophic lateral sclerosis, multiple sclerosis). One method that can be used to assess presynaptic inhibition is to apply a conditioning stimulus to a heteronymous muscle and then evoke H reflexes in a test muscle (Morin et al., 1984; Nielsen & Petersen, 1994). For example, a brief train of vibration (three pulses at 200 Hz) applied to the tendon of tibialis anterior will elicit action potentials in the Ia afferents arising from the muscle, and this input will activate the interneurons that mediate presynaptic inhibition of Ia terminals onto motor neurons of heteronymous muscles (figure 7.19). This type of conditioning stimulus will evoke presynaptic inhibition of Ia afferents from soleus and quadriceps femoris that project to soleus motor neurons for about

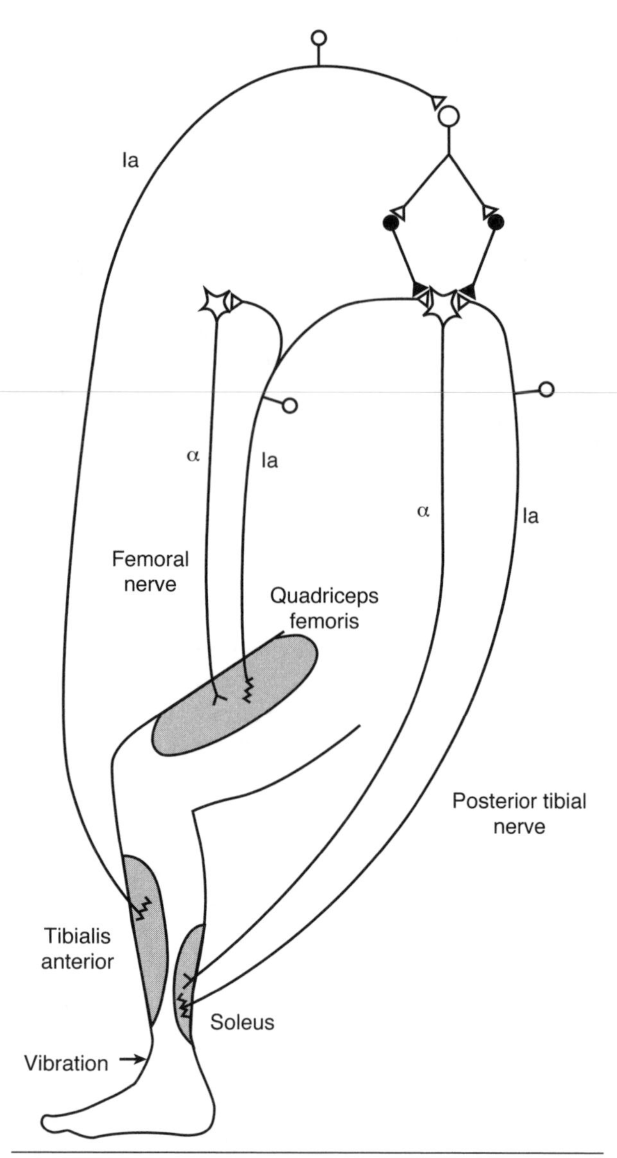

Figure 7.19 Presynaptic inhibition elicited by brief vibration applied to the tendon of a heteronymous muscle.

250 ms. Pierrot-Deseilligny (1990) found that such a conditioning stimulus depressed the soleus H reflex less in patients with amyotrophic lateral sclerosis than in healthy subjects. The reduction in H-reflex depression in patients with spinal cord lesions is probably due to an interruption of the descending pathways that maintain resting levels of presynaptic inhibition in healthy individuals.

Cutaneomuscular and Withdrawal Responses

The afferent pathways involving cutaneous receptors include specialized cutaneous pathways and a group of pathways in which many different afferents converge on common interneurons. Because this group of pathways can evoke the flexion reflex of Sherrington (1910), they are known as the **flexion reflex afferent** pathways. The motor neurons of fast-twitch motor units are excited by cutaneous afferents but inhibited by other flexor reflex afferents, such as joint and group III muscle afferents (Burke et al., 1970*a*). The flexor reflex afferents provide excitation that supports voluntary actions.

Withdrawal responses, which can be observed in most leg muscles under resting conditions, are elicited by painful stimuli applied to the skin or to a nerve. The intensity of the stimulus must be sufficient to activate small-diameter afferents. The muscles recruited for a withdrawal reflex depend on the location of the stimulus (Andersen et al., 1999; Hagbarth, 1960; Kugelberg, 1962). In general, stimuli applied to the leg to evoke withdrawal reflexes involve activation of flexor muscles, especially those proximal to the stimulus, and inhibition of extensor muscles. Similarly, activation of cutaneous receptors in the index finger evokes EMG activity in proximal arm muscles and inhibition of hand muscles (Floeter et al., 1998). The intensity of the withdrawal response can vary with activity, depending on the role of the muscle in the task; it is depressed when the muscle is essential for the task.

The technique often used to assess **cutaneomuscular reflexes** in humans is to determine the influence of activating cutaneous receptors on the ongoing EMG activity during a voluntary contraction. For example, a single shock (two times perception threshold) delivered to the digital nerves of the index finger produced a triphasic response in the first dorsal interosseus muscle: an early excitation (E1), an inhibition (I1), and a late excitation (E2) (Jenner & Stephens, 1982). This response indicates that the cutaneous afferents produce both excitation and inhibition of the motor neurons supplying the hand muscle. The projection of the cutaneous afferents seems to vary across the motor neuron pool as it increases the recruitment threshold of low-threshold motor units and decreases the recruitment threshold of high-threshold motor units (Garnett & Stephens, 1981; Kanda & Desmedt, 1983). Cutaneous afferents are also able to increase the amount of facilitation produced by monosynaptic Ia input to heteronymous muscles (Nielsen & Kagamihara, 1993). As with the withdrawal responses, the cutaneomuscular reflexes, especially the E2 component, are modulated during performance; the E2 component likely involves a transcortical pathway (Gibbs et al., 1995; Turner et al., 2002; Van Wezel et al., 1997). The main role of cutaneous afferents during movement is to modulate the transmission in pathways that provide feedback to the CNS about the sense of position and movement.

Propriospinal Pathways

One of the primary functions of the spinal cord is to transmit the command for a movement from higher centers to the motor neurons. Although there are some direct connections from the motor cortex to the motor neurons in humans, most pathways include multiple neurons. One type of neuron involved in these pathways is the **propriospinal neuron,** which is a spinal neuron with an axon that traverses several spinal cord segments. Clusters of propriospinal neurons in the cervical and lumbar regions of the spinal cord receive both excitatory and inhibitory input from peripheral and descending sources and transmit output to motor neurons. In this capacity, the propriospinal neurons can integrate a variety of inputs and deliver the result to the motor neurons.

Figure 7.20 shows some of the many inputs and outputs of a cervical propriospinal neuron in the cat (Pierrot-Deseilligny & Burke, 2005). There are, however, significant differences in these connections across species (Lemon & Griffiths, 2005). In the human, an electrical stimulus (0.5 times motor threshold) applied to the median nerve at the wrist has been shown to facilitate the discharge of a motor unit in the flexor carpi radialis muscle at a nonmonosynaptic latency (Malmgren & Pierrot-Deseilligny, 1987). The facilitation disappears when the stimulus intensity is increased, presumably due to activation of inhibitory inputs to the propriospinal neurons. A stronger stimulus presumably activates cutaneous afferents that can act through an inhibitory interneuron and inhibit the propriospinal excitation to the motor neurons (figure 7.20). A similar pattern of excitation and inhibition has been observed when EMG responses are preceded by a conditioning stimulus applied to the nerves of heteronymous muscles (Burke et al., 1992; Nicolas et al., 2001).

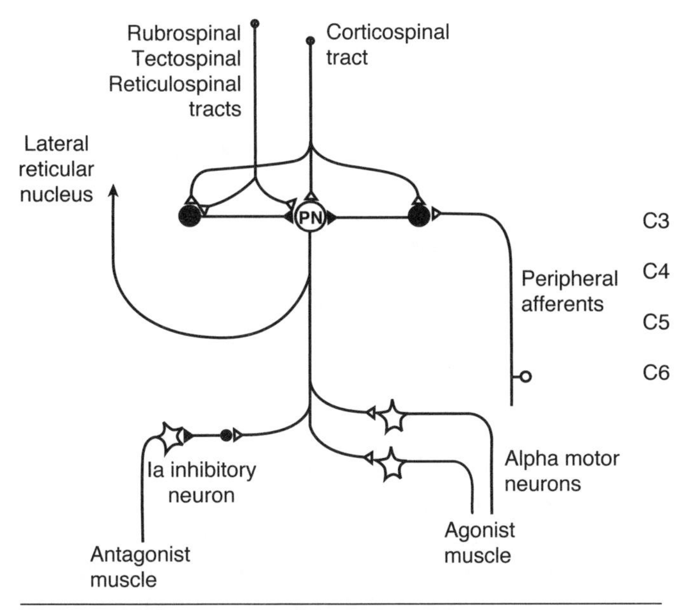

Figure 7.20 Some of the connections of a C3-C4 propriospinal neuron in the cat. The propriospinal neuron (PN) projects to alpha motor neurons that innervate forelimb muscles and receives monosynaptic excitation from descending pathways and, to a lesser extent, peripheral afferents. The descending pathways include the corticospinal, rubrospinal, tectospinal, and reticulospinal tracts. The propriospinal neuron receives inhibitory input from interneurons that receive a mixture of descending and peripheral input. The propriospinal neuron sends a recurrent collateral to the lateral reticular nucleus.

From E. Pierrot-Deseilligny and D. Burke, 2005, Propriospinal relay for descending motor commands. In *The circuitry of the human spinal cord. Its role in motor control and movement disorders* (New York: Cambridge University Press), 453. Adapted by the permission of Cambridge University Press.

These results indicate that propriospinal neurons receive monosynaptic excitation from peripheral and corticospinal inputs and disynaptic inhibition from the same pathways. However, the inputs are not spread evenly across all propriospinal neurons, as they are organized into subsets based on the excitatory muscle afferent input they receive. For example, propriospinal neurons excited by muscle afferents from the wrist extensors are inhibited by cutaneous afferents from the back of the hand and not by those from the palm (Nielsen & Pierrot-Deseilligny, 1991). Thus, a descending command to extend the wrist could be terminated when the back of the hand contacts an object. Furthermore, excitation of the inhibitory neurons by descending input was shown to be greater at the end of a movement, which is consistent with the pathway being able to contribute to movement termination (Pierrot-Deseilligny et al., 1995). Such results indicate that at least part of the descending command for movement is transmitted through propriospinal neurons.

The lumbar propriospinal neurons project to motor neurons that innervate leg muscles. As with the cervical propriospinal neurons, a low-intensity stimulus applied to a peripheral nerve can facilitate the discharge of single motor units, increase the amplitude of an H reflex, and modulate ongoing EMG activity at nonmonosynaptic latencies. In contrast to what occurs in the cervical system, the lumbar propriospinal neurons receive strong input from group I and II afferents, with each neuron receiving afferent input from many muscles (Forget et al., 1989). Furthermore, propriospinal neurons can be inhibited by conditioning stimuli applied to the nerves of heteronymous muscles, but not by cutaneous afferents (Chaix et al., 1997). However, corticospinal inputs can facilitate both the lumbar propriospinal neurons and the inhibitory neurons that receive input from peripheral afferents.

Transmission through the lumbar propriospinal neurons has been demonstrated at the onset of voluntary contractions with quadriceps femoris (Forget et al., 1989). The test involved comparing the effect of a conditioning stimulus applied to the common peroneal nerve (tibialis anterior) on the soleus H reflex. The facilitation caused by the conditioning stimulus was decreased during the contraction at latencies consistent with the convergence of the afferent feedback and the descending command onto the same propriospinal neurons. This observation suggested that the corticospinal input excited the inhibitory interneurons at the onset of the voluntary contraction. Both cervical and lumbar propriospinal systems, therefore, function as major integrating centers for descending motor commands and afferent feedback from the periphery.

Spinal Pathways and Movement

Movements are accomplished by the nervous system activating the motor neurons that innervate the muscles responsible for the action. Voluntary contractions are made possible by motor neurons that receive input from both descending systems and afferent pathways. Because afferent pathways can provide excitatory and inhibitory input to motor neurons, the execution of a movement generally involves the facilitation of excitatory pathways and the suppression of inhibitory pathways by descending systems.

Figure 7.21 provides an example of how input from corticospinal pathways controls the input received by a pool of motor neurons to perform a flexion movement. For clarity, only some of the afferent pathways are shown. Figure 7.21 shows a symmetrical set of afferent pathways for the flexor (agonist) and extensor (antagonist) muscles. These pathways include monosynaptic Ia excitation, reciprocal Ia inhibition, recurrent inhibition and recurrent facilitation, and presynaptic inhibition of Ia terminals. Superimposed on these afferent pathways are inputs from the corticospinal tract that promote excitation of the motor neurons supplying the flexor muscle. The corticospinal input provides direct excitation of the alpha and gamma motor neurons innervating the flexor muscle and suppresses input that could diminish the excitation of the alpha motor neurons. This includes inhibition of the Renshaw cell and an interneuron in the pathway that mediates presynaptic inhibition of Ia terminals. To support the activity of the alpha motor neurons supplying the flexor muscle, the corticospinal input reduces the excitation of the motor neurons that innervate the extensor muscle by facilitating recurrent inhibition, suppressing reciprocal Ia inhibition, and reducing presynaptic inhibition of the Ia terminals. This scheme underscores the essential role of afferent pathways in supporting the activation of motor neurons by descending systems during voluntary contractions (Stein & Thompson, 2006).

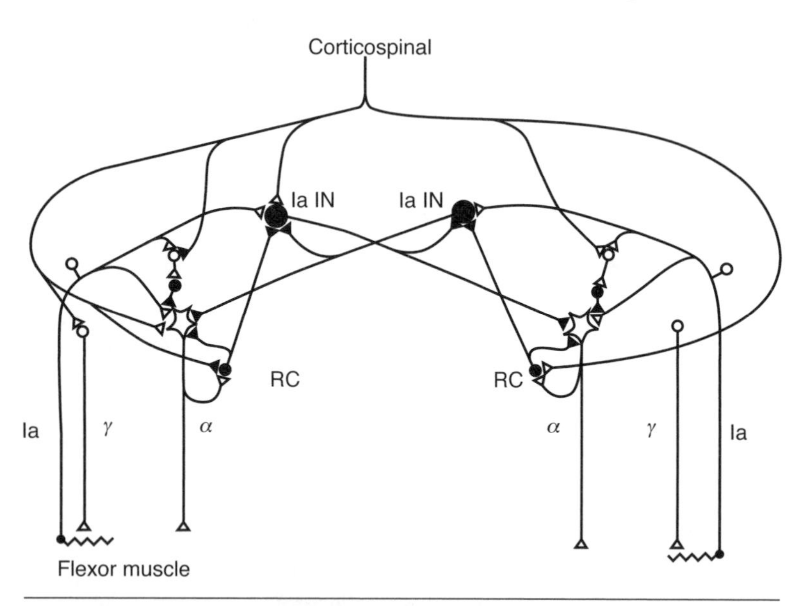

Figure 7.21 Control of afferent pathways by the corticospinal pathways during a voluntary contraction to perform a flexion movement. The flexor and extensor muscles are innervated by alpha (α) and gamma (γ) motor neurons, and a branch of the axon from the alpha motor neuron connects to Renshaw cells (RC). The Ia afferent from the muscle spindle delivers input to the alpha motor neurons, to the Ia inhibitory interneurons (Ia IN), and to the interneurons that mediate presynaptic inhibition of the Ia terminals.

The spinal pathways that project to the motor neurons innervating sets of muscles can vary. The wrist flexor and extensor muscles provide an example of the differences that can exist in these connections (figure 7.22). Flexor and extensor carpi radialis act as antagonists for flexion and extension but function as synergists for wrist abduction. The key differences in the spinal pathways for these muscles include mutual recurrent inhibition to both sets of motor neurons and the projection of Ia and Ib afferents from both sets of muscles to interneurons that mediate nonreciprocal group I inhibition. Nonreciprocal group I inhibition differs from reciprocal Ia inhibition in that the reciprocal inhibition between muscles cannot be evoked by Ia input alone and the interneurons are not inhibited by Renshaw cells. Corticospinal input to the group I interneurons can facilitate or suppress nonreciprocal group I inhibition to meet the needs of the task being performed. For example, activation of the flexor carpi radialis muscle to perform wrist flexion will involve descending facilitation of the group I interneurons projecting to the antagonist muscle (extensor carpi radialis), whereas the group I interneurons supplying the flexor muscle will be inhibited. Once the contraction is under way, feedback from the Ia and Ib afferents arising from the contracting flexor muscle will provide additional excitation to the group I interneurons that are delivering nonreciprocal group I inhibition to the extensor motor neurons.

Not all movements, however, involve excitation of an agonist muscle and inhibition of the

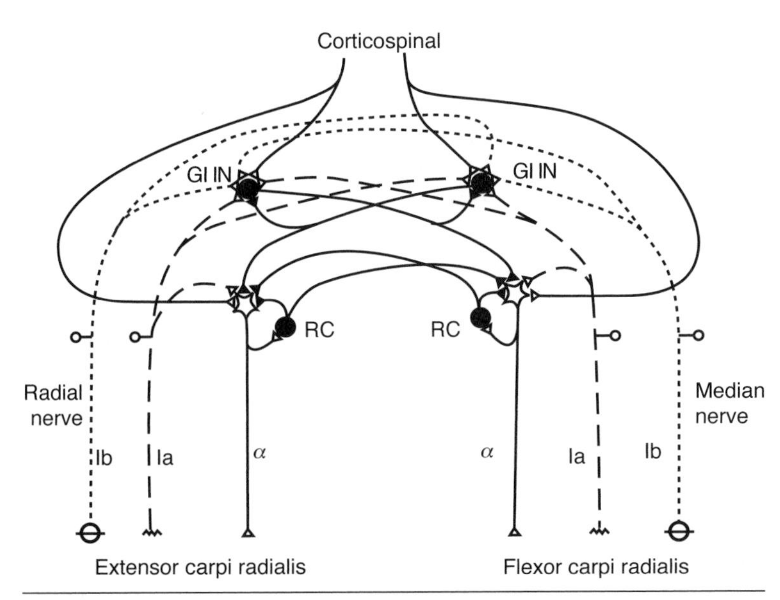

Figure 7.22 Some spinal pathways associated with wrist flexor and extensor muscles. The two muscles are innervated by alpha (α) motor neurons, and a branch of the axon from the alpha motor neuron connects to Renshaw cells (RC) that provide inhibitory feedback to both sets of motor neurons. The Ia afferent from the muscle spindle and the Ib afferent from the tendon organ deliver excitatory input to group I interneurons (GI IN) that mutually inhibit each other and also inhibit the alpha motor neuron innervating the other muscle. Corticospinal input projects to both the alpha motor neurons and the group I interneurons.

antagonist muscle. Many tasks involve the coactivation of agonist and antagonist muscles, which requires suppression of reciprocal inhibition between muscles. This can be accomplished by descending modulation of reciprocal Ia inhibition and recurrent inhibition (Hansen et al., 2002; Lévénez et al., 2005; Nielsen et al., 1993). As shown in figure 7.21, this would involve withdrawal of corticospinal facilitation of the Ia inhibitory interneuron and the Renshaw cell. Furthermore, coactivation is accompanied by an increase in presynaptic inhibition of the Ia terminals yet an increase in the activation of gamma motor neurons. The contribution of these different mechanisms in supporting coactivation differs with contraction force: Depression of reciprocal Ia inhibition is greatest at low forces, whereas recurrent and presynaptic inhibition increase with the strength of the contraction.

EXAMPLE 7.6

Activation of Spinal Pathways During Walking

In addition to supporting the ongoing activation of muscles during walking, spinal pathways must be ready to react to disturbances that could disrupt the gait (Capaday, 2002; Lam & Pearson, 2002; Misiaszek, 2006; Nielsen & Sinkjær, 2002). Measurement of the H reflex in soleus during walking has demonstrated that its amplitude is depressed compared with that during standing, and that it is greatest during the stance phase and absent during the swing phase (Capaday & Stein, 1986; Ferris et al., 2001; Simonsen & Dyhre-Poulson, 1999). Modulation of the H reflex is accomplished by descending control of presynaptic inhibition of the Ia terminals (Schneider et al., 2000). Although it might seem contrary to suppress Ia feedback that could be used to react to disturbances, too much feedback could cause instability of the segmental stretch reflex by activating cycles of muscle stretch and muscle contraction (Windhorst, 2007). Rather, presynaptic inhibition reduces monosynaptic Ia excitation of homonymous motor neurons, but the Ia afferents still provide information to other spinal and ascending pathways. Also, excitation of the soleus motor neurons by group II afferents increases during the stance phase to support the descending activation (Sinkjær et al., 2000).

Despite presynaptic inhibition of Ia terminals during the stance phase, a rapid stretch of the calf muscles can evoke a stretch reflex in the early part of the stance phase (Sinkjær et al., 1996; Yang et al., 1991). An unexpected stretch of the soleus muscle during the stance phase will also evoke medium-latency (M2) and long-latency (M3) responses (Grey et al., 2001; Sinkjær et al., 1999). These results indicate that stretch responses contribute to the stability of the support limb during the stance phase of walking (Stein & Thompson, 2006).

Cutaneous afferents can also evoke rapid responses that help to sustain the gait cycle. For example, stimulation of the superficial peroneal nerve that activated cutaneous afferents depressed the EMG activity in tibialis anterior during the swing phase, whereas stimulation of the tibial nerve increased the EMG activity of the biceps femoris and vastus lateralis muscles during the swing phase (Zehr et al., 1997). In contrast, stimulation of the sural nerve increases the EMG activity in tibialis anterior early in swing phase and decreases it later in the swing phase (Van Wezel et al., 1997). These results indicate that cutaneous reflexes can have a significant effect on muscle activation during walking, but that the response depends on the location of the cutaneous receptors and on the phase of the gait cycle. Similar modulation of cutaneous reflexes occurs during cycling (Zehr et al., 2001*b*).

The afferent signals arising from sensory receptors also contribute to perceptions about the positions and actions of the limbs. The term **proprioception** refers to perceived sensations about the position and velocity of a movement and the muscular forces generated to perform the task (Gandevia, 1996). Inputs from cutaneous, joint, and muscle receptors provide the CNS with information about the movements we perform and those that are imposed on us (Collins et al., 2005; Windhorst, 2007; Winter et al., 2005). For example, when descending systems provide a motor command to the spinal centers to produce a specific movement, proprioceptive feedback informs the CNS about the position of the limb and whether or not the movement was performed as expected. Our ability to detect movement varies with the speed of the movement, the joint being tested, and the contractile state of the involved muscles (Allen & Proske, 2006; Cordo et al., 2000; Goodwin et al., 1972; Refshauge et al., 1998; Taylor & McCloskey, 1992). Experiments have shown that humans can detect movements as slow as 2°/min at the knee, ankle, and finger joints.

Proprioceptive acuity depends on the integration of afferent input with descending motor commands (Gandevia et al., 2006; Winter et al., 2005). In combination, these two signals can be used to develop perceptions about limb position and velocity, the force exerted during the task, and the heaviness of an object that is lifted (McCloskey et al., 1983). Experimenters often demonstrate position sense by asking a subject to match with one limb the position to which the other limb was moved by the experimenter. For example, Gandevia and colleagues (2006) found that subjects could replicate the final position of the wrist angle with an accuracy of about 1° when the forearm muscles were passive and about 2° to 4° during moderate contractions with the forearm muscles. The ability to match the position of wrist angle was lost when the arm moved by the experimenter was anesthetized, and the error increased with attempts to contract the anesthetized muscles; this finding indicates that the ability to match the two wrist angles relied on an intact motor command to the muscles of the limb that was moved. The perception of movement involves alpha-gamma coactivation (Windhorst, 2007; Vallbo, 1971) and the capacity of the CNS to use feedback from muscle spindles to differentiate between intended and imposed displacements (Gandevia et al., 1993; McCloskey et al., 1983; Proske, 2006). The judgment of force and heaviness refers to the estimated relative muscle force that is used to perform a task. When a muscle is weakened or fatigued, for example, an object seems heavier because a greater proportion of the available force is required to lift it (Gandevia, 1987).

Information on joint position and movement is probably combined with knowledge of limb lengths to develop a perceptual map of the body segments, which is necessary for achieving absolute positions in three-dimensional space (Gandevia, 1996). These internal representations are essential for the control of posture and movement (Bosco & Poppele, 2001; Wolpert & Ghahramani, 2000).

AUTOMATIC RESPONSES

Feedback from sensory receptors is also important for a class of behaviors known as automatic responses. These behaviors are produced by pathways that are more complex than those associated with reflexes (Prochazka et al., 2000). The behaviors range from fight-or-flight responses associated with the preservation of life through to postural adjustments that precede the performance of movement (Melvill Jones, 2000). As with reflexes, the details of automatic responses depend on the afferent feedback that is provided to the CNS. As examples of automatic responses, we will discuss postural contractions and rhythmic activation patterns produced by networks of neurons.

Postural Contractions

Postural activity comprises muscle contractions that place the body in the necessary location from which a movement is performed. The contractions are enabled by feedback from somatosensory, vestibular, and visual sensors and can involve responses to disturbances as well as activity that precedes a movement (Bacsi & Colebatch, 2005; Bent et al., 2005; Buchanan & Horak, 2003; Dietz, 1992). In general, these automatic responses serve postural orientation and postural equilibrium functions (Horak, 2006; Horak & Macpherson, 1996). **Postural orientation** entails positioning the body relative to its surroundings, such as the line of gravity relative to the base of support, and locating the body segments with respect to one another. To understand the neural mechanisms involved in the control of postural orientation, researchers have attempted to identify the body segment that remains in a relatively fixed location during a movement. For many tasks, such as those involving movement of the arms and legs, muscles are activated automatically to maintain the trunk in a vertical orientation. In other tasks, however, such as those involving whole-body movement, muscle activity associated with postural orientation is more focused on maintaining the position of the head. Because the magnitude and sequence of muscle contractions participating in these automatic responses can vary, the focal point chosen for postural orientation varies across individuals and movements. The postural expectations related to a movement cause an individual to prepare a set of muscles to provide the automatic response; the muscles so prepared are known as the **postural set** (Gurfinkel et al., 2006; Schieppati & Nardone, 1995; Winstein et al., 2000).

In addition to maintaining the position of the body, postural orientation involves aligning the body segments both within and between limbs (Chabran et al., 2001; Hoy et al., 1985; J. P. Hunter et al., 2004; Patla & Prentice, 1995; Zernicke & Smith, 1996). Consider, for example, the movement shown in figure 7.23*a*. The subject is asked to shake the forearm rapidly in a forward-backward motion while keeping the upper arm horizontal. This movement is accomplished by alternately activating the flexor and extensor muscles that cross the elbow joint. The movement also requires postural activation of shoulder muscles to stabilize the upper arm and to minimize the inertial effects of the forearm motion on other body segments. Furthermore, the muscles that cross the wrist joint need to be activated to control the motion of the hand, which could vary from an uncontrolled flail to a slow wave to no relative motion between the hand and the forearm.

The muscle contractions that contribute to **postural equilibrium** are an attempt to maintain the balance of the individual. These responses involve controlling the small displacements that occur during a steady posture, reacting to perturbations that disturb the position of the body, and anticipating a movement-related disturbance of balance. In an upright standing posture, the base of support is determined by the position of the feet and includes the area underneath and between the feet. A person in an upright standing posture is in equilibrium as long as the line of action of the weight vector remains within the boundaries of the base of support, and the person is stable as long as the musculoskeletal system can accommodate perturbations and return to an equilibrium position (Henry et al., 2001; Schieppati et al., 2002; van Emmerik & van Wegen, 2002). When we stand in an upright posture, our bodies sway back and forth, but we do not fall because of the automatic postural activity that maintains equilibrium (Krishnamoorthy & Latash, 2005; Latash et al., 2003; Tokuno et al., 2006; Wang et al., 2006).

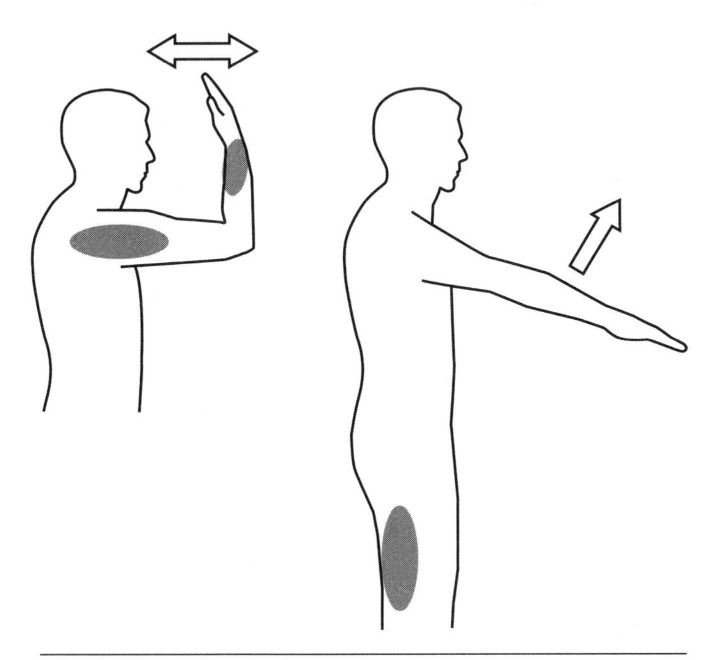

Figure 7.23 Distribution of postural activity with rapid movements. *(a)* Rapid alternating flexion-extension of the elbow joint requires postural activation of the shoulder and wrist muscles to control the inertial effect due to forearm motion. *(b)* Rapid elevation of the arm to an extended horizontal position is preceded by postural activity in the legs.

Contrary to the common explanation, however, the automatic postural activity that maintains equilibrium during upright standing does not comprise reflexes evoked by the back-and-forth sway of the body. By using an ultrasound scanner to measure small changes in the length of the calf muscles during normal standing, Loram and colleagues (2005*a, b*) found that the muscles shortened during forward sway and lengthened during backward sway. Thus, the muscle contracted during the forward sway to lengthen the tendon and use series elasticity to resist the forward motion. Furthermore, the EMG records comprised brief bursts of activity (~2.7 times per second) that displaced the center of mass by small distances (30 to 300 µm) as the body swayed in each direction. The control of equilibrium during standing, therefore, appears to be based on predicted displacements of the center of mass that are learned rather than on responses evoked by afferent feedback. Instead, the sensory information is integrated to enable the prediction of the next required EMG activity.

Nonetheless, reflex responses are likely necessary when the disturbance to the standing posture exceeds the ability of the prediction strategy to maintain equilibrium (Misiaszek, 2006; Stein & Thompson, 2006). A method commonly used to study the automatic responses to large postural disturbances is to have subjects or patients stand on a platform that can be moved suddenly in several different directions (Nashner, 1971, 1972). With the use of this technology, investigators have identified several response strategies that depend on the type of perturbation experienced by the individual. When the support surface is moved suddenly, muscle contractions are evoked at a latency of 70 to 100 ms after the onset of the disturbance. These automatic responses can be modified by afferent and descending inputs that provide information on the direction and velocity of the disturbance, by the initial position of the subject, by the prior experience and expectations of the subject, and by the task being performed (Chabran et al., 2001; Horak, 2006; Horak & Macpherson, 1996; Hughey & Fung, 2005; Schieppati et al., 2002).

Two examples underscore the specificity of these automatic responses. When the support surface was perturbed so that the ankle was dorsiflexed and the calf muscles were stretched, the intensity of the automatic responses depended on whether or not the muscle activity was needed to maintain balance. For example, backward translation of the platform evoked greater EMG activity in posterior leg muscles (soleus, hamstrings, and gastrocnemius) than did toes-up rotation of the platform (figure 7.24*a*). The second example involves the influence of initial conditions on the muscles that participate in the automatic response. When a subject was in an upright standing posture on the movable platform, the response to the perturbation was initiated in the leg muscles (figure 7.25). However, the response began in the arm muscles when the subject was able to hold a handle.

Not only are the automatic responses associated with postural equilibrium specific for each condition; they are also readily adaptable to meet changing needs (Benvenuti et al., 1997; Cordo & Nashner, 1982; Gruneberg et al., 2004). This property is apparent in the comparison of EMG responses of the calf muscles to perturbations involving backward translation and toes-up rotation of the support surface (figure 7.24*b*). Although the muscle spindles likely experienced similar stretches with the two disturbances, subjects knew immediately that it was necessary to activate the gastrocnemius muscle to maintain stability in the translation condition and to activate tibialis anterior for the rotation condition (Gollhofer et al., 1989; Hansen et al., 1988). The ability to switch between the two activation strategies may depend on our ability to sense the location of the total body weight vector relative to the base of support (Dietz, 1997).

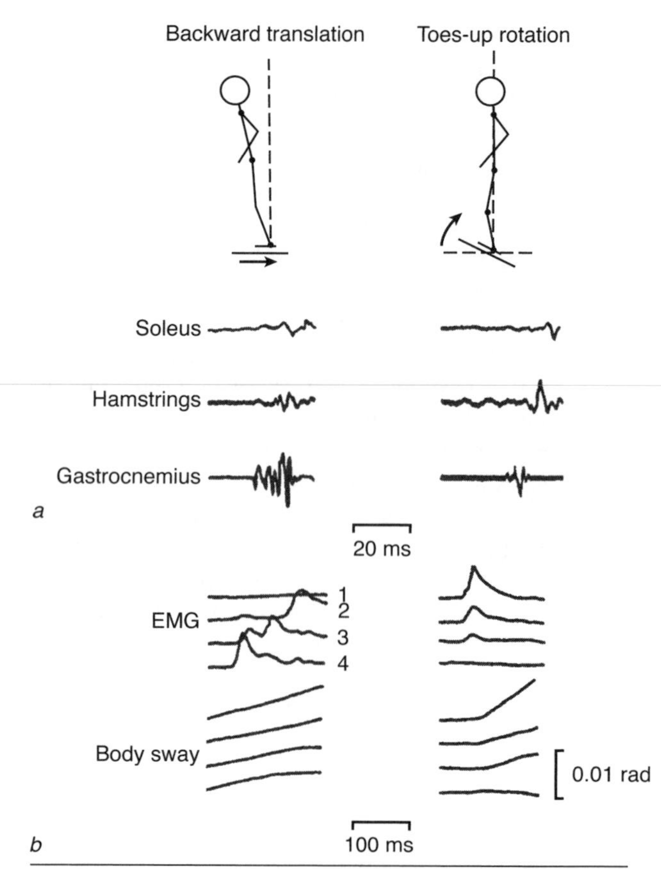

Figure 7.24 Postural responses of a subject when the support surface was translated backward (left column) or rotated in the toes-up direction (right column). *(a)* Although both perturbations induced dorsiflexion of the ankle and a comparable stretch of the calf muscles, the intensity of the EMG activity in the posterior leg muscles, especially gastrocnemius, varied depending on their role in maintaining balance. *(b)* The rectified EMG of gastrocnemius and body sway during four trials for the two conditions. For both the EMG and body sway traces, Trial 1 is the top line and Trial 4 is the bottom line. Because the posterior leg muscles are required to maintain balance for the backward translation but not for the toes-up rotation, subjects learn to increase EMG activity across trials when the disturbance involves a backward translation and to depress it when experiencing an upward rotation of the toes. The change in EMG activity across trials reduces the accompanying body sway.

Postural Contractions That Precede Movement

The maintenance of equilibrium can also include postural contractions that are performed before a movement is begun. These actions are known as **anticipatory postural adjustments.** Suppose an individual standing upright is asked to raise an arm as rapidly as possible to a horizontal position (figure 7.23*b*). This is a reaction-time task in which the movement is done as quickly as possible after a "go" signal. The anterior deltoid muscle is mainly responsible for raising the arm. The briefest reaction time between a "go" signal and the onset of muscle activity is about 120 ms. In this task, however, the hamstring muscles on the same side of the body are activated about 50 ms before the activation of the arm muscles (Belen'kii et al., 1967). The activation of the leg muscles serves at least two purposes: It provides anticipatory stabilization against the inertial effects of the subsequent arm movement, and it establishes a rigid connection between the limb motion and the associated ground reaction force (Stapley et al., 1999). By increasing the rigidity of the body segments not directly involved in the movement, the anticipatory adjustments can probably facilitate the subsequent movement, even to the extent of transferring energy through intersegmental dynamics. The presence of anticipatory postural adjustments depends on the task, the support provided by the surroundings, and the

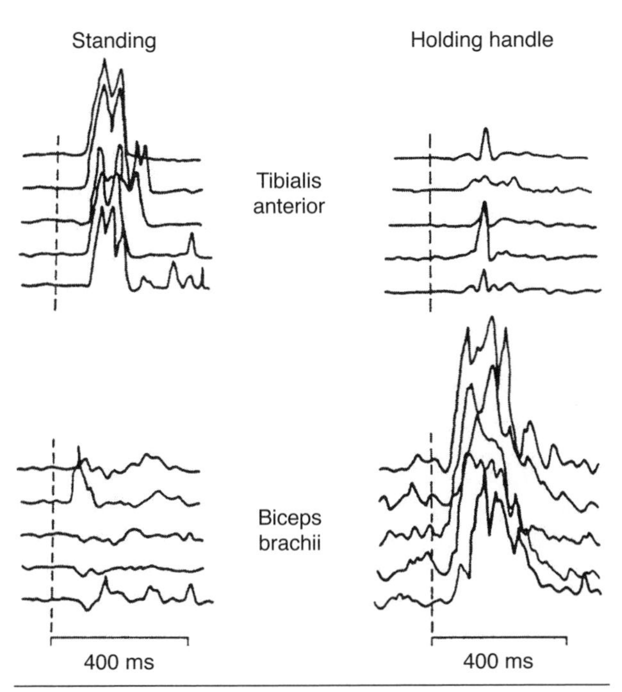

Figure 7.25 Electromyogram responses to postural disturbances. The vertical line indicates the moment when a constant forward translation was applied to the support surface. There were five trials for each condition (standing and holding a handle). The responses of tibialis anterior (rectified EMG) were greatest for the standing condition, whereas the biceps brachii EMG was greatest when the subject held a handle.

Modified with permission, from P.J. Cordo and L.M. Nashner, 1982, "Properties of postural adjustments associated with rapid arm movements," *Journal of Neurophysiology* 47: 296. Used with permission.

neurological status of the subject (Aruin et al., 1998, 2001; Benvenuti et al., 1997; De Wolf et al., 1998; Dietz et al., 2000). For example, patients with Parkinson's disease have difficulty combining anticipatory postural adjustments with intended (voluntary) movements (Rogers, 1991). The anticipatory postural adjustments are controlled independently of the voluntary action by the cerebral cortex (Taylor, 2005).

Distribution of Postural Responses

The automatic responses evoked by a disturbance of one limb often occur in both limbs, just as it is possible for a stimulus applied to one limb to evoke a reflex response in another limb (Baldissera et al., 1998; Delwaide et al., 1988; Koceja, 1995; Tax et al., 1995; Zehr et al., 2001*a*). For example, displacement perturbations of a single leg during stance, balancing, or gait typically evoke a bilateral response with a similar latency in the two legs (Dietz, 1992). The bilateral response probably provides a more stable base from which to compensate for the perturbation. Similarly, when a hand holding a support is perturbed, this can elicit an automatic postural response, such as an increased grip force, in the contralateral arm and hand (Marsden et al., 1983*b*). Such responses are automatic and are superimposed on other movements being performed by the nonperturbed limb. The appearance of automatic postural adjustments in muscles that are not perturbed suggests that these responses represent a coordinated and predetermined motor pattern.

Perhaps the most extreme example of a distributed set of automatic responses to a disturbance is the **startle reaction.** This is the response of the body to an unexpected auditory stimulus; sometimes it can be evoked with visual, vestibular, or somesthetic stimuli (Bisdorff et al., 1994, 1999; Hawk & Cook, 1997). It is typically described as a generalized flexion response that is most prominent in the face, neck, shoulders, and arms (Blouin et al., 2006; Brown, 1995; Landis & Hunt, 1939), although it can also be evoked in the leg muscles of humans (Delwaide & Schepens, 1995). The startle response belongs to a class of behaviors known as escape responses (Ritzmann & Eaton, 1997) that prepare the body for subsequent action after an intense and unexpected sensory stimulus.

The startle reaction appears to originate in the caudal brain stem and to involve a subcortical reflex loop (Brown, 1995; Davis, 1984). An early study showed that this reaction involves rapid responses that begin with an eye blink (40 ms) and can progress to include flexion at the neck (75-120 ms), trunk and shoulders (100-120 ms), elbows (125-195 ms), fingers (145-195 ms), and legs (145-395 ms) (Landis & Hunt, 1939). Subsequent studies have suggested that the eye blink response has two parts, an initial auditory blink reflex and the subsequent generalized startle reaction (Brown, 1995). The magnitude of the startle reaction is variable; it habituates with repeat stimuli, decreases in a dose-dependent way with ethanol, increases with fear and arousal, and can be modulated by supraspinal centers (Davis, 1984). The most effective stimulus is a brief, loud noise, such as 90 dB for $\leq$30 ms. The startle reaction is exaggerated in some clinical conditions, such as hyperekplexia.

The startle reaction can have a marked effect on voluntary movements (Nieuwenhuijzen et al., 2000; Siegmund et al., 2001). For example, when a loud acoustic stimulus was randomly superimposed on reaction-time movements, the reaction time was shortened by about one-half (Carlsen et al., 2004*b;* Valls-Solé et al., 1999). The reaction-time task involved flexion and extension of the wrist or rising up onto the toes in a standing posture. Under control conditions, the time from the "go" signal to the beginning of the response was 204 ms for the wrist movement and 244 ms for the foot movement. With the superimposition of the startle stimulus, the reaction time decreased to 104 ms for the wrist movement and 123 ms for the foot movement. Because the EMG pattern was similar for trials with and without the startle stimulus, it was concluded that each movement is likely triggered by activity at subcortical levels (Carlsen et al., 2004*a;* Valls-Solé et al., 1999).

EXAMPLE 7.7

Automatic Responses During Hand Movements

In the spectrum of movement capabilities, a qualitatively different type of movement involves the exploration of our environment. This distinction is typified by the functions performed by the hands as compared with those of the arms and legs. The human hand and brain are close partners in our ability to explore our physical world and to reshape it. Both of these functions depend on accurate descriptions of mechanical events when objects come in close contact with the hand. Much of this information is provided by the mechanoreceptive afferents that innervate the hairless skin of the hand. These sensory receptors participate in a behavior we call **active touch** (Flanagan et al., 2006; Johansson, 1996, 2002; Overduin et al., 2008; Phillips, 1986; Wing et al., 1996).

The role of cutaneous mechanoreceptors in the control of motor output has been explored in the study of grip-force responses to unexpected changes in load (Flanagan et al., 1999; Johansson et al., 1992*a, b, c;* Macefield & Johansson, 2003). When an object is squeezed between the index finger and thumb in a pinch grip, an unexpected change in the pulling load elicits an automatic adjustment in the grip force after a brief delay. This response, which is sometimes referred to as the **reactive grip response,** occurs when the control of a handheld object appears uncertain, as when it begins slipping or when its balance is disturbed. The latency of the grip-force response decreases (174 to 80 ms) as the rate of change in the pulling load increases. The automatic response consists of two parts: The initial phase comprises a bell-shaped rate of change in force, and the second phase includes a slow increase in the rate of change in force (figure 7.26). Similar biphasic responses have been reported for isometric tasks, compensation for body sway elicited by translation of the support surface, and eye movements during smooth pursuit. The initial phase is a standard element that has a constant duration, but its amplitude varies with the rate of change in the pulling load. The second phase appears for longer-duration increases in grip force and is abolished if the index finger and thumb are anesthetized. Thus, the second phase depends on afferent feedback from mechanoreceptors in the hand.

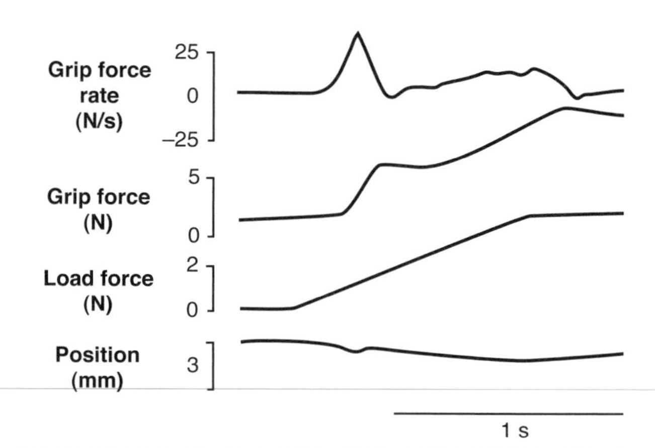

Figure 7.26 Automatic grip-force responses to an unexpected change in the load force. A motor applied the load force, and the subject was required to prevent movement (or change in position). The rate of change in grip force (top trace) consists of two phases, an initial bell-shaped response followed by a more gradual change.

Data from Johansson et al. 1992*c*.

Central Pattern Generators

The second example of automatic responses we will consider is the output of **central pattern generators,** which are neuronal circuits that produce coordinated motor patterns in response to central inputs (Grillner & Wallen, 1985; Kiehn, 2006; Rossignol et al., 2006; Stein et al., 1997). Central pattern generators (CPGs) produce automatic movements, such as locomotion, respiration, swallowing, and defense reactions (Edgerton & Roy, 2006; Pearson & Gordon, 2000). The motor pattern generated by a CPG is specific to the behavior it controls. To accommodate the diversity of behaviors that can be produced, CPGs vary in the organization and properties of the constituent neurons and the interactions among these neurons; this enables the network to use one of several mechanisms to produce its characteristic bursts of action potentials (Calabrese, 1998; Edgerton et al., 2004; Kiehn et al., 1997; Rossignol et al., 2006)

The flow of information through the nervous system that produces motor behavior is shown in figure 7.27. The CPG is located between the higher centers and the motor neurons. It is typically activated by signals from command and modulatory neurons located in higher centers. The activity of the CPG, however, is also influenced by sensory feedback and by hormones (neuromodulation). Most is known about the structure and function of GPGs that control rhythmical movements, the classic example being locomotion. For such movements, the CPG must produce a rhythm that results in alternating activation of antagonistic muscles; it must be able to vary the frequency of the output; and it must shape the pattern of the output

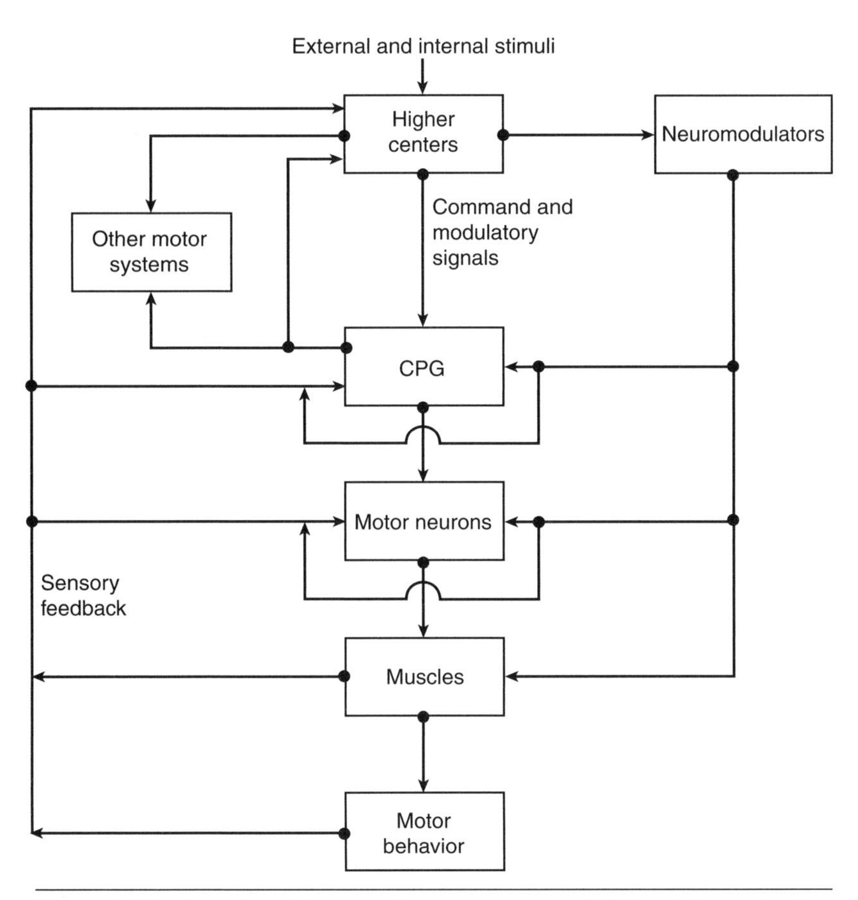

Figure 7.27 Location of the central pattern generator (CPG) relative to the flow of information through the nervous system to produce a movement.

Stein, Paul S.G., Sten Grillner, Allen I. Selverston, and Douglas G. Stuart, eds., *Neurons, networks, and motor behavior,* figure on page 106. © 1997 Massachusetts Institute of Technology, published by The MIT Press.

(Edgerton & Roy, 2006). One of the first models of a CPG was proposed by Brown (1911) to explain the motor pattern observed in the leg muscles of the cat during locomotion. To explain the observation that an isolated spinal cord could produce alternating activation of flexor and extensor muscles, Brown proposed the **half-center model,** which comprises one set of CPG neurons that project to motor neurons innervating extensor muscles and another set that project to motor neurons innervating flexor muscles (figure 7.28*a*). These two sets of neurons, or half-centers, inhibit each other reciprocally so that when one half-center is active the other is inhibited. The periods of excitation and inhibition of each half-center are limited by the intrinsic properties of the neurons and the properties of the associated synapses (McCrea & Rybak, 2008).

Three examples of mechanisms that can produce changes in the response of one half-center over time are adaptation, postinhibitory rebound, and depression of synaptic efficacy. **Adaptation** is the progressive increase in the time between successive action potentials discharged by a neuron despite the neuron receiving constant excitatory input—that is, a slowing of discharge rate. **Postinhibitory rebound** refers to an increase in the excitability of a neuron after it has experienced a period of inhibition. **Depression of synaptic efficacy** can be produced by a decline in the availability of a neurotransmitter or the accumulation of a limiting factor. Thus, the decline in the discharge of a neuron over time due to the depression of synaptic efficacy or adaptation will release the neuron in the other half-center from the inhibition it receives from the interneuron; and, due to postinhibitory rebound, it will respond more vigorously to any excitatory input that it receives. By such mechanisms, continuous input to the CPGs from higher centers results in alternating output from the two half-centers.

Since Brown's proposal, we have learned many details about the CPG components, such as the source of the rhythm, the types of neurons involved in the CPG, and the cellular interactions that define the motor pattern (Edgerton & Roy, 2006; Hooper & DiCaprio, 2004; Selverston, 2005; Stein et al., 1997). Because of the complexity of the CPG, most of the details have emerged from studies on vertebrate CPGs with fewer elements. These have included such systems as the heartbeat CPG in crustacea, the feeding and swimming CPGs in the mollusk, the pyloric and gastric CPGs in the crustacean stomatogastric system, the flight CPG in the locust, the swimming CPGs in the lamprey and leech, and the respiratory CPGs in mammals. As an example, the CPG that produces a scratch behavior by one hindlimb of the turtle is shown in figure 7.28*b*. This behavior involves alternating activation of the hip flexor and extensor muscles, which requires interactions among interneurons (excitatory and inhibitory) and motor neurons that control both legs. Thus, when one leg is activated, the other is inhibited.

The existence of CPGs in humans seems certain on the basis of observations of rhythmic reciprocal contractions and long-latency reflex responses in paraplegic patients, as well as the locomotor patterns that can be produced by isolated spinal cords of primates (Dimitrijevic et al., 1998; Kiehn & Kullander, 2004). The evidence from such studies suggests that the CPG controlling limb movement is

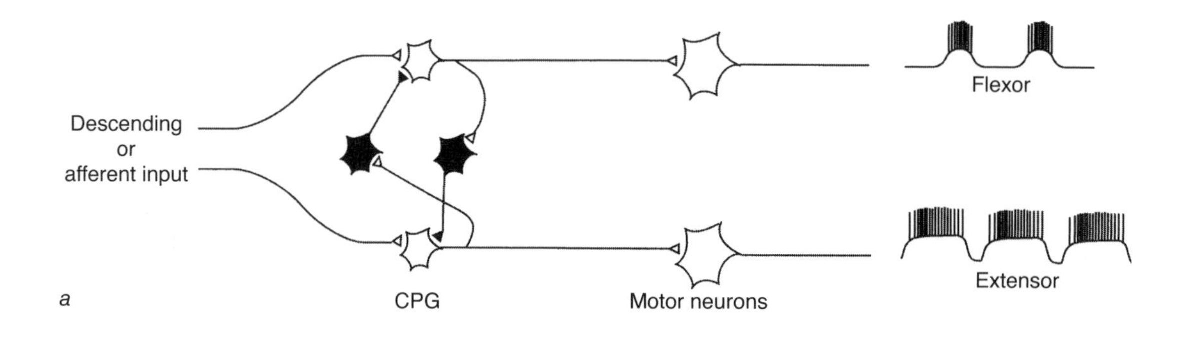

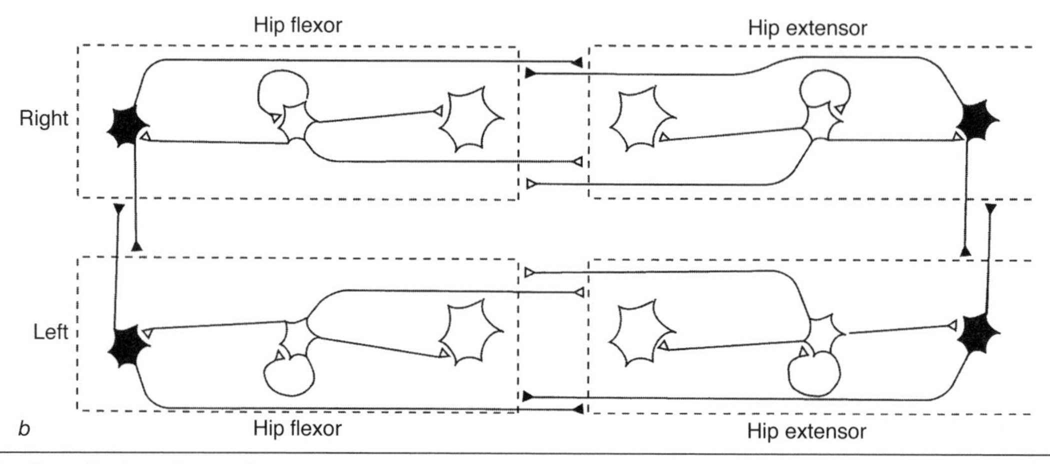

Figure 7.28 Organization of central pattern generators (CPGs). *(a)* The half-center model of Brown (1911), which represents reciprocal inhibition (filled-in neurons) between neurons that project to motor neurons innervating flexor and extensor muscles. The motor rhythm is shown as alternating bursts of action potentials in the nerves projecting to the antagonist muscles. *(b)* The CPG network that controls rostral scratching in the turtle. The CPG for each leg (right and left) and muscle group (flexor and extensor) is shown within the dashed lines. According to this scheme, the scratch behavior by one leg involves rhythmic activation of the hip flexor and extensor muscles, and interacts with the neurons controlling the other leg.

distributed along the spinal cord and is not a single entity (Gelfand et al., 1988; Ivanenko et al., 2006*b*; Kiehn & Kjærulff, 1998; Salteil et al., 1998; Zehr et al., 2003). For example, spinal cord–injured patients who trained to step on a treadmill were able to recover the capacity to produce relatively normal foot kinematics, but they achieved this by developing new patterns of muscle activity that involved a reorganization of the locomotor network from the cervical to the sacral segments of the spinal cord (Grasso et al., 2004).

EXAMPLE 7.8
Capabilities of Spinal Pathways

The existence of CPGs in the spinal cord that can be activated to produce locomotion has considerable significance for individuals with spinal cord injuries (Edgerton et al., 2004, 2006). Numerous studies have been performed to characterize the capabilities of these spinal networks. The observations of Brown (1911) have been extended to demonstrate, for example, that a decerebrated cat can walk on a treadmill when a specific region in the brain stem is stimulated electrically (Shik et al., 1966), that spinalized cats can be trained to walk (Eidelberg et al., 1980; Forssberg et al., 1980), that electrical stimulation of the lumbosacral cord in paraplegic mammals can produce a locomotor rhythm (Dimitrijevic et al., 1998), and that a human with no supraspinal control below a low thoracic lesion can learn to stand and initiate steps (Harkema, 2001, 2008). These achievements are possible due to the interactions between the spinal locomotor system, including the CPGs that produce the rhythm, and afferent feedback from sensory receptors (Edgerton & Roy,

2006). However, the recovery of stepping in spinalized mice was enhanced by administration of a serotonin agonist to the spinal cord (Fong et al., 2005).

One of the more remarkable demonstrations of the capabilities of the spinal cord has been a series of studies on the speed and the specificity of the learning response (Edgerton et al., 2008). For example, a spinal cat stepping on a treadmill learned within one step cycle to flex the limb during the swing phase to a greater extent to avoid an obstacle (Edgerton et al., 2001*a;* Forssberg, 1970). This adjustment produced a stumbling reaction in subsequent steps until the gait cycle returned to normal. Similarly, spinalized rats quickly learned to alter the locomotor rhythm to accommodate a disturbing force applied to the hindlimbs (de Leon et al., 2002; Timoszyk et al., 2002). The specificity of the adaptation was demonstrated by the improvement in the stepping ability of spinalized cats trained to step, whereas those trained to stand experienced only an enhanced capacity to stand (de Leon et al., 1998a, 1999). These capabilities suggest that it will be possible to identify appropriate rehabilitation strategies for individuals with spinal cord injuries (Behrman et al., 2006).

Structure of a Central Pattern Generator

Although the oscillatory output of a CPG can be produced by the endogenous pacemaker properties of generator neurons, most CPGs are based on the properties of a network of neurons (Feldman et al., 2003; Kiehn et al., 2008). Three features of neurons that comprise a rhythm-generating network can be manipulated to produce a CPG: (1) cellular properties, (2) synaptic properties, and (3) connections among the neurons (Rossignol et al., 2006; Stein et al., 1997). The cellular properties include threshold, the relation between input current and discharge rate, adaptation, afterhyperpolarization, delayed excitation, postinhibitory rebound, and plateau potentials. The synaptic properties that can vary are the sign of the postsynaptic potential (excitation or inhibition), the form of the postsynaptic potential (action potential or local potential), the strength of the connection and its time course, the type of connection (electrical or chemical), and the amount of facilitation or depression. The connections can include reciprocal inhibition, recurrent inhibition, and mutual excitation.

As an example of the combination of properties that can exist in a CPG, we consider the organization of the CPG that produces swimming in the lamprey (Parker & Grillner, 2000; Cangiano & Grillner, 2005). Lampreys swim by alternating activation of muscles on each side of the body, which progresses from head to tail. A CPG at each spinal segment controls the activation of the associated muscles (figure 7.29). Descending input from the reticular formation in the brain stem excites the network by activating metabotropic receptors in neurons on one side of the network to produce plateau potentials. The depolarization of the membrane associated with the plateau potentials activates low-voltage Ca^{2+} channels, and the influx of Ca^{2+} increases the depolarization and activates calcium-dependent K^+ channels. The efflux of K^+ repolarizes the membrane potential and contributes to terminating the plateau potential. The effect of the K^+ efflux on the membrane potential is enhanced by the summation of a slow afterhyperpolarization after successive action potentials.

The network on each side comprises two types of inhibitory interneurons (I and L) and a number of excitatory interneurons (E) that project to the motor neurons. Activation of the network by the reticular formation causes the excitatory interneurons to evoke excitatory postsynaptic potentials in the motor neurons and both types of inhibitory interneurons. Activation of the I-type interneuron produces reciprocal inhibition in the network on the other side. The L-type interneuron, however, experiences delayed excitation, which causes it to eventually inhibit the I-type interneuron and to remove the reciprocal inhibition. With the removal of reciprocal inhibition, the interneurons on the other side of the network (E, I, and L) experience postinhibitory rebound and respond to the descending excitation from the reticular formation.

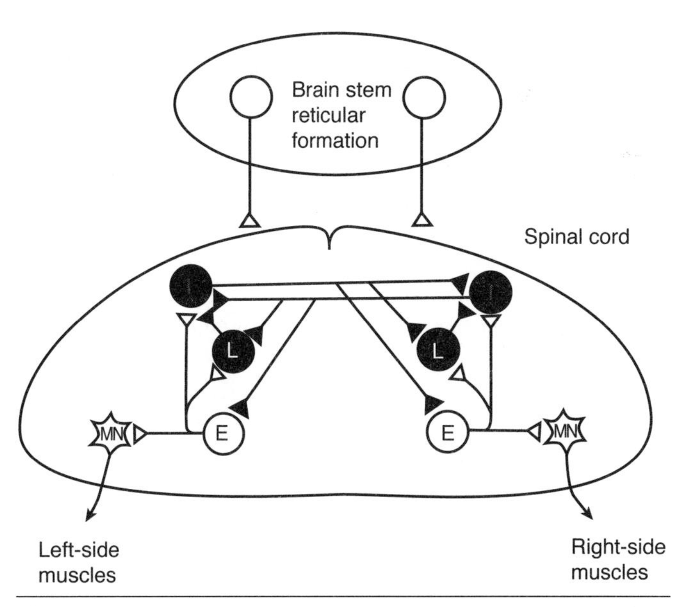

Figure 7.29 The network of neurons at a spinal segment of the lamprey that can produce the locomotor rhythm for swimming. E = excitatory interneurons; I and L = inhibitory interneurons; MN = motor neurons.

Adapted, by permission, from K. Pearson and J. Gordon, 2000, Locomotion. In *Principles of neural science,* 4th ed., edited by E.R. Kandel, J.H. Schwartz, and T.M. Jessel (New York: McGraw-Hill Companies), 745. Copyright 2000 by the McGraw-Hill Companies, Inc.

To produce swimming, the actions of each segmental network must also be coupled to those of its neighbors to coordinate the progression of the alternating muscle activity along the body of the lamprey.

Sensory Modulation

Although CPGs are capable of generating a specific motor pattern, feedback from sensory receptors is absolutely necessary for a normal pattern in an intact behaving animal (Büschges & El Manira, 1998; Edgerton & Roy, 2006; Pearson, 2008; Rossignol et al., 2006). From studies on the integration of sensory feedback into CPG function, four principles have emerged: (1) Sensory feedback contributes to the generation and maintenance of rhythmic activity; (2) phasic sensory signals initiate the major phase transitions in intact motor systems; (3) sensory signals regulate the magnitude of ongoing motor activity; and (4) transmission in reflex pathways can vary during a movement (Pearson & Ramirez, 1997).

Generation and Maintenance of Rhythm When the CPG is disconnected from the muscles it controls, the output produced by the CPG is referred to as a **fictive** pattern; that is, no actual motor pattern is expressed in the muscles (figure 7.28*a*). In such preparations, the CPG can also be isolated from sensory feedback (deafferented) and can be separated from high centers by decerebration or spinalization. The critical test of a CPG is that it must be able to generate a rhythm under these conditions. However, the CPG has to be activated artificially in these preparations, for example by electrical stimulation, by the introduction of pharmacologic agents, or by phasic afferent input. The ability to activate the isolated CPG with afferent input underscores the contribution that sensory feedback can make to the function of the CPG. Although the rhythm that can be observed in a fictive preparation can be similar to that found in normal behavior, the motor pattern is typically quite different due to the absence of afferent input.

In intact and behaving systems, there are a number of examples of the significant role of sensory feedback in shaping a motor rhythm. One example is chewing. Three jaw-closing muscles (temporalis, masseter, and pterygoid) and one jaw-opening muscle (digastric) control mastication in mammals. The motor pattern used during mastication depends on the type of food that is being chewed. The motor pattern is controlled by a CPG located in the brain stem and involves the intrinsic properties of the trigeminal motor system (Lund et al., 1998). Feedback from muscle spindles in jaw-closing muscles and pressure receptors in the periodontum provides excitatory support during mastication, which increases cycle duration and the bursts of activity in the motor neurons. In contrast, the spindles are active during resting conditions when the jaw-closing muscles are stretched, not when the jaw is closed. Thus, the muscle spindles and pressure sensors enhance and modify the motor pattern during the jaw-closing phase of mastication. Moreover, chronic musculoskeletal pain slows the rate of chewing movements and alters the associated motor patterns (Svensson et al., 1997).

Another example comes from a study involving the performance of locomotor-like muscle activity in the legs of patients with complete spinal cord injuries (Kawashima et al., 2005). The investigators compared the EMG activity in the leg muscles of 10 patients as their legs were moved using three types of straight-leg movements: the movement of one leg, an alternating movement with both legs, and a synchronous (in-phase) movement with both legs. All three passive movements evoked EMG activity in soleus and medial gastrocnemius muscles of all patients and in biceps femoris of eight patients. The amplitude of the EMG activity was greatest for the task that imposed alternating movements on the two legs. Thus, sensory feedback when the legs were moved in a gaitlike pattern (alternating movement of both legs) amplified the muscle activity that was observed with all three movements.

Phase Transitions One observation made by Sherrington (1910) was that afferent input appears to be critical in controlling the switch from the stance to the swing phase in locomotion. When cats walked on a treadmill, he noted that the switch from the stance phase to the swing phase depended on the amount of extension at the hip joint. Later, Grillner and Rossignol (1978) found in spinalized cats (CPG disconnected from higher centers at the spinal level) that blocking extension of the hip joint during walking on a treadmill prevented the hindlimb from initiating the swing phase. As a result of stretching and vibration of various muscles during the step cycle, it seems that feedback from muscle spindles located in hip and ankle flexor muscles (pathway 5 in figure 7.30) is capable of resetting the locomotor rhythm (Hiebert et al., 1996; McVea et al., 2005; Yakovenko et al., 2005). In general, muscle afferents provide input that acts as on-off switches to set the range of motion about the involved joints. In contrast, cutaneous inputs are involved in the placement of the limb during normal locomotion or after a disturbance (Rossignol et al., 2006).

The stance-to-swing transition during walking also depends on feedback from tendon organs (Pearson & Ramirez, 1997). Tendon organs provide information on muscle force. Under typical quiescent conditions, activation of tendon organs results in disynaptic inhibition of the motor neurons innervating the muscle in which the tendon organs reside (pathway 2 in figure 7.30). During locomotion, however, input from Ib afferents of extensor muscles produces excitation of the motor neurons that project to the extensor muscles (pathway 3 in figure 7.30)

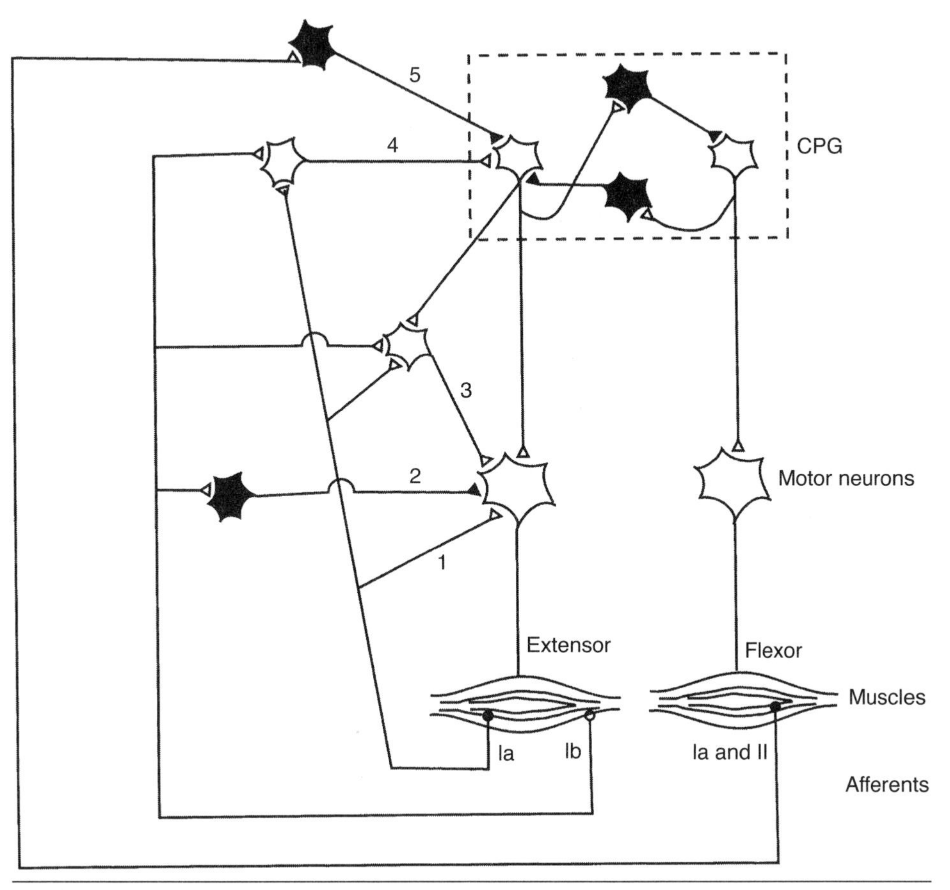

Figure 7.30 Some effects of afferent feedback from muscle spindles (Ia and II) and tendon organ (Ib) to motor neurons and locomotor central pattern generator (CPG).

Data from Pearson, 1993; and Pearson and Ramirez, 1997.

(Donelan & Pearson, 2004). This input has the effect of increasing the extensor muscle activity during the stance phase. Electrical stimulation of Ib afferents from knee and ankle extensor muscles prolongs the stance phase in decerebrate cats (CPG disconnected from higher centers at the level of the midbrain) walking on a treadmill (Whelan et al., 1995). These results indicate that the force in the extensor muscles must decline before the CPG will switch from stance to the swing phase.

Magnitude of Motor Activity One of the functions of feedback from large-diameter afferents is to enhance muscle activation during various phases of the rhythm, such as the stance phase of walking. Three pathways reinforce the activation of the extensor muscles during the stance phase of walking (Pearson & Ramirez, 1997). First, there is the monosynaptic excitation from Ia afferents (pathway 1 in figure 7.30). Second, there is disynaptic excitation from Ia and Ib afferents to the extensor motor neurons (pathway 3 in figure 7.30). This pathway is also activated by the CPG during locomotion. Third, there is polysynaptic excitation of the extensor half-center of the CPG by feedback from Ia and Ib afferents (pathway 4 in figure 7.30). This reinforcement of muscle activity during the locomotor rhythm might be necessary to accommodate changes in the interaction between the limb and its surroundings, as in walking up an incline, walking backward, carrying a heavy load, and walking into a head wind (Ferris et al., 2004; Lamont & Zehr, 2006; Maegele et al., 2002).

Investigators have demonstrated the influence of afferent feedback on the intensity of the locomotor rhythm by recording EMG activity in subjects as they walked on a treadmill with various levels of body weight support. Harkema and colleagues (1997) recorded the EMG activity of leg muscles in four individuals with complete spinal cord lesions at the thoracic level and two able-bodied subjects. The average EMG activity in the soleus, medial gastrocnemius, and tibialis anterior muscles varied with the peak load experienced by the leg. This effect was most pronounced in soleus and medial gastrocnemius when a harness supported at least 50% of body weight. A similar effect of loading on EMG activity was observed in another group of individuals with clinically complete spinal cord injuries when one leg stepped on a treadmill and the other leg experienced various loading conditions (Ferris et al., 2004). These findings indicate that the sensory feedback about the load on the legs can influence the amplitude of the locomotor rhythm (Duysens et al., 2000).

Modulation of Transmission in Reflex Pathways As we discussed in the first part of this chapter, the magnitude of a reflex is not fixed but instead varies across tasks and within the different phases of a single task. When the locomotor CPGs are activated, for example, reflex pathways switch from a posture control mode to a movement control mode (McCrea, 2001; Sinkjær et al., 1996; Stephens & Yang, 1996). This change in control strategy can even lead to a reflex reversal, whereby the same afferent input can evoke the opposite motor response (Stecina et al., 2005). For example, displacement of a limb can elicit a reflex that resists the disturbance when a posture is maintained but a reflex that enhances the displacement during movement (Büschges & El Manira, 1998). Similarly, feedback by Ib afferents inhibits motor neurons during holding of a posture (pathway 2 in figure 7.30) but provides excitation during locomotion.

There are three mechanisms that can change the gain and direction (resistance or assistance) of reflex pathways. One mechanism is efferent modulation of proprioceptors, such as fusimotor control of the sensitivity of muscle spindles (Prochazka, 1989; A. Taylor et al., 2000). Another mechanism is presynaptic modulation of afferent pathways. Variations in the amplitude of monosynaptic reflexes in cats and H reflexes in humans during locomotion are mainly due to presynaptic inhibition of afferent feedback in the spinal cord (Gossard, 1996; McCrea, 2001; Menard et al., 2003). Furthermore, the modulation of H and cutaneous reflexes differs during rhythmic activity (Zehr et al., 2001*b*). The third of these mechanisms involves neuromodulatory substances that are released during a particular state, which can alter transmission in reflex pathways (Burke, 1999; McCrea et al., 1995). Neuromodulators, which are normally amines or peptides, exert an effect through second messengers that can occur rapidly (within 10 ms) and produce a long-term change in the properties of the neuron (Faumont et al., 2005; Fenelon et al., 2003; Thoby-Brisson & Simmers, 2002). For example, substance P can alter transmission in the pathways of pulmonary reflexes during hypoxia, which results in an increase in the ventilatory drive to the respiratory muscles (Pearson & Ramirez, 1997). These mechanisms enable the system to match the efficacy of reflex pathways to its needs during rhythmic pattern generation.

Modulation by Chemicals and Steroids

One of the principles to emerge from the study of CPGs is that they are not fixed circuits but are able to change and thereby modify the behavior produced by the CPG. In addition to modulation by sensory feedback, the intrinsic properties of neurons and the synaptic connections between them are influenced by inputs from modulatory neurons (Ramirez et al., 2004; Wood et al., 2004). For example, neuromodulators can change both the frequency of the locomotor pattern and the intensity of muscle contractions (Sillar et al., 1997). The modulators include such chemicals as amino acids, amines, and peptides, which can alter the electrical properties of neurons in the CPG and can cause the release of neurotransmitters at presynaptic terminals within the CPG (Johnson et al., 2005).

Two of the better-known modulators are serotonin (5-HT) and gamma-aminobutyric acid (GABA). The actions of the amine 5-HT include control of cellular conductances, transmitter receptors, and synaptic connections (Hsiao et al., 2005; Ladewig et al., 2004). For example, 5-HT can alter some of the voltage-gated conductances of motor neurons and thereby influence the timing and intensity of the discharge of motor neurons during locomotion (Sillar et al., 1997). The raphe region in the brain stem controls the release of 5-HT that acts on the locomotor CPG (Hornung, 2003); the activity of these neurons increases during motor behavior (Jacobs et al., 2002). Similarly, GABA acts through metabotropic receptors ($GABA_B$) to modulate the synaptic transmission (Cazalets et al., 1998). It can presynaptically modulate synaptic transmission from inhibitory and excitatory interneurons, and postsynaptically it can reduce Ca^{2+} currents. GABA is delivered by neurons that are either intrinsic or extrinsic to the spinal cord.

Another class of neurons that can influence CPGs and behavior are those that release steroid hormones (Weeks & McEwen, 1997). Steroids are able to induce neuronal plasticity, such as dendritic remodeling, neurogenesis, neuronal death, and alterations in excitability and neuropeptide expression. Among vertebrates, the rat is the species in which the effects of steroids are most commonly studied (Kiehn & Butt, 2003). Sex and steroid hormones, along with influencing development, can modify the structure and characteristics of neurons in the brain. For example, both sex and stress hormones can alter the structure and function of the hippocampus, which participates in contextual memory, spatial memory, and declarative memory (Normann & Clark, 2005; Weeks & McEwen, 1997). The importance of these mechanisms is underscored by findings that many of the molecular and cellular processes involved in steroidal modulation are conserved across species. These hormones play key roles in the construction, expression, and modulation of the neural networks that produce motor behavior (Nusbaum et al., 2001).

Descending Pathways

Although the CPGs that produce the motor pattern for locomotion are located in the spinal cord, the control of locomotion involves descending pathways to the spinal cord from the brain. In general, the descending pathways activate the spinal locomotor system, adjust the motor pattern based on sensory feedback from the limbs, and guide limb movement in response to visual and vestibular input (Drew et al., 2008; Grillner et al., 2008; Jordan et al., 2008; Rossignol et al., 2006).

The activation signal for locomotion arises from the medial reticular formation in the brain stem and is transmitted to the spinal cord via the reticulospinal tract (figure 7.31). Neurons in the medial reticular formation receive input from several sources, including the brain stem areas known as the **mesencephalic locomotor region** and the **diencephalic locomotor region** (lateral hypothalamus). One of the classic studies on locomotion demonstrated

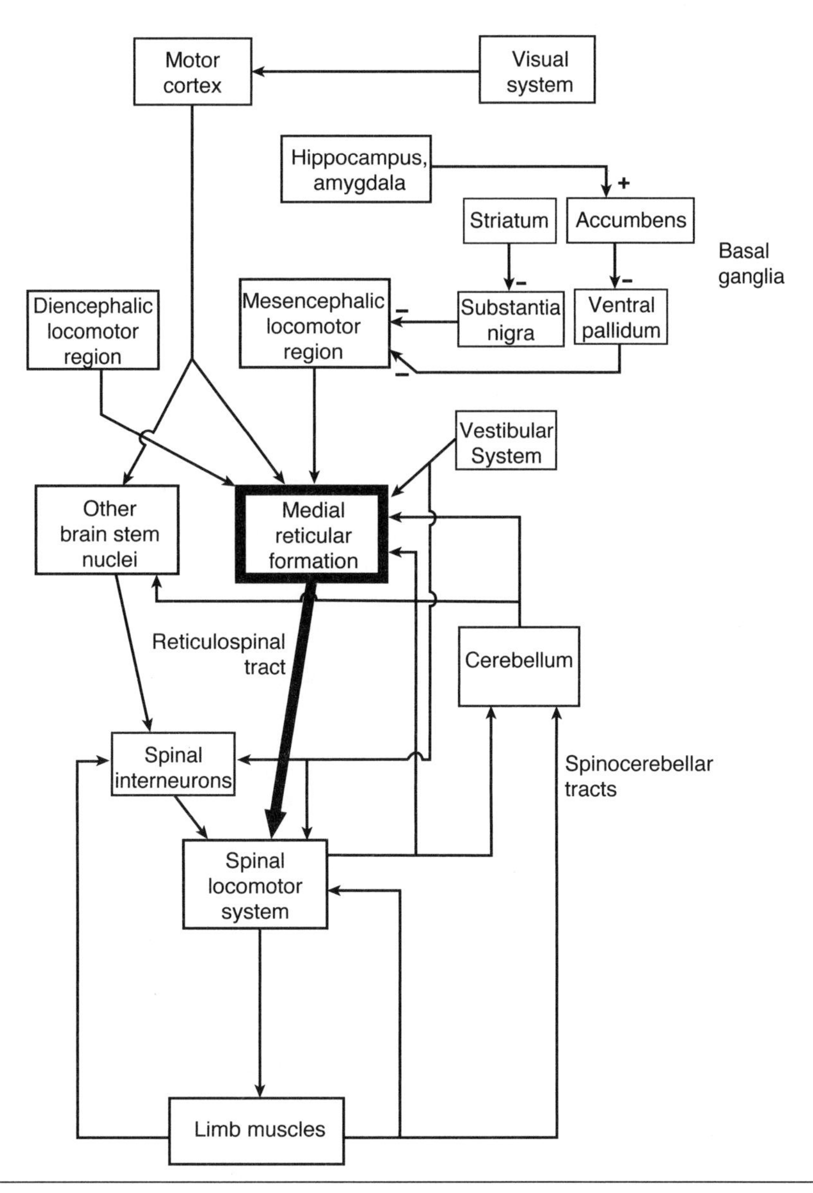

Figure 7.31 Descending pathways influencing the motor output from the spinal locomotor system that includes the rhythm-producing CPGs. All connections are excitatory, except those within and from the basal ganglia to the mesencephalic locomotor region.

that electrical stimulation of the mesencephalic locomotor region could enable decerebrate cats to walk, trot, or gallop, depending on the intensity of the stimulation (Shik et al., 1966). This observation indicated that the activation signal delivered by the reticulospinal tract from the mesencephalic locomotor region could be converted into the relevant patterns of muscle activity by the spinal locomotor system. The activation signal, therefore, controls both the initiation and speed of locomotion.

The mesencephalic locomotor region can be activated by the **basal ganglia,** which comprise several nuclei with major projections to the cerebral cortex, thalamus, and some brain stem nuclei (figure 7.32). The input received by the basal ganglia arrives in the three divisions of the **striatum** (caudate nucleus, the putamen, and the ventral striatum, which includes the nucleus accumbens) and the **subthalamic nucleus.** The outputs arise from the **pallidum,** which comprises the globus pallidus pars interna, substantia nigra pars reticulata, and ventral pallidum. The output of the basal ganglia tonically inhibits the motor commands for rapid eye movements (saccades), locomotion, and posture. In order for these actions to be evoked, therefore, the output of the basal ganglia must be suppressed. As indicated in figure 7.32, the connections within the basal ganglia can either facilitate or depress their output. When the output is depressed (gray lines), the tonic inhibition is reduced and the motor commands can generate the various actions. When the output is

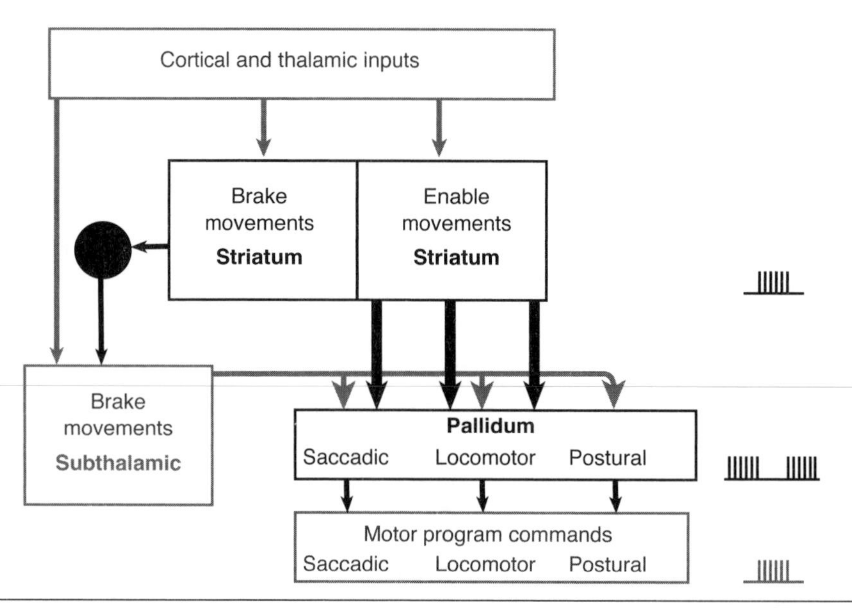

Figure 7.32 Schematic representation of the functional pathways involving the basal ganglia. The input from the cerebral cortex and thalamus arrives in the pallidum and subthalamic nucleus. The input received by the striatum projects either directly or indirectly (through the subthalamic nucleus) to the output nuclei, which are shown as the pallidum. The output from the basal ganglia tonically inhibits the motor centers that produce many automatic behaviors, such as saccades, locomotion, and the control of posture. The inhibitory output is suppressed by the direct pathway from the striatum, and is facilitated by the indirect pathway from the striatum through the subthalamic nucleus. The enabling of movement is indicated by the trains of action potentials on the right; when neurons in the part of the striatum that enables movement discharge action potentials, the tonic inhibition is suppressed and the motor program can produce a train of action potentials.

Adapted from *Trends in Neurosciences*, Vol. 28, S. Grillner, J. Hellgren, A. Ménard, K. Saitoh, and M.A. Wikström, "Mechanisms for selection of basic motor programs—roles for the striatum and pallidum," pg. 7, Copyright 2005, with permission from Elsevier.

enhanced (black lines), however, the inhibitory effects of the basal ganglia are further increased (Grillner et al., 2005; Rossignol et al., 2006; Takakusaki et al., 2004). The different pathways available to activate the reticulospinal neurons (figure 7.31) are associated with the various behavioral contexts in which locomotion may be necessary, such as exploratory, appetitive, and defensive locomotion (Jordan, 1998).

Sensory feedback from the limbs can influence the activity of both the spinal rhythm generator and the descending signals. This input adjusts the motor pattern to the surroundings in which the locomotion is being performed. Sensory inputs can be strong enough to stop and start locomotion. For example, mechanical stimulation of the skin or electrical stimulation of cutaneous afferents can stop locomotion (Viala et al., 1978). More commonly, however, afferent inputs impinge on supraspinal structures involved in the control of locomotion and modulate the discharge of neurons in the medial reticular formation (Drew et al., 2008). Furthermore, perturbations applied during locomotion can even influence the discharge of action potentials by neurons in the motor cortex, presumably as a response to ensure appropriate foot placement during stepping (Christensen et al., 2000; Grillner et al., 1997; Matsuyama et al., 2004). The involvement of neurons in the motor cortex during locomotion is greatest when the surroundings require greater precision of limb kinematics (Armstrong & Marple-Horvat, 1996).

Most importantly, clinical observations underscore the significant role of the **cerebellum** in adjusting the locomotor rhythm. Patients with damage to the cerebellum exhibit an abnormal gait that includes variation in the speed and range of motion in single limbs and the coordination between limbs (Morton & Bastian, 2004; Timmann & Horak, 2001). In general, the cerebellum influences the activity in the motor system by adjusting the performance to reduce differences between the intended and the actual motor behavior. In the context of locomotion, the cerebellum receives input from proprioceptors in the involved muscles and from the rhythm generators in the spinal cord. This input is transmitted over two different spinocerebellar pathways. The cerebellum compares these two inputs and distributes outputs to several supraspinal structures, including the medial reticular formation (figure 7.33). This output, however, appears to contribute predictive (feedforward) rather than reactive (feedback) adjustments to the locomotor cycle (Armstrong & Marple-Horvat, 1996; Morton & Bastian, 2006).

In addition to the role of sensory feedback in adjusting the motor patterns during locomotion, the trajectory of the

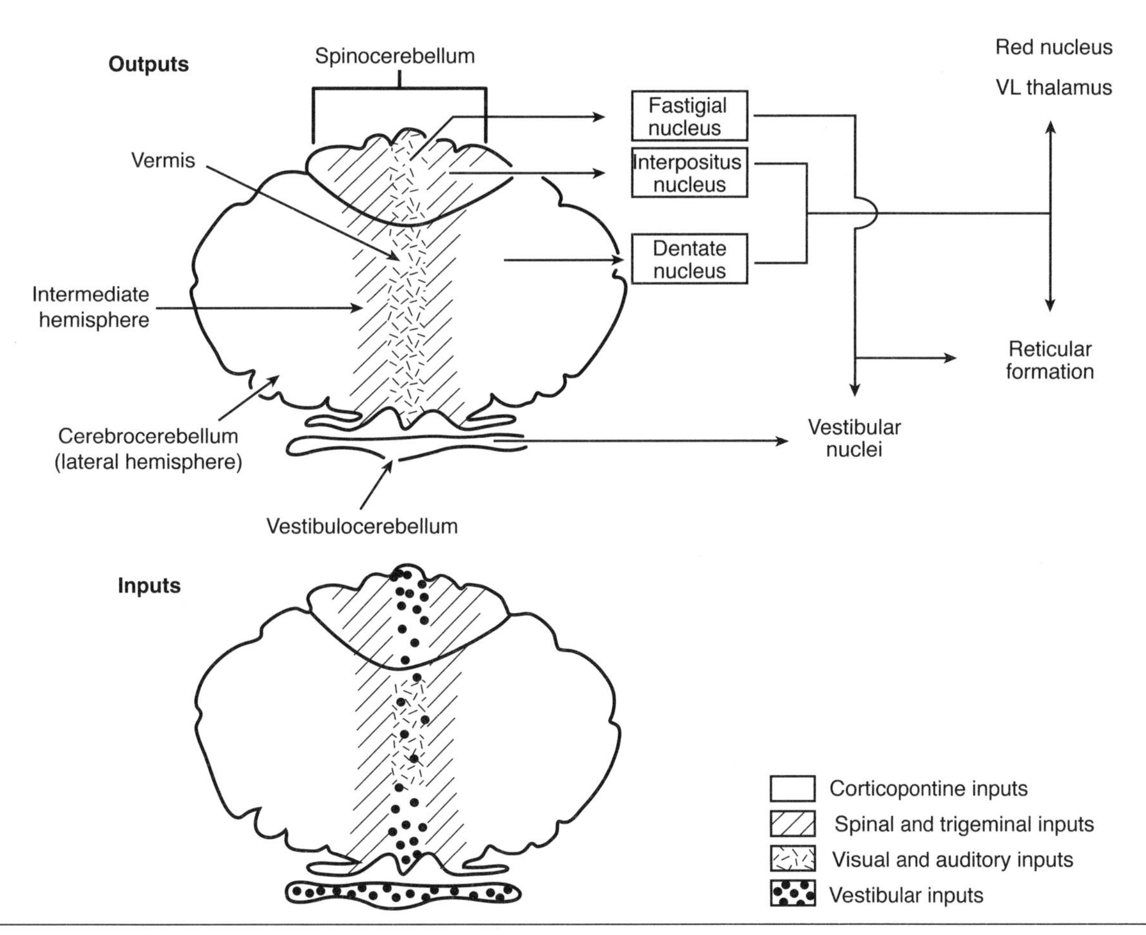

Figure 7.33 Three functional subdivisions of the cerebellum, each of which receives input from and sends output to distinct parts of the CNS. The three divisions are the spinocerebellum, cerebrocerebellum, and vestibulocerebellum. The **spinocerebellum** receives most of its input from the spinal cord and sends its output to the fastigial nucleus and the interpositus nucleus. The **cerebrocerebellum** receives input from the cerebral cortex via the pons and sends its output through the dentate nucleus to the thalamus. The **vestibulocerebellum** receives its input from the vestibular nuclei in the medulla and sends its output back to the same nuclei.

limbs is guided by input from the visual and vestibular systems (Drew et al., 2008; Grillner et al., 2008). Visual input is used to choose a direction and to avoid obstacles and, together with vestibular input, to establish the orientation of the body (Bent et al., 2005; Rossignol et al., 2006). For example, visual information enables an individual to judge the distance to an obstacle and to prepare the appropriate avoidance strategy (Mohagheghi et al., 2004; Patla & Greig, 2006). These adjustments are accomplished by output from the visual cortex being directed through the motor cortex to working memory, which likely involves the posterior parietal cortex, and to descending brain stem pathways (Drew et al., 2004; McVea & Pearson, 2006). In the lamprey, visual and vestibular inputs converge on the same reticulospinal neurons, which enable the vertebrate to compensate for imbalances in the sensory feedback it receives (Rossignol et al., 2006). Visual input likely dominates during normal locomotion, but the influence of vestibular input increases when visual information is insufficient (Deshpande & Patla, 2005).

EXAMPLE 7.9

Symptoms and Signs of Cerebellar Diseases

The most sensitive indicators of cerebellar disease in humans are disturbances of upright stance and gait (Thach & Bastian, 2004). Disorders of the human cerebellum can be grouped into three categories:

1. **Hypotonia**—a lack of muscle tone that results in a diminished resistance to passive limb displacements. One manifestation of hypotonia is the continued oscillation (about seven times) of the leg after a single tap of the patellar tendon has been used to evoke the tendon jerk reflex.
2. **Ataxia**—an impaired ability to execute voluntary movements, sometimes described as a

lack of coordination. The functional deficits include a delay in initiating responses, errors in the range of movement (dysmetria), an impairment in the ability to match a prescribed rate and regularity of a movement, and difficulty in producing the appropriate timing of the components in a multijoint movement.

3. **Tremor**—an action (or intention) tremor (oscillation) that is expressed when a limb approaches a target. The tremor represents erroneous corrections to the range of motion due to poor adaptive control by the cerebellum.

Individuals with cerebellar damage exhibit difficulty in anticipating and adjusting for variation in the mechanical demands associated with movement. Bastian (2002) reported two difficulties experienced by these individuals: (1) a reduction in the accuracy of multijoint movements due to the influence of interaction torques between the involved body segments, and (2) an impaired ability to adapt a catching movement to a change in the mass of the ball. These results suggested that the cerebellum is involved in the rapid adaptation to variation in the load encountered during a movement. Similarly, subjects with cerebellar dysfunction had a reduced ability to scale anticipatory postural adjustments to an impending perturbation during step initiation (Timmann & Horak, 2001).

EXAMPLE 7.10

Assessing Vestibular Input During Movement

Two strategies are available to study the influence of vestibular inputs on the control of movement: assessing the capabilities of patients with dysfunction of the vestibular system and evaluating the effect of a transient perturbation of the vestibular system on movements performed by healthy persons. A transient disturbance of vestibular feedback can be achieved with **galvanic vestibular stimulation,** which involves applying a direct current for several seconds to the eighth cranial nerve through electrodes placed over the mastoid processes (Fitzpatrick & Day, 2004). The current alters the discharge rate of action potentials along the vestibular afferents (Goldberg et al., 1984). When the current (~1 mA) is passed through the anode (negative terminal of a galvanic cell) to the cathode, the discharge rate of vestibular afferents decreases, and subjects tilt the head toward the electrode when standing and follow a curved path when walking. Vestibular afferents discharge spontaneously at rest, although the regularity of the discharge varies across neurons; and changes in discharge rate indicate acceleration of the head in a specific direction. Galvanic vestibular stimulation can influence afferents arising from both the otolith organs and the semicircular canals and has a greater effect on neurons that discharge irregularly (Fitzpatrick & Day, 2004).

Galvanic vestibular stimulation can evoke reflex responses in the EMG signals of limb and trunk muscles when they are involved in balance activities (Fitzpatrick et al., 1994). The onset of the current elicits short-latency (~40 ms) and medium-latency (55-65 ms in the arms and 110-120 ms in the legs) responses; the converse effects are observed when the current is turned off. The directions of the two responses are opposite such that a short-latency increase in EMG is followed by a medium-latency decrease in EMG. The medium-latency response is more readily influenced by other sensory input and can be attenuated by visual input (Britton et al., 1993). The amplitude of the responses varies with the intensity of the stimulation. These responses are complicated, however, because reflexes evoked by galvanic vestibular stimulation represent the response to a perceived threat to balance that is not real. To maintain balance, therefore, other sensory inputs must provide compensatory reactions (Fitzpatrick & Day, 2004).

Galvanic vestibular stimulation has been used to examine the influence of vestibular input during tasks that involve acceleration of the body, such as walking and stepping forward (Bent et al., 2005). For example, the influence of galvanic vestibular stimulation on the placement of the foot in the medial-lateral direction varies across consecutive stance phases during walking and is greatest at heel contact compared with midstance and toe-off (Bent et al., 2004). When the vestibular input was perturbed, therefore, the medial-lateral placement of the feet, which assists the individual in maintaining balance, varied across the gait cycle. The interpretation of this finding was that vestibular input is most important during the period of double support (stance phase) when the trajectory for the next step is being planned.

Motor Programs for Locomotion

An individual who has learned to perform a particular movement can accomplish it without having to think about the details of what has to be done. For example, we rarely have to pay attention to the execution of the stance and swing phases of the gait cycle when we walk from one location to another. This ability is often explained by invoking the concept of a **motor program,** which is defined as a characteristic timing of muscle activations that produce specific kinematic events. According to this scheme, the act of learning a movement involves establishing the muscle activations that are required to perform the action. Once the movement has been learned, all that is required is an activation signal to trigger the set of established muscle activations.

The most common experimental approach used to study motor programs is to identify stereotypical patterns of muscle activation that are associated with a movement and infer that these correspond to the control signals dispatched from the spinal cord (d'Avella & Bizzi, 2005; Krouchev et al., 2006; Lamb & Yang, 2000; Merkle et al., 1998). Seminal work by Lacquaniti and colleagues has used this approach to study the motor patterns employed by humans to walk and run (Cappellini et al., 2006; Ivanenko et al., 2004, 2006*a*). A factor analysis of EMG recordings from 32 muscles on the same side of the body during walking (figure 4.13) indicated that the muscle activations corresponded to five specific events: (1) weight acceptance; (2) propulsion; (3) trunk stabilization during double support; (4) liftoff; and (5) touchdown (figure 7.34). The same pattern is produced for both legs, but shifted in time, and the duration of each set of activations expands and contracts with changes in the speed of walking (Ivanenko et al., 2006*a*). Although the EMG activity of individual muscles can vary for persons with a spinal cord injury compared with able-bodied individuals during weight-supported walking, a similar set of five muscle activations is also observed in those with spinal cord injury (Ivanenko et al., 2003). Furthermore, a similar set of muscle activations can be identified in running, although the timing of the second component occurs earlier in the gait cycle (Cappellini et al., 2006) (figure 7.34).

The motor program associated with walking is not invariant, however, as the different clusters of activity can be modified to accommodate superimposed tasks,

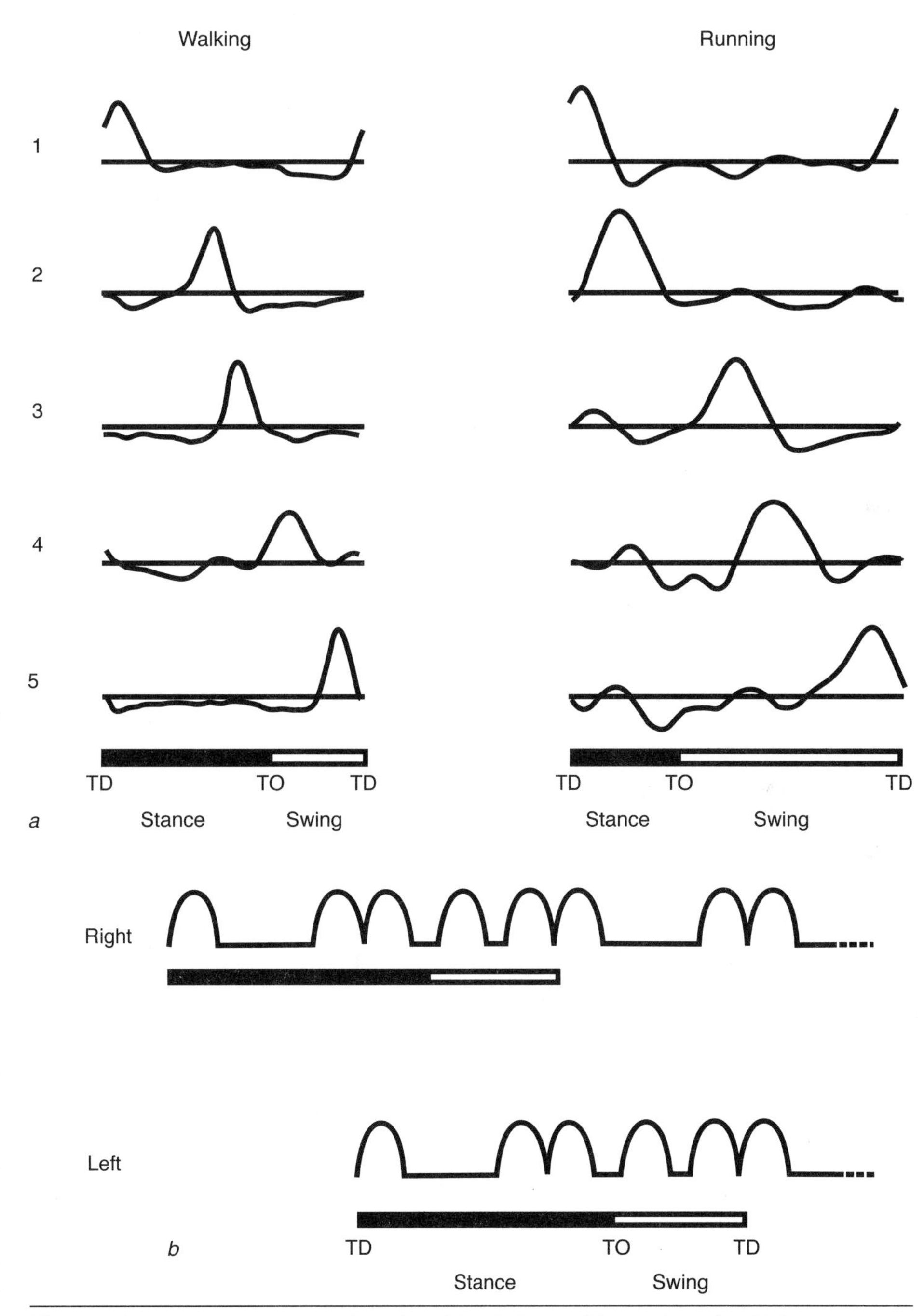

Figure 7.34 The five patterns of EMG activity during walking and running by humans. *(a)* A statistical analysis of rectified and filtered EMG recordings from muscles extending from the neck to the foot identified five clusters of activity (1 through 5). The patterns of activity are indicated relative to the touchdown (TD) and toe-off (TO) events of the stance and swing phases of walking and running. *(b)* The timing of activity on each side of the body during walking.

Adapted from Cappellini et al. 2006; Ivanenko et al. 2006a.

such as stepping over an obstacle, turning a corner, stooping to pick something up off the ground, and kicking a ball (Ivanenko et al., 2005). However, the individual muscles that contribute to each group of muscle activations will vary across conditions. For example, the addition of a kick involves greater activation by muscles innervated by motor neurons in the lumbar spinal cord, whereas stooping involves more activity in proximal muscles and delayed activation of trunk muscles. Nonetheless, the five basic muscle activations are present when voluntary actions are performed during walking, although the relative timing of each pattern can be altered.

VOLUNTARY ACTIONS

The motor system comprises three main levels of control within the CNS: the spinal cord, the descending systems of the brain stem, and the motor areas of the cerebral cortex (figure 7.35). Reflexes and automatic behaviors are mediated by the spinal cord and brain stem, whereas actions produced in response to a perceived need are initiated and controlled by the cortical motor centers.

Organization of the Motor System

Although the spinal cord contains neuronal networks that can produce reflexes and automatic behaviors independently of input from the brain stem and the cerebral cortex, the function of these spinal networks can be modified and controlled by descending input from these higher centers. The reticulospinal system of the brain stem, for example, distributes its projections to the spinal cord widely and can influence the function of many networks, including those that produce the locomotor rhythm. The interactions among the three levels of the motor system are extensive, and no action involves exclusively only one level (Graziano, 2006; Scott, 2008).

The brain stem comprises the medulla oblongata, pons, and mesencephalon (figure 7.36). It integrates information provided by the equilibrium organ (vestibular apparatus) and sensory receptors in the neck region, along with input from the cerebral cortex and cerebellum. The brain stem has four motor centers that send efferent fibers to the spinal cord: red nucleus, lateral vestibular nucleus, tectum, and reticular formation. These motor centers give rise to the **rubrospinal, vestibulospinal, tectospinal,** and **reticulospinal tracts,** respectively. These brain stem pathways are under the control of the sensorimotor cortex and the cerebellum. The cerebellum, which receives input from ascending sensory pathways and the cerebral cortex, has no direct connection to either the spinal cord or the cerebral cortex.

The highest level in the motor system includes the basal ganglia, thalamus, and cerebral cortex. The basal ganglia and thalamus comprise subcortical loops that interact with the cerebral cortex. The cortical areas involved in voluntary actions are the premotor cortex, motor cortex, and parietal cortex (figure 7.37). The motor cortex includes the primary motor cortex and the secondary motor cortex. The more recently evolved part of the cerebral cortex is known as the **neocortex.** It consists of the outer shell of each cerebral hemisphere and is typically organized into six layers; layer 1 corresponds to the surface layer of neurons, and layer 6 denotes the deepest layer of

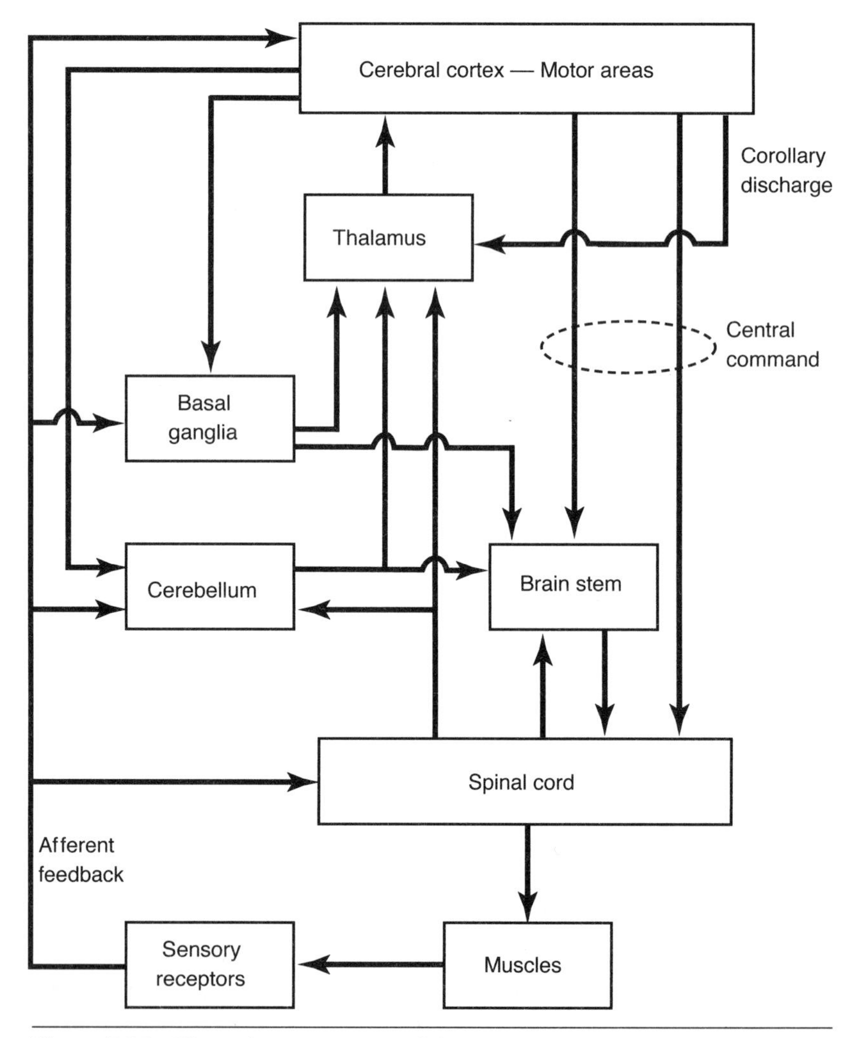

Figure 7.35 The major components of the motor system.

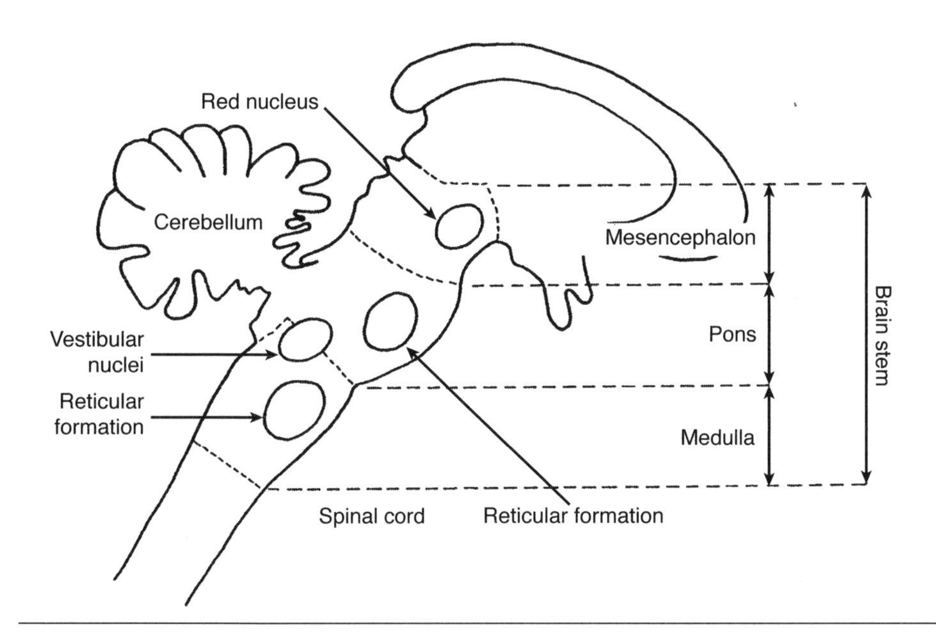

Figure 7.36 Components of the brain stem and the locations of the motor centers.

Springer *Human Physiology,* edited by R.F. Schmidt and G. Thews, 1983, pg. 93, "Motor systems," R.F. Schmidt, ©1983. With permission from Springer Science and Business Media.

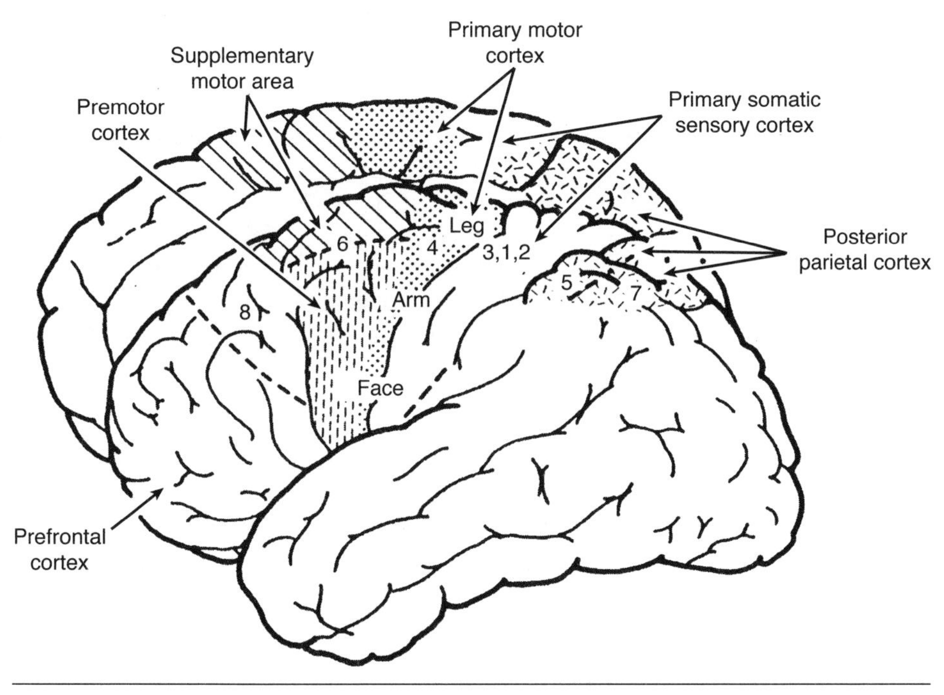

Figure 7.37 Regions of the left and right halves of the cerebral cortex.

neurons. Inputs often arrive in layers 1 and 4; connections between cortical areas occur in layers 2 and 3; outputs to most cortical targets arise from layer 5; and projections to the thalamus occur in layers 5 and 6. The specific inputs and outputs vary across the neocortex and involve the processing of visual, auditory, somatosensory, and motor information. The thalamus provides the main excitatory input to the cerebral cortex; and several sources in the midbrain, including the locus coeruleus, raphe nucleus, and dopaminergic cells, modulate the excitability of the cortex. The motor cortex sends outputs to the basal ganglia, cerebellum, brain stem, and spinal cord.

The primary motor cortex has a **somatotopic** organization: The medial part contains the leg and foot representations, and the most lateral part has the face, tongue, and mouth representations (figure 7.37). The projections from the primary motor cortex to the spinal cord, which comprise the **corticospinal tract,** involve some monosynaptic connections to motor neurons; but mostly the cortical axons terminate on interneurons. Neurons in the primary motor cortex typically synapse on motor neurons belonging to several motor pools; that is, each projection usually influences more than one muscle (Bennett & Lemon, 1994; Cheney & Fetz, 1985; Shinoda et al., 1981). Other areas of the motor cortex, including the premotor cortex and the supplementary motor area, also project to the spinal cord. Thus, there are parallel outputs from the cerebral cortex to the spinal cord. These outputs are often referred to as the **central command** or motor command. A copy of the outgoing central command, known as the **corollary discharge** or efference copy, is returned to cortical centers and compared with incoming sensory information to determine differences between the intended and actual performance of a movement.

Estimating Locations for Reaching

To enable an appreciation of the interactions among the various elements of the motor system in the performance of a movement, this section focuses on the actions required to perform a specific task rather than providing a detailed account of the anatomical connections between the structures. The anatomy is presented when it is relevant to the discussion. The task selected for this purpose is reaching and pointing, which encompasses such actions as reaching to grasp an object, directing the beam of a laser pointer, placing the head of a hammer on a nail, and moving a cursor on a computer screen. For performance of these actions, the CNS must control the location of the hand, laser beam, hammer, and cursor, which are collectively referred to as **end effectors.**

Before one of these actions can be initiated, the CNS must determine the locations of the end effector and the target and then compute a **difference vector** that specifies the amplitude and direction of a movement that would cause the two locations to coincide. The CNS uses the difference vector to develop the motor plan. How does the CNS determine the location of an end effector, such as the hand? It could be determined with vision, but we rarely look at our hand when performing a reaching movement. Alternatively, it could be determined from proprioceptive feedback that provides information about muscle lengths and joint angles. Conversely, the location of the target is most often determined with vision. The CNS, therefore, uses both vision and proprioception to determine the locations of the end effector and the target.

The integration of sensory information to define the configuration of the limb is distributed throughout the CNS. The integration begins early along the ascending pathways as single spinal neurons that project to the cerebellum in the dorsal spinocerebellar tract already convey information from several muscles. About half of the neurons that contribute to the dorsal spinocerebellar tract vary discharge rate in a way that reflects the location of the end effector, whereas the other half of the neurons in this tract encode the configuration of the limb in joint coordinates (Bosco & Poppele, 2001; Bosco et al., 2000). When cats walked on a treadmill, however, the discharge of these neurons was related more to the orientation of the leg and the load that it experienced (Bosco et al., 2006). When the sensory input reaches the cerebral cortex, some of the involved neurons display a discharge pattern that varies with the displacement of the end effector. For example, many neurons in the primary somatosensory cortex (figure 7.37) increase discharge rate more gradually when the hand is moved away from the body compared with when it is moved in the left-right and up-down directions (Tillery et al., 1996). Similarly, neurons in both the posterior parietal cortex and the primary motor cortex modulate discharge rate with variation in hand location and joint angles (Kalaska et al., 1983; Scott & Kalaska, 1997; Scott et al., 1997) and the orientation of the limb (Lacquaniti et al., 1995).

But how is this information used to calculate the difference vector? The difference vector, as the term implies, is the difference between the vectors that specify the locations of the end effector and the target relative to the origin of a coordinate system. Subtracting one of these vectors from the other to derive the difference vector, however, requires that both be expressed in the same coordinate system (Beurze et al., 2006). If the target is located with vision, its position vector will originate from the retina. Conversely, the use of proprioception to specify the location of the end effector will result in a position vector relative to a proximal location on the body, such as the shoulder or the head. Scientists have not yet determined whether the common coordinate system for visually guided reaching and pointing movements is based in the head, shoulder, or an external reference point (Ghafouri & Lestienne, 2006; Henriques et al., 2002; Lemay & Stelmach, 2005). For the purposes of discussion, however, we will focus on a **fixation-centered coordinate system** in which the two locations are specified relative to an external point that is being fixated by the eyes (Shadmehr & Wise, 2005).

A scheme for computing the difference vector in a fixation-centered coordinate system is shown in figure 7.38. The eyes fixate on a point in the surroundings that enables the location of the target to be determined with vision and its position vector ($\mathbf{r}_t$) to be specified in fixation coordinates. The location of the end effector ($\mathbf{r}_{ee}$), however, is based on visual and proprioceptive information about the arm, hand, and eyes (Battaglia-Mayer et al., 2007 ; Graziano et al., 2000). The movement required for the subject to reach the target is simply the difference between these two vectors ($\mathbf{r}_{dv}$). Neurons in area 5 of the posterior parietal cortex appear to represent target and hand locations in fixation-centered coordinates (Buneo & Andersen, 2006; Buneo et al., 2002; Grefkes et al., 2004).

A typical experiment to study the contribution of different regions in the cerebral cortex to a reaching movement involves recording the discharge of neurons in a monkey when it reaches to a specified target. To distinguish between the generation of the difference vector, which corresponds to the development of a motor plan, and the command to perform the reaching movement, the task involves several phases: (1) presentation of the target, (2) a delay before the instruction to perform the movements, and (3) a "go" signal. The results of such experiments have shown that neurons in both the posterior parietal cortex (especially area 5d) and the premotor cortex are active during the delay period, presumably contributing to the development of the motor plan (Crammond & Kalaska, 1989, 2000; Gail & Andersen, 2006). Furthermore, if the task involved either a push or a pull, the discharge of these neurons was modulated more for one direction than the other, which is referred to as the **preferred direction** for the neuron (figure 7.39*a*). When

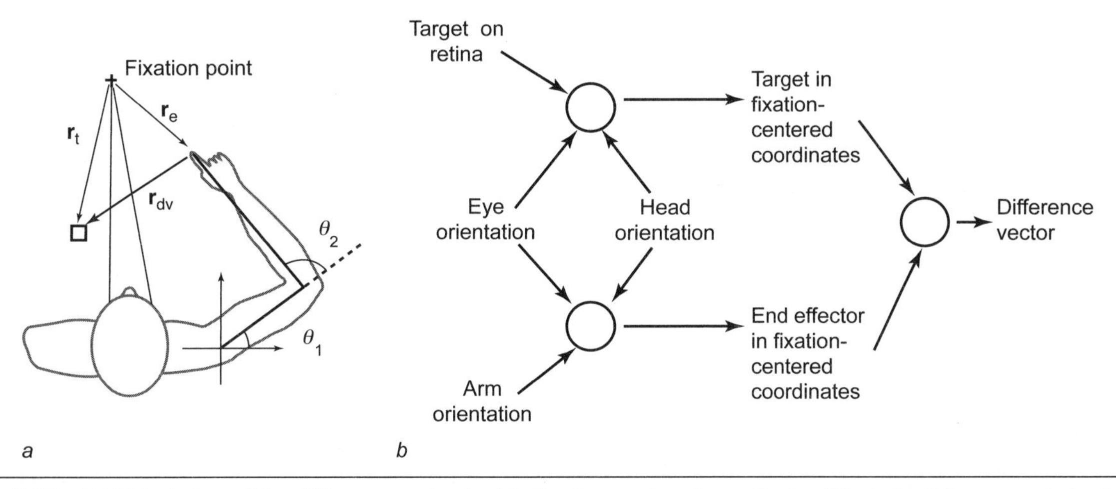

Figure 7.38 *(a)* Calculation of the difference vector ($\mathbf{r}_{dv}$) in fixation-centered coordinates. The task of the person is to place the index finger on the target. *(b)* From the locations of the fixation point and the target on the retina, the position vector to the target ($\mathbf{r}_t$) can be specified. Based on visual and proprioceptive information on arm, head, and eye orientation, the position vector to the end effector (index finger) ($\mathbf{r}_{ee}$) can be expressed in fixation-centered coordinates.

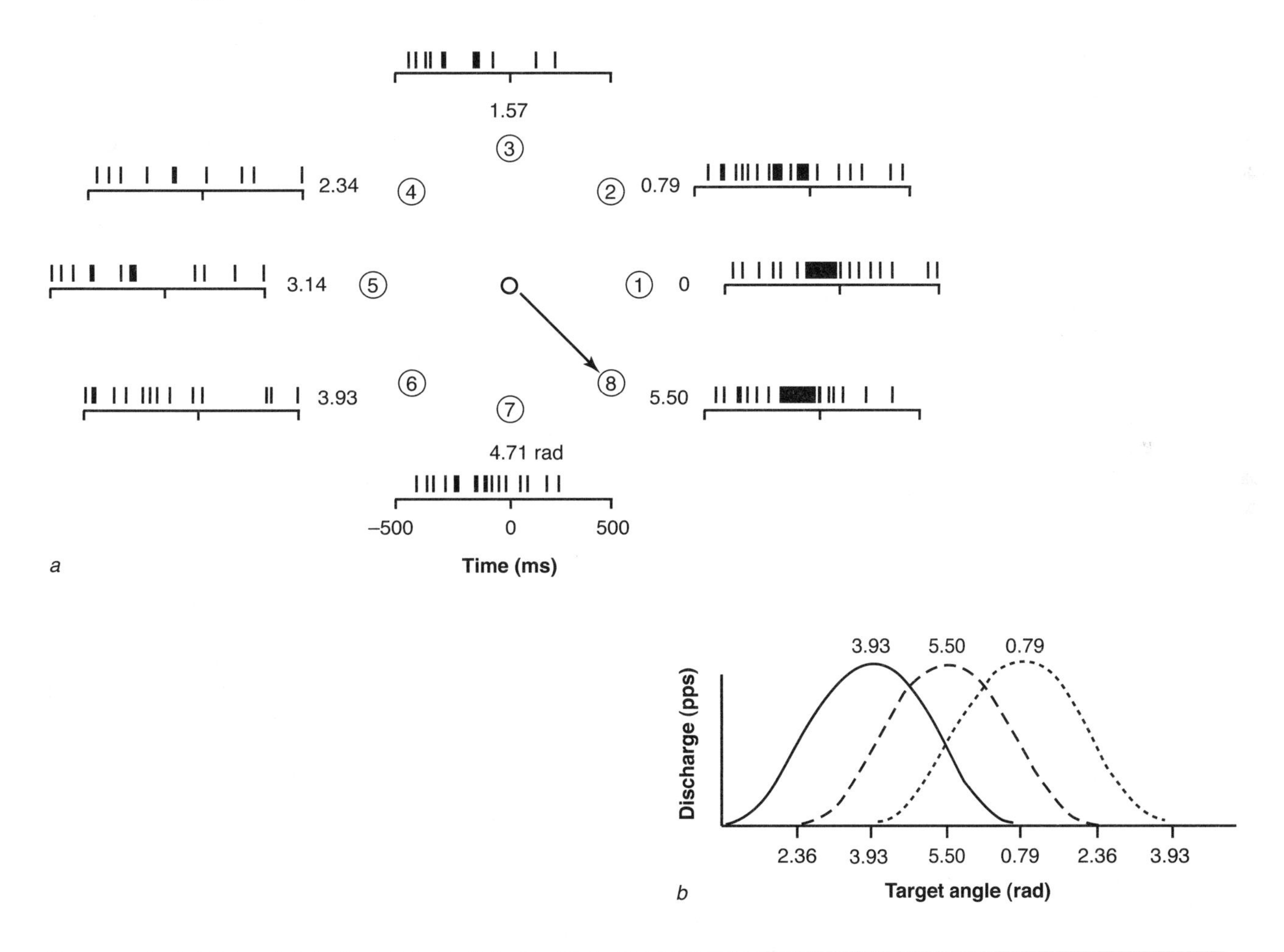

Figure 7.39 Preferred directions of neurons in the primary motor cortex. *(a)* The discharge of a single neuron in the primary motor cortex of a monkey changed the most when the monkey reached toward target 8, which indicated its preferred direction. The discharge of the neuron was also elevated, however, when the reach was toward targets 7, 1, and 2. *(b)* When this distribution of activity was represented with equation 7.1, the resulting function indicated the directional tuning of the neuron. Tuning curves are shown for three neurons with preferred directions of 3.93, 5.50, and 0.79 rad, with the middle curve (5.50 rad) corresponding to the neuron shown in panel *a.*

Data from Scott et al. 2001; Shadmehr and Wise 2005.

EXAMPLE 7.11

Preferred Direction and Directional Tuning

The discharge of many neurons in the posterior parietal, premotor, and primary motor cortices is modulated more strongly when movements are either planned or performed in one particular direction compared with other directions. The direction that is associated with the greatest change in discharge rate is known as the preferred direction for the neuron. Figure 7.39 shows experimental recordings that were used to determine the preferred direction of a neuron in the primary motor cortex. The experiment involved a monkey that performed reaching movements in a horizontal plane from a central location to one of eight targets. Representative trains of action potentials discharged by the neuron are shown beside each target; the records indicate the discharge times of action potentials 500 ms before and after the onset of the reaching movement (0 ms). The change in discharge rate was greatest when the movement was toward target 8, as indicated by the arrow.

The trains of action potentials indicate that the neuron modulated its discharge rate not only when the movement was toward target 8, but also when it was toward neighboring targets. The change in discharge rate for neighboring targets, however, was less than that for target 8. This distribution of activity is quantified with a cosine function (equation 7.1) that represents the difference in discharge rate during a control period and the peak discharge rate during either the preparation or movement time:

$$a(\omega) = b_0 + b_1 \cos(\omega - \omega_{pd}) \tag{7.1}$$

where $a(\omega)$ represents the discharge rate of the neuron when the monkey reaches in direction ω, b_0 indicates the resting discharge rate, b_1 is a scalar term, and ω_{pd} is the preferred direction (Georgopoulos et al., 1982). The middle curve in figure 7.39*b* indicates the cosine function for the neuron shown in figure 7.39*a*. The cosine function corresponds to the directional tuning of the neuron. Because the tuning curves of individual neurons overlap, a reaching movement to a specific target involves the activation of many neurons. For example, a reach to target 7 (4.71 rad) would involve a change in discharge rate for all three neurons shown in figure 7.39*b,* with the relative change in activity for each neuron indicated by the amplitude of its cosine function in the movement direction.

the task involved reaching in a horizontal plane from a central location to one target in a circle of potential targets, discharge was modulated for a few neighboring targets; this activity is known as **directional tuning** (figure 7.39*b*). Accordingly, if the discharge of a neuron was modulated for a target presented to the left, a change in the target to the right during the delay period would cause the neuron to cease the modulation of its discharge (Wise & Mauritz, 1985). In general, it takes neurons about 140 ms to respond once a target has been presented. The directional tuning curves of many neurons correspond to a cosine function.

Experiments on primates indicate that neurons in the premotor cortex and posterior parietal cortex contribute to different aspects of the motor plan. For example, although neurons in both area 5d (posterior parietal cortex) and the dorsal premotor cortex reflect potential targets and movements, those in the dorsal premotor cortex represent the motor plan that has been selected for execution (Cisek & Kalaska, 2002; Kalaska & Crammond, 1995; Koch et al., 2006). Experimenters have shown that when the task involved several reaching movements, the discharge of neurons in the posterior parietal cortex coded the location of the target for the next reaching movement (Batista & Andersen, 2001). When the end effector was a cursor on a computer screen, movement of the cursor to the target involved the same modulation of neurons in ventral and dorsal premotor cortex whether the mouse was moved with the right or the left hand (Cisek et al., 2003; Hoshi & Tanji, 2002). The discharge of these neurons presumably contributes to the development of the difference vector.

The Motor Plan

The successful completion of a reaching movement requires displacements that reduce the difference vector to zero. Because the limb displacements involved in this action are produced by muscle torques, the difference vector must be transformed from fixation-centered coordinates to intrinsic-limb coordinates. Figure 7.40 provides a general scheme of the interactions that are required to define the difference vector ($\mathbf{r}_{dv}$) and to estimate the necessary muscle torque (τ). A central feature of this scheme is the neural networks that transform a set of inputs to a desired output. Each network likely corresponds to a unique set of interconnected neurons that receive one or more inputs and convert them to an output in either the same or a different coordinate system (Battaglia et al., 2003; Burnod et al., 1999).

Network 1 receives proprioceptive input in intrinsic-body coordinates that can be used to *estimate* joint angles ($\hat{\mu}$) and the location of the end effector ($\hat{\mathbf{r}}_{ee}$). Network 1 is known as a **location map** because it specifies the location of the end effector in fixation-centered coordinates. When the end effector is visible while the target is being fixated, visual feedback can also contribute to identifying the location of the end effector in fixation-centered coordinates. Network 2 receives visual input and provides an estimate of target location ($\hat{\mathbf{r}}_t$) in fixation-centered coordinates. Recall that it is necessary for the two position vectors ($\hat{\mathbf{r}}_{ee}$ and $\hat{\mathbf{r}}_t$) to be expressed in the same coordinate system so that simple subtraction will yield the difference vector. Estimates of the two position vectors are used as input to Network 3; and the outputs, in fixation-centered coordinates, are joint angles ($\hat{\theta}$) and the difference vector ($\mathbf{r}_{dv}$). These two outputs then pass to Network 4 and are transformed into estimated displacements ($\Delta\hat{\theta}$) in intrinsic-body coordinates. Network 4 is known as a **displacement map.** The estimated displacements then pass to Network 5, which estimates the muscle torques ($\hat{\tau}$) required to achieve these displacements. Network 5 corresponds to a **dynamics map** as it transforms displacements into torques. In the performance of a reaching movement, Networks 1 and 2 receive continuous inputs that are used to update $\mathbf{r}_{dv}$ and $\hat{\tau}$ throughout the movement.

Experimenters have identified several locations in the frontal lobe, largely the premotor cortex and the primary motor cortex, and the cerebellum where the discharge of neurons is modulated in accordance with the displacement and dynamics maps. For example, neurons included in Network 4 exhibit activity that is related to joint angles, joint displacements, or the direction and magnitude of the difference vector. Furthermore, the activity of some neurons in Network 4 is related to fixation-centered coordinates, whereas the activity of other neurons is more related to intrinsic-body coordinates.

Experiments that have varied gaze location and the location and orientation of the hand have indicated that neurons in the ventral premotor cortex encode the location of the target relative to the end effector and not either joint displacements or patterns of muscle activity (Graziano et al., 1997; Kakei et al., 2001). Similarly, neurons in the dorsal premotor cortex displayed activity related to the difference vector in extrinsic coordinates rather than intrinsic-body coordinates when visual feedback was varied during a reaching task (Ochiai et al., 2002; Shen & Alexander, 1997). Furthermore, the discharge of most neurons in the dorsal premotor cortex is related more to the direction of the difference vector than to its amplitude (Messier & Kalaska, 2000). The activity of neurons in the dorsal premotor cortex, however, represents more than just the difference vector. In an experiment that involved a get-ready signal, a delay period, and then a "go" signal, the discharge of neurons in the dorsal premotor cortex was greatest when the tuning curves for the locations of the get-ready signal and the target coincided (di Pellegrino & Wise, 1993). These results suggest that neurons in the premotor cortex contribute to the transformations for Networks 2, 3, and 4 (Kakei et al., 2003; Kurata & Hoshi, 2002). Neurons on the dorsal premotor cortex may be more involved in the dynamics of the task (Hoshi & Tanji, 2004; Xiao et al., 2006), whereas those in the ventral premotor cortex appear to contribute more to target location (Hoshi & Tanji, 2002; Ochiai et al., 2005).

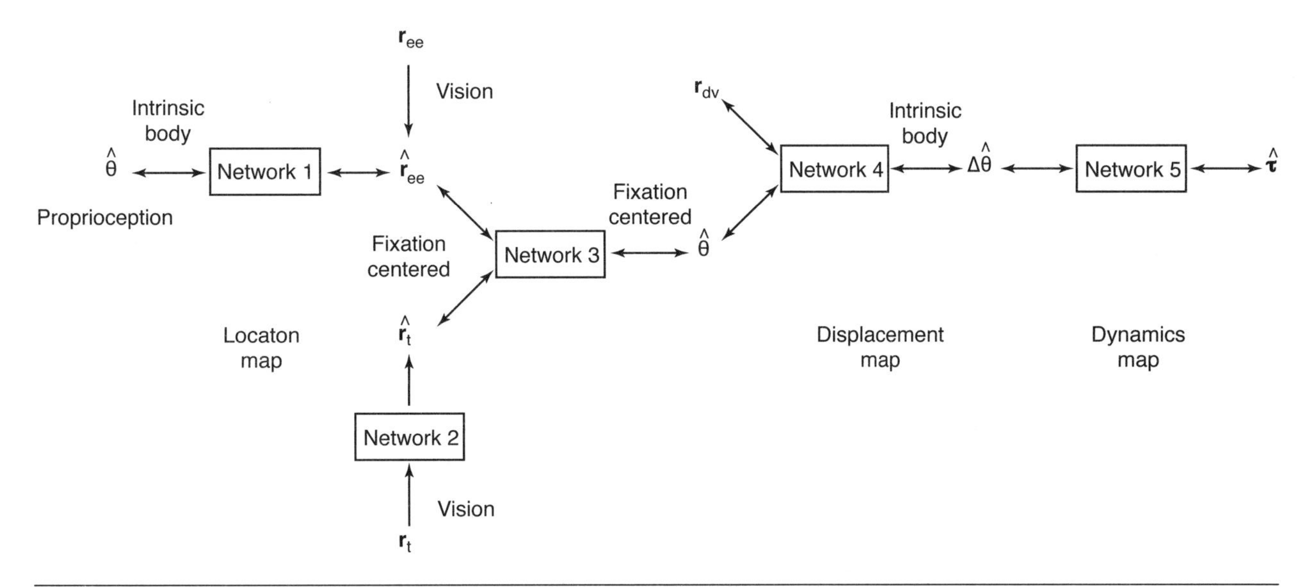

Figure 7.40 Sensorimotor transformations required to perform a reaching task. The hat symbol (∧) denotes an estimate of a variable, such as a position vector ($\hat{\mathbf{r}}$), a joint angle ($\hat{\theta}$), and a muscle torque ($\hat{\tau}$).

Shadmehr, Reza, and Steven P. Wise, *The computational neurobiology of reaching and pointing: A foundation for motor learning,* figure 14.1A, © 2004 Massachusetts Institute of Technology, published by The MIT Press.

The final transformation is performed in Network 5 and involves neurons in the primary motor cortex more than those in the premotor cortex (Hamel-Paquet et al., 2006). According to the scheme in figure 7.40, Network 5 receives information about the required angular displacements for the reaching movement and produces an output that is related to the dynamics of the task (Sergio et al., 2005). Consistent with this expectation, Sergio and Kalaska (2003) found that neurons in the primary motor cortex exhibited tuning curves that were related to the force exerted during an isometric contraction but that varied when the position of the arm was changed. The results from other studies, however, suggest that the change in discharge rate of neurons in the primary motor cortex during voluntary contractions is related to both the magnitude of force and the rate of change in force (Hepp-Reymond et al., 1999). Still other studies suggest that the tuning curves of neurons in the primary motor cortex are related to peak angular power, which is the product of dynamic (torque) and kinematic (angular velocity) variables (Scott et al., 2001). Although there is no consensus on the movement parameters encoded by neurons in the primary motor cortex, some of the divergent findings may relate to whether these neurons are part of Network 5 or both Networks 4 and 5. It seems that there are multiple levels of representation in the motor cortex (Sergio et al., 2005).

Changes in Location During Movement

As indicated in figure 7.40, a reaching task begins with visual and proprioceptive input that is used to estimate the locations of the target and the end effector. The CNS must revise these estimated locations when the eyes, target, or end effector moves during the reaching action. In the simplest reach, the individual remains fixated on the same extrinsic point, and only the location of the end effector changes as the difference vector is reduced to zero. Even in this instance, however, the location of the end effector is updated periodically during the movement. In many reaching movements, both the target and the eyes move; this requires a continuous revision of these locations. For example, when one is playing a video game that involves moving a cursor to targets on a screen, the fixation points and the target change frequently, and the difference vector must be updated with each change in location.

The revision of a location is not based solely on feedback about observed changes; it can begin before the change occurs. This predictive revision was observed in an experiment that recorded the discharge of neurons in the posterior parietal cortex in response to a visual stimulus and superimposed rapid eye movements (Duhamel et al., 1992). Rapid eye movements to a target, which are known as saccades, are controlled by the superior colliculus that sends projections to the posterior parietal cortex. Neurons in the posterior parietal cortex respond to specific areas on the visual field; these are known as **activity fields.** When a monkey fixated on a point and a visual stimulus was presented in the activity field of the neuron, the neuron responded to the stimulus by increasing its discharge rate. When the monkey next fixated on a point that was distant from the activity field of the neuron, presentation of the visual stimulus did not alter the discharge rate of the neuron. However, when the fixation point changed to a location close to the activity field of the neuron and the monkey performed a saccade to the new fixation point, the discharge of the neuron changed 80 ms before the saccade and 150 ms before the visual stimulus came into the activity field of the neuron. This indicates that the neuron knew about the change in location before it occurred; hence it received input that predicted the change in location.

The source of the prediction for target location involves a corollary discharge (efference copy) of the motor commands that move the eyes (Colby et al., 2005; Wurtz & Sommer, 2004). Six muscles control the position of the eyes, and these comprise the oculomotor system. Because the corollary discharge pathway from the oculomotor center passes through the thalamus and back to the cerebral cortex (Bellebaum et al., 2005), the increase in discharge rate of neurons in the dorsomedial nucleus of the thalamus 140 ms before the onset of the saccade suggests that this pathway contributed to the prediction of the new location (Sommer & Wurtz, 2002). Studies on humans have demonstrated that the posterior parietal cortex, which receives input from the thalamus, is involved in updating target location during movements (Medendorp et al., 2003; Prablanc et al., 2003). For example, the application of **transcranial magnetic stimulation** to disrupt the function of the posterior parietal cortex impaired the ability of individuals to accommodate a change in target location that occurred during a movement (Desmurget et al., 1999).

Similarly, the location of the end effector must be updated during a reach. End-effector location can be estimated quite easily if it can be seen (Saunders & Knill, 2003), but most reaching movements involve fixating on the target and not on the end effector. In the absence of vision, therefore, it is reasonable to assume that the location of the end effector is estimated from proprioceptive feedback. However, reaching movements performed when subjects were asked to look in the direction of the hand when it could not be seen suggest that corollary discharge also contributes to the estimation of end-effector location. In these experiments, subjects made saccades along a series of self-selected fixation points that predicted the location of the hand 150 ms later (Ariff et al., 2002). Thus, the eye muscles (oculomotor system) knew where the hand would be in 150 ms—knowledge that must have been based on corollary discharge and not on proprioceptive feedback.

These findings suggest that knowledge of the planned motor commands, which is provided by the corollary discharge, can contribute to estimating target and end-effector locations. When the reaching task involves the displacement of a limb, however, the ability to predict the location of the end effector, as indicated by the saccade data, requires that the CNS also have knowledge about the biomechanics of the limb. It would be necessary to know, for example, the configuration of the limb and at least its mass properties and lengths. Thus, the CNS must have an internal model of how the motion of the limb will be altered by the planned motor commands (Hwang & Shadmehr, 2005; Ren et al., 2006; Tong & Flanagan, 2003; Wolpert & Ghahramani, 2000). Internal models can predict the forces that are needed to reach a target. The type of internal model that can be used to predict future locations, for an end effector for example, is known as a **forward model** (Kawato et al., 2003; Mehta & Schaal, 2002). Figure 7.41 provides an example of a forward model that could generate predictive saccades indicating the location of the end effector at some future time. In this scheme, the corollary discharge of the motor command to the end effector (cd_{ee}) is led back to Network 1, where it combines with proprioceptive feedback from the limb and the corollary discharges to the eye (cd_i) and head (cd_h) muscles to estimate the future (Δt) location of the end effector [$\hat{\mathbf{r}}_{ee}(t + \Delta t)$]. Based on the time it takes proprioceptive signals to reach the CNS, the proprioceptive feedback that is integrated with the corollary discharges provides information about previous limb motion.

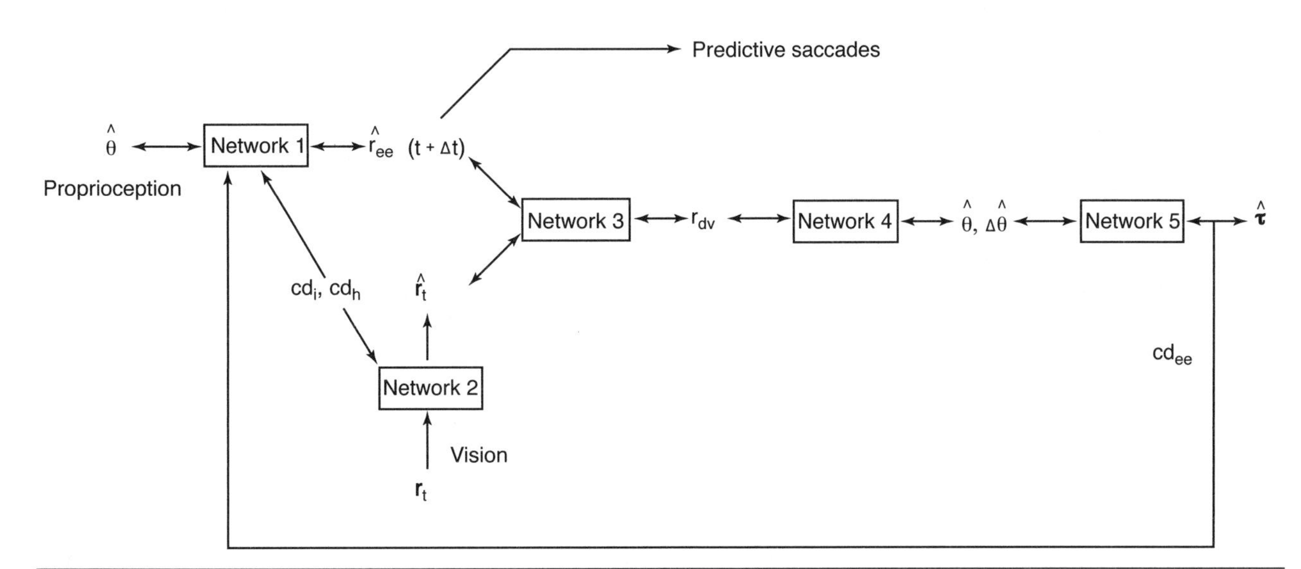

Figure 7.41 A forward model that combines corollary discharges and proprioceptive feedback to predict the future location of an end effector.

EXAMPLE 7.12

Shocks to the Brain Evoke Muscle Responses

One way to study transmission along neural pathways is to generate action potentials artificially at one location and record the response evoked at another location. The responses are known as **evoked potentials.** For example, an electric stimulus applied in the periphery evokes small brain potentials that can be averaged and then measured. These potentials can also be measured along the spinal cord with surface electrodes placed on the back. Similarly, electric or magnetic stimuli applied to the brain evoke muscle contractions, as occurs with transcranial magnetic stimulation (Mills, 1991; Reis et al., 2008; Taylor & Gandevia, 2004; Terao & Ugawa, 2002). This technique involves energizing a coil of wire encased in plastic by the rapid discharge of a large capacitor, which results in rapidly changing currents in the windings of the coil that produce a magnetic field perpendicular to the plane of the coil. The magnetic field passes through the scalp and skull and induces an electric current in the brain that flows in the opposite direction tangential to the skull (figure 7.42). The induced current activates presynaptic elements that

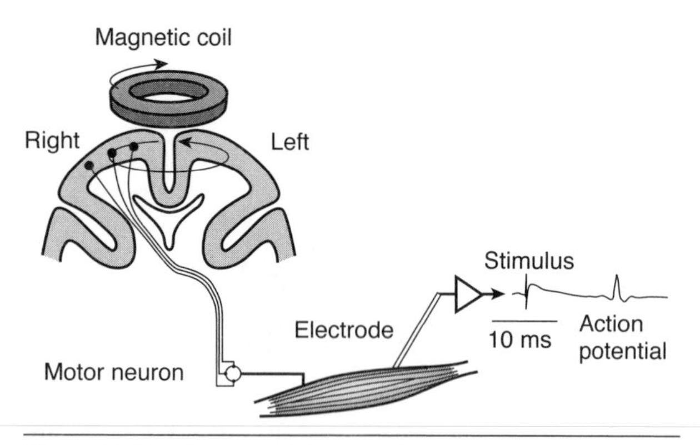

Figure 7.42 Transcranial magnetic stimulation can evoke an action potential in a single motor unit.

project to the output neurons, which then activate muscle (Edgley et al., 1997). By moving the coil to different locations over the scalp, it is possible to evoke a response in various muscles throughout the body (Schieppati et al., 1996).

By measuring the time it takes for the stimulus to evoke an action potential in a single motor unit, it is possible to identify the pathway activated by transcranial magnetic stimulation. The latency from the stimulus to the response in a hand muscle is about 25 ms, which means that the stimulus must have activated a monosynaptic pathway (Garland & Miles, 1997). Such measurements can provide useful information about various neurological conditions and neural circuits (Mills, 1991; Petersen et al., 2003; Taylor & Gandevia, 2004). An increase in the latency, for example, can suggest delays in central conduction, as experienced by patients with multiple sclerosis, motor neuron disease, cervical myelopathy, and Friedreich's ataxia. Delays in central conduction can be caused by a decrease in the number of axons in the corticospinal tract, demyelination of these axons, or a block of action potential propagation in these axons. Since the development of the magnetic stimulator in the early 1980s, the technique has been used to study projections from the motor cortex to the spinal motor neurons and to identify motor impairments such as weakness and spasticity in patients (McKay et al., 2005; Mills, 1991).

Motor Output

When an individual performs a reaching movement, the path traveled by the end effector is usually straight unless there is an obstacle that must be avoided. Figure 7.43 shows an example of point-to-point reaching movements in a horizontal plane (Morasso, 1981). The subject was instructed to move from one target to another; figure 7.43*b* shows the displacement of the hand in moving from Target 2 to Target 5 and from Target 1 to Target 4. The displacement of the hand between the targets was linear with a single peak in the velocity–time graph. This meant that each reaching movement was accomplished by a continual decrease in the difference vector. The straight-line trajectory for the hand, however, required angular displacements about the shoulder and elbow joints in both directions, as indicated by the positive and negative angular velocities for both the shoulder and elbow (figure 7.43*f*). When this observation was combined with the results of experiments that involved distorting the feedback on hand location during a reach (Wolpert et al., 1995), the findings indicated that the CNS chooses visually straight displacements for such reaching movements. Furthermore, these data indicate that the reaching tasks involved minimizing a difference vector computed in visual coordinates (Flanagan & Rao, 1995).

Because reaching trajectories, especially those that have been practiced, are so consistent, it is reasonable to assume that the motor plan specifies a desired trajectory for a specific task. When a reaching movement is perturbed or the target changes, however, the continuation of the reach to the target is uninterrupted, which suggests that the CNS does not have to stop and determine a new trajectory (Hoff & Arbib, 1993; Nijhof, 2003; Scheidt et al., 2005; Ulloa & Bullock, 2003). Rather, trajectories are likely determined by a system known as a **next-state planner** (Shadmehr & Wise, 2005). This system combines feedback to estimate the current state of the limb (position, velocity, and acceleration) with the estimate of target location to determine the next step in displacement; that is, the next-state planner determines what to do next. With this strategy, trajectories emerge as a series of small displacements, and the motor output can produce a range of arbitrary trajectories (Schaal & Schweighofer, 2005). Patients with Huntington's disease, which involves a progressive atrophy of the striatum in the basal ganglia, have difficulty modifying trajectories during a reach; it seems that the basal ganglia are involved in next-state planning (Novak et al., 2002, 2003; Nowak & Hermsdörfer, 2006; Smith et al., 2000). The problem appears to be that proprioceptive feedback is diminished in these patients (Abbruzzese et al., 1990; Berardelli et al., 1999; Meyer et al., 1992) and this reduces their ability to correct errors that occur in a trajectory.

In healthy individuals, the CNS exhibits considerable plasticity in being able to adapt reaching and pointing movements to changing conditions, such as variable force fields applied to the end effector and rotations of the whole body (Lackner & DiZio, 2005; Lai et al., 2003; Shadmehr & Mussa-Ivaldi, 1994). This adaptability is

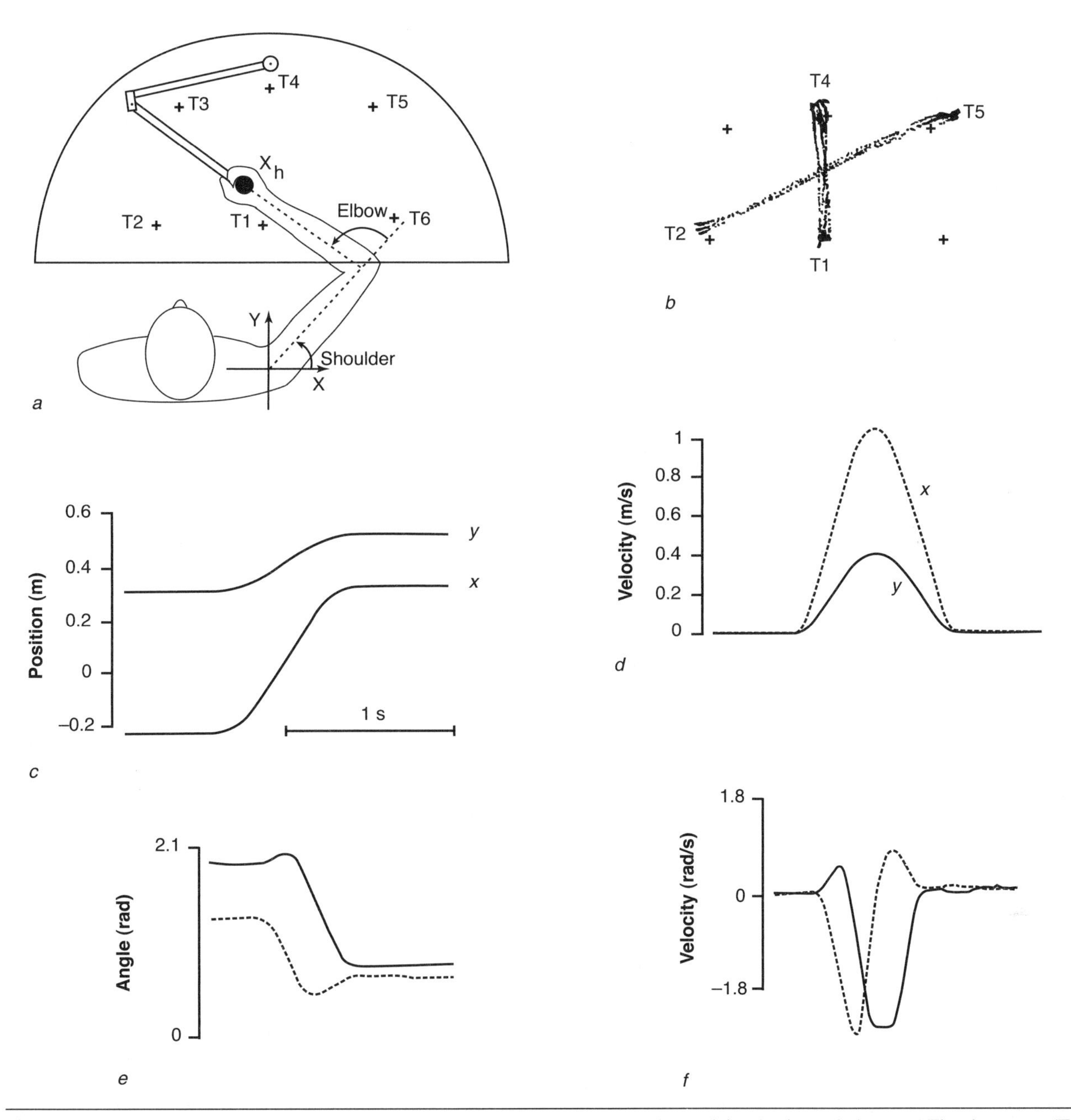

Figure 7.43 Hand and arm trajectories associated with a point-to-point reaching task in a horizontal plane. *(a)* The six targets (T1-T6) used for the reaching movements. *(b)* Hand path during reaching from T2 to T5 and from T1 to T4. *(c)* Position–time graph for the hand in *x-y* coordinates for the reach from T2 to T5. *(d)*. Velocity–time graph for the hand in *x-y* coordinates for the reach from T2 to T5. *(e)* Angle–time graph for the elbow (solid line) and shoulder (dashed line) joints for the reach from T2 to T5. *(f)* Angular velocity–time graph for the elbow (solid line) and shoulder (dashed line) joints for the reach from T2 to T5.

Shadmehr, Reza, and Steven P. Wise, *The computational neurobiology of reaching and pointing: A foundation for motor learning,* figure 18.2, © 2004 Massachusetts Institute of Technology, published by The MIT Press.

presumed to rely on an internal model of dynamics in which networks are able to predict the forces required to achieve a specific state of the limb given its current state. In this scheme, the internal model maps limb states to forces and not to the timing details of the task (Conditt & Mussa-Ivaldi, 1999; Karniel & Mussa-Ivaldi, 2003). When circumstances disrupt the accuracy of the forces predicted by the internal model, the muscles involved in the task will experience unexpected changes in length that can evoke long-latency reflex responses. With sufficient exposure to the altered condition, the internal model will adapt and change the relative activation of the muscle to accommodate the new state (Thoroughman & Shadmehr, 1999; Wang et al., 2001). Individuals with damage to the cerebellum have difficulty in adapting to conditions that disturb the accuracy of an internal model (Maschke et al., 2004; Rost et al., 2005), which underscores the observation that the cerebellum is involved in the predictive control of movements (Bastian, 2006).

The capacity of muscle to exert only a pulling force means that each degree of freedom about a joint is controlled by at least two muscles that provide opposing actions. As a consequence of this organization, a rapid pointing task that requires displacement about a single degree of freedom is accomplished as the two muscles produce a series of EMG bursts that accelerate the limb toward the target and then slow it down to arrive at the target. These two muscles are often referred to as the agonist and antagonist; the **agonist** muscle provides the acceleration toward the target and the **antagonist** muscle generates the acceleration in the opposite direction. When the agonist and antagonist muscles are activated at the same time, an occurrence known as **coactivation,** the actions oppose one another and increase the stiffness about the joint (slope of the torque–angle relation). The initial adjustment of an individual when new conditions are encountered during the performance of a reaching movement is to increase the level of coactivation, thereby increasing the stiffness about the involved joints (Hinder & Milner, 2005; Osu et al., 2002). The advantage of using this strategy is that perturbations cause less of a displacement away from the intended trajectory. As the internal model adapts to the new conditions, the heightened levels of muscle activation, and hence joint stiffness, are reduced back to normal levels.

Due to the off-axis attachments of most muscles and the number of two-joint muscles in the human body (figure 6.39), there are relatively few pairs of muscles that exert only opposing actions. Rather, actions are produced by the concurrent activation of a group of muscles. A particular balance of muscle activations that changes across time is known as a **muscle synergy,** and movement is produced through the coordinated activation of these synergies (d'Avella et al., 2003; Tresch et al., 2006). For example, d'Avella and colleagues (2006) recorded the EMG activity of 19 shoulder and arm muscles as subjects performed point-to-point reaches to eight targets in either a sagittal or frontal plane; they found that the movements involved only four or five muscle synergies. Each synergy comprised the coordinated activations of specific muscle groups that included synchronized bursts of EMG for some muscles and asynchronous increases and decreases in EMG activity for other muscles. Furthermore, the synergies extracted from the point-to-point movements explained much of the variability in EMG activity for other reaching tasks. Hence, the mapping of motor plans into muscle contractions can be accomplished by the flexible and task-dependent combination of a few patterns of muscle activity. A number of motor behaviors, such as grasping, reaching, walking, and running, involve only a few synergies, which presumably simplifies the control strategy used by the CNS for these actions (Cappellini et al., 2006; Hart & Giszter, 2004; Krishnamoorthy et al., 2003; Overduin et al., 2008; Weiss & Flanders, 2004).

EXAMPLE 7.13

Coactivation Does Not Limit Performance in Stroke Patients

Individuals who experience a stroke exhibit varying degrees of residual motor function following the cerebrovascular accident. The remaining capabilities in the arm can be assessed through evaluation of the ability of patients to perform such tasks as touching the opposite knee, placing the hand on the chin, and raising the arm overhead (Gowland et al., 1992). The inability of a patient to perform one of these tasks is a consequence of the generation of an inadequate net muscle torque. Neurodevelopmental theory (Bobath, 1978) suggests that this impairment is due to an inappropriate coactivation of muscle because of a failure to inhibit antagonist activity. However, the problem has more to do with muscle weakness and insufficient motor unit activity than with heightened coactivation (Burke, 1988; Fellows et al., 1994; Gowland et al., 1992; Kamper et al., 2006). Moreover, persons with poststroke hemiplegia benefit from physical training during rehabilitation (Brown & Kautz, 1998; Brown et al., 2005; Kluding & Bilinger, 2005; Ouellette et al., 2004).

Grasping

The goal of a reaching movement is often to grasp, lift, and manipulate an object. These actions are also based on predictive control but include a greater reliance on sensory feedback both to assess the accuracy of the predictions and to evaluate the characteristics of the object being manipulated (Castiello, 2005; Flanagan et al., 2006; Henriques & Soechting, 2005; Overduin et al., 2008). Once the target has been identified, the planning of the reach-to-grasp movement must include a decision on how the object is to be grasped, and its mass properties must be estimated. Although the hand can form many different shapes (Brochier et al., 2004; Weiss & Flanders, 2004), most studies distinguish between about four types of grips: pinch grip, key grip, power grip, and precision grip (Mason et al., 2001; McDonnell et al., 2005; Napier, 1960; Perkins et al., 1994). The pinch grip involves holding the object between the tips of the thumb and index finger. The key grip corresponds to squeezing the object between the sides of the thumb and index finger, as occurs when one inserts a key into a lock. The power grip, which places all fingers and the thumb around the object, is used when one grasps a barbell or a tennis racket. With the precision grip, all five fingertips contact the object. The common feature of all grips is that the force exerted by the thumb is opposed by the actions of one or more fingers, which establishes biomechanical constraints on the applied forces (Dumont et al., 2006; Zatsiorsky & Latash, 2004).

When the reach involves grasping an object, the output from the motor cortex prior to initiation of the movement already includes information about the object. Experimenters in one study demonstrated this by imposing pairs of transcranial magnetic stimulation pulses to assess the excitability of corticocortical inputs to the motor cortex as subjects prepared to grasp different objects (Cattaneo et al., 2005). There was an increase in excitability of the corticospinal neurons that projected to muscles involved in the action at least 600 ms before the movement. Furthermore, the change in excitability was related to the geometrical properties of the object that was about to be grasped. Some of the input received by the neurons in the motor cortex likely arose from the posterior parietal cortex, as it has been shown with functional magnetic resonance imaging that activity in the regions increased when individuals were required to coordinate the fingertip forces exerted by the thumb and index finger in a lifting task (Ehrsson et al., 2003). Direct recordings have indicated that the discharge rate of neurons in the posterior parietal cortex (areas 5 and 7b) of monkeys increased 200 to 500 ms before contact, peaked at contact, and declined after the grasp was secured (Gardner et al., 2007*a*). Area 5 is presumed to be involved in sensorimotor transformations that coordinate the reach and grasp behaviors. The actions of grasping and lifting the object likely involve separate internal models (Flanagan et al., 2003; Quaney et al., 2005; Salimi et al., 2003).

Once the object has been grasped, subsequent actions can involve moving the object to another location. The contact, lift, arrival at the new location, and release of the object give rise to distinct sensory events, which are referred to as **contact events** and can be used to predict expected and actual afferent feedback (Flanagan et al., 2006). The feedback comprises tactile signals, such as from mechanoreceptors in the glabrous skin (Johansson & Cole, 1992), but also involves proprioception, vision, and audition. The feedback contributes to the increase in discharge rate of neurons in the primary somatosensory cortex during the contact and grasp phases of the task and is sent both to the primary motor cortex to adjust the current grip and to the posterior parietal cortex to update plans for future grasp-and-lift actions (Gardner et al., 2007*b*). Tactile afferents can provide information about the friction of the contact; the shape of the contact surface; and the direction of the fingertip forces that can be used to generate rapid reactive grip responses to correct errors in the estimates of mass distribution, weight, and stable fingertip force vectors that were based on visual feedback. Significantly, the critical information is conveyed by the timing of multiple afferent signals rather than by the rate coding of action potentials from single sensory receptors (Johansson & Birznieks, 2004). The sensory feedback is used to develop and update internal models of the dynamic properties of the object (Flanagan et al., 2006).

SUMMARY

This chapter is the third of three on the properties of the motor system. It examines the control of three classes of actions produced by the CNS: spinal reflexes, automatic responses, and voluntary actions. The chapter is organized into three sections, each addressing one of the classes of action. The chapter begins by characterizing the properties of the sensory receptors that provide afferent feedback for the control of movement. After entering the spinal cord, the afferent axons from the different receptors are distributed widely and contribute to all three classes of actions. Although the pathways that mediate spinal reflexes typically contain few neurons, the responses are highly adaptable depending on the context in which the reflex is evoked. As discussed in the second section of the chapter, the afferent signals can also trigger and modify automatic responses—both postural adjustments and the output of neuronal circuits (central pattern generators) that produce coordinated patterns in response to descending inputs. The second section includes a description of the organization and properties of the constituent neurons and the interactions among the neurons that comprise central pattern generators and introduces the concept of descending pathways and the modulation of spinal cord activity by higher centers. As an introduction to the control issues and strategies used by the higher centers in the performance of voluntary actions, the third section of the chapter discusses reaching and pointing movements. The issues considered are how the CNS estimates locations during reaching tasks; the interactions that contribute to the development of a motor plan; how the CNS deals with changes in location during a movement; the characteristics of the motor output; and the control of grasping actions.

SUGGESTED READINGS

Biewener, A.A. (2003). *Animal Locomotion.* Oxford, UK: Oxford University Press, chapters 1-3.

Enoka, R.M., & Pearson, K.G. (in press). The motor unit and muscle action. In E.R. Kandel, J.H. Schwartz, & T.M. Jessell (Eds.), *Principles of Neural Science* (5th ed.). New York: McGraw Hill, in press.

Heckman, C.J., & Enoka, R.M. (2004). Physiology of the motor neuron and the motor unit. In A. Eisen (Ed.), *Clinical Neurophysiology of Motor Neuron Diseases. Handbook of Clinical Neurophysiology* (Vol. 4, pp. 119-147). Amsterdam: Elsevier.

Jones, D.A., & Round, J.M. (1990). *Skeletal Muscle in Health and Disease.* Manchester: Manchester University Press, chapters 1-4.

Lieber, R.L. (1992). *Skeletal Muscle Structure and Function.* Baltimore: Williams & Wilkins, chapters 1-3.

Pierrot-Deseilligny, E., & Burke, D. (2005). *The Circuitry of the Human Spinal Cord.* Cambridge, UK: Cambridge University Press.

Shadmehr, R., & Wise, S.P. (2005). *The Computational Neurobiology of Reaching and Pointing.* Cambridge, MA: MIT Press.

Vogel, S. (2001). *Prime Mover.* New York: Horton, chapters 1-3 and 6.

Part II Summary

The goal of part II (chapters 5-7) has been to define the structure and function of the motor system. The motor system comprises those parts of the nervous system and muscle that are responsible for movement. After reading these three chapters, you should be able to do the following:

- Appreciate the role of electricity in establishing the properties of excitable membranes
- Identify the paths along which current will flow in an electrical circuit
- Understand the mechanisms that produce the resting potential across an excitable membrane
- Note the properties of neurons that enable them to generate action potentials
- Conceive of the means by which information is transmitted rapidly throughout the motor system
- Recognize the direct and indirect means by which incoming signals can evoke responses in target neurons
- Realize the bidirectional effects of neurons and muscle fibers on each other
- Know the appropriate methods for the recording, measurement, and interpretation of muscle fiber potentials
- Comprehend the structure of muscle down to the molecules that form the contractile proteins and the associated connective tissues
- Acknowledge the critical role of calcium in connecting the excitation from the nervous system to the contraction of muscle
- Understand the interaction between contractile proteins that produces muscle force
- Realize that the motor unit is the basic functional unit of the motor system and describe its neuronal and muscular features
- Perceive the differences between the various types of muscle fibers and motor units
- Know that the nervous system controls muscle force by varying the number of activated motor units and the rate at which each motor neuron discharges action potentials
- Distinguish the influence of muscle length and rate of change in length on the force that can be exerted by a single muscle fiber
- Differentiate the effects of muscle and joint architecture on the force exerted by muscle
- Appreciate the influence of the mechanical properties of muscle on the force exerted, the work performed, and power produced by activated muscle fibers
- Comprehend the basic features of afferent feedback and the properties of the sensory receptors that contribute to the control of movement
- Understand a spinal reflex as a fast response that involves an afferent signal into the spinal cord and an efferent signal out to the muscle
- Elaborate on how the study of reflex pathways provides information about the contribution of sensory feedback to the control of movement
- Discern the reflex pathways and the connections between neurons in the spinal cord
- Describe examples of automatic responses that are triggered by afferent inputs and descending pathways

- Distinguish the two functions served by postural contractions
- Appreciate the organization and properties of the neurons that comprise central pattern generators, as well as the interactions among them
- Observe the descending pathways from higher centers that are involved in the control of central pattern generators
- Conceive of the connections between the various components of the motor system, including those in the cerebral cortex and brain stem
- Grasp the transformations that must be performed by the CNS to accomplish pointing and reaching tasks
- Acknowledge the importance of coordinating visual feedback with CNS activity in achieving motor goals
- Delineate the roles served by different parts of the cerebral cortex to perform voluntary actions
- Understand the control issues that confront the CNS during the performance of a motor task
- Recognize the significance of predictions within the strategies used by the CNS to control movement
- Consider internal models to denote the CNS representation of the kinematics and dynamics of a motor task
- Appreciate that critical sensory information can be conveyed by the relative timing of multiple afferent signals
- Comprehend that sensory feedback is used to develop and update internal models of the dynamic properties of objects

part III

Adaptability of the Motor System

In this text, we have characterized human movement as an interaction between the motor system and its surroundings. In part I, we discussed the concepts and principles that have been derived from physics for the study of motion. In part II, we examined the motor system, which comprises those parts of the nervous system and muscle that are responsible for movement. Our approach was to begin with the principles of electricity to explain the properties of excitable membranes; such properties provide the means for the rapid communication that can extend from the cerebral cortex to the muscle fibers. We also considered how the nervous system activates muscle and how the CNS organizes the activation signal. Part III of the text describes the acute adjustments (chapter 8) and chronic adaptations (chapter 9) that are exhibited by the motor system in response to physical activity.

OBJECTIVES

To conclude our study of the neuromechanics of human movement, we examine the adaptive capabilities of the motor system. The goal of part III is to describe the ways in which the motor system adapts to various types of physical stress. Specific objectives include the following:

- To explain the effect of altering core temperature on performance capabilities
- To indicate the techniques that have been developed to improve flexibility
- To outline the mechanisms that can cause muscle fatigue and discuss a strategy to identify the functionally significant adjustments that occur during fatiguing contractions
- To describe the potentiating capabilities of muscle
- To review some principles of exercise prescription
- To define the performance characteristics of strength and power and identify the mechanisms that mediate changes in these capabilities

- To document the adaptations that occur after periods of reduced activity
- To evaluate the changes that occur with aging and discuss the physiological basis of these changes

chapter 8

Acute Adjustments

One prominent characteristic of the motor system is its adaptability. When subjected to an acute or chronic stress, it can adapt to the altered demands of usage. These adaptations can be extensive, and they have been shown to influence all aspects of the motor system, both morphological and functional. In this chapter, we consider the immediate (acute) response of the motor system to the stress associated with a single bout of physical activity. We will examine the effects of a warm-up, the techniques and mechanisms underlying changes in flexibility, muscle soreness and damage, the mechanisms that cause muscle fatigue and task failure, the phenomenon of muscle potentiation, and the effects of arousal on performance.

WARM-UP EFFECTS

Often when an individual undertakes a bout of physical activity, this is preceded by light exercise to prepare the body for the ensuing stress. The purpose of the light exercises is to elicit a warm-up effect, which includes increasing core temperature and disrupting transient connective tissue bonds. The increase in core temperature will improve the biomechanical performance of the motor system, and the stretch may reduce the possibility of a muscle strain (Amako et al., 2003; Garrett, 1990). However, a regular program of stretching exercises does not produce a clinically relevant reduction in the incidence of lower limb injury (Pope et al., 2000; Thacker et al., 2004; Yeung & Yeung, 2001). Nevertheless, stretching may be beneficial prior to the performance of activities that involve stretch-shortening cycles (Witvrouw et al., 2004). The effects of a warm-up are different from those achieved with exercise designed to increase flexibility—that is, those intended to induce long-term increases in the range of motion about a joint.

Temperature

The warm-up has a significant effect on the physiological processes that are influenced by temperature. The elevation of core temperature can increase the dissociation of oxygen from hemoglobin and myoglobin, enhance metabolic reactions, facilitate muscle blood flow, reduce muscle viscosity, increase the extensibility of connective tissue, increase the conduction velocity of action potentials, and increase the twitch force of motor units (Farina et al., 2005; Rall & Woledge, 1990). Because our focus is on human movement, we are most interested in the effects of a warm-up on the ability of muscle to exert force, perform work, and produce power.

Muscle Function

Performance in an activity such as a vertical jump or a throw is generally enhanced after a warm-up. The improvement occurs because the warm-up increases the maximal power that a muscle can produce (figure 8.1) and these activities depend on the power produced by the involved muscles. The enhancement of jump performance following a warm-up is caused by the effect of muscle temperature on contraction speed. For example, changes in temperature within the physiological range alter the maximal velocity of muscle shortening (the *x*-axis intercepts in figure 8.1) but not the maximal isometric force (Binkhorst et al., 1977; Cheung & Sleivert, 2004; de Ruiter et al., 1999). The increase in contraction speed produces an increase in peak muscle power (de Ruiter & de Haan, 2000). However, passive heating by 2° C can impair the voluntary activation of muscle and thereby reduce the maximal voluntary contraction (MVC) force (Morrison et al., 2004; Todd et al., 2005). Conversely, a reduction in muscle temperature decreases its work capacity (Wade et al., 2000) but does not alter the force

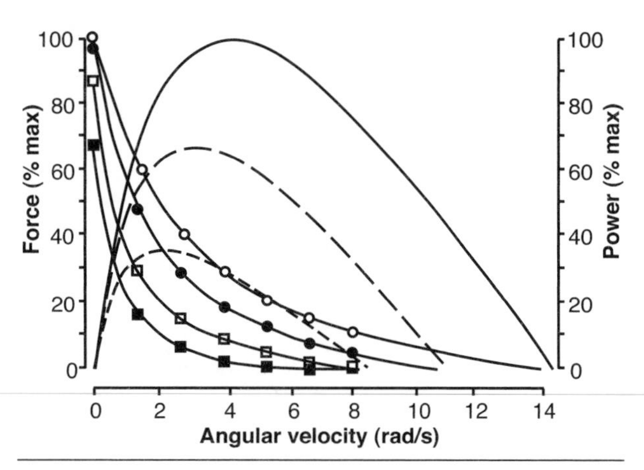

Figure 8.1 The influence of muscle temperature on the force (measured) and power (estimated) produced by the adductor pollicis muscle at selected shortening speeds (angular velocity). Muscle temperature was reduced by immersion of the hand in a water bath for 20 min before each measurement of the force exerted by the thumb; the temperatures were 37.1° (open circles), 31.4° (filled circles), 25.6° (open squares), and 22.2° C (filled squares). The muscle was activated by electrical stimulation of the ulnar nerve. Force is expressed relative to the maximal isometric value (0 rad/s) measured at a muscle temperature of 37.1° C. The power curves were calculated from the Hill model (Example 6.2): The solid line corresponds to the open circles, the long-dash line to the filled circles, and the short-dash line to the open squares.

Data from De Ruiter and De Haan 2000.

fluctuations during steady submaximal contractions (Geurts et al., 2004).

Muscle force can be altered when there is a substantial change in muscle temperature (Ranatunga et al., 1987; Steinen et al., 1996). For example, de Ruiter and de Haan (2000) found that the maximal force that could be evoked in a hand muscle with electrical stimulation did not change until muscle temperature decreased from 25.6° to 22.2° C. Conversely, Bergh and Ekblom (1979) observed an increase in the maximal isometric torque of the knee extensor muscles from 262 N•m at 30.4° C to 312 N•m at 38.5° C (2.4%•$°C^{-1}$), with the temperature measured in the vastus lateralis muscle. Furthermore, these changes increased vertical jump height by 44% (17 cm) and maximal power production in cycling by 32% (316 W).

The influence of temperature on muscle force is much less than the effect on the consumption of adenosine triphosphate (ATP) by muscle (figure 8.2). On the basis of measurements made on muscle fiber segments obtained by biopsy from the rectus abdominis and vastus lateralis muscles, Steinen and colleagues (1996) determined myofibrillar adenosine triphosphatase (ATPase) activity (an index of ATP consumption) and force for various muscle fiber types. Measurements were made at four different temperatures and normalized to the values obtained at 20° C. As indicated by the slopes of the two lines in figure 8.2, the myofibrillar ATPase activity varied more than specific force over the temperature range examined. Although the data in figure 8.2 are for type IIa fibers, similar relations were observed for type I and type IIx fibers. The cost of force generation, however, when expressed as the ratio of ATP consumption to specific force, was least for type I fibers and greatest for type IIx fibers. Furthermore, the cost of force generation changed with temperature, as indicated by the difference in the slope for the two lines in figure 8.2.

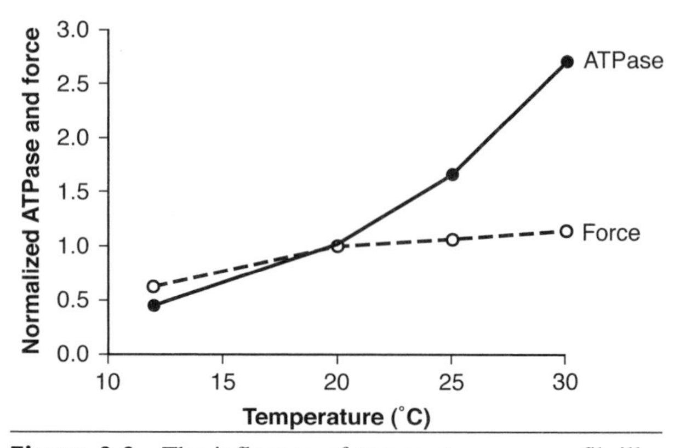

Figure 8.2 The influence of temperature on myofibrillar adenosine triphosphatase (ATPase) activity (mmol/L•s^{-1}) and force (N/mm^2) for type IIa fibers.

Data from Steinen et al. 1996.

EXAMPLE 8.1

Warm-Up Effect Independent of Temperature

A preceding activity, such as that performed during a warm-up, can have an effect on the force capacity of muscle independent of changes in temperature. When single muscle fibers were stimulated to produce ten 400 ms tetani (stimulation at 70 Hz) with 4 s between each tetanus, the peak force increased by 10% (Bruton et al., 1996). The increase in force lasted for about 15 min. The potentiation could not be attributed to augmentation of Ca^{2+} release or to changes in intracellular pH. The most likely explanation was the 40% reduction in inorganic phosphate (P_i), which can produce an 8% to 12% increase in tetanic force (Steinen et al., 1990). Conversely, some of the decline in force after cast immobilization is probably caused by an elevation of P_i (Cooke et al., 1988; Pathare et al., 2005).

Warm-Up Techniques

The strategies that can be used during a warm-up can be classified as either passive or active heating techniques (Bishop, 2003*a*). An active warm-up is one in which the change in temperature results from muscle activity, whereas a passive warm-up involves heating with an external source, such as a warm bath or a heating pad. The increase in muscle temperature during an active warm-up depends on the intensity of the muscle contraction. When subjects performed isometric contractions (8-20 s in duration) with the quadriceps femoris muscles, Saugen and Vøllestad (1995) found that the increase in temperature in vastus lateralis was greatest for forces in the range of 30% to 70% of MVC force. The increase in temperature ranged from 3.1 mK/s at 10% MVC to 14 mK/s at 70% MVC (mK = millikelvin). The increase in temperature was not greater with stronger contractions; this suggests that warm-up exercises should not exceed moderate intensities.

Ingjer and Strømme (1979) were able to raise the intramuscular temperature of the lateral part of quadriceps femoris from 35.9° C to about 38.4° C with both active and passive techniques. Nonetheless, they favored an active warm-up because the result was that aerobic processes provided a greater proportion of the energy expenditure during a treadmill run. Active warm-up improves the execution of a brief performance (<10 s) more than does a passive warm-up (Bishop, 2003*b*) and can attenuate the accumulation of lactate in a subsequent intense activity (Gray et al., 2002). The duration of the warm-up should be greater than 5 min, and the intensity should be equivalent to a 7.5 min/mile pace for a trained athlete or sufficient to cause perspiration and to increase heart rate in an untrained individual (Bishop, 2003*b;* Ingjer & Strømme, 1979). Stewart and Sleivert (1998) found that a warm-up at ~65% of maximal oxygen consumption for 15 min produces the greatest improvement in the range of motion and the subsequent anaerobic performance. Similarly, a warm-up at a moderate intensity and of long duration is optimal for swimmers (Houmard et al., 1991). An overly intense warm-up or insufficient recovery time prior to the event can impair performance (Bishop, 2003*b;* Drust et al., 2005). The increase in muscle temperature from warm-up activities is lost by about 15 min after the warm-up; therefore, the time between the warm-up and the event should be no longer than 15 min. Passive techniques can be used to maintain the increase in temperature acquired with an active warm-up (Bishop, 2003*b*).

Stiffness of Passive Muscle

In addition to elevating muscle temperature, warm-up exercises are used to increase the range of motion about a joint (Bishop, 2003*b;* Smith, 1994; Stewart & Sleivert, 1998). Increases in core temperature, whether due to muscle contractions or a passive heat source, enhance the extensibility of the tissues around the joint. This effect is evident only while temperature is elevated.

It is the structural elements of muscle and tendon that resist the imposed increase in length during stretching exercises. As indicated by the force–length relation of muscle (figure 6.33), the passive resisting force of muscle increases with the amount of stretch. The resistance of passive muscle to stretch can be visualized on a torque–angle graph (figure 8.3*c*). For example, Hufschmidt and Mauritz (1985) measured the passive resistance of the plantarflexor muscles in healthy subjects and in patients with varying degrees of spasticity. They did this by using a torque motor to displace the ankle joint through a 0.35 rad range of motion (figure 8.3*a*), from 0.175 rad (10°) of plantarflexion to 0.175 rad of dorsiflexion, and measuring

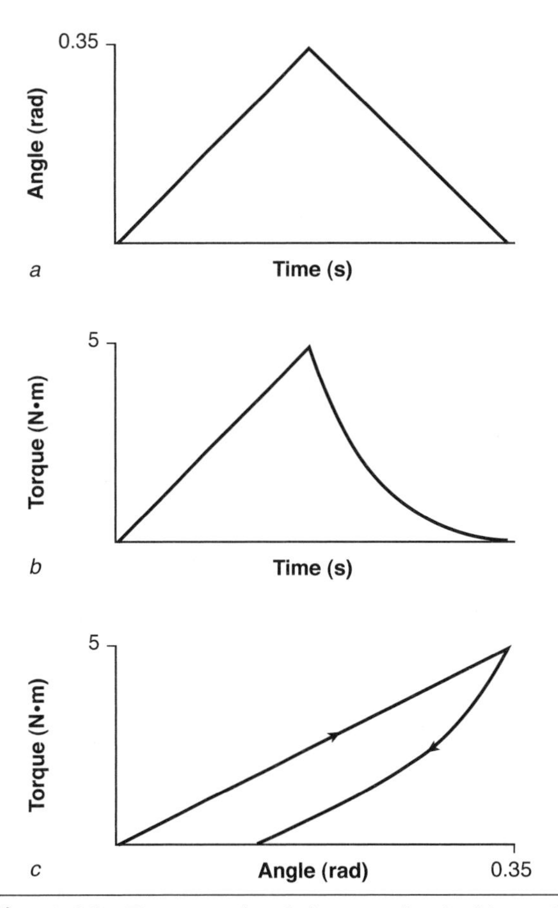

Figure 8.3 Torque–angle relation associated with passive stretch and release of the plantarflexor muscles: *(a)* the increase (stretch) and decrease (release) in ankle angle during the stretch; *(b)* the passive torque exerted by the muscles during the stretch; *(c)* the resulting torque–angle relation, which does not overlap during the stretch and release phases.

Data from A. Hufschmidt and K.H. Mauritz, 1985, "Chronic transformation of muscle in spasticity: A peripheral contribution to increased tone," *Journal of Neurology, Neurosurgery, and Psychiatry* 48: 678.

the resistance exerted by the plantarflexor muscles through the foot (figure 8.3*b*). The area enclosed between the stretch (up arrow) and release (down arrow) indicates the energy absorbed by the plantarflexor muscles during the task (figure 8.3*c*). Because spasticity involves changes in muscle that increase the passive resistance to angular displacement about a joint, patients who had exhibited spastic symptoms for more than one year had stiffer muscles (Hufschmidt & Mauritz, 1985). This increased stiffness was evident as an increase in the energy absorbed during the stretch and release, which corresponds to the area of the loop in figure 8.3*c*. This effect is greater for the extensor (antigravity) muscles (Dietz, 1992).

The resistance that a passive muscle offers to an imposed stretch is indicated by the slope of the torque–angle relation, which corresponds to the stiffness of the muscle. Muscle stiffness increases as the duration between the consecutive stretches gets longer. For example, when the time between stretches of the passive plantarflexor muscles increased from 0 to 30 s, Hufschmidt and Mauritz (1985) found that the stiffness of the muscles increased. This is evident in figure 8.4 by the greater slope during the stretch, especially the initial part of the stretch. This effect—which has been observed in single muscle fibers, finger muscles, and the plantarflexor muscles—involves an increase in muscle stiffness that occurs over a 30 min rest interval following muscle activation (Hufschmidt & Mauritz, 1985; Kilgore & Mobley, 1991; Lakie & Robson, 1988*a*). The increase in stiffness is greatest immediately after the activity and then becomes more gradual. The increased stiffness can be eliminated by active or passive movements but not by isometric contractions (Lakie & Robson, 1988*a*).

The dependence of muscle stiffness on its immediate history of activity is known as thixotropy (Campbell & Lakie, 1988; Proske et al., 1993; Walsh, 1992). It is a property exhibited by various gels, including muscle. In general, the gel becomes more fluid when shaken, stirred,

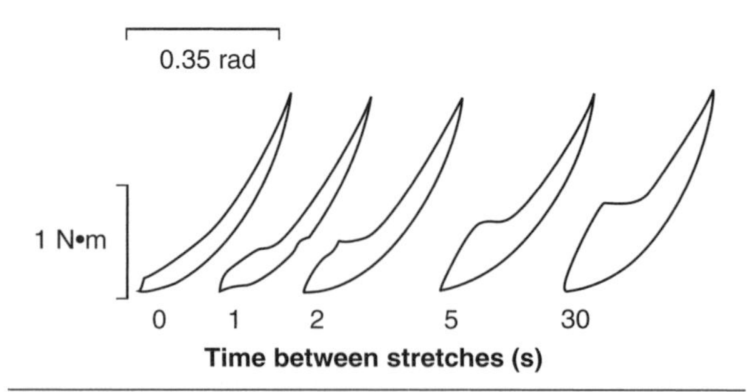

Figure 8.4 Effect of variable rest intervals on the torque–angle relation associated with passive stretch and release of the plantarflexor muscles.

Adapted, by permission, from A. Hufschmidt and K.H. Mauritz, 1985, "Chronic transformation of muscle in spasticity: A peripheral contribution to increased tone," *Journal of Neurology, Neurosurgery, and Psychiatry* 48: 678. Copyright 1985 by *Journal of Neurology, Neurosurgery, and Psychiatry.*

or otherwise disturbed, and it sets again when allowed to stand. The molecular rearrangement in muscle underlying thixotropy probably involves the development of stable bonds between actin and myosin filaments (Campbell & Lakie, 1998; Proske & Morgan, 1999). With inactivity, the number of bonds increases, making the muscle stiffer. With a brief stretch or period of physical activity, however, most of the bonds are broken and muscle stiffness decreases (Hagbarth et al., 1985; Lakie & Robson, 1988*b;* Wiegner, 1987). A means of accomplishing this is to move most major muscle groups through a complete range of motion (Wiktorsson-Möller et al., 1983).

The stiffness characteristics shown in figures 8.3 and 8.4, however, depend on the properties of both muscle and the connective tissues that attach the muscle to the skeleton. The mechanical properties of tendon and aponeurosis can be estimated with ultrasonography (Maganaris & Paul, 2000*a;* Magnusson et al., 2001). By imaging the displacement of a landmark, such as an attachment location, and measuring the cross-sectional area, it is possible to determine the strain and stress exhibited by the various tissues. For example, Maganaris and Paul (1999) found that the force in the tibialis anterior tendon reached 530 N during a maximal isometric contraction and that the tendon elongated by 4.1 mm, which corresponded to a stress of 25 MPa and a strain of 2.5%. They subsequently found that strain in the aponeurosis during an MVC was greater than that of the tendon and that it varied along the aponeurosis: tendon = 3.1%, distal aponeurosis = 3.5%, proximal aponeurosis = 9.2% (Maganaris & Paul, 2000*b*). Strain differences have also been reported between the tendon and aponeurosis for the knee extensors and plantarflexors and between synergist muscles (Arampatzis et al., 2005; Bojsen-Møller et al., 2004; Kubo et al., 2005; Magnusson et al., 2003*a*). When the plantarflexor muscles were passive and the ankle joint was rotated by 0.8 rad, there was also a difference in the strain experienced by the tendon and aponeurosis (Muraoka et al., 2002). Because the elastic properties of tendon are correlated with muscle strength (Muraoka et al., 2005; Narici & Maganaris, 2006), the amount of strain experienced by the tissues during passive stretching varies across individuals.

The time course of muscle and connective tissue thixotropy should define the protocol for passive range of motion activities in patients who are immobile. However, measurements indicate that it takes 16 h of passive motion per day to prevent joint stiffness and that even this amount of activity does not maintain bone density (Gebhard et al., 1993). Such durations are impractical for a therapist who typically uses several stretches, with each stretch lasting about 15 s, to maintain the range of motion about various joints; this is similar to the duration necessary to counteract the effects of inactivity (Roy et al., 1998).

Muscle Tone

The thixotropic property of muscle and its associated connective tissues endows it with a passive stiffness that resists changes in length. Clinicians refer to this resistance to stretch by a relaxed muscle as **muscle tone.** The resistance to stretch can, of course, be supplemented by the stretch reflex, but this does not occur in a relaxed subject at a low rate of stretch. Apparently, muscle tone is similar in relaxed healthy individuals and totally anesthetized patients (Rothwell, 1994).

Clinicians can use alterations in muscle tone to identify underlying pathologies. For example, reduced levels of muscle tone, known as **hypotonia,** are exhibited by patients who present with lesions in the cerebellar hemispheres and in persons who experience a spinal transection. Rothwell (1994) suggests that hypotonia is probably due to decreased excitability of the stretch reflex. Interestingly, muscle tone is reduced after low-frequency vibration (Walsh, 1992).

In contrast, an increase in muscle tone, referred to as **hypertonia,** is probably caused by low levels of motor unit activity despite an attempt to relax. The two most common forms of hypertonia are spasticity and rigidity. **Spasticity** describes a pathologically induced state of heightened excitability of the stretch reflex (Condliffe et al., 2005; Lin & Sabbahi, 1999). It occurs as a consequence of several different motor disorders, including brain trauma, spinal cord injury, and some systemic degenerative processes such as multiple sclerosis. Spasticity can be induced by a lesion in the central nervous system (CNS), primarily the corticospinal tract, and by transection of the spinal cord. When a spastic muscle is stretched, it responds with a more vigorous stretch reflex than normal muscle. Furthermore, the exaggerated stretch reflex increases with the velocity of the stretch. Many mechanisms underlie spasticity, including changes in the excitability of motor neurons, postsynaptic hypersensitivity to a neurotransmitter, enlargement of motor unit territories by sprouting, and increases in the passive thixotropic properties of muscle. The symptoms associated with spasticity include increased passive resistance to movement in one direction, a hyperactive tendon tap reflex, the adoption of a characteristic posture by the involved limb, an apparent inability to relax the involved muscle, and an inability to move the involved joint quickly or in alternating directions (Burne et al., 2005; Hidler et al., 2002; Levin et al., 2000; Singer et al., 2003).

One misconception associated with spasticity is that the changes in muscle tone impair movement capabilities. This is not correct. Spasticity in an antagonist muscle is not the primary factor that impairs the ability of an agonist muscle to perform a movement. Rather, the impairment is attributable to an inability of the agonist muscle to recruit a sufficient number of motor units (Ada et al., 2006; Kamper et al., 2006; McComas et al., 1973; Sahrmann & Norton, 1977). Consequently, the appropriate clinical intervention is to improve the activation patterns of the agonist rather than attempting to reduce the spasticity in the antagonist muscle. For example, the long-term application of transcutaneous electrical nerve stimulation to the common peroneal nerve significantly increased the dorsiflexion force (agonist), but not the plantarflexion force, and reduced clinical spasticity (Dewald et al., 1996; Levin & Hui-Chan, 1992).

The other form of hypertonia is rigidity. Rigidity and spasticity have different symptoms. The symptoms associated with **rigidity** include a bidirectional resistance to passive movement that is independent of the movement velocity and that occurs in the absence of an exaggerated tendon tap reflex (Xia et al., 2006). The most common occurrence of rigidity is in Parkinson's disease and involves a persistent muscle contraction that appears during passive manipulation and during a series of interrupted jerks (cogwheel rigidity).

FLEXIBILITY

Frequently no distinction is made between warm-up exercises and those designed to increase flexibility. Stretching exercises should be used with caution in a warm-up routine because long-duration stretches can impair maximal performance (Avela et al., 2004; Behm et al., 2004; Cramer et al., 2005; Weir et al., 2005). However, the increase in core temperature that occurs during a warm-up enhances the passive extensibility of the tissues around the joint and increases the range of motion (Magnusson et al., 1998; Zakas et al., 2006). Nonetheless, flexibility corresponds to the range of motion about a joint, and differences in flexibility between joints and individuals (figure 8.5) are due to long-term adaptations and not the changes that occur after a set of warm-up exercises (Halbertsma et al., 1996; Magnusson, 1998).

Stretching Techniques

The research on flexibility training has focused on developing effective strategies to increase the range of motion and on identifying the factors that limit flexibility (Bandy et al., 1998; Guissard & Duchateau, 2006; Hutton, 1992; Sharman et al., 2006). The two basic stretches are the static stretch and the ballistic (dynamic) stretch. The **static stretch** involves lengthening the muscle to the limit of its range of motion and holding this position for several seconds, whereas the **ballistic stretch** comprises repetitive bouncing movements to the limit of the range of motion. The two types of stretches can produce similar

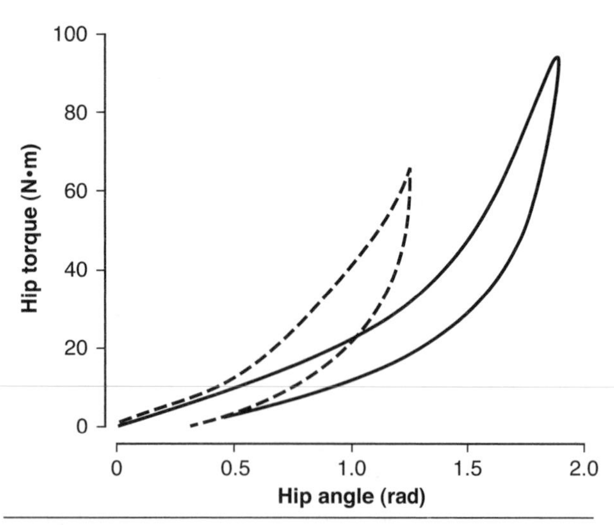

Figure 8.5 The increase in torque about the hip joint as hip angle was increased for two individuals during the straight-leg raise test. The measurements represent flexion of the hip and an extensor torque about the hip joint due to the passive properties of the joint tissues. The more flexible individual (solid line) was able to tolerate greater discomfort.

Data from McHugh et al. 1998.

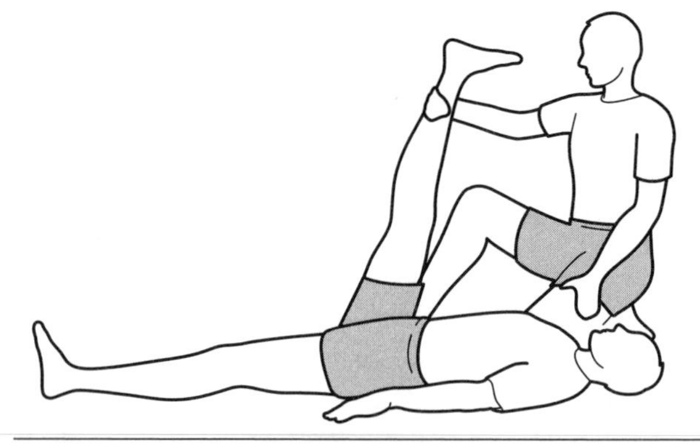

Figure 8.6 Partner-assisted flexibility training. An example of the contract-relax, antagonist-contract technique in which your partner stretches your hamstrings maximally (hip flexion) while you perform an isometric contraction with your quadriceps femoris muscles.

increases in the range of motion (Mahieu et al., 2007). Because stretching exercises are often associated with sensations of tightness in the muscle, some stretching techniques have been designed to improve relaxation and thereby increase flexibility. Three of these types of exercise have been derived from a rehabilitation procedure known as *proprioceptive neuromuscular facilitation* (PNF) (Kabat & Knott, 1953; Knott & Voss, 1968). The **contract-relax** technique consists of an initial maximal isometric contraction of the muscle to be stretched (target muscle) followed by relaxation and stretch of the target muscle to the limit of its range of motion. The **antagonist-contract stretch** requires the assistance of a partner or therapist (figure 8.6). After your partner moves your limb so that the joint is at the limit of its range of motion, you contract the opposing muscle (e.g., quadriceps femoris) while your partner pulls on your limb to stretch the target muscle (e.g., hamstrings). The **contract-relax, antagonist-contract** stretch is a combination of the contract-relax and antagonist-contract techniques. For the example shown in figure 8.6, the contract-relax, antagonist-contract stretch involves an initial maximal isometric contraction of the hamstrings (target muscle) followed by relaxation and stretch of the hamstrings; the hamstrings stretch is accomplished by manual assistance from your partner and by contraction of quadriceps femoris (opposing muscle).

In general, static, ballistic, and PNF stretches are all effective at increasing the range of motion about a joint (Bandy et al., 1998; Condon & Hutton, 1987; Etnyre & Abraham, 1986; Lucas & Koslow, 1984; Wallin et al., 1985). The PNF stretches, however, typically produce greater gains in flexibility (Etnyre & Lee, 1988; Ferber et al., 2002; Funk et al., 2003; Guissard et al., 1988). Although a single stretch can increase the range of motion about a joint (Etnyre & Lee, 1988; Lucas & Koslow, 1984; Moore & Hutton, 1980; Sullivan et al., 1992), the increase in muscle length can dissipate rapidly after a single session that includes only a few stretches (Magnusson, 1998; Spernoga et al., 2001). Increases in flexibility can be achieved only by the inclusion of stretches in a regular program of physical activity. For example, Guissard and Duchateau (2004) examined the effects of a six-week training program performed five times per week on the dorsiflexion range of motion about the ankle joint. The range of motion increased by 31% (0.14 rad) after 30 training sessions. It is necessary to perform stretching exercises once or twice a week to retain the desired range of motion (Sharman et al., 2006).

Mechanisms That Limit Flexibility

The viscous and elastic properties of muscle and the connective tissues associated with the muscle and the joint that resist stretches and limit the range of motion (figure 8.3) are plastic and will adapt to the demands of usage, both within a single session and with training (Stromberg & Wiederhielm, 1969; Taylor et al., 1990). Three classes of factors have been identified as contributing to gains in flexibility: passive tissue properties, spinal reflex excitability, and tolerance of discomfort (Magnusson et al., 1997).

Changes Within a Single Session

The passive properties of muscle and its connective tissues can change during both a single stretch and a series of stretches. For example, a constant force applied to tendon causes a gradual elongation of muscle and tendon; this is known as **creep** (figure 8.7*a*). Conversely, repeated stretches to a specific joint angle will result in a lower resistance force across stretches; this is known as **stress relaxation** (figure 8.7*b*). Within a single stretch, changes in the passive properties of the tissues are characterized with measures of stiffness (slope of the torque–angle curve) and the area enclosed between the stretch and release phases of the stretch (figure 8.3*c*). The area, which corresponds to a hysteresis, denotes the amount of energy dissipated as heat due to damping within the tissues during the maneuver.

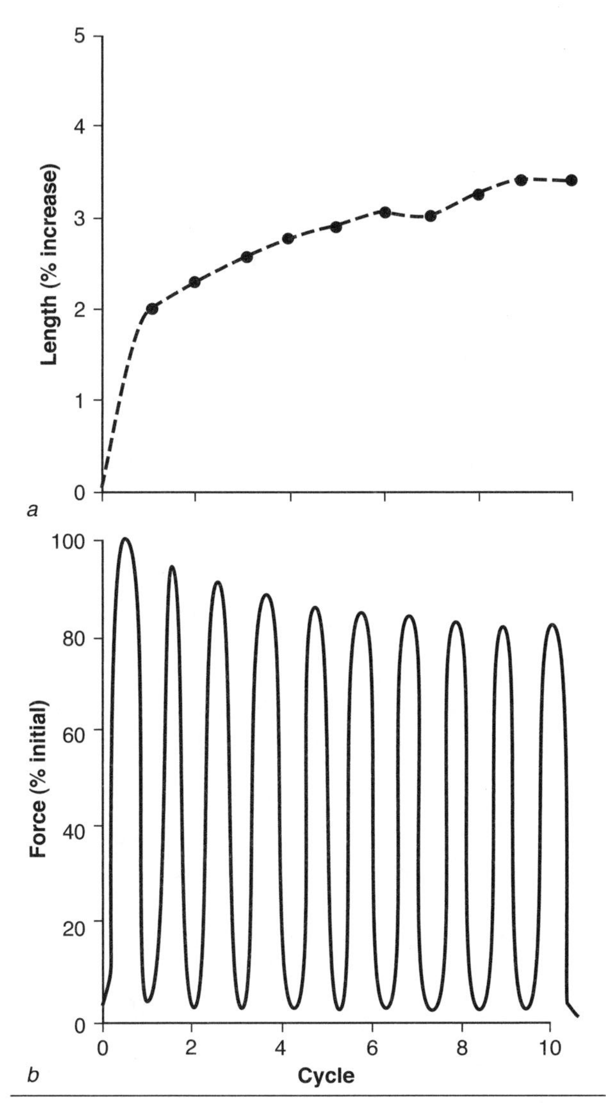

Figure 8.7 Stretch-induced changes in force and length of a muscle-tendon unit. *(a)* To maintain a peak force of 78 N on an isolated muscle (extensor digitorum longus) during 10 stretches applied over 30 s, its length had to be increased gradually in successive stretches. *(b)* The peak force achieved in each of 10 stretches decreased when the muscle was stretched by 10% of its initial length each time.

Springer, *American Journal of Sports Medicine,* Vol. 18, 1990, p. 4303, 4304, "The viscoelastic properties of muscle-tendon units" by D.C. Taylor, J. Dalton, A.V. Seaber, and W.E. Garrett, 304. Copyright 1990 by the American Orthopaedic Society for Sports Medicine. With permission from Sage Publications.

Within a single session, continuous passive motion produced a greater reduction in stiffness of the passive dorsiflexor muscles compared with a sustained hold, whereas the decrease in peak force during the stretch was greater for the hold task (McNair et al., 2001). Because stress relaxation has been observed in the absence of electromyographic (EMG) activity, these adjustments likely involve the passive properties of the tissues (McHugh et al., 1992). Furthermore, a static stretch of the plantarflexor muscles for 10 min reduced both the stiffness and the hysteresis of the muscles and connective tissues (Kubo et al., 2001*b*).

However, the observation that vibration, which depresses the Hoffmann reflex (Shinohara et al., 2005), can augment the increase in range of motion during stretching exercises underscores a significant role for afferent feedback in limiting flexibility (Sands et al., 2006; van den Tillaar, 2006). To evaluate the contribution of spinal reflex excitability to gains in flexibility, it is necessary to examine the reflex pathways directly (Guissard & Duchateau, 2006). By comparing the change in amplitude of reflexes and evoked responses, it is possible to estimate the involvement of various mechanisms in the adjustments that occur during stretching: H reflex, tendon tap reflex, and motor-evoked potentials. The H and tendon tap reflexes correspond to the responses of motor neurons to input from Ia afferents of muscle spindles, whereas the motor-evoked potentials elicited by transcranial magnetic stimulation (TMS) indicate the efficacy of transmission in descending pathways (example 7.12). The amplitude of the H and tendon tap reflexes depends on both the excitability of the motor neuron pool and the amount of presynaptic inhibition of the Ia afferents (figure 7.18). In addition, the amplitude of the tendon tap reflex is influenced by the responsiveness of the muscle spindles to the percussion of the tendon. Transmission in the descending pathways that contribute to the motor-evoked potentials is not modulated by presynaptic inhibition.

Guissard and colleagues (1988) found that the decline in the amplitude of the tendon tap reflex in soleus was greater than that for the H reflex during the static stretch of the plantarflexor muscles, and that the H reflex, but not the tendon tap reflex, returned to control values when the stretch was terminated (figure 8.8). In contrast, the amplitude of the motor-evoked potentials did not decrease until the static stretch reached >0.35 rad of dorsiflexion (Guissard et al., 2001). The reduction in the size of the reflexes during small-amplitude stretches (0.17 rad), therefore, was attributable to presynaptic inhibition, whereas the

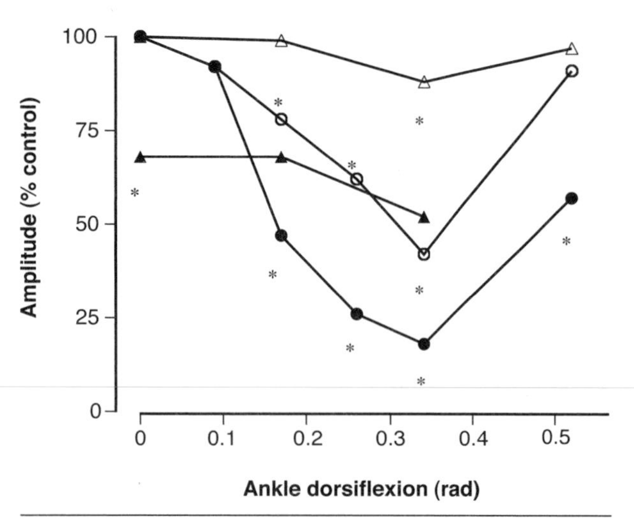

Figure 8.8 Decrease in the amplitudes of the H (open circles) and tendon tap (filled circles) reflexes and motor-evoked potentials (open triangles) elicited with transcranial magnetic stimulation in the soleus muscle as a function of ankle joint angle. The stretch of soleus increased with dorsiflexion angle. The amplitude of the conditioned H reflex (filled triangles), which is compared with that of normal H reflexes, was depressed at the initial length and did not change with an increase in the stretch. The asterisk indicates a significant depression of amplitude.

Data from Guissard and Duchateau 2006.

decrease in motor neuron excitability contributed to the depression of the reflexes as the stretch increased. The origin of the input that reduces motor neuron excitability can be estimated with a conditioned H-reflex test that involves activation of heteronymous Ib afferents (nerve to medial gastrocnemius) prior to the stimulus delivered to the soleus motor neurons (Pierrot-Deseilligny et al., 1979). Input by Ib afferents causes postsynaptic inhibition of motor neurons. Figure 8.8 shows that there was less of a difference in the amplitude of the H and conditioned H reflexes with an increase in the stretch of soleus, which indicates that the Ib afferent input to the motor neurons could not be augmented as the stretch increased. Thus, the reduction in motor neuron excitability during a static stretch is caused, at least in part, by greater input delivered to the motor neurons by Ib afferents from tendon organs (Guissard & Duchateau, 2006).

In addition to the adjustments that occur in spinal reflex pathways during a static stretch, the PNF procedures further reduce the likelihood that a stretch will evoke a reflex. When the stretch is preceded by a voluntary contraction with the target muscle (contract–relax technique), the amplitudes of the H and tendon tap reflexes are depressed for about 10 s (Crone & Nielsen, 1989; Enoka et al., 1980; Moore & Kukulka, 1991). The depression is presumably caused by presynaptic inhibition and not postsynaptic inhibition, such as can be produced by tendon organs (Guissard & Duchateau, 2006; Sharman et al., 2006). When the stretch involves concurrent contraction of the opposing muscle (antagonist-contract technique), the descending command activates the alpha and gamma motor neurons that innervate the opposing muscle and the interneuron that mediates reciprocal inhibition to the target muscle (figure 7.14). Consistent with this rationale, the amplitude of the H reflex was depressed more during the contract-relax and antagonist-contract stretches compared with a static stretch of the plantarflexor muscles, and presumably these mechanisms contributed to the greater stretch achieved with the PNF techniques (Guissard et al., 1988).

Stretch Training

The adjustments that occur during a single session of stretches provide the foundation for the adaptations associated with stretch training. The increase in flexibility after a single session is compounded with several weeks of training. For example, the resistance force generated by the passive plantarflexor muscle from a neutral ankle angle to maximal dorsiflexion was reduced by 18% after a single session of contract-relax stretches, and by 36% after three weeks of twice-a-day stretch training (Toft et al., 1989). Similarly, Guissard and Duchateau (2004) found that 10 min of static stretching of the plantarflexor muscles (four exercises with five 30 s stretches each), five times a week for six weeks, increased the ankle dorsiflexion range of motion by 31%, and that 56% of the gain in flexibility was achieved after two weeks. Furthermore, 74% of the increase in the range of motion was retained 30 days after the stretch training program. In contrast, 2 min of static stretching with the plantarflexor muscles once a day for six weeks did not alter the range of motion for ankle dorsiflexion (Youdas et al., 2003).

Guissard and Duchateau (2004) found a strong correlation between the increase in the range of motion and the reduction in passive resistance to stretch provided by the plantarflexor muscles and the associated connective tissues. Kubo and colleagues (2002) reported a similar increase in compliance with stretch training (five 45 s stretches twice a day for 20 days), and by measuring tendon length with ultrasonography they concluded that stretch training reduced the viscosity of tendon but not its elasticity. Such training programs, however, can evoke adaptations in spinal reflex pathways in addition to the changes in passive properties. For example, Guissard and Duchateau (2004) found that the amplitude of the H reflex decreased by 36% and that the amplitude of the tendon tap reflex was reduced by 14% after 30 stretch training sessions. Because the change in the ratio of the tendon tap and H reflexes, which provides an index of muscle spindle responsiveness and tendon stiffness, was not correlated with the change in passive stiffness during stretch training, the gain in flexibility was not entirely explained by the decrease in passive stiffness (Guissard & Duchateau, 2006).

Some studies have shown, however, that it is possible to increase the range of motion for hip flexion after stretch training of the hamstring muscles without adaptations in either passive properties or EMG activity (Magnusson et al., 1996*a*; Reid & McNair, 2004). For example, Magnusson and colleagues (1996*a*) had subjects perform five stretches of the hip extensor muscles once every 12 h for 20 consecutive days, with each stretch held for 45 s. This stretch has been shown to produce a 29% stress relaxation (Magnusson et al., 1995). After the 20 training sessions, the hip flexion range of motion of 0.3 rad (17°) increased; but there were no changes in the passive tissue properties (stiffness or energy dissipated), and the EMG activity was negligible. Accordingly, Magnusson and colleagues concluded that the increase in flexibility was attributable to an improvement in the tolerance to the stretch. Others have also found that stretch training can increase the ability of an individual to tolerate the discomfort associated with stretching a muscle to its limit (Guissard & Duchateau, 2004; Halbertsma et al., 1996; McHugh et al., 1998).

Research on the mechanisms that limit flexibility suggest that adjustments in both the passive properties and the spinal reflex excitability can contribute to increases in the range of motion about a joint. However, the relative contributions of these mechanisms may differ across joints. Although several observations also suggest a significant role for a greater tolerance of the associated discomfort, the underlying mechanisms remain to be identified.

MUSCLE SORENESS AND DAMAGE

Strenuous physical activity can have diverse effects on muscle, ranging from subcellular damage of muscle fibers to stretch-induced muscle strains. The subcellular damage, which most active individuals experience, frequently produces an inflammatory response and is associated with muscle soreness that begins hours after completion of the exercise. In contrast, strain injuries typically occur as an acute painful injury during intense activity and require clinical intervention.

Muscle Soreness

Because the perception of soreness is not evident until 24 to 48 h after the exercise, postexercise muscle soreness is called **delayed-onset muscle soreness.** This term distinguishes postexercise soreness from the exertional pain that occurs during exercise (Asmussen, 1952, 1956). The clinical symptoms associated with delayed-onset muscle soreness include an increase in plasma enzymes (e.g., creatine kinase), myoglobin, and protein metabolites from injured muscles; structural damage to subcellular components of muscle fibers, as seen with light and electron microscopy; and temporary impairment of muscle function (Armstrong, 1990). The major sensation associated with delayed-onset muscle soreness is tenderness to applied pressure (Howell et al., 1993; Jones & Round, 1997; Walsh et al., 2004). Although the sensation is often attributed to feedback delivered by group III and IV afferents, the heightened discomfort associated with the application of vibration to the sore muscle suggests that it involves muscle spindle afferents (Weerakkody et al., 2003). Presumably, the feedback delivered by vibration-sensitive afferents gains access to the pain pathway after the exercise that produces the soreness (Gregory et al., 2003; Proske & Allen, 2005).

Delayed-onset muscle soreness occurs after an individual performs unaccustomed physical activity, especially if it involves lengthening contractions (Fridén & Lieber, 2001; Clarkson & Hubal, 2002; Proske & Allen, 2005). Most studies of muscle soreness, therefore, typically focus on the consequences of performing an intense bout of lengthening contractions. Two major theories have been proposed to explain the origin of muscle soreness: One asserts that signaling factors are responsible, whereas the other proposes a mechanical origin. The signaling theory is based on a disruption of excitation-contraction coupling that results in an increase in the intracellular concentration of Ca^{2+} (Warren et al., 2001). According to this scheme, the prolonged decrease in strength after an intense bout of lengthening contractions is largely attributable to an inability to activate intact force-generating structures within muscle fibers. The responsible mechanisms, at least for the first three days after the exercise, do not involve either the function of the sarcoplasmic reticulum or the propagation of muscle fiber potentials, but rather a disruption of signaling between receptors in the T tubule and the sarcoplasmic reticulum (figure 6.5). After the first three days of recovery, Warren and colleagues suggest that the lingering decline in strength is caused by a loss of contractile proteins. This theory, however, cannot explain all of the subsequent changes that are observed in muscle function.

The mechanical theory proposes that muscle soreness is caused by mechanical damage of the muscle fibers (Fridén & Lieber, 2001). Recall that muscle fibers are composed of myofibrils and that each myofibril consists of many (often thousands) of sarcomeres in series (figures 6.1 and 6.2). The force produced by each muscle fiber during a contraction is due to the interaction of the contractile proteins in each sarcomere, and the amount of force depends on the overlap between the thick and thin filaments (figure 6.26). The amount of overlap during a contraction, however, varies among the sarcomeres in a single myofibril, especially with performance of lengthening contractions (Patel et al., 2004; Proske & Morgan, 2001). As a consequence, those sarcomeres with less overlap have a reduced capacity to resist the stretch that occurs during a lengthening contraction and will become overstretched and dysfunctional. Repeat contractions will increase the number of damaged sarcomeres, both along the myofibril and to adjacent myofibrils, and the change in the distribution of strain within the fiber will damage membranes in the sarcoplasmic reticulum, T tubule, and sarcolemma (Fridén & Lieber, 1998; Takekura et al., 2001). Damage to the membranes will disrupt their capacity to maintain ionic gradients, which leads to an influx of Na^+ and Ca^{2+} into the sarcoplasm (Allen et al., 2005). These events alter the mechanical properties of muscle and lead to delayed soreness, tenderness, and swelling, which may be necessary to induce an increase in the number of sarcomeres in series (Yu et al., 2004).

Muscle Weakness

One of the most consistent outcomes of an exercise protocol that induces delayed-onset muscle soreness is prolonged muscle weakness. For example, Howell and colleagues (1993) found that 5 to 15 repetitions of slowly (5-9 s) lowering a heavy load with the elbow flexor muscles resulted in a 35% decrease in strength that was half recovered in about six weeks (figure 8.9*a*). Furthermore, Walsh and colleagues (2004) found that the time course of recovery in MVC force after five sets of 10 repetitions at about 30% MVC force with the elbow flexor muscles on an isokinetic dynamometer (1.05 rad/s) was slower for lengthening contractions compared with shortening contractions. Maximal voluntary contraction force decreased by 26% immediately after the shortening contractions and had returned to control values 2 h later. In contrast, MVC force declined by 46% immediately after the lengthening contractions, was reduced by 30% 2 h later, and did not return to control values until 72 h after the exercise

The site of the impairment that produces the muscle weakness can be identified, in broad terms, through comparison of the force that can be exerted during an MVC with that evoked by electrical stimulation using the twitch interpolation, paired-stimulus, and low-frequency methods. The **twitch interpolation technique** involves superimposing an electric shock on a muscle during a voluntary contraction and measuring the amount of extra force evoked by the stimulus. As indicated in figure 8.10, the amplitude of the evoked force varies with the intensity of the muscle contraction and provides an index of the extent to which the activation elicits the force that the muscle is capable of producing. The level of voluntary activation is quantified as

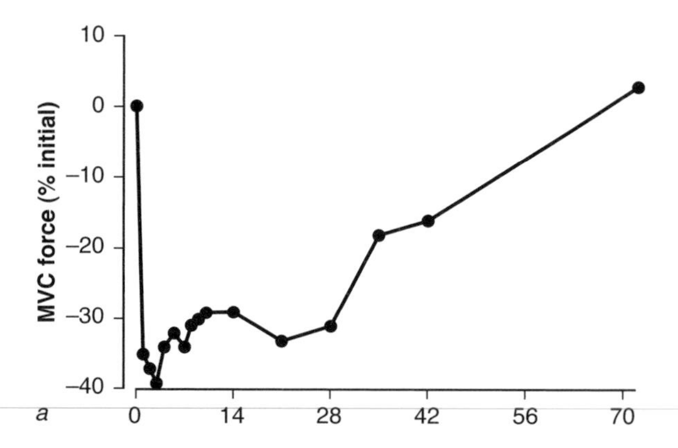

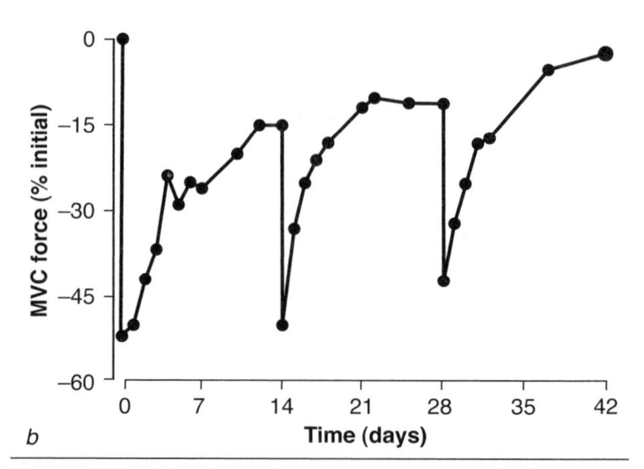

Figure 8.9 Muscle weakness after a protocol of eccentric contractions with the elbow flexor muscles. Weakness is expressed as the percentage reduction in the maximal voluntary contraction (MVC) force. *(a)* Weakness after a single bout of eccentric contractions; *(b)* weakness after exercise at days 0, 14, and 28.

(a) Data from Howell et al., 1993; *(b)* Newham et al., 1987.

$$\text{Voluntary activation (\%)} = \left(1 - \frac{\text{Superimposed twitch}}{\text{Resting twitch}}\right) \times 100 \quad (8.1)$$

where *superimposed twitch* corresponds to the amplitude of the extra force evoked during the voluntary contraction and *resting twitch* refers to the amplitude of the force produced by the same stimulus when the muscle is at rest (Allen et al., 1998). The **paired-stimulus technique** is used to measure the peak force that can be evoked in a resting muscle by a pair of electric shocks (10 ms apart); the amplitude of the response provides an index of the force capacity of muscle (Klass et al., 2005). The third method involves comparing the forces evoked by trains of stimuli delivered at low (20 Hz) and high (80 Hz) frequencies. When an intervention causes a greater relative decrease in the force elicited by the low-frequency stimulation, this is referred to as **low-frequency depression of force** and indicates an impairment of

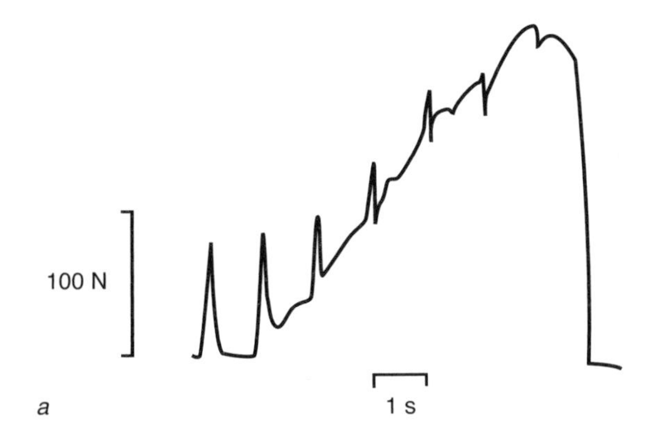

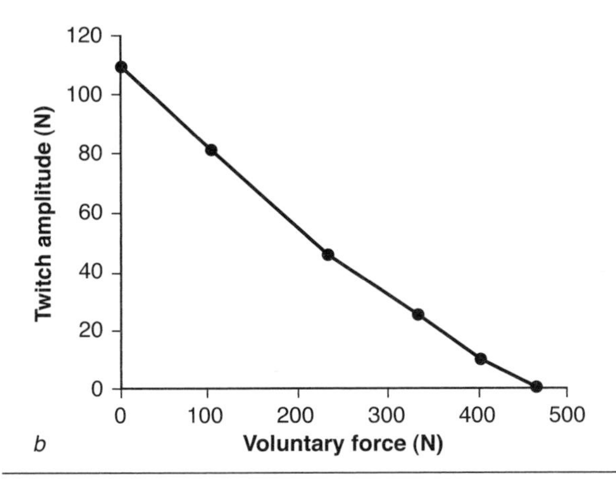

Figure 8.10 The twitch interpolation method. *(a)* Decrease in the amplitude of the twitch force elicited by an electric shock as the voluntary force exerted by the quadriceps femoris increased; *(b)* decline in twitch force as a function of voluntary force. The twitch was elicited at various levels of force during voluntary contractions. The maximal voluntary contraction (MVC) force was about 470 N.

Adapted, by permission, from B. Bigland-Ritchie, E. Cafarelli, and N.K. Vøllestad, 1986, "Fatigue of submaximal static contractions," *Acta Physiological Scandinavia* Suppl. 556:138.

excitation-contraction coupling (Edwards et al., 1977; Parikh et al., 2004).

With the use of these two techniques, Prasartwuth and colleagues (2006) examined the changes present in the biceps brachii and brachialis muscles after the elbow flexor muscles performed lengthening contractions until the MVC torque declined by 40%. When the tests were performed with the elbow joint at a right angle, voluntary activation declined from 97% before the exercise to 90% when measured at 2 and 24 h after the exercise. By day 8 after the exercise, voluntary activation had returned to control levels, but MVC torque was still reduced by 25%. The force evoked with the paired-stimulus method was reduced by 70% at 2 and 24 h after the exercise and remained depressed by 40% at day 8. Hence, much of the weakness observed by Prasartwuth and colleagues (2006) was associated with a decline in the force capacity of the muscles rather than insufficient activation by the nervous system. Other investigators, however, have also reported low-frequency depression of force after lengthening contractions, presumably indicating a contribution by impairment of excitation-contraction coupling to the postexercise muscle weakness (Balnave & Allen, 1995; Ingalls et al., 1998; Parikh et al., 2004).

Muscle Length

Consistent with the mechanical damage theory of muscle soreness, the length at which a muscle fiber or whole muscle produces its peak force increases after a bout of lengthening contractions (Jones et al., 1997; Philippou et al., 2004; Prasartwuth et al., 2006). Through comparison of the forces evoked at several muscle lengths (joint angles) before and after the exercise, the elongation is quantified as the shift in the length at which the peak force occurs. The length at which the peak force occurs is known as the optimal length (L_o). The shift in L_o, which has been observed in both isolated muscle and in vivo (Prasartwuth et al., 2006; Whitehead et al., 2003), is attributed to an increase in the length of myofibrils and muscle fibers due to the overstretch of some of the sarcomeres (Proske & Allen, 2005). Because the dysfunctional sarcomeres essentially represent passive tissue, there is an increase in the amount of stretch that must be applied to the muscle fibers before the active force is transmitted to the skeleton. This corresponds to an increase in the compliance (mm/N) of the damaged muscle fibers.

As a consequence of the shift in L_o, Prasartwuth and colleagues (2006) found that the optimal angle for peak MVC torque for the elbow flexor muscles increased by 0.3 rad (17°) at 2 h and by 0.24 rad at 24 h after a series of lengthening contractions had reduced MVC force by 40%. The optimal angle of the elbow joint for peak MVC torque was similar at day 8 to that measured before exercise, yet the MVC torque remained depressed by 25%. The shift in the optimal angle for the force evoked by paired stimuli was much less: 0.07 rad (4°) at 2 h and 0.14 rad at 24 h. These shifts in L_o, however, had only a minor influence on peak torque produced during an MVC or evoked with paired stimuli.

Lengthening contractions that produce muscle damage also cause passive tension to increase immediately after the exercise. When a resting muscle is stretched gradually, the tissue resists the lengthening with a force that is known as passive tension (figure 6.33). Passive tension at L_o increased by 170% immediately after a set of lengthening contractions in one experiment (Whitehead et al., 2003). A series of stretches, however, reduces the passive tension back to baseline levels, although it will gradually increase again. The increase in passive tension, which is likely caused by an influx of Ca^{2+} into damaged muscle fibers that produces an injury contracture (Proske & Allen, 2005), increases the activation of tendon organs and the feedback delivered to the spinal cord by group Ib afferents (Gregory et al., 2003).

In addition to shifting L_0 to longer lengths and increasing passive tension, lengthening contractions impair the performance of muscle at short muscle lengths. The impairments involve both the activation of muscle and its contractile capabilities. For example, voluntary activation of the elbow flexor muscles at an elbow angle of 1.05 rad (60°) declined from 94% before exercise to 67% at 2 h after lengthening contractions and remained depressed (85%) at day 8 (Prasartwuth et al., 2006). Similarly, there was greater low-frequency depression of force for the knee extensor muscles at a short muscle length compared with a long muscle length after exercise-induced muscle damage (Skurvydas et al., 2006); this was likely due to an influence on the length dependence of calcium sensitivity (Balnave & Allen, 1996).

Repeated-Bout Effect

One of the characteristics of the impairment produced by an intense bout of lengthening contractions is that the exercise induces less weakness in subsequent sessions (figure 8.9*b*) and there is less evidence of muscle damage (Stupka et al., 2001; Thompson et al., 2002). This is known as the repeated-bout effect (Clarkson & Hubal, 2002; McHugh, 2003). The protection does not arise when the initial activity involves shortening contractions or lengthening contractions performed at a different muscle length (Gleeson et al., 2003; McHugh & Pasiakos, 2004; Whitehead et al., 1998). Furthermore, the amount of protection increases with the load (% maximum) during the initial set of lengthening contractions (Chen et al., 2006).

Based on the mechanical theory of muscle damage for delayed-onset muscle soreness, the repeated-bout effect is attributed to the addition of new sarcomeres to the damaged myofibrils so that subsequent lengthening contractions are less likely to stretch sarcomeres to a length that damages them (Proske & Allen, 2005). There is evidence that the number of sarcomeres in series can change with physical activity. For example, Butterfield and colleagues (2005) observed that the number of sarcomeres in series in the muscle fibers of the vastus intermedius and vastus lateralis muscles in rats decreased after 10 days (15-35 min/day) of walking uphill on a treadmill (shortening contractions), whereas the number of sarcomeres in vastus intermedius increased when the training program involved walking downhill (lengthening contractions). Presumably, the number of sarcomeres changes so that the displacement of the thick and thin filaments occurs within the range around maximal overlap of the filaments. The cellular mechanisms that signal and organize the addition and deletion of sarcomeres are not yet known (Hentzen et al., 2006).

Muscle Strain Injury

In contrast to the microinjuries that often accompany delayed-onset muscle soreness, muscle can experience acute, painful events of greater magnitude, such as cramps and strains. **Muscle cramp** is a painful, involuntary shortening of muscle that is triggered by a variety of mechanisms, including both neural and muscular factors (Bentley, 1996; Bertolasi et al., 1993; Miller & Layzer, 2005; Ross & Thomas, 1995). Most evidence suggests that the involuntary contractions associated with a muscle cramp arise from the motor nerve in the periphery and not from the CNS. For example, although muscle cramps can be triggered by a forceful voluntary contraction at a short muscle length, they can also be evoked by electrical stimulation applied distal to a nerve block (Bertolasi et al., 1993). A muscle cramp can involve discharge rates of action potentials up to 150 pps, is usually confined to one muscle or one part of a muscle, can begin and end with muscle twitches in various parts of the muscle, and is usually terminated by a muscle stretch. Muscle cramps can occur in individuals with motor neuron disease, metabolic disorders, or inherited syndromes; as a side effect of medications; after an acute depletion of extracellular volume; and for no apparent reason in elderly persons (Miller & Layzer, 2005).

Types of Strains

A *muscle strain* injury occurs because of an unexpected and substantial stretch of a muscle (Best et al., 1997; Brockett et al., 2004). Muscle strain injuries, which are also referred to as *pulls* and *tears,* can be categorized as mild, moderate, or severe. Mild strains involve minor structural disruption, local tenderness, and minimal deficits in function. Moderate strains comprise some structural damage, visible swelling, marked tenderness, and some impairment of function. Severe strains exhibit substantial structural damage that usually requires surgical intervention. Best and colleagues (1997) report that gradual increases in strain (from 13% to 23%) were accompanied by progressive impairment of contractile function, reduction in EMG amplitude, and tissue damage. They also noted that the disruption of muscle fibers preceded that of connective tissue.

Clinical reports indicate that muscle strain injuries invariably occur at the muscle-tendon or the muscle-bone junction; muscle strain injuries have been reported for medial gastrocnemius, rectus femoris, triceps brachii, adductor longus, pectoralis major, and semimembranosus (Best et al., 1997; Garrett, 1996). Muscles most prone to strain injuries are two-joint muscles (because they can be stretched more), muscles that limit the range of motion about a joint, and muscles that have a high proportion of type II muscle fibers (Noonan & Garrett, 1999; Prior et al., 2001). Because of these factors, there is often one muscle within a synergistic group that is more prone than others to injury—for example, adductor longus in the hip adductors and rectus femoris in the knee extensors. Moreover, this type of injury most often occurs during powerful lengthening contractions, when the force can be several times greater than the maximal isometric force. The injury frequently involves bleeding with the subsequent accumulation of blood in the subcutaneous tissues. The most appropriate immediate treatment is to rest the muscle and to apply ice and compression (Best et al., 1997). Rehabilitation should include physical therapy to improve the range of motion and to prescribe *functional* strengthening exercises.

Mechanism of Injury

Although muscle fibers experience strain (change in length relative to resting length) during both voluntary contractions and passive stretch, the strain usually progresses to an injury only during the forced lengthening of an active muscle, that is, during a lengthening contraction (Butterfield & Herzog, 2006; Stauber, 2004). As an example of the strain experienced by muscle fibers during voluntary contractions, Pasquet and colleagues (2006) found a 21% change in fascicle length for tibialis anterior during shortening and lengthening contractions over a 0.35 rad (20°) range of motion performed on an isokinetic dynamometer. Muraoka and colleagues (2001) observed a strain of 40% in the fascicles of vastus lateralis during slow pedaling on an ergometer that involved a 1.31 rad range of motion at the knee joint. In contrast, the fascicle strain experienced in leg muscles during stretch-shortening cycles, as occur during walking, running, and jumping, is less than 10% (Butterfield & Herzog, 2006; Ishikawa et al., 2005; Hoyt et al., 2005).

A key question regarding the mechanism of injury is whether the muscle strain injuries experienced in vivo are caused by excessive strains or by a progression of the microinjuries associated with muscle soreness. Although tendons can rupture during physical activity, the range of symptoms associated with strain injuries suggests a more progressive impairment, and much evidence favors the microinjury explanation. For example, Butterfield and Herzog (2006) found that the changes in muscle torque caused by repeated stretch-shortening cycles varied with the initial muscle length and the timing of the muscle activation, even though a 5% strain was used for all conditions. The reduction in peak torque was greater for stretch-shortening cycles that began from longer muscle lengths. According to the mechanical damage theory of muscle soreness, the weakness produced by the stretch-shortening cycles would have damaged some sarcomeres and initiated cellular events that altered the number of sarcomeres in series. Whether the adaptation involves an increase or a decrease in the number of sarcomeres depends on the subsequent physical activity, including that performed during rehabilitation. In individuals with a history of unilateral hamstring strain injuries, Brockett and colleagues (2004) found that the knee joint angle at which the hamstring muscles exerted peak torque was less than that for the uninjured leg and for control subjects. They suggested that the rehabilitation after each injury inappropriately reduced the muscle length at which peak torque occurred and predisposed the individual to a subsequent injury when the muscle was stretched to a long length during a lengthening contraction. Consistent with this explanation, the incidence of strain injuries increases from 16% in previously uninjured athletes to 34% in those with a history of strain injuries (Orchard & Seward, 2002; Seward et al., 1993). Brockett and colleagues (2004) suggest that rehabilitation activities should include lengthening contractions and measurement of the torque–angle relation to ensure that the injured muscle is not being reduced to a shorter operating length.

MUSCLE FATIGUE

In everyday language, the term *muscle fatigue* is used to denote a variety of conditions that range from an exercise-induced impairment of motor performance through to the sensations of tiredness and weakness that accompany some neurological disorders. Although convenient for casual conversation, such broad usage precludes the systematic study of the underlying mechanisms. Accordingly, many investigators who study the physiology of muscle fatigue prefer a more restricted definition that considers fatigue to be an exercise-induced reduction in the ability of muscle to produce force or power, whether or not the task can be sustained (Enoka & Stuart, 1992; Fitts, 2006; Gandevia, 2001; Meeusen et al., 2006; Nybo & Secher, 2004). Importantly, it is recognized that the reduction begins soon after the onset of sustained physical activity (figure 8.11*a*).

Sites of Impairment

Since the seminal work of Angelo Mosso (1846-1910) (Di Giulio et al., 2006), the typical strategy used by experimentalists in the study of fatigue has been to determine whether the mechanism responsible for fatigue was located in the exercising muscles or in those parts of the nervous system that activated the involved muscles. The approach has often been to distinguish between an impairment of contractile protein function and a decrease in the magnitude of the activation signal. This work has

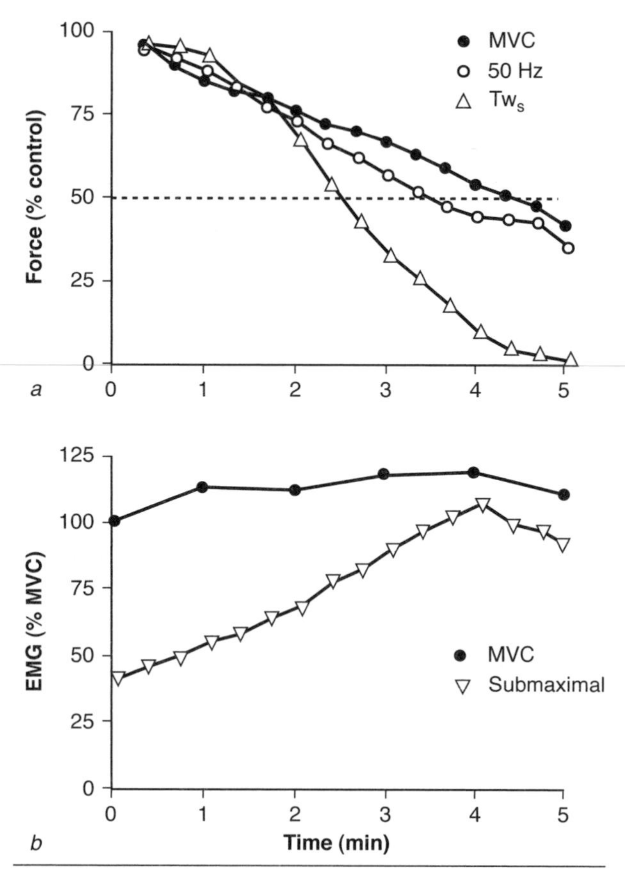

Figure 8.11 Changes in force and EMG during a fatiguing contraction that involved intermittent contractions (6 s contraction, 4 s rest) with the quadriceps femoris muscle to a target force that was 50% of the maximal voluntary contraction (MVC) force. The task was terminated when the MVC force declined to the target force. *(a)* MVC force decreased in parallel with an evoked tetanic force (50 Hz stimulation), and the superimposed twitch force (Tw_s) declined to zero when the task could not be performed any longer. *(b)* The integrated EMG for quadriceps femoris during an MVC did not change during the task, whereas the EMG amplitude during the intermittent contractions to the target force (50% of maximum) first increased and then decreased.

Adapted, by permission, from B. Bigland-Ritche, F. Furbush, and J.J. Woods, 1986, "Fatigue of intermittent submaximal voluntary contractions: Central and peripheral factors," *Journal of Applied Physiology* 61: 424. Used with permission.

demonstrated that the function of the contractile proteins and the capacity of the myofibrils to produce force can be impaired by a decline in the efficacy of excitation-contraction coupling, metabolic changes that influence the supply of ATP and impede the function of cross-bridges, and an inadequate perfusion of the active muscles (figure 8.12). In contrast, activation of the contractile proteins can be reduced by a decline in neuromuscular propagation and the inward spread of action potentials down the transverse tubules, a reduction in the motor output from the spinal cord, a decline in the afferent support of motor neuron discharge, an increase in afferent feedback that reduces motor neuron excitability, a decrease in the activation of the motor unit population by descending inputs, and an insufficient drive to the neurons in the motor cortex. The next few sections of this chapter provide examples of situations in which these different physiological processes can be impaired during sustained activity and thereby potentially contribute to muscle fatigue.

Cerebral Cortex

Just as muscle fibers require energy, produce metabolites, generate heat, and need a neurotransmitter to enable activation, so too do the neurons in the cerebral cortex. These functions can be disturbed, especially during prolonged exercise, and contribute to the development of fatigue (Meeusen et al., 2006; Nybo & Secher, 2004). In addition to impairing the output of the cerebral cortex to the active muscles, sustained activity impedes the ability of the cerebral cortex to perform other functions, such as cognitive tasks. Performance on a choice reaction task, for example, worsened when subjects concurrently sustained a fatiguing contraction (Lorist et al., 2002; Zijdewind et al., 2006*b*).

One of the prevailing mechanisms thought to impair the motor output of the cerebral cortex during prolonged exercise is a disturbance in the balance of neurotransmitter levels. The candidates that have received the most attention are serotonin, dopamine, and norepinephrine. Because the serotonergic system influences such behaviors as appetite, emotion, mood, and sleep, Newsholme and colleagues (1987) proposed that the extracellular accumulation of serotonin during prolonged contractions can reduce mental alertness and the capacity of the individual to sustain a task. This **serotonin-fatigue hypothesis** has been evaluated with studies that have imposed nutritional manipulation of neurotransmission (branched-chain amino acids, tryptophan, and carbohydrate supplementation), evoked pharmacological alterations of neurotransmission (selective serotonin reuptake inhibitors, combined reuptake inhibitors), and administered serotonin receptor agonists and antagonists (Meeusen et al., 2006). The results of these interventions indicate that an increase in serotonin levels in the brain does not consistently reduce the exercise capacity of an individual.

Rather, impairment due to neurotransmitter disturbances likely involves multiple neurotransmitters, such as serotonin, catecholamines (dopamine, norepinephrine), amino acid neurotransmitters (glutamate, gamma-aminobutyric acid [GABA]), and acetylcholine. These neurotransmitters influence a number of attributes that are essential for sustained activity: arousal, anxiety, mood, motivation, reward, and vigilance. The potential role of

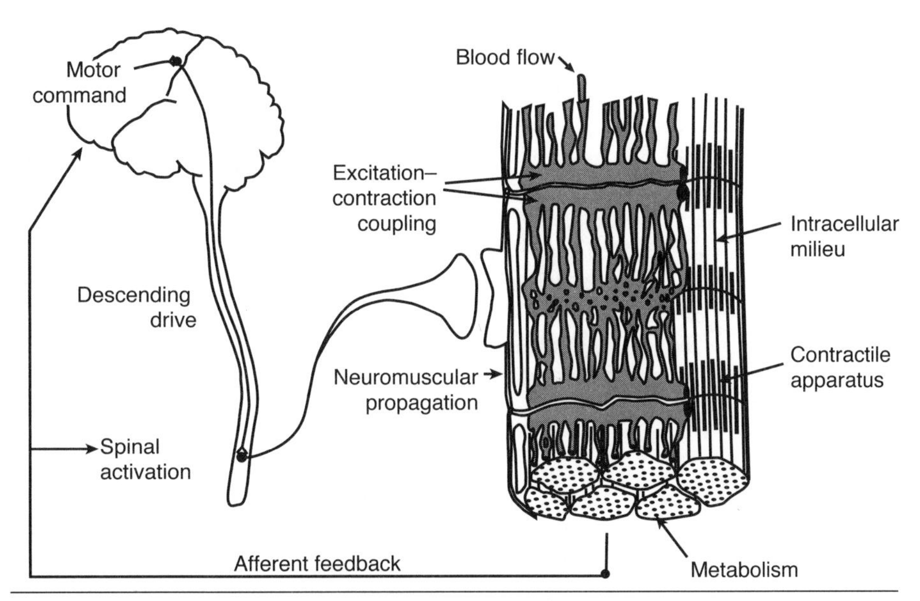

Figure 8.12 Potential processes that can be impaired during sustained activity and can contribute to muscle fatigue.

Adapted from *The Journal of Cell Biology*, 1969, 42: 46-67. Copyright 1969 by The Rockefeller University Press.

these various neurotransmitters in contributing to fatigue has been examined by administering reuptake inhibitors that influence the actions of two or more neurotransmitters. Because a reuptake inhibitor prolongs the action of a neurotransmitter, the comparison of different inhibitors should identify those that are critical for a particular task. Although such interventions do evoke the expected neuroendocrine effects, they have not influenced the time it takes athletes to perform a prescribed amount of work (Piacentini et al., 2002*a, b,* 2004). Alternatively, a number of experimental approaches on animals demonstrate that depletion of dopamine can be associated with a decline in the capacity to sustain a performance (Bailey et al., 1993). However, the potential role of dopamine depletion in humans remains uncertain (Meeusen et al., 2006; Nybo & Secher, 2004).

In contrast to the uncertainty over the role of neurotransmitters in contributing to fatigue, some exercise conditions can disrupt the processes that generate the motor output of the cerebral cortex. Two such conditions are an elevation of core temperature **(hyperthermia)** and a reduction in the level of glucose in the blood **(hypoglycemia)**. The capacity of individuals to sustain prolonged activity is reduced in hot environments, and this cannot be explained by the depletion of muscle glycogen, the elevation of muscle and blood concentrations of lactate, or the extracellular accumulation of potassium (González-Alonso et al., 1999; Nielsen et al., 1997; Nielsen & Nybo, 2003). Rather, hyperthermia impairs the ability of the cerebral cortex to provide the required level of muscle activation (Todd et al., 2005). The inability to sustain muscle activation in the heat was demonstrated by Nybo and Nielsen (2001) when they compared the adjustments that occurred during a 2 min MVC with the knee extensor muscles after subjects exercised on a cycle ergometer in hot and in normal environments. Subjects were exhausted after 50 min of cycling in the hot environment, but they were able to perform the exercise for 60 min without reaching exhaustion in the normal environment; core temperatures for the two conditions were 40° C and 38° C, respectively. At the completion of the exercise, peak torque was similar at the onset of the 2 min MVC, but force and muscle activation declined more rapidly after exercise in the hot environment (figure 8.13). This suggests that the earlier termination of the cycling exercise in the hot environment was likely caused by an inability to sustain the activation of the involved muscles.

The other exercise condition known to influence the motor output of the cerebral cortex is hypoglycemia. It has been known for some time that carbohydrate supplementation during prolonged exercise can prevent hypoglycemia and prolong the time to exhaustion (Coggan & Coyle, 1987; Coyle et al., 1983, 1986). The improvement in performance is typically attributed to an increase in the uptake of blood glucose by the exercising muscle. However, the greater availability of substrate also improves metabolism within the cerebral cortex. The benefit of carbohydrate supplementation on the function of the cerebral cortex was demonstrated by an evaluation of the performance of subjects on a 2 min MVC after they had cycled for 3 h while receiving either glucose or placebo supplementation (Nybo, 2003). The subjects became hypoglycemic with the placebo supplementation but maintained normal levels of blood glucose (euglycemic) with glucose supplementation. The torque exerted by the knee extensor muscles at the onset of the 2 min MVC was the same after both conditions, but both the torque and muscle activation declined more rapidly when the subjects did not receive the glucose supplement. Furthermore, glucose uptake by the cerebral cortex was reduced more during the placebo condition, which indicates that there was less energy turnover in the cerebral cortex when the subjects were hypoglycemic (Nybo et al., 2003*a,b*). As with the hyperthermia study (Nybo & Nielsen, 2001), these results indicate that the improvement in performance with carbohydrate feeding during

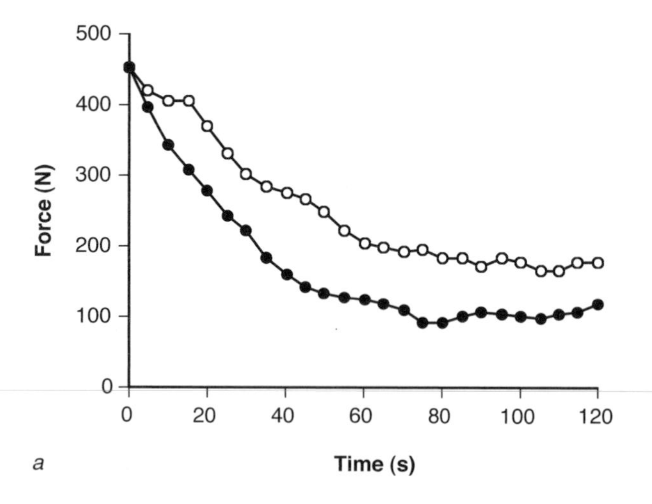

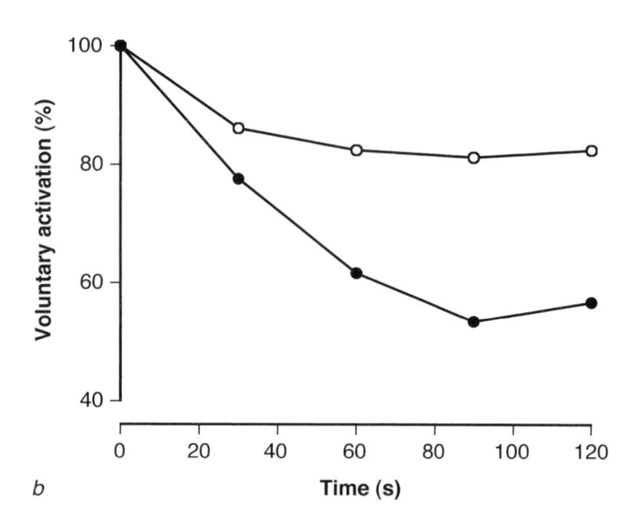

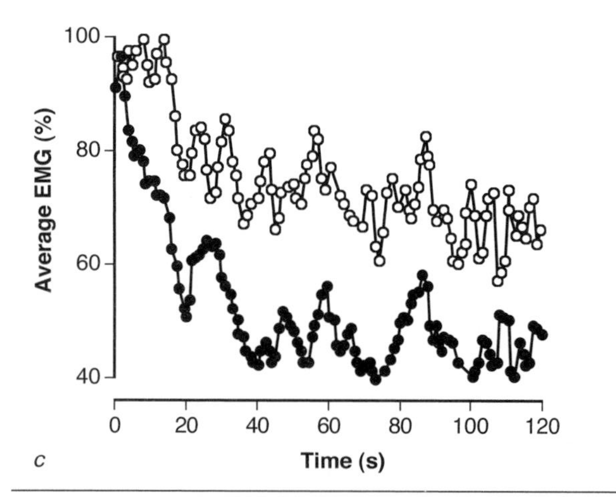

Figure 8.13 The decline in *(a)* force, *(b)* voluntary activation, and *(c)* rectified EMG for vastus lateralis during a sustained 2 min MVC with the knee extensor muscles after exercising in normal (open circles; 18° C) and hot (filled circles; 40° C) environments for about 1 h.

Data from Nybo and Nielsen, 2001.

prolonged exercise involves a preservation of function in the cerebral cortex.

Variation in the levels of some neurotransmitters, such as glutamate, acetylcholine, and GABA, may be influenced by the excessive accumulation of ammonia (NH_3) in the cerebral cortex (Davis & Bailey, 1997; Sahlin et al., 1999; Snow et al., 2000). Ammonia is released by active muscles into the blood supply and can cross the blood-brain barrier. **Hyperammonemia** can disturb cerebral blood flow, energy metabolism, synaptic transmission, and the regulation of some neurotransmitters (Felipo & Butterworth, 2002). Because the plasma concentration and the cerebral uptake of ammonia increase after prolonged exercise in the heat and without glucose supplementation, its accumulation in the cerebral cortex may disturb the function of some neurotransmitters and thereby impair the ability of an individual to sustain muscle activation (Blomstrand et al., 2005; Mohr et al., 2006; Nybo et al., 2005).

Descending Drive

The excitation delivered from supraspinal centers to motor neurons is not impaired during high-force fatiguing contractions, but it can be during prolonged contractions (Gandevia, 2001; Nybo & Secher, 2004). This limitation is expressed as an increase in the effort associated with the task, a decline in voluntary activation of the involved muscles, and the spread of activation to accessory muscles in an attempt to maintain the desired motor output.

Perceived Effort Recall from chapter 7 that in the performance of a voluntary contraction, a copy of the motor command, known as the corollary discharge (figure 7.35), is returned to cortical centers and enables an individual to judge the effort associated with performing the contraction. This judgment is known as the **sense of effort** (Carson et al., 2002; McCloskey et al., 1974; Winter et al., 2005). The sense of effort is a subjective judgment of the intensity of the outgoing motor command associated with performing a task at a specific instant in time. This sensation is independent of the mechanisms that impair the ability of muscle to exert a desired force. For example, imagine holding a heavy briefcase in one hand while you are waiting for a bus. After a few minutes, although you are still able to hold the briefcase, you notice that the task requires an increase in the effort necessary to do so. To continue holding the briefcase in the same hand, it is necessary to increase the descending drive, which can be detected as an increase in the sense of effort, to compensate for the decline in the force produced by the motor units that were activated at the onset of the task. For sustained submaximal contractions performed by a motivated individual, the effort will always increase before the force begins to decline (Jones, 1995; Jones & Hunter, 1983).

One approach that is used to monitor the change in effort during a fatiguing contraction is to have a subject perform a contralateral-matching experiment (Allen & Proske, 2006; McCloskey et al., 1974). In this protocol, the subject exerts a sustained force with one limb (the test limb) and periodically indicates the effort associated with the sustained contraction by performing a matching contraction with the other limb. The subject accomplishes this by activating the muscles in the matching limb until the senses of effort for the two limbs are equal. The results of such an experiment are shown in figure 8.14. The force sustained by the test limb for 10 min remained constant (filled circles), whereas the force exerted by the matching limb increased progressively to indicate a gradual increase in the effort during the contraction. Furthermore, the ability to match the position of the test limb and to detect the speed at which it moves is impaired after the muscles of the test limb have been fatigued (Allen & Proske, 2006).

Voluntary Activation In addition to estimating changes in effort during fatiguing contractions, investigators assess the adequacy of the descending drive to the involved muscles by estimating the level of muscle activation. A common approach is to provide electric or magnetic stimuli at various locations between the motor cortex and the neuromuscular junction to determine if the force being exerted during the voluntary contraction can be supplemented by artificial stimulation (Gandevia, 2001; Taylor & Gandevia, 2008). This approach can involve applying single shocks or a brief train of shocks to the nerve during a fatiguing contraction. When subjects performed a sustained 60 s MVC with a thumb muscle (adductor pollicis), for example, the force declined by 30% to 50%, but an electric shock applied to the ulnar nerve could not increase the voluntary force (Bigland-Ritchie et al., 1982). Similarly, the maximal voluntary and electrically elicited (50 Hz) forces declined in parallel (figure 8.11*a*) when subjects performed intermittent, submaximal contractions (50% MVC force) with quadriceps femoris (Bigland-Ritchie et al., 1986*c*). The parallel decline in the voluntary and evoked forces and the decrease in the superimposed twitch indicate that the descending drive remained adequate for muscle activation during these two tasks.

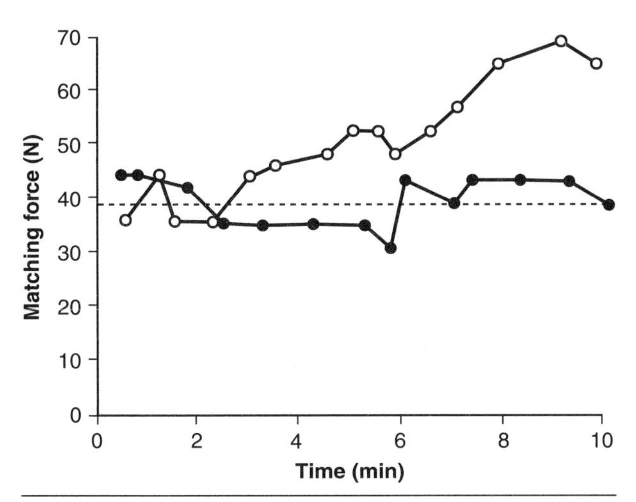

Figure 8.14 A contralateral-matching experiment in which one arm sustained a force of 40 N for 10 min (filled circles) and the other arm (open circles) performed intermittent contractions to the same level of effort associated with the sustained contractions being performed by the other arm. The increase in the force exerted by the matching arm indicates that the effort associated with the fatiguing contraction increased progressively during the task

Adapted from *Experimental Neurology*, Vol. 42, D.I. McCloskey, P. Ebeling, and G.M. Goodwin, "Estimation of weights and tensions and apparent involvement of a sense of effort," pg. 226. Copyright 1974, with permission from Elsevier.

In contrast, the voluntary activation provided by the descending drive became inadequate during a sustained submaximal contraction (30% MVC force) with the plantarflexor muscles. Löscher and colleagues (1996) found that when subjects could no longer achieve the target force after 400 s with a voluntary contraction, the required force could be elicited by electrical stimulation applied over the muscles. After 60 s of electrical stimulation, during which time the central processes could recover, the subjects were able to continue the task with a voluntary contraction for another 85 s. This study suggests that the adequacy of the descending drive during a fatiguing contraction may depend on the involved muscles and the target force.

The extent to which the descending drive is inadequate and the locus of the impairment can be examined through measurement of voluntary activation (equation 8.1) in response to stimuli delivered at various locations along the pathway from the motor cortex to the muscle (Andersen et al., 2003; Löscher & Nordlund, 2002; Taylor et al., 2006; Todd et al., 2004). For example, studies on the fatigue experienced by the elbow flexor muscles can involve a comparison of responses evoked by transcranial stimulation of the motor cortex, stimulation of the corticospinal tract at the level of the cervicomedullary junction (transmastoid processes), stimulation of the brachial plexus at the supraclavicular fossa, and stimulation of the intramuscular branches of the motor nerve (figure 8.15). Stimulation of the motor cortex, corticospinal tract, and brachial plexus evokes measurable EMG and force responses, whereas stimulation applied over the muscle elicits only a measurable force response. Because TMS generates action potentials in the axons of cortical neurons that impinge on the output neurons of the motor cortex, a difference in the responses evoked by cortical and corticospinal stimulation provides information on changes in the cortical neurons that transmit the descending

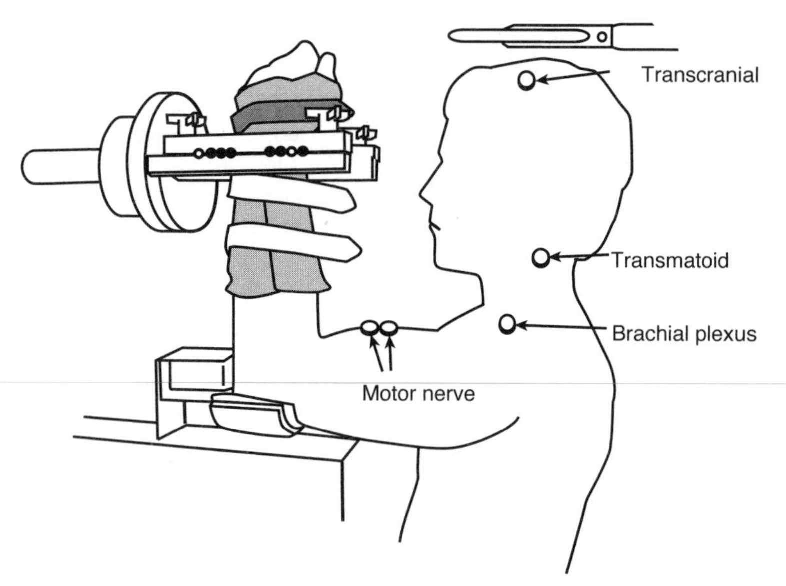

Figure 8.15 Changes that occur along the pathway from the motor cortex to the muscle fibers of the elbow flexor muscles can be examined through comparison of responses evoked by stimulation of the motor cortex (transcranial stimulation), the corticospinal tract (transmastoid stimulation), the peripheral nerve (brachial plexus), and the intramuscular branches of the motor axons (motor nerve).

drive. Similarly, a difference in the responses between the corticospinal and brachial plexus stimulation informs the investigator about changes in the excitability of the motor neurons.

The distributed-stimulation approach has provided a range of observations on the adjustments that can occur during fatiguing contractions. For example, Søgaard and colleagues (2006) examined the adjustments that occurred when subjects sustained an isometric contraction with the elbow flexor muscles at a target force of 15% MVC force for 43 min. To measure voluntary activation during MVCs performed once every 3 min, transcranial stimulation followed by either brachial plexus or motor nerve (paired-stimulus method) stimulation was applied. Maximal voluntary contraction torque declined to 58% of the initial value and the paired-stimulus force decreased to 59% at the end of the fatiguing contraction. The progressive increase in perceived effort was accompanied by a decline in voluntary activation from 98% at the start of the task to 72% for transcranial stimulation and 77% for brachial plexus stimulation at the end of the contraction. On the basis of these data, Søgaard and colleagues concluded that approximately 40% of the decline in MVC torque during this task was caused by an inadequate output from the motor cortex.

Another study used transcranial stimulation to assess the contribution of voluntary activation to the difference in fatigability exhibited by men and women (Hunter et al., 2006). Subjects performed six 22 s MVCs with the elbow flexor muscles; there was a 10 s rest between each MVC. Voluntary activation prior to beginning the task was 96% for the men and 93% for the women. The men were more fatigable and experienced a 65% decrease in MVC torque compared with a 52% reduction for the women. At the end of the sixth MVC, voluntary activation decreased to similar values: 77% for the men and 73% for the women. In contrast, the men experienced a greater decrease in the estimated twitch torque (59%) compared with the women (27%). These data were interpreted to indicate that the sex difference in fatigability (Hunter et al., 2004*b;* Russ & Kent-Braun, 2003; West et al., 1995), at least for this task, is not attributable to a difference in voluntary activation.

Patterns of Muscle Activity A resultant muscle torque about a joint can be achieved by a variety of muscle activation patterns. This flexibility certainly exists among a group of synergist muscles, but it can also involve accessory muscles and antagonist muscles. Because of this possibility, the motor system can delay the progression of fatigue by varying the contribution of synergist and accessory muscles to the resultant muscle torque and by minimizing the activation of antagonist muscles. Although varying the contribution of synergist and accessory muscles is available only when the task requires submaximal torques, most activities of daily living involve such torques.

Four examples describe the range of adjustments in the pattern of muscle activity that can occur during a fatiguing contraction: alternation among synergist muscles, coactivation of antagonist muscles, activation of contralateral muscles, and the specificity of the activation. When low forces (≤5% MVC force) are sustained for an hour or so, several studies have shown that the EMG activity of the involved muscles can vary despite a constant resultant muscle torque (Sjøgaard et al., 1986, 1988). The alternating activity has been observed most often among the quadriceps femoris muscles. For example, Kouzaki and Shinohara (2006) quantified the alternating activity during a 60 min isometric contraction at 2.5% MVC force with the knee extensor muscles. They found that the EMG alternated between rectus femoris and vastus lateralis and between rectus femoris and vastus medialis seven times for each pair during the contraction. In contrast, the EMG activity alternated only twice between vastus lateralis and vastus medialis. The number of times the EMG activity alternated varied among the 41 subjects, and those subjects who exhibited the greatest amount of alternating activity experienced the least fatigue (decline in MVC force). Consistent with this observation of a functional significance to the alternating activity, there is a strong association between the variation in EMG activity and perfusion of the involved muscles (Kouzaki et al., 2003; Laaksonen et al., 2006) and magnetic resonance imaging (MRI) measures of muscle activation (Akima et al., 2004).

A common observation during fatiguing contractions is that activation of antagonist muscles increases progressively, which reduces the net muscle torque (Ebenbichler et al., 1998; Hunter et al., 2003; Psek & Cafarelli, 1993). For example, Lévénez and colleagues (2005) found that the EMG activity of the agonist (tibialis anterior) and antagonist (soleus and lateral gastrocnemius) muscles increased progressively when subjects sustained a dorsiflexion torque at 50% of maximum for as long as possible. The maximal dorsiflexion torque decreased by 41% and the associated EMG amplitude of tibialis anterior declined by 37% at the end of the task. Despite the gradual increase in EMG of the antagonist muscles, they did not exhibit any fatigue at the conclusion of the task. The amplitude of the H reflex evoked in soleus and lateral gastrocnemius (antagonists), however, exhibited a biphasic pattern: It increased to about 150% of the initial values during the first 20% of the contraction and then declined to around 70%. The decline in H-reflex amplitude was likely caused by a gradual increase in presynaptic inhibition of the group Ia afferents that connected to the motor neurons innervating the antagonist muscles. Lévénez and colleagues (2008) subsequently found that the biphasic pattern of evoked responses was only evident for H reflexes and not for those elicited with transcranial or transmastoid stimulation. Thus, although the EMG activity increased in parallel for the agonist and antagonist muscles, the biphasic adjustment in spinal reflex excitability for the antagonist muscles suggests that the two sets of muscles are controlled by independent descending inputs.

As a fatiguing contraction progresses, activation spreads to muscles not directly involved in the task, including muscles in the contralateral limb. For example, Zijdewind and Kernell (2001) found that intermittent contractions to a target force of 30% MVC force performed with a muscle in one hand were accompanied by a gradual increase in EMG activity up to 8% of maximum in the same muscle of the other hand. Nonetheless, the contralateral activity was relatively minor and did not influence the performance of a subsequent fatiguing contraction by the muscle in the contralateral hand (Todd et al., 2003*a*; Zijdewind et al., 1998). However, when the contralateral limb was required to participate in a pointing task with the limb in which a test muscle had been fatigued, the performance of the unfatigued contralateral limb deteriorated also (increased EMG activity and tremor) (Morrison et al., 2005). A similar effect was observed among intrinsic hand muscles (Danion et al., 2000). By comparing potentials evoked with transcranial and transmastoid stimulation during contractions performed with the left and right elbow flexors, Zijdewind and colleagues (2006*a*) found that the contralateral activity (left arm) arose from motor areas in the appropriate hemisphere (right), which meant that both the right and left hemispheres were active during a voluntary contraction performed with the elbow flexors in the right arm. Furthermore, they observed that cortical excitability in the right hemisphere was greater during the contralateral activity compared with when the muscles in the left arm themselves performed the voluntary contraction to the same EMG level. These types of fatiguing contractions are capable of reducing intracortical facilitation (Baumer et al., 2002).

The study of contralateral interactions has also indicated that the amount of fatigue experienced during a task depends on the familiarity of the individual with the task. For example, Rube and Secher (1990) examined the ability of subjects to perform an isometric leg extension task (concurrent hip and knee extension) with either one or two legs. Subjects performed 150 MVCs for both conditions (one and two legs) before and after five weeks of strength training. The subjects were assigned to one of three groups: control group, one-legged training group, or two-legged training group. Both the one- and two-legged groups increased strength after five weeks of training. Furthermore, the rate of decline in force during the 150 contractions was less after training (figure 8.16). But the effect was specific to the training mode. The one-legged training group was less fatigable only during the one-leg task, and the two-legged training group was less fatigable only during the two-leg task. Thus, even though the subjects in the two-legged group trained both legs, the improvement in fatigability was observed only when both legs performed the task. These data indicate that the neural strategy adopted during the one-leg task differed from that used during the two-leg task, and that this difference influenced the fatigue experienced in each task. Similarly, the selective fatigue of two out of many muscles involved in a throwing action significantly impaired the subsequent performance of the action (Forestier & Nougier, 1998), and the adjustments depend on the location of the muscles that are fatigued (Huffenus et al., 2006).

Another frequently cited example of the specificity of muscle activation is the difference in fatigability of shortening and lengthening contractions (Babault et al., 2006; Baudry et al., 2007; Molinari et al., 2006; Tesch et al., 1990). For example, Pasquet and colleagues (2000) compared the adjustments that occurred in the dorsiflexor muscles when subjects performed five sets of 30 MVCs at 0.87 rad/s over a 0.52 rad (30°) range of motion. Examples of the isokinetic contractions at the beginning and end of the protocol are shown for one subject in figure 8.17. The decline in maximal torque during

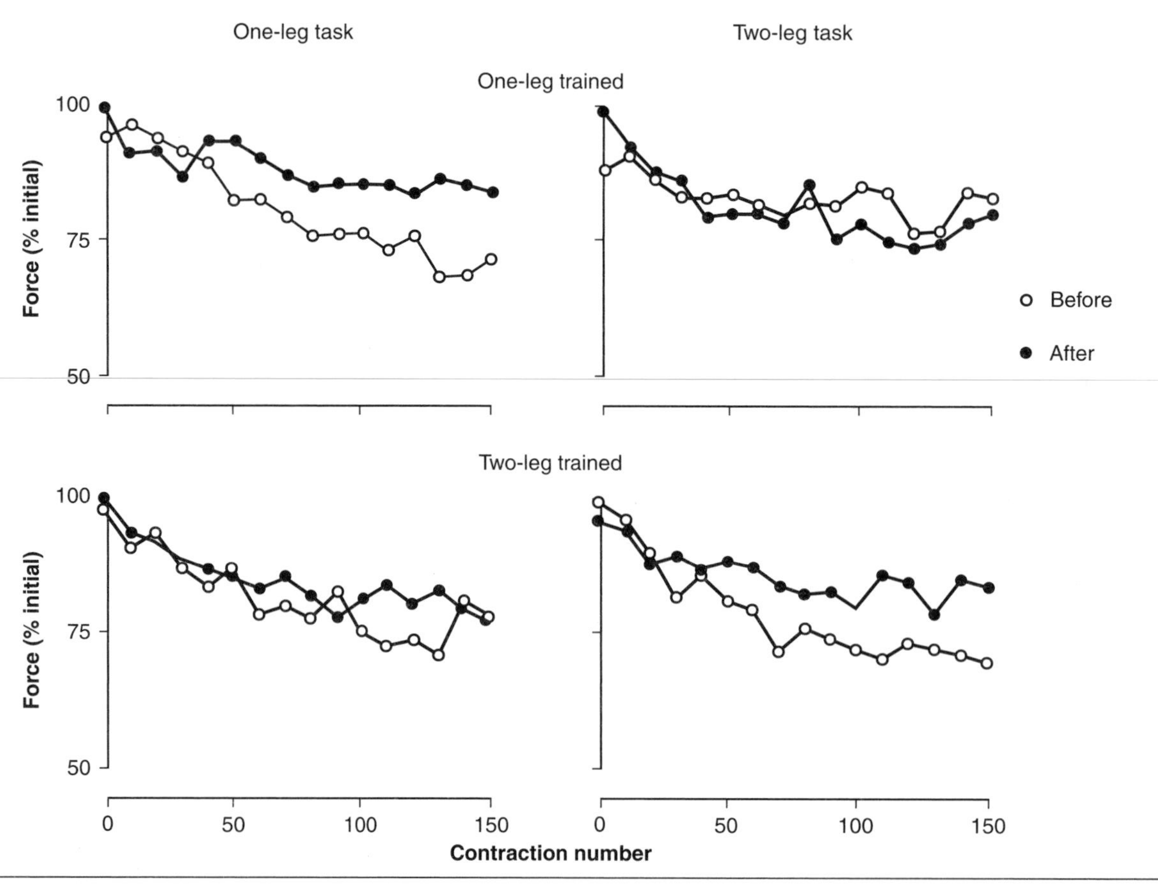

Figure 8.16 The decline in force during one- and two-legged performances of a leg extension task before and after five weeks of training. The upper row shows the results for the one-legged training group; the bottom row indicates the data for the two-legged training group. The rate of decline in force was less after training, but for the one-legged group during the one-leg task only and for the two-legged group during the two-leg task only.

Data from Rube and Secher 1990.

the 150 contractions for the group of subjects was greater for the shortening contractions (–32%) compared with the lengthening contractions (–24%). Similarly, the decrease in peak EMG was greater for the shortening contractions (–26% vs. –18%). During the 60 s pause between each set of 30 MVCs, Pasquet and colleagues delivered single and paired stimuli to the peroneal nerve to assess the M-wave and twitch responses in tibialis anterior. The peak twitch torque and the rate of change in torque during the twitch were greater during the protocol with lengthening contractions; in contrast, the one-half relaxation time and M-wave duration were more prolonged and twitch potentiation was less depressed during the series of shortening contractions (figure 8.18). Because there was no difference in either the superimposed twitch response or the decline in EMG normalized to M-wave amplitude, the greater fatigue experienced during the shortening contractions was not caused by a difference in voluntary activation or neuromuscular propagation. Rather, the adjustments during the 150 contractions and the subsequent recovery period suggest a greater impairment of excitation-contraction coupling during the shortening contractions.

EXAMPLE 8.2

Identifying the Cause of Fatigue in Patients

To identify the physiological mechanisms that underlie the fatigue experienced by patients presenting with a variety of conditions, clinical investigators often perform a battery of tests and compare the outcomes with those for healthy individuals. One example is multiple sclerosis, in which fatigue is a common and disabling symptom. When patients with multiple sclerosis and healthy control subjects performed a 45 s maximal contraction with a hand muscle, the force declined by 45% in the patients but only by 20% in the control subjects (Sheean et al., 1997). The strength of the hand muscle prior to the fatiguing contraction was

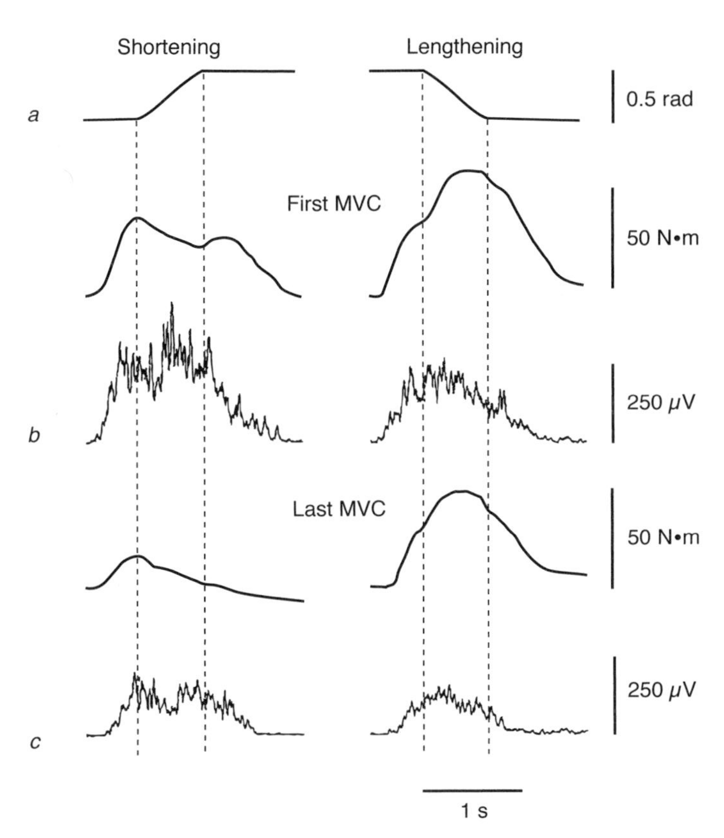

Figure 8.17 Maximal shortening and lengthening contractions performed with the ankle dorsiflexor muscles on an isokinetic dynamometer. *(a)* The change in ankle joint angle. *(b)* Peak torque during the first MVC was greater during the lengthening contraction, whereas the peak surface EMG for tibialis anterior was greater during the shortening contraction. *(c)* The decline in peak torque and EMG by the 150th MVC was greater for the shortening contractions.

Muscle & Nerve, Vol. 23, 2000, pg. 1729. "Muscle fatigue during concentric and eccentric contractions" by B. Pasquet, A. Carpentier, J. Duchateau, and K. Hainaut. Copyright © 2000 John Willey & Sons, Inc. Adapted with permission of Wiley-Liss Inc., a subsidiary of John Wiley & Sons, Inc., 2000.

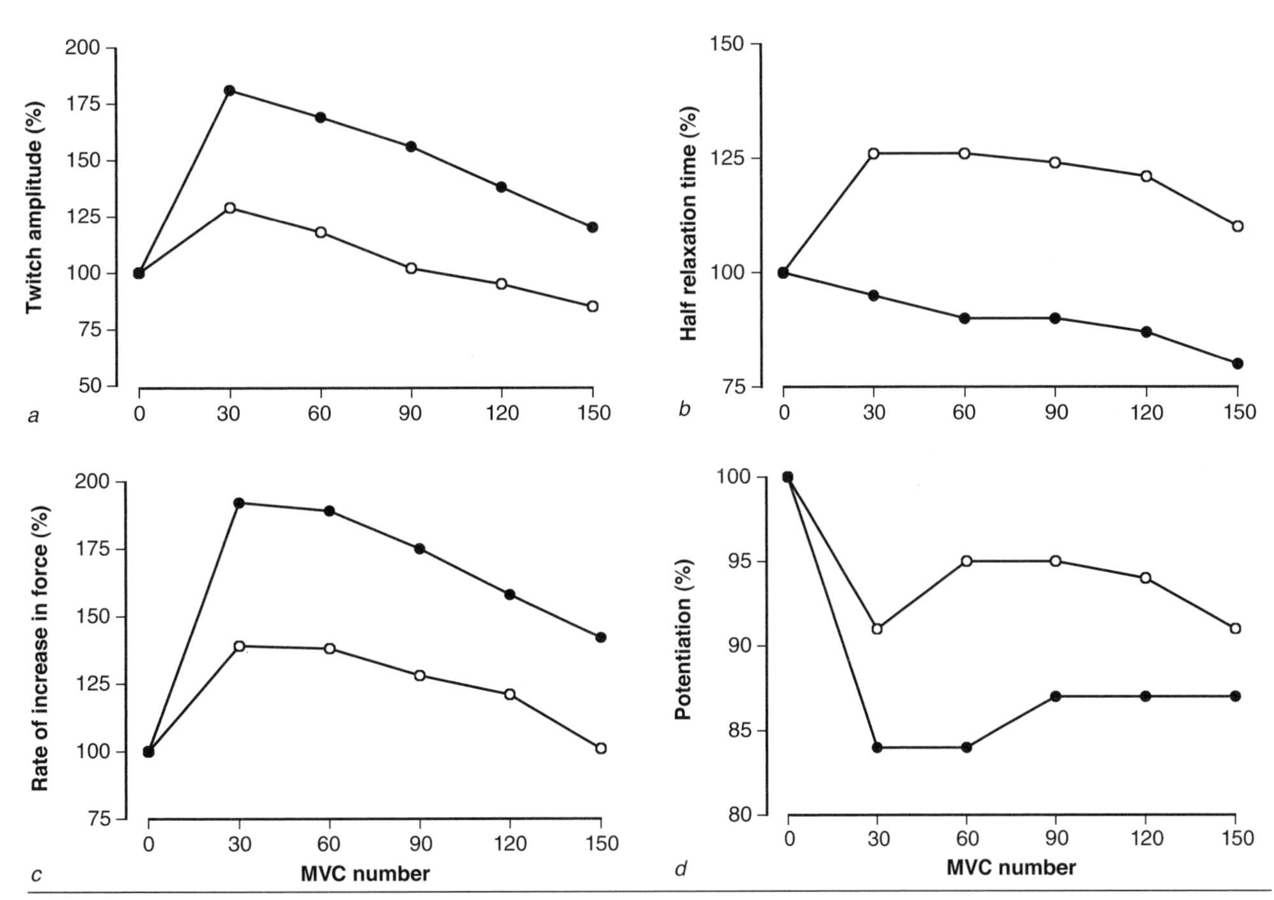

Figure 8.18 *(a)* Changes in twitch amplitude, *(b)* one-half relaxation time of the twitch, *(c)* rate of change in twitch torque, and *(d)* twitch potentiation during 150 maximal shortening (open circles) and lengthening (filled circles) contractions with the ankle dorsiflexor muscles.

Data from Pasquet et al. 2000.

similar for the two groups of subjects, and there was no difference in the influence of fatigue on contractile function. Furthermore, based on potentials evoked by TMS, there was no impairment of the primary motor pathways or the excitability of the motor cortex in the patients. These results led Sheean and colleagues (1997) to conclude that the fatigue experienced by patients with this task was due to a decline in the activation of the primary motor cortex. This conclusion was confirmed in other studies on patients with multiple sclerosis (Liepert et al., 2005; Sandroni et al., 1992; Thickbroom et al., 2006) and in a study of patients with chronic fatigue syndrome (Lloyd et al., 1991). Consistent with these observations, Leocani and colleagues (2001) found more widespread activity in the electroencephalogram recorded over the sensorimotor cortex in multiple sclerosis patients who complained of fatigue compared with healthy individuals when they performed a thumb movement. Furthermore, another study showed no association between cerebral abnormalities, as determined by MRI, and neurological disability in patients with multiple sclerosis (van der Werf et al., 1998).

The magnitude of the decline in force during a fatiguing contraction experienced by patients with multiple sclerosis is often not related to their clinical symptoms. Part of the difficulty faced by clinicians who attempt to associate signs and symptoms of multiple sclerosis with laboratory measures of fatigue is the many meanings of the word "fatigue." It is important to distinguish muscle fatigue from other signs of impairment, such as weakness and tiredness. Muscle fatigue refers to the decline in the force or power capacity of muscle due to exercise, whereas **muscle weakness** indicates a reduction in the peak force for a given volume of muscle mass, **tiredness** denotes sensations of lethargy, and **lassitude** corresponds to a subjective sense of reduced energy (Schwid et al., 2002). With these distinctions in mind, Ng and colleagues (2004) found that the impairment in voluntary activation exhibited by patients with multiple sclerosis was related to weakness and deficits in walking, but not to fatigue. By focusing on these differences, clinicians may be able to offer more effective interventions.

Spinal Activation

When the force exerted by activated muscle fibers is insufficient to sustain a submaximal contraction, the CNS can adjust the motor output from the spinal cord to compensate for the deficit. Although the adjustments can involve either the activation of more motor units or an increase in the discharge rate of the units that are already active, the options depend on the intensity of the muscle contraction. When the task requires a force that exceeds the upper limit of motor unit recruitment, for example, the adjustment cannot include the activation of additional motor units. Furthermore, discharge rate tends to decrease during high-force contractions (figure 8.19) even though the rates during voluntary contractions do not reach the plateau of the force–frequency relation (figure 6.16). Consequently, it is not possible to attenuate the decrease in muscle force during strong fatiguing contractions by altering motor unit activity.

In contrast, it is possible to prolong the duration over which a force less than the upper limit of motor unit recruitment can be sustained by adjusting motor unit activity. The adjustments include the recruitment of motor units that were not active at the beginning of the contraction and changes in the rate at which the motor units discharge action potentials (Adam & De Luca, 2003; Christova & Kossev, 1998; Enoka et al., 1989; Jensen et al., 2000; Miller et al., 1996). The activation of additional motor units is relatively straightforward, as the order in which motor units are recruited during a sustained contraction is similar to that observed during control contractions (Adam & De Luca, 2003; Carpentier et al., 2001). The change in discharge rate, however, depends on both the recruitment threshold of the motor unit and whether or not it was active from the beginning of the contraction (Riley et al., 2008). For example, Carpentier and colleagues (2001) found that the change in discharge rate of motor units in a hand muscle during intermittent, isometric contractions performed to a target force of 50% MVC force varied with recruitment threshold. Those units with low recruitment thresholds (<25% MVC force), which were recruited from the onset of the task, experienced a decrease in discharge rate, whereas discharge

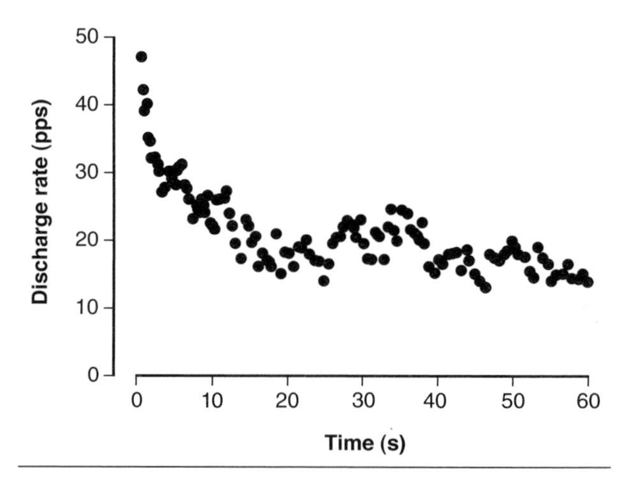

Figure 8.19 Decrease in the discharge rate of a single motor unit in a hand muscle (adductor pollicis) during a 60 s maximal contraction.

Data from Marsden et al., 1983*a*.

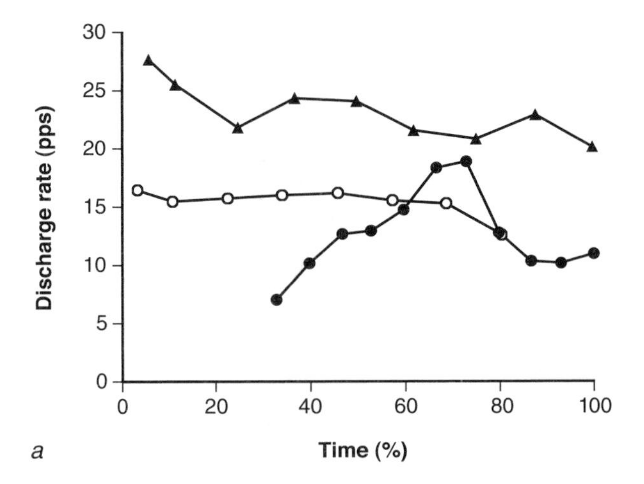

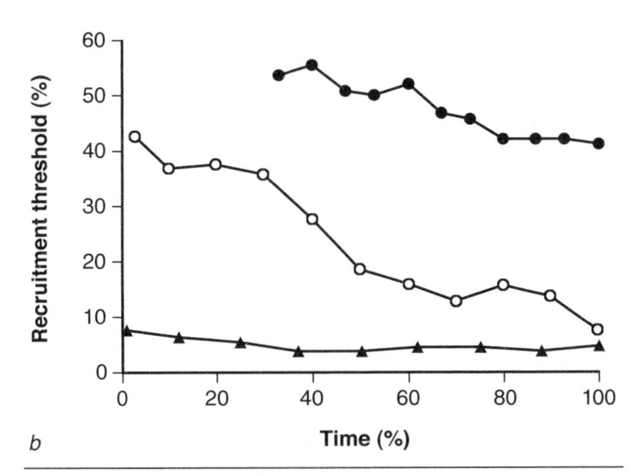

Figure 8.20 Change in *(a)* discharge rate and *(b)* recruitment threshold of three concurrently active motor units in a hand muscle during a series of intermittent, isometric contractions to a target force of 50% MVC force. Each contraction comprised a 3 s increase to the target force, 10 s at the target force, and then a 3 s decrease to zero. Two motor units were active from the beginning of a fatiguing contraction, and one was recruited at about 30% of the task duration. The unit with the lowest recruitment threshold (7.4%) had the greatest initial discharge rate (28 pps). The unit with the intermediate recruitment threshold (42%) stopped discharging during the plateau phase of each intermittent contraction at about 80% of task duration. The unit recruited during the task had a recruitment threshold of 54%.

Adapted, by permission, from A. Carpentier, J. Duchateau, and K. Hainaut, 2001, "Motor unit behaviour and contractile changes during fatigue in the human first dorsal interosseus." *Journal of Physiology* 534: 907.

rate did not change for units with higher thresholds, although some of them stopped discharging action potentials (figure 8.20*a*). In addition, units that were recruited during the fatiguing contraction displayed an initial increase in discharge rate followed by a decrease. Because the descending drive increased progressively during the intermittent contractions, the decrease in discharge rate was likely caused by changes in either the intrinsic properties of motor neurons, the afferent feedback received by the motor neurons, or both mechanisms. Furthermore, only the motor units with recruitment thresholds >25% MVC force exhibited a decrease in recruitment threshold during the fatiguing contraction (figure 8.20*b*), which suggests either that these motor neurons received greater net excitatory input or that they experienced a reduction in the voltage threshold.

The change in discharge rate during fatiguing contractions also depends on the type of muscle contraction performed during the task. Griffin and colleagues (1998) recorded the discharge of motor units in triceps brachii when subjects performed 50 repetitions of lifting and lowering a load (20% of maximum) over a 0.7 rad displacement about the elbow joint. The 50 repetitions fatigued the muscle, and MVC torque decreased by 29% at the end of the task. To compensate for the decline in force capacity, additional motor units were recruited during the task, and there was no change in average discharge rate for the motor units that were active from the beginning of the task. For comparison, Griffin and colleagues (2000) performed the same protocol with isometric contractions and found that the discharge rate of most motor units declined over the course of the 50 repetitions. These two studies suggest that feedback related to the type of muscle contraction can influence the adjustments in discharge rate that occur during fatiguing contractions.

The decline in discharge rate that occurs when the fatigue task involves isometric contractions has been regarded as an adjustment designed to match the change in the mechanical state of the muscle (Bigland-Ritchie et al., 1983*a, b*); Binder-Macleod & Guerin, 1990; Jones et al., 1979; Gordon et al., 1990a). Because the relaxation rate of the twitch decreases and the duration of the twitch increases with the development of fatigue, the same degree of fusion in the force during a tetanus can be achieved with a lower rate of activation (figure 8.21). Therefore, the reduction in discharge rate may represent an adjustment that makes the activation signal more economical. The reduction in the discharge rate of motor units to match the change in the mechanical state of the muscle fibers is known as **muscle wisdom** (Marsden et al., 1983*a*). The potential mechanisms that might contribute to the slowing of discharge rate during a prolonged contraction include reflexes, motor neuron properties, and descending drive. Two lines of evidence, however, suggest that muscle wisdom is not a global strategy during fatiguing contractions. First, electrical stimulation of a hand muscle with a rate that declined from 30 Hz to 15 Hz evoked a more rapid decline in force than that elicited with a constant rate of 30 Hz (Fuglevand & Keen, 2003). This result indicates that the decline in discharge rate did not optimize muscle activation and

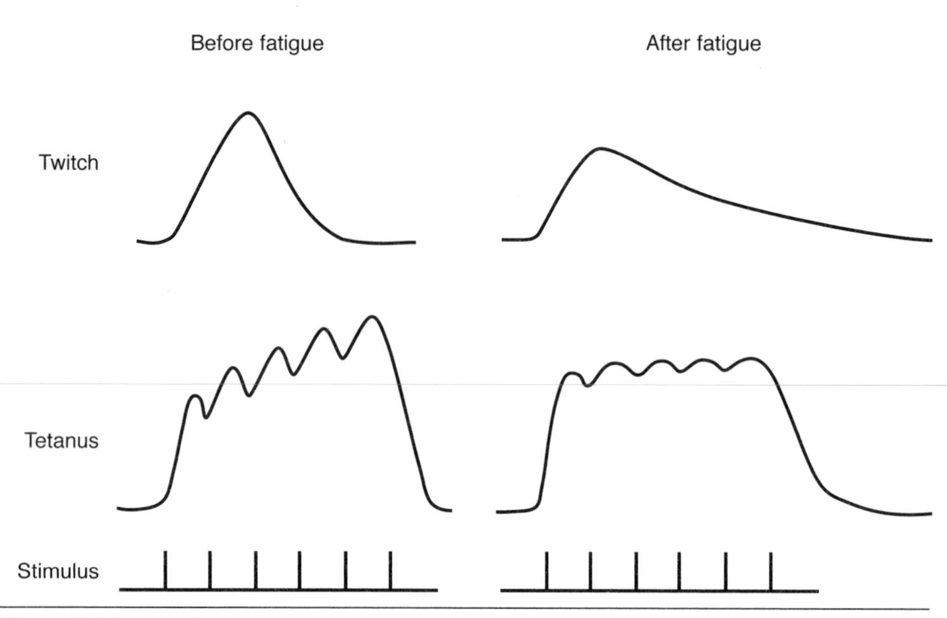

Figure 8.21 A fatiguing contraction is usually associated with an increase in the relaxation time of a twitch and an increase in the degree of fusion of an unfused tetanus for the same frequency of stimulation.

actually contributed to the decrease in force during a fatiguing contraction. Second, the discharge rate of motor units during submaximal fatiguing contractions does not always decrease, and those units that are recruited during the contraction invariably exhibit an increase in discharge rate (Carpentier et al., 2001; Garland et al., 1994; Kuchinad et al., 2004; Nordstrom & Miles, 1991).

Afferent Feedback

Two approaches have been used to identify the changes that occur in afferent feedback during fatiguing contractions: measuring the adjustments in the transduction properties of the sensory receptors and recording the changes in reflex pathways. The findings on the changes in the sensory receptors are often mixed and depend on the experimental preparation used in the study. For example, studies on experimental animals have generally shown that fatigue enhances the sensitivity of muscle spindle afferents (group Ia and II) to contractions by single motor units (Christakos & Windhorst, 1986; Nelson & Hutton, 1985; Zytnicki et al., 1990), presumably due to an increase in the discharge rate of gamma motor neurons (Ljubisavljevic et al., 1992). However, results from studies on humans indicate that fatigue is associated with a decline in gamma motor neuron activation of muscle spindles (Bongiovanni & Hagbarth, 1990; Macefield et al., 1991) and a reduction in the discharge of muscle spindle afferents during sustained isometric contractions (Macefield et al., 1991). Feedback from tendon organs (group Ib afferents) is also depressed by fatigue, as indicated by a reduction in the response of tendon organs to whole-muscle stretch (Hutton & Nelson, 1986) and a decrease in the inhibitory effect of Ib afferents on motor neurons (Zytnicki et al., 1990).

One can estimate the functional significance of these adjustments by comparing the changes that occur in reflex responses (Hunter et al., 2004*c;* Nicol et al., 2006). An example is the measurement of changes in short- and long-latency reflexes evoked by application of either a brief electrical stimulation to the group Ia fibers or a quick stretch of the muscle to activate the muscle spindle afferents. The short-latency reflex largely comprises monosynaptic input to the spinal motor neurons, whereas the long-latency reflex involves transmission of the elicited activity traveling to supraspinal centers before being received by the motor neurons. When subjects sustained a maximal contraction with a hand muscle (first dorsal interosseus) until the force declined to 50% of maximum, Duchateau and Hainaut (1993) found that the amplitude of the short-latency reflex (H reflex) decreased by 30% and that there was no change in the amplitude of the long-latency reflex. In contrast, there was an increase in the amplitude of the long-latency reflex in a hand muscle (abductor pollicis brevis) not involved in the fatigue task, which indicates an increase in the descending excitatory drive to the motor neuron pool of first dorsal interosseus to compensate for the changes in synaptic input and intrinsic properties of the motor neurons.

To evaluate the magnitude of the adjustments, Duchateau and colleagues (2002) compared the change in amplitude of the H and long-latency reflexes when subjects sustained isometric contractions with abductor pollicis brevis at target forces of 25% and 50% of maximum for as long as possible. The decrease in H-reflex amplitude was similar for the two contractions (25% and 50%) and comparable to that observed by Duchateau and Hainaut (1993) for the sustained maximal contraction (figure 8.22*a*). Because the amplitude of the H reflex did not decrease when subjects performed intermittent contractions (less occlusion of blood flow) to the 25% target force for a similar duration, Duchateau and colleagues (2002) concluded that the depression of the H reflex was likely caused by feedback from group III-IV afferents. In contrast to the consistent reduction in H-reflex amplitude, the

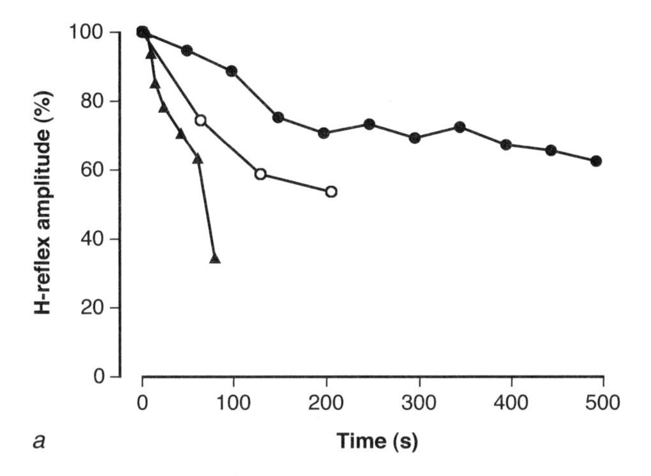

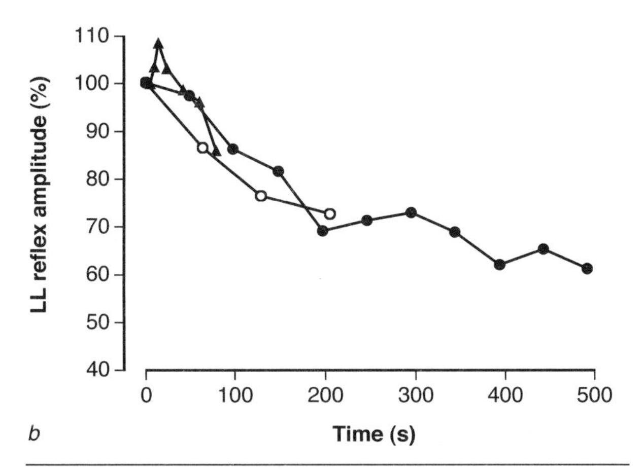

Figure 8.22 Changes in reflex amplitude of *(a)* the H reflex and *(b)* the long-latency (LL) reflex as a function of contraction force during sustained fatiguing contractions with a hand muscle. The target forces were 25% (filled circles), 50% (open circles), and 100% (filled triangles) of MVC force.

Data from Duchateau and Hainaut 1993; Duchateau et al. 2002.

decline in the amplitude of the long-latency component varied with the target force and, therefore, the duration of the contraction (figure 8.22*b*). The fatiguing contraction at 100% MVC force had the briefest duration and exhibited the smallest decline in the amplitude of the long-latency reflex. These results indicate that the decrease in excitatory input to the motor neurons from spinal or peripheral sources was constant for all three contractions but that the compensation by an increase in the descending excitatory drive varied with the duration of the contraction. This finding is consistent with the observation that the decline in voluntary activation is greater for long-duration contractions (Taylor et al., 2006).

Investigators have also examined transmission in reflex pathways by comparing two short-latency reflexes: H reflex and tendon tap reflex. Both reflexes involve the group Ia afferents, but they are activated by electrical stimulation of the peripheral nerve for the H reflex and by a stretch of the muscle spindles for the tendon tap reflex. Thus, a comparison of the two reflexes provides information about the sensitivity of the muscle spindles (figure 7.5). Klass and colleagues (2004) recorded the two reflexes in the soleus muscle before and after subjects performed a series of plantarflexion movements (anisometric contractions) against a load of 50% of maximum. At the end of the task, MVC torque during an isometric contraction decreased by 23%. The fatiguing contractions caused the amplitude of the H reflex to decrease by 10% and that for the tendon tap reflex to decline by 13%. The absence of a difference in the reduction of the two reflexes indicates that the depression of the reflexes was not related to changes in the sensitivity of the muscle spindle. A similar conclusion was reached in another study when the fatigue task involved a sustained MVC (Balestra e al., 1992). However, the decrease in the amplitude of these two reflexes after the anisometric contractions was much less than that observed after sustained isometric contractions (figure 8.22*a*).

In addition to modulating the activity transmitted by large-diameter afferents, fatiguing contractions activate the sensory receptors served by group III-IV afferents (Garland & Kaufman, 1995). The receptors innervated by group III afferents are sensitive to changes in both the mechanical state and the metabolic environment of the muscle, whereas group IV afferents are most responsive to the chemical milieu in the muscle. Because the receptors innervated by these afferents are so numerous, small variations in activity are likely to have a large effect in the CNS. A common strategy used to study the influence of fatigue on the feedback delivered by group III-IV afferents is to compare the recovery of function when blood flow is normal and when it is impeded. When blood flow is occluded, the metabolites that accumulate in the fatigued muscle continue to provide a stimulus that sustains the discharge of group III-IV afferents (Hayes et al., 2006). With this approach, Bigland-Ritchie and colleagues (1986*b*) found that the depression of discharge rate for motor units in biceps brachii after a sustained MVC did not recover during the 3 min when blood flow was occluded but did recover to control values within 3 min when blood flow was restored. This result led to the conclusion that a peripheral reflex mediated by group III-IV afferents from the fatigued muscle contributed to the decrease in discharge rate during this protocol. Conversely, feedback from the afferents that are sensitive to an occlusion of blood flow did not contribute to the decrease in voluntary activation after an MVC sustained by the elbow flexor muscles for 2 min (Butler et al., 2003; J.L. Taylor et al., 2000).

The central connections of group III-IV afferents, however, are not as direct as those for the large-diameter afferents and can evoke diverse responses (Duchateau & Hainaut, 1993; Pettorossi et al., 1999). Martin and colleagues (2006) examined the contribution of feedback by group III-IV afferents to the fatigue experienced during sustained 2 min MVCs with the elbow flexor and extensor muscles. The protocols involved comparing the amplitude of potentials evoked in muscle with transmastoid stimulation of the corticospinal tract (figure 8.15) during the MVCs and during recovery when blood flow to the muscle was occluded (ischemia) and when it was not; transmastoid stimulation is used to probe the excitability of the motor neurons. When a muscle is kept ischemic, the accumulated metabolites enhance the feedback delivered by group III-IV afferents (Hayes et al., 2006; Kaufman et al., 1984; Rotto & Kaufman, 1988). By comparing the amplitude of the evoked potentials in the presence and absence of enhanced group III-IV feedback, it is possible to estimate the influence of these afferents on the excitability of motor neurons. When the fatiguing contraction was performed with the triceps brachii muscle, the amplitude of the evoked response decreased by 35% during the MVC and remained depressed by 28% during ischemia but recovered within 15 s with the removal of ischemia. The amplitude of the evoked potentials in triceps brachii decreased by 20% after the fatiguing contraction was performed with the elbow flexor muscles. In contrast, the amplitude of the evoked potentials in biceps brachii increased by 25% after a fatiguing contraction with triceps brachii. These results indicate that group III-IV afferents depressed the excitability of triceps brachii motor neurons but facilitated those that innervate biceps brachii. These findings are consistent with direct measurements of chemical activation of group III-IV afferents in extensor muscles causing hyperpolarization of extensor motor neurons and depolarization of flexor motor neurons (Kniffki et al., 1979, 1980, 1981*a, b*). Thus, the contribution of feedback from group III-IV afferents to the decline in motor unit activity during a fatiguing contraction probably differs for flexor and extensor muscles.

Neuromuscular Propagation

Several processes are involved in converting an axonal action potential into a sarcolemmal action potential. Collectively, these processes are referred to as neuromuscular propagation. Sustained activity can impair some of the processes involved in neuromuscular propagation, and this can contribute to the decline in force associated with fatigue. Potential impairments include a failure of the axonal action potential to invade all the branches of the axon (branch-point failure), a reduction in the ability of the action potential to mobilize vesicles in the presynaptic terminal, a depletion of neurotransmitter, and a decrease in the sensitivity of postsynaptic receptors and membrane (Krnjevic & Miledi, 1958; Kugelberg & Lindegren, 1979; Kuwabara et al., 2002; Sieck & Prakash, 1995; Spira et al., 1976).

The most common way to test for impairment of neuromuscular propagation in humans is to elicit M waves before, during, and after a fatiguing contraction. Recall that an M wave is an EMG response caused by action potentials that are evoked in axons of alpha motor neurons in response to an electric (or magnetic) shock applied to the nerve (figure 7.10). A decline in M-wave amplitude is interpreted as an impairment of one or more of the processes involved in converting the axonal action potential (initiated by the electric shock) into the muscle fiber potentials. Figure 8.23 shows an example of a decline in M-wave amplitude immediately after a fatiguing contraction, as well as the eventual recovery of the M wave after 10 min of rest. This type of decline in M-wave amplitude tends to occur in long-duration, low-force contractions (Bellemare & Garzaniti, 1988; Fuglevand et al., 1993; Kranz et al., 1983; Linnamo et al., 2002; Millet & Lepers, 2004; Milner-Brown & Miller, 1986). Although modeling studies suggest that impairment of neuromuscular propagation is one of several mechanisms that can contribute to the decline in force during long-duration fatiguing contractions (Fuglevand et al., 1993), the M wave provides a rather coarse index of the changes that can occur in neuromuscular propagation (Keenan et al., 2006).

Excitation–Contraction Coupling

Under normal conditions, excitation by the nervous system results in the activation of muscle and the

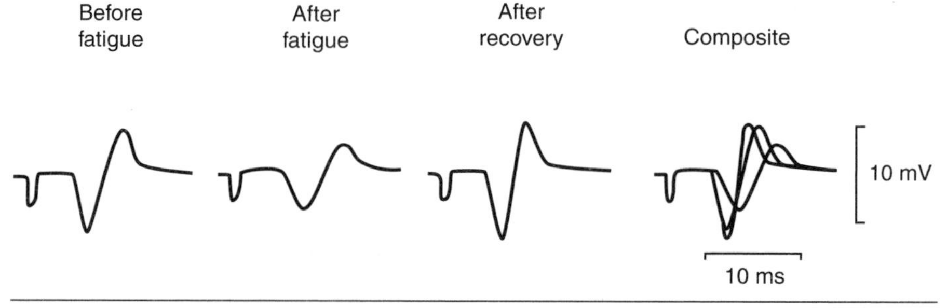

Figure 8.23 M waves elicited in a hand muscle (first dorsal interosseus) by stimulation of the ulnar nerve. M waves were evoked before and after a fatiguing contraction and after 10 min of recovery. The fatiguing contraction involved sustaining an isometric force at 35% of the maximal voluntary contraction force for as long as possible. M-wave amplitude declined immediately after the fatigue task but recovered quickly.

Adapted, by permission, from A.J. Fuglevand, K.M. Zackowski, K.A. Huey, and R.M. Enoka, 1993, "Impairment of neuromuscular propagation during human fatiguing contractions at submaximal forces," *Journal of Physiology* 460: 556.

subsequent cycling of cross-bridges. Seven major processes (Fitts, 2006) are involved in the conversion of the excitation (muscle fiber action potential) into a muscle fiber force (figure 6.5): (1) propagation of the action potential along the sarcolemma, (2) propagation of the action potential down the T tubule, (3) change in the Ca^{2+} conductance of the sarcoplasmic reticulum, (4) movement of Ca^{2+} down its concentration gradient into the sarcoplasm, (5) beginning of reuptake of Ca^{2+} by the sarcoplasmic reticulum, (6) binding of Ca^{2+} to troponin, and (7) interaction of actin and myosin and the work done by the cross-bridge.

Impairments Observed Each of these processes can be influenced by many different factors. For example, opening of the Ca^{2+} release channels in the sarcoplasmic reticulum (figure 6.5) is facilitated by ATP, inhibited by Mg^{2+}, and altered by P_i and pH. Impairment of excitation–contraction coupling, however, does not contribute to the initial decline in force during a fatiguing contraction (D.G. Allen et al., 1995; Edman, 1995). Rather, as fatigue progresses, the availability and efficacy of Ca^{2+} as an activation signal decrease and the force exerted by individual cross-bridges declines. These effects are distinguished as **failure of activation** and **myofibrillar fatigue** (Edman, 1995).

Activation failure generally occurs later than myofibrillar fatigue during high-force contractions. It is caused by a decrease in the sensitivity of the myofibrils to Ca^{2+} and a reduction in the release of Ca^{2+} from the sarcoplasmic reticulum. Although the reduction in the sensitivity of the myofibrils to Ca^{2+} can be caused by the development of acidosis, it is possible to evoke the decrease in sensitivity in the absence of acidosis (Allen, 2004; Stephenson et al., 1995; Westerblad et al., 1998); and the development of acidosis may preserve action potential propagation during fatiguing contractions (Pedersen et al., 2004). As an example of a condition in which fatigue was caused by a decrease in the myofibrillar sensitivity to Ca^{2+}, Moopanar and Allen (2005) compared the fatigue experienced by muscle fibers when activated with a series of maximal tetanic contractions in the presence and absence of a reactive oxygen species scavenger. The antioxidant agent reduced the rate of fatigue (decline in tetanic force) when the muscle fibers were tested at body temperature (37° C), but not when fibers were tested at a lower temperature (22° C). The investigators found no change in the intracellular (muscle fiber) concentration of Ca^{2+} during the fatigue protocol at body temperature and a similar reduction in the force that could be evoked by maximal levels of intracellular Ca^{2+} at the two temperatures. These results led them to conclude that the more rapid rate of fatigue in the absence of the antioxidant agent was due to a quicker decrease in the sensitivity of the myofibrils to Ca^{2+}. Although the mechanisms are not yet known, the accumulation of reactive oxygen species in muscle can facilitate the development of fatigue (Cooke, 2007; Essig & Nosek, 1997; Reid, 2008).

The other impairment that can cause activation failure involves a reduction in the release of Ca^{2+} from the sarcoplasmic reticulum, which likely involves an ATP-dependent mechanism (Westerblad et al., 1998). Apart from the direct inhibition of the Ca^{2+} release channels by low concentrations of ATP, there are at least four ATP-sensitive sites that could contribute to the impairment of Ca^{2+} release: (1) sarcolemmal Na^+-K^+ ATPase pumps, which are required to prevent the gradual depolarization of the membrane; (2) sarcolemmal K^+ channels that open when the ATP concentration is low, which may reduce the duration of the action potential; (3) sarcoplasmic reticulum Ca^{2+} ATPase, which is responsible for returning Ca^{2+} to the sarcoplasmic reticulum; and (4) processes that connect the T-tubule activation to the terminal cisternae of the sarcoplasmic reticulum (Lamb, 2002; Leppik et al., 2004; Renaud, 2002; Westerblad et al., 2000). Alternatively, the extracellular accumulation of K^+ due to action potential activity is capable of impeding the inward propagation of action potentials in some fatiguing contractions (Edman, 1996; Nielsen et al., 2004; Sejersted & Sjøgaard, 2000), and the precipitation of calcium and phosphate in the sarcoplasmic reticulum may reduce the release of calcium (Allen, 2004; Steele & Duke, 2003).

Myofibrillar fatigue corresponds to an impairment of cross-bridge function that is evident as a decrease in both the isometric force and the shortening velocity (Edman, 1995). Experiments on single muscle fibers have indicated that the decline in cross-bridge function during fatigue at physiological temperatures is not due to acidification (Cooke, 2007; Westerblad et al., 1998). Rather, myofibrillar fatigue is attributed to the accumulation of P_i ions (Cooke, 2007; Dahlstedt et al., 2001; Edman, 1995; Westerblad et al., 2002). Similarly, the decrease in shortening velocity likely involves other factors besides H^+, perhaps the accumulation of adenosine diphosphate (ADP) (Fitts, 2006).

Identifying the Impairment One way to demonstrate a role for impairment of excitation-contraction coupling in the fatigue experienced by humans is to show that the decline in force cannot be ascribed to either a decrease in the activation delivered to muscle or a change in metabolic factors. This approach was used by Bigland-Ritchie and colleagues (1986*a, b*) for a task in which subjects performed a series of 6 s submaximal contractions ($\leq$50% of maximum). There was a 4 s rest between each 6 s contraction. The subjects performed this sequence for 30 min. The force exerted during an MVC declined in parallel to the electrically elicited force and the superimposed twitch decreased to zero (figure 8.11*a*), which indicates that the subjects voluntarily exerted as much force as the muscle was capable of producing. There were no significant changes in muscle lactate, ATP, or phosphocreatine, and the depletion of glycogen was minimal and confined to type I and IIa muscle fibers. Consequently, the decline in MVC force could not be explained by inadequate muscle activation, acidosis, or lack of metabolic substrates. Additionally, the amplitude of the twitch response evoked during the rest periods declined more rapidly than the MVC force; this is regarded as evidence of impaired excitation-contraction coupling. For these reasons, the decline in force during this protocol was attributed to the impairment of one or more processes associated with excitation-contraction coupling (Bigland-Ritchie et al., 1986*a, b;* Cheng & Rice, 2005; de Ruiter et al., 2005; Klass et al., 2004; Saugen et al., 1997).

Another experimental approach that has been used to demonstrate a role for impairment of excitation-contraction coupling in muscle fatigue is to monitor the recovery after a fatiguing contraction (figure 8.24). The experimenters compare the recovery of the force evoked with high (80 Hz) and low (20 and 1 Hz) frequencies of stimulation (Martin et al., 2004). Electric shocks delivered at 80 Hz provide a measure of the maximal force that can be evoked from the muscle independent of activation by the CNS. In contrast, the force evoked with lower rates of stimulation is influenced by the efficacy of excitation-contraction coupling. Figure 8.24 shows the time course of recovery for forces evoked in a muscle after it was fatigued with a series of intermittent trains of stimuli at 20 Hz. The prolonged recovery of the forces evoked with low-frequency stimulation (1 and 20 Hz) has been termed **low-frequency fatigue** (de Ruiter et al., 2005; Edwards et al., 1977; Fitts, 2006). Because the force depression persists in the absence of activation and metabolic disturbances, it is attributed to an impairment of excitation-contraction coupling. Many different types of exercises can cause low-frequency fatigue (Blangsted et al., 2005*a*; Jones, 1996; Klass et al., 2004; Ratkevicius et al., 1998*b*; Rijkelijkhuizen et al., 2005).

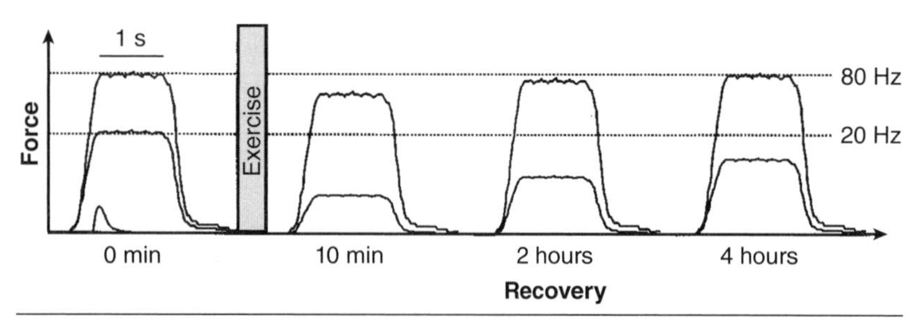

Figure 8.24 Recovery of muscle force after a fatiguing contraction. The maximal force capacity of muscle (adductor pollicis and quadriceps femoris) recovered much more rapidly than the force associated with submaximal contractions.

Adapted from Edwards et al. 1977.

Metabolic Pathways

Muscle contractions require the high-energy intermediate ATP to enable such processes as cross-bridge cycling, Ca^{2+} release and reuptake into the sarcoplasmic reticulum, and Na^+-K^+ pump activity. The availability of ATP, however, does not decline to such a level that it contributes to fatigue (Cooke, 2007; Fitts, 2006). Furthermore, although the levels of the immediate source of ATP rephosphorylation (phosphocreatine) decline with intense exercise, the differences in the time courses of the decrease in phosphocreatine and muscle force suggest that phosphocreatine depletion does not contribute to muscle fatigue. Rather, the metabolic factors that can contribute to fatigue include the products of the pathways that supply ATP and the availability of substrate for these pathways.

The activation of glycolysis results in the production of lactic acid, which dissociates into lactate and free H^+ that causes the pH of the muscle fiber to decline. Although H^+ accumulation can inhibit glycolysis, this interaction does not appear to be a major mechanism causing the decline in force at physiological temperatures. For example, reducing intracellular pH (7.0 to 6.6) in single, intact muscle fibers by increasing the CO_2 in the extracellular medium caused only moderate declines in the number of attached cross-bridges, the force exerted by each cross-bridge, and the speed of cross-bridge cycling during shortening contractions (Edman & Lou, 1990; Lännergren & Westerblad, 1989; Westerblad et al., 1998). Similarly, high-intensity leg exercises that elevated the concentration of blood lactate did not reduce muscle glycogenolysis or glycolysis (Bangsbo et al., 1996). Rather, it has been suggested that the severe plasma acidosis that can impair whole-body exercise may do so by impairing the voluntary activation of muscle (Cairns, 2006). It is probable, however, that other products of ATP hydrolysis (e.g., Mg-ADP, P_i) contribute to the decline in force. An increase in the concentration of P_i, for example, can reduce the maximal isometric force but does not influence the maximal speed of shortening. Conversely, an increase in the concentration of Mg-ADP can cause a small increase in the maximal isometric force and a modest decline in the maximal speed of shortening (Chase & Kushmerick, 1988; Cooke et al., 1988). The task conditions under which these products might be important in muscle fatigue are not yet known.

Nonetheless, the metabolic consequences of physical activity do appear to influence performance and to differ across contraction types. Jones and colleagues compared muscle performance in isometric and anisometric contractions evoked in humans by electrical stimulation. In one study, the quadriceps femoris muscle was activated by

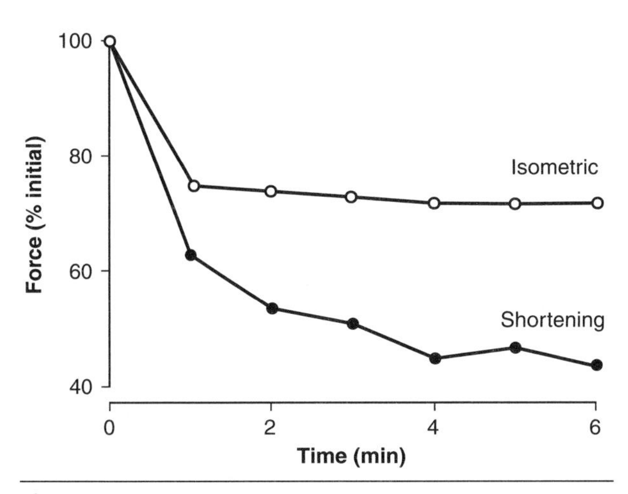

Figure 8.25 The force evoked by electrical stimulation of the quadriceps femoris muscle, first held isometric and then allowed to shorten against an isokinetic dynamometer (1.57 rad/s).

Data from Jones 1993.

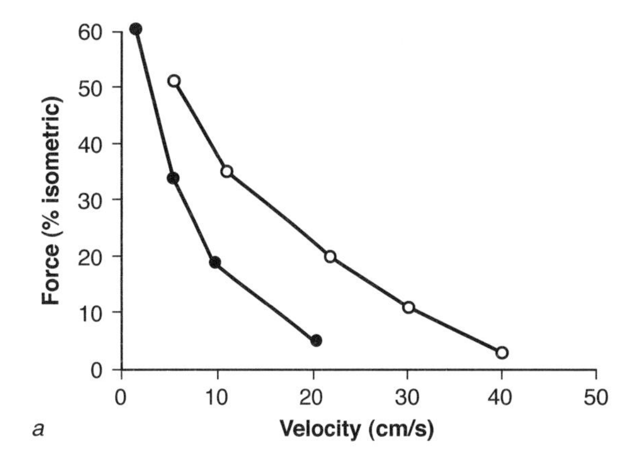

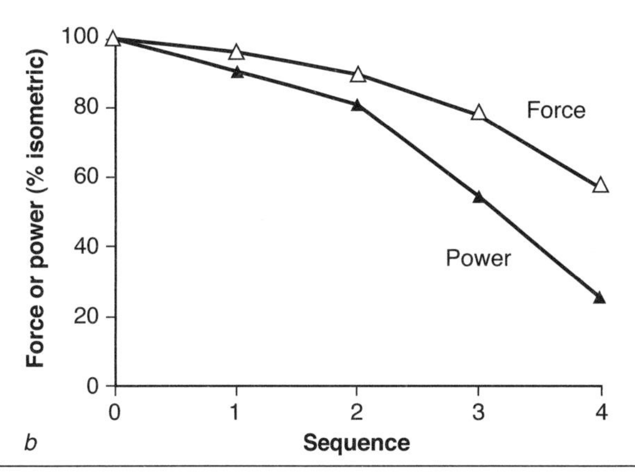

Figure 8.26 The influence of fatigue on the force and power capabilities of a hand muscle (adductor pollicis). Fatigue was induced by electrical stimulation of the muscle. The protocol involved four sequences in which the muscle was activated to produce first five isometric contractions and then four shortening contractions against different loads. *(a)* The force–velocity relation shifted to the left from control (open circles) to the fatigued condition (filled circles). The loads (*y*-axis) are expressed relative to the force evoked during the preceding isometric contraction. *(b)* The decrease in peak power during the shortening contractions (filled triangles) was greater than the decline in force during the isometric contractions (open triangles).

Data from Jones 1993.

stimuli applied to the femoral nerve to perform first an isometric contraction and then a shortening contraction (Jones, 1993). Although the contractions were performed one after the other, the decline in force was much greater for the shortening contraction (figure 8.25). In another study, a hand muscle was stimulated electrically, and the phosphorus metabolites were measured by magnetic resonance spectroscopy (Cady et al., 1989) and compared with the change in force and power (Jones, 1993). Blood flow to the arm was occluded during the sequence of evoked contractions. As with the study on quadriceps femoris, the isometric force evoked in the hand muscle declined less than the peak power produced during the shortening contraction (figure 8.26). Consistent with this difference, more phosphocreatine was utilized and more lactate accumulated during the shortening contraction (table 8.1). The metabolic cost was estimated at 9.3 mM ATP/s for the shortening contraction compared with 4.7 mM ATP/s for the isometric contraction. This difference in energy cost probably explains much of the difference in the fatigability of the two types of contractions.

The ability to sustain a task can be limited by the availability of substrate for the metabolic pathways (Sahlin et al., 1998). For example, when subjects exercised on a cycle ergometer at a rate of 70% to 80% of maximal aerobic power, exhaustion and the inability to sustain the forces necessary for the task coincided with the depletion of glycogen from the muscle fibers of vastus lateralis (Hermansen et al., 1967). Similarly, when exercising subjects were fed glucose or had it infused intravenously, they were able to exercise longer (Coggan & Coyle, 1987; Coyle et al., 1986). Consequently, the availability of carbohydrates determines how long motivated subjects can ride a cycle ergometer at 65% to 85% of maximal aerobic power (Broberg & Sahlin, 1989; Costill & Hargreaves, 1992; Hultman et al., 1990; Sahlin et al., 1998). However, exhaustion from performing intermittent submaximal contractions (30% MVC) is not associated with either substrate depletion or the accumulation of P_i and H^+ (Vøllestad, 1995).

The hypoglycemia that occurs during prolonged exercise—due to the failure of hepatic glucose production to satisfy the increased demands for glucose—can have widespread consequences. Nybo (2003) found that subjects experienced less fatigue when the prescribed task was preceded by activity in which subjects received either glucose supplementation or a placebo. The fatiguing contraction, which involved an MVC that was sustained for 2

Table 8.1 Metabolite Contents of Adductor Pollicis at Rest and After Isometric and Shortening Contractions

	Rest	Isometric	Shortening
Inorganic phosphate	8.2 ± 0.3	19.9 ± 1.0	25.1 ± 1.0
Phosphocreatine	30.0 ± 0.6	18.3 ± 0.4	11.6 ± 0.9
ATP	8.2 ± 1.1	7.9 ± 0.4	7.5 ± 0.4
pH	7.18 ± 0.04	7.06 ± 0.01	6.78 ± 0.04
Δ lactate	—	±7.5	±18.3

Data from Cady et al. 1989.

min with the knee extensor muscles, was preceded by 3 h of cycling on an ergometer (60% of maximal oxygen consumption). Both the decrease in voluntary activation and the decline in force during the fatiguing contraction were less when the subjects received glucose supplementation (figure 8.27). The glucose supplementation maintained blood glucose levels during the 3 h of cycling, whereas subjects experienced a decrease in blood glucose from 4.5 to 3.0 mM during the placebo condition. Because the subjects were able to exert the same net torque at the onset of the fatiguing contraction, Nybo (2003) concluded that decline in substrate impaired the ability of the subjects to provide an adequate activation signal to the muscles during the fatiguing contraction.

Blood Flow

Among the mechanisms that can contribute to fatigue, the impairment of blood flow to active muscles was one of the first to be identified. An increase in muscle blood flow with motor activity is necessary for the supply of substrate, the removal of metabolites, and the dissipation of heat (Hamann et al., 2005). When a muscle is active, however, there is an increase in intramuscular pressure that compresses blood vessels and occludes blood flow when it exceeds systolic pressure. For example, when the knee extensor muscles sustain an isometric contraction for as long as possible at a submaximal force (5% to 50% of MVC force), blood flow decreases with an increase in the level of the sustained force (Sjøgaard et al., 1988). In brief isometric contractions, de Ruiter and colleagues (2007) found that maximal deoxygenation (all oxygen consumed) occurred at 35% MVC force for rectus femoris and at 25% MVC force for vastus medialis and lateralis. Blood flow is probably not impaired for tasks that involve sustained forces of less than 15% MVC force (Blangsted et al., 2005*b*; Gaffney et al., 1990; Sjøgaard et al., 1988) and when the task involves intermittent or dynamic contractions (Sjøgaard et al., 2004; Wigmore et al., 2006). Similarly, when blood pressure is increased

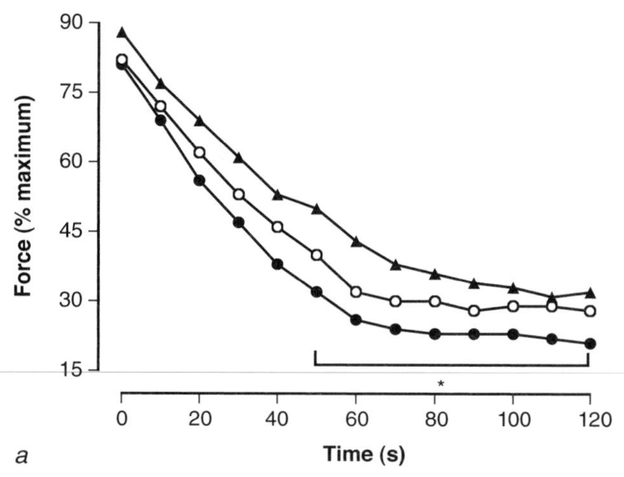

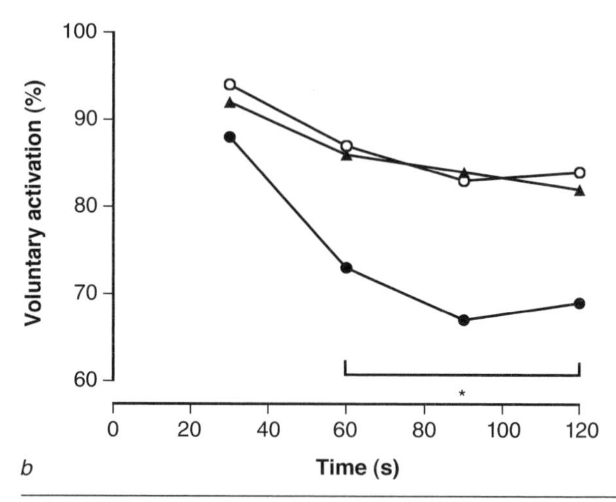

Figure 8.27 The force exerted by the knee extensor muscles (top) and the associated levels of voluntary activation (bottom) during maximal contractions sustained for 120 s. Three conditions were examined: baseline (filled triangles), glucose trial (open circles), and placebo trial (filled circles). The baseline condition represents the decreases in force and voluntary activation when the task was performed after a 15 min warm-up. The glucose and placebo trials were performed after 3 h of cycling either with (glucose trial) or without (placebo trial) glucose supplementation. The force data (top) are reported as a percentage of the peak values recorded during a maximal voluntary contraction. *$p < 0.05$ compared with the baseline and glucose trial.

Data from Nybo 2003.

above intramuscular pressure, the endurance time for contractions in which the force is less than 60% of maximum is increased (Butler et al., 2004; Fitzpatrick et al., 1996; Petrofsky & Hendershot, 1984).

In addition to the influence of intramuscular pressure on muscle blood flow, an elevated blood flow (hyperfusion) can decrease fatigue by a mechanism that is independent of oxygen or substrate delivery. A reduction in blood flow exerts an effect on mechanical performance that is independent of a reduction in the oxygen content of the blood (hypoxemia). The increase in blood flow associated with hyperfusion may improve the removal of metabolites and thereby diminish the inhibitory effect of metabolite accumulation (Barclay, 1986; Stainsby et al., 1990). These findings indicate that the force capacity of muscle is sensitive to changes in arterial blood pressure (Cole & Brown, 2000; Wright et al., 2000).

Task Failure

As suggested by the scheme shown in figure 8.12 and substantiated by the preceding discussion on the sites of impairment, the fatigue that occurs during voluntary contractions performed by humans can be caused by many different factors. The dominant mechanism depends on the characteristics of the task being performed (Asmussen, 1979; Bigland-Ritchie et al., 1995; Enoka & Stuart, 1992; Gandevia, 2001). The task details that influence the contributing mechanisms include the type and intensity of exercise, the muscle groups involved in the activity, and the physical environment in which the task is performed. These findings have led to a principle known as the **task dependency of muscle fatigue,** which states that there is no single cause of muscle fatigue and that the dominant mechanism depends on the details of the task being performed. Furthermore, the decline in force during a fatiguing contraction often involves multiple mechanisms.

Variation in the dominant mechanism across tasks has confused the literature on muscle fatigue. For example, compare the results of two studies aimed at determining if old adults are more fatigable than young adults. In one study, Baudry and colleagues (2007) had young and old adults perform a series of maximal shortening and lengthening contractions with the dorsiflexor muscles. The decline in peak torque during both types of contractions was greater for the old adults, which led to the conclusion that old adults are more fatigable than young adults. In the other study, Hunter and colleagues (2005) found that old men could sustain an isometric contraction with the elbow flexor muscles at a force that was 20% of maximum for a longer duration than young men. At the end of the submaximal contraction, the two groups of men exhibited a similar amount of fatigue, as indicated by equivalent declines in MVC force. Hence, the old men were less fatigable when they performed this task. Furthermore, Yassierli and colleagues (2007) found that the relative durations over which young and old adults could sustain submaximal isometric contractions (30-70% MVC force) differed for shoulder and torso muscles. Consistent with the principle of task dependency, this comparison indicates that the influence of age on muscle fatigue depends on the task being performed (Bilodeau, 2006).

Because of the task dependency of muscle fatigue, it is not possible to answer the question "What causes muscle fatigue?" An alternative strategy in the study of muscle fatigue, therefore, is to identify the mechanisms responsible for the failure of specific tasks (Enoka & Duchateau, 2008; Maluf & Enoka, 2005). Such an approach, which is known as the **task-failure approach,** emphasizes the functional significance of the changes that occur during fatiguing contractions. The relative role of the adjustments that take place during fatiguing contractions can be assessed through identification of those changes that are associated with differences in the time to failure of two performances. Comparisons can include, for example, one group of subjects performing two similar tasks, two groups of subjects sustaining the same task, or one group of subjects performing the same task before and after an intervention. The task-failure approach is illustrated with studies that have addressed the influence of load type, limb posture, and sex differences on the fatigue experienced by humans.

Load Type

When muscles contract and a limb exerts a force against its surroundings, it can encounter four types of loads: elastic, immovable, inertial, and viscous. An elastic load provides a resistance that is proportional to the stretch it experiences; an immovable load corresponds to an object that cannot be moved by the force of the limb; an inertial load is a mass that resists a change in its velocity; and a viscous load offers a resistance that is proportional to its velocity. The task-failure approach has been used to compare the adjustments that occur when subjects sustain isometric contractions against an immovable load and an inertial load. In both conditions, subjects were required to sustain a submaximal contraction (15-20% of MVC force) for as long as possible. When they were acting against an immovable load, the task was to exert a force that matched a target displayed on a monitor; this is known as a **force task.** When they were supporting an inertial load, the task was to match a joint angle to a target displayed on a monitor; this is known as a **position task.**

When the force and position tasks were performed with the elbow flexor muscles (figure 8.28, *b* and *c*) and a hand muscle (figure 8.28*a*), the time to task failure was briefer for the position task (table 8.2). The difference in the time to failure was caused by a more rapid recruitment of the motor units during the position task as indicated by a faster increase in the surface EMG (figure 8.29), more frequent bursts of activity in the EMG signal, and no difference in the time to failure for the two tasks when the load exceeded the upper limit of motor unit

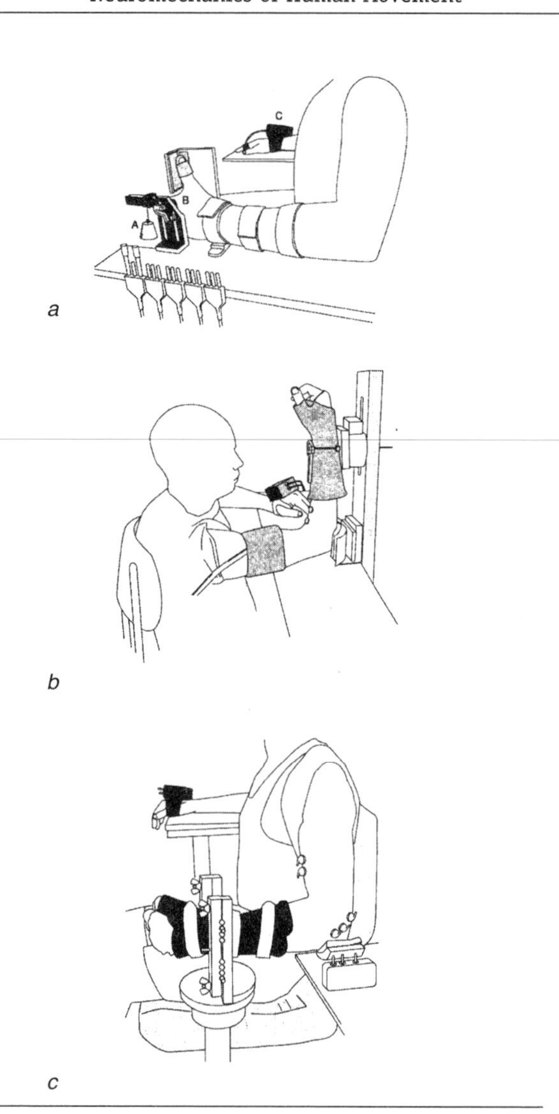

Figure 8.28 Limb position and test muscle for task-failure studies. Experimental arrangement for the *(a, c)* force and *(b)* position tasks as performed by *(b, c)* the elbow flexor and *(a)* the first dorsal interosseus muscles.

Adapted, by permission, from K. Maluf and R.M. Enoka, 2005, "Task failure during fatiguing contractions," *Journal of Applied Physiology* 99: 390. Copyright 2005 by The American Physiological Society.

Table 8.2 Time to Failure for the Force and Position Tasks Shown in Figure 8.28

	Force task (s)	Position task (s)	Ratio (%)
First dorsal interosseus (figure 8.28*a*)	983 ± 1328	593 ± 212	63 ± 28
Elbow flexors (figure 8.28*b*)	609 ± 250	477 ± 276	77 ± 21
Elbow flexors (figure 8.28*c*)	1402 ± 728	702 ± 582	51 ± 26

Note. The ratio (%) indicates the duration of the position task relative to the force task.

Data (mean ± SD) from Hunter et al. 2002; Rudroff et al. 2005; Maluf et al. 2005.

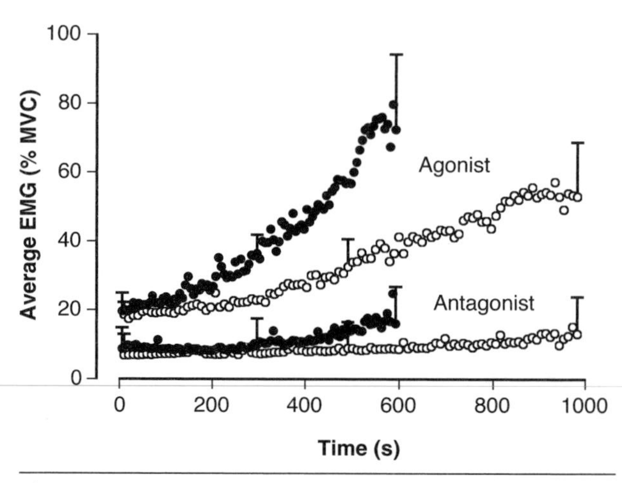

Figure 8.29 EMG activity during force (open circles) and position (filled circles) tasks performed with the first dorsal interosseus muscle (agonist). The antagonist muscle was second palmar interosseus.

Data from Maluf et al. 2005.

recruitment (Hunter et al., 2002; Maluf et al., 2005). The antagonist muscle increased its EMG only slightly, and similarly, during the two tasks (figure 8.29). Because the net muscle torque exerted by each subject was similar for the two tasks, the load on the muscle fibers did not differ at the beginning of the two fatiguing contractions. As the fatiguing contractions progressed, however, more motor units were recruited during the position task, which would have reduced the load experienced by each active muscle fiber. Consequently, the more rapid recruitment of motor units during the position task must have been caused by a difference in the control strategy used by the CNS and not due to a greater load placed on the muscle fibers (Klass et al., 2008).

The behavior of single motor units during the force and position tasks was examined directly by Mottram and colleagues (2005). They recorded the discharge of the same motor unit in biceps brachii when subjects sustained the two contractions with the same load (~20% MVC force) for around 3 min. The average discharge rate (13 pps) and coefficient of variation for discharge times (22.8%) were similar at the beginning of the two contractions, but discharge rate decreased more and the coefficient of variation increased more during the position task (figure 8.30). Furthermore, Mottram and colleagues observed the recruitment of the same 26 units during the two tasks, of 6 different motor units during the force task, and of 20 different motor units during the position task. Thus, the synaptic inputs received by the motor unit pool differed during the two tasks, presumably due to the difference in load type, even though the magnitude of the load was the same. Consequently, the briefer duration for the position task was attributable to a more rapid activation of motor units in the muscle.

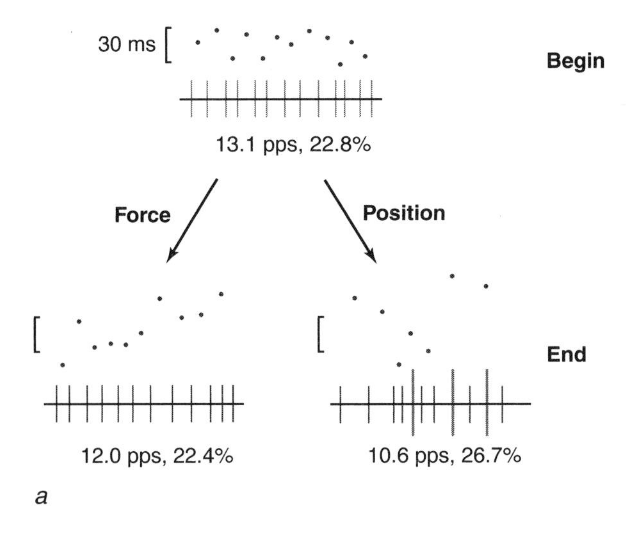

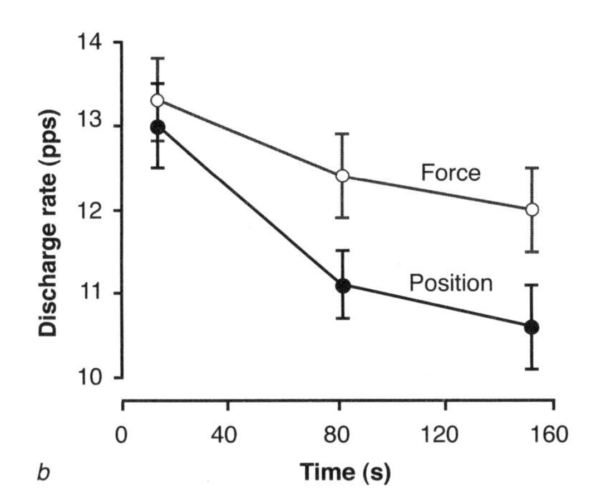

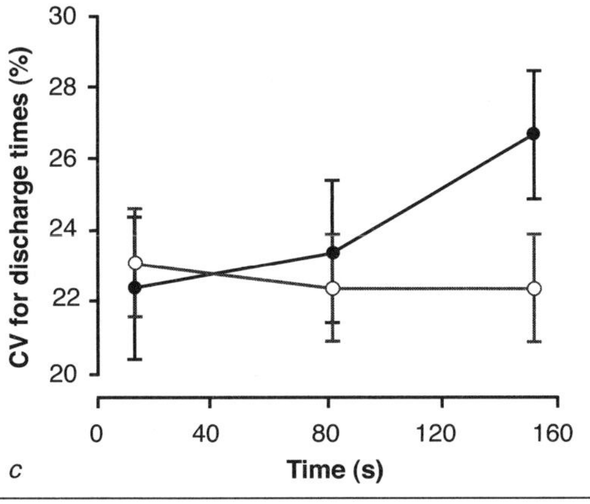

Figure 8.30 The discharge of single motor units (n = 32) in biceps brachii during force and position tasks. *(a)* At the beginning of the two tasks, average discharge rate (13.1 pps) and coefficient of variation for discharge times (22.8%) were similar. *(b)* Even though the two tasks were sustained for an identical duration, discharge rate decreased more during the position task. *(c)* Coefficient of variation for discharge times did not change during the force task but increased during the position task.

Data from Mottram et al. 2005.

Limb Posture

In addition to the influence of load type, time to task failure can depend on the posture of the limb due to differences in the load placed on accessory muscles and changes in the involvement of the synergist muscles. When the force and position tasks were performed by the elbow flexor muscles with the arm in the positions shown in figure 8.28, *a* and *b,* the duration of the position task relative to the force task was 51% when the forearm was horizontal compared with 77% when it was vertical (table 8.2). This difference indicates that the influence of load type on the time to failure was supplemented by another factor. When the upper arm was vertical (figure 8.28*a*), it was abducted from the trunk by about 0.4 rad, and subjects had to activate the external rotator muscles at the shoulder (supraspinatus, infraspinatus, teres minor) to prevent internal rotation of the arm about the shoulder joint during the position task. These muscles were activated to a much lesser extent during the force task when the arm was attached to the apparatus. Accordingly, Rudroff and colleagues (2007) found that the rate of increase in average EMG of the rotator cuff and posterior deltoid muscles was greater during the position task compared with the force task, and that the load experienced by the accessory muscles contributed to the much greater reduction in the time to failure for the position task (figure 8.31). This effect was not observed when the forearm was vertical (figure 8.28*b*) and there was no difference in the load placed on the accessory muscles during the two tasks. This finding indicates that the demands experienced by postural muscles can limit the duration over which a task can be sustained, which has implications for exercises prescribed during rehabilitation and for the ergonomic design of workstations.

Variation in limb posture can also change the relative activation of the synergist muscles due to reflex connections between the muscles. Sensory receptors send afferent feedback both to the motor neurons that innervate the muscle in which they reside and to other muscles, synergists as well as antagonists. The afferent feedback can evoke excitatory or inhibitory potentials in the motor neurons; in general, the responses are excitatory when the muscles contribute to the same function and inhibitory for opposing actions. When the forearm is in a neutral position midway between pronation and supination (figure 8.28*c*), the muscles biceps brachii, brachialis, and brachioradialis all contribute to the net flexion torque about the elbow joint. When the forearm is supinated, however, biceps brachii contributes to the supination torque and brachioradialis contributes to both the supination and pronation torques. Because of these opposing actions, afferent feedback from brachioradialis can evoke inhibitory effects in the motor neurons that innervate biceps brachii (Barry et al., 2007; Naito et al., 1996, 1998). The change in sign

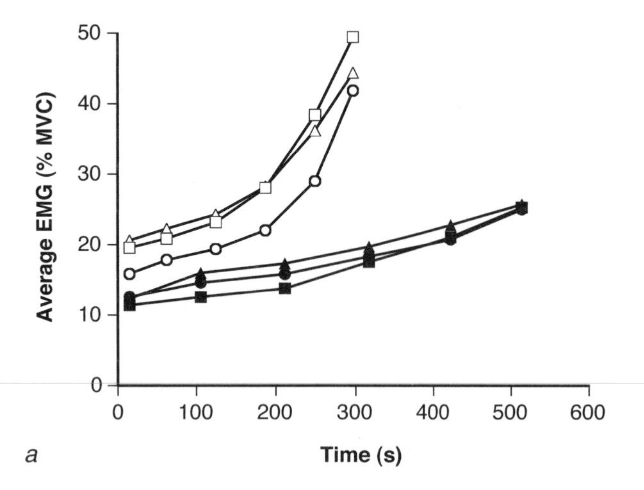

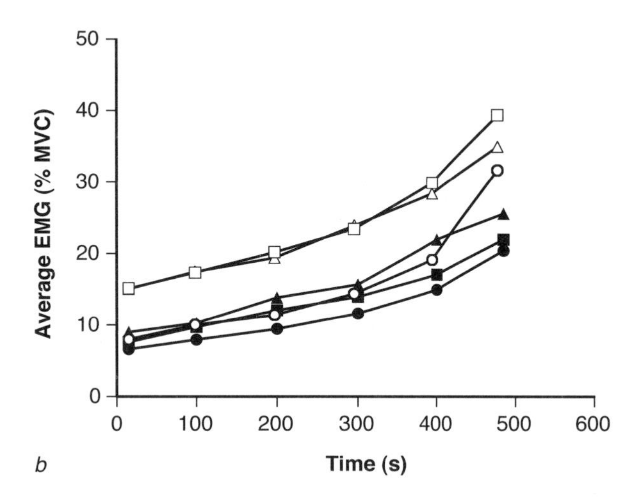

Figure 8.31 The average rectified EMG for three rotator cuff muscles when the force (filled symbols) and position (open symbols) tasks were performed with the forearm *(a)* horizontal and *(b)* vertical. The data are for supraspinatus (circles), infraspinatus (triangles), and teres minor (squares).

Data from Rudroff et al. 2007.

(excitatory to inhibitory) of the afferent feedback from brachioradialis to biceps brachii likely causes the time to task failure to differ for the two forearm positions.

Sex Differences

When men and women perform submaximal fatiguing contractions under normal conditions, women are usually capable of a longer time to task failure (Clark et al., 2005; Hicks et al., 2001; Hunter & Enoka, 2001; Hunter et al., 2006). The typical explanation for this sex difference is that because men are often stronger than women, they experience a greater occlusion of blood flow and different metabolic activity when the task is performed at the same relative intensity (% of maximum). Consistent with this explanation, the time to failure when the elbow flexor and plantarflexor muscles performed the force task (20% MVC force) was similar for men and women who were matched for strength (Hatzikotoulas et al., 2004; Hunter et al., 2004*a*), and there was no difference in the time to failure on the force task (25% MVC force) between the sexes for the knee extensor muscles when blood flow was occluded (Clark et al., 2005).

In contrast, when men and women of equal strength performed intermittent contractions (6 s contraction, 4 s rest) to a target force of 50% MVC force with the elbow flexor muscles, the time to task failure was longer for the women (Hunter et al., 2004*b*). Even though the two groups of subjects exerted a similar net muscle torque, EMG activity increased more rapidly for the men and they reached task failure sooner than the women (1133 s vs. 1408 s). Similarly, the decline in MVC torque of the elbow flexor muscle after a series of MVCs was greater for men than for women, and this difference was not attributable to a greater impairment of voluntary activation during the fatiguing contractions (Hunter et al., 2006). Furthermore, because the amount of fatigue experienced by men and women during intermittent contractions is similar when blood flow is occluded (Russ & Kent-Braun, 2003), the sex difference is likely caused by one or more factors located within the muscle. One possibility is that the relative contributions of the metabolic pathways used to supply ATP during a muscle contraction can differ for men and women during fatiguing contractions and contribute to differences in the time to task failure (Russ et al., 2005).

MUSCLE POTENTIATION

In contrast to the negative effects of fatigue on performance, several mechanisms can enhance the output of the neuromuscular system during a performance (Hutton, 1984; McComas, 1996). After a brief period of activity, these mechanisms can increase both the electrical and mechanical output above resting values. Examples of these capabilities include the potentiation of monosynaptic responses, miniature end-plate potentials, M waves, twitch force, and the discharge of muscle spindle receptors.

Monosynaptic Responses

It seems reasonable to assume that most processes in the motor system can be augmented by brief periods of activity. This is even evident at the level of input-output relations for the spinal cord. Lloyd (1949) delivered single electric shocks to the muscle nerve of an experimental animal and measured the output in the ventral root. The experimental preparation is schematized in figure 8.32*a*. Because the ventral root was cut, the electric shock (stimulus) generated action potentials that were transmitted along the afferent axons and into the spinal cord. The synaptic input delivered by the afferent axons activated motor neurons, and the monosynaptic response was measured in efferent axons (figure 8.32*a*). The input-output relation involved populations of afferent and efferent axons. The monosynaptic responses were measured before and after tetanic stimulation of the nerve (12 s duration at 555 Hz), shown at time zero in figure 8.32*b*. The tetanic stimulation increased the amplitude of the monosynaptic response by seven times compared with the values measured before the tetanus (control). The potentiation of the monosynaptic response decayed over a 3 min interval (figure 8.32*b*). A similar effect has been reported for potentiation of the H and tendon tap reflexes after a brief period of high-frequency electrical stimulation (Hagbarth, 1962).

When Lloyd (1949) activated different afferent pathways after the tetanic stimulation, the potentiation was limited to the afferent pathway that received tetanic stimulation. Therefore, the mechanism underlying the potentiation was presynaptic; that is, it was located before the synaptic contact with the motor neurons. Given the type of stimuli that elicit potentiation of the monosynaptic response, the mechanism probably involves group Ia afferents (Hutton, 1984; Koerber & Mendell, 1991). Furthermore, the amount of potentiation differs across the group Ia synapses onto motor neurons (Davis et al., 1985). The effects could include an increase in the quantity of neurotransmitter released, an increase in the efficacy of the neurotransmitter, or a reduction in axonal branch-point failure along the group Ia afferents (Clamman et al., 1989; Kuno, 1964; Lüscher et al., 1983).

Miniature End-Plate Potentials

The spontaneous release of neurotransmitter (acetylcholine) at the neuromuscular junction elicits miniature end-plate potentials in the muscle fiber membrane. The miniature end-plate potentials are not constant events (Fahim, 1992); the amplitude is greater for fast-twitch compared with slow-twitch muscle fibers, and the frequency can decline as a function of age (Alshuaib & Fahim, 1991; Lømo & Waerhaug, 1985). Similarly, a brief period of high-frequency stimulation can increase the amplitude and frequency of the spontaneously released miniature end-plate potentials (Pawson & Grinnell, 1990; Vrbová & Wareham, 1976). The increase in frequency lasts for a few minutes, whereas the increase in amplitude has been reported to last for several hours. Furthermore, the effect is greater at neuromuscular junctions where more neurotransmitter is released per unit length of the junction. Heightened levels of miniature end-plate potentials presumably keep the potential of the muscle fiber membrane closer to its voltage threshold and more responsive to incoming action potentials.

The mechanisms seem to involve activity-dependent increases in the sensitivity of the postsynaptic membrane and in the influx of Ca^{2+} into the presynaptic terminal (Elrick & Charlton, 1999; Robitaille & Tremblay, 1991). The increase in postsynaptic membrane sensitivity means that a quantum of neurotransmitter will elicit a greater

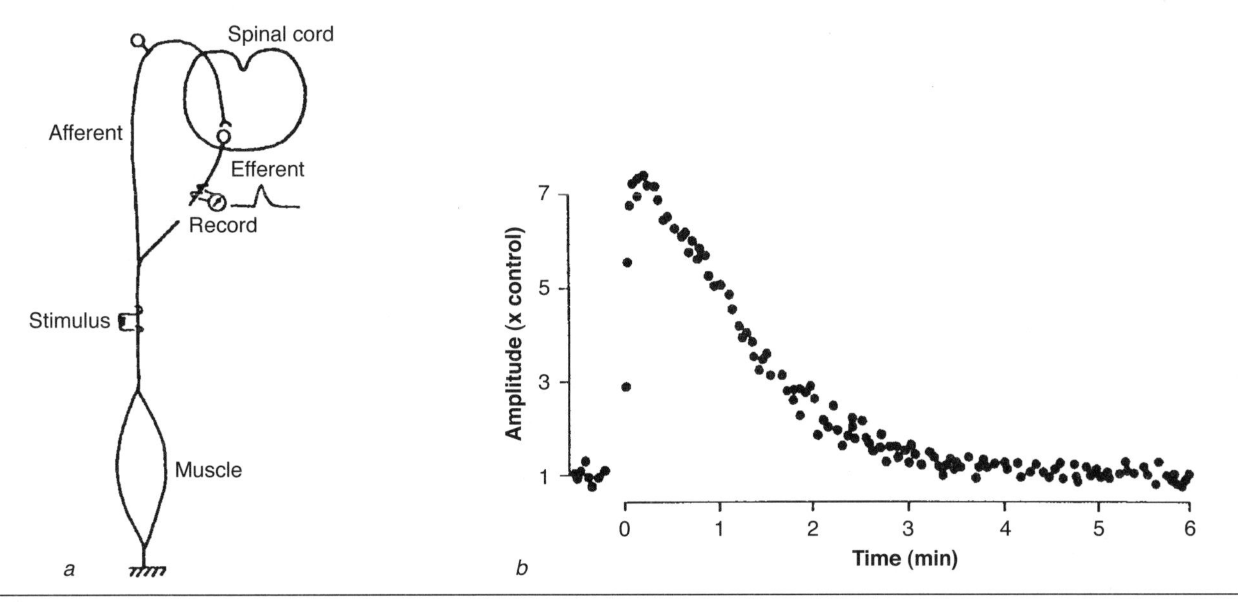

Figure 8.32 Potentiation of monosynaptic responses: *(a)* the experimental model and *(b)* amplitude of the monosynaptic response.

From "Post-tetanic potentiation of response in monosynaptic reflex pathways of the spinal cord" by D.P.C. Lloyd. Reproduced from *The Journal of General Physiology*, 1949, vol. 33, p.149 by copyright permission of The Rockefeller University Press.

response (amplitude of the synaptic potential) in the muscle fiber membrane. In addition, because Ca^{2+} is necessary for the fusion of vesicles to the presynaptic membrane and the subsequent release of neurotransmitter, a greater influx of Ca^{2+} will lead to an enhanced frequency of spontaneous release of neurotransmitter. The influx of Ca^{2+} can be modulated by second messengers and by the presence of different types of Ca^{2+} channels in the presynaptic terminal (Rathmayer et al., 2002; Yoshihara et al., 2000), which contributes to the variation in amplitude and frequency of the miniature end-plate potentials.

M-Wave Amplitude

The processes involved in converting an axonal action potential into a muscle fiber action potential are collectively referred to as neuromuscular propagation. One way to test the efficacy of neuromuscular propagation is to elicit M waves (figure 7.10). This test involves stimulating the muscle nerve and measuring the evoked muscle action potential, which comprises the sum of the muscle fiber potentials within the recording volume of the electrodes (Keenan et al., 2006; Zehr, 2002). When a human subject performs a voluntary contraction or when a muscle is activated with electrical stimulation, there is often an initial transient increase in M-wave amplitude (figure 8.33). For example, when subjects performed 20 maximal contractions (3 s duration for each contraction, 1.5 s between contractions) with the thenar muscles, M-wave amplitude increased immediately after the first contraction and then reached a plateau at an average increase of about 24% (Hicks et al., 1989). The potentiation of M-wave amplitude is greater for high-force contractions (Nagata & Christianson, 1995), for contractions elicited with electrical stimulation at rates of <30 Hz (Cupido et al., 1996; Harrison & Flatman, 1999), and for fatigable motor units (Enoka et al., 1992; Hamada et al., 2003).

Although M-wave amplitude can be reduced by presynaptic factors (e.g., branch-point failure, neurotransmitter depletion, decrease in availability of synaptic vesicles), only postsynaptic factors can increase M-wave amplitude. These factors include a reduction in the temporal dispersion of the muscle fiber potentials and an increase in the amplitude of individual muscle fiber potentials. The temporal distribution of the individual fiber potentials influences the amount of cancellation that occurs between the positive and negative phases of the potentials (Dimitrova & Dimitrov, 2002; Farina et al., 2004*a*). Different motor unit properties influence timing variability between potentials, including motor neuron activation times and axonal and muscle fiber conduction velocities (Burke et al., 2001; Magistris et al., 1998; Ståalberg & Karlsson, 2001). For example, Keenan and colleagues (2006) found that a decrease in the mean conduction velocity of muscle fiber potentials from 5.0 to 2.5 m/s, which can occur during fatiguing contractions (Cupido et al., 1996), decreased the amplitude of the simulated M wave (–13%) but increased its area (+73%). In contrast, they found that an increase in the standard deviation of conduction velocity for the muscle fiber potentials from 0.1 to 0.6 m/s decreased both the amplitude (–13%) and the area (–14%) of the simulated M waves. Conversely, a reduction in the standard deviation of conduction velocity, such as could occur during a high-force contraction when the conduction velocity of the largest motor units declines (Farina et al., 2000, 2002*a*), would be expected to increase the amplitude and area of the M waves.

Studies on single muscle fibers that were subjected to repetitive stimulation suggest that the increase in M-wave amplitude was at least partially due to an activity-dependent increase in the amplitude of muscle fiber potentials (Hicks & McComas, 1989). Activation increases the activity of the Na^+-K^+ pump, which lowers (hyperpolarizes) the resting membrane potential and produces a greater change in voltage across the membrane during an action potential. According to Cupido and colleagues (1996), the reason for the potentiation of M-wave size is that the muscle fiber potentials become less biphasic and therefore experience less cancellation.

A key functional question is whether the potentiation of M-wave amplitude has any effect on muscle force. The potentiation of M-wave area observed by Enoka and colleagues (figure 8.33) was not associated with the increase in force experienced by the fatigable motor units during the contractions evoked with electrical stimulation. Conversely, Gong and colleagues (2003) found that the efflux of K^+ through ATP-dependent channels depressed the amplitude of muscle fiber potentials and reduced the rate at which force declined during a fatiguing contraction. They observed that a decrease in force occurred when the amplitude of the overshoot for action potentials recorded in the fibers from leg muscles of mice declined from above 20 mV to less than 5 mV. Thus, either the increase in muscle fiber potentials during M-wave potentiation was not substantial enough or the increase in size of the muscle fiber potentials does not enhance muscle force.

Posttetanic Potentiation

Perhaps the best known of the potentiation responses is the effect of prior activity on twitch force. The magnitude of the twitch force is extremely variable and depends on the activation history of the muscle. A twitch elicited in a resting muscle does not produce the maximal twitch force. Rather, twitch force is maximal following a brief

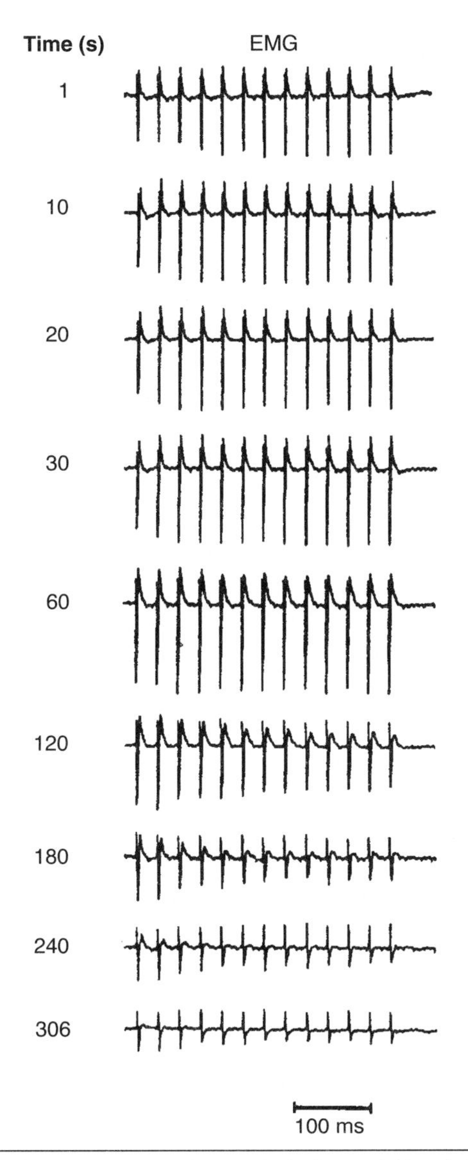

Figure 8.33 Changes in EMG during a 360 s fatiguing contraction involving electrical stimulation of a cat hindlimb muscle. The stimulus regimen consisted of 330 ms trains of 13 stimuli given once each second. The EMG represents the summed muscle potentials (M waves) that were elicited by the electric shocks. M-wave amplitude increased during the first 60 s of the test and then declined.

Adapted from Enoka et al. 1992.

tetanus; this effect is known as **posttetanic potentiation** of twitch force (Baudry & Duchateau, 2004; Hodgson et al., 2005; Sale, 2002).

The posttetanic potentiation of twitch force can be substantial and can be elicited by either voluntary contractions or electrical stimulation. For example, the potentiation of twitch force in the ankle dorsiflexor muscles of human volunteers ranged from 29% to 150% after 20 to 40 s of intermittent electrical stimulation applied to the nerve (Garner et al., 1989), by 5% to 140% after a 7 s tetanus at 100 Hz (O'Leary et al., 1997), and by 150% to 180% after a 6 s MVC (Baudry & Duchateau, 2004). Posttetanic potentiation is greatest after contractions that last 5 s to 10 s and decreases with longer-duration contractions. The amount of twitch potentiation is similar for isometric, shortening, and lengthening MVCs and does not substantially change either the contraction time or half-relaxation time of the twitch or the M wave (Baudry & Duchateau, 2004). However, twitch potentiation in the knee extensor muscles after a fatiguing contraction (20% MVC force) was greater at a short muscle length and was associated with a longer time to task failure (Place et al., 2005), and was reduced in the ankle plantarflexor muscles after a series of anisometric contractions that fatigued the muscles (Klass et al., 2004).

At least two processes are involved in posttetanic twitch potentiation (Grange & Houston, 1991). An early potentiation occurs after brief contractions and decays relatively quickly. After a delay of about 60 s, a late potentiating process emerges, which reaches a peak at about 200 s and then decays to control levels after 8 to 12 min of recovery. The mechanisms underlying these potentiation processes may involve an alteration in calcium kinetics (Duchateau & Hainaut, 1986; O'Leary et al., 1997), the phosphorylation of myosin light chains (Grange et al., 1998; Sweeney & Stull, 1990; Sweeney et al., 1993; Zhi et al., 2005), and the force–velocity characteristics of the cross-bridges (Edman et al., 1997; MacIntosh & Willis, 2000). The most commonly accepted explanation for posttetanic potentiation is an increase in the sensitivity of the contractile proteins to activation by Ca^{2+} due to phosphorylation of the myosin light chains.

Potentiation of the submaximal force occurs in all three types of motor units (types S, FR, and FF). When motor units were activated with a stimulus that elicited a submaximal tetanic force, the potentiation (increase in peak force) was greater for the fast-twitch motor units (50-60% of control in types FR and FF) than for the slow-twitch motor units (20% of control in type S). However, the incidence of potentiation among the motor units was greater for the fatigue-resistant motor units (60-75% for types S and FR) compared with the fatigable units (40% for type FF). Because the occurrence of potentiation was distributed across all three types of motor units, the mechanisms underlying potentiation differ from those that define motor unit type (Gordon et al., 1990*b*).

The study of posttetanic twitch potentiation has emphasized that the processes of potentiation and fatigue occur concurrently, beginning from the onset of activation (Hamada et al, 2003; Place et al., 2005). For example, when the extensor digitorum longus muscle of rats was stimulated with a protocol that reduced the submaximal tetanic force to an average peak force of 36% of the control value, 50% of the muscles exhibited posttetanic twitch potentiation (Rankin et al., 1988). This effect was not observed in soleus, which largely consists of slow-twitch fibers. The coexistence of potentiation and fatigue has also been observed in the quadriceps femoris muscles of humans after a 60 s MVC (Grange & Houston, 1991).

A scheme for the interaction of these two processes and the net effect on twitch force during a specific protocol is shown in figure 8.34. In the experiment from which the scheme was devised, the ankle dorsiflexor muscles of human volunteers were electrically stimulated to elicit a 3 s submaximal tetanus once every 5 s for 180 s. Between the tetani, a twitch was elicited by a single electrical shock. For this particular protocol, twitch force increased and then decreased during the stimulation period, and decreased during the recovery period. Garner and colleagues (1989) proposed that the time course of the change in twitch force was due to the interaction of the processes that mediate potentiation and fatigue.

Because posttetanic potentiation is typically demonstrated for the twitch response and a twitch occurs rarely during voluntary contractions, the functional significance of twitch potentiation seems uncertain. Baudry and Duchateau (2007) examined this issue by comparing the potentiation of the twitch, the force evoked by a train of 15 electrical stimuli at 250 Hz, and the force associated with rapid voluntary contractions in hand muscles after a 6 s MVC. The potentiation was greatest for the twitch (200%) and much less for the electrically stimulated (17%) and the rapid voluntary contractions (9-24%). Potentiation was maximal immediately after the MVC for the twitch, but the maximum did not occur until 60 s later for the other two contractions. Because the rate of increase in torque was similar during the evoked and voluntary contractions, the mechanisms that caused potentiation did not depend on the activation characteristics. These results suggest that posttetanic potentiation may be one of the positive outcomes of warm-up exercises that can enhance subsequent physical activity.

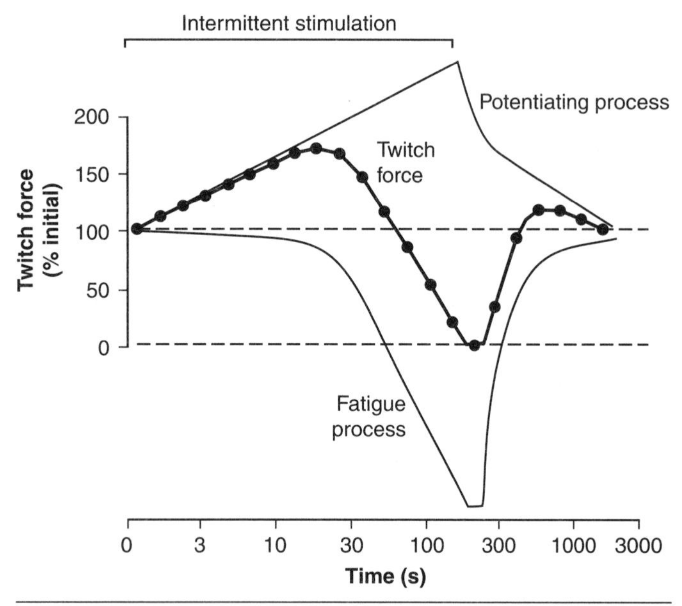

Figure 8.34 Coexistence of potentiation and fatigue and the net effect on the twitch force. When a muscle is stimulated electrically, the contraction activates processes that both diminish (fatigue) and enhance (potentiate) the muscle force. The result is a nonlinear change in the amplitude of the twitch force.

Postcontraction Sensory Discharge

In addition to the effects on motor processes, a brief period of intense activity can also influence sensory processes. One example of this effect is the increased neural activity that has been recorded in the dorsal roots of experimental animals after a contraction; this phenomenon has been termed **postcontraction sensory discharge** (Hutton et al., 1973). The increase in the dorsal root activity is primarily due to an increase in the muscle spindle discharge, mainly from group Ia afferents. Moreover, the postcontraction sensory discharge is abolished if the muscle is stretched immediately after the contraction. The mechanism responsible for this phenomenon is the development of stable cross-bridges in the intrafusal fibers. These cross-bridges develop during the muscle contraction and persist after the extrafusal fibers relax so that the muscle spindle is in a state of increased tension and the resting discharge is higher than before the contraction (Emonet-Denand et al., 1985; Gregory et al., 1986; Hutton et al., 1973). Consequently, a stretch of the muscle will break the cross-bridge bonds and abolish the postcontraction sensory discharge.

Because postcontraction sensory discharge increases the excitatory input to the motor neuron pool, it influences subsequent activity for up to 15 min, with a peak effect at 5 to 20 s after the contraction. The strength of postcontraction sensory discharge can be sufficient to increase the resting discharge of motor neurons (Suzuki & Hutton, 1976). As a consequence, the heightened excitatory feedback can enable the system to respond more rapidly and forcefully to subsequent perturbations of muscle length, influence kinesthetic sensations, and oppose the relaxation of muscle that may be desired during stretching maneuvers (Gregory et al., 1998; Proske et al., 1993).

In general, this section of the chapter describes several examples of changes in input-output relations, in which the magnitude of a response increases despite a constant input signal. This indicates an improvement in the efficacy of the processes that couple the input and the output.

Some phenomena, such as posttetanic potentiation, have an obvious effect on muscle force, whereas the association is less obvious for other potentiated responses.

EXAMPLE 8.3

Aftercontractions

The sensory inflow after a brief bout of intense voluntary contraction produces an aftercontraction known as the **Kohnstamm effect** (Kohnstamm, 1915). We can demonstrate this by performing an MVC and then observing the postcontraction effects. For example, stand in a doorway with your arms extended from your trunk so that the backs of your hands can push against the door frame. After a 15 s MVC, step forward and relax your arms. For most individuals, the arms will rise slightly due to postcontraction activation of the motor neurons innervating proximal arm muscles (Kozhina et al., 1996). The magnitude and direction of the Kohnstamm effect depend on the intensity and duration of the preceding voluntary activity (Parkinson & McDonagh, 2006; Sapirstein et al., 1937), and the effect is more prominent in proximal muscles (Adamson & McDonagh, 2004). The maximal EMG during the aftercontraction can reach up to 50% of the value recorded during an MVC (Adamson & McDonagh, 2004). Functional MRI indicates that aftercontractions involve activation of the same brain areas that control voluntary activation (primary sensory and motor cortices, premotor cortex, and anterior and posterior cingulated gyrus) and structures (posterior parietal cortex) that contribute to sensorimotor integration (Duclos et al., 2007).

AROUSAL

Arousal is an internal state of alertness. It is a component of several emotional responses, including fear and anxiety, and is mediated by the neuroendocrine system. Its physiological manifestations commonly include increases in blood pressure and heart rate; sweating; dryness of the mouth; hyperventilation; and musculoskeletal disturbances such as restlessness, tremor, and feelings of weakness (Bonnet et al., 1995; Hoehn-Saric et al., 2004; Spielberger & Rickman, 1990).

The level of arousal varies along a continuum from deep sleep to the fight-or-fight response. It can have a substantial effect on movement. For example, a moderate amount of arousal maximizes performance both in laboratory tasks, such as reaction-time tests and movement accuracy tasks, and in competitive sport (Arent & Landers, 2003; Millalieu et al., 2004). This effect has been explained by the **inverted-U hypothesis** (Arent & Landers, 2003; Raglin, 1992), which suggests that moderate levels of arousal (~65% of maximum) enhance performance. Due to variability among individuals, however, the desired level of arousal can vary, and this has produced the **Zone of Optimal Function** theory (Jokela & Hanin, 1999; Turner & Raglin, 1996), which also suggests that performance is optimal at an intermediate level of arousal.

Assessment of Arousal

Because arousal induces changes in several autonomic functions, one strategy to assess arousal is to measure the changes in selected physiological variables or the neuroendocrine factors that mediate the responses. The commonly measured physiological variables include heart rate, blood pressure, pupil dilation, and skin conductance. The association between arousal and changes in these physiological variables, however, is not direct. Typically, there is a low correlation between changes in the physiological variables and variations in arousal. Furthermore, the physiological changes differ with the stressor used to manipulate arousal (Gavrilovic & Dronjak, 2005; Kopin, 1995; Pacak et al., 1998; Raglin, 1992).

The neuroendocrine factors commonly associated with elevations in arousal include the catecholamines, adrenocorticotropic hormone, cortisol, growth hormone, and prolactin (Campeau et al., 1997; Pancheri & Biondi, 1990). There is a strong association between changes in arousal and modulation of these factors. Although an arousal response can begin within milliseconds of the presentation of the appropriate stimulus, the time before changes in the levels of the neuroendocrine factors can be detected is usually too long for an application to movement analysis. Most movements are completed within seconds, whereas the time course of detectable changes in the circulating levels of the neuroendocrine factors is in the order of minutes (Kirschbaum & Hellhammer, 1994; Roy, 2004; Van Eck et al., 1996).

As a supplement to measurement of the physiological and neuroendocrine variables, arousal can also be quantified by self-reports of the perceived level of arousal. Two common approaches are used: One determines an individual's anxiety, and the other estimates the moment-to-moment level of arousal. The assessment of anxiety usually involves having an individual complete a questionnaire that assesses both the average (trait anxiety) and the current (state anxiety) levels of anxiety (Spielberger et al., 1983). The scores indicate the extent to which the individual might be aroused by stress. The moment-to-moment level of arousal is measured with the visual analog scale, a 10 cm line anchored at each end with descriptive polar phrases such as "Not at all anxious" on the left end and "Very anxious" on the right end (Cella & Perry, 1986; O'Connor & Cook, 1999). The comprehensive assessment of arousal in humans, therefore, requires the concurrent measurement of physiological variables, neuroendocrine factors, and perceived levels of anxiety.

Mechanism of Action

Although there is a consensus that arousal can influence motor performance, the mechanisms that mediate the effect are uncertain. One explanation suggests that as arousal varies, so does the attention afforded by the individual to the task, which could produce a corresponding variation in performance (Raglin, 1992). Whatever the cognitive consequences of arousal, however, there must be some variation in the neural commands sent to muscle that cause changes in the performance. Little is known about the nature of these changes.

Because arousal involves an increase in the circulating levels of various neuroendocrine factors, it is likely that the neural activation patterns vary with the level of arousal. Two examples support this possibility. First, the time to task failure during fatiguing contractions depends on the level of such factors as epinephrine and 5-hydroxytryptamine (5-HT, serotonin) (Blomstrand et al., 1988; Heyes et al., 1988; Nybo & Secher, 2004). For example, the duration over which subjects can cycle at 80% of the maximal rate of oxygen uptake is decreased after oral administration of an agent that enhances serotonergic activity (Marvin et al., 1997). This is not too surprising, as serotonin levels in the CNS have been implicated in the regulation of sleep, depression, anxiety, aggression, appetite, temperature, sexual behavior, and pain sensation (Birdsall, 1998). Furthermore, therapeutic administration of 5-hydroxytryptophan, an intermediate metabolite in the synthesis of serotonin, is effective in treating depression, fibromyalgia, binge eating with obesity, chronic headaches, and insomnia (Birdsall, 1998; Halford et al., 2005; Ribeiro, 2000; Spath, 2002; Turner et al., 2006).

Second, some of the neurotransmitters that modulate the activity of the autonomic nervous system can act as neuromodulators and modify the function of the spinal circuits underlying motor performance (Heckman & Enoka, 2004; Marder, 1998; Rekling et al., 2000). For example, numerous stressors can increase the activity of serotonergic neurons and thereby raise the extracellular levels of 5-HT (Chaouloff et al., 1999), which can alter evoked metabotropic effects and the response of a neural network to synaptic drive (Heckman et al., 2003; Nistri et al., 2006). Variation in the level of serotonin can alter the activity of spinal motor neurons (Liu & Jordan, 2005; Li Volsi et al., 1998), EMG activity during the performance of a choice reaction-time task (Rihet et al., 1999), and even the central pattern generators associated with locomotion (Feraboli-Lohnherr et al., 1999; Jovanovic et al., 1996; Pearlstein et al., 2005). These interactions indicate that it is necessary to examine the effects of arousal on motor performance in order to obtain a more complete understanding of the function of the nervous system.

Neuromuscular Function

The functional significance of variations in arousal has been demonstrated by studies that have examined its influence on maximal force, steadiness of submaximal contractions, fatiguing contractions, and the distribution of muscle activity.

Maximal Force

Findings from a study by Ikai and Steinhaus (1961) are often cited as exemplifying the potentiating effect that arousal can have on the maximal force. In that study, subjects performed brief maximal isometric contractions with the elbow flexor muscles once each minute for 30 min. Occasionally, and unexpectedly, one of the investigators discharged a firearm prior to an MVC. In addition, each subject was asked to shout loudly prior to the last MVC. For most subjects, the MVC force was greater after the loud noise associated with the discharge of the gun and with the shout. The effect of the loud noise, however, provides evidence of the potentiating effect of a startle response (Siegmund et al., 2001; Valls-Solé et al., 1999, 2005) and not an enhancement due to arousal. In contrast, the maximal force that individuals can exert during a handgrip task did not increase after they participated in arousal-enhancing activities that included doing mental math problems and receiving electric shock (Noteboom & Enoka, 2001*a, b*).

If arousal does have an effect on muscle strength, three distinct mechanisms might be involved: changes in voluntary activation, the contractility of muscle, or the coordination of the involved muscles. Because motivated individuals can maximally activate muscle by voluntary command under standard laboratory conditions (Allen et al., 1998; Gandevia et al., 1998; Williams & Bilodeau, 2004), it is unlikely that arousal can increase the activation of muscle. Similarly, the catecholamine epinephrine, which is secreted in greater amounts with increases in arousal, can potentiate twitch responses but not tetanic force (Marsden & Meadows, 1970; Williams & Barnes, 1989). In contrast, the modulatory effects of neuroendocrine factors on spinal networks might optimize coordination and increase the load that could be lifted with heightened arousal.

Steadiness

When an individual performs a steady contraction with hand, arm, or leg muscles, the force exerted by the limb is not constant but rather fluctuates about an average value (Christou et al., 2002; Enoka et al., 2003; Slifkin & Newell, 1999). The variability in force can be quantified in absolute terms as the standard deviation or in relative terms as the coefficient of variation. Studies on steady contractions with the knee extensor muscles indicated that both the standard deviation of force during isometric contractions (figure 8.35*a*) and the standard deviation of acceleration during shortening and lengthening contractions (figure 8.35*b*) increase with stronger muscle contractions. The variation in force fluctuations during isometric contractions across the operating range of a hand muscle is attributable to the distribution of discharge rate variability among the active motor units (Barry et al., 2007; Moritz et al., 2005*a*).

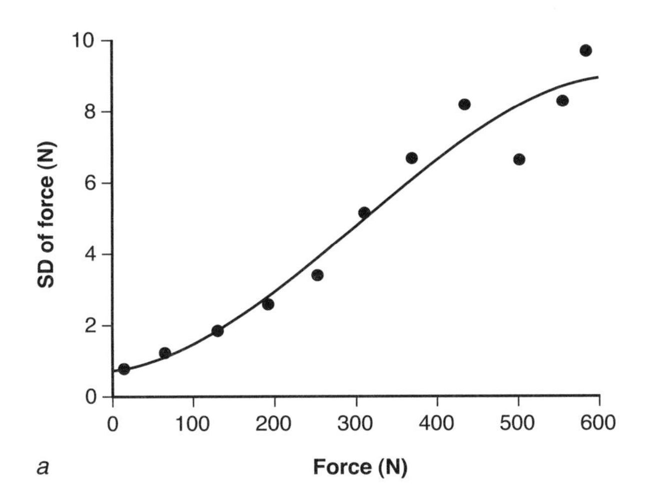

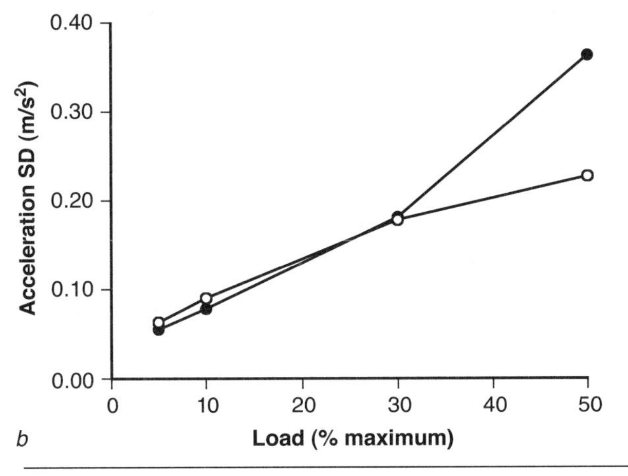

Figure 8.35 Fluctuations in force and acceleration during steady contractions with the knee extensor muscles. *(a)* The standard deviation of force during brief isometric contractions at target forces that ranged from 2% to 95% of maximum. The task was to match as closely as possible the force exerted by the leg to a target line on a monitor. *(b)* The standard deviation of ankle acceleration during shortening (filled circles) and lengthening (open circles) contractions with loads ranging from 5% to 50% of the maximal load that could be lifted once.

Data for part *a* from Christou et al. 2002 ; part *b* from Enoka et al. 2003.

When arousal is manipulated by the application of a stressor (mental math or electric shock), the steadiness of a submaximal pinch grip decreases (Noteboom & Enoka, 2001*a, b*). This effect was demonstrated by Christou and colleagues (2004) when subjects performed a pinch grip at 2% MVC force during a 70 min protocol that involved three phases: anticipatory, stressor, and recovery. Subjects received noxious electrical stimuli intermittently on the back of the hand that was not performing the pinch grip during the stressor phase. Subjects reported an increased level of arousal during the stressor phase (figure 8.36*a*) that was accompanied by a marked increase in the standard deviation of force (figure 8.36*b*), and this effect was most pronounced in old adults (figure 8.36*c*). Although women reported greater levels of arousal (figure 8.35*a*), there was no sex difference in the influence of the stressor on steadiness.

Fatiguing Contractions

The amount of work that can be performed during fatiguing activity can be enhanced by an elevation of arousal. Asmussen and Mazin (1978*a, b*) compared the amount of work that could be performed by elbow and finger flexor muscles when subjects were exposed to actions that increased arousal. Subjects performed a series of contractions to task failure with 2 min rest periods between each bout of exercise. The amount of work that could be performed in successive bouts was greater when subjects were active during the rest periods (physical and mental activity) and when the work was performed with the eyes open compared with eyes closed. These effects were not attributable to differences in blood flow to the muscles performing the fatiguing contraction and were associated with a more vigorous tendon tap reflex. The authors concluded that the improvement in performance was attributable to an increase in arousal.

The influence of visual feedback on fatiguing contractions was also demonstrated by Mottram and colleagues (2006*a*) when they compared the time to failure of the position task (15% MVC force) as the sensitivity of the displayed signal was varied. The task was to keep the elbow angle constant; the desired elbow angle was displayed on a monitor. The task was performed on two occasions, and the sensitivity of the angle signal was either high or low. When the sensitivity was high, the movement of the elbow angle displayed on the monitor was large for a small change in elbow angle. The time to task failure in men was similar for the low (6.0 min)- and

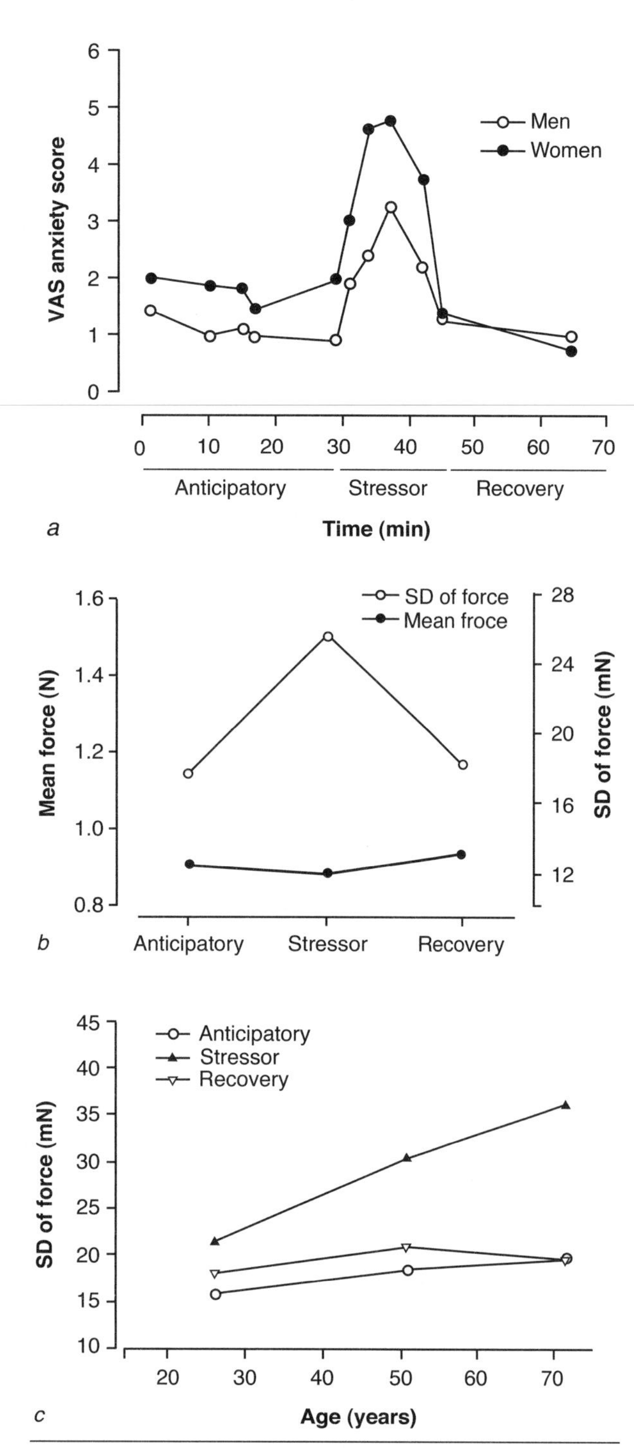

Figure 8.36 The influence of electric shock on the steadiness of a submaximal pinch grip. *(a)* The scores on the visual analog scale (VAS) increased during the 15 min stressor phase and were greater for the women than for the men during both the anticipatory (30 min) and stressor phases. *(b)* The standard deviation of force increased during the stressor phase even though the average force during the pinch grip remained constant during all three phases. *(c)* The standard deviation of force increased during the stressor phase for young, middle-aged, and old adults, with the greatest increase being exhibited by the old adults.

Data from Christou et al. 2004.

high (5.9 min)-sensitivity conditions, but it was less in the women for the high-sensitivity condition (8.7 min) compared with the low-sensitivity condition (11.9 min). The greater attentiveness required during the high-sensitivity condition, therefore, reduced the time to task failure for the women but not the men. Furthermore, the physiological variables that predicted the time to failure for this task differed for the men and women.

Spread of Muscle Activity

When a person performs a task that requires a high level of effort, there is a spread of activation to other muscles besides those principally responsible for the task. For example, other muscles are typically activated during the performance of a strength task and in the course of a fatiguing contraction (Howard & Enoka, 1991; Rudroff et al., 2007; Zijdewind et al., 2006*a*). This spread of activation probably enhances the postural stability of the individual and enables the transfer of power across joints by two-joint muscles.

Increased levels of arousal also enhance the amount of muscle activation used to perform a task. Weinberg and Hunt (1976) found, for example, that subjects with high levels of anxiety (an index of arousability; Wilken et al., 2000), as compared with less anxious subjects, responded to negative feedback on performance by increasing the amount of muscle EMG during a throwing task. The change in EMG activity caused a decline in performance for the highly anxious subjects but an improvement in performance of the less anxious subjects. Conversely, training with EMG biofeedback can reduce the levels of state and trait anxiety, and the effect is greater in more anxious individuals (Hurley & Meminger, 1992).

The effect of arousal on the spread of muscle activation has been used as the basis for some therapeutic interventions with patients. For example, many of the techniques utilized in PNF are based on the spread of muscle activation, which is often referred to as **irradiation** (Zijdewind & Kernell, 2001). In a typical application, a patient is encouraged to perform a particular movement pattern against a maximal resistance, which spreads the activation to synergist muscles. Moreover, the basic PNF procedures recommend the use of strong verbal commands "to simulate a stress situation" and thereby enhance the spread of activation to the synergist muscles (Knott & Voss, 1968). Clinical experience suggests that patients have a greater capacity to involve impaired muscles under such conditions of heightened arousal (Kofotolis & Kellis, 2006; Noth et al., 2006; Ward et al., 2006).

SUMMARY

This chapter is the first of two to examine the effects of physical activity on the motor system. The focus of this chapter is the immediate (acute) response of the system to the stress associated with a single bout of physical activity. The chapter addresses six topics that characterize the adjustments exhibited by the motor system: warm-up effects, flexibility, muscle soreness and damage, muscle fatigue and task failure, muscle potentiation, and arousal. First, we discuss warm-up effects and consider the influence of changes in temperature on the mechanical output of the system and on the passive stiffness of muscle and connective tissue. Second, we distinguish warm-up activities from those related to flexibility and consider the techniques used to increase flexibility and the factors that limit joint range of motion. Third, we examine the muscle soreness that occurs after strenuous activity. Delayed-onset muscle soreness occurs more frequently after lengthening than after shortening contractions. The consequences of such activity include an increase in the tenderness of muscle, elevated levels of plasma creatine kinase, and a temporary impairment of muscle function. Fourth, we consider the effects of sustained activity on the ability to exert force (muscle fatigue). Because the mechanisms that cause muscle fatigue vary with the details of the task, it is suggested that the functional significance of the adjustments that occur during fatiguing contractions can be determined through identification of the mechanisms that limit the duration over which a task can be sustained. Fifth, we characterize the potentiating effects of brief contractions on monosynaptic responses, miniature end-plate potentials, M-wave amplitude, posttetanic twitch potentiation, and postcontraction sensory discharge. Sixth, we explore the effect of arousal on motor performance, including how it can be measured, why it is significant, and what is known about its effect on neuromuscular performance. These examples, however, are not intended to represent the complete set of adjustments that can be produced by the motor system.

SUGGESTED READINGS

Kernell, D. (2006). *The Motoneurone and Its Muscle Fibres*. Oxford, Great Britain: Oxford University Press, chapter 9.

MacIntosh, B.R., Gardiner, P.F., and McComas, A.J. (2006). *Skeletal Muscle: Form and Function* (2nd ed.). Champaign, IL: Human Kinetics, chapter 15.

chapter 9

Chronic Adaptations

In chapter 8, we examined the acute adjustments of the motor system in response to a single bout of physical activity. We learned that the adjustments can be extensive and that the response involves the processes and components of the system that are stressed by the activity. Chapter 9 addresses the cumulative (chronic) response of the motor system to the stress associated with long-term physical activity. We will examine the adaptations associated with muscle strength and muscle power, adaptation to reduced use, motor recovery from injury, and adaptations with age.

MUSCLE STRENGTH

The measurement of strength is used as an index of the force-generating capacity of muscle. In clinical and experimental settings, strength is commonly measured in one of three ways: as the maximal force that can be exerted during an isometric contraction, the maximal load that can be lifted once, or the peak torque during an isokinetic (shortening or lengthening) contraction (figure 9.1). The isometric contraction task is usually referred to as a maximal voluntary contraction (MVC) and the load that can be lifted once is known as the **1-repetition maximum load** (1-RM load). While the magnitude of the MVC force depends primarily on the size of the involved muscle, the 1-RM load and the peak isokinetic torque depend on both muscle size and the coordinated activation of muscles by the nervous system (Folland & Williams, 2007; Herzog, 2004*a;* Rutherford and Jones, 1986; Semmler & Enoka, 2000).

The strength gains experienced by an individual depend on how strength is measured and on the exercises performed in the training program (Barry & Carson, 2004; Dvir, 2004; Gallagher et al., 2004). Because of this association, we will discuss the training and loading techniques used in strength training before considering the adaptations that produce the increases in strength.

Training Techniques

The adaptations experienced by the motor system with repeated bouts of exercise depend on the specific tasks performed in the training program. As described in chapters 6 and 7, variation in an exercise task influences the motor units that are activated, the coordination among synergist muscles, the amount of postural support needed for the movement, and the type of sensory feedback received by the central nervous system (CNS). Because of these effects, the most effective training strategy is to match the exercise to the intended outcome (Aagaard et al., 1996). This section describes six training modalities that one should consider when designing a training program to achieve specific goals.

Isometric Contractions

An isometric contraction (*iso* = constant, *metric* = whole-muscle length) was previously defined as a condition in which the torque due to the load is matched by a muscle torque that has an equal magnitude but an opposite direction. This occurs, for example, when a limb pushes against an immovable object in the surroundings. Although there is no change in whole-muscle length during an isometric contraction, the muscle fibers shorten (Griffiths, 1991; Ito et al., 1998; Kawakami & Fukunaga, 2006).

The strength gains achieved with isometric contractions usually peak after about six to eight weeks of training (Davies et al., 1988; Duchateau & Hainaut, 1984; Folland et al., 2005; Garfinkel & Cafarelli, 1992; Kubo et al., 2001*a*). For example, Kitai and Sale (1989) measured the gains achieved by six subjects who trained

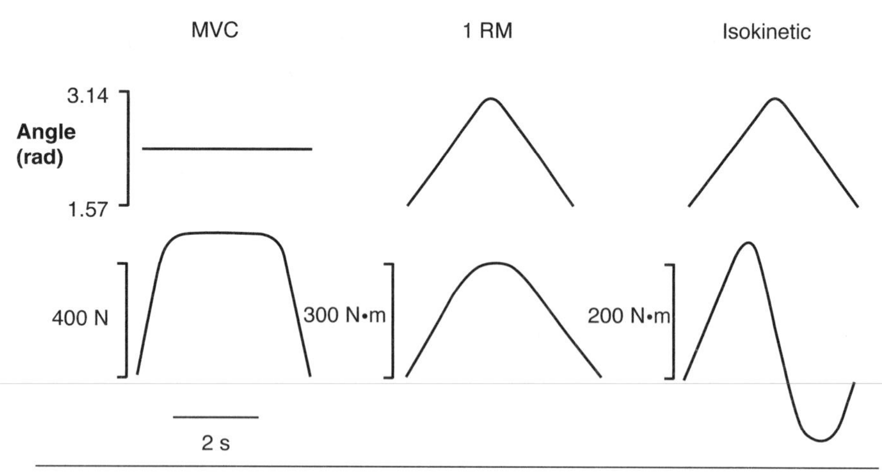

Figure 9.1 Idealized performances of a maximal voluntary contraction (MVC), a 1-repetition maximum (1-RM) lift, and an isokinetic contraction with the knee extensor muscles. The second half of the isokinetic task involves the knee flexor muscles. The upper row indicates knee angle (3.14 rad = complete extension). The lower row shows force (N) for the MVC and the resultant muscle torque (N•m) for the 1-RM and isokinetic tasks.

with isometric exercises for six weeks. The subjects trained three times a week by performing two sets of 10 repetitions. Each repetition was a 5 s maximal (100%) isometric contraction of the plantarflexor muscles with the ankle joint fixed at a right angle between the shank and the foot. Two contractions were performed each minute, and there was a 2 min rest between sets. The training resulted in an increase in strength (MVC torque) at the training angle of 18% (25.5 N•m) and increases of 17% and 14% at adjacent (0.17 rad) plantarflexion and dorsiflexion angles. With a 16 week intervention, Alway and colleagues (1989) found that isometric exercises increased maximal plantarflexor torque by 44%. The influence of isometric training on muscle volume and dynamic strength differs for high-force, short-duration and moderate-force, long-duration contractions (Kanehisa et al., 2002). However, there is generally a poor correlation between increases in isometric strength and performance or the gains achieved with anisometric contractions (Wilson & Murphy, 1996). Accordingly, isometric contractions are used infrequently in training programs, but more often in rehabilitation activities (Duncan et al., 1989; Symons et al., 2005; Williams et al., 2003).

Although most of the research on isometric training has focused on the gains achieved by the prime mover muscles, many other muscles perform isometric contractions during strength training exercises to provide postural support for the task. The significant role of the isometric activity is underscored by the effect of posture on the measured outcome of a strength training program (Bouisset et al., 2002; Hortobágyi & Katch, 1990; Roy et al., 2003; Rutherford & Jones, 1986). For example, the improvement in performance after eight weeks of weight training with the squat and bench press lifts was evident only when the training and testing procedures were similar (Wilson et al., 1996). The improvements in performance included an increase in the maximal bench press load (12%), the bench press throw (8%), and the isokinetic bench press (13%), but not the ground reaction force during a push-up test or the peak torque during horizontal arm adduction. These results indicate that the strength gains were specific to the training exercises, resulting in a poor correlation between MVC force and performance but a moderate association between the rate of increase in MVC force and performance (Morrisey et al., 1995; Wilson & Murphy, 1996).

Lifting Weights

When an exercise involves lifting a load, the muscle torque is greater than the load torque when the load is lifted and less than the load torque when the load is lowered. These actions require shortening and lengthening muscle contractions, respectively. Because most movements involve both shortening and lengthening contractions, the greatest strength gains are achieved with training programs that involve both actions (Gur et al., 2002).

As proposed by DeLorme (1945), strength training that involves lifting weights should be based on a scheme in which the load is increased progressively across the training session. DeLorme proposed that each exercise should comprise three sets of 10 repetitions in which the load is increased for each set. The load is based on the amount that can be lifted 10 times, which is referred to as the 10-RM load (RM = repetition maximum). The first set of 10 repetitions is done with one-half of the 10-RM load, the second set with three-quarters of the 10-RM load, and the final set with the full 10-RM load. Because most exercises include both shortening and lengthening contractions, the 10-RM load does not provide a consistent stress for all phases of each repetition, and the capabilities of the shortening action typically limit performance and thus determine the 10-RM load. Because the amount of exercise stress depends on the magnitude of the load relative to maximal capabilities, the shortening contraction experiences the greater stress and subsequent adaptation (Hortobágyi & Katch, 1990).

Dudley and colleagues (1991) compared the efficacy of training with shortening and lengthening contractions in middle-aged men who trained for 19 weeks. Subjects performed four to five sets of 6 to 12 repetitions with a leg press and knee extension exercise. Two groups of men performed the exercises with shortening contractions only, whereas another group used both shortening and lengthening contractions. The maximal load that could be lifted three times (3-RM load) during the leg press exercise increased by 26% for the group that trained with both contractions, whereas it increased by 8% to 15% for the two groups that trained with only shortening contractions. Similarly, the 3-RM load for the knee extension exercise increased by 29% for the group that used both types of contractions and by 16% for the subjects who trained with shortening contractions only. Spurway and colleagues (2000) observed similar specificity in young men and women after they trained one leg with shortening contractions and the other with lengthening contractions for six weeks on a leg extension machine. The peak force exerted against the dynamometer after training increased during lengthening contractions; the increase ranged from 18% for the leg of the women who trained with shortening contractions up to 31% for the leg of the men who trained with lengthening contractions. The changes in peak force during the shortening contractions were much less.

The strength gains achieved by training with only shortening or lengthening contractions can be relatively specific to the training mode. For example, Vikne and colleagues (2006) compared the influence of 12 weeks of training with shortening and lengthening contractions on the performance of the elbow flexor muscles in strength-trained men. Both types of training increased (14% and 18%) peak force during shortening contractions, whereas the increase in peak force during the lengthening contractions was greater for the group that trained with lengthening contractions (26%) compared with the group that trained with shortening contractions (9%). A likely explanation for this difference is that the adaptations evoked by the training with lengthening contractions involve changes in both muscle fiber size and the activation of muscle by the nervous system (Colson et al., 1999; Spurway et al., 2000; Valour et al., 2004; Vikne et al., 2006). Accordingly, training programs that include both shortening and lengthening contractions produce greater increases in muscle fiber size than programs that involve only shortening contractions (Hather et al., 1991).

Another feature of training with lengthening contractions compared with shortening contractions is that a greater load can be imposed on the exercising muscle for a comparable demand on the cardiovascular system. In patients whose exercise tolerance is limited by cardiovascular stress, lengthening contractions can provide a useful adjunct to standard rehabilitation (Dufour et al., 2004; Meyer et al., 2003).

Accommodation Devices

Exercise machines in which the load is controlled by gear or friction systems (e.g., Cybex, Biodex), by hydraulic cylinders (e.g., Ariel, Kincom, Lido, Omnitron), or by pneumatic systems (e.g., Keiser) provide an accommodating resistance. These systems can generate a load equal in magnitude but opposite in direction to the force exerted by the subject. One consequence of the gear systems and of some hydraulic devices is a movement in which the angular velocity of the displaced body segment is constant, which results in an isokinetic contraction (Ichinose et al., 2000; Kaufman et al., 1995; Kawakami & Fukunaga, 2006).

Isokinetic devices can be used to perform shortening and lengthening contractions (figure 9.2) (Chino et al., 2007). When an individual pushes against the device and performs a shortening contraction, power flows from the individual as positive work is done on the device. When the individual resists the load imposed on the limb by the device, the action requires a lengthening contraction and power flows from the device as it does negative work on the individual. Many investigators use isokinetic devices (figure 9.3*a*) to quantify the torque, work, and power output of muscle. Typically, they accomplish this by measuring the resistance provided by the machine ($\boldsymbol{F}_l$ in figure 9.3*b*) and expressing the effort in the appropriate units. A free body diagram of the system (figure 9.3*b*), however, indicates that the load torque includes both the machine load and the weight of the limb. Accordingly, the acceleration of the limb at the beginning and end of a repetition will influence the estimated muscle torque (Iossifidou & Baltzopoulos, 1998; Winter et al., 1981). However, the error due to these acceleration effects can be determined and incorporated into the appropriate calculations to

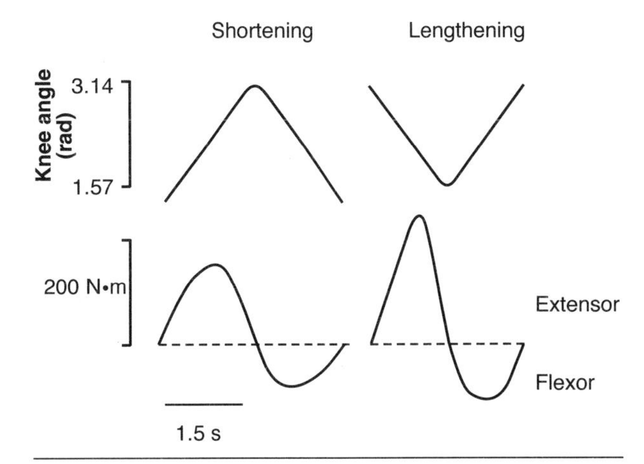

Figure 9.2 Shortening and lengthening isokinetic contractions with the knee extensor and flexor muscles. The upper row indicates knee angle (3.14 rad = complete extension) whereas the lower row shows the resultant muscle torque exerted by the knee extensor (positive) and flexor (negative) muscles.

Adapted from R.M. Enoka 1988.

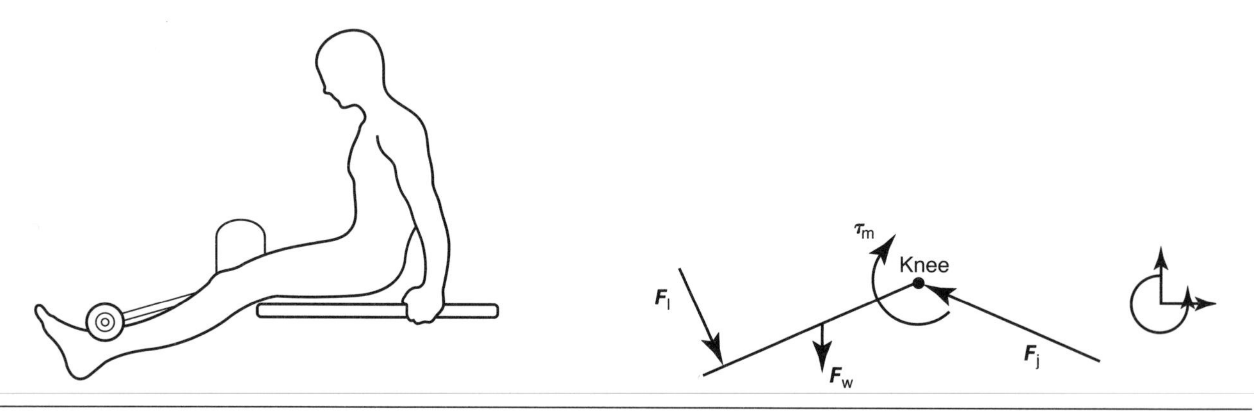

Figure 9.3 A subject performing an isokinetic knee extension exercise: *(a)* the isokinetic dynamometer; *(b)* a free body diagram of the forces acting on the shank. F_j = joint reaction force; F_w = limb weight; F_l = machine load; τ_m = resultant muscle torque.

allow more accurate measurements (Aagaard et al., 1994*b*; Kaufman et al., 1995; Kellis & Baltzopoulos, 1995). Without this correction, the torque exerted by the machine ($F_l \times$ moment arm) is not a measure of the resultant muscle torque (τ_m).

One disadvantage of isokinetic devices is that they do not mimic natural movements (Iossifidou et al., 2005). For example, unconstrained movements rarely involve a constant angular velocity of a limb; isokinetic contractions can involve greater coactivation of antagonist muscles (Aagaard et al., 2000*a*; Kellis & Baltzopoulos, 1997); and the maximal speed on many isokinetic devices (5 rad/s) is much less than the maximal angular velocity achieved during movements such as running, jumping, and throwing. These characteristics may explain why the peak torques exerted during isokinetic contractions are usually less than those achieved during less constrained movements. For example, the peak torque that the knee extensors can exert during a shortening contraction at 1.05 rad/s is about 160 N•m (Crenshaw et al., 1995) compared with a peak torque of 300 N•m during the shortening contraction of a vertical jump (Bobbert & Ingen Schenau, 1988). Similarly, a strength training program that included the squat lift resulted in significant gains in maximal squat load (21%), 40 m sprint time (2%), peak power during a 6 s cycle (10%), and vertical jump height (21%), but not peak torque during an isokinetic contraction (Wilson et al., 1996). Additionally, strength training with isokinetic contractions increases the peak torque during an isokinetic contraction but does not alter maximal kicking performance (Aagaard et al., 1996).

Despite these limitations, isokinetic training can produce adaptations comparable to those achieved with less constrained techniques, although these changes may be specific to the training speed and type of contraction (Aagaard et al., 1996; Higbie et al., 1996; Morrissey et al., 1995; Seger et al., 1998; Seger & Thorstensson, 2005). For example, when subjects trained the knee extensors for seven weeks on an isokinetic device at 3.14 rad/s, there were significant increases in peak torque (about 14%) for contractions from 0 to 3.14 rad/s but not at 4.2 and 5.2 rad/s (Lesmes et al., 1978). Similarly, subjects who trained for 10 weeks at 1.1 rad/s experienced a 9% increase (25 N•m) in peak torque, whereas those who trained at 4.2 rad/s had a 20% increase (30 N•m) in peak torque; each of these increases was specific to the training speed (Ewing et al., 1990). Furthermore, the relative increase in peak torque during an isokinetic contraction is greater after training with lengthening contractions compared with shortening contractions (Farthing & Chilibeck, 2003; Higbie et al., 1996), even when training involves the same absolute load (Hortobágyi et al., 1996*a*). For example, subjects who trained the knee extensor muscles with lengthening contractions for 12 weeks experienced a 46% increase in the peak torque compared with a 13% increase in peak torque exhibited by subjects who trained with shortening contractions (Hortobágyi et al., 1996*b*). Furthermore, the strength gains achieved with either shortening or lengthening contractions are specific to the training mode (Aagaard et al., 1996; Morrissey et al., 1995).

Isokinetic devices offer several advantages. One significant advantage is that a muscle group may be stressed differently throughout its range of motion. This accommodation is useful in rehabilitation settings (Kellis & Baltzopoulos, 1995). When one specific locus in a range of motion is painful, the patient can reduce the effort at this point yet exercise the joint system in the other nonpainful regions (Shirakura et al., 1992). Furthermore, the patient can simply stop in the middle of an exercise without having to worry about controlling the load. Because of the accommodation property, the resistance provided by the device varies in proportion to the capabilities of the user over the range of motion. Isokinetic devices also provide substantial support for the user, removing the need to offer some of the stabilizing support that must be generated with other rehabilitation exercises. As with training programs that involve lifting loads, an isokinetic training program should incorporate both shortening and lengthening contractions to maximize the transfer to functional tasks (Gur et al., 2002).

In contrast to isokinetic devices, hydraulic and pneumatic systems provide an accommodating resistance but do not control movement speed. Hydraulic devices provide resistance as individuals exert a force against a leaky, fluid-filled chamber (Fothergill et al., 1996; Pinder & Grieve, 1997). The muscles perform shortening contractions against this type of resistance. The resistance is manipulated by variation in the diameter of the valve through which the fluid passes from one chamber to another; the resistance is high when the diameter of the valve is small. The hydraulic system offers an accommodating resistance. The greater the force exerted by the individual, the greater the resistance provided by the device. However, movement speed increases as the applied force increases. Nonetheless, hydraulic devices can elicit strength gains comparable to those achieved with free weights (O'Hagan et al., 1995). For example, Hortobágyi and Katch (1990) trained one group of subjects with free weights and another group with a hydraulic device for 12 weeks. The average improvements in the maximal bench press and squat loads were 23% for subjects who trained with free weights and 20% for those who trained with a hydraulic device. Furthermore, changes in force, velocity, and power evaluated by isokinetic and hydraulic dynamometers were not different (~8%) for the two groups. Pneumatic systems also provide a variable resistance by using a compressor to alter the pressure and hence the resistance against which the individual works. The user can change the pressure rapidly, simply by depressing a button. Although the pressure in a pneumatic system can be increased sufficiently to enable the performance of lengthening contractions, control of the device is minimal under these conditions.

Plyometric Training

In contrast to the isokinetic device, **plyometric** exercises were designed to train a specific movement pattern, the lengthening and shortening actions associated with the stretch–shorten cycle. Most human movements involve a stretch–shorten cycle in which a shortening contraction is preceded by an increase in whole-muscle length. The initial stretch may or may not involve a lengthening of active muscle fibers (Griffiths, 1991; Hoyt et al., 2005; Ishikawa et al., 2007; Lichtwark et al., 2007; Sousa et al., 2007). Because the initial stretch during a stretch–shorten cycle is usually so brief, the CNS likely establishes a relatively constant level of activation and the action of the surroundings determines the extent to which the whole muscle (fibers + tendon) is stretched (Finni et al., 2003; Ishikawa et al., 2006; Lindstedt et al., 2002).

Plyometric exercises are able to induce a training effect (Cutlip et al., 2006; Herrero et al., 2006; Pousson et al., 1991; Reich et al., 2000). For example, Blattner and Noble (1979) trained two groups of subjects for eight weeks, one group on an isokinetic device and the other with plyometric exercises. The two groups increased vertical jump height by about the same amount (5 cm) as a result of the training. Similarly, Häkkinen and colleagues (1985) found that jump training caused a minor increase in the maximal knee extensor force during an isometric contraction but a substantial increase in the rate of force development and an increase in the area of fast-twitch fibers. Although it is difficult to train experimental animals to perform jump training, Dooley and colleagues (1990) trained rats to perform 30 jumps a day, five days a week for at least eight weeks. The jumps were estimated to be between 30% and 67% of the maximal height that the rats could achieve. The training produced some changes in the medial gastrocnemius muscle but not in the soleus. There was a 15% increase in the maximal tetanic force, a 3% increase in the maximal rate of force development, a 15% increase in fatigability, and a 4% decrease in the percentage of type IIa muscle fibers. Artificially evoked stretch–shorten cycles, however, can induce similar changes in the soleus muscle of rats (Almeida-Silveira et al., 1996). These observations indicate that regular vertical jump exercises performed at a moderate intensity are able to induce adaptations in the involved muscles.

The advantage of plyometric exercises is that they enable an individual to train the stretch–shorten component of a movement, which is difficult to accomplish with any other technique. Nonetheless, plyometric exercises should not be the sole component of a training program but rather should be part of a more diverse program that addresses the strength, speed, endurance, and flexibility needs of the individual (Zatsiorsky, 1995).

Neuromuscular Electrical Stimulation

Electrical stimulation applied over the skin has been used since the 18th century as a rehabilitation tool (Hainaut & Duchateau, 1992; Liberson et al., 1961), but only since the 1970s has it been applied to noninjured active athletes as a supplement to conventional training (Paillard, 2008). The electric current that passes through a muscle to evoke a muscle contraction generates action potentials in the intramuscular nerve branches rather than directly exciting muscle fibers (Hultman et al., 1983). Because the technique involves artificial activation of both the nerve and the muscle, it is known as **neuromuscular electrical stimulation.**

Stimulation Protocols Neuromuscular electrical stimulation can be applied with a variety of protocols. The parameters include stimulus frequency, stimulus waveform, stimulus intensity, electrode size, and electrode type. The simplest stimulus protocol is to apply a train of rectangular pulses, typically at constant frequencies (usually 15-40 Hz) with narrow pulse widths (50-400 μs) (figure 9.4*a*). One limitation of this protocol, however, is that it requires a frequency of about 100 Hz to elicit the maximal force in a muscle, which is much more painful than stimuli at <50 Hz (Barr et al., 1986). The discomfort associated with neuromuscular electrical stimulation can be circumvented with a protocol of high-frequency stimulation (10 kHz) that is modulated (i.e., turned on and off) at a lower frequency (50-100 Hz) (figure 9.4*b*). Moreno-Aranda and Seireg (1981*a, b, c*) found that the optimal regimen involved the application of the stimulus for 1.5 s once every 6 s for 60 s, followed by a 60 s rest. This type of protocol minimizes the pain associated with the procedure and can elicit a force equivalent to the MVC force (Delitto et al., 1989).

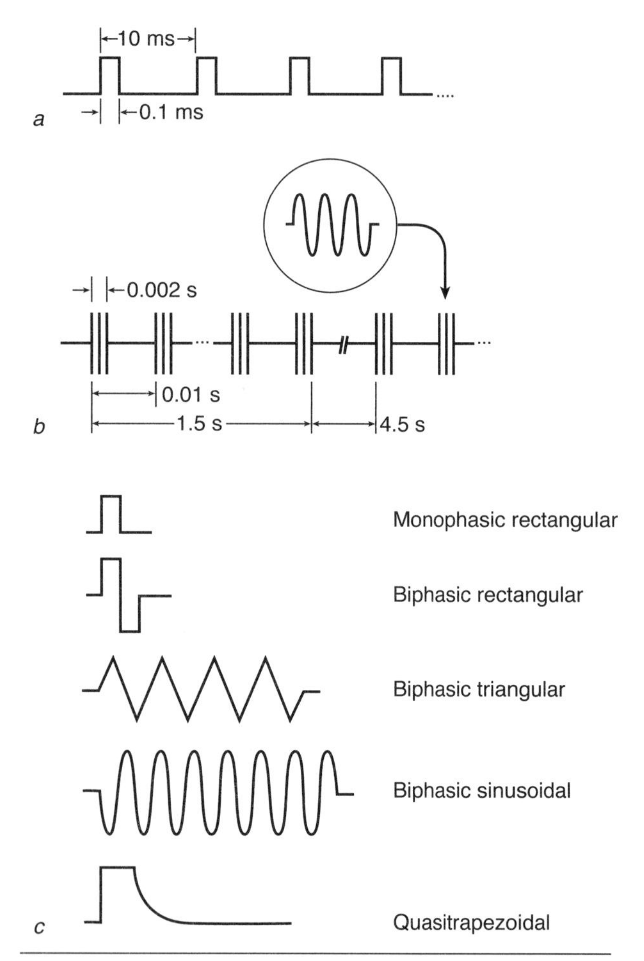

Figure 9.4 Selected stimulus protocols used in neuromuscular electrical stimulation: *(a)* a conventional train of low-frequency (100 Hz) rectangular stimuli with a pulse width of 0.1 ms; *(b)* a pattern of high-frequency stimulation (1 kHz sine wave) that is modulated at a low frequency (100 Hz) with 0.01 s between trains of stimuli; *(c)* waveform shapes. The biphasic waveforms vary about a zero line.

Adapted, by permission, from R.M. Enoka, 1988, "Muscle strength and its development: New perspectives," *Sports Medicine* 6:149. Copyright 1988 by ADIS International Ltd.

In addition to varying stimulus frequency, it is also possible to use stimulus waveforms that have different shapes (figure 9.4*c*). The shape of the stimulus waveform can affect the *comfort* associated with neuromuscular electrical stimulation. Commercially available clinical stimulators provide a variety of waveform shapes (e.g., rectangular, triangular, sinusoidal) that can deliver electric current in either a positive (monophasic) or positive-negative (biphasic) pulse. Although no waveform shape is universally preferred, subjects and patients do have individual preferences (Baker et al., 1988; Delitto & Rose, 1986).

Both the size and material of the electrode influence the efficacy of the stimulation. Large electrodes enable the current to be more readily disbursed throughout the muscle, which is essential for large muscles such as quadriceps femoris. Furthermore, the current density (nA/cm^2) is less with large electrodes; this permits a greater amount of current to be passed without damaging the underlying tissue and thus minimizes the discomfort associated with the procedure. In addition to a large surface area, the electrode should have a low impedance—that is, a low resistance to the flow of electric current. Large carbonized-rubber electrodes are particularly effective at meeting these needs (Lieber & Kelly, 1991).

As with neuromuscular electrical stimulation, the artificial generation of action potentials in the peripheral motor system can be achieved with **magnetic stimulation** (Ellaway et al., 1997; Lotz et al., 1989). Commercially available magnetic stimulators can induce an electric field and the flow of current, which will depolarize excitable membranes and generate action potentials. In contrast to the situation with neuromuscular electrical stimulation, however, the threshold for a motor response with magnetic stimulation is lower than that for a sensory response (Panizza et al., 1992). Therefore, magnetic stimulation is less painful than electrical stimulation. Furthermore, the stimulus declines less over distance than with electrical stimulation and thus is more widespread.

Training Effects A training program with neuromuscular electrical stimulation can increase muscle strength in healthy individuals and in patients (Fitzgerald & Delitto, 2006; Gondon et al., 2006; Paillard, 2008; Stevens et al., 2004). For example, Duchateau and Hainaut (1992) examined the adaptations evoked in a hand muscle by a six-week program of neuromuscular electrical stimulation at 100 Hz (figure 9.4*a*). The daily training program comprised 10 sets of twenty 1 s contractions in which subjects acted against a load of about 65% of maximum. The increase in MVC force after the stimulation training (15.5%) was less than that achieved with a comparable amount of training using voluntary contractions (22.2%).

With a substantially different protocol, Banerjee and colleagues (2005) trained old adults five times a week for six weeks with 60 min of neuromuscular electrical stimulation. The stimulation (4 Hz) was administered through large electrodes placed over the quadriceps, gluteal, hamstring, and calf muscles (600 cm^2 per leg). In addition to an increase in MVC of the knee extensor muscles (361 N to 448 N), subjects improved treadmill walking time, 6 min walking distance, and peak oxygen consumption.

In many training studies using neuromuscular electrical stimulation, the maximal intensity of current tolerated by the subjects is about 60 mA (Laughman et al., 1983). In a case study, however, Delitto and colleagues (1989) examined the enhancement of performance when training was supplemented with high-intensity stimulation (200 mA). The subject was a highly motivated and experienced weightlifter. He participated in a 140-day study that was divided into four phases of about one month each. The dependent variable was the maximal load that could be lifted in the clean and jerk, snatch, and squat lifts. In the second and fourth periods of training, the voluntary training regimen (3 h/day) was supplemented with neuromuscular electrical stimulation of the quadriceps femoris (2.5 kHz modulated at 75 Hz was delivered to evoke ten 11 s contractions per thigh in each session, with three sessions each week). The results indicate that performance in all three lifts improved substantially with the intervention (figure 9.5). This study suggests that outcomes depend on the intensity of stimulation used in the protocol (Miller & Thépaut-Mathieu, 1993). However, athletes are generally reluctant to use neuromuscular electrical stimulation as a supplement to training because the imposed activation is applied during isometric contractions rather than during the movements for which they are training (Zatsiorsky, 1995).

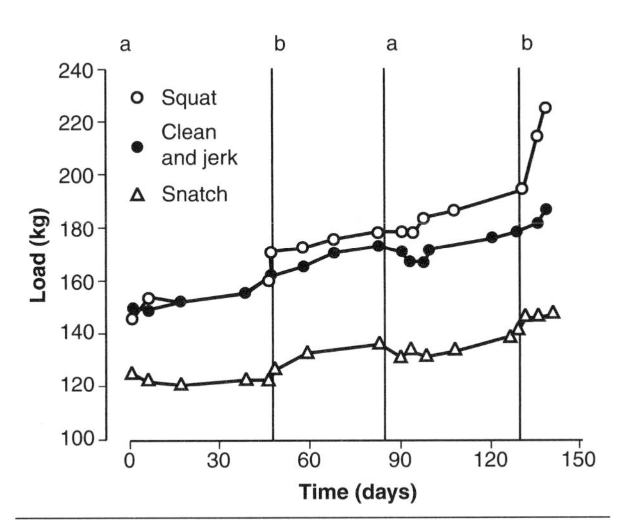

Figure 9.5 Changes in the maximal loads lifted by an experienced weightlifter for three lifts. The weightlifter performed regular training during the *a* periods, and this was supplemented with neuromuscular electrical stimulation of the quadriceps femoris during the *b* periods.

Adapted, by permission, from A. Delitto et al., 1989, "Electrical stimulation of quadriceps femoris in an elite weight lifter: A single-subject experiment," *International Journal of Sports Medicine* 10: 187-197.

One of the negative attributes of neuromuscular electrical stimulation is that it hastens the onset of muscle fatigue. This is a major limitation in the artificial activation of muscle that has lost its innervation, as in individuals with spinal cord injury (Butler et al., 2004; Olive et al., 2003; Scott et al., 2005). Because neuromuscular electrical stimulation can activate only muscle fibers close to the electrodes (Adams et al., 1993*a;* Vanderthommen et al., 1997), the stimulus intensity has to be greater than that associated with producing a comparable force by voluntary activation. Furthermore, the artificial stimulation activates the muscle fibers synchronously, in contrast to the asynchronous pattern that occurs during voluntary contractions. Because the synchronous activation is confined to one part of the involved muscle, the activated fibers must consume more energy to produce the force required for a specific voluntary task. Nuclear magnetic studies have shown that energy metabolism differs when a task is accomplished with voluntary contractions compared with electrically evoked contractions (Ratkevicius et al., 1998*a*; Vanderthommen et al., 1999). When moderate forces (<40% MVC) were sustained with leg muscles, adenosine triphosphate turnover was greater, reliance on glycolysis was greater, intracellular pH was lower, and muscle blood flow was increased when the force was sustained by electrical stimulation (Vanderthommen et al., 2006). The greater energetic demand associated with neuromuscular electrical stimulation causes an earlier onset of muscle fatigue.

Despite the greater energetic demands of neuromuscular electrical stimulation, it is a useful modality in rehabilitation, especially when there is an impairment of muscle activation. Examples are rehabilitation after a surgical procedure on a joint, such as reconstruction of the anterior cruciate ligament or knee joint arthroplasty, and maintenance of function in individuals with spinal cord injury (Fitzgerald et al., 2003; Jacobs & Nash, 2004; Thrasher et al., 2006). After total knee arthroplasty, patients experience a deficit in voluntary activation and pronounced muscle atrophy of quadriceps femoris that can take months to resolve (Mizner et al., 2005*a*). It has been shown that the decreases in voluntary activation (–17%) and muscle cross-sectional area (–10%) explain 85% of the decline in quadriceps strength (Mizner et al., 2005*b*). There is a strong association between recovery of quadriceps strength and functional performance. Neuromuscular electrical stimulation is effective at reducing the deficit in voluntary activation and increasing the rate at which quadriceps strength is regained (Petterson & Snyder-Mackler, 2006; Stevens et al., 2004).

Neuromuscular electrical stimulation is used on individuals with a spinal cord injury for at least two different purposes: to restore function and to reduce the deconditioning that occurs as a result of inactivity. When the stimulation is used to restore function, it is known as **functional electrical stimulation.** A common application of functional electrical stimulation is the restoration of gait in individuals with an incomplete spinal cord injury, where the challenges are to devise a stimulation protocol that minimizes fatigue and elicits the coordinated activation of the required muscles (Field-Fote & Tepavac, 2002). Nonetheless, functional electrical stimulation can be an effective intervention in this population (Hesse et al., 2004; Thrasher et al., 2006).

As a countermeasure to physical deconditioning, neuromuscular electrical stimulation is used to minimize muscle atrophy, the increase in fatigability, the loss of bone mass, and the decline in the cardiovascular system (Fornusek & Davis, 2004; Shields et al., 2006*a, b;* Stoner et al., 2007). For example, stimulation (20 Hz) of tibialis anterior for 1 to 2 h over a six-week interval was sufficient to reduce the fatigability of the muscle, by increasing the proportion of type I muscle fibers, and to increase contraction and relaxation times, but not to increase muscle fiber size or strength (Stein et al., 1992). The adaptations, however, depend on the type of load encountered by the activated muscle. Crameri and colleagues (2004*a*) trained six paraplegic individuals with neuromuscular electrical stimulation (35 Hz) 45 min per day, three days per week for 10 weeks. The quadriceps femoris and hamstring muscles performed an isometric contraction with one leg and lifted a light load with the other leg. Both legs experienced an increase in quadriceps strength, but the increase was greater for the leg that performed the isometric contractions (122% vs. 51%). Furthermore, the fibers in the vastus lateralis muscle of the leg that performed isometric contractions exhibited greater increases in the proportion of type I fibers, fiber cross-sectional area, capillary-to-fiber ratio, and citrate synthase activity. Such results indicate that neuromuscular electrical stimulation can increase muscle size and reduce fatigability in individuals with a spinal cord injury (Sabatier et al., 2006).

Mechanisms of Action The strength gains produced by neuromuscular electrical stimulation involve both peripheral and central adaptations (Enoka, 2006). The peripheral adaptations correspond to the changes experienced by the activated muscle fibers, whereas the central adaptations are related to the sensory consequences of the imposed stimulation. One issue associated with the peripheral adaptations has to do with the motor units that are activated by neuromuscular electrical stimulation. Because large-diameter axons are more easily excited by imposed electric fields (Baratta et al., 1989; Fang & Mortimer, 1991), the classic view is that recruitment order is reversed during electrical stimulation compared with voluntary activation. When the current source is delivered through electrodes attached to the skin, however, the distance between the electrodes and the intramuscular axons reduces the influence of axon diameter on recruitment order. As a result, the activation order of motor units with neuromuscular electrical stimulation is more similar to the normal small-to-large order than to a reversed order (Thomas et al., 2002). For example, Feiereisen and colleagues (1997) compared the recruitment order of pairs of motor units in tibialis anterior during voluntary contractions and contractions evoked with electrical stimulation. Recruitment order for the 249 pairs observed during voluntary contractions was reversed for 5.7% of the pairs; that is, the unit with the higher recruitment threshold was activated first. During the electrically evoked contractions, however, the recruitment order for 35% of 231 pairs was reversed. These results indicate that activation order is relatively random during neuromuscular electrical stimulation.

Neuromuscular electrical stimulation also generates action potentials in the axons of sensory receptors that are propagated to the CNS. The afferent feedback associated with neuromuscular electrical stimulation has been shown to reach the primary somatosensory cortex and is presumed to contribute to the neural adaptations that have been observed after a training program (Gondon et al., 2006; Han et al., 2003; Jubeau et al., 2006; Kimberley et al., 2004). As an example of the central effects of neuromuscular electrical stimulation, Howard and Enoka (1991) compared the influence of stimulation applied to the right quadriceps femoris on the MVC force exerted by the left leg in recreationally active young men and in weightlifters. The stimulation (2.5 kHz modulated at 50 Hz) was delivered at the maximal tolerable intensity through large electrodes (7.5 × 15.5 cm) placed on the proximal and distal ends of the right quadriceps femoris. The stimulation applied to the right thigh increased the left-leg MVC force in both groups of subjects: The increase was 16.2% in the weightlifters and 6.2% in the recreationally active group. The increase in MVC force was attributed to the sensory feedback associated with neuromuscular electrical stimulation of the contralateral homologous muscle. Similarly, Hortobágyi and colleagues (1999) found that six weeks of neuromuscular electrical stimulation training of quadriceps femoris in one leg increased MVC force in the contralateral (untrained) leg by 66%.

When the width of the stimulation pulses is increased to 1 ms, the contribution of the sensory feedback to the force evoked by neuromuscular electrical stimulation is increased (Collins, 2007). The magnitude of this effect was demonstrated by comparing the force evoked with

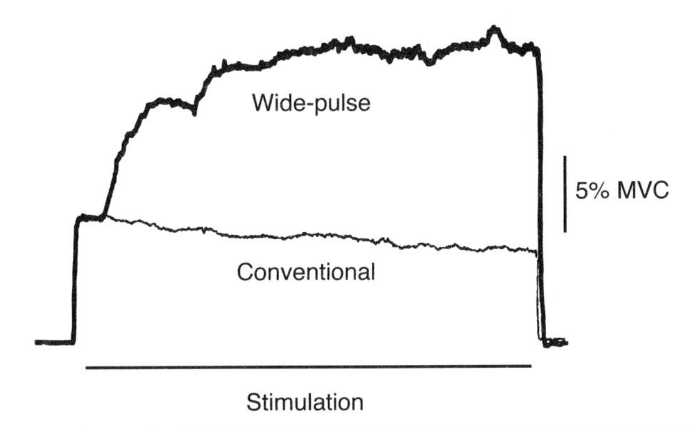

Figure 9.6 Torque evoked in the calf muscles of one subject with normal (25 Hz, 0.05 ms) and wide-pulse (100 Hz, 1 ms) neuromuscular electrical stimulation. Stimulation frequency was constant during each contraction.

Data from Collins et al. 2002.

conventional stimulus parameters (pulse width, 0.05 ms; rate, 25 Hz) and wide-pulse stimulation (1 ms and 100 Hz). The stimulus intensity was adjusted to match the torque produced by both protocols (figure 9.6), and the greater torque elicited with the wide-pulse stimulation was attributed to sensory-activated central mechanisms (Collins et al., 2002). The central origin of the extra force was confirmed by an anesthetic block of the peripheral nerves proximal to the site of stimulation, which prevented the propagation of afferent signals back to the CNS. Collins (2007) suggests that one of the mechanisms likely activated by the afferent feedback is persistent inward current in the motor neurons (figure 5.21). Because the afferent feedback activates different motor units than those recruited directly by the stimulation, wide-pulse stimulation may reduce the fatigue that occurs during neuromuscular electrical stimulation (Baldwin et al., 2006; Klakowicz et al., 2006).

EXAMPLE 9.1

The Russian Protocol

As described by Zatsiorsky (1995), the Russian protocol for neuromuscular electrical stimulation comprised the following routine:

- Carrier signal—sinusoidal or triangular
- Frequency—2.5 kHz
- Modulation—50 Hz
- Duty cycle—50%, with 10 ms on and 10 ms off
- Stimulus amplitude—at least sufficient to evoke a force equal to MVC force, or up to the level of tolerance
- Contraction time—10 s
- Rest between contractions—50 s
- Number of contractions—10 per day
- Training frequency—five days per week for up to 25 training days

When such a protocol was applied to both calf muscles of eight subjects (16-17 years) for 19 days, there were significant improvements in MVC force, vertical jump height, and calf circumference (figure 9.7).

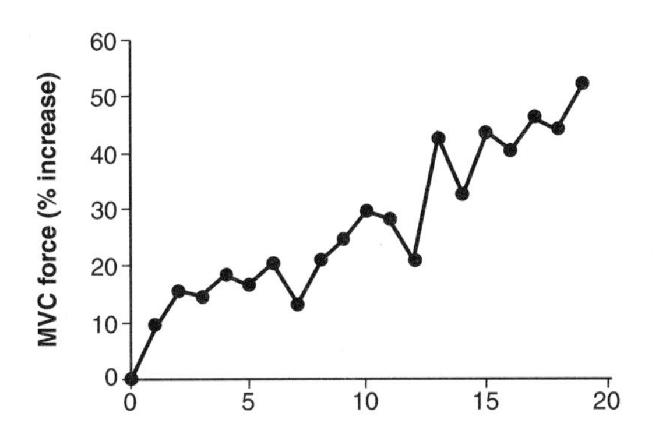

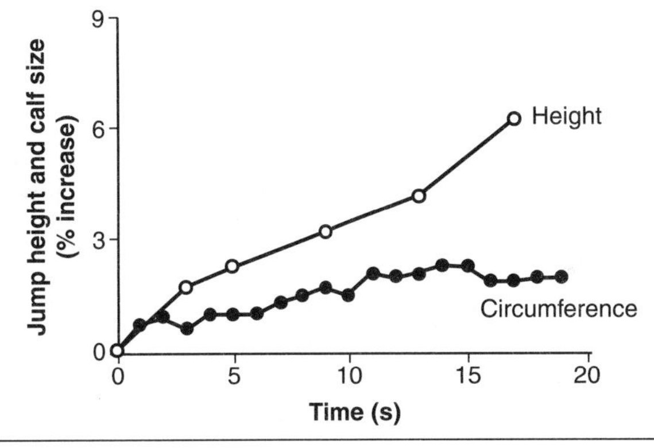

Figure 9.7 Increases in maximal voluntary contraction (MVC) force, vertical jump height, and calf circumference with neuromuscular electrical stimulation.

Data from Zatsiorsky and Kraemer 2006.

Vibration Training

Because the application of vibration to a muscle or its tendon can evoke a reflex that increases muscle force (Burke et al., 1976; Eklund & Hagbarth, 1966), vibration has been incorporated into some training programs (Cardinale & Bosco, 2003; Jordan et al., 2005; Luo et al., 2005). The vibration can be applied either directly to a muscle with a handheld device or indirectly by an individual standing on or holding on to a vibration source. The amplitude and frequency of the vibration experienced by the muscle are more similar to those of the source with the direct-application technique. Although prolonged exposure to vibration can cause pathology, short-term application can enhance neuromuscular performance (Cardinale & Wakeling, 2005).

A few seconds of vibration applied directly to the muscle has only a minor effect on MVC force (Bongiovanni & Hagbarth, 1990; Ribot-Ciscar et al., 2003). In contrast, minutes of vibration depresses MVC force and requires greater motor unit activity to achieve a specific target force, but increases the amplitude of the short-latency stretch reflex during a submaximal contraction (Shinohara et al., 2005; Steyvers et al., 2003; Yoshitake et al., 2004). Although vibration can increase motor unit activity in a fatigued muscle, it reduces the time a limb position can be maintained with a submaximal contraction (Bongiovanni & Hagbarth, 1990; Griffin et al., 2001; Mottram et al., 2006*b*). In contrast, vibration applied to a limb during a movement can increase the 1-RM load and the maximal power produced by the elbow flexor muscles, and the effects have been shown to be greater in elite athletes (Issurin & Tenenbaum, 1999; Liebermann & Issurin, 1997).

Vibration training has produced a range of effects that are likely due to differences in the training programs (Cardinale & Wakeling, 2005; Jordan et al., 2005; Moran et al., 2007; Poston et al., 2007). For example, de Ruiter and colleagues (2003) examined the adaptations induced in the knee extensor muscles by 11 weeks of training (three times per week) with whole-body vibration (30 Hz, 8 mm amplitude). The training involved five to eight sets of 60 s vibration with a 60 s rest between sets. In comparison with values for a control group, whole-body vibration did not increase MVC force, the maximal rate of force development, or the peak height reached during a countermovement jump. However, the training did augment the rate of increase in force evoked with neuromuscular electrical stimulation. Similarly, Kvorning and colleagues (2006) found that improvements in performance after nine weeks of training with squat lifts were not further increased when the lifts were performed on a vibration platform (20-25 Hz).

In contrast, Delecluse and colleagues (2003) found that whole-body vibration can produce strength gains comparable to those achieved with a weightlifting program in previously untrained women. One of the unique features of this study was that the intensity of the vibration training increased across the 12-week program (table 9.1). The improvements in performance of the knee extensor muscles were similar for the subjects who trained with vibration and those who performed weightlifting exercises: MVC force increased by 16.6% and 14.4%, respectively, and peak torque during four isokinetic contractions increased by 9.0% and 6.2%. The vibration group also experienced a 7.6% increase in the peak height reached during a countermovement jump, whereas jump height did not change for the weightlifting group. Roelants and colleagues (2004) obtained similar results for older women (58 to 74 years) who trained with either whole-body vibration or weightlifting exercises. Similarly, whole-body vibration can improve strength and function in adults with cerebral palsy, with some effects superior to those from strength training (Ahlborg et al., 2006).

Table 9.1 Volume and Intensity of Whole Body Vibration at the Start and End of a 12-Week Training Program

	Start	End
Volume		
Longest vibration without rest (s)	30	60
Total duration of vibration (min)	3	20
Sets of each exercise	1	3
Number of exercises	2	6
Intensity		
Rest interval between exercises (s)	60	5
Vibration amplitude (mm)	2.5	5.0
Vibration frequency (Hz)	35	40

Data from Delecluse et al., 2003.

Studies on vibration training indicate that it can improve neuromuscular performance. However, much remains to be learned about the vibration characteristics (e.g., amplitude, frequency, duration of exposure) that are most effective at eliciting performance gains. Also, vibration training may be more appropriate for some types of actions than others.

Loading Techniques

Several issues are important to consider when one is deciding how to vary the load in order to induce strength gains; these include progressive-resistance exercises, the magnitude of the load, and the way in which the load varies over the range of motion.

Progressive-Resistance Exercises

The load used in strength training can also be manipulated in a number of ways. DeLorme (1945) proposed varying the load systematically from one set of repetitions to another; this technique is known as **progressive-resistance exercises.** DeLorme's technique involves three sets of repetitions of an exercise in which the load increases with each set. In conventional weight training programs, this typically means increasing the weight of the barbell from one set to another. However, the torque exerted by muscle depends on the size of the moment arm and the speed of the movement in addition to the magnitude of the external load (the amount of the barbell weight). As a consequence, one can accomplish progressive-resistance exercises without changing the size of the load (barbell weight) simply by varying moment arm length or the speed of the movement. For example, figure 9.8*a* shows an individual in the middle of a bent-knee sit-up. This exercise entails use of the abdominal and hip flexor muscles (Andersson et al., 1996) to raise and lower the upper body. As the four positions shown in figure 9.8*b* indicate, there are four variations of the exercise that involve placing the arms in different positions. The change in the arm position does not alter the weight of the upper body ($\boldsymbol{F}_w$), but it does shift the center of gravity of the upper body toward the head and thus increases the moment arm of the upper body weight relative to the hip joint (dotted line in figure 9.8*b*). The net effect is an increase in the force that the muscles must exert ($\boldsymbol{F}_m$) to accomplish the movement (Cordo et al., 2006).

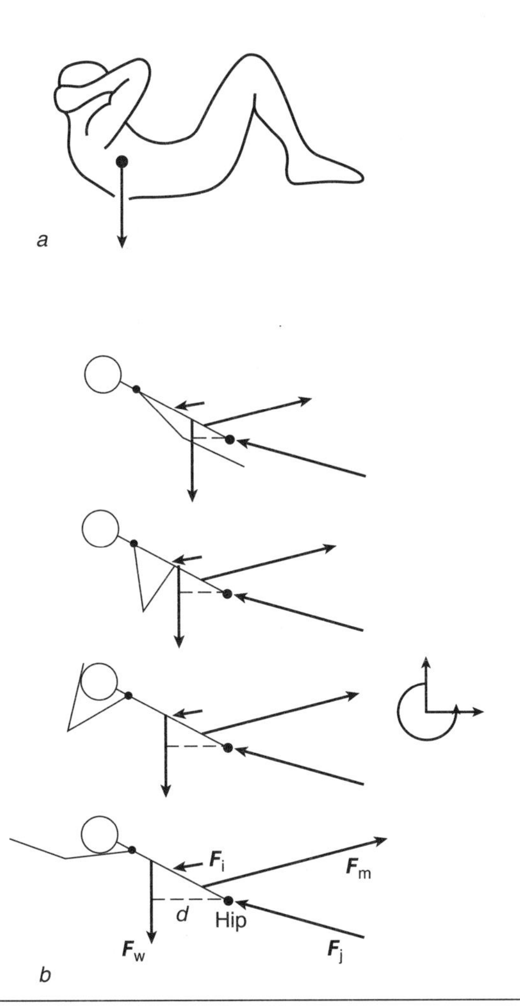

Figure 9.8 Subject performing a bent-knee sit-up. *(a)* Whole-body figure. *(b)* Changing arm position increases the moment arm (*d*) of the upper body weight vector ($\boldsymbol{F}_w$) with respect to the hip joint, and thus a greater muscle force ($\boldsymbol{F}_m$) is required to perform the task. $\boldsymbol{F}_i$ and $\boldsymbol{F}_j$ represent the force due to the intra-abdominal pressure and the joint reaction force, respectively.

This example (figure 9.8) underscores the general point that the demands imposed on a muscle vary as the posture assumed by the individual changes. This includes changing the relative angles between adjacent body segments and altering the orientation of the body relative to the force of gravity. For example, Zatsiorsky (1995) found that varying the forward lean of the trunk during the squat exercise had a substantial effect on the resultant muscle torque about the knee joint. Similarly, the location in the range of motion at which the load torque is maximum differs when an exercise is performed in a supine compared with a standing position. This feature of progressive-resistance exercises can be used to enhance the diversity of a training program.

Magnitude of the Load

When defining the concept of progressive-resistance exercises, DeLorme (1945) proposed that each of the three sets should include 10 repetitions. Currently, most strength training programs advocate using from one to eight repetitions in a set. The number of repetitions performed in a set generally determines the load that is lifted; as the load becomes heavier, fewer repetitions are performed. Heavier loads are assumed to increase the stress imposed on the involved muscles. However, this occurs only if the kinematics of the movement remain constant across loads. If an individual performs two squats, for example, one with a barbell weight of 500 N and the other with a weight of 1000 N, then the torque exerted by the knee extensors with the heavier load will be about twice that for the lighter load if the movement is performed in exactly the same way each time. When subjects lifted loads of 40%, 60%, and 80% of the 4-RM load, the kinematics of the movement were altered (e.g., through flexing more at the hip during a squat), and the resultant muscle torque did not increase in proportion to the load (Hay et al., 1980, 1983). In weightlifting and in rehabilitation, therefore, it is necessary to focus on the form of the movement to ensure that the appropriate muscles are involved in the task.

The American College of Sports Medicine recommends that an exercise session be organized so that large muscles are exercised before small muscles, exercises involving multiple joints be performed before those that focus on a single joint, and high-intensity exercises precede those involving a lesser intensity (Kraemer et al., 2002). The loads used in each exercise are specified in terms of the maximal number of repetitions that can be completed. According to the scheme proposed by Sale and MacDougall (1981), a 5- to 6-RM load corresponds to about 85% to 90% of the maximum load (figure 9.9).

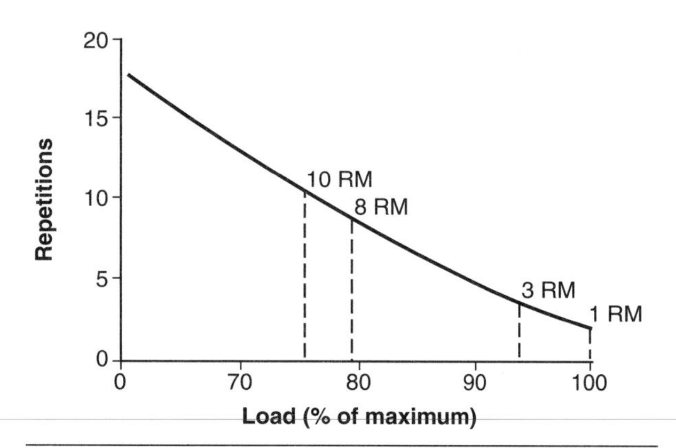

Figure 9.9 The relation between the number of repetitions in an exercise set (RM = repetition maximum) and the magnitude of the load as a percentage of maximum.

Adapted, by permission, from D.G. Sale and D. MacDougall, 1981, "Specificity in strength training: A review for the coach and athlete," *Journal of Applied Sports Sciences* 6: 90. Copyright 1981 by the Canadian Association of Sport Sciences.

Novices should begin with loads of 8- to 12-RM and perform 1 to 2 s shortening and lengthening contractions. When the individual can perform one or two more repetitions of an exercise than the target number, the load should be increased by 2% to 10%. Intermediate and advanced lifters use a wider range of loads (1- to 12-RM), with a focus on heavy loads (1- to 6-RM) and 3 min between sets. Moderate-volume training with various combinations of repetitions and sets produces greater improvements in strength activities compared with low- and high-volume training (Gonzalez-Badillo et al., 2005). When the goal of a lifting program is to maximize muscle hypertrophy, the program should comprise greater numbers of repetitions and sets of exercises.

Meta-analyses confirm that the training program to achieve maximal strength gains depends on the training history of an individual (Peterson et al., 2005). (1) Untrained individuals should use loads of 60% of maximum, perform four sets of exercises per muscle group, and train three days per week; (2) recreational lifters should use loads that correspond to 80% of maximum, perform four sets of exercises per muscle group, and train two days per week; (3) athletes are advised to use loads of 85% of maximum, perform eight sets of exercises per muscle group, and train two days a week. However, the frequency recommended by the American College of Sports Medicine is two to three days per week for novice and intermediate lifters and four to five days per week for advanced lifters (Kraemer et al., 2002). Furthermore, the number of sets that should be performed in a session remains controversial (Carpinelli, 2002; Galvao & Taaffe, 2005; Hass et al., 2000).

An alternative to manipulating the magnitude of the load during weightlifting is to vary the speed of the movement. Some investigators have compared the strength gains achieved with conventional weightlifting exercises to those that result from rapid movements (often called explosive exercises, ballistic contractions, or dynamic training). For example, Duchateau and Hainaut (1984) compared the adaptations exhibited by a hand muscle after training for 12 weeks with either isometric MVCs or dynamic contractions with a load that was one-third of maximum. Both training programs increased the maximal force evoked with electrical stimulation (100 Hz), but the gain was greater with the isometric contractions (20% vs. 11%). The increase in the rate of force development during the electrically evoked contractions, however, was greater after training with the dynamic contractions (31% vs. 18%). The dynamic training improved the speed of shortening contractions against light loads, whereas the improvement after isometric training was with heavy loads. Similarly, strength training with a heavy load increases the rate of force development during a maximal isometric contraction (Aagaard et al., 2002*a*). Due to the different adaptations evoked by training with slow and rapid contractions, programs that incorporate both modalities have been shown to be effective (Bastiaans et al., 2001; Häkkinen et al., 1998*a*).

Constant and Variable Loads

With the exception of isometric exercises, strength training activities involve changing the length of an active muscle over a prescribed range of motion. Because the torque that a muscle can exert changes with joint angle (figures 6.38 and 6.43), it is necessary to indicate not only the size of the load but also the way in which the load changes over the range of motion. The main distinction is whether the load applied to the limb remains constant or varies over the range of motion; these loading strategies are known as **constant-load** and **variable-load** training, respectively. In general, free weights (i.e., barbells and dumbbells) provide constant loads whereas machines (e.g., Nautilus, Universal) provide loads that vary over a range of motion.

When lifting and lowering a load held in the hands by flexing and then extending the elbow joint (figure 9.10*a*), the magnitude and direction of the load vector ($\mathbf{F}_l$) remain constant over the range of motion. With some machines, however, the load held in the hands varies over the range of motion (figure 9.10*b*) because the load provided by the machine is a torque that corresponds to the product of the weight in the weight stack and the moment arm of the cam. Although the weight remains constant for the exercise, the moment arm varies over the range of motion because the cam is not circular. The moment arm is the perpendicular distance from the axis of rotation of the cam to the cable that connects to the weight stack (dashed arrow in the upper right diagram of figure 9.10*b*).

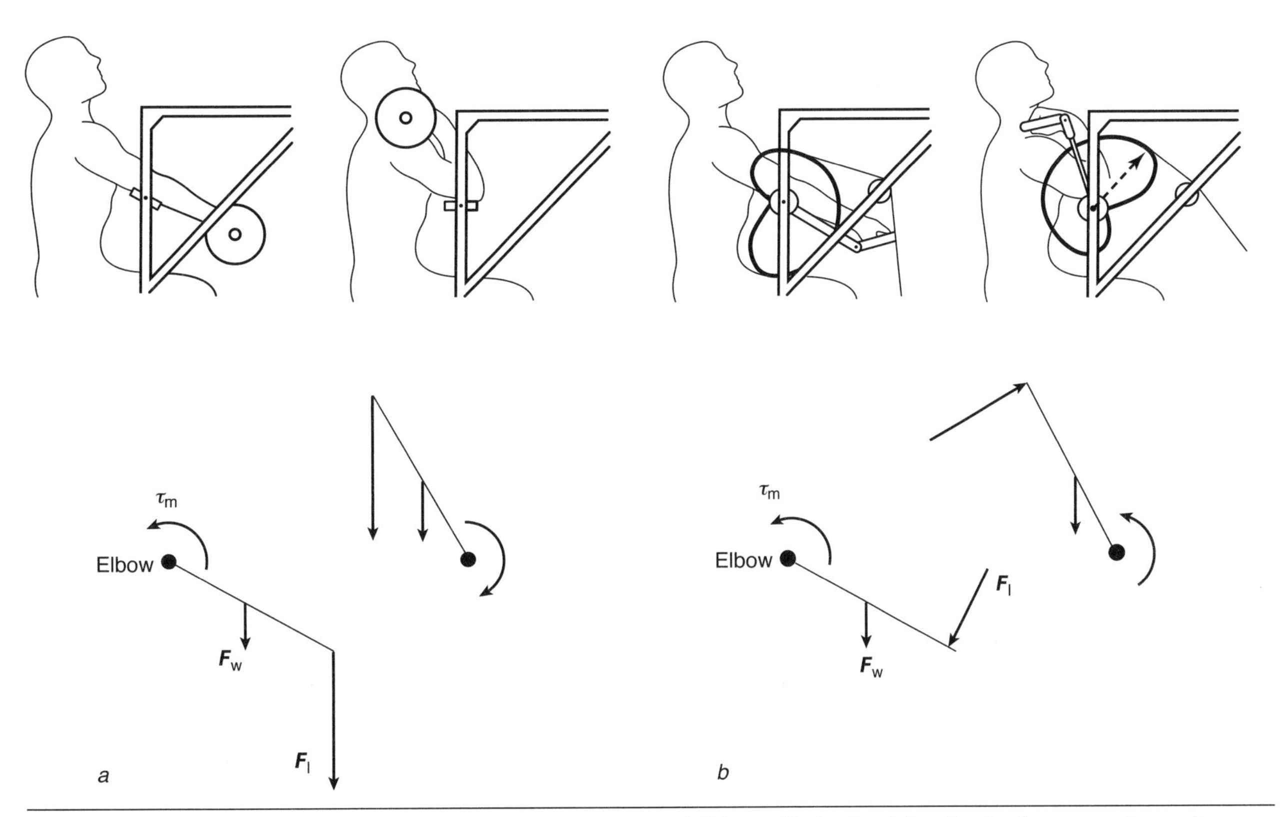

Figure 9.10 Loading conditions associated with *(a)* constant- and *(b)* variable-load training for the forearm curl exercise. τ_m = resultant muscle torque; F_w = weight of the forearm and hand; F_l = magnitude of the load.

The variable load is shown in the free body diagram as a change in the length of the load vector.

The effect of the two loading techniques on the resultant muscle torque over the range of motion is shown in figure 9.11. The resultant muscle torque about the elbow joint (τ_m) was determined for the constant- and variable-load tasks (figure 9.10) and compared with the MVC force determined with isometric contractions at several joint angles. The load used for the constant- and variable-load tasks was 60% of 4-RM. The comparison indicates that the resultant muscle torque during the variable-load exercise paralleled the change in MVC force, whereas that for the constant load did not. Hence, the involved muscle experienced a similar relative stress throughout the range of motion with the variable load.

The reason for the difference between the two loading conditions is that the load applied to the hands is not the same as the load about the elbow joint. Specifically, the load torque about the elbow joint is the product of the load at the hands and the moment arm from the hands to the joint (figure 9.12). For the constant-load exercise, the moment arm varies from zero when the forearm is vertical to a maximum when the forearm is horizontal. In contrast, the direction of the variable load is approximately perpendicular to the arm over the range of motion, which means that the moment arm for the load from the hands to

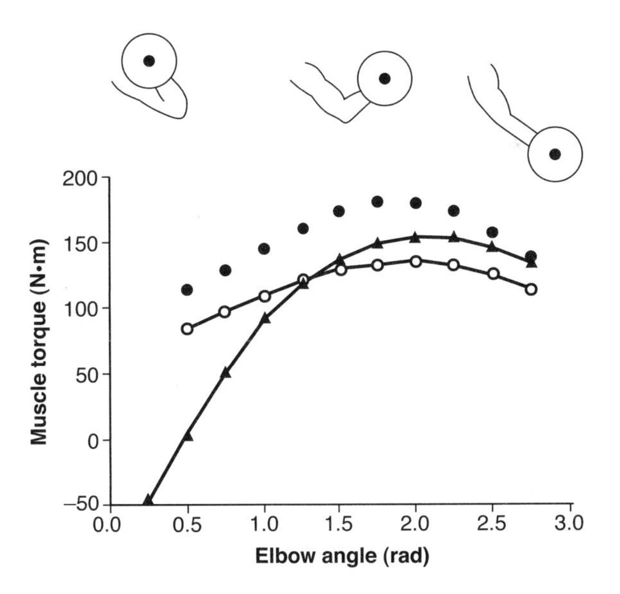

Figure 9.11 Resultant muscle torque about the elbow joint during a forearm curl exercise with a barbell (filled triangles) and a Nautilus machine (open circles) in comparison with the MVC torque (filled circles) that can be exerted over the range of motion.

Adapted from F. Smith, 1982, "Dynamic variable resistance and the universal system," *National Strength & Conditioning Association Journal* 4:14-19.

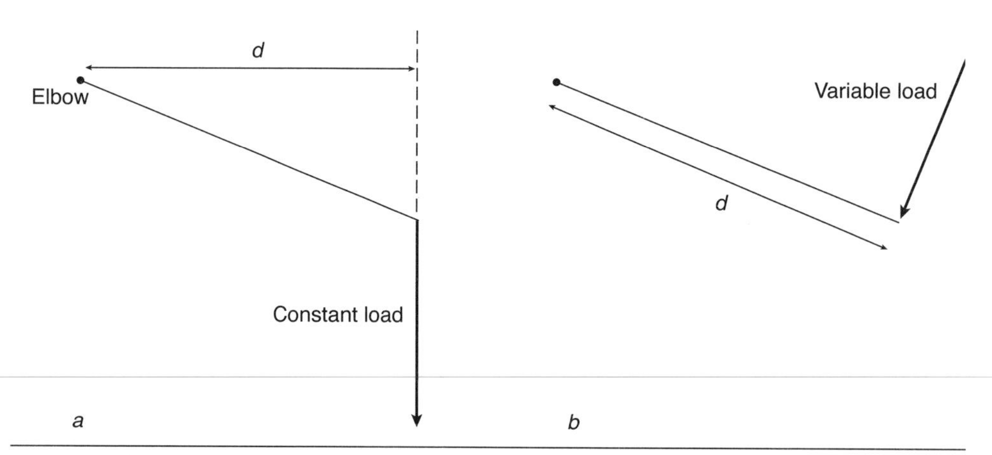

Figure 9.12 The load torque about the elbow joint with *(a)* a constant load or *(b)* a variable load held in the hands.

the joint changes very little. Thus, contrary to the names, the load torque about the elbow joint varies more with the constant load than it does for the variable load.

EXAMPLE 9.2
A Cam With Two Moment Arms

An alternative design for variable-load devices involves two moment arms, one related to the cable that goes to the weight stack and another related to the cable that goes to the lifter. In such designs, the two moment arms vary in a reciprocal manner, as dictated by the shape of the cam (figure 9.13). The variation in moment arms depends on the strength capability of an average adult throughout a given range of motion. At the location of greatest strength (figure 9.13*a*), the lifter's moment arm (*d*) is less than the load's moment arm (*e*). Conversely, the lengths of the two moment arms are reversed in regions of least strength (figure 9.13*b*). With this scheme, the variable load (weight stack × moment arm) is greatest where the lifter is strongest. This enables the machine to provide a variable load that matches the strength capability of the average adult throughout a prescribed range of motion.

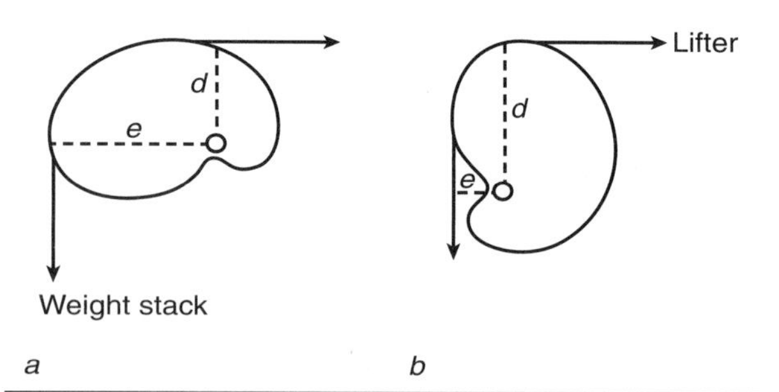

Figure 9.13 A variable-load cam that controls two moment arms. *(a)* Due to the shape of the cam, the moment arm for the cable going to the lifter (*d*) is less than that for the cable to the weight stack (*e*) in this position. *(b)* The converse relation occurs in a different location in the range of motion.

Data from Stone and O'Bryant 1987.

EXAMPLE 9.3
Beginning Movement Load Training

Motivated by the concept of specificity of training and the poor correspondence between standard weightlifting exercises and natural movements, Yasushi Koyama has designed a set of exercise equipment that has two unique features: (1) a gear system that varies the load throughout the range of motion so that the load is maximal when the relaxed muscle begins a shortening contraction and (2) multiple degrees of freedom to facilitate nonplanar motion. Beginning Movement Load (BML) training on these machines involves exercises in which the movement requires rotations about multiple axes, an activation sequence that progresses from proximal to distal muscles, stretch of the relaxed principal muscles involved in the action before they perform a shortening contraction, and minimal amounts of coactivation. Most BML exercises include a dodge movement, which comprises a longitudinal rotation about the twist axis of the involved body segment. Similar to the diagonal movement patterns used in proprioceptive neuromuscular facilitation exercises, the dodge movement is used to facilitate the actions of the principal muscles.

Some of these features are evident in the lat pull-down exercise shown in figure 9.14. In contrast to the conventional machine that permits displacement only in a frontal plane, the BML machine also allows both pronation-supination of the forearm (dodge movement) and forward-backward displacement of the arm in a horizontal plane. Figure 9.14*a* and *b* shows the position of the arms at the beginning of the exercise as performed on the conventional and BML machines. Compared to what occurs with the conventional machine, the additional degrees of freedom on the BML machine increase the vertical displacement of the wrist, permit medial-lateral displacement of the wrist, and involve a reduced rotation about the elbow joint (figure 9.14*c*). Although not shown, the electromyographic (EMG) activity of the latissimus dorsi, anterior and posterior deltoid, serratus anterior, trapezius, and biceps and triceps brachii differs when the exercise is performed on the two machines. Beginning Movement Load training offers the possibility of training with exercises that may transfer more effectively to activities of daily living.

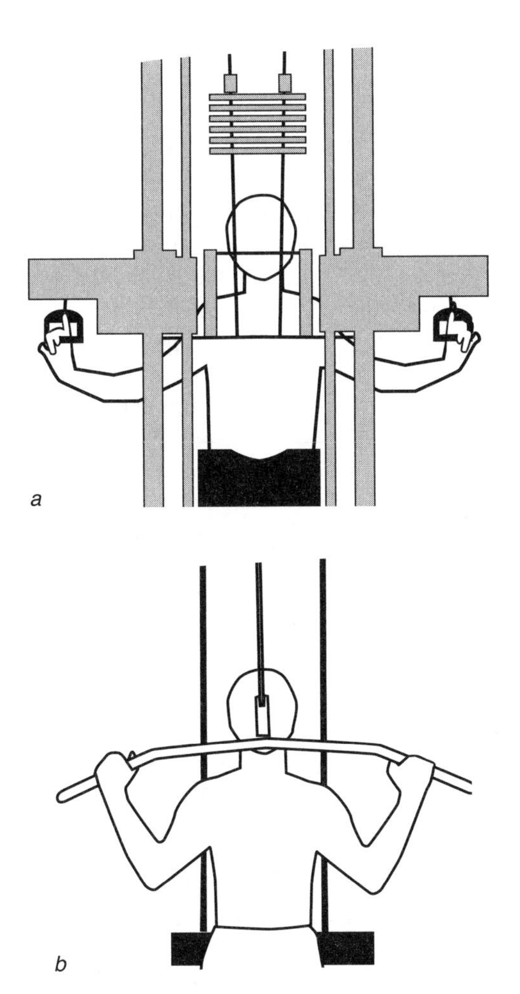

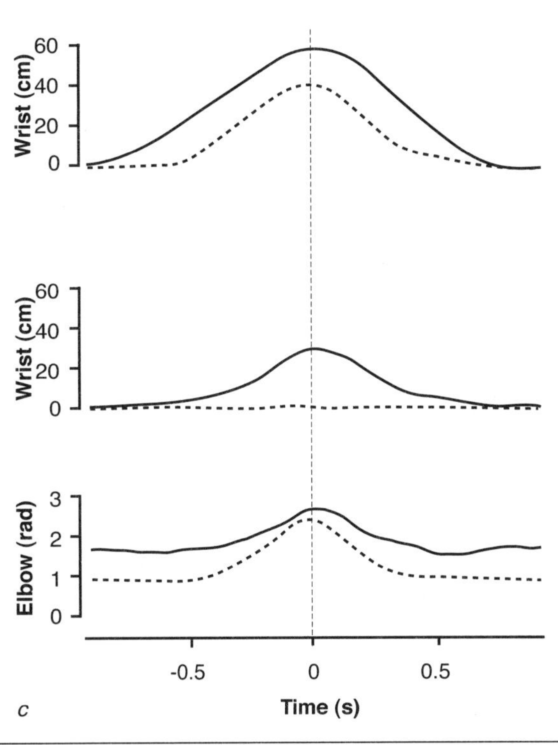

Figure 9.14 Comparison of the lat pull-down exercise performed on two machines. The position of the arms at the beginning of the exercise as performed on *(a)* a conventional machine and *(b)* a Beginning Movement Load (BML) machine. Both parts of the figure show a back view of the weightlifter. The palms are facing backward on the BML machine. *(c)* Selected kinematics of the arm during one repetition of the exercise performed on the conventional machine (dotted line) and the BML machine (solid line). The data represent the vertical displacement of the wrist (top panel), the medial-lateral displacement of the wrist (middle panel), and the angular displacement about the elbow joint (bottom panel).

Data provided by Yasushi Koyama, M.S.

Neuromuscular Adaptations

Changes in muscle strength are generally attributed to either neural or muscular factors. The principal muscular factor is muscle size (Lieber & Fridén, 2000). As indicated in figure 3.4, variation in the cross-sectional area of muscle between individuals can account for about 50% of the differences in strength. In addition, the efficacy of force transmission from the sarcomeres to the skeleton varies among individuals due to differences in the cytoskeleton and associated connective tissue structures (Magnusson et al., 2003*b;* Narici & Maganaris, 2006). Two lines of evidence, however, suggest that neural factors are also important (Semmler & Enoka, 2000). These are the dissociation between changes in muscle size and strength and the specificity of the strength gains.

Dissociated Changes in Muscle Size and Strength

Two examples of dissociated changes are the timing of the adaptations and the strength gains that can be achieved without muscle activation. The timing example occurs during the early stages of a strength training program when the increase in EMG precedes the muscle hypertrophy (Akima et al., 1999; Häkkinen et al., 2001*a, b;* Rutherford & Jones, 1986). For example, figure 9.15 shows the time course of changes in the cross-sectional area of quadriceps femoris (as determined by magnetic resonance imaging, MRI), integrated EMG, and the maximal isometric force in response to 60 days of training and 40 days of detraining (Narici et al., 1989). Similarly, eight weeks of training increased the load that subjects could lift by 100% to 200%, but there were no changes in the cross-sectional areas of muscle fibers (Staron et al., 1994); and knee extensor strength increased before changes occurred in either the cytoskeletal proteins or fiber cross-sectional area (Woolstenhulme et al., 2006).

Dissociated changes in muscle size and strength have also been demonstrated with protocols that evoke an increase in muscle strength without even subjecting the muscle to physical training. This occurs, for example, when subjects train one limb and experience a strength gain in the contralateral limb. This effect is known as **cross-education** (Hortobágyi, 2005; Lee & Carroll, 2007; Semmler & Enoka, 2000; Zhou, 2000). Most studies on cross-education have shown that when the muscles in one limb participate in a strength training program, the homologous muscles in the contralateral limb experience an increase in strength despite the absence of activation during the training program and no change in muscle fiber characteristics (figure 9.16). A cross-education effect occurs even when the training program involves neuromuscular electrical stimulation (Hortobágyi et al., 1999). After a survey of this literature, Munn and colleagues (2004) concluded that the strength gain in the contralateral limb averaged 7.8%, which corresponded to 35.1% of the increase achieved in the trained limb. Furthermore, they found that cross-education was present only after six weeks of training with three sets at a load of 6- to 8-RM, and there was a trend for fast contractions to produce a greater effect than slow contractions (Munn et al., 2005). Cross-education is not accompanied by a change in spinal reflex excitability as tested with the Hoffmann (H) reflex (Hortobágyi et al., 2003; Lagerquist et al., 2006). Rather, cross-education is likely mediated either by adaptations in cortical and descending pathways (Hortobágyi, 2005) or by adaptations in cortical, subcortical, and spinal levels (Carroll et al., 2006).

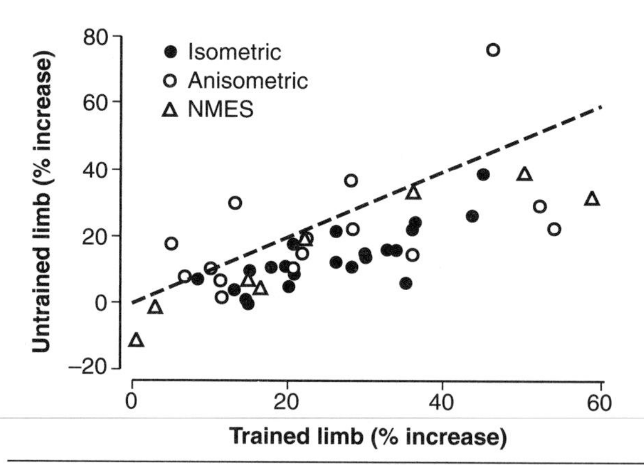

Figure 9.16 Results from studies of cross-education show the range of associations between increases in strength for trained and untrained limbs. Strength, measured as the MVC force, was increased by voluntary contractions (isometric or anisometric) or by neuromuscular electrical stimulation (NMES) in several different muscles. Most values lie below the line of identity (dashed line), which indicates that the increase in strength in the trained limb was greater than that in the untrained limb.

Adapted, by permission, from R.M. Enoka, 1988, "Muscle strength and its development: New perspectives," *Sports Medicine* 6:155. Copyright 1988 by ADIS International Ltd.

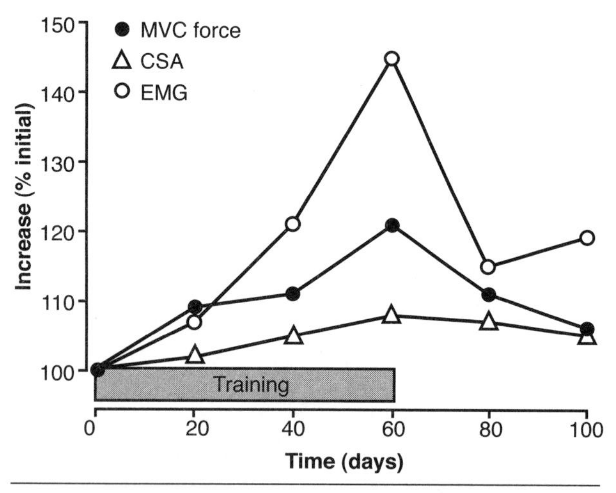

Figure 9.15 Changes in cross-sectional area (CSA) of the quadriceps femoris, integrated EMG of vastus lateralis during a maximal contraction, and the maximal voluntary contraction (MVC) force during isokinetic training and detraining.

Data from Narici et al., 1989, "Changes in force, cross-sectional area, and neural activation during strength training and detraining of the human quadriceps," *European Journal of Applied Physiology* 59: 313.

A related effect can occur with imagined contractions. A four-week program of training with imagined MVCs for a hand muscle (abductor digiti minimi) increased the MVC force by 22% compared with a 30% increase for subjects who actually performed contractions during training (Yue & Cole, 1992). Furthermore, the cross-education effect was 11% for the subjects who performed imagined contractions and 14% for those who performed actual contractions. Similarly, Zijdewind and colleagues

(2003) reported a 36% increase in MVC torque of the plantarflexor muscles after seven weeks of training with imagined contractions. When a similar protocol was performed on the elbow flexor muscles with subjects who were able to activate the muscles maximally by voluntary command, however, the imagined contractions did not increase strength (Herbert et al., 1998). The efficacy of imagined contractions, therefore, appears to differ across muscles, but can have a potent effect on the excitability of spinal reflex pathways (Li et al., 2004).

Specificity of Strength Gains

Although strength training can increase the size of a muscle, the maximal force that is exerted varies across tasks (Aagaard et al., 1996; Almåsbakk & Hoff, 1996; Folland et al., 2005; Wilson et al., 1996). In general, the improvement in strength is greatest when the training and testing modalities are the same. For example, 12 weeks of training that involved raising and lowering a load with the knee extensor muscles produced ~200% increases in 1-RM load but only ~20% increases in the isometric MVC force (Rutherford & Jones, 1986). Similarly, 20 weeks of training the elbow flexor muscles with a hydraulic resistance device increased the cross-sectional area of the involved muscles, but there were task and sex differences in strength gains (O'Hagan et al., 1995). The increases in peak force on the hydraulic device at the speed used in training, as well as the increases in 1-RM load, were about 50% for men and 120% for women. In contrast, the peak torque exerted on an isokinetic dynamometer at four angular velocities did not change by much (<25% increase).

The specificity of the strength gains is most pronounced for tasks that require more learning, such as less constrained movements (Chilibeck et al., 1998; Rutherford & Jones, 1986; Wilson et al., 1996), those involving voluntary activation compared with electrical stimulation (McDonagh et al., 1983; Young et al., 1985), and those involving lengthening contractions (Higbie et al., 1996). For example, 12 weeks of training the knee extensor muscles on an isokinetic dynamometer produced greater specificity in the subjects who trained with lengthening contractions than in those who trained with shortening contractions (figure 9.17). In addition, cross-education was greater in subjects who used lengthening contractions (Farthing & Chilibeck, 2003; Hortobágyi et al., 1997) and even greater when the lengthening contractions were evoked with neuromuscular electrical stimulation (Hortobágyi et al., 1999).

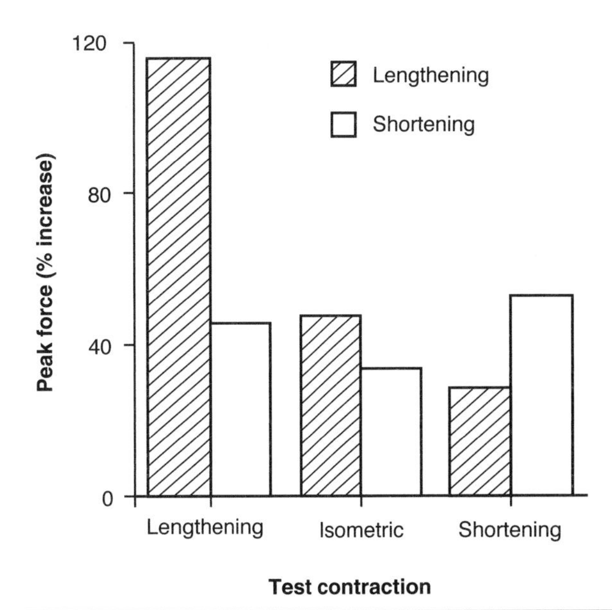

Figure 9.17 Specificity of the increases in peak force exerted by the knee extensor muscles on an isokinetic dynamometer. Some subjects trained with shortening contractions (open columns), whereas others trained with lengthening contractions (hatched columns). Peak forces were measured for both groups of subjects while they performed lengthening, isometric, and shortening contractions.

Data from Hortobágyi et al. 1996*b*.

Neural Factors

Based on the time course of the adaptations and the dissociation between changes in muscle size and strength, it appears possible to obtain an increase in strength without an adaptation in the muscle, but not without an adaptation in the nervous system. Nonetheless, it has proved difficult to identify specific mechanisms that underlie increases in strength (Carroll et al., 2002; Jensen et al., 2005). Figure 9.18 indicates some potential sites within the nervous system where the adaptations may occur (Adkins et al., 2006). The adaptations can be classified as changes in activation maximality, coordination of the muscles involved in the task, and plasticity in the spinal cord.

Activation Maximality As represented by sites 1, 6, and 7 in figure 9.18, one neural adaptation that might contribute to strength gains is an increase in the neural drive to the muscle during an MVC (Duchateau et al., 2006; Fisher & White, 1999). Researchers have examined this possibility by measuring changes in the EMG, by testing voluntary activation with the twitch interpolation technique (figure 8.10), and by determining the ratio of the evoked tetanic force to the MVC force. These methods have produced mixed results. For example, although the surface EMG during an MVC often increases after strength training (Aagaard et al., 2000*b;* Häkkinen et al., 1998*b*; Moritani & de Vries, 1979), this has not been a consistent result (Carolan & Cafarelli, 1992; Keen et al., 1994). Given what we know about the influence of amplitude cancellation on the surface EMG (Keenan

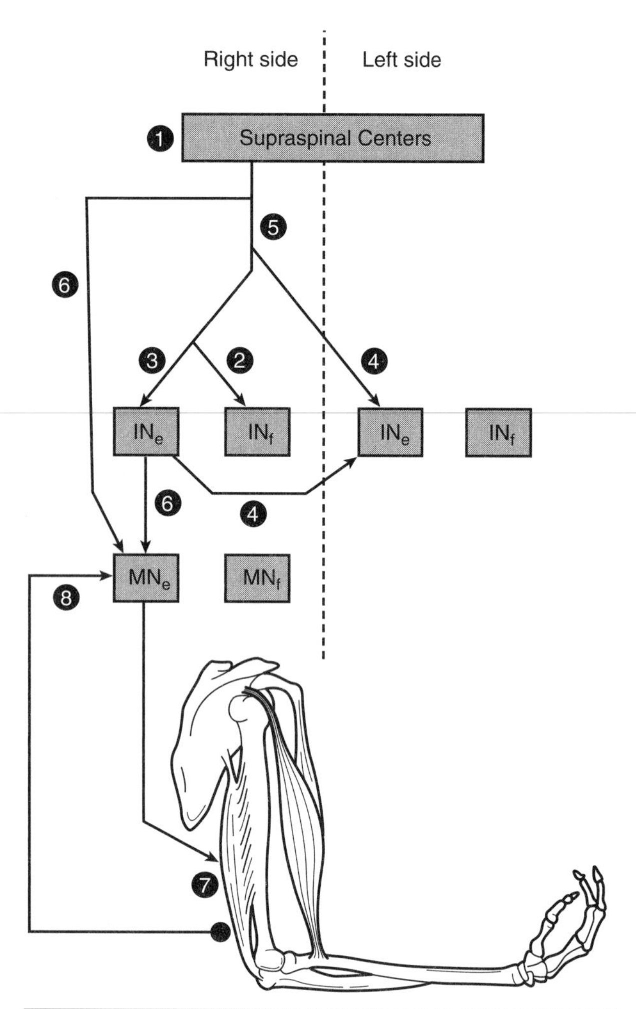

Figure 9.18 Potential sites of adaptations in the nervous system that might contribute to increases in strength: (1) enhanced output from supraspinal centers, as suggested by findings with imagined contractions; (2) reduced coactivation of antagonist muscles; (3) greater activation of agonist and synergist muscles; (4) enhanced coupling among spinal interneurons (IN) that produces cross-education; (5) changes in descending drive that reduce the bilateral deficit; (6) shared input to motor neurons that increases motor unit synchronization; (7) greater muscle activation (EMG); and (8) heightened excitability and altered connections onto motor neurons (MN). The scheme indicates potential interactions between limbs (right and left sides) and between extensor (e) and flexor (f) muscles when the elbow extensor muscles are the agonists.

Data from Semmler and Enoka 2000.

et al., 2005), it is not surprising that there is a variable association between the amplitude of the surface EMG and strength gains.

The assessment of voluntary activation with the twitch interpolation technique indicates that most individuals are able to achieve full activation of a muscle in about one out of four attempts during maximal isometric and shortening contractions (G.M. Allen et al., 1995; De Serres & Enoka, 1998; Gandevia et al., 1998). A similar finding was obtained when responses were evoked in a muscle with transcranial magnetic stimulation during an MVC (Todd et al., 2003*b*). In contrast, voluntary activation is less than maximal when the superimposed stimulus involves brief trains of shocks (Kent-Braun & Le Blanc, 1996; Stevens et al., 2003) and during lengthening contractions (Amiridis et al., 1996; Pinninger et al., 2000; Westing et al., 1990). There are minimal changes in voluntary activation after strength training when it is assessed with the twitch interpolation technique (Harridge et al., 1999; Knight & Kamen, 2001).

An alternative approach is to compare the force exerted during an MVC with that evoked by maximal electric stimulation. Duchateau and Hainaut (1988) found that six weeks of strength training with a hand muscle (adductor pollicis) increased MVC force (22%) more than it increased tetanic force (15%). The 7% difference suggests that the training increased the capacity of the nervous system to activate the motor unit pool for the hand muscle. However, the training may also have improved the contribution of the synergist muscles to supporting the hand muscle during the MVC. There remains some uncertainty, therefore, as to whether activation is maximal during the measurement of strength, and it likely depends on the task being performed.

Coordination Of all the potential neural factors, perhaps the most significant is the coordination of activity among the muscles involved in the strength task (Carson, 2006; Carroll et al., 2001*a*). This is the most likely explanation for the specificity of strength gains. For example, Rutherford and Jones (1986) trained a group of subjects for 12 weeks with a bilateral knee extension task. The subjects used a 6-RM load. Both the MVC force and the training load were recorded over the course of the training program. The relative changes in MVC force and training load are shown in figure 9.19. The increase in MVC force, which is evident as a shift along the *x*-axis, was 20% for the men and 4% for the women. In contrast, the training load increased, as shown as a shift along the *y*-axis, by 200% for the men and by 240% for the women. Because MVC force was tested with an isometric contraction whereas the training involved anisometric contractions, the greater increase in training load must have been due to an improvement in the coordination associated with the knee extension task. In a similar example, subjects who strength trained a hand muscle for eight weeks experienced a 33% increase in MVC force but only an 11% increase in the tetanic force evoked by electrical stimulation (Davies et al., 1985). Furthermore, when another group of subjects trained the muscle with electrical stimulation for eight weeks, there was no change in the evoked tetanic force whereas the MVC force declined by 11%.

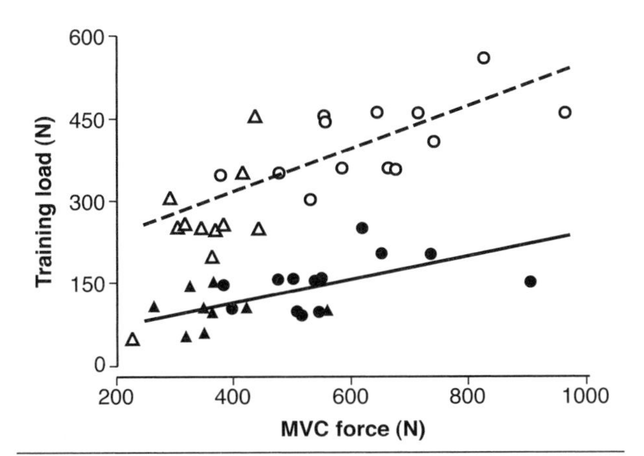

Figure 9.19 The relation between maximal voluntary contraction (MVC) force and training load in men (circles) and women (triangles) before (filled symbols) and after (open symbols) a 12-week training program.

Data from R.M. Rutherford and D.A. Jones, 1986, "The role of learning and coordination in strength training," *European Journal of Applied Physiology* 55: 105.

Strength can improve the ability of individuals to perform difficult coordination tasks. For example, Carroll and colleagues (2001*b*) compared the ability of individuals to coordinate flexion or extension movements of the index finger with the beat of a metronome after four weeks of strength training the muscles that control these movements. The subjects performed four 66 s tasks: extend on the beat, flex on the beat, extend between the beats, flex between the beats. The pacing of the metronome increased across each task. Performing a movement (flexion or extension) on a beat is easier than performing it between beats; as a consequence, an increase in the pacing beat causes the between-beat movements to transition to on-the-beat movements. Training involved lifting and lowering a load over a 45° range of motion with the index finger extensors, which increased in strength by 26%. Strength training increased the duration over which the subjects could sustain the extend-between-the-beats task before it transitioned to the extend-on-the-beat task. Electromyogram recordings indicated a more consistent activation of the muscle during the task after training.

Another example of the adaptability of coordination is the influence of training history on the reduction in strength that is typically observed when homologous muscles in two limbs are activated concurrently. The decline in force that occurs during MVCs is referred to as a **bilateral deficit** (Howard & Enoka, 1991; Jakobi & Chilibeck, 2001; Li et al., 1998). The bilateral deficit (site 5 in figure 9.18) occurs when homologous muscles in two limbs or multiple fingers in the same hand exert an MVC. An example of a bilateral deficit is shown in figure 9.20 for a task that involved isometric contractions of the triceps brachii muscles with the elbow at about a right angle. Each column in figure 9.20 corresponds to one trial. In Trial 1 (left column), the subject performed a maximal isometric contraction with the right arm. In Trial 2 (middle column), the subject concurrently activated the elbow extensors of both arms maximally. In Trial 3 (right column), the subject performed a maximal isometric contraction with the left arm. Both the maximum force exerted by each arm and the quantity of EMG were less during the bilateral contraction compared to the force and EMG during the single-limb trial. The reduction in force with a bilateral deficit averages about 5% to 10% of maximum but can be substantial (25-54%), especially during rapid contractions (Koh et al., 1993). Bilateral interactions, however, are modifiable with training (Secher, 1975; Taniguchi, 1998). For example, Howard and Enoka (1991) compared the maximal isometric force exerted during one- and two-limb knee extensor contractions and found a bilateral deficit for untrained subjects (–9.5%) and elite cyclists (–6.6%) but a **bilateral facilitation** for weightlifters (+6.2%). A bilateral facilitation means that the maximal isometric force occurred during the bilateral contraction rather than the unilateral contraction.

Changes in coordination can also involve altering the relative activation of agonist and antagonist muscles. The resultant muscle torque about a joint depends on the difference in the torque exerted by opposing muscles. Consequently, one way to increase strength is to reduce the size of the opposing muscle torque (site 2 in figure 9.18). Although activation of the antagonist muscle can decline with strength training and is less in elite athletes,

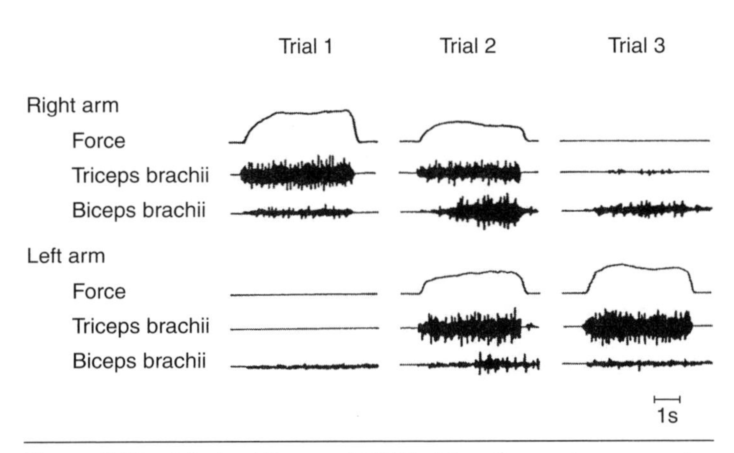

Figure 9.20 Maximal force and EMG of the elbow extensor muscles during maximal voluntary contractions performed with the right arm (Trial 1), both arms (Trial 2), and the left arm (Trial 3).

Adapted from *Behavioral Brain Research*, Vol. 7, T. Ohtsuki, "Decrease in human voluntary isometric arm strength induced by simultaneous bilateral exertion" p.169. Copyright 1983 with permission from Elsevier Science.

the magnitude of the reduction is relatively minor compared with the improvements in strength (Amiridis et al., 1996; Carolan & Cafarelli, 1992; Häkkinen et al., 1998*b*; Kawakami et al., 1995). Thus a reduction in coactivation of the antagonist muscles does not contribute significantly to the strength gains of agonist muscles.

Spinal Cord Endurance training produces dendrite restructuring, increases in protein synthesis, an elevation of axonal transport of proteins, enhanced neuromuscular propagation, and changes in the electrophysiological properties of motor neurons (Gardiner, 2006; Gardiner et al., 2006). Presumably, strength training is likely to cause comparable adaptations. For example, Remple and colleagues (2001) found that rats trained to perform a strong contraction during a reaching task exhibited a higher density of excitatory, but not inhibitory, synapses onto motor neurons in cervical segments of the spinal cord. However, most of our knowledge of the spinal adaptations in response to strength training is confined to measures of motor unit discharge and the responsiveness of specific pathways.

Two studies demonstrate that the discharge rate of motor units can change with strength training. One study showed that strength gains were accompanied by increases in average discharge rate of motor units. Knight and Kamen (2001) trained the knee extensor muscles of young and old adults for six weeks with a standard heavy-load program. Before training, MVC force was less for the old adults (396 vs. 520 N), and average discharge rates of motor units in vastus lateralis during an MVC were greater for young adults (24.7 vs. 17.8 pps). Maximal voluntary contraction torque increased for both the young (29%) and old (36%) adults and remained less for the old adults (514 vs. 713 N) at the end of training. There was a corresponding increase in the average discharge rate of 15% for the young adults and 49% for the old adults, which resulted in similar values at the end of training.

The other study examined the influence of training on the rate at which motor units could discharge action potentials during a rapid contraction. Van Cutsem and colleagues (1998) compared initial discharge rates of motor units in tibialis anterior before and after 12 weeks of training with rapid contractions against a moderate load (~35% of maximum). Training augmented the rate of torque development and the EMG increase during rapid submaximal contractions (figure 9.21*a*). The underlying change in motor unit activity responsible for the more rapid increase in the surface EMG was assessed through measurement of the instantaneous discharge rate of the first four action potentials (figure 9.21*b*). The average instantaneous discharge rate increased from 69 to 96 pps with training, and there was a significant increase in the number of motor units (5-33%) with brief times (<5 ms) between consecutive action potentials. Thus, the capacity

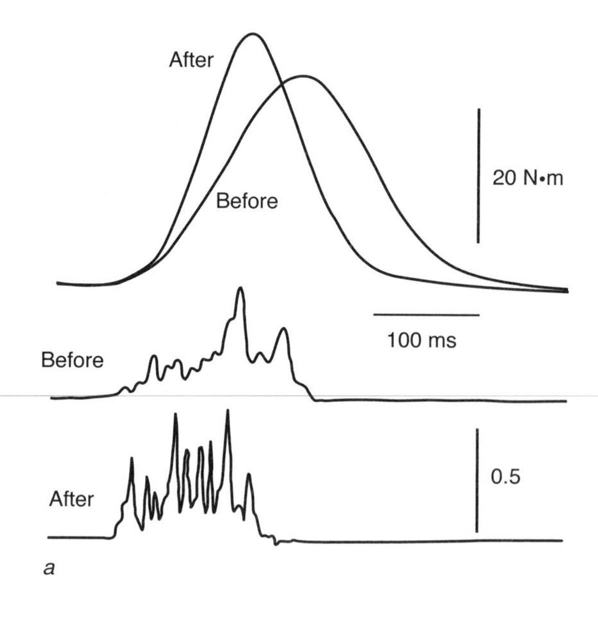

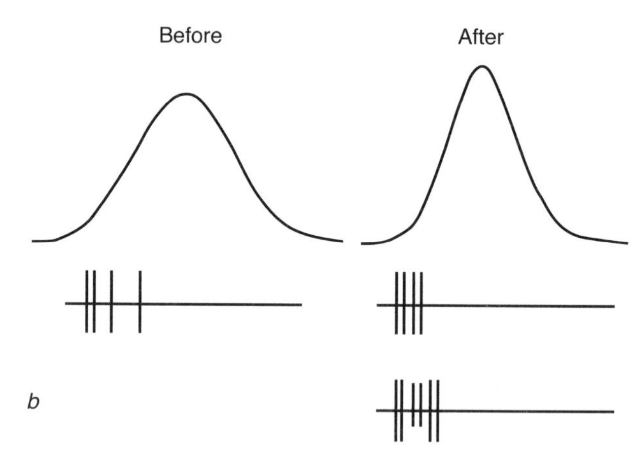

Figure 9.21 Changes in the rate of torque development, surface EMG, and motor unit discharge rate during rapid submaximal contractions. *(a)* Torque exerted by the dorsiflexor muscles and associated surface EMG activity for tibialis anterior during typical rapid contractions before and after 12 weeks of training. *(b)* Schematic illustrations of the corresponding changes in discharge rate, which included a greater incidence of brief interspike intervals (bottom trace) and a higher average discharge rate (second-to-bottom trace) after training.

Reprinted, by permission, from J. Duchateau, J.G. Semmler and R.M. Enoka, 2006, "Training adaptations in the behavior of human motor units," *Journal of Applied Physiology* 101:1770. Used with permission.

to produce more rapid increases in torque was associated with an adaptation in motor unit discharge rate (Van Cutsem et al., 1998) and an increase in the amplitude of the H reflex (Holtermann et al., 2007).

One of the most frequently cited examples of neural adaptations that accompany strength training is a change in the amount of motor unit synchronization (site 6 in figure 9.18). As described in chapter 6, synchronization is quantified as the coincidental discharge of action potentials by pairs of motor units (figure 6.25) and is attributed to the shared input that is delivered by branched axons of last-order neurons (Nordstrom et al., 1992; Schmied et al., 2000). The level of synchronization between a pair of motor units is determined from the peak in a cross-correlation histogram. The amount of motor unit synchronization can vary with contraction type (Semmler et al., 2002) and the type of habitual activity performed by the individual (Semmler & Nordstrom, 1998; Semmler et al., 2004). However, four weeks of training increased the strength of a hand muscle, but this was not accompanied by any change in the amount of synchronization between pairs of motor units (Kidgell et al., 2006), which is consistent with the prediction of a computer model that changes in synchronization do not influence maximal force (Yao et al., 2000). These findings suggest that motor unit synchronization is more related to fine motor skills than to strength.

Spinal cord adaptations have also been observed in the testing of reflexes (site 8 in figure 9.18). Electrical stimulation of a muscle nerve during a voluntary contraction (V) can evoke two reflex responses (V1 and V2) in addition to the M wave. Changes in the amplitude of V1 and V2 relative to the maximal M wave have been used as an index of reflex potentiation (Sale et al., 1983). The relative amplitude of the V1 response evoked by supramaximal electric stimulation increases after strength training. Aagaard and colleagues (2002*b*) compared the amplitudes of the V1 response and H reflex relative to the maximal M wave before and after 14 weeks of strength training. The measurements were made on the soleus muscle during a maximal isometric contraction. Training increased MVC torque by 20%, the amplitude of the V1 response by ~50%, and the amplitude of the H reflex by ~20%. Aagaard and colleagues (2002*b*) interpreted these results as indicating an increase in the efferent drive to soleus after strength training of the plantarflexor muscles. Studies on operant conditioning of the stretch reflex and the H reflex suggest that much of the plasticity appears to be located in the spinal cord, to involve the motor neurons, and also to be expressed in the contralateral, untrained limb (Chen et al, 2006; Wolpaw, 1994; Wolpaw et al., 1994).

Much of the adaptation that occurs in the spinal cord with training likely involves the integrative actions of the interneurons (Nielsen, 2004). For example, Carroll and colleagues (2002) examined the influence of strength training on the responses evoked in a hand muscle (first dorsal interosseus) by transcranial magnetic stimulation (TMS) and transcranial electric stimulation (TES). Transcranial magnetic stimulation generates action potentials in axons that synapse onto cortical neurons, whereas TES evokes action potentials in the axons of neurons that contribute to the corticospinal tract (Di Lazzaro et al., 1998). Four weeks of training increased MVC torque by 33%. Responses were evoked by TMS and TES at a range of contraction intensities (5-60% of MVC torque). The size of the responses evoked by both TMS and TES at the various target torques was reduced after training, which Carroll and colleagues (2002) interpreted as indicating that strength training did not alter the organization of the motor cortex but did influence the functional properties of the spinal cord circuitry.

Although a compelling case can be made for a significant role of neural factors in strength gains, the specific mechanisms remain elusive. There is neither a consensus on individual mechanisms nor evidence on the relative significance of the various mechanisms.

Muscular Factors

In contrast to the uncertainties regarding the neural factors, there is no doubt that specific muscular factors contribute to differences in strength among individuals and across time (Folland & Williams, 2007). As we discussed in chapter 3, there is a strong correlation between MVC force and cross-sectional area (figure 3.4). This same relation applies at the level of single motor units and muscle fibers. An increase in cross-sectional area of muscle can be caused by **hypertrophy,** an increase in the cross-sectional area of individual muscle fibers, or by **hyperplasia,** an increase in the number of muscle fibers. Most experimental evidence suggests that the typical response of human subjects to strength training involves hypertrophy but that hyperplasia may occur under some conditions (Antonio & Gonyea, 1993; Campos et al., 2002; Gillies et al., 2006; Higbie et al., 1996; McCall et al., 1996).

Hypertrophy The magnitude of the increase in cross-sectional area with training depends on several factors, including the initial strength of the individual, the duration of the training program, and the training technique used. In novice subjects, six weeks of isometric training increased the cross-sectional area of the elbow flexor muscles (biceps brachii, brachioradialis) by about 5% (Davies et al., 1988), whereas eight weeks of isometric training increased the cross-sectional area of the quadriceps femoris by 15% (Garfinkel & Cafarelli, 1992). Similarly, 60 days of isokinetic training at 2.1 rad/s with the knee extensors increased the cross-sectional area of the quadriceps femoris by 9% (Narici et al., 1989). Furthermore, 19 weeks of shortening and lengthening contractions with the knee extensor muscle produced a greater increase in cross-sectional area than exercises that involved only shortening contractions (Hather et al., 1991). In contrast, 24 weeks of dynamic training by experienced bodybuilders failed to elicit an increase in the cross-sectional area of muscle fibers in biceps brachii (Alway et al., 1992).

The effect of strength training on the different muscle fiber types can also be diverse. For example, 16 weeks of isometric training with the triceps surae muscles did not alter the fiber type proportions in either soleus or lateral gastrocnemius despite a 45% increase in MVC force (Alway et al., 1989). In contrast, 19 weeks of a knee extension exercise increased the proportion of type IIa muscle fibers and decreased the proportion of type IIb (IIx) muscle fibers in vastus lateralis (Hather et al., 1991). This adaptation even included a reduction in the type IIx (IIb) myosin heavy chains (Adams et al., 1993*b*), which appears to involve a change in genetic expression (Caiozzo et al., 1996). Similarly, the 16-week isometric program increased the cross-sectional area of type I (20%) and II (27%) muscle fibers in soleus and the type II fibers (50%), but not the type I fibers, in lateral gastrocnemius. Furthermore, the increase in cross-sectional area of muscle fibers in vastus lateralis is greater for programs that involve shortening and lengthening contractions (type I, 14% increase; type II, 32% increase) than for shortening contractions alone (type II, 27% increase) (Hather et al., 1991). Taken together, these results indicate that much of the hypertrophy produced by strength training occurs in type II fibers (Folland & Williams, 2007).

In a comprehensive study of the morphological changes that occur in muscle with strength training, Aagaard and colleagues (2001) measured muscle cross-sectional area and volume with MRI, the pennation angle of muscle fibers in vastus lateralis by ultrasound, and the cross-sectional area of muscle fibers obtained by muscle biopsy from vastus lateralis. The 14 weeks of training increased the MVC torque of the knee extensors by 16% (282 to 327 N•m). The morphological changes included a 10% increase in the volume of quadriceps femoris (1676 to 1841 cm^3), a 36% increase in the pennation angle of fibers in vastus lateralis (8° to 11°), an 18% increase in the cross-sectional area of type II fibers (3952 to 4572 μm^2), and no statistically significant change in the area of type I fibers (3582 to 3910 μm^2). The fiber type proportions did not change with training. The greater increase in MVC torque (16%) compared with muscle volume (10%) was attributed to the increase in pennation angle, which increased the force capacity of the muscle per unit volume.

As discussed in chapter 3, the force capacity of muscle also depends on its specific tension and the ability of the connective tissues to transmit the force to the skeleton. Specific tension corresponds to the intrinsic force-generating capacity of individual muscle fibers. Specific tension, which has an average value of 22.5 N/cm^2, has been reported to decline with bed rest and aging (Larsson et al., 1996, 1997) and to change with training (Harber et al., 2004). The measurement of whole-muscle force relative to its size, however, represents normalized force because it depends on both specific tension and the properties of the connective tissues. Aagaard and colleagues (2001) calculated normalized force for quadriceps femoris from the cross-sectional area, pennation angle, and length of fibers in vastus lateralis relative to the volume of quadriceps femoris. They estimated that normalized force increased from 42.6 N/cm^3 before training to 45.3 N/cm^3 after training, which indicates that some of the increase in strength was the result of changes in specific tension, the properties of connective tissues, or both.

EXAMPLE 9.4
Strength Training Strategies

We can appreciate the practical issues related to the development of strength compared with muscle mass if we compare the training strategies used by weightlifters and bodybuilders. The goal of weightlifters is to increase the maximal load that can be lifted in prescribed events, whereas the goal of bodybuilders is to increase muscle mass. As described by Garhammer and Takano (1992), weightlifting programs are based on the concept of periodization, which means that the training program is divided into several phases. The original model of periodization comprised a preparation phase and a competition phase. The preparation phase is characterized by high volume and low intensity, which means many repetitions and relatively light loads—for example, 6 to 15 training sessions a week, three to six exercises each session, with each exercise comprising four to eight sets of four to six repetitions. The competition phase, in contrast, would involve 5 to 12 training sessions a week, one to four exercises each session, with each exercise comprising three to five sets of one to three repetitions. A weightlifter may train five or six days a week with one to three sessions each day. The duration of each phase may last from several weeks to several months, and two or more complete cycles of preparation and competition may fit into a single training year.

In contrast, bodybuilders tend to use lighter loads (6- to 12-RM) and to reach task failure in each set of repetitions (Tesch, 1992). Elite bodybuilders use a split system with four consecutive days of training followed by a day off. The split system focuses on two or three major muscle groups in a training session (table 9.2). Each muscle group performs 20 to 25 sets of 6 to 12 repetitions for a total of 40 to 70 sets in a single session. Bodybuilders use brief intervals (1-2 min) between sets of exercises. This training approach, which appears to maximize muscle hypertrophy, focuses on high volume (sets × repetitions × load) with each muscle group worked to failure. The physiological mechanisms responsible for these different effects are unknown.

Table 9.2 Split System Used by Bodybuilders

Day 1	Day 2	Day 3	Day 4	Day 5
Chest	Quadriceps	Back	Hamstrings	Rest
Triceps brachii	Calves	Biceps brachii Abdominals	Shoulders	

Note. Scheme taken from Tesch (1992).

Hypertrophy Mechanisms The hypertrophy of muscle requires both a change in the ratio of protein synthesis to degradation and an increase in the number of myonuclei to manage the increase in the volume of contractile proteins. New myonuclei are differentiated from satellite cells, which are quiescent mononucleated cells located between the basal lamina and the plasmalemma (figure 9.22). The number of satellite cells is similar in type I and II fibers in the vastus lateralis muscle of young adults and is reduced in the muscles of old adults (Kadi et al., 2006; Thornell et al., 2003). Satellite cells are activated by such stimuli as injury and an increase in muscle force. Once activated, they replicate and control the repair of injured fibers and the growth of stressed fibers (Anderson, 2006; Kadi et al., 2005).

Satellite cell content is higher in the muscles of athletes who have performed strength training for several years, and the number of satellite cells increases after strength training (Eriksson et al., 2005; Kadi et al., 1999; Kadi & Thornell, 2000). Satellite cells can be activated by a single bout of high-intensity exercise, but this stimulus is insufficient for the cells to undergo terminal differentiation (Crameri et al., 2004; Dreyer et al., 2006). Differentiation of satellite cells into myonuclei requires a significant increase in fiber size. For example, the number of myonuclei does not change when strength training increases fiber area by 17%, whereas there is an increase in number when the fibers hypertrophy by 36% (Kadi et al., 2004; Mackey et al., 2007). These results suggest that when the transcriptional activity of the myonuclei reaches some upper level, the progeny of the satellite cells become involved in protein synthesis by enhancing the number of nuclear domains. There is a strong association between the number of myonuclei added after strength training and the magnitude of muscle fiber hypertrophy (Petrella et al., 2006).

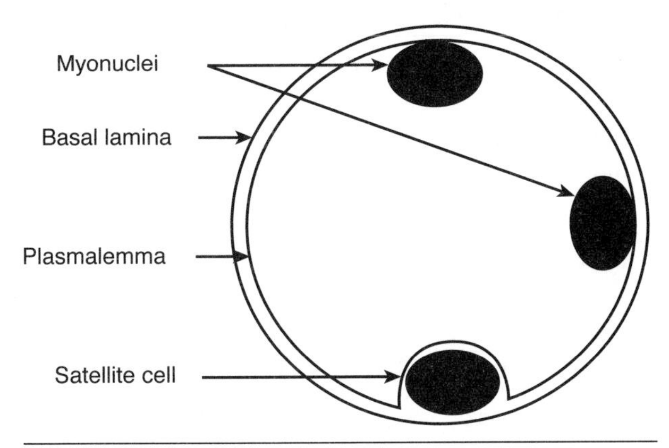

Figure 9.22 Locations of the myonuclei and satellite cells in a cross-sectional view of a muscle fiber.

Reprinted, by permission, from F. Kadi, 2000, "Adaptation of human skeletal muscle to training and anabolic steroids," *Acta Physiologica Scandinavica* 168, Suppl 646: 20.

Kadi and colleagues (2005) hypothesize that exercise, including strength training, can have four effects on satellite cells: (1) activation without proliferation; (2) proliferation and withdrawal from differentiation, which increases the number of satellite cells; (3) proliferation and differentiation into myonuclei; and (4) proliferation and differentiation to develop new muscle fibers or to repair injured segments of muscle fibers. The number of times a satellite cell can replicate is limited by the length of the telomeres at the ends of the chromosomes. Telomeres can be shortened by oxidative stress, which limits cell replication; but strength training does not contribute to this process, and satellite cells have sufficient cell divisions to ensure repair and growth across the human life span (Collins et al., 2003; Thornell et al., 2003).

Muscle fiber hypertrophy is produced by a net increase in protein synthesis. There are a number of intracellular signaling pathways that can mediate this adaptation (Tsika, 2006). One of the critical transduction pathways involves protein kinase B (Akt) and its activation of the mammalian target of rapamycin (mTOR) (Bodine, 2006). Akt signaling is activated by an isoform of insulin-like growth factor-1 (IGF-1) that is produced by skeletal muscle in response to an exercise stress. This isoform of IGF-1, called a mechano growth factor (MGF), contributes to the upregulation of protein synthesis (Goldspink & Harridge, 2004). After binding to a transmembrane receptor, MGF initiates a series of interactions that activate Akt, which increases protein synthesis through mTOR pathways and decreases protein degradation by phosphorylating a transcription factor that is expelled from the nuclei and reduces the production of proteins involved in degradation (Tidball, 2005). It seems likely, however, that other exercise signals besides MGF are also able to activate the Akt pathway. Many of the underlying details involved in muscle fiber hypertrophy, including the relative contributions of changes in translation and the increase in myonuclei number, remain to be determined.

MUSCLE POWER

The power that a muscle can produce depends on the product of muscle force and the velocity of shortening (figures 6.35 and 8.1). Power production is maximal when the muscle acts against a load that is one-third of MVC force or when the muscle shortens at one-third of the maximal shortening velocity (chapter 6). Consequently any adaptation that increases the maximal force a muscle can exert or its maximal shortening velocity will increase the maximal power that it can produce. In this section, we consider the measurement of power production at various levels of the motor system and the chronic adaptations that occur with training.

Power Production and Movement

Success in many athletic endeavors depends on the ability of the performer to sustain power production at the highest level possible for the duration of the event. The maximum power that can be sustained is known as the **critical power** (Walsh, 2000). The maximal sustainable power is inversely related to the duration of the event. Despite this significant role for power production, it is one of the least-examined biomechanical parameters in the analysis of human movement. Some of this lack of attention can be attributed to the difficulty associated with measuring power and the abstract nature of the parameter. The next few sections provide examples that address these two issues.

Power Production and Whole-Body Tasks

The power produced by the motor system can be determined by a task performance (e.g., vertical jump, weightlifting), with the use of an ergometer, or by an isolated-muscle experiment. The evaluation of a task performance provides an index of whole-body power. The vertical jump is commonly used for this purpose. The average power produced during a squat jump, which involves only isometric and shortening contractions, can be estimated by kinematic or kinetic measurements as we discussed in chapter 2. These procedures can be performed quite quickly and are useful for monitoring the progress of athletes during a training program or comparing the efficacy of various training strategies.

The kinematic and kinetic approaches involve estimating the average power and average velocity during the movement. The kinematic approach entails calculating the average power during the descent of the flight phase in the vertical jump, which must equal the power provided to the body during the takeoff phase. As indicated by equation 2.35, the calculation requires only the mass of the individual (m) and the height that the center of mass was raised (r):

$$\bar{P} = (9.81 \bullet m) \bullet \sqrt{r \bullet 4.9}$$

The first term ($9.81 \bullet m$) corresponds to the average force, and the second term ($\sqrt{r \bullet 4.9}$) indicates the average velocity during the descent phase.

Alternatively, we can estimate the average power by performing a kinetic analysis, which is based on the vertical component of the ground reaction force (figure 9.23). As indicated in equation 2.36, the calculation requires knowledge of the net impulse (shaded area in figure 9.23; $\int F_{net}\,dt$), the absolute impulse ($\int F_{g,y}\,dt$), the duration of the propulsive phase of the impulse (t), and the mass of the individual.

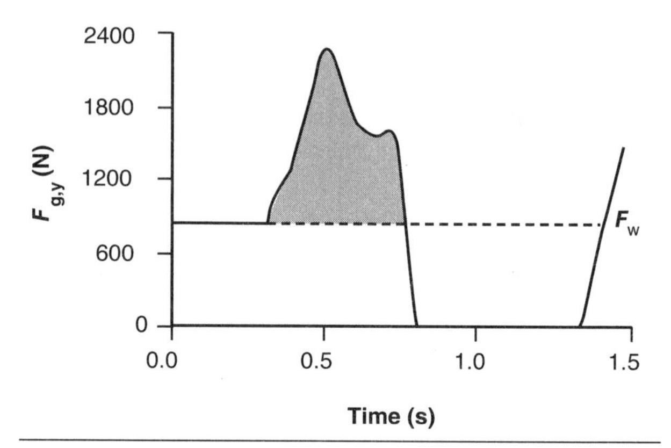

Figure 9.23 Vertical component of the ground reaction force during a squat jump for maximal height. The net impulse is shown as the shaded area.

$$\bar{P} = \frac{\int F_{g,y}\,dt}{t} \bullet \frac{\int F_{net}\,dt}{2\,m}$$

The first term indicates the average force and the second term the average velocity. Another kinetic approach is to use equation 2.37, which is based on the change in mechanical energy (potential and kinetic energy).

$$\bar{P} = \frac{mg(r+h)}{t}$$

where r = jump height, h = vertical displacement of the jumper's center of mass during the takeoff phase, and t = the duration of the propulsive phase of the vertical impulse.

EXAMPLE 9.5

A Comparison of the Three Techniques to Estimate Power

When an individual with a mass (m) of 87.2 kg performed a squat jump for maximal height, he was able to raise his center of mass (r) by 42.3 cm by producing an absolute impulse ($\int F_{g,y}\,dt$) of 510 N•s and a net impulse ($\int F_{net} dt$) of 236 N•s that was applied for a duration (t) of 0.32 s to raise his center of mass during the takeoff phase (h) by 56.7 cm. Let us calculate the average power ($\bar{P}$) produced by the jumper using the three approaches:

First, the kinematic approach:

$$\begin{aligned}\bar{P} &= (9.81 \bullet \text{m}) \bullet \sqrt{r \bullet 4.9} \\ &= (9.81 \bullet 87.2) \bullet \sqrt{0.423 \bullet 4.9} \\ &= 1232 \text{ W}\end{aligned}$$

Second, the impulse-momentum (kinetic) approach:

$$\begin{aligned}\bar{P} &= \frac{\int F_{g,y}\, dt}{t} \bullet \frac{\int F_{net}\, dt}{2\,m} \\ &= \frac{510}{0.32} \bullet \frac{236}{(2 \times 87.2)} \\ &= 2156 \text{ W}\end{aligned}$$

Third, the mechanical energy (kinetic) approach:

$$\begin{aligned}\bar{P} &= \frac{mg(r+h)}{t} \\ &= \frac{855(0.423+0.567)}{0.32} \\ &= 2645 \text{ W}\end{aligned}$$

These estimates vary because of the assumptions associated with each procedure. The mechanical energy approach is the most accurate (Hatze, 1998).

For comparison, based on measurements of the torque–velocity relation, the maximal power produced by the plantarflexor muscles during the squat jump was calculated as 2499 W (Ingen Schenau et al., 1985). Similarly, Josephson (1993) estimated the maximal power for muscle to range from 5 to 500 W/kg, with the sustainable power ranging from 5 to 150 W/kg.

Another approach that can be used to determine power production, for weightlifting events for example, is to calculate the rate of change in mechanical energy (Garhammer, 1980). Because the quantity of work done can be determined from the change in energy, and power equals the rate of doing work, power can be calculated from the rate of change in energy.

$$\begin{aligned}P &= \frac{\Delta \text{ energy}}{\Delta \text{ time}} \\ &= \frac{E_{k,t} + E_{p,g}}{\Delta t} \\ &= \frac{0.5mv^2 + mgh}{\Delta t} \qquad (9.1)\end{aligned}$$

where $E_{k,t}$ refers to the maximal kinetic energy (the initial value was zero), $E_{p,g}$ represents change in potential energy from the initial position, and time indicates the time to the maximal kinetic energy.

An example of this approach is provided in table 9.3, which indicates the power delivered to a barbell by a weightlifter during three lifts: (1) the clean, a movement that requires the displacement of the barbell from the floor to the lifter's chest in one continuous rapid motion; (2) the squat, a movement in which the lifter supports a barbell on the shoulders and goes from a standing erect position to a knee-flexed position (thighs parallel to the floor) and then returns to standing; and (3) the bench press, a movement in which the lifter assumes a supine position and lowers the barbell from a straight-arm location above the chest down to touch the chest, then raises it back to the initial position. Data have been taken from the literature (clean—Enoka, 1979; squat—McLaughlin et al., 1977; bench press—Madsen & McLaughlin, 1984) to illustrate the differences in power production for the three lifts (table 9.3).

These data suggest that the peak power delivered to the barbell is greatest during the clean and least for the bench press. This is interesting, because the squat and bench press are two of the three lifts that comprise the sport of powerlifting. Indeed, these data suggest that success in the squat and bench press lifts is not determined solely by power production but also depends on muscle strength.

Production of Power About a Single Joint

In contrast to the estimate of whole-body or whole-limb average power, an ergometer enables us to focus more specifically on one muscle group (Grabiner & Jeziorowski, 1992). An ergometer permits measurement of the relation between torque and angular velocity about

Table 9.3 Power Delivered to a Barbell During the Three Types of Lifts

	Clean	Squat	Bench press
Barbell weight (N)	1226	3694	1815
Peak barbell velocity (m/s)	2	0.30	1.54
Height (m)	0.30	0.30	0.06
Time to $E_{k,t}$ (s)	0.35	1.30	0.70
$E_{k,t}$ (J)	250	17	219
$E_{p,g}$ (J)	368	1108	109
Power (W)	1766	865	469

$E_{k,t}$ = maximal kinetic energy; $E_{p,g}$ = potential energy.

a joint, and the data can then be used to calculate the associated power (Aagaard et al., 1994*a,b;* De Looze et al., 1992) or even to determine a power loop (Stevens, 1993). For example, De Koning and colleagues (1985) measured the torque and angular velocity about the elbow joint for three groups of subjects: untrained women, untrained men, and arm-trained males (i.e., track and field athletes, rowers, weightlifters, bodybuilders, handball players, karate practitioners, and tug-of-war competitors). The MVC torque (P_0) varied in the order that you would expect: arm-trained men, untrained men, and untrained women. There was, however, only a small difference in the maximal angular velocity (ω_{max}) among the groups (table 9.4). Nonetheless, there was a significant difference among the groups in the peak power that the elbow flexor muscles could produce.

Table 9.4 Elbow Flexor Strength (P_0), Maximal Angular Velocity (ω_{max}), and Peak Power for Arm-Trained Men and Untrained Men and Women

	Arm-trained men	Untrained men	Untrained women
P_0 (N•m)	90.9 ± 15.6	68.5 ± 11.0	42.7 ± 6.9
ω_{max} (rad/s)	17.0 ± 1.6	16.6 ± 1.5	14.9 ± 1.3
Power (W)	253.0 ± 58.0	195.0 ± 46.0	111.0 ± 24.0

Values are given as mean ± SD.

Another approach we can use to determine power production about a single joint is to perform a biomechanical analysis and calculate the product of torque and angular velocity about a joint (Enoka, 1988; Winter, 1983). This approach requires a description of limb kinematics and data on a contact force, such as the ground reaction force or a pedal force. For example, we can determine the power that a subject can apply to the pedal of a cycle ergometer by calculating the product of pedal force and pedal velocity or by summing the joint power (torque × angular velocity) for the hip, knee, and ankle joints (Ingen Schenau et al., 1990*b*). These two methods produce similar results and indicate the power produced by the leg throughout one complete pedal revolution (figure 9.24*a*). However, the individual joint powers provide additional information on the contribution of the three joints to the total leg power (figure 9.24*b*).

Power Production for Isolated Muscle

The power production capabilities of the motor system can also be measured at the level of the single muscle or muscle fiber. Brooks and Faulkner (1991), for example, compared the ability of slow-twitch muscle (soleus) and fast-twitch muscle (extensor digitorum longus) of mice to sustain isometric force and power production. The peak isometric force that extensor digitorum longus could exert was greater than that for soleus (363 vs. 273 mN). When force was normalized relative to muscle size (cross-sectional area), however, peak force was similar for the two muscles (24.5 N/cm^2 for extensor digitorum longus compared with 23.7 N/cm^2 for soleus). Nonetheless, extensor digitorum longus was more fatigable and experienced a more rapid decline in isometric force with sustained activation; the maximal sustainable force was 1.38 N/cm^2 for extensor digitorum longus and 4.58 N/cm^2 for soleus. In contrast, when power production was expressed relative to muscle mass, the peak sustainable power was greater for extensor digitorum longus (9.1 W/kg) compared with soleus (7.4 W/kg). Although force declined more for extensor digitorum longus with sustained activation, the greater velocity of shortening resulted in a greater sustainable power. This observation suggests that type S and type FR motor units (see chapter 6) are important for the ability to sustain isometric force but that type FR and type FF motor units are more important for sustaining power production (see also Rome et al., 1988; Swoap et al., 1997).

Peak power production has also been measured for segments of single muscle fibers taken from primates (Fitts et al., 1998). Muscle samples were obtained by biopsy from soleus and medial gastrocnemius, and the contractile properties were measured in vitro. From measurements of the force–velocity relation, the peak power was calculated for type I and type II fiber segments from each muscle. The peak power was greater for the type II fibers compared with the type I fibers; it was 5 times greater for the soleus fibers and 8.5 times greater for the medial gastrocnemius fibers. Fitts and colleagues found that the greater peak power of the type II fibers was due to a greater maximal shortening velocity and a lower curvature of the force–velocity relation. In addition, the maximal force of the type II fibers in medial gastrocnemius was greater than that for type I fibers.

Power Production and Training

Because of the technical demands associated with measuring joint power, there are few studies in the literature on the effects of training on power production at this level. Duchateau and Hainaut (1984) performed one such study on a hand muscle (adductor pollicis) of human volunteers. Two groups of subjects trained the muscle at a moderate intensity (10 repetitions) every day for three months. One group used maximal isometric contractions (5 s duration, once every minute), and the other group did rapid shortening contractions against a load of ~35%

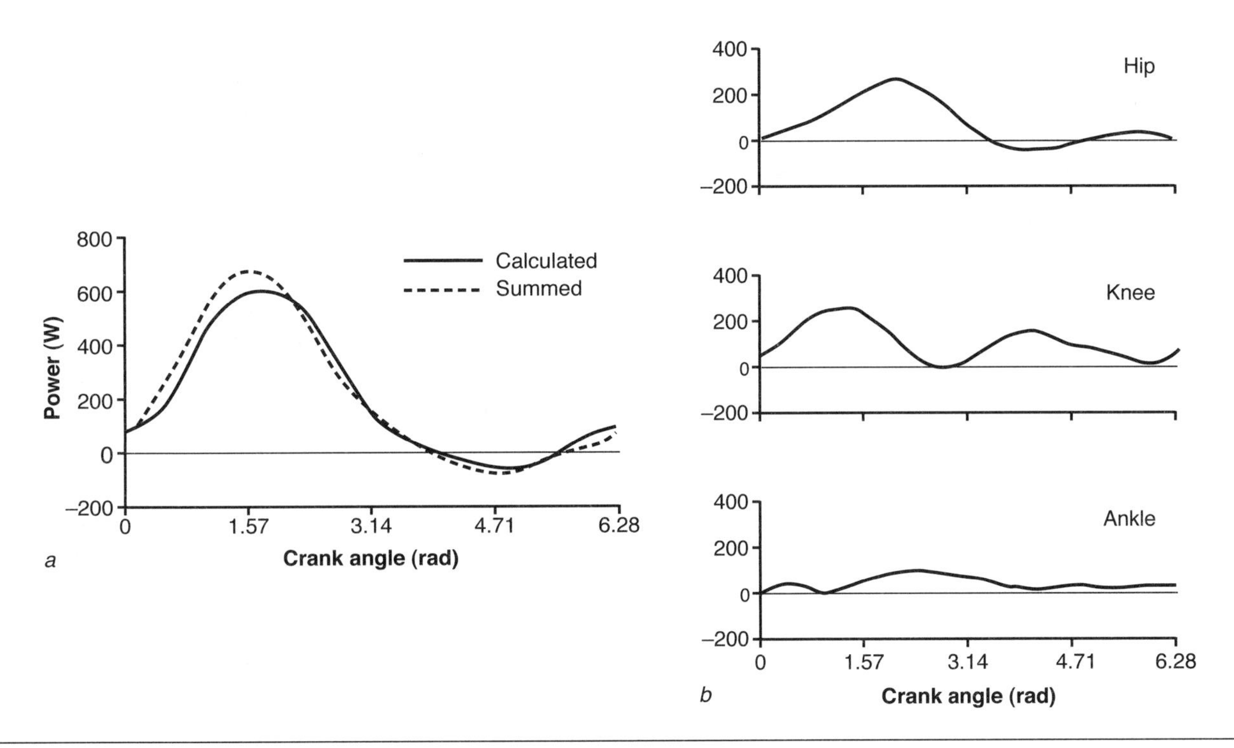

Figure 9.24 Power production during a single pedal cycle on an ergometer. *(a)* Power produced as a function of crank angle. The "Summed" power was determined by summing the joint powers (ankle, knee, and hip), whereas the "Calculated" power was estimated from the rate of change in segmental energy plus the power transferred to the pedal. *(b)* Joint power, which was calculated as the product of torque and angular velocity for the hip, knee, and ankle as a function of crank angle.

of maximum. The moderate load for the dynamic contractions was chosen because muscle produces maximal power when the contraction force is about one-third of maximum. Isometric training produced a greater increase in the maximal power and a shift in the force at which the maximum occurred.

The group that trained with the greater loads (MVC force) experienced a significant increase in the MVC force, whereas the other group did not (figure 9.25). Surprisingly, the isometric group experienced a 51% increase in maximal power compared with a 19% increase for the dynamic group. A greater increase in power was obtained with high-force rather than high-velocity training. A similar effect was seen with leg muscles (Aagaard et al., 1994*a;* de Vos et al., 2005). After 12 weeks of strength training, Aagaard and colleagues (1994*a*) found that subjects who used a high load experienced increases in both strength and power production, whereas subjects who used a low load did not. In comparison, subjects who trained with loaded kicking movements experienced modest increases in net muscle torque and peak power at intermediate velocities.

The effect of training load was examined more systematically through a comparison of the effects of training the elbow flexor muscles with loads that were 15%, 35%, or 90% of 1-RM load (Moss et al., 1997). The subjects, assigned to one of the three groups, trained three times a week for nine weeks. The outcome was expressed as the increase in peak power when lifting loads that ranged from 15% to 90% 1-RM as fast as possible. The 1-RM load increased by 6.6%, 10.1%, and 15.2% for the three groups (15%, 35%, and 90% 1-RM). The increase in peak power was modest with the lightest load, whereas it was greatest, but with a high degree of specificity, for the group that trained with the heaviest load (figure 9.26). In contrast, the increase in peak power was most consistent for the subjects who trained with a 35% 1-RM load. Similarly, Moritani (1992) found that training loads of 30% 1-RM produced greater increases in power production than loads of either 0% or 100% 1-RM; these increases were accompanied by significant increases in the quantity of EMG and a decrease in the mean power frequency. These data suggest that the strategy an individual should choose depends on the goal of the training program (Cronin & Sleivert, 2005; Kawamori & Haff, 2004; Wilson et al., 1993).

Despite the significance of the training program, the ability of muscle fibers to produce power can be increased

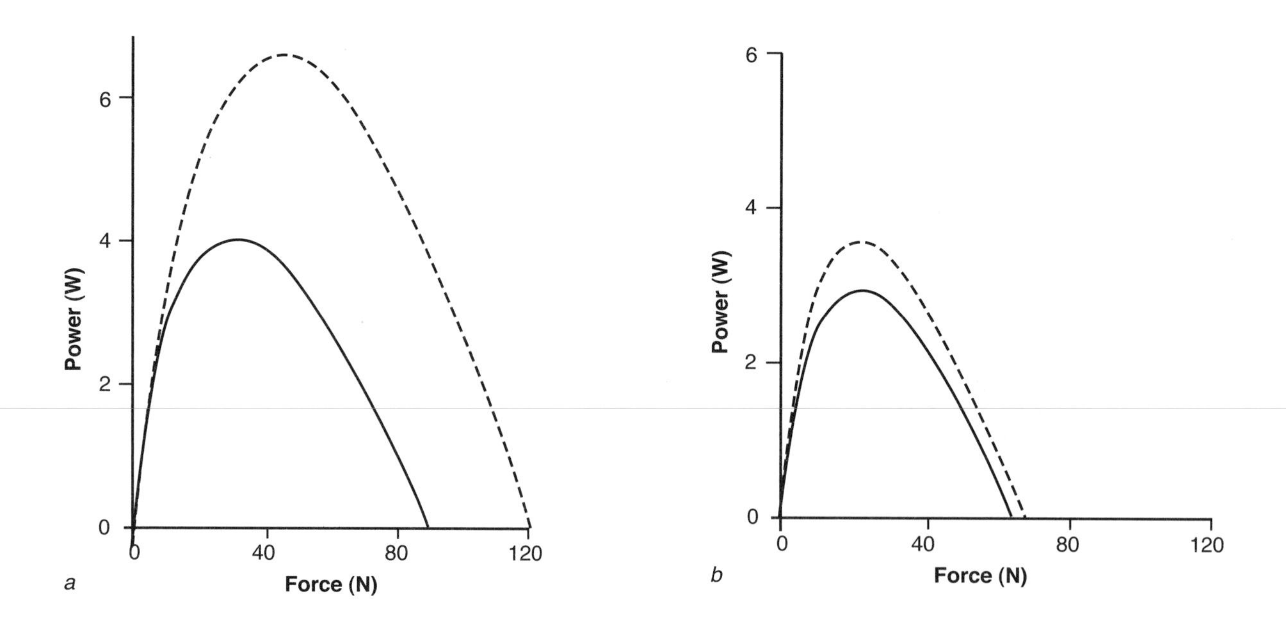

Figure 9.25 The effects of *(a)* isometric and *(b)* dynamic training on muscle power production by a hand muscle. The solid lines denote the values before training, and the dashed lines indicate the changes produced by the two training programs.

Data from Duchateau and Hainaut 1984.

with quite different types of training. For example, Trappe and colleagues (2000) measured the mechanical properties of skinned fiber segments obtained by muscle biopsy from the vastus lateralis muscle before and after 12 weeks of strength training (80% 1-RM load) by older men. Myosin heavy chain I and IIa (MHC-I and MHC-IIa) fibers increased in diameter (20% and 13%, respectively), peak tetanic force (55% and 25%), maximal shortening velocity (75% and 45%), and peak power (129% and 61%). Although Widrick and colleagues (2002*a*) found

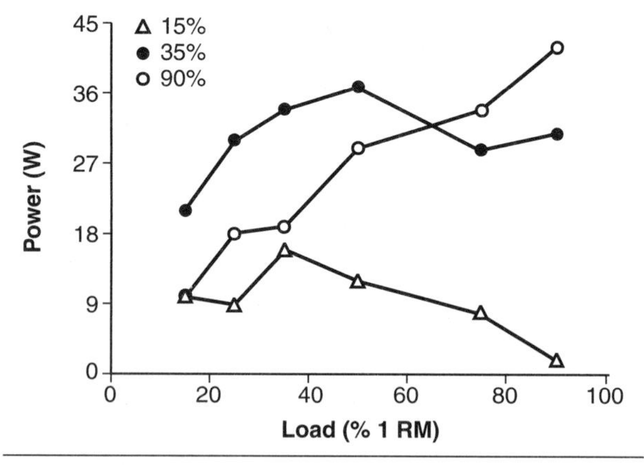

Figure 9.26 Increases in peak power produced by the elbow flexor muscles when lifting a range of loads after training with 15% (open triangles), 35% (filled circles), or 90% (open circles) of the 1-RM load.

Data from Moss et al. 1997.

that the maximal shortening velocity of fiber segments did not change after strength training in young men, there was a 30% increase in cross-sectional area for both MHC-I and MHC-IIa fibers, as well as 30% and 42% increases in peak power for MHC-I and MHC-IIa fibers, respectively.

Malisoux and colleagues (2006) made similar measurements on fiber segments from vastus lateralis before and after eight weeks of training with maximal-effort stretch-shorten cycles during jumping exercises. Training increased 1-RM load for the leg press by 12% and vertical jump height by 13%. MHC-I and MHC-IIa fibers experienced increases in cross-sectional area (23% and 22%), peak force (19% and 15%), maximal shortening velocity (18% and 29%), and peak power (25% and 34%). In contrast to the similar results with strength training and jump training, Trappe and colleagues (2006) found that eight weeks of training for a marathon produced quite different results for fiber segments taken from gastrocnemius. The size of MHC-I and MHC-IIa fibers decreased by 20% but peak tetanic force was unchanged, which meant that specific tension increased by ~60%. Although maximal shortening velocity increased by 28% in MHC-I fibers and did not change in MHC-IIa fibers, peak power increased by 56% and 16%, respectively.

Given these reasonably similar adaptations in muscle fiber properties, the functional differences observed with different types of training programs suggest that the activation of the involved muscles contributes significantly to the various outcomes. Little is known about the adaptations in muscle activation with power training. We do know that training with fast contractions can increase the rate of rise in force, and this appears to be achieved by adaptations in the discharge rate of motor units (figure 9.21) (Aagaard et al., 2002*a;* Van Cutsem et al., 1998). However, we do not yet know how the behavior of a motor unit pool or the distribution of activity across multiple pools changes with power training.

ADAPTATION TO REDUCED USE

Just as increases in physical activity can enhance the properties of the motor system, reductions in activity are accompanied by corresponding adaptations. Consequently, the adaptability of the motor system can be studied with experimental models that involve either an increase or a decrease in the amount of activity. To assess the role of the nervous system in the adaptations experienced by muscle, a number of experimental models have been developed to alter the connection between the nervous system and muscle and, by subtraction, to determine the role of the nervous system in defining muscle properties (table 9.5). In this section, we consider three commonly used models of reduced use and examine the types of adaptations that have been observed with each model. Information on other models is available elsewhere (Roy et al., 2007; Tsika, 2006; Wackerhage & Rennie, 2006).

Table 9.5 Experimental Models Used to Study the Effects of Changes in Activity

	Neural	Muscle
Enhanced use	Electrical stimulation	Stretch
	Exercise	Exercise
		Compensatory hypertrophy
Reduced use	Denervation	Tenotomy
	Tetrodotoxin	Limb immobilization
	Curare	Hindlimb unloading
	Colchicine	Barbiturate sleep
	Rhizotomy	Bed rest
	Spinal transection	Water immersion
	Spinal isolation	

Limb Immobilization

People who have an injured limb immobilized in a cast for a few weeks often experience a loss of muscle mass and function that is apparent upon removal of the cast. This adaptation is of obvious concern to the clinician, who would like to know how to minimize the loss of mass and function (Labarque et al., 2002; Mattiello-Sverzut et al., 2006; Stevens et al., 2006; Vandenborne et al., 1998; Veldhuizen et al., 1993). However, the scientist regards this adaptation as an opportunity to characterize the changes that occur in the motor system with this type of stress (reduction in activity) and to identify the mechanisms that mediate these changes.

Reduction in Activity

Many studies have been performed on both animals and humans to examine the adaptations that occur with limb immobilization. A critical feature of protocols that address the effects of reduced use is the extent to which the intervention halts activation of the involved muscles (Cruz-Martinez et al., 2000; Hodgson et al., 2005). When healthy humans had an arm immobilized in a cast, the activity of the elbow flexor muscles decreased modestly (Semmler et al., 2000*a*). This effect was quantified by measurement of bursts of EMG activity for 24 h periods before and during immobilization (figure 9.27). In this study, the EMG activity of biceps brachii declined by 38% and that for brachioradialis decreased by 29% during four weeks of immobilization. Interestingly, this effect differed for men and women. The women had greater levels of EMG prior to immobilization and experienced a greater reduction in EMG during immobilization. Similarly, Yasuda and colleagues (2005) found that 14 days of knee immobilization produced a greater decline in the strength of the knee extensors in women despite comparable reductions in quadriceps and muscle fiber size in men and women.

Multiple measures of EMG during immobilization (figure 9.28*a*) indicate that it declines steadily and then remains at the reduced level (Fournier et al., 1983). However, the amount of the decline in EMG depends on the length at which the muscle is immobilized; the decline is greatest for short muscle lengths and negligible for long lengths. The integrated EMG decreased by 77% for soleus and 50% for medial gastrocnemius at the shortest lengths. These reductions in EMG were accompanied by a 36% loss of muscle mass for soleus and a 47% loss of muscle mass for medial gastrocnemius at the shortest lengths.

The amount of atrophy (loss of muscle mass) due to limb immobilization varies across studies. A reduction in muscle mass (wet weight) as high as 50% has been recorded (Fournier et al., 1983), and the decrease in muscle fiber cross-sectional area has been found to be

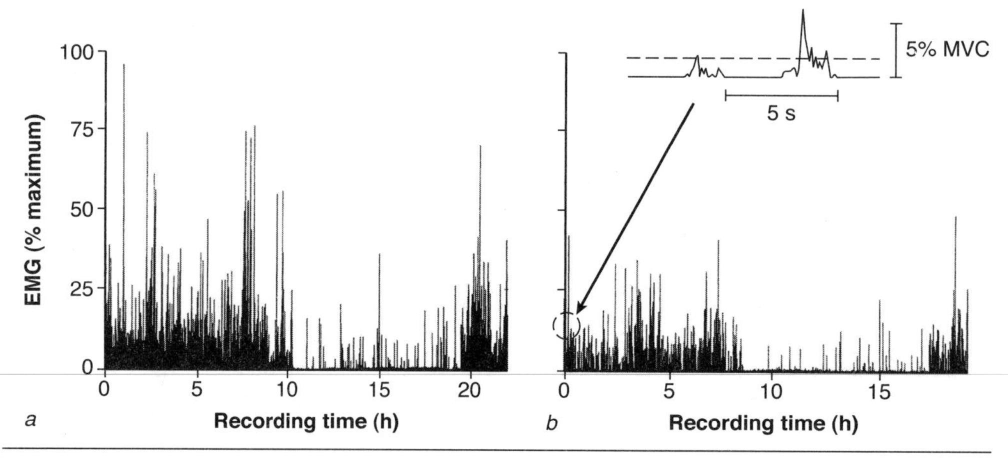

Figure 9.27 Electromyographic (EMG) activity in biceps brachii *(a)* before and *(b)* at week 3 during four weeks of arm immobilization. The recordings were based on bursts of EMG activity, which are shown in the inset.

Data from Semmler et al. 2000*a*.

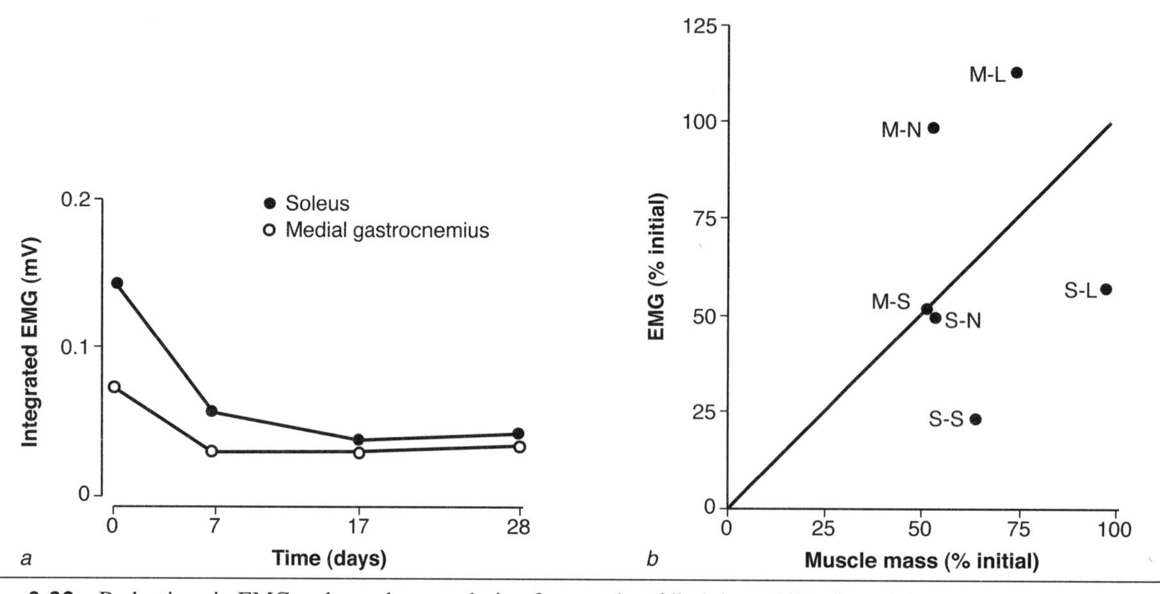

Figure 9.28 Reductions in EMG and muscle mass during four weeks of limb immobilization of the rat hindlimb: *(a)* decline in integrated EMG for the soleus and medial gastrocnemius muscles constrained to be at a short length; *(b)* relation between the decline in EMG and reduction in muscle mass. S-S = soleus, short length; S-N = soleus, neutral length; S-L = soleus, long length; M-S = medial gastrocnemius, short length; M-N = medial gastrocnemius, neutral length; M-L = medial gastrocnemius, long length.

as much as 42% (Nicks et al., 1989) with a few weeks of limb immobilization. Others, however, have reported a modest reduction of 17% in wet weight (Robinson et al., 1991) and a 14% decrease in the mean diameter of type I but not type II muscle fibers (Gibson et al., 1987) for a similar duration of immobilization. The atrophy is probably due both to a decline in the rate of protein synthesis (Gibson et al., 1987) and to a loss of muscle fibers (Oishi et al., 1992).

Despite the reduction in EMG and the muscle atrophy that have been reported to occur with limb immobilization, these results are difficult to interpret because of the dissociations between the decline in EMG and muscle atrophy and between the muscle atrophy and loss of function. These effects are apparent in figure 9.28*b* as a deviation of the data points from the line of identity. Fournier and colleagues (1983) found no reduction in EMG for the medial gastrocnemius muscle when it was immobilized at a neutral length, but the muscle experienced a 54% decline in mass. Similarly, rat soleus and medial gastrocnemius muscles experienced similar declines in wet weight (43-52%) at short and neutral lengths, whereas the decline in maximal isometric force at the short length was 72% to 77% compared with 45% at the neutral length (Simard et al., 1982; Spector et al., 1982). The decrease in maximal isometric force did not parallel the reduction in wet weight.

Some of the dissociation between muscle atrophy and the decline in force can be attributed to the definition of atrophy as the loss of mass. As discussed previously, the maximal force a muscle can exert is closely related to its cross-sectional area, not the quantity of muscle mass. When muscle atrophy is expressed in terms of cross-sectional area, there is a much tighter correlation with the decline in force (Hortobágyi et al., 2000; Lieber, 1992). However, another study showed that the cross-sectional areas of type I and IIa muscle fibers declined by about 25% with immobilization, whereas the maximal isometric force decreased by 40% for type FR and 52% for type S motor units (Nordstrom et al., 1995). Clearly, the remodeling that occurs in the neuromuscular apparatus during short-term immobilization is more complex than can be predicted by a linear relation between the decline in EMG, loss of muscle mass, and impairment of performance. It is likely that the altered neuromechanical conditions (e.g., fixed muscle length, continued isometric contractions, altered sensory feedback) are just as important as the reduction in neuromuscular activity in determining the nature of the adaptations.

Neuromuscular Adaptations

Given the absence of a simple relation between the decrease in activity and impairment of performance with limb immobilization, it has been difficult to develop a coherent view of the adaptations that occur with this intervention. The adaptations appear to extend from the genetic regulation of muscle mass to the descending control of muscle activation.

Muscle Fibers and Motor Units Seven days of immobilization of the rat soleus muscle at a short length reduced muscle mass by 37%, depolarized muscle fiber membranes by 5 mV, decreased the frequency of miniature end-plate potentials by 60%, and reduced Na^+-K^+ transport across the membrane by 25% (Zemková et al., 1990). Similarly, three weeks of immobilization of the rat plantaris muscle at a long length increased the postsynaptic areas of junctional folds and clefts at the neuromuscular junctions of type I and type II muscle fibers (Pachter & Eberstein, 1986).

One observation that has been reported in a number of studies is the slow-to-fast conversion of muscle fiber types. There is a decline in the proportion of slow-twitch (type I) muscle fibers and an increase in the proportion of fast-twitch, fatigue-resistant (type IIa) fibers (Fitts et al., 1989; Hortobágyi et al., 2000; Lieber et al., 1988; Oishi et al., 1992). For example, after one week of limb immobilization there was an increase in the quantities of messenger ribonucleic acid (mRNA) for faster isoforms of the MHCs in the gastrocnemius, plantaris, and soleus muscles of the rat hindlimb (Jänkälä et al., 1997). A similar effect has been observed in the vastus lateralis muscle of humans after three weeks of leg immobilization (Hortobágyi et al., 2000), although not after two weeks of immobilization (Labarque et al., 2002; Yasuda et al., 2005). The typical explanation for this adaptation is that the muscle fibers most affected by the immobilization are those whose activity is reduced the most, namely, the type I muscle fibers.

Despite the appeal of this rationale and the observations in muscles of the rat, dog, and human, studies on motor units in a cat hindlimb muscle have not shown a similar change in motor unit proportions or a differential reduction in cross-sectional area after several weeks of immobilization (Mayer et al., 1981; Nordstrom et al., 1995; Robinson et al., 1991). Nonetheless, the decline in force seems to be greatest in type S and FR motor units of cat hindlimb muscle (Petit & Gioux, 1993), which is consistent with the slow-to-fast hypothesis.

Six to eight weeks of immobilization alters both the properties and behavior of motor units in human hand muscles (Duchateau & Hainaut, 1990). When recruitment threshold was expressed relative to maximal force, there was an increase in the number of high-threshold motor units in the immobilized muscle. However, the average force exerted by these units was less (figure 9.29*a*), and there was a decline in the peak-to-peak amplitude of the motor unit action potentials. Although recruitment order was not altered by the immobilization, there was an increase in the range of recruitment and a decrease in the range of discharge rate modulation (figure 9.29*b*). This indicates a change in the activation strategy used to grade muscle force.

Muscle Activation A common finding in studies of limb immobilization is that the loss of strength often exceeds the decrease in muscle atrophy (Kitahara et al., 2003; Miles et al., 1994; Yasuda et al., 2005). For example, Hortobágyi and colleagues (2000) found that the strength of the quadriceps femoris muscles decreased by 47% after three weeks of immobilization, whereas the average decrease in the cross-sectional area of the muscle fibers was 11%. The dissociation between strength and muscle size could be caused by either a decrease in the intrinsic force-generating capacity of the muscle fibers or a decline in the activation of muscle by the nervous system. Consistent with a role for a change in muscle fiber force, Prathare and colleagues (2006) found that the

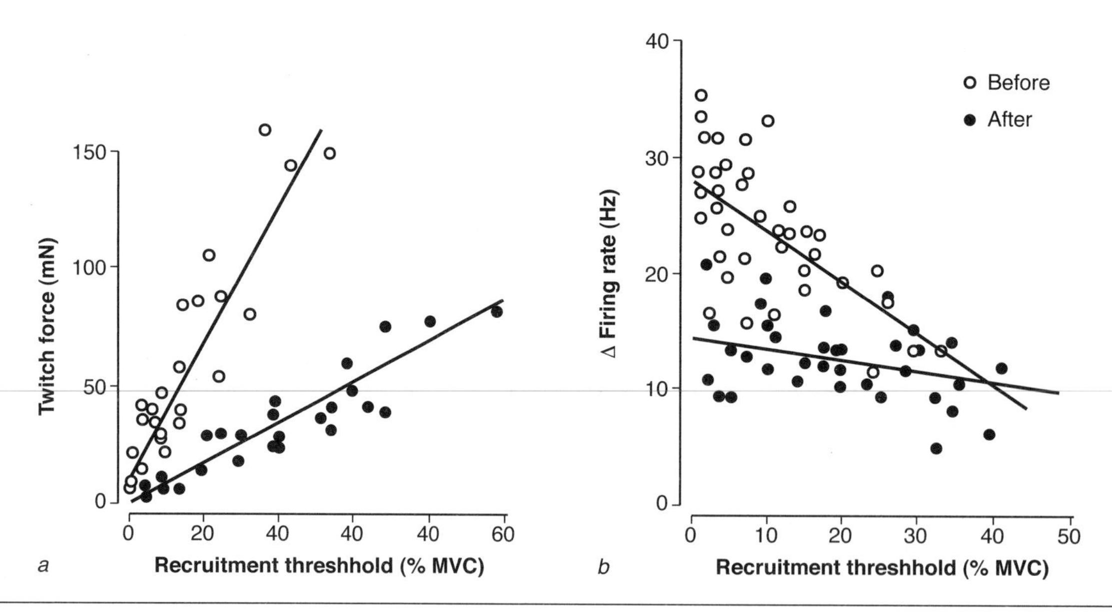

Figure 9.29 Changes in motor unit *(a)* twitch force and *(b)* range of discharge rate modulation as a function of recruitment threshold after several weeks of joint immobilization. Each data point corresponds to a single motor unit either before (open circles) or after (filled circles) limb immobilization.

Reprinted, by permission, from J. Duchateau and K. Hainaut, 1990, "Effects of immobilization on contractile properties, recruitment, and firing rates of human motor units," *Journal of Physiology* 422: 60-62.

deficits in strength and size of the plantarflexor muscles after six weeks of immobilization for an ankle fracture were accompanied by an increase in the resting levels of inorganic phosphate (P_i) and the ratio of P_i to phosphocreatine. Because elevated levels of P_i have been shown to inhibit the development of force in skinned muscle fibers, Prathare and colleagues found that the decline in P_i to resting levels was associated with the recovery of muscle strength. Furthermore, when muscles, motor units, or muscle fibers are activated with electrical stimulation, there remains a discrepancy between the decline in force and the decrease in cross-sectional area.

Nonetheless, a decline in activation of muscle by the nervous system also contributes to the loss of strength after immobilization, especially when a limb is immobilized due to an injury. The decrease has been measured as a reduction in voluntary activation and a decline in motor unit discharge rate. Stevens and colleagues (2006) found that patients with ankle fractures who were immobilized in a cast for seven weeks experienced decreases of 42% in voluntary activation, 26% in maximal cross-sectional area, and 75% in peak torque. The magnitude of voluntary activation accounted for 56% of the variability in peak plantarflexor torque during the 10 weeks of rehabilitation. A reduction in the average discharge rate of motor units was observed in a hand muscle after six weeks of joint immobilization (Seki et al., 2001). The average discharge rate was less after immobilization at both relative (% MVC force) and absolute forces (N). The adaptation in discharge rate is rapid, as it can occur after one week of immobilization and return to initial values after one week of recovery (Seki et al., 2007). These results indicate that some of the reduction in strength after immobilization is due to an impairment of activation by the nervous system.

Performance The muscle atrophy associated with limb immobilization results in a substantial loss of strength and an impairment of most activities of daily living (Imms et al., 1977; Imms & MacDonald, 1978). For example, six weeks of immobilization due to a fracture produced a 55% decline in the force and a 45% reduction in the EMG during an MVC with a hand muscle (Duchateau & Hainaut, 1991). There was also a reduction in the ability of subjects to generate maximal voluntary activation of the hand muscle. Nonetheless, these changes did not impair the ability of the subjects to sustain the maximal force for 60 s. Similarly, others have found that three weeks of immobilization in a plaster cast reduced the electrically evoked force (10%) and the MVC force (23%), but did not influence the fatigability of the triceps surae muscle in human volunteers (Davies et al., 1987).

The effect of immobilization on the fatigability of muscle differs for high- and low-force contractions. For example, the time to failure for an isometric contraction sustained at 65% MVC force by the elbow flexor muscles did not change after four weeks of limb immobilization, whereas the time to failure for a contraction at 20% MVC increased by 59% (Yue et al., 1997). Furthermore, this effect differs with the sex of the individual (Miles et al., 2005). Semmler and colleagues (1999, 2000*a*) found that four weeks of limb immobilization increased the time to failure for an isometric contraction at 15% MVC force with the elbow flexor muscles increased by 220% for one group of subjects (mainly women) but not change for the other group (men).

Recovery of strength after immobilization appears to depend on the health of the person exposed to the intervention (Miles et al., 2005). Studies performed on healthy subjects indicate a complete recovery of muscle strength within a few weeks (Duchateau & Hainaut, 1987; Semmler et al., 2000; Yue et al., 1997). Although recovery can occur in these individuals as a consequence of normal activities of daily living, it is facilitated by exercises that include lengthening contractions (Hortobágyi et al., 2000). In contrast, measurements performed on persons who were immobilized due to an injury or a surgical procedure indicate that complete recovery can take months (Stevens et al., 2006; Vandenborne et al., 1998).

Limb Unloading

When humans were first sent into space, several models were developed that mimicked many of the changes known to occur during spaceflight (Adams et al., 2003). One model involves suspending the hindlimbs of an animal, usually a rat, off the ground for a few weeks so that the limbs are free to move but cannot touch the ground or any support surface (figure 9.30*a*). This experimental technique is known as **hindlimb suspension** (Morey-Holton et al., 2005). The animal can perform many of its daily functions but experiences some stress (Sonnenfeld, 2005). Similarly, one leg of a human can be placed in a sling to reduce its level of activity (figure 9.31); this technique is known as **leg unloading** (Berg & Tesch, 1996; Berg et al., 1991; Dudley et al., 1992). The individual ambulates using crutches and the nonsuspended leg; the thickness of the sole of the shoe worn on the nonsuspended leg is increased by about 10 cm to reduce the possibility that the suspended foot will strike the ground during ambulation. Alternatively, the limbs of humans can be unloaded by prolonged **bed rest,** which is more commonly used than leg unloading as a ground-based model of the adaptations that occur in microgravity conditions (Alkner & Tesch, 2004; Trappe et al., 2004; Widrick et al., 2002*b*).

Reduction in Activity

Four important ways in which the hindlimb suspension model mimics spaceflight are a cephalic shift in fluid, loss of bone mineral content, decreased growth, and removal of the need for postural activity in the leg muscles. Despite the reduced need for hindlimb postural support, the EMG activity recorded in ankle muscles is not substantially depressed for the duration of the suspension period (figure 9.32*a*). The activity of soleus and medial gastrocnemius muscles is depressed for about the first three days of suspension but then recovers to control levels (Alford et al., 1987; Riley et al, 1990). Paradoxically, the muscles atrophy during the period when there seem to be normal levels of EMG (figure 9.32*b*), meaning that there is a dissociation between the levels of activity

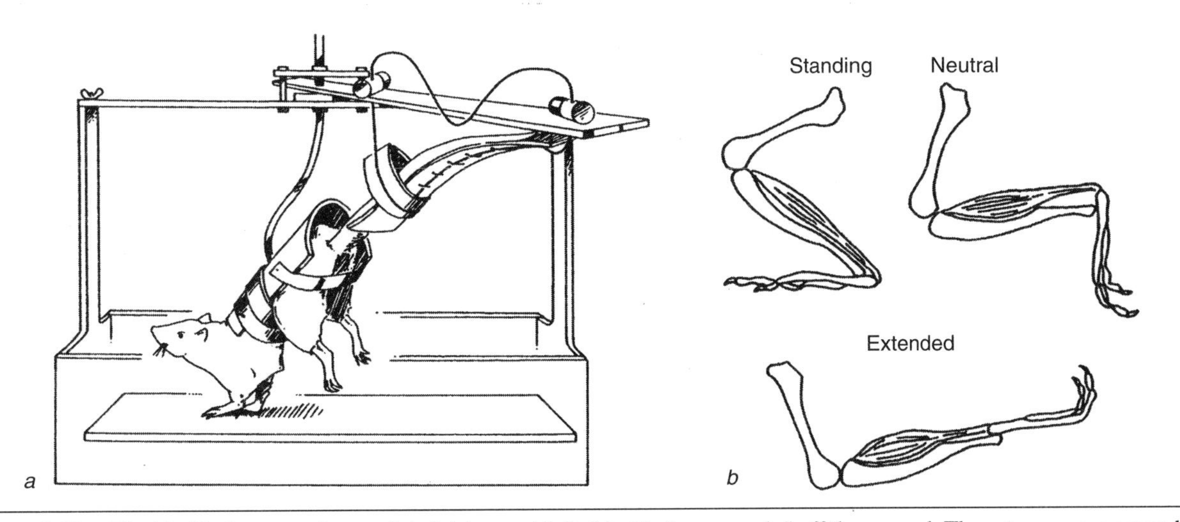

Figure 9.30 The hindlimb suspension model. *(a)* A rat with its hindlimbs suspended off the ground. The rat can move around its cage by using its front legs. *(b)* Change in the range of motion for the soleus muscle during hindlimb suspension; the top left image shows the minimal angle at the ankle joint during normal standing; the top right image indicates the neutral position during hindlimb suspension; and the bottom image represents the maximal angle for both conditions.

(9.30*a*) Provided by Charles M. Tipton, Ph.D. (9.30*b*) Adapted, by permission, from D.A. Riley et al., 1990, "Rat hindlimb unloading: Soleus histochemistry, ultrastructure, and electromyography" *Journal of Applied Physiology* 68: 58-66. Used with permission.

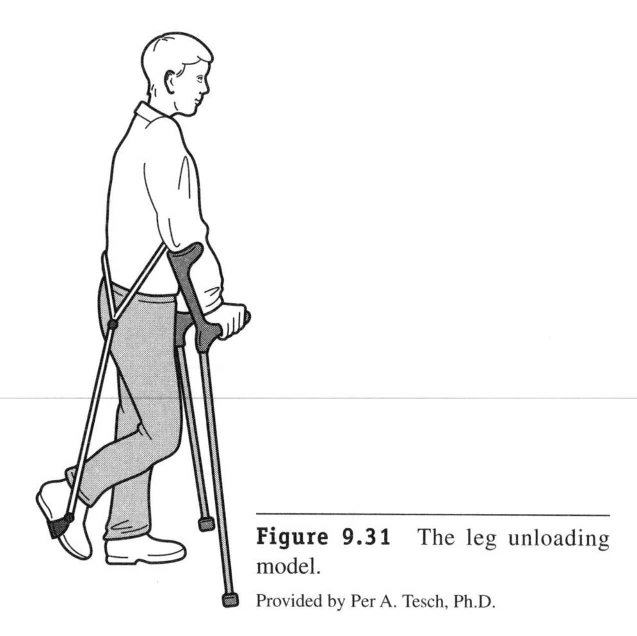

Figure 9.31 The leg unloading model.

Provided by Per A. Tesch, Ph.D.

and the loss of muscle mass (figure 9.33). Numerous studies have shown that the soleus muscle (slow-twitch fibers) experiences greater atrophy than either its fast-twitch synergist (medial gastrocnemius, plantaris) or its antagonist (tibialis anterior, extensor digitorum longus) muscles (Thomason & Booth, 1990). Furthermore, forelimb muscles experience heightened activity during hindlimb suspension (Allaf & Goubel, 1999).

The dissociation between EMG and muscle atrophy is probably attributable to an altered leg posture (figure 9.30*b*). In a weight-bearing stance, the range of motion at the ankle joint extends from about 0.5 rad to 3.14 rad. After a few days of suspension, however, the ankle adopts a neutral angle of about 1.57 rad, which reduces the range of motion substantially. Also, recall that the EMG is influenced by muscle length; for the same muscle force, the EMG is greater at shorter muscle lengths. Nonetheless, the extent to which the muscles are compromised by hindlimb suspension is much greater for the slow-twitch soleus compared with its fast-twitch synergists (Michel & Gardiner, 1990). By comparing the combined and individual effects of hindlimb suspension, tenotomy, and denervation on the atrophy experienced by single muscle fibers, Ohira and colleagues (2006) found no difference for fibers from soleus but observed that the atrophy of fibers from plantaris was further increased by removal of the innervation. These results suggest that an intact nerve–muscle connection protects fast-twitch muscle to some extent. However, Kyparos and colleagues (2005) found that mechanical stimulation of the plantar surface of the foot during hindlimb suspension can reduce the atrophy of soleus substantially, with a lesser effect on medial gastrocnemius. This indicates that the absence of sensory feedback contributes to the greater atrophy experienced by soleus.

A number of human studies have shown significant reductions in neuromuscular function after short-term exposure to leg unloading. For example, six weeks of leg unloading reduces whole-muscle and fiber cross-sectional area (Dudley et al., 1992; Hather et al., 1992). The cross-sectional areas of the vasti were reduced by 16%, that of the soleus by 17%, and that of gastrocnemius by 26%. Biopsies of the vastus lateralis muscle indicated reductions in cross-sectional area of 12% for the type I fibers and 15% for the type II muscle fibers. Furthermore, skin temperature over the calf muscles was 4° C cooler for the suspended leg, presumably due to decreased blood flow through the less active limb (Berg & Tesch, 1996).

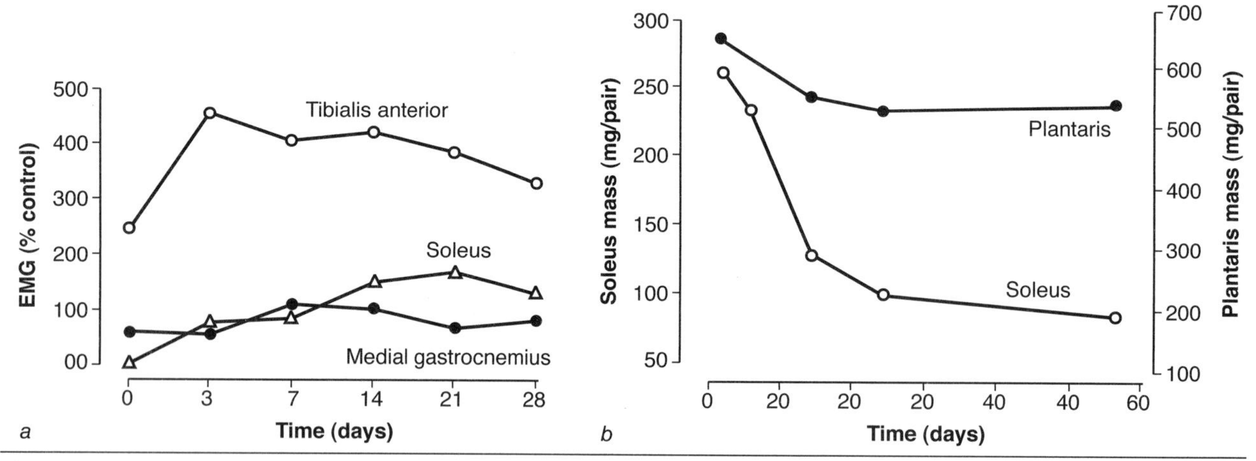

Figure 9.32 Results of 28 days of hindlimb suspension: *(a)* integrated EMG activity normalized to control values for three rat hindlimb muscles; *(b)* loss of mass in two muscles during 28 days of hindlimb suspension. Muscle mass is expressed in milligrams (mg) for the two hindlimb muscles of each animal.

Adapted, by permission, from D.B. Thomason and F.W. Booth, 1990, "Atrophy of the soleus muscle by hindlimb unweighting," *Journal of Applied Physiology* 68: 2, 4. Used with permission.

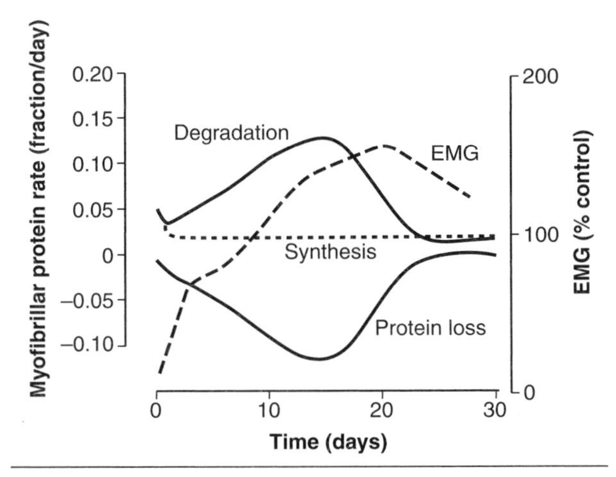

Figure 9.33 Time course of protein degradation, protein synthesis, protein loss, and integrated EMG activity in the rat soleus muscle during 28 days of hindlimb suspension. The EMG values are normalized to control measurements prior to the intervention. The net loss of protein is largely due to an increase in protein degradation despite an increase in EMG.

Reprinted, by permission, from D.B.Thomason and F.W. Booth, 1990, "Atrophy of the soleus muscle by hindlimb unweighing," *Journal of Applied Physiology* 68: 8. Used with permission.

Neuromuscular Adaptations

Although hindlimb suspension and leg unloading both reduce the availability of limbs to participate in activities of daily living, the two interventions produce different adaptations (Widrick et al., 2002*b*). Accordingly, this section briefly describes some of the adaptations produced by each intervention and compares these with the adaptations that have been observed after spaceflight.

Hindlimb Suspension Many animal studies have shown a preferential effect of hindlimb suspension on slow-twitch muscle (Esser & Hardeman, 1995; Fitts et al., 1986; Ohira et al., 2002; Thomason et al., 1987). Consequently, studies often focus on the effects elicited in soleus (Thomason & Booth, 1990). A few weeks of suspension produces a decline in the proportion of type I fibers in soleus with little effect on synergist (medial gastrocnemius) or antagonist muscles (tibialis anterior). The suspension produces a decrease in the concentration of myofibrillar and myosin proteins, with an increased appearance of the fast myosin isoforms. However, this effect occurs in some, but not all, type I muscle fibers of soleus (Gardetto et al., 1989). There is a reduction in the proportion of muscle fibers that comprise only the MHC for type I fibers and an increase in the proportion that contain multiple isoforms of the MHCs for type II fibers (Oishi et al., 1998; Toursel et al., 2002). As a consequence there is a decrease in the proportion of slow-twitch motor units in the soleus muscle after a few weeks of hindlimb suspension (Leterme & Falempin, 1996). These adaptations, however, do not involve changes in the soma size or oxidative capacity of the innervating motor neurons (Ishihara et al., 1997).

Accompanying these changes in protein content and myosin isoforms are some changes in the metabolic capabilities of the muscle fibers (Roy et al., 1991). The concentrations of enzymes involved in oxidative metabolism (succinate dehydrogenase and citrate synthase) increase in soleus with hindlimb suspension. After several weeks of suspension, however, predominantly fast-twitch muscles (medial gastrocnemius, tibialis anterior) exhibit lower levels of some of the enzymes (succinate dehydrogenase) that characterize oxidative metabolism but not others (citrate synthase). In contrast, both type I and type II muscle fibers in hindlimb extensor and flexor muscles either maintain or elevate the levels of enzymes (e.g., alpha glycerophosphate dehydrogenase) associated with glycolytic metabolism.

The changes in protein and enzyme content with hindlimb suspension produce several adaptations in the mechanical properties of muscle. The maximal force capacity of the soleus muscle is reduced, but the decline is greater than would be anticipated based on the loss of muscle mass. This indicates a reduction in normalized force (Herbert et al., 1988; McDonald & Fitts, 1995; Thomason & Booth, 1990), which involves a decrease in specific tension (Larsson et al., 1996; Riley et al., 2000; Thompson et al., 1998) that is likely due to a decrease in thin filament density and changes in the structural distribution of myosin (Riley et al., 2005; Zhong et al., 2006). At the muscle fiber level, type I muscle fibers from both soleus and medial gastrocnemius show a significant reduction in diameter and peak force, whereas the type IIa fibers from medial gastrocnemius exhibit a decline in diameter but no change in peak force (Gardetto et al., 1989). In contrast, Leterme and Casasnovas (1999) found a decrease in the peak tetanic force of motor units in lateral gastrocnemius after 14 days of hindlimb suspension. In addition to the influence of hindlimb suspension on the contractile proteins, there is a decrease in the amount of some cytoskeletal proteins, such as titin (Toursel et al., 2002).

One functional outcome of these adaptations is an increase in contraction speed (Roy et al., 1991; Thomason & Booth, 1990). The soleus muscle becomes faster, which should be expected due to the shift in myosin isoforms and the reduction in the proportion of type I muscle fibers (Riley et al., 2005; Stelzer & Widrick, 2003). This was evident by an increase in the maximal velocity of shortening (V_{max}) for the soleus muscle and some of its type I muscle fibers (Thompson et al., 1998). Not all of the muscle fibers indicated an increase in V_{max}, which is consistent with a limitation of the change in myosin isoforms to a subpopulation of these muscle fibers (Gardetto et al., 1989). The increase in V_{max} shifts the force–velocity relation to the right. Other effects on contraction speed include a reduction in both contraction time and one-half relaxation time; these reductions are probably related to changes in Ca^{2+} kinetics as hindlimb suspension increases the Ca^{2+} concentration necessary to activate the contractile apparatus (Gardetto et al., 1989; McDonald & Fitts, 1995).

The influence of hindlimb suspension on the fatigability of the soleus muscle depends on the intensity of the imposed fatigue test. There is no change in the fatigability of soleus with moderate stimulation, but fatigability with high-intensity stimulation increases markedly (100 Hz, 100 ms trains, 120/min) (McDonald et al., 1992; Winiarski et al., 1987). In contrast, the gastrocnemius-plantaris-soleus complex exhibited less fatigue than soleus after 15 days of suspension with the high-intensity protocol (McDonald et al., 1992). The rat medial gastrocnemius muscle displays a more complicated adaptation that involves no effect on fatigability after seven days of suspension but an increase in fatigability after 28 days of suspension (Winiarski et al., 1987). Similarly, the fatigability of motor units in rat lateral gastrocnemius did not change after 14 days of suspension (Leterme & Casasnovas, 1999).

Leg Unloading There are some similarities but some important differences between the changes that occur with hindlimb suspension and the adaptations exhibited by fibers from human leg muscles after leg unloading. Widrick and colleagues (2002*b*) measured the contractile properties of MHC-type fiber segments from soleus and gastrocnemius after 12 days of leg unloading. The only fiber type to experience a decrease in diameter was the type I fibers from soleus (7%). However, the type I fibers from both muscles displayed a reduction in peak force: 18% for soleus and 14% for gastrocnemius. Some of the decrease in peak force for the soleus fibers was caused by a reduction in specific tension, and it was greater than that observed with hindlimb suspension. In contrast to findings for hindlimb suspension, maximal shortening velocity was reduced by leg unloading in type I fibers from soleus (10%) and increased in type IIa fibers from gastrocnemius (12%). Furthermore, 6 h after the subjects began reambulating, the capabilities of the type I fibers from soleus were further reduced and there was evidence of fiber damage. The muscle atrophy related to leg unloading is associated with a decrease in the capacity for protein synthesis (Gamrin et al., 1998).

Because the decrease in muscle strength with limb unloading is greater than the amount of muscle atrophy (Dudley et al., 1992; Hather et al., 1992), the intervention seems to impair the activation of muscle. For example, exercise-induced shifts in the signal intensity of magnetic resonance images (figure 6.9) indicate that greater volumes (50-130%) of the knee extensor muscles are activated to lift submaximal loads after five weeks of suspension (Ploutz-Snyder et al., 1995). Similarly, the EMG of the knee extensor muscles associated with exerting a force of 30% to 45% MVC increased by 25% after 10 days of unloading (Berg & Tesch, 1996). This adaptation may be due to an impairment of excitation-contraction coupling rather than a reduced ability to maximally activate the muscle. Furthermore, four weeks of leg unloading reduced the work capacity of the knee extensor muscles (Berg et al., 1993).

Spaceflight Most organ systems experience some adaptation when an organism is exposed to spaceflight (Antonutto & di Prampero, 2003; Clement et al., 2005; Heer & Paloski, 2006; Lang et al., 2006; Tesch et al., 2005). The adaptations are pronounced in the neuromuscular system (Fitts et al., 2000). Some of the adaptations are similar to those observed with ground-based models (e.g., hindlimb suspension, leg unloading, and bed rest), but others are not. For example, the atrophy experienced by single muscle fibers differs for rats and humans and between spaceflight and ground-based models (figure 9.34); in the human vastus lateralis muscle, the atrophy was greatest for type IIx fibers and least for type I fibers. Furthermore, rats experience a decrease in the number of fibers in antigravity muscles and an increase in the number

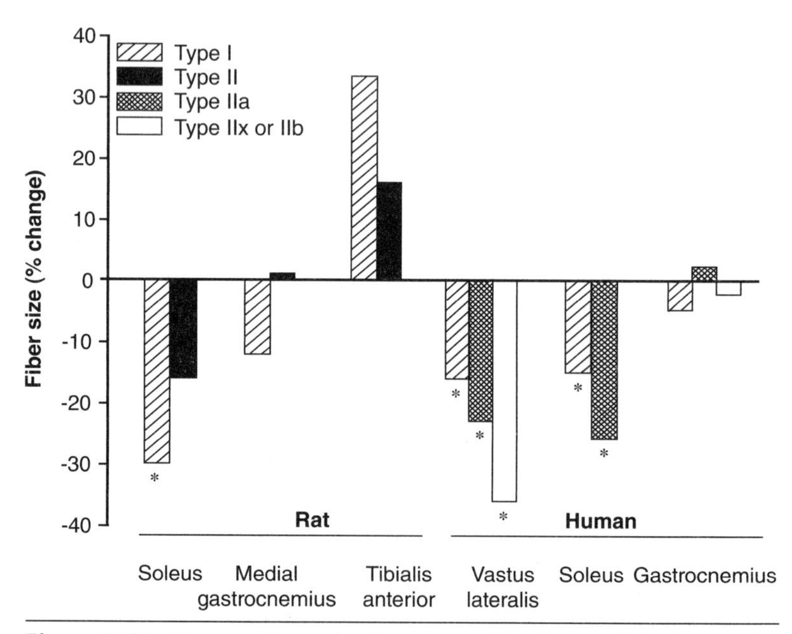

Figure 9.34 Percent change in the cross-sectional area of single muscles sampled from rats and humans after 11 to 17 days of spaceflight. $*p < 0.05$ compared with baseline values.

Reprinted, by permission, from R.H. Fitts, D.R. Riley, and J.J. Widrick, 2000, "Physiology of a microgravity environment. Invited review: Microgravity and skeletal muscle," *Journal of Applied Physiology* 89: 824. Used by permission.

of fibers containing fast-type myosin, whereas humans exhibited a similar pattern in soleus but an increase in myosin adenosine triphosphatase (ATPase) activity for type IIa fibers but not type I fibers in vastus lateralis. The duration of spaceflight influences the decline in protein content; 12.5 days is sufficient to decrease myofibrillar proteins in the slow-twitch vastus intermedius but not the fast-twitch vastus lateralis. The data from rats and humans suggest that the atrophy involves a selective loss of slow-type myosin.

Microgravity induces declines in muscle strength that are significant after 17 days in space. The decrease in strength for humans is greater for thigh muscles than for arm muscles and for extensors than for flexors. For example, a 28-day flight produced a 20% decrease in peak torque for the knee extensors compared with a 10% decline for the thigh flexor and arm extensor muscles. As with the muscle atrophy, there is substantial variability between individuals. The decrease in strength appears to reach a new steady-state level, as the strength of the plantarflexor muscle after 110 days in space remained the same after 237 days. However, the strength of the dorsiflexor muscles declined over the entire 237 days until they exhibited the same percent change as the plantarflexor muscles (Fitts et al., 2000).

At the single-fiber level in humans, the peak force of type I fiber segments from soleus decreased by 21% and that for type IIa fibers declined by 25% after a 17-day flight (Widrick et al., 1999). The reduction in peak force for the type I fibers was not proportional to the atrophy they experienced, which suggested a loss of myofibril content. The maximal shortening velocity was greater after the flight for both type I and type IIa fibers from soleus, and to a lesser extent for type I fibers from gastrocnemius (Widrick et al., 1999, 2001). Although peak force declined, the increase in shortening velocity resulted in relatively minor reductions in peak power production for both fiber types after the 17-day flight. Peak power produced by the leg muscles, however, decreased by 32% after 31 days, which suggests a likely contribution from impairment of muscle activation (Antonutto et al., 1999).

As with hindlimb suspension, Caiozzo and colleagues (1994) found that the soleus muscles of rats are more fatigable (40 Hz, 300 ms trains, 60/min) than other muscles after a six-day spaceflight. The tetanic force evoked by the electrical stimulation declined by 36% in control soleus muscles after 2 min of stimulation and by 66% in the muscles from rats that had flown in space. The greater fatigability was attributed to a more rapid utilization of glycogen and a reduced capacity to oxidize fats. Presumably, additional impairments will emerge when humans performing voluntary contractions are studied (Antonutto et al., 1999; Edgerton et al., 2001*b*) as the proprioceptive and vestibular systems of humans exhibit adaptations after exposure to microgravity (Lambertz et al., 2003; McCall et al., 2003; Reschke et al., 1998; Roll et al., 1993, 1998).

In general, these data indicate that hindlimb suspension, leg unloading, and spaceflight all reduce the mass and strength of muscle. Nonetheless, there are some significant differences due to species and type of intervention that preclude a general description of the effects of limb unloading.

Spinal Transection

In contrast to limb immobilization and limb unloading, which represent models of restraint, **spinal transection** imposes a reduction in use by disconnecting parts of the nervous system from muscle. The separation is imposed in experimental animals at the level of the spinal cord, usually T12-T13, and is referred to as an **upper motor neuron** lesion because it eliminates supraspinal control of the hindlimbs. This lesion is distinct from a **lower motor neuron** lesion whereby the muscle is separated from all components of the nervous system (i.e., denervation).

Reduction in Activity

Transection of the spinal cord, which is also known as **spinalization,** produces an immediate flaccid paralysis such that the hindlimbs are dragged by the experimental animal (Lieber, 1992). About three to four weeks after the spinalization, the muscles develop spasticity, which eventually leads to sustained extensor activity with no apparent voluntary activation of the muscles. In this paralyzed state, however, the neuromuscular system can still be electrically activated; and with appropriate support and afferent feedback, the animals can be trained to perform hindlimb locomotion on a treadmill (Belanger et al., 1996; Gregor et al., 1988; Lovely et al., 1990). Spinalization reduced the daily EMG in rats to <1% of control measures in soleus, <2% in medial gastrocnemius and vastus lateralis, and <8% in tibialis anterior (Roy et al., 2007). The mean amplitudes of the EMG bursts of activity were normal in soleus but increased in medial gastrocnemius (two times), vastus lateralis (two times), and tibialis anterior (four times).

In humans who experience a spinal cord injury, many motor units in muscles that are deprived of descending control are active spontaneously. For example, Thomas (1997) found that motor units in thenar muscles of patients who had an injury at C7 or higher for more than a year exhibited spontaneous activity, which the patients were unable to control. The motor units discharged at low rates, in the range of 1 to 8 pps, with occasional rapid bursts of activity. The involuntary activity can be depressed by vibration of the muscles (Butler et al., 2006). Furthermore, motor units in an individual with an incomplete spinal cord transection can continue to discharge action potentials after a voluntary contraction has concluded (Zijdewind & Thomas, 2003). Thus, chronic paralysis does not eliminate activity in paralyzed muscle.

Neuromuscular Adaptations

Because spinal transection eliminates input from supraspinal centers to the spinal cord, the motor neuron pools receive less synaptic input and the muscles are accordingly more quiescent than in an intact system. The ensuing adaptations influence the properties of muscle fibers and motor units, as well as the performance capabilities of the residual motor system.

Muscle Fibers and Motor Units As with limb immobilization and limb unloading, spinalization produces muscle atrophy with a preferential effect on slow-twitch muscle and a slow-to-fast conversion of muscle fiber and motor unit types (Celichowski et al., 2006; West et al., 1986). The conversion of fiber types and loss of muscle mass are most pronounced in single-joint muscles involved in postural support. Soleus and medial gastrocnemius atrophy by about 45% and 30%, respectively, by two weeks after spinalization. There is a decrease in the proportion of type I muscle fibers and an increase in type II fibers; these adaptations result in an increase in the proportion of type FR motor units in soleus and an increase of type FF and FI (fast-twitch, intermediate fatigability) motor units in medial gastrocnemius (Munson et al., 1986). These changes include an increase in the expression of fast myosin isoforms, a decrease in the number of fibers in soleus that react with a slow MHC antibody, an increase in myosin ATPase of 50% for soleus and 30% for medial gastrocnemius, and an increase in the proportion of fibers from the rat soleus muscle that include multiple MHC isoforms (Jiang et al., 1990; Roy et al., 1984; Talmadge et al., 1995). Such adaptations were found to be less pronounced in the tibialis anterior muscle (Pierotti et al., 1994).

The slow-to-fast conversion of muscle fiber types also influences the activities of enzymes associated with oxidative and glycolytic metabolism (Castro et al., 1999; Otis et al., 2004; Roy et al., 1991). These effects, however, are complex because the magnitude of the metabolic adaptations at the single-fiber level depends on both the age at spinalization and the muscle fiber type as defined by myosin properties. For example, there appear to be different effects on citrate synthase and succinate dehydrogenase, two enzymes involved in the citric acid cycle. In contrast, spinalization does not seem to alter the relation between the glycolytic potential of a muscle fiber and its type as defined by myosin ATPase.

Because the spinal transection model involves the removal of supraspinal control over spinal networks, changes occur in the properties and connectivity of spinal cord neurons. For example, motor neurons that innervate the soleus muscle become less excitable and come to resemble motor neurons that innervate fast-twitch muscle. These changes include a decrease in the afterhyperpolarization duration and an increase in rheobase (Cope et al., 1986). There are, however, mixed reports on changes in the motor neurons that innervate the medial gastrocnemius muscle (Czéh et al., 1978; Foehring & Munson, 1990; Foehring et al., 1987*a, b*), which appear to differ with motor unit type (Hochman & McCrea, 1994*c*). In contrast, paralysis that is induced by the application of tetrodotoxin, which does not involve sectioning the spinal cord, evokes changes in the twitch and tetanic responses of motor units in soleus but minimal changes in the motor neurons (Gardiner & Seburn, 1997).

Performance Capabilities The physiological adaptations exhibited by spinalized muscles are consistent with the changes in contractile and metabolic properties (figure 9.35). The slow-to-fast conversion in the cat soleus muscle is associated with a decrease in both time to peak force and half-relaxation time and an increase in the maximal velocity of shortening (V_{max}). The increase in V_{max} is to be expected given the increase in myosin ATPase, the increase in proportion of type II muscle fibers, and the increased incidence of fast myosin isoforms (Roy et al., 1984). The change in the time course of the twitch (time to peak force and half-relaxation time) probably reflects changes in Ca^{2+} kinetics. In the cat medial gastrocnemius muscle, there is no change in the time course of the twitch, but V_{max} increases due to the increase in myosin ATPase. Both slow- and fast-twitch muscle manifest an increase in glycolytic enzyme activity after spinal cord transection.

Somewhat different adaptations have been observed in the vastus lateralis muscle of humans who have had a spinal cord injury for more than three years. Malisoux and colleagues (2007) compared measurements made on fiber segments obtained by biopsy from five spinal cord–injured individuals with data from 10 able-bodied persons. They found that the cross-sectional area and peak force of MHC-typed fibers were similar for type I, IIa, and IIx fibers between the two groups, whereas the maximal shortening velocity (26% and 47%, respectively) and peak power (46% and 118%) were greater for type IIa and IIx fibers of the able-bodied persons. Malisoux and colleagues (2007) also found a decrease in the proportions of type I and IIa fibers, as well as an increase in the proportion of type IIx fibers from 7% to 40%. Gregory and colleagues (2003) reported a similar adaptation for human vastus lateralis, but a much lesser switch in rat vastus lateralis.

The fatigability of the soleus muscle in animals is not affected by spinalization (Gordon & Pattullo, 1993). For

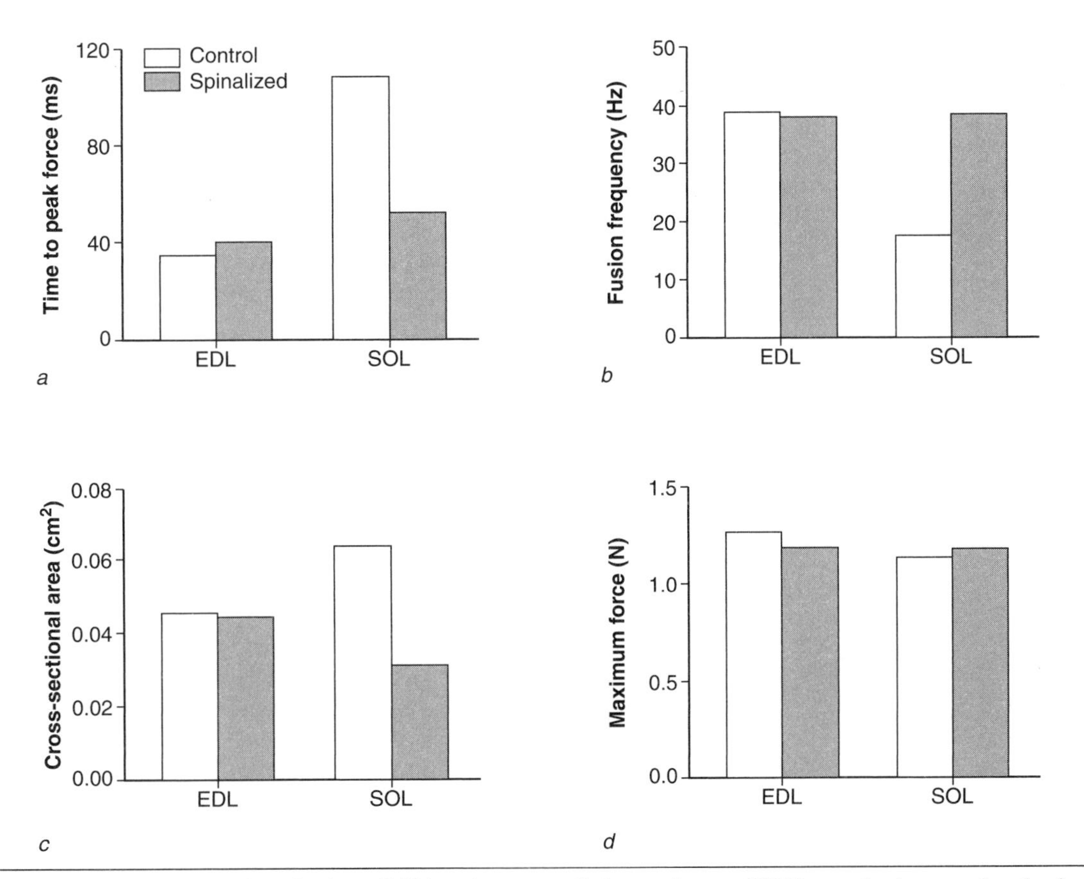

Figure 9.35 Contractile properties of soleus (SOL) and extensor digitorum longus (EDL) muscles in normal and spinalized rat hindlimbs: *(a)* time to reach the peak twitch force; *(b)* stimulus frequency at which the tetanic force became fused (smooth); *(c)* cross-sectional area; *(d)* maximal tetanic force. The greater effects occurred in the soleus muscle.

Adapted, by permission, from R.L. Lieber, 1992, *Skeletal and muscle structure and function* (Baltimore, MD: Lippincott, Williams, and Wilkins), 229.

medial gastrocnemius, however, fatigability can change depending on the age at spinal transection. There is no effect on the fatigability of medial gastrocnemius in young animals, but the muscle becomes more fatigable when an adult animal experiences a spinal transection (Celichowski et al., 2006). In humans, the effect of spinalization on fatigability changes with time. Soon after spinalization (six weeks or less), there are minimal changes in the fatigability of the soleus or quadriceps femoris muscles (Castro et al., 1999; Gaviria & Ohanna, 1999; Gerrits et al., 1999; Shields, 1995). Patients who have been spinalized for more than a year, however, exhibit a marked increase in the fatigability of these muscles. Motor units in thenar muscles of individuals who have been paralyzed for about 10 years are also more fatigable than usual, likely due to adaptations that have occurred in the muscles (Klein et al., 2006).

The changes in motor neuron properties are accompanied by changes in segmental reflexes. In humans with cervical spinal cord lesions, it is possible to elicit interlimb responses that are not normally present in able-bodied individuals. For example, brief electric shocks to the lower extremities are sufficient to elicit unusual ipsilateral and contralateral upper extremity responses (Calancie, 1991). The contralateral responses, which predominate, involve the response of motor units to light touch, individual hair movement, and thermal stimulation (Calancie et al., 1996). The amplitude and spread of tendon tap reflexes can be used to distinguish complete from incomplete lesions of descending motor pathways (Calancie et al., 2004). Furthermore, the excitatory postsynaptic potentials recorded in motor neurons that innervate ankle extensor muscles after stretch are enhanced after six weeks of spinal cord transection (Hochman & McCrea, 1994*a,b,c*). Such adaptations may contribute to the development of spinal spasticity.

Incomplete Transection Muscle and motor unit properties after spinal cord injury in humans have been examined in muscles that are completely paralyzed and in those that retain some voluntary activity. An incomplete spinal cord injury causes atrophy of leg muscles, with average reductions in cross-sectional area of 30% for thigh muscles and 25% for shank muscles (Shah et al., 2006). Among 17 individuals who had an incomplete spinal cord injury at C7 or higher for more than a year, the strength of the paralyzed thenar muscles for half the subjects was similar to that of able-bodied subjects (Thomas, 1997). The contraction time of the evoked twitch response did not differ between the injured and able-bodied subjects. However, the properties of the twitch and tetanic responses were much more variable in the spinal cord–injured subjects. Furthermore, the muscles of these individuals comprised fewer motor units, which ranged in strength, from exerting no detectable force to being five times normal strength (Thomas, 1997; Yang et al., 1990).

Similar studies have been performed on the triceps brachii muscle in spinal cord–injured subjects who retain some voluntary control of the muscle. Maximal voluntary contraction force is significantly less in these individuals compared with able-bodied persons (Thomas et al., 1998, 1997*b*). Although these muscles exhibit marked atrophy, they are also weak due to an impairment of the neural drive to the motor neurons (Thomas et al., 1997*b*). As observed in paralyzed muscle, the forces exerted by motor units in partially paralyzed muscle range from near zero to much greater than normal. For example, 11% of the motor units in partially paralyzed triceps brachii generated normal EMGs but no measurable force; 65% of the units were similar to control units; and 24% were stronger than usual (Thomas et al., 1997*a*). Nonetheless, these motor units were activated voluntarily in order of increasing force, which is consistent with the orderly recruitment of motor units.

MOTOR RECOVERY FROM INJURY

Although the neuromuscular system has remarkable adaptive capabilities, there are limits to the adaptations that can be achieved. There are limits, for example, to muscle plasticity (Pette & Vrbová, 1999). Moreover, the nervous system has minimal regenerative capabilities (Bhardwaj et al., 2006; Ramirez-Amaya et al., 2006; van Pragg et al., 2005). To describe this feature of chronic adaptations, we consider the capabilities of the neuromuscular system to recover motor function after an injury to the peripheral nerve or a lesion in the CNS.

Peripheral Nervous System

The neuromuscular system is capable of some recovery of function after an injury to a peripheral nerve. Because neurogenesis does not occur in the adult CNS (Bhardwaj et al., 2006), recovery depends on the ability of the neurons to reinnervate appropriate targets. To accomplish this, either the injured axons regenerate or the surviving axons develop sprouts and reinnervate the abandoned targets.

Axon Regeneration

When an axon is cut, in a procedure known as **axotomy,** degenerative changes occur both in the axon distal to the lesion and in the neuron (figure 9.36). About two to three days after the axotomy, the soma begins to swell and may double in size, and the rough endoplasmic reticulum (stained as Nissl substance) breaks apart and moves to the periphery of the soma (figure 9.36, *b* & *c*). The dissolution of the Nissl substance is referred to as **chromatolysis.** The process of chromatolysis, which lasts one to three weeks, involves a massive resynthesis of the proteins necessary for regeneration of the axon and the formation of a growth cone that enables the development of sprouts. If a sprout is able to invade the remaining myelin fragments (figure 9.36*c*) or newly generated Schwann cells, then it is likely that reinnervation of the original target will occur. This is possible in the motor, sensory, and autonomic divisions of the peripheral nervous system. Invasion of the appropriate myelin fragment is the most critical step in the recovery of function. Failure of reinnervation, however, results in degeneration of the axotomized neuron, the distal segment of the axon, and the target cell (MacIntosh et al., 2006).

Because recovery depends on the ability of sprouts to reinnervate appropriate targets, the nature of the injury to a peripheral nerve is an important determinant in the extent of the recovery of function. In general, a complete transection of the nerve offers the worst prognosis; a partial denervation (sparing of some axons) is less severe; and a crush injury has only minimal long-term effects (Leterme & Tyc, 2004; Ruijs et al., 2005). Injuries that result in a transection of peripheral nerve trunks are usually accompanied by lasting motor and sensory deficits, which represent poor functional recovery (Cope et al., 1994; Haftel et al., 2005; Scott, 1996). There are surgical techniques for resuturing a severed nerve with the aim of permitting regenerating axons to develop sprouts and reinnervate original targets (Lanzetta et al., 2005). These procedures, however, typically result in a significant number of misdirected reinnervations (table 9.6). When this occurs, motor neurons that originally innervated one muscle now innervate another muscle. It is possible to demonstrate this by finding a motor unit in a test muscle and then determining which muscle the individual must contract to activate the motor unit. In a study on the first dorsal interosseus muscle, 39% of the motor units were reinnervated correctly; 11% of the motor units were reinnervated by axons that previously went to abductor digiti minimi (abducts the fifth finger), 22% by axons that previously went to adductor pollicis (adducts the thumb), and 28% by axons that previously went to other muscles (Thomas et al., 1987).

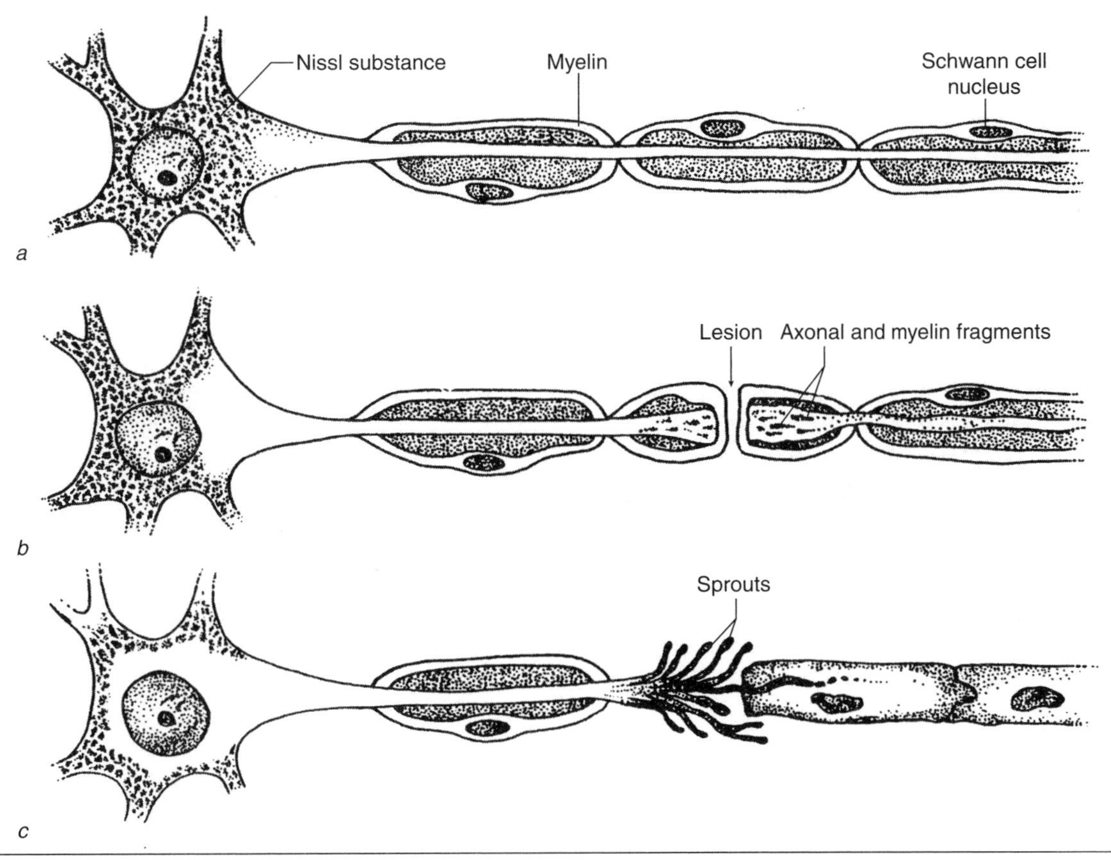

Figure 9.36 Changes in the distal axon and in the neuron after axotomy: *(a)* a normal soma and axon of a neuron; *(b)* two to three days after axotomy; *(c)* one to three weeks after axotomy.

Adapted, by permission, from J.P. Kelly, 1985, Reactions of neurons to injury. In *Principles of neural science,* 2nd ed., edited by E.R. Kandel and J.H. Schwartz (New York: McGraw-Hill Companies), 187. Adapted with permission from The McGraw-Hill Companies.

Table 9.6 Origin of Motor Axons in Reinnervated Muscles After Complete Transection of the Ulnar Nerve at the Wrist

	Abductor digiti minimi	First dorsal interosseus	Adductor pollicis	Others
Abductor digiti minimi	16 (34%)	15 (31%)	9 (18%)	8 (17%)
First dorsal interosseus	6 (11%)	22 (39%)	12 (22%)	16 (28%)

The rows indicate motor unit location (abductor digiti minimi or first dorsal interosseus), whereas the four columns denote the muscles in which voluntary contractions were needed to activate the motor unit.

Note. Data are from Thomas et al., 1987.

An obvious consequence of these misdirected reinnervations is that the activation of a motor neuron pool does not lead to the selective activation of one muscle. For example, table 9.6 indicates that activating the motor neuron pool for first dorsal interosseus will result in activation of at least the first dorsal interosseus (index finger abduction) and abductor digiti minimi (fifth finger abduction) muscles. Similarly, activation of the motor neuron pool for adductor pollicis (thumb adduction) will activate both first dorsal interosseus and abductor digiti minimi. This reorganization obviously impairs motor coordination for fine movements. With time, however, the regenerated axons establish the size-dependent order observed in healthy muscle so that the first-recruited motor units comprise the lowest innervation numbers and recruitment proceeds in the order of increasing size (Gordon et al., 2004). Nonetheless, recovery is less impressive for sensory axons, which makes it difficult for patients to localize a sensory stimulus, although they can detect it.

Collateral Sprouting

Compared with the recovery associated with a complete nerve transection, there is less impairment following recovery from a partial transection of the nerve close to the muscle. The principal recovery mechanism following a partial lesion is somewhat different from that associated with a complete transection; it still involves axonal growth and reinnervation, but the axons of surviving motor units develop sprouts to reinnervate the muscle fibers that have been denervated (Dengler et al., 1989; Luff et al., 1988; Slawinska et al., 1998). Sprouting by the surviving axons, referred to as **collateral sprouting,** is confined to the distal region of the motor axon and occurs close to the target. Apparently all the motor units within a pool compensate for these types of nerve injuries by collateral sprouting (Rafuse et al., 1992; Rafuse & Gordon, 1996). The number of additional fibers reinnervated depends on the size of the motor unit; larger motor units are capable of supporting a larger number of muscle fibers. Furthermore, the extent of the sprouting depends on the number of fibers that were denervated (figure 9.37). The greater the extent of the partial lesion and thus the number of muscle fibers that are denervated, the greater the amount of collateral sprouting and the greater the size of the surviving motor units. Infusion of a drug (tetrodotoxin) that blocks axonal action potentials can limit collateral sprouting in partially denervated muscles (Tam et al., 2002). Moreover, the performance of physical activity, such as running, can further increase the tetanic force of already enlarged motor units (Seburn & Gardiner, 1996), although increases in physical activity are detrimental in muscles that have been extensively denervated (Tam et al., 2001; Tam & Gordon, 2003).

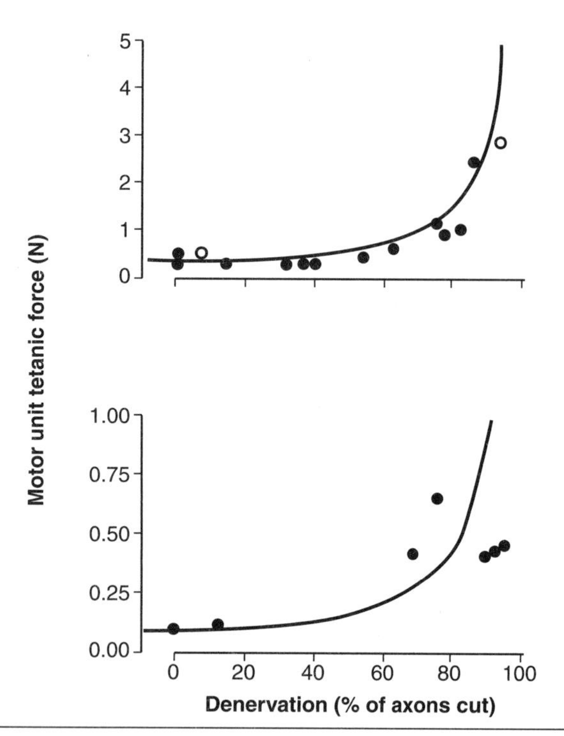

Figure 9.37 Tetanic force of motor units as a function of the degree of partial denervation (percentage of the nerve that was cut): *(a)* motor units in medial (filled circles) and lateral (open circles) gastrocnemius; *(b)* motor units in soleus.

Adapted, by permission, from V.F. Rafuse, T. Gordon, and R. Orozco, 1992, "Proportional enlargement of motor units after partial denervation of cat triceps," *Journal of Neurophysiology* 68:1266. Used with permission.

Collateral sprouting can account for up to about 80% of the lost force, with motor units capable of enlarging to about five times the original size (Gordon et al., 1993). These observations indicate that when the peripheral nerve is cut, the neuromuscular system can recover the maximal force capability up to about 80%. This recovery could result from three mechanisms: an increase in the innervation number, an increase in the average cross-sectional area of muscle fibers, or an increase in the specific tension. The most important mechanism is an increase in the innervation number, which must be due to collateral sprouting (Rafuse & Gordon, 1998; Tötösy de Zepetnek et al., 1992). The potential for collateral sprouting does not depend on descending drive, because it occurs to the same extent whether or not the spinal cord is transected above the level of the motor neuron pool.

EXAMPLE 9.6

Denervation Changes Muscle Fiber Properties

When the nerve to a muscle is cut, the muscle loses its innervation; this is referred to as **denervation.** In addition to the changes that occur in the proximal and distal segments of the severed axons, the target cells of the motor axons—the muscle fibers—undergo a number of changes. As described in more detail by MacIntosh and colleagues (2006), these changes include the following:

- Denervation atrophy: A decrease in fiber size, which begins about three days after denervation, occurs in all fibers.
- Displacement of nuclei: The nuclei change shape and migrate from the outer membranes to the center of the fiber.
- Necrosis: Some fibers degenerate and die after several months.
- Reduction in size: As the fibers atrophy, many organelles (e.g., mitochondria, sarcoplasmic reticulum) also decrease in size.
- Lowered enzyme activity: The activities of many, but not all, enzymes decrease so that the fibers appear more homogeneous.
- Decline in contractile properties: As a consequence of the fiber atrophy and decrease in enzyme activities, the amplitude of the twitch

and tetanus decline and the twitch becomes slower.

- Sarcolemmal adaptations: The resting membrane potential of the sarcolemma is hyperpolarized; the permeability to K^+ and Cl^- declines; the duration of the action potential increases; the sensitivity to acetylcholine is heightened; fibrillation potentials appear; the concentration of acetylcholinesterase decreases; and the neuromuscular junction releases factors that promote sprouts in nearby motor axons.

The connection between motor neurons and muscle fibers can be disrupted by **neuropathies,** which are disorders of peripheral nerves. These disorders can be acute or chronic and can involve the myelin sheath or the axon. Guillain-Barré syndrome, which is an autoimmune disorder that disrupts the myelination of peripheral nerves, is an example of an acute neuropathy. Chronic neuropathies can result from genetic diseases, metabolic disorders, intoxication, nutritional disorders, carcinomas, and immunological disorders.

Central Nervous System

Recovery in the CNS does not involve **neurogenesis** (generation of new neurons), but it does involve **synaptogenesis** (formation of new synapses) (Bhardwaj et al., 2006). Synaptogenesis is enhanced with activity (Adkins et al., 2006; Jones et al., 1999; Sakata & Jones, 2003; van Pragg et al., 2005). For example, when intracortical connections in the cat brain were subjected to long-term stimulation (four days), the results were an increased density of specific classes of synapses in layers II to III in the motor cortex, an increase in certain structural features of synapses, and alterations in the patterns of synaptic activity (Keller et al., 1992). It has been suggested that this reorganization underlies processes associated with motor learning and memory.

Other attempts to examine sprouting in the CNS have focused on the ability of the spinal cord to compensate for removal of selected inputs or pathways. Three classical preparations are used to examine CNS sprouting: the spared-root preparation, a hemisection model, and a deafferentation preparation. The **spared-root** preparation involves transection of all the dorsal roots supplying a hindlimb except one (usually L6). The animal recovers from the surgery and learns to reuse the limb. Once the motor recovery has been measured (weeks to months later), experiments are performed to determine the extent of the sprouting in the spinal cord by the spared dorsal root. The **hemisection model** involves the removal of one half (left or right) of the spinal cord at some appropriate level, and then monitoring both the recovery of function and subsequently the extent of sprouting in the spinal cord. The **deafferentation** preparation follows the same principle and measures the recovery of function and sprouting after transection of the dorsal (afferent) roots.

There is no consensus on the results from these experiments; some scientists conclude that sprouting does occur and others suggest that it does not (Goldberger et al., 1993). Distinguishing between sprouting and unmasking is one difficulty, and accounting for interanimal variability is another. **Unmasking** refers to the activation of a dormant neural pathway (or set of synapses) that is not needed until another (primary) pathway has been interrupted. As is the case for the peripheral nervous system, there is some evidence that collateral sprouting occurs in the spinal cord with the spared-root preparation. There are also clear examples in which the recovery of function can be explained by unmasking (O'Hara & Goshgarian, 1991). Furthermore, because the evidence for sprouting relies on the measurement of synaptic density, and because there is considerable variability in this parameter between animals, it is possible that evidence of sprouting is missed when data from different animals are combined rather than examined individually (Goldberger et al., 1993). Nevertheless, it has been demonstrated that ischemic stroke, for example, results in axonal sprouting both locally and over distances, such as interhemispheric projections (Carmichael, 2003).

More success in examining reorganization within the human brain has been achieved with imaging technologies (Rossini & Dal Forno, 2004). A common approach in the study of CNS plasticity is to document the adaptations that occur after either focal injury in the CNS (e.g., stroke) or alterations in the periphery (e.g., amputation, immobilization) (Chen et al., 2002; Elbert & Rockstroh, 2004; Jones, 2000). This can be accomplished with positron emission tomography (PET), which images the synthesis of compounds that contain a labeled isotope (Curt et al., 2002; Powers & Zazulia, 2003). For example, PET has been used to image the turnover of an agent that modulates synaptic transmission in the intact brain (Imahori et al., 1999). The investigators compared turnover of the neuromodulator in patients after a focal lesion (due to either a stroke or a brain tumor) with activity in healthy subjects. They found that the early events in reorganization of the neural connections began in remote regions of the brain rather than in the areas close to the lesions. Presumably this activity indicates the activation of alternative pathways in an attempt to compensate for the damage (Butefisch, 2004; Danause et al., 2005).

One of the critical factors in CNS regeneration is the presence or absence of scar tissue. Crushing of the spinal cord causes massive glial and connective tissue scarring from which recovery is minimal (Freed et al., 1985). Sprouts cannot cross physical barriers such as scars, and this severely limits regeneration. Furthermore, the sprouts from damaged axons in the CNS are usually short, and this limits the range of regeneration. Scientists do not yet know why the regenerative capacity is so different in the peripheral and central nervous systems. Key determinants appear to be the abundance of factors that promote sprouting; the presence of axon elongation inhibitors; the expression of proteins that facilitate axon elongation; and the magnitude of sequelae that can include glial cell formation, scar tissue, inflammation, and invasion by immune cells.

Rehabilitation Strategies

Given its capacity for reorganization, there is considerable practical interest in the ability of the injured CNS to recover motor function and in the interventions that can facilitate this recovery. As examples of this work, we will discuss current rehabilitation strategies used with individuals who have a spinal cord injury and those who have experienced an ischemic lesion in the cerebral cortex due to a stroke.

Spinal Cord Injury

An injury to the spinal cord can cause either a complete or an incomplete disruption of the descending motor pathways (Calancie et al., 2004). For both categories of injury, a principal concern is to maintain the physiological integrity of the paralyzed limbs. Individuals who experience an incomplete motor lesion retain the possibility of recovery of some motor function, which can be facilitated with functional electrical stimulation and assisted-locomotion training. As discussed at the beginning of this chapter, functional electrical stimulation refers to the artificial activation of muscle with a protocol of electric shocks that enhances the function of muscle. Within the context of this chapter, the primary purpose of this procedure is to counteract the effects of inactivity (Gordon & Mao, 1994; Shields & Dudley-Javoroski, 2006). The three major consequences of reduced activity for patients who experience a spinal cord injury are disuse atrophy, increased fatigability, and reduced bone density (Shields, 2002; Jayaraman et al., 2006; Stein et al., 2002). Functional electrical stimulation can reduce these negative adaptations (Belanger et al., 2000; Kern et al., 2004; Shields et al., 2006*a, b*).

A variety of functional electrical stimulation protocols have been used on paralyzed muscle. These have evoked adaptations ranging from a change in the myosin heavy chain composition of muscle fibers to increases in bone density, muscle force, and endurance. The adaptations evoked with electrical stimulation depend on the details of the stimulation protocol. For example, maintenance of the endurance capability of the tibialis anterior muscle in spinal cord–injured patients requires 1 to 2 h of stimulation (20 Hz) per day, but this amount does not alter muscle force (Stein et al., 1992). The efficacy of electrical stimulation was demonstrated in a two-year study of seven men that was begun within six weeks of a spinal cord injury (Shields & Dudley-Javoroski, 2006). The training program focused on the plantarflexor muscles of one leg and comprised four bouts of stimulation with 125 trains of 10 pulses (15 Hz over 667 ms) applied once every 2 s. The training took about 35 min per day and was performed five days a week for two years. Due to the position and fixation of the leg during stimulation, the mean compressive load experienced by the tibia was estimated at ~1.25 times body weight. Compared with values for the untrained leg, training increased the peak torque by 24%, the rate of torque development by 45%, a fatigue index by 50%, and bone mineral density in the trabecular bone of the distal tibia by 31%. These outcomes indicate that this approach represents an effective intervention.

As with able-bodied individuals, there is also a need to identify the most effective loading strategy. For example, should the muscles be activated to perform static or dynamic contractions, and how much load should oppose the contraction? Crameri and colleagues (2004*a*) trained one leg in six individuals with a complete spinal cord injury with static contractions against a load and the other leg with dynamic contractions using a minimal load. The training involved electrical stimulation of quadriceps femoris and lasted for 10 weeks (three days each week, 45 min each session). The force evoked with 35 Hz stimulation of quadriceps femoris increased by 122% for the leg trained with static contractions and by 52% for the leg trained with dynamic contractions. The two legs experienced a similar decrease in the proportion of type IIx fibers in vastus lateralis and increase in the proportions of type IIa and type IIc fibers, as well as increase in the capillary-to-fiber ratio. In contrast to the dynamic contractions, the static contractions increased the cross-sectional area of muscle fibers (119%), the activity level of citrate synthase (43%), and relative oxygenation during a 5 min isometric contraction. The adaptations were clearly greater for the leg that experienced static contractions against a load.

The inactivity imposed by spinal cord injury also causes significant deconditioning of the cardiovascular system (Jacobs & Nash, 2004). To counteract this adaptation, people need to perform activities that provide a cardiovascular stress, such as cycle ergometry (Barstow et al., 2000). To address this need, Scremin and colleagues (1999) found that strength training with electrical stimulation increased the cross-sectional area of many thigh muscles and enabled the patients to progress to a cycling program. Even six weeks of cycle ergometry in which the legs were activated with functional electrical stimulation is sufficient to reduce the fatigability of leg muscles (Gerrits et al., 2000).

Individuals with a spinal cord injury that spares some of the motor pathways can benefit from **assisted-locomotion training** (Stewart et al., 2004; Wirz et al., 2005). In the most common protocol, the person walks on a treadmill while wearing a harness that provides partial support of body weight. The individual, however, requires assistance with the swing phase of each step; this is provided either by therapists or by a robotic device. This type of training improves walking speed, endurance, and performance on functional tasks but does not alter the need for walking aids, orthosis, or external physical assistance (Behrman et al., 2006; Dietz & Harkema, 2004; Field-Fote, 2001; Wirz et al., 2005). Furthermore, the treadmill protocol provides outcomes that are similar to those achieved with overground mobility therapy (Dobkin et al., 2006). The training has been shown to confer a positive outcome in individuals with acute and chronic spinal cord injuries. Dobkin and colleagues (2006) provided 12 weeks of training for 146 subjects within eight weeks of an injury that resulted in their being graded on the American Spinal Injury Association (ASIA) scale as grades B, C, or D with levels from C5 to L3; Wirz and colleagues (2005) provided training for 20 individuals with injuries for at least two years with ASIA grades of C or D. ASIA grades of B, C, and D indicate an incomplete spinal cord injury, with a B grade denoting the greatest deficit.

In a landmark study, Dietz and colleagues (1994) provided nine individuals with 12 weeks of assisted-locomotion training on a treadmill; five of the patients had complete paraplegia and four incomplete. Figure 9.38 shows the adaptations in muscle activation that one of the incomplete patients achieved with the training. At the beginning of training, a harness supported 37% of body weight, and this declined to no support by the end of training. The patient was able to walk ~25 steps unsupported by the end of training. Although there was

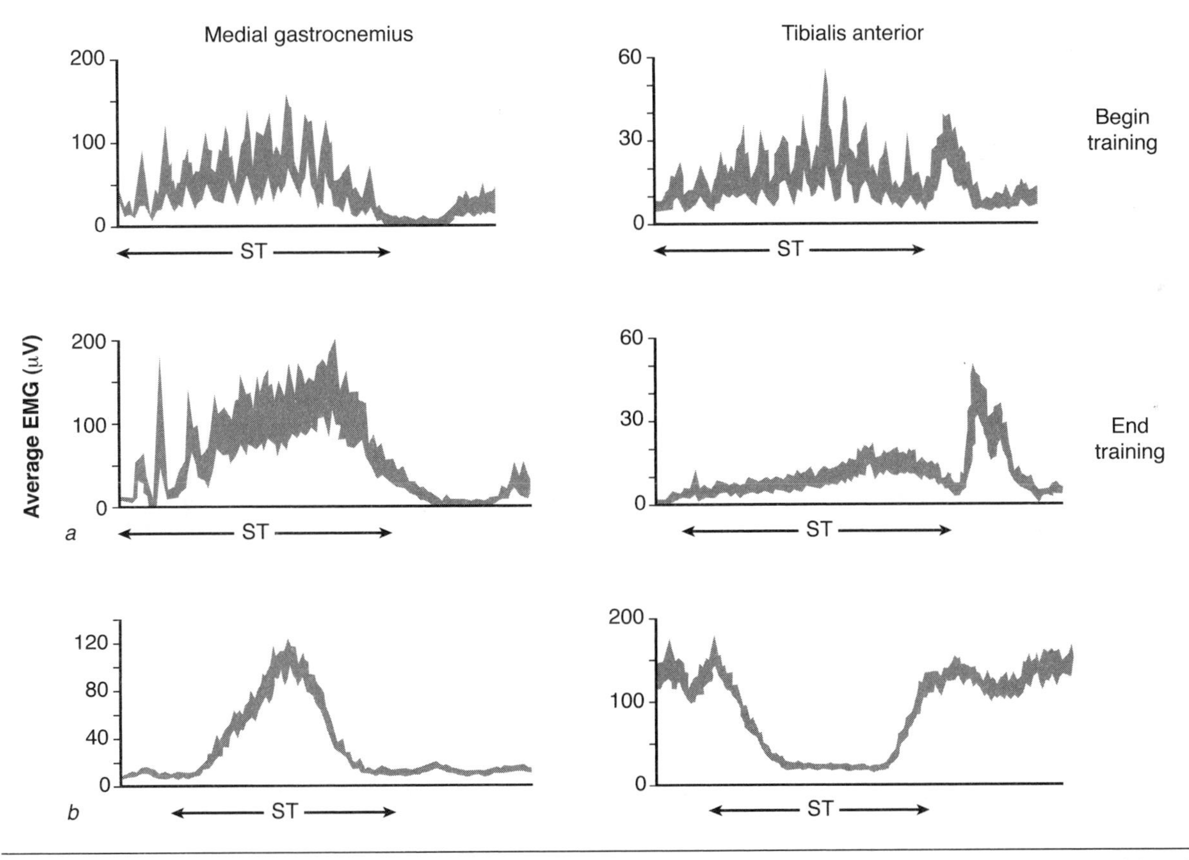

Figure 9.38 Amplitude of the electromyogram (EMG) for medial gastrocnemius (left column) and tibialis anterior (right column) in *(a)* a patient with incomplete paraplegia and *(b)* a healthy subject while walking on a treadmill. EMGs are shown for the patient at the beginning (top row) and end of training (middle row). ST indicates the stance phase.

no change in voluntary strength, the quality of the EMG activity for medial gastrocnemius and tibialis anterior became more similar to that observed in an able-bodied control subject. The magnitude of the training effect is greatest in persons with partial lesions that result in incomplete paralysis (Dietz, 1997).

Although individuals with a chronic injury can benefit from assisted-locomotion training, they exhibit a more rapid decline in EMG activity during a training session compared with those who have an acute injury. Because the amplitude of the action potentials evoked by electrical stimulation can remain unchanged with repetitive activation in these individuals, Dietz and Muller (2004) concluded that the decline in the EMG observed during a training session was caused by an impairment of input to the motor neurons. Furthermore, the authors reported that the reduction in EMG was not improved by 13 weeks of training, which suggests that the properties of the spinal neurons had adapted to the inactivity and could not be rescued. Thus the training of individuals with a spinal cord injury should begin as soon after the injury as possible (Dietz & Muller, 2004; Norrie et al., 2005).

A number of studies by Edgerton and colleagues have demonstrated that the outcomes exhibited by animals with spinal cord injury depend on the type of training they receive (Edgerton et al., 2006). For example, spinalized adult cats can be trained to stand or to perform stepping movements (de Leon et al., 1998*a, b;* Roy et al., 1998). The training regimen involves about 30 min per day, five days a week, for 8 to 12 weeks. The adaptations exhibited by the involved muscles, including the fiber type and MHC composition and the contractile properties, differ for the two training tasks (Roy et al., 1999). More importantly, the reorganization of the pathways within the spinal cord differs for the two types of training (Edgerton et al., 2001*a*). In the lumbosacral cord, for example, transection of the spinal cord results in an upregulation of the GABAergic and glycinergic inhibitory systems, and step training can reduce this adaptation back to control levels (figure 9.39). Consequently, an animal can learn to stand if it is trained to stand, whereas stand training has a minimal effect on the ability to step. These results underscore the specificity of the adaptations and suggest that the therapeutic strategies should be specific to the motor task that needs to be improved.

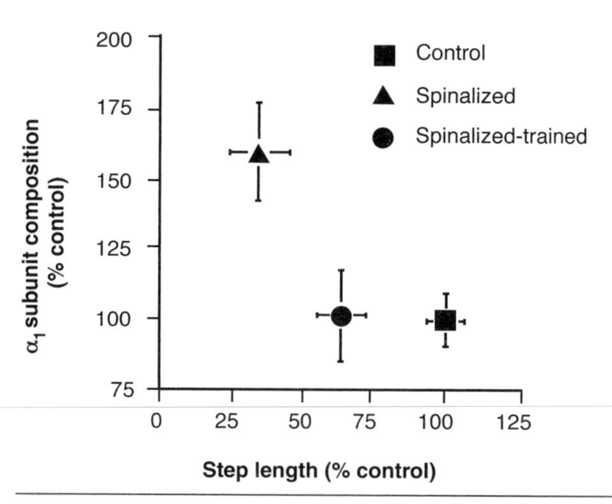

Figure 9.39 Step training of rats with a spinal cord transection normalizes step length and the α_1 subunit of the glycine receptor.

Reprinted, by permission, from Edgerton et al. 2001*a* "Retraining the injured spinal cord," *Journal of Physiology* 533: 20.

Stroke

Another major challenge the CNS can encounter is the damage that occurs after disruption of its blood supply, which can impair the function of the motor system. Because such events are often localized to one side of the brain, only one side of the body is compromised. Consequently, therapy can focus on either training the less-affected limb or restraining the less-affected limb and forcing the individual to use the more-affected limb; the latter is known as **constraint-induced movement therapy** (Wolf et al., 2006). A typical protocol involves training the more-affected arm for 6 h a day for 10 consecutive days and restraining the less-affected arm for about 90% of waking hours for two weeks. When compared with individuals who receive more conventional rehabilitation, the patients who perform constraint-induced movement therapy typically experience greater gains in motor performance (Platz et al., 2005; Suputtitada et al., 2004; Taub et al., 2006; Wolf et al., 2006; Wu et al., 2007).

The efficacy of this therapy is presumably related to the interactions that can occur between the two sides of the cerebral cortex. For example, the sensorimotor cortex that was not damaged and that controls the less-affected limb experiences adaptations that enhance its ability to learn new tasks, and depresses the capabilities of the contralateral side that controls the more-affected limb (Allred et al., 2005; Bury & Jones, 2004). Nonetheless, some evidence indicates that enhanced plasticity in the unimpaired cortex provides an environment that facilitates projections to the damaged hemisphere (Schallert et al., 2003). Rehabilitation, therefore, should balance the activities performed by the two sides, as is done with constraint-induced movement therapy.

ADAPTATIONS WITH AGE

Senescence is generally accompanied by a marked decline in the capabilities of the motor system. Although these adaptations can often be attributed to pathological processes, even healthy, active elderly individuals experience reductions in performance capabilities that represent the natural consequences of aging. In this section, we begin by describing some of the remodeling that occurs in the motor system with advancing age and then characterize the functional consequences of these changes.

Remodeling of the Motor System

Throughout this text, the concept has been emphasized that the motor unit represents the functional unit of the motor system. It represents the link that connects the nervous system and muscle. Aging is accompanied by adaptations in the properties of motor units, muscle, and the neural pathways that provide input to the motor neurons.

Motor Unit Properties

Counts of motor neuron numbers in the lumbosacral cord of human cadavers indicate a progressive loss of motor neurons in older individuals (Tomlinson & Irving, 1977) (figure 9.40*a*). A consequence of the decline in motor neuron number is a decrease in the number of functioning motor units in a muscle (Campbell et al., 1973) as assessed with the Motor Unit Number Estimation technique (Example 6.1) (figure 9.40*b*). Based on an average innervation number for tibialis anterior of 562 (table 6.5), a decline in the number of functioning motor units from 150 for young men to 59 for the old men (80-89 years) would correspond to a loss of innervation for 50,960 muscle fibers.

It is likely that muscle fibers deprived of an innervation release an agent (e.g., insulin-like growth factor, glial cell-derived neurotrophic factor, neurocrescin) that promotes collateral sprouting by axons of surviving motor neurons and that some of the denervated muscle fibers are rescued from degeneration (Keller-Peck et al., 2001; Payne et al., 2006; Wehrwein et al., 2002). The release of the agent is transient and presumably reduced in the muscles of older individuals. Nonetheless, the reinnervation of some abandoned muscle fibers results in a motor unit population that has a higher than normal average innervation number and larger twitch forces (Masakado et al., 1994; Semmler et al., 2000*b*). The average force exerted during a weak contraction, for example, was 293 mN for motor units in a hand muscle of old adults compared with 17.4 mN for motor units in young adults, even though the old adults had a lesser MVC force (24.1 N vs. 29.4 N) (Galganski et al., 1993).

Accompanying the remodeling of motor unit territories, the motor neurons of old adults exhibit a reduced capacity for rate coding (Erim et al., 1999). For example, average discharge rates during MVCs are less for old adults in both tibialis anterior (22.3 pps vs. 28.1 pps) and vastus lateralis (17.8 pps vs. 24.7 pps) (Kamen & Knight, 2004; Rubenstein & Kamen, 2005). Similarly, single

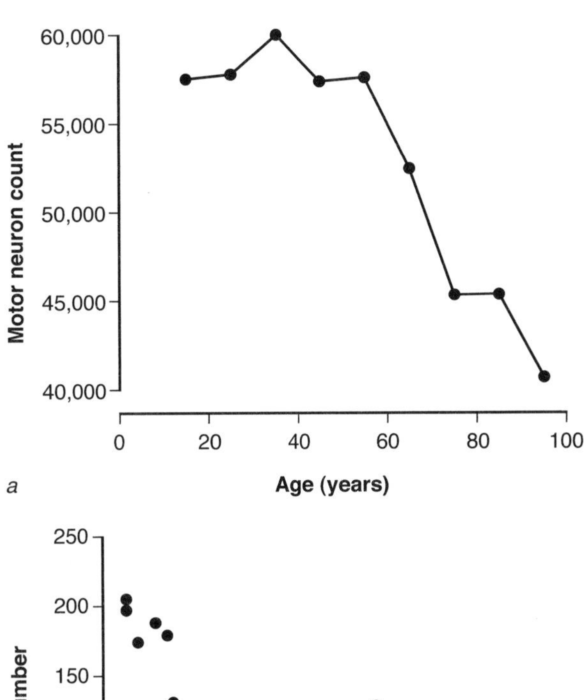

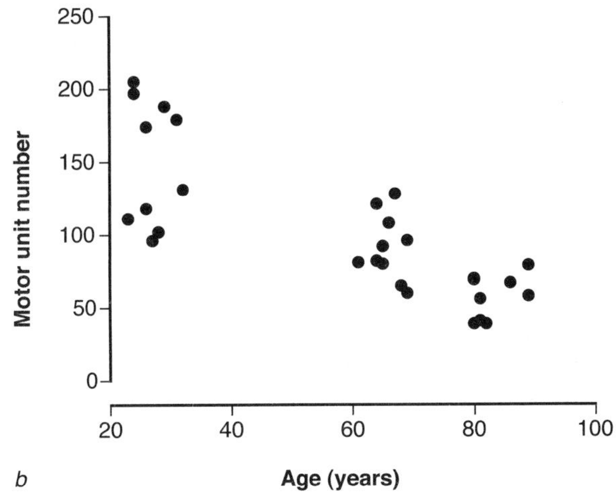

Figure 9.40 Decreases in *(a)* the number of motor neurons in the lumbosacral cord of human cadavers and *(b)* the number of functioning motor units in a foot muscle (extensor digitorum brevis) of human volunteers.

Data from Tomlinson and Irving, 1977 (9.40*a*) and McNeil et al., 2005 (9.40*b*).

motor units in a hand muscle of old adults exhibit less of a range from minimal to peak values during brief isometric contractions (Barry et al., 2007) and less modulation during a force-tracking task (Knight & Kamen, 2007). Nonetheless, the coefficient of variation for discharge times of motor units in a hand muscle is similar for young and old adults (Barry et al., 2007).

Some measures of the relative timing of motor unit discharges change with age, whereas others do not. The amount of motor unit synchronization, which is a measure of the near-simultaneous discharge of action potentials by two motor neurons, is similar in a hand muscle for young and old adults (Kamen & Roy, 2000; Semmler et al., 2000*b*, 2006). This result indicates that the proportion of common input received by pairs of motor neurons does not change with age. In contrast, common oscillatory inputs to motor neurons are greater for the motor units of a hand muscle in old adults, but they do not modulate the correlated discharge across isometric, lengthening, and shortening contractions as occurs in young adults (Semmler et al., 2003, 2006). The failure of old adults to exhibit this strategy indicates that they must rely on other mechanisms to control muscle force across these various tasks.

Muscle Properties

The motor unit remodeling that occurs with aging results in at least three significant changes in the muscle fibers. First, the muscle fibers deprived of an innervation degenerate, contributing to a decrease in muscle mass. The decrease in muscle mass and accompanying decline in strength are known as **sarcopenia** (Evans, 1995; Morse et al., 2005*a;* Sowers et al., 2005). There is a strong association between the loss of muscle mass and the decrease in strength (Frontera et al., 2000*b*). For example, Frontera and colleagues (2000*a*) reported that most of the decreases in the strength of the knee extensor and flexor muscles in 12 men over a 12-year period were explained by the decrease in muscle mass. Using computed tomography, they found that the cross-sectional area of quadriceps femoris decreased by 16.1% and the hamstrings by 14.9%, and these adaptations were accompanied by 20% to 30% decreases in peak torque during isokinetic contractions. The loss of muscle mass is often accompanied by a decrease in the pennation angle of muscle fascicles in the muscles of old adults (Morse et al., 2005*b*; Narici et al., 2003).

Second, as indicated by the examples shown in figure 9.41*a* for vastus lateralis of two women, the proportion of muscle occupied by type I muscle fibers increases with aging (Klein et al., 2003; Lexell et al., 1988). Third, the cross-sectional area of all fiber types decreases. For example, Hunter and colleagues (1999) found average decreases in cross-sectional area of 31% for muscle fibers in vastus lateralis of women, ranging from 19% for type I fibers to 39% for type II fibers (figure 9.41*c*). In humans, aging is not associated with a preferential loss of type II fibers (Klein et al., 2003), but there is a greater amount of atrophy in these fibers and therefore less of a contribution to muscle volume.

Consistent with an increase in the relative volume of muscle occupied by type I fibers, the muscles of old adults have slower contractile properties (D'Antona et al., 2003). For example, Baudry and colleagues (2005) found that the maximal rate of torque development during the response evoked in tibialis anterior by two electric stimuli (10 ms apart) was 209 N•m/s for young adults compared with 106 N•m/s for old adults. Similarly, the maximal rate of decrease in torque was less for the old adults (132 vs. 68 N•m/s). The potentiation in the rate of torque development by a preceding MVC, however, was similar for young (33%) and old (28%) adults. In contrast, the potentiation of the force in response to a single stimulus (twitch) was greater for young adults, which is consistent with an adaptation in excitation–contraction coupling in the muscles of older adults (Fulle et al., 2005).

The slowing of the intrinsic contractile speed of muscle, as demonstrated with evoked contractions, is accompanied by a reduction in the maximal rate at which old adults can discharge action potentials during rapid contractions. When subjects performed submaximal isometric voluntary contractions as fast as possible with the dorsiflexor muscles, Klass and colleagues (2007*a*) found that the maximal rate of torque development was 48% less for old adults compared with young adults and that this was associated with an increase of about 27% in the intervals between the first four action potentials. Because the difference in MVC torque between the young and old adults was only 28%, the adaptations that limited the maximal speed of a voluntary contraction differed from those that were responsible for the decrease in torque capacity (Barry et al., 2005*b*; Kent-Braun & Ng, 1999).

The adaptations at the single-fiber level appear to be complicated by sex differences and variation across muscles. On the basis of MRI and ultrasound measurements of lateral gastrocnemius, Morse and colleagues (2005*b*) estimated physiological cross-sectional area (muscle size) and the fascicle force produced by young and old men during an MVC. They found that the 47% reduction in MVC torque produced by the old men was partly due to a 30% decrease in the fascicle force normalized to cross-sectional area. In contrast, measurements of fiber force from segments obtained by muscle biopsy suggest no change in specific force with aging. For example, Trappe and colleagues (2003) reported that the lesser peak force for fibers from the vastus lateralis muscle of young and old women compared with men disappeared when expressed relative to cell size. However, the peak power produced by type I fibers of young women was 25% less than for the other three groups, and that produced by type IIa fibers was ~30% less in old women compared with the other three groups. In contrast, Krivickas and colleagues (2006) found no difference in either specific force or specific power of type I and type IIa fiber segments from the vastus lateralis muscle of old men and women.

One consistent finding at the single-fiber level is the decrease in the number of satellite cells in the fibers of old adults, which reduces the regenerative capacity of these muscles (Goldspink & Harridge, 2004). The decline has been demonstrated by a lesser increase in the number of satellite cells and expression of myogenic regulatory factors and IGF-1-related genes after a single bout of

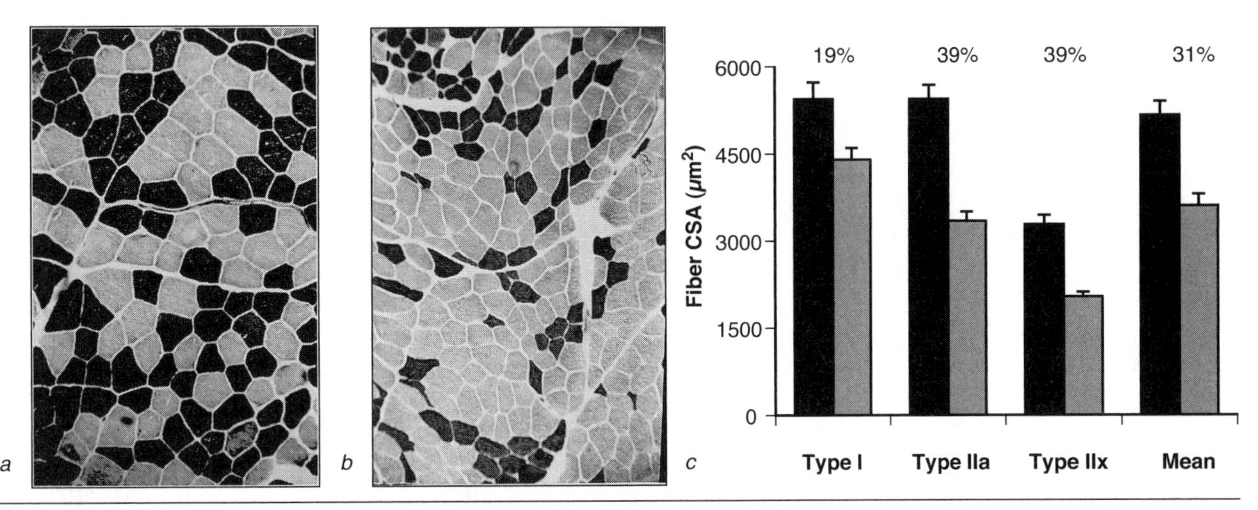

Figure 9.41 Distribution of type I (light stain) and type II (dark stain) fibers in the vastus lateralis of *(a)* a young woman (25 years) and *(b)* an old woman (67 years). *(c)* The cross-sectional area (CSA) of muscle fibers in vastus lateralis for young (n = 17; black bars) and old (n = 11; gray bars) women. The numbers at the top indicate the percentage decrease in fiber CSA for the old women.

Muscle fiber images (9.41*a-b*) were provided by Sandra K. Hunter, Ph.D. Data from Hunter et al. 1999 (9.41*c*).

exercise (Dreyer et al., 2006; Kim et al., 2005). Nonetheless, the number of satellite cells in muscle fibers does increase with strength training (Mackey et al., 2007); and these cells are capable of a sufficient number of cell divisions to ensure repair (Thornell et al., 2003), although the regeneration is compromised after periods of reduced use (Pattison et al., 2003; Suetta et al., 2007).

The decline in muscle capabilities with advancing age is further compounded by the adaptations that occur in tendon. Kubo and colleagues (2007*a*) found that the strain of the Achilles tendon during an MVC was less for older compared with younger adults: 5.1% for 20-year-olds, 4.4% for 30-year-olds, 4.0% for 50-year-olds, and 3.1% for 70-year-olds. Conversely, Onambele and colleagues (2006) reported strains in the Achilles tendon of 6.8% for young adults (24 years), 8.5% for middle-aged adults (46 years), and 8.8% for old adults (68 years) during an MVC with the plantarflexor muscles. These strains corresponded to decreases in both stiffness (22.4, 15.4, and 15.1 MPa, respectively) and the modulus of elasticity (0.36, 0.26, and 0.26 GPa) with increasing age. Onambele and colleagues (2006) compared muscle-tendon properties with the ability of the participants to maintain balance during standing tasks and found that decrements in balance performance were associated with a reduction in the stiffness of the Achilles tendon.

Neural Pathways

As suggested by the reduction in motor neuron number (figure 9.40*a*), the number of neurons decreases and the functional capabilities of the CNS decline as we age (Kramer et al., 2006; Raz & Rodrigue, 2006). The adaptations that influence movement most directly extend from spinal reflexes to the coordination of multiple muscles in the performance of voluntary actions. Several studies have demonstrated that modulation of reflex pathways can differ between young and old adults during some tasks. For example, Tsuruike and colleagues (2003) found that the amplitude of the soleus H reflex was increased by a Jendrassik maneuver in all standing positions tested for young adults, but only in some standing positions for old adults. Similarly, old adults modulate presynaptic inhibition of Ia afferent feedback, the short-latency component of the stretch reflex (M1), and reciprocal inhibition less than young adults during low-force contractions compared with resting levels (Earles et al., 2001; Kawashima et al., 2004; Kido et al., 2004). However, there is no age effect in the modulation of the H reflex during walking, in the ability to reduce the amplitude of the H reflex with training, or in the capacity to depress the H reflex during a challenging standing posture (Chalmers & Knutzen, 2002; Kido et al., 2004; Mynark & Koceja, 2002).

In contrast to the lesser modulation of some spinal reflexes during voluntary contractions, old adults tend to exhibit greater levels of activity in the cerebral cortex than young adults during simple motor tasks. With the use of functional MRI, investigators have demonstrated that old adults have slower reaction times and recruit additional cortical and subcortical areas, especially in ipsilateral sensorimotor and premotor cortex, when performing finger and hand movements (Heuninckx et al., 2005; Hutchinson et al., 2002; Mattay et al., 2002; Riecker et al., 2006). The greater ipsilateral distribution of activation in old adults results in their displaying less functional lateralization during the control of simple motor tasks, which suggests that the control of these actions is less automatic (Naccarato et al., 2006; Shinohara et al., 2003). Adaptations in the motor cortex have been probed with

TMS; old adults have displayed smaller motor-evoked potentials in limb muscles, lesser intracortical inhibition, higher stimulus intensities for achieving maximal motor output, and greater differences between hands (Eisen et al., 1996; Kossev et al., 2002; Matsunaga et al., 1998; Pitcher et al., 2003; Sale & Semmler, 2005).

Researchers have tested the functional significance of the cortical adaptations by estimating their contribution to the decline in strength that occurs with aging. At least two cortical mechanisms could contribute to the decrease in strength: a reduction in voluntary activation and an increase in coactivation of antagonistic muscles. Although most studies have shown that old adults are able to achieve levels of voluntary activation similar to those of young adults during isometric and anisometric contractions, some have reported differences between the two groups (Klass et al., 2007*b*). When a difference is detected, it is usually associated with a larger proximal muscle, such as biceps brachii or quadriceps femoris, during an isometric contraction. However, these differences are usually not substantial, although exceptions have been reported (Kubo et al., 2007*a*).

Although the level of coactivation during an MVC is not consistently greater in old adults (Klass et al., 2007*b*), it is often higher in old adults during submaximal contractions (Barry & Carson, 2004). The amount of coactivation, however, depends on the muscle group performing the task and the type of muscle contraction. Old adults, for example, use greater coactivation when performing steady, submaximal contractions with hand muscles (Burnett et al., 2000), when stepping down from a platform (Hortobágyi & DeVita, 2000), and when tracking a target force with the dorsiflexor muscles (Patten & Kamen, 2000). Because descending pathways control coactivation of the antagonist muscle independently of the drive to the agonist muscle (Hortobágyi & DeVita, 2006; Lévénez et al., 2005, 2007), the greater coactivation that is sometimes used by old adults is consistent with the more widespread distribution of cortical activity observed in these individuals.

A difference in the descending projections to the spinal neurons may also be responsible for the greater response of old adults to increases in physiological arousal. When exposed to an electric shock stressor, an individual shows a markedly impaired ability to match a target force of 2% MVC force with a pinch grip (Christou et al., 2004). The increase in the force fluctuations is greatest for old adults compared with young and middle-aged adults (figure 8.36*c*). Presumably, descending projections that control the level of arousal, such as those originating in the brain stem that provide monoaminergic input to the spinal neurons (Aston-Jones et al., 2000; Jacobs et al., 2002), responded more vigorously to the intervention in old adults.

Movement Capabilities of Older Adults

The adaptations that occur in the motor system with advancing age cause a range of changes in motor performance that include a decline in strength, changes in fatigability, a slowing of rapid reactions, an increased postural instability and diminished postural control, and decreased control of submaximal force. Even an active lifestyle is not sufficient to prevent these declines in performance (Korhonen et al., 2006; Marcell, 2003).

Muscle Strength and Power

The decline in strength has been observed in many different muscles in humans (Doherty, 2003; Reimers et al., 1998; Vandervoort, 2002). In some muscles, the decrease in strength seems to begin at about age 60 years (figure 9.42*a*), whereas other muscles show a constant decline with advancing age (figure 9.42*b*). For example,

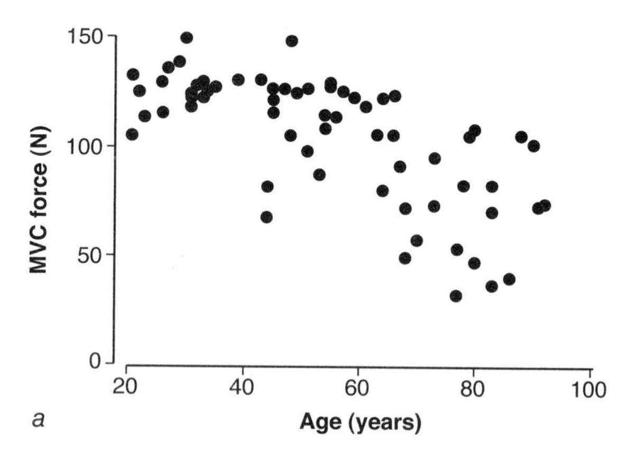

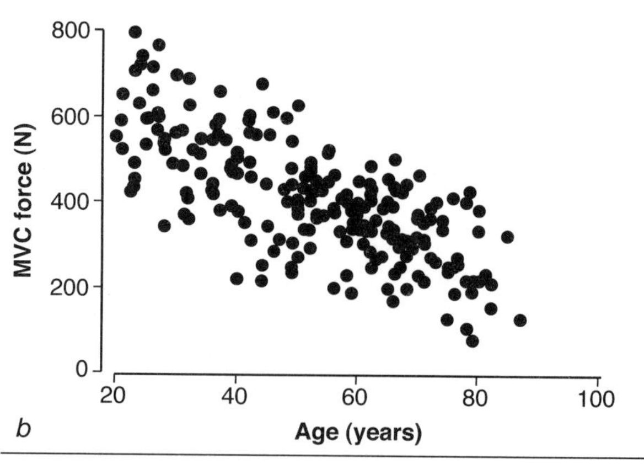

Figure 9.42 The decrease in maximal voluntary contraction (MVC) force for individuals of different ages for *(a)* the hand muscle adductor pollicis and *(b)* quadriceps femoris.

a longitudinal study on men indicated that MVC torque decreased progressively by 16% to 30% for the elbow and knee flexor and extensor muscles over 12 years, except at fast speeds for the elbow extensors (Frontera et al., 2000*a*). The magnitude of the decline in strength, however, varies among individuals. This is obvious in figure 9.42*a* by the increased scatter of the data points after age 60 years. The variable loss of strength suggests that the responsible mechanisms are activated to different degrees among individuals, but in healthy persons does not involve a substantial reduction in the ability to achieve maximal activation of muscle during a voluntary contraction (Klass et al., 2007*b*).

The decrease in strength is usually greater than the loss of muscle mass. For example, the cross-sectional area in a longitudinal study (Frontera et al., 2000*a*) declined by 16.1% for the quadriceps femoris and by 14.9% for the knee flexor muscles compared with average strength reductions of 20% to 30%. Furthermore, the percentage decrease in strength (peak torque) was similar for the knee flexor and extensor muscles but not for the elbow flexor and extensor muscles. As a result of this dissociation, muscle strength at the onset of a 12-year study and the changes in cross-sectional area were independent predictors of muscle strength at the end of the 12 years.

Despite the presence of sarcopenia, numerous studies have demonstrated that muscle strength can be enhanced in older adults with training (Barry & Carson, 2004; G.R. Hunter et al., 2004). The strength gains can be substantial, especially in sedentary persons (Fiatarone et al., 1990; Seynnes et al., 2004; Sullivan et al., 2007). Such adaptations, however, typically involve minimal changes in muscle size. For example, Lexell and colleagues (1995) found that 11 weeks of strength training by old men and women increased 1-RM load for the knee extensor muscles by ~160% but that the cross-sectional area of the fibers in the vastus lateralis muscle showed no change. In contrast, an active lifestyle appears to reduce the amount of sarcopenia that is evident in muscles of older adults (Reimers et al., 1998) and to enhance physiological status (Nakamura et al., 1996).

Traditional strength training programs, as discussed at the beginning of this chapter, typically involve lifting heavy loads in several sets of repetitions about three times a week. However, older adults can achieve impressive strength gains with variations in the traditional program. For example, twice-weekly training sessions enhance exercise adherence yet are sufficient to produce substantial increases in 1-RM load (figure 9.43) in both arm (arm curl, military press, and bench press) and leg (leg press) muscles (McCartney et al., 1996). Alternatively, a few weeks of training with light loads and steady contractions is sufficient to increase 1-RM load, albeit less than with heavy loads (Laidlaw et al., 1999; Tracy & Enoka, 2006). Older adults even experience an increase in muscle strength with tai chi training (Lan et al., 1998).

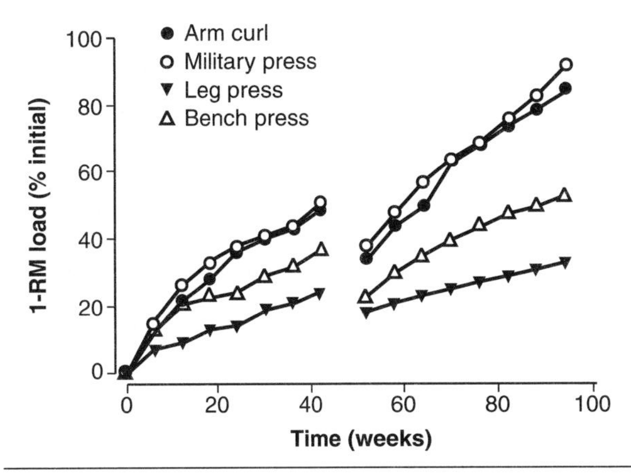

Figure 9.43 Increases in 1-repetition maximum (1-RM) load over a two-year strength training program. There was an eight-week break in the middle of the program.

Data from McCartney et al. 1996.

The improvements in strength achieved with these alternative protocols likely involve adaptations in the nervous system, because the changes in muscle size are minimal (Barry & Carson, 2004). One mechanism that can mediate such adaptations is an improved coordination among the muscles involved in the task (Capranica et al., 2004). Old adults, for example, take longer to achieve a target torque with the elbow flexor muscles during rapid contractions, but this can be improved with strength training that emphasizes the rate of force development (Barry et al., 2005*a, b*). Observed improvements in performance involved an increase in the initial EMG activity of the brachialis and brachioradialis muscles, but not the biceps brachii muscle. In contrast, young adults decreased the time to target and increased the initial EMG activity for all of these muscles. Although the old adults did improve performance of rapid contractions with training, the adaptation was less than that achieved by the young adults.

The concurrent declines in MVC torque and maximal rate of torque development can reduce the amount of power that can be produced and alter its distribution across the joints involved in the actions of old adults (Candow & Chilibeck, 2005; DeVita & Hortobágyi, 2000; Lanza et al., 2003; McNeil et al., 2007; Petrella et al., 2005; Yanagiya et al., 2004). Curiously, the peak torque that can be achieved by old adults during lengthening contractions (power absorption) is reduced less than those produced during shortening and isometric contractions (Hortobágyi et al., 1995; Klass et al., 2005; Porter et al., 1997). Some investigators suggest that the decrease in power production capacity is more critical than the loss of strength. For example, there are strong associations between the peak power that can be produced by the dorsiflexor and plantarflexor muscles and both chair rise time and stair climb time, which indicates that ankle muscle power is a significant contributor to functional mobility (Suzuki et al., 2001). Furthermore, the ability of older adults to recover from tripping depends on the coordination and power that can be produced by the leg muscles (Madigan, 2006; Robinovitch et al., 2000; Thelen et al., 2000). As with strength training, however, power production in old adults can be improved with training (Connelly & Vandervoort, 2000; de Vos et al., 2005; Ferri et al., 2003; Kraemer et al., 2002).

Fatigability

As described in chapter 8, there is no single cause of muscle fatigue, and the impairment that dominates during a fatiguing contraction varies across tasks. It should not be surprising, therefore, to learn that old adults are more fatigable than young adults on some tasks but less fatigable on other tasks. Two studies illustrate this difference.

Baudry and colleagues (2007) compared the decline in peak torque during shortening and lengthening contractions performed on an isokinetic dynamometer by young (20-35 years) and old (70-87 years) adults. The task was to perform five sets of 30 maximal shortening or lengthening contractions with the dorsiflexor muscles. Each maximal contraction was performed over a 30° range of motion at a speed of 50°/s, and one contraction was performed every 3.5 s with a 60 s rest between sets. The peak torque declined in each set of 30 contractions and across each of the five sets for both the shortening and lengthening contractions (figure 9.44*a*). The decrease in peak torque at the end of the 150 contractions was greater for the old adults for both the shortening (young = –41%; old = –50%) and lengthening (young = –27%; old = –42%) contractions. The old adults experienced a similar decrease in peak torque for the two types of contractions, whereas for the young adults the reduction during the shortening contractions was greater than that during the lengthening contractions. The fatigue was caused by an impairment of excitation-contraction coupling in both groups of subjects and by a decline in the efficacy of neuromuscular propagation for the old adults.

Hunter and colleagues (2005) measured the time during which young and old men could sustain a force at 20% of maximum with the elbow flexor muscles. The study involved eight pairs of men who were matched for strength so that it was possible to exclude differences in muscle perfusion as influencing performance. The average torque exerted by the young (18-31 years) and old (67-76 years) men was 65 N•m. The sustained contraction was terminated either when the torque declined by at least 10% from the target value or when the subject lifted the elbow off the support surface. Although each pair of subjects exerted the same net muscle torque, the time to task failure was 22.6 min for the old adults and 13.0 min for the young adults (figure 9.44*b*). The decrease in peak torque at task failure was similar for the two groups of men (–31% MVC torque). The rates of increase in perceived exertion, heart rate, fluctuations (SD) in torque, and average EMG activity were greater for the young men. Despite similar target torques for the

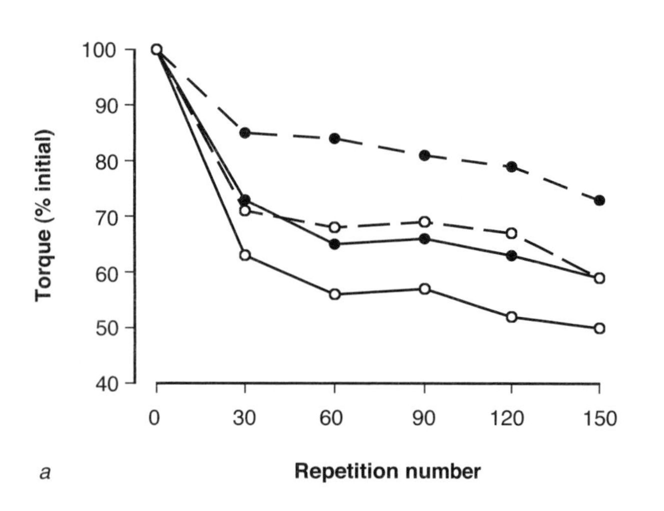

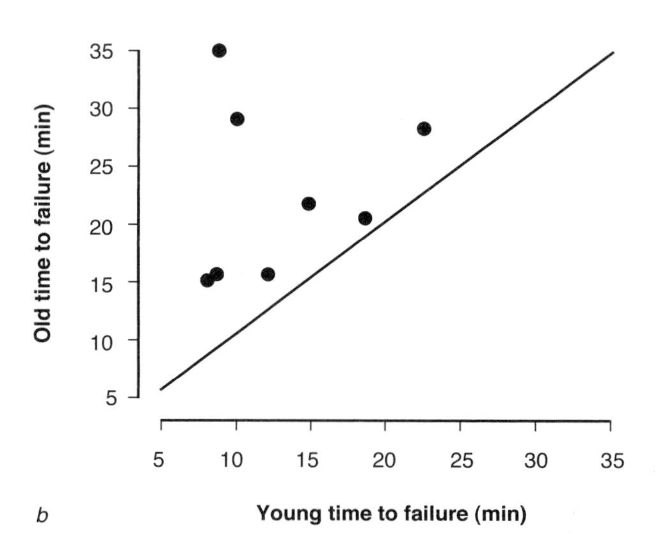

Figure 9.44 Fatigability of young and old adults. *(a)* The decline in peak torque of young (filled circles) and old (open circles) adults during 150 maximal shortening (dashed lines) and lengthening (solid lines) contractions with the dorsiflexor muscles. *(b)* The time to task failure for eight pairs of strength-matched young and old men during performance of an isometric contraction sustained at 20% MVC force with the elbow flexor muscles.

Data from Baudry et al. 2007 (Part *a*); Hunter et al. 2005 (Part *b*).

two groups of men, the young men recruited the motor unit pools of the elbow flexor muscles more rapidly and reached task failure sooner.

The results of these two studies demonstrate that old adults are not generally more fatigable than young adults, and that the physiological processes that can contribute to fatigue do not always decline in parallel with advancing age.

Rapid Reactions

The study of rapid reactions to visual cues or sudden movements provides some further insight on the adaptations that occur with aging. One can quantify this effect by determining the **reaction time,** which can be measured as the time from target displacement to the beginning of the response by the subject. For example, Warabi and colleagues (1986) compared the response time of eye and hand muscles to various target displacements. The response of the eyes was measured with an electrooculogram, and that of the hand was measured as the movement of a joystick. Both reaction times increased by similar amounts with age and were greatest for the larger target displacements. Because the motor components of these movements (e.g., peak velocity, duration, amplitude) were not altered with age, Warabi and colleagues (1986) concluded that the principal effect of age on the rapid reaction task was an impairment of sensory processing. This conclusion is consistent with other reports of a decline in sensory capabilities with age (Sarlegna, 2006; Stevens et al., 1998; Thelen et al., 1998; Wishart et al., 2002). Furthermore, the differential effects of age on the various components of the stretch reflex argue against a significant role for impairment of central processing in changes during relatively simple tasks (Corden & Lippold, 1996).

When tasks involve multiple components, however, old adults do require longer times to generate the appropriate motor response. For example, electroencephalogram recordings during various reaction-time tasks indicate that the longer reaction times of old adults when multiple choices are available are not related to either the stimulus processing or the response selection, but to slower activation over the contralateral motor cortex (Kolev et al., 2006; Yordanova et al., 2004). Similarly, the response time to make a sudden turn during walking is greater in old adults (Cao et al., 1997), and women needed longer to make the turn compared with men of the same age. When individuals perform two tasks concurrently, such as walking while spelling five-letter words backward, performance is typically more compromised in old adults. Hollman and colleagues (2007) found that walking speed was slower and the variability in speed across strides was greater in old adults compared with young adults when they performed the walking–spelling task. The impairment exhibited by old adults is related to a reduction in their cognitive abilities to perform two tasks concurrently (Holtzer et al., 2005; Logie et al., 2007). These studies indicate that all movements, from reflexes to multitask activities, become slower with advancing age.

Control of Posture

Aging is typically accompanied by reductions in the ability to control automatic movements, including posture and gait (Hortobágyi & DeVita, 2000; Tirosh & Sparrow, 2005; Wu & Hallett, 2005). These effects are most readily documented as a decline in the amount of walking and in the performance of the daily activities of living among elderly individuals. The adaptations that underlie these reductions are extensive, and they involve both motor and sensory processes. Alterations in the control of posture are significant because they influence the balance of an individual and the likelihood of accidental falls and injuries (Kannus et al., 1999). Furthermore, the concurrent loss of strength and reduction in balance exaggerates the functional impairments (Rantanen et al., 1999).

Maintenance of an upright posture requires an individual to keep the projection of the weight vector within the base of support. This requires the involvement of several different processes: (a) sensory information to detect the person's orientation and motion, (b) selection of an appropriate response strategy to maintain balance, and (c) activation of the muscles that can overcome any postural imbalance. The sensory information can be derived from visual, somatosensory, and vestibular sources. Not all of this information is necessary, and an individual can choose from among these sensory signals. Furthermore, patients with deficits in one of these sensory systems can readily learn to rely on the other two. However, aging diminishes the ability of an individual to select the appropriate sensory information and results in a change in the strategy for maintaining balance (Allum et al., 2002; Amiridis et al., 2003; Benjuya et al., 2004; Fransson et al., 2004).

Because of the diminished control of posture, considerable attention has focused on the association between balance and falls in old adults. The ability to maintain balance is often quantified from the displacement of the center of pressure during standing. The measures include the average displacement and the range of displacement during leaning and swaying tasks. Laughton and colleagues (2003) found that elderly fallers displayed both greater anteroposterior sway and EMG activity in leg muscles than young adults during quiet standing. Elderly nonfallers also used greater levels of EMG activity than young adults, but the amount of sway was similar for the two groups. In each group, the amount of postural sway was correlated with muscle activity. Similarly, the response of less stable old adults to a balance perturbation was slower and involved a greater proportion of muscle capacity than in stable old adults and young adults (Lin & Woollacott, 2002). These studies demonstrate a strong association between the functional capabilities of old adults and their ability to maintain balance, and both can be improved with training (Faber et al., 2006; Gatts & Woollacott, 2006; Hess & Woollacott, 2005).

Control of Submaximal Force

The preceding sections describe the adaptations that occur in the nervous system of the aging human as ranging from the remodeling of motor unit territories to a greater involvement of cortical and subcortical structures in the performance of voluntary contractions. A means of demonstrating the significance of these changes for activities of daily living is to compare the ability of young and old adults to control force during a submaximal contraction.

The force exerted by a muscle during a voluntary contraction is never constant, even if the individual attempts to keep it constant. Rather, the force will fluctuate about an average value. The magnitude of the fluctuations is often expressed as the standard deviation of force or acceleration and used as a measure of the **steadiness** of the contraction (Christou et al., 2002; Galganski et al., 1993). Old adults are often, but not always, less steady than young adults, as has been observed for hand, arm, and leg muscles performing isometric and anisometric contractions (Enoka et al., 2003; Vaillancourt et al., 2003). Differences in steadiness between young and old adults are least during brief (<10 s) isometric contractions to target forces and are greatest when a force is sustained without visual feedback in a state of heightened arousal (Barry et al., 2007; Christou et al., 2004). The impaired steadiness of older adults is worsened when they have to track a target force based on visual feedback (Sosnoff & Newell, 2006; Vaillancourt & Newell, 2003). Steadiness has been used as a predictor of performance during brief, stressful tasks and as a discriminator among fallers and nonfallers (Carville et al., 2007; Seynnes et al., 2005). The two features of motor unit activity most responsible for differences in steadiness are the variability in motor unit discharge and the low-frequency common input received by the motor neuron pool (Barry et al., 2007; Erim et al., 1999; Laidlaw et al., 2000; Sosnoff et al., 2004; Welsh et al., 2007).

Variability in motor output, such as the measurement of steadiness, can also increase across repeat performances of the same task (Christou & Carlton, 2002*b;* Jones et al., 2002). The functional consequences of greater motor output variability across trials are that the individual is less likely to produce the same force or trajectory each time the task is performed. The difference in motor output variability between young and old adults is greatest during rapid contractions, whether the task involves isometric, shortening, or lengthening contractions (Christou & Carlton, 2002*a*; Poston et al., 2008). For example, Christou and Carlton (2001) compared the ability of young (25 years) and old (73 years) adults to reproduce a force–time parabola with a rapid contraction of the quadriceps femoris muscles. The target forces ranged from 5% to 90% of MVC force, and the time to the peak force was 200 ms. The old adults displayed greater variability in the time to peak force, the magnitude of the peak force, and the force–time integral (impulse), especially at lower target forces. These data indicate that old adults have greater motor output variability during rapid, low-force contractions.

The decline in steadiness with aging also influences grip force during the manipulation of small objects (Cole, 1991). Subjects were instructed to lift a small object (1.7 cm width, 1.6 N weight) in a vertical direction about 5 cm, and the pinch force exerted by the thumb and index finger was measured. The slipperiness of the sides of the object could be altered. The elderly subjects exerted a force that was about two times greater than the force used by the young subjects and 2.5 times greater than the force needed to prevent the object from slipping. The excessive grip forces by the older subjects reflect an adaptation in tactile afferents that reduce hand sensibilities and to alterations in the glabrous skin that increase hand slipperiness (Cole et al., 1999). Furthermore, old adults exhibit a reduced ability to coordinate the actions of multiple fingers to achieve specific target forces with the hand (Koegh et al., 2006; Shim et al., 2004; Shinohara et al., 2004; Voelcker-Rehage & Alberts, 2005).

SUMMARY

This is the second of two chapters examining the effects of physical activity on the motor system. This chapter focuses on the long-term (chronic) response of the system to the stress associated with altered levels of physical activity, presenting five topics that characterize the chronic adaptive capabilities of the system: muscle strength, muscle power, adaptation to reduced use, motor recovery from injury, and adaptations with age. First, the concept of strength is defined; the training and loading techniques used to increase strength are presented; and the neural and muscular adaptations that accompany strength training are examined. Second, the concept of power production and its importance for movement are discussed, and the training techniques that can be used to increase power production are described. Third, we review the neuromuscular adaptations that occur when the level of physical activity is reduced, considering the examples of limb immobilization, limb unloading, and spinal transection. Fourth, we discuss the ability of the neuromuscular system to recover motor function from an injury to the nervous system. The capabilities of the peripheral and central nervous systems are distinguished, and two common rehabilitation strategies are described. Fifth, the movement capabilities of older adults are characterized, and the adaptations that accompany advancing age are discussed. These examples are not intended to represent a complete summary of the chronic adaptive responses, but rather to illustrate the capabilities of the system.

SUGGESTED READINGS

Kernell, D. (2006). *The Motoneurone and Its Muscle Fibres*. Oxford, Great Britain: Oxford University Press, chapters 10-12.

MacIntosh, B.R., Gardiner, P.F., and McComas, A.J. (2006). *Skeletal Muscle: Form and Function* (2nd ed.). Champaign, IL: Human Kinetics, chapters 17 and 19-22.

Part III Summary

The goal of part III (chapters 8 and 9) has been to describe the adaptive capabilities of the motor system. This description has focused on the acute (chapter 8) and chronic (chapter 9) adaptive capabilities of the system. As a result of reading these chapters, you should be familiar with the following features and concepts related to the motor system and be able to do the following:

- Realize the effect of altering core temperature on performance capabilities
- Conceive of the rationale for the techniques that have been developed to alter flexibility
- Identify the many factors that can contribute to muscle fatigue and acknowledge that performance can be impaired by different processes depending on the details of the task
- Understand that one can identify the rate-limiting adjustments during fatiguing contractions by comparing the rate of change in physiological processes during tasks that are sustained to failure
- Comprehend the various forms of muscle potentiation
- Know the performance characteristics of strength and power and the mechanisms that mediate changes in these capabilities
- Realize the extent of the adaptations that occur in the neuromuscular system with a reduction in the level of physical activity
- Acknowledge the ability of the motor system to recover some function following an injury to the nervous system
- Understand the procedures and limitations associated with two common rehabilitation techniques used with patients who have a spinal cord injury
- Identify the adaptations in motor function that occur with advancing age and the mechanisms that mediate these changes

appendix A

SI Units

The abbreviation *SI* is derived from *Le Systeme Internationale d'Unités,* which represents the modern metric system. There are seven base units in this measurement system, from which the other units of measurement are derived.

BASE

1. **length**—meter (m): Defined as the length of the path traveled by light in a vacuum during 1/299,792,458th of a second.

   1 in. = 2.54 cm

   1 ft = 30.48 cm = 0.3048 m

   1 yd = 0.9144 m

   1 mile = 1609 m = 1.609 km

2. **mass**—kilogram (kg): Defined as the mass of a platinum iridium cylinder preserved in Sevres, France.

   1 lb = 0.454 kg

   2.2 lb = 1 kg

   1 kg-force = 9.81 N

   1 lb-force = 4.45 N

DERIVED

area—square meters (m^2): Two-dimensional measure of length.

$1\ ft^2 = 0.0929\ m^2$

1 acre = 0.4047 hectares (ha)

volume—cubic meters (m^3): Three-dimensional measure of length. Although not an SI unit of measurement, volume is often measured in liters (L), where

1 mL of H_2O = 1 cm^3

1 L = 0.001 m^3

density—kilogram/cubic meters (kg/m^3): Mass per unit volume.

energy, work—joule (J): Energy denotes the capacity to perform work, and work refers to the application of a force over a distance.

1 J = 1 N•m

1 kcal = 4.183 kJ

1 kpm = 9.807 J

force—newton (N): One newton is a force that accelerates a 1 kg mass at a rate of 1 m/s^2.

1 N = 1 kg•m/s^2

1 kg-force = 9.81 N

1 lb-force = 4.45 N

impulse—newton•second (N•s): The application of a force over an interval of time; the area under a force–time curve.

moment of inertia (kg•m^2): The resistance that an object offers to a change in its state of angular motion; a measure of the proximity of the mass of an object to an axis of rotation.

BASE	DERIVED
mass *(continued)*	**power—**watt (W): The rate of performing work. 1 W = 1 J/s 1 horsepower = 736 W **pressure—**pascal (Pa): The force applied per unit area. 1 Pa = 1 N/m^2 1 mmHg = 133.3 Pa **torque—**moment of force (N•m): The rotary effect of a force. 1 ft-lb = 1.356 N•m
3. **time—**second (s): Defined in terms of one characteristic frequency of a cesium clock (9,192,631,770 cycles of radiation associated with a specified transition of the cesium-133 atom).	**acceleration** (m/s^2): Time rate of change in velocity. Gravity produces an acceleration of 9.807 m/s^2. 1 ft/s^2 = 0.3048 m/s^2 **frequency—**hertz (Hz): The number of cycles per second. 1 Hz = 1 cycle/s **momentum** (kg•m/s): Quantity of motion. 1 slug•ft/s = 4.447 kg•m/s **speed, velocity** (m/s): Time rate of change in position, where speed refers to the size of the change and velocity indicates its size and direction. 1 ft/s = 0.3048 m/s 1 mph = 0.447 m/s = 1.609 km/h
4. **electric current—**ampere (A): Rate of flow of charged particles. The ampere is the constant current that would produce a force of 2×10^{-7} N/m between two conductors of infinite length, negligible circular cross section, and placed 1 m apart in a vacuum.	**capacitance—**farad (F): The property of an electrical system of conductors and insulators that enables it to store electric charge when a potential difference exists between the conductors. **conductance—**siemen (S): The reciprocal of resistance, thus the ease with which charged particles move through an object. **resistance—**ohm (Ω): The difficulty with which charged particles flow through an object. **voltage—**volt (V): The difference in net distribution of charged particles between two locations.
5. **temperature—**kelvin (K): A measure of the velocity of vibration of the molecules of a body, which is the thermodynamic temperature. There is 100 K from ice point to steam point. 0 K = absolute zero	**Celsius** (°C) 0° C = 273.15 K **Fahrenheit** (°F) 32° F = 0° C = 273.15 K °F = 1.8° C + 32
6. **amount of substance—**mole (mol): The amount of substance containing the same number of particles as there are in 12 g (1 mol) of the nuclide ^{12}C.	**concentration** (mol/m^3): Amount of substance per unit volume.

BASE

7. **luminous intensity—**candela (cd): The radiant intensity of 1/683 watt per steradian from a source that emits monochromatic radiation of frequency 540×10^{12} Hz.

Supplementary Unit

8. **angle—**radian (rad): Measurement of an angle in a plane (i.e., two-dimensional angle).

   1 rad = 57.3 degrees

DERIVED

lumen (lm): Measure of light flux.

steradian (sr): Three-dimensional angle, also known as a solid angle.

appendix B

Conversion Factors

Acceleration

1 centimeter/second/second (cm/s^2)	=	0.036 kilometers/hour/second (km/h/s)
1 centimeter/second/second (cm/s^2)	=	0.01 meters/second/second (m/s^2)
1 foot/second/second (ft/s^2)	=	30.48 centimeters/second/second (cm/s^2)
1 foot/second/second (ft/s^2)	=	1.097 kilometers/hour/second (km/h/s)
1 foot/second/second (ft/s^2)	=	0.3048 meters/second/second (m/s^2)
1 kilometer/hour/second (km/h/s)	=	27.78 centimeters/second/second (cm/s^2)
1 kilometer/hour/second (km/h/s)	=	0.2778 meters/second/second (m/s^2)
1 meter/second/second (m/s^2)	=	100 centimeters/second/second (cm/s^2)
1 mile/hour/second	=	0.447 meters/second/second (m/s^2)
1 revolution/minute/minute	=	0.001745 radians/second/second (rad/s^2)
1 revolution/second/second	=	6.283 radians/second/second (rad/s^2)

Angle

1 circumference	=	6.283 radians (rad)
1 degree (°)	=	0.01745 radians (rad)
1 degree (°)	=	$\frac{\pi}{180}$ radians (rad)
1 minute (min)	=	0.0002909 radians (rad)
1 radian (rad)	=	57.3 degrees (°)
1 revolution (rev)	=	360 degrees (°)
1 revolution (rev)	=	2π radians (rad)

Area

1 acre	=	4047 square meters (m^2)
1 acre	=	0.4047 hectares (ha)
1 ares	=	100 square meters (m^2)
1 barn	=	10^{-28} square meters (m^2)
1 centare (ca)	=	1 square meter (m^2)
1 hectare (ha)	=	10,000 square meters (m^2)
1 shed	=	10^{-30} square meters (m^2)

1 square centimeter (cm^2)	=	0.0001 square meters (m^2)
1 square centimeter (cm^2)	=	100 square millimeters (mm^2)
1 square degree	=	0.00030462 steradians (sr)
1 square foot (ft^2)	=	929 square centimeters (cm^2)
1 square foot (ft^2)	=	0.092903 square meters (m^2)
1 square foot (ft^2)	=	92,900 square millimeters (mm^2)
1 square inch ($in.^2$)	=	6.4516 square centimeters (cm^2)
1 square inch ($in.^2$)	=	645.16 square millimeters (mm^2)
1 square inch ($in.^2$)	=	0.0006452 square meters (m^2)
1 square inch ($in.^2$)	=	0.006944 square feet (ft^2)
1 square kilometer (km^2)	=	1,000,000 square meters (m^2)
1 square meter (m^2)	=	0.0001 hectare (ha)
1 square meter (m^2)	=	10,000 square centimeters (cm^2)
1 square mile ($mile^2$)	=	2.590 square kilometers (km^2)
1 square mile ($mile^2$)	=	2,590,000 square meters (m^2)
1 square millimeter (mm^2)	=	10^8 square micrometers (μm^2)
1 square millimeter (mm^2)	=	0.01 square centimeters (cm^2)
1 square yard (yd^2)	=	8361 square centimeters (cm^2)
1 square yard (yd^2)	=	0.836127 square meters (m^2)

Density

1 pound/cubic foot (lb/ft^3)	=	16.01846 kilograms/cubic meter (kg/m^3)
1 slug/cubic foot ($slug/ft^3$)	=	515.3788 kilograms/cubic meter (kg/m^3)
1 pound/gallon (UK)	=	99.77633 kilograms/cubic meter (kg/m^3)
1 pound/gallon (USA)	=	119.8264 kilograms/cubic meter (kg/m^3)

Electricity

1 biot (Bi)	=	10 amperes (A)
1 ampere/square inch ($A/in.^2$)	=	1550 amperes/square meter (A/m^2)
1 faraday/second	=	96,490 amperes (A)
1 faraday	=	96,490 coulombs (coulomb = ampere-second)
1 mho	=	1 siemen (S)

Energy and Work

1 btu (British thermal unit)	=	1055 joules (J)
1 btu	=	1.0548 kilojoules (kJ)
1 btu	=	0.0002928 kilowatt-hours (kW-h)
1 calorie	=	4.184 joules (J)
1 erg	=	0.0001 millijoules (mJ)
1 foot-poundal	=	0.04214 joules (J)
1 foot-pound force	=	1.355818 joules (J)
1 gram-centimeter	=	0.09807 millijoules (mJ)
1 horsepower-hour	=	2684 kilojoules (kJ)

1 horsepower-hour	=	0.7457 kilowatt-hours (kW-h)
1 kilocalorie (International)	=	4.1868 kilojoules (kJ)
1 kilocalorie	=	4183 joules (J)
1 kilopond-meter (kp•m)	=	9.807 joules (J)
1 kilowatt-hour	=	3600 kilojoules (kJ)

Force

1 dyne	=	0.01 millinewtons (mN)
1 foot-pound (ft-lb)	=	1.356 joules (J)
1 foot-pound/second (ft-lb/s)	=	0.001356 kilowatts (kW)
1 gram (g)	=	9.807 millinewtons (mN)
1 kilogram-force (kg-f)	=	9.807 newtons (N)
1 kilopond (kp)	=	9.807 newtons (N)
1 poundal	=	0.138255 newtons (N)
1 pound-force (lb-f)	=	4.448222 newtons (N)
1 stone (weight)	=	62.275 newtons (N)
1 ton (long)	=	9964 newtons (N)
1 ton (metric)	=	9807 newtons (N)

Length

1 angstrom (Å)	=	0.0001 micrometers (μm)
1 angstrom (Å)	=	10^{-10} meters (m)
1 bolt	=	36.576 meters (m)
1 centimeter (cm)	=	0.00001 kilometers (km)
1 centimeter (cm)	=	0.01 meters (m)
1 centimeter (cm)	=	10 millimeters (mm)
1 chain	=	20.12 meters (m)
1 fathom	=	1.8288 meters (m)
1 foot (ft)	=	30.48 centimeters (cm)
1 foot (ft)	=	0.3048 meters (m)
1 foot (ft)	=	304.8 millimeters (mm)
1 furlong	=	201.17 meters (m)
1 hand	=	10.16 centimeters (cm)
1 inch (in.)	=	2.54 centimeters (cm)
1 inch (in.)	=	0.0254 meters (m)
1 inch (in.)	=	25.4 millimeters (mm)
1 light year	=	9,460,910,000,000 kilometers (9.46×10^{12} km)
1 mile (nautical)	=	1.852 kilometers (km)
1 mile (nautical)	=	1852 meters (m)
1 mile (statute)	=	5280 feet (ft)
1 mile (statute)	=	1.609 kilometers (km)
1 mile (statute)	=	1609 meters (m)
1 mile (statute)	=	1760 yards (yd)

1 rod	=	5.029 meters (m)
1 sphere (solid angle)	=	12.57 steradians (sr)
1 yard (yd)	=	91.44 centimeters (cm)
1 yard (yd)	=	0.9144 meters (m)

Luminous Intensity

1 candela/square centimeter	=	10,000 candela/square meter (cd/m^2)
1 candela/square foot	=	10.76 candela/square meter (cd/m^2)
1 foot-candle	=	10.764 lumen/square meter (lm/m^2)
1 foot-lambert (fL)	=	3.426 candela/square meter (cd/m^2)
1 lambert (L)	=	3183 candela/square meter (cd/m^2)
1 phot	=	10,000 lumen/square meter (lm/m^2)

Moment of Inertia

1 slug-foot squared (slug-ft^2)	=	1.35582 kilogram-square meters (kg•m^2)
1 pound-foot squared (lb-ft^2)	=	0.04214 kilogram-square meters (kg•m^2)

Mass

1 hundredweight	=	50.8 kilograms (kg)
1 metric ton (tonne; T)	=	1000 kilograms (kg)
1 ounce (oz)	=	28.3495 grams (g)
1 pound (lb)	=	0.453592 kilograms (kg)
1 slug	=	14.59 kilograms (kg)
1 ton	=	1016 kilograms (kg)

Power

1 btu/hour	=	0.2931 watts (W)
1 btu/minute	=	17.57 watts (W)
1 calorie/second	=	4.187 watts (W)
1 erg/second	=	10^{-7} watts (W)
1 foot-pound/second	=	1.356 watts (W)
1 foot-poundal/second	=	0.04214 watts (W)
1 gram-calorie	=	0.001162 watt-hours
1 horsepower (UK)	=	745.7 watts (W)
1 horsepower (metric)	=	735.5 watts (W)
1 kilocalorie/minute (kcal/min)	=	69.767 watts (W)
1 kilogram-meter/second	=	9.807 watts (W)
1 kilopond-meter/minute (kpm/min)	=	0.1634 watts (W)

Pressure

1 atmosphere	=	760 millimeters of mercury (at 0° C)
1 atmosphere	=	101,340 pascals (Pa)

1 bar	=	100,031 pascals (Pa)
1 centimeter of mercury	=	1333.224 pascals (Pa)
1 centimeter of water	=	0.738 millimeters of mercury
1 centimeter of water	=	98.3919 pascals (Pa)
1 dyne/square centimeter	=	0.10 pascals (Pa)
1 foot of water (4° C)	=	2989 pascals (Pa)
1 inch of mercury	=	3386 pascals (Pa)
1 inch of water (4° C)	=	249 pascals (Pa)
1 kilogram-force/square meter	=	9.807 pascals (Pa)
1 millibar	=	100 pascals (Pa)
1 millimeter of mercury (1 torr)	=	133.322387 pascals (Pa)
1 pound-force/square foot (lb/ft^2)	=	47.88026 pascals (Pa)
1 pound-force/square inch ($lb/in.^2$)	=	6.8948 kilopascals (kPa)
1 poundal/square foot	=	1.488 pascal (Pa)
1 torr	=	1 mmHg at 0° C
1 torr	=	133.3 pascals (Pa)

Speed, Velocity

1 centimeter/second (cm/s)	=	0.036 kilometers/hour (km/h, kph)
1 centimeter/second (cm/s)	=	0.6 meters/minute (m/min)
1 foot/minute (ft/min)	=	0.508 centimeters/second (cm/s)
1 foot/minute (ft/min)	=	0.01829 kilometers/hour (km/h, kph)
1 foot/minute (ft/min)	=	0.3048 meters/minute (m/min)
1 foot/second (ft/s)	=	30.48 centimeters/second (cm/s)
1 foot/second (ft/s)	=	1.097 kilometers/hour (km/h, kph)
1 foot/second (ft/s)	=	18.29 meters/minute (m/min)
1 foot/second (ft/s)	=	0.3048 meters/second (m/s)
1 kilometer/hour (km/h, kph)	=	16.67 meters/minute (m/min)
1 kilometer/hour (km/h, kph)	=	0.2778 meters/second (m/s)
1 knot	=	1.8532 kilometers/hour (km/h, kph)
1 knot	=	51.48 centimeters/second (cm/s)
1 knot	=	0.5155 meters/second (m/s)
1 meter/minute (m/min)	=	1.667 centimeters/second (cm/s)
1 meter/minute (m/min)	=	0.06 kilometers/hour (km/h, kph)
1 meter/second (m/s)	=	3.6 kilometers/hour (km/h, kph)
1 meter/second (m/s)	=	0.06 kilometers/minute (km/min)
1 mile/hour (mph)	=	44.7 centimeters/second (cm/s)
1 mile/hour (mph)	=	1.6093 kilometers/hour (km/h, kph)
1 mile/hour (mph)	=	0.447 meters/second (m/s)
1 mile/minute	=	2682 centimeters/second (cm/s)
1 mile/minute	=	1.6093 kilometers/minute (km/min)
1 revolution/minute	=	0.1047 radians/second (rad/s)
1 revolution/second	=	6.283 radians/second (rad/s)

Temperature

1 degree centigrade (°C)	=	(°C × 9/5) + 32 degrees Fahrenheit (°F)
1 degree centigrade (°C)	=	°C + 273.18 degrees kelvin

Time

1 minute (min)	=	60 seconds (s)
1 hour (h)	=	60 minutes (min)
1 hour (h)	=	3600 seconds (s)
1 day (d)	=	24 hours (h)
1 day (d)	=	86,400 seconds (s)
1 year (y)	=	365.2422 days (d)
1 year (y)	=	3.156×10^7 seconds (s)

Torque

1 foot-pound (ft-lb)	=	1.356 newton-meters (N•m)
1 kilopond-meter (kpm)	=	9.807 newton-meters (N•m)

Volume

1 barrel (UK)	=	0.1637 cubic meters (m^3)
1 barrel (USA)	=	0.11921 cubic meters (m^3)
1 bushel (UK)	=	0.03637 cubic meters (m^3)
1 bushel (USA)	=	0.03524 cubic meters (m^3)
1 bushel	=	35.24 liters (L)
1 cubic centimeter (cm^3)	=	0.000001 cubic meters (m^3)
1 cubic centimeter (cm^3)	=	0.001 liters (L)
1 cubic foot (ft^3)	=	0.02832 cubic meters (m^3)
1 cubic foot (ft^3)	=	28.32 liters (L)
1 cubic foot/minute (ft^3/min)	=	472 cubic centimeters/second (cm^3/s)
1 cubic foot/minute (ft^3/min)	=	0.472 liters/second (L/s)
1 cubic inch ($in.^3$)	=	16.387 cubic centimeters (cm^3)
1 cubic inch ($in.^3$)	=	0.0000164 cubic meters (m^3)
1 cubic inch ($in.^3$)	=	0.0164 liters (L)
1 cubic meter (m^3)	=	1,000,000 cubic centimeters (cm^3)
1 cubic meter (m^3)	=	1000 liters (L)
1 cubic yard (yd^3)	=	0.7646 cubic meters (m^3)
1 cubic yard (yd^3)	=	764.6 liters (L)
1 cubic yard/minute (yd/min)	=	12.74 liters/second (L/s)
1 dram	=	3.6967 cubic centimeters (cm^3)
1 gallon (UK)	=	0.004546 cubic meters (m^3)
1 gallon (USA)	=	3785 cubic centimeters (cm^3)
1 gallon (USA)	=	0.003785 cubic meters (m^3)
1 gallon	=	3.785 liters (L)

1 gallon/minute	=	0.06308 liters/second (L/s)
1 gill (UK)	=	142.07 cubic centimeters (cm^3)
1 gill (USA)	=	118.295 cubic centimeters (cm^3)
1 gill (USA)	=	0.1183 liters (L)
1 liter (L)	=	1000 cubic centimeters (cm^3)
1 liter (L)	=	0.001 cubic meters (m^3)
1 ounce (fluid)	=	0.02957 liters (L)
1 ounce (fluid—UK)	=	28.413 cubic centimeters (cm^3)
1 ounce (fluid—USA)	=	29.573 cubic centimeters (cm^3)
1 peck (UK)	=	9.0919 liters (L)
1 peck (USA)	=	8.8096 liters (L)
1 pint	=	473.2 cubic centimeters (cm^3)
1 pint	=	0.4732 liters (L)
1 quart	=	946.4 cubic centimeters (cm^3)
1 quart	=	0.9463 liters (L)

appendix C

Equations

$$\text{Velocity} = \frac{\Delta \text{ position}}{\Delta \text{ time}} \quad (1.1)$$

$$\text{Acceleration} = \frac{\Delta \text{ velocity}}{\Delta \text{ time}} \quad (1.2)$$

$$v_f = v_i + at \quad (1.3)$$

$$r_f - r_i = v_i t + \tfrac{1}{2}at^2 \quad (1.4)$$

$$v_f^2 = v_i^2 + 2a(r_f - r_i) \quad (1.5)$$

$$\boldsymbol{d} \bullet \boldsymbol{F} = dF \cos\theta \quad (1.6)$$

$$\boldsymbol{r} \times \boldsymbol{F} = rF \sin\theta \quad (1.7)$$

$$s = r\theta \quad (1.8)$$

$$v = r\omega \quad (1.9)$$

$$\boldsymbol{v} = \boldsymbol{\omega} \times \boldsymbol{r} \quad (1.10)$$

$$a = \sqrt{(r\omega^2)^2 + (r\alpha)^2} \quad (1.11)$$

$$a = \frac{v^2}{r} \quad (1.12)$$

$$x(t) = a_0 + a_1 t + a_2 t^2 + \ldots + a_n t^n \quad (1.13)$$

$$x(t) = a_0 + \sum_{i=1}^{n}\left[a_i \cos\left(\frac{2\pi t}{T} \bullet i\right) + b_i \sin\left(\frac{2\pi t}{T} \bullet i\right)\right] \quad (1.14)$$

$$x'(i) = a_0 x_i + a_1 x_{i-1} + a_2 x_{i-2} + b_1 x'_{i-1} + b_2 x'_{i-2} \quad (1.15)$$

$$\dot{x}(i) = \frac{x_{i+1} - x_{i-1}}{2(\Delta t)} \quad (1.16)$$

$$\ddot{x}(i) = \frac{x_{i+1} - 2x_i + x_{i-1}}{(\Delta t)^2} \quad (1.17)$$

$$F_c = \frac{mv^2}{r} \quad (2.1)$$

$$G = mv \quad (2.2)$$

$$F = ma \quad (2.3)$$

$$\boldsymbol{\tau} = \boldsymbol{r} \times \boldsymbol{F} \quad (2.4)$$

$$F \propto \frac{m_1 m_2}{r^2} \quad (2.5)$$

$$I = \sum_{i=1}^{n} m_i r_i^2 \quad (2.6)$$

$$I = \int r^2 dm \quad (2.7)$$

$$I_o = I_g + md^2 \quad (2.8)$$

$$I_o = \frac{F_w \bullet h \bullet T^2}{4\pi^2} \quad (2.9)$$

$$F_{\text{s,max}} = \mu F_{\text{g,y}} \quad (2.10)$$

$$F_{\text{f}} = kAv^2 \quad (2.11)$$

$$F_{\text{b}} = V_{\text{o}}\gamma \quad (2.12)$$

$$\text{Impulse} = \int_{t_1}^{t_2} \boldsymbol{F} dt \quad (2.13)$$

$$\int_{t_1}^{t_2} \boldsymbol{F} dt = \Delta \mathbf{G} \quad (2.14)$$

$$\sum \boldsymbol{F} t = \Delta m\text{v or } \overline{F} \bullet t = \Delta mv \quad (2.15)$$

$$(m_{\text{A}} v_{\text{A}})_{\text{before}} + (m_{\text{B}} v_{\text{B}})_{\text{before}} = (m_{\text{A}} v_{\text{A}})_{\text{after}} + (m_{\text{B}} v_{\text{B}})_{\text{after}} \quad (2.16)$$

$$m_{\text{A}} \Delta v_{\text{A}} = m_{\text{B}} \Delta v_{\text{B}} \quad (2.17)$$

$$\frac{\Delta v_A}{\Delta v_B} = \frac{m_B}{m_A} \quad (2.18)$$

$$v_{\text{B,a}} - v_{\text{b,a}} = -e\,(v_{\text{B,b}} - v_{\text{b,b}}) \quad (2.19)$$

$$v_{\text{b,a}} = v_{\text{B,a}} + e\,(v_{\text{B,b}} - v_{\text{b,b}}) \quad (2.20)$$

$$v_{\text{B,a}} = v_{\text{b,a}} - e\,(v_{\text{B,b}} - v_{\text{b,b}}) \quad (2.21)$$

$$v_{\text{B,a}} = \frac{m_{\text{b}} v_{\text{bb}}(1+e) + v_{\text{Bb}}(m_{\text{B}} - m_{\text{b}} e)}{m_{\text{b}} + m_{\text{B}}} \quad (2.22)$$

$$v_{\text{b,a}} = \frac{m_{\text{B}} v_{\text{Bb}}(1+e) + v_{\text{bb}}(m_{\text{b}} - m_{\text{B}} e)}{m_{\text{B}} + m_{\text{b}}} \quad (2.23)$$

$$\int_{t_1}^{t_2} (\boldsymbol{r} \times \boldsymbol{F})\, dt = \Delta \boldsymbol{H} \quad (2.24)$$

$$\boldsymbol{H}^{S/CS} = \boldsymbol{H}^{B1/C1} + \boldsymbol{H}^{B2/C2} + \boldsymbol{H}^{B3/C3} + \ldots \text{(local terms)}$$
$$\boldsymbol{H}^{C1/CS} + \boldsymbol{H}^{C2/CS} + \boldsymbol{H}^{C3/CS} + \ldots \text{(remote terms)} \quad (2.25)$$

$$U = \int F\, dr \quad (2.26)$$

$$U = \sum_{i=1}^{N} F_i\, \Delta x \quad (2.27)$$

$$U = \int \boldsymbol{F} \bullet \boldsymbol{dr} \quad (2.28)$$

$$E_k = \frac{1}{2} m v^2 + \frac{1}{2} I \omega^2 \quad (2.29)$$

$$\Delta E_k = U \quad (2.30)$$

$$E_{p,g} = mgh \quad (2.31)$$

$$E_{p,s} = \frac{1}{2} k x^2 \quad (2.32)$$

$$\Delta E_k + \Delta E_p = U_{nc} \quad (2.33)$$

$$E_k + E_p = \text{Constant} \quad (2.34)$$

$$\bar{P} = (9.81 \bullet m) \bullet \sqrt{r \bullet 4.9} \quad (2.35)$$

$$\bar{P} = \frac{\int F_{g,y}\, dt}{t} \bullet \frac{\int F_{net}\, dt}{2m} \quad (2.36)$$

$$\bar{P} = \frac{mg(r+h)}{t} \quad (2.37)$$

$$F_m = \text{Specific tension} \times \text{Cross-sectional area} \quad (3.1)$$

$$F_e = kx \quad (3.2)$$

$$E = \frac{\sigma}{\epsilon} \quad (3.3)$$

$$\sum F_x = 0 \quad (3.4)$$

$$\sum F_y = 0 \quad (3.5)$$

$$\sum \tau_o = 0 \quad (3.6)$$

$$\sum F_x = m a_x \quad (3.7)$$

$$\sum F_y = m a_y \quad (3.8)$$

$$\sum \tau_o = I_o \alpha + mad \quad (3.9)$$

$$\sum F_n = m r \omega^2 \quad (3.10)$$

$$\sum F_t = m r \alpha \quad (3.11)$$

$$\sum \tau_g = I_g \alpha \quad (3.12)$$

$$s_M = s_{M/A} + s_{A/K} + s_{K/H} + s_H \quad (3.13)$$

$$v_M = r_f \omega_f + r_s \omega_s + r_t \omega_t + v_H \quad (3.14)$$

$$a_M = \sqrt{(r_f \omega_f^{\,2})^2 + (r_f \alpha_f)^2} + \sqrt{(r_s \omega_s^{\,2})^2 + (r_s \alpha_s)^2} + \sqrt{(r_t \omega_t^{\,2})^2 + (r_t \alpha_t)^2} + a_H \quad (3.15)$$

$$\boldsymbol{F}_{j,k} + \boldsymbol{F}_{w,s} = m_s \bullet \mathbf{a}_s \quad (3.16)$$

$$\boldsymbol{a}_s = \boldsymbol{a}_h + (\boldsymbol{\alpha}_t \times \boldsymbol{r}_{k/h}) + (\boldsymbol{\omega}_t \times \boldsymbol{\omega}_t \times \boldsymbol{r}_{k/h}) + (\boldsymbol{\alpha}_s \times \boldsymbol{r}_{s/k}) + (\boldsymbol{\omega}_s \times \boldsymbol{\omega}_s \times \boldsymbol{r}_{s/k}) \quad (3.17)$$

$$\boldsymbol{F}_{j,k} = m_s \boldsymbol{a}_h + m_s(\boldsymbol{\alpha}_t \times \boldsymbol{r}_{k/h}) + m_s(\boldsymbol{\omega}_t \times \boldsymbol{\omega}_t \times \boldsymbol{r}_{k/h}) + m_s(\boldsymbol{\alpha}_s \times \boldsymbol{r}_{s/k}) + m_s(\boldsymbol{\omega}_s \times \boldsymbol{\omega}_s \times \boldsymbol{r}_{s/k}) - \boldsymbol{F}_{w,s} \quad (3.18)$$

$$\boldsymbol{\tau}_{m,k} + (\boldsymbol{r}_{k/s} \times \boldsymbol{F}_{j,k}) = I_{g,s}\, \boldsymbol{\alpha}_s \quad (3.19)$$

$$\boldsymbol{\tau}_{m,h} - \boldsymbol{\tau}_{m,k} + (\boldsymbol{r}_{h/t} \times \boldsymbol{F}_{j,h}) - (\boldsymbol{r}_{k/s} \times \boldsymbol{F}_{j,k}) = I_{g,t}\, \boldsymbol{\alpha}_t \quad (3.20)$$

$$\text{Peak } F_{g,y}\text{(BW)} = 0.856 \frac{T}{t_c} + 0.089 \quad (4.1)$$

$$\text{Centripetal force} = \frac{mv^2}{l} \quad (4.2)$$

$$\text{Froude number} = \frac{v^2}{gl} \quad (4.3)$$

$$\kappa = \frac{\Delta \tau_m}{\Delta \theta} \quad (4.4)$$

$$k_{leg} = \frac{\text{Peak } F_{g,y}}{\Delta l} \quad (4.5)$$

$$\%\ \text{recovery} = \frac{U_v + U_f - U_e}{U_v + U_f} \bullet 100 \quad (4.6)$$

$$v_x(t) = \int \frac{F_{g,x}}{m} dt + c \quad (4.7)$$

$$E_{k,f}(t) = \frac{m}{2} [v_x(t)]^2 \quad (4.8)$$

$$v_y(t) = \int \frac{F_{g,y} - F_w}{m} dt + c_1 \quad (4.9)$$

$$h(t) = \int v_y\, dt + c_2 \quad (4.10)$$

$$E_{p,g}(t) = mgh(t) \quad (4.11)$$

$$V = \frac{J}{C} \quad (5.1)$$

$$I = \frac{Q}{t} \quad (5.2)$$

$$g = \sigma \frac{A}{L} \quad (5.3)$$

$$I = gV \quad (5.4)$$

$$C = \frac{Q}{V} \quad (5.5)$$

$$V_{ab} = E \frac{R_1}{R_1 + R_2} \quad (5.6)$$

$$I = g(E - V_C) \tag{5.7}$$

$$E_X = \frac{58 \text{ mV}}{z} \log \frac{[X]_o}{[X]_i} \tag{5.8}$$

$$\Delta V_m = \frac{I_m}{g_{in}} \tag{5.9}$$

$$g_K = \frac{I_K}{(V_m - E_K)} \text{ and } g_{Na} = \frac{I_{Na}}{(V_m - E_{Na})} \tag{5.10}$$

$$\text{EMG}_{rms} = \sqrt{\frac{1}{N}\sum_{i=1}^{N} x_i^2} \tag{5.11}$$

$$\text{Coefficient of variation (\%)} = \left(\frac{\text{Standard deviation}}{\text{Mean}}\right) \times 100 \tag{6.1}$$

$$\text{Standard deviation (pps)} = \sqrt{\frac{(\text{SD of interspike interval})^2}{(\text{Mean of interspike interval})^3}} \tag{6.2}$$

$$\text{fCSA} = \frac{\text{Volume (cm}^3) \times \cos\beta}{\text{Fiber length (cm)}} \tag{6.3}$$

$$a(\omega) = b_0 + b_1 \cos(\omega - \omega_{pd}) \tag{7.1}$$

$$\text{Voluntary activation (\%)} = \left(1 - \frac{\text{Superimposed twitch}}{\text{Resting twitch}}\right) \times 100 \tag{8.1}$$

$$P = \frac{0.5mv^2 + mgh}{\Delta t} \tag{9.1}$$

Glossary

A band—Anisotropic or dark striation in skeletal muscle due to the double refraction of light rays from a single source.

abduction—Movement away from the midline of the body or a body part.

absolute angle—The angle of a body segment relative to a fixed reference, such as a horizontal or vertical line.

acceleration—Rate of change in velocity with respect to time (m/s^2); the derivative of velocity with respect to time; the slope of a velocity–time graph.

accommodation device—A training device in which the resistance varies to match the force exerted by the user.

accuracy—The closeness of an estimate to the true value.

acetylcholine (ACh)—A small-molecule neurotransmitter used in both the brain and the peripheral nervous system.

acetylcholine receptor—An integral protein in a membrane to which the neurotransmitter ACh can bind to evoke an effector action.

acetylcholinesterase—An enzyme that degrades acetylcholine.

actin—A globular protein that combines to form filaments, such as the thin filament of the muscle fiber.

action potential—A transient reversal of the potential difference across a membrane that is transmitted rapidly along the excitable membrane.

active touch—The exploration of the surroundings through reliance on the sensory receptors associated with the hand.

active zone—The specialized site for release of neurotransmitter from a presynaptic terminal.

activity field—A region of space for which a neuron is responsive. Activity fields are known as receptive fields for sensory neurons and as motor fields for motor neurons.

adaptation—A decline in the discharge rate of a neuron despite a constant excitatory input. The decline is due to the biophysical properties of the neuron.

adduction—Movement toward the midline of the body or a body part.

ADP—Adenosine diphosphate.

afferent—An axon that transmits signals from sensory receptors.

aftercontraction—An involuntary muscle contraction that follows a strong contraction.

afterhyperpolarization—The last phase of an action potential when the membrane potential is more negative on the inside compared with the steady-state voltage.

agonist—(1) A muscle whose activation produces the acceleration required for a movement. (2) A neurotransmitter, drug, or molecule that stimulates receptors to produce a desired action.

air-righting reaction—The ability of cats (and other animals) when upside down to land on their feet from a low height.

alpha chain—A sequence of about 1,000 amino acids in a polypeptide chain of collagen.

alpha-gamma coactivation—Concurrent activation of alpha and gamma motor neurons during voluntary activation of a muscle.

alpha motor neuron—A motor neuron that innervates skeletal muscle fibers.

AMP—Adenosine monophosphate.

amplitude cancellation—The reduction in EMG amplitude due to the summation of overlapping positive and negative phases of muscle fiber action potentials.

angle–angle diagram—A graph that shows the changes in one angle (y-axis) as a function of another angle (x-axis).

angle of pull—The angle between a muscle force and a body segment.

angular impulse—The area under a torque–time curve. An angular impulse changes the angular momentum of a body.

angular momentum—The quantity of angular motion.

angular motion—Rotation.

anion—A negatively charged atom or molecule in solution.

anisometric—Characterizing a contraction that is not isometric, that is, one in which whole-muscle length changes.

antagonist—(1) A muscle whose activation produces an acceleration in the direction opposite to that required for a movement. (2) A neurotransmitter, drug, or molecule that blocks receptors.

antagonist–contract stretch—A PNF-based stretch in which an assistant places the joint at the limit of rotation and then has the individual contract the opposing muscle (e.g., quadriceps femoris) to assist in further increasing the range of motion by stretching the agonist muscle (e.g., hamstrings).

anthropometric—Referring to measurements of the human body.

anticipatory postural adjustment—A change in the activity of muscles prior to a movement to establish a stable position of the body relative to the surroundings.

antidromic—Conduction of action potentials along an axon in the backward direction.

aponeurosis—The connective tissue extension of the tendon into muscle.

arousal—An internal state of alertness.

assisted-locomotion training—The training of patients with spinal cord injuries to walk on a treadmill by partially supporting body weight and providing passive assistance with the leg displacements.

ataxia—An inability to perform voluntary movements.

ATP—Adenosine triphosphate.

ATP hydrolysis—The breakdown of ATP to ADP, P_i, and energy.

atrophy—Loss of muscle mass.

automatic response—An afferent–efferent response involving neural pathways that are more complex than those responsible for spinal reflexes.

axial conductance—The capacity of a dendrite to transmit current along its length.

axon—A nerve fiber; a tubular process that arises from the soma of a neuron and functions as a cable for transmitting the electrical signals generated by the neuron.

axon hillock—The anatomical region where the axon leaves the soma.

axoplasmic transport—The movement of proteins and organelles on the inside of an axon. This material can be moved in both the orthograde (soma-to-synapse) and retrograde (synapse-to-soma) directions.

axotomy—Transection of an axon.

ballistic stretch—A technique to increase the range of motion about a joint by performing a series of rapid, bouncing stretches of the muscle.

band-pass filter—A filter that preserves the amplitude of the frequencies in a selected range and attenuates the other frequencies.

basal ganglia—Closely related nuclei (caudate, putamen, globus pallidus, subthalamic nuclei, and substantia nigra) that receive input from the cerebral cortex and send their output either to the brain stem or back to the cerebral cortex via the thalamus.

battery (*E*)—A device that contains a potential difference between two terminals due to the potential energy that can be released from a chemical reaction.

bed rest—An experimental technique of reduced use in which a human is confined to a bed for several weeks, often with the head slightly lower than the feet.

Bernoulli's principle—Principle that fluid pressure is inversely related to fluid velocity.

bilateral deficit—The reduction in force that occurs during a concurrent bilateral contraction compared with the force exerted during a unilateral contraction.

bilateral facilitation—The increase in force that occurs during a concurrent bilateral contraction compared with the force exerted during a unilateral contraction.

biomechanics—The use of physics to study biological systems.

bipolar recording—The difference in the signals detected by a pair of electrodes.

boundary layer—The streamline in fluid flow that is closest to an object.

brain stem—The components of the CNS between the brain and the spinal cord, which include the medulla oblongata, pons, and mesencephalon; the major route of information flow to and from the spinal cord and peripheral nerves.

buoyancy force—An upward-directed force that depends on the weight of a fluid that has been displaced.

Butterworth filter—A type of digital filter.

Ca^{2+}—Calcium ion.

cAMP—Cyclic AMP (adenosine monophosphate): a messenger molecule in the cell interior.

capacitance—The amount of charge that is stored on the two plates of a capacitor.

capacitive current (I_C)—The current that flows into and out of a capacitor. Positive charges accumulate on one plate of the capacitor and leave from the other plate.

capacitor (*C*)—A device that comprises two conductors separated by a layer of insulation.

cartwheel axis—An axis of rotation that passes through the human body from front to back.

cation—A positively charged atom or molecule in solution.

caudate nucleus—One of the basal ganglia nuclei.

center of mass—The point about which the mass of a system is evenly distributed; sometimes referred to as the center of gravity.

center of percussion—The location along a bat where no reaction force is felt at the hands when the bat strikes a ball.

center of pressure—The point of application of the ground reaction force.

central command—The neural signal transmitted by the motor cortex to lower motor centers (brain stem, spinal cord) for the execution of a movement.

central pattern generator—A neural network that is capable of generating behaviorally relevant patterns of motor output in the absence of sensory input.

centripetal force—A force directed toward the axis of rotation that is responsible for changing linear motion into angular motion.

cerebellar nuclei—Brain stem structures (dentate, fastigial, interpositus) that mediate most of the output from the cerebellum.

cerebellar peduncles—Six neural tracts connecting the cerebellum with the rest of the CNS.

cerebellum—A brain structure involved in the control of movement. It is located inferior to the cerebral hemispheres and posterior to the brain stem.

cerebral cortex—The outer layer of the cerebral hemispheres of the brain.

cerebral hemispheres—The two specialized halves of the brain.

cerebrocerebellum—The lateral hemispheres of the cerebellum, which receive input from the cerebral cortex and send output to the dentate nucleus.

chemical synapse—A synapse at which the communication involves the release of a neurotransmitter and its attachment to ligand-gated channels in the membrane of a postsynaptic cell.

chromatolysis—Dissolution of the Nissl substance (rough endoplasmic reticulum) and resynthesis of proteins that occur after axotomy.

Cl^-—Chloride ion.

climbing fibers—One of the afferent pathways into the cerebellum.

CNS—Central nervous system.

coactivation—Concurrent activation of the muscles around a joint, usually involving the agonist and antagonist muscles.

coefficient of restitution—A ratio that describes the speed of an object after a collision compared with its speed before the collision.

collateral—A branch of an axon.

collateral sprouting—The development of sprouts (neurites) by distal segments of an axon for the purpose of reinnervating a denervated target. This usually occurs after damage or an injury to a target cell (e.g., muscle fiber, neuron).

collinear—Lying on the same line.

collision—A brief contact between objects in which the force associated with the contact is much larger than other forces acting on the objects.

common drive—The modulation of average discharge rate of concurrently active motor units at 1 to 2 Hz.

common mode rejection—The elimination of signals that are common to a pair of electrodes.

compliance—The amount of a material that can be forcibly stretched. This property is expressed as the amount of change in length per unit of force used to stretch the material (mm/N).

composition—The process of determining the resultant of several vectors.

compressive force—A pushing force.

computed tomography (CT)—A method of reconstructing three-dimensional images based on a series of two-dimensional X-ray images.

concentration gradient—The diffusion force that causes particles to move from an area of high concentration to one that has a lower concentration.

concurrent—Pertaining to vectors whose lines of action converge and intersect at a point.

conditioning stimulus—A stimulus applied to afferent axons to test the functional state of a reflex pathway.

conductance (*g*)—The capacity of an object to transmit current. The reciprocal of resistance.

conduction velocity—The speed at which an action potential is propagated along an excitable membrane.

conductivity—The ability of an excitable membrane to propagate a wave of excitation.

connexins—The subunits of a gap junction.

conservation of mechanical energy—Idea that when there are no nonconstant opposing forces, such as friction, the sum of the changes in kinetic and potential energy is equal to zero.

conservation of momentum—A state in which the quantity of motion remains constant.

constant load—A training load that remains constant (e.g., barbell) during the performance of an exercise.

constraint-induced movement therapy—A therapy in which the less-affected limb of a patient is restrained so that the individual is forced to use the more-affected limb during activities of daily living.

contact event—Four moments that occur during a grasping action to move an object: contact with the fingertips, lifting of the object, arrival at the new location, and release of the object.

contractile element—The component of the Hill model that accounts for the contractility of muscle.

contractility—The ability of a tissue to modify its length.

contraction—A state of activation in which the cross-bridges of a muscle cycle in response to a muscle fiber action potential. Muscle length may decrease (shorten), stay the same, or increase (lengthen) during this state of activation.

contraction time—The time from force onset to the peak force during a twitch.

contract–relax—A PNF-based stretching technique in which the muscle to be stretched first performs a brief maximal contraction.

contract–relax, antagonist–contract—A PNF-based stretching technique that combines the contract-relax and antagonist–contract techniques.

corollary discharge—A copy of the outgoing central command that is returned to the cerebral cortex through the thalamus. It is also known as *efference copy*.

corpus callosum—A major neural tract that connects the two cortical hemispheres.

cortex (cerebral, cerebellar)—The external layer of thin and densely packed neuron bodies.

corticobulbar tract—Axons in the corticospinal tract that terminate in the cranial nuclei and innervate facial muscles.

corticospinal tract—Projections from the primary motor cortex to the spinal cord.

cost coefficient—The ratio of the rate of oxygen consumption relative to the speed of locomotion.

cotransmitters—Different neurotransmitters that are released from the same presynaptic terminal.

countermovement jump—A vertical jump that involves a lowering of the center of mass prior to the upward propulsive phase.

creep—The gradual increase in tissue length that is necessary to maintain a constant stress.

critical power—The maximal power that can be sustained for the duration of a task.

cross-bridge—The extension of the myosin molecule that can interact with actin and perform work.

cross-bridge cycle—The interaction of actin and myosin that occurs as a result of Ca^{2+} disinhibition.

cross-bridge theory of muscle contraction—A theory that explains the force exerted during a muscle contraction as due to the action of cross-bridges.

cross-education—An adaptation in motor capabilities that occurs in one limb as a consequence of physical training by the contralateral limb.

cross-links—Biochemical bonds that hold the collagen molecules together in connective tissue.

cross-sectional area—The area of the end-on view of an object (e.g., muscle) when it has been sectioned (cut) at right angles to its long axis.

cross-talk—A signal that originates in one muscle but is recorded by electrodes placed over another nearby muscle.

cross (vector) product—The procedure used to multiply two vectors to produce a vector product.

cubic—Relating to a third-degree polynomial.

current (I)—The flow of positive charges in an electrical circuit.

curve fitting—The derivation of a mathematical function to represent a data set.

cutaneomuscular reflexes—Responses evoked by activation of cutaneous receptors in muscle.

cutaneous mechanoreceptor—A sensory receptor that is sensitive to such stimuli as skin displacement, pressure on the skin, and temperature. These include Pacinian corpuscles, Merkel disks, Meissner corpuscles, Ruffini endings, and free nerve endings.

cutoff frequency—A frequency that is specified for a filter (digital or electronic) to indicate the location in the frequency where the input signal should be reduced to one-half of its power. The rate at which the frequency spectrum is modified depends on the type of filter used.

cytoskeleton—The structural proteins that provide the physical framework for the organization and interaction of contractile proteins.

deafferentation—An experimental model of CNS function and recovery in which all the afferent axons supplying one or more muscles are cut.

decerebrate preparation—An animal model in which the cerebral cortex has been disconnected from the brain stem and spinal cord.

decomposition—A technique that extracts identified action potentials from a multiunit recording.

degree of freedom at a joint—An axis of rotation. A joint with three degrees of freedom permits rotation about three different axes.

delayed-onset muscle soreness—The perception of muscle soreness (tenderness) that is associated with subcellular damage and occurs 24 to 48 h after an exercise. The soreness may be due to the inflammatory response consequent to the damage.

dendrite—A branch, other than the axon, that extends from the soma of a neuron.

denervation—A state in which the nerve innervating a muscle has been cut so that the muscle is without its neural input.

dentate nucleus—One of the deep subcortical nuclei of the cerebellum.

depolarization—A reduction in the potential difference across a membrane so that the inside becomes less negative relative to the outside.

depression of synaptic efficacy—A decrease in transmission across a synapse due to a decline in the availability of neurotransmitter or the accumulation of a limiting factor.

derecruitment—The cessation of discharge of action potentials by a motor unit during a voluntary contraction.

descending pathways—Neural pathways that transmit information from the brain to the brain stem and spinal cord. The major tracts include the corticospinal tract, corticobulbar tract, pyramidal tract, rubrospinal tract, vestibulospinal tract, reticulospinal tracts, and tectospinal tract.

desmin—A protein in the cytoskeleton.

diencephalic locomotor region—An area in the lateral hypothalamus that activates neurons in the medial reticular formation, which project to the spinal locomotor system.

difference vector—The difference between the vectors that define the positions of a target and an end effector.

digital filter—A numerical procedure that manipulates the frequency spectrum of an input signal to produce an output signal that contains a substantially attenuated frequency above the cutoff frequency.

digitizer—An instrument that determines the spatial coordinates *(xyz)* of selected landmarks.

directional tuning—The range of force or displacement directions over which the discharge of a neuron increases from its resting rate.

discharge rate—The rate at which neurons discharge action potentials. The unit of measurement is pulses per second (pps).

disinhibition—Inhibition of an inhibitor. For example, the Renshaw cell disinhibits the Ia inhibitory interneuron, and Ca^{2+} disinhibits the inhibitory effect of the regulatory proteins (troponin and tropomyosin).

displacement—A change in position. It is measured in meters (m) or radians (rad).

displacement map—A set of neurons that estimates angular displacements about the involved joints during reaching and pointing tasks.

DNA—Deoxyribonucleic acid, the genetic material of the cell nucleus.

dopamine—A neurotransmitter whose deficiency leads to Parkinson's disease.

dorsal roots—Structures on the back side of the spinal cord by which afferent axons enter the spinal cord.

double discharge—The occurrence of two action potentials discharged by a single motor neuron within 10 ms.

drag force—The component of a fluid resistance vector that acts parallel to the direction of fluid flow.

driving force—The difference between two potential differences, such as those for a battery and a capacitor, in an electrical circuit.

dynamic analysis—Mechanical analysis in which the forces acting on a system are not balanced and hence the system experiences an acceleration. The right-hand side of Newton's law of acceleration is nonzero.

dynamics map—A set of neurons that estimates the torques required to achieve specific angular displacements.

E–C coupling—Excitation–contraction coupling. The processes that link activation of the contractile proteins by an action potential.

economy—The amount of energy needed to perform a prescribed quantity of work. A performance is economical when the energy use is minimal.

effective synaptic current—The net effect of synaptic input to a neuron as measured by an intracellular microelectrode, which presumably represents the input that reaches the axon hillock.

efference copy—Internal, command-related signals that cancel sensory discharges due to motor commands (reafferent signals) and leave unaffected the sensory signals (exafferent signals) caused by external influences.

efferent—An axon that transmits signals from a neuron to an effector organ.

efficiency—The ratio of the work done to the energy used. The more work performed per unit of energy expenditure, the more efficient the system.

elastic collision—A collision in which the objects that collide bounce off one another.

elastic energy—The potential energy stored in a stretched spring.

elastic force—The passive property of a stretched material that tends to return it toward its original length.

elastic region—The initial linear part of a stress–strain or force–length graph.

electrical potential—Voltage.

electrical synapse—A synapse at which the communication involves the direct flow of current.

electrode—A probe that can be used to measure a physical quantity, such as the concentration of a chemical (ion, metabolite, pH), the potential difference between two locations, or the flow of current.

electroencephalography (EEG)—A method of recording the electrical waves of brain activity with electrodes placed over the skull.

electrogenic—Characterizing the contribution of Na^+-K^+ pump activity to the electrical potential across an excitable membrane.

electromyogram (EMG)—A recording of muscle fiber action potentials.

electrotonic conduction—The passive spread of a change in membrane potential.

end effector—The distal element of a limb (e.g., hand, laser beam, tennis racket, or cursor on a computer screen) that is being controlled during reaching and pointing tasks.

endomysium—A connective tissue matrix that surrounds individual muscle fibers.

endosarcomeric—Relating to the cytoskeletal proteins that maintain the orientation of the thick and thin filaments within the sarcomere.

end plate—The specialized region of muscle membrane included in the nerve-muscle synapse.

end-plate potential—The synaptic potential that is generated in a muscle fiber in response to the release of acetylcholine.

energy cost—The energy required to perform the work associated with a specific task.

epimysium—The outer layer of connective tissue that ensheathes an entire muscle.

equilibrium potential—The membrane potential at which the chemical and electrical forces acting on an ion are equal in magnitude but opposite in direction, that is, at which the net force is zero.

evoked potential—A recording of action potentials that are generated artificially by an external stimulus.

excitability—Responsiveness of an excitable membrane (e.g., axolemma, sarcolemma) to input signals.

excitation–contraction coupling—The electrochemical processes involved in converting a muscle action potential into mechanical work performed by the cross-bridges.

excitatory postsynaptic potential (EPSP)—A depolarizing (excitatory) synaptic potential that is elicited in the postsynaptic membrane (distal to the synapse).

exocytosis—The process of fusion by a vesicle to the presynaptic membrane and the subsequent release of its neurotransmitter.

exosarcomeric—Relating to the cytoskeletal components that maintain the lateral alignment of sarcomeres.

extension—An increase in the angle between two adjacent body segments.

external work—The work done to displace the center of mass in locomotion.

extracellular—Outside the cell.

extrafusal fiber—A skeletal muscle fiber that is not part of the muscle spindle.

F actin—Fibrous actin; a strand of a few hundred G-actin molecules.

failure of activation—An exercise-induced decrease in the availability and efficacy of Ca^{2+} as an activation signal for muscle.

fast axonal transport—A mechanism whereby an intra-axonal cytoskeleton transports enzymes and precursors between a cell body and its presynaptic terminals at rates up to 400 mm/day.

fastigial nucleus—One of the deep subcortical nuclei of the cerebellum.

fatigue—An exercise-induced reduction in the ability of muscle to produce force or power, whether or not the task can be sustained.

feedback—Signals arising from various peripheral receptors that provide information to the nervous system on the mechanical state of the neuromuscular system.

feedback control—A control system that generates forcing functions by comparing a desired performance (as dictated by command signals) with actual performance (sensed by feedback sensors).

feedforward control—A control system that generates command signals independently of the outcome.

FFT—Fast Fourier transform; the procedure used to determine the Fourier series for a given signal.

fictive—Referring to the output produced by a central pattern generator when the nerves have been disconnected from the muscles.

filtering—In signal processing, an alteration of the frequency content of a signal. For example, an EMG signal can be filtered to remove the high frequencies, and the resulting EMG signal is much smoother.

final common pathway—A description of the motor neuron as the route by which the nervous system sends control signals to muscle.

finite differences—A numerical procedure used to calculate the first and second derivatives, such as velocity and acceleration from position.

first law of thermodynamics—Law stating that the performance of work requires the expenditure of energy; work = Δ energy.

fixation-centered coordinate system—A set of coordinates based at an external point that is being fixated by the eyes.

flexibility—The range of motion about a joint.

flexion—A decrease in the angle between two adjacent body segments.

flexion reflex afferents—A group of afferent pathways that converge on common interneurons and are capable of evoking a flexion reflex.

fluid resistance—The resistance that a fluid offers to any object that passes through it. The magnitude of the resistance depends on the physical characteristics of the fluid and the extent to which the motion of the object disturbs the fluid.

focal adhesion—A structure that connects intracellular proteins to extracellular space.

force—A mechanical interaction between an object and its surroundings. The SI unit of measurement for force is the newton (N).

force enhancement—The increase in force at a given muscle length when an active muscle is stretched to that length.

force–frequency relation—The association between the rate of activation (e.g., discharge rate, electrical stimulation) and muscle force.

force–length relation—The association between maximal force and the length of a muscle. Also referred to as the length–tension relation.

force relaxation—The gradual decline in tissue stress while the length of the tissue remains constant.

force sharing—The relative contribution of each muscle to the net force exerted by a group of muscles.

force task—A fatiguing contraction in which the task of the subject is to sustain a prescribed target force.

force–velocity relation—The effect that the rate of change in muscle length has on the maximal force a muscle can exert.

form drag—The drag due to a difference in the pressure between the front and back of an object, which increases as the flow becomes more turbulent.

forward dynamics—A dynamic analysis to determine the kinematics that will be exhibited by a system based on the forces and torques that it experiences.

forward model—The computation of a future state (position, velocity, and acceleration) based on current feedback and motor commands.

Fourier analysis—The derivation of a series of sine and cosine terms to represent a signal.

Fourier series—The set of sine and cosine terms that describe a variable that changes over time.

free body diagram—A graphic-analysis technique that defines a system and indicates how the system interacts with its surroundings. A free body diagram represents a graphic version of the left-hand side of Newton's law of acceleration.

free nerve endings—Terminations of small-diameter axons that sense abnormal mechanical stress and chemical agents.

frequency domain—A representation of a signal that has frequency (Hz) as the independent variable (x-axis).

friction—Contact resistance due to the relative motion of one body sliding, rolling, or flowing over another. In movements involving contacts with a support surface, friction is the resultant of the two horizontal components of the ground reaction force.

friction drag—The drag due to the friction between the boundary layer and an object.

frontal plane—A plane that divides the body into front and back.

Froude number—The ratio of centripetal force to gravitational force, which is used as an index of dimensionless speed.

functional cross-sectional area (fCSA)—The measure of muscle cross-sectional area that takes into account muscle fiber pennation. The measurement is made perpendicular to the long axis of the muscle fibers and is proportional to the maximal force that the fibers can exert.

functional electrical stimulation—The artificial activation of muscle with a protocol of electric shocks that enhances muscle function.

fundamental frequency—The single sine and cosine term that best describes how a signal varies during one cycle.

fused tetanus—The force evoked in muscle fibers, motor units, or muscle in response to activation at a high frequency.

fusimotor—Because the muscle spindle has a fusiform shape, gamma motor neurons are sometimes called fusimotor neurons and their axons are referred to as fusimotor fibers.

F wave—A response in muscle that is evoked by stimulation of the axon to produce first an antidromic potential back to the soma and then an orthodromic potential out to the muscle.

G actin—Globular actin molecule.

gain—The amount by which the amplitude of an input signal is increased. In a control system, this is calculated as the ratio of the change in the control signal relative to the change in the controlled variable.

galvanic vestibular stimulation—Excitation of afferents from the vestibular apparatus by a direct current applied behind the ears.

gamma-aminobutyric acid (GABA)—A neurotransmitter.

gamma motor neuron—A motor neuron that innervates intrafusal muscle fibers in the muscle spindle.

gap junctions—Specialized proteins that conduct the flow of current from a presynaptic cell to a postsynaptic cell.

gating—The opening and closing of ion channels.

globus pallidus—A nucleus in the basal ganglia that comprises internal and external segments and the ventral pallidum.

Golgi ending—A thinly encapsulated, fusiform corpuscle that is similar to a tendon organ and may sense tension in a ligament.

gravitational potential energy—The energy possessed by a system due to its location in a gravitational field above a baseline.

gravity—The force of attraction between an object and a planet, which is an object with large mass. The force of gravity causes an acceleration of 9.81 m/s^2 on Earth.

ground reaction force—The reaction force provided by a horizontal support surface.

group Ia afferent—A muscle spindle afferent.

group Ib afferent—The tendon organ afferent.

group II afferent—Afferent with a diameter that is one class lower than that of the group I afferents. The muscle spindle has group II afferents.

H^+—Hydrogen ion.

half-center model—A model of a central pattern generator in which two sets of neurons inhibit each other reciprocally.

half-relaxation time—The time it takes for twitch force to decline from the peak to one-half of the peak force.

harmonic—Referring to a multiple of the fundamental frequency.

H band—Hellerscheibe; the region of the A band that is devoid of thin filaments.

heavy chain—A protein with a high molecular weight, such as those in myosin.

heavy meromyosin—The head-end fragment of the myosin molecule. This fragment can be further subdivided into subfragments 1 and 2 (S1 and S2).

hemisection model—An experimental model of CNS recovery in which one half of the spinal cord (left or right) is removed somewhere along the cord.

heteronymous—Referring to an anatomical relation that identifies sensory receptors and motor neurons supplying synergist muscles.

hindlimb suspension—An experimental protocol in which the hindlimbs of an experimental animal are lifted up off the ground for a few weeks so that they can still move but cannot contact the ground or any support surface.

Hoffmann reflex (H)—A muscle response evoked by electrical stimulation of the Ia afferents in a peripheral nerve. The afferent volley activates homonymous motor neurons and elicits an EMG and force response.

homogeneous—Of a similar kind or nature.

homonymous—Referring to an anatomical relation that identifies sensory receptors and motor neurons supplying the same muscle.

hyperammonemia—An elevation in the extracellular levels of ammonia, including those in the plasma and brain.

hypermobile joint—A joint about which the range of motion is exaggerated; individuals with such joints are sometimes described as double-jointed.

hyperplasia—An increase in muscle mass due to an increase in the number of muscle fibers.

hyperpolarization—An increase in the potential difference across a membrane such that the inside becomes more negative than the typical resting membrane potential.

hyperthermia—An elevation of core temperature.

hypertonia—An increase in muscle tone, such as spasticity and rigidity.

hypertrophy—An increase in muscle mass due to an increase in the cross-sectional area of the muscle fibers.

hypoglycemia—A decrease in the level of glucose in the blood.

hypothalamus—A structure in the diencephalon, important in autonomic control and in the expression of emotions.

hypotonia—A decrease in muscle tone, which is exhibited as a decreased resistance to a passive stretch.

hysteresis loop—A history-dependent relation in which the loading and unloading phases of a load–deformation graph do not coincide. For example, the change in length (deformation) of a tissue when a force is released does not retrace the curve obtained when the force was first applied.

H zone—Pale area in the center of the A band in the muscle fiber.

Ia inhibitory interneuron—An interneuron in the pathway from Ia afferents of one muscle to the motor neurons that innervate an antagonist muscle. This interneuron evokes inhibitory postsynaptic potentials in other neurons, including the motor neurons of the antagonist muscle.

I band—Isotropic or light band of skeletal muscle, named for the single refraction of light rays from a single source.

impedance—The apparent opposition in an electrical circuit to the flow of alternating current.

impulse—The area under a force–time graph (N•s), which corresponds to the force–time integral.

inertia—The resistance that an object offers to any changes in its motion.

inertial force—The force that an object exerts due to its motion.

inhibitory postsynaptic potential (IPSP)—A hyperpolarizing (inhibitory) synaptic potential that is elicited in the postsynaptic membrane (distal to the synapse).

innervation number—The number of muscle fibers innervated by a single motor neuron.

innervation zone—The area of a muscle that contains the neuromuscular junctions of its muscle fibers.

in-parallel—Arranged side by side.

input conductance (g_{in})—The ability of a neuron to conduct an applied current.

input resistance (R_{in})—The electrical resistance of a neuron to an applied input.

in-series—Arranged end to end.

integration—A mathematical procedure for measuring the area under a curve, such as a voltage–time or force–time relation. This procedure is often applied to a rectified EMG and results in a measure of the area under the EMG–time signal. Sometimes, however, a procedure that smoothes (filters out the high frequencies of) the rectified EMG is also called integration. This technique is more accurately referred to as "leaky integration" but is the procedure used most frequently to integrate the EMG signal and is commonly described as integration.

integrin—Transmembrane protein that may serve to connect myofibrils to the extracellular matrix of connective tissue.

interaction torque—The inertial force that a moving body segment exerts on its neighbors.

interelectrode distance—The space between a pair of electrodes.

interference EMG—An EMG recording that comprises the overlapping positive and negative phases of many muscle fiber action potentials. This is sometimes referred to as the "raw" EMG.

intermediate fiber—One of the components of the exosarcomeric cytoskeleton, arranged longitudinally along and transversely across sarcomeres to maintain the alignment of the sarcomeres.

internal model—A CNS representation of the kinematics and dynamics of a motor task.

interneuron—A neuron that receives information from and transmits it to other neurons.

interpositus nucleus—One of the deep subcortical nuclei of the cerebellum.

intersegmental dynamics—The inertial forces that a moving body segment exerts on its neighbors.

interspike interval—The time between consecutive action potentials discharged by the same motor neuron.

intra-abdominal pressure—The pressure (Pa) inside the abdominal cavity; the intra-abdominal pressure acts on the diaphragm and vertebral column to cause the trunk to extend.

intracellular—Inside the cell.

intradiscal pressure—The pressure inside intervertebral disks.

intrafusal fiber—A miniature skeletal muscle fiber that forms part of the muscle spindle.

intramuscular pressure—The pressure inside muscle, which increases during a contraction.

intrathoracic pressure—The pressure (Pa) inside the thoracic cavity.

inverse dynamics—A dynamic analysis to determine the forces and torques acting on a system based on the kinematics of the motion.

inverted-pendulum model—A model of the stance phase of walking in which the leg is represented as a strut that supports body mass and rotates about the ankle joint.

inverted-U hypothesis—The idea that maximal performance is associated with moderate levels of arousal.

ion—Electrically charged atom or molecule.

ion channel—A protein in a cell membrane that is capable of allowing one or more types of ions to pass through its pore.

ionic current—A current that is carried by ions.

ionotropic receptor—A membrane protein that contains both attachment sites for a neurotransmitter and an ion channel.

irradiation—The spread of activation to components of the motor system not directly involved in a task.

irritability—The ability to respond to a stimulus.

ischemia—A block of the blood flow to an area, which disrupts the transmission of action potentials along axons.

isoform—A protein that is synthesized with a slightly different amino acid composition to another isoform of the same protein. Also referred to as an isoenzyme.

isokinetic contraction—A movement in which the angular velocity of a displaced body segment is kept constant by an external device.

isometric contraction—A muscle contraction in which the muscle torque is equal to the load torque and as a consequence whole-muscle length does not change.

isotonic contraction—A muscle contraction in which a muscle contracts and does work against a constant load.

Jendrassik maneuver—The use of remote muscle activity to increase the excitability of a motor neuron pool and the size of the Hoffmann (H) reflex.

jerk—The derivative of acceleration with respect to time.

jitter—Variation in the arrival times of action potentials in muscle fibers innervated by the same motor neuron.

joint reaction force—The net force transmitted from one segment to another due to muscle, ligament, and bony contacts that are exerted across a joint.

joint receptor—A class of sensory receptors that sense joint-related events.

joint stiffness—The ratio of the change in resultant muscle torque to the change in angular displacement.

K^+—Potassium ion.

kilogram—The metric measurement for mass.

kinematic—Referring to a description of motion in terms of position, velocity, and acceleration.

kinesiology—The study of movement.

kinesthesia—The ability of the system to use information derived from sensory receptors to determine the position of the limbs, to identify the agent (itself or something else) that causes it to move, to distinguish the senses of effort and heaviness, and to perceive the timing of movements.

kinetic diagram—A diagram that indicates the effects (mass × acceleration) of the forces acting on a body; a diagram of the right-hand side of the law of acceleration.

kinetic energy—The capacity of an object to perform work because of its motion.

kinetics—Relating to motion, a description that includes consideration of force as the cause of motion.

Kohnstamm effect—The postcontraction activation of motor neurons that evokes an aftercontraction in an apparently relaxed muscle.

Lambert-Eaton myasthenic syndrome—An immune disorder that results in the destruction of voltage-gated Ca^{2+} channels in the presynaptic motor nerve terminal.

laminar flow—Uniform flow of fluid (streamlines) around an object.

lassitude—The subjective sense of reduced energy levels.

latency—The time delay between a stimulus and a response.

lateral sac—The enlargement of the sarcoplasmic reticulum that is adjacent to the transverse tubules; also referred to as the terminal cisternae.

law of acceleration—F = ma.

law of action–reaction—Law stating that for every action there is an equal and opposite reaction.

law of gravitation—Law stating that all bodies attract one another with a force that is proportional to the product of their masses and inversely proportional to the square of the distance between them.

law of inertia—Law stating that a force is required to stop, start, or alter motion.

leak current—The flux of ions through resting channels.

leg spring—A model of the leg muscles as behaving like springs during the stance phase of running.

leg unloading—An experimental technique of reduced use in which one leg is placed in a sling and not used during activities of daily living.

length—Distance; measured in meters (m).

length constant (λ)—The distance over which current will spread passively to change the membrane potential by 37% of the change that occurs at the current source.

lift force—The component of the fluid resistance vector that acts perpendicular to the direction of the fluid flow.

light chain—A protein with a low molecular weight, such as those on myosin.

light meromyosin—The tail-end fragment of the myosin molecule.

linear—Pertaining to a straight-line relation.

linear momentum—The quantity of linear motion.

line of action—The direction of a vector.

linked system—A system that is represented as a series of rigid links. The human body is regarded as a linked system when a biomechanical analysis is performed.

location map—A set of neurons that defines the location of an end effector during reaching and pointing tasks.

lower motor neuron—A motor neuron in the spinal cord.

low-frequency depression of force—A relative reduction in the force that can be evoked from muscle with low frequencies of stimulation.

low-frequency fatigue—An exercise-induced depression of the force evoked by low frequencies of activation.

M1—The short-latency component (~30 ms) of the stretch reflex.

M2—A long-latency component (50-60 ms) of the stretch reflex.

macro EMG—A needle electrode with a large recording area.

magnetic resonance imaging (MRI)—An imaging technique that involves the use of magnetic fields to determine the spatial localization of protons; can be used to study the structure and function of the motor system.

magnetic stimulation—The use of a stimulator to generate a magnetic field that induces an electric field and elicits axonal action potentials.

magnitude—Size or amplitude.

Magnus force—The sideways pressure gradient across an object.

mass—The amount of matter (kg).

maxima—The high points or peaks in a graph (e.g., position–time graph); the slope of a graph is zero at a maximum.

M band—Mittelscheibe; the band that connects two sets of thick filaments in a sarcomere.

mean frequency—The central value of a frequency spectrum.

mechanomyogram (MMG)—Surface recording of the transverse displacement of a muscle during a contraction.

median frequency—The frequency that divides a spectrum into two equal halves based on the energy content of the signal.

medulla oblongata—The lower portion of the brain stem.

Meissner corpuscle—A cutaneous mechanoreceptor that is sensitive to local, maintained pressure.

Merkel disk—A cutaneous mechanoreceptor that is sensitive to local vertical pressure.

mesencephalic locomotor region—An area in the brain stem that activates neurons in the medial reticular formation, which project to the spinal locomotor system.

mesencephalon—The rostral component of the brain stem; merges anteriorly into the thalamus and hypothalamus.

metabotropic receptor—A receptor activated by a neurotransmitter to evoke various intracellular actions in a postsynaptic neuron, including the activation of ion channels.

meter—The SI unit of measurement for length.

MHC—Myosin heavy chain, which is part of the myosin molecule.

microneurography—The recording of axonal action potentials with an electrode inserted into a peripheral nerve.

miniature end-plate potentials—Small synaptic potentials at the neuromuscular junction that occur in response to the spontaneous release of neurotransmitter.

minima—The low points or valleys in a graph (e.g., position–time graph); the slope of a graph is zero at a minimum.

M line—A narrow, dark line in the center of the H zone.

modulatory synaptic actions—The influence of metabotropic receptors on the actions that occur at a synapse.

modulus of elasticity—The slope of the elastic region of a stress–strain relation; the ratio of stress to strain.

moment arm—The shortest distance (perpendicular) from the line of action of a force vector to the axis of rotation.

moment of force—The rotary effect of a force; torque.

moment of inertia—The resistance that an object offers to any change in its angular motion. It represents the distribution of the mass of the object about the axis of rotation. The symbol for moment of inertia is I, and the SI unit of measurement is $kg \cdot m^2$.

moment of momentum—Angular momentum.

momentum—The quantity of motion possessed by an object; a vector quantity. The SI units of measurement are $kg \cdot m/s$ for linear momentum (G) and $kg \cdot m^2/s$ for angular momentum (H).

monopolar recording—A recording made with a single electrode.

monosynaptic—Pertaining to a neural circuit that involves a single synapse.

monosynaptic reflex—A pathway between an afferent input and an efferent output that involves a single synapse.

morphology—A branch of biology that deals with the form and structure of animals and plants.

motion—A change in position (m) that occurs over an interval of time.

motion-dependent effects—Interactive forces between body segments during human movement.

motor control—The control of movement by the nervous system.

motor neuron—A neuron that innervates muscle fibers.

motor neuron pool—The group of motor neurons that innervate a single muscle.

motor program—A characteristic timing of muscle activations that produce specific kinematic events.

motor system—The neuromuscular elements involved in the production of movement.

motor unit—The cell body and dendrites of a motor neuron, the multiple branches of its axon, and the muscle fibers that it innervates.

motor unit synchronization—An increased coincidence in the timing of action potentials discharged by different motor units.

motor unit territory—The subvolume of muscle in which the muscle fibers belonging to a single motor unit are located.

multiple sclerosis—A systemic disease that leads to a loss of the myelin sheath around axons in the CNS.

muscle—A tissue that contains contractile cells capable of converting chemical energy into mechanical energy and that has the properties of irritability, conductivity, contractility, and a limited growth and regenerative capacity.

muscle architecture—The design of muscle including such factors as length, cross-sectional area, pennation, and the attachment points on the skeleton.

muscle cramp—A painful, involuntary shortening of muscle that appears to be triggered by peripheral stimuli.

muscle fatigue—An exercise-induced reduction in the ability of muscle to produce force or power.

muscle force—The force exerted by structural (passive) and active (cross-bridges) elements of muscle.

muscle mechanics—The study of the mechanical properties of the force-generating units of muscle, whether or not the task can be continued.

muscle potentiation—An increase in the force evoked in muscle by a constant activation signal.

muscle power—The rate at which muscle can do work.

muscle soreness—*See* delayed-onset muscle soreness.

muscle spindle—An intramuscular sensory receptor that monitors unexpected changes in muscle length. It is arranged in parallel with skeletal muscle fibers.

muscle strain—An unexpected and substantial strain of muscle that occurs as an acute and painful event and is immediately recognized as an injury. Muscle strains are also referred to as "pulls" and "tears."

muscle synergy—A particular balance of muscle activations that change across time to assure a particular movement or preserve a specific posture.

muscle tone—The passive resistance of muscle to a change in its length.

muscle unit—All the muscle fibers innervated by the same motor neuron.

muscle weakness—A reduction in the force that a muscle can exert relative to its size.

muscle wisdom—The reduction in discharge rate of motor neurons during a fatiguing contraction to match the change in the biochemically mediated reduction in relaxation rate.

musculoskeletal force—A force inside the human body that contributes to movement.

musculotendinous unit—The combination of muscle and associated connective tissue structures that are involved in transmitting the force exerted by muscle fibers to the skeleton.

M wave—An EMG response that is evoked by electrical stimulation of the peripheral nerve to generate action potentials in axons belonging to alpha motor neurons.

myasthenia gravis—An immunological disorder in which the acetylcholine receptors are targeted by the immune system.

myelin—A fatty, insulating sheath that surrounds axons.

myofibril—A subunit of a muscle fiber that comprises several rows of sarcomeres arranged in series. The cross-sectional area of a myofibril is about 1 to 2 μm^2.

myofibrillar fatigue—An exercise-induced decline in the force exerted by individual cross-bridges.

myofilaments—The thick and thin filaments of a muscle fiber that contain the contractile apparatus.

myogenic—Pertaining to an effect attributable to muscle.

myonucleus—A nucleus in a muscle fiber.

myosin—The major protein in the thick filament of the muscle fiber. It includes the cross-bridge.

myosin ATPase—The enzyme that catalyzes the actomyosin reaction associated with the cross-bridge cycle.

myotendinous junction—The connection between the myofibrils and the tendon.

Na^+—Sodium ion.

Na^+-K^+ pump—A membrane-bound protein that transports Na^+ from the intracellular to the extracellular fluid and transports K^+ in the reverse direction.

nebulin—A protein in the endosarcomeric cytoskeleton that may maintain the lattice array of actin.

negative feedback—Outcome information that decreases the magnitude of the command controlling the outcome.

negative work—The work done by the surroundings on a system. Energy is absorbed by the system from the surroundings during negative work. Muscle does negative work when the resultant muscle torque is less than the load torque.

neocortex—The more recently evolved regions of the cerebral cortex.

Nernst equation—An equation to determine the equilibrium potential of an ion in the presence of an electrical field and a gradient in its concentrations inside and outside a cell.

neurofilament—One of the cytoskeletal structures involved in axoplasmic transport.

neurogenesis—The generation of new neurons.

neurogenic—Relating to an effect attributable to the nervous system.

neuroglia—One of two cell types in the nervous system. They provide structural, metabolic, and protective support for neurons.

neuromechanical—Referring to the integration of biomechanics and neurophysiology in the study of human movement.

neuromodulator—An agent that evokes a metabotropic effect.

neuromuscular compartment—The part of a muscle that is innervated by a primary branch of the muscle nerve.

neuromuscular electrical stimulation—A clinical or experimental procedure in which electric shocks are used to artificially generate axonal action potentials that elicit a muscle contraction.

neuromuscular junction—Nerve–muscle synapse.

neuromuscular propagation—The physiological processes that transform an axonal action potential into a muscle fiber action potential.

neuron—One of two cell types in the nervous system; capable of generating and transmitting an electrical signal.

neuropathy—A disease of the nerves, which can prevent transmission of action potentials.

neurotransmitter—A chemical substance that is released from a presynaptic terminal of a nerve cell and binds to specific receptors in the membrane of a postsynaptic cell.

neurotrophism—The sustaining influence that one biological element exerts on another.

next-state planner—A system that combines feedback on the current state of the limb with an estimate of target location to determine what to do next during a reaching and pointing task.

nociceptors—Small sensory endings that generate action potentials in response to potentially damaging stimuli (temperature, pressure, and certain chemicals). These sensations contribute to the sense of pain.

nonreciprocal group I inhibitory pathway—A pathway that transmits inhibitory input to homonymous and heteronymous motor neurons. Both Ia and Ib afferents provide input to the pathway.

normal component—A component that acts at a right angle to a surface.

normalization—The expression of an outcome measure relative to a factor that contributes to differences in a measure among individuals.

normalized muscle force—The net force relative to the cross-sectional area of the muscle.

nuclear bag fiber—An intrafusal fiber in which the nuclei cluster in a group.

nuclear chain fiber—An intrafusal fiber in which the nuclei are arranged end to end.

Ohm's law—$I = gV$, where I = current, g = conductance, and V = potential difference.

omphalion—The center of the navel.

1-repetition maximum load—The maximal load that can be lifted once.

optimum—The value of an independent variable that provides the best outcome for the dependent variable.

orderly recruitment—The repeatable sequence of motor unit activation for a specific task.

orthodromic—The conduction of action potentials in the forward direction, for example from the soma to the neuromuscular junction.

orthogonal—Perpendicular; independent.

orthograde transport—Axonal transport away from the cell body.

overlap of myofilaments—Interdigitation of thick and thin filaments that allows the cross-bridges to cycle and to exert a force. The number of cross-bridges that can be formed, and hence the force that can be exerted, depends on the amount of overlap.

overshoot—The phase of an action potential when the inside of a membrane is positive with respect to the outside.

Pacinian corpuscle—Thickly encapsulated receptor with a parent axon of 8 to 12 μm, a low threshold to mechanical stress, and the ability to detect acceleration of a joint.

paired-stimulus technique—The use of a pair of electric shocks (~10 ms apart) to test the force capacity of muscle.

pallidum—The output region of the basal ganglia, which comprises the globus pallidus pars interna, substantia nigra pars reticulata, and ventral pallidum.

parallel axis theorem—A theory that explains how to determine the moment of inertia for a system about an axis that is parallel to the one for which the moment of inertia is known.

parallel elastic element—The component of the Hill model that accounts for the elasticity of muscle in parallel with the contractile element.

parallelogram law of combinations—A theory that explains how to indicate the magnitude and direction of the resultant of two vectors or, conversely, two components of one vector.

Parkinson's disease—A disorder that involves an impairment of the basal ganglia.

PCr—Phosphocreatine.

pedunculopontine nucleus—A nucleus located in the midbrain tegmentum; functions as an ancillary loop of the basal ganglia.

pendulum method—A method to determine the moment of inertia for a system.

pennation—The angle between the orientation of muscle fibers and the line of pull of a muscle.

perception—The recognition and interpretation of afferent input from sensory receptors.

perimysium—The connective tissue matrix that collects bundles of muscle fibers into fascicles.

persistent inward current—Long-lasting current due to the activation of metabotropic receptors.

phase transition—The moment when a movement changes from one phase to another, for example from the stance phase to the swing phase in locomotion.

phasic—Intermittent.

piezoelectric—Pertaining to the generation of electric potentials due to pressure.

pinocytosis—The closure and release of a vesicle from the membrane of a presynaptic terminal after exocytosis.

planar—Existing in a single plane.

plasma membrane—A lipid bilayer that is semipermeable and excitable; also known as the plasmalemma.

plastic region—The part of a stress–strain or force–length graph where the tissue will not return to its original length when the pulling force is removed.

plateau potential—The sustained depolarization of a neuron produced by a persistent inward current.

plyometric—Referring to exercises for the stretch-shorten cycle.

polynomial—Referring to the sum of terms that are powers of a variable.

pons—The intermediate section of the brain stem, between the medulla and the mesencephalon.

position—The location of an object relative to some reference.

position task—A fatiguing contraction in which the task of the subject is to sustain a prescribed limb position.

positive feedback—Outcome information that increases the magnitude of the command controlling the outcome.

positive work—The work done by a system on its surroundings. Energy flows from the system to the surroundings during positive work. Muscle does positive work when the resultant muscle torque exceeds the load torque.

postcontraction sensory discharge—The increase in neural activity, primarily due to increased muscle spindle discharge, in the dorsal roots of the spinal cord after a muscle contraction.

posterior parietal cortex—A component of the sensorimotor cortex that is posterior to the central sulcus; Brodman's areas 5 and 7.

postinhibitory rebound—The increase in the excitability of a neuron after it has experienced a period of inhibition.

poststimulus time histogram—A histogram of the interspike intervals of a tonically discharging motor unit before and after a stimulus is applied to a sensory receptor or an afferent axon.

postsynaptic—On the distal side of a synapse.

postsynaptic inhibition—An inhibitory response on the distal side of a synapse that is evoked by the release of a neurotransmitter.

posttetanic potentiation—A transient increase in twitch force after a brief tetanus.

postural equilibrium—Maintenance of balance.

postural orientation—Positioning of the body relative to its surroundings.

postural set—The group of muscles activated before a movement to counteract the expected disturbance in posture.

posture—Orientation of the body for a task and the maintenance of equilibrium.

potential difference—A difference in the net electrical charge between two locations. It is measured in volts (V).

potential energy—The energy that a system possesses due to its location away from a more stable location. There are two forms of potential energy, gravitational and strain.

potentiation—An augmentation of a response (e.g., force, EMG) despite a constant input.

power—The rate of doing work; the rate of change in energy; the product of force and velocity.

power absorption—The flow of mechanical energy from the surroundings to a system. A system absorbs power when it does negative work.

power density spectrum—A graph of the relative contribution (y-axis) of different frequencies (x-axis) in a signal. The relative contribution is expressed as the power or amplitude of each frequency.

power production—The flow of mechanical energy from a system to its surroundings. A system produces power when it does positive work.

preferred direction—The direction of force or displacement in which the discharge of a neuron exhibits its greatest change from its resting rate.

premotor cortex—A component of the sensorimotor cortex that is anterior to the central sulcus; Brodman's area 6.

pressure drag—The drag due to a difference in the pressure between the front and back of an object, which increases as the flow becomes more turbulent.

presynaptic—On the proximal (incoming) side of a synapse.

presynaptic facilitation—An increase in the amount of neurotransmitter released by a neuron due to modulation that it receives at a synapse near its presynaptic terminal.

presynaptic inhibition—A decrease in the amount of neurotransmitter released by a neuron due to modulation that it receives at a synapse near its presynaptic terminal.

presynaptic terminal—A specialized swelling of an axon.

primary motor area—A component of the sensorimotor cortex that is anterior to the central sulcus; Brodman's area 4.

primary somatosensory cortex—A component of the sensorimotor cortex that is posterior to the central sulcus; Brodman's areas 1-3.

principle of transmissibility—The theory that explains why it is possible to slide a vector along its line of action without altering its mechanical effect on a rigid body.

progressive-resistance exercise—A training strategy in which the load increases progressively from one set of repetitions to the next.

projectile—An object that is displaced so that it has a phase of nonsupport and the only forces it experiences are those due to gravity and air resistance.

propagation—The conduction of an action potential by active, regenerative processes that tend to preserve the amplitude of the potential.

proprioception—Perceived sensations about the position and velocity of a movement and the muscular forces generated to perform a task.

proprioceptor—A sensory receptor (e.g., muscle spindle, tendon organ, joint receptor, cutaneous receptor) that provides information used to develop perceptions about movement and muscle contractions.

propriospinal neuron—A spinal neuron with an axon that traverses several spinal cord segments.

psychophysics—The quantification of sensory experience. Study of the relationship between a physical variable and its corresponding sensation.

Purkinje cells—Large inhibitory cells in the cerebellum; they provide the output of the cerebellum.

putamen nucleus—One of the basal ganglia nuclei.

pyramidal cells—Large neurons in the cerebral cortex.

Pythagorean relationship—$a^2 = b^2 + c^2$.

qualitative analysis—Description of the type or kind.

quantitative analysis—Description of the quantity or how much.

quantum—The quantity of neurotransmitter released by a vesicle.

quasistatic—Pertaining to a mechanical state in which the acceleration experienced by a system is small enough that it can be assumed to be zero.

radian—An angle defined as the quotient of a distance on the circumference of the circle relative to its radius. It is a dimensionless unit.

range of motion—The maximal angular displacement about a joint.

rate coding—Control of the rate at which motor neurons discharge action potentials.

reaction time—The minimum time from the presentation of a stimulus to the onset of a response.

reactive grip response—An automatic increase in grip force when a handheld object begins to slip or its balance is disturbed.

readiness potential—A slow, negative shift in the EEG that begins about 1.5 s prior to a voluntary movement.

receptor potential—The electrical potential generated in a sensory receptor in response to a stimulus.

reciprocal Ia inhibition—A pathway that includes projections from Ia afferents of one muscle onto interneurons that evoke inhibitory postsynaptic potentials in motor neurons supplying an antagonist muscle.

recruitment—The process of motor unit activation.

recruitment curve—A graph of the relation between stimulus intensity and the amplitudes of the M wave and H reflex.

recruitment threshold—The force at which a motor unit is recruited (activated) during a voluntary contraction.

rectification—A process that involves eliminating or inverting the negative phases in a signal. This process is frequently applied to the EMG interference pattern. Half-wave rectification refers to eliminating the negative phase, whereas full-wave rectification involves inverting the negative phases. Rectification may be accomplished numerically or electronically.

recurrent facilitation—Inhibition of an Ia inhibitory interneuron by Renshaw cells to depress reciprocal Ia inhibition.

recurrent inhibition—A local reflex circuit whereby action potentials discharged by the motor neuron lead to activation, via axon collaterals, of the Renshaw cell and the subsequent generation of an inhibitory postsynaptic potential in the motor neuron.

red nucleus—A cluster of cells located in the mesencephalon that give rise to the rubrospinal tract.

reflex—A short-latency connection between an input signal (afferent feedback) and an output response (muscle activation).

refractory—Referring to a period during which an excitable membrane is less excitable than normal.

regression analysis—The statistical process of finding a simple function, such as a line, to describe the relation between two variables.

regulation—The process of maintaining a constant state within a system that can accommodate disturbances.

reinnervation—The reconnection of a nerve to its muscle.

relative angle—The angle between two adjacent body segments.

Renshaw cell—An interneuron in the spinal cord that elicits inhibitory postsynaptic potentials in other neurons. The Renshaw cell receives input from collateral branches of motor neuron axons and from descending pathways.

repeated-bout effect—A reduction in the amount of delayed-onset muscle soreness upon repeat performances of the activity that originally caused the soreness.

repolarization—The phase of an action potential when the membrane potential is being returned from a reversal in the polarity (overshoot) to the resting membrane potential.

residual analysis—A technique used in curve fitting to determine the difference between the filtered and unfiltered signals.

residual moment of force—The term in an equation of motion that remains when all the other quantities have been determined. Resolving for this term is the most common procedure for calculating the resultant muscle force.

resistance (*R*)—The opposition that an object presents to the transmission of current.

resistance to fatigue—The ability of muscle (muscle fiber, motor unit, whole muscle) to sustain force with repetitive activation.

resolution—The process of breaking a vector down into several components.

resting channels—Ion channels that are open during steady-state conditions.

resting length of muscle—The length at which a passive force is first detected when a muscle is stretched.

resting membrane potential—The transmembrane voltage during steady-state conditions.

resultant muscle force (or torque)—The net force (torque) exerted by a group of muscles about a joint.

reticular formation—A diffuse cluster of cells in the brain stem with concentrations in the pons and the medulla that give rise to two reticulospinal tracts.

reticulospinal tract—Descending pathway from the reticular formation to the spinal cord.

retrograde transport—Axonal transport toward the cell body.

reversal potential—The membrane potential at which the flux of an ion across a membrane changes direction.

rheobase current—A measure of neuron excitability as the amount of current that has to be injected into the neuron in order for it to generate an action potential.

rheological models—Models that are used to study the deformation and flow of matter.

right-hand coordinate system—Standard positive directions for *x*-, *y*-, and *z*-axes (to the right, up, and out of page toward the reader, respectively).

right-hand-thumb rule—A technique for determining the direction of a vector product. The direction of the product is always perpendicular to the plane in which the other two vectors lie.

rigidity—A bidirectional resistance to passive movement that is independent of the movement velocity and that occurs in the absence of an exaggerated tendon tap reflex.

RNA—Ribonucleic acid. Genetic material formed from DNA and used to construct proteins in the cytoplasm.

root mean square—A numerical process used to estimate the amplitude of such signals as the electromyogram and mechanomyogram.

rotation—A motion in which not all parts of the system are displaced by a similar amount.

rubrospinal tract—A descending pathway from the red nucleus to motor neurons in the spinal cord.

Ruffini endings—Two to six thinly encapsulated, globular corpuscles with a single myelinated parent axon that has a diameter of 5 to 9 μm. These mechanoreceptors are capable of signaling joint position and displacement, angular velocity, and intra-articular pressure.

running—A mode of human locomotion that includes a phase when neither foot is on the ground.

ryanodine receptor—Calcium-releasing channel in the sarcoplasmic reticulum of the muscle fiber.

sag—A tetanus in which the force declines after the initial four to eight action potentials and then increases again.

sagittal plane—A plane that divides the body into right and left.

saltatory conduction—Propagation of an action potential along a myelinated axon by the depolarization of the axolemma exposed at each node of Ranvier.

sarcolemma—The excitable plasma membrane of a muscle fiber.

sarcomere—The zone of a myofibril from one Z band to the next Z band.

sarcopenia—The decline in muscle mass and strength with advancing age.

sarcoplasm—The fluid enclosed within a muscle fiber by the sarcolemma.

sarcoplasmic reticulum—A hollow membranous system within a muscle fiber that bulges into lateral sacs in the vicinity of the transverse tubules.

scalar—Referring to a variable that is defined by magnitude only.

scalar product—The procedure used to multiply vectors to produce a scalar quantity (product).

segmental analysis—A biomechanical analysis that involves graphically separating body segments and drawing a free body diagram to derive the equations of motion for each segment.

self-sustained firing—The maintained discharge of action potentials by a neuron due to a persistent inward current.

sense of effort—A perception of the effort associated with performing a muscle contraction.

series elastic element—The component of the Hill model that accounts for the elasticity of muscle in series with the contractile element.

serotonin-fatigue hypothesis—Hypothesis that the reduction in the ability of the cerebral cortex to provide an adequate descending drive during prolonged exercise is caused by a disturbance of the serotonergic neurotransmitter system.

servo—A feedback control system that enables a controller to achieve a desired output.

servo hypothesis—Notion that muscle length is maintained at a centrally specified value by means of the tonic stretch reflex.

shear force—A force that acts parallel to a contact surface.

shortening deactivation—Depression of muscle force after a shortening contraction.

shortening velocity—The rate at which a muscle shortens its length.

short-range stiffness—A mechanical property of muscle that produces a high initial stiffness during a stretch.

significant figures—The number of digits in a number that are used to indicate the accuracy of a measurement.

SI system—The international metric system, Le Systeme Internationale d'Unités.

size principle—The concept that the recruitment of motor units is based on motor neuron size, proceeding from the smallest to the largest.

skelemin—A protein that is part of the cytoskeleton.

sliding filament theory—A concept describing the sliding of thick and thin myofilaments past one another during a muscle contraction.

slope—The rate of change in one variable (dependent variable; *y*-axis variable) relative to another variable (independent variable; *x*-axis variable); rise over the run.

slow axonal transport—An intra-axonal system of cytoskeletal structures that can transport enzymes and precursors between a cell body and its presynaptic terminals at 0.5 to 5 mm/day.

smoothing—The process of reducing the fluctuations in a signal, which has the effect of decreasing the higher-frequency components in the signal.

sodium–potassium pump—An integral membrane protein that translocates Na^+ and K^+ against their concentration gradients across an excitable membrane.

soma—The cell body of a neuron.

somatosensory cortical areas—Areas 1, 2, 3a, and 3b in the parietal cortex, which receive inputs from the thalamus.

somatosensory receptors—The six sensory receptors that provide feedback from the periphery: muscle spindle, tendon organs, joint receptors, cutaneous mechanoreceptors, nociceptors, and thermal sensors.

somatotopic—Relating to representations in the nervous system, a body map.

somersault axis—An axis that passes through the human body from side to side.

space diagram—A schematic of a free body diagram that does not include the correct magnitudes of the force vectors.

spared-root preparation—An experimental procedure of CNS recovery in which all the dorsal roots supplying a hindlimb except one are transected.

spasticity—Pathological disruption of transmission along descending pathways, which is manifested as uncontrolled spasms, increased muscle tone, and heightened excitability of the stretch reflex.

spatial summation—The sum of inputs from multiple synapses.

specific membrane capacitance—The amount of capacitance that can be stored on 1 cm^2 of excitable membrane.

specific membrane conductance—The amount of current that can cross 1 cm^2 of an excitable membrane.

specific tension—The intrinsic capacity of muscle to exert force (N/cm^2).

speed—The size of the velocity vector (tells how fast).

spike-triggered stimulation—A technique used to assess the influence of a feedback pathway on the discharge of an isolated motor unit. The stimulus is delivered to afferent axons either before or after an action potential is discharged by the motor unit, and the time to the next action potential is recorded.

spinalization—An experimental procedure that disconnects the spinal cord from supraspinal centers.

spinal reflexes—Fast responses that involve an afferent signal into the spinal cord and an efferent signal out to the muscle.

spinal transection—Cutting of the spinal cord so that it is disconnected from the brain.

spinocerebellum—The vermis and intermediate hemispheres of the cerebellum. Receives input from the spinal cord and sends output to the fastigial and interposed nuclei.

spinocervical tract—An ascending pathway that ends in the cervical cord and contributes to the descending control of pain.

spinoreticular tract—An ascending pathway that conveys noxious and thermal information from the spinal cord to the reticular formation.

spinothalamic tract—An ascending pathway that conveys noxious and thermal information from the spinal cord to the thalamus.

spring–mass system—A simple model used to represent the elastic (spring) and inertial (mass) properties of a system.

stability—A state of equilibrium to which a system returns after it has been perturbed.

staggered muscle fibers—Serially attached muscle fibers that are shorter than the length of the whole muscle.

stance phase—The phase of the gait cycle when the foot is on the ground.

startle reaction—The response of the body to an unexpected loud auditory stimulus; sometimes it can be evoked with visual, vestibular, or somesthetic stimuli.

state-dependent response—A response that varies with the state of a system. Reflexes are described as state-dependent responses, which means that their amplitude varies depending on the activity of the individual (e.g., walking vs. standing, stance vs. swing).

static analysis—Mechanical analysis in which the forces acting on the system are balanced and thus the system is not accelerating. The system is either stationary or moving at a constant velocity.

static stretch—A technique aimed at increasing the range of motion about a joint by sustaining a muscle stretch for 15 to 30 s.

steadiness—The fluctuations in force or acceleration when an individual attempts to keep force constant.

step—One-half of a stride, for example from left foot toe-off to right foot toe-off. There are two steps in one complete gait cycle (stride).

stiffness—The change in force per unit change in length (N/mm); the slope of a force–length graph.

strain—The change in length relative to initial length (%).

strain energy—The potential energy (J) stored by a system when it is stretched from a resting position.

streamline—The schematized lines of fluid flow around an object.

stress—Force applied per unit area (Pa).

stress relaxation—A decrease in the stress experienced by tissue when repeatedly stretched to a specific length.

stretch reflex—The response of a muscle to a sudden, unexpected increase in its length. The response activates the stretched muscle to minimize the length increase.

stretch–shorten cycle—A muscle activation scheme in which an activated muscle lengthens before it shortens.

striatum—The major input region of the basal ganglia comprising the caudate nucleus, putamen, and ventral striatum.

stride—One complete cycle of human gait, for example from left foot toe-off to left foot toe-off. A stride contains two steps.

stride length—The distance covered in a single stride.

stride rate—The frequency (Hz) at which strides are performed. The inverse of stride rate is the time it takes to complete a stride.

subfragment 1—The two globular heads of the myosin molecule. One globular head contains an ATP-binding site and the other an actin-binding site.

subfragment 2—The nonglobular head region of heavy meromyosin.

substantia nigra—One of the output nuclei of the basal ganglia. It consists of the pars compacta and the pars reticulata.

subthalamic nucleus—One of the basal ganglia nuclei.

superior colliculus—Located on the dorsal surface of the midbrain; coordinates head and eye movement to visual targets.

supplementary motor area—A component of the sensorimotor cortex (area 6) that is anterior to the central sulcus.

supraspinal—The part of the CNS above (rostral to) the spinal cord.

suprasternale—The most caudal point on the margin of the jugular notch of the sternum.

surface drag—Drag due to the friction between the boundary layer of a fluid and an object.

surface mechanomyogram (MMG)—A measure of the transverse displacement or acceleration of the skin over a muscle during a contraction.

swing phase—The period of a gait cycle when the foot is not on the ground.

synapse—The specialized zone of contact at which a neuron communicates with another cell.

synaptic bouton—The swelling of a motor neuron axon with the capacity to release neurotransmitter at the nerve-muscle synapse.

synaptic cleft—The space (20-40 nm) at a chemical synapse between the pre- and postsynaptic membranes.

synaptic delay—The time it takes for an action potential to cross a synapse.

synaptic potential—A conducted (nonpropagated) excitatory or inhibitory electrical potential that is elicited in neurons and muscle fibers in response to a neurotransmitter. The amplitude of a synaptic potential is variable, depending on the excitability of the cell and the quantity of neurotransmitter released.

synaptic vesicle—An organelle that stores thousands of molecules of a neurotransmitter.

synaptogenesis—Formation of new synapses.

T_1 relaxation time—An MRI measure of the time constant for the oscillations of atomic nuclei to return to an initial level of longitudinal magnetization after exposure to a radio-frequency pulse. The recovery rates vary across tissues, which enables MRI to differentiate between different types of tissue.

T_2 relaxation time—An MRI measure of the time constant for the oscillations of atomic nuclei to become random again after exposure to a radio-frequency pulse. The T_2 for a tissue largely depends on its chemical environment, which can change with exercise.

tangential component—A component that acts parallel to and along a surface.

task dependency of muscle fatigue—The association between the demands of a task and the physiological processes that are impaired during a fatiguing contraction.

task-failure approach—An experimental strategy that seeks to identify the physiological adjustments that are responsible for an individual not being able to continue performing a prescribed task.

tectospinal tract—A descending pathway that arises from neurons in the superior colliculus, which is a mesencephalic area concerned with orientation to visual inputs.

temporal summation—The sum of successive inputs at a given synapse.

tendon organ—An intramuscular sensory receptor that monitors muscle force. It is arranged in series with skeletal muscle fibers.

tendon tap reflex—A monosynaptic reflex evoked by a mechanical stimulus, such as a brief tap applied to the muscle tendon.

tensile force—A pulling force.

terminal velocity—The velocity of a falling object when the forces due to gravity and air resistance are equal in magnitude.

tetanus—A force evoked in muscle (single fiber, motor unit, or whole muscle) in response to a series of action potentials that produces a summation of twitch responses.

tetrodotoxin—A drug that blocks the propagation of action potentials by deactivating Na^+ channels.

thalamus—A brain structure that integrates and distributes most of the sensory and motor information going to the cerebral cortex. It is important in coordination.

thixotropy—The property exhibited by various gels, such as muscle, of becoming fluid when shaken, stirred, or otherwise disturbed and of setting again when allowed to stand.

time—Duration; measured in seconds (s).

time constant—The time it takes an exponential function to reach 63% of its final value.

time domain—A graph that has time as the independent variable (x-axis).

tiredness—A sensation of lethargy.

titin—A protein in the endosarcomeric cytoskeleton that may be responsible for resting muscle elasticity.

TN-C—The component of troponin that binds calcium.

TN-I—The component of troponin that inhibits G actin from binding to myosin.

TN-T—The component of troponin that binds to tropomyosin.

tonic—Continuous, sustained.

tonic stretch reflex—A polysynaptic reflex that increases the level of muscle activation in response to a slow muscle stretch.

tonic vibration reflex—A polysynaptic reflex that is elicited by small-amplitude, high-frequency (50-150 Hz) vibration of a muscle. The vibration activates the muscle spindles, which provide reflex activation of the motor neurons.

torque (τ)—The rotary effect of a force, which is quantified as the product of force and moment arm ($N \cdot m$).

torque–angle relation—The variation in muscle torque as a function of joint angle.

torque–velocity relation—The variation in muscle torque as a function of joint angular velocity.

torsional spring—An angular spring, such as a snapping mouse trap.

tractive resistance—The combined effect of air resistance and rolling resistance.

trajectory—A position–time record for a movement.

transcranial magnetic stimulation—A technique that activates brain cells by creating a magnetic field through the skull.

transduction—A process by which energy is converted from one form to another.

translation—A motion in which all parts of the system are displaced by a similar amount.

transverse plane—A plane that divides the body into upper and lower.

transverse (T) tubules—Invaginations of the sarcolemma that facilitate a rapid communication between sarcolemmal events (action potentials) and myofilaments located in the interior of the muscle fiber.

tremor—An oscillation of a body part that is produced by alternating muscle activity. The oscillations can occur at about 3 to 5 Hz in patients with cerebellar disorders, ~6 Hz in patients with Parkinson's disease, and at 8 to 12 Hz in healthy persons (physiological tremor).

tropomyosin—A thin-filament protein involved in regulating the interaction between actin and myosin.

troponin—A three-component molecule that forms part of the thin filament and is involved in the regulation of the interaction between actin and myosin.

turbulent flow—Nonuniform flow of fluid (streamlines) around an object.

twist axis—An axis that passes through the human body from head to toe.

twitch—The force response of muscle (single fiber, motor unit, or whole muscle) to a single action potential.

twitch interpolation technique—A technique used to test the maximality of voluntary activation. The technique involves delivering one to three supramaximal electric shocks to the muscle nerve during a voluntary contraction to determine if the muscle force can be increased. The technique is also known as twitch superimposition and twitch occlusion.

two-joint muscle—A muscle whose attachment sites span two joints.

type FF motor unit—A fast-twitch, fatigable motor unit. This type of unit is characterized by a sag response and a decline in force during a standard fatigue test.

type FR motor unit—A fast-twitch, fatigue-resistant motor unit. This type of unit is characterized by a sag response and a minimal decline in force during a standard fatigue test.

type I muscle fiber—A slow-twitch muscle fiber, as defined by a low level of myosin ATPase.

type II muscle fiber—A fast-twitch muscle fiber, as defined by a high level of myosin ATPase.

type S motor unit—A slow-twitch motor unit. This type of unit is characterized by a zero sag response and the absence of a decline in force during a standard fatigue test.

unfused tetanus—An irregular muscle force (in a muscle fiber, motor unit, whole muscle) elicited by a low frequency of activation. The peak forces of the individual twitch responses are evident.

unmasking—The uncovering of a previously dormant pathway that enables the recovery of function following an injury to the nervous system.

upper motor neuron—A cortical neuron that projects to the spinal cord.

Valsalva maneuver—Voluntary pressurization of the abdominal cavity.

variable load—A training load that varies over a range of motion.

vector—Quantity that conveys both magnitude and direction.

vector product—The procedure used to multiply vectors to produce a vector quantity (product).

velocity—The rate of change in position with respect to time (m/s); the derivative of position with respect to time; the slope of a position–time graph.

ventral roots—Structures on the front side of the spinal cord from which efferent axons exit the spinal cord.

vermis—Middle region of the cerebellum.

vestibular nuclei—Nuclei located in the medulla that give rise to the vestibulospinal tracts.

vestibulocerebellum—A structure that occupies the flocculonodular lobe of the cerebellum. Receives input from the vestibular nuclei and sends output back to the same nuclei.

vestibulospinal tract—A descending pathway from the lateral vestibular nucleus to ipsilateral motor neurons.

vimentin—A protein that is part of the cytoskeleton.

viscoelastic—Characterizing a material that has both viscous and elastic properties.

viscosity—A measure of the shear stress that must be applied to a fluid to obtain a rate of deformation. Viscosity varies with temperature and depends on the cohesive forces between molecules and the momentum interchange between colliding molecules. The SI unit of measurement is $N \cdot s/cm^2$. Oil has a greater viscosity than water.

voltage clamp—A device that holds membrane potential at a user-set value by passing current across the membrane.

voltage-divider circuit—The distribution of a battery's potential difference to two resistors, connected in series, in proportion to their resistance.

voltage-gated channel—A pathway through an excitable membrane that is opened by a decrease in the electrical potential across a membrane.

voltage threshold—The amount of depolarization required from the resting membrane potential to cause a neuron to discharge an action potential.

volume conduction—The spread of electrical signals in the body.

voluntary actions —Contractions generated by the cerebral cortex in response to a perceived need.

voluntary activation—A measure of the extent to which the nervous system can elicit the maximal force a muscle can produce (equation 8.1).

voluntary contraction—An action generated by the cerebral cortex in response to a perceived need.

walking—A mode of human locomotion that is defined in either of two ways: (1) At least one foot is always in contact with the ground or (2) the center of mass reaches a maximum height in the middle of the stance phase.

warm-up—An intervention that increases muscle temperature.

wave resistance—The resistive effects of waves encountered by a swimmer. The resistance is probably due to the difference between the density of water and that of air.

weight—The amount of gravitational attraction between an object and Earth.

weighting coefficient—A coefficient used to modulate the amplitude of a variable and its contribution to a compound expression, for example the harmonic terms of a Fourier analysis.

Westphal phenomenon—An abrupt reflex excitation of muscle in response to an externally imposed shortening of the muscle.

withdrawal response—The withdrawal of a limb away from a noxious stimulus.

wobbling mass—The soft tissue component of a body segment.

work—A scalar quantity that describes the extent to which a force can move an object in a specified direction. The symbol for work is U, and its SI unit of measurement is the joule (J).

work loop—A graph that shows the net work performed by a muscle as the difference between the work input (lengthening contraction) and the work output (shortening contraction). The method is most often applied to cyclic locomotor activities.

xyphion—The midpoint of the sulcus between the body of the sternum and the xyphoid process.

Z band—Zwischenscheibe; intrasarcomere connection of the two sets of thin filaments in adjacent sarcomeres.

Zone of Optimal Function theory—Theory regarding the level of arousal that produces maximal performance, which varies across individuals.

References

Aagaard, P., Andersen, J.L., Dyhre-Poulsen, P., Leffers, A.M., Wagner, A., Magnusson, S.P., Halkjaer-Kristensen, J., & Simonsen, E.B. (2001). A mechanism for increased contractile strength in response to strength training: Changes in muscle architecture. *Journal of Physiology, 534,* 613-623.

Aagaard, P., & Bangsbo, J. (2006). The muscular system: Design, function, and performance relationships. In C.M. Tipton (Ed.), *ACSM's Advanced Exercise Physiology* (pp. 144-160). Philadelphia: Lippincott Williams & Wilkins.

Aagaard, P., Simonsen, E.B., Andersen, J.L., Magnusson, S.P., Bojsen-Møller, F., & Dyhre-Poulsen, P. (2000*a*). Antagonist muscle coactivation during isokinetic knee extension. *Scandinavian Journal of Medicine and Science in Sports, 10,* 58-67.

Aagaard, P., Simonsen, E.B., Andersen, J.L., Magnusson, S.P., & Dyhre-Poulsen, P. (2002*a*). Increased rate of force development and neural drive of human skeletal muscle following resistance training. *Journal of Applied Physiology, 93,* 1318-1326.

Aagaard, P., Simonsen, E.B., Andersen, J.L., Magnusson, S.P., & Dyhre-Poulsen, P. (2002*b*). Neural adaptation to resistance training: Changes in evoked V-wave and H-reflex responses. *Journal of Applied Physiology, 92,* 2309-2318.

Aagaard, P., Simonsen, E.B., Andersen, J.L., Magnusson, S.P., Halkjaer-Kristensen, J., & Dyhre-Poulsen, P. (2000*b*). Neural inhibition during maximal eccentric and concentric quadriceps contraction: Effects of resistance training. *Journal of Applied Physiology, 89,* 2249-2257.

Aagaard, P., Simonsen, E.B., Trolle, M., Bangsbo, J., & Klausen, K. (1994*a*). Effects of different strength training regimes on moment and power generation during dynamic knee extensions. *European Journal of Applied Physiology, 69,* 382-386.

Aagaard, P., Simonsen, E.B., Trolle, M., Bangsbo, J., & Klausen, K. (1994*b*). Moment and power generation during maximal knee extensions performed at low and high speeds. *European Journal of Applied Physiology, 69,* 376-381.

Aagaard, P., Simonsen, E.B., Trolle, M., Bangsbo, J., & Klausen, K. (1996). Specificity of training velocity and training load on gains in isokinetic knee joint strength. *Acta Physiologica Scandinavica, 156,* 123-129.

Abbruzzese, G., Dall'Agata, D., Morena, M., Reni, L., & Favale, E. (1990). Abnormalities of parietal and prerolandic somatosensory evoked potentials in Huntington's disease. *Electroencephalography and Clinical Neurophysiology, 77,* 340-346.

Ada, L., O'Dwyer, N., & O'Neill, E. (2006). Relation between spasticity, weakness and contracture of the elbow flexors and upper limb activity after stroke: An observational study. *Disability and Rehabilitation, 28,* 891-897.

Adam, A., & De Luca, J.C. (2003). Recruitment order of motor units in human vastus lateralis muscle is maintained during fatiguing contractions. *Journal of Neurophysiology, 90,* 2919-2927.

Adams, G.R., Caiozzo, V.J., & Baldwin, K.M. (2003). Skeletal muscle unweighting: Spaceflight and ground-based models. *Journal of Applied Physiology, 95,* 2185-2201.

Adams, G.R., Duvoisin, M.R., & Dudley, G.A. (1992). Magnetic resonance imaging and electromyography as indexes of muscle function. *Journal of Applied Physiology, 73,* 1578-1583.

Adams, G.R., Harris, R.T., Woodard, D., & Dudley, G.A. (1993*a*). Mapping of electrical muscle stimulation using MRI. *Journal of Applied Physiology, 74,* 532-537.

Adams, G.R., Hather, B.M., Baldwin, K.M., & Dudley, G.A. (1993*b*). Skeletal muscle myosin heavy chain composition and resistance training. *Journal of Applied Physiology, 74,* 911-915.

Adamson, G., & McDonagh, M. (2004). Human involuntary postural aftercontractions are strongly modulated by limb position. *European Journal of Applied Physiology, 92,* 343-351.

Adkins, D.L., Boychuk, J., Remple, M.S., & Kleim, J.A. (2006). Motor training induces experience-specific patterns of plasticity across motor cortex and spinal cord. *Journal of Applied Physiology, 101,* 1776-1782.

Adrian, E.D. (1925). Interpretation of the electromyogram. *Lancet,* June 13, 1229-1233.

Adrian, E., & Bronk, D. (1929). The discharge of impulses in motor nerve fibres. Part II. The frequency of discharges in reflex and voluntary contractions. *Journal of Physiology, 67,* 119-151.

Ahlborg, L., Andersson, C., & Julin, P. (2006). Whole-body vibration training compared with resistance training: Effect on spasticitiy, muscle strength, and motor performance in adults with cerebral palsy. *Journal of Rehabilitation Medicine, 38,* 302-308.

Akima, H., Takahashi, H., Kuno, S.Y., & Katsuta, S. (2004). Coactivation pattern in human quadriceps during isokinetic knee-extension by muscle functional MRI. *European Journal of Applied Physiology, 91,* 7-14.

Akima, H., Takahashi, H., Kuno, S.-Y., Masuda, K., Masuda, T., Shimojo, H., Anno, I., Itai, Y., & Katsuta, S. (1999). Early phase adaptations of muscle use and strength to isokinetic training. *Medicine and Science in Sports and Exercise, 31,* 588-594.

Albani, G., Sandrini, G., Kunig, G., Martin-Soelch, C., Mauro, A., Pignatti, R., Pacchetti, C., Dietz, V., & Leenders, K.L. (2003). Differences in the EMG pattern of leg muscle activation during locomotion in Parkinson's disease. *Functional Neurology, 18,* 165-170.

Alexander, R.M. (1984). Walking and running. *American Scientist, 72,* 348-354.

Al-Falahe, N.A., Nagaoka, M., & Vallbo, Å.B. (1990). Response profiles of human muscle afferents during active finger movements. *Brain, 113,* 325-346.

Alford, E.K., Roy, R.R., Hodgson, J.A., & Edgerton, V.R. (1987). Electromyography of rat soleus, medial gastrocnemius, and tibialis anterior during hind limb suspension. *Experimental Neurology, 96,* 635-649.

Alkjaer, T., Simonsen, E.B., Jorgensen, U., & Dyhre-Poulsen, P. (2003). Evaluation of the walking pattern in two types of patients with anterior cruciate ligament deficiency: Copers and non-copers. *European Journal of Applied Physiology, 89,* 301-308.

Alkner, B.A., & Tesch, P.A. (2004). Knee extensor and plantar flexor muscle size and function following 90 days of bed rest with or without resistance exercise. *European Journal of Applied Physiology, 93,* 294-305.

Allaf, O., & Goubel, F. (1999). The rat suspension model is also a good tool for inducing muscle hyperactivity. *Pflügers Archiv, 437,* 504-507.

Allen, D.G. (2004). Skeletal muscle function: Role of ionic changes in fatigue, damage, and disease. *Clinical and Experimental Pharmacology and Physiology, 31,* 485-493.

Allen, D.G., Lee, J.A., & Westerblad, H. (1989). Intracellular calcium and tension during fatigue in isolated single muscle fibres from *Xenopus Laevis. Journal of Physiology, 415,* 433-458.

Allen, D.G., Westerblad, H., & Lännergren, J. (1995). The role of intracellular acidosis in muscle fatigue. In S.C. Gandevia, R.M. Enoka, A.J. McComas, D.G. Stuart, & C.K. Thomas (Eds.), *Fatigue: Neural and Muscular Mechanisms* (pp. 57-68). New York: Plenum Press.

Allen, D.G., Whitehead, N.P., & Yeung, E.W. (2005). Mechanisms of stretch-induced muscle damage in normal and dystrophic muscle: Role of ionic changes. *Journal of Physiology, 567,* 723-735.

Allen, G.M., Gandevia, S.C., & McKenzie, D.K. (1995). Reliability of measurements of muscle strength and voluntary activation using twitch interpolation. *Muscle & Nerve, 18,* 593-600.

Allen, G.M., McKenzie, D.K., & Gandevia, S.C. (1998). Twitch interpolation of the elbow flexor muscles at high forces. *Muscle & Nerve, 21,* 318-328.

Allen, T.J., & Proske, U. (2006). Effect of muscle fatigue on the sense of limb position and movement. *Experimental Brain Research, 170,* 30-38.

Allred, R.P., Maldonado, M.A., Hsu And, J.E., & Jones, T.A. (2005). Training the "less-affected" forelimb after unilateral cortical infarct interferes with functional recovery of the impaired forelimb in rats. *Restorative Neurology and Neuroscience, 23,* 297-302.

Allum, J.H., Carpenter, M.G., Honegger, F., Adkin, A.L., & Bloem, B.R. (2002). Age-dependent variations in the directional sensitivity of balance corrections and compensatory arm movements in man. *Journal of Physiology, 542,* 643-663.

Almåsbakk, B., & Hoff, J. (1996). Coordination, the determinant of velocity specificity? *Journal of Applied Physiology, 80,* 2046-2052.

Almeida-Silveira, M.-I., Lambertz, D., Pérot, C., & Goubel, F. (2000). Changes in stiffness induced by hindlimb suspension in rat Achilles tendon. *European Journal of Applied Physiology, 81,* 252-257.

Almeida-Silveira, M.-I., Pérot, C., & Goubel, F. (1996). Neuromuscular adaptations in rats trained by muscle stretch-shortening. *European Journal of Applied Physiology, 72,* 261-266.

Alshuaib, W.B., & Fahim, M.A. (1991). Depolarization reverses age-related decrease of spontaneous transmitter release. *Journal of Applied Physiology, 70,* 2066-2071.

Alway, S.E., Grumbt, W.H., Stray-Gundersen, J., & Gonyea, W.J. (1992). Effects of resistance training on elbow flexors of highly competitive bodybuilders. *Journal of Applied Physiology, 72,* 1512-1521.

Alway, S.E., MacDougall, J.D., & Sale, D.G. (1989). Contractile adaptations in the human triceps surae after isometric exercise. *Journal of Applied Physiology, 66,* 2725-2732.

Amako, M., Oda, T., Masuoka, K., Yokoi, H., & Campisi, P. (2003). Effect of static stretching on prevention of injuries for military recruits. *Military Medicine, 168,* 442-446.

Amar, J. (1920). *The Human Motor.* London: Routledge.

Amiridis, I.G., Hatzitaki, V., & Arabatzi, F. (2003). Age-induced modifications of static postural control in humans. *Neuroscience Letters, 35,* 137-140.

Amiridis, I.G., Martin, A., Morlon, B., Martin, L., Cometti, G., Pousson, M., & van Hoecke, J. (1996). Co-activation and tension-regulating phenomena during isokinetic knee extension in sedentary and highly skilled humans. *European Journal of Applied Physiology, 73,* 149-156.

An, K.N., Hui, F.C., Morrey, B.F., Linscheid, R.L., & Chao, E.Y. (1981). Muscles across the elbow joint: A biomechanical analysis. *Journal of Biomechanics, 10,* 659-669.

An, K.N., Kwak, B.M., Chao, E.Y., & Morrey, B.F. (1984). Determination of muscle and joint forces: A new technique to solve the indeterminate problem. *Journal of Biomechanical Engineering, 106,* 364-367.

Andersen, B., Westlund, B., & Krarup, C. (2003). Failure of activation of spinal motoneurones after muscle fatigue in healthy subjects studied by transcranial magnetic stimulation. *Journal of Physiology, 551,* 345-356.

Andersen, O.K., Sonnenberg, F.A., & Arendt-Nielsen, L. (1999). Modular organization of human leg withdrawal reflexes elicited by electrical stimulation of the foot sole. *Muscle & Nerve, 22,* 1520-1530.

Anderson, J.E. (2006). The satellite cell as a companion in skeletal muscle plasticity: Currency, conveyance, clue, connector and colander. *Journal of Experimental Biology, 209,* 2276-2292.

Andersson, E.A., Nilsson, J., Ma, Z., & Thorstensson, A. (1996). Abdominal and hip flexor muscle activation during various training exercises. *European Journal of Applied Physiology, 75,* 115-123.

Andersson, G.B.J., Örtengren, R., & Nachemson, A. (1977). Intradiscal pressure, intra-abdominal pressure, and myoelectric back muscle activity related to posture and loading. *Clinical Orthopedics Research, 129,* 156-164.

Andruchov, O., Andruchova, O., Wang, Y., & Galler, S. (2006). Dependence of cross-bridge kinetics on myosin light chain isoforms in rabbit and rat skeletal muscle fibres. *Journal of Physiology, 571,* 231-242.

Antonio, J., & Gonyea, W.J. (1993). Skeletal muscle fiber hyperplasia. *Medicine and Science in Sports and Exercise, 25,* 1333-1345.

Antonutto, G., Capelli, C., Girardis, M., Zamparo, P., & di Prampero, P.E. (1999). Effects of microgravity on maximal power of lower limbs during very short efforts in humans. *Journal of Applied Physiology, 86,* 85-92.

Antonutto, G., & di Prampero, P.E. (2003). Cardiovascular deconditioning in microgravity: Some possible countermeasures. *European Journal of Applied Physiology, 90,* 283-291.

Arampatzis, A., Brüggemann, G.-P., & Metzler, V. (1999). The effect of speed on leg stiffness and joint kinetics in human running. *Journal of Biomechanics, 32,* 1349-1353.

Arampatzis, A., Knicker, A., Metzler, V., & Brüggemann, G.-P. (2000). Mechanical power in running: A comparison of different approaches. *Journal of Biomechanics, 33,* 457-463.

Arampatzis, A., Stafilidis, S., DeMonte, G., Karamanidis, K., Morey-Klapsing, G., & Bruggemann, G.P. (2005). Strain and elongation of the human gastrocnemius tendon and aponeurosis during maximal plantarflexion effort. *Journal of Biomechanics, 38,* 833-841.

Arendt-Nielsen, L., Mills, K.R., & Forster, A. (1989). Changes in muscle fiber conduction velocity, mean power frequency, and mean EMG voltage during prolonged submaximal contractions. *Muscle & Nerve, 12,* 493-497.

Arent, S.M., & Landers, D.M. (2003). Arousal, anxiety, and performance: A reexamination of the inverted-U hypothesis. *Research Quarterly for Exercise and Sport, 74,* 436-444.

Ariff, G., Donchin, O., Nanayakkara, T., & Shadmehr, R. (2002). A real-time state predictor in motor control: Study of saccadic eye movements during unseen reaching movements. *Journal of Neuroscience, 22,* 7721-7729.

Arjmand, N., & Shirazi-Adl, A. (2006). Role of intra-abdominal pressure in the unloading and stabilization of the human spine during static lifting tasks. *European Spine Journal, 15,* 1265-1275.

Arjmand, N., Shirazi-Adl, A., & Bazrgari, B. (2006). Wrapping of trunk thoracic extensor muscles influences muscle forces and spinal loads in lifting tasks. *Clinical Biomechanics, 21,* 668-675.

Armstrong, D.M., & Marple-Horvat, D.E. (1996). Role of the cerebellum and motor cortex in the regulation of visually controlled locomotion. *Canadian Journal of Physiology and Pharmacology, 74,* 443-455.

Armstrong, R.B. (1990). Initial events in exercise-induced muscular injury. *Medicine and Science in Sports and Exercise, 22,* 429-435.

Arsenault, A.B., & Chapman, A.E. (1974). An electromyographic investigation of the individual recruitment of the quadriceps muscles during isometric contraction of the knee extensors in different patterns of movement. *Physiotherapy of Canada, 26,* 253-261.

Aruin, A.S., Forrest, W.R., & Latash, M.L. (1998). Anticipatory postural adjustments in conditions of postural instability. *Electroencephalography and Clinical Neurophysiology, 109,* 350-359.

Aruin, A.S., Ota, T., & Latash, M.L. (2001). Anticipatory postural adjustments associated with lateral and rotational perturbations during standing. *Journal of Electromyography and Kinesiology, 11,* 39-51.

Asami, T., & Nolte, V. (1983). Analysis of powerful ball kicking. In H. Matsui & K. Kobayashi (Eds.), *Biomechanics VIII-B* (pp. 695-700). Champaign, IL: Human Kinetics.

Ashley, C.C., & Ridgway, E.B. (1968). Simultaneous recording of membrane potential, calcium transient, and tension in single muscle fibres. *Nature, 219,* 1168-1169.

Ashley, C.C., & Ridgway, E.B. (1970). On the relationships between membrane potential, calcium transient, and tension in single barnacle muscle fibres. *Journal of Physiology, 209,* 105-130.

Askew, G.N., & Marsh, R.L. (2002). Muscle designed for maximum short-term power output: Quail flight muscle. *Journal of Experimental Biology, 205,* 2153-2160.

Asmussen, E. (1952). Positive and negative muscular work. *Acta Physiologica Scandinavica, 28,* 365-382.

Asmussen, E. (1956). Observations on experimental muscle soreness. *Acta Rheumatologica Scandinavica, 1,* 109-116.

Asmussen, E. (1979). Muscle fatigue. *Medicine and Science in Sports, 11,* 313-321.

Asmussen, E., & Mazin, B. (1978*a*). A central nervous component in local muscular fatigue. *European Journal of Applied Physiology, 38,* 9-15.

Asmussen, E., & Mazin, B. (1978*b*). Recuperation after muscular fatigue by "diverting activities." *European Journal of Applied Physiology, 38,* 1-7.

Aston-Jones, G., Rajkowski, J., & Cohen, J. (2000). Locus coeruleus and regulation of behavioral flexibility and attention. *Progress in Brain Research, 126,* 165-182.

Atwater, A.E. (1977). Biomechanics of throwing: Correction of common misconceptions. Paper presented at the Joint Meeting of the National College Physical Education Association for Men and the National Association for Physical Education of College Women, Orlando, FL.

Atwater, A.E. (1979). Biomechanics of overarm throwing movements and of throwing injuries. *Exercise and Sport Sciences Reviews, 7,* 43-85.

Avela, J., Finni, T., & Komi, P.V. (2006). Excitability of the soleus reflex arc during intensive stretch–shortening cycle exercise in two power-trained athlete groups. *European Journal of Applied Physiology, 97,* 486-493.

Avela, J., Finni, T., Liikavainio, T., Niemelä, E., & Komi, P.V. (2004). Neural and mechanical responses of the triceps surae muscle group after 1 h of repeated fast passive stretches. *Journal of Applied Physiology, 96,* 2325-2332.

Avela, J., Kryolainen, H., Komi, P.V., & Rama, D. (1999). Reduced stretch reflex sensitivity persists several days after long-lasting stretch–shortening cycle exercise. *Journal of Applied Physiology, 86,* 1292-1300.

Aymard, C., Chia, L., Katz, R., Lafitte, C., & Pénicaud, A. (1995). Reciprocal inhibition between wrist flexors and extensors in man: A new set of interneurons? *Journal of Physiology, 487,* 221-235.

Babault, N., Desbrosses, K., Fabre, M.S., Michaut, A., & Pousson, M. (2006). Neuromuscular fatigue development during maximal concentric and isometric knee extensions. *Journal of Applied Physiology, 100,* 780-785.

Babij, P., & Booth, F.W. (1988). Sculpturing new muscle phenotypes. *News in Physiological Sciences, 3,* 100-102.

Bacsi, A.M., & Colebatch, J.G. (2005). Evidence for reflex and perceptual vestibular contributions to postural control. *Experimental Brain Research, 160,* 22-28.

Bailey, S.P., Davis, J.M., & Ahlborn, E.N. (1993). Neuroendocrine and substrate responses to altered brain 5-HT activity during prolonged exercise to fatigue. *Journal of Applied Physiology, 74,* 3006-3012.

Baker, J.R., Davey, N.J., Ellaway, P.H., & Friedland, C.L. (1992). Short-term synchrony of motor unit discharge during weak isometric contraction in Parkinson's disease. *Brain, 115,* 137-154.

Baker, L.L., Bowman, B.R., & McNeal, D.R. (1988). Effects of waveform on comfort during neuromuscular electrical stimulation. *Clinical Orthopaedics and Related Research, 233,* 75-85.

Baldissera, F., Cavallari, P., & Leocani, L. (1998). Cyclic modulation of the H-reflex in a wrist flexor during rhythmic flexion–extension movements of the ipsilateral foot. *Experimental Brain Research, 118,* 427-430.

Baldissera, F., Hultborn, H., & Illert, M. (1981). Integration of spinal neuronal systems. In V.B. Brooks (Ed.), *Handbook of Physiology: Sec. I. The Nervous System, Vol. 2. Motor Control* (Pt 1, pp. 509-595). Bethesda, MD: American Physiological Society.

Baldwin, E.R., Klakowicz, P.M., & Collins, D.F. (2006). Wide-pulse-width, high-frequency neuromuscular stimulation: Implications for functional electrical stimulation. *Journal of Applied Physiology, 101,* 228-240.

Balestra, C., Duchateau, J., & Hainaut, K. (1992). Effects of fatigue on the stretch reflex in a human muscle. *Electroencephalography and Clinical Neurophysiology, 85,* 46-52.

Balnave, C.D., & Allen, D.G. (1995). Intracellular calcium and force in single mouse fibres followed by repeated contractions with stretch. *Journal of Physiology, 488,* 25-36.

Balnave, C.D., & Allen, D.G. (1996). The effect of muscle length on intracellular calcium and force in single fibres from mouse skeletal muscle. *Journal of Physiology, 492,* 705-713.

Bamman, M.M., Newcomer, B.R., Larson-Meyer, D.E., Weinsier, R.L., & Hunter, G.R. (2000). Evaluation of the strength–size relationship in vivo using various muscle size indices. *Medicine and Science in Sports and Exercise, 32,* 1307-1313.

Bandy, W.D., Irion, J.M., & Briggler, M. (1998). The effect of static stretch and dynamic range of motion training on the flexibility of the hamstring muscles. *Journal of Orthopedic Sports Physical Therapy, 27,* 295-300.

Banerjee, P., Caulfield, B., Crowe, L., & Clark, A. (2005). Prolonged electrical muscle stimulation exercise improves strength and aerobic capacity in healthy sedentary adults. *Journal of Applied Physiology, 99,* 2307-2311.

Bangsbo, J., Madsen, K., Kiens, B., & Richter, E.A. (1996). Effect of muscle acidity on muscle metabolism and fatigue during intense exercise in man. *Journal of Physiology, 495,* 587-596.

Bárány, M. (1967). ATPase activity of myosin correlated with speed of muscle shortening. *Journal of General Physiology, 50,* 197-218.

Baratta, R.V., Solomonow, M., Best, R., & D'Ambrosia, R. (1993). Isotonic length/force models of nine different skeletal muscles. *Medical & Biological Engineering & Computing, 31,* 449-458.

Baratta, R., Solomonow, M., Zhou, B.H., Letson, D., Chuinard, R., & D'Ambrosia, R. (1988). Muscular coactivation. The role of the antagonist musculature in maintaining knee stability. *American Journal of Sports Medicine, 16,* 113-122.

Barbeau, H., Marchand-Pauvert, V., Meunier, S., Nicolas, G., & Pierrot-Deseilligny, E. (2000). Posture-related changes in heteronymous recurrent inhibition from quadriceps to ankle muscles in humans. *Experimental Brain Research, 130,* 345-361.

Barclay, J.K. (1986). A delivery-independent blood flow effect on skeletal muscle fatigue. *Journal of Applied Physiology, 61,* 1084-1090.

Baret, M., Katz, R., Lamy, J.C., Pénicaud, A., & Wargon, I. (2003). Evidence for recurrent inhibition of reciprocal inhibition from soleus to tibialis anterior in man. *Experimental Brain Research, 152,* 133-136.

Barr, J.O., Nielsen, D.H., & Soderberg, G.L. (1986). Transcutaneous electrical nerve stimulation characteristics for altering pain perception. *Physical Therapy, 66,* 1515-1521.

Barrance, P.J., Williams, G.N., Snyder-Mackler, L., & Buchanan, T.S. (2006). Altered knee kinematics in ACL-deficient non-copers: A comparison using dynamic MRI. *Journal of Orthopedic Research, 24,* 132-140.

Barrance, P.J., Williams, G.N., Snyder-Mackler, L., & Buchanan, T.S. (2007). Do ACL-injured copers exhibit differences in knee kinematics? An MRI study. *Clinical Orthopedics and Research, 454,* 74-80.

Barry, B.K., & Carson, R.G. (2004). The consequences of resistance training for movement control in older adults. *Journal of Gerontology, 59,* 730-754.

Barry, B.K., Pascoe, M.A., Jesunathadas, M., & Enoka, R.M. (2007). Rate coding is compressed but variability is unaltered for motor units in a hand muscle of old adults. *Journal of Neurophysiology, 97,* 3206-3218.

Barry, B.K., Riek, S., & Carson, R.G. (2005*a*). Muscle coordination during rapid force production by young and older adults. *Journal of Gerontology, 60A,* 232-240.

Barry, B.K., Warman, G.E., & Carson, R.G. (2005*b*). Age-related differences in rapid muscle activation after rate of force development training of the elbow flexors. *Experimental Brain Research, 162,* 122-132.

Barstow, T.J., Scremin, A.M., Mutton, D.L.L., Kunkel, C.F., Cagle, T.G., & Whipp, B.J. (2000). Peak and kinetic cardiorespiratory responses during arm and leg exercise in patients with spinal cord injury. *Spinal Cord, 38,* 340-345.

Bartee, H., & Dowell, L. (1982). A cinematographical analysis of twisting about the longitudinal axis when performers are free of support. *Journal of Human Movement Studies, 8,* 41-54.

Bastiaans, J.J., van Diemen, A.B., Veneberg, T., & Jeukendrup, A.E. (2001). The effects of replacing a portion of endurance training by explosive strength training on performance in trained cyclists. *European Journal of Applied Physiology, 86,* 79-84.

Bastian, A.J. (2002). Cerebellar limb ataxia: Abnormal control of self-generated and external forces. *Annals of the New York Academy of Sciences, 978,* 16-27.

Bastian, A.J. (2006). Learning to predict the future: The cerebellum adapts feedforward movement control. *Current Opinion in Neurobiology, 16,* 1-5.

Bastian, A.J., & Nakajima, S. (1974). Action potential in the transverse tubules and its role in the activation of skeletal muscle. *Journal of General Physiology, 63,* 257-278.

Bastien, G.J., Willems, P.A., Schepens, A., & Heglund, N.C. (2005). Effect of load and speed on the energetic cost of human walking. *European Journal of Applied Physiology, 94,* 76-83.

Batista, A.P., & Andersen, R.A. (2001). The parietal reach region codes the next planned movement in a sequential reach task. *Journal of Neurophysiology, 85,* 539-544.

Battaglia-Mayer, A., Caminiti, R., Lacquaniti, F., & Zago, M. (2003). Multiple levels of representation of reaching in the parieto-frontal network. *Cerebral Cortex, 13,* 1009-1022.

Battaglia-Mayer, A., Mascaro, M., & Caminiti, R. (2007). Temporal evolution and strength of neural activity in parietal cortex during eye and hand movements. *Cerebral Cortex, 17,* 1350-1363.

Baudry, S., & Duchateau, J. (2004). Postactivation potentiation in human muscle is not related to the type of maximal conditioning contraction. *Muscle & Nerve, 30,* 328-336.

Baudry, S., & Duchateau, J. (2007). Postactivation potentiation in human muscle: Effect on the rate of torque development of tetanic and voluntary isometric contractions. *Journal of Applied Physiology, 103,* 1318-1325.

Baudry, S., Klass, M., & Duchateau, J. (2005). Postactivation potentiation influences differently the nonlinear summation of contraction in young and elderly adults. *Journal of Applied Physiology, 98,* 1243-1250.

Baudry, S., Klass, M., Pasquet, B., & Duchateau, J. (2007). Age-related fatigability of the ankle dorsiflexor muscles during concentric and eccentric contractions. *European Journal of Applied Physiology, 100,* 543-551.

Baumer, T., Munchau, A., Weiller, C., & Liepert, J. (2002). Fatigue suppresses ipsilateral intracortical facilitation. *Experimental Brain Research, 146,* 467-473.

Bawa, P., & Calancie, B. (1983). Repetitive doublets in human flexor carpi radialis muscle. *Journal of Physiology, 339,* 123-132.

Behm, D.G., Bambury, A., Cahill, F., & Power, K. (2004). Effect of acute static stretching on force, balance, reaction time, and movement time. *Medicine and Science in Sports and Exercise, 36,* 1397-1402.

Behrman, A.L., Bowden, M.G., & Nair, P.M. (2006). Neuroplasticity after spinal cord injury and training: An emerging paradigm shift in rehabilitation and walking recovery. *Physical Therapy, 86,* 1406-1425.

Belanger, M., Drew, T., Provencher, J., & Rossignol, S. (1996). A comparison of treadmill locomotion in adult cats before and after spinal transection. *Journal of Neurophysiology, 76,* 471-491.

Belanger, M., Stein, R.B., Wheeler, G.D., Gordon, T., & Leduc, B. (2000). Electrical stimulation: Can it increase muscle strength and reverse osteopenia in spinal cord injured individuals? *Archives of Physical Medicine and Rehabilitation, 81,* 1090-1098.

Belen'kii, V.Ye., Gurfinkel, V.S., & Pal'tsev, Ye.I. (1967). Elements of control of voluntary movements. *Biofizika, 12,* 135-141.

Bell, J., Bolanowski, S., & Holmes, M.H. (1994). The structure and function of Pacinian corpuscles: A review. *Progress in Neurobiology, 42,* 79-128.

Bellebaum, C., Daum, I., Koch, B., Schwarz, M., & Hoffmann, K.P. (2005). The role of the human thalamus in processing corollary discharge. *Brain, 128,* 1139-1154.

Bellemare, F., & Garzaniti, N. (1988). Failure of neuromuscular propagation during human maximal voluntary contraction. *Journal of Applied Physiology, 64,* 1084-1093.

Bellemare, F., Woods, J.J., Johansson, R.S., & Bigland-Ritchie, B. (1983). Motor-unit discharge rates in maximal voluntary contractions of three human muscles. *Journal of Neurophysiology, 50,* 1380-1392.

Benjuya, N., Melzer, I., & Kaplanski, J. (2004). Aging-induced shifts from a reliance on sensory input to muscle cocontraction during balanced standing. *Journal of Gerontology, 59A,* 155-171.

Bennett, K.M.B., & Lemon, R.N. (1994). The influence of single monkey cortico-motoneuronal cells. *Journal of Physiology, 477,* 291-307.

Bennett, M.B., Ker, R.F., Dimery, N.J., & Alexander, R.M. (1986). Mechanical properties of various mammalian tendons. *Journal of Zoology, 209A,* 537-548.

Bent, L.R., Inglis, J.T., & McFadyen, B.J. (2004). When is vestibular information important during walking? *Journal of Neurophysiology, 92,* 1269-1275.

Bent, L.R., McFadyen, B.J., & Inglis, J.T. (2005). Vestibular contributions during human locomotor tasks. *Exercise and Sport Sciences Reviews, 33,* 107-113.

Bentley, S. (1996). Exercise-induced muscle cramp. Proposed mechanisms and management. *Sports Medicine, 21,* 409-420.

Benvenuti, F., Stanhope, S.J., Thomas, S.L., Panzer, V.P., & Hallett, M. (1997). Flexibility of anticipatory postural adjustments revealed by self-paced and reaction-time arm movements. *Brain Research, 761,* 59-70.

Berardelli, A., Noth, J., Thompson, P.D., Bollen, E.L., Curra, A., Deuschl, G., van Dijk, J.G., Topper, R., Schwarz, M., & Roos, R.A. (1999). Pathophysiology of chorea and bradykinesia in Huntington's disease. *Movement Disorders, 14,* 398-403.

Berg, H.E., Dudley, G.A., Häggmark, T., Ohlsén, H., & Tesch, P.A. (1991). Effects of lower limb unloading on skeletal muscle mass and function in humans. *Journal of Applied Physiology, 70,* 1882-1885.

Berg, H.E., Dudley, G.A., Hather, B., & Tesch, P.A. (1993). Work capacity and metabolic and morphologic characteristics of the human quadriceps muscle in response to unloading. *Clinical Physiology, 13,* 337-347.

Berg, H.E., & Tesch, P.A. (1996). Changes in muscle function in response to 10 days of lower limb unloading in humans. *Acta Physiologica Scandinavica, 157,* 63-70.

Berger, W., Quintern, J., & Dietz, V. (1984). Tension development and muscle activation in the leg during gait in spastic hemiparesis: The independence of muscle hypertonia and exaggerated stretch reflexes. *Journal of Neurology, Neurosurgery, and Psychiatry, 47,* 1029-1033.

Bergh, U., & Ekblom, B. (1979). Influence of muscle temperature on maximal muscle strength and power output in human skeletal muscles. *Acta Physiologica Scandinavica, 107,* 33-37.

Bergmann, G., Graichen, F., & Rohlmann, A. (1993). Hip joint loading during walking and running, measured in two patients. *Journal of Biomechanics, 26,* 969-990.

Bertolasi, L., De Grandis, D., Bongiovanni, L.G., Zanette, G.P., & Gasperini, M. (1993). The influence of muscular lengthening on cramps. *Annals of Neurology, 33,* 176-180.

Best, T.M., Hasselman, C.T., & Garrett, W.E., Jr. (1997). Muscle strain injuries: Biomechanical and structural studies. In S. Salmon (Ed.), *Muscle Damage* (pp. 145-167). Oxford, UK: Oxford University Press.

Beurze, S.M., Van Pelt, S., & Medendorp, W.P. (2006). Behavioral reference frames for planning human reaching movements. *Journal of Neurophysiology, 96,* 352-362.

Bhardwaj, R.D., Curtis, M.A., Spalding, K.L., Buchholz, B.A., Fink, D., Bjork-Eriksson, T., Nordborg, C., Gage, F.H., Druid, H., Eriksson, P.S., & Frisen, J. (2006). Neocortical neurogenesis in humans is restricted to development. *Proceedings of the National Academy of Sciences USA, 103,* 12564-12568.

Biewener, A.A. (1998). Muscle function in vivo: A comparison of muscles used as springs for elastic energy savings versus muscles used to generate mechanical power. *American Zoologist, 38,* 703-717.

Biewener, A.A., Farely, C.T., Roberts, T.J., & Temaner, M. (2004). Muscle mechanical advantage of human walking and running: Implications for energy cost. *Journal of Applied Physiology, 97,* 2266-2274.

Biewener, A.A., & Roberts, T.J. (2000). Muscle and tendon contributions to force, work, and elastic savings: A comparative perspective. *Exercise and Sport Sciences Reviews, 28,* 99-107.

Bigland-Ritchie, B. (1981). EMG/force relations and fatigue of human voluntary contractions. *Exercise and Sport Sciences Reviews, 9,* 75-117.

Bigland-Ritchie, B., Cafarelli, E., & Vøllestad, N.K. (1986*a*). Fatigue of submaximal static contractions. *Acta Physiologica Scandinavica, 556,* 137-148.

Bigland-Ritchie, B., Dawson, N.J., Johansson, R.S., & Lippold, O.C.J. (1986*b*). Reflex origin for the slowing of motoneurone firing rates in fatigue of human voluntary contractions. *Journal of Physiology, 379,* 451-459.

Bigland-Ritchie, B., Fuglevand, A.J., & Thomas, C.K. (1998). Contractile properties of human motor units: Is man a cat? *Neuroscientist, 4,* 240-249.

Bigland-Ritchie, B., Furbush, F., & Woods, J.J. (1986*c*). Fatigue of intermittent submaximal voluntary contractions: Central and peripheral factors. *Journal of Applied Physiology, 61,* 421-429.

Bigland-Ritchie, B., Johansson, R., Lippold, O.C.J., Smith, S., & Woods, J.J. (1983*a*). Changes in motoneurone firing rates during sustained maximal voluntary contractions. *Journal of Physiology, 340,* 335-346.

Bigland-Ritchie, B., Johansson, R., Lippold, O.C.J., & Woods, J.J. (1983*b*). Contractile speed and EMG changes during fatigue of sustained maximal voluntary contractions. *Journal of Neurophysiology, 50,* 313-324.

Bigland-Ritchie, B., Kukulka, C.G., Lippold, O.C.G., & Woods, J.J. (1982). The absence of neuromuscular transmission failure in sustained maximal voluntary contractions. *Journal of Physiology, 330,* 265-278.

Bigland-Ritchie, B., Rice, C.L., Garland, S.J., & Walsh, M.L. (1995). Task-dependent factors in fatigue of human voluntary contractions. In S.C. Gandevia, R.M. Enoka, A.J. McComas, D.G. Stuart, & C.K. Thomas (Eds.), *Fatigue: Neural and Muscular Mechanisms* (pp. 361-380). New York: Plenum Press.

Bilodeau, M. (2006). Central fatigue in continuous and intermittent contractions of triceps brachii. *Muscle & Nerve, 34,* 205-213.

Binder, M.D., Heckman, C.J., & Powers, R.K. (1996). The physiological control of motoneuron activity. In L.B. Rowell & J.T. Shepherd (Eds.), *Handbook of Physiology: Sec. 12. Exercise: Regulation and Integration of Multiple Systems* (pp. 3-53). New York: Oxford University Press.

Binder, M.D., Heckman, C.J., & Powers, R.K. (2002). Relative strengths and distributions of different sources of synaptic input to the motoneurone pool. *Advances in Experimental Medicine and Biology, 508,* 207-212.

Binder, M.D., Kroin, J.S., Moore, G.P., & Stuart, D.G. (1977). The response of Golgi tendon organs to single motor unit contractions. *Journal of Physiology (London), 271,* 337-349.

Binder-Macleod, S.A., & Guerin, T. (1990). Preservation of force output through progressive reduction of stimulation frequency in human quadriceps femoris muscle. *Physical Therapy, 70,* 619-625.

Binkhorst, R.A., Hoofd, L., & Vissers, A.C.A. (1977). Temperature and force-velocity relationship of human muscles. *Journal of Applied Physiology, 42,* 471-475.

Birdsall, T.C. (1998). 5-hydroxytryptophan: A clinically-effective serotonin precursor. *Alternative Medicine Reviews, 3,* 271-280.

Bisdorff, A.R., Bronstein, A.M., & Gretsy, M.A. (1994). Responses in neck and facial muscles to sudden free fall and a startling auditory stimulus. *Electromyography and Clinical Neurophysiology, 93,* 409-416.

Bisdorff, A.R., Bronstein, A.M., Wolsey, C., Gretsy, M.A., Davies, A., & Young, A. (1999). EMG responses to free fall in elderly subjects and akinetic rigid patients. *Journal of Neurology, Neurosurgery, and Psychiatry, 66,* 447-455.

Bishop, D. (2003*a*). Warm-up I: Potential mechanisms and the effects of passive warm-up on exercise performance. *Sports Medicine, 33,* 439-454.

Bishop, D. (2003*b*). Warm-up II: Performance changes following active warm-up and how to structure the warm-up. *Sports Medicine, 33,* 483-498.

Bizzi, E., & Abend, W. (1983). Posture control and trajectory formation in single- and multi-joint arm movements. In J.E. Desmedt (Ed.), *Motor Control Mechanisms in Health and Disease* (pp. 31-45). New York: Raven Press.

Blackard, D.O., Jensen, R.L., & Ebben, W.P. (1999). Use of EMG analysis in challenging kinetic chain terminology. *Medicine and Science in Sports and Exercise, 31,* 443-448.

Blangsted, A.K., Sjøgaard, G., Madeleine, P., Olsen, H.B., & Søgaard, K. (2005*a*). Voluntary low-force contraction elicits prolonged low-frequency fatigue and changes in surface electromyography and mechanomyography. *Journal of Electromyography and Kinesiology, 15,* 138-148.

Blangsted, A.K., Vedsted, P., Sjøgaard, G., & Søgaard, K. (2005*b*). Intramuscular pressure and tissue oxygenation during low-force static contraction do not underlie muscle fatigue. *Acta Physiologica Scandinavica, 183,* 379-388.

Blattner, S.E., & Noble, L. (1979). Relative effects of isokinetic and plyometric training on vertical jumping performance. *Research Quarterly, 50,* 583-588.

Blevins, C.E. (1967). Innervation patterns of the human stapedius muscle. *Archives of Otolaryngology, 86,* 136-142.

Blin, O., Ferrandez, A.M., Pailhouse, J., & Serratrice, G. (1991). Dopa-sensitive and dopa-resistant gait parameters in Parkinson's disease. *Journal of Neurological Sciences, 103,* 51-54.

Blinks, J.R., Rüdel, R., & Taylor, S.R. (1978). Calcium transients in isolated amphibian skeletal muscle fibres: Detection with aequorin. *Journal of Physiology, 277,* 291-323.

Bloch, R.J., & Gonzalez-Serratos, J. (2003). Lateral force transmission across costameres in skeletal muscle. *Exercise and Sport Sciences Reviews, 31,* 73-78.

Bloom, W., & Fawcett, D.W. (1968). *A Textbook of Histology* (9th ed.). Philadelphia: Saunders.

Blomstrand, E., Celsing, F., & Newsholme, E.A. (1988). Changes in plasma concentrations of aromatic and branched-chain amino acids during sustained exercise in man and their possible role in fatigue. *Acta Physiologica Scandinavica, 133,* 115-121.

Blomstrand, E., Møller, K., Secher, N.H., & Nybo, L. (2005). Effect of carbohydrate ingestion on brain exchange of amino acids during exercise in human subjects. *Acta Physiologica Scandinavica, 185,* 203-209.

Blouin, J.S., Inglis, J.T., & Siegmund, G.P. (2006). Startle responses elicited by whiplash perturbations. *Journal of Physiology, 573,* 857-867.

Bobath, B. (1978). *Adult hemiplegia: Evaluation and treatment.* London: Heinemann Medical Books.

Bobbert, M.F., & Casius, L.J. (2005). Is the effect of a countermovement on jump height due to active state development? *Medicine and Science in Sports and Exercise, 37,* 440-446.

Bobbert, M.F., Gerritsen, K.G.M., Litjens, M.C.A., & Soest, A.J. van. (1996). Why is the countermovement jump height greater than squat jump height? *Medicine and Science in Sports and Exercise, 28,* 1402-1412.

Bobbert, M.F., & Ingen Schenau, G.J. van. (1988). Coordination in vertical jumping. *Journal of Biomechanics, 21,* 249-262.

Bobbert, M.F., Schamhardt, H.C., & Nigg, B.M. (1991). Calculation of vertical ground reaction force estimates during running from positional data. *Journal of Biomechanics, 24,* 1095-1105.

Bobbert, M.F., & van Soest, A.J. (2001). Why do people jump the way they do? *Exercise and Sport Sciences Reviews, 29,* 95-102.

Bodine, S.C. (2006). mTOR signaling and the molecular adaptation to resistance exercise. *Medicine and Science in Sports and Exercise, 38,* 1950-1957.

Bodine, S.C., Garfinkel, A., Roy, R.R., & Edgerton, V.R. (1988). Spatial distribution of motor unit fibers in the cat soleus and tibialis anterior muscles: Local interactions. *Journal of Neuroscience, 8,* 2142-2152.

Bodine, S.C., Roy, R.R., Eldred, E., & Edgerton, V.R. (1987). Maximal force as a function of anatomical features of motor units in the cat tibialis anterior. *Journal of Neurophysiology, 57,* 1730-1745.

Boe, S.G., Stashuk, D.W., & Doherty, T.J. (2006). Within-subject reliability of motor unit number estimates and quantitative motor unit analysis in a distal and proximal upper limb muscle. *Clinical Neurophysiology, 117,* 596-603.

Bogert, A.J. van den, Read, L., & Nigg, B.M. (1996). A method for inverse dynamic analysis using accelerometry. *Journal of Biomechanics, 29,* 949-954.

Bogert, A.J. van den, Read, L., & Nigg, B.M. (1999). An analysis of hip joint loading during walking, running, and skiing. *Medicine and Science in Sports and Exercise, 31,* 131-142.

Bojsen-Møller, J., Hansen, P., Aagaard, P., Svantesson, U., Kjær, M., & Magnusson, S.P. (2004). Differential displacement of the human soleus and medial gastrocnemius aponeuroses during isometric plantar flexor contractions in vivo. *Journal of Applied Physiology, 97,* 1908-1914.

Bojsen-Møller, J., Magnusson, S.P., Rasmussen, L.R., Kjær, M., & Aagaard, P. (2005). Muscle performance during maximal isometric and dynamic contractions is influenced by the stiffness of the tendinous structures. *Journal of Applied Physiology, 99,* 986-994.

Bongiovanni, L.G., & Hagbarth, K.-E. (1990). Tonic vibration reflexes elicited during fatigue from maximal voluntary contractions in man. *Journal of Physiology, 423,* 1-14.

Bonnet, M., Bradley, M.M., Lang, P.J., & Requin, J. (1995). Modulation of spinal reflexes: Arousal, pleasure, action. *Psychophysiology, 32,* 367-372.

Bosco, G., Eian, J., & Poppele, R.E. (2006). Phase-specific sensory representations in spinocerebellar activity during stepping: Evidence for a hybrid kinematic/kinetic framework. *Experimental Brain Research, 175,* 83-96.

Bosco, G., & Poppele, R.E. (2001). Proprioception from a spinocerebellar perspective. *Physiological Reviews, 81,* 539-568.

Bosco, G., Poppele, R.E., & Eian, J. (2000). Reference frames for spinal proprioception: Limb endpoint based or joint-level based? *Journal of Neurophysiology, 83,* 2931-2945.

Botterman, B.R., Binder, M.D., & Stuart, D.G. (1978). Functional anatomy of the association between motor units and muscle receptors. *American Zoologist, 18,* 135-152.

Botterman, B.R., Iwamoto, G.A., & Gonyea, W.J. (1986). Gradation of isometric tension by different activation rates in motor units of cat flexor carpi radialis muscle. *Journal of Neurophysiology, 56,* 494-506.

Bottinelli, R., & Reggiani, C. (2000). Human skeletal muscle fibers: Molecular and functional diversity. *Progress in Biophysics and Molecular Biology, 73,* 195-262.

Bouisset, S., Le Bozec, S., & Ribreau, C. (2002). Postural dynamics in maximal isometric ramp efforts. *Biological Cybernetics, 87,* 211-219.

Boyd, I.A., & Gladden, M. (Eds.). (1985). *The Muscle Spindle.* London: Macmillan.

Boyer, K.A., & Nigg, B.M. (2006). Soft tissue vibrations within one soft tissue compartment. *Journal of Biomechanics, 39,* 645-651.

Brancazio, P.J. (1984). *Sport Science: Physical Laws and Optimum Performance.* New York: Simon & Schuster.

Brand, R.A., Crowninshield, R.D., Wittstock, C.E., Pedersen, D.R., Clark, C.R., & van Krieken, F.M. (1982). A model of lower extremity muscular anatomy. *Journal of Biomechanical Engineering, 104,* 304-310.

Brand, R.A., Pedersen, D.R., & Friederich, J.A. (1986). The sensitivity of muscle force predictions to changes in physiologic cross-sectional area. *Journal of Biomechanics, 19,* 589-596.

Breit, G.A., & Whalen, R.T. (1997). Prediction of human gait parameters from temporal measures of foot–ground contact. *Medicine and Science in Sports and Exercise, 29,* 540-547.

Bremer, A.K., Sennwald, G.R., Favre, P., & Jacob, H.A. (2006). Moment arms of forearm rotators. *Clinical Biomechanics, 21,* 683-691.

Bremner, F.D., Baker, J.R., & Stephens, J.A. (1991*a*). Correlation between the discharges of motor units recorded from the same and from different finger muscles in man. *Journal of Physiology, 432,* 355-380.

Bremner, F.D., Baker, J.R., & Stephens, J.A. (1991*b*). Effect of task on the degree of synchronization of intrinsic hand muscle motor units in man. *Journal of Neurophysiology, 66,* 2072-2083.

Bremner, F.D., Baker, J.R., & Stephens, J.A. (1991*c*). Variation in the degree of synchrony exhibited by motor units lying in different finger muscles in man. *Journal of Physiology, 432,* 381-399.

Britton, T.C., Day, B.L., Brown, P., Rothwell, J.C., Thompson, P.D., & Marsden, C.D. (1993). Postural electromyographic responses in the arm and leg following galvanic vestibular stimulation in man. *Experimental Brain Research, 94,* 143-151.

Broberg, S., & Sahlin, K. (1989). Adenine nucleotide degradation in human skeletal muscle during prolonged exercise. *Journal of Applied Physiology, 67,* 116-122.

Brochier, T., Spinks, R.L., Umilta, M.A., & Lemon, R.N. (2004). Patterns of muscle activity underlying object-specific grasp by the macaque monkey. *Journal of Neurophysiology, 92,* 1770-1782.

Brockett, C.L., Morgan, D.L., & Proske, U. (2004). Predicting hamstring strain injury in elite athletes. *Medicine and Science in Sports and Exercise, 36,* 379-387.

Brooke, J.D., & Zehr, E.P. (2006). Limits to fast-conducting somatosensory feedback in movement control. *Exercise and Sport Sciences Reviews, 34,* 22-28.

Brooke, M.H., & Kaiser, K.K. (1974). The use and abuse of muscle histochemistry. *Annals of the New York Academy of Sciences, 228,* 121-144.

Brooks, S.V., & Faulkner, J.A. (1991). Forces and powers of slow and fast skeletal muscle in mice during repeated contractions. *Journal of Physiology, 436,* 701-710.

Brown, D.A., & Kautz, S.A. (1998). Increased workload enhances force output during pedaling exercise in persons with poststroke hemiplegia. *Stroke, 29,* 598-606.

Brown, D.A., Nagpal, S., & Chi, S. (2005). Limb-loaded cycling program for locomotor intervention following stroke. *Physical Therapy, 85,* 159-168.

Brown, P. (1995). Physiology of startle phenomena. *Advances in Neurology, 67,* 273-287.

Brown, T.G. (1911). The intrinsic factors in the act of progression in the mammal. *Proceedings of the Royal Society of London B, 84,* 308-319.

Bruton, J.D., Westerblad, H., Katz, A., & Lännergren, J. (1996). Augmented force output in skeletal muscle fibres of *Xenopus* following a preceding bout of activity. *Journal of Physiology, 493,* 211-217.

Buchanan, J.J., & Horak, F.B. (2003). Voluntary control of postural equilibrium patterns. *Behavioral Brain Research, 143,* 121-140.

Buchanan, T.S., Lloyd, D.G., Manal, K., & Besier, T.F. (2005). Estimation of muscle forces and joint moments using a forward-inverse dynamics model. *Medicine and Science in Sports and Exercise, 37,* 1911-1916.

Buchanan, T.S., Rovai, G.P., & Rymer, W.Z. (1989). Strategies for muscle activation during isometric torque generation at the human elbow. *Journal of Neurophysiology, 62,* 1201-1212.

Buchthal, F. (1961). The general concept of the motor unit. Neuromuscular disorders. *Research Publication of the Association for Research on Nervous and Mental Disorders, 38,* 3-30.

Buchthal, F., & Schmalbruch, H. (1970). Contraction times of twitches evoked by the H-reflexes. *Acta Physiologica Scandinavica, 80,* 378-382.

Buneo, C.A., & Andersen, R.A. (2006). The posterior parietal cortex: Sensorimotor interface for the planning and online control of visually guided movements. *Neuropsychologia, 44,* 2594-2606.

Buneo, C.A., Jarvis, M.R., Batista, A.P., & Andersen, R.A. (2002). Direct visuomotor transformations for reaching. *Nature, 416,* 632-636.

Burden, A.M., Trew, M., & Baltzopoulos, V. (2003). Normalisation of gait EMGs: A re-examination. *Journal of Electromyography and Kinesiology, 13,* 519-532.

Burgess, P.R., Horch, K.W., & Tuckett, R.P. (1987). Mechanoreceptors. In G. Adelman (Ed.), *Encyclopedia of Neuroscience* (Vol. II, pp. 620-621). Boston: Birkhäuser.

Burke, D. (1988). Spasticity as an adaptation to pyramidal tract injury. *Advances in Neurology, 47,* 401-422.

Burke, D., Gracies, J.M., Mazevet, D., Meunier, S., & Pierrot-Deseilligny, E. (1992). Convergence of descending and various peripheral inputs onto common propriospinal-like neurones in man. *Journal of Physiology, 449,* 655-671.

Burke, D., Hagbarth, K.E., & Löfstedt, L. (1978). Muscle spindle activity in man during shortening and lengthening contractions. *Journal of Physiology, 277,* 131-142.

Burke, D., Hagbarth, K.E., Lofstedt, L., & Walling, B.G. (1976). The responses of human muscle spindle endings to vibration of non-contracting muscles. *Journal of Physiology, 261,* 673-693.

Burke, D., Kiernan, M.C., & Bostock, H. (2001). Excitability of human axons. *Clinical Neurophysiology, 112,* 1575-1585.

Burke, D., McKeon, B., & Skuse, N.F. (1981). The irrelevance of fusimotor activity to the Achilles tendon jerk of relaxed humans. *Annals of Neurology, 10,* 547-550.

Burke, R.E. (1981). Motor units: Anatomy, physiology, and functional organization. In V.B. Brooks (Ed.), *Handbook of Physiology: Sec. 1. The Nervous System, Vol. 2. Motor Control* (Pt. 1, pp. 345-422). Bethesda, MD: American Physiological Society.

Burke, R.E. (1999). The use of state-dependent modulation of spinal reflexes as a tool to investigate the organization of spinal interneurons. *Experimental Brain Research, 128,* 263-277.

Burke, R.E., Dum, R., Fleshman, J., Glenn, L., Lev-Tov, A., O'Donovan, M., & Pinter, M. (1982). An HRP study of the relation between cell size and motor unit type in cat ankle extensor motoneurons. *Journal of Comparative Neurology, 209,* 17-28.

Burke, R.E., Fedina, L., & Lundberg, A. (1971). Spatial synaptic distribution of recurrent and group Ia inhibitory systems in cat spinal motoneurones. *Journal of Physiology, 214,* 305-326.

Burke, R.E., Jankowska, E., & Ten Bruggencate, G. (1970*a*). A comparison of peripheral and rubrospinal synaptic input to slow and fast twitch motor units of triceps surae. *Journal of Physiology, 207,* 709-732.

Burke, R.E., Levine, D.N., Tsairis, P., & Zajac, F.E. (1973). Physiological types and histochemical profiles in motor units of the cat gastrocnemius. *Journal of Physiology, 234,* 723-748.

Burke, R.E., Rudomin, P., & Zajac, F.E. III. (1970*b*). Catch property in single mammalian motor units. *Science, 168,* 122-124.

Burke, R.E., & Tsairis, P. (1973). Anatomy and innervation ratios in motor units of cat gastrocnemius. *Journal of Physiology, 234,* 749-765.

Burne, J.A., Carleton, V.L., & O'Dwyer, N.J. (2005). The spasticity paradox: Movement disorder or disorder of resting limbs? *Journal of Neurology, Neurosurgery, and Psychiatry, 76,* 47-54.

Burnett, R.A., Laidlaw, D.H., & Enoka, R.M. (2000). Coactivation of the antagonist muscle does not covary with steadiness in old adults. *Journal of Applied Physiology, 89,* 61-71.

Burnfield, J.M., & Powers, C.M. (2006). Prediction of slips: An evaluation of utilized coefficients of friction and available slip resistance. *Ergonomics, 49,* 982-995.

Burnod, Y., Baraduc, P., Battaglia-Mayer, A., Guigon, E., Koechlin, E., Ferraina, S., Lacquaniti, F., & Caminiti, R. (1999). Parieto-frontal coding of reaching: An integrated framework. *Experimental Brain Research, 129,* 325-346.

Bury, S.D., & Jones, T.A. (2004). Facilitation of motor skill learning by callosal intervention or forced forelimb use in adult rats. *Behavioral Brain Research, 150,* 43-53.

Büschges, A., & El Manira, A. (1998). Sensory pathways and their modulation in the central control of locomotion. *Current opinion in Neurobiology, 8,* 733-739.

Butefisch, C.M. (2004). Plasticity in the human cerebral cortex: Lessons from the normal brain and from stroke. *Neuroscientist, 10,* 163-173.

Butler, J.E., Godfrey, S., & Thomas, C.K. (2006). Depression of involuntary activity in muscles paralyzed by spinal cord injury. *Muscle & Nerve, 33,* 637-644.

Butler, J.E., Ribot-Ciscar, E., Zijdewind, I., & Thomas, C.K. (2004). Increased blood pressure can reduce fatigue of thenar muscles paralyzed after spinal cord injury. *Muscle & Nerve, 29,* 575-584.

Butler, J.E., Taylor, J.L., & Gandevia, S.C. (2003). Responses of human motoneurons to corticospinal stimulation during maximal voluntary contractions and ischemia. *Journal of Neuroscience, 23,* 10224-10230.

Butterfield, T.A., & Herzog, W. (2006). Effect of altering length and activation timing of muscle on fiber strain and muscle damage. *Journal of Applied Physiology, 100,* 1489-1498.

Butterfield, T.A., Leonard, T.R., & Herzog, W. (2005). Differential serial sarcomere number adaptations in knee extensor muscles of rats is contraction type dependent. *Journal of Applied Physiology, 99,* 1352-1358.

Cady, E.B., Jones, D.A., Lynn, J., & Newham, D.J. (1989). Changes in force and intracellular metabolites during fatigue of human skeletal muscle. *Journal of Physiology, 418,* 311-325.

Cahouet, V., Luc, M., & David, A. (2002). Static optimal estimation of joint accelerations for inverse dynamics problem solution. *Journal of Biomechanics, 35,* 1507-1513.

Caiozzo, V.J., Baker, M.J., Herrick, R.E., Tao, M., & Baldwin, K.M. (1994). Effect of spaceflight on skeletal muscle: Mechanical properties and myosin isoform content of a slow muscle. *Journal of Applied Physiology, 80,* 1503-1512.

Caiozzo, V.J., Baker, M.J., Huang, K., Chou, H., Wu, Y.Z., & Baldwin, K.M. (2003). Single-fiber myosin heavy chain polymorphism: How many patterns and what proportions? *American Journal of Physiology, 285,* R570-R580.

Caiozzo, V.J., Haddad, F., Baker, M.J., & Baldwin, K.M. (1996). Influence of mechanical loading on myosin heavy-chain protein and mRNA isoform expression. *Journal of Applied Physiology, 80,* 1503-1512.

Caiozzo, V.J., & Rourke, B. (2006). The muscular system: Structural and functional plasticity. In C.M. Tipton (Ed.), *ACSM's Advanced Exercise Physiology* (pp. 112-143). Philadelphia: Lippincott Williams & Wilkins.

Cairns, S.P. (2006). Lactic acid and exercise performance. Culprit or friend? *Sports Medicine, 36,* 279-291.

Calabrese, R.L. (1998). Cellular, synaptic network, and modulatory mechanisms involved in rhythm generation. *Current Opinion in Neurobiology, 8,* 710-717.

Calancie, B. (1991). Interlimb reflexes following cervical spinal cord injury in man. *Experimental Brain Research, 85,* 458-469.

Calancie, B., Lutton, S., & Lutton, J.G. (1996). Central nervous system plasticity after spinal cord injury in man: Interlimb reflexes and the influence of cutaneous stimulation. *Electroencephalography and Clinical Neurophysiology, 101,* 304-315.

Calancie, B., Molano, M.R., & Broton, J.G. (2004). Tendon reflexes for predicting movement recovery after acute spinal cord injury in humans. *Clinical Neurophysiology, 115,* 2350-2363.

Calvert, T.W., & Chapman, A.E. (1977). Relationship between the surface EMG and force transients in muscle: Simulation and experimental studies. *Proceedings of the IEEE, 65,* 682-6889.

Camhi, J.M. (1984). *Neuroethology: Nerve Cells and the Natural Behavior of Animals.* Sunderland, MA: Sinauer Associates.

Campbell, K.S., & Lakie, M. (1998). A cross-bridge mechanism can explain the thixotropic short-range elastic component of relaxed frog skeletal muscle. *Journal of Physiology, 510,* 941-962.

Campbell, M.J., McComas, A.J., & Petito, F. (1973). Physiological changes in ageing muscles. *Journal of Neurology, Neurosurgery, and Psychiatry, 36,* 174-182.

Campeau, S., Falls, W., Cullinan, W., Helmreich, D., Davis, M., & Watson, S. (1997). Elicitation and reduction of fear: Behavioral and neuroendocrine indices and brain induction of the immediate-early gene c-fos. *Neuroscience, 78,* 1087-1104.

Campos, G.E., Luecke, T.J., Wendeln, H.K., Toma, K., Hagerman, F.C., Murray, T.F., Ragg, K.E., Ratamess, N.A., Kraemer, W.J., & Staron, R.S. (2002). Muscular adaptations in response to three different resistance-training regimens: Specificity of repetition maximum training zones. *European Journal of Applied Physiology, 88,* 50-60.

Candow, D.G., & Chilibeck, P.D. (2005). Differences in size, strength, and power of upper and lower body muscle groups in young and older men. *Journal of Gerontology, 60A,* 148-156.

Cangiano, L., & Grillner, S. (2005). Mechanisms of rhythm generation in a spinal locomotor network deprived of crossed connections: The lamprey hemicord. *Journal of Neuroscience, 25,* 923-935.

Cao, C., Ashton-Miller, J.A., Schultz, A.B., & Alexander, N.B. (1997). Abilities to turn suddenly while walking: Effects of age, gender, and available response time. *Journal of Gerontology, 52A,* M88-M93.

Capaday, C. (2002). The special nature of human walking and its neural control. *Trends in Neurosciences, 25,* 370-376.

Capaday, C., & Stein, R.B. (1986). Amplitude modulation of the soleus H-reflex in the human during walking and standing. *Journal of Neuroscience, 6,* 1308-1313.

Cappellini, G., Ivanenko, Y.P., Poppele, R.E., & Lacquaniti, F. (2006). Motor patterns in human walking and running. *Journal of Neurophysiology, 95,* 3426-3437.

Capranica, L., Tessitore, A., Olivieri, B., Minganti, C., & Pesce, C. (2004). Field evaluation of cycled coupled movements of hand and foot in older individuals. *Gerontology, 50,* 399-406.

Cardinale, M., & Bosco, C. (2003). The use of vibration as an exercise intervention. *Exercise and Sport Sciences Reviews, 31,* 3-7.

Cardinale, M., & Wakeling, J. (2005). Whole body vibration exercise: Are vibrations good for you? *British Journal of Sports Medicine, 39,* 585-589.

Carlsen, A.N., Chua, R., Inglis, J.T., Sanderson, D., & Franks, I.M. (2004*a*). Can prepared responses be stored subcortically? *Experimental Brain Research, 159,* 301-309.

Carlsen, A.N., Chua, R., Inglis, J.T., Sanderson, D., & Franks, I.M. (2004*b*). Prepared movements are elicited early by startle. *Journal of Motor Behavior, 36,* 253-264.

Carlsöö, S. (1958). Motor units and action potentials in masticatory muscles. An electromyographic study of the form and duration of the action potentials and an anatomic study of the size of the motor units. *Acta Morphologica Neerlando-Scandinavica, 2,* 13-19.

Carmichael, S.T. (2003). Plasticity of cortical projections after stroke. *Neuroscientist, 9,* 64-75.

Carolan, B., & Cafarelli, E. (1992). Adaptations in coactivation after isometric resistance training. *Journal of Applied Physiology, 73,* 911-917.

Carp, J.S., Herchenroder, P.A., Chen, X.Y., & Wolpaw, J.R. (1999). Sag during unfused tetanic contractions in rat triceps surae motor units. *Journal of Neurophysiology, 81,* 2647-2661.

Carpentier, A., Duchateau, J., & Hainaut, K. (2001). Motor unit behaviour and contractile changes during fatigue in the human first dorsal interosseus. *Journal of Physiology, 534,* 903-912.

Carpinelli, R.N. (2002). Berger in retrospect: Effect of varied weight training programmes on strength. *British Journal of Sports Medicine, 36,* 319-324.

Carroll, T.J., Herbert, R.D., Munn, J., Lee, M., & Gandevia, S.C. (2006). Contralateral effects of unilateral strength training: Evidence and possible mechanisms. *Journal of Applied Physiology, 101,* 1514-1522.

Carroll, T.J., Riek, S., & Carson, R.G. (2001*a*). Corticospinal responses to motor training revealed by transcranial magnetic stimulation. *Exercise and Sport Sciences Reviews, 29,* 54-59.

Carroll, T.J., Riek, S., & Carson, R.G. (2001*b*). Resistance training enhances the stability of sensorimotor coordination. *Proceedings of the Royal Society of London B, 268,* 221-227.

Carroll, T.J., Riek, S., & Carson, R.G. (2002). The sites of neural adaptation induced by resistance training in humans. *Journal of Physiology, 544,* 641-652.

Carson, R.G. (2006). Changes in muscle coordination with training. *Journal of Applied Physiology, 101,* 1506-1513.

Carson, R.G., Riek, S., & Shahbazpour, N. (2002). Central and peripheral mediation of human force sensation following eccentric or concentric contractions. *Journal of Physiology, 539,* 913-925.

Carville, S.F., Perry, M.C., Rutherford, O.M., Smith, I.C., & Newham, D.J. (2007). Steadiness of quadriceps contractions in young and older adults with and without a history of falling. *European Journal of Applied Physiology, 100,* 527-533.

Castiello, U. (2005). The neuroscience of grasping. *Nature Neuroscience, 6,* 726-736.

Castro, M.J., Apple, D.F., Staron, R.S., Campos, G.E.R., & Dudley, G.A. (1999). Influence of complete spinal cord injury on skeletal muscle within 6 mo of injury. *Journal of Applied Physiology, 86,* 350-358.

Cattaneo, L., Voss, M., Brochier, T., Prabhu, G., Wolpert, D.M., & Lemon, R.N. (2005). A cortico-cortical mechanism mediating object-driven grasp in humans. *Proceedings of the National Academy of Science USA, 102,* 898-903.

Cavagna, G.A. (1975). Force platforms as ergometers. *Journal of Applied Physiology, 39,* 174-179.

Cavagna, G.A., & Citterio, G. (1974). Effect of stretching on the elastic characteristics of the contractile component of the frog striated muscle. *Journal of Physiology, 239,* 1-14.

Cavagna, G.A., & Franzetti, P. (1986). The determinants of the step frequency in walking in humans. *Journal of Physiology, 373,* 235-242.

Cavagna, G.A., Heglund, B.C., & Taylor, C.R. (1977). Mechanical work in terrestrial locomotion: Two basic mechanisms for minimizing energy expenditure. *American Journal of Physiology, 233,* R243-R261.

Cavagna, G.A., Mantovani, M., Willems, P.A., & Musch, G. (1997). The resonant step frequency in human running. *Pflügers Archiv, 434,* 678-684.

Cavagna, G.A., Saibene, F.P., & Margaria, R. (1964). Mechanical work in running. *Journal of Applied Physiology, 19,* 249-256.

Cavagna, G.A., Thys, H., & Zamboni, A. (1976). The sources of external work in level walking and running. *Journal of Physiology, 262,* 639-657.

Cavagna, G.A., Willems, P.A., & Heglund, N.C. (2000). The role of gravity in human walking: Pendular energy exchange, external work and optimal speed. *Journal of Physiology, 528,* 657-668.

Cavagna, G.A., Willems, P.A., Legramandi, M.A., & Heglund, N.C. (2002). Pendular energy transduction within the step in human walking. *Journal of Experimental Biology, 205,* 3413-3422.

Cavanagh, P.R., & Ae, M. (1980). A technique for the display of pressure distributions beneath the foot. *Journal of Biomechanics, 13,* 69-76.

Cavanagh, P.R., & Grieve, D.W. (1973). The graphical display of angular movement of the body. *British Journal of Sports Medicine, 7,* 129-133.

Cavanagh, P.R., & Kram, R. (1989). Stride length in distance running: Velocity, body dimensions, and added mass effects. *Medicine and Science in Sports and Exercise, 21,* 476-479.

Cavanagh, P.R., & Lafortune, M.A. (1980). Ground reaction forces in distance running. *Journal of Biomechanics, 13,* 397-406.

Cazalets, J.R., Bertrand, S., Sqalli-Houssaini, Y., & Clarac, F. (1998). GABAergic control of spinal locomotor networks in the neonatal rat. *Annals of the New York Academy of Sciences, 16,* 168-180.

Celichowski, J., Mrowczynski, W., Kruti, P., Gorska, T., Majczynski, H., & Slawinska, U. (2006). Changes in contractile properties of motor units of the rat medial gastrocnemius muscle after spinal cord transection. *Experimental Physiology, 91,* 887-895.

Cella, D.F., & Perry, S.W. (1986). Reliability and concurrent validity of three visual-analogue mood scales. *Psychological Reports, 59,* 827-833.

Cescon, C., Sguazzi, E., Merletti, R., & Farina, D. (2006). Non-invasive characterization of single motor unit electromyographic and mechanomyographic activities in the biceps brachii muscle. *Journal of Electromyography and Kinesiology, 16,* 17-24.

Chabran, E., Maton, B., Ribreau, C., & Fourment, A. (2001). Electromyographic and biomechanical characteristics of segmental body adjustments associated with voluntary wrist movements. Influence of an elbow support. *Experimental Brain Research, 141,* 133-145.

Chaix, Y., Marque, P., Meunier, S., Pierrot-Deseilligny, E., & Simonetta-Moreau, M. (1997). Further evidence for non-monosynaptic group I excitation of motoneurones in the human lower limb. *Experimental Brain Research, 115,* 35-46.

Challis, J.H. (1996). Accuracy of human limb moment of inertia estimations and their influence on resultant joint moments. *Journal of Applied Biomechanics, 12,* 517-530.

Challis, J.H., & Pain, M.T.G. (2008). Soft tissue motion influences skeletal loads during impacts. *Exercise and Sport Sciences Reviews, 36,* 71-75.

Chalmers, G.R., & Knutzen, K.M. (2002). Soleus H-reflex gain in healthy elderly and young adults when lying, standing, and balancing. *Journal of Gerontology A, 57,* B321-B329.

Chandler, R.F., Clauser, C.E., McConville, J.T., Reynolds, H.M., & Young, J.W. (1975). *Investigation of Inertial Properties of the Human Body* (AMRL-TR-74-137). Wright-Patterson Air Force Base, OH: Aerospace Medical Research Laboratories, Aerospace Medical Division (NTIS No. AD-A016 485).

Chang, Y.-H., & Kram, R. (1999). Metabolic cost of generating horizontal forces during human running. *Journal of Applied Physiology*, *86*, 1657-1662.

Chang, Y.H., Huang, H.W., Hamerski, C.M., & Kram, R. (2000). The independent effects of gravity and inertia on running mechanics. *Journal of Experimental Biology, 203,* 229-238.

Chaouloff, F., Berton, O., & Mormede, P. (1999). Serotonin and stress. *Neuropsychopharmacology*, *21* (Suppl. 2), 28S-35S.

Chase, P.B., & Kushmerick, M.J. (1988). Effects of pH on contraction of rabbit fast muscle slow skeletal muscle fibers. *Biophysical Journal, 53,* 935-946.

Chen, R., Cohen, L.G., & Hallett, M. (2002). Nervous system reorganization following injury. *Neuroscience, 111,* 761-773.

Chen, T.C., Nosaka, K., & Sacco, P. (2006). Intensity of eccentric exercise, shift of optimum angle, and the magnitude of repeated bout effect. *Journal of Applied Physiology, 102,* 992-999.

Chen, X.Y., Chen, L., Chen, Y., & Wolpaw, J.R. (2006). Operant conditioning of reciprocal inhibition in rat soleus muscle. *Journal of Neurophysiology, 96,* 2144-2150.

Cheney, P.D., & Fetz, E.E. (1985). Comparable patterns of muscle facilitation evoked by individual corticomotoneuronal (CM) cells and by single intracortical microstimuli in primates: Evidence for functional groups of CM cells. *Journal of Neurophysiology, 53,* 786-804.

Cheng, A.J., & Rice, C.L. (2005). Fatigue and recovery of power and isometric torque following isotonic knee extensions. *Journal of Applied Physiology, 99,* 1446-1452.

Cheng, K.B., & Hubbard, M. (2005). Optimal compliant-surface jumping: A multi-segment model of springboard standing jumps. *Journal of Biomechanics, 38,* 1822-1829.

Chester, V.L., & Jensen, R.K. (2005). Changes in infant segment inertias during the first three months of independent walking. *Dynamic Medicine, 28,* 4-9.

Cheung, S.S., & Sleivert, G.G. (2004). Lowering of skin temperature decreases isokinetic maximal force production independent of core temperature. *European Journal of Applied Physiology, 91,* 723-728.

Chilibeck, P.D., Calder, A.W., Sale, D.G., & Webber, C.E. (1998). A comparison of strength and muscle mass increases during resistance training in young women. *European Journal of Applied Physiology, 77,* 170-175.

Chino, K., Oda, T., Kurihara, T., Nagayoshi, T., Kanehisa, H., Fukashiro, S., Fukunaga, T., & Kawakami, Y. (2003). In vivo muscle fiber behavior of the triceps surae muscles during isokinetic concentric and eccentric plantar flexions. *Japanese Journal of Biomechanics, Sports, and Exercise, 7,* 206-213.

Chino, K., Oda, T., Kurihara, T., Nagayoshi, T., Yoshikawa, K., Kanehisa, H., Fukunaga, T., Fukashiro, S., & Kawakami, Y. (2007). In vivo fascicle behavior of synergistic muscles in concentric and eccentric plantar flexions in humans. *Journal of Electromyography and Kinesiology, 32,* 00-00.

Cholewicki, J., Juluru, K., & McGill, S.M. (1999). Intra-abdominal pressure mechanism for stabilizing the lumbar spine. *Journal of Biomechanics, 32,* 13-17.

Cholewicki, J., McGill, S.M., & Norman, R.W. (1991). Lumbar spine loads during the lifting of extremely heavy weights. *Medicine and Science in Sports and Exercise, 23,* 1179-1186.

Christakos, C.N., & Windhorst, U. (1986). Spindle gain increase during muscle unit fatigue. *Brain Research, 365,* 388-392.

Christensen, E. (1959). Topography of terminal motor innervation in striated muscles from stillborn infants. *American Journal of Physical Medicine, 38,* 65-78.

Christensen, L.O.D., Petersen, N., Andersen, J.B., Sinkjær, T., & Nielsen, J.B. (2000). Evidence for transcortical reflex pathways in the lower limb of man. *Progress in Neurobiology, 62,* 251-272.

Christie, A., & Kamen, G. (2006). Doublet discharges in motoneurons of young and older adults. *Journal of Neurophysiology*, *95*, 2787-2795.

Christou, E.A., & Carlton, L.G. (2001). Old adults exhibit greater motor output variability than young adults only during rapid discrete isometric contractions. *Journal of Gerontology, 56A,* B524-B532.

Christou, E.A., & Carlton, L.G. (2002*a*). Age and contraction type influence motor output variability in rapid discrete tasks. *Journal of Applied Physiology, 93,* 489-498.

Christou, E.A., & Carlton, L.G. (2002*b*). Motor output is more variable during eccentric compared with concentric contractions. *Medicine and Science in Sports and Exercise, 34,* 1773-1778.

Christou, E.A., Grossman, M., & Carlton, L.G. (2002). Modeling variability of force during isometric contractions of the quadriceps femoris. *Journal of Motor Behavior, 34,* 67-81.

Christou, E.A., Jakobi, J.M., Critchlow, A., Fleshner, M., & Enoka, R.M. (2004). The 1- to 2-Hz oscillations in muscle force are exacerbated by stress, especially in older adults. *Journal of Applied Physiology, 97,* 225-235.

Christova, P., & Kossev, A. (1998). Motor unit activity during long-lasting intermittent muscle contractions in humans. *European Journal of Applied Physiology, 77,* 379-387.

Cisek, P., Crammond, D.J., & Kalaska, J.F. (2003). Neural activity in primary motor and dorsal premotor cortex in reaching tasks with the contralateral versus the ipsilateral arm. *Journal of Neurophysiology, 89,* 922-942.

Cisek, P., & Kalaska, J.F. (2002). Simultaneous encoding of multiple potential reach directions in dorsal premotor cortex. *Journal of Neurophysiology, 87,* 1149-1154.

Claasen, H., Gerber, C., Hoppeler, H., Lüthi, J.-M., & Vock, P. (1989). Muscle filament spacing and short-term heavy-resistance exercise in humans. *Journal of Physiology, 409,* 491-495.

Clamann, H.P., Mathis, J., & Lüscher, H.R. (1989). Variance analysis of excitatory postsynaptic potentials in cat spinal motoneurons during posttetanic potentiation. *Journal of Neurophysiology, 61,* 403-416.

Clark, B.C., Collier, S.R., Manini, T.M., & Ploutz-Snyder, L.L. (2005). Sex differences in muscle fatigability and activation patterns of the human quadriceps femoris. *European Journal of Applied Physiology, 94,* 196-206.

Clark, J.M., & Haynor, D.R. (1987). Anatomy of the abductor muscles of the hip as studied by computed tomography. *Journal of Bone and Joint Surgery, 69-A,* 1021-1031.

Clark, K.A., McElhinny, A.S., Beckerie, M.C., & Gregorio, C.C. (2002). Striated muscle cytoarchitecture: An intricate web of form and function. *Annual Reviews of Cell and Developmental Biology, 18,* 637-706.

Clarkson, P.M., & Hubal, M.J. (2002). Exercise-induced muscle damage in humans. *American Journal of Physical Medicine and Rehabilitation, 81* (Suppl. 11), S52-S69.

Clarys, J.P., & Marfell-Jones, M.J. (1986). Anthropometric prediction of component tissue masses in the minor limbs of the human body. *Human Biology, 58,* 761-769.

Clarys, J.P., Martin, A.D., & Drinkwater, D.T. (1984). Gross tissue weights in the human body by cadaver dissection. *Human Biology, 54,* 459-473.

Clausen, T., Nielsen, O.B., Harrison, A.P., Flatman, J.A., & Overgaard, E. (1998). The Na^+-K^+ pump and muscle excitability. *Acta Physiologica Scandinavica, 162,* 183-190.

Clauser, C. E., McConville, J. T., & Young, J. W. (1969). *Weight, Volume, and Center of Mass of Segments of the Human Body* (AMRL-TR-69-70). Wright-Patterson Air Force Base, OH: Aerospace Medical Research Laboratories, Aerospace Medical Division.

Clement, G., Reschke, M., & Wood, S. (2005). Neurovestibular and sensorimotor studies in space and Earth benefits. *Current Pharmacology and Biotechnology, 6,* 267-283.

Coggan, A.R., & Coyle, E.F. (1987). Reversal of fatigue during prolonged exercise by carbohydrate infusion or ingestion. *Journal of Applied Physiology, 63,* 2388-2395.

Colby, C.L., Berman, R.A., Heiser, L.M., & Saunders, R.C. (2005). Corollary discharge and spatial updating: When the brain is split, is space still unified? *Progress in Brain Research, 149,* 187-205.

Cole, K.J. (1991). Grasp force control in older adults. *Journal of Motor Behavior, 23,* 251-258.

Cole, K.J., Rotella, D.L., & Harper, J.G. (1999). Mechanisms for age-related changes of fingertip forces during precision gripping and lifting in adults. *Journal of Neuroscience, 19,* 3238-3247.

Cole, M.A., & Brown, M.D. (2000). Response of the human triceps surae muscle to electrical stimulation during varying levels of blood flow restriction. *European Journal of Applied Physiology, 82,* 39-44.

Collins, D.F. (2007). Central contributions to contractions evoked by tetanic neuromuscular electrical stimulation. *Exercise and Sport Sciences Reviews, 35,* 102-109.

Collins, D.F., Burke, D., & Gandevia, S.C. (2002). Sustained contractions produced by plateau-like behavior in human motoneurones. *Journal of Physiology, 538,* 289-301.

Collins, D.F., Refshauge, K.M., Todd, G., & Gandevia, S.C. (2005). Cutaneous receptors contribute to kinesthesia at the index finger, elbow, and knee. *Journal of Neurophysiology, 94,* 1699-1706.

Collins, M., Renault, V., Grobler, L.A., St. Clair Gibson, A., Lambert, M.I., Wayne Derman, E., Butler-Browne, G.S., Noakes, T.D., & Mouly, V. (2003). Athletes with exercise-associated fatigue have abnormally short muscle DNA telomeres. *Medicine and Science in Sports and Exercise, 35,* 1524-1528.

Colson, S., Pousson, M., Martin, A., & Van Hoecke, J. (1999). Isokinetic elbow flexion and coactivation following eccentric training. *Journal of Electromyography and Kinesiology, 9,* 13-20.

Conditt, M.A., & Mussa-Ivaldi, F.A. (1999). Central representation of time during motor learning. *Proceedings of the National Academy of Sciences USA, 96,* 11625-11630.

Condliffe, E.G., Clark, D.J., & Patten, C. (2005). Reliability of elbow stretch reflex assessment in chronic post-stroke hemiparesis. *Clinical Neurophysiology, 116,* 1870-1878.

Condon, S.M., & Hutton, R.S. (1987). Soleus muscle electromyographic activity and ankle dorsiflexion range of motion during four stretching procedures. *Physical Therapy, 67,* 24-30.

Connelly, D.M., & Vandervoort, A.A. (2000). Effects of isokinetic strength training on concentric and eccentric torque development in the ankle dorsiflexors of older adults. *Journal of Gerontology, 55A,* B465-B472.

Cooke, R. (2007). Modulation of the actomyosin interaction during fatigue of skeletal muscle. *Muscle & Nerve, 36,* 756-777.

Cooke, R., Franks, K., Luciani, G.B., & Pate, E. (1988). The inhibition of rabbit skeletal muscle contraction by hydrogen ions and phosphate. *Journal of Physiology, 395,* 77-97.

Cope, T.C., Bodine, S.C., Fournier, M., & Edgerton, V.R. (1986). Soleus motor units in chronic spinal transected cat: Physiological and morphological alterations. *Journal of Neurophysiology, 55,* 1202-1220.

Cope, T.C., Bonasera, S.J., & Nichols, T.R. (1994). Reinnervated muscles fail to produce stretch reflexes. *Journal of Neurophysiology, 71,* 817-820.

Corden, D.M., & Lippold, O.C.J. (1996). Age-related impaired reflex sensitivity in a human hand muscle. *Journal of Physiology, 76,* 2701-2706.

Cordo, P.J., Flores-Vieira, C., Verschueren, S.M.P., Inglis, J.T., & Gurfinkel, V. (2002). Position sensitivity of human muscle spindles: Single afferent and population representations. *Journal of Neurophysiology, 87,* 1186-1195.

Cordo, P.J., Gurfinkel, V.S., & Levik, Y. (2000). Position sense during imperceptibly slow movements. *Experimental Brain Research, 132,* 1-9.

Cordo, P.J., Hodges, P.W., Smith, T.C., Brumagne, S., & Gurfinkel, V.S. (2006). Scaling and non-scaling of muscle activity, kinematics, and dynamics in sit-ups with different degrees of difficulty. *Journal of Electromyography and Kinesiology, 16,* 506-521.

Cordo, P.J., & Nashner, L.M. (1982). Properties of postural adjustments associated with rapid arm movements. *Journal of Neurophysiology, 47,* 287-302.

Cormery, B., Marini, J.F., & Gardiner, P.F. (2000). Changes in electrophysiological properties of tibial motoneurones in the rat following 4 weeks of tetrodotoxin-induced paralysis. *Neuroscience Letters, 287,* 21-24.

Costill, D.L., & Hargreaves, M. (1992). Carbohydrate nutrition and fatigue. *Sports Medicine, 13,* 86-92.

Coyle, E.F., Coggan, A.R., Hemmert, M.K., & Ivy, J.L. (1986). Muscle glycogen utilization during prolonged strenuous exercise when fed carbohydrate. *Journal of Applied Physiology, 61,* 165-172.

Coyle, E.F., Hagberg, J.M., Hurley, B.F., Martin, W.H., Ehsani, A.A., & Holloszy, J.O. (1983). Carbohydrate feeding during prolonged strenuous exercise can delay fatigue. *Journal of Applied Physiology, 55,* 230-235.

Crago, P.E., Houk, J.C., & Rymer, W.Z. (1982). Sampling of total muscle force by tendon organs. *Journal of Neurophysiology, 47,* 1069-1083.

Cramer, J.T., Housh, T.J., Weir, J.P., Johnson, G.O., Coburn, J.W., & Beck, T.W. (2005). The acute effects of static stretching on peak torque, mean power output, electromyography, and mechanomyography. *European Journal of Applied Physiology, 93,* 530-539.

Crameri, R.M., Cooper, P., Sinclair, P.J., Bryant, G., & Weston, A. (2004*a*). Effect of load during stimulation training in spinal cord injury. *Muscle & Nerve, 29,* 104-111.

Crameri, R.M., Langberg, H., Magnusson, P., Jensen, C.H., Schroder, H.D., Olesen, J.L., Suetta, C., Teisner, B., & Kjaer, M. (2004*b*). Changes in satellite cells in human skeletal muscle after a single bout of high intensity exercise. *Journal of Physiology, 558,* 333-340.

Crammond, D.J., & Kalaska, J.F. (1989). Neuronal activity in primate parietal cortex area 5 varies with intended movement direction during an instructed-delay period. *Experimental Brain Research, 76,* 458-462.

Crammond, D.J., & Kalaska, J.F. (2000). Prior information in motor and premotor cortex: Activity during the delay period and effect on pre-movement activity. *Journal of Neurophysiology, 84,* 986-1006.

Crenshaw, A.G., Karlsson, S., Styf, J., Bäcklund, T., & Fridén, J. (1995). Knee extension torque and intramuscular pressure of the vastus lateralis muscle during eccentric and concentric activities. *European Journal of Applied Physiology, 70,* 13-19.

Cresswell, A.G., Grundström, H., & Thorstensson, A. (1992). Observations on intra-abdominal pressure and patterns of abdominal intra-muscular activity in man. *Acta Physiologica Scandinavica, 144,* 409-418.

Crone, C., Hultborn, H., Jespersen, B., & Nielsen, J. (1987). Reciprocal Ia inhibition between ankle flexors and extensors in man. *Journal of Physiology, 389,* 163-185.

Crone, C., Hultborn, H., Maziéres, L., Morin, C., Nielsen, J., & Pierrot-Deseilligny, E. (1990). Sensitivity of monosynaptic test reflexes to facilitation and inhibition as a function of the test reflex size: A study in man and the cat. *Experimental Brain Research, 81,* 35-45.

Crone, C., & Nielsen, J. (1989). Spinal mechanisms in man contributing to reciprocal inhibition during voluntary dorsiflexion of the foot. *Journal of Physiology, 416,* 255-272.

Cronin, J., & Sleivert, G. (2005). Challenges in understanding the influence of maximal power training on improving athletic performance. *Sports Medicine, 35,* 213-234.

Cruz-Martinez, A., Ramirez, A., & Arpa, J. (2000). Quadriceps atrophy after knee traumatisms and immobilization: Electrophysiological assessment. *European Neurology, 43,* 110-114.

Cullheim, S., Fleshman, J.W., Glenn, L.L., & Burke, R.E. (1987). Membrane area and dendritic structure in type-identified triceps alpha motoneurons. *Journal of Comparative Neurology, 255,* 68-81.

Cullheim, S., & Kellerth, J-O. (1978). A morphological study of the axons and recurrent axon collaterals of cat α-motoneurones supplying different hind-limb muscles. *Journal of Physiology, 281,* 285-299.

Cupido, C.M., Galea, V., & McComas, A.J. (1996). Potentiation and depression of the M wave in human biceps brachii. *Journal of Physiology, 491,* 541-550.

Curt, A., Bruehlmeier, M., Leenders, K.L., Roelcke, U., & Dietz, V. (2002). Differential effect of spinal cord injury and functional impairment on human brain activation. *Journal of Neurotrama, 19,* 43-51.

Cutlip, R.G., Baker, B.A., Geronilla, K.B., Mercer, R.R., Kashon, M.L., Miller, G.R., Murlastis, Z., & Alway, S.E. (2006). Chronic exposure to stretch-shortening contractions results in skeletal muscle adaptation in young rats and maladaptation in old rats. *Applied Physiology, Nutrition, and Metabolism, 31,* 573-587.

Czéh, G., Gallego, R., Kudo, N., & Kuno, M. (1978). Evidence for the maintenance of motoneurone properties by muscle activity. *Journal of Physiology, 281,* 239-252.

Daggfeldt, K., & Thorstensson, A. (1997). The role of intra-abdominal pressure in spinal unloading. *Journal of Biomechanics, 30,* 1149-1155.

Dahlkvist, N.J., Mayo, P., & Seedhom, B.B. (1982). Forces during squatting and rising from a deep squat. *Engineering in Medicine, 11,* 69-76.

Dahlstedt, A.J., Katz, A., & Westerblad, H. (2001). Role of myoplasmic phosphate in contractile function of skeletal muscle: studies on creatine kinase–deficient mice. *Journal of Physiology, 533,* 379-388.

Dainis, A. (1981). A model for gymnastics vaulting. *Medicine and Science in Sports and Exercise, 13,* 34-43.

Damiani, E., & Margreth, A. (1994). Characterization study of the ryanodine receptor and of calsequesterin isoforms of mammalian skeletal muscles in relation to fibre types. *Journal of Muscle Research and Cell Motility, 15,* 86-101.

Damiano, D.L., Laws, E., Carmines, D.V., & Abel, M.F. (2006). Relationship of spasticity to knee angular velocity and motion during gait in cerebral palsy. *Gait & Posture, 23,* 1-8.

Danause, N., Barbay, S., Frost, S.B., Plautz, E.J., Chen, D., Zoubina, E.V., Stowe, A.M., & Nudo, R.J. (2005). Extensive cortical rewiring after brain injury. *Journal of Neuroscience, 25,* 10167-10179.

Danion, F., Latash, M.L., Li, Z.M., & Zatsiorsky, V.M. (2000). The effect of fatigue on multifinger co-ordination in force production tasks in humans. *Journal of Physiology, 523,* 523-532.

D'Antona, G., Pelligrino, M.A., Adami, R., Rossi, R., Carlizzi, C.N., Canepari, M., Saltin, B., & Bottinelli, R. (2003). The effect of ageing and immobilization on structure and function of human skeletal muscle fibres. *Journal of Physiology, 552,* 499-511.

Darling, W.G., & Cole, K.J. (1990). Muscle activation patterns and kinetics of human index finger movements. *Journal of Neurophysiology, 63,* 1098-1108.

Datta, A.K., Farmer, S.F., & Stephens, J.A. (1991). Central nervous pathways underlying synchronization of human motor unit firing studied during voluntary contraction. *Journal of Physiology, 432,* 401-425.

d'Avella, A., & Bizz, E. (2005). Shared and specific muscle synergies in natural motor behaviors. *Proceedings of the National Academy of Sciences USA, 102,* 3076-3081.

d'Avella, A., Portone, A., Fernandez, L., & Lacquaniti, F. (2006). Control of fast-reaching movements by muscle synergy combinations. *Journal of Neuroscience, 26,* 7791-7810.

d'Avella, A., Saltiel, P., & Bizzi, E. (2003). Combinations of muscle synergies in the construction of a natural motor behavior. *Nature Neuroscience, 6,* 300-308.

Davey, N.J., Ellaway, P.H., Baker, J.R., & Friedland, C.L. (1993). Rhythmicity associated with a high degree of short-term synchrony of motor unit discharge in man. *Experimental Physiology, 78,* 649-661.

Davies, C.T.M., Dooley, P., McDonagh, M.J.N., & White, M.J. (1985). Adaptation of mechanical properties of muscle to high force training in man. *Journal of Physiology, 365,* 277-284.

Davies, C.T.M., Rutherford, I.C., & Thomas, D.O. (1987). Electrically evoked contractions of the triceps surae during and following 21 days of voluntary leg immobilization. *European Journal of Applied Physiology, 56,* 306-312.

Davies, J., Parker, D.F., Rutherford, O.M., & Jones, D.A. (1988). Changes in strength and cross sectional area of elbow flexors as a result of isometric strength training. *European Journal of Applied Physiology, 57,* 667-670.

Davis, B.M., Collins, W.F. 3rd, & Mendell, L.M. (1985). Potentiation of transmission at Ia-motoneuron connections induced by repeated short bursts of afferent activity. *Journal of Neurophysiology, 54,* 1541-1552.

Davis, B.L., & Vaughan, C.L. (1993). Phasic behavior of EMG signals during gait: Use of multivariate statistics. *Journal of Electromyography and Kinesiology, 3,* 51-60.

Davis, J.M., & Bailey, S.P. (1997). Possible mechanisms of central nervous system fatigue during exercise. *Medicine and Science in Sports and Exercise, 29,* 45-57.

Davis, J.R., & Mirka, G.A. (2000). Transverse-contour modeling of trunk muscle-distributed forces and spinal loads during lifting and twisting. *Spine, 25,* 180-189.

Davis, M. (1984). The mammalian startle response. In R.C. Eaton (Ed.), *Neural Mechanisms of Startle Behavior* (pp. 287-351). New York: Plenum Press.

Day, S.J., & Hulliger, M. (2001). Experimental simulation of the cat electromyogram: Evidence for algebraic summation of motor-unit action-potential trains. *Journal of Neurophysiology, 86,* 2144-2158.

Debicki, D.B., & Gribble, P.L. (2005). Persistence of inter-joint coupling during single-joint elbow flexions after shoulder fixation. *Experimental Brain Research, 163,* 252-257.

de Carvalho, V.C. (1976). Study of motor units and arrangement of myons of human musculus plantaris. *Acta Anatomica, 96,* 444-448.

de Haan, A. (1998). The influence of stimulation frequency on force–velocity characteristics of in situ rat medial gastrocnemius muscle. *Experimental Physiology, 83,* 77-84.

De Koning, F.L., Binkhorst, R.A., Vos, J.A., & van't Hof, M.A. (1985). The force–velocity relationship of arm flexion in untrained males and females and arm-trained athletes. *European Journal of Applied Physiology, 54,* 89-94.

De Koning, J.J., de Groot, G., & van Ingen Schenau, G.J. (1992). Ice friction during speed skating. *Journal of Biomechanics, 25,* 565-571.

Delecluse, C., Roelants, M., & Verschueren, S. (2003). Strength increase after whole-body vibration compared with resistance training. *Medicine and Science in Sports and Exercise, 35,* 1033-1041.

de Leon, R.D., Hodgson, J.A., Roy, R.R., & Edgerton, V.R. (1998*a*). Full weight-bearing hindlimb standing following stand training in the adult spinal cat. *Journal of Neurophysiology, 79,* 83-91.

de Leon, R.D., Hodgson, J.A., Roy, R.R., & Edgerton, V.R. (1998*b*). Locomotor capacity attributable to step training versus spontaneous recovery after spinalization in adult cats. *Journal of Neurophysiology, 79,* 1329-1340.

de Leon, R.D., Hodgson, J.A., Roy, R.R., & Edgerton, V.R. (1999). Retention of hindlimb stepping ability in adult spinal cats after the cessation of training. *Journal of Neurophysiology, 81,* 85-94.

de Leon, R.D., Reinkensmeyer, D.J., Timoszyk, W.K., London, N.J., & Edgerton, V.R. (2002). Use of robotics in assessing the adaptive capacity of the rat lumbar spinal cord. *Progress in Brain Research, 137,* 141-149.

de Leva, P. (1996). Adjustments to Zatsiorsky-Seluyanov's segment inertia parameters. *Journal of Biomechanics, 29,* 1223-1230.

Delitto, A., Brown, M., Strube, M.J., Rose, S.J., & Lehman, R.C. (1989). Electrical stimulation of quadriceps femoris in an elite weight lifter: A single subject experiment. *International Journal of Sports Medicine, 10,* 187-191.

Delitto, A., & Rose, S.J. (1986). Comparative comfort of three waveforms used in electrically eliciting quadriceps femoris muscle contractions. *Physical Therapy, 66,* 1704-1707.

De Looze, M.P., Bussmann, J.B.J., Kingma, I., & Toussaint, H.M. (1992). Different methods to estimate total power and its components during lifting. *Journal of Biomechanics, 25,* 1089-1095.

DeLorme, T.L. (1945). Restoration of muscle power by heavy-resistance exercises. *Journal of Bone and Joint Surgery, 27,* 645-667.

De Luca, C.J., & Erim, Z. (1994). Common drive of motor units in regulation of muscle force. *Trends in Neuroscience, 17,* 299-305.

De Luca, C.J., & Erim, Z. (2002). Common drive in motor units of a synergistic muscle pair. *Journal of Neurophysiology, 87,* 2200-2204.

De Luca, C.J., LeFever, R.S., McCue, M.P., & Xenakis, A.P. (1982). Behaviour of human motor units in different muscles during linearly varying contractions. *Journal of Physiology, 329,* 113-128.

De Luca, C.J., & Mambrito, B. (1987). Voluntary control of motor units in human antagonist muscles: Coactivation and reciprocal activation. *Journal of Neurophysiology, 58,* 525-542.

Delwaide, P.J., Sabatino, M., Pepin, J.L., & La Grutta, V. (1988). Reinforcement of reciprocal inhibition by contralateral movements in man. *Experimental Neurology, 99,* 10-16.

Delwaide, P.J., & Schepens, B. (1995). Auditory startle (audio-spinal) reaction in normal man: EMG responses and H reflex changes in antagonistic lower limb muscles. *Electroencephalography and Clinical Neurophysiology, 97,* 416-423.

Dempster, W.T. (1955). *Space Requirements of the Seated Operator* (WADC-TR-55-159). Wright-Patterson Air Force Base, OH: Aerospace Medical Research Laboratory (NTIS No. AD-87892).

Dempster, W.T. (1961). Free-body diagrams as an approach to the mechanics of human posture and motion. In G.F. Evans (Ed.), *Biomechanical Studies of the Musculo-Skeletal System* (pp. 81-135). Springfield, IL: Charles C Thomas.

Dengler, R., Konstabzer, A., Hesse, S., Schubert, M., & Wolf, W. (1989). Collateral nerve sprouting and twitch forces of single motor units in conditions with partial denervation in man. *Neuroscience Letters, 97,* 118-122.

Denny-Brown, D., & Pennybacker, J.B. (1938). Fibrillation and fasciculation in voluntary muscle. *Brain, 61,* 311-334.

de Ruiter, C.J., & de Haan, A. (2000). Temperature effect on the force/velocity relationship of the fresh and fatigued human adductor pollicis muscle. *Pflügers Archiv, 440,* 163-170.

de Ruiter, C.J., de Haan, A., Jones, D.A., & Sargeant, A.J. (1998). Shortening-induced force depression in human adductor pollicis muscle. *Journal of Physiology, 507,* 583-591.

de Ruiter, C.J., Elzinga, M.J., Verdijk, P.W., van Mechelen, W., & de Haan, A. (2005). Changes in force, surface and motor unit EMG during post-exercise development of low-frequency fatigue in vastus lateralis muscle. *European Journal of Applied Physiology, 94,* 659-669.

de Ruiter, C.J., Goudsmit, F.A., van Tricht, J.A., & de Haan, A. (2007). The isometric torque at which knee-extensor muscle reoxygenation stops. *Medicine and Science in Sports and Exercise, 39,* 443-452.

de Ruiter, C.J., Jones, D.A., Sargeant, A.J., & de Haan, A. (1999). Temperature effect on the rates of isometric force development and relaxation in the fresh and fatigued human adductor pollicis muscle. *Experimental Physiology, 84,* 1137-1150.

de Ruiter, C.J., van Raak, S.M., Schilperoort, J.V., Hollander, A.P., & de Haan, A. (2003). The effects of 11 weeks' whole body vibration training on jump height, contractile properties, and activation of human knee extensors. *European Journal of Applied Physiology, 90,* 595-600.

De Serres, S.J., & Enoka, R.M. (1998). Older adults can maximally activate the biceps brachii muscle by voluntary command. *Journal of Applied Physiology, 84,* 284-291.

Deshpande, N., & Patla, A.E. (2005). Dynamic visual-vestibular integration during goal directed human locomotion. *Experimental Brain Research, 166,* 237-247.

Desmedt, J.E., & Godaux, E. (1977). Fast motor units are not preferentially activated in rapid voluntary contractions in man. *Nature, 267,* 717-719.

Desmurget, M., Epstein, C.M., Turner, R.S., Prablanc, C., Alexander, G.E., & Grafton, S.T. (1999). Role of the posterior parietal cortex in updating reaching movements to a visual target. *Nature Neuroscience*, *2*, 563-567.

DeVita, P. (2005). Musculoskeletal modeling and the prediction of in vivo muscle and joint forces. *Medicine and Science in Sports and Exercise, 37,* 1909-1910.

DeVita, P., & Hortobágyi, T. (2000). Age causes a redistribution of joint torques and powers during gait. *Journal of Applied Physiology, 88,* 1804-1811.

DeVita, P., Hortobágyi, T., & Barrier, J. (1998). Gait biomechanics are not normal after anterior cruciate ligament reconstruction and accelerated rehabilitation. *Medicine and Science in Sports and Exercise, 30,* 1481-1488.

DeVita, P., Lassiter, T., Jr., Hortobágyi, T., & Torry, M. (1998). Functional knee brace effects during walking in patients with anterior cruciate ligament reconstruction. *American Journal of Sports Medicine, 26,* 778-784.

DeVita, P., Torry, M., Glover, K.L., & Speroni, D.L. (1996). A functional knee brace alters joint torque and power patterns during walking and running. *Journal of Biomechanics, 29,* 583-588.

de Vos, N.J., Singh, N.A., Ross, D.A., Stavrinos, T.M., Orr, R., & Fiatarone Singh, M.A. (2005). Optimal load for increasing muscle power during explosive resistance training in older adults. *Journal of Gerontology, 60A,* 638-647.

Dewald, J.P., Given, J.D., & Rymer, W.Z. (1996). Long-lasting reductions of spasticity induced by skin electrical stimulation. *IEEE Transactions on Rehabilitation Engineering, 4,* 231-242.

De Wit, B., De Clercq, D., & Aert, P. (2000). Biomechanical analysis of the stance phase during barefoot and shod running. *Journal of Biomechanics, 33,* 269-278.

De Wolf, S., Slijper, H., & Latash, M.L. (1998). Anticipatory postural adjustments during self-paced and reaction-time movements. *Experimental Brain Research, 121,* 7-19.

Diener, H.C., Dichgans, J., Bootz, E., & Bacher, M. (1984). Early stabilization of human posture after a sudden disturbance: Influence of rate and amplitude of displacement. *Experimental Brain Research, 56,* 126-134.

Dietz, V. (1992). Human neuronal control of automatic functional movements: Interaction between central programs and afferent input. *Physiological Reviews, 72,* 33-69.

Dietz, V. (1997). Neurophysiology of gait disorders: Present and future applications. *Electroencephalography and Clinical Neurophysiology, 103,* 333-355.

Dietz, V., Colombo, G., & Jensen, L. (1994). Locomotor activity in spinal man. *Lancet, 344,* 1260-1263.

Dietz, V., & Harkema, S.J. (2004). Locomotor activity in spinal cord-injured persons. *Journal of Applied Physiology, 96,* 1954-1960.

Dietz, V., Kowalewski, R., Nakazawa, K., & Colombo, G. (2000). Effects of changing stance conditions on anticipatory postural adjustment and reaction time to voluntary arm movement in humans. *Journal of Physiology, 524,* 617-627.

Dietz, V., & Muller, R. (2004). Degradation of neuronal function following spinal cord injury: Mechanisms and countermeasures. *Brain, 127,* 2221-2231.

Dietz, V., Quintern, J.B., & Sillem, M. (1987). Stumbling reactions in man: Significance of proprioceptive and preprogrammed mechanisms. *Journal of Physiology, 368,* 149-163.

Dietz, V., Schmidtbleicher, D., & Noth, J. (1979). Neuronal mechanisms of human locomotion. *Journal of Neurophysiology, 42,* 1212-1222.

Di Giulio, C., Daniele, F., & Tipton, C.M. (2006). Angelo Mosso and muscular fatigue: 116 years after the first congress of physiologists: IUPS commemoration. *Advances in Physiology Education, 30,* 51-57.

Di Lazzaro, V., Oliviero, A., Profice, P., Saturno, E., Pilato, F., Insola, A., Mazzone, P., Tonali, P., & Rothwell, J.C. (1998). Comparison of descending volleys evoked by transcranial magnetic and electric stimulation in conscious humans. *Electroencephalography and Clinical Neurophysiology, 109,* 397-401.

Dimitrijevic, M.R., Gersaimenko, Y., & Pinter, M.M. (1998). Evidence for a spinal pattern generator in humans. *Annals of the New York Academy of Sciences, 860,* 360-376.

Dimitrova, N., & Dimitrov, G. (2002). Amplitude-related characteristics of motor unit and M-wave potentials during fatigue. A simulation study using literature data on intracellular potential changes found in vitro. *Journal of Electromyography and Kinesiology, 12,* 339-349.

di Pellegrino, G., & Wise, S.P. (1993). Visuospatial vs. visuomotor activity in the premotor and prefrontal cortex of a primate. *Journal of Neuroscience, 13,* 1227-1243.

di Prampero, P. (2000). Cycling on earth, in space, and on the moon. *European Journal of Applied Physiology, 82,* 345-350.

Dixon, S.J., Collop, A.C., & Batt, M.E. (2000). Surface effects on ground reaction forces and lower extremity kinematics in running. *Medicine and Science in Sports and Exercise, 32,* 1919-1926.

D'Lima, D.D., Patil, S., Steklov, N., Slamin, J.E., & Colwell, C.W. Jr. (2005). The Chitranjan Ranawat Award: In vivo knee forces after total knee arthroplasty. *Clinical Orthopedics and Related Research, 440,* 45-49.

Dobkin, B., Apple, D., Barbeau, H., Basso, M., Behrman, A., Deforge, D., Ditunno, J., Dudley, G., Elashoff, R., Fugate, L., Harkema, S., Saulino, M., & Scott, M. (2006). Weight-supported treadmill vs. over-ground training for walking after acute incomplete SCI. *Neurology, 66,* 484-493.

Doherty, T.J. (2003). Aging and sarcopenia. *Journal of Applied Physiology, 95,* 1717-1727.

Donelan, J.M., & Kram, R. (2000). Exploring dynamic similarity in human running using simulated reduced gravity. *Journal of Experimental Biology, 203,* 2405-2415.

Donelan, J.M., Kram, R., & Kuo, A.D. (2002). Mechanical work for step-to-step transitions is a major determinant of the metabolic cost of human walking. *Journal of Experimental Biology, 205,* 3717-3727.

Donelan, J.M., & Pearson, K.G. (2004). Contribution of force feedback to ankle extensor activity in decerebrate walking cats. *Journal of Neurophysiology, 92,* 2093-2104.

Donelan, J.M., Shipman, D.W., Kram, R., & Kuo, A.D. (2004). Mechanical and metabolic requirements for active lateral stabilization in human walking. *Journal of Biomechanics, 37,* 827-835.

Donkers, M.J., An, K.N., Chao, E.Y., & Morrey, B.F. (1993). Hand position affects elbow joint load during push-up exercise. *Journal of Biomechanics, 26,* 625-632.

Dooley, P.C., Bach, T.M., & Luff, A.R. (1990). Effect of vertical jumping on the medial gastrocnemius and soleus muscles of rats. *Journal of Applied Physiology, 69,* 2004-2011.

Doorenbosch, C.A., Joosten, A., & Haarlaar, J. (2005). Calibration of EMG to force for knee muscles is applicable with submaximal voluntary contractions. *Journal of Electromyography and Kinesiology, 15,* 429-435.

Dörge, H.C., Andersen, T.B., Sørensen, H., & Simonsen, E.B. (2002). Biomechanical differences in soccer kicking with the preferred and the non-preferred leg. *Journal of Sports Science, 20,* 293-299.

Dörge, H.C., Andersen, T.B., Sørensen, H., Simonsen, E.B., Dyhre-Poulsen, P., & Klausen, K. (1999). EMG activity of the iliopsoas muscle and leg kinetics during the soccer place kick. *Scandinavian Journal of Sports Medicine, 9,* 195-200.

Dounskaia, N. (2005). The internal model and the leading joint hypothesis: Implications for the control of multi-joint movements. *Experimental Brain Research, 166,* 1-16.

Drew, T., Andujar, J.-E., Lajoie, K., & Yakovenko, S.(2008). Cortical mechanisms involved in visuomotor coordination during precision walking. *Brain Research Reviews, 57,* 199-211.

Drew, T., Prentice, S., & Schepens, B. (2004). Cortical and brainstem control of locomotion. *Progress in Brain Research, 143,* 251-261.

Dreyer, H.C., Blanco, C.E., Sattler, F.R., Schroeder, E.T., & Wiswell, R.A. (2006). Satellite cell numbers in young and older men 24 hours after eccentric exercise. *Muscle & Nerve, 33,* 242-253.

Drust, B., Rasmussen, P., Mohr, M., Nielsen, B., & Nybo, L. (2005). Elevations in core and muscle temperature impairs repeated spring performance. *Acta Physiologica Scandinavica, 183,* 181-190.

Duchateau, J., Balestra, C., Carpentier, A., & Hainaut, K. (2002). Reflex regulation during sustained and intermittent submaximal contractions in humans. *Journal of Physiology, 541,* 959-967.

Duchateau, J., & Hainaut, K. (1984). Isometric or dynamic training: Differential effects on mechanical properties of a human muscle. *Journal of Applied Physiology, 56,* 296-301.

Duchateau, J., & Hainaut, K. (1986). Nonlinear summation of contractions in striated muscle. I. Twitch potentiation in human muscle. *Journal of Muscle Research and Cell Motility, 7,* 11-17.

Duchateau, J., & Hainaut, K. (1987). Electrical and mechanical changes in immbolized human muscle. *Journal of Applied Physiology, 62,* 2168-2173.

Duchateau, J., & Hainaut, K. (1988). Training effects of submaximal electrostimulation in a hand muscle. *Medicine and Science in Sports and Exercise, 20,* 99-104.

Duchateau, J., & Hainaut, K. (1990). Effects of immobilization on contractile properties, recruitment, and firing rates of human motor units. *Journal of Physiology, 422,* 55-65.

Duchateau, J., & Hainaut, K. (1991). Effects of immobilization on electromyogram power spectrum changes during fatigue. *European Journal of Applied Physiology, 63,* 458-462.

Duchateau, J., & Hainaut, K. (1992). Neuromuscular electrical stimulation and voluntary exercise. *Sports Medicine, 14,* 100-113.

Duchateau, J., & Hainaut, K. (1993). Behaviour of short and long latency reflexes in fatigued human muscles. *Journal of Physiology, 471,* 787-799.

Duchateau, J., Semmler, J.G., & Enoka, R.M. (2006). Training adaptations in the behavior of human motor units. *Journal of Applied Physiology, 101,* 1766-1775.

Duclos, C., Roll, R., Kavounoudias, A., & Roll, J.P. (2007). Cerebral correlates of the "Kohnstamm phenomenon": An fMRI study. *NeuroImage, 34,* 774-783.

Duda, G.N., Schneider, E., & Chao, E.Y.S. (1997). Internal forces and moments in the femur during walking. *Journal of Biomechanics, 30,* 933-941.

Dudley, G.A., Duvoisin, M.R., Adams, G.R., Meyer, R.A., Belew, A.H., & Buchanan, P. (1992). Adaptations to unilateral lower limb suspension in humans. *Aviation, Space, and Environmental Medicine, 63,* 678-683.

Dudley, G.A., Tesch, P.A., Miller, B.J., & Buchanan, P. (1991). Importance of eccentric muscle actions in performance adaptations to resistance training. *Aviation, Space, and Environmental Medicine, 62,* 543-550.

Dufour, S.P., Lampert, E., Doutreleau, S., Lonsdorfer-Wolf, E., Billat, V.L., Piquard, F., & Richard, R. (2004). Eccentric cycle exercise: Training application of specific circulatory adjustments. *Medicine and Science in Sports and Exercise, 36,* 1900-1906.

Duhamel, J.R., Colby, C.L., & Goldberg, M.E. (1992). The updating of the representation of visual space in parietal cortex by intended eye movements. *Science, 255,* 90-92.

Dum, R.P., & Kennedy, T.T. (1980). Physiological and histochemical characteristics of motor units in cat tibialis anterior and extensor digitorum longus muscles. *Journal of Neurophysiology, 43,* 1615-1630.

Dumas, R., Cheze, L., & Verriest, J.P. (2007). Adjustments to McConville et al. and Young et al. body segment inertial parameters. *Journal of Biomechanics, 40,* 543-553.

Dumont, C.E., Popovic, M.R., Keller, T., & Sheikh, R. (2006). Dynamic force-sharing in multi-digit task. *Clinical Biomechanics, 21,* 138-146.

Duncan, A., & McDonagh, M.J.N. (2000). Stretch reflex distinguished from pre-programmed muscle activations following landing impacts in man. *Journal of Physiology, 526,* 456-468.

Duncan, P.W., Chandler, J.M., Cavanaugh, D.K., Johnson, K.R., & Buehler, A.G. (1989). Mode and speed specificity of eccentric and concentric exercise training. *Journal of Orthopedic and Sports Physical Therapy, 11,* 70-75.

Durkin, J.L., & Dowling, J.J. (2003). Analysis of body segment parameter differences between four human populations and the estimation errors of four popular mathematical models. *Journal of Biomechanical Engineering, 125,* 515-522.

Duysens, J., Clarac, F., & Cruse, H. (2000). Load-regulating mechanisms in gait and posture. *Physiological Reviews, 80,* 83-133.

Dvir, Z. (2004). Isokinetic strength testing: Devices and protocols. In S. Kumar (Ed.), *Muscle Strength* (pp. 157-176). Boca Raton, FL: CR Press LLC.

Dyhre-Poulsen, P., & Krogsgaard, M.R. (2000). Muscular reflexes elicited by electrical stimulation of the anterior cruciate ligament in humans. *Journal of Applied Physiology, 89,* 2191-2195.

Earles, D., Vardaxis, V., & Koceja, D. (2001). Regulation of motor output between young and old adults. *Clinical Neurophysiology, 112,* 1273-1279.

Ebenbichler, G.R., Kollmitzer, J., Glöckler, L., Bochdansky, T., Kopf, A., & Fialka, V. (1998). The role of the biarticular agonist and cocontracting antagonist pair in isometric muscle fatigue. *Muscle & Nerve, 21,* 1706-1713.

Eccles, J.C., Eccles, R.M., Iggo, A., & Lundberg, A. (1961). Electrophysiological investigations on Renshaw cells. *Journal of Physiology, 159,* 461-478.

Eccles, J.C., Eccles, R.M., & Lundberg, A. (1957). The convergence of monosynaptic excitatory afferents onto many different species of alpha motoneurones. *Journal of Physiology, 137,* 22-50.

Eccles, J.C., & Lundberg, A. (1958). Integrative patterns of Ia synaptic actions on motoneurones of hip and knee muscles. *Journal of Physiology, 144,* 271-298.

Edgerton, V.R., Apor, P., & Roy, R.R. (1990). Specific tension of human elbow flexor muscles. *Acta Physiologica Hungarica, 75,* 205-216.

Edgerton, V.R., Courtine, G., Gerasimenko, Y.P, Lavrov, I., Ichiyama, R.M., Fong, A.J., Cai, L.L., Otoshi, C.K., Tillakaratne, N.J.K., Burdick, J.W., & Roy, R.R. (2008). Training locomotor networks. *Brain Research Reviews, 57,* 241-254.

Edgerton, V.R., de Leon, R.D., Harkema, S.J., Hodgson, J.A., London, N., Reinkensmeyer, D.J., Roy, R.R., Talmadge, R.J., Tillakaratne, N.J., Timoszyk, W., & Tobin, A. (2001*a*). Retraining the injured spinal cord. *Journal of Physiology, 533,* 15-22.

Edgerton, V.R., Kim, S.J., Ichiyama, R.M., Gerasimenko, Y.P., & Roy, R.R. (2006). Rehabilitative therapies after spinal cord injury. *Journal of Neurotrama, 23,* 560-570.

Edgerton, V.R., McCall, G.E., Hodgson, J.A., Gotto, J., Goulet, C., Fleischmann, K., & Roy, R.R. (2001*b*). Sensorimotor adaptations to microgravity in humans. *Journal of Experimental Biology, 204,* 3217-3224.

Edgerton, V.R., & Roy, R.R. (2006). The nervous system and movement. In C.M. Tipton (Ed.), *ACSM's Advanced Exercise Physiology* (pp. 41-94). Philadelphia: Lippincott Williams & Wilkins.

Edgerton, V.R., Tillakaratne, N., Bigbee, A., de Leon, R., & Roy, R.R. (2004). Plasticity of spinal circuitry after injury. *Annual Reviews of Neuroscience, 27,* 145-167.

Edgley, S.A., Eyre, J.A., Lemon, R.N., & Miller, S. (1997). Comparison of activation of corticospinal neurons and spinal motor neurons by magnetic and electrical transcranial stimulation in the lumbosacral cord of the anesthetized monkey. *Brain, 120,* 839-853.

Edin, B.B. (2001). Cutaneous afferents provide information about knee joint movements in humans. *Journal of Physiology, 531,* 289-297.

Edman, K.A.P. (1988). Double-hyperbolic force-velocity relation in frog muscle fibres. *Journal of Physiology, 404,* 301-321.

Edman, K.A.P. (1992). Contractile performance of skeletal muscle fibres. In P.V. Komi (Ed.), *Strength and Power in Sport* (pp. 96-114). Champaign, IL: Human Kinetics.

Edman, K.A.P. (1995). Myofibrillar fatigue versus failure of activation. In S.C. Gandevia, R.M. Enoka, A.J. McComas, D.G. Stuart, & C.K. Thomas (Eds.), *Fatigue: Neural and Muscular Mechanisms* (pp. 29-43). New York: Plenum Press.

Edman, K.A.P. (1996). Fatigue vs. shortening-induced deactivation in striated muscle. *Acta Physiologica Scandinavica, 156,* 183-192.

Edman, K.A.P., Elzinga, G., & Noble, M.I.M. (1978). Enhancement of mechanical performance by stretch during tetanic contractions of vertebrate skeletal muscle fibres. *Journal of Physiology, 281,* 139-155.

Edman, K.A.P., & Lou, F. (1990). Changes in force and stiffness induced by fatigue and intracellular acidification in frog muscle fibres. *Journal of Physiology, 424,* 133-149.

Edman, K.A.P., Månsson, A., & Caputo, C. (1997). The biphasic force-velocity relationship in frog muscle fibres and its evaluation in terms of cross-bridge function. *Journal of Physiology, 503,* 141-156.

Edman, K.A.P., Reggiani, C.,. Schiaffino, S., & te Kronnie, G. (1988). Maximum velocity of shortening related to myosin isoform composition in frog skeletal muscle fibres. *Journal of Physiology, 95,* 679-694.

Edman, K.A.P., & Tsuchiya, T. (1996). Strain of passive elements during force enhancement by stretch in frog muscle fibres. *Journal of Physiology, 490,* 191-205.

Edwards, A.G., & Byrnes, W.C. (2007). Aerodynamic characteristics as determinants of the drafting effect in cycling. *Medicine and Science in Sports and Exercise, 39,* 170-176.

Edwards, R.H.T., Hill, D.K., Jones, D.A., & Merton, P.A. (1977). Fatigue of long duration in human skeletal muscle after exercise. *Journal of Physiology, 272,* 769-778.

Ehrsson, H.H., Fagergren, A., Johansson, R.S., & Forssberg, H. (2003). Evidence for the involvement of the posterior parietal cortex in coordination of fingertip forces for grasp stability in manipulation. *Journal of Neurophysiology, 90,* 2978-2986.

Eidelberg, E., Story, J.L., Meyer, B.L., & Nystel, J. (1980). Stepping by chronic spinal cats. *Experimental Brain Research, 40,* 241-246.

Eie, N., & Wehn, P. (1962). Measurements of the intra-abdominal pressure in relation to weight bearing of the lumbosacral spine. *Journal of Oslo City Hospitals, 12,* 205-217.

Eisen, A., Entezari-Taher, M., & Stewart, H. (1996). Cortical projections to spinal motoneurons: Changes with aging and amyotrophic lateral sclerosis. *Neurology, 46,* 1396-1404.

Eklund, G., & Hagbarth, K.E. (1966). Normal variability of tonic vibration reflexes in man. *Experimental Neurology, 16,* 80-92.

Elbert, T., & Rockstroh, B. (2004). Reorganization of human cerebral cortex: The range of changes following use and injury. *Neuroscientist, 10,* 129-141.

Elftman, H. (1938). The measurement of the external force in walking. *Science, 88,* 152-153.

Elftman, H. (1939). Forces and energy changes in the leg during walking. *American Journal of Physiology, 125,* 339-356.

Ellaway, P.H., Davey, N.J., Ferrell, W.R., & Baxendale, R.H. (1996). The action of knee joint afferents and the concomitant influence of cutaneous (sural) afferents on the discharge of triceps surae γ-motoneurones in the cat. *Experimental Physiology, 81,* 45-66.

Ellaway, P.H., Rawlinson, S.R., Lewis, H.S., Davey, N.J., & Maskill, D.W. (1997). Magnetic stimulation excites skeletal muscle via motor nerve axons in the cat. *Muscle & Nerve, 20,* 1108-1114.

Ellerby, D.J., Henry, H.T., Carr, J.A., Buchanan, C.I., & Marsh, R.L. (2005). Blood flow in guinea fowl Numida meleagris as an indicator of energy expenditure by individual muscles during walking and running. *Journal of Physiology, 564,* 631-648.

Elliott, B.C. (2000). Hitting and kicking. In V.M. Zatsiorsky (Ed.), *Biomechanics in Sport* (pp. 487-504). Oxford, UK: Blackwell Science.

Elliott, B.C., Alderson, J.A., & Denver, E.R. (2007). System and modeling errors in motion analysis: Implications for the measurement of the elbow angle in cricket bowling. *Journal of Biomechanics, 40,* 2679-2685.

Elrick, D.B., & Charlton, M.P. (1999). Alpha-latrocrustatoxin increases neurotransmitter release by activating a calcium influx pathway at crayfish neurotransmitter junction. *Journal of Neurophysiology, 82,* 3550-3562.

Emonet-Denand, F., Hunt, C.C., & Laporte, Y. (1985). Fusimotor after-effects on responses of primary endings to test dynamic stimuli in cat muscle spindles. *Journal of Physiology, 360,* 187-200.

English, A.W. (1984). An electromyographic analysis of compartments in cat lateral gastrocnemius during unrestrained locomotion. *Journal of Neurophysiology, 52,* 114-125.

English, A.W., & Ledbetter, W.D. (1982). Anatomy and innervation patterns of cat lateral gastrocnemius and plantaris muscles. *American Journal of Anatomy, 164,* 67-77.

Engstrom, C.M., Loeb, G.E., Reid, J.G., Forrest, W.J., & Avruch, L. (1991). Morphometry of the human thigh muscles. A comparison between anatomical sections and computer tomographic and magnetic resonance images. *Journal of Anatomy, 176,* 139-156.

Enocson, A.G., Berg, H.E., Vargas, R., Jenner, G., & Tesch, P.A. (2005). Signal intensity of MR-images of thigh muscles following acute open- and closed chain kinetic knee extensor exercise—index of muscle use. *European Journal of Applied Physiology, 94,* 357-363.

Enoka, R.M. (1979). The pull in Olympic weightlifting. *Medicine and Science in Sports, 11,* 131-137.

Enoka, R.M. (1983). Muscular control of a learned movement: The speed control system hypothesis. *Experimental Brain Research, 51,* 135-145.

Enoka, R.M. (1988). Load- and skill-related changes in segmental contributions to a weightlifting movement. *Medicine and Science in Sports and Exercise, 20,* 178-187.

Enoka, R.M. (2006). Neuromuscular electrical stimulation: What is activated? In A. Rainoldi, M.A. Minetto, & R. Merletti (Eds.), *Biomedical Engineering in Exercise and Sports* (pp. 181-186). Torino, Italy: Edizioni Minerva Medica.

Enoka, R.M., Christou, E.A., Hunter, S.K., Kornatz, K.W., Semmler, J.G., Taylor, A.M., & Tracy, B.L. (2003). Mechanisms that contribute to differences in motor performance between young and old adults. *Journal of Electromyography and Kinesiology, 13,* 1-12.

Enoka, R.M., & Duchateau, J. (2008). Muscle fatigue: What, why, and how it influences muscle function. *Journal of Physiology, 586,* 11-23.

Enoka, R.M., & Fuglevand, A.J. (2001). Motor unit physiology: Some unresolved issues. *Muscle & Nerve, 24,* 4-17.

Enoka, R.M., Hutton, R.S., & Eldred, E. (1980). Changes in excitability of tendon tap and Hoffmann reflexes following voluntary contractions. *Electroencephalography and Clinical Neurophysiology, 48,* 664-672.

Enoka, R.M., Miller, D.I., & Burgess, E.M. (1982). Below-knee amputee running gait. *American Journal of Physical Medicine, 61,* 66-84.

Enoka, R.M., Robinson, G.A., & Kossev, A.R. (1989). Task and fatigue effects on low-threshold motor units in human hand muscle. *Journal of Neurophysiology, 62,* 1344-1359.

Enoka, R.M., & Stuart, D.G. (1992). Neurobiology of muscle fatigue. *Journal of Applied Physiology, 72,* 1631-1648.

Enoka, R.M., Trayanova, N., Laouris, Y., Bevan, L., Reinking, R.M., & Stuart, D.G. (1992). Fatigue-related changes in motor unit action potentials of adult cats. *Muscle & Nerve, 14,* 138-150.

Eriksson, A., Kadi, F., Malm, C., & Thornell, L.E. (2005). Skeletal muscle morphology in power-lifters with and without anabolic steroids. *Histochemistry and Cell Biology, 124,* 167-175.

Erim, Z., Beg, M.F., Burke, D.T., & De Luca, C.J. (1999). Effects of aging on motor-unit control properties. *Journal of Neurophysiology, 82,* 2081-2091.

Escamilla, R.F., Fleisig, G.S., Barrentine, S.W., Zheng, N., & Andrews, J.E. (1998*a*). Kinematic comparisons of throwing different types of baseball pitches. *Journal of Applied Biomechanics, 14,* 1-23.

Escamilla, R.F., Fleisig, G.S., Zheng, N., Barrentine, S.W., Wilk, K.E., & Andrews, J.R. (1998*b*). Biomechanics of the knee during closed kinetic chain and open kinetic chain exercises. *Medicine and Science in Sports and Exercise, 30,* 556-569.

Essendrop, M., Andersen, T.B., & Schibye, B. (2002). Increase in spinal stability obtained at levels of intra-abdominal pressure and back muscle activity realistic to work situations. *Applied Ergonomics, 33,* 471-476.

Esser, K.A., & Hardeman, E.C. (1995). Changes in contractile protein mRNA accumulation in response to spaceflight. *American Journal of Physiology, 268,* C466-C471.

Essig, D.A., & Nosek, T.M. (1997). Muscle fatigue and induction of stress protein genes: A dual function of reactive oxygen species? *Canadian Journal of Applied Physiology, 22,* 409-428.

Etnyre, B.R., & Abraham, L.D. (1986). Gains in range of ankle dorsiflexion using three popular stretching techniques. *American Journal of Physical Medicine, 65,* 189-196.

Etnyre, B.R., & Lee, E.J. (1988). Chronic and acute flexibility of men and women using three different stretching techniques. *Research Quarterly, 59,* 222-228.

Ettema, G.J.C., Styles, G., & Kippers, V. (1998). The moment arms of 23 muscle segments of the upper limb with varying elbow and forearm positions: Implications for motor control. *Human Movement Science, 17,* 201-220.

Evans, W.J. (1995). What is sarcopenia? *Journal of Gerontology A, 50,* Spec. No., 5-8.

Ewing, J.L., Wolfe, D.R., Rogers, M.A., Amundson, M.L., & Stull, G.A. (1990). Effects of velocity of isokinetic training on strength, power, and quadriceps muscle fibre characteristics. *European Journal of Applied Physiology, 61,* 159-162.

Faaborg-Andersen, K. (1957). Electromyographic investigation of intrinsic laryngeal muscles in humans. *Acta Physiologica Scandinavica, 41* (Suppl. 140), 1-149.

Faber, M.J., Bosscher, R.J., Chin A Paw, M.J., & van Wieringen, P.C. (2006). Effects of exercise programs on falls and mobility in frail and pre-frail older adults: A multicenter randomized controlled trial. *Archives of Physical Medicine and Rehabilitation, 87,* 885-896.

Fahim, M.A. (1992). An analysis of miniature synaptic events recorded from locust muscle—variations in amplitude and time course. *Cellular and Molecular Biology, 38,* 231-241.

Faist, M., Dietz, V., & Pierrot-Deseilligny, E. (1996). Modulation, probably presynaptic in origin, of monosynaptic Ia excitation during human gait. *Experimental Brain Research, 109,* 441-449.

Fallon, J.B., Bent, L.R., McNulty, P.A., & Macefield, V.G. (2005). Evidence for strong synaptic coupling between single tactile afferents from the sole of the foot and motoneurons supplying leg muscles. *Journal of Neurophysiology, 94,* 3795-3804.

Fang, Z.-P., & Mortimer, J.T. (1991). Selective activation of small motor axons by quasitrapezoidal current pulses. *IEEE Transactions on Biomedical Engineering, 38,* 168-174.

Farina, D. (2006). Interpretation of the surface electromyogram in dynamic contractions. *Exercise and Sport Sciences Reviews, 35,* 121-127.

Farina, D., Arendt-Nielsen, L., & Graven-Nielsen, T. (2005). Effect of temperature on spike-triggered average torque and electrophysiological properties of low-threshold motor units. *Journal of Applied Physiology, 99,* 197-203.

Farina, D., Arendt-Nielsen, L., Merletti, R., & Graven-Nielsen, T. (2002*a*). Assessment of single motor unit conduction velocity during sustained contractions of the tibialis anterior muscle with advanced spike triggered averaging. *Journal of Neuroscience Methods, 115,* 1-12.

Farina, D., Fattorini, L., Felici, F., & Filligoi, G. (2002*b*). Nonlinear surface EMG analysis to detect changes of motor unit conduction velocity and synchronization. *Journal of Applied Physiology, 93,* 1753-1763.

Farina, D., Fortunato, E., & Merletti, R. (2000). Noninvasive estimation of motor unit conduction velocity distribution using linear electrode arrays. *IEEE Transactions on Biomedical Engineering, 47,* 380-388.

Farina, D., Gazzoni, M., & Merletti, R. (2003). Assessment of low back muscle fatigue by surface EMG signal analysis: Methodological aspects. *Journal of Electromyography and Kinesiology, 13,* 319-332.

Farina, D., Madeleine, P., Graven-Nielsen, T., Merletti, R., & Arendt-Nielsen, L. (2002*c*). Standardising surface electromyogram recordings for assessment of activity and fatigue in the upper human trapezius muscle. *European Journal of Applied Physiology, 86,* 469-478.

Farina, D., Mertilli, R., Indino, B., Nazzaro, M., & Pozzo, M. (2002*d*). Surface EMG crosstalk between knee extensor muscles: Experimental and model results. *Muscle & Nerve, 26,* 681-695.

Farina, D., Merletti, R., & Enoka, R.M. (2004*a*). The extraction of neural strategies from the surface EMG. *Journal of Applied Physiology, 96,* 1486-1495.

Farina, D., Merletti, R., Indino, B., & Graven-Nielsen, T. (2004*b*). Surface EMG crosstalk evaluated from experimental recordings and simulated signals. *Methods of Information in Medicine, 43,* 30-35.

Farley, C.T., & Ferris, D.P. (1998). Biomechanics of walking and running: Center of mass movements to muscle action. *Exercise and Sport Sciences Reviews, 26,* 253-285.

Farley, C.T., & Gonzalez, O. (1996). Leg stiffness and stride frequency in human running. *Journal of Biomechanics, 29,* 181-186.

Farley, C.T., Houdijk, H.H.P., van Strien, C., & Louie, M. (1998). Mechanism of leg stiffness adjustment for hopping on surfaces of different stiffnesses. *Journal of Applied Physiology, 85,* 1044-1055.

Farley, C.T., & McMahon, T.A. (1992). Energetics of walking and running: Insights from simulated reduced-gravity experiments. *Journal of Applied Physiology*, *73*, 2709-2712.

Farmer, S.F. (1998). Rhythmicity, synchronization, and binding in human and primate motor systems. *Journal of Physiology, 509,* 3-14.

Farmer, S.F., Bremner, F.D., Halliday, D.M., Rosenberg, J.R., & Stephens, J.A. (1993*a*). The frequency content of common synaptic inputs to motoneurones studied during voluntary isometric contraction in man. *Journal of Physiology, 470,* 127-155.

Farmer, S.F., Sheean, G.L., Mayson, M.J., Rothwell, J.C., Marsden, C.D., Conway, B.A., Halliday, D.M., Rosenberg, J.R., & Stephens, J.A. (1998). Abnormal motor unit synchronization of antagonist muscles underlies pathological co-contraction in upper limb dystonia. *Brain, 121,* 801-814.

Farmer, S.F., Swash, M., Ingram, D.A., & Stephens, J.A. (1993*b*). Changes in motor unit synchronization following central nervous system lesions in man. *Journal of Physiology, 463,* 83-105.

Farthing, J.P., & Chilibeck, P.D. (2003). The effects of eccentric and concentric training at different velocities of muscle hypertrophy. *European Journal of Applied Physiology, 89,* 578-586.

Fattorini, L., Felici, F., Filligoi, G.C., Traballesi, M., & Farina, D. (2005). Influence of high motor unit synchronization levels on non-linear and spectral variables of the surface EMG. *Journal of Neuroscience Methods, 143,* 133-139.

Faumont, S., Combes, D., Meyrand, P., & Simmers, J. (2005). Reconfiguration of multiple motor networks by short- and long-term actions of an identified modulatory neuron. *European Journal of Neuroscience, 22,* 2489-2502.

Feiereisen, P., Duchateau, J., & Hainaut, K. (1997). Motor unit recruitment order during voluntary and electrically induced contractions in the tibialis anterior. *Experimental Brain Research, 114,* 117-123.

Feinstein, B., Lindegård, B., Nyman, E., & Wohlfart, G. (1955). Morphologic studies of motor units in normal human muscles. *Acta Anatomica, 23,* 127-142.

Feldman, J.L., Mitchell, G.S., & Nattie, E.E. (2003). Breathing: Rhythmicity, plasticity, chemosensitivity. *Annual Reviews of Neuroscience, 26,* 239-266.

Felipo, V., & Butterworth, R.F. (2002). Neurobiology of ammonia. *Progress in Neurobiology, 67,* 259-279.

Fellows, S.J., Kaus, C., Ross, H.F., & Thilmann, A.F. (1994). Agonist and antagonist EMG activation during isometric torque development at the elbow in spastic hemiparesis. *Electroencephalography and Clinical Neurophysiology, 93,* 106-112.

Fellows, S.J., & Rack, P.M.H. (1987). Changes in the length of the human biceps brachii muscle during elbow movements. *Journal of Physiology, 383,* 405-412.

Feltner, M.E. (1989). Three-dimensional interactions in a two-segment kinetic chain. Part II: Application to the throwing arm in baseball pitching. *International Journal of Sport Biomechanics, 5,* 420-450.

Feltner, M., & Dapena, J. (1986). Dynamics of the shoulder and elbow joints of the throwing arm during a baseball pitch. *International Journal of Sport Biomechanics, 2,* 235-259.

Feltner, M.E., & Dapena, J. (1989). Three-dimensional interactions in a two-segment kinetic chain. Part I: General model. *International Journal of Sport Biomechanics, 5,* 403-419.

Feltner, M.E., & Taylor, G. (1997). Three-dimensional kinetics of the shoulder, elbow, and wrist during a penalty throw in water polo. *Journal of Applied Biomechanics, 13,* 347-372.

Fenelon, V., Le Feuvre, Y., Bem, T., & Meyrand, P. (2003). Maturation of rhythmic neural network: Role of central modulatory inputs. *Journal of Physiology (Paris), 97,* 59-68.

Fenn, W.O. (1924). The relation between the work performed and the energy liberated in muscular contraction. *Journal of Physiology, 58,* 373-395.

Fenn, W.O. (1930). Work against gravity and work due to velocity changes in running. *American Journal of Physiology, 93,* 433-462.

Feraboli-Lohnherr, D., Barthe, J.Y., & Orsal, D. (1999). Serotonin-induced activation of the network for locomotion in adult spinal rats. *Journal of Neuroscience Research, 55,* 87-98.

Ferber, R., Gravelle, D.C., & Osternig, L.R. (2002). Effect of proprioceptive neuromuscular facilitation stretch techniques on trained and untrained older adults. *Journal of Aging and Physical Activity, 10,* 132-142.

Ferri, A., Scaglioni, G., Pousson, M., Capodaglio, P., Van Hoecke, J., & Narici, M.V. (2003). Strength and power changes of the human plantar flexors and knee extensors in response to resistance training in old age. *Acta Physiologica Scandinavica, 177,* 69-78.

Ferris, D.P., Aagaard, P., Simonsen, E.B., Farley, C.T., & Dyrhe-Poulsen, P. (2001). Soleus H-reflex gain in humans walking and running under simulated reduced gravity. *Journal of Physiology, 530,* 167-180.

Ferris, D.P., Bohra, Z.A., Lukos, J.R., & Kinnaird, C.R. (2006). Neuromechanical adaptation to hopping with an elastic ankle-foot orthosis. *Journal of Applied Physiology, 100,* 163-170.

Ferris, D.P., Gordon, K.E., Beres-Jones, J.A., & Harkema, S.J. (2004). Muscle activation during unilateral stepping occurs in the nonstepping limb of humans with clinically complete spinal cord. *Spinal Cord, 42,* 14-23.

Ferris, D.P., Liang, K., & Farley, C.T. (1999). Runners adjust leg stiffness for their first step on a new running surface. *Journal of Biomechanics, 32,* 787-794.

Ferris, D.P., Louie, M., & Farley, C.T. (1998). Running in the real world: Adjusting leg stiffness for different surfaces. *Proceedings of the Royal Society of London B, 265,* 989-994.

Fiatarone, M.A., Marks, E.C., Ryan, N.D., Meredith, C.N., Lipsitz, L.A., & Evans, W.J. (1990). High-intensity strength training in nonagenarians. *Journal of the American Medical Association, 263,* 3029-3034.

Fick, R. (1904). *Handbuch der anatomie des menschen* (Vol. 2). Stuttgart: Gustav Fischer Verlag.

Field-Fote, E.C. (2001). Combined use of body weight support, functional electrical stimulation, and treadmill training to improve walking ability in individuals with chronic incomplete spinal cord injury. *Archives of Physical Medicine and Rehabilitation, 82,* 818-824.

Field-Fote, E.C., Lindley, S.D., & Sherman, A.L. (2005). Locomotor training approaches for individuals with spinal cord injury: A preliminary report of walking-related outcomes. *Journal of Neurological Physical Therapy, 29,* 127-137.

Field-Fote, E.C., & Tepavac, D. (2002). Improved intralimb coordination in people with incomplete spinal cord injury following training with body weight support and electrical stimulation. *Physical Therapy, 82,* 707-715.

Finni, T., Ikegawa, S., Lepola, V., & Komi, P.V. (2003). Comparison of force-velocity relationships of vastus lateralis muscle in isokinetic and in stretch-shortening cycle exercises. *Acta Physiologica Scandinavica, 177,* 483-491.

Finni, T., Komi, P.V., & Lepola, V. (2000). Mechanical output of the triceps surae and quadriceps femoris muscles during jumping in humans. *European Journal of Applied Physiology, 83,* 416-426.

Finni, T., Komi, P.V., & Lukkariniemi, J. (1998). Achilles tendon loading during walking: Application of a novel optic fiber technique. *European Journal of Applied Physiology, 77,* 289-291.

Fischer-Rasmussen, T., Krogsgaard, M., Jensen, D.B., & Dyhre-Poulsen, P. (2001). Inhibition of dynamic thigh muscle contraction by electrical stimulation of the posterior cruciate ligament in humans. *Muscle & Nerve, 24,* 1482-1488.

Fisher, M.J., Meyer, R.A., Adams, G.R., Foley, J.M., & Potchen, E.J. (1990). Direct relationship between proton T2 and exercise intensity in skeletal muscle MR images. *Investigative Radiology, 25,* 480-485.

Fisher, W.J., & White, M.J. (1999). Training-induced adaptations in the central command and peripheral reflex components of the pressor response to isometric exercise of the human triceps surae. *Journal of Physiology, 520,* 621-628.

Fitts, R.H. (2006). The muscular system: Fatigue processes. In C.M. Tipton (Ed.), *ACSM's Advanced Exercise Physiology* (pp. 178-196). Philadelphia: Lippincott Williams & Wilkins.

Fitts, R.H., Bodine, S.C., Romatowski, J.G., & Widrick, J.J. (1998). Velocity, force, power, and Ca^{2+} sensitivity of fast and slow monkey skeletal muscle fibers. *Journal of Applied Physiology, 84,* 1776-1787.

Fitts, R.H., Brimmer, C.J., Heywood-Cooksey, A., & Timmerman, R.J. (1989). Single muscle fiber enzyme shifts with hindlimb suspension and immobilization. *American Journal of Physiology, 256,* C1082-C1091.

Fitts, R.H., & Metzger, J.M. (1993). Mechanisms of muscular fatigue. In J.R. Poortmans (Ed.), *Principles of Exercise Biochemistry* (pp. 248-268). Basel: Karger.

Fitts, R.H., Metzger, J.M., Riley, D.A., & Unsworth, B.R. (1986). Models of disuse: A comparison of hindlimb suspension and immobilization. *Journal of Applied Physiology, 60,* 1946-1953.

Fitts, R.H., Riley, D.A., & Widrick, J.J. (2000). Microgravity and skeletal muscle. *Journal of Applied Physiology, 89,* 823-839.

Fitzgerald, G.K., & Delitto, A. (2006). Neuromuscular electrical stimulation for muscle strength training. In A. Rainoldi, M.A. Minetto, & R. Merletti (Eds.), *Biomedical Engineering in Exercise and Sports* (pp. 199-207). Torino, Italy: Edizioni Minerva Medica.

Fitzgerald, G.K., Piva, S.R., & Irrgang, J.J. (2003). A modified neuromuscular electrical stimulation protocol for quadriceps strength training following anterior cruciate ligament reconstruction. *Journal of Orthopedic Sports Physical Therapy, 33,* 492-501.

Fitzpatrick, R.C., & Day, B.L. (2004). Probing the human vestibular system with galvanic stimulation. *Journal of Applied Physiology, 96,* 2301-2316.

Fitzpatrick, R.C., Burke, D., & Gandevia, S.C. (1994). Task-dependent reflex responses and movement illusions evoked by galvanic vestibular stimulation in standing humans. *Journal of Physiology, 478,* 363-372.

Fitzpatrick, R.C., Taylor, J.L., & McCloskey, D.I. (1996). Effects of arterial perfusion pressure on force production in working hand muscles. *Journal of Physiology, 495,* 885-891.

Flanagan, J.R., Bowman, M.C., & Johansson, R.S. (2006). Control strategies in object manipulation tasks. *Current Opinion in Neurobiology, 16,* 650-659.

Flanagan, J.R., Burstedt, M.K., & Johansson, R.S. (1999). Control of fingertip forces in multidigit manipulation. *Journal of Neurophysiology, 81,* 1706-1717.

Flanagan, J.R., & Rao, A.K. (1995). Trajectory adaptation to a nonlinear visuomotor transformation: Evidence of motion planning in visually perceived space. *Journal of Neurophysiology, 74,* 2174-2178.

Flanagan, J.R., Vetter, P., Johansson, R.S., & Wolpert, D.M. (2003). Prediction precedes control in motor learning. *Current Biology, 13,* 146-150.

Fleckenstein, J.L., Canby, R.C., Parkey, R.W., & Peshock, R.M. (1988). Acute effects of exercise on MR imaging of skeletal muscle in normal volunteers. *American Journal of Roentgenology, 151,* 231-237.

Fleckenstein, J.L., & Shellock, F.G. (1991). Exertional muscle injuries: Magnetic resonance imaging evaluation. *Topics in Magnetic Resonance Imaging, 3,* 50-70.

Fleckenstein, J.L., Watumull, D., Bertocci, L.A., Parkey, R.W., & Peshock, R.M. (1992). Finger-specific flexor recruitment in humans: Depiction of exercise-enhanced MRI. *Journal of Applied Physiology, 72,* 1974-1977.

Fleisig, G.S., Kingsley, D.S., Loftice, J.W., Dinnen, K.P., Ranganathan, R., Dun, S., Escamilla, R.F., & Andrews, J.R. (2006). Kinetic comparison among the fastball, curveball, change-up, and slide in collegiate baseball players. *American Journal of Sports Medicine, 34,* 423-430.

Fleshman, J.W., Munson, J.B., & Sypert, G.W. (1981*a*). Homonymous projection of individual group Ia-fibers to physiologically characterized medial gastrocnemius motoneurons in the cat. *Journal of Neurophysiology, 46,* 1339-1348.

Fleshman, J.W., Munson, J.B., Sypert, G.W., & Friedman, W.A. (1981*b*). Rheobase, input resistance, and motor-unit type in medial gastrocnemius motoneurons in the cat. *Journal of Neurophysiology, 46,* 1326-1338.

Floeter, M.K., Gerloff, C., Kouri, J., & Hallett, M. (1998). Cutaneous withdrawal reflexes of the upper extremity. *Muscle & Nerve, 21,* 591-598.

Foehring, R.C., & Munson, J.B. (1990). Motoneuron and muscle-unit properties after long-term direct innervation of soleus muscle by medial gastrocnemius nerve in cat. *Journal of Neurophysiology, 64,* 847-861.

Foehring, R.C., Sypert, G.W., & Munson, J.B. (1987*a*). Motor unit properties following cross-reinnervation of cat lateral gastrocnemius and soleus muscles with medial gastrocnemius nerve. I. Influence of motoneurons on muscle. *Journal of Neurophysiology, 57,* 1210-1226.

Foehring, R.C., Sypert, G.W., & Munson, J.B. (1987*b*). Motor unit properties following cross-reinnervation of cat lateral gastrocnemius and soleus muscles with medial gastrocnemius nerve. II. Influence of muscle on motoneurons. *Journal of Neurophysiology, 57,* 1227-1245.

Folland, J.P., Hawker, K., Leach, B., Little, T., & Jones, D.A. (2005). Strength training: Isometric training at a range of joint angles versus dynamic training. *Journal of Sports Science, 23,* 817-824.

Folland, J.P., & Williams, A.G. (2007). The adaptations to strength training. Morphological and neurological contributions to increased strength. *Sports Medicine, 37,* 145-168.

Fong, A.J., Cai, L.L., Otoshi, C.K., Reinkensmeyer, D.J., Burdick, J.W., Roy, R.R., & Edgerton, V.R. (2005). Spinal cord-transected mice learn to step in response to quipazine treatment and robotic training. *Journal of Neuroscience, 25,* 11738-11747.

Forestier, N., & Nougier, V. (1998). The effects of muscular fatigue on the coordination of multijoint movement in human. *Neuroscience Letters, 252,* 187-190.

Forget, R., Hultborn, H., Meunier, S., Pantieri, R., & Pierrot-Deseilligny, E. (1989). Facilitation of quadriceps motoneurones by group I afferents from pretibial flexors in man. 2. Changes occurring during voluntary contraction. *Experimental Brain Research, 78,* 21-27.

Forner Cordero, A., Koopman, H.J., & van der Helm, F.C. (2004). Use of pressure insoles to calculate the complete ground reaction forces. *Journal of Biomechanics, 37,* 1427-1432.

Forner Cordero, A., Koopman, H.J., & van der Helm, F.C. (2006). Inverse dynamics calculations during gait with restricted ground reaction force information from pressure insoles. *Gait & Posture, 23,* 189-199.

Fornusek, C., & Davis, G.M. (2004). Maximizing muscle force via low-cadence functional stimulation cycling. *Journal of Rehabilitation Medicine, 36,* 232-237.

Forssberg, H. (1970). Stumbling corrective reaction: A phase-dependent compensatory reaction during locomotion. *Journal of Neurophysiology, 42,* 936-953.

Forssberg, H., Grillner, S., Halbertsma, J., & Rossignol, S. (1980). The locomotion of the low spinal cat. II. Interlimb coordination. *Acta Physiologica Scandinavica, 108,* 283-295.

Fothergill, D.M., Grieve, D.W., & Pinder, A.D.J. (1996). The influence of task resistance on the characteristics of maximal one- and two-handed lifting exertions in men and women. *European Journal of Applied Physiology, 72,* 430-439.

Fournier, E., Katz, R., & Pierrot-Deseilligny, E. (1983). Descending control of reflex pathways in the production of voluntary isolated movements in man. *Brain Research, 288,* 375-377.

Fournier, E., Meunier, S., Pierrot-Deseilligny, E., & Shindo, M. (1986). Evidence for interneuronally mediated Ia excitatory effects to human quadriceps motoneurones. *Journal of Physiology, 377,* 143-169.

Fournier, M., Roy, R.R., Perham, H., Simard, C.P., & Edgerton, V.R. (1983). Is limb immobilization a model of muscle disuse? *Experimental Neurology, 80,* 147-156.

Fransson, P.A., Kristinsdottir, E.K., Hafstrom, A., Magnusson, M., & Johansson, R. (2004). Balance control and adaptations during vibratory perturbations in middle-aged and elderly humans. *European Journal of Applied Physiology, 91,* 595-603.

Freed, W.J., de Medinaceli, L., & Wyatt, R.J. (1985). Promoting functional plasticity in the damaged nervous system. *Science, 227,* 1544-1552.

Freriks, B., Hermens, H., Disselhorst-Klug, C., & Rau, G. (1999). The recommendations for sensors and sensor placement procedures for surface electromyography. In H. Hermens, B. Freriks, R. Merletti, D. Stegeman, J. Blok, G. Rau, C. Disselhorst-Klug, & G. Hägg (Eds.), *European Recommendations for Surface Electromyography* (pp. 15-53). Enschede, The Netherlands; Roessingh Research and Development b.v.

Fridén, J., & Lieber, R.L. (1998). Segmental muscle fiber lesions after repetitive eccentric contractions. *Cell and Tissue Research, 293,* 165-171.

Fridén, J., & Lieber, R.L. (2001). Eccentric exercise-induced injuries to contractile and cytoskeletal muscle fibre components. *Acta Physiologica Scandinavica, 171,* 321-326.

Friederich, J.A., & Brand, R.A. (1990). Muscle fiber architecture in the human lower limb. *Journal of Biomechanics, 23,* 91-95.

Friedman, W.A., Sypert, G.W., Munson, J.B., & Fleshman, J.W. (1981). Recurrent inhibition in type-identified motoneurons. *Journal of Neurophysiology, 46,* 1349-1359.

Frohlich, C. (1980, March). The physics of somersaulting and twisting. *Scientific American,* pp. 154-164.

Frolkis, V.V., Tanin, S.A., Marcinko, V.I., Kulchitsky, O.K., & Yasechko, A.V. (1985). Axoplasmic transport of substances in motoneuronal axons of the spinal cord in old age. *Mechanisms of Ageing and Development, 29,* 19-28.

Frontera, W.R., Hughes, V.A., Fielding, R.A., Fiatarone, M.A., Evans, W.J., & Roubenoff, R. (2000*a*). Aging of skeletal muscle: A 12-yr longitudinal study. *Journal of Applied Physiology, 88,* 1321-1326.

Frontera, W.R., Hughes, V.A., Krivickas, L.S., Kim, S.K., Foldvari, M., & Roubenoff, R. (2003). Strength training in older women: Early and late changes in whole muscle and single cells. *Muscle & Nerve, 28,* 601-608.

Frontera, W.R., Suh, D., Krivickas, L.S., Hughes, V.A., Goldstein, R., & Roubenoff, R. (2000*b*). Skeletal muscle fiber quality in older men and women. *American Journal of Physiology, 279,* C611-C618.

Fuglevand, A.J., & Keen, D.A. (2003). Re-evaluation of muscle wisdom in the human adductor pollicis using physiological rates of stimulation. *Journal of Physiology, 549,* 865-875.

Fuglevand, A.J., Macefield, V.G., & Bigland-Ritchie, B. (1999). Force-frequency and fatigue properties of motor units in muscles that control digits of the human hand. *Journal of Neurophysiology, 81,* 1718-1729.

Fuglevand, A.J., Winter, D.A., & Patla, A.E. (1993). Models of recruitment and rate coding organization in motor-unit pools. *Journal of Neurophysiology, 70,* 2470-2488.

Fuglevand, A.J., Zackowski, K.M., Huey, K.A., & Enoka, R.M. (1993). Impairment of neuromuscular propagation during human fatiguing contractions at submaximal forces. *Journal of Physiology, 460,* 549-572.

Fuglsang-Frederiksen, A. (2000). The utility of interference pattern analysis. *Muscle & Nerve, 23,* 18-36.

Fujiwara, M., & Basmajian, J.V. (1975). Electromyographic study of two-joint muscles. *American Journal of Physical Medicine, 54,* 234-242.

Fukanaga, T., Kubo, K., Kawakami, Y., Fukashiro, S., Kanehisa, H., and Maganaris, C.N. (2001). In vivo behaviour of human muscle tendon during walking. *Proceedings of the Royal Society of London B, 268,* 229-233.

Fukunaga, T., Roy, R.R., Shellock, F.G., Hodgson, J.A., & Edgerton, V.R. (1996). Specific tension of plantar flexors and dorsiflexors. *Journal of Applied Physiology, 80,* 158-165.

Fukashiro, S., & Komi, P.V. (1987). Joint moment and mechanical power flow of the lower limb during vertical jump. *International Journal of Sports Medicine, 8,* 15-21.

Fukashiro, S., Komi, P.V., Jarvinen, M., & Miyashita, M. (1995). In vivo Achilles tendon loading during jumping in humans. *European Journal of Applied Physiology, 71,* 453-458.

Fulle, S., Belia, S., & Di Tano, G. (2005). Sarcopenia is more than a muscular deficit. *Archives of Italian Biology, 143,* 229-234.

Funk, D.A., An, K.N., Morrey, B.F., & Daube, J.R. (1987). Electromyographic analysis of muscles across the elbow joint. *Journal of Orthopaedic Research, 5,* 529-538.

Funk, D.C., Swank, A.M., Mikla, B.M., Fagan, T.A., & Farr, B.K. (2003). Impact of prior exercise on hamstring flexibility: A comparison of proprioceptive neuromuscular facilitation and static stretching. *Journal of Strength and Conditioning Research, 17,* 489-492.

Gabaldon, A.M., Nelson, F.E., & Roberts, T.J. (2004). Mechanical function of two ankle extensors in wild turkeys: Shifts from energy production to energy absorption during incline versus decline running. *Journal of Experimental Biology, 207,* 2277-2288.

Gaffney, F.A., Sjøgaard, G., & Saltin, B. (1990). Cardiovascular and metabolic responses to static contraction in man. *Acta Physiologica Scandinavica, 138,* 249-258.

Gail, A., & Andersen, R.A. (2006). Neural dynamics in monkey parietal reach region reflect context-specific sensorimotor transformations. *Journal of Neuroscience, 26,* 9376-9384.

Galea, V., & Norman, R.W. (1985). Bone-on-bone forces at the ankle joint during a rapid dynamic movement. In D.A. Winter, R.W. Norman, R.P. Wells, K.C. Hayes, & A.E. Patla (Eds.), *Biomechanics IX-A* (pp. 71-76). Champaign, IL: Human Kinetics.

Galganski, M.E., Fuglevand, A.J., & Enoka, R.M. (1993). Reduced control of motor output in a human hand muscle of elderly subjects during submaximal contractions. *Journal of Neurophysiology, 69,* 2108-2115.

Gallagher, S., Moore, J.S., & Stobbe, T.J. (2004). Isometric, isoinertial, and psychophysical strength testing: Devices and protocols. In S. Kumar (Ed.), *Muscle Strength* (pp. 129-156). Boca Raton, FL: CR Press LLC.

Galvao, D.A., & Taaffe, D.R. (2005). Resistance exercise dosage in older adults: Single- versus multiset effects on physical performance and body composition. *Journal of the American Geriatric Society, 53,* 2090-2097.

Gamrin, L., Berg, H.E., Essen, P., Tesch, P.A., Hultman, E., Garlick, P.J., McNurlan, M.A., & Wernerman, J. (1998). The effect of unloading on protein synthesis in human skeletal muscle. *Acta Physiologica Scandinavica, 163,* 369-377.

Gandevia, S.C. (1987). Roles for perceived voluntary motor commands in motor control. *Trends in Neuroscience, 10,* 81-85.

Gandevia, S.C. (1996). Kinesthesia: Roles for afferent signals and motor commands. In L.B. Rowell & J.T. Shepherd (Eds.), *Handbook of Physiology: Sec. 12. Exercise: Regulation and Integration of Multiple Systems* (pp. 128-172). New York: Oxford University Press.

Gandevia, S.C. (2001). Spinal and supraspinal factors in human muscle fatigue. *Physiological Reviews, 81,* 1725-1789.

Gandevia, S.C., Herbert, R.D., & Leeper, J.B. (1998). Voluntary activation of human elbow flexor muscles during maximal concentric contractions. *Journal of Physiology, 512,* 595-602.

Gandevia, S.C., Killian, K., McKenzie, D.K., Crawford, M., Allen, G.M., Gorman, R.B., & Hales, J.P. (1993). Respiratory sensations, cardiovascular control, kinaesthesia and transcranial stimulation during paralysis in humans. *Journal of Physiology, 470,* 85-107.

Gandevia, S.C., Smith, J.L., Crawford, M., Proske, U., & Taylor, J.L. (2006). Motor commands contribute to human position sense. *Journal of Physiology, 571,* 703-710.

Ganley, K.J., & Powers, C.M. (2006). Intersegmental dynamics during the swing phase of gait: A comparison of knee kinetics between 7 year-old children and adults. *Gait & Posture, 23,* 499-504.

Gao, C., Abeysekera, J., Hirvonen, M., & Gronqvist, R. (2004). Slip resistant properties of footwear on ice. *Ergonomics, 47,* 710-716.

Gardetto, P.R., Schluter, J.M., & Fitts, R.H. (1989). Contractile function of single muscle fibers after hindlimb suspension. *Journal of Applied Physiology, 66,* 2739-2749.

Gardiner, P.F. (2006). Changes in α-motoneurone properties with altered physical activity levels. *Exercise and Sport Sciences Reviews, 34,* 54-58.

Gardiner, P.F., Dai, Y., & Heckman, C.J. (2006). Effects of exercise on α-motoneurons. *Journal of Applied Physiology, 101,* 1228-1236.

Gardiner, P.F., & Seburn, K.L. (1997). The effects of tetrodotoxin-induced muscle paralysis on the physiological properties of muscle units and their innervating motoneurons in rat. *Journal of Physiology, 499,* 207-216.

Gardner, E.P., Babu, K.S., Reitzen, S.D., Ghosh, S., Brown, A.M., Chen, J., Hall, A.L., Herzlinger, M., Kohlenstein, J.B., & Ro, J.Y. (2007*a*). Neurophysiology of prehension. I. Posterior parietal cortex and object-oriented hand behaviors. *Journal of Neurophysiology, 97,* 387-406.

Gardner, E.P., Ro, J.Y., Babu, K.S., & Ghosh, S. (2007*b*). Neurophysiology of prehension. II. Response diversity in primary somatosensory (S-I) and motor (M-I) cortex. *Journal of Neurophysiology, 97,* 1656-1670.

Garfinkel, S., & Cafarelli, E. (1992). Relative changes in maximal force, EMG, and muscle cross-sectional area after isometric training. *Medicine and Science in Sports and Exercise, 24,* 1220-1227.

Garhammer, J. (1980). Power production by Olympic weightlifters. *Medicine and Science in Sports and Exercise, 12,* 54-60.

Garhammer, J., & Takano, B. (1992). Training for weightlifting. In P.V. Komi (Ed.), *Strength and Power in Sport* (pp. 357-369). Oxford, UK: Blackwell Scientific.

Garland, S.J., Enoka, R.M., Serrano, L.P., & Robinson, G.A. (1994). Behavior of motor units in human biceps brachii during a submaximal fatiguing contraction. *Journal of Applied Physiology, 76,* 2411-2419.

Garland, S.J., & Griffin, L. (1999). Motor unit double discharges: Statistical anomaly or functional entity? *Canadian Journal of Applied Physiology, 24,* 113-130.

Garland, S.J., & Kaufman, M.P. (1995). Role of muscle afferents in the inhibition of motoneurons during fatigue. In S.C. Gandevia, R.M. Enoka, A.J. McComas, D.G. Stuart, & C.K. Thomas (Eds.), *Fatigue: Neural and Muscular Mechanisms* (pp. 271-278). New York: Plenum Press.

Garland, S.J., & Miles, T.S. (1997). Responses of human single motor units to transcranial magnetic stimulation. *Electroencephalography and Clinical Neurophysiology, 105,* 94-101.

Garner, S.H., Hicks, A.L., & McComas, A.J. (1989). Prolongation of twitch potentiating mechanism throughout muscle fatigue and recovery. *Experimental Neurology, 103,* 277-281.

Garnett, R., & Stephens, J.A. (1981). Changes in the recruitment threshold of motor units produced by cutaneous stimulation in man. *Journal of Physiology, 311,* 463-473.

Garrett, W.E. (1990). Muscle strain injuries: Clinical and basic aspects. *Medicine and Science in Sports and Exercise, 22,* 436-443.

Garrett, W.E. Jr. (1996). Muscle strain injuries. *American Journal of Sports Medicine, 24* (Suppl. 6), S2-S8.

Gatesy, S.M., & Biewener, A.A. (1991). Bipedal locomotion—effects of speed, size and limb posture in birds and humans. *Journal of Zoology, 224,* 127-147.

Gatts, S.K., & Woollacott, M.H. (2006). Neural mechanisms underlying balance control improvement with short term tai chi training. *Aging Clinical and Experimental Research, 18,* 7-19.

Gaviria, M., & Ohanna, F. (1999). Variability of the fatigue response of paralyzed skeletal muscle in relation to the time after spinal cord injury: Mechanical and electrophysiological characteristics. *European Journal of Applied Physiology, 80,* 145-153.

Gavrilovic, L., & Dronjak, S. (2005). Activation of rat pituitary-adrenocortical and sympatho-adrenomedullary system in response to different stressors. *Neuro Endocrinology Letters, 26,* 515-520.

Gebhard, J.S., Kabo, J.M., & Meals, R.A. (1993). Passive motion: The dose effects on joint stiffness, muscle mass, bone density, and regional swelling. *Journal of Bone and Joint Surgery, 75-A,* 1636-1647.

Gelfand, I.M., Orlovsky, G.N., & Shik, M.L. (1988). Locomotion and scratching in tetrapods. In A.H. Cohen, S. Rossignol, & S. Grillner (Eds.), *Neural Control of Rhythmic Movements in Vertebrates* (pp. 167-199). New York: Wiley.

Georgopoulos, A.P., Kalaska, J.F., Caminiti, R., & Massey, J.T. (1982). On the relations between the direction of two-dimensional arm movements and cell discharge in primate motor cortex. *Journal of Neuroscience, 2,* 1527-1537.

Gerdle, B., Eriksson, N.E., & Brundin, L. (1990). The behaviour of the mean power frequency of the surface electromyogram in biceps brachii with increasing force and during fatigue. With special regard to the electrode distance. *Electromyography and Clinical Neurophysiology, 30,* 483-489.

Gerilovsky, L., Tsvetinov, P., & Trenkova, G. (1989). Peripheral effects on the amplitude of monopolar and bipolar H-reflex potentials from the soleus muscle. *Experimental Brain Research, 76,* 173-181.

Gerrits, H.L., de Haan, A., Hopman, M.T.E., Van der Woude, L.H.V., Jones, D.A., & Sargeant, A.J. (1999). Contractile properties of the quadriceps muscle in individuals with spinal cord injury. *Muscle & Nerve, 22,* 1249-1256.

Gerrits, H.L., de Haan, A., Sargeant, A.J., Dallmeijer, A., & Hopman, M.T. (2000). Altered contractile properties of the quadriceps muscle in people with spinal cord injury following functional electrical stimulated cycle training. *Spinal Cord, 38,* 214-223.

Geurts, C., Sleivert, G.G., & Cheung, S.S. (2004). Temperature effects on the contractile characteristics and submaximal voluntary isometric force production of the first dorsal interosseus muscle. *European Journal of Applied Physiology, 91,* 41-45.

Geyer, H., Seyfarth, A., & Blickhan, R. (2006). Compliant leg behavior explains basic dynamics of walking and running. *Proceedings of Biological Sciences, 273,* 2861-2867.

Ghafouri, M., & Lestienne, F.G. (2006). Contribution of reference frames for movement planning in peripersonal space representation. *Experimental Brain Research, 169,* 24-36.

Gibbs, J., Harrison, L.M., & Stephens, J.A. (1995). Cutaneomuscular reflexes recorded from the lower limb in man during different tasks. *Journal of Physiology, 487,* 237-242.

Gibson, J.N.A., Halliday, D., Morrison, W.L., Stoward, P.J., Hornsby, G.A., Watt, P.W., Murdoch, G., & Rennie, M.J. (1987). Decrease in human quadriceps muscle protein turnover consequent upon leg immobilization. *Clinical Science, 72,* 503-509.

Gielen, C.C.A.M., Ingen Schenau, G.J. van, Tax, T., & Theeuwen, M. (1990). The activation of mono- and bi-articular muscles in multi-joint movements. In J.M. Winters & S.L-Y. Woo (Eds.), *Multiple Muscle Systems: Biomechanics and Movement Organization* (pp. 302-311). New York: Springer-Verlag.

Gillard, D.M., Yakovenko, S., Cameron, T., & Prochazka, A. (2000). Isometric muscle length-tension curves do not predict angle-torque curves of human wrist in continuous active movements. *Journal of Biomechanics, 33,* 1341-1348.

Gillies, E.M., Putnam, C.T., & Bell, G.J. (2006). The effect of varying the time of concentric and eccentric muscle actions during resistance training on skeletal muscle adaptations in women. *European Journal of Applied Physiology, 97,* 443-453.

Gittoes, M.J., Brewin, M.A., & Kerwin, D.G. (2006). Soft tissue contributions to impact forces simulated using a four-segment wobbling mass model of forefoot-heel landings. *Human Movement Sciences, 25,* 775-787.

Gladden, M.H., Jankowska, E., & Czarkowska-Bauch, J. (1998). New observations on coupling between group II muscle afferents and feline γ-motoneurones. *Journal of Physiology, 512,* 507-520.

Gleeson, N., Eston, R., Marginson, V., & McHugh, M.P. (2003). Effects of prior concentric training on eccentric exercise induced muscle damage. *British Journal of Sports Medicine, 37,* 119-125.

Glitsch, U., & Baumann, W. (1997). The three-dimensional determination of internal loads in the lower extremity. *Journal of Biomechanics, 30,* 1123-1131.

Goldberg, J.M., Smith, C.E., & Fernandez, C. (1984). Relation between discharge regularity and responses to externally applied galvanic currents in vestibular nerve afferents of the squirrel monkey. *Journal of Neurophysiology, 51,* 1236-1256.

Goldberger, M., Murray, M., & Tessler, A. (1993). Sprouting and regeneration in the spinal cord: Their roles in recovery of function after spinal injury. In A. Gorio (Ed.), *Neuroregeneration* (pp. 241-264). New York: Raven Press.

Goldspink, G., & Harridge, S.D. (2004). Growth factors and muscle ageing. *Experimental Gerontology, 39,* 1433-1438.

Gollhofer, A., Horstmann, G.A., Berger, W., & Dietz, V. (1989). Compensation for translational and rotational perturbations in human posture: Stabilization of the centre of gravity. *Neuroscience Letters, 105,* 73-78.

Gondon, J., Guette, M., Ballay, Y., & Martin, A. (2006). Neural and muscular changes to detraining after electrostimulation training. *European Journal of Applied Physiology, 97,* 165-173.

Gong, B., Legault, D., Miki, T., Seino, S., & Renaud, J.M. (2003). KATP channels depress force by reducing action potential amplitude in mouse EDL and soleus muscle. *American Journal of Physiology, 285,* C1464-C1474.

González-Alonso, J., Calbet, J., & Nielsen, B. (1999). Metabolic and thermodynamic responses to dehydration-induced reductions in muscle blood flow in exercising muscle. *Journal of Physiology, 520,* 577-589.

Gonzalez-Badillo, J.J., Gorostiaga, E.M., Arellano, R., & Izquierdo, M. (2005). Moderate resistance training volume produces more favorable strength gains than high or low volumes during a short-term training cycle. *Journal of Strength and Conditioning Research, 19,* 689-697.

Goodwin, G.M., McCloskey, D.I., & Matthews, P.B. (1972). The contribution of muscle afferents to kinaesthesia shown by vibration induced illusions of movement and by the effects of paralyzing joint afferents. *Brain, 95,* 705-748.

Gordon, A.M., Huxley, A.F., & Julian, F.J. (1966). The variation in isometric tension with sarcomere length in vertebrate muscle fibres. *Journal of Physiology, 184,* 170-192.

Gordon, A.M., Regnier, M., & Homsher, R. (2001). Skeletal and cardiac muscle contractile activation: Tropomyosin "rocks and rolls." *News in Physiological Sciences, 16,* 49-55.

Gordon, D.A., Enoka, R.M., Karst, G.M., & Stuart, D.G. (1990*a*). Force development and relaxation in single motor units of adult cats during a standard fatigue test. *Journal of Physiology, 421,* 583-594.

Gordon, D.A., Enoka, R.M., & Stuart, D.G. (1990*b*). Motor-unit force potentiation in adult cats during a standard fatigue test. *Journal of Physiology, 421,* 569-582.

Gordon, T., & Mao, J. (1994). Muscle atrophy and procedures for training after spinal cord injury. *Physical Therapy, 74,* 50-60.

Gordon, T., & Pattullo, M.C. (1993). Plasticity of muscle fiber and motor unit types. *Exercise and Sport Sciences Reviews, 21,* 331-362.

Gordon, T., Thomas, C.K., Munson, J.B., & Stein, R.B. (2004). The resilience of the size principle in the organization of motor unit properties in normal and reinnervated adult skeletal muscles. *Canadian Journal of Physiology and Pharmacology, 82,* 645-661.

Gordon, T., Yang, J.F., Ayer, K., Stein, R.B., & Tyreman, N. (1993). Recovery potential of muscle after partial denervation: A comparison between rats and humans. *Brain Research Bulletin, 30,* 477-482.

Goske, S., Erdemir, A., Petre, M., Budhabhatti, S., & Cavanagh, P.R. (2006). Reduction of plantar heel pressure: Insole design using finite element analysis. *Journal of Biomechanics, 39,* 2363-2370.

Gossard, J.-P. (1996). Control of transmission in muscle group Ia afferents during fictive locomotion in the cat. *Journal of Neurophysiology, 76,* 4104-4112.

Gossard, J.P., Brownstone, R.M., Barajon, I., & Hultborn, H. (1994). Transmission in a locomotor-related group Ib pathway from hindlimb extensor muscles in the cat. *Experimental Brain Research, 98,* 213-228.

Gottschall, J.S., & Kram, R. (2005). Ground reaction forces during downhill and uphill running. *Journal of Biomechanics, 38,* 445-452.

Gowland, C., deBruin, H., Basmajian, J.V., Plews, N., & Burcea, I. (1992). Agonist and antagonist activity during voluntary upper-limb movement in patients with stroke. *Physical Therapy, 72,* 624-633.

Grabiner, M.D., & Jeziorowski, J.J. (1992). Isokinetic trunk extension discriminates uninjured subjects from subjects with previous low back pain. *Clinical Biomechanics, 7,* 195-200.

Grabiner, M.D., Koh, T.J., & Andrish, J.T. (1992). Decreased excitation of vastus medialis oblique and vastus lateralis in patellofemoral pain. *European Journal of Experimental Musculoskeletal Research, 1,* 33-39.

Grabowski, A., Farley, C.T., & Kram, R. (2005). Independent metabolic costs of supporting body weight and accelerating body mass during walking. *Journal of Applied Physiology, 98,* 579-583.

Graf, A., Judge, J.O., Ounpuu, S., & Thelen, D.G. (2005). The effect of walking speed on lower-extremity joint powers among elderly adults who exhibit low physical performance. *Archives of Physical Medicine and Rehabilitation, 86,* 2177-2183.

Graham-Smith, P., & Lees, A. (2005). A three-dimensional kinematic analysis of the long jump take-off. *Journal of Sports Science, 23,* 891-903.

Granata, K.P., Abel, M.F., & Damiano, D.L. (2000). Joint angular velocity in spastic gait and the influence of muscle-tendon lengthening. *Journal of Bone and Joint Surgery (American), 82,* 174-186.

Grange, R.W., & Houston, M.E. (1991). Simultaneous potentiation and fatigue in quadriceps after a 60-second maximal voluntary isometric contraction. *Journal of Applied Physiology, 70,* 726-731.

Grange, R.W., Vandenboom, R., Xeni, J., & Houston, M.E. (1998). Potentiation of in vitro concentric work in mouse fast muscle. *Journal of Applied Physiology, 84,* 236-243.

Granzier, H.L., & Labeit, S. (2006). The giant muscle protein titin is an adjustable molecular spring. *Exercise and Sport Sciences Reviews, 34,* 50-53.

Grasso, R., Ivanenko, Y.P., Zago, M., Molinari, M., Scivoletto, G., Castellano, V., Macellari, V., & Lacquaniti, F. (2004). Distributed plasticity of locomotor pattern generators in spinal cord injured patients. *Brain, 127,* 1019-1034.

Grasso, R., Zago, M., & Lacquaniti, F. (2000). Interactions between posture and locomotion: Motor patterns in humans walking with bent posture versus erect posture. *Journal of Neurophysiology, 83,* 288-300.

Gray, S.C., Devito, G., & Nimmo, M.A. (2002). Effect of active warm-up on metabolism prior to and during intense dynamic exercise. *Medicine and Science in Sports and Exercise, 34,* 2091-2096.

Graziano, M.S. (2006). The organization of behavioral repertoire in motor cortex. *Annual Reviews of Neuroscience, 29,* 105-134.

Graziano, M.S., Cooke, D.F., & Taylor, C.S. (2000). Coding the location of the arm by sight. *Science, 290,* 1782-1786.

Graziano, M.S., Hu, X.T., & Gross, C.G. (1997). Visuospatial properties of ventral premotor cortex. *Journal of Neurophysiology, 77,* 2268-2292.

Green, H., Dahly, A., Shoemaker, K., Goreham, C., Bombardier, E., & Ball-Burnett, M. (1999*a*). Serial effects of high-resistance and prolonged endurance training on Na^+-K^+ pump concentration and enzymatic activities in human vastus lateralis. *Acta Physiologica Scandinavica, 165,* 177-184.

Green, H., MacDougall, J., Tarnopolsky, M., & Melissa, N.L. (1999*b*). Downregulation of Na^+-K^+-ATPase pumps in skeletal muscle with training in normobaric hypoxia. *Journal of Applied Physiology, 86,* 1745-1748.

Greensmith, L., & Vrbová, G. (1996). Motoneurone survival: A functional approach. *Trends in Neurosciences, 19,* 450-455.

Grefkes, C., Ritzl, A., Zilles, K., & Fink, G.R. (2004). Human medial intraparietal cortex subserves visuomotor coordinate transformation. *NeuroImage, 23,* 1494-1506.

Gregor, R.J., Cavanagh, P.R., & LaFortune, M. (1985). Knee flexor moments during propulsion in cycling—a creative solution to Lombard's paradox. *Journal of Biomechanics, 18,* 307-316.

Gregor, R.J., Komi, P.V., Browning, R.C., & Jarvinen, M. (1991). A comparison of the triceps surae and residual muscle moments at the ankle during cycling. *Journal of Biomechanics, 24,* 287-297.

Gregor, R.J., Komi, P.V., & Järvinen, M. (1987). Achilles tendon forces during cycling. *International Journal of Sports Medicine, 8,* 9-14.

Gregor, R.J., Roy, R.R., Whiting, W.C., Lovely, R.G., Hodgson, J.A., & Edgerton, V.R. (1988). Mechanical output of the cat soleus during treadmill locomotion: In vivo vs in situ characteristics. *Journal of Biomechanics, 21,* 721-732.

Gregory, C.M., Vandenborne, K., Castro, M.J., & Dudley, G.A. (2003). Human and rat skeletal muscle adaptations to spinal cord injury. *Canadian Journal of Applied Physiology, 28,* 491-500.

Gregory, J.E., Brockett, C.L., Morgan, D.L., Whitehead, N.P., & Proske, U. (2002). Effect of eccentric muscle contractions on Golgi tendon organ responses to passive and active tension in the cat. *Journal of Physiology, 538,* 210-218.

Gregory, J.E., Morgan, D.L., & Proske, U. (1986). Aftereffects in the responses of cat muscle spindles. *Journal of Neurophysiology, 56,* 451-461.

Gregory, J.E., Morgan, D.L., & Proske, U. (2003). Tendon organs as monitors of muscle damage from eccentric contractions. *Experimental Brain Research, 151,* 346-355.

Gregory, J.E., Wise, A.K., Wood, S.A., Prochazka, A., & Proske, U. (1998). Muscle history, fusimotor activity and the human stretch reflex. *Journal of Physiology, 513,* 927-934.

Grey, M.J., Ladouceur, M., Andersen, J.B., Nielsen, J.B., & Sinkjær, T. (2001). Group II muscle afferents probably contribute to the medium latency soleus stretch reflex during walking in humans. *Journal of Physiology, 534,* 925-933.

Griffin, L., Garland, S.J., & Ivanova, T. (1998). Discharge patterns in human motor units during fatiguing arm movements. *Journal of Applied Physiology, 85,* 1684-1692.

Griffin, L., Garland, S.J., & Ivanova, T. (2000). Role of limb movement in the modulation of motor unit discharge rate during fatiguing contractions. *Experimental Brain Research, 130,* 392-400.

Griffin, L., Garland, S.J., Ivanova, T., & Gossen, E. (2001). Muscle vibration sustains motor unit firing rate during submaximal isometric fatigue in humans. *Journal of Physiology, 535,* 929-936.

Griffin, T.M., Roberts, T.J., & Kram, R. (2003). Metabolic cost of generating muscular force in human walking: Insights from load-carrying and speed experiments. *Journal of Applied Physiology, 95,* 172-183.

Griffin, T.M., Tolani, N.A., & Kram, R. (1999). Walking in simulated reduced gravity: Mechanical energy fluctuations and exchange. *Journal of Applied Physiology, 86,* 383-390.

Griffiths, R.I. (1991). Shortening of muscle fibres during stretch of the active cat medial gastrocnemius muscle: The role of tendon compliance. *Journal of Physiology, 436,* 219-236.

Grigg, P. (2001). Properties of sensory neurons innervating synovial joints. *Cells, Tissues, and Organs, 169,* 218-225.

Grillner, S., Georgopoulos, A.P., & Jordan, L.M. (1997). Selection and initiation of motor behavior. In P.S.G. Stein, S. Grillner, A.I. Selverston, & D.G. Stuart (Eds.), *Neurons, Networks, and Motor Behavior* (pp. 3-19). Cambridge, MA: MIT Press.

Grillner, S., Hellgren, J., Ménard, A., Saitoh, K., & Wikström, M.A. (2005). Mechanisms for selection of basic motor programs—roles for the striatum and pallidum. *Trends in Neurosciences, 28,* 364-370.

Grillner, S., & Rossignol, S. (1978). On the initiation of the swing phase of locomotion in chronic spinal cats. *Brain Research, 146,* 269-277.

Grillner, S., & Wallén, P. (1985). Central pattern generators for locomotion, with special reference to vertebrates. *Annual Reviews of Neuroscience, 8,* 233-261.

Grillner, S., Wallén, P., Saitoh, K., Kozlov, A., & Robertson, B. (2008). Neural bases of goal-directed locomotion in vertebrates—an overview. *Brain Research Reviews, 57,* 2-12.

Gruneberg, C., Bloem, B.R., Honegger, F., & Allum, J.H. (2004). The influence of artificially increased hip and trunk stiffness on balance control in man. *Experimental Brain Research, 157,* 472-485.

Guissard, N., & Duchateau, J. (2004). Effect of static training on neural and mechanical properties of the human plantar-flexor muscles. *Muscle & Nerve, 29,* 248-255.

Guissard, N., & Duchateau, J. (2006). Neural aspects of muscle stretching. *Exercise and Sport Sciences Reviews, 34,* 154-158.

Guissard, N., Duchateau, J., & Hainaut, K. (1988). Muscle stretching and motoneuron excitability. *European Journal of Applied Physiology, 58,* 47-52.

Guissard, N., Duchateau, J., & Hainaut, K. (2001). Mechanisms of decreased motoneurone excitation during passive muscle stretching. *Experimental Brain Research, 137,* 163-169.

Gunning, P., & Hardeman, E. (1991). Multiple mechanisms regulate muscle fiber diversity. *FASEB Journal, 5,* 3064-3070.

Gur, H., Cakin, N., Akova, B., Okay, E., & Kucukoglu, S. (2002). Concentric versus combined concentric-eccentric isokinetic training: Effects on functional capacity and symptoms in patients with osteoarthrosis of the knee. *Archives of Physical Medicine and Rehabilitation, 83,* 308-316.

Gurfinkel, V., Cacciatore, T., Cordo, P.J., Horak, F., Nutt, J., & Skoss, R. (2006). Postural muscle tone in the body axis of healthy humans. *Journal of Neurophysiology, 96,* 2678-2687.

Gustafsson, B., & Pinter, M.J. (1984). An investigation of threshold properties among cat spinal alpha-motoneurons. *Journal of Physiology, 357,* 453-483.

Gydikov, A., & Kosarov, D. (1974). Some features of different motor units in human biceps brachii. *Pflügers Archiv, 347,* 75-88.

Gydikov, A.A., Kossev, A.R., Kosarov, D.S., & Kostov, K.G. (1987). Investigations of single motor units firing during movements against elastic resistance. In B. Jonsson (Ed.), *Biomechanics X-A* (pp. 227-232). Champaign, IL: Human Kinetics.

Haftel, V.K., Bichler, E.K., Wang, Q.B., Prather, J.F., Pinter, M.J., & Cope, T.C. (2005). Central suppression of regenerated proprioceptive afferents. *Journal of Neuroscience, 25,* 4733-4742.

Hagbarth, K.-E. (1960). Spinal withdrawal reflexes in the human lower limbs. *Journal of Neurology, Neurosurgery, and Psychiatry, 23,* 222-227.

Hagbarth, K.-E. (1962). Post-tetanic potentiation of myotatic reflexes in man. *Journal of Neurology, Neurosurgery, and Psychiatry, 25,* 1-10.

Hagbarth, K.-E., Hagglund, J.V., Nordin, M., & Wallin, E.U. (1985). Thixotropic behaviour of human finger flexor muscles with accompanying changes in spindle and reflex responses to stretch. *Journal of Physiology, 368,* 323-342.

Hagins, M., Pietrek, M., Sheikhzadeh, A., & Nordin, M. (2006). The effects of breath control on maximum force and IAP during a maximum isometric lifting task. *Clinical Biomechanics, 21,* 775-780.

Hainaut, K., & Duchateau, J. (1992). Neuromuscular electrical stimulation and voluntary exercise. *Sports Medicine, 14,* 100-113.

Häkkinen, K., Kallinen, M., Izquierdo, M., Jokelainen, K., Lassila, H., Mälkiä, E., Kraemer, W.J., Newton, R.U., & Alén, M. (1998*a*). Changes in agonist–antagonist EMG, muscle CSA, and force during strength training in middle-aged and older people. *Journal of Applied Physiology, 84,* 1341-1349.

Häkkinen, K., & Keskinen, K.L. (1989). Muscle cross-sectional area and voluntary force production characteristics in elite strength- and endurance-trained athletes and sprinters. *European Journal of Applied Physiology, 59,* 215-220.

Häkkinen, K., Komi, P.V., & Alén, M. (1985). Effect of explosive type strength training on isometric force- and relaxation-time, electromyographic and muscle fibre characteristics of leg extensor muscles. *Acta Physiologica Scandinavica, 125,* 587-600.

Häkkinen, K., Kraemer, W.J., Newton, R.U., & Alén, M. (2001*a*). Changes in electromyographic activity, muscle fibre and force production characteristics during heavy resistance/power strength training in middle-aged and older men and women. *Acta Physiologica Scandinavica, 171,* 51-62.

Häkkinen, K., Newton, R.U., Gordon, S.E., McCormick, M., Volek, J.S., Nindl, B.C., Gotshalk, L.A., Campbell, W.W., Evans, W.J., Häkkinen, A., Humphries, B.J., & Kraemer, W.J. (1998*b*). Changes in muscle morphology, electromyographic activity, and force production characteristics during progressive strength training in young and older men. *Journal of Gerontology, 53,* B415-B423.

Häkkinen, K., Parkarinen, A., Kraemer, W.J., Häkkinen, A., Valkeinen, H., & Alén, M. (2001*b*). Selective muscle hypertrophy, changes in EMG and force, and serum hormones during strength training in older women. *Journal of Applied Physiology, 91,* 569-580.

Halbertsma, J.P.K., van Bolhuis, A.I., & Göeken, L.N.H. (1996). Sport stretching: Effect on passive muscle stiffness of short hamstrings. *Archives of Physical Medicine and Rehabilitation, 77,* 688-692.

Halford, J.C., Harrold, J.A., Lawton, C.L., & Blundell, J.E. (2005). Serotonin (5-HT) drugs: Effects on appetite expression and use for the treatment of obesity. *Current Drug Targets, 6,* 201-213.

Hallett, M., Berardelli, A., Delwaide, P., Freund, H.-J., Kimura, K., Lücking, C., Rothwell, J.C., Shahani, B.T., & Yanagisawa, N. (1994). Central EMG and tests of motor control. Report of an IFCN committee. *Electroencephalography and Clinical Neurophysiology, 90,* 404-432.

Hamada, T., Sale, D.G., MacDougall, J.D., & Tarnopolsky, M.A. (2003). Interaction of fibre type, potentiation and fatigue in human knee extensor muscles. *Acta Physiologica Scandinavica, 178,* 165-173.

Hamann, J.J., Kluess, H.A., Buckwalter, J.B., & Clifford, P.S. (2005). Blood flow response to muscle contractions is more closely related to metabolic rate than contractile work. *Journal of Applied Physiology, 98,* 2096-2100.

Hamel-Paquet, C., Sergio, L.E., & Kalaska, J.F. (2006). Parietal area 5 activity does not reflect the differential time-course of motor output kinetics during arm-reaching and isometric-force tasks. *Journal of Neurophysiology, 95,* 3353-3370.

Han, B.S., Jang, S.H., Chang, Y., Byun, W.M., Lim, S.K., & Kang, D.S. (2003). Functional magnetic resonance image finding of cortical activation by neuromuscular electrical stimulation on wrist extensor muscles. *American Journal of Physical Medicine and Rehabilitation, 82,* 17-20.

Hanavan, E.P. (1964). *A Mathematical Model of the Human Body* (AMRL-TR-64-102). Wright Patterson Air Force Base, OH: Aerospace Medical Research Laboratories (NTIS No. AD-608463).

Hanavan, E.P. (1966). A personalized mathematical model of the human body. *Journal of Spacecraft and Rockets, 3,* 446-448.

Hannerz, J. (1974). Discharge properties of motor units in relation to recruitment order in voluntary contractions. *Acta Physiologica Scandinavica, 91,* 374-384.

Hansen, P., Bojsen-Møller, J., Aagaard, P., Kjaer, M., & Magnusson, S.P. (2006). Mechanical properties of the human patellar tendon, in vivo. *Clinical Biomechanics, 21,* 54-58.

Hansen, P.D., Woollacott, M.H., & Debu, B. (1988). Postural responses to changing task conditions. *Experimental Brain Research, 73,* 627-636.

Hansen, S., Hansen, N.L., Christensen, L.O., Petersen, N.T., & Nielsen, J.B. (2002). Coupling of antagonistic ankle muscles during co-contraction in humans. *Experimental Brain Research, 146,* 282-292.

Harber, M.P., Gallagher, P.M., Creer, A.R., Minchev, K.M., & Trappe, S.W. (2004). Single muscle fiber properties during a competitive season in male runners. *American Journal of Physiology, 287,* R1124-R1131.

Harkema, S.J. (2001). Neural plasticity after human spinal cord injury: Application of locomotor training to the rehabilitation of walking. *Neuroscientist, 7,* 455-468.

Harkema, S.J. (2008). Plasticity of interneuronal networks of the functionally isolated human spinal cord. *Brain Research Reviews, 57,* 255-264.

Harkema, S.J., Hurley, S.L., Patel, U.K., Requejo, P.S., Dobkin, B.H., & Edgerton, V.R. (1997). Human lumbosacral spinal cord interprets loading during stepping. *Journal of Neurophysiology, 77,* 797-811.

Harman, E.A., Frykman, P.N., Clagett, E.R., & Kraemer, W.J. (1988). Intra-abdominal and intra-thoracic pressures during lifting and jumping. *Medicine and Science in Sports and Exercise, 20,* 195-201.

Harman, E.A., Rosenstein, R.M., Frykman, P.N., & Nigro, G.A. (1989). Effects of a belt on intra-abdominal pressure during weight lifting. *Medicine and Science in Sports and Exercise, 21,* 186-190.

Harridge, S.D.R., Bottinelli, R., Canepari, M., Pellegrino, M.A., Reggiani, C., Esbjörnsson, M., & Saltin, B. (1996). Whole-muscle and single-fibre contractile properties and myosin heavy chain isoforms in humans. *Pflügers Archiv, 432,* 913-920.

Harridge, S.D.R., Kryger, A., & Stensgaard, A. (1999). Knee extensor strength, activation, and size in very elderly people following strength training. *Muscle & Nerve, 22,* 831-839.

Harris, A.J., Duxson, M.J., Butler, J.E., Hodges, P.W., Taylor, J.L., & Gandevia, S.C. (2005). Muscle fiber and motor unit behavior in the longest human skeletal muscle. *Journal of Neuroscience, 25,* 8528-8533.

Harrison, A.P., & Flatman, J.A. (1999). Measurement of force and both surface and deep M wave properties in isolated rat soleus muscles. *American Journal of Physiology, 277,* R1646-R1653.

Harrison, R.N., Lees, A., McCullagh, P.J.J., & Rowe, W.B. (1986). A bioengineering analysis of human muscle and joint forces in the lower limbs during running. *Journal of Sports Sciences, 4,* 201-218.

Hart, C.B., & Giszter, S.F. (2004). Modular premotor drives and unit bursts as primitives for frog motor behaviors. *Journal of Neuroscience, 24,* 5269-5282.

Hasan, Z. (1991). Biomechanics and the study of multi-joint movements. In D.R. Humphrey & H.-J. Freund (Eds.), *Motor Control: Concepts and Issues* (pp. 75-84). Chichester: Wiley.

Hasan, Z., & Stuart, D.G. (1984). Mammalian muscle receptors. In R.A. Davidoff (Ed.), *Handbook of the spinal cord* (pp. 559-607). New York: Dekker.

Hass, C.J., Garzarella, L., De Hoyos, D., & Pollock, M.L. (2000). Single versus multiple sets in long-term recreational weightlifters. *Medicine and Science in Sports and Exercise, 32,* 235-242.

Hasson, C.J., Dugan, E.L., Doyle, T.L., Humphries, B., & Newton, R.U. (2004). Neuromechanical strategies employed to increase jump height during the initiation of the squat jump. *Journal of Electromyography and Kinesiology, 14,* 515-521.

Hather, B.M., Adams, G.R., Tesch, P.A., & Dudley, G.A. (1992). Skeletal muscle responses to lower limb suspension in humans. *Journal of Applied Physiology, 72,* 1493-1498.

Hather, B.M., Tesch, P.A., Buchanan, P., & Dudley, G.A. (1991). Influence of eccentric actions on skeletal muscle adaptations to resistance training. *Acta Physiologica Scandinavica, 143,* 177-185.

Hatze, H. (1976). Forces and duration of impact, and grip tightness during the tennis stroke. *Medicine and Science in Sports, 8,* 88-95.

Hatze, H. (1980). A mathematical model for the computational determination of parameter values of anthropomorphic segments. *Journal of Biomechanics, 13,* 833-843.

Hatze, H. (1981*a*). A comprehensive model for human motion simulation and its application to the take-off phase of the long jump. *Journal of Biomechanics, 14,* 135-141.

Hatze, H. (1981*b*). Estimation of myodynamic parameter values from observations on isometrically contracting muscle groups. *European Journal of Applied Physiology, 46,* 325-338.

Hatze, H. (1998). Validity and reliability of methods for testing vertical jumping performance. *Journal of Applied Biomechanics, 14,* 127-140.

Hatze, H. (2002). The fundamental problem of myoskeletal inverse dynamics and its implications. *Journal of Biomechanics, 35,* 109-115.

Hatzikotoulas, K., Siatras, T., Spyropoulou, E., Paraschos, I., & Patikas, D. (2004). Muscle fatigue and electromyographic changes are not different in women and men matched for strength. *European Journal of Applied Physiology, 92,* 298-304.

Hawk, L.W., & Cook, E.W. (1997). Affective modulation of tactile startle. *Psychophysiology, 34,* 23-31.

Hawkins, D., & Hull, M.L. (1990). A method for determining lower extremity muscle-tendon lengths during flexion/extension movements. *Journal of Biomechaincs, 23,* 487-494.

Hay, J.G. (1993). *The Biomechanics of Sports Techniques* (4th ed.). Englewood Cliffs, NJ: Prentice-Hall.

Hay, J.G., Andrews, J.G., & Vaughan, C.L. (1980). The influence of external load on the joint torques exerted in a squat exercise. In J.M. Cooper & B. Haven (Eds.), *Proceedings of the Biomechanics Symposium* (pp. 286-293). Bloomington, IN: Indiana State Board of Health.

Hay, J.G., Andrews, J.G., Vaughan, C.L., & Ueya, K. (1983). Load, speed, and equipment effects in strength-training exercises. In H. Matsui & K. Kobayashi (Eds.), *Biomechanics VIII-B* (pp. 939-950). Champaign, IL: Human Kinetics.

Hay, J.G., Miller, J.A., & Cantera, R.W. (1986). The techniques of elite male long jumpers. *Journal of Biomechanics, 19,* 855-866.

Hay, J.G., Wilson, B.D., Dapena, J., & Woodworth, G.G. (1977). A computational technique to determine the angular momentum of the human body. *Journal of Biomechanics, 10,* 269-277.

Hayes, S.G., Kindig, A.E., & Kaufman, M.P. (2006). Cyclooxygenase blockade attenuates responses of group III and IV muscle afferents to dynamic exercise in cats. *American Journal of Physiology, 290,* H2239-H2246.

Heckathorne, C.W., & Childress, D.S. (1981). Relationships of the surface electromyogram to the force, length, velocity, and contraction rate of the cineplastic human biceps. *American Journal of Physical Medicine, 60,* 1-19.

Heckman, C.J. (2003). Active conductances in motoneuron dendrites enhance movement capabilities. *Exercise and Sport Sciences Reviews, 31,* 96-101.

Heckman, C.J., & Binder, M.D. (1990). Neural mechanisms underlying the orderly recruitment of motoneurons. In M.D. Binder & L.M. Mendell (Eds.), *The Segmental Motor System* (pp. 182-204). New York: Oxford University Press.

Heckman, C.J., & Binder, M.D. (1991). Computer simulation of the steady-state input-output function of the cat medial gastrocnemius motoneuron pool. *Journal of Neurophysiology, 65,* 952-967.

Heckman, C.J., & Enoka, R.M. (2004). Physiology of the motor neuron and the motor unit. In A. Eisen (Ed.), *Clinical Neurophysiology of Motor Neuron Diseases. Handbook of Clinical Neurophysiology* (Vol. 4, pp. 119-147). Amsterdam: Elsevier.

Heckman, C.J., Lee, R.H., & Brownstone, R.M. (2003). Hyperexcitable dendrites in motoneurons and their neuromodulatory control during motor behavior. *Trends in Neuroscience, 26,* 688-695.

Heer, M., & Palowski, W.H. (2006). Space motion sickness: Incidence, etiology, and countermeasures. *Autonomic Neuroscience, 129,* 77-79.

Heller, M.O., Bergmann, G., Kassi, J.P., Claes, L., Haas, N.P., & Duda, G.N. (2005). Determination of muscle loading at the hip joint for use in pre-clinical testing. *Journal of Biomechanics, 38,* 1155-1163.

Henneman, E. (1957). Relation between size of neurons and their susceptibility to discharge. *Science, 126,* 1345-1347.

Henneman, E. (1979). Functional organization of motoneuron pools: The size principle. In H. Asanuma & V.J. Wilson (Eds.), *Integration in the Nervous System* (pp. 13-25). Tokyo: Igaku-Shoin.

Hennig, E.M. (2007). Influence of racket properties and performance in tennis. *Exercise and Sports Sciences Reviews, 35,* 62-66.

Hennig, E.M., Cavanagh, P.R., Albert, H.T., & Macmillan, N.H. (1982). A piezoelectric method of measuring the vertical contact stress beneath the human foot. *Journal of Biomedical Engineering, 4,* 213-222.

Hennig, E.M., & Milani, T.L. (1995). In-shoe pressure distribution for running in various types of footwear. *Journal of Applied Biomechanics, 11,* 299-310.

Hennig, E.M., Valiant, G.A., & Liu, Q. (1996). Biomechanical variables and the perception of cushioning for running in various types of footwear. *Journal of Applied Biomechanics, 12,* 143-150.

Henriques, D.Y., Medendorp, W.P., Khan, A.Z., & Crawford, J.D. (2002). Visuomotor transformations for eye-hand coordination. *Progress in Brain Research, 140,* 329-340.

Henriques, D.Y.P., & Soechting, J.F. (2005). Approaches to the study of haptic sensing. *Journal of Neurophysiology, 93,* 3036-3043.

Henry, H.T., Ellerby, D.J., & Marsh, R.L. (2005). Performance of guinea fowl Numida meleagris during jumping requires storage and release of elastic energy. *Journal of Experimental Biology, 208,* 3293-3302.

Henry, S.M., Fung, J., & Horak, F.B. (2001). Effect of stance width on multidirectional postural responses. *Journal of Neurophysiology, 85,* 559-570.

Hentzen, E.R., Lahey, M., Peters, D., Mathew, L., Barash, I.A., Fridén, J., & Lieber, R.L. (2006). Stress-dependent and -independent expression of the myogenic regulatory factors and the MARP genes after eccentric contractions in rats. *Journal of Physiology, 570,* 157-167.

Hepp-Reymond, M-C., Kirkpatrick-Tanner, M., Gabernet, L., Qi, H-X., & Weber, B. (1999). Context-dependent force coding in motor and premotor cortical areas. *Experimental Brain Research, 128,* 123-133.

Herbert, M.E., Roy, R.R., & Edgerton, V.R. (1988). Influence of one week of hindlimb suspension and intermittent high load exercise on rat muscles. *Experimental Neurology, 102,* 190-198.

Herbert, R.D., Dean, C., & Gandevia, S.C. (1998). Effects of real and imagined training on voluntary muscle activation during maximal isometric contractions. *Acta Physiologica Scandinavica, 163,* 361-368.

Hermansen, L., Hultman, E., & Saltin, B. (1967). Muscle glycogen during prolonged severe exercise. *Acta Physiologica Scandinavica, 71,* 129-139.

Hernandez, O.M., Jones, M., Guzman, G., & Szczesna-Cordary, D. (2007). Myosin essential light chain in health and disease. *American Journal of Physiology, 292,* H1643-H1654.

Herrero, J.A., Izquierdo, M., Maffiuletti, N.A., & Garcia-Lopez, J. (2006). Electromyostimulation and plyometric training effects on jumping and sprint time. *International Journal of Sports Medicine, 27,* 533-539.

Hershler, C., & Milner, M. (1980*a*). Angle–angle diagrams in the assessment of locomotion. *American Journal of Physical Medicine, 59,* 109-125.

Hershler, C., & Milner, M. (1980*b*). Angle–angle diagrams in above-knee amputee and cerebral palsy gait. *American Journal of Physical Medicine, 59,* 165-183.

Herzog, W. (1996). Force-sharing among synergistic muscles: Theoretical considerations and experimental approaches. *Exercise and Sport Sciences Reviews, 24,* 173-202.

Herzog, W. (1998). History dependence of force production in skeletal muscle: A proposal for mechanisms. *Journal of Electromyography and Kinesiology, 8,* 111-117.

Herzog, W. (2004*a*). Determinants of muscle strength. In S. Kumar (Ed.), *Muscle Strength* (pp. 45-103). Boca Raton, FL: CR Press LLC.

Herzog, W. (2004*b*). History dependence of skeletal muscle force production: Implications for movement control. *Human Movement Sciences, 23,* 591-604.

Herzog, W., Lee, E.J., & Rassier, D.E. (2006). Residual force enhancement in skeletal muscle. *Journal of Physiology, 574,* 635-642.

Herzog, W., & Read, L.J. (1993). Lines of action and moment arms of the major force carrying structures that cross the human knee joint. *Journal of Anatomy, 182,* 213-230.

Hess, J.A., & Woollacott, M.H. (2005). Effect of high-intensity strength-training on functional measures of balance ability in balance-impaired older adults. *Journal of Manipulative and Physiological Therapeutics, 28,* 582-590.

Hesse, S., Werner, C., & Bardeleben, A. (2004). Electromechanical gait training with functional electrical stimulation: Case studies in spinal cord injury. *Spinal Cord, 42,* 346-352.

Heuninckx, S., Wenderoth, N., Debaere, F., Peeters, R., & Swimmen, S.P. (2005). Neural basis of aging: The penetration of cognition into action control. *Journal of Neuroscience, 25,* 6787-6796.

Heyes, M.P., Garnett, E.S., & Coates, G. (1988). Nigrostriatal dopaminergic activity is increased during exhaustive exercise stress in rats. *Life Sciences, 42,* 1537-1542.

Hicks, A., Adams, M.M., Martin-Ginis, K., Giangregorio, L., Latimer, A., Phillips, S.M., & McCartney, N. (2005). Long-term body-weight supported treadmill training and subsequent follow-up in persons with chronic SCI: Effects on functional walking ability and measures of subjective well-being. *Spinal Cord, 43,* 291-298.

Hicks, A., Fenton, J., Garner, S., & McComas, A.J. (1989). M wave potentiation during and after muscle activity. *Journal of Applied Physiology, 66,* 2606-2610.

Hicks, A.L., Kent-Braun, J., & Ditor, D.S. (2001). Sex differences in human skeletal muscle fatigue. *Exercise and Sport Sciences Reviews, 29,* 109-112.

Hicks, A., & McComas, A.J. (1989). Increased sodium pump activity following repetitive stimulation of rat soleus muscles. *Journal of Physiology, 414,* 337-349.

Hidler, J.M., Harvey, R.L., & Rymer, W.Z. (2002). Frequency response characteristics of ankle plantar flexors in humans following spinal cord injury: Relation to degree of spasticity. *Annals of Biomedical Engineering, 30,* 969-981.

Hiebert, G.W., Whelan, R.J., Prochazka, A., & Pearson, K.G. (1996). Contribution of hind limb flexor muscle afferents to the timing of phase transitions in the cat step cycle. *Journal of Neurophysiology, 75,* 1126-1137.

Higbie, E.J., Cureton, K.J., Warren, G.L., & Prior, B.M. (1996). Effects of concentric and eccentric training on muscle strength, cross-sectional area, and neural activation. *Journal of Applied Physiology, 81,* 2173-2181.

Hill, A.V. (1928). The air-resistance to a runner. *Proceedings of the Royal Society of London, B102,* 380-385.

Hill, A.V. (1938). The heat of shortening and the dynamic constraints of muscle. *Proceedings of the Royal Society of London B, 126,* 136-195.

Hinder, M.R., & Milner, T.E. (2005). Novel strategies in feed-forward adaptation to a position-dependent perturbation. *Experimental Brain Research, 165,* 239-249.

Hinrichs, R.N. (1987). Upper extremity function in running. II: Angular momentum considerations. *International Journal of Sport Biomechanics, 3,* 242-263.

Hinrichs, R.N., Cavanagh, P.R., & Williams, K.R. (1987). Upper extremity function in running. I: Center of mass and propulsion considerations. *International Journal of Sport Biomechanics, 3,* 222-241.

Hirashima, M., Kudo, K., & Ohtsuki, T. (2003). Utilization and compensation of interaction torques during ball-throwing movements. *Journal of Neurophysiology, 89,* 1784-1796.

Hirashima, M., Kudo, K., & Ohtsuki, T. (2007*a*). A new non-orthogonal decomposition method to determine effective torques for three-dimensional joint rotation. *Journal of Biomechanics, 40,* 137-145.

Hirashima, M., Kudo, K., Watarai, K., & Ohtsuki, T. (2007*b*). Control of 3D limb dynamics in unconstrained overarm throws of different speeds performed by skilled baseball players. *Journal of Neurophysiology, 97,* 680-691.

Hochman, S., & McCrea, D.A. (1994*a*). Effects of chronic spinalization on ankle extensor motoneurons. I. Composite monosynaptic Ia EPSPs in four motoneuron pools. *Journal of Neurophysiology, 71,* 1452-1467.

Hochman, S., & McCrea, D.A. (1994*b*). Effects of chronic spinalization on ankle extensor motoneurons. II. Motoneuron electrical properties. *Journal of Neurophysiology, 71,* 1468-1479.

Hochman, S., & McCrea, D.A. (1994*c*). Effects of chronic spinalization on ankle extensor motoneurons. I. Composite Ia EPSPs in motoneurons separated into motor unit types. *Journal of Neurophysiology, 71,* 1480-1490.

Hodges, P.W., Cresswell, A.G., Daggfeldt, K., & Thorstensson, A. (2001). In vivo measurement of the effect of intra-abdominal pressure on the human spine. *Journal of Biomechanics, 34,* 347-353.

Hodges, P.W., Cresswell, A.G., & Thorstensson, A. (2004). Intra-abdominal pressure response to multidirectional support-surface translation. *Gait & Posture, 20,* 163-170.

Hodges, P.W., Eriksson, A.E.M., Shirley, D., & Gandevia, S.C. (2005). Intra-abdominal pressure increases stiffness of the lumbar spine. *Clinical Biomechanics, 38,* 1873-1880.

Hodges, P.W., & Gandevia, S.C. (2000). Changes in intra-abdominal pressure during postural and respiratory activation of the human diaphragm. *Journal of Applied Physiology, 89,* 967-976.

Hodgson, J.A., Roy, R.R., Higuchi, N., Monti, R.J., Zhong, H., Grossman, E., & Edgerton, V.R. (2005). Does daily activity level determine muscle phenotype? *Journal of Experimental Biology, 208,* 3761-3770.

Hodgson, M., Docherty, D., & Robbins, D. (2005). Post-activation potentiation: Underlying physiology and implications for motor performance. *Sports Medicine, 35,* 585-595.

Hoehn-Saric, R., McLeod, D.R., Funderburk, F., & Kowalski, P. (2004). Somatic symptoms and physiologic responses in generalized anxiety disorder: An ambulatory monitor study. *Archives of General Psychiatry, 61,* 913-921.

Hof, A.L. (1984). EMG and muscle force: An introduction. *Human Movement Sciences, 3,* 119-153.

Hof, A.L. (1998). In vivo measurement of the series elasticity release curve of human triceps surae muscle. *Journal of Biomechanics, 31,* 793-800.

Hof, A.L. (2003). Muscle mechanics and neuromotor control. *Journal of Biomechanics, 36,* 1031-1038.

Hof, A.L., Elzinga, H., Grimmius, W., & Halbertsma, J.P. (2002). Speed dependence of averaged EMG profiles in walking. *Gait & Posture, 16,* 78-86.

Hof, A.L., & van den Berg, J. (1981*a*). EMG to force processing I: An electrical analogue of the Hill muscle model. *Journal of Biomechanics, 14,* 747-758.

Hof, A.L., & van den Berg, J. (1981*b*). EMG to force processing II: Estimation of parameters of the Hill muscle model for the human triceps surae by means of a calf ergometer. *Journal of Biomechanics, 14,* 759-770.

Hof, A.L., & van den Berg, J. (1981*c*). EMG to force processing III: Estimation of model parameters for the human triceps surae muscle and assessment of the accuracy by means of a torque plate. *Journal of Biomechanics, 14,* 771-785.

Hof, A.L., & van den Berg, J. (1981*d*). EMG to force processing IV: Eccentric-concentric contractions on a spring-flywheel setup. *Journal of Biomechanics, 14,* 787-792.

Hoff, B., & Arbib, M.A. (1993). Models of trajectory formation and temporal interaction of reach and grasp. *Journal of Motor Behavior, 25,* 175-192.

Hoffmann, P. (1918). Über die Beziehungen der Sehnenreflexe zur willkürlichen Bewegung und Zum Tonus. *Zeitschrift für Biologie, 68,* 351-370.

Hoffmann, P. (1922). *Untersuchungen über die Eigenreflexe (Sehnenreflexe) menschlischer Muskeln.* Berlin: Springer.

Hogervorst, T., & Brand, R.A. (1998). Mechanoreceptors in joint function. *Journal of Bone and Joint Surgery American, 80,* 1365-1378.

Holder-Powell, H., Di Matteo, G., & Rutherford, O.M. (2001). Do knee injuries have long-term consequences for isometric and dynamic muscle strength? *European Journal of Applied Physiology, 85,* 310-316.

Hollerbach, J.M., & Flash, T. (1982). Dynamic interactions between limb segments during planar arm movement. *Biological Cybernetics, 44,* 67-77.

Hollman, J.H., Koyash, F.M., Kubik, J.J., & Linbo, R.A. (2007). Age-related differences in spatiotemporal markers of gait stability during dual task walking. *Gait & Posture, 26,* 113-119.

Holtermann, A., Roeleveld, F., Engstrøm, M., & Sand, T. (2007). Enhance H-reflex with resistance training is related to increased rate of force development. *European Journal of Applied Physiology, 101,* 301-312.

Holtzer, R., Stern, Y., & Rakitin, B.C. (2005). Predicting age-related dual-task effects with individual differences on neuropsychological tests. *Neuropsychology, 19,* 18-27.

Hongo, T., Lundberg, A., Phillips, C.G., & Thompson, R.F. (1984). The pattern of monosynaptic Ia-connections to hindlimb motor nuclei in the baboon: A comparison with the cat. *Proceedings of the Royal Society of London B, 221,* 261-289.

Hooper, S.L., & DiCaprio, R.A. (2004). Crustacean motor pattern generator networks. *Neurosignals, 13,* 50-69.

Hopper, B.J. (1973). *The Mechanics of Human Movement.* New York: Elsevier.

Horak, F.B. (2006). Postural orientation and equilibrium: What do we need to know about neural control of balance to prevent falls? *Age and Ageing, 35* Suppl. 2, ii7-ii11.

Horak, F.B., & Macpherson, J.M. (1996). Postural orientation and equilibrium. In L.B. Rowell & J.T. Shepherd (Eds.), *Handbook of Physiology: Sec. 12. Exercise: Regulation and Integration of Multiple Systems* (pp. 255-292). New York: Oxford University Press.

Horch, K.W., Tuckett, R.P., & Burgess, P.R. (1977). A key to the classification of cutaneous mechanoreceptors. *Journal of Investigative Dermatology, 69,* 75-82.

Hore, J., O'Brien, M., & Watts, S. (2005). Control of joint rotations in overarm throws of different speeds made by dominant and nondominant arms. *Journal of Neurophysiology, 94,* 3975-3986.

Horita, T., Kitamura, K., & Kohno, N. (1991). Body configuration and joint moment analysis during standing long jump in 6-yr-old children and adult males. *Medicine and Science in Sports and Exercise, 23,* 1068-1077.

Hornung, J.P. (2003). The human raphe nuclei and the serotonergic system. *Journal of Chemical Neuroanatomy, 26,* 331-343.

Hortobágyi, T. (2005). Cross education and the human central nervous system. *IEEE Engineering in Medicine and Biology Magazine,* January/February, 22-28.

Hortobágyi, T., Barrier, J., Beard, D., Braspennincx, J., Koens, P., DeVita, P., Dempsey, L., & Lambert, N.J. (1996*a*). Greater initial adaptations to submaximal muscle lengthening than maximal shortening. *Journal of Applied Physiology, 81,* 1677-1682.

Hortobágyi, T., Dempsey, L., Fraser, D., Zheng, D., Hamilton, G., Lambert, J., & Dohm, L. (2000). Changes in muscle strength, muscle fibre size and myofibrillar gene expression after immobilization and retraining in humans. *Journal of Physiology, 524,* 293-304.

Hortobágyi, T., & DeVita, P. (2000). Altered movement strategy increases lower extremity stiffness during stepping down in the aged. *Journal of Gerontology, 54A,* B63-B70.

Hortobágyi, T., & DeVita, P. (2006). Mechanisms responsible for the age-associated increase in coactivation of antagonist muscles. *Exercise and Sport Sciences Reviews, 34,* 29-35.

Hortobágyi, T., Hill, J.P., Houmard, J.A., Fraser, D.D., Lambert, N.J., & Israel, R.G. (1996*b*). Adaptive responses to muscle lengthening and shortening in humans. *Journal of Applied Physiology, 80,* 765-772.

Hortobágyi, T., & Katch, F.I. (1990). Role of concentric force in limiting improvement in muscular strength. *Journal of Applied Physiology, 68,* 650-658.

Hortobágyi, T., Lambert, J.N., & Hill, J.P. (1997). Greater cross education following strength training with muscle lengthening than shortening, *Medicine and Science in Sports and Exercise, 29,* 107-112.

Hortobágyi, T., Scott, K., Lambert, J., Hamilton, G., & Tracy, J. (1999). Cross-education of muscle strength is greater with stimulated than voluntary contractions. *Motor Control, 3,* 205-219.

Hortobágyi, T., Taylor, J.E., Petersen, N.T., Russell, G., & Gandevia, S.C. (2003). Changes in segmental and motor cortical output with contralateral muscle contractions and altered sensory inputs in humans. *Journal of Neurophysiology, 90,* 2451-2459.

Hortobágyi, T., Zheng, D., Weidner, M., Lambert, N.J., Westbrook, S., & Houmard, J.A. (1995). The influence of aging on muscle strength and muscle fiber characteristics with special reference to eccentric strength. *Journal of Gerontology, 50A,* B399-B406.

Hoshi, E., & Tanji, J. (2002). Contrasting neuronal activity in the dorsal and ventral premotor areas during preparation to reach. *Journal of Neurophysiology, 87,* 1123-1128.

Hoshi, E., & Tanji, J. (2004). Functional specialization in dorsal and ventral premotor areas. *Progress in Brain Research, 143,* 507-511.

Houmard, J.A., Johns, R.A., Smith, L.L., Wells, J.M., Kobe, R.W., & McGoogan, S.A. (1991). The effect of warm-up on responses to intense exercise. *International Journal of Sports Medicine, 12,* 480-483.

Howard, J.D., & Enoka, R.M. (1991). Maximum bilateral contractions are modified by neurally mediated interlimb effects. *Journal of Applied Physiology, 70,* 306-316.

Howell, J.N., Chleboun, G., & Conaster, R. (1993). Muscle stiffness, strength loss, swelling and soreness following exercise-induced injury in humans. *Journal of Physiology, 464,* 183-196.

Howells, J., Trevillion, L., Jankelowitz, S., & Burke, D. (2006). Augmentation of the contraction force of human thenar muscles by and during brief discharge trains. *Muscle & Nerve, 33,* 384-392.

Hoy, M.G., Zernicke, R.J., & Smith, J.L. (1985). Contrasting roles of inertial and muscle moments at knee and ankle during paw-shake response. *Journal of Neurophysiology, 54,* 1282-1295.

Hoyt, D.F., Wickler, S.J., Biewener, A.A., Cogger, E.A., & De La Paz, K.L. (2005). In vivo muscle function vs speed. I. Muscle strain in relation to length change of the muscle-tendon unit. *Journal of Experimental Biology, 208,* 1175-1190.

Hsiao, C.F., Wu, N., & Chandler, S.H. (2005). Voltage-dependent calcium currents in trigeminal motoneurons of early postnatal rats: Modulation by 5-HT receptors. *Journal of Neurophysiology, 94,* 2063-2072.

Hubbard, M. (2000). The flight of sports projectiles. In V.M. Zatsiorsky (Ed.), *Biomechanics in Sport* (pp. 381-400). Oxford, UK: Blackwell Science.

Huffenus, A-F., Amarantini, D., & Forestier, N. (2006). Effects of distal and proximal arm muscles fatigue on multi-joint movement organization. *Experimental Brain Research, 170,* 438-447.

Hufschmidt, A., & Mauritz, K.-H. (1985). Chronic transformation of muscle in spasticity: A peripheral contribution to increased tone. *Journal of Neurology, Neurosurgery, and Psychiatry, 48,* 676-685.

Hughey, L.K., & Fung, J. (2005). Postural responses triggered by multidirectional leg lifts and surface tilts. *Experimental Brain Research, 165,* 152-166.

Huijing, P.A. (2003). Muscular force transmission necessitates a multilevel integrative approach to the analysis of function of skeletal muscle. *Exercise and Sport Sciences Reviews, 31,* 167-175.

Huizar, P., Kuno, M., Kudo, N., & Miyata, Y. (1978). Reaction of intact spinal motoneurones to partial denervation of the muscle. *Journal of Physiology, 265,* 175-193.

Hultborn, H. (2001). State-dependent modulation of sensory feedback. *Journal of Physiology, 533,* 5-13.

Hultborn, H. (2006). Spinal reflexes, mechanisms and concepts: From Eccles to Lundberg and beyond. *Progress in Neurobiology, 78,* 125-232.

Hultborn, H., Jankowska, E., & Lindström, S. (1971). Recurrent inhibition from motor axon collaterals of transmission in the Ia inhibitory pathway to motoneurones. *Journal of Physiology, 215,* 591-612.

Hultborn, H., Meunier, S., Morin, C., & Pierrot-Deseilligny, E. (1987*a*). Assessing changes in presynaptic inhibition of Ia fibres: A study in man and the cat. *Journal of Physiology, 389,* 729-756.

Hultborn, H., Meunier, S., Pierrot-Deseilligny, E., & Shindo, M. (1987*b*). Changes in presynaptic inhibition of Ia fibres at the onset of voluntary contractions in man. *Journal of Physiology, 389,* 757-772.

Hultborn, H., & Pierrot-Deseilligny, E. (1979). Changes in recurrent inhibition during voluntary soleus contractions in man studied by an H-reflex technique. *Journal of Physiology, 297,* 229-251.

Hultman, E., Bergström, M., Spriet, L.L., & Söderlund, K. (1990). Energy metabolism and fatigue. In A.W. Taylor, P.D. Gollnick, H.J. Green, C.D. Ianuzzo, E.G. Noble, G. Métivier, & J.R. Sutton (Eds.), *Biochemistry of Exercise VII* (pp. 73-92). Champaign, IL: Human Kinetics.

Hultman, E., Sjöholm, H., Jäderhaolm-Ek, I., & Krynicki, J. (1983). Evaluation of methods for electrical stimulation of human skeletal muscle in situ. *Pflügers Archiv, 398,* 139-141.

Hunter, G.R., McCarthy, J.P., & Bamman, M.M. (2004). Effects of resistance training on older adults. *Sports Medicine, 34,* 329-348.

Hunter, J.P., Marshall, R.N., & McNair, P.J. (2004). Segment-interaction analysis of the stance limb in sprint running. *Journal of Biomechanics, 37,* 1439-1446.

Hunter, S., White, M., & Thompson, M. (1998). Techniques to evaluate elderly human muscle function: A physiological basis. *Journal of Gerontology, 53A,* B204-B216.

Hunter, S.K., Butler, J.E., Todd, G., Gandevia, S.C., & Taylor, J.L. (2006). Supraspinal fatigue does not explain the sex difference in muscle fatigue of maximal contractions. *Journal of Applied Physiology, 101,* 1036-1044.

Hunter, S.K., Critchlow, A., & Enoka, R.M. (2005). Muscle endurance is greater for old men compared with strength-matched young men. *Journal of Applied Physiology, 99,* 890-897.

Hunter, S.K., Critchlow, A., Shin, I.S., & Enoka, R.M. (2004*a*). Fatigability of the elbow flexor muscles for a sustained submaximal contraction is similar in men and women matched for strength. *Journal of Applied Physiology, 96,* 195-202.

Hunter, S.K., Critchlow, A., Shin, I.S., & Enoka, R.M. (2004*b*). Men are more fatigable than strength-matched women when performing intermittent submaximal contractions. *Journal of Applied Physiology, 96,* 2125-2132.

Hunter, S.K., Duchateau, J., & Enoka, R.M. (2004*c*). Muscle fatigue and the mechanisms of task failure. *Exercise and Sport Sciences Reviews, 32,* 44-49.

Hunter, S.K., & Enoka, R.M. (2001). Sex differences in the fatigability of arm muscles depends on absolute force during isometric contractions. *Journal of Applied Physiology, 91,* 2686-2694.

Hunter, S.K., Lepers, R., MacGillis, C.J., & Enoka, R.M. (2003). Activation among the elbow flexor muscles differs when maintaining arm position during a fatiguing contraction. *Journal of Applied Physiology, 94,* 2439-2447.

Hunter, S.K., Ryan, D.L., Ortega, J.D., & Enoka, R.M. (2002). Task differences with the same load torque alter the endurance time of submaximal fatiguing contractions in humans. *Journal of Neurophysiology, 88,* 3087-3096.

Hunter, S.K., Thompson, M.W., Ruell, P.A., Harmer, A.R., Thom, J.M., Gwinn, T.H., & Adams, R.D. (1999). Human skeletal sarcoplasmic reticulum Ca^{2+} uptake and muscle function with aging and strength training. *Journal of Applied Physiology, 86,* 1858-1865.

Hurley, J.D., & Meminger, S.R. (1992). A relapse-prevention program: Effects of electromyographic training on high and low levels of state and trait anxiety. *Perceptual and Motor Skills, 74,* 699-705.

Hutchison, S., Kobayashi, M., Horkan, C.M., Pascual-Leone, A., Alexander, M.P., & Schlaug, G. (2002). Age-related differences in movement representation. *NeuroImage, 17,* 1720-1728.

Hutton, R.S. (1984). Acute plasticity in spinal segmental pathways with use: Implications for training. In M. Kumamoto (Ed.), *Neural and Mechanical Control of Movement* (pp. 90-112). Kyoto: Yamaguchi Shoten.

Hutton, R.S. (1992). Neuromuscular basis of stretching exercises. In P.V. Komi (Ed.), *Strength and Power in Sport* (pp. 29-38). Oxford, UK: Blackwell Scientific.

Hutton, R.S., & Nelson, D.L. (1986). Stretch sensitivity of Golgi tendon organs in fatigued gastrocnemius muscle. *Medicine and Science in Sports and Exercise, 18,* 69-74.

Hutton, R.S., Smith, J.L., & Eldred, E. (1973). Postcontraction sensory discharge from muscle and its source. *Journal of Neurophysiology, 36,* 1090-1103.

Huxley, A.F. (1957). Muscle structure and theories of contraction. *Progress in Biophysics and Molecular Biology, 7,* 255-318.

Huxley, A.F. (2000). Mechanics and models of the myosin motor. *Philosophical Transactions of the Royal Society, B355,* 433-440.

Huxley, A.F., & Niedergerke, R. (1954). Structural changes in muscle during contraction. Interference microscopy of living muscle fibres. *Nature, 173,* 971-973.

Huxley, A.F., & Simmons, R.M. (1971). Proposed mechanism of force generation in striated muscle. *Nature, 233,* 533-538.

Huxley, H.E., & Hanson, J. (1954). Changes in cross-striations of muscle during contraction and stretch and their structural interpretation. *Nature, 173,* 973-976.

Hwang, E.J., & Shadmehr, R. (2005). Internal models of limb dynamics and the encoding of limb state. *Journal of Neural Engineering, 2,* S266-S278.

Hyatt, J.P., Roy, R.R., Baldwin, K.K., Wernig, A., & Edergton, V.R. (2006). Activity-unrelated neural control of myogenic factors in a slow muscle. *Muscle & Nerve, 33,* 49-60.

Ichinose, Y., Kawakami, Y., Ito, M., Kanehisa, H., & Fukunaga, T. (2000). In vivo estimation of contraction velocity of human vastus lateralis muscle during "isokinetic" action. *Journal of Applied Physiology, 88,* 851-856.

Iggo, A., & Andres, K.H. (1982). Morphology of cutaneous receptors. *Annual Reviews of Neuroscience, 5,* 1-31.

Ikai, M., & Steinhaus, A.H. (1961). Some factors modifying the expression of human strength. *Journal of Applied Physiology, 16,* 157-163.

Iles, J.E., & Pardoe, J. (1999). Changes in transmission in the pathway of heteronymous spinal recurrent inhibition from soleus to quadriceps motor neurons during movement in man. *Brain, 122,* 1757-1764.

Imahori, Y., Fujii, R., Kondo, M., Ohmori, Y., & Nakajima, K. (1999). Neural features of recovery from CNS injury revealed by PET in human brain. *NeuroReport, 10,* 117-121.

Imms, F.J., Hackett, A.J., Prestidge, S.P., & Fox, R.H. (1977). Voluntary isometric muscle strength of patients undergoing rehabilitation following fractures of the lower limb. *Rheumatology and Rehabilitation, 16,* 162-171.

Imms, F.J., & MacDonald, I.C. (1978). Abnormalities of the gait occurring during recovery from fractures of the lower limb and their improvement during rehabilitation. *Scandinavian Journal of Rehabilitation Medicine, 10,* 193-199.

Ingalls, C.P., Warren, G.L., Williams, J.H., Ward, C.W., & Armstrong, R.B. (1998). E-C coupling failure in mouse EDL muscle after in vivo eccentric contractions. *Journal of Applied Physiology, 85,* 58-67.

Ingen Schenau, G.J. van (1990). On the action of bi-articular muscles, a review. *Netherlands Journal of Zoology, 40,* 521-540.

Ingen Schenau, G.J. van, Bobbert, M.F., & de Haan, A. (1997). Does elastic energy enhance work and efficiency in the stretch-shortening cycle? *Journal of Applied Biomechanics, 13,* 389-415.

Ingen Schenau, G.J. van, Bobbert, M.F., Huijing, P.A., & Woittiez, R.D. (1985). The instantaneous torque-angular velocity relation in plantar flexion during jumping. *Medicine and Science in Sports and Exercise, 17,* 422-426.

Ingen Schenau, G.J. van, Bobbert, M.F., & Soest, A.J. van. (1990*a*). The unique action of bi-articular muscles in leg extensions. In J.M. Winters & S.L.-Y. Woo (Eds.), *Multiple Muscle Systems: Biomechanics and Movement Organization* (pp. 639-652). New York: Springer-Verlag.

Ingen Schenau, G.J. van, & Cavanagh, P.R. (1990). Power equations in endurance sports. *Journal of Biomechanics, 23,* 865-881.

Ingen Schenau, G.J. van, Woensel, W.W.L.M. van, Boots, P.J.M., Snackers, R.W., & de Groot, G. (1990*b*). Determination and interpretation of mechanical power in human movement: Application to ergometer cycling. *European Journal of Applied Physiology, 61,* 11-19.

Ingjer, F., & Strømme, S.B. (1979). Effects of active, passive, or no warm-up on the physiological response to heavy exercise. *European Journal of Applied Physiology, 40,* 273-282.

Inman, V.T., Ralston, H.J., Saunders, J.B., Feinstein, B., & Wright, E.W. (1952). Relation of human electromyogram to muscular tension. *Electroencephalography and Clinical Neurophysiology, 4,* 187-194.

Iossifidou, A.N., & Baltzopoulos, V. (1998). Inertial effects on the assessment of performance in isokinetic dynamometry. *International Journal of Sports Medicine, 19,* 567-573.

Iossifidou, A.N., Baltzopoulos, V., & Giakas, G. (2005). Isokinetic knee extension and vertical jumping: Are they related? *Journal of Sports Science, 23,* 1121-1127.

Ishihara, A., Oishi, Y., Roy, R.R., & Edgerton, V.R. (1997). Influence of two weeks of non-weight bearing on rat soleus motoneurons and muscle fibers. *Aviation, Space, and Environmental Medicine, 68,* 421-425.

Ishikawa, M., Komi, P.V., Finni, T., & Kuitunen, S. (2006). Contribution of the tendinous tissue to force enhancement during stretch-shortening cycle exercise depends on the prestretch and concentric phase intensities. *Journal of Electromyography and Kinesiology, 16,* 423-431.

Ishikawa, M., Niemelä, E., & Komi, P.V. (2005). Interaction between fascicle and tendinous tissues in short-contact stretch-shortening cycle exercise with varying eccentric intensities. *Journal of Applied Physiology, 99,* 217-223.

Ishikawa, M., Pakaslahti, J., & Komi, P.V. (2007). Medial gastrocnemius muscle behavior during human running and walking. *Gait & Posture, 25,* 380-384.

Issurin, V.B., & Tenenbaum, G. (1999). Acute and residual effects of vibratory stimulation on explosive strength in elite and amateur athletes. *Journal of Sports Sciences, 17,* 177-182.

Ito, M., Akima, H., & Fukunaga, T. (2000). In vivo moment arm determination using B-mode ultrasonography. *Journal of Biomechanics, 33,* 215-218.

Ito, M., Kawakami, Y., Ichinose, Y., Fukashiro, S., & Fukunaga, T. (1998). Nonisometric behavior of fascicles during isometric contractions of a human muscle. *Journal of Applied Physiology, 85,* 1230-1235.

Ivancic, P.C., Cholewicki, J., & Radebold, A. (2002). Effects of the abdominal belt on muscle-generated spinal stability and L4/L5 joint compression force. *Ergonomics, 45,* 501-513.

Ivanenko, Y.P., Cappellini, G., Dominici, N., Poppele, R.E., & Lacquaniti, F. (2005). Coordination of locomotion with voluntary movements in humans. *Journal of Neuroscience, 25,* 7238-7253.

Ivanenko, Y.P., Grasso, R., Zago, M., Molinari, M., Scivoletto, G., Castellano, V., Macellari, V., & Lacquaniti, F. (2003). Temporal components of the motor patterns expressed by the human spinal cord reflect foot kinematics. *Journal of Neurophysiology, 90,* 3555-3565.

Ivanenko, Y.P., Poppele, R.E., & Lacquaniti, F. (2004). Five basic muscle activation patterns account for muscle activity during human locomotion. *Journal of Physiology, 556,* 267-282.

Ivanenko, Y.P., Poppele, R.E., & Lacquaniti, F. (2006*a*). Motor control programs and walking. *Neuroscientist, 12,* 339-348.

Ivanenko, Y.P., Poppele, R.E., & Lacquaniti, F. (2006*b*). Spinal cord maps of spatiotemporal alpha-motoneuron activation in humans walking at different speeds. *Journal of Neurophysiology, 95,* 602-618.

Jacobs, B.L., Martin-Cora, F.J., & Fornal, C.A. (2002). Activity of medullary serotonergic neurons in freely moving animals. *Brain Research and Brain Research Reviews, 40,* 45-52.

Jacobs, P.L., & Nash, M.S. (2004). Exercise recommendations for individuals with spinal cord injury. *Sports Medicine, 34,* 727-751.

Jakobi, J.M., & Chilibeck, P.D. (2001). Bilateral and unilateral contractions: Possible differences in maximal voluntary force. *Canadian Journal of Applied Physiology, 26,* 12-22.

Jami, L. (1992). Golgi tendon organs in mammalian skeletal muscle: Functional properties and central actions. *Physiological Reviews, 72,* 623-666.

Jami, L., Murthy, K.S.K., Petit, J., & Zytnicki, D. (1983). After-effects of repetitive stimulation at low frequency on fast-contracting motor units of cat muscle. *Journal of Physiology, 340,* 129-143.

Jänkälä, H., Harjola, V.-P., Petersen, N.E., & Härkönen, M. (1997). Myosin heavy chain mRNA transform to faster isoforms in immobilized skeletal muscle: A quantitative PCR study. *Journal of Applied Physiology, 82,* 977-982.

Jankowska, E. (1992). Interneuronal relay in spinal pathways from proprioceptors. *Progress in Neurobiology, 38,* 335-378.

Jankowska, E., & Lundberg, A. (1981). Interneurones in the spinal cord. *Trends in Neurosciences, 4,* 230-233.

Jasmin, B.J., Lavoie, P.-A., & Gardiner, P.F. (1987). Fast axonal transport of acetylcholine in rat sciatic motoneurons is enhanced following prolonged daily running, but not following swimming. *Neuroscience Letters, 78,* 156-160.

Jayaraman, A., Gregory, C.M., Bowden, M., Stevens, J.E., Shah, P., Behrman, A.L., & Vandenborne, K. (2006). Lower extremity muscle function in persons with incomplete spinal cord injury. *Spinal Cord, 44,* 680-687.

Jenner, G., Foley, J.M., Cooper, T.G., Potchen, E.J., & Meyer, R.A. (1994). Changes in magnetic resonance images of muscle depend on exercise intensity and duration, not work. *Journal of Applied Physiology, 76,* 2119-2124.

Jenner, J.R., & Stephens, J.A. (1982). Cutaneous responses and their central nervous pathways studied in man. *Journal of Physiology, 333,* 405-419.

Jenny, A.B., & Inukai, J. (1983). Principles of motor organization of the monkey cervical spinal cord. *Journal of Neuroscience, 3,* 567-575.

Jensen, B.R., Pilegaard, M., & Sjøgaard, G. (2000). Motor unit recruitment and rate coding in response to fatiguing shoulder abductions and subsequent recovery. *European Journal of Applied Physiology, 83,* 190-199.

Jensen, J.L., Marstrand, P.C., & Nielsen, J.B. (2005). Motor skill training and strength training are associated with different plastic changes in the central nervous system. *Journal of Applied Physiology, 99,* 1558-1568.

Jensen, R.K., Doucet, S., & Treitz, T. (1996). Changes in segment mass and mass distribution during pregnancy. *Journal of Biomechanics, 29,* 251-256.

Jiang, B., Roy, R.R., & Edgerton, V.R. (1990). Expression of a fast fiber enzyme profile in the cat soleus after spinalization. *Muscle & Nerve, 13,* 1037-1049.

Johansson, H., Sjölander, P., & Sojka, P. (1991). Receptors in the knee joint ligaments and their role in the biomechanics of the joint. *CRC Critical Reviews in Biomedical Engineering, 18,* 341-368.

Johansson, R.S. (1996). Sensory and memory information in the control of dextrous manipulation. In F. Lacquaniti & P. Viviani (Eds.), *Neural Bases of Motor Behavior* (pp. 205-260). Netherlands: Kluwer Academic.

Johansson, R.S. (2002). Dynamic use of tactile afferent signals in control of dexterous manipulation. *Advances in Experimental and Medical Biology, 508,* 397-410.

Johansson, R.S., & Birznieks, I. (2004). First spikes in ensembles of human tactile afferents code complex spatial fingertip events. *Nature Neuroscience, 7,* 170-177.

Johansson, R.S., & Cole, K.J. (1992). Sensory-motor coordination during grasping and manipulative actions. *Current Opinion in Neurobiology, 2,* 815-823.

Johansson, R.S., Häger, C., & Bäckström, L. (1992*a*). Somatosensory control of precision grip during unpredictable pulling loads. III. Impairments during digital anesthesia. *Experimental Brain Research, 89,* 204-213.

Johansson, R.S., Häger, C., & Riso, R. (1992*b*). Somatosensory control of precision grip during unpredictable pulling loads. II. Changes in load force rate. *Experimental Brain Research, 89,* 192-203.

Johansson, R.S., Riso, R., Häger, C., & Bäckström, L. (1992*c*). Somatosensory control of precision grip during unpredictable pulling loads. I. Changes in load force amplitude. *Experimental Brain Research, 89,* 181-191.

Johnson, B.R., Schneider, L.R., Nadim, F., & Harris-Warrick, R.M. (2005). Dopamine modulation of phasing of activity in a rhythmic motor network: Contribution of synaptic and intrinsic modulatory actions. *Journal of Neurophysiology, 94,* 3101-3111.

Jokela, M., & Hanin, Y.L. (1999). Does the individual zones of optimal functioning model discriminate between successful and less successful athletes? A meta-analysis. *Journal of Sports Science, 17,* 873-887.

Jones, C., Allen, T., Talbot, J., Morgan, D.L., & Proske, U. (1997). Changes in the mechanical properties of human and amphibian muscle after eccentric exercise. *European Journal of Applied Physiology, 76,* 21-31.

Jones, D.A. (1993). How far can experiments in the laboratory explain the fatigue of athletes in the field? In A.J. Sargeant & D. Kernell (Eds.), *Neuromuscular Fatigue* (pp. 100-108). Amsterdam: North-Holland.

Jones, D.A. (1996). High- and low-frequency fatigue revisited. *Acta Physiologica Scandinavica, 156,* 265-270.

Jones, D.A., Bigland-Ritchie, B., & Edwards, R.H.T. (1979). Excitation frequency and muscle fatigue: Mechanical responses during voluntary and stimulated contractions. *Experimental Neurology, 64,* 401-413.

Jones, D.A., & Round, J.M. (1997). Human muscle damage induced by eccentric exercise or reperfusion injury: A common mechanism? In S. Salmon (Ed.), *Muscle Damage* (pp. 64-75). Oxford, UK: Oxford University Press.

Jones, D.A., Rutherford, O.M., & Parker, D.F. (1989). Physiological changes in skeletal muscle as a result of strength training. *Quarterly Journal of Experimental Physiology, 74,* 233-256.

Jones, E.G. (2000). Cortical and subcortical contributions to activity-dependent plasticity in primate somatosensory cortex. *Annual Reviews of Neuroscience, 23,* 1-37.

Jones, K.E., Hamilton, A.F., & Wolpert, D.M. (2002). Sources of signal-dependent noise during isometric force production. *Journal of Neurophysiology, 88,* 1533-1544.

Jones, K.E., Wessberg, J., & Vallbo, Å.B. (2001). Directional tuning of human forearm muscle afferents during voluntary wrist movements. *Journal of Physiology, 536,* 635-647.

Jones, L.A. (1995). The senses of effort and force during fatiguing contractions. In S.C. Gandevia, R.M. Enoka, A.J. McComas, D.G. Stuart, & C.K. Thomas (Eds.), *Fatigue: Neural and Muscular Mechanisms* (pp. 305-313). New York: Plenum Press.

Jones, L.A., & Hunter, I.W. (1983). Effect of fatigue on force sensation. *Experimental Neurology, 81,* 640-650.

Jones, T.A., Chu, C.J., Grande, L.A., & Gregory, A.D. (1999). Motor skills training enhances lesion-induced structural plasticity in the motor cortex of adult rats. *Journal of Neuroscience, 19,* 10153-10163.

Jordan, L.M. (1998). Initiation of locomotion in mammals. *Annals of the New York Academy of Sciences, 860,* 83-93.

Jordan, L.M., Liu, J., Hedlund, P.B., Akay, T., & Pearson, K.G. (2008). Descending command systems for the initiation of locomotion in mammals. *Brain Research Reviews, 57,* 183-191.

Jordan, M.J. Norris, S.R., Smith, D.J., & Herzog, W. (2005). Vibration training: An overview of the area, training consequences, and future considerations. *Journal of Strength and Conditioning Research, 19,* 459-466.

Josephson, R.K. (1985). Mechanical power output from striated muscle during cyclic contraction. *Journal of Experimental Biology, 114,* 493-512.

Josephson, R.K. (1993). Contraction dynamics and power output of skeletal muscle. *Annual Reviews in Physiology, 55,* 527-546.

Josephson, R.K. (1999). Dissecting muscle power. *Journal of Experimental Biology, 202,* 3369-3375.

Jovanovic, K., Petrov, T., Greer, J.J., & Stein, R.B. (1996). Serotonergic modulation of the mudpuppy (Necturus maculatus) locomotor pattern in vitro. *Experimental Brain Research, 111,* 57-67.

Joyce, G.C., Rack, P.M.H., & Westbury, D.R. (1969). The mechanical properties of cat soleus muscle during controlled lengthening and shortening movements. *Journal of Physiology, 204,* 461-474.

Jubeau, M., Zory, R., Gondin, J., Martin, A., & Maffiuletti, N.A. (2006). Late neural adaptations to electrostimulation resistance training of the plantar flexor muscles. *European Journal of Applied Physiology, 98,* 202-211.

Kabat, H., & Knott, M. (1953). Proprioceptive facilitation techniques for treatment of paralysis. *Physical Therapy Reviews, 33,* 53-64.

Kadi, F. (2000). Adaptation of human skeletal muscle to training and anabolic steroids. *Acta Physiologica Scandinavica, 168* (Suppl. 646), 1-52.

Kadi, F., Charifi, N., Denis, C., Lexell, J., Andersen, J.L., Schjerling, P., Olsen, C., & Kjaer, M. (2005). The behaviour of satellite cells in response to exercise: What have we learned from human studies? *Pflügers Archiv, 41,* 319-327.

Kadi, F., Charifi, H., & Henriksson, J. (2006). The number of satellite cells in slow and fast fibres from human vastus lateralis muscle. *Histochemistry and Cell Biology, 126,* 83-87.

Kadi, F., Eriksson, A., Holmner, S., Butler-Browne, G.S., & Thornell, L.E. (1999). Cellular adaptation of the trapezius muscle in strength-trained athletes. *Histochemistry and Cell Biology, 111,* 189-195.

Kadi, F., Schjerling, P., Andersen, L.L., Charifi, N., Madsen, J.L., Christensen, L.R., & Andersen, J.L. (2004). The effects of heavy resistance training and detraining on satellite cells in human skeletal muscles. *Journal of Physiology, 558,* 1005-1012.

Kadi, F., & Thornell, L.E. (2000). Concomitant increases in myonuclear and satellite cell content in female trapezius muscle following strength training. *Histochemistry and Cell Biology, 113,* 99-103.

Kadota, K., Matsuo, T., Hashizume, K., & Tezuka, K. (2004). Practice changes the usage of moment components in executing a multijoint task. *Research Quarterly for Exercise and Sport, 75,* 138-147.

Kakei, S., Hoffman, D.S., & Strick, P.L. (2001). Direction of action is represented in the ventral premotor cortex. *Nature Neuroscience, 4,* 1020-1025.

Kakei, S., Hoffman, D.S., & Strick, P.L. (2003). Sensorimotor transformations in cortical motor areas. *Neuroscience Research, 46,* 1-10.

Kakihana, W., & Suzuki, S. (2001). The EMG activity and mechanics of the running jump as a function of takeoff angle. *Journal of Electromyography and Kinesiology, 11,* 365-372.

Kakuda, N., & Nagaoka, M. (2000). Dynamic response of human muscle spindle afferents to stretch during voluntary contraction. *Journal of Physiology, 527,* 397-628.

Kakuda, N., Vallbo, Å.B., & Wessberg, J. (1996). Fusimotor and skeletomotor activities are increased with precision finger movement in man. *Journal of Physiology, 492,* 921-929.

Kalaska, J.F., Caminiti, R., & Georgopoulos, A.P. (1983). Cortical mechanisms related to the direction of two-dimensional arm movements: Relations in parietal area 5 and comparison with motor cortex. *Experimental Brain Research, 51,* 247-260.

Kalaska, J.F., & Crammond, D.J. (1995). Deciding not to GO: Neuronal correlates of response selection in a GO/NOGO task in primate premotor and parietal cortex. *Cerebral Cortex, 5,* 410-428.

Kamen, G., & Knight, C.A. (2004). Training-related adaptations in motor unit discharge rate in young and older adults. *Journal of Gerontology, 59,* 1334-1338.

Kamen, G., & Roy, A. (2000). Motor unit synchronization in young and elderly adults. *European Journal of Applied Physiology, 81,* 403-410.

Kamper, D.G., Fischer, H.C., Cruz, E.G., & Rymer, W.Z. (2006). Weakness is the primary contributor to finger impairment in chronic stroke. *Archives in Physical Medicine and Rehabilitation, 87,* 1262-1269.

Kanda, K., Burke, R.E., & Walmsley, B. (1977). Differential control of fast and slow twitch motor units in the decerebrate cat. *Experimental Brain Research, 29,* 57-74.

Kanda, K., & Desmedt, J.E. (1983). Cutaneous facilitation of large motor units and motor control in human fingers in precision grip. *Advances in Neurology, 39,* 253-261.

Kandarian, S.C., & White, T.P. (1990). Mechanical deficit persists during long-term muscle hypertrophy. *Journal of Applied Physiology, 69,* 861-867.

Kandarian, S.C., & Williams, J.H. (1993). Contractile properties of skinned fibers from hypertrophied muscle. *Medicine and Science in Sports and Exercise, 25,* 999-1004.

Kandel, E.R., & Schwartz, J.H. (1985). *Principles of neural science.* New York: Elsevier.

Kane, T.R., & Scher, M.P. (1969). A dynamical explanation of the falling cat phenomenon. *International Journal of Solids and Structures, 5,* 663-670.

Kanehisa, H., Nagareda, H., Kawakami, Y., Akima, H., Masani, K., Kouzaki, M., & Fukunaga, T. (2002). Effects of equivolume isometric training programs comprising medium or high resistance on muscle size and strength. *European Journal of Applied Physiology, 87,* 112-119.

Kannus, P., Parkkari, J., Koskinen, S., Niemi, S., Palvanen, M., Järvinen, M., & Vuori, I. (1999). Fall-induced injuries and deaths among older individuals. *Journal of the American Medical Association, 281,* 1895-1899.

Karamanidis, K., & Arampatzis, A. (2005). Mechanical and morphological properties of different muscle-tendon units in the lower extremity and running mechanics: Effect of aging and physical activity. *Journal of Experimental Biology, 208,* 3907-3923.

Karniel, A., & Mussa-Ivaldi, F.A. (2003). Sequence, time, or state representation: How does the motor control system adapt to variable environments? *Biological Cybernetics, 89,* 10-21.

Katz, R., Meunier, S., & Pierrot-Deseilligny, E. (1988). Changes in presynaptic inhibition of Ia fibres in man while standing. *Brain, 111,* 417-437.

Katz, R., Penicaud, A., & Rossi, A. (1991). Reciprocal Ia inhibition between elbow flexors and extensors in the human. *Journal of Physiology, 437,* 269-286.

Kaufman, K.R., An, K.-N., & Chao, E.Y.S. (1995). A comparison of intersegmental joint dynamics to isokinetic dynamometer measurements. *Journal of Biomechanics, 28,* 1243-1256.

Kaufman, M.P., Rybicki, K.J., Waldrop, T.G., & Ordway, G.A. (1984). Effect of ischemia on responses of group III and IV afferents to contraction. *Journal of Applied Physiology, 57,* 644-650.

Kawakami, Y., Abe, T., Kuno, S.Y., & Fukunaga, T. (1995). Training-induced changes in muscle architecture and specific tension. *European Journal of Applied Physiology, 72,* 37-43.

Kawakami, Y., & Fukunaga, T. (2006). New insights into in vivo human skeletal muscle function. *Exercise and Sport Sciences Reviews, 34,* 16-21.

Kawakami, Y., Ichinose, Y., & Fukunaga, T. (1998). Architectural and functional features of human triceps surae muscles during contraction. *Journal of Applied Physiology, 85,* 398-404.

Kawakami, Y., Kubo, K., Kanehisa, H., & Fukunaga, T. (2002*a*). Effect of series elasticity on isokinetic torque-angle relationship in humans. *European Journal of Applied Physiology, 87,* 381-387.

Kawakami, Y., Kumagai, K., Huijing, P.A., Hijung, T., & Fukunaga, T. (2000). The length-force characteristics of human gastrocnemius and soleus muscle in vivo. In W. Herzog (Ed.), *Skeletal Muscle Mechanics: From Mechanisms to Function* (pp. 327-341). Chichester: Wiley.

Kawakami, Y., & Lieber, R.L. (2000). Interaction between series compliance and sarcomere kinetics determines internal sarcomere shortening during fixed-end contractions. *Journal of Biomechanics, 33,* 1249-1255.

Kawakami, Y., Muraoka, T., Ito, S., Kanehisa, H., & Fukunaga, T. (2002*b*). In vivo muscle-fibre behaviour during counter-movement exercise in human reveals significant role for tendon elasticity. *Journal of Physiology, 540,* 635-646.

Kawakami, Y., Nakazawa, K., Fujimoto, T., Nozaki, D., Miyashita, M., & Fukunaga, T. (1994). Specific tension of elbow flexor and extensor muscles based on magnetic resonance imaging. *European Journal of Applied Physiology, 68,* 139-147.

Kawamori, N., & Haff, G.G. (2004). The optimal training load for the development of muscular power. *Journal of Strength and Conditioning Research, 18,* 675-684.

Kawashima, N., Nakazawa, K., Yamamoto, S.I., Nozaki, D., Akai, M., & Yano, H. (2004). Stretch reflex excitability of the anti-gravity ankle extensor muscle in elderly humans. *Acta Physiologica Scandinavica, 180,* 99-105.

Kawashima, N., Nozaki, D., Abe, M.O., Akai, M., & Nakazawa, K. (2005). Alternate leg movement amplifies locomotor-like muscle activity in spinal cord injured persons. *Journal of Neurophysiology, 93,* 777-785.

Kawato, M., Kuroda, T., Imamizu, H., Nakano, E., Miyauchi, S., & Yoshioka, T. (2003). Internal forward models in the cerebellum: fMRI study on grip force and load force coupling. *Progress in Brain Research, 142,* 171-188.

Keen, D.A., & Fuglevand, A.J. (2004). Distribution of motor unit force in human extensor digitorum assessed by spike-triggered averaging and intraneural microstimulation. *Journal of Neurophysiology, 91,* 2515-2523.

Keen, D.A., Yue, G.H., & Enoka, R.M. (1994). Training-related enhancement in the control of motor output in elderly humans. *Journal of Applied Physiology, 77,* 2648-2658.

Keenan, K.G., Farina, D., Maluf, K.S., Merletti, R., & Enoka, R.M. (2005). Influence of amplitude cancellation on the simulated surface electromyogram. *Journal of Applied Physiology, 98,* 120-131.

Keenan, K.G., Farina, D., Merletti, R., & Enoka, R.M. (2006). Influence of motor unit properties on the size of the simulated evoked surface EMG potential. *Experimental Brain Research, 169,* 37-49.

Keller, A., Arissian, K., & Asanuma, H. (1992). Synaptic proliferation in the motor cortex of adult cats after long-term thalamic stimulation. *Journal of Neurophysiology, 68,* 295-308.

Keller-Peck, C.R., Feng, G., Sanes, J.R., Yan, Q., Lichtman, J.W., & Snider, W.D. (2001). Glial cell line-derived neurotrophic factor administration in postnatal life results in motor unit enlargement and continuous synaptic remodeling at the neuromuscular junction. *Journal of Neuroscience, 21,* 6136-6146.

Kellis, E., & Baltzopoulos, V. (1995). Isokinetic eccentric exercise. *Sports Medicine, 19,* 202-222.

Kellis, E., & Baltzopoulos, V. (1997). The effects of antagonist moment on the resultant knee joint moment during isokinetic testing of the knee extensors. *European Journal of Applied Physiology, 76,* 253-259.

Kellis, E., & Baltzopoulos, V. (1999). In vivo determination of the patella tendon and hamstrings moment arm in adult males using videofluoroscopy during submaximal knee extension and flexion. *Clinical Biomechanics, 14,* 118-124.

Kent-Braun, J.A., & Le Blanc, R. (1996). Quantitation of central activation failure during maximal voluntary contractions in humans. *Muscle & Nerve, 19,* 861-869.

Kent-Braun, J.A., & Ng, A.V. (1999). Specific strength and voluntary muscle activation in young and elderly women and men. *Journal of Applied Physiology, 87,* 22-29.

Kerdok, A.E., Biewener, A.A., McMahon, T.A., Weyand, P.G., & Herr, H.M. (2002). Energetics and mechanics of human running on surfaces of different stiffnesses. *Journal of Applied Physiology, 92,* 469-478.

Kern, H., Boncompagni, S., Rossini, K., Mayr, W., Fano, G., Zanin, M.E., Podhorska-Okolow, M., Protasi, F., & Carraro, U. (2004). Long-term denervation in humans causes degeneration of both contractile and excitation-contraction coupling apparatus, which is reversible by functional electrical stimulation (FES): A role for myofiber regeneration? *Journal of Neuropathology and Experimental Neurology, 63,* 919-931.

Kernell, D. (1965). The adaptation and the relation between discharge frequency and current strength of cat lumbosacral motoneurones stimulated by long-lasting injected currents. *Acta Physiologica Scandinavica, 65,* 65-73.

Kernell, D. (1992). Organized variability in the neuromuscular system: A survey of task-related adaptations. *Archives Italiennes de biologie, 130,* 19-66.

Kernell, D. (2006). *The Motoneurone and Its Muscle Fibres.* Oxford, Great Britain: Oxford University Press.

Kernell, D., & Hultborn, H. (1990). Synaptic effects on recruitment gain: A mechanism of importance for the input-output relations of motoneurone pools? *Brain Research, 507,* 176-179.

Kidgell, D.J., Sale, M.V., & Semmler, J.G. (2006). Motor unit synchronization measured by cross-correlation is not influenced by short-term strength training of a hand muscle. *Experimental Brain Research, 175,* 745-753.

Kido, A., Tanaka, N., & Stein, R.B. (2004). Spinal excitation and inhibition decrease as humans age. *Canadian Journal of Physiology and Pharmacology, 82,* 238-248.

Kiehn, O. (2006). Locomotor circuits in the mammalian spinal cord. *Annual Reviews of Neuroscience, 29,* 279-306.

Kiehn, O., & Butt, S.J. (2003). Physiological, anatomical and genetic identification of CPG neurons in the developing spinal cord. *Progress in Neurobiology, 70,* 347-361.

Kiehn, O., Hounsgaard, J., & Sillar, K.T. (1997). Basic building blocks of vertebrate spinal central pattern generators. In P.S.G. Stein, S. Grillner, A.I. Selverston, & D.G. Stuart (Eds.), *Neurons, Networks, and Motor Behavior* (pp. 47-59). Cambridge, MA: MIT Press.

Kiehn, O., & Kjærulff, O. (1998). Distribution of central pattern generators for rhythmic motor outputs in the spinal cord of limbed vertebrates. *Annals of the New York Academy of Sciences, 860,* 110-129.

Kiehn, O., & Kullander, K. (2004). Central pattern generators deciphered by molecular genetics. *Neuron, 41,* 317-321.

Kiehn, O., Quinlan, K.A., Restrepo, C.E., Lundfald, L., Borgius, L., Talpalar, A.E., & Endo, T. (2008). Excitatory components of the mammalian locomotor CPG. *Brain Research Reviews, 57,* 55-63.

Kilbreath, S.L., Gorman, R.B., Raymond, J., & Gandevia, S.C. (2002). Distribution of forces produced by motor unit activity in the human flexor digitorum profundus. *Journal of Physiology, 543,* 289-296.

Kilgore, J.B., & Mobley, B.A. (1991). Additional force during stretch of single frog muscle fibres following tetanus. *Experimental Physiology, 76,* 579-588.

Kim, J.S., Kosek, D.J., Petrella, J.K., Cross, J.M., & Bamman, M.M. (2005). Resting and load-induced levels of myogenic gene transcripts differ between older adults with demonstrable sarcopenia and young men and women. *Journal of Applied Physiology, 99,* 2149-2158.

Kimberley, T.J., Lewis, S.M., Auerbach, E.J., Dorsey, L.L., Lojovich, J.M., & Carey, J.R. (2004). Electrical stimulation driving functional improvements and cortical changes in subjects with stroke. *Experimental Brain Research, 154,* 450-460.

Kinugasa, R., Kawakami, Y., & Fukunaga, T. (2005). Muscle activation and its distribution within human triceps surae muscles. *Journal of Applied Physiology, 99,* 1149-1156.

Kirkwood, P.A. (1979). On the use and interpretation of cross-correlation measurements in the mammalian central nervous system. *Journal of Neuroscience Methods, 1,* 107-132.

Kirsch, R.F., Boskov, D., & Rymer, W.Z. (1994). Muscle stiffness during transient and continuous movements of cat muscle: Perturbation characteristics and physiological relevance. *IEEE Transactions on Biomedical Engineering, 41,* 758-770.

Kirsch, R.F., & Kearney, R.E. (1997). Identification of time-varying stiffness dynamics of the human ankle joint during an imposed movement. *Experimental Brain Research, 114,* 71-85.

Kirschbaum, C., & Hellhammer, D.H. (1994). Salivary cortisol in psychoneuroendocrine research: Recent developments and applications. *Pscyhoneuroendocrinology, 19,* 313-333.

Kitahara, A., Hamaoka, T., Murase, N., Homma, T., Kurosawa, Y., Ueda, C., Nagasawa, T., Ichimura, S., Motobe, M., Yashiro, K., Nakano, S., & Katsumura, T. (2003). Deterioration of muscle function after 21-day forearm immobilization. *Medicine and Science in Sports and Exercise, 35,* 1697-1702.

Kitai, T.A., & Sale, D.G. (1989). Specificity of joint angle in isometric training. *European Journal of Applied Physiology, 58,* 744-748.

Kitamura, K., Tokunaga, M., Iwane, A., H., & Yanagida, T. (1999). A single myosin head moves along an actin filament with regular steps of 5.3 nanometres. *Nature, 397,* 129-134.

Klakowicz, P.M., Baldwin, E.R., & Collins, D.F. (2006). Contribution of M-waves and H-reflexes to contractions evoked by tetanic nerve stimulation in humans. *Journal of Neurophysiology, 96,* 1293-1302.

Klass, M., Baudry, S., & Duchateau, J. (2005). Aging does not affect voluntary activation of the ankle dorsiflexors during isometric, concentric, and eccentric contractions. *Journal of Applied Physiology, 99,* 31-38.

Klass, M., Baudry, S., & Duchateau, J. (2007*a*). Age-related decline in rate of torque development is accompanied by lower maximal motor unit discharge frequency during ballistic contractions. *Journal of Neurophysiology, 100,* 515-525.

Klass, M., Baudry, S., & Duchateau, J. (2007*b*). Voluntary activation during maximal contraction with advancing age: A brief review. *European Journal of Applied Physiology, 100,* 543-551.

Klass, M., Guissard, N., & Duchateau, J. (2004). Limiting mechanisms of force production after repetitive dynamic contractions in human triceps surae. *Journal of Applied Physiology, 96,* 1516-1521.

Klass, M., Lévénez, M., Enoka, R.M., & Duchateau, J. (2008). Spinal mechanisms contribute to differences in time to failure of submaximal fatiguing contractions performed with different loads. *Journal of Neurophysiology, 99,* 1096-1104.

Klein, C.S., Hage-Ross, C.K., & Thomas, C.K. (2006). Fatigue properties of human thenar motor units paralysed by chronic spinal cord injury. *Journal of Physiology, 573,* 161-171.

Klein, C.S., Marsh, G.D., Petrella, R.J., & Rice, C.L. (2003). Muscle fiber number in the biceps brachii muscle of young and old men. *Muscle & Nerve, 28,* 62-68.

Klein, C.S., Rice, C.L., & Marsh, G.D. (2001). Normalized force, activation, and coactivation in the arm muscles of young and old men. *Journal of Applied Physiology, 91,* 1341-1349.

Kleine, B., Stegeman, D.F., Mund, D., & Anders, C. (2001). Influence of motoneuron firing synchronization on SEMG characteristics in dependence of electrode position. *Journal of Applied Physiology, 91,* 1588-1599.

Kluding, P., & Billinger, S.A. (2005). Exercise-induced changes of the upper extremity in chronic stroke survivors. *Topics in Stroke Rehabilitation, 12,* 58-68.

Kniffki, K.D., Schomburg, E.D., & Steffens, H. (1979). Synaptic responses of lumbar alpha-motoneurones to chemical algesic stimulation of skeletal muscle in spinal cats. *Brain Research, 160,* 549-552.

Kniffki, K.D., Schomburg, E.D., & Steffens, H. (1980). Action of muscular group III and IV afferents on spinal locomotor activity in cat. *Brain Research, 186,* 445-447.

Kniffki, K.D., Schomburg, E.D., & Steffens, H. (1981*a*). Convergence in segmental reflex pathways from fine muscle afferents and cutaneous or group II muscle afferents to alpha-motoneurones. *Brain Research, 218,* 342-346.

Kniffki, K.D., Schomburg, E.D., & Steffens, H. (1981*b*). Synaptic effects from chemically activated fine muscle afferents upon alpha-motoneurones in decerebrate and spinal cats. *Brain Research, 206,* 361-370.

Knight, C.A., & Kamen, G. (2001). Adaptations in muscle activation of the knee extensor muscles with strength training in young and older adults. *Journal of Electromyography and Kinesiology, 11,* 405-412.

Knight, C.A., & Kamen, G. (2007). Modulation of motor unit firing rates during a complex sinusoidal force task in young and older adults. *Journal of Applied Physiology, 102,* 122-129.

Knott, M., & Voss, D.E. (1968). *Proprioceptive Neuromuscular Facilitation: Patterns and Techniques* (2nd ed.). New York: Hoeber Medical Division, Harper & Row.

Koceja, D.M. (1995). Quadriceps mediated changes in soleus motoneuron excitability. *Electromyography and Clinical Neurophysiology, 35,* 25-30.

Koch, G., Franca, M., Del Olmo, M.F., Cheeran, B., Milton, R., Alvarez-Sauco, M.A., & Rothwell, J.C. (2006). Time course of functional connectivity between dorsal premotor and contralateral motor cortex during movement selection. *Journal of Neuroscience, 26,* 7452-7459.

Koegh, J., Morrison, S., & Barrett, R. (2006). Age-related differences in inter-digit coupling during finger pinching. *European Journal of Applied Physiology, 97,* 76-88.

Koerber, H.R., & Mendell, L.M. (1991). Modulation of synaptic transmission at Ia-afferent fiber connections on motoneurons during high-frequency stimulation: Role of postsynaptic target. *Journal of Neurophysiology, 65,* 590-597.

Kofotolis, N., & Kellis, E. (2006). Effects of two 4-week proprioceptive neuromuscular facilitation programs on muscle endurance, flexibility, and functional performance in women with chronic low back pain. *Physical Therapy, 86,* 1001-1012.

Koh, T.J., Grabiner, M.D., & Clough, C.A. (1993). Bilateral deficit is larger for step than for ramp isometric contractions. *Journal of Applied Physiology, 74,* 1200-1205.

Kohnstamm, O. (1915). Demonstration einer katatonieartigen Erscheinung beim esunden (Katatonusversuch). *Neurologisches Centralblatt, 34,* 290-291.

Kolev, V., Falkenstein, M., & Yordanova, J. (2006). Motor-response generation as a source of aging-related behavioural slowing in choice-reaction tasks. *Neurobiology of Aging, 27,* 1719-1730.

Komi, P.V. (1990). Relevance of in vivo force measurements to human biomechanics. *Journal of Biomechanics, 23* (Suppl. 1), 23-34.

Komi, P.V. (1992). Stretch-shortening cycle. In P. V. Komi (Ed.), *Strength and Power in Sport* (pp. 169-179). Champaign, IL: Human Kinetics.

Komi, P.V., Belli, A., Huttunen, V., Bonnefroy, R., Greyssant, A., & Lacour, J.R. (1996). Optic fibre as a transducer of tendomuscular forces. *European Journal of Applied Physiology, 72,* 278-280.

Komi, P.V., & Nicol, C. (2000). Stretch-shortening cycle of muscle function. In V.M. Zatsiorsky (Ed.), *Biomechanics in Sport* (pp. 87-102). Oxford, UK: Blackwell Science.

Komistek, R.D., Kane, T.R., Mahfouz, M., Ochoa, J.A., & Dennis, D.A. (2005). Knee mechanics: A review of past and present techniques to determine in vivo loads. *Journal of Biomechanics, 38,* 215-228.

Kopin, I.J. (1995). Definitions of stress and sympathetic neuronal responses. *Stress, 771,* 19-30.

Korhonen, M.T., Cristea, A., Alén, M., Häkkinen, K., Sipila, S., Mero, A., Viitasalo, J.T., Larsson, L., & Suominen, H. (2006). Aging, muscle fiber type, and contractile function in sprint-trained athletes. *Journal of Applied Physiology, 101,* 906-917.

Koshland, G.F., Galloway, J.C., & Nevoret-Bell, C.J. (2000). Control of the wrist in three-joint arm movements to multiple directions in the horizontal plane. *Journal of Neurophysiology, 83,* 3188-3195.

Kossev, A.R., Schrader, C., Dauper, J., Dengler, R., & Rollnik, J.D. (2002). Increased intracortical inhibition in middle-aged humans: A study using paired-pulse transcranial magnetic stimulation. *Neuroscience Letters, 333,* 83-86.

Kouzaki, M., & Shinohara, M. (2006). The frequency of alternate muscle activity is associated with the attenuation of muscle fatigue. *Journal of Applied Physiology, 101,* 715-720.

Kouzaki, M., Shinohara, M., Masani, K., Tachi, M., Kanehisa, H., & Fukunaga, T. (2003). Local blood circulation among knee extensor synergists in relation to alternate muscle activity during low-level sustained contraction. *Journal of Applied Physiology, 95,* 49-56.

Kovanen, V. (2002). Intramuscular extracellular matrix: Complex environment of muscle cells. *Exercise and Sport Sciences Reviews, 30,* 20-25.

Kozhina, G.V., Person, R.S., Popov, K.E., Smetanin, B.N., & Shlikov, V.Y. (1996). Motor unit discharge during muscular after-contraction. *Journal of Electromyography and Kinesiology, 6,* 169-175.

Kraemer, W.J., Adams, K., Cafarelli, E., Dudley, G.A., Dooly, C., Feigenbaum, M.S., Fleck, S.J., Franklin, B., Fry, A.C., Hoffman, J.R., Newton, R.U., Potteiger, J., Stone, M.H., Ratamess, N.A., & Tripplett-McBride, T. (2002). American College of Sports Medicine position stand. Progression models in resistance training for healthy adults. *Medicine and Science in Sports and Exercise, 34,* 364-380.

Kram, R. (2000). Muscular force or work: What determines the metabolic energy cost of running? *Exercise and Sport Sciences Reviews, 28,* 138-142.

Kram, R., Domingo, A., & Ferris, D.P. (1997). Effect of reduced gravity on the preferred walk-run transition speed. *Journal of Experimental Biology, 200,* 821-826.

Kram, R., & Taylor, C.R. (1990). Energetics of running: A new perspective. *Nature, 346,* 265-267.

Kramer, A.F., Erickson, K.I., & Colcombe, S.J. (2006). Exercise, cognition, and the aging brain. *Journal of Applied Physiology, 101,* 1237-1242.

Kranz, H., Williams, A.M., Cassell, J., Caddy, D.J., & Silberstein, R.B. (1983). Factors determining the frequency content of the electromyogram. *Journal of Applied Physiology, 55,* 392-399.

Krevolin, J.L., Pandy, M.G., & Pearce, J.C. (2004). Moment arm of the patellar tendon in the human knee. *Journal of Biomechanics, 37,* 785-788.

Krishnamoorthy, V., Goodman, S., Zatsiorsky, V.M., & Latash, M.L. (2003). Muscle synergies during shifts of the center of pressure by standing persons: Identification of muscle modes. *Biological Cybernetics, 89,* 152-161.

Krishnamoorthy, V., & Latash, M.L. (2005). Reversals of anticipatory postural adjustments during voluntary sway in humans. *Journal of Physiology, 565,* 675-684.

Krivickas, L.S., Fielding, R.A., Murray, A., Callahan, D., Johansson, A., Dorer, D.J., & Frontera, W.R. (2006). Sex differences in single muscle fiber power in older adults. *Medicine and Science in Sports and Exercise, 38,* 57-63.

Krnjevic, K., & Miledi, R. (1958). Failure of neuromuscular propagation in rats. *Journal of Physiology, 140,* 440-461.

Krouchev, N., Kalaska, J.F., & Drew, T. (2006). Sequential activation of muscle synergies during locomotion in the intact cat as revealed by cluster analysis and direct decomposition. *Journal of Neurophysiology, 96,* 1991-2010.

Kruidhof, J., & Pandy, M.G. (2006). Effect of muscle wrapping on model estimates of neck muscle strength. *Computer Methods in Biomechanics and Biomedical Engineering, 9,* 343-352.

Kubo, K., Kanehisa, H., & Fukunaga, T. (2001*a*). Effects of different isometric contractions on tendon elasticity in human quadriceps femoris. *Journal of Physiology, 536,* 649-655.

Kubo, K., Kanehisa, H., & Fukunaga, T. (2002). Effect of stretching training on the viscoelastic properties of human tendon structures in vivo. *Journal of Applied Physiology, 92,* 595-601.

Kubo, K., Kanehisa, H., & Fukunaga, T. (2005). Comparison of elasticity of human tendon and aponeurosis in knee extensors and ankle plantar flexors in vivo. *Journal of Applied Biomechanics, 21,* 129-142.

Kubo, K., Kanehisa, H., Kawakami, Y., & Fukunaga, T. (2001*b*). Influence of static stretching on viscoelastic properties of human tendon structures in vivo. *Journal of Applied Physiology, 90,* 520-527.

Kubo, K., Kanehisa, H., Miyatani, M., Tachi, M., & Fukunaga, T. (2003). Effect of low-load resistance training on the tendon properties in middle-aged and elderly women. *Acta Physiologica Scandinavica, 178,* 25-32.

Kubo, K., Morimoto, M., Komuro, T., Tsunoda, N., Kanehisa, H., & Fukunaga, T. (2007*a*). Age-related differences in the properties of the plantar flexor muscles and tendons. *Medicine and Science in Sports and Exercise, 39,* 541-547.

Kubo, K., Morimoto, M., Komuro, T., Tsunoda, N., Kanehisa, H., & Fukunaga, T. (2007*b*). Influences of tendon stiffness, joint stiffness, and electromyographic activity on jump performance using single joint. *European Journal of Applied Physiology, 99,* 235-243.

Kuchinad, R.A., Ivanova, T.D., & Garland, S.J. (2004). Modulation of motor unit discharge rate and H-reflex amplitude during submaximal fatigue of the human soleus muscle. *Experimental Brain Research, 158,* 345-355.

Kuechle, D.K., Newman, S.R., Itoi, E., Niebur, G.L., Morrey, B.F., & An, K.N. (2000). The relevance of the moment arm of shoulder muscles with respect to axial rotation of the glenohumeral joint in four positions. *Clinical Biomechanics, 15,* 322-329.

Kugelberg, E. (1962). Polysynaptic reflexes of clinical importance. *Electroencephalography and Clinical Neurophysiology,* Suppl. 22, 111.

Kugelberg, E., & Lindegren, B. (1979). Transmission and contraction fatigue of rat motor units in relation to succinate dehydrogenase activity of motor unit fibres. *Journal of Physiology, 288,* 285-300.

Kuitunen, S., Komi, P.V., & Kryolainen, H. (2002). Knee and ankle joint stiffness in spring running. *Medicine and Science in Sports and Exercise, 34,* 166-173.

Kukulka, C.G., & Clamann, H.P. (1981). Comparison of the recruitment and discharge properties of motor units in human brachial biceps and adductor pollicis during isometric contractions. *Brain Research, 219,* 45-55.

Kuno, M. (1964). Mechanism of facilitation and depression of the excitatory synaptic potential in spinal motoneurons. *Journal of Physiology, 175,* 100-112.

Kurata, K., & Hoshi, E. (2002). Movement-related neuronal activity reflecting the transformation of coordinates in the ventral premotor cortex of monkeys. *Journal of Neurophysiology, 88,* 3118-3132.

Kurokawa, S., Fukunaga, T., Nagano, A., & Fukashiro, S. (2003). Interaction between fascicles and tendinous structures during counter movement jumping investigated in vivo. *Journal of Applied Physiology, 95,* 2306-2314.

Kuwabara, S., Cappelen-Smith, C., Lin, C.S., Mogyoros, I., & Burke, D. (2002). Effects of voluntary activity on the excitability of motor axons in the peroneal nerve. *Muscle & Nerve, 25,* 176-184.

Kvorning, T., Bagger, M., Caserotti, P., & Madsen, K. (2006). Effects of vibration and resistance training on neuromuscular and hormonal measures. *European Journal of Applied Physiology, 96,* 615-625.

Kwon, O., & Lee, K.W. (2004). Reproducibility of statistical motor unit number estimates in amyotrophic lateral sclerosis: Comparisons between size- and number-weighted modifications. *Muscle & Nerve, 29,* 211-217.

Kyparos, A., Feeback, D.L., Layne, C.S., Martinez, D.A., & Clarke, M.S. (2005). Mechanical stimulation of the plantar foot surface attenuates soleus muscle atrophy induced by hindlimb unloading in rats. *Journal of Applied Physiology, 99,* 739-746.

Laaksonen, M.S., Kyrolainen, H., Kalliokoski, K.K., Nuutila, P., & Knuuti, J. (2006). The association between muscle EMG and perfusion in knee extensor muscles. *Clinical Physiology and Functional Imaging, 26,* 99-105.

Labarque, V.L., Op 't Eijnde, B., & Van Leemputte, M. (2002). Effect of immobilization and retraining on torque-velocity relationship of human knee flexor and extensor muscles. *European Journal of Applied Physiology, 86,* 251-257.

Lackner, J.R., & DiZio, P. (2005). Motor control and learning in altered dynamic environments. *Current Opinion in Neurobiology, 15,* 653-659.

Lacquaniti, F., Guigon, E., Bianchi, L., Ferraina, S., & Caminiti, R. (1995). Representing spatial information for limb movement: Role of area 5 in the monkey. *Cerebral Cortex, 5,* 391-409.

Ladewig, T., Lalley, P.M., & Keller, B.U. (2004). Serotonergic modulation of intracellular calcium dynamics in neonatal hypoglossal motoneurons from mouse. *Brain Research, 1001,* 1-12.

Ladin, Z., & Wu, G. (1991). Combining position and acceleration measurements for joint force estimation. *Journal of Biomechanics, 24,* 1173-1187.

Lafleur, J., Zytnicki, D., Horcholle-Bossavit, G., & Jami, L. (1993). Declining inhibition in ipsi- and contralateral lumbar motoneurons during contractions of an ankle extensor muscle in the cat. *Journal of Neurophysiology, 70,* 1797-1804.

Lagerquist, O., Zehr, E.P., & Docherty, D. (2006). Increased spinal reflex excitability is not associated with neural plasticity underlying the cross-education effect. *Journal of Applied Physiology, 100,* 83-90.

Lago, P., & Jones, N.D. (1977). Effect of motor unit firing time statistics on EMG spectra. *Medical and Biological Engineering and Computing, 15,* 648-655.

Lai, E.J., Hodgson, A.J., & Milner, T.E. (2003). Influence of interaction force levels on degree of motor adaptation in a stable dynamic force field. *Experimental Brain Research, 153,* 76-83.

Laidlaw, D.H., Bilodeau, M., & Enoka, R.M. (2000). Steadiness is reduced and motor unit discharge is more variable in old adults. *Muscle & Nerve, 23,* 600-612.

Laidlaw, D.H., Kornatz, K.W., Keen, D.A., Suzuki, S., & Enoka, R.M. (1999). Strength training improves the steadiness of slow lengthening contractions performed by old adults. *Journal of Applied Physiology, 87,* 1786-1795.

Lakie, M., & Robson, L.G. (1988*a*). Thixotropic changes in human muscle stiffness and the effects of fatigue. *Quarterly Journal of Experimental Physiology, 73,* 487-500.

Lakie, M., & Robson, L.G. (1988*b*). Thixotropy: The effect of stretch size in relaxed frog muscle. *Quarterly Journal of Experimental Physiology, 73,* 127-129.

Lakomy, H.K.A. (1987). Measurement of human power output in high intensity exercise. In B. Van Gheluwe & J. Atha (Eds.), *Current Research in Sport Biomechanics* (pp. 46-57). Basel: Karger.

Lam, T., & Pearson, K.G. (2002). The role of proprioceptive feedback in the regulation and adaptation of locomotor activity. *Advances in Experimental and Medical Biology, 508,* 343-355.

Lamb, G.D. (2002). Excitation–contraction coupling and fatigue mechanisms in skeletal muscle: Studies with mechanically skinned fibres. *Journal of Muscle Research and Cell Motility, 23,* 81-91.

Lamb, T., & Yang, J.F. (2000). Could different directions of infant stepping be controlled by the same locomotor central pattern generator? *Journal of Neurophysiology, 83,* 2814-2824.

Lambertz, D., Goubel, F., Kaspranski, R., & Perot, C. (2003). Influence of long-term spaceflight on neuromechanical properties of muscles in humans. *Journal of Applied Physiology, 94,* 490-498.

Lamont, E.V., & Zehr, E.P. (2006). Task-specific modulation of cutaneous reflexes expressed at functionally relevant gait cycle phases during level and incline walking and stair climbing. *Experimental Brain Research, 173,* 185-192.

Lan, C., Lai, J-S., Chen, S-Y., & Wong, M-K. (1998). 12-month Tai Chi training in the elderly: Its effect on health fitness. *Medicine and Science in Sports and Exercise, 30,* 345-351.

Lander, J.E., Bates, B.T., & DeVita, P. (1986). Biomechanics of the squat exercise using a modified center of mass bar. *Medicine and Science in Sports and Exercise, 18,* 468-478.

Lander, J.E., Hundley, J.R., & Simonton, R.L. (1992). The effectiveness of weight-belts during multiple repetitions of the squat exercise. *Medicine and Science in Sports and Exercise, 24,* 603-609.

Landis, C., & Hunt, W.A. (1939). *The Startle Pattern.* New York: Farrar and Rinehart.

Lang, T.F., Leblanc, A.D., Evans, H.J., & Lu, Y. (2006). Adaptation of the proximal femur to skeletal loading after long-duration spaceflight. *Journal of Bone and Mineral Research, 21,* 1224-1230.

Lännergren, J., & Westerblad, H. (1989). Maximum tension and force-velocity properties of fatigued, single *Xenopus* muscle fibres studied by caffeine and high K^+. *Journal of Physiology, 409,* 473-490.

Lanza, I.R., Towse, T.F., Caldwell, G.E., Wigmore, G.E., & Kent-Braun, J.A. (2003). Effects of age on human muscle torque, velocity, and power in two muscle groups. *Journal of Applied Physiology, 95,* 2361-2369.

Lanzetta, M., Pozzo, M., Bottin, A., Merletti, R., & Farina, D. (2005). Reinnervation of motor units in intrinsic muscles of a transplanted hand. *Neuroscience Letters, 373,* 138-143.

Laouris, Y., Kalli-Laouri, J., & Schwartze, P. (1990). The postnatal development of the air-righting reaction in albino rats. Quantitative analysis of normal development and the effect of preventing neck-torso and torso-pelvis rotations. *Behavioral Brain Research, 37,* 37-44.

Larsson, L., Li, X., Berg, H.E., & Frontera, W.R. (1996). Effects of removal of weight-bearing function on contractility and myosin isoform in single human skeletal muscle cells. *Pflügers Archiv, 432,* 320-328.

Larsson, L., Li, X., & Frontera, W.R. (1997). Effects of aging on shortening velocity and myosin isoform composition in single human skeletal muscle cells. *American Journal of Physiology, 272,* C638-C649.

Latash, M.L., Ferreira, S.S., Wieczorek, S.A., & Duarte, M. (2003). Movement sway: Changes in postural sway during voluntary shifts in the center of pressure. *Experimental Brain Research, 150,* 314-324.

Laughman, R.K., Youdas, J.W., Garrett, T.R., & Chao, E.Y.S. (1983). Strength changes in the normal quadriceps femoris muscle as a result of electrical stimulation. *Physical Therapy, 63,* 494-499.

Laughton, C.A., Slavin, M., Katdare, K., Nolan, L., Bean, J.F., Kerrigan, D.C., Phillips, E., Lipsitz, L.A., & Collins, J.J. (2003). Aging, muscle activity, and balance control: Physiologic changes associated with balance control. *Gait & Posture, 18,* 101-108.

Lawrence, J.H., & De Luca, C.J. (1983). Myoelectric signal versus force relationship in different human muscles. *Journal of Applied Physiology, 54,* 1653-1659.

Lawrence, J.H., Nichols, T.R., & English, A.W. (1993). Cat hindlimb muscles exert substantial torques outside the sagittal plane. *Journal of Neurophysiology, 69,* 282-285.

Lee, C.R., & Farley, C.T. (1998). Determinants of the center of mass trajectory in human walking and running. *Journal of Experimental Biology, 201,* 2935-2944.

Lee, M., & Carroll, T.J. (2007). Cross education: Possible mechanisms for the contralateral effects of unilateral resistance training. *Sports Medicine, 37,* 1-14.

Lee, W., Karwowski, W., Marras, W.S., & Rodrick, D. (2003). A neuro-fuzzy model for estimating electromyographical activity of trunk muscles due to manual lifting. *Ergonomics, 46,* 285-309.

Lemay, M., & Stelmach, G.E. (2005). Multiple frames of reference for pointing to a remembered target. *Experimental Brain Research, 164,* 301-310.

Lemon, R.N., & Griffiths, J. (2005). Comparing the function of the corticospinal system in different species: Organizational differences for motor specialization? *Muscle & Nerve, 32,* 261-279.

Leocani, L., Colombo, B., Magnani, G., Martinelli-Boneschi, F., Cursi, M., Rossi, P., Martinelli, V., & Comi, G. (2001). Fatigue in multiple sclerosis is associated with abnormal cortical activation to voluntary movement—EEG evidence. *NeuroImage, 13,* 1186-1192.

Leonard, T.R., & Herzog, W. (2005). Does the speed of shortening affect steady-state force depression in cat soleus muscle? *Journal of Biomechanics, 38,* 2190-2197.

Leppik, J.A., Aughey, R.J., Medved, I., Fairweather, I., Carey, M.F., & McKenna, M.J. (2004). Prolonged exercise to fatigue in humans impairs skeletal muscle Na^+-K^+-ATPase activity, sarcoplasmic reticulum Ca^{2+} release, and Ca^{2+} uptake. *Journal of Applied Physiology, 97,* 1414-1423.

Lesmes, G.R., Costill, D.L., Coyle, E.F., & Fink, W.J. (1978). Muscle strength and power changes during maximal isokinetic training. *Medicine and Science in Sports, 10,* 266-269.

Leterme, D., & Casasnovas, B. (1999). Adaptation of rat lateral gastrocnemius muscle motor units during hindlimb unloading. *European Journal of Applied Physiology, 79,* 312-317.

Leterme, D., & Falempin, M. (1996). Contractile properties of rat soleus motor units following 14 days of hindlimb unloading. *Pflügers Archive, 432,* 313-319.

Leterme, D., & Tyc, F. (2004). Re-innervation and recovery of rat soleus muscle and motor unit function after nerve crush. *Experimental Physiology, 89,* 353-361.

Lévénez, M., Garland, S.J., Klass, M., & Duchateau, J. (2008). Cortical and spinal modulatioin of antagonist coactivation during a submaximal fatiguing contraction in humans. *Journal of Neurophysiology, 99,* 554-563.

Lévénez, M., Kotzamanidis, C., Carpentier, A., & Duchateau, J. (2005). Spinal reflexes and coactivation of ankle muscles during a submaximal fatiguing contraction. *Journal of Applied Physiology, 99,* 1182-1188.

Levin, M.F., & Hui-Chan, C.W.Y. (1992). Relief of hemiparetic spasticity by TENS is associated with improvement in reflex and voluntary motor functions. *Electroencephalography and Clinical Neurophysiology, 85,* 131-142.

Levin, M.F., Selles, R.W., Verheul, M.H., & Meijer, O.G. (2000). Deficits in the coordination of agonist and antagonist muscles in stroke patients: Implications for normal motor control. *Brain Research, 853,* 352-369.

Lewis, G.N., MacKinnon, C.D., & Perreault, E.J. (2006). The effect of task instruction on the excitability of spinal and supraspinal reflex pathways projecting to the biceps muscle. *Experimental Brain Research, 174,* 413-425.

Lexell, J., Downham, D.Y., Larsson, Y., Bruhn, E., & Morsing, B. (1995). Heavy-resistance training in older Scandinavian men and women: Short-and long-term effects on arm and leg muscles. *Scandinavian Journal of Medicine and Science in Sports, 5,* 329-341.

Lexell, J., Taylor, C.C., & Sjöstrom, M. (1988). What is the cause of ageing atrophy? Total number, size and proportion of different fiber types studies in whole vastus lateralis muscle from 15- to 83-year-old men. *Journal of the Neurological Sciences, 84,* 275-294.

Li, C., & Atwater, A.E. (1984). Temporal and kinematic analysis of arm motion in sprinters. Paper presented at the Olympic Scientific Congress, Eugene, OR.

Li, L., & Caldwell, G.E. (1998). Muscle coordination in cycling: Effect of surface incline and posture. *Journal of Applied Physiology, 85,* 927-934.

Li, S., Kamper, D.G., Stevens, J.A., & Rymer, W.Z. (2004). The effect of motor imagery on spinal segmental excitability. *Journal of Neuroscience, 24,* 9674-9680.

Li, Z.-M., Latash, M.L., & Zatsiorsky, V.M. (1998). Force sharing among fingers as a model of the redundancy problem. *Experimental Brain Research, 119,* 276-286.

Liao, H., & Belkoff, S.M. (1999). A failure model for ligaments. *Journal of Biomechanics, 32,* 183-188.

Liberson, W.T., Holmquest, H.J., Scot, D., & Dow, M. (1961). Functional electrotherapy: Stimulation of the peroneal nerve synchronized with the swing phase of the gait of hemiplegic patients. *Archives of Physical Medicine and Rehabilitation, 42,* 101-105.

Lichtwark, G.A., Bougoulias, K., & Wilson, A.M. (2007). Muscle fascicle and series elastic element length changes along the length of the human gastrocnemius during walking and running. *Journal of Biomechanics, 40,* 157-164.

Liddell, E.G.T., & Sherrington, C.S. (1924). Reflexes in response to stretch (myotatic reflexes). *Proceedings of the Royal Society of London B, 96,* 212-242.

Liddell, E.G.T., & Sherrington, C.S. (1925). Recruitment and some other factors of reflex inhibition. *Proceedings of the Royal Society of London B, 97,* 488-518.

Lieber, R.L. (1992). *Skeletal Muscle Structure and Function.* Baltimore: Williams & Wilkins.

Lieber, R.L., Fazeli, B.M., & Botte, M.J. (1990). Architecture of selected wrist flexor and extensor muscles. *Journal of Hand Surgery, 15A,* 244-250.

Lieber, R.L., & Fridén, J. (2000). Functional and clinical significance of skeletal muscle architecture. *Muscle & Nerve, 23,* 1647-1666.

Lieber, R.L., Fridén, J.O., Hargens, A.R., Danzig, L.A., & Gershuni, D.H. (1988). Differential response of the dog quadriceps muscle to external skeletal fixation of the knee. *Muscle & Nerve, 11,* 193-201.

Lieber, R.L., & Kelly, M.J. (1991). Factors influencing quadriceps femoris muscle torque using transcutaneous neuromuscular electrical stimulation. *Physical Therapy, 71,* 715-723.

Lieber, R.L., Loren, G.J., & Fridén, J. (1994). In vivo measurement of human wrist extensor muscle sarcomere length changes. *Journal of Neurophysiology, 71,* 874-881.

Liebermann, D.G., & Issurin, V.B. (1997). Effort perception during isotonic muscle contractions with superimposed mechanical vibration stimulation. *Journal of Human Movement Studies, 32,* 171-186.

Liepert, J., Mingers, D., Heesen, C., Baumer, T., & Weiller, C. (2005). Motor cortex excitability and fatigue in multiple sclerosis: A transcranial magnetic stimulation study. *Multiple Sclerosis, 11,* 316-321.

Lin, F.M., & Sabbahi, M. (1999). Correlation of spasticity with hyperactive stretch reflexes and motor dysfunction in hemiplegia. *Archives of Physical Medicine and Rehabilitation, 80,* 526-530.

Lin, S.I., & Woollacott, M. (2002). Postural muscle responses following changing balance threats in young, stable older, and unstable older adults. *Journal of Motor Behavior, 34,* 37-44.

Linnamo, V., Strojnik, V., & Komi, P.V. (2002). EMG power spectrum and features of the superimposed M-wave during voluntary eccentric and concentric actions at different activation levels. *European Journal of Applied Physiology, 86,* 534-540.

Linnamo, V., Strojnik, V., & Komi, P.V. (2006). Maximal force during eccentric and isometric actions at different elbow angles. *European Journal of Applied Physiology, 96,* 672-678.

Lindstedt, S.L., Reich, T.E., Keim, P., & LaStayo, P.C. (2002). Do muscles function as adaptable locomotor springs? *Journal of Experimental Biology, 205,* 2211-2216.

Liu, J., & Jordan, L.M. (2005). Stimulation of the parapyramidal region of the neonatal rat brain stem produces locomotor-like activity involving spinal 5-HT7 and 5-HT2A receptors. *Journal of Neurophysiology, 94,* 1392-1404.

Liu, W., & Nigg, B.M. (2000). A mechanical model to determine the influence of masses and mass distribution on the impact force during running. *Journal of Biomechanics, 33,* 219-224.

Li Volsi, G., Lacata, F., Ciranna, L., Caserta, C., & Santangelo, F. (1998). Electromyographic effects of serotonin application into the lateral vestibular nucleus. *NeuroReport, 3,* 2539-2543.

Ljubisavljevic, M., Jovanovic, K., & Anastasikevic, R. (1992). Changes in discharge rate of fusimotor neurones provoked by fatiguing contractions of cat triceps surae muscles. *Journal of Physiology, 445,* 499-513.

Lloyd, A.R., Gandevia, S.C., & Hales, J.P. (1991). Muscle performance, voluntary activation, twitch properties and perceived effort in normal subjects and patients with chronic fatigue syndrome. *Brain, 114,* 85-98.

Lloyd, D.P.C. (1949). Post-tetanic potentiation of response in monosynaptic reflex pathways of the spinal cord. *Journal of General Physiology, 33,* 147-170.

Lloyd, D.P.C. (1943). Conduction and synaptic transmission of the reflex response to stretch in spinal cats. *Journal of Neurophysiology, 6,* 317-326.

Lloyd, D.G., & Besier, T.F. (2003). An EMG-driven musculoskeletal model to estimate muscle forces and knee joint moments in vivo. *Journal of Biomechanics, 36,* 765-776.

Lloyd, D.G., Buchanan, T.S., & Besier, T.F. (2005). Neuromuscular biomechanical modeling to understand knee ligament loading. *Medicine and Science in Sports and Exercise, 37,* 1939-1947.

Logie, R.H., Della Sala, S., MacPherson, S.E., & Cooper, J. (2007). Dual task demands on encoding and retrieval processes: Evidence from healthy adult ageing. *Cortex, 43,* 159-169.

Lombardi, V., & Piazzesi, G. (1990). The contractile response during lengthening of stimulated frog muscle fibres. *Journal of Physiology, 431,* 141-171.

Lømo, T., & Waerhaug, O. (1985). Motor endplates in fast and slow muscles of the rat: What determines their differences? *Journal of Physiology (Paris), 80,* 290-297.

Loram, I.D., Maganaris, C.N., & Lakie, M. (2005*a*). Active, non-spring-like muscle movements in human postural sway: How might paradoxical changes in muscle length be produced? *Journal of Physiology, 564,* 281-293.

Loram, I.D., Maganaris, C.N., & Lakie, M. (2005*b*). Human postural sway results from frequent, ballistic bias impulses by soleus and gastrocnemius. *Journal of Physiology, 564,* 295-311.

Loren, G.J., & Lieber, R.L. (1995). Tendon biomechanical properties enhance human wrist muscle specialization. *Journal of Biomechanics, 28,* 791-799.

Loren, G.J., Shoemaker, S.D., Burkholder, T.J., Jacobson, M.D., Fridén, J., & Lieber, R.L. (1996). Influences of human wrist motor design on joint torque. *Journal of Biomechanics, 29,* 331-342.

Lorist, M.M., Kernell, D., Meijman, T.F., & Zijdewind, I. (2002). Motor fatigue and cognitive task performance in humans. *Journal of Physiology, 545,* 313-319.

Löscher, W.N., Cresswell, A.G., & Thorstensson, A. (1996). Central fatigue during a long-lasting submaximal contraction of the triceps surae. *Experimental Brain Research, 108,* 305-314.

Löscher, W.N., & Nordlund, M.M. (2002). Central fatigue and motor cortical excitability during repeated shortening and lengthening actions. *Muscle & Nerve, 25,* 864-872.

Lotz, B.P., Dunne, J.W., & Daube, J.R. (1989). Preferential activation of muscle fibers with peripheral magnetic stimulation of the limb. *Muscle & Nerve, 12,* 636-639.

Lovely, R.G., Gregor, R.J., Roy, R.R., & Edgerton, V.R. (1990). Weight-bearing hindlimb stepping in treadmill-exercised adult spinal cats. *Brain Research, 514,* 206-218.

Lowery, M.M., Stoykov, N.S., & Kuiken, T.A. (2003). A simulation study to examine the use of cross-correlation as an estimate of surface EMG cross talk. *Journal of Applied Physiology, 94,* 1324-1334.

Lowey, S., Waller, G.S., & Trbus, K.M. (1993). Skeletal muscle myosin light chains are essential for physiological speeds of shortening. *Nature, 365,* 454-456.

Lowrie, M.B., & Vrbová, G. (1992). Dependence of postnatal motoneurones on their targets: Review and hypothesis. *Trends in Neurosciences, 15,* 80-84.

Lu, T.W., Lin, H.C., & Hsu, H.C. (2006). Influence of functional bracing on the kinetics of anterior cruciate ligament-injured knees during level walking. *Clinical Biomechanics, 21,* 517-524.

Lu, T.W., Taylor, S.J.G., O'Connor, J.J., & Walker, P.S. (1997). Influence of muscle activity on the forces in the femur: An in vivo study. *Journal of Biomechanics, 30,* 1101-1106.

Lucas, R.C., & Koslow, R. (1984). Comparative study of static, dynamic and proprioceptive neuromuscular facilitation stretching techniques on flexibility. *Perceptual and Motor Skills, 58,* 615-618.

Luff, A.R., Hatcher, D.D., & Torkko, K. (1988). Enlarged motor units resulting from partial denervation of cat hindlimb muscles. *Journal of Neurophysiology, 59,* 1377-1394.

Luhtanen, P., & Komi, P.V. (1978). Mechanical factors influencing running speed. In E. Asmussen & K. Jørgensen (Eds.), *Biomechanics VI-B* (pp. 23-29). Baltimore: University Park Press.

Lund, J.P., Kolta, A., Westberg, K.-G., & Scott, G. (1998). Brainstem mechanisms underlying feeding behaviors. *Current Opinions in Neurobiology, 8,* 718-724.

Lundberg, A., Malmgren, K., & Schomburg, E.D. (1977). Cutaneous facilitation of transmission in reflex pathways from Ib afferents to motoneurones. *Journal of Physiology, 265,* 763-780.

Lundberg, A., Malmgren, K., & Schomburg, E.D. (1987). Reflex pathways from group II muscle afferents. 3. Secondary spindle afferents and the FRA: A new hypothesis. *Experimental Brain Research, 65,* 294-306.

Luo, J., McNamara, B., & Moran, K. (2005). The use of vibration training to enhance muscle strength and power. *Sports Medicine, 35,* 23-41.

Lusby, L.A., & Atwater, A.E. (1983). Speed-related position-time profiles of arm motion in trained women distance runners. *Medicine and Science in Sports and Exercise, 15,* 171.

Lüscher, H.-R., Ruenzel, P., & Henneman, E. (1983). Effects of impulse frequency, PTP, and temperature on responses elicited in large populations of motoneurons by impulses in single Ia-fibers. *Journal of Neurophysiology, 50,* 1045-1058.

Lutz, G.J., & Lieber, R.L. (1999). Skeletal muscle myosin II structure and function. *Exercise and Sport Sciences Reviews, 27,* 63-77.

Macefield, V.G., Fuglevand, A.J., & Bigland-Ritchie, B. (1996). Contractile properties of single motor units in human toe extensors assessed by intraneural motor axon stimulation. *Journal of Neurophysiology, 75,* 2509-2519.

Macefield, V.G., Hagbarth, K.-E., Gorman, R., Gandevia, S.C., & Burke, D. (1991). Decline in spindle support to α-motoneurones during sustained voluntary contractions. *Journal of Physiology, 440,* 497-512.

Macefield, V.G., & Johansson, R.S. (2003). Loads applied tangential to a fingertip during an object restraint task can trigger short-latency as well as long-latency EMG responses in hand muscles. *Experimental Brain Research, 152,* 143-149.

MacIntosh, B.R., Gardiner, P.F., & McComas, A.J. (2006). *Skeletal Muscle: Form and Function.* Champaign, IL: Human Kinetics.

MacIntosh, B.R., & Willis, J.C. (2000). Force-frequency relationship and potentiation in mammalian skeletal muscle. *Journal of Applied Physiology, 88,* 2088-2096.

Mackey, A.L., Esmarck, B., Kadi, F., Koskinen, S.O., Kongsgaard, M., Sylerstersen, A., Hansen, J.J., Larsen, G., & Kjaer, M. (2007). Enhanced satellite cell proliferation with resistance training in elderly men and women. *Scandinavian Journal of Medicine and Science in Sports, 17,* 34-42.

Madigan, M.L. (2006). Age-related differences in muscle power during single-step balance recovery. *Journal of Applied Biomechanics, 22,* 186-193.

Madsen, N., & McLaughlin, T. (1984). Kinematic factors influencing performance and injury risk in the bench press exercise. *Medicine and Science in Sports and Exercise, 16,* 376-381.

Maegele, M., Muller, S., Wernig, A., Edgerton, V.R., & Harkema, S.J. (2002). Recruitment of spinal motor pools during voluntary movements versus stepping after human spinal cord injury. *Journal of Neurotrama, 19,* 1217-1229.

Maertens de Noordhout, A., Rothwell, J.C., Day, B.L., Nakashima, D.K., Thompson, P.D., & Marsden, C.D. (1992). Effect of digital nerve stimuli on responses to electrical or magnetic stimulation of the human brain. *Journal of Physiology, 447,* 535-548.

Maganaris, C.N. (2004). Imaging-based estimates of moment arm length in intact human muscle-tendons. *European Journal of Applied Physiology, 91,* 130-139.

Maganaris, C.N., Baltzopoulos, V., Ball, D., & Sargeant, A.J. (2001). In vivo specific tension of human skeletal muscle. *Journal of Applied Physiology, 90,* 865-872.

Maganaris, C.N., Baltzopoulos, V., & Sargeant, A.J. (1999). Changes in the tibialis anterior tendon moment arm from rest to maximum isometric dorsiflexion: In vivo observations in man. *Clinical Biomechanics, 14,* 661-666.

Maganaris, C.N., & Paul, J.P. (1999). In vivo human tendon mechanical properties. *Journal of Physiology, 521,* 307-313.

Maganaris, C.N., & Paul, J.P. (2000*a*). Hysteresis measurements in intact human tendon. *Journal of Biomechanics, 33,* 1723-1727.

Maganaris, C.N., & Paul, J.P. (2000*b*). In vivo human tendinous tissue stretch upon maximum muscle force generation. *Journal of Biomechanics, 33,* 1453-1459.

Magistris, M.R., Rösler, K.M., Truffet, A., & Myers, J.P. (1998). Transcranial stimulation excites virtually all motor neurons supplying the target muscle. A demonstration and a method improving the study of motor evoked potentials. *Brain, 121,* 437-450.

Magladery, J.W., Porter, W.E., Park, A.M., & Teasdall, R.D. (1951). Electrophysiological studies of nerve and reflex activity in normal man. IV. Two-neurone reflex and identification of certain action potentials from spinal roots and cord. *Bulletin of Johns Hopkins Hospital, 88,* 499-519.

Magnus, R. (1922). Wie sich die fallende katze in der luft umdrecht. *Archives Neerlandaises de Physiologie de l'Homme et des Animaux, 7,* 218-222.

Magnusson, S.P. (1998). Passive properties of human skeletal muscle during stretch maneuvers. A review. *Scandinavian Journal of Medicine and Science in Sports, 8,* 65-71.

Magnusson, S.P., Aagaard, P., Dyhre-Poulson, P., & Kjær, M. (2001). Load-displacement properties of the human triceps surae aponeurosis in vivo. *Journal of Physiology, 531,* 277-288.

Magnusson, S.P., Aagard, P., Simonsen, E., & Bojsen-Møller, F. (1998). A biomechanical evaluation of cyclic and static stretch in human skeletal muscle. *Internatonal Journal of Sports Medicine, 19,* 310-316.

Magnusson, S.P., Hansen, P., Aagaard, P., Brond, J., Dyhre-Poulson, P., Bojsen-Møller, J., & Kjær, M. (2003*a*). Differential strain patterns of the human gastrocnemius aponeurosis and free tendon, in vivo. *Acta Physiologica Scandinavica, 177,* 185-195.

Magnusson, S.P., Hansen, P., & Kjær, M. (2003*b*). Tendon properties in relation to muscular activity and physical training. *Scandinavian Journal of Medicine and Science in Sports, 13,* 211-223.

Magnusson, S.P., Simonsen, E.B., Aagaard, P., Boesen, J., Johannsen, F., & Kjær, M. (1997). Determinants of musculoskeletal flexibility: Viscoelastic properties, cross-sectional area, EMG and stretch tolerance. *Scandinavian Journal of Medicine and Science in Sports, 7,* 195-202.

Magnusson, S.P., Simonsen, E.B., Aagaard, P., Gleim, P., McHugh, G.W., & Kjær, M. (1995). Viscoelastic response to repeated static stretching in the human hamstring muscle. *Scandinavian Journal of Medicine and Science in Sports, 5,* 342-347.

Magnusson, S.P., Simonsen, E.B., Aagaard, P., Gleim, P., McHugh, G.W., & Kjær, M. (1996*a*). Viscoelastic stress relaxation during static stretch in human skeletal muscle in the absence of EMG activity. *Scandinavian Journal of Medicine and Science in Sports, 6,* 323-328.

Magnusson, S.P., Simonsen, E.B., Aagaard, P., Strensen, H., & Kjær, M. (1996*b*). A mechanism for altered flexibility in human skeletal muscle. *Journal of Physiology, 497,* 291-298.

Mahieu, N.N., McNair, P., De Muynck, M., Stevens, V., Blanckaert, I., Smits, N., & Witrouw, E. (2007). Effect of static and ballistic stretching on the muscle-tendon tissue properties. *Medicine and Science in Sports and Exercise, 39,* 494-501.

Malamud, J.G., Godt, R.E., & Nichols, T.R. (1996). Relationship between short-range stiffness and yielding in type-identified, chemically skinned muscle fibers from the cat triceps surae muscles. *Journal of Neurophysiology, 76,* 2280-2289.

Malisoux, L., Francaux, M., Nielens, H., & Theisen, D. (2006). Stretch-shortening exercises: An effective training paradigm to enhance power output of human single muscle fibers. *Journal of Applied Physiology, 100,* 771-779.

Malisoux, L., Jamart, C., Delplace, K., Nielens, H., Francaux, M., & Theisen, D. (2007). Effect of long-term muscle paralysis on human single fiber mechanics. *Journal of Applied Physiology, 102,* 340-349.

Malmgren, K., & Pierrot-Deseilligny, E. (1987). Evidence that low threshold afferents both evoke and depress polysynaptic excitation of wrist flexor motoneurones in man. *Experimental Brain Research, 67,* 429-432.

Maluf, K.S., & Enoka, R.M. (2005). Task failure during fatiguing contractions performed by humans. *Journal of Applied Physiology, 99,* 389-396.

Maluf, K.S., Shinohara, M., Stephenson, J.L., & Enoka, R.M. (2005). Muscle activation and time to task failure differ with load type and contraction intensity for a human hand muscle. *Experimental Brain Research, 167,* 165-177.

Manter, J.T. (1938). The dynamics of quadrupedal walking. *Journal of Experimental Biology, 15,* 522-540.

Mao, C.C., Ashby, P., Wang, M., & McCrea, D. (1984). Synaptic connections from large muscle afferents to the motoneurones of various leg muscles in man. *Experimental Brain Research, 56,* 341-350.

Marcell, T.J. (2003). Sarcopenia: Causes, consequences, and preventions. *Journal of Gerontology, 58A,* M911-M916.

Marchand-Pauvert, V., & Nielsen, J.B. (2002). Modulation of non-monosynaptic excitation from ankle dorsiflexor afferents to quadriceps motoneurones during human walking. *Journal of Physiology, 538,* 647-657.

Marder, E. (1998). From biophysics to models of network function. *Annual Reviews of Neuroscience, 21,* 25-45.

Maréchal, G., & Plaghki, L. (1979). The deficit of the isometric tetanic tension redeveloped after a release of frog muscle at a constant velocity. *Journal of General Physiology, 73,* 453-467.

Marey, E.-J. (1879). *Animal Mechanism: A Treatise on Terrestrial and Aerial Locomotion.* New York: D. Appleton and Co.

Marey, E.J. (1894). Des mouvements que certains animaux executent pour retomber sur leurs pieds, lorsqu'ils sont precipites d'un lieu eleve. *Academie des Sciences, 119,* 714-718.

Marras, W.S., Joynt, R.L., & King, A.I. (1985). The force-velocity relation and intra-abdominal pressure during lifting activities. *Ergonomics, 28,* 603-613.

Marras, W.S., & Mirka, G.A. (1992). A comprehensive evaluation of trunk response to asymmetric trunk motion. *Spine, 17,* 318-326.

Marsden, C.D., & Meadows, J.C. (1970). The effect of adrenaline on the contraction of human muscle. *Journal of Physiology, 207,* 429-448.

Marsden, C.D., Meadows, J.C., & Merton, P.A. (1983*a*). "Muscular wisdom" that minimized fatigue during prolonged effort in man: Peak rates of motoneuron discharge and slowing of discharge during fatigue. In J.E. Desmedt (Ed.), *Motor Control Mechanisms in Health and Disease* (pp. 169-211). New York: Raven Press.

Marsden, C.D., Merton, P.A., & Morton, H.B. (1983*b*). Rapid postural reactions to mechanical displacement of the hand in man. In J.E. Desmedt (Ed.), *Motor Control Mechanisms in Health and Disease* (pp. 645-659). New York: Raven Press.

Marsden, C.D., Rothwell, J.C., & Day, B.L. (1983*c*). Long-latency automatic responses to muscle stretch in man: Origin and function. In J.E. Desmedt (Ed.), *Motor Control Mechanisms in Health and Disease* (pp. 509-539). New York: Raven Press.

Marsden, J.F., Farmer, S.F., Halliday, D.M., Rosenberg, J.R., & Brown, P. (1999). The unilateral and bilateral control of motor unit pairs in the first dorsal interosseous and paraspinal muscles in man. *Journal of Physiology, 521,* 553-564.

Marsh, E., Sale, D., McComas, A.J., & Quinlan, J. (1981). The influence of joint position on ankle dorsiflexion in humans. *Journal of Applied Physiology, 51,* 160-167.

Marsh, R.L. (1999). How muscles deal with real-world loads: The influence of length trajectory on muscle performance. *Journal of Experimental Biology, 202,* 3377-3385.

Martin, P.E., Mungiole, M., Marzke, M.W., & Longhill, J.M. (1989). The use of magnetic resonance imaging for measuring segment inertial properties. *Journal of Biomechanics, 22,* 367-376.

Martin, P.E., Sanderson, D.J., & Umberger, B.R. (2000). Factors affecting preferred rates of movement in cyclic activities. In V.M. Zatsiorsky (Ed.), *Biomechanics of Sport* (pp. 143-160). Oxford, UK: Blackwell Science.

Martin, P.G., Smith, J.L., Butler, J.E., Gandevia, S.C., & Taylor, J.L. (2006). Fatigue-sensitive afferents inhibit extensor but not flexor motoneurons in humans. *Journal of Neuroscience, 26,* 4796-4802.

Martin, V., Millet, G.Y., Martin, A., Deley, G., & Lattier, G. (2004). Assessment of low-frequency fatigue with two methods of electrical stimulation. *Journal of Applied Physiology, 97,* 1923-1929.

Marvin, G., Sharma, A., Aston, W., Field, C., Kendall, M.J., & Jones, D.A. (1997). The effects of buspirone on perceived exertion and time to fatigue in man. *Experimental Physiology, 82,* 1057-1060.

Masakado, Y., Noda, Y., Nagata, M., Kimura, A., Chino, N., & Akaboshi, K. (1994). Macro-EMG and motor unit recruitment threshold: Differences between the young and aged. *Neuroscience Letters, 179,* 1-4.

Maschke, M., Gomez, C.M., Ebner, T.J., & Konczak, J. (2004). Hereditary cerebellar ataxia progressively impairs force adaptation during goal-directed arm movements. *Journal of Neurophysiology, 91,* 230-238.

Mason, C.R., Gomez, J.E., & Ebner, T.J. (2001). Hand synergies during reach-to-grasp. *Journal of Neurophysiology, 86,* 2896-2910.

Matsunaga, K., Uozumi, T., Tsuji, S., & Murai, Y. (1998). Age-dependent changes in physiological threshold asymmetries for the motor evoked potential and silent period following transcranial magnetic stimulation. *Electroencephalography and Clinical Neurophysiology, 109,* 502-507.

Matsuo, A., Ozawa, H., Goda, K., & Fukunaga, T. (1995). Moment of inertia of whole body using an oscillating table in adolescent boys. *Journal of Biomechanics, 28,* 219-223.

Matsuyama, K., Mori, F., Nakajima, K., Drew, T., Aoki, M., & Mori, S. (2004). Locomotor role of the corticoreticular-reticulospinal-spinal interneuronal system. *Progress in Brain Research, 143,* 239-249.

Mattay, V.S., Fera, F., Tessitore, A., Hariri, A.R., Das, S., Callicott, J.H., & Weinberger, D.R. (2002). Neurophysiological correlates of age-related changes in human motor function. *Neurology, 58,* 630-635.

Matthews, P.B.C. (1972). *Mammalian Muscle Receptors and Their Central Actions.* Baltimore: Williams & Wilkins.

Matthews, P.B.C. (1991). The human stretch reflex and the motor cortex. *Trends in Neurosciences, 14,* 87-91.

Mattiello-Sverzut, A.C., Carvalho, L.C., Cornachione, A., Nagashima, M., Neder, L., & Shimano, A.C. (2006). Morphological effects of electrical stimulation and intermittent muscle stretch after immobilization in soleus muscle. *Histology and Histopathology, 21,* 957-964.

Mayer, R.F., Burke, R.E., Toop, J., Hodgson, J.A., Kanda, K., & Walmsley, B. (1981). The effect of long-term immobilization on the motor unit population of the cat medial gastrocnemius muscle. *Neuroscience, 6,* 725-739.

Mazzocchio, R., Rossi, A., & Rothwell, J.C. (1994). Depression of Renshaw recurrent inhibition by activation of corticospinal fibres in human upper and lower limb. *Journal of Physiology, 481,* 487-498.

McCall, G.E., Byrnes, W.C., Dickinson, A., Pattany, P.M., & Fleck, S.J. (1996). Muscle fiber hypertrophy, hyperplasia, and capillary density in college men after resistance training. *Journal of Applied Physiology, 81,* 2004-2012.

McCall, G.E., Goulet, C., Boorman, G.I., Roy, R.R., & Edgerton, V.R. (2003). Flexor bias of joint position in humans during spaceflight. *Experimental Brain Research, 152,* 87-94.

McCartney, N., Hicks, A.L., Martin, J., & Webber, C.E. (1996). A longitudinal trial of weight training in the elderly: Continued improvements in year 2. *Journal of Gerontology, 51A,* B425-B433.

McCloskey, D.I., Ebeling, P., & Goodwin, G.M. (1974). Estimation of weights and apparent involvement of a "sense of effort." *Experimental Neurology, 42,* 220-232.

McCloskey, D.I., Gandevia, S., Poter, E.K., & Colebatch, J.G. (1983). Muscle sense and effort: Motor commands and judgements about muscular contractions. *Advances in Neurology, 39,* 151-167.

McComas, A.J. (1995). Motor-unit estimation: The beginning. *Journal of Clinical Neurophysiology, 12,* 560-564.

McComas, A.J. (1996). *Skeletal Muscle: Form and Function.* Champaign, IL: Human Kinetics.

McComas, A.J. (1998). Motor units: How many, how large, what kind? *Journal of Electromyography and Kinesiology, 8,* 391-402.

McComas, A.J., Sica, R.E., Upton, A.R.M., & Aguilera, G.C. (1973). Functional changes in motoneurons of hemiparetic patients. *Journal of Neurology, Neurosurgery, and Psychiatry, 36,* 183-193.

McCrea, D. (1998). Neuronal basis of afferent-evoked enhancement of locomotor activity. *Annals of the New York Academy of Sciences, 802,* 216-225.

McCrea, D.A. (2001). Spinal circuitry of sensorimotor control of locomotion. *Journal of Physiology, 533,* 41-50.

McCrea, D.A., & Rybak, I.A. (2008). Organization of mammalian locomotor rhythm and pattern generation. *Brain Research Reviews, 57,* 134-146.

McCrea, D.A., Shefchyk, S.J., Stephens, M.J., & Pearson, K.G. (1995). Disynaptic group I excitation of synergist ankle extensor motoneurones during fictive locomotion. *Journal of Physiology, 487,* 527-539.

McCully, K.K., & Faulkner, J.A. (1985). Injury to skeletal muscle fibers of mice following lengthening contractions. *Journal of Applied Physiology, 59,* 119-126.

McDonagh, M.J.N., & Davies, C.T.M. (1984). Adaptive response of mammalian skeletal muscle to exercise with high loads. *European Journal of Applied Physiology, 52,* 139-155.

McDonagh, M.J.N., Hayward, C.M., & Davies, C.T.M. (1983). Isometric training in human elbow flexor muscles: The effects on voluntary and electrically evoked forces. *Journal of Bone and Joint Surgery, 65B,* 355-358.

McDonald, C., & Dapena, J. (1991). Angular momentum in the men's 110-m and women's 110-m hurdle races. *Medicine and Science in Sports and Exercise, 23,* 1392-1402.

McDonald, K.S., Delp, M.D., & Fitts, R.H. (1992). Fatigability and blood flow in the rat gastrocnemius-plantaris-soleus after hindlimb suspension. *Journal of Applied Physiology, 73,* 1135-1140.

McDonald, K.S., & Fitts, R.H. (1995). Effect of hindlimb unloading on rat soleus fiber force, stiffness, and calcium sensitivity. *Journal of Applied Physiology, 79,* 1796-1802.

McDonnell, M.N., Ridding, M.C., Flavel, S.C., & Miles, T.S. (2005). Effect of human grip strategy on force control in precision tasks. *Experimental Brain Research, 161,* 368-373.

McGill, S.M., Norman, R.W., & Sharratt, M.T. (1990). The effect of an abdominal belt on trunk muscle activity and intra-abdominal pressure during squat lifts. *Ergonomics, 33,* 147-160.

McHugh, M.P. (2003). Recent advances in the understanding of the repeated bout effect: The protective effect against muscle damage from a single bout of eccentric exercise. *Scandinavian Journal of Medicine and Science in Sports, 13,* 88-97.

McHugh, M.P., Kremenic, I.J., Fox, M.B., & Gleim, G.W. (1998). The role of mechanical and neural restraints to joint range of motion during passive stretch. *Medicine and Science in Sports and Exercise, 30,* 928-932.

McHugh, M.P., Magnusson, S.P., Gleim, G.W., & Nicholas, J.A. (1992). Viscoelastic stress relaxation in human skeletal muscle. *Medicine and Science in Sports and Exercise, 24,* 1375-1382.

McHugh, M.P., & Pasiakos, S. (2004). The role of exercising muscle length in the protective adaptation to a single bout of eccentric exercise. *European Journal of Applied Physiology, 93,* 286-293.

McKay, W.B., Lee, D.C., Lim, H.K., Holmes, S.A., & Sherwood, A.M. (2005). Neurophysiological examination of the corticospinal system and voluntary motor control in motor-incomplete human spinal cord injury. *Experimental Brain Research, 163,* 379-387.

McKenna, M.J., Medved, I., Goodman, C.A., Brown, M.J., Bjorksten, A.R., Murphy, K.T., Petersen, A.C., Sostaric, S., & Gong, X. (2006). N-acetylcysteine attenuates the decline in muscle Na^+-K^+–pump activity and delays fatigue during prolonged exercise in humans. *Journal of Physiology, 576,* 279-288.

McLaughlin, T.M., Dillman, C.J., & Lardner, T.J. (1977). A kinematic model of performance in the parallel squat by champion powerlifters. *Medicine and Science in Sports, 9,* 128-133.

McLean, S.P., & Hinrichs, R.N. (1998). Sex differences in the centre of buoyancy location of competitive swimmers. *Journal of Sports Science, 16,* 373-383.

McMahon, T.A., & Cheng, G.C. (1990). The mechanics of running: How does stiffness couple with speed? *Journal of Biomechanics, S1,* 65-78.

McMahon, T.A., Valiant, G., & Frederick, E.C. (1987). Groucho running. *Journal of Applied Physiology, 62,* 2326-2337.

McNair, P.J., Dombroski, E.W., Hewson, D.J., & Stanley, S.N. (2001). Stretching at the ankle joint: Viscoelastic responses to holds and continuous passive motion. *Medicine and Science in Sports and Exercise, 33,* 354-358.

McNeil, C.J., Doherty, T.J., Stashuk, D.W., & Rice, C.L. (2005). Motor unit number estimates in the tibialis anterior muscle of young, old, and very old men. *Muscle & Nerve, 31,* 461-467.

McNeil, C.J., Vandervoort, A.A., & Rice, C.L. (2007). Peripheral impairments cause a progressive age-related loss of strength and velocity-dependent power in the dorsiflexors. *Journal of Applied Physiology, 102,* 1962-1968.

McNulty, P.A., & Macefield, V.G. (2001). Modulation of ongoing EMG by different classes of low-threshold mechanoreceptors in the human hand. *Journal of Physiology, 537,* 1021-1032.

McNulty, P.A., & Macefield, V.G. (2005). Intraneural microstimulation of motor axons in the study of human single motor units. *Muscle & Nerve, 32,* 119-139.

McVea, D.A., Donelan, J.M., Tachibana, A., & Pearson, K.G. (2005). A role for hip position in initiating the swing-to-stance transition in walking cats. *Journal of Neurophysiology, 94,* 3497-3508.

McVea, D.A., & Pearson, K.G. (2006). Long-lasting, context-dependent modification of stepping in the cat after repeated stumbling-corrective responses. *Journal of Neurophysiology, 97,* 659-669.

Medendorp, W.P., Goltz, H.C., Vilis, T., & Crawford, J.D. (2003). Gaze-centered updating of visual space in human parietal cortex. *Journal of Neuroscience, 23,* 6209-6214.

Meeusen, R., Watson, P., Hasegawa, H., Roelands, B., & Piacentini, M.F. (2006). Central fatigue. The serotonin hypothesis and beyond. *Sports Medicine, 36,* 881-909.

Mehta, B., & Schaal, S. (2002). Forward models in visuomotor control. *Journal of Neurophysiology, 88,* 942-953.

Melvill Jones, G. (2000). Posture. In E.R. Kandel, J.H. Schwartz, & T.M. Jessell (Eds.), *Principles of Neural Science* (4th ed., pp. 816-831). New York: Elsevier.

Menard, A., Leblond, H., & Gossard, J.P. (2003). Modulation of monosynaptic transmission by presynaptic inhibition during fictive locomotion in the cat. *Brain Research, 964,* 67-82.

Menz, H.B., & Morris, M.E. (2006). Clinical determinants of plantar forces and pressures during walking in older people. *Gait & Posture, 24,* 229-236.

Meriam, J.L., & Kraige, L.G. (1987). *Engineering Mechanics.* New York: Wiley.

Merkle, L.A., Layne, C.S., Bloomberg, J.J., & Zhang, J.J. (1998). Using factor analysis to identify neuromuscular synergies during treadmill walking. *Journal of Neuroscience Methods, 82,* 207-214.

Merletti, R., Knaflitz, M., & De Luca, C.J. (1990). Myoelectric manifestations of fatigue in voluntary and electrically elicited contractions. *Journal of Applied Physiology, 69,* 1810-1820.

Merletti, R., & Parker, P.A. (2004). *Electromyography. Physiology, Engineering, and Noninvasive Applications.* Hoboken, NJ: Wiley.

Merletti, R., Rainoldi, A., & Farina, D. (2001). Surface electromyography for noninvasive characterization of muscle. *Exercise and Sport Sciences Reviews, 29,* 20-25.

Mero, A., & Komi, P.V. (1987). Electromyographic activity in sprinting at speeds ranging from sub-maximal to supramaximal. *Medicine and Science in Sports and Exercise, 19,* 266-274.

Messier, J., & Kalaska, J.F. (2000). Covariation of primate dorsal premotor cell activity with direction and amplitude during a memorized-delay reaching task. *Journal of Neurophysiology, 84,* 152-165.

Meunier, S., & Pierrot-Deseilligny, E. (1989). Gating of the afferent volley of the monosynaptic stretch reflex during movement in man. *Journal of Physiology, 419,* 753-763.

Meunier, S., Pierrot-Deseilligny, E., & Simonetta, M. (1993). Pattern of monosynaptic heteronymous Ia connections in the human lower limb. *Experimental Brain Research, 96,* 534-544.

Meyer, B.U., Noth, J., Lange, H.W., Bischoff, C., Machetanz, J., Weindl, A., Roricht, S., Benecke, R., & Conrad, B. (1992). Motor responses evoked by magnetic brain stimulation in Huntington's disease. *Electroencephalography and Clinical Neurophysiology, 85,* 197-208.

Meyer, K., Steiner, R., Lastayo, P., Lippuner, K., Allemann, Y., Eberli, F., Schmid, J., Saner, H., & Hoppeler, H. (2003). Eccentric exercise in coronary patients: Central hemodynamic and metabolic responses. *Medicine and Science in Sports and Exercise, 35,* 1076-1082.

Meyer, R.A., & Prior, B.M. (2000). Functional magnetic resonance imaging of muscle. *Exercise and Sport Sciences Reviews, 28,* 89-92.

Michel, R.N., & Gardiner, P.F. (1990). To what extent is hindlimb suspension a model of disuse? *Muscle & Nerve, 13,* 646-653.

Miles, M.P., Clarkson, P.M., Bean, M., Ambach, K., Mulroy, J., & Vincent, K. (1994). Muscle function at the wrist following 9 d of immobilization and suspension. *Medicine and Science in Sports and Exercise, 26,* 615-623.

Miles, M.P., Heil, D.P., Larson, K.R. Conant, S.B., & Schneider, S.M. (2005). Prior resistance training and sex influence muscle responses to arm suspension. *Medicine and Science in Sports and Exercise, 37,* 1983-1989.

Millalieu, S.D., Hanton, S., & O'Brien, M. (2004). Intensity and direction of competitive anxiety as a function of sport type and experience. *Scandinavian Journal of Medicine and Science in Sports, 14,* 326-334.

Miller, C., & Thépaut-Mathieu, C. (1993). Strength training by electrostimulation conditions for efficacy. *International Journal of Sports Medicine, 14,* 20-28.

Miller, D.I. (1976). A biomechanical analysis of the contribution of the trunk to standing vertical jump take-offs. In J. Broekhoff (Ed.), *Physical Education, Sports, and the Sciences* (pp. 355-374). Eugene, OR: Microform.

Miller, D.I. (1978). Biomechanics of running—what should the future hold? *Canadian Journal of Applied Sport Sciences, 3,* 229-236.

Miller, D.I. (1979). Modelling in biomechanics: An overview. *Medicine and Science in Sports, 11,* 115-122.

Miller, D.I. (1981). *Biomechanics of Diving.* Report to the Canadian Amateur Diving Association.Miller, D.I. (1990). Ground reaction forces in distance running. In P.R. Cavanagh (Ed.), *Biomechanics of Distance Running* (pp. 203-224). Champaign, IL: Human Kinetics.

Miller, D.I. (2000). Springboard and platform diving. In V.M. Zatsiorsky (Ed.), *Biomechanics in Sport* (pp. 325-348). Oxford, UK: Blackwell Science.

Miller, D.I., Enoka, R.M., McCulloch, R.G., Burgess, E.M., Hutton, R.S., & Frankel, V.H. (1979). *Biomechanical Analysis of Lower Extremity Amputee Extra-Ambulatory Activities* (Contract No. V5244P-1540/VA). New York: Veterans Administration.

Miller, D.I., & Nelson, R.C. (1973). *Biomechanics of Sport.* London: Henry Kimpton.

Miller, K.J., Garland, S.J., Ivanova, T., & Ohtsuki, T. (1996). Motor-unit behavior in humans during fatiguing arm movements. *Journal of Neurophysiology, 75,* 1629-1636.

Miller, T.M., & Layzer, R.B. (2005). Muscle cramps. *Muscle & Nerve, 32,* 431-442.

Millet, G.Y., & Lepers, R. (2004). Alterations of neuromuscular function after prolonged running, cycling and skiing exercises. *Sports Medicine, 34,* 105-116.

Mills, K.R. (1991). Magnetic brain stimulation: A tool to explore the action of the motor cortex on single human spinal motoneurones. *Trends in Neuroscience, 14,* 401-405.

Milner-Brown, H.S., & Miller, R.G. (1986). Muscle membrane excitation and impulse propagation velocity are reduced during muscle fatigue. *Muscle & Nerve, 9,* 367-374.

Milner-Brown, H., & Stein, R.B. (1975). The relation between the surface electromyogram and muscular force. *Journal of Physiology, 246,* 549-569.

Misiaszek, J.E. (2006). Neural control of walking balance: IF falling THEN react ELSE continue. *Exercise and Sport Sciences Reviews, 34,* 128-134.

Mitoma, H., Hayashi, R., Yanagisawa, N., & Tsukagoshi, H. (2000). Characteristics of parkinsonian and ataxic gaits: A study using surface electromyograms, angular displacements and floor reaction forces. *Journal of Neurological Sciences, 174,* 22-39.

Miyamoto, K., Iinuma, N., Maeda, M., Wada, E., & Shimizu, K. (1999). Effects of abdominal belts on intra-abdominal pressure, intra-muscular pressure in the erector spinae muscles and myoelectrical activities of trunk muscles. *Clinical Biomechanics, 14,* 79-87.

Mizner, R.L., Petterson, S.C., & Snyder-Mackler, L. (2005*a*). Quadriceps strength and the time course of functional recovery after total knee arthroplasty. *Journal of Orthopedic Sports Physical Therapy, 35,* 424-436.

Mizner, R.L., Petterson, S.C., Stevens, J.E., Vandenborne, K., & Snyder-Mackler, L. (2005*b*). Early quadriceps strength loss after total knee arthroplasty. The contributions of muscle atrophy and failure of voluntary muscle activation. *Journal of Bone and Joint Surgery, 87,* 1047-1053.

Mohagheghi, A.A., Moraes, R., & Patla, A.E. (2004). The effects of distant and on-line visual information on the control of approach phase and step over an obstacle during locomotion. *Experimental Brain Research, 155,* 459-468.

Mohamed, O., Cerny, K., Jones, W., & Burnfield, J.M. (2005). The effect of terrain on foot pressure during walking. *Foot & Ankle International, 26,* 859-869.

Mohr, M., Rasmussen, P., Drust, B., Nielsen, B., & Nybo, L. (2006). Environmental heat stress, hyperammonemia and nucleotide metabolism during intermittent exercise. *European Journal of Applied Physiology, 97,* 89-95.

Molinari, F., Knaflitz, M., Bonato, P., & Actis, M.V. (2006). Electrical manifestations of muscle fatigue during concentric and eccentric isokinetic knee flexion-extension movements. *IEEE Transactions on Biomedical Engineering, 53,* 1309-1316.

Mommersteeg, T.J.A., Huiskes, R., Blankevoort, L., Kooloos, J.G.M., & Kauer, J.M.G. (1997). An inverse dynamics modeling approach to determine the restraining function of human knee ligament bundles. *Journal of Biomechanics, 30,* 139-146.

Monster, A.W., & Chan, H.C. (1977). Isometric force production by motor units of extensor digitorum communis muscle in man. *Journal of Neurophysiology, 40,* 1432-1443.

Monster, A.W., Chan, H.C., & O'Connor, D. (1978). Activity patterns of human skeletal muscle: Relation to muscle fiber type composition. *Science, 200,* 314-317.

Monti, R.J., Roy, R.R., Hodgson, J.A., & Edgerton, V.R. (1999). Transmission of forces within mammalian skeletal muscles. *Journal of Biomechanics, 32,* 371-380.

Moopanar, T.R., & Allen, D.G. (2005). Reactive oxygen species reduce myofibrillar Ca^{2+} sensitivity in fatiguing mouse skeletal muscle at 37° C. *Journal of Physiology, 564,* 189-199.

Moore, J.A., & Appenteng, K. (1991). The morphology and electrical geometry of rat jaw-elevator motoneurones. *Journal of Physiology, 440,* 325-343.

Moore, M.A., & Hutton, R.S. (1980). Electromyographic investigation of muscle stretching techniques. *Medicine and Science in Sports and Exercise, 12,* 322-329.

Moore, M.A., & Kukulka, C.G. (1991). Depression of Hoffmann reflexes following voluntary contraction and implications for proprioceptive neuromuscular facilitation therapy. *Physical Therapy, 71,* 321-333.

Morag, E., & Cavanagh, P.R. (1999). Structural and functional predictors of regional peak pressures under the foot during walking. *Journal of Biomechanics, 32,* 359-370.

Moran, K., McNamara, B., & Luo, J. (2007). Effect of vibration training in maximal effort (70% 1RM) dynamic bicep curl. *Medicine and Science in Sports and Exercise, 39,* 526-533.

Morasso, P. (1981). Spatial control of arm movements. *Experimental Brain Research, 42,* 223-227.

Moreno-Aranda, J., & Seireg, A. (1981*a*). Electrical parameters for over-the-skin muscle stimulation. *Journal of Biomechanics, 14,* 579-585.

Moreno-Aranda, J., & Seireg, A. (1981*b*). Force response to electrical stimulation of canine skeletal muscles. *Journal of Biomechanics, 14,* 595-599.

Moreno-Aranda, J., & Seireg, A. (1981*c*). Investigation of over-the-skin electrical stimulation parameters for different normal muscles and subjects. *Journal of Biomechanics, 14,* 587-593.

Morey-Holton, E., Globus, R.K., Kaplansky, A., & Durnova, G. (2005). The hindlimb unloading rat model: Literature overview, technique update and comparison with space flight data. *Advances in Space Biology and Medicine, 10,* 7-40.

Morin, C., & Pierrot-Deseilligny, E. (1977). Role of Ia afferents in the soleus motoneurones. Inhibition during a tibialis anterior voluntary contraction in man. *Experimental Brain Research, 27,* 509-522.

Morin, C., Pierrot-Deseilligny, E., & Hultborn, H. (1984). Evidence for presynaptic inhibition of muscle spindle Ia afferents in man. *Neuroscience Letters, 44,* 137-142.

Morita, H., Crone, C., Christenhuis, D., Petersen, N.T., & Nielsen, J.B. (2001). Modulation of presynaptic inhibition and disynaptic reciprocal Ia inhibition during voluntary movement in spasticity. *Brain, 124,* 826-837.

Moritani, T. (1992). Time course of adaptations during strength and power training. In P.V. Komi (Ed.), *Strength and Power in Sport* (pp. 266-278). Champaign, IL: Human Kinetics.

Moritani, T., & de Vries, H.A. (1979). Neural factors vs hypertrophy in the time course of muscle strength gain. *American Journal of Physical Medicine, 58,* 115-130.

Moritani, T., & Muro, M. (1987). Motor unit activity and surface electromyogram power spectrum during increasing force of contraction. *European Journal of Applied Physiology, 56,* 260-265.

Moritani, T., Stegeman, D., & Merletti, R. (2004). Basic physiology and biophysics of EMG signal generation. In R. Merletti & P.A. Parker (Eds.), *Electromyography. Physiology, Engineering, and Noninvasive Applications* (pp. 1-25). Hoboken, NJ: Wiley.

Moritz, C.T., Barry, B.K., Pascoe, M.A., & Enoka, R.M. (2005*a*). Discharge rate variability influences the variation in force fluctuations across the working range of a hand muscle. *Journal of Neurophysiology, 93,* 2449-2459.

Moritz, C.T., Christou, E.A., Meyer, F.G., & Enoka, R.M. (2005*b*). Coherence at 16-32 Hz can be caused by short-term synchrony of motor units. *Journal of Neurophysiology, 94,* 105-118.

Moritz, C.T., & Farley, C.T. (2006). Human hoppers compensate for simultaneous changes in surface compression and damping. *Journal of Biomechanics, 39,* 1030-1038.

Moritz, C.T., Greene, S.M., & Farley, C.T. (2004). Neuromuscular changes for hopping on a range of damped surfaces. *Journal of Applied Physiology, 96,* 1996-2004.

Morris, J.M., Lucas, D.B., & Bressler, B. (1961). Role of the trunk in stability of the spine. *Journal of Bone and Joint Surgery, 43A,* 327-351.

Morrissey, M.C., Harman, E.A., & Johnson, M.J. (1995). Resistance training modes: Specificity and effectiveness. *Medicine and Science in Sports and Exercise, 27,* 648-660.

Morrison, S., Kavanagh, J., Obst, S.J., Irwin, J., & Haseler, L.J. (2005). The effects of unilateral muscle fatigue on bilateral physiological tremor. *Experimental Brain Research, 167,* 609-621.

Morrison, S., Sleivert, G.G., & Cheung, S.S. (2004). Passive hyperthermia reduces voluntary activation and isometric force production. *European Journal of Applied Physiology, 91,* 729-736.

Morse, C.I., Thom, J.M., Birch, K.M., & Narici, M.V. (2005*a*). Changes in triceps surae muscle architecture with sarcopenia. *Acta Physiologica Scandinavica, 183,* 291-298.

Morse, C.I., Thom, J.M., Mian, O.S., Birch, K.M., & Narici, M.V. (2007). Gastrocnemius specific force is increased in elderly males following a 12-month physical training programme. *European Journal of Applied Physiology, 100,* 563-570.

Morse, C.I., Thom, J.M., Reeves, N.D., Birch, K.M., & Narici, M.V. (2005*b*). In vivo physiological cross-sectional area and specific force are reduced in the gastrocnemius of elderly men. *Journal of Applied Physiology, 99,* 1050-1055.

Morton, S.M., & Bastian, A.J. (2004). Cerebellar control of balance and locomotion. *Neuroscientist, 10,* 247-259.

Morton, S.M., & Bastian, A.J. (2006). Cerebellar contributions to locomotor adaptations during splitbelt treadmill walking. *Journal of Neuroscience, 26,* 9107-9116.

Moss, B.M., Refsnes, P.E., Abildgaard, A., Nicolaysen, K., & Jensen, J. (1997). Effects of maximal effort strength training with different loads on dynamic strength, cross-sectional area, load-power and load-velocity relationships. *European Journal of Applied Physiology, 75,* 193-199.

Mottram, C.J., Hunter, S.K., Rochette, L., Anderson, M.K., & Enoka, R.M. (2006*a*). Time to task failure varies with the gain of the feedback signal for women, but not for men. *Experimental Brain Research, 174,* 575-587.

Mottram, C.J., Jakobi, J.M., Semmler, J.G., & Enoka, R.M. (2005). Motor-unit activity differs with load type during a fatiguing contraction. *Journal of Neurophysiology, 93,* 1381-1392.

Mottram, C.J., Maluf, K.S., Stephenson, J.L., Anderson, M.K., & Enoka, R.M. (2006*b*). Prolonged vibration of the biceps brachii tendon reduces time to failure when maintaining arm position with a submaximal load. *Journal of Neurophysiology, 95,* 1185-1193.

Munn, J., Herbert, R.D., & Gandevia, S.C. (2004). Contralateral effects of unilateral resistance training: A meta-analysis. *Journal of Applied Physiology, 96,* 1861-1866.

Munn, J., Herbert, R.D., Hancock, M.J., & Gandevia, S.C. (2005). Training with unilateral resistance exercise increases contralateral strength. *Journal of Applied Physiology, 99,* 1880-1884.

Munro, C.F., Miller, D.I., & Fuglevand, A.J. (1987). Ground reaction forces in running: A reexamination. *Journal of Biomechanics, 20,* 147-155.

Munson, J.B., Foehring, R.C., Lofton, S.A., Zengel, J.E., & Sypert, G.W. (1986). Plasticity of medial gastrocnemius motor units following cordotomy in the cat. *Journal of Neurophysiology, 55,* 619-633.

Muraoka, T., Kawakami, Y., Tachi, M., & Fukunaga, T. (2001). Muscle fiber and tendon length changes in the human vastus lateralis during slow pedaling. *Journal of Applied Physiology, 91,* 2035-2040.

Muraoka, T., Muramatsu, T., Fukunaga, T., & Kanehisa, H. (2005). Elastic properties of human Achilles tendon are correlated with muscle strength. *Journal of Applied Physiology, 99,* 665-669.

Muraoka, T., Muramatsu, T., Takeshita, D., Kawakami, Y., & Fukunaga, T. (2002). Length change of human gastrocnemius aponeurosis and tendon during passive joint motion. *Cells, Tissues, and Organs, 171,* 260-268.

Murray, W.M., Buchanan, T.S., & Delp, S.L. (2000). The isometric functional capacity of muscles that cross the elbow. *Journal of Biomechanics, 33,* 943-952.

Murray, W.M., Buchanan, T.S., & Delp, S.L. (2002). Scaling of peak moment arms of elbow muscles with upper extremity bone dimensions. *Journal of Biomechanics, 33,* 1-26.

Murray, W.M., Delp, S.L., & Buchanan, T.S. (1995). Variation of muscle moment arms with elbow and forearm position. *Journal of Biomechanics, 28,* 513-525.

Mynark, R.G., & Koceja, D.M. (2002). Down training of the elderly soleus H reflex with the use of a spinally induced balance perturbation. *Journal of Applied Physiology, 93,* 127-133.

Naccarato, M., Calauitti, C., Jones, P.S., Day, D.J., Carpenter, T.A., & Baron, J.C. (2006). Does healthy aging affect the hemispheric activation balance during paced index-to-thumb opposition task? An fMRI study. *NeuroImage, 32,* 1250-1256.

Nachemson, A.L., Andersson, G.B.J., & Schultz, A.B. (1986). Valsalva maneuver biomechanics. *Spine, 11,* 476-479.

Nagata, A., & Christianson, J.C. (1995). M-wave modulation at relative levels of maximal voluntary contraction. *European Journal of Applied Physiology, 71,* 77-86.

Naito, A., Shindo, M., Miyasaka, T., Sun, Y-J., Momoi, H., & Chishima, M. (1996). Inhibitory projection from brachioradialis to biceps brachii motoneurones in humans. *Experimental Brain Research, 111,* 483-486.

Naito, A., Shindo, M., Miyasaka, T., Sun, Y-J., Momoi, H., & Chishima, M. (1998). Inhibitory projections from pronator teres to biceps brachii motoneurones in humans. *Experimental Brain Research, 121,* 99-102.

Nakamura, E., Moritani, T., & Kanetaka, A. (1996). Effects of habitual physical exercise on physiological age in men aged 20-85 years as estimated using principal component analysis. *European Journal of Applied Physiology, 73,* 410-418.

Napier, J.R. (1960). Studies of the hands of living primates. *Proceedings of the Zoological Society London, 134,* 647-657.

Narici, M. (1999). Human muscle skeletal architecture studied in vivo by non-invasive imaging techniques: Functional significance and applications. *Journal of Electromyography and Kinesiology, 9,* 97-103.

Narici, M.V., Bordini, M., & Cerretelli, P. (1991). Effect of aging on human adductor pollicis muscle function. *Journal of Applied Physiology, 71,* 1277-1281.

Narici, M.V., Landoni, L., & Minetti, A.E. (1992). Assessment of human knee extensor muscle stress from in vivo physiological cross-sectional area and strength measurements. *European Journal of Applied Physiology, 65,* 438-444.

Narici, M.V., & Maganaris, C.N. (2006). Adaptability of elderly human muscles and tendons to increased loading. *Journal of Anatomy, 208,* 433-443.

Narici, M.V., Maganaris, C.N., Reeves, N.D., & Capodaglio, P. (2003). Effect of aging on human muscle architecture. *Journal of Applied Physiology, 95,* 2229-2234.

Narici, M.V., Roi, G.S., & Landoni, L. (1988). Force of knee extensor and flexor muscles and cross-sectional area determined by nuclear magnetic resonance imaging. *European Journal of Applied Physiology, 57,* 39-44.

Narici, M.V., Roi, G.S., Landoni, L., Minetti, A.E., & Ceretelli, P. (1989). Changes in force, cross-sectional area and neural activation during strength training and detraining of the human quadriceps. *European Journal of Applied Physiology, 59,* 310-319.

Nashner, L.M. (1971). A model describing vestibular detection of body sway motion. *Acta Otolaryngology, 72,* 429-436.

Nashner, L.M. (1972). Vestibular postural control model. *Kybernetic, 10,* 106-110.

Nashner, L.M. (1976). Adapting reflexes controlling the human posture. *Experimental Brain Research, 26,* 59-72.

Nashner, L.M. (1977). Fixed patterns of rapid postural responses among leg muscles during stance. *Experimental Brain Research, 30,* 13-24.

Naunheim, R.S., Bayly, P.V., Standeven, J., Neubauer, J.S., Lewis, L.M., & Genin, G.M. (2003). Linear and angular head accelerations during heading of a soccer ball. *Medicine and Science in Sports and Exercise, 35,* 1406-1412.

Nelson, D.L., & Hutton, R.S. (1985). Dynamic and static stretch responses in muscle spindle receptors in fatigued muscle. *Medicine and Science in Sports and Exercise, 17,* 445-450.

Németh, G., & Ohlsén, H. (1986). Moment arm lengths of trunk muscles to the lumbosacral joint obtained in vivo with computed tomography. *Spine, 11,* 158-160.

Neptune, R.R., & Sasaki, K. (2005). Ankle plantar flexor force production is an important determinant of the preferred walk-to-run transition speed. *Journal of Experimental Biology, 208,* 799-808.

Newham, D.J., Jones, D.A., & Clarkson, P.M. (1987). Repeated high-force eccentric exercise: Effects on muscle pain and damage. *Journal of Applied Physiology, 63,* 1381-1386.

Newsholme, E.A., Acworth, I., & Blomstrand, E. (1987). Amino acids, brain neurotransmitters and a functional link between muscle and brain that is important in sustained exercise. In G. Benzi (Ed.), *Advances in Myochemistry* (pp. 127-133). London: John Libbey Eurotext.

Ng, A.V., Miller, R.G., Gelinas, D., & Kent-Braun, J.A. (2004). Functional relationships of central and peripheral muscle alterations in multiple sclerosis. *Muscle & Nerve, 29,* 843-852.

Nicholls, R.L., Miller, K., & Elliott, B.C. (2006). Numerical analysis of maximal bat performance in baseball. *Journal of Biomechanics, 39,* 1001-1009.

Nicks, D.K., Beneke, W.M., Key, R.M., & Timson, B.F. (1989). Muscle fibre size and number following immobilisation atrophy. *Journal of Anatomy, 163,* 1-5.

Nicol, C., Avela, J., & Komi, P.V. (2006). The stretch-shortening cycle: A model to study naturally occurring neuromuscular fatigue. *Sports Medicine, 36,* 977-999.

Nicol, C., & Komi, P.V. (1998). Significance of passively induced stretch reflexes on Achilles tendon force enhancement. *Muscle & Nerve, 21,* 1546-1548.

Nicolas, G., Marchand-Pauvert, V., Burke, D., & Pierrot-Deseilligny, E. (2001). Corticospinal excitation of presumed cervical propriospinal neurones and its reversal to inhibition in humans. *Journal of Physiology, 533,* 903-919.

Nielsen, B., & Nybo, L. (2003). Cerebral changes during exercise in the heat. *Sports Medicine, 33,* 1-11.

Nielsen, B., Strange, S., Christensen, N.J., & Saltin, B. (1997). Acute and adaptive responses in humans to exercise in a warm, humid environment. *Pflügers Archiv, 434,* 49-56.

Nielsen, J.B. (2004). Sensorimotor integration at spinal levels as a basis for muscle coordination during voluntary movement in humans. *Journal of Applied Physiology, 96,* 1961-1967.

Nielsen, J., & Kagamihara, Y. (1993). Differential projection of the sural nerve to early and late recruited human tibialis anterior motor units: Change of recruitment gain. *Acta Physiologica Scandinavica, 147,* 385-401.

Nielsen, J.J., Mohr, M., Klarskov, C., Kristensen, M., Krustrup, P., Juel, C., & Bangsbo, J. (2004). Effects of high-intensity intermittent training on potassium kinetics and performance in human skeletal muscle. *Journal of Physiology, 554,* 857-870.

Nielsen, J., & Petersen, N. (1994). Is presynaptic inhibition distributed to corticospinal fibres in man? *Journal of Physiology, 477,* 47-58.

Nielsen, J., Petersen, N., Deuschl, G., & Ballegaard, M. (1993). Task-related changes in the effect of magnetic brain stimulation on spinal neurones in man. *Journal of Physiology, 471,* 223-243.

Nielsen, J., & Pierrot-Deseilligny, E. (1991). Pattern of cutaneous inhibition of the propriospinal-like excitation to human upper limb motoneurones. *Journal of Physiology, 434,* 169-182.

Nielsen, J., & Pierrot-Deseilligny, E. (1996). Evidence of facilitation of soleus-coupled Renshaw cells during voluntary co-contraction of antagonistic ankle muscles in man. *Journal of Physiology, 493,* 603-611.

Nielsen, J.B., & Sinkjær, T. (2002). Reflex excitation of muscles during human walking. *Advances in Experimental and Medical Biology, 508,* 369-375.

Nielsen, O.B., & Harrison, A.P. (1998). The regulation of the Na^+-K^+ pump in contracting skeletal muscle. *Acta Physiologica Scandinavica, 162,* 191-200.

Nieuwenhuijzen, P.H.J.A., Schillings, A.M., Van Galen, G.P., & Duysens, J. (2000). Modulation of the startle response during human gait. *Journal of Neurophysiology, 84,* 65-74.

Nigg, B.M. (Ed.). (1986). *Biomechanics of Running Shoes.* Champaign, IL: Human Kinetics.

Nigg, B.M., & Herzog, W. (Eds.) (1994). *Biomechanics of the Musculo-Skeletal System.* New York: Wiley.

Nijhof, E.J. (2003). On-line trajectory modifications of planar, goal-directed arm movements. *Human Movement Sciences, 22,* 13-36.

Nilsson, J., Thorstensson, A., & Halbertsma, J. (1985). Changes in leg movements and muscle activity with speed of locomotion and mode of progression in humans. *Acta Physiologica Scandinavica, 123,* 457-475.

Nissinen, M., Preiss, R., & Brüggemann, P. (1985). Simulation of human airborne movements on the horizontal bar. In D.A. Winter, R.W. Norman, R.P. Wells, K.C. Hayes, & A.E. Patla (Eds.), *Biomechanics IX-B* (pp. 373-376). Champaign, IL: Human Kinetics.

Nistri, A., Ostroumov, K., Sharifullina, E., & Taccola, G. (2006). Tuning and playing a motor rhythm: How metabotropic glutamate receptors orchestrate generation of motor patterns in the mammalian central nervous system. *Journal of Physiology, 572,* 323-334.

Noonan, T.J., & Garrett, W.E. Jr. (1999). Muscle strain injury: Diagnosis and treatment. *Journal of the American Academy of Orthopedic Surgery, 7,* 262-269.

Nordstrom, M.A., Enoka, R.M., Callister, R.J., Reinking, R.M., & Stuart, D.G. (1995). Effects of six weeks of limb immobilization on the cat tibialis posterior: 1. Motor units. *Journal of Applied Physiology, 78,* 901-913.

Nordstrom, M.A., Fuglevand, A.J., & Enoka, R.M. (1992). Estimating the strength of common input to motoneurons from the cross-correlogram. *Journal of Physiology, 453,* 547-574.

Nordstrom, M.A., Gorman, R.B., Laouris, Y., Spielmann, J.M., & Stuart, D.G. (2007). Does motoneuron adaptation contribute to muscle fatigue? *Muscle & Nerve, 35,* 135-158.

Nordstrom, M.A., & Miles, T.S. (1991). Instability of motor unit firing rates during prolonged isometric contractions in human masseter. *Brain Research, 549,* 268-274.

Normann, C., & Clark, K. (2005). Selective modulation of Ca^{2+} influx pathways by 5-HT regulates synaptic long-term plasticity in the hippocampus. *Brain Research, 1037,* 187-193.

Norrie, B.A., Nevett-Duchcherer, J.M., & Gorassini, M.A. (2005). Reduced functional recovery by delaying motor training after spinal cord injury. *Journal of Neurophysiology, 94,* 255-264.

Noteboom, J.T., Barnholt, K.R., & Enoka, R.M. (2001*a*). Activation of the arousal response and impairment of performance increase with anxiety and stressor intensity. *Journal of Applied Physiology, 91,* 2093-2101.

Noteboom, J.T., Fleshner, M., & Enoka, R.M. (2001*b*). Activation of the arousal response can impair performance on a simple motor task. *Journal of Applied Physiology, 91,* 821-831.

Noth, J., Fromm, C., & Weiss, P.H. (2006). Unusual form of proprioceptive facilitation during recovery from hemiplegia. *Movement Disorders, 21,* 722-725.

Nourbakhsh, M.R., & Kukulka, C.G. (2004). Relationship between muscle length and moment arm on EMG activity of human triceps surae muscle. *Journal of Electromyography and Kinesiology, 14,* 263-273.

Novak, K.E., Miller, L.E., & Houk, J.C. (2002). The use of overlapping submovements in the control of rapid hand movements. *Experimental Brain Research, 144,* 351-364.

Novak, K.E., Miller, L.E., & Houk, J.C. (2003). Features of motor performance that drive adaptation in rapid hand movements. *Experimental Brain Research, 148,* 388-400.

Nowak, D.A., & Hermsdörfer, J. (2006). Predictive and reactive control of grasping forces: On the role of the basal ganglia and sensory feedback. *Experimental Brain Research, 173,* 650-660.

Nunome, H., Ikegami, Y., Kozakai, R., Apriantono, T., & Sano, S. (2006). Segmental dynamics of soccer instep kicking with the preferred and non-preferred leg. *Journal of Sports Science, 24,* 529-541.

Nusbaum, M.P., Blitz, D.M., Swensen, A.M., Wood, D., & Marder, E. (2001). The roles of co-transmission in neural network modulation. *Trends in Neuroscience, 24,* 146-154.

Nussbaum, M.A., Chaffin, D.B., & Rechtien, C.J. (1995). Muscle lines-of-action affect predicted forces in optimization-based spine muscle modeling. *Journal of Biomechanics, 28,* 401-409.

Nybo, L. (2003). CNS fatigue and prolonged exercise: Effect of glucose supplementation. *Medicine and Science in Sports and Exercise, 35,* 589-594.

Nybo, L., Dalsgaard, M.K., Steensberg, A., Møller, K., & Secher, N.H. (2005). Cerebral ammonia uptake and accumulation during prolonged exercise in humans. *Journal of Physiology, 563,* 285-290.

Nybo, L., Møller, K., Pedersen, B., Nielsen, B., & Secher, N.H. (2003*a*). Association between fatigue and failure to preserve cerebral energy turnover during prolonged exercise. *Acta Physiologica Scandinavica, 179,* 67-74.

Nybo, L., & Nielsen, B. (2001). Hyperthermia and central fatigue during prolonged exercise in humans. *Journal of Applied Physiology, 91,* 1055-1060.

Nybo, L., Nielsen, B., Blomstrand, E., Møller, K., & Secher, N.H. (2003*b*). Neurohumoral responses during prolonged exercise in humans. *Journal of Applied Physiology, 95,* 1125-1131.

Nybo, L., & Secher, N.H. (2004). Cerebral perturbations provoked by prolonged exercise. *Progress in Neurobiology, 72,* 223-261.

Ochiai, T., Mushiake, H., & Tanji, J. (2002). Effects of image motion in the dorsal premotor cortex during planning of an arm movement. *Journal of Neurophysiology, 88,* 2167-2171.

Ochiai, T., Mushiake, H., & Tanji, J. (2005). Involvement of the ventral premotor cortex in controlling image motion of the hand during performance of a target-capturing task. *Cerebral Cortex, 15,* 929-937.

O'Connor, P.J., & Cook, D.B. (1999). Exercise and pain: The neurobiology, measurement, and laboratory study of pain in relation to exercise in humans. *Exercise and Sport Sciences Reviews, 27,* 119-166.

Ogawa, K., & Yoshida, A. (1998). Throwing fracture of the humeral shaft. An analysis of 90 patients. *American Journal of Sports Medicine, 26,* 242-246.

O'Hagan, F.T., Sale, D.G., MacDougall, J.D., & Garner, S.H. (1995). Comparative effectiveness of accommodating and weight resistance training modes. *Medicine and Science in Sports and Exercise, 27,* 1210-1219.

O'Hara, T.E., & Goshgarian, H.G. (1991). Quantitative assessment of phrenic nerve functional recovery mediated by the crossed phrenic reflex at various time intervals after spinal cord injury. *Experimental Neurology, 111,* 244-250.

Ohira, Y., Yoshinaga, T., Nomura, T., Kawano, F., Ishihara, A., Nonaka, I., Roy, R.R., & Edgerton, V.R. (2002). Gravitational unloading effects on muscle fiber size, phenotype and myonuclear number. *Advances in Space Research, 30,* 777-781.

Ohira, Y., Yoshinaga, T., Ohara, M., Kawano, F., Wang, X.D., Higo, Y., Terada, M., Matsuoka, Y., Roy, R.R., & Edgerton, V.R. (2006). The role of neural and mechanical influences in maintaining normal fast and slow muscle properties. *Cells, Tissues, and Organs, 182,* 129-142.

Ohtsuki, T. (1983). Decrease in human voluntary isometric arm strength induced by simultaneous bilateral exertion. *Behavioural Brain Research, 7,* 165-178.

Oishi, Y., Ishihara, A., & Katsuta, S. (1992). Muscle fibre number following hindlimb immobilization. *Acta Physiologica Scandinavica, 146,* 281-282.

Oishi, Y., Ishihara, A., Yamamoto, H., & Miyamoto, E. (1998). Hindlimb suspension induces the expression of multiple myosin heavy chain isoforms in single fibres of the rat soleus muscle. *Acta Physiologica Scandinavica, 162,* 127-134.

O'Leary, D.D., Hope, K., & Sale, D.G. (1997). Posttetanic potentiation of human dorsiflexors. *Journal of Applied Physiology, 83,* 2131-2138.

Olive, J.L., Slade, J.M., Dudley, G.A., & McCully, K.K. (2003). Blood flow and muscle fatigue in SCI individuals during electrical stimulation. *Journal of Applied Physiology, 94,* 701-708.

Onambele, G.L., Narici, M.V., & Maganaris, C.N. (2006). Calf muscle–tendon properties and postural balance in old age. *Journal of Applied Physiology, 100,* 2048-2056.

Orchard, J., & Seward, H. (2002). Epidemiology of injuries in the Australian Football League, seasons 1997-2000. *British Journal of Sports Medicine, 36,* 39-44.

Orizio, C., & Gobbo, M. (2006). Mechanomyography. In M. Akay (Ed.), *Wiley Encyclopedia of Biomedical Engineering.* Hoboken, NJ: Wiley.

Ortega, J.D., & Farley, C.T. (2005). Minimizing center of mass vertical movement increases metabolic cost in walking. *Journal of Applied Physiology, 99,* 2099-2107.

Osu, R., Franklin, D.W., Kato, H., Gomi, H., Domen, K., Yoshioka, T., & Kawato, M. (2002). Short- and long-term changes in joint co-contraction associated with motor learning as revealed from surface EMG. *Journal of Neurophysiology, 88,* 991-1004.

Otis, J.S., Roy, R.R., Edgerton, V.R., & Talmadge, R.J. (2004). Adaptations in metabolic capacity of rat soleus after paralysis. *Journal of Applied Physiology, 96,* 584-596.

Otten, E. (1988). Concepts and models of functional architecture in skeletal muscle. *Exercise and Sport Sciences Reviews, 16,* 89-137.

Otten, E. (2003). Inverse and forward dynamics: Models of multi-body systems. *Philosophical Transactions of the Royal Society of London B Biological Sciences, 358,* 1493-1500.

Ouellette, M.M., LeBrasseur, N.K., Bean, J.F., Phillips, E., Stein, J., Frontera, W.R., & Fielding, R.A. (2004). High-intensity resistance training improves muscle strength, self-reported function, and disability in long-term stroke survivors. *Stroke, 35,* 1404-1409.

Ounjian, M., Roy, R.R., Eldred, E., Garfinkel, A., Payne, J.R., Armstrong, A., Toga, A.W., & Edgerton, V.R. (1991). Physiological and developmental implications of motor unit anatomy. *Journal of Neurobiology, 22,* 547-559.

Overduin, S.A., d'Avella, A., Roh, J., & Bizzi, E. (2008). Modulation of muscle synergy recruitment in primate grasping. *Journal of Neuroscience, 28,* 880-892.

Pacak, K., Palkovits, M., Yadid, G., Kvetnansky, R., Kopin, I.J., & Goldstein, D.S. (1998). Heterogeneous neurochemical responses to different stressors: A test of Selye's doctrine of nonspecificity. *American Journal of Physiology, 275,* R1247-R1255.

Pachter, B.R., & Eberstein, A. (1986). The effect of limb immobilization and stretch on the fine structure of the neuromuscular junction in rat muscle. *Experimental Neurology, 92,* 13-19.

Pai, Y.-C., Rogers, M.W., Patton, J., Cain, T.D., & Hanke, T.A. (1998). Static versus dynamic predictions of protective stepping following waist-pull perturbations in young and older adults. *Journal of Biomechanics, 31,* 1111-1118.

Paillard, T. (2008). Combined application of neuromuscular electrical stimulation and voluntary muscular contractions. *Sports Medicine, 38,* 161-177.

Pain, M.T.G., & Challis, J.H. (2004). Simulation of a wobbling mass model during an impact: A sensitivity analysis. *Journal of Applied Biomechanics, 20,* 309-316.

Pain, M.T.G., & Challis, J.H. (2006). The influence of soft tissue movement on ground reaction forces, joint torques, and joint reaction forces in drop landings. *Journal of Biomechanics, 39,* 119-124.

Palliyath, S., Hallett, M., Thomas, S.L., & Lebiedowska, M.K. (1998). Gait in patients with cerebellar ataxia. *Movement Disorders, 13,* 958-964.

Pancheri, P., & Biondi, M. (1990). Biological and psychological correlates of anxiety: The role of psychoendocrine reactivity as a marker of the anxiety response. In N. Sartorius (Ed.), *Anxiety: Psychobiological and Clinical Perspectives* (pp. 101-113). New York: Hemisphere.

Pandy, M.G., & Shelburne, K.B. (1997). Dependence of cruciate-ligament loading on muscle forces and external load. *Journal of Biomechanics, 30,* 1015-1024.

Panizza, M., Nilsson, J., Roth, B.J., Basser, P.J., & Hallett, M. (1992). Relevance of stimulus duration for activation of motor and sensory fibers: Implications for the study of H reflexes and magnetic stimulation. *Electroencephalography and Clinical Neurophysiology, 85,* 22-29.

Parikh, S., Morgan, D.L., Gregory, J.E., & Proske, U. (2004). Low-frequency depression of tension in the cat gastrocnemius muscle after eccentric exercise. *Journal of Applied Physiology, 97,* 1195-1202.

Parker, D., & Grillner, S. (2000). Neuronal mechanisms of synaptic and network plasticity in the lamprey spinal cord. *Progress in Brain Research, 125,* 381-398.

Parker, P.A., Englehart, K.B., & Hudgins, B.S. (2004). Control of powered upper limb prostheses. In R. Merletti & P.A. Parker (Eds.), *Electromyography. Physiology, Engineering, and Noninvasive Applications* (pp. 453-475). Hoboken, NJ: Wiley.

Parkinson, A., & McDonagh, M. (2006). Evidence for position feedback during involuntary aftercontractions. *Experimental Brain Research, 171,* 516-523.

Parry, D.J. (2001). Myosin heavy chain expression and plasticity: Role of myoblast diversity. *Exercise and Sport Sciences Reviews, 29,* 175-179.

Pasquet, B., Carpentier, A., & Duchateau, J. (2006). Specific modulation of motor unit discharge for a similar change in fascicle length during shortening and lengthening contractions in humans. *Journal of Physiology, 577,* 753-765.

Pasquet, B., Carpentier, A., Duchateau, J., & Hainaut, K. (2000). Muscle fatigue during concentric and eccentric contractions. *Muscle & Nerve, 23,* 1727-1735.

Patel, T.H., Das, R., Fridén, J., Lutz, G.J., & Lieber, R.L. (2004). Sarcomere strain and heterogeneity correlate with injury to frog skeletal muscle fiber bundles. *Journal of Applied Physiology, 97,* 1803-1813.

Patel, T.J., & Lieber, R.L. (1997). Force transmission in skeletal muscle: From actomyosin to external tendons. *Exercise and Sport Sciences Reviews, 25,* 321-363.

Pathare, N., Walter, G.A., Stevens, J.E., Yang, Z., Okerke, E., Gibbs, J.D., Esterhai, J.L., Scarborough, M.T., Gibbs, C.P., Sweeney, H.L., & Vandenborne, K. (2005). Changes in inorganic phosphate and force production in human skeletal muscle after cast immobilization. *Journal of Applied Physiology, 98,* 307-314.

Patla, A.E. (1985). Some characteristics of EMG patterns during locomotion: Implications for the locomotor control process. *Journal of Motor Behavior, 17,* 443-461.

Patla, A.E., & Greig, M. (2006). Any way you look at it, successful obstacle negotiation needs visually guided on-line foot placement regulation during the approach phase. *Neuroscience Letters, 397,* 110-114.

Patla, A.E., & Prentice, S.D. (1995). The role of active forces and intersegmental dynamics in the control of limb trajectory over obstacles during locomotion in humans. *Experimental Brain Research, 106,* 499-504.

Patten, C., & Kamen, G. (2000). Adaptations in motor unit discharge activity with force control training in young and older human adults. *European Journal of Applied Physiology, 83,* 128-143.

Patten, C., Meyer, R.A., & Fleckenstein, J.L. (2003). T2 mapping of muscle. *Seminars in Musculoskeletal Radiology, 7,* 297-305.

Pattison, J.S., Folk, L.C., Madsen, R.W., & Booth, F.W. (2003). Identification of differentially expressed genes between young and old rat soleus muscle during recovery from immobilization-induced atrophy. *Journal of Applied Physiology, 95,* 2171-2179.

Pavol, M.J., Owings, T.M., & Grabiner, M.D. (2002). Body segment inertial parameter estimation for the general population of older adults. *Journal of Biomechanics, 35,* 707-712.

Pawson, P.A., & Grinnell, A.D. (1990). Physiological differences between strong and weak frog neuromuscular junctions: A study involving tetanic and posttetanic potentiation. *Journal of Neuroscience, 10,* 1769-1778.

Payne, A.M., & Delbono, O. (2004). Neurogenesis of excitation-contraction uncoupling in aging skeletal muscle. *Exercise and Sport Sciences Reviews, 32,* 36-40.

Payne, A.M., Zheng, Z., Messi, M.L., Milligan, C.E., Gonzalez, E., & Delbono, O. (2006). Motor neurone targeting of IGF-1 prevents specific force decline in ageing mouse muscle. *Journal of Physiology, 570,* 283-294.

Pearlstein, E., Mabrouk, F.B., Pflieger, J.F., & Vinay, L. (2005). Serotonin refines the locomotor-related alternations in the in vitro neonatal rat spinal cord. *European Journal of Neuroscience, 21,* 1338-1346.

Pearson, K. (1993). Common principles of motor control in vertebrates and invertebrates. *Annual Reviews in Neuroscience, 16,* 265-297.

Pearson, K.G. (2000). Spinal reflexes. In E.R. Kandel, J.H. Schwartz, & T.M. Jessell (Eds.), *Principles of Neural Science* (4th ed., pp. 713-736). New York: Elsevier.

Pearson, K.G. (2004). Generating the walking gait: Role of sensory feedback. *Progress in Brain Research, 143,* 123-129.

Pearson, K.G. (2008). Role of sensory feedback in the control of stance duration in walking cats. *Brain Research Reviews, 57,* 222-227.

Pearson, K.G., & Gordon, J. (2000). Locomotion. In E.R. Kandel, J.H. Schwartz, & T.M. Jessell (Eds.), *Principles of Neural Science* (4th ed., pp. 737-755). New York: Elsevier.

Pearson, K.G., & Ramirez, J.-M. (1997). Sensory modulation of pattern-generating circuits. In P.S.G. Stein, S. Grillner, A.I. Selverston, & D.G. Stuart (Eds.), *Neurons, Networks, and Motor Behavior* (pp. 225-235). Cambridge, MA: MIT Press.

Pedersen, T.H., Nielsen, O.B., Lamb, G.D., & Stephenson, D.G. (2004). Intracellular acidosis enhances the excitability of working muscle. *Science, 305,* 1144-1147.

Pepe, F.A., & Drucker, B. (1979). The myosin filament IV: Myosin content. *Journal of Molecular Biology, 130,* 379-393.

Pereon, Y., Genet, R., & Guiheneuc, P. (1995). Facilitation of motor evoked potentials: Timing of Jendrassik maneuver effects. *Muscle & Nerve, 18,* 1427-1432.

Perez, M.A., Field-Fote, E.C., & Floeter, M.K. (2003). Patterned sensory stimulation induces plasticity in reciprocal Ia inhibition in humans. *Journal of Neuroscience, 23,* 2014-2018.

Perez, M.A., Lungholt, B.K., & Nielsen, J.B. (2005). Presynaptic control of group Ia afferents in relation to acquisition of a visuo-motor skill in healthy humans. *Journal of Physiology, 568,* 343-354.

Perkins, T.A., Brindley, G.S., Donaldson, N.D., Polkey, C.E., & Rushton, D.N. (1994). Implant provision of key, pinch and power grips in a C6 tetraplegic. *Medical and Biological Engineering and Computing, 32,* 367-372.

Perry, J. (1993). Determinants of muscle function in the spastic lower extremity. *Clinical Orthopaedics and Related Research, 288,* 10-26.

Person, R.S., & Kudina, L.P. (1972). Discharge frequency and discharge pattern of human motor units during voluntary contraction of muscle. *Electroencephalography and Clinical Neurophysiology, 32,* 471-483.

Peters, S.E. (1989). Structure and function in vertebrate skeletal muscle. *American Zoologist, 29,* 221-234.

Petersen, N.T., Pyndt, H.S., & Nielsen, J.B. (2003). Investigating human motor control by transcranial magnetic stimulation. *Experimental Brain Research, 152,* 1-16.

Peterson, M.D., Rhea, M.R., & Alvar, B.A. (2005). Applications of the dose-response for muscular strength development: A review of meta-analytic efficacy and reliability for designing training prescription. *Journal of Strength and Conditioning Research, 19,* 950-958.

Petit, J., Filippi, G.M., Gioux, M., Hunt, C.C., & Laporte, Y. (1990). Effects of tetanic contraction of motor units of similar type on the initial stiffness to ramp stretch of the cat peroneus longus muscle. *Journal of Neurophysiology, 64,* 1724-1732.

Petit, J., & Gioux, M. (1993). Properties of motor units after immobilization of cat peroneus longus muscle. *Journal of Applied Physiology, 74,* 1131-1139.

Petit, J., Giroux-Metges, M-A., & Gioux, M. (2003). Power developed by motor units of the peroneus tertius muscle of the cat. *Journal of Neurophysiology, 90,* 3095-3104.

Petit, J., Scott, J.J.A., & Reynolds, K.J. (1997). Tendon organ sensitivity to steady-state isotonic contraction of in-series motor units in feline peroneus tertius muscle. *Journal of Physiology, 500,* 227-233.

Petrella, J.K., Kim, J.S., Cross, J.M., Kosek, D.J., & Bamman, M.M. (2006). Efficacy of myonuclear addition may explain differential myofiber growth among resistance-trained young and older men and women. *American Journal of Physiology, 291,* E937-E946.

Petrella, J.K., Kim, J.S., Tuggle, S.C., Hall, S.R., & Bamman, M.M. (2005). Age differences in knee extension power, contractile velocity, and fatigability. *Journal of Applied Physiology, 98,* 211-220.

Petrof, B.J., Shrager, J.B., Stedman, H.H., Kelly, A.M., & Sweeney, H.L. (1993). Dystrophin protects the sarcolemma from stresses developed during muscle contraction. *Proceedings of the National Academy of Sciences, 90,* 3710-3714.

Petrofsky, J.S., & Hendershot, D.M. (1984). The interrelationship between blood pressure, intramuscular pressure, and isometric endurance in fast and slow twitch skeletal muscle in the cat. *European Journal of Applied Physiology, 53,* 106-111.

Petrofsky, J.S., & Lind, A.R. (1980). Frequency analysis of the surface EMG during sustained isometric contractions. *European Journal of Applied Physiology, 43,* 173-182.

Pette, D., Peuker, H., & Staron, R.S. (1999). The impact of biochemical methods for single muscle fibre analysis. *Acta Physiologica Scandinavica, 166,* 261-277.

Pette, D., & Vrbová, G. (1999). What does chronic electrical stimulation teach us about muscle plasticity? *Muscle & Nerve*, *22,* 666-677.

Petterson, S., & Snyder-Mackler, L. (2006). The use of neuromuscular electrical stimulation to improve activation in a patient with chronic quadriceps strength impairments following total knee arthroplasty. *Journal of Orthopedic Sports Physical Therapy, 36,* 678-685.

Pettorossi, V.E., Della Torre, G., Bortolami, R., & Brunetti, O. (1999). The role of capsaicin-sensitive muscle afferents in fatigue-induced modulation of the monosynaptic reflex in the rat. *Journal of Physiology, 515,* 599-607.

Philippou, A., Bogdanis, G.C., Nevill, A.M., & Maridaki, M. (2004). Changes in the angle-force curve of human elbow flexors following eccentric and isometric exercise. *European Journal of Applied Physiology, 93,* 237-244.

Phillips, C.G. (1986). *Movements of the Hand.* Liverpool: Liverpool University Press.

Phillips, S.J., & Roberts, E.M. (1980). Muscular and non-muscular moments of force in the swing limb of Masters runners. In J.M. Cooper & B. Haven (Eds.), *Proceedings of the Biomechanics Symposium* (pp. 256-274). Bloomington, IN: Indiana State Board of Health.

Phillips, S.J., Roberts, E.M., & Huang, T.C. (1983). Quantification of intersegmental reactions during rapid swing motion. *Journal of Biomechanics, 16,* 411-418.

Piacentini, M.F., Meeusen, R., Buyse, L., de Schutter, G., Kempemaers, F., & de Meirleir, K. (2002*a*). No effect of a selective serotonergic/noradrenergic reuptake inhibitor on endurance performance. *European Journal of Sports Science, 2,* 1-10.

Piacentini, M.F., Meeusen, R., Buyse, L., de Schutter, G., Kempemaers, F., van Nijvel, J., & de Meirleir, K. (2002*b*). No effect of a noradrenergic reuptake inhibitor on performance in trained cyclists. *Medicine and Science in Sports and Exercise, 34,* 1189-1193.

Piacentini, M.F., Meeusen, R., Buyse, L., de Schutter, G., & de Meirleir, K. (2004). Hormonal responses during prolonged exercise are influenced by a selective DA/NA reuptake inhibitor. *British Journal of Sports Medicine, 38,* 129-133.

Piazza, S.J., & Delp, S.L. (1996). The influence of muscles on knee flexion during the swing phase of gait. *Journal of Biomechanics, 29,* 723-733.

Piazzesi, G., Reconditi, M., Dobbie, I., Linari, M., Boesecke, P., Diat, O., Irving, M., & Lombardi, V. (1999). Changes in conformation of myosin heads during the development of isometric contraction and rapid shortening in single frog muscle fibres. *Journal of Physiology, 514,* 305-312.

Pierrot-Deseilligny, E. (1990). Electrophysiological assessment of the spinal mechanisms underlying spasticity. In P.M. Rossini & F. Mauguiére (Eds.), *New Trends and Advanced Techniques in Clinical Neurophysiology* (pp. 364-373). Amsterdam: Elsevier.

Pierrot-Deseilligny, E., & Burke, D. (2005). *The Circuitry of the Human Spinal Cord.* Cambridge, UK: Cambridge University Press.

Pierrot-Deseilligny, E., Katz, R., & Morin, C. (1979). Evidence for Ib inhibition in human subjects. *Experimental Brain Research, 166,* 176-179.

Pierrot-Deseilligny, E., Mazevet, D., & Meunier, S. (1995). Cutaneous inhibition of the descending command passing through the propriospinal relay might contribute to curtail human movements. In A. Taylor, M.H. Gladden, & R. Durbaba (Eds.), *Alpha and Gamma Motor System* (pp. 607-615). New York: Plenum Press.

Pierrot-Deseilligny, E., Morin, C., Bergego, C., & Tankov, N. (1981). Pattern of group I fibre projections from ankle flexor and extensor muscles in man. *Experimental Brain Research, 42,* 337-350.

Pierotti, D.J., Roy, R.R., Hodgson, J.A., & Edgerton, V.R. (1994). Level of independence of motor unit properties from neuromuscular activity. *Muscle & Nerve, 17,* 1324-1335.

Pigeon, P., Yahia, H., & Feldman, A.G. (1996). Moment arms and lengths of human upper limb muscles as functions of joint angles. *Journal of Biomechanics, 29,* 1365-1370.

Pinder, A.D.J., & Grieve, D.W. (1997). Hydro-resistive measurement of dynamic lifting strength. *Journal of Biomechanics, 30,* 399-402.

Pinninger, G.J., Steele, J.R., Thorstensson, A., & Cresswell, A.G. (2000). Tension regulation during lengthening and shortening actions of the human soleus muscle. *European Journal of Applied Physiology, 81,* 375-383.

Pitcher, J.B., Ogston, K.M., & Miles, T.S. (2003). Age and sex differences in human motor cortex input-output characteristics. *Journal of Physiology, 546,* 605-613.

Pitman, M.I., Nainzadeh, N., Menche, D., Gasalberti, R., & Song, E.K. (1992). The intraoperative evaluation of the neurosensory function of the anterior cruciate ligament in human using somatosensory evoked potentials. *Arthroscopy, 8,* 442-447.

Pitman, M.I., & Peterson, L. (1989). Biomechanics of skeletal muscle. In M. Nordin & V.H. Frankel (Eds.), *Basic Biomechanics of the Musculoskeletal System* (pp. 89-111). Philadelphia: Lea & Febiger.

Place, N., Maffiuletti, N.A., Ballay, Y., & Lepers, R. (2005). Twitch potentiation is greater after a fatiguing submaximal isometric contraction performed at short vs. long quadriceps muscle length. *Journal of Applied Physiology, 98,* 429-436.

Platz, T., Eickhof, C., van Kaick, S., Engel, U., Pinkowski, C., Kalok, S., & Pause, M. (2005). Impairment-oriented training or Bobath therapy for severe arm paresis after stroke: A single-blind, multicentre randomized controlled trial. *Clinical Rehabilitation, 19,* 714-724.

Ploutz-Snyder, L.L., Tesch, P.A., Crittenden, D.J., & Dudley, G.A. (1995). Effect of unweighting on skeletal muscle use during exercise. *Journal of Applied Physiology, 79,* 168-175.

Pope, R.P., Herbert, R.D., Kirwan, J.D., & Graham, B.J. (2000). A randomized trial of preexercise stretching for prevention of lower limb injury. *Medicine and Science in Sports and Exercise, 32,* 271-277.

Porter, M.M., Vandervoort, A.A., & Kramer, J.F. (1997). Eccentric peak torque of the plantar and dorsiflexors is maintained in older women. *Journal of Gerontology, 52A,* B125-B131.

Poston, B., Enoka, J.A., & Enoka, R.M. (2008). Practice and endpoint accuracy with the left and right hands of old adults: The right-hemisphere aging model. *Muscle & Nerve, 37,* 376-386.

Poston, B., Holcomb, W.R., Guadagnoli, M.A., & Linn, L.L. (2007). The acute effects of mechanical vibration on power output in the bench press. *Journal of Strength and Conditioning Research, 21,* 199-203.

Potvin, J.R., & Brown, S.H. (2004). Less is more: High pass filtering, to remove up to 99% of the surface EMG signal power, improves EMG-based biceps brachii muscle force estimates. *Journal of Electromyography and Kinesiology, 14,* 389-399.

Pousson, M., Perot, C., & Goubel, F. (1991). Stiffness changes and fibre type transitions in rat soleus muscle produced by jumping training. *Pflügers Archiv, 419,* 127-130.

Powell, P.L., Roy, R.R., Kanim, P., Bello, M.A., & Edgerton, V.R. (1984). Predictability of skeletal muscle tension from architectural determinations in guinea pig hindlimbs. *Journal of Applied Physiology, 57,* 1715-1721.

Powers, R.K., Marder-Meyer, J., & Rymer, W.Z. (1989). Quantitative relations between hypertonia and stretch reflex threshold in spastic hemiparesis. *Annals of Neurology, 23,* 115-124.

Powers, W.J., & Zazulia, A.R. (2003). The use of positron emission tomography in cerebrovascular disease. *Neuroimaging Clinics of North America, 13,* 741-758.

Prablanc, C., Desmurget, M., & Grea, H. (2003). Neural control of on-line guidance of hand reaching movements. *Progress in Brain Research, 142,* 155-170.

Prado, L.G., Makarenko, I., Andresen, C., Krüger, M., Opitz, C.A., & Linke, W.A. (2005). Isoform diversity of giant proteins in relation to passive and active contractile properties of rabbit skeletal muscles. *Journal of General Physiology, 126,* 461-480.

Prasartwuth, O., Allen, T.J., Butler, J.E., Gandevia, S.C., & Taylor, J.L. (2006). Length-dependent changes in voluntary activation, maximum voluntary torque and twitch responses after eccentric damage in humans. *Journal of Physiology, 571*, 243-252.

Prathare, N.C., Stevens, J.E., Walter, G.A., Shah, P., Jayaraman, A., Tillman, S.M., Scarborough, M.T., Parker Gibbs, C., & Vandenborne, K. (2006). Deficit in human muscle strength with cast immobilization: Contribution of inorganic phosphate. *European Journal of Applied Physiology, 98,* 71-78.

Price, T.B., Kamen, G., Damon, B.M., Knight, C.A., Applegate, B., Gore, J.C., Eward, K., & Signorile, J.F. (2003). Comparison of MRI with EMG to study muscle activity associated with dynamic plantar flexion. *Magnetic Resonance Imaging, 21,* 853-861.

Prilutsky, B.I. (2000). Eccentric muscle action in sport exercise. In V.M. Zatsiorsky (Ed.), *Biomechanics in Sport* (pp. 56-86). Oxford, UK: Blackwell Science.

Prilutsky, B.I., & Gregor, R.J. (2001). Swing- and support-related muscle actions differentially trigger human walk–run and run–walk transitions. *Journal of Experimental Biology, 204,* 2277-2287.

Prilutsky, B.I., Herzog, W., Leonard, T.R., & Allinger, T.L. (1996*a*). Role of the muscle belly and tendon of soleus, gastrocnemius, and plantaris in mechanical energy absorption and generation during cat locomotion. *Journal of Biomechanics, 29,* 417-434.

Prilutsky, B.I., Petrova, L.N., & Raitsin, L.M. (1996*b*). Comparison of mechanical energy expenditure of joint moments and muscle forces during human locomotion. *Journal of Biomechanics, 29,* 405-415.

Prilutsky, B.I., & Zatsiorsky, V.M. (1994). Tendon action of two-joint muscles: Transfer of mechanical energy between joints during jumping, landing, and running. *Journal of Biomechanics, 27,* 25-34.

Prior, B.M., Foley, J.M., Jayaraman, R.C., & Meyer, R.A. (1999). Pixel T2 distribution in functional magnetic resonance images of muscle. *Journal of Applied Physiology, 87,* 2107-2114.

Prior, B.M., Jayaraman, R.C., Reid, R.W., Cooper, T.G., Foley, J.M., Dudley, G.A., & Meyer, R.A. (2001). Biarticular and monoarticular muscle activation and injury in human quadriceps muscle. *European Journal of Applied Physiology, 85,* 185-190.

Prochazka, A. (1989). Sensorimotor gain control: A basic strategy of motor systems? *Progress in Neurobiology, 33,* 281-307.

Prochazka, A. (1996). Proprioceptive feedback and movement regulation. In L.B. Rowell & J.T. Shepherd (Eds.), *Handbook of Physiology: Sec. 12. Exercise: Regulation and Integration of Multiple Systems* (pp. 89-127). New York: Oxford University Press.

Prochazka, A., Clarac, F., Loeb, G.E., Rothwell, J.C., & Wolpaw, J.R. (2000). What do reflex and voluntary mean? Modern views on an ancient debate. *Experimental Brain Research, 130,* 417-432.

Prochazka, A., & Gorassini, M. (1998). Ensemble firing of muscle afferents recorded during normal locomotion in cats. *Journal of Physiology, 507,* 293-304.

Proske, U. (1997). The mammalian muscle spindle. *News in Physiological Sciences, 12*, 37-42.

Proske, U. (2006). Kinesthesia: The role of muscle receptors. *Muscle & Nerve, 34,* 545-558.

Proske, U., & Allen, T.J. (2005). Damage to skeletal muscle from eccentric exercise. *Exercise and Sport Sciences Reviews, 33*, 98-104.

Proske, U., & Morgan, D.L. (1999). Do cross-bridges contribute to the tension during stretch of passive muscle? *Journal of Muscle Research and Cell Motility, 20,* 433-442.

Proske, U., & Morgan, D.L. (2001). Muscle damage from eccentric exercise: Mechanism, mechanical signs, adaptation and clinical applications. *Journal of Physiology, 537,* 333-345.

Proske, U., Morgan, D.L., & Gregory, J.E. (1993). Thixotropy in skeletal muscle and in muscle spindles: A review. *Progress in Neurobiology, 41,* 705-721.

Psek, J.A., & Cafarelli, E. (1993). Behavior of coactive muscles during fatigue. *Journal of Applied Physiology, 74,* 170-175.

Putnam, C.A. (1983). Interaction between segments during a kicking motion. In H. Matsui & K. Kobayashi (Eds.), *Biomechanics VIII-B* (pp. 688-694). Champaign, IL: Human Kinetics.

Putnam, C.A. (1991). A segment interaction analysis of proximal-to-distal sequential segment motion patterns. *Medicine and Science in Sports and Exercise, 23,* 130-144.

Quaney, B.M., Nudo, R.J., & Cole, K.J. (2005). Can internal models of objects be utilized for different prehension tasks? *Journal of Neurophysiology, 93,* 2021-2027.

Quevedo, J., Feirchuk, B., Gosnach, S., & McCrea, D. (2000). Group I disynaptic excitation of cat hindlimb flexor and bifunctional motoneurones during fictive locomotion. *Journal of Physiology, 525,* 549-564.

Rack, P.M.H., & Westbury, D.R. (1969). The effects of length and stimulus rate on tension in the isometric cat soleus muscle. *Journal of Physiology, 204,* 443-460.

Rack, P.M.H., & Westbury, D.R. (1974). The short range stiffness of active mammalian muscle and its effect on mechanical properties. *Journal of Physiology, 240,* 331-350.

Rafuse, V.F., & Gordon, T. (1996). Self-reinnervated cat medial gastrocnemius muscles. I. Comparisons of the capacity for regenerating nerves to form enlarged motor units after extensive peripheral nerve injuries. *Journal of Neurophysiology, 75,* 268-281.

Rafuse, V.F., & Gordon, T. (1998). Incomplete matching of nerve and muscle properties in motor units after extensive nerve injuries in cat hindlimb muscle. *Journal of Physiology, 509,* 909-926.

Rafuse, V.F., Gordon, T., & Orozco, R. (1992). Proportional enlargement of motor units after partial denervation of cat triceps surae muscles. *Journal of Neurophysiology, 68,* 1261-1276.

Raglin, J.S. (1992). Anxiety and sport performance. *Exercise and Sport Sciences Reviews, 20,* 243-274.

Rainoldi, A., Garlardi, G., Maderna, L., Comi, G., Lo Conte, L., & Merletti, R. (1999). Repeatability of surface EMG variables during voluntary isometric contractions of the biceps brachii muscle. *Journal of Electromyography and Kinesiology, 9,* 105-119.

Rainoldi, A., Melchiorri, G., & Caruso, I. (2004). A method for positioning electrodes during surface EMG recordings in lower limb muscles. *Journal of Neuroscience Methods, 15,* 37-43.

Rall, J.A., & Woledge, R.C. (1990). Influence of temperature on mechanics and energetics of muscle contraction. *American Journal of Physiology, 259,* R197-R203.

Ralston, H.J., Inman, V.T., Strait, L.A., & Shaffrath, M.D. (1947). Mechanics of human isolated voluntary muscle. *American Journal of Physiology, 151,* 612-620.

Ramirez, J.M., Tryba, A.K., & Pena, F. (2004). Pacemaker neurons and neuronal networks: An integrative view. *Current Opinion in Neurobiology, 14,* 665-674.

Ramirez-Amaya, V., Marrone, D.F., Gage, F.H., Worley, P.F., & Barnes, C.A. (2006). Integration of new neurons into functional neural networks. *Journal of Neuroscience, 26,* 12237-12241.

Ranatunga, K.W., Sharpe, B., & Turnbull, B. (1987). Contractions of a human skeletal muscle at different temperatures. *Journal of Physiology, 390,* 383-395.

Rankin, L.L., Enoka, R.M., Volz, K.A., & Stuart, D.G. (1988). Coexistence of twitch potentiation and tetanic force decline in rat hindlimb muscle. *Journal of Applied Physiology, 65,* 2687-2695.

Rantanen, T., Guralnik, J.M., Ferrucci, L., Leveille, S., & Fried, L.P. (1999). Coimpairments: Strength and balance as predictors of severe walking disability. *Journal of Gerontology, 54A,* M172-M176.

Rassier, D.E., & Herzog, W. (2004). Considerations on the history dependence of muscle contraction. *Journal of Applied Physiology, 96,* 419-427.

Rassier, D.E., Herzog, W., Wakeling, J., & Syme, D. (2003). Stretch-induced, steady-state force enhancement in single skeletal muscle fibres exceeds the isometric force at optimum fiber length. *Journal of Biomechanics, 36,* 1309-1316.

Rassier, D.E., MacIntosh, B.R., & Herzog, W. (1999). Length dependence of active force production in skeletal muscle. *Journal of Applied Physiology, 86,* 1445-1457.

Rathmayer, W., Djokaj, S., Gaydukov, A., & Kreissl, S. (2002). The neuromuscular junctions of the slow and the fast excitatory axon in the closer of the crab Eriphia spinifrons are endowed with different Ca^{2+} channel types and allow neuron-specific modulation of transmitter release by two neuropeptides. *Journal of Neuroscience, 22,* 708-717.

Ratkevicius, A., Mizuno, M., Povilonis, E., & Quistorff, B. (1998*a*). Energy metabolism of the gastrocnemius and soleus muscles during isometric voluntary and electrically induced contractions in man. *Journal of Physiology, 507,* 593-602.

Ratkevicius, A., Skurvydas, A., Povilonis, E., Quistorff, B., & Lexell, J. (1998*b*). Effects of contraction duration on low-frequency fatigue in voluntary and electrically induced exercise of quadriceps femoris muscle in humans. *European Journal of Applied Physiology, 77,* 462-468.

Ravn, S., Voigt, M., Simonsen, E.B., Alkjær, T., Bojsen-Møller, F., & Klausen, K. (1999). Choice of jumping strategy in two standard jumps, squat and countermovement jump—effect of training background or inherited preference? *Scandinavian Journal of Medicine and Science in Sports, 9,* 201-208.

Rayment, I., Holden, H.M., Whittaker, M., Yohn, C.B., Lorenz, M., Holmes, K.C., & Milligan, R.A. (1993). Structures of the actin-myosin complex and its implications for muscle contraction. *Science, 261,* 58-65.

Raynor, A.J., Yi, C.J., Abernethy, B., & Jong, Q.J. (2002). Are transitions in human gait determined by mechanical, kinetic, or energetic factors? *Human Movement Sciences, 21,* 785-805.

Raz, N., & Rodrigue, K.M. (2006). Differential aging of the brain: Patterns, cognitive correlates and modifiers. *Neuroscience and Biobehavioral Reviews, 30,* 730-748.

Redfern, M.S., Cham, R., Gielo-Perczak, K., Gronqvist, R., Hirvonen, M., Lanshammar, H., Marpet, M., Pai, C.Y., & Powers, C. (2001). Biomechanics of slips. *Ergonomics, 44,* 1138-1166.

Reeves, N.D., Maganaris, C.N., & Narici, M.V. (2004). Ultrasonographic assessment of human skeletal muscle size. *European Journal of Applied Physiology, 91,* 116-118.

Reeves, N.D., Narici, M.V., & Maganaris, C.N. (2006). Myotendinous plasticity to ageing and resistance exercise in humans. *Experimental Physiology, 91,* 483-498.

Refshauge, K.M., Kilbreath, S.M., & Gandevia, S.C. (1998). Movement detection at the distal joint of the human thumb and fingers. *Experimental Brain Research, 122,* 85-92.

Reich, T.E., Lindstedt, S.L., LaStayo, P.C., & Pierotti, D.J. (2000). Is the spring quality of muscle plastic? *American Journal of Physiology, 278,* R1661-R1666.

Reid, D.A., & McNair, P.J. (2004). Passive force, angle, and stiffness changes after stretching of hamstring muscles.

Medicine and Science in Sports and Exercise, 36, 1944-1948.

Reid, M.B. (2008). Free radicals and muscle fatigue: Of ROS, canaries, and the IOC. *Free Radical Biology & Medicine, 44,* 169-179.

Reimers, C.D., Harder, T., & Saxe, H. (1998). Age-related muscle atrophy does not affect all muscles and can partly be compensated by physical activity: An ultrasound study. *Journal of the Neurological Sciences, 159,* 60-66.

Reis, J., Swayne, O.B., Vandermeeren, Y., Camus, M., Dimyan, M.A., Harris-Love, M., Perez, M.A., Ragert, P., Rothwell, J.C., & Cohen, L.G. (2008). Contribution of transcranial magnetic stimulation to the understanding of cortical mechanisms involved in motor control. *Journal of Physiology, 586,* 325-351.

Rekling, J.C., Funk, G.D., Bayliss, D.A., Dong, X.W., & Feldman, J.L. (2000). Synaptic control of motoneuronal excitability. *Physiological Reviews, 80,* 767-852.

Remple, M.S., Bruneau, R.M., VandenBerg, P.M., Goertzen, C., & Kleim, J.A. (2001). Sensitivity of cortical movement representations to motor experience: Evidence that skill training but not strength training induces cortical reorganization. *Behavioral Brain Research, 123,* 133-141.

Remy-Neris, O., Tiffreau, V., Bouilland, S., & Bussel, B. (2003). Intrathecal Baclofen in subjects with spastic hemiplegia: Assessment of the antispastic effect during gait. *Archives of Physical Medicine and Rehabilitation, 84,* 643-650.

Ren, L., Khan, A.Z., Blohm, G., Henriques, D.Y., Sergio, L.E., & Crawford, J.D. (2006). Proprioceptive guidance of saccades in eye-hand coordination. *Journal of Neurophysiology, 96,* 1464-1477.

Renaud, J.M. (2002). Modulation of force development by Na^+, K^+, Na^+-K^+ pump, and KATP channel during muscular activity. *Canadian Journal of Applied Physiology, 27,* 296-315.

Renshaw, B. (1941). Influence of discharge of motoneurons upon excitation of neighboring motoneurons. *Journal of Neurophysiology, 4,* 167-183.

Reschke, M.F., Bloomberg, J.J., Harm, D.L., Paloski, W.H., Layne, C.S., & McDonald, P.V. (1998). Posture, locomotion, spatial orientation, and motion sickness as a function of space flight. *Brain Research Reviews, 28,* 102-117.

Ribeiro, C. (2000). L-5-Hydroxytryptophan in the prophylaxis of chronic tension-type headache: A double-blind, randomized, placebo-controlled study. For the Portuguese Head Society. *Headache, 40,* 451-456.

Ribot-Ciscar, R., Butler, J.E., & Thomas, C.K. (2003). Facilitation of triceps brachii muscle contraction by tendon vibration after chronic cervical spinal cord injury. *Journal of Applied Physiology, 94,* 2358-2367.

Richardson, R.S., Frank, L.R., & Haseler, L.J. (1998). Dynamic knee-extensor and cycle exercise: Functional MRI of muscular activity. *International Journal of Sports Medicine, 19,* 182-187.

Riecker, A., Groschel, K., Ackermann, H., Steinbrink, C., Witte, O., & Kastrup, A. (2006). Functional significance of age-related differences in motor activation patterns. *NeuroImage, 32,* 1345-1354.

Rihet, P., Hasbroucq, T., Blin, O., & Possamai, C.A. (1999). Serotonin and human information processing: An electromyographic study of the effects of fluvoxamine on choice reaction time. *Neuroscience Letters, 16,* 143-146.

Rijkelijkhuizen, J.M., de Ruiter, C.J., Huijing, P.A., & de Haan, A. (2005). Low-frequency fatigue, post-tetanic potentiation and their interaction at different muscle lengths following eccentric exercise. *Journal of Experimental Biology, 208,* 55-63.

Riley, D.A., Bain, J.L., Romatowski, J.G., & Fitts, R.H. (2005). Skeletal muscle fiber atrophy: Altered thin filament density changes slow fiber force and shortening velocity. *American Journal of Physiology, 288,* C360-C365.

Riley, D.A., Bain, J.L.W., Thompson, J.L., Fitts, R.H., Widrick, J.J., Trappe, S.W., Trappe, T.A., & Costill, D.L. (2000). Increased thin filament density and length in human atrophic soleus muscle fibers after spaceflight. *Journal of Applied Physiology, 88,* 567-572.

Riley, D.A., Slocum, G.R., Bain, J.L.W., Sedlak, F.R., Sowa, T.E., & Mellender, J.W. (1990). Rat hindlimb unloading: Soleus histochemistry, ultrastructure, and electromyography. *Journal of Applied Physiology, 69,* 58-66.

Riley, Z.A., Maerz, A.H., Litsey, J.C., & Enoka, R.M. (2008). Motor unit recruitment in human biceps brachii during sustained voluntary contractions. *Journal of Physiology, 586,* 2183-2193.

Ríos, E., Ma, J., & González, A. (1991). The mechanical hypothesis of excitation-contraction (EC) coupling in skeletal muscle. *Journal of Muscle Research and Cell Motility, 12,* 127-135.

Ríos, E., & Pizarró, G. (1988). Voltage sensors and calcium channels of excitation-contraction coupling. *News in Physiological Sciences, 3,* 223-227.

Ritzmann, R.E., & Eaton, R.C. (1997). Neural substrates for the initiation of startle responses. In P.S.G. Stein, S. Grillner, A.I. Selverston, & D.G. Stuart (Eds.), *Neurons, Networks, and Motor Behavior* (pp. 33-44). Cambridge, MA: MIT Press.

Roatta, S., & Passatore, M. (2006). Muscle sensory receptors. In M. Akay (Ed.), *Wiley Encyclopedia of Biomedical Engineering.* Hoboken, NJ: Wiley.

Roberts, T.J., & Marsh, R.L. (2003). Probing the limits to muscle-powered accelerations: Lessons from jumping bullfrogs. *Journal of Experimental Biology, 206,* 2567-2580.

Robinovitch, S.N., Hsiao, E.T., Sandler, R., Cortez, J., Liu, Q., & Paiement, G.D. (2000). Prevention of falls and fall-related fractures through biomechanics. *Exercise and Sport Sciences Reviews, 28,* 74-79.

Robinson, G.A., Enoka, R.M., & Stuart, D.G. (1991). Immobilization-induced changes in motor unit force and fatigability in the cat. *Muscle & Nerve, 14,* 563-573.

Robitaille, R., & Tremblay, J.P. (1991). Non-uniform responses to Ca^{2+} along the frog neuromuscular junction: Effects on the probability of spontaneous and evoked transmitter release. *Neuroscience, 40,* 571-585.

Rodacki, A.L., Fowler, N.E., & Bennett, S.J. (2002). Vertical jump coordination fatigue effects. *Medicine and Science in Sports and Exercise, 34,* 105-116.

Roelants, M., Delecluse, C., & Verschueren, S.M. (2004). Whole-body-vibration increases knee-extension strength and speed of movement in older women. *Journal of the American Geriatric Society, 52,* 901-908.

Rogers, M.W. (1991). Motor control problems in Parkinson's disease. In M.J. Lister (Ed.), *Contemporary Management of Motor Control Problems* (pp. 195-208). Alexandria, VA: Foundation for Physical Therapy.

Rogers, M.W., & Pai, Y.-C. (1990). Dynamic transitions in stance support accompanying leg flexion movements in man. *Experimental Brain Research, 81,* 398-402.

Roll, J.P., Popov, K., Gurfinkel, V., Lipshits, M., Deshays, A.C., Gilhodes, J.C., & Quoniam, C. (1993). Sensorimotor and perceptual function of muscle proprioception in microgravity. *Journal of Vestibular Research, 3,* 259-273.

Roll, R., Gilhodes, J.C., Roll, J.P., Popov, K., Charade, O., & Gurfinkel, V. (1998). Proprioceptive information processing in weightlessness. *Experimental Brain Research, 122,* 393-402.

Rome, L.C., Funke, R.P., Alexander, R.M., Lutz, G., Aldridge, H., Scott, F., & Freadman, M. (1988). Why animals have different muscle fibre types. *Nature, 335,* 824-827.

Rose, P.K., Keirstead, S.A., & Vanner, S.J. (1985). A quantitative analysis of the geometry of cat motoneurons innervating neck and shoulder muscles. *Journal of Comparative Neurology, 239,* 89-107.

Ross, B.H., & Thomas, C.K. (1995). Human motor unit activity during induced muscle cramp. *Brain, 118,* 983-993.

Rossi, A.E., & Dirksen, R.T. (2006). Sarcoplasmic reticulum: The dynamic calcium governor of muscle. *Muscle & Nerve, 33,* 715-731.

Rossignol, S., Dubic, R., & Gossard, J.-P. (2006). Dynamic sensorimotor interactions in locomotion. *Physiological Reviews, 86,* 89-154.

Rossini, P.M., & Dal Forno, G. (2004). Integrated technology for evaluation of brain function and neural plasticity. *Physical Medicine and Rehabilitation Clinics of North America, 15,* 263-306.

Rost, K., Nowak, D.A., Timmann, D., & Hermsdorfer, J. (2005). Preserved and impaired aspects of predictive grip force control in cerebellar patients. *Clinical Neurophysiology, 116,* 1405-1414.

Rothwell, J.C. (1994). *Control of Human Voluntary Movement* (2nd ed.). Kent, UK: Croom Helm.

Rotto, D.M., & Kaufman, M.P. (1988). Effect of metabolic products of muscular contraction on discharge of group III and IV afferents. *Journal of Applied Physiology, 64,* 2306-2313.

Rousanoglou, E.N., Oskouei, A.E., & Herzog, W. (2006). Force depression following muscle shortening in sub-maximal voluntary contractions of human adductor pollicis. *Journal of Biomechanics, 40,* 1-8.

Roy, A.L., Keller, T.S., & Colloca, C.J. (2003). Posture-dependent trunk extensor EMG activity during maximum isometrics exertions in normal male and female subjects. *Journal of Electromyography and Kinesiology, 13,* 469-476.

Roy, M.P. (2004). Patterns of cortisol reactivity to laboratory stress. *Hormones and Behavior, 46,* 618-627.

Roy, R.R., Baldwin, K.M., & Edgerton, V.R. (1991). The plasticity of skeletal muscle: Effects of neuromuscular activity. *Exercise and Sport Sciences Reviews, 19,* 269-312.

Roy, R.R., & Edgerton, V.R. (1992). Skeletal muscle architecture and performance. In P.V. Komi (Ed.), *Strength and Power in Sport* (pp. 115-129). Champaign, IL: Human Kinetics.

Roy, R.R., Pierotti, D.J., Baldwin, K.M., Zhong, H., Hodgson, J.A., & Edgerton, V.R. (1998). Cyclical passive stretch influences the mechanical properties of the inactive cat soleus. *Experimental Physiology, 83,* 377-385.

Roy, R.R., Sacks, R.D., Baldwin, K.M., Short, M., & Edgerton, V.R. (1984). Interrelationships of contraction time, V_{max} and myosin ATPase after spinal transection. *Journal of Applied Physiology, 56,* 1594-1601.

Roy, R.R., Talmadge, R.J., Hodgson, J.A., Oishi, Y., Baldwin, K.M., & Edgerton, V.R. (1999). Differential response of fast hindlimb extensor and flexor muscles to exercise in adult spinalized cats. *Muscle & Nerve, 22,* 230-241.

Roy, R.R., Talmadge, R.J., Hodgson, J.A., Zhong, H., Baldwin, K.M., & Edgerton, V.R. (1998). Training effects on soleus of cats spinal cord transected (T12-13) as adults. *Muscle & Nerve, 21,* 63-71.

Roy, R.R., Zhong, H., Khalili, N., Kim, S.J., Higuchi, N., Monti, R.J., Grossman, E., Hodgson, J.A., & Edgerton, V.R. (2007). Is spinal cord isolation a good model of muscle disuse? *Muscle & Nerve, 35,* 312-321.

Rube, N., & Secher, N.H. (1990). Effect of training on central factors in fatigue following two- and one-leg static exercise in man. *Acta Physiologica Scandinavica, 141,* 87-95.

Rubenson, J., Henry, H.T., Dimoulas, P.M., & Marsh, R.L. (2006). The cost of uphill running: Linking organismal and muscle energy use in guinea fowl (Numida meleagris). *Journal of Experimental Biology, 209,* 2395-2408.

Rubenstein, S., & Kamen, G. (2005). Decreases in motor unit firing rate during sustained maximal-effort contractions in young and older adults. *Journal of Electromyography and Kinesiology, 15,* 536-543.

Rudomin, P., & Schmidt, R.F. (1999). Presynaptic inhibition in the vertebrate spinal cord revisited. *Experimental Brain Research, 129,* 1-37.

Rudroff, T., Barry, B.K., Stone, A.L., Barry, C.J., & Enoka, R.M. (2007). Accessory muscle activity influences variation in time to task failure for different arm postures and loads. *Journal of Applied Physiology, 102,* 1000-1006.

Rudroff, T., Poston, B., Shin, I.S., Bojsen-Møller, J., & Enoka, R.M. (2005). Net excitation of the motor unit pool varied with load type during fatiguing contractions. *Muscle & Nerve, 31,* 78-87.

Rüegg, J.C. (1983). Muscle. In R.F. Schmidt & F. Thews (Eds.), *Human Physiology* (pp. 32-50). Berlin: Springer-Verlag.

Rugg, S.G., Gregor, R.J., Mandelbaum, B.R., & Chiu, L. (1990). In vivo moment arm calculations at the ankle using magnetic resonance imaging (MRI). *Journal of Biomechanics, 23,* 495-501.

Ruijs, A.C., Jaquet, J.B., Kalmijn, S., Giele, H., & Hovius, S.E. (2005). Median and ulnar nerve injuries: A meta-analysis of predictors of motor and sensory recovery after modern microsurgical nerve repair. *Plastic and Reconstructive Surgery, 116,* 484-494.

Russ, D.W., & Kent-Braun, J.A. (2003). Sex differences in human skeletal muscle fatigue are eliminated under ischemic conditions. *Journal of Applied Physiology, 94,* 2414-2422.

Russ, D.W., Lanza, I.R., Rothman, D., & Kent-Braun, J.A. (2005). Sex differences in glycolysis during brief, intense isometric contractions. *Muscle & Nerve, 32,* 647-655.

Russell, B., Motlagh, D., & Ashley, W.W. (2000). Form follows function: How muscle shape is regulated by work. *Journal of Applied Physiology, 88,* 1127-1132.

Rutherford, O.M., & Jones, D.A. (1986). The role of learning and coordination in strength training. *European Journal of Applied Physiology, 55,* 100-105.

Sabatier, M.J., Stoner, L., Mahoney, E.T., Black, C., Elder, C., Dudley, G.A., & McCully, K. (2006). Electrically stimulated resistance training in SCI individuals increases muscle fatigue resistance but not femoral artery size or blood flow. *Spinal Cord, 44,* 227-233.

Sabick, M.B., Kim, Y.K., Torry, M.R., Keirns, M.A., & Hawkins, R.J. (2005). Biomechanics of the shoulder in youth baseball pitchers: Implications for the development of proximal humeral epiphysiolysis and humeral retrotorsion. *American Journal of Sports Medicine, 33,* 1716-1722.

Sabick, M.B., Torry, M.R., Kim, Y.K., & Hawkins, R.J. (2004). Humeral torque in professional pitchers. *American Journal of Sports Medicine, 32,* 892-898.

Sahlin, K., Tonkonogi, M., & Söderlund, K. (1998). Energy supply and muscle fatigue in humans. *Acta Physiologica Scandinavica, 162,* 261-266.

Sahlin, K., Tonkonogi, M., & Söderlund, K. (1999). Plasma hypoxanthine and ammonia in humans during prolonged exercise. *European Journal of Applied Physiology, 80,* 417-422.

Sahrmann, S.A., & Norton, B.J. (1977). The relationship of voluntary movement to spasticity in the upper motor neuron syndrome. *Annals of Neurology, 2,* 460-464.

Saibene, F., & Minetti, A.E. (2003). Biomechanical and physiological aspects of legged locomotion in humans. *European Journal of Applied Physiology, 88,* 297-316.

Sainburg, R.L., Ghez, C., & Kalakanis, D. (1999). Intersegmental dynamics are controlled by sequential anticipatory, error correction, and postural mechanisms. *Journal of Neurophysiology, 81,* 1045-1056.

Saito, M., Kobayashi, K., Miyashita, M., & Hoshikawa, T. (1974). Temporal patterns in running. In R.C. Nelson & C.A. Morehouse (Eds.), *Biomechanics IV* (pp. 106-111). Baltimore: University Park Press.

Sakata, J.T., & Jones, T.A. (2003). Synaptic mitochondrial changes in the motor cortex following unilateral cortical lesions and motor skills training in adult male rats. *Neuroscience Letters, 337,* 159-162.

Sale, D.G. (2002). Postactivation potentiation: Role in human performance. *Exercise and Sport Sciences Reviews, 30,* 138-143.

Sale, D.G., & MacDougall, J.D. (1981). Specificity in strength training: A review for the coach and athlete. *Canadian Journal of Applied Sport Sciences, 6,* 87-92.

Sale, D.G., MacDougall, J.D., Upton, A.R.M., & McComas, A.J. (1983). Effect of strength training upon motoneuron excitability in man. *Medicine and Science in Sports and Exercise, 125,* 57-62.

Sale, M.V., & Semmler, J.G. (2005). Age-related differences in corticospinal control during functional isometric contractions in left and right hands. *Journal of Applied Physiology, 99,* 1483-1493.

Salimi, I., Frazier, W., Reilmann, R., & Gordon, A.M. (2003). Selective use of visual information signaling objects' center of mass for anticipatory control of manipulative fingertip forces. *Experimental Brain Research, 150,* 9-18.

Saltiel, P., Tresch, M.C., & Bizzi, E. (1998). Spinal cord modular organization and rhythm generation: An NMDA iontophoretic study in the frog. *Journal of Neurophysiology, 80,* 2323-2339.

Sanders, D.B., Stålberg, E.V., & Nandedkar, S.D. (1996). Analysis of the electromyographic interference pattern. *Journal of Clinical Neurophysiology, 13,* 385-400.

Sandroni, P., Walker, C., & Starr, A. (1992). "Fatigue" in patients with multiple sclerosis. Motor pathway conduction and event-related potentials. *Archives of Neurology, 49,* 517-524.

Sands, W.A., McNeal, J.R., Stone, M.H., Russell, E.M., & Jemni, M. (2006). Flexibility enhancement with vibration: Acute and long-term. *Medicine and Science in Sports and Exercise, 38,* 720-725.

Sant'ana Pereira, J.A.A., Wessels, A., Nijtmans, L., Moorman, A.F.M., & Sargeant, A.J. (1994). New method for the accurate characterization of single human skeletal muscle fibres demonstrates a relation between mATPase and MyHC expression in pure and hybrid fibre types. *Journal of Muscle Research and Cell Motility, 16,* 21-24.

Santello, M., & Fuglevand, A.J. (2004). Role of across-muscle motor unit synchrony for the coordination of forces. *Experimental Brain Research, 159,* 501-508.

Santo Neto, H.S., de Carvalho, V.C., & Peneteado, C.V. (1985). Motor units of the human abductor digiti mnimi. *Archives of Italian Anatomy and Embriology, XC,* 47-51.

Santo Neto, H.S., Filho, J.M., Passini, R., & Marques, M.J. (2004). Number and size of motor units in thenar muscles. *Clinical Anatomy, 17,* 308-311.

Sapirstein, M.R., Herman, R.C., & Wallace, G.B. (1937). A study of after-contraction. *American Journal of Physiology, 119,* 549-556.

Sarlegna, F.R. (2006). Impairment of online control of reaching movements with aging: A double-step study. *Neuroscience Letters, 403,* 309-314.

Sarre, G., Lepers, R., Maffiuletti, N., Millet, G., & Martin, A. (2003). Influence of cycling cadence on neuromuscular activity of the knee extensors in humans. *European Journal of Applied Physiology, 88,* 476-479.

Saugen, E., & Vøllestad, N.K. (1995). Nonlinear relationship between heat production and force during voluntary contractions in humans. *Journal of Applied Physiology, 79,* 2043-2049.

Saugen, E., Vøllestad, N.K., Gibson, H., Martin, P.A., & Edwards, R.H.T. (1997). Dissociation between metabolic and contractile responses during intermittent isometric exercise in man. *Experimental Physiology, 82,* 213-226.

Saunders, J.A., & Knill, D.C. (2003). Humans use continuous visual feedback from the hand to control fast reaching movements. *Experimental Brain Research, 152,* 341-352.

Sayers, S.P., Harackiewicz, D.V., Harman, E.A., Frykman, P.N., & Rosenstein, M.T. (1999). Cross-validation of three jump power equations. *Medicine and Science in Sports and Exercise, 31,* 572-577.

Schaal, S., & Schweighofer, N. (2005). Computational motor control in humans and robots. *Current Opinion in Neurobiology, 15,* 675-682.

Schallert, T., Fleming, S.M., & Woodlee, M.T. (2003). Should the injured and intact hemispheres be treated differently during the early phases of physical restorative therapy in experimental stroke or parkinsonism? *Physical Medicine and Rehabilitation Clinics of North America, 14,* S27-S46.

Schechtman, H., & Bader, D.L. (1997). In vitro fatigue of human tendons. *Journal of Biomechanics, 30,* 829-835.

Scheidt, R.A., Conditt, M.A., Secco, E.L., & Mussa-Ivaldi, F.A. (2005). Interaction of visual and proprioceptive feedback during adaptation of human reaching movements. *Journal of Neurophysiology, 93,* 3200-3213.

Schiaffino, S., & Reggiani, C. (1996). Molecular diversity of myofibrillar proteins: Gene regulation and functional significance. *Physiological Reviews, 76,* 371-423.

Schieppati, M., Giordano, A., & Nardone, A. (2002). Variability in a dynamic postural task attests ample flexibility in balance control mechanisms. *Experimental Brain Research, 144,* 200-210.

Schieppati, M., & Nardone, A. (1991). Free and supported stance in Parkinson's disease. The effect of posture and postural set on leg muscle responses to perturbation, and its relation to the severity of the disease. *Brain, 114,* 1227-1244.

Schieppati, M., & Nardone, A. (1995). Time course of "set"-related changes in muscle responses to stance perturbation in humans. *Journal of Physiology, 487,* 787-796.

Schieppati, M., & Nardone, A. (1999). Group II spindle afferent fibers in humans: Their possible role in the reflex control of stance. In M.D. Binder (Ed.), *Progress in Brain Research* (Vol. 123, pp. 461-472). Amsterdam: Elsevier Science.

Schieppati, M., Nardone, A., Silotto, R., & Grasso, M. (1995). Early and last stretch responses of human foot muscles induced by perturbation of stance. *Experimental Brain Research, 105,* 411-422.

Schieppati, M., Trompetto, C., & Abbruzzese, G. (1996). Selective facilitation of responses to cortical stimulation of proximal and distal arm muscles by precision tasks in man. *Journal of Physiology, 491,* 551-562.

Schmied, A., Pagni, S., Sturm, H., & Vedel, J.-P. (2000). Selective enhancement of motoneurone short-term synchrony during an attention-demanding task. *Experimental Brain Research, 133,* 377-390.

Schmied, A., Pouget, J., & Vedel, J.-P. (1999). Electromechanical coupling and synchronous firing of single wrist extensor motor units in sporadic amyotrophic lateral sclerosis. *Clinical Neurophysiology, 110,* 960-977.

Schmied, A., Vedel, J.-P., & Pagni, S. (1994). Human spinal lateralization assessed from motoneurone synchronization: Dependence on handedness and motor unit type. *Journal of Neurophysiology, 480,* 369-387.

Schmied, A., Vedel, J.-P., Pouget, J., Forget, R., Lamarre, Y., & Paillard, J. (1995). Changes in motoneurone connectivity evaluated from neuronal synchronization analysis. In A. Taylor, M.H. Gladden, & R. Durbaba (Eds.), *Alpha and Gamma Motor Systems* (pp. 469-477). New York: Plenum Press.

Schneider, C., Lavoie, B., & Capaday, C. (2000). On the origin of the soleus H-reflex modulation pattern during human walking and its task-dependent differences. *Journal of Neurophysiology, 83,* 2881-2890.

Schneider, K., & Zernicke, R.F. (1992). Mass, center of mass, and moment of inertia estimates for infant limb segments. *Journal of Biomechanics, 25,* 145-148.

Schroeter, J.P., Bretaudiere, J.P., Sass, R.L., & Goldstein, M.A. (1996). Three-dimensional structure of the Z band in a normal mammalian skeletal muscle. *Journal of Cell Biology, 133,* 571-583.

Schwid, S.R., Covington, M., Segal, B.M., & Goodman, A.D. (2002). Fatigue in multiple sclerosis: Current understanding and future directions. *Journal of Rehabilitation Research and Development, 39,* 211-224.

Schwindt, P.C. (1973). Membrane-potential trajectories underlying motoneuron rhythmic firing at high rates. *Journal of Neurophysiology, 36,* 434-439.

Schwindt, P.C., & Calvin, W.H. (1972). Membrane potential trajectories between spikes underlying motoneuron rhythmic firing. *Journal of Neurophysiology, 35,* 311-325.

Scott, G., Menz, H.B., & Newcombe, L. (2007). Age-related differences in foot structure and function. *Gait & Posture, 26,* 68-75.

Scott, J.J.A. (1996). The functional recovery of muscle proprioceptors after peripheral nerve lesions. *Journal of the Peripheral Nervous System, 1,* 19-27.

Scott, S.H. (2008). Inconvenient truths about neural processing in primary motor cortex. *Journal of Physiology, 586,* 1217-1224.

Scott, S.H., Gribble, P.L., Graham, K.M., & Cabel, D.W. (2001). Dissociation between hand motion and population vectors from neural activity in motor cortex. *Nature, 413,* 161-165.

Scott, S.H., & Kalaska, J.F. (1997). Reaching movements with similar hand paths but different arm orientations. I. Activity of individual cells in motor cortex. *Journal of Neurophysiology, 77,* 826-852.

Scott, S.H., Sergio, L.E., & Kalaska, J.F. (1997). Reaching movements with similar hand paths but different arm orientations. II. Activity of individual cells in dorsal premotor cortex and parietal area 5. *Journal of Neurophysiology, 78,* 2413-2426.

Scott, W.B., Lee, S.C., Johnston, T.E., & Binder-Macleod, S.A. (2005). Switching stimulation patterns improves performance of paralyzed human quadriceps muscle. *Muscle & Nerve, 31,* 581-588.

Scremin, A.M.E., Kurta, L., Gentili, A., Wiseman, B., Perell, K., Kunkel, C., & Scremin, O.U. (1999). Increasing muscle mass in spinal cord injured persons with a functional electrical stimulation protocol. *Archives of Physical Medicine and Rehabilitation, 80,* 1531-1536.

Sears, T.A., & Stagg, D. (1976). Short-term synchronization of intercostal motoneurone activity. *Journal of Physiology, 263,* 357-381.

Seburn, K.L., & Gardiner, P.F. (1996). Properties of sprouted rat motor units: Effects of period of enlargement and activity level. *Muscle & Nerve, 19,* 1100-1109.

Secher, N.H. (1975). Isometric rowing strength of experienced and inexperienced oarsmen. *Medicine and Science in Sports and Exercise, 7,* 280-283.

Seger, J.Y., Arvidsson, B., & Thorstensson, A. (1998). Specific effects of eccentric and concentric training on muscle strength and morphology in humans. *European Journal of Applied Physiology, 79,* 49-57.

Seger, J.Y., & Thorstensson, A. (2005). Effects of eccentric versus concentric training on thigh muscle strength and EMG. *International Journal of Sports Medicine, 26,* 45-52.

Segers, V., Aerts, P., Lenoir, M., & De Clerq, D. (2007). Dynamics of the body centre of mass during actual acceleration across transition speed. *Journal of Experimental Biology, 15,* 578-585.

Sejersted, O.M., & Sjøgaard, G. Dynamics and consequences of potassium shifts in skeletal muscle and heart during exercise. *Physiological Reviews, 80,* 1411-1481.

Seki, K., Kizuka, T., & Yamada, H. (2007). Reduction in maximal firing rate of motoneurons after 1-week immobilization of finger muscle in human subjects. *Journal of Electromyography and Kinesiology, 17,* 113-120.

Seki, K., Perlmutter, S.I., & Fetz, E.E. (2003). Sensory input to primate spinal cord is presynaptically inhibited during voluntary movements. *Nature Neuroscience, 6,* 1309-1316.

Seki, K., Taniguchi, Y., & Narusawa, M. (2001). Effects of joint immobilization on firing rate modulation of human motor units. *Journal of Physiology, 530,* 507-519.

Selverston, A.I. (2005). A neural infrastructure for rhythmic motor patterns. *Cellular and Molecular Neurobiology, 25,* 223-244.

Selverston, A.I., Panchin, Y.V., Arshavsky, Y.I., & Orlovsky, G.N. (1997). Shared features of invertebrate central pattern generators. In P.S.G. Stein, S. Grillner, A.I. Selverston, & D.G. Stuart (Eds.), *Neurons, Networks, and Motor Behavior* (pp. 105-117). Cambridge, MA: MIT Press.

Semmler, J.G., & Enoka, R.M. (2000). Neural contributions to changes in muscle strength. In V.M. Zatsiorsky (Ed.), *Biomechanics in Sport* (pp. 3-20). Oxford, UK: Blackwell Science.

Semmler, J.G., Kornatz, K.W., & Enoka, R.M. (2003). Motor-unit coherence during isometric contractions is greater in a hand muscle of older adults. *Journal of Neurophysiology, 90,* 1346-1349.

Semmler, J.G., Kornatz, K.W., Kutzscher, D.V., Zhou, S., & Enoka, R.M. (2002). Motor unit synchronization is enhanced during slow lengthening contractions of a hand muscle. *Journal of Physiology, 545,* 681-695.

Semmler, J.G., Kornatz, K.W., Meyer, F.G., & Enoka, R.M. (2006). Diminished task-related adjustments of common inputs to hand muscle motor neurons in older adults. *Experimental Brain Research, 172,* 507-518.

Semmler, J.G., Kutzscher, D.V., & Enoka, R.M. (1999). Gender differences in the fatigability of human skeletal muscle. *Journal of Neurophysiology, 82,* 3590-3593.

Semmler, J.G., Kutzscher, D.V., & Enoka, R.M. (2000*a*). Limb immobilization alters muscle activation patterns during a fatiguing isometric contraction. *Muscle & Nerve, 23,* 1381-1392.

Semmler, J.G., & Nordstrom, M.A. (1998). Motor unit discharge and force tremor in skill- and strength-trained individuals. *Experimental Brain Research, 119,* 27-38.

Semmler, J.G., Nordstrom, M.A., & Wallace, C.J. (1997). Relationship between motor unit short-term synchronization and common drive in human first dorsal interosseus muscle. *Brain Research, 767,* 314-320.

Semmler, J.G., Sale, M.V., Meyer, F.G., & Nordstrom, M.A. (2004). Motor-unit coherence and its relation with synchrony are influenced by training. *Journal of Neurophysiology, 92,* 3320-3331.

Semmler, J.G., Steege, J.W., Kornatz, K.W., & Enoka, R.M. (2000*b*). Motor-unit synchronization is not responsible for larger motor-unit forces in old adults. *Journal of Neurophysiology, 84,* 358-366.

Sergio, L.E., Hamel-Pâquet, C., & Kalaska, J.F. (2005). Motor cortex neural correlates of output kinematics and kinetics during isometric-force and arm-reaching tasks. *Journal of Neurophysiology, 94,* 2353-2378.

Sergio, L.E., & Kalaska, J.F. (2003). Systematic changes in motor cortex cell activity with arm posture during directional isometric force generation. *Journal of Neurophysiology, 89,* 212-228.

Seth, A., & Pandy, M.G. (2007). A neuromusculoskeletal tracking method for estimating individual muscle forces in human movement. *Journal of Biomechanics, 40,* 356-366.

Seward, H., Orchard, J., Hazard, H., & Collinson, D. (1993). Football injuries in Australia at the elite level. *Medical Journal of Australia, 159,* 298-301.

Seyfarth, A., Geyer, H., Gunther, M., & Blickhan, R. (2002). A movement criterion for running. *Journal of Biomechanics, 35,* 649-655.

Seynnes, O., Fiatarone Singh, M.A., Hue, O., Pras, P., Legros, P., & Bernard, P.L. (2004). Physiological and functional responses to low-moderate versus high-intensity progressive resistance training in frail elders. *Journal of Gerontology, 59A,* 503-509.

Seynnes, O., Hue, O.A., Garrandes, F., Colson, S.S., Bernard, P.L., Legros, P., & Fiatarone Singh, M.A. (2005). Force steadiness in the lower extremities as an independent predictor of functional performance in older women. *Journal of Aging and Physical Activity, 13,* 395-408.

Shadmehr, R., & Mussa-Ivaldi, F.A. (1994). Adaptive representation of dynamics during learning of a motor task. *Journal of Neuroscience, 14,* 3208-3224.

Shadmehr, R., & Wise, S.P. (2005). *The Computational Neurobiology of Reaching and Pointing.* Cambridge, MA: MIT Press.

Shah, P.K., Stevens, J.E., Gregory, C.M., Pathare, N.C., Jayaraman, A., Bickel, S.C., Bowden, M., Behrman, A.L., Walter, G.A., Dudley, G.A., & Vandenborne, K. (2006). Lower-extremity muscle cross-sectional area after incomplete spinal cord injury. *Archives of Physical Medicine and Rehabilitation, 87,* 772-778.

Shan, G., & Bohn, C. (2003). Anthropometrical data and coefficients of regressions related to gender and race. *Applied Ergonomics, 34,* 327-337.

Shanebrook, J.R., & Jaszczak, R.D. (1976). Aerodynamic drag analysis of runners. *Medicine and Science in Sports, 8,* 43-45.

Sharman, M.J., Cresswell, A.G., & Riek, S. (2006). Proprioceptive neuromuscular facilitation stretching: Mechanisms and clinical implications. *Sports Medicine, 36,* 929-939.

Sheard, P.W. (2000). Tension delivery from short fibers in long muscles. *Exercise and Sport Sciences Reviews, 28,* 51-56.

Sheean, G.L., Murray, N.M.F., Rothwell, J.C., Miller, D.H., & Thompson, A.J. (1997). An electrophysiological study of the mechanism of fatigue in multiple sclerosis. *Brain, 120,* 299-315.

Shefner, J.M. (2004). Motor unit number estimates. In A. Eisen (Ed.), *Clinical Neurophysiology of Motor Neuron Diseases. Handbook of Clinical Neurophysiology* (Vol. 4, pp. 271-281). Amsterdam: Elsevier.

Shelburne, K.B., & Pandy, M.G. (1997). A musculoskeletal model of the knee for evaluating ligament forces during isometric contractions. *Journal of Biomechanics, 30,* 163-176.

Shelburne, K.B., Torry, M.R., & Pandy, M.G. (2005). Muscle, ligament, and joint-contact forces at the knee during walking. *Medicine and Science in Sports and Exercise, 37,* 1948-1956.

Shellock, F.G., Fukunaga, T., Mink, J.H., & Edgerton, V.R. (1991). Exertional muscle injury: Evaluation of concentric versus eccentric actions with serial MRI imaging. *Radiology, 179,* 659-664.

Shen, L., & Alexander, G.E. (1997). Preferential representation of instructed target location versus limb trajectory in dorsal premotor area. *Journal of Neurophysiology, 77,* 1195-1212.

Sherrington, C.S. (1897). On reciprocal innervation of antagonist muscles. Third note. *Proceedings of the Royal Society, 60,* 408-417.

Sherrington, C.S. (1910). Flexion-reflex of the limb, crossed extension-reflex, and reflex stepping and standing. *Journal of Physiology, 40,* 28-121.

Sherrington, C.S. (1925). Remarks on some aspects of reflex inhibition. *Proceedings of the Royal Society of London B, 97,* 19-45.

Shields, R.K. (1995). Fatigability, relaxation properties, and electromyographic responses of the human paralyzed soleus muscle. *Journal of Neurophysiology, 73,* 2195-2206.

Shields, R.K. (2002). Muscular, skeletal, and neural adaptations following spinal cord injury. *Journal of Orthopedic Sports Physical Therapy, 32,* 65-74.

Shields, R.K., & Dudley-Javoroski, S. (2006). Musculoskeletal plasticity after acute spinal cord injury: Effects of long-term neuromuscular electrical stimulation training. *Journal of Neurophysiology, 95,* 2380-2390.

Shields, R.K., Dudley-Javoroski, S., & Law, L.A. (2006*a*). Electrically induced muscle contractions influence bone density decline after spinal cord injury. *Spine, 31,* 548-553.

Shields, R.K., Dudley-Javoroski, S., & Littmann, A.E. (2006*b*). Postfatigue potentiation of the paralyzed soleus muscle: Evidence for adaptation with long-term electrical stimulation training. *Journal of Applied Physiology, 101,* 556-565.

Shik, M.L., Severin, F.V., & Orlovsky, G.N. (1966). Control of walking and running by means of electrical stimulation of the mid-brain. *Biophysics, 11,* 756-765.

Shim, J.K., Lay, B.S., Zatsiorsky, V.M., & Latash, M.L. (2004). Age-related changes in finger coordination in static prehension tasks. *Journal of Applied Physiology, 97,* 213-224.

Shinoda, Y., Yokata, J-L., & Futami, T. (1981). Divergent projection of individual corticospinal axons to motoneurons of multiple muscles in the monkey. *Neuroscience Letters, 23,* 7-12.

Shinohara, M., Keenan, K.G., & Enoka, R.M. (2003). Contralateral activity in a homologous hand muscle during voluntary contractions is greater in old adults. *Journal of Applied Physiology, 94,* 966-974.

Shinohara, M., Keenan, K.G., & Enoka, R.M. (2008). Fluctuations in motor output of a hand muscle can be altered by the mechanical properties of the position sensor. *Journal of Neuroscience Methods, 168,* 164-173.

Shinohara, M., Moritz, C.T., Pascoe, M.A., & Enoka, R.M. (2005). Prolonged muscle vibration increases stretch reflex amplitude, motor unit discharge rate, and force fluctuations in a hand muscle. *Journal of Applied Physiology, 99,* 1835-1842.

Shinohara, M., Scholz, J.P., Zatsiorsky, V.M., & Latash, M.L. (2004). Finger interaction during accurate multi-finger force production tasks in young and elderly persons. *Experimental Brain Research, 156,* 282-292.

Shinohara, M., & Søgaard, K. (2006). Mechanomyography for studying force fluctuations and muscle fatigue. *Exercise and Sport Sciences Reviews, 34,* 59-64.

Shirakura, K., Kato, K., & Udagawa, E. (1992). Characteristics of the isokinetic performance of patients with injured cruciate ligaments. *American Journal of Sports Medicine, 20,* 755-760.

Shorten, M.R. (1987). Muscle elasticity and human performance. In B. Van Gheluwe & J. Atha (Eds.), *Medicine and*

Sport Science: Current Research in Sport Biomechanics (Vol. 25, pp. 1-18). Basel: Karger.

Sieck, G.C., & Prakash, Y.S. (1995). Fatigue at the neuromuscular junction. In S.C. Gandevia, R.M. Enoka, A.J. McComas, D.G. Stuart, & C.K. Thomas (Eds.), *Fatigue: Neural and Muscular Mechanisms* (pp. 83-100). New York: Plenum Press.

Siegmund, G.P., Inglis, J.T., & Sanderson, D.J. (2001). Startle response of human neck muscles sculpted by readiness to perform ballistic head movements. *Journal of Physiology, 535,* 289-300.

Sillar, K.T., Kiehn, O., & Kudo, N. (1997). Chemical modulation of vertebrate circuits. In P.S.G. Stein, S. Grillner, A.I. Selverston, & D.G. Stuart (Eds.), *Neurons, Networks, and Motor Behavior* (pp. 183-193). Cambridge, MA: MIT Press.

Silva, M.P., & Ambrosio, J.A. (2004). Sensitivity of the results produced by the inverse dynamic analysis of a human stride to perturbed input data. *Gait & Posture, 19,* 35-49.

Simard, C.P., Spector, S.A., & Edgerton, V.R. (1982). Contractile properties of rat hind limb muscles immobilized at different lengths. *Experimental Neurology, 77,* 467-482.

Simonetta-Moreau, M., Marque, P., Marchand-Pauvert, V., & Pierrot-Deseilligny, E. (1999). The pattern of excitation of human lower limb motoneurones by probable group II muscle afferents. *Journal of Physiology, 517,* 287-300.

Simonsen, E.B., & Dyhre-Poulsen, P. (1999). Amplitude of the human soleus H reflex during walking and running. *Journal of Physiology, 515,* 929-939.

Simonsen, E.B., Dyhre-Poulsen, P., Voigt, M., Aagaard, P., Sjøgaard, G., & Bojsen-Møller, F. (1995). Bone-on-bone forces during loaded and unloaded walking. *Acta Anatomica, 152,* 133-142.

Singer, B.J., Dunne, J.W., Singer, K.P., & Allison, G.T. (2003). Velocity dependent passive plantarflexor resistive torque in patients with acquired brain injury. *Clinical Biomechanics, 18,* 157-165.

Sinkjær, T. (1997). Muscle, reflex, and central components in the control of the ankle joint in healthy and spastic man. *Acta Neurologica Scandinavica, 170,* 1-28.

Sinkjær, T., Andersen, J.B., Ladouceur, M., Christensen, L.O.D., & Nielsen, J.B. (2000). Major role for sensory feedback in soleus EMG activity in the stance phase of walking in man. *Journal of Physiology, 523,* 817-827.

Sinkjær, T., Andersen, J.B., & Larsen, B. (1996). Soleus stretch reflex modulation during gait in humans. *Journal of Neurophysiology, 76,* 1112-1120.

Sinkjær, T., Andersen, J.B., Nielsen, J.F., & Hansen, H.J. (1999). Soleus long-latency stretch reflexes during walking in healthy and spastic humans. *Clinical Neurophysiology, 110,* 951-959.

Sjøgaard, G. (1996). Potassium and fatigue: The pros and cons. *Acta Physiologica Scandinavica, 156,* 257-264.

Sjøgaard, G., Jensen, B.R., Hargens, A.R., & Sogaard, K. (2004). Intramuscular pressure and EMG relate during static contractions but dissociate with movement and fatigue. *Journal of Applied Physiology, 96,* 1522-1529.

Sjøgaard, G., Kiens, B., Jørgensen, K., & Saltin, B. (1986). Intramuscular pressure, EMG and blood flow during low-level prolonged static contraction in man. *Acta Physiologica Scandinavica, 128,* 475-484.

Sjøgaard, G., Savard, G., & Juel, C. (1988). Muscle blood flow during isometric activity and its relation to muscle fatigue. *European Journal of Applied Physiology, 57,* 327-335.

Sjölander, P., Johansson, H., & Djupsjöbacka, M. (2002). Spinal and supraspinal effects of activity in ligament afferents. *Journal of Electromyography and Kinesiology, 12,* 167-176.

Sjöström, M., Ängquist, K.-A., Bylund, A.-C., Fridén, J., Gustavsson, L., & Scherstén, T. (1982). Morphometric analyses of human muscle fiber types. *Muscle & Nerve, 5,* 538-553.

Skurvydas, A., Sipaviciene, S., Krutulyte, G., Gailiuniene, A., Stasisulis, A., Mamkus, G., & Stanislovaitis, A. (2006). Dynamics of indirect symptoms of skeletal muscle damage after stretch-shortening exercise. *Journal of Electromyography and Kinesiology, 16,* 629-636.

Slawinska, U., Tyc, F., Kasicki, S., Navarrete, R., & Vrbová, G. (1998). Time course of changes in EMG activity of fast muscles after partial denervation. *Experimental Brain Research, 120,* 193-201.

Slifkin, A.B., & Newell, K.M. (1999). Noise, information transmission, and force variability. *Journal of Experimental Psychology, 25,* 837-851.

Smith, C.A. (1994). The warm-up procedure: To stretch or not to stretch. A brief review. *Journal of Orthopedic and Sports Physical Therapy, 19,* 12-17.

Smith, J.L., & Zernicke, R.F. (1987). Predictions for neural control based limb dynamics. *Trends in Neurosciences, 10,* 123-128.

Smith, M.A., Brandt, J., & Shadmehr, R. (2000). The motor dysfunction in Huntington's disease begins as a disorder in error feedback control. *Nature, 403,* 544-549.

Snow, R., Carey, M., Stathis, C., Febbraio, M.A., & Hargreaves, M. (2000). Effect of carbohydrate ingestion on ammonia metabolism during exercise in humans. *Journal of Applied Physiology, 88,* 1576-1580.

Søgaard, K., Gandevia, S.C., Todd, G., Petersen, N.T., & Taylor, J.L. (2006). The effect of sustained low-intensity contractions on supramaximal fatigue in human elbow flexor muscles. *Journal of Physiology, 573,* 511-523.

Sommer, M.A., & Wurtz, R.H. (2002). A pathway in primate brain for internal monitoring of movements. *Science, 296,* 1480-1482.

Sonnenfeld, G. (2005). Use of animal models for space flight physiology studies, with special focus on the immune system. *Gravitational Space Biology Bulletin, 18,* 31-35.

Sosnoff, J.J., & Newell, K.M. (2006). Aging, visual intermittency, and variability in isometric force output. *Journal of Gerontology, 61B,* P117-P124.

Sosnoff, J.J., Vaillancourt, D.E., & Newell, K.M. (2004). Aging and rhythmical force output: Loss of adaptive control of multiple neural oscillators. *Journal of Neurophysiology, 91,* 172-181.

Sousa, F., Ishikawa, M., Vilas-Boas, J.P., & Komi, P.V. (2007). Intensity- and muscle-specific fascicle behavior during human drop jumps. *Journal of Applied Physiology, 102,* 382-389.

Sowers, M.R., Crutchfield, M., Richards, K., Wilkin, M.K., Furniss, A., Jannausch, M., Zhang, D., & Gross, M. (2005). Sarcopenia is related to physical functioning and leg strength in middle-aged women. *Journal of Gerontology, 60A,* 486-490.

Spath, M. (2002). Current experience with 5-HT3 receptor antagonists in fibromyalgia. *Rheumatic Diseases Clinics of North America, 28,* 319-328.

Spector, S.A., Simard, C.P., Fournier, M., Sternlicht, E., & Edgerton, V.R. (1982). Architectural alterations of rat hindlimb skeletal muscles immobilized at different lengths. *Experimental Neurology, 76,* 94-110.

Spernoga, S.G., Uhl, T.L., Arnold, B.L., & Gansneder, B.M. (2001). Duration of maintained hamstring flexibility after a one-time, modified hold-relax stretching protocol. *Journal of Athletic Training, 36,* 44-48.

Spielberger, C.D., Gorsuch, R.L., Lushene, R.E., Vagg, P.R., & Jacobs, G.A. (1983). *Manual for the State–Trait Anxiety Inventory STAI (Form Y).* Palo Alto, CA: Consulting Psychologists Press.

Spielberger, C.D., & Rickman, R.L. (1990). Assessment of state and trait anxiety. In N. Sartorius (Ed.), *Anxiety: Psychological and clinical perspectives* (pp. 69-83). New York: Hemisphere Books.

Spira, M.E., Yarom, Y., & Parnas, I. (1976). Modulation of spike frequency by regions of special axonal geometry and by synaptic inputs. *Journal of Neurophysiology, 39,* 882-899.

Spoor, C.W., van Leeuwen, J.L., Meskers, C.G.M., Titulaer, A.F., & Huson, A. (1990). Estimation of instantaneous moment arms of lower-leg muscles. *Journal of Biomechanics, 23,* 1247-1259.

Spring, E., Savolainen, S., Erkkilä, J., Hämäläinen, T., & Pihkala, P. (1988). Drag area of a cross-country skier. *International Journal of Sport Biomechanics, 4,* 103-113.

Spurway, N.C., Watson, H., McMillan, K., & Connolly, G. (2000). The effect of strength training on the apparent inhibition of eccentric force production in voluntarily activated human quadriceps. *European Journal of Applied Physiology, 82,* 374-380.

Stafilidis, S., Karamanidis, K., Morey-Klapsing, G., Demonte, G., Brüggemann, G.P., & Arampatzis, A. (2005). Strain and elongation of the vastus lateralis aponeurosis and tendon in vivo during maximal isometric contraction. *European Journal of Applied Physiology, 94,* 317-322.

Stainsby, W.N., Brechue, W.F., O'Drobinak, D.M., & Barclay, J.K. (1990). Effects of ischemic and hypoxic hypoxia on $\dot{V}O_2$ and lactic acid output during tetanic contractions. *Journal of Applied Physiology, 68,* 574-579.

Stålberg, E., & Karlsson, L. (2001). The motor nerve simulator. *Clinical Neurophysiology, 112,* 2118-2132.

Stapley, P.J., Pozzo, T., Cheron, G., & Grishin, A. (1999). Does the coordination between posture and movement during human whole-body reaching ensure center of mass stabilization? *Experimental Brain Research, 129,* 134-146.

Staron, R.S., Karapondo, D.L., Kraemer, W.J., Fry, A.C., Gordon, S.E., Falkel, J.E., Hagerman, F.C., & Hikida, R.S. (1994). Skeletal muscle adaptations during early phase of heavy-resistance training in men and women. *Journal of Applied Physiology, 76,* 1247-1255.

Staron, R.S., & Pette, D. (1986). Correlation between myofibrillar ATPase activity and myosin heavy chain composition in rabbit muscle fibres. *Histochemistry, 86,* 19-23.

Stauber, W.T. (2004). Factors involved in strain-induced injury in skeletal muscles and outcomes of prolonged exposures. *Journal of Electromyography and Kinesiology, 14,* 61-70.

Staudenmann, D., Kingma, I., Daffertshofer, A., Stegeman, D.F., & van Dieën, J.H. (2006). Improving EMG-based muscle force estimation by using a high-density EMG grid and principal component analysis. *IEEE Transactions on Biomedical Engineering, 53,* 712-719.

Stecina, K., Quevedo, J., & McCrea, D.A. (2005). Parallel reflex pathways from flexor muscle afferents evoking resetting and flexion enhancement during fictive locomotion and scratch in the cat. *Journal of Physiology, 569,* 275-290.

Steele, D.S., & Duke, A.M. (2003). Metabolic factors contributing to altered Ca^{2+} regulation in skeletal muscle fatigue. *Acta Physiologica Scandinavica, 179,* 39-48.

Stein, P.S.G., Grillner, S., Selverston, A.I., & Stuart, D.G. (Eds.). (1997). *Neruons, Networks, and Motor Behavior.* Cambridge, MA: MIT Press.

Stein, P.S.G., & Smith, J.L. (1997). Neural and biomechanical control strategies for different forms of vertebrate hindlimb motor tasks. In P.S.G. Stein, S. Grillner, A.I. Selverston, & D.G. Stuart (Eds.), *Neurons, Networks, and Motor Behavior* (pp. 61-73). Cambridge, MA: MIT Press.

Stein, R.B. (1995). Presynaptic inhibition in humans. *Progress in Neurobiology, 47,* 533-544.

Stein, R.B., Chong, S.L., James, K.B., Kido, A., Bell, G.J., Tubman, L.A., & Belanger, M. (2002). Electrical stimulation for therapy and mobility after spinal cord injury. *Progress in Brain Research, 137,* 27-34.

Stein, R.B., Gordon, T., Jefferson, J., Sharfenberger, A., Yang, J.F., Tötösy de Zepetnek, J., & Belanger, M. (1992). Optimal stimulation of paralyzed muscle after human spinal cord injury. *Journal of Applied Physiology, 72,* 1393-1400.

Stein, R.B., Gossen, E.R., & Jones, K.E. (2005). Neuronal variability: Noise or part of the signal. *Nature Reviews Neuroscience, 6,* 389-398.

Stein, R.B., & Thompson, A.K. (2006). Muscle reflexes in motion: How, what, and why? *Exercise and Sport Sciences Reviews, 34,* 145-153.

Stein, R.B., & Yang, J.F. (1990). Methods for estimating the number of motor units in human muscles. *Annals of Neurology, 28,* 487-495.

Steinen, G.J.M., Kiers, J.L., Bottinelli, R., & Reggiani, C. (1996). Myofibrillar ATPase activity in skinned human skeletal muscle fibres: Fibre type and temperature dependence. *Journal of Physiology, 493,* 299-307.

Steinen, G.J.M., Roosemalen, M.C.M., Wilson, M.G.A., & Elzinga, G. (1990). Depression of force by phosphate in skinned skeletal muscle fibers of the frog. *American Journal of Physiology, 259,* C349-C357.

Stelzer, J.E., & Widrick, J.J. (2003). Effect of hindlimb suspension on the functional properties of slow and fast soleus fibers from three strains of mice. *Journal of Applied Physiology, 95,* 2425-2433.

Stensdotter, A.K., Hodges, P.W., Mellor, R., Sundelin, G., & Hager-Ross, C. (2003). Quadriceps activation in closed and in open kinetic chain exercise. *Medicine and Science in Sports and Exercise, 35,* 2043-2047.

Stephens, J.A., Usherwood, T.P., & Garnett, R. (1976). Technique for studying synaptic connections in single motoneurones in man. *Nature, 263,* 343-344.

Stephens, M.J., & Yang, J.F. (1996). Short latency, non-reciprocal group I inhibition is reduced during the stance phase of walking in humans. *Brain Research, 743,* 24-31.

Stephenson, D.G., Lamb, G.D., Stephenson, G.M.M., & Fryer, M.W. (1995). Mechanisms of excitation–contraction coupling relevant to skeletal muscle fatigue. In S.C. Gandevia, R.M. Enoka, A.J. McComas, D.G. Stuart, & C.K. Thomas (Eds.), *Fatigue: Neural and Muscular Mechanisms* (pp. 45-56). New York: Plenum Press.

Sterzing, T., & Hennig, E.M. (2008). The influence of soccer shoes on kicking velocity in full-instep kicks. *Exercise and Sport Sciences Reviews, 36,* 91-97.

Stevens, E.D. (1993). Relation between work and power calculated from force–velocity curves to that done during oscillatory work. *Journal of Muscle Research and Cell Motility, 14,* 518-526.

Stevens, J.C., Cruz, L.A., Marks, L.E., & Lakatos, S. (1998). A multimodal assessment of sensory thresholds in aging. *Journal of Gerontology, 53A,* B263-B272.

Stevens, J.E., Mizner, R.L., & Snyder-Mackler, L. (2004). Neuromuscular electrical stimulation for quadriceps muscle strengthening after bilateral total knee arthroplasty: A case series. *Journal of Orthopedic Sports Physical Therapy, 34,* 21-29.

Stevens, J.E., Pathare, N.C., Tillman, S.M., Scarborough, M.T., Gibbs, C.P., Shah, P., Jayaraman, A., Walter, G.A., & Vandenborne, K. (2006). Relative contributions of muscle activation and muscle size to plantarflexor torque during rehabilitation after immobilization. *Journal of Orthopaedic Research, 24,* 1729-1736.

Stevens, J.E., Stackhouse, S.K., Binder-Macleod, S.A., & Snyder-Mackler, L. (2003). Are voluntary muscle activation deficits in older adults meaningful? *Muscle & Nerve, 27,* 99-101.

Stewart, B.G., Tarnopolsky, M.A., Hicks, A.L., McCartney, B., Mahoney, D.J., Staron, R.S., & Phillips, S.M. (2004). Treadmill training-induced adaptations in muscle phenotype in persons with incomplete spinal cord injury. *Muscle & Nerve, 30,* 61-68.

Stewart, I.B., & Sleivert, G.G. (1998). The effect of warm-up intensity on range of motion and anaerobic performance. *Journal of Orthopedic and Sports Physical Therapy, 27,* 154-161.

Steyvers, M., Levin, O., Van Baelen, M., & Swimmen, S.P. (2003). Corticospinal excitability changes following prolonged muscle tendon vibration. *NeuroReport, 14,* 1901-1905.

Stodden, D.F., Fleisig, G.S., McLean, S.P., & Andrews, J.R. (2005). Relationship of biomechanical factors to baseball pitching velocity: Within pitcher variation. *Journal of Applied Biomechanics, 21,* 44-56.

Stokes, M., & Young, A. (1984). The contribution of reflex inhibition to arthrogenous muscle weakness. *Clinical Science, 67,* 7-14.

Stoloff, R.H., Zehr, E.P., & Ferris, D.P. (2007). Recumbent stepping has similar but simpler neural control compared to walking. *Experimental Brain Research, 178,* 427-438.

Stoner, L., Sabatier, M.J., Mahoney, E.T., Dudley, G.A., & McCully, K.K. (2007). Electrical stimulation-evoked resistance exercise therapy improves arterial health after chronic spinal cord injury. *Spinal Cord, 45,* 49-56.

Street, S.F. (1983). Lateral transmission of tension in frog myofibers: A myofibrillar network and transverse cytoskeletal connections are possible transmitters. *Journal of Cell Physiology, 114,* 346-364.

Stromberg, D.D., & Wiederhielm, C.A. (1969). Viscoelastic description of a collagenous tissue in simple elongation. *Journal of Applied Physiology, 26,* 857-862.

Stone, M.H., & O'Bryant, H.S. (1987). *Weight Training: A Scientific Approach.* Minneapolis: Bellweather Press.

Stroup, F., & Bushnell, D.L. (1970). Rotation, translation, and trajectory in diving. *Research Quarterly, 40,* 812-817.

Stucke, H., Baudzus, W., & Baumann, W. (1984). On friction characteristics of playing surfaces. In E.C. Frederick (Ed.), *Sport Shoes and Playing Surfaces* (pp. 87-97). Champaign, IL: Human Kinetics.

Stupka, N., Tarnopolsky, M.A., Yardley, N.J., & Phillips, S.M. (2001). Cellular adaptation to repeated eccentric exercise-induced muscle damage. *Journal of Applied Physiology, 91,* 1669-1678.

Suetta, C., Aagaard, P., Magnusson, S.P., Andersen, L.L., Sipila, S., Rosted, A., Jakobsen, A.K., Duus, B., & Kjaer, M. (2007). Muscle size, neuromuscular activation, and rapid force characteristics in elderly men and women: Effects of unilateral long-term disuse due to hip-osteoarthritis. *Journal of Applied Physiology, 102,* 942-948.

Sullivan, D.H., Roberson, P.K., Smith, E.S., Price, J.A., & Bopp, M.M. (2007). Effects of muscle strength training and megestrol acetate on strength, muscle mass, and function in frail older people. *Journal of the American Geriatric Society, 55,* 20-28.

Sullivan, M.K., Dejulia, J.J., & Worrell, T.W. (1992). Effect of pelvic position and stretching method on hamstring muscle flexibility. *Medicine and Science in Sports and Exercise, 24,* 1383-1389.

Suputtitada, A., Suwanwela, N.C., & Tumvitee, S. (2004). Effectiveness of constraint-induced movement therapy in chronic stroke patients. *Journal of the Medical Association of Thailand, 87,* 1482-1490.

Suzuki, S., & Hutton, R.S. (1976). Postcontractile motoneuronal discharge produced by muscle afferent activation. *Medicine and Science in Sports, 8,* 258-264.

Suzuki, S., Watanabe, S., & Homma, S. (1982). EMG activity and kinematics of human cycling movements at different constant velocities. *Brain Research, 240,* 245-258.

Suzuki, T., Bean, J.F., & Fielding, R.A. (2001). Muscle power of the ankle flexors predicts functional performance in community-dwelling older women. *Journal of the American Geriatrics Society*, *49*, 1161-1167.

Svensson, P., Houe, L., & Arendt-Nielsen, L. (1997). Bilateral experimental muscle pain changes electromyographic activity of human jaw-closing muscles during mastication. *Experimental Brain Research, 116,* 182-185.

Sweeney, H.L., Bowman, B.F., & Stull, J.T. (1993). Myosin light chain phosphorylation in vertebrate striated muscle: Regulation and function. *American Journal of Physiology, 264,* C1085-C1095.

Sweeney, H.L., & Stull, J.T. (1990). Alteration of cross-bridge kinetics by myosin light chain phosphorylation in rabbit skeletal muscle: Implications for regulation of actin-myosin interaction. *Proceedings of the National Academy of Sciences, 87,* 414-418.

Swoap, S.J., Caiozzo, V.J., & Baldwin, K.M. (1997). Optimal shortening velocities for in situ power production of rat soleus and plantaris muscles. *American Journal of Physiology, 273,* C1057-C1063.

Symons, T.B., Vandervoort, A.A., Rice, C.L., Overend, T.J., & Marsh, G.D. (2005). Effects of maximal isometric and isokinetic resistance training on strength and functional mobility in older adults. *Journal of Gerontology, 60A,* 777-781.

Takakusaki, K., Saitoh, K., Harada, H., & Kashiwayanagi, M. (2004). Role of basal ganglia–brainstem pathways in the control of motor behaviors. *Neuroscience Research, 50,* 137-151.

Takekura, H., Fujinami, N., Nishizawa, T., Ogasawara, H., & Kasuga, N. (2001). Eccentric exercise–induced morphological changes in the membrane systems involved in excitation–contraction coupling in rat skeletal muscle. *Journal of Physiology, 533,* 571-583.

Talmadge, R.J., Roy, R.R., & Edgerton, V.R. (1995). Prominence of myosin heavy chain hybrid fibers in soleus muscle of spinal cord–transected rats. *Journal of Applied Physiology, 78,* 1256-1265.

Talmadge, R.J., Roy, R.R., & Edgerton, V.R. (1999). Persistence of hybrid fibers in rat soleus after spinal cord transection. *Anatomical Record, 255,* 188-201.

Tam, S.L., Archibald, V., Jassr, B., Tyreman, N., & Gordon, T. (2001). Increased neuromuscular activity reduces sprouting in partially denervated muscles. *Journal of Neuroscience, 21,* 654-667.

Tam, S.L., Archibald, V., Tyreman, N., & Gordon, T. (2002). Tetrodotoxin prevents motor unit enlargement after partial denervation in rat hindlimb muscles. *Journal of Physiology, 543,* 655-663.

Tam, S.L., & Gordon, T. (2003). Neuromuscular activity impairs axonal sprouting in partially denervated muscles by inhibiting bridge formation of perisynaptic Schwann cells. *Journal of Neurobiology, 57,* 221-234.

Tanaka, R. (1974). Reciprocal Ia inhibition during voluntary movements in man. *Experimental Brain Research, 21,* 529-540.

Taniguchi, Y. (1998). Relationship between the modifications of bilateral deficit in upper and lower limbs by resistance training in humans. *European Journal of Applied Physiology, 78,* 226-230.

Tate, C.M., Williams, G.N., Barrance, P.J., & Buchanan, T.S. (2006). Lower extremity muscle morphology in young athletes: An MRI-based analysis. *Medicine and Science in Sports and Exercise, 38,* 122-128.

Taub, E., Uswatte, G., King, D.K., Morris, D., Crago, J.E., & Chatterjee, A. (2006). A placebo-controlled trial of constraint-induced movement therapy for upper extremity after stroke. *Stroke, 37,* 1045-1049.

Tax, A.A.M., Wezel, B.M.H. van, & Dietz, V. (1995). Bipedal reflex coordination to tactile stimulation of the sural nerve during human running. *Journal of Neurophysiology, 73,* 1947-1964.

Taylor, A. (2002). Give proprioceptors a chance. *Advances in Experimental and Medical Biology, 508,* 327-334.

Taylor, A., Durbaba, R., Ellaway, P.H., & Rawlinson, S. (2000). Patterns of fusimotor activity during locomotion in the decerebrate cat deduced from recordings from hindlimb muscle spindles. *Journal of Physiology, 522,* 515-532.

Taylor, C.R., Heglund, N.C., McMahon, T.A., & Looney, T.R. (1980). Energetic cost of generating muscular force during running. *Journal of Experimental Biology*, *86*, 9-18.

Taylor, D.C., Dalton, J., Seaber, A.V., & Garrett, W.E. (1990). The viscoelastic properties of muscle-tendon units. *American Journal of Sports Medicine, 18,* 300-309.

Taylor, J.L. (2005). Independent control of voluntary movements and associated anticipatory postural responses in a bimanual task. *Clinical Neurophysiology, 116,* 2083-2090.

Taylor, J.L., & Gandevia, S.C. (2004). Noninvasive stimulation of the human corticospinal tract. *Journal of Applied Physiology, 96,* 1496-1503.

Taylor, J.L., & Gandevia, S.C. (2008). A comparison of central aspects of fatigue in submaximal and maximal voluntary contractions. *Journal of Applied Physiology*, *104,* 542-550.

Taylor, J.L., & McCloskey, D.I. (1992). Detection of slow movements imposed at the elbow during active flexion in man. *Journal of Physiology, 457,* 503-513.

Taylor, J.L., Petersen, N., Butler, J.E., & Gandevia, S.C. (2000). Ischaemia after exercise does not reduce responses of human motoneurones to cortical or corticospinal tract stimulation. *Journal of Physiology, 525,* 793-801.

Taylor, J.L., Todd, G., & Gandevia, S.C. (2006). Evidence for a supraspinal contribution to human muscle fatigue. *Clinical and Experimental Pharmacology and Physiology, 33,* 400-405.

Taylor, W.R., Heller, M.O., Bergmann, G., & Duda, G.N. (2004). Tibio-femoral loading during human gait and stair climbing. *Journal of Orthopaedic Research, 22,* 625-632.

ter Haar Romeny, B.M., Denier van der Gon, J.J., & Gielen, C.C.A.M. (1982). Changes in recruitment order of motor units in the human biceps brachii muscle. *Experimental Neurology, 78,* 360-368.

Terao, Y., & Ugawa, Y. (2002). Basic mechanisms of TMS. *Journal of Clinical Neurophysiology, 19,* 322-343.

Tesch, P.A. (1992). Training for bodybuilding. In P.V. Komi (Ed.), *Strength and Power in Sport* (pp. 370-380). Oxford, UK: Blackwell Scientific.

Tesch, P.A., Berg, H.E., Bring, D., Evans, H.J., & LeBlanc, A.D. (2005). Effects of 17-day spaceflight on knee extensor muscle function and size. *European Journal of Applied Physiology, 93,* 463-468.

Tesch, P.A., Dudley, G.A., Duvoisin, M.R., Hather, B.M., & Harris, R.T. (1990). Force and EMG signal patterns during repeated bouts of concentric and eccentric muscle actions. *Acta Physiologica Scandinavica, 138,* 263-271.

Thach, W.T., & Bastian, A.J. (2004). Role of the cerebellum in the control and adaptation of gait in health and disease. *Progress in Brain Research, 143,* 353-366.

Thacker, S.B., Gilchrist, J., Stroup, D.F., & Kimsey, C.D. Jr. (2004). The impact of stretching on sports injury risk: A systematic review of the literature. *Medicine and Science in Sports and Exercise, 36,* 371-378.

Thambyah, A., Pereira, B.P., & Wyss, U. (2005). Estimation of bone-on-bone contact forces in the tibiofemoral joint during walking. *Knee, 12,* 383-388.

Thelen, D.G., Brockmiller, C., Ashton-Miller, J.A., Schultz, A.B., & Alexander, N.B. (1998). Thresholds for detecting foot dorsi- and plantarflexion during upright stance: Effects of age and velocity. *Journal of Gerontology, 53A,* M33-M38.

Thelen, D.G., Chumanov, E.S., Best, T.M., Swanson, S.C., & Heiderscheit, B.C. (2005). Simulation of biceps femoris musculotendon mechanics during the swing phase of sprinting. *Medicine and Science in Sports and Exercise, 37,* 1931-1938.

Thelen, D.G., Muriuki, M., James, J., Schultz, A.B., Ashton-Miller, J.A., & Alexander, N.B. (2000). Muscle activities used by young and old adults when stepping to regain balance during a forward fall. *Journal of Electromyography and Kinesiology, 10,* 93-101.

Thickbroom, G.W., Sacco, P., Kermode, A.G., Archer, S.A., Byrnes, M.L., Guilfoyle, A., & Mastaglia, F.L. (2006). Central motor drive and perception of effort during fatigue in multiple sclerosis. *Journal of Neurology, 253,* 1048-1053.

Thoby-Brisson, M., & Simmers, J. (2002). Long-term neuromodulatory regulation of a motor-pattern generating network: Maintenance of synaptic efficacy and oscillatory properties. *Journal of Neurophysiology, 88,* 2942-2953.

Thomas, C.K. (1997). Contractile properties of human thenar muscles paralyzed by spinal cord injury. *Muscle & Nerve, 20,* 788-799.

Thomas, C.K., Broton, J.G., & Calancie, B. (1997*a*). Motor unit forces and recruitment patterns after cervical spinal cord injury. *Muscle & Nerve, 20,* 212-220.

Thomas, C.K., Johansson, R.S., & Bigland-Ritchie, B. (1991). Attempts to physiologically classify human thenar motor units. *Journal of Neurophysiology, 65,* 1501-1508.

Thomas, C.K., Johansson, R.S., & Bigland-Ritchie, B. (1999). Pattern of pulses that maximize force output from single human thenar motor units. *Journal of Neurophysiology, 82,* 3188-3195.

Thomas, C.K., Nelson, G., Than, I., & Zijdewind, I. (2002). Motor unit activation order during electrically evoked contractions of paralyzed or partially paralyzed muscles. *Muscle & Nerve, 25,* 797-804.

Thomas, C.K., Stein, R.B., Gordon, T., Lee, R.G., & Elleker, M.G. (1987). Patterns of reinnervation and motor unit recruitment in human hand muscles after complete ulnar and median nerve section and resuture. *Journal of Neurology, Neurosurgery, and Psychiatry, 50,* 259-268.

Thomas, C.K., Tucker, M.E., & Bigland-Ritchie, B. (1998). Voluntary muscle weakness and co-activation after chronic cervical spinal cord injury. *Journal of Neurotrauma, 15,* 149-161.

Thomas, C.K., Zaidner, E.Y., Calancie, B., Broton, J.G., & Bigland-Ritchie, B.R. (1997*b*). Muscle weakness, paralysis, and atrophy after human cervical spinal cord injury. *Experimental Neurology, 148,* 414-423.

Thomason, D.B., & Booth, F.W. (1990). Atrophy of the soleus muscle by hindlimb unweighting. *Journal of Applied Physiology, 68,* 1-12.

Thomason, D.B., Herrick, R.E., & Baldwin, K.M. (1987). Activity influences on soleus muscle myosin during rodent hindlimb suspension. *Journal of Applied Physiology, 63,* 138-144.

Thompson, L.V., Johnson, S.A., & Shoeman, J.A. (1998). Single soleus muscle fiber function after hindlimb unweighting in adult and aged rats. *Journal of Applied Physiology, 84,* 1936-1942.

Thompson, H.S., Clarkson, P.M., & Scordilis, S.P. (2002). The repeated bout effect and heat shock proteins: Intramuscular HSP27 and HSP70 expression following two bouts of eccentric exercise in humans. *Acta Physiologica Scandinavica, 174,* 47-56.

Thornell, L.E., Carlsson, E., Kugelberg, E., & Grove, B.K. (1987). Myofibrillar M-band structure and composition of physiologically defined rat motor units. *American Journal of Physiology, 253,* C456-C468.

Thornell, L.E., Lindstrom, M., Renault, V., Mouly, V., & Butler-Browne, G.S. (2003). Satellite cells and training in the elderly. *Scandinavian Journal of Medicine and Science in Sports, 13,* 48-55.

Thoroughman, K.A., & Shadmehr, R. (1999). Electromyographic correlates of learning an internal model of reaching movements. *Journal of Neuroscience, 19,* 8573-8588.

Thrasher, T.A., Flett, H.M., & Popovic, M.R. (2006). Gait training regimen for incomplete spinal cord injury using functional electrical stimulation. *Spinal Cord, 44,* 357-361.

Tidball, J.G. (2005). Mechanical signal transduction in skeletal muscle growth and adaptation. *Journal of Applied Physiology, 98,* 1900-1908.

Tillery, S.I., Soechting, J.F., & Ebner, T.J. (1996). Somatosensory cortical activity in relation to arm posture: Nonuniform spatial tuning. *Journal of Neurophysiology, 76,* 2423-2438.

Timmann, D., & Horak, F.B. (2001). Perturbed step initiation in cerebellar subjects: 2. Modification of anticipatory postural adjustments. *Experimental Brain Research, 141,* 110-120.

Timoszyk, W.K., de Leon, R.D., London, N., Roy, R.R., Edgerton, V.R., & Reinkensmeyer, D.J. (2002). The rat lumbosacral spinal cord adapts to robotic loading applied during stance. *Journal of Neurophysiology, 88,* 3108-3117.

Tirosh, O., & Sparrow, W.A. (2005). Age and walking speed effects on muscle recruitment in gait termination. *Gait & Posture, 21,* 279-288.

Todd, G., Butler, J.E., Taylor, J.L., & Gandevia, S.C. (2005). Hyperthermia: A failure of the motor cortex and the muscle. *Journal of Physiology, 563,* 621-631.

Todd, G., Petersen, N.T., Taylor, J.L., & Gandevia, S.C. (2003*a*). The effect of a contralateral contraction on maximal voluntary activation and central fatigue in elbow flexor muscles. *Experimental Brain Research, 150,* 308-313.

Todd, G., Taylor, J.L., & Gandevia, S.C. (2003*b*). Measurement of voluntary activation of fresh and fatigued human muscles using transcranial magnetic stimulation. *Journal of Physiology, 551,* 661-671.

Todd, G., Taylor, J.L., & Gandevia, S.C. (2004). Reproducible measurement of voluntary activation of human elbow flexors with motor cortical stimulation. *Journal of Applied Physiology, 97,* 236-242.

Toft, E., Espersen, C.T., Kalund, S., Sinkjær, T., & Hornemann, B.C. (1989). Passive tension of the ankle before and after stretching. *American Journal of Sports Medicine, 17,* 489-494.

Tokuno, C.D., Carpenter, M.G., Thorstensson, A., & Cresswell, A.G. (2006). The influence of natural body sway on neuromuscular responses to an unpredictable surface translation. *Experimental Brain Research, 174,* 19-28.

Tomlinson, B.E., & Irving, D. (1977). The number of limb motor neurons in the human lumbosacral cord throughout life. *Journal of Neurological Sciences, 34,* 213-219.

Tong, C., & Flanagan, J.R. (2003). Task-specific internal models for kinematic transformations. *Journal of Neurophysiology, 90,* 578-585.

Torre, M. (1953). Nombre et dimensions des unités motrices dans les muscles extrinsêques de l'œil et, en général, dans les muscles squélettiques reliés à des organes de sens. *Archives Suisses de Neurologie et de Psychiatrie, 72,* 362-376.

Tötösy de Zepetnek, J.E., Zung, H.V., Erdebil, S., & Gordon, T. (1992). Innervation ratio is an important determinant of force in normal and reinnervated rat tibialis anterior muscles. *Journal of Neurophysiology, 67,* 1385-1403.

Toursel, T., Stevens, L., Granzier, H., & Mounier, Y. (2002). Passive tension of rat skeletal soleus muscle fibers: Effects of unloading conditions. *Journal of Applied Physiology, 92,* 1465-1472.

Toussaint, H.M., van Baar, C.E., van Langen, P.P., de Looze, M.P., & van Dieën, J.H. (1992). Coordination of the leg muscles in backlift and leglift. *Journal of Biomechanics, 25,* 1279-1289.

Townend, M.S. (1984). *Mathematics in Sport.* New York: Halstead.

Tracy, B.L., & Enoka, R.M. (2006). Steadiness training with light loads in the knee extensors of elderly adults. *Medicine and Science in Sports and Exercise, 38,* 735-745.

Trappe, S., Gallagher, P., Harber, M., Carrithers, J., Fluckey, J., & Trappe, T. (2003). Single muscle fiber contractile properties in young and old men and women. *Journal of Physiology, 552,* 47-58.

Trappe, S., Harber, M., Creer, A., Gallagher, P., Slivka, D., Minchev, K., & Whitsett, D. (2006). Single muscle fiber adaptations with marathon training. *Journal of Applied Physiology, 101,* 721-727.

Trappe, S., Trappe, T., Gallagher, P., Harber, M., Alkner, B., & Tesch, P. (2004). Human single muscle fibre function with 84 day bed-rest and resistance exercise. *Journal of Physiology, 557,* 501-513.

Trappe, S., Williamson, D., Godard, M., Porter, D., Rowden, G., & Costill, D. (2000). Effect of resistance training on single muscle fiber contractile function in older men. *Journal of Applied Physiology, 89,* 143-152.

Tresch, M.C., Cheung, V.C.K., & d'Avella, A. (2006). Matrix factorization algorithms for the identification of muscle synergies: Evaluation on simulated and experimental data sets. *Journal of Neurophysiology, 95,* 2199-2212.

Trimble, M.H., & Enoka, R.M. (1991). Mechanisms underlying the training effects associated with neuromuscular electrical stimulation. *Physical Therapy, 71,* 273-282.

Trotter, J.A. (1990). Interfiber tension transmission in series-fibered muscles of the cat hindlimb. *Journal of Morphology, 206,* 351-361.

Trotter, J.A. (1993). Functional morphology of force transmission in skeletal muscle. *Acta Anatomica, 146,* 205-222.

Tsaopoulos, D.E., Baltzopoulos, V., & Maganaris, C.N. (2006). Human patellar tendon moment arm length: Measurement considerations and clinical implications for joint loading assessment. *Clinical Biomechanics, 21,* 657-667.

Tsika, R. (2006). The muscular system: The control of muscle mass. In C.M. Tipton (Ed.), *ACSM's Advanced Exercise Physiology* (pp. 161-177). Philadelphia: Lippincott Williams & Wilkins.

Tsika, R.W., Herrick, R.E., & Baldwin, K.M. (1987). Subunit composition of rodent isomyosins and their distribution in hindlimb skeletal muscles. *Journal of Applied Physiology, 63,* 2101-2110.

Tsuruike, M., Koceja, D.M., Yabe, K., & Shima, N. (2003). Age comparison of H-reflex modulation with the Jendrassik maneuver and postural complexity. *Clinical Neurophysiology, 114,* 945-953.

Turner, E.H., Loftis, J.M., & Blackwell, A.D. (2006). Serotonin a la carte: Supplementation with the serotonin precursor

5-hydroxytryptophan. *Pharmacology Therapy, 109,* 325-338.

Turner, L.C., Harrison, L.M., & Stephens, J.A. (2002). Finger movement is associated with attenuated cutaneous reflexes recorded from human first dorsal interosseus. *Journal of Physiology, 542,* 559-566.

Turner, P.E., & Raglin, J.S. (1996). Variability in precompetition anxiety and performance in college track and field athletes. *Medicine and Science in Sports and Exercise, 28,* 378-385.

Ulfhake, B., & Cullheim, S. (1988). Postnatal development of cat hind limb motoneurons. III: Changes in size of motoneurons supplying the triceps surae muscle. *Journal of Comparative Neurology, 278,* 103-120.

Ulfhake, B., & Kellerth, J-O. (1983). A quantitative morphological study of HRP-labelled cat α-motoneurones supplying different hindlimb muscles. *Brain Research, 264,* 1-19.

Ulfhake, B., & Kellerth, J-O. (1982). Does α-motoneurone size correlate with motor unit type in cat triceps surae? *Brain Research, 251,* 201-209.

Ulloa, A., & Bullock, D. (2003). A neural network simulating human reach–grasp coordination by continuous updating of vector positioning commands. *Neural Networks, 16,* 1141-1160.

Vaillancourt, D.E., Larsson, L., & Newell, K.M. (2003). Effects of aging on force variability, single motor unit discharge patterns, and the structure of 10, 20, and 40 Hz EMG activity. *Neurobiology of Aging, 24,* 25-35.

Vaillancourt, D.E., & Newell, K.M. (2003). Aging and the time and frequency structure of force output variability. *Journal of Applied Physiology, 94,* 903-912.

Valero-Cuevas, F.J., Johanson, M.E., & Towles, J.D. (2003). Towards a realistic biomechanical model of the thumb: The choice of kinematic description may be more critical than the solution method or the variability/uncertainty of musculoskeletal parameters. *Journal of Biomechanics, 36,* 1019-1030.

Vallbo, Å.B. (1971). Muscle spindle response at the onset of isometric voluntary contractions in man. Time difference between fusimotor and skeletomotor effects. *Journal of Physiology, 218,* 405-431.

Vallbo, Å.B., Hagbarth, K-E., & Wallin, G.B. (2004). Microneurography: How the technique developed and its role in the investigation of the sympathetic nervous system. *Journal of Applied Physiology, 96,* 1261-1269.

Valls-Solé, J., Kofler, M., Kumru, H., Castellote, J.M., & Sanegre, M.T. (2005). Startle-induced reaction time shortening is not modified by prepulse inhibition. *Experimental Brain Research, 165,* 541-548.

Valls-Solé, J., Rothwell, J.C., Goulart, F., Cossou, G., & Muñoz, E. (1999). Patterned ballistic movements triggered by a startle in healthy humans. *Journal of Physiology, 516,* 931-938.

Valour, D., Rouji, M., & Pousson, M. (2004). Effects of eccentric training on torque-angular velocity-power characteristics of elbow flexor muscles in older women. *Experimental Gerontology, 39,* 359-368.

Van Cutsem, M., & Duchateau, J. (2005). Preceding muscle activity influences motor unit discharge and rate of torque development during ballistic contractions in humans. *Journal of Physiology, 562,* 635-644.

Van Cutsem, M., Duchateau, J., & Hainaut, K. (1998). Neural adaptations mediate increase in muscle contraction speed and change in motor unit behaviour after dynamic training. *Journal of Physiology, 513,* 295-305.

Van Cutsem, M., Feiereisen, P., Duchateau, J., & Hainaut, K. (1997). Mechanical properties and behaviour of motor units in the tibialis anterior during voluntary contractions. *Canadian Journal of Applied Physiology, 22,* 585-597.

Vandenborne, K.A., Elliott, M.A., Walter, G.A., Abdus, S., Okereke, E., Shaffer, M., Tahernia, D., & Esterhai, J.L. (1998). Longitudinal study of skeletal muscle adaptations during immobilization and rehabilitation. *Muscle & Nerve, 21,* 1006-1012.

van den Tillaar, R. (2006). Will whole-body vibration training help increase the range of motion of the hamstrings? *Journal of Strength and Conditioning Research, 20,* 192-196.

van der Helm, F.C.T., & Veenbaas, R. (1991). Modelling the mechanical effect of muscles with large attachment sites: Application to the shoulder mechanism. *Journal of Biomechanics, 24,* 1151-1163.

Vanderthommen, M., Depresseux, J.C., Bauvir, P., Degueldre, C., Delfiore, G., Peters, J.M., Sluse, F., & Crielaard, J.M. (1997). A positron emission tomography study of voluntarily and electrically contracted human quadriceps. *Muscle & Nerve, 20,* 505-507.

Vanderthommen, M., Duteil, S., Wary, C., Raynaud, J.S., Leroy-Willig, A., & Carlier, P.G. (2006). Comparison of muscle energetics during voluntary and electrically induced contractions in humans. In A. Rainoldi, M.A. Minetto, & R. Merletti (Eds.), *Biomedical Engineering in Exercise and Sports* (pp. 187-198). Torino, Italy: Edizioni Minerva Medica.

Vanderthommen, M., Gilles, R., Carlier, P., Ciancabilla, F., Zahlan, O., Sluse, F., & Crielaard, J.M. (1999). Human muscle energetics during voluntary and electrically induced isometric contractions as measured by ^{31}P NMR spectroscopy. *International Journal of Sports Medicine, 20,* 279-283.

van der Vaart, A.J.M., Savelberg, H.H.C.M., de Groot, G., Hollander, A.P., Toussaint, H.M., & Ingen Schenau, G.J. van (1987). An estimation of drag in front crawl swimming. *Journal of Biomechanics, 20,* 543-546.

Vandervoort, A.A. (2002). Aging of the human neuromuscular system. *Muscle & Nerve, 25,* 17-25.

van der Werf, S.P., Jongen, P.J.H., Lycklama à Nijeholt, G.J., Barkhof, F., Hommes, O.R., & Bleijenberg, G. (1998). Fatigue in multiple sclerosis: Interrelations between fatigue complaints, cerebral MRI abnormalities and neurological disability. *Journal of the Neurological Sciences, 160,* 164-170.

Van Eck, M.M.M., Nicolson, N.A., Berkhof, H., & Sulon, J. (1996). Individual differences in cortisol responses to a laboratory speech task and their relationship to responses to stressful daily events. *Biological Psychology, 43,* 69-84.

van Emmerik, R.E.A., & van Wegen, E.E.H. (2002). On the functional aspects of variability in postural control. *Exercise and Sport Sciences Reviews, 30,* 177-183.

van Mameren, H., & Drukker, J. (1979). Attachment and composition of skeletal muscles in relation to their function. *Journal of Biomechanics, 12,* 859-867.

van Pragg, H., Shubert, T., Zhao, C., & Gage, F.H. (2005). Exercise enhances learning and hippocampal neurogenesis in aged mice. *Journal of Neuroscience, 25,* 8680-8685.

Van Wezel, B.M.H., Ottenhoff, F.A., & Duysens, J. (1997). Dynamic control of location-specific information in tactile cutaneous reflexes from the foot during human walking. *Journal of Neuroscience, 17,* 3804-3814.

van Zuylen, E.J., Gielen, C.C.A.M., & Denier van der Gon, J.J. (1988). Coordination and inhomogeneous activation of human arm muscles during isometric torques. *Journal of Neurophysiology, 60,* 1523-1548.

Vaughan, C.L. (1984). Biomechanics of running gait. *CRC Critical Reviews in Biomedical Engineering, 12,* 1-48.

Veldhuizen, J.W., Verstappen, F.T.J., Vroemen, J.P.A.M., Kuipers, H., & Greep, J.M. (1993). Functional and morphological adaptations following four weeks of knee immobilization. *International Journal of Sports Medicine, 14,* 283-287.

Verburg, E., Hallén, J., Sejersted, O.M., & Vøllestad, N. (1999). Loss of potassium from muscle during moderate exercise in humans: A result of insufficient activation of the Na^+-K^+ pump? *Acta Physiologica Scandinavica, 165,* 357-367.

Viala, G., Orsal, D., & Buser, P. (1978). Cutaneous fiber groups involved in the inhibition of fictive locomotion in the rabbit. *Experimental Brain Research, 33,* 257-267.

Viitasalo, J.T., & Komi, P.V. (1978). Interrelationships of EMG signal characteristics at different levels of muscle tension during fatigue. *Electromyography and Clinical Neurophysiology, 18,* 167-178.

Vikne, H., Refsnes, P.E., Ekmark, M., Medbo, J.I., Gundersen, V., & Gundersen, K. (2006). Muscular performance after concentric and eccentric exercise in trained men. *Medicine and Science in Sports in Exercise, 38,* 1770-1781.

Visser, J.J., Hoogkamer, J.E., Bobbert, M.F., & Huijing, P.A. (1990). Length and moment arm of human leg muscles as a function of knee and hip-joint angles. *European Journal of Applied Physiology, 61,* 453-460.

Voelcker-Rehage, C., & Alberts, J.L. (2005). Age-related changes in grasping force modulation. *Experimental Brain Research, 166,* 61-70.

Voigt, M., Chelli, F., & Frigo, C. (1998*a*). Changes in the excitability of soleus muscle short latency stretch reflexes during human hopping after 4 weeks of hopping training. *European Journal of Applied Physiology, 78,* 522-532.

Voigt, M., Dyhre-Poulsen, P., & Simonsen, E.B. (1998*b*). Modulation of short latency stretch reflexes during human hopping. *Acta Physiologica Scandinavica, 163,* 181-194.

Vøllestad, N.K. (1995). Metabolic correlates of fatigue from different types of exercise in man. In S.C. Gandevia, R.M. Enoka, A.J. McComas, D.G. Stuart, & C.K. Thomas (Eds.). *Fatigue: Neural and Muscular Mechanisms* (pp. 185-194). New York: Plenum Press.

Volkmann, N., Hanein, D., Ouyang, G., Trybus, K.M., DeRosier, D.J., & Lowey, S. (2000). Evidence for cleft closure in actomyosin upon ADP release. *Nature Structural Biology, 7,* 1147-1155.

Vorontsov, A.R., & Rumyantsev, V.A. (2000). Resistive forces in swimming. In V.M. Zatsiorsky (Ed.), *Biomechanics in Sport* (pp. 184-204). Oxford, UK: Blackwell Science.

Vrbová, G., & Wareham, A.C. (1976). Effects of nerve activity on the postsynaptic membrane of skeletal muscle. *Brain Research, 118,* 371-382.

Wackerhage, H., & Rennie, M.J. (2006). How nutrition and exercise maintain the human musculoskeletal mass. *Journal of Anatomy, 208,* 451-458.

Wade, A.J., Broadhead, M.W., Cady, E.B., Llewelyn, M.E., Tong, H.N., & Newham, D.J. (2000). Influence of muscle temperature during fatiguing work with the first dorsal interosseous muscle in man: A ^{31}P-NMR study. *European Journal of Applied Physiology, 81,* 203-209.

Wallin, D., Ekblom, B., Grahn, R., & Nordenborg, T. (1985). Improvement of muscle flexibility. A comparison between two techniques. *American Journal of Sports Medicine, 13,* 263-268.

Walmsley, B., Hodgson, J.A., & Burke, R.E. (1978). Forces produced by medial gastrocnemius and soleus muscles during locomotion in freely moving cats. *Journal of Neurophysiology, 41,* 1203-1216.

Walmsley, B., & Proske, U. (1981). Comparison of stiffness of soleus and medial gastrocnemius muscles in cats. *Journal of Neurophysiology, 46,* 250-259.

Walsh, E.G. (1992). *Muscles, Masses and Motion.* London: Mac Keith Press.

Walsh, L.D., Hess, C.W., Morgan, D.L., & Proske, U. (2004). Human forearm position sense after fatigue of elbow flexor muscles. *Journal of Physiology, 558,* 705-715.

Walsh, M.L. (2000). Whole body fatigue and critical power. *Sports Medicine, 29,* 153-166.

Wang, K., McCarter, R., Wright, J., Beverly, J., & Ramirez-Mitchell, R. (1993). Viscoelasticity of the sarcomere matrix of skeletal muscles: The titin–myosin composite filament is a dual-stage molecular spring. *Biophysical Journal, 64,* 1161-1177.

Wang, L., & Buchanan, T.S. (2002). Prediction of joint moments using a neural network model of muscle activations from EMG signals. *IEEE Transactions on Neural and Systems Rehabilitation Engineering, 10,* 30-37.

Wang, T., Dordevic, G.S., & Shadmehr, R. (2001). Learning the dynamics of reaching movements results in the modification of arm impedance and long-latency perturbation responses. *Biological Cybernetics, 85,* 437-448.

Wang, Y., Asaka, T., Zatsiorsky, V.M., & Latash, M.L. (2006). Muscle synergies during voluntary body sway: Combining across-trials and within-a-trial analyses. *Experimental Brain Research, 174,* 679-693.

Warabi, T., Noda, H., & Kato, T. (1986). Effect of aging on sensorimotor functions of eye and hand movements. *Experimental Neurology, 92,* 686-697.

Ward, N.S., Brown, M.M., Thompson, A.J., & Frackowiak, R.S. (2006). Longitudinal changes in cerebral response to proprioceptive input in individual patients after stroke: An FMRI study. *Neurorehabilitation and Neural Repair, 20,* 398-405.

Ward-Smith, A.J. (1985). A mathematical analysis of the influence of adverse and favourable winds on sprinting. *Journal of Biomechanics, 18,* 351-357.

Warren, G.L., Ingalls, C.P., Lowe, D.A., & Armstrong, R.B. (2001). Excitation–contraction uncoupling: major role in contraction-induced muscle injury. *Exercise and Sport Sciences Reviews, 29,* 82-87.

Warshaw, D.M. (1996). The in vitro motility assay: A window into the myosin molecular motor. *News in Physiological Sciences, 11,* 1-6.

Waterman-Storer, C.M. (1991). The cytoskeleton of skeletal muscle: Is it affected by exercise? A brief review. *Medicine and Science in Sports and Exercise, 23,* 1240-1249.

Webber, S., & Kriellaars, D. (1997). Neuromuscular factors contributing to in vivo eccentric moment generation. *Journal of Applied Physiology, 83,* 40-45.

Weeks, J.C., & McEwen, B.S. (1997). Modulation of neural circuits by steroid hormones in rodent and insect model systems. In P.S.G. Stein, S. Grillner, A.I. Selverston, & D.G. Stuart (Eds.), *Neurons, Networks, and Motor Behavior* (pp. 195-207). Cambridge, MA: MIT Press.

Weerakkody, N.S., Percival, P., Hickey, M.W., Morgan, D.L., Gregory, J.E., Canny, B.J., & Proske, U. (2003). Effects of local pressure and vibration on muscle pain from eccentric exercise and hypertonic saline. *Pain, 105,* 425-435.

Wehrwein, E.A., Roskelley, E.M., & Spitsbergen, J.M. (2002). GDNF is regulated in an activity-dependent manner in rat skeletal muscle. *Muscle & Nerve, 26,* 206-211.

Weinberg, R.S., & Hunt, V.V. (1976). The interrelationships between anxiety, motor performance and electromyography. *Journal of Motor Behavior, 8,* 219-224.

Weir, D.E., Tingley, J., & Elder, G.C.B. (2005). Acute passive stretching alters the mechanical properties of human plantar flexors and the optimal angle for maximal voluntary contraction. *European Journal of Applied Physiology, 93,* 614-623.

Weiss, E.J., & Flanders, M. (2004). Muscular and postural synergies of the human hand. *Journal of Neurophysiology, 92,* 523-535.

Welsh, S.J., Dinenno, D.V., & Tracy, B.L. (2007). Variability of quadriceps femoris motor neuron discharge and muscle force in human aging. *Experimental Brain Research, 179,* 219-233.

West, S.P., Roy, R.R., & Edgerton, V.R. (1986). Fiber type and fiber size of cat ankle, knee and hip extensors and flexors following low thoracic spinal cord transection at an early age. *Experimental Neurology, 91,* 174-182.

West, W., Hicks, A., Clements, L., & Dowling, J. (1995). The relationship between voluntary electromyogram, endurance time and intensity of effort in isometric handgrip exercise. *European Journal of Applied Physiology, 71,* 301-305.

Westbury, J.R., & Shaughnessy, T.G. (1987). Associations between spectral representation of the surface electromyogram and fiber type distribution and size in human masseter muscle. *Electromyography and Clinical Neurophysiology, 27,* 427-435.

Westerblad, H., Allen, D.G., Bruton, J.D., Andrade, F.H., & Lännergren, J. (1998). Mechanisms underlying the reduction of isometric force in skeletal muscle fatigue. *Acta Physiologica Scandinavica, 162,* 253-260.

Westerblad, H., Allen, D.G., & Lannergren, J. (2002). Muscle fatigue: Lactic acid or inorganic phosphate the major cause? *News in Physiological Sciences, 17,* 17-21.

Westerblad, H., Bruton, J.D., Allen, D.G., & Lännergren, J. (2000). Functional significance of Ca^{2+} in long-lasting fatigue of skeletal muscle. *European Journal of Applied Physiology, 83,* 166-174.

Westing, S.H., Seger, J.Y., & Thorstensson, A. (1990). Effects of electrical stimulation on eccentric and concentric torque-velocity relationships during knee extension in man. *Acta Physiologica Scandinavica, 140,* 17-22.

Weyand, P.G., & Davis, J.A. (2005). Running performance has a structural basis. *Journal of Experimental Biology, 208,* 2625-2631.

Weyand, P.G., Sternlight, D.B., Bellizzi, M.J., & Wright, S. (2000). Faster top running speeds are achieved with greater ground forces not more rapid leg movements. *Journal of Applied Physiology, 89,* 1991-1999.

Weytjens, J.L., & van Steenberghe, D. (1984). The effects of motor unit synchronization on the power spectrum of the electromyogram. *Biological Cybernetics, 51,* 71-77.

Whelan, P.J., Hiebert, G.W., & Pearson, K.G. (1995). Stimulation of the group I extensor afferents prolongs the stance phase in walking cats. *Experimental Brain Research, 103,* 20-30.

Whitehead, N.P., Allen, T.J., Morgan, D.L., & Proske, U. (1998). Damage to human muscle from eccentric exercise after training with concentric exercise. *Journal of Physiology, 512,* 615-620.

Whitehead, N.P., Morgan, D.L., Gregory, J.E., & Proske, U. (2003). Rises in whole muscle passive tension of mammalian muscle after eccentric contractions at different lengths. *Journal of Applied Physiology*, *95*, 1224-1234.

Wickiewicz, T.L., Roy, R.R., Powell, P.L., & Edgerton, V.R. (1983). Muscle architecture of the human lower limb. *Clinical Orthopaedics and Related Research, 179,* 275-283.

Widrick, J.J., Knuth, S.T., Norenberg, K.M., Romatowski, J.G., Bain, J.L., Riley, D.A., Karhanek, M., Trappe, S.W., Trappe, T.A., Costill, D.L., & Fitts, R.H. (1999). Effect of a 17 day spaceflight on contractile properties of human soleus muscle fibres. *Journal of Physiology, 516,* 915-930.

Widrick, J.J., Romatowski, J.G., Norenberg, K.M., Knuth, S.T., Bain, J.L., Riley, D.A., Trappe, S.W., Trappe, T.A., Costill, D.L., & Fitts, R.H. (2001). Functional properties of slow and fast gastrocnemius muscle fibers after a 17-day spaceflight. *Journal of Applied Physiology, 90,* 2203-2211.

Widrick, J.J., Steizer, J.E., Shoepe, T.C., & Garner, D.P. (2002*a*). Functional properties of human muscle fibers after short-term resistance exercise training. *American Journal of Physiology, 283,* R408-R416.

Widrick, J.J., Trappe, S.W., Romatowski, J.G., Riley, D.A., Costill, D.L., & Fitts, R.H. (2002*b*). Unilateral lower limb suspension does not mimic bed rest or spaceflight effects on human muscle fiber function. *Journal of Applied Physiology, 93,* 354-360.

Wiegner, A.W. (1987). Mechanism of thixotropic behaviour at relaxed joints in the rat. *Journal of Applied Physiology, 62,* 1615-1621.

Wight, J., Richards, J., & Hall, S. (2004). Influence of pelvis rotation styles on baseball pitching mechanics. *Sports Biomechanics, 3,* 67-83.

Wigmore, D.M., Propert, K., & Kent-Braun, J.A. (2006). Blood flow does not limit skeletal muscle force production during incremental isometric contractions. *European Journal of Applied Physiology, 96,* 370-378.

Wiktorsson-Möller, M., Öberg, B., Ekstrand, J., & Gillquist, J. (1983). Effects of warming up, massage, and stretching on range of motion and muscle strength in the lower extremity. *American Journal of Sports Medicine, 11,* 249-252.

Williams, D.M., & Bilodeau, M. (2004). Assessment of voluntary activation by stimulation of one muscle or two synergistic muscles. *Muscle & Nerve, 29,* 112-119.

Williams, G.N., Barrance, P.J., Snyder-Mackler, L., Axe, M.J., & Buchanan, T.S. (2003). Specificity of muscle action after anterior cruciate ligament injury. *Journal of Orthopedic Research, 21,* 1131-1137.

Williams, J.H., & Barnes, W.S. (1989). The positive inotropic effect of epinephrine on skeletal muscle: A brief review. *Muscle & Nerve, 12,* 968-975.

Williams, K.R. (1985). Biomechanics of running. *Exercise and Sport Sciences Reviews, 13,* 389-441.

Wilken, J.A., Smith, B.D., Tola, K., & Mann, M. (2000). Trait anxiety and prior exposure to non-stressful stimuli: Effects on psychophysiological arousal and anxiety. *International Journal of Psychophysiology, 37,* 233-242.

Wilson, B.D. (1977). Toppling techniques in diving. *Research Quarterly, 48,* 800-804.

Wilson, G.J., & Murphy, A.J. (1996). The use of isometric tests of muscular function in athletic assessment. *Sports Medicine, 22,* 19-37.

Wilson, G.J., Murphy, A.J., & Walshe, A. (1996). The specificity of strength training: The effect of posture. *European Journal of Applied Physiology, 73,* 346-352.

Wilson, G.J., Newton, R.U., Murphy, A.J., & Humphries, B.J. (1993). The optimal training load for the development of dynamic athletic performance. *Medicine and Science in Sports and Exercise, 25,* 1279-1286.

Winarski, A.M., Roy, R.R., Alford, E.K., Chiang, P.C., & Edgerton, V.R. (1987). Mechanical properties of rat skeletal muscle after hindlimb suspension. *Experimental Neurology, 96,* 650-660.

Windhorst, U. (1996). On the role of recurrent inhibitory feedback in motor control. *Progress in Neurobiology, 49,* 517-587.

Windhorst, U. (2007). Muscle proprioceptive feedback and spinal networks. *Brain Research Bulletin, 73,* 155-202.

Windhorst, U., Hamm, T.M., & Stuart, D.G. (1989). On the function of muscle and reflex partitioning. *Behavioral and Brain Sciences, 12,* 629-682.

Wing, A.M., Haggard, P., & Flanagan, J.R. (Eds.). (1996). *Hand and Brain. The Neurophysiology and Psychology of Hand Movements.* New York: Academic Press.

Winstein, C.J., Horak, F.B., & Fisher, B.E. (2000). Influence of central set on anticipatory and triggered grip-force adjustments. *Experimental Brain Research, 130,* 298-308.

Winter, D.A. (1983). Moments of force and mechanical power in jogging. *Journal of Biomechanics, 16,* 91-97.

Winter, D.A. (1990). *Biomechanics and Motor Control of Human Movement* (2nd ed.). New York: Wiley.

Winter, D.A., Wells, R.P., & Orr, G.W. (1981). Errors in the use of isokinetic dynamometers. *European Journal of Applied Physiology, 46,* 397-408.

Winter, D.A., & Yack, H.J. (1987). EMG profiles during normal human walking: Stride-to-stride and inter-subject variability. *Electroencephalography and Clinical Neurophysiology, 67,* 402-411.

Winter, J.A., Allen, T.J., & Proske, U. (2005). Muscle spindle signals combine with the sense of effort to indicate limb position. *Journal of Physiology, 568,* 1035-1046.

Wirz, M., Zemon, D.H., Rupp, R., Scheel, A., Colombo, G., Dietz, V., & Hornby, T.G. (2005). Effectiveness of automated locomotor training in patients with chronic incomplete spinal cord injury: A multicenter trial. *Archives of Physical Medicine and Rehabilitation, 86,* 672-680.

Wise, S.P., & Mauritz, K-H. (1985). Set-related neuronal activity in the premotor cortex of rhesus monkeys: Effects of changes in motor set. *Proceedings of the Royal Society of London B, 223,* 331-354.

Wishart, L.R., Lee, T.D., Cunningham, S.J., & Murdoch, J.E. (2002). Age-related differences and the role of augmented visual feedback in learning a bimanual coordination pattern. *Acta Psychologica, 110,* 247-263.

Witvrouw, E., Mahieu, N., Danneels, L., & McNair, P. (2004). Stretching and injury prevention: An obscure relationship. *Sports Medicine, 34,* 443-449.

Wolf, S.L., Winstein, C.J., Miller, J.P., Taub, E., Uswatte, G., Morris, D., Giuliani, C., Light, K.E., Nichols-Larsen, D., & EXCITE Investigators. (2006). Effect of constraint-induced movement therapy on upper extremity function 3 to 9 months after stroke: The EXCITE randomized clinical trial. *Journal of the American Medical Association, 296,* 2095-2104.

Wolpaw, J.R. (1994). Acquisition and maintenance of the simplest motor skill: Investigation of CNS mechanisms. *Medicine and Science in Sports and Exercise, 26,* 1475-1479.

Wolpaw, J.R., Maniccia, D.M., & Elia, T. (1994). Operant conditioning of primate H-reflex: Phases of development. *Neuroscience Letters, 170,* 203-207.

Wolpert, D.M., & Ghahramani, Z. (2000). Computational principles of movement neuroscience. *Nature Neuroscience,* 3 Suppl., 1212-1217.

Wolpert, D.M., Ghahramani, Z., & Jordan, M.I. (1995). Are arm trajectories planned in kinematic or dynamic coordinates? An adaptation study. *Experimental Brain Research, 103,* 460-470.

Wood, D.E., Manor, Y., Nadim, F., & Nusbaum, M.P. (2004). Intercircuit control via rhythmic regulation of projection neuron activity. *Journal of Neuroscience, 24,* 7455-7463.

Wood, G.A. (1982). Data smoothing and differentiation procedures in biomechanics. *Exercise and Sport Sciences Reviews, 10,* 308-362.

Woolstenhulme, M.T., Conlee, R.K., Drummond, M.J., Stites, A.W., & Parcell, A.C. (2006). Temporal response of desmin and dystrophin proteins to progressive resistance exercise in human skeletal muscle. *Journal of Applied Physiology, 100,* 1876-1882.

Wright, J.R., McCloskey, D.I., & Fitzpatrick, R.C. (2000). Effects of systemic arterial blood pressure on the contractile force of a human hand muscle. *Journal of Applied Physiology, 88,* 1390-1396.

Wright, S., & Weyand, P.G. (2001). The application of ground force explains the energetic cost of running backward and forward. *Journal of Experimental Biology, 204,* 1805-1815.

Wu, C.Y., Chen, C.L., Tsai, W.C., Lin, K.C., & Chou, S.H. (2007). A randomized controlled trial of modified constraint-induced movement therapy for elderly stroke survivors: Changes in motor impairment, daily functioning, and quality of life. *Archives of Physical Medicine and Rehabilitation, 88,* 273-278.

Wu, G., & Ladin, Z. (1993). The kinematometer—an integrated kinematic sensor for kinesiologic measurements. *Journal of Biomechanical Engineering, 115,* 53-62.

Wu, T., & Hallett, M. (2005). The influence of normal human ageing on automatic movements. *Journal of Physiology, 562,* 605-615.

Wurtz, R.H., & Sommer, M.A. (2004). Identifying corollary discharges for movement in the primate brain. *Progress in Brain Research, 144,* 47-60.

Xia, R., Bush, B.M.H., & Karst, G.M. (2005). Phase-dependent and task-dependent modulation of stretch reflexes during rhythmical hand tasks in humans. *Journal of Physiology, 564,* 941-951.

Xia, R., Markopoulou, K., Puumala, S.E., & Rymer, W.Z. (2006). A comparison of the effects of imposed extension and flexion movements on Parkinsonian rigidity. *Clinical Neurophysiology, 117,* 2302-2307.

Xiao, J., Padoa-Schioppa, C., & Bizzi, E. (2006). Neuronal correlates of movement dynamics in the dorsal and ventral premotor area of the monkey. *Experimental Brain Research, 168,* 106-119.

Xu, H., Akai, M., Kakurai, S., Yokota, K., & Kaneko, H. (1999). Effect of shoe modifications on center of pressure and in-shoe plantar pressures. *American Journal of Physical Medicine and Rehabilitation, 78,* 516-524.

Yakovenko, S., McCrea, D.A., Stecina, K., & Prochazka, A. (2005). Control of locomotor cycle durations. *Journal of Neurophysiology, 94,* 1057-1065.

Yamaguchi, G.T., Sawa, A.G.U., Moran, D.W., Fessler, M.J., & Winters, J.M. (1990). A survey of human musculotendon actuator parameters. In J.M. Winters & S.L-Y. Woo (Eds.), *Multiple Muscle Systems: Biomechanics and Movement Organization* (pp. 717-773). New York: Springer-Verlag.

Yamashita, N. (1988). EMG activities in mono- and bi-articular thigh muscles in combined hip and knee extension. *European Journal of Applied Physiology, 58,* 274-277.

Yanagiya, T., Kanehisa, H., Tachi, M., Kuno, S., & Fukunaga, T. (2004). Mechanical power during maximal treadmill walking and running in young and elderly men. *European Journal of Applied Physiology, 92,* 33-38.

Yang, J.F., Stein, R.B., & James, K.B. (1991). Contribution of peripheral afferents to the activation of the soleus muscle during walking in humans. *Experimental Brain Research, 87,* 679-687.

Yang, J.F., Stein, R.B., Jhamandas, J., & Gordon, T. (1990). Motor unit numbers and contractile properties after spinal cord injury. *Annals of Neurology, 28,* 496-502.

Yao, W., Fuglevand, A.J., & Enoka, R.M. (2000). Motor-unit synchronization increases EMG amplitude and decreases force steadiness of simulated contractions. *Journal of Applied Physiology, 83,* 441-452.

Yassierli, Nussbaum, M.A., Iridiastadi, H., & Wojcik, L.A. (2007). The influence of age on isometric endurance and fatigue is muscle dependent: A study of shoulder abduction and torso extension. *Ergonomics, 50,* 26-45.

Yasuda, N., Glover, E.I., Phillips, S.M., Isfort, R.J., & Tarnopolsky, M.A. (2005). Sex-based differences in skeletal muscle function and morphology with short-term limb immobilization. *Journal of Applied Physiology, 99,* 1085-1092.

Yeadon, M.R. (1990). The simulation of aerial movement—II. A mathematical inertia model of the human body. *Journal of Biomechanics, 23,* 67-74.

Yeadon, M.R. (1993*a*). The biomechanics of twisting somersaults part I: Rigid body motions. *Journal of Sports Sciences, 11,* 187-198.

Yeadon, M.R. (1993*b*). The biomechanics of twisting somersaults part II: Contact twist. *Journal of Sports Sciences, 11,* 199-208.

Yeadon, M.R. (1993*c*). The biomechanics of twisting somersaults part III: Aerial twist. *Journal of Sports Sciences, 11,* 209-218.

Yeadon, M.R. (1993*d*). The biomechanics of twisting somersaults part IV: Partitioning performances using the tilt angle. *Journal of Sports Sciences, 11,* 219-225.

Yeadon, M.R. (1993*e*). Twisting techniques used by competitive divers. *Journal of Sports Sciences, 11,* 337-342.

Yeadon, M.R., Kong, P.W., & King, M.A. (2006). Parameter determination for a computer simulation model of a diver and a springboard. *Journal of Applied Biomechanics, 22,* 167-176.

Yeung, E.W., & Yeung, S.S. (2001). Interventions for preventing lower limb soft-tissue injuries in runners. *Cochrane Database Systematic Reviews, 3,* CD001256.

Yordanova, J., Kolev, V., Hohnsbein, J., & Falkenstein, M. (2004). Sensorimotor slowing with ageing is mediated by a functional dysregulation of motor-generation processes: Evidence from high-resolution event-related potentials. *Brain, 127,* 351-362.

Yoshihara, M., Suzuki, K., & Kidokoro, Y. (2000). Two independent pathways mediated by cAMP and protein kinase A enhance spontaneous transmitter release at Drosophila neuromuscular junctions. *Journal of Neuroscience, 20,* 8315-8322.

Yoshitake, Y., Shinohara, M., Kouzaki, M., & Fukunaga, T. (2004). Fluctuations in plantar flexion force are reduced after prolonged tendon vibration. *Journal of Applied Physiology, 97,* 2090-2097.

Youdas, J.W., Krause, D.A., Egan, K.S., Therneau, T.M., & Laskowski, E.R. (2003). The effect of static stretching of the calf muscle-tendon unit on active ankle dorsiflexion range of motion. *Journal of Orthopedic Sports Physical Therapy, 33,* 408-417.

Young, A., Stokes, M., & Iles, J.F. (1987). Effects of joint pathology on muscle. *Clinical Orthopaedics and Related Research, 219,* 21-27.

Young, K., McDonagh, M.J.N., & Davies, C.T.M. (1985). The effects of two forms of isometric training on the mechanical properties of the triceps surae in man. *Pflügers Archiv, 405,* 384-388.

Young, R.R. (1994). Spasticity: A review. *Neurology, 44* (Suppl. 9), S12-S20.

Yu, J.G., Carlsson, L., & Thornell, L.E. (2004). Evidence for myofibril remodeling as opposed to myofibril damage in human muscles with DOMS: An ultrastructural and immunoelectron microscopic study. *Histochemistry and Cell Biology, 121,* 219-227.

Yue, G.H., Bilodeau, M., Hardy, P.A., & Enoka, R.M. (1997). Task-dependent effect of limb immobilization on the fatigability of the elbow flexor muscles in humans. *Experimental Physiology, 82,* 567-592.

Yue, G., & Cole, K.J. (1992). Strength increases from the motor program: A comparison of training with maximal voluntary and imagined muscle contractions. *Journal of Neurophysiology, 67,* 1114-1123.

Zadpoor, A.A., Asadi Nikooyan, A., & Reza Arshi, A. (2007). A model-based parametric study of impact force during running. *Journal of Biomechanics, 40,* 2012-2021.

Zajac, F.E. (1993). Muscle coordination of movement: A perspective. *Journal of Biomechanics, 26,* 109-124.

Zakas, A., Grammatikopoulou, M.G., Zakas, N., Zahariadis, P., & Vamvakoudis, E. (2006). The effect of active warm-up and stretching on the flexibility of adolescent soccer players. *Journal of Sports Medicine and Physical Fitness, 46,* 57-61.

Zatsiorsky, V.M. (1995). *Science and Practice of Strength Training*. Champaign, IL: Human Kinetics.

Zatsiorsky, V.M., & Kraemer, W.J. (2006). *Science and Practice of Strength Training* (2nd ed.). Champaign, IL: Human Kinetics.

Zatsiorsky, V.M., & Latash, M.L. (2004). Prehension synergies. *Exercise and Sport Sciences Reviews, 32,* 75-80.

Zatsiorsky, V., & Seluyanov, V. (1983). The mass and inertia characteristics of the main segments of the human body. In H. Matsui & K. Kobayashi (Eds.), *Biomechanics VIII-B* (pp. 1152-1159). Champaign, IL: Human Kinetics.

Zatsiorsky, V., Seluyanov, V., & Chugunova, L. (1990*a*). In vivo body segment inertial parameters determination using a gamma-scanner method. In N. Berme & A. Cappozzo (Eds.), *Biomechanics of Human Movement: Applications in Rehabilitation, Sports and Ergonomics* (pp. 186-202). Worthington, OH: Bertec.

Zatsiorsky, V.M., Seluyanov, V.N., & Chugunova, L.G. (1990*b*). Methods of determining mass-inertial characteristics of human body segments. In G.G. Chernyi & S.A. Regirer (Eds.), *Contemporary Problems of Biomechanics* (pp. 272-291). Boca Raton, FL: CRC Press.

Zehr, E.P. (2002). Considerations for use of the Hoffmann reflex in exercise studies. *European Journal of Applied Physiology, 86,* 455-468.

Zehr, E.P. (2006). Training-induced adaptive plasticity in human somatosensory reflex pathways. *Journal of Applied Physiology, 101,* 1783-1794.

Zehr, E.P., Collins, D.F., & Chua, R. (2001*a*). Human interlimb reflexes evoked by electrical stimulation of cutaneous nerves innervating the hand and foot. *Experimental Brain Research, 140,* 495-504.

Zehr, E.P., Collins, D.F., Frigon, A., & Hoogenboom, N. (2003). Neural control of rhythmic human arm movement: Phase dependence and task modulation of Hoffmann reflexes in forearm muscles. *Journal of Neurophysiology, 89,* 12-21.

Zehr, E.P., Hesketh, K.L., & Chua, R. (2001*b*). Differential regulation of cutaneous and H-reflexes during leg cycling in humans. *Journal of Neurophysiology, 85,* 1178-1184.

Zehr, E.P., Komiyama, T., & Stein, R.B. (1997). Cutaneous reflexes during human gait: Electromyographic and kinematic responses to electrical stimulation. *Journal of Neurophysiology, 77,* 3311-3325.

Zehr, E.P., & Stein, R.B. (1999). Interaction of the Jendrássik maneuver with segmental presynaptic inhibition. *Experimental Brain Research, 124,* 474-480.

Zemková, H., Teisinger, J., Almon, R.R., Vejsada, R., Hník, P., & Vyskocil, F. (1990). Immobilization atrophy and membrane properties in rat skeletal muscle fibres. *Pflügers Archiv, 416,* 126-129.

Zengel, J., Reid, S., Sypert, G., & Munson, J. (1985). Membrane electrical properties and prediction of motor-unit type of medial gastrocnemius motoneurons in the cat. *Journal of Neurophysiology, 53,* 1323-1344.

Zernicke, R.F., & Smith, J.L. (1996). Biomechanical insights into neural control of movement. In L.B. Rowell and J.T. Shepherd (eds.), *Handbook of Physiology: Sec. 12, Exercise: Regulation and Integration of Multiple Systems* (pp. 293-330), New York: Oxford University Press.

Zhang, L., Butler, J., Nishida, T., Nuber, G., Huang, H., & Rymer, W.Z. (1998). In vivo determination of the direction of rotation and moment-angle relationship of individual elbow muscles. *Journal of Biomechanical Engineering, 120,* 625-633.

Zheng, N., Fleisig, G.S., Escamilla, R.F., & Barrentine, S.W. (1998). An analytical model of the knee for estimation of internal forces during exercise. *Journal of Biomechanics, 31,* 963-967.

Zhi, G., Ryder, J.W., Huang, J., Ding, P., Chen, Y., Zhao, Y., Kamm, K.E., & Stull, J.T. (2005). Myosin light chain kinase and myosin phosphorylation effect frequency-dependent potentiation of skeletal muscle contraction. *Proceedings of the National Academy of Sciences, 102,* 17519-17524.

Zhong, S., Lowe, D.A., & Thompson, L.V. (2006). Effects of hindlimb unweighting and aging on rat semimembranosus muscle and myosin. *Journal of Applied Physiology, 101,* 873-880.

Zhou, S. (2000). Chronic neural adaptations to unilateral exercise: Mechanisms of cross education. *Exercise and Sport Sciences Reviews, 28,* 177-184.

Zijdewind, I., Butler, J.E., Gandevia, S.C., & Taylor, J.L. (2006*a*). The origin of activity in the biceps brachii muscle during voluntary contractions of the contralateral elbow flexor muscles. *Experimental Brain Research, 175,* 526-535.

Zijdewind, I., & Kernell, D. (2001). Bilateral interactions during contractions of intrinsic hand muscles. *Journal of Neurophysiology, 85,* 1907-1913.

Zijdewind, I., & Thomas, C.K. (2003). Motor unit firing during and after voluntary contractions of human thenar muscles weakened by spinal cord injury. *Journal of Neurophysiology, 89,* 2065-2071.

Zijdewind, I., Toering, S.T., Bessem, B., van der Laan, O., & Diercks, R.L. (2003). Effects of imagery motor training on torque production of ankle plantar flexor muscles. *Muscle & Nerve, 28,* 168-173.

Zijdewind, I., van Duinen, H., Zielman, R., & Lorist, M.M. (2006*b*). Interaction between force production and cognitive performance in humans. *Clinical Neurophysiology, 117,* 660-667.

Zijdewind, I., Zwarts, M.J., & Kernell, D. (1998). Influence of a voluntary fatigue test on the contralateral homologous muscle in humans. *Neuroscience Letters, 253,* 41-44.

Zimny, M.L., & Wink, C.S. (1991). Neuroreceptors in the tissues of the knee joint. *Journal of Electromyography and Kinesiology, 3,* 148-157.

Zytnicki, D., Lafleur, J., Horcholle-Bossavit, G., Lamy, F., & Jami, L. (1990). Reduction of Ib autogenetic inhibition in motoneurons during contractions of an ankle extensor muscle in the cat. *Journal of Neurophysiology, 64,* 1380-1389.

Index

Note: The italicized *f* and *t* following page numbers refer to figures and tables, respectively.

Q

R

U

V

About the Author

Roger M. Enoka, PhD, is a professor and chair in the department of integrative physiology at the University of Colorado at Boulder. He is also a professor in the Health Sciences Center, department of medicine, geriatrics, at the University of Colorado. Previously, Enoka was a biomechanist in the department of biomedical engineering at the Cleveland Clinic Foundation and a professor in the department of physiology at the University of Arizona.

For more than 30 years, Enoka has focused his research and teaching on the combination of biomechanics and neurophysiology of movement. He conducts an interdisciplinary research program, which has received continuous funding by the National Institutes of Health for almost two decades. Internationally known for his achievements as a teacher and researcher, Enoka is the author of about 350 journal articles, books, chapters, reviews, and abstracts related to his research. He is also a reviewer for numerous journals and serves on the editorial board for the *Journal of Applied Physiology, Journal of Electromyography and Kinesiology*, *Motor Control*, *Sports Medicine*, *Muscle and Nerve*, and the *Scandinavian Journal of Medicine and Science in Sports*.

Enoka's professional affiliations include the American College of Sports Medicine, the American Physiological Society, International Society of Biomechanics, and the Society for Neuroscience. He is a former member of the Advisory Panel on Research for the American Physical Therapy Association and the Respiratory and Applied Physiology Study Section of the National Institutes of Health. Enoka is also a past president and current member of the American Society of Biomechanics.